Methods in Enzymology

Volume XIX
PROTEOLYTIC ENZYMES

METHODS IN ENZYMOLOGY

EDITORS-IN-CHIEF

Sidney P. Colowick   Nathan O. Kaplan

QP601
M49
V.19

Methods in Enzymology

Volume XIX

Proteolytic Enzymes

EDITED BY

Gertrude E. Perlmann

THE ROCKEFELLER UNIVERSITY
NEW YORK, NEW YORK

and

Laszlo Lorand

DIVISION OF BIOCHEMISTRY
DEPARTMENT OF CHEMISTRY
NORTHWESTERN UNIVERSITY
EVANSTON, ILLINOIS

1970

ACADEMIC PRESS   New York and London

ACADEMIC PRESS, INC.
111 Fifth Avenue, New York, New York 10003

United Kingdom Edition published by
ACADEMIC PRESS, INC. (LONDON) LTD.
Berkeley Square House, London W1X 6BA

LIBRARY OF CONGRESS CATALOG CARD NUMBER: 54-9110

PRINTED IN THE UNITED STATES OF AMERICA

Table of Contents

Section I. Titration Methods

Section II. Individual Proteolytic Enzymes

A. The Serine Proteases

Section III. Enzymes Primarily Considered as Transpeptidases

Section IV. Naturally Occurring Activators and Inhibitors of Proteolytic Enzymes

Section V. Water-Insoluble Proteolytic Enzymes

Contributors to Volume XIX

Article numbers are shown in parentheses following the names of contributors.
Affiliations listed are current.

E. D. ADAMSON (60, 61), *Department of Biochemistry, University of Toronto, Toronto, Ontario, Canada*

BENJAMIN ALEXANDER (72), *The New York Blood Center, New York, New York*

S. Y. ALI (65), *Department of Pathology, Institute of Orthopaedics (University of London), Royal National Orthopaedics Hospital, Stanmore, Middlesex, England*

KEI ARIMA (30), *Department of Agricultural Chemistry, The University of Tokyo, Bunkyo-ku, Tokyo, Japan*

RUTH ARNON (14), *Department of Chemical Immunology, The Weizmann Institute of Science, Rehovot, Israel*

TAGE ASTRUP (64), *The James F. Mitchell Foundation, Institute for Medical Research, Washington, D.C.*

ESTELLE BAR-ELI (24), *Department of Biochemistry, Tel Aviv University, Tel Aviv, Israel*

GRANT H. BARLOW (48), *Molecular Biology Department, Abbott Laboratories, North Chicago, Illinois*

D. JOE BAUGHMAN (9), *Ortho Pharmaceutical Corporation, Raritan, New Jersey*

MAKONNEN BELEW (40), *Ethiopian Nutrition Institute, Children's Nutrition Unit, Addis Ababa, Ethiopia*

MYRON L. BENDER (12), *Department of Chemistry, Northwestern University, Evanston, Illinois*

A. BERGER (35), *Department of Biophysics, The Weizmann Institute of Science, Rehovot, Israel*

ZVI BOHAK (22), *Department of Biophysics, The Weizmann Institute of Science, Rehovot, Israel*

K. BROCKLEHURST (71), *Department of Biochemistry and Chemistry, Medical College of St. Bartholomew's Hospital, Charterhouse Square, London, England*

JOYCE BRUNER-LORAND (59), *Division of Biochemistry, Department of Chemistry, Northwestern University, Evanston, Illinois*

F. M. BUMPUS (51), *Cleveland Clinic Foundation, Cleveland, Ohio*

PHILIP J. BURCK (67), *Biochemical Research, Eli Lilly and Company, Indianapolis, Indiana*

BENEDICT J. CAMPBELL (54), *Department of Biochemistry, University of Missouri, Columbia, Missouri*

R. CHAPUIS (37b), *Laboratorium für Molekular-Biologie, Chemischer Richtung, Eidgenössische Technische Hochschule, Zürich, Switzerland*

THEODORE CHASE, JR. (2), *Department of Biochemistry and Microbiology, College of Agriculture and Environmental Science, Rutgers University, New Brunswick, New Jersey*

ICHIRO CHIBATA (58), *Research Laboratory, Tanabe Seiyaku Company, Ltd., Higashiyodogawa-ku, Osaka, Japan*

G. E. CONNELL (60, 61), *Department of Biochemistry, University of Toronto, Toronto, Canada*

E. M. CROOK (71), *Department of Biochemistry and Chemistry, Medical College of St. Bartholomew's Hospital, Charterhouse Square, London, England*

SAM T. DONTA (23), *Department of Medicine, Boston University School of Medicine, Boston, Massachusetts*

R. F. DOOLITTLE (38), *Department of Chemistry, University of California, San Diego, La Jolla, California*

S. D. ELLIOTT (16), *Department of Path-*

ology, *Corpus Christi College, Cambridge, England*

ARACELI M. ENGEL (72), *The New York Blood Center, New York, New York*

M. P. ESNOUF (53), *Nuffield Department of Clinical Biochemistry, Radcliffe Infirmary, Oxford, England*

L. FISHMAN (49), *Department of Biochemistry, College of Dentistry, New York University, New York, New York*

J. E. FOLK (6, 32), *Section on Enzyme Chemistry, Laboratory of Biochemistry, National Institute of Dental Research, National Institutes of Health, Bethesda, Maryland*

B. FOLTMANN (28), *University of Copenhagen, Institute of Genetics B, (Biochemical Genetics), Copenhagen, Denmark*

BERNARD FRIEDENSON (17), *Department of Biochemistry Research, University of Buffalo, Buffalo, New York*

KITTY L. A. GLANVILLE (52), *Department of Pathology, Institute of Orthopaedics, Royal National Orthopaedic Hospital, Brockley Hill, Stanmore, Middlesex, England*

LEON GOLDSTEIN (70), *Department of Biophysics, The Weizmann Institute of Science, Rehovot, Israel*

T. GOTOH (59a), *Division of Biochemistry, Department of Chemistry, Northwestern University, Evanston, Illinois*

EDWARD E. HALEY (55, 56), *Veterans Administration Hospital, West Haven, Connecticut*

ELVIN HARPER (44), *Developmental Biology Laboratory, Department of Medicine, Massachusetts General Hospital, Boston, Massachusetts*

WILLIAM F. HARRINGTON (45), *McCollum Pratt Institute, The Johns Hopkins University, Baltimore, Maryland*

S. RALPH HIMMELHOCH (33), *Laboratory of Biochemistry, National Cancer Institute, National Institutes of Health, Department of Health, Education and Welfare, Bethesda, Maryland*

T. HOFMANN (25), *Department of Bio-*

chemistry, *University of Toronto, Toronto, Ontario, Canada*

JEROME F. HRUSKA (13), *University of Minnesota, Minneapolis, Minnesota*

EIJI ICHISHIMA (26), *Laboratory of Microbiology and Enzymology, Tokyo Noko University, Fuchu, Tokyo, Japan*

TSUTOMU ISHIKAWA (58), *Research Laboratories, Tanabe Seiyaku Company, Ltd., Higashiyodogawa-ku, Osaka, Japan*

SHINJIRO IWASAKI (30), *Meito Sangijo Company, Ltd., Chiao-ku, Tokyo, Japan*

RICHARD L. JACKSON (42), *Experimental Therapeutics Branch, National Health Institute, National Institutes of Health, Bethesda, Maryland*

E. T. KAISER (1), *Departments of Chemistry and Biochemistry, University of Chicago, Chicago, Illinois*

BEATRICE KASSELL (21, 66, 66a, 66b, 66c, 66d, 66e, 66f, 66g), *Department of Biochemistry, Marquette School of Medicine, Milwaukee, Wisconsin*

F. J. KÉZDY (1), *Department of Biochemistry, University of Chicago, Chicago, Illinois*

PREBEN KOK (64), *The James F. Mitchell Foundation, Institute for Medical Research, Washington, D.C.*

DONALD K. KUNIMITSU (15), *Department of Chemistry, Fresno State College, Fresno, California*

CHARLES H. LACK (52), *Department of Pathology, Institute of Orthopaedics, Royal National Orthopaedic Hospital, Stanmore, Middlesex, England*

ABEL LAJTHA (36), *New York State Institute of Neurochemistry and Drug Addiction, Ward's Island, New York*

JOHN H. LAW (13), *Department of Biochemistry, University of Chicago, Chicago, Illinois*

MILTON LEVY (49), *Department of Biochemistry, College of Dentistry, New York University, New York, New York*

IRVIN E. LIENER (17), *Department of Biochemistry, College of Biological*

Sciences, University of Minnesota, St. Paul, Minnesota

TEH-YUNG LIU (16), Department of Biology, Brookhaven National Laboratory, Upton, New York

L. LORAND (59a), Division of Biochemistry, Department of Chemistry, Northwestern University, Evanston, Illinois

JAMES McCONN (39), Division of Biochemistry, Department of Chemistry, Northwestern University, Evanston, Illinois

STAFFAN MAGNUSSON (9a), Institut für Molekylar Biologi, Aarhus Universitet, Aarhus, Denmark

NEVILLE MARKS (36), New York State Institute of Neurochemistry and Drug Addiction, Ward's Island, New York

FRITZ MARKWARDT (69), Pharmakologisches Institut, Medizinische Akademie, Erfurt, Germany (D.D.R.)

HIROSHI MATSUBARA (46), The Central Research Institute, Sunory Limited, Kita-ku, Osaka, Japan

GARY R. MATSUEDA (42), Department of Biochemistry and Biophysics, University of Hawaii, Honolulu, Hawaii

PATRICIA A. MEITNER (21), Department of Biochemistry, Marquette School of Medicine, Milwaukee, Wisconsin

TERENCE G. MERRETT (24), The Radiochemical Centre, Amersham, Buckinghamshire, England

KENT D. MILLER (8), Department of Medicine, Division of Laboratory Medicine, University of Miami School of Medicine, Miami, Florida

WILLIAM M. MITCHELL (45), Department of Microbiology, Vanderbilt University, Nashville, Tennessee

FRANK C. MONKHOUSE (68), Department of Physiology, University of Toronto, Medical Sciences Building, Toronto, Ontario, Canada

ROBIN E. MONRO (62), Instituto de Biologia Celular, Centro de Investigaciones Biologicas, Madrid, Spain

TAKASHI MURACHI (18), Department of Biochemistry, Nagoya City University, School of Medicine, Nagoya, Japan

MARY J. MYCEK (19), New Jersey College of Medicine and Dentistry, Newark, New Jersey

ROLF MYHRMAN (59), Division of Biochemistry, Department of Chemistry, Northwestern University, Evanston, Illinois

YASUSHI NAKAGAWA (41), The Rockefeller University, New York, New York

YOSHIKO NARAHASHI (47), Polymer Chemistry Laboratory, Institute of Physical and Chemical Research, Yamato-Machi, Saitama Prefecture, Japan

MARTIN OTTESEN (11, 30a), Chemical Department, Carlsberg Laboratory, Copenhagen-Valby, Denmark

PHILIP H. PÉTRA (31), Departments of Obstetrics, Gynecology, and Biochemistry, University of Washington, Seattle, Washington

G. PFLEIDERER (34), Lehrstuhl für Biochemie, Abteilung für Chemie, Ruhr-Universität Bochum, Bochum, Germany

MANFRED PHILIPP (12), Department of Chemistry, Northwestern University, Evanston, Illinois

LASZLO POLGAR (12), Institute of Biochemistry, Hungarian Academy of Sciences, Budapest, Hungary

JERKER PORATH (40), Institute of Biochemistry, Uppsala, Sweden

ELINE S. PRADO (50), Departamento de Bioquimica e Farmacologia, Escola Paulista de Medicina, Sao Paulo, Brasil

W. RICKERT (30a), Department of Statistics, University of Waterloo, Waterloo, Ontario, Canada

KENNETH C. ROBBINS (10), Biochemistry Section, Michael Reese Blood Center, Chicago, Illinois

G. RONCARI (37a), Laboratorium für Molekular-Biologie, Chemischer Richtung, Eidgenössische Technische Hochschule, Zürich, Switzerland

C. A. RYAN (66f), Department of Agricultural Chemistry, Washington State University, Pullman, Washington

A. P. RYLE (20), *Department of Biochemistry, University of Edinburgh Medical School, Edinburgh, Scotland*

I. SCHENKEIN (49), *Irvington House Laboratories, Department of Medicine, New York University School of Medicine, New York, New York*

CHRISTIAN SCHWABE (57), *Department of Biological Chemistry, Harvard School of Dental Medicine, Boston, Massachusetts*

SAM SEIFTER (44), *Department of Biochemistry, Albert Einstein College of Medicine, Yeshiva University, Bronx, New York*

ELLIOTT SHAW (2), *Biology Department, Brookhaven National Laboratory, Upton, New York*

DAVID M. SHOTTON (7), *Department of Biochemistry, University of Bristol, Bristol, England*

R. R. SMEBY (51), *Cleveland Clinic Foundation, Cleveland, Ohio*

J. ŠODEK (25), *Department of Biochemistry, University of Alberta, Edmonton, Alberta, Canada*

LOUIS SUMMARIA (10), *Biochemistry Section, Michael Reese Research Foundation, Chicago, Illinois*

IB SVENDSEN (11), *Carlsberg Laboratory, Copenhagen, Denmark*

A. SZEWCZUK (61), *Institute of Immunology and Experimental Therapy, Wroclaw, Poland*

JORDAN TANG (27), *Neurosciences Section, Oklahoma Medical Research Foundation, Oklahoma City, Oklahoma*

FLETCHER B. TAYLOR, JR. (63), *Department of Medicine, Allergy and Immunology Division, University of Pennsylvania School of Medicine, Philadelphia, Pennsylvania*

RUSSELL H. TOMAR (63), *Department of Medicine, Allergy and Immunology Division, University of Pennsylvania*

School of Medicine, Philadelphia, Pennsylvania

TETSUYA TOSA (58), *Research Laboratories, Tanabe Seiyaku Company, Ltd., Higashiyodogawa-ku, Osaka, Japan*

HELEN VAN VUNAKIS (23), *Graduate Department of Biochemistry, Brandeis University, Waltham, Massachusetts*

K. A. WALSH (3, 4), *Department of Biochemistry, University of Washington, Seattle, Washington*

MARION E. WEBSTER (50), *Experimental Therapeutics Branch, National Heart and Lung Institute, National Institute of Health, Bethesda, Maryland*

C. W. WHARTON (71), *Department of Biochemistry, The University of Birmingham, Birmingham, London, England*

D. R. WHITAKER (43), *Biochemistry Laboratory, National Research Council of Canada, Ottawa, Ontario, Canada*

JOHN R. WHITAKER (29), *Department of Food Science and Technology, University of California, Davis, California*

WILFRID F. WHITE (48), *Biochemical Research Department, Abbott Laboratories, North Chicago, Illinois*

P. E. WILCOX (3, 5), *Department of Chemistry, Harvard University, Cambridge, Massachusetts*

A. YARON (35), *Department of Biophysics, The Weizmann Institute of Science, Rehovot, Israel*

KERRY T. YASUNOBU (15, 39), *Department of Biochemistry and Biophysics, University of Hawaii, Honolulu, Hawaii*

JUHYUN YU (30), *Department of Food Technology, College of Science and Engineering, Yonsei University, Seudaimun-ku, Seoul, Korea*

H. ZUBER (37a, 37b), *Laboratorium für Molekular-Biologie, Chemischer Richtung, Eidgenössische Technische Hochschule, Zürich, Switzerland*

Preface

The "Proteolytic Enzymes" like other volumes of "Methods in Enzymology" provides information on the purification and mode of assay of enzymes. It is meant to be of value to the graduate student as well as to the more advanced investigator. The authors have been asked to give an up-to-date synopsis of other knowledge relating to physical properties (molecular weight, electrophoretic behavior, optical rotatory dispersion) and chemical structure (amino acid composition and sequences) as well as of some details on the kinetics of the various reactions catalyzed by the enzymes discussed. We, thus, believe that this volume will fill some of the needs of a primary reference work in this area.

Classification of enzymes is a difficult task, and proteolytic enzymes are no exception. Regardless of the criteria adopted for their grouping, one soon notices the arbitrary nature of imposed systematization, and contradictions become obvious. While it may be convenient to group a number of enzymes together on the basis of a catalytic functionality common to the enzyme proteins themselves (e.g., serine proteases, cf. Section II [3]) it is also of some advantage to have sections based on entirely different criteria such as the pH optimum of activity (acidic proteases) or substrate specificities (carboxypeptidases and aminopeptidases, etc.). Other difficulties are also obvious; nevertheless, we felt that the presentation of a volume of this sort would gain even by such ambiguous subdivisions into various sections.

Similarity of the catalytic pathway for a number of estero-proteolytic enzymes provided the idea for including methods on enzyme titrations. Apart from being of immediate use for the standardization of the particular enzymes discussed in the special chapters of Section I, hopefully they will provide an impetus, in general, for designing appropriate methodology for a variety of other proteases as well. Eventually, these titration methods are bound to be of prime usefulness in enzyme standardization for a variety of purposes (e.g., World Health Organization).

Although there are several enzymes which could be assayed by the rate of hydrolysis of given common substrates (be it casein, hemoglobin, or tosylarginine methyl ester), optimal conditions for the assay of each enzyme will vary appreciably and, in spite of possible duplications, these rate assays are better dealt with in connection with the descriptions of individual enzymes.

In addition to proteases and peptidases, this volume includes a number of transpeptidating (transamidating) enzymes. Of course, several of these also catalyze hydrolytic reactions just as many of the enzymes conventionally regarded as proteases function in transpeptidating reactions. Special reasons could be offered for including a few other miscellaneous enzymes, too. Because water-insoluble enzymes hold great promise, detailed coverage was given to some solid-phase enzyme derivatives. The section on the naturally occurring activators and inhibitors of proteolytic enzymes will, no doubt, enhance the value of this book.

Some duplications in a task of this nature are unavoidable. However, we are somewhat more unhappy about possible omissions, a few of which we are already aware (serum trypsin inhibitors; the complement system), and hope that the reader will not hesitate to draw our attention to these. The proteolytic enzyme field appears to be in a state of rapid expansion, and it is conceivable that some of the omissions could be included in a later volume.

Although we did recommend certain guidelines for presentation, it was decided that we would not try to mold contributions of the different authors into a uniform format. It is doubtful that we could have accomplished this even if we attempted it. Such is the inherent diversity of the subject matter of proteolytic enzymes—and of those who write about them.

Special thanks are due to Dr. Joyce Bruner-Lorand for valuable help.

METHODS IN ENZYMOLOGY

EDITED BY

Sidney P. Colowick and Nathan O. Kaplan

VANDERBILT UNIVERSITY
SCHOOL OF MEDICINE
NASHVILLE, TENNESSEE

DEPARTMENT OF CHEMISTRY
UNIVERSITY OF CALIFORNIA
AT SAN DIEGO
LA JOLLA, CALIFORNIA

I. Preparation and Assay of Enzymes
II. Preparation and Assay of Enzymes
III. Preparation and Assay of Substrates
IV. Special Techniques for the Enzymologist
V. Preparation and Assay of Enzymes
VI. Preparation and Assay of Enzymes (*Continued*)
Preparation and Assay of Substrates
Special Techniques
VII. Cumulative Subject Index

METHODS IN ENZYMOLOGY

EDITORS-IN-CHIEF

Sidney P. Colowick   Nathan O. Kaplan

Section I

Titration Methods

[1] Principles of Active Site Titration of Proteolytic Enzymes[1]

By F. J. Kézdy and E. T. Kaiser

Introduction

The purpose of this chapter is to discuss methods which have been developed for the measurement of the active site molarities of enzyme solutions. Rather than presenting an exhaustive survey of the active site titration techniques now available, we will consider first the basic principles involved in such measurements, and then we will describe a few representative titration methods for α-chymotrypsin.

Any quantitative chemical study has to rely ultimately on the knowledge of the concentration of each reacting species in the reaction mixture. The measurement of concentrations—the titration—ideally would require a method which should be accurate, precise, and specific for each compound. The development and perfection of such methods is still one of the major preoccupations of analytical chemistry. In the field of enzymology, accurate titration methods only recently became a realistic goal, after the remarkable progress achieved in the past decades concerning the purification, structural analysis, and understanding of the mechanism of a multitude of enzymes.

Conceptually, the process of establishing a method of titration could be divided into the following elementary steps: (1) definition of the function(s) or the molecular species to be measured; (2) selection of an experimentally measurable physical or chemical property which is common and proper to these compounds or functions; and (3) the development of theoretical equations permitting the translation of the experimental results into concentrations. The methods of titration of proteolytic enzymes are no exception to this rule although the concepts involved are further complicated by the complex chemical nature of the enzymes.

Enzymes are proteins of well-defined chemical composition and theoretically the molecular species to be titrated could be defined as the protein molecules having amino acid sequences identical to an established standard. For the present, the lack of adequate criteria for protein purity renders impossible the attainment of this goal. Furthermore, the most important property of an enzyme is its catalytic activity,

[1] Supported in part by grants from the National Institute of Arthritis and Metabolic Diseases and the National Institute of General Medical Sciences.

which is determined not only by the primary sequence, but also depends critically on the secondary and tertiary structure of the protein. Finally, the catalysis occurs at a limited area of the protein surface—the active site or active center of the enzyme—and the enzymatic activity is in many cases independent of slight chemical or structural changes undergone by other parts of the protein molecule. For these reasons it would be impractical to define the goal of enzyme titrations as the determination of the number of protein molecules of identical chemical structure and of identical catalytic activity. In the present discussion, therefore, we would like to limit ourselves to the titration of active sites, i.e., the determination of the concentration of catalytic centers possessing identical enzymatic properties.

The selection of the method should take into account the fact that the enzyme to be titrated will be contaminated by inactive protein, most likely of identical or similar amino acid composition. Thus, even if the active center contains a unique chemical function—such as the sulfhydryl group of several proteolytic enzymes of plant origin—a simple determination of that particular function will not necessarily yield the concentration of active centers. The titrating agent should be chosen in such a way that its reaction would be dependent upon the integrity of the active center. The enzyme should either catalyze the reaction in some way, or at least the environment of the active site should modify the reactivity of the particular chemical function in order to distinguish it from a similar group on enzymatically inactive fragments.[1a] In other words, in designing titrating reagents, one should exploit fully the specificity of the enzyme.

In general, the reagents suitable for titration will be small molecules of high purity possessing a pronounced chemical or physical affinity toward the active site. The reaction of the reagent with the enzyme should be accompanied by an easily and accurately measurable change in a physical property. The reaction should have a one-to-one stoichiometry and optimally should consist of the irreversible formation of a covalent bond. Reagents which combine reversibly with the active site or which bind to it noncovalently are less satisfactory from the experimental point of view. Covalent bond-forming reagents can be further classified according to whether they react with groups involved in the catalytic process or whether they react with other "structural" groups of the active site.

Several criteria will help to establish that the titration occurs at the active center:

[1a] This modification of the reactivity might sometimes be obtained by adsorbing substrate or competitive inhibitor molecules to the active site.

1. Comparison of the rate constant of the enzyme reaction with that of an appropriate model system will show the presence of catalysis.

2. The kinetics of the reaction should be simple first or second order—depending on the relative concentration of the reagent and the enzyme—or it should show a saturation phenomenon if the titrant binds to the active site before reaction can occur. Substrates and competitive inhibitors specific to the enzyme should normally inhibit the titration reaction. The presence of such an inhibition is a good indication—but not a rigorous proof—that the reaction occurs at the active site. In the same vein, the absence of inhibition does not rule out the participation of the active center, especially when the group to be titrated is not one of the catalytic functions.

3. The combination of the titrant with catalytic groups should yield an inactive enzyme. Even if the reaction involves only "structural" groups, a considerable modification of the activity should result. As a corollary, reversing the reaction of titration should restore fully the enzymatic activity.

4. In the case of a titrating agent possessing an asymmetric center, a large difference in activity of the antipodes is again excellent evidence for the involvement of the active site in the reaction. This difference should be more pronounced when the structure of the titrating agent better approximates the structure of substrates specific for the enzyme.

5. Finally, the use of experimental conditions which lead to the suppression of enzymatic catalysis—such as extreme pH's, high temperature, or $8 M$ urea—should also result in the disappearance of the reaction of titration.

The establishment of theoretical equations allowing one to convert the experimental data into concentrations of active centers is a relatively simple process only in the case of titrations accompanied by the irreversible formation of a covalent bond. The use of titrants based on the formation of reversible products and transient intermediates necessitates a more complicated mathematical treatment, the validity of which has to be ascertained by a detailed kinetic study of the phenomenon in each case. A kinetic analysis is also necessary in order to reveal any eventual heterogeneity of the enzyme which might arise from the presence of isozymes, partially active degradation products, or even other enzymes of similar specificity. In all cases, the purity, the yield, and the chemical identity of the product of titration should be carefully determined.

Finally, it should be emphasized that titration procedures, although highly accurate and theoretically straightforward, have a rather low sensitivity inherent to the method of detection. For example, most spectrophotometric titrations will require enzyme concentrations of the order of $10^{-6} M$ at least. Furthermore, even the simplest and most selective titrations will be reliable only with relatively pure protein preparations, when the concentration of interfering impurities is at a reasonably low level. Therefore it is unlikely that titration procedures will ever completely replace rate assays which can easily determine relative enzyme concentrations of the order of $10^{-9} M$ by the use of appropriate specific substrates and extremely simple experimental procedures. Titration procedures will be of practical application only with enzymes which are readily obtained in large quantities and with high purity. For other enzymes the titration of one preparation will allow one to standardize the rate assay in absolute concentrations, thereby eliminating the use of cumbersome and ill-defined enzyme units.

The result of active site titrations is expressed in units of molarity of active sites. This value is independent of the purity of the protein preparation, and thus it renders possible quantitative mechanistic studies on unstable or imperfectly purified enzymes. Combination of titration data with other determinations can also yield valuable conclusions about enzyme purity and about the number of active sites and subunits per molecule. Finally, titration of pure enzymes is one of the most accurate ways of determining their molecular weights.

In the examples of active site titrations which we will discuss now the emphasis will be on the use of spectrophotometric methods. Numerous other techniques have been developed for the titration of active sites, but we feel that the spectrophotometric methods are unexcelled because of their convenience, accuracy, and rapidity. In the first experimental method we will consider the use of a slowly reversible, covalently bound inhibitor. Next, we will consider the use of rapidly reversible, covalently bound substrates as titrants.

The Use of Slowly Reversible, Covalently Bound Inhibitors

Titration of the Active Site of α-Chymotrypsin with
2-Hydroxy-5-nitro-α-toluenesulfonic Acid Sultone

The titration procedure described below appears to be the most convenient method at the present time for the determination of the operational molarity of the active sites in α-chymotrypsin solutions. The reaction utilized is the extremely rapid formation of 2-hydroxy-5-nitro-α-toluenesulfonyl-α-chymotrypsin from the attack of the serine hydroxyl at the active site of chymotrypsin on 2-hydroxy-5-nitro-α-toluenesulfonic acid sultone (Eq. 1).[2] The sulfonyl enzyme decomposes slowly, regenerat-

---

[2] J. H. Heidema and E. T. Kaiser, *J. Am. Chem. Soc.* **90**, 1860 (1968); *ibid.*, **89**, 460 (1967).

ing active enzyme, and the slowness of this desulfonylation reaction as compared to the sulfonylation process makes it possible to employ the sultone as an active site titrant.

$$
\text{S} + \text{E} \underset{K_S}{\longrightarrow} \text{ES} \quad \begin{array}{c} k_2 \uparrow\downarrow k_{-2} \end{array} \quad \text{P} + \text{E} \xleftarrow{k_3} \text{ES}' \tag{1}
$$

S   E   ES

P   E   ES'

Site of Reaction

Experiments on the degradation of the sulfonyl enzyme, 2-hydroxy-5-nitro-α-toluenesulfonyl-α-chymotrypsin, which were done in an attempt to demonstrate the site of attachment of the sulfonyl group to the enzyme, were unsuccessful. However, other kinds of evidence for the location of the sulfonyl group on the enzyme were obtained.

The evidence that the sulfonyl group of 2-hydroxy-5-nitro-α-toluene-sulfonyl-α-chymotrypsin is located on Ser_{195} is 3-fold. First, the sulfonyl enzyme is inactive toward both the active site titrant cinnamoyl imidazole[3] and the specific ester substrate N-acetyl-L-tryptophan methyl ester. Also, the inhibition constant obtained for the inhibition by N-acetyl-L-tryptophan of the sulfonylation of chymotrypsin by 2-hydroxy-5-nitro-α-toluenesulfonic acid sultone agrees well with inhibition constants found using other chymotrypsin substrates known to react at the active site of the enzyme.[2] A second type of evidence implicating Ser_{195} rather than His_{57} as the active site residue which is sulfonylated by the sultone comes from spectrophotometric titration of the phenolic hydroxyl group and from the kinetics of desulfonylation of the sulfonyl enzyme (steps $k_{-2} + k_3$ of Eq. 1). The titration behavior of the phenolic hydroxyl indicates that the ionization of this residue is perturbed by the ionization of another residue on the enzyme which is almost certainly the imidazole moiety of His_{57}. Furthermore, the pH dependence of the desulfonylation reaction clearly implicates the participation of a histidine residue in this process. In addition to these observations supporting the assignment of

[3] G. R. Schonbaum, B. Zerner, and M. L. Bender, J. Biol. Chem. 236, 2930 (1961).

Ser$_{195}$ as the site of covalent binding of the sulfonyl group, there is a third kind of evidence based on a comparison of the behavior in acidic solution of 2-hydroxy-5-nitro-α-toluenesulfonyl-α-chymotrypsin with that of α-toluenesulfonyl-α-chymotrypsin.[4] In both of these cases when the sulfonyl enzymes were desulfonylated at low pH and elevated temperature (pH 1.9 at 40°, for example) and the pH was quickly adjusted to 7 and the temperature to 25°, the desulfonylated enzyme species obtained were nearly inactive initially against the specific substrate N-acetyl-L-tryptophan methyl ester, and they slowly regained activity. Whatever the mechanism of these reactions is, this parallel behavior of the sulfonyl enzymes offers convincing evidence that the enzyme is sulfonylated at the same site in both sulfonyl chymotrypsins. Since a variety of evidence (including X-ray crystallographic data[5]) exists demonstrating the attachment of the sulfonyl group in α-toluenesulfonyl-α-chymotrypsin to Ser$_{195}$, we conclude that the sulfonyl group in 2-hydroxy-5-nitro-α-toluenesulfonyl-α-chymotrypsin is attached to the same site in the enzyme.

Kinetic Considerations

To establish thoroughly the validity of an active site titration, it is necessary to determine the kinetics of the reaction between the titration reagent and the enzyme. This point will now be illustrated for the titration of chymotrypsin with 2-hydroxy-5-nitro-α-toluenesulfonic acid sultone.

Under conditions of $S_0 \gg E_0$, assuming the reaction scheme of Eq. (1), the relationship between ES′ and E_0 (the total concentration of active enzyme in solution) is given by Eq. (2).[6] When titrations with 2-hydroxy-5-nitro-α-toluenesulfonic acid

---

[4] A. M. Gold and D. E. Fahrney, *Biochemistry* **3**, 783 (1964); A. M. Gold, *Biochemistry* **4**, 897 (1965); D. E. Fahrney, Ph.D. thesis, Columbia University, New York, 1963.

[5] B. W. Matthews, P. B. Sigler, R. Henderson, and D. M. Blow, *Nature* **214**, 652 (1967).

[6] Equation (2) can be derived as follows. From Eq. (1), $E = ES(K_s/S)$. The total enzyme concentration $E_0 = E + ES + ES'$. Thus,

$$ES = \frac{E_0 - ES'}{1 + K_s/S}$$

Making the stationary state assumption $(dES'/dt) = 0 = k_2ES - ES'(k_{-2} + k_3)$, the expression

$$ES' = \frac{k_2(E_0 - ES')}{(k_{-2} + k_3)(1 + K_s/S)}$$

can be obtained. Rearrangement of this equation leads to Eq. (2).

$$\frac{(ES')}{(E_0)} = \frac{\dfrac{k_2}{k_{-2} + k_3}}{\left(1 + \dfrac{K_s}{S}\right) + \dfrac{k_2}{k_{-2} + k_3}} \tag{2}$$

sultone are done according to the titration procedure which will be described later, the ratio $(ES')/(E_0)$ is very close to 1. In other words, the titration reagent measures the active site concentrations of chymotrypsin solutions with great accuracy. Since the kinetic parameters of Eq. (2) are known, we can be confident of this point as can be shown by a sample calculation. At pH 5.9, for instance, the value of $k_{-2} + k_3 \approx$ 2.5×10^{-3} sec^{-1}, $k_2 \approx 10$ sec^{-1} and $K_s \approx 5 \times 10^{-4}\,M$.[7] At a typical substrate concentration of $5 \times 10^{-5}\,M$ used in a titration, the ratio $k_2/(k_{-2} + k_3)$ is about 4×10^3 which is enormously larger than the quantity $1 + (K_s/S)$. Thus, the titration efficiency is considerably better than one part in a hundred.

Comparison of 2-Hydroxy-5-nitro-α-toluenesulfonic Acid Sultone with trans-Cinnamoyl Imidazole as an α-Chymotrypsin Titrant

To justify the use of any new titrant, a comparison should be made to the use of other titrants, if they exist. Currently, *trans*-cinnamoyl imidazole which was introduced as a chymotrypsin active site titrant by Bender in 1961[3] is the most widely used reagent for this purpose. The availability of the *trans*-cinnamoyl imidazole titration method has contributed greatly to the rapid progress made in understanding the chemistry of chymotrypsin-catalyzed reactions. A calculation similar to that which we have performed for the sultone can be done to show that at acidic pH values (pH 5, for example) *trans*-cinnamoyl imidazole is a very effective titrant for chymotrypsin active sites. What then are the advantages to using the sultone?

First, the sultone can be employed as a titrant over a much wider pH range than *trans*-cinnamoyl imidazole. Chymotrypsin can be titrated with the sultone over the pH range 5–7.5 with no difficulty since the sulfonyl enzyme is quite stable. At pH 7.27 (0.01 M phosphate buffer), for example, the first-order rate constant[7] observed for the formation of the hydrolysis product 2-hydroxy-5-nitro-α-toluenesulfonic acid from the decomposition of 2-hydroxy-5-nitro-α-toluenesulfonyl-α-chymotrypsin is 8.1×10^{-4} sec^{-1}. On the other hand, *trans*-cinnamoyl-α-chymotrypsin is considerably less stable in the neutral pH range. At pH 7.39 in phosphate buffer the first-order rate constant for the formation of *trans*-cinnamic

[7] J. H. Heidema, Ph.D. thesis, Department of Chemistry, University of Chicago, Chicago, Illinois, 1969.

acid from *trans*-cinnamoyl-α-chymotrypsin[8] is 7.8×10^{-3} sec^{-1}, a value about ten times greater than that for the production of the sulfonic acid quoted above. Even above pH 7.5 it is possible to use the sultone, although the rapidity of its nonenzymatic hydrolysis begins to be a problem.

A second advantage to using the sultone is that it is easier to purify and store than *trans*-cinnamoyl imidazole. The sultone is readily purified by crystallization from ethanol and can be stored indefinitely. However, it is hard to remove the last traces of impurity from *trans*-cinnamoyl imidazole which can be crystallized from cyclohexane or hexane. Furthermore, the stability of this compound on storage is not good.

Finally, titrations with 2-hydroxy-5-nitro-α-toluenesulfonic acid sultone can be done at 390 nm, a wavelength where enzyme absorption is negligible. In the case of *trans*-cinnamoyl imidazole, if titrations are done at 310 nm, the enzyme absorption cannot be neglected (although it is not very significant at 335 nm, another recommended wavelength for titrations with *trans*-cinnamoyl imidazole).

TABLE I

CONCENTRATION OF A CHYMOTRYPSIN STOCK SOLUTION AS DETERMINED BY TITRATION WITH DIFFERENT REAGENTS AT VARIOUS pH VALUES

Reagent	pH	Wave-length (nm)	Buffer	Enzyme conc.[a] $M_E \times 10^4$ (M)
trans-Cinnamoyl	5.05	335	0.10 M Acetate	9.87[b]
imidazole	5.05	335	0.10 M Acetate	9.96[b]
	6.23	335	0.08 M 2,6-Lutidine (NO$_3$$^-$)	9.85
2-Hydroxy-5-nitro-α-	5.05	321	0.10 M Acetate	9.44[c]
toluenesulfonic	6.84	390	0.08 M 2,6-Lutidine (NO$_3$$^-$)	9.61
acid sultone	7.48	390	0.10 M Tris (NO$_3$$^-$)	9.58

[a] Average of duplicate determinations.
[b] These titrations at pH 5.05 were performed a week apart on a CT solution stored in a refrigerator at 2°.
[c] Average of triplicate determinations.

Results of titrations of chymotrypsin solutions with 2-hydroxy-5-nitro-α-toluenesulfonic acid sultone and *trans*-cinnamoyl imidazole are compared in Table I. The agreement between the active site molarities determined by the two methods is excellent.

[8] M. L. Bender, G. R. Schonbaum, and B. Zerner, *J. Am. Chem. Soc.* **84**, 2562 (1962).

Titration Procedure[9]

The chymotrypsin (Worthington Biochemical Corp., Freehold, N.J., thrice crystallized) stock solution[10] employed was about $1.0 \times 10^{-3} M$ in active enzyme. The stock solution of the substrate was $3.0 \times 10^{-3} M$ in 2-hydroxy-5-nitro-α-toluenesulfonic acid sultone and was prepared in acetonitrile solvent.

To 3.00 ml of the desired buffer in a 1-cm pathlength rectangular spectrophotometer cuvette was added 100 μl of chymotrypsin stock solution. The absorbance A_1 of the resultant solution at the appropriate wavelength served as a baseline. Then 50 μl of the reagent stock solution was added with stirring. The new absorbance of the solution, A_2, was determined by linear extrapolation of the observed absorbance back to the time of reagent addition.[11] When this procedure is used, Eq. (3)[12] gives the molarity of enzyme active sites in the stock solution, M_E. $\epsilon_{ES'}$

[9] This titration procedure has been described by Dr. J. H. Heidema. See footnote 7. See also J. H. Heidema and E. T. Kaiser, *Chem. Commun.* p. 300 (1968).

[10] Stock solutions of chymotrypsin were usually prepared by dissolving 0.300 g of the commercially obtained, three times crystallized and lyophilized enzyme in 10.0 ml of pH 5.05 acetate buffer. Such a solution had a chymotrypsin active site molarity of about $1.0 \times 10^{-3} M$. Stock solutions of chymotrypsin were always stored in a refrigerator at $2°$.

[11] Absorbance measurements were made on a Cary 15 spectrophotometer. A chart speed of 10 seconds per division (2 inches/minute) was used. After the immediate absorbance burst there is a slow nearly linear change in absorbance with time because of the slow solvent-catalyzed desulfonylation of the sulfonyl enzyme. On the time scale employed for the titrations these slow reactions cause a virtually constant $\Delta A/\Delta t$ and thus justify a simple linear extrapolation of the absorbance to zero time to obtain A_2 ($A_2 = A_1$ plus the absorbance burst). At pH 5 these subsequent reactions are so slow that $\Delta A/\Delta t$ is essentially negligible. At pH values above 7.5 they become sufficiently fast that $\Delta A/\Delta t$ is no longer virtually constant and extrapolation to zero time must therefore be nonlinear.

[12] Equation (3) results from the following considerations. Immediately after the absorbance burst the solution contains only the sulfonyl enzyme, ES' and the excess sultone S. Thus $A = \epsilon_{ES'} C_{ES'} + \epsilon_S C_S + \epsilon_E C_E$ where C_X represents the concentration of X in the solution. The absorbance caused by the enzyme protein must be included since $\epsilon_{ES'}$ actually represents the difference in extinction coefficients between ES' and E. (At the wavelengths chosen $\epsilon_E C_E$ was very small, always less than 0.01.) Then, considering the various dilutions of the stock solutions in the procedure employed

$$A_2 = \epsilon_{ES'} \times 10 M_E/315 + \epsilon_S(M_S/63 - 10 M_E/315) + 310 A_1/315$$
$$A_2 = M_E \times 10(\epsilon_{ES'} - \epsilon_S)/315 + 310 A_1/315 + \epsilon_S M_S/63$$

Therefore

$$M_E = \frac{A_2 - 310 A_1/315 - \epsilon_S M_S/63}{10(\epsilon_{ES} - \epsilon_S)/315} = \frac{A_2 - 62 A_1/63 - \epsilon_S M_S/63}{2(\epsilon_{ES} - \epsilon_S)/63}$$

and ϵ_{S} are extinction coefficients of the sulfonyl enzyme, 2-hydroxy-5-nitro-α-toluenesulfonyl chymotrypsin, and the titration reagent, respectively, and M_{S} is the molarity of the reagent stock solution. M_{S} is known from the method of preparation of the reagent stock solution. $\epsilon_{ES'}$ may be found by adding a known volume of the reagent stock solution to buffer in a cuvette at the desired pH containing a molar excess of enzyme. In the presence of excess enzyme the reagent will be converted entirely to the sulfonyl enzyme. From the absorbance burst ΔA observed, $\epsilon_{ES'}$ may be calculated simply by the use of Eq. (4). ϵ_{S} may be determined likewise, but employing buffer which does not contain enzyme. The value of ϵ_{S} observed at any given wavelength does not depend on pH. However, the absorbance of 2-hydroxy-5-nitro-α-toluenesulfonyl-α-chymotrypsin depends upon the extent of ionization of a nitrophenol chromophore which changes with pH. Thus $\epsilon_{ES'}$ must be determined at the exact pH employed for the titration. This can be accomplished simply by repeating the titration procedure in the same buffer with the addition of only 25 μl of the reagent stock solution. Under these conditions the enzyme remains in excess and $\epsilon_{ES'}$ can be calculated using Eq. (5).

$$M_{E} = \frac{A_2 - 62A_1/63 - (\epsilon_{S}M_{S}/63)}{2(\epsilon_{ES'} - \epsilon_{S})/63} \tag{3}$$

$$\epsilon_{ES'} = \Delta A/(M_{S} \times \text{dilution factor}) \tag{4}$$

$$\epsilon_{ES'} = \frac{A_2 - 124A_1/125}{M_{S}/125} \tag{5}$$

In practice some simplification of Eqs. (3) and (5) can be made. At all of the wavelengths used, A_1, the absorbance caused by the enzyme may be so small that the small adjustments in A_1, theoretically necessary because of dilution, can be neglected. Also, in titrations done at 390 nm (near neutral pH) ϵ_{S} can generally be neglected. Thus, under these conditions Eq. (3) can be simplified to Eq. (6) and Eq. (5) becomes Eq. (7).

$$M_{E} = 31.5(A_2 - A_1)/\epsilon_{ES'} \tag{6}$$

$$\epsilon_{ES'} = 125(A_2 - A_1)/M_{S} \tag{7}$$

$$M_{E} = (A_2 - A_1 - 33.2\,M_{S})/149.7 \tag{8}$$

When titrations were performed at pH 5.05, the reaction was observed at 321 nm, and Eq. (8) was employed. At this pH the nitrophenol chromophore of the sulfonyl enzyme is almost entirely protonated, and the value of $\epsilon_{ES'}$ at 321 nm, the absorption maximum for the nitrophenol chromophore, has been found to be rather insensitive to small changes in pH.

In titrations with 2-hydroxy-5-nitro-α-toluenesulfonic acid sultone at 390 nm, an alternative method for the determination of $\epsilon_{ES'}$ can be

TABLE II
Effect of pH on the Extinction Coefficient of
2-Hydroxy-5-nitro-α-toluenesulfonyl
Chymotrypsin at 390 nm[a]

pH	$\epsilon_{ES'}$	pH	$\epsilon_{ES'}$
6.37	3,492	7.62	9,276
6.55	4,300	7.76	9,734
6.87	5,849	7.95	10,720
7.15	7,246	8.29	11,850
7.41	8,315	8.71	12,960

[a] Dr. John Heidema has found $\epsilon_{ES'}$ to be independent of the various buffer systems used in his work.

used. If the pH of the titration medium is accurately determined, $\epsilon_{ES'}$ may be found from the data of Table II. The method using Eq. (7) is expected to be more accurate since it does not require a precise pH measurement, and it is unaffected by any differences in absorbance as measured on different spectrophotometers.[13]

Syntheses

Preparation of 2-Hydroxy-α-toluenesulfonic Acid Sultone.[14] Sodium 2-hydroxy-α-toluenesulfonate was prepared first. To 41.3 g (0.397 mole) of sodium bisulfite was added 49.2 g (0.397 mole) of 2-hydroxybenzyl alcohol (Eastman) and enough water to dissolve the mixture (ca. 900 ml). The solution was refluxed for 8 hours, most of the excess water was removed by distillation, and the remaining solution was evaporated to dryness on a rotary evaporator, yielding a white residue which was air dried. Continuous extraction of this residue with ethanol using a Soxhlet extractor gave 78.3 g (94% yield) of sodium 2-hydroxy-α-toluenesulfonate which was isolated as a flaky, white product.

The sultone was synthesized by the cyclization of sodium 2-hydroxy-α-toluenesulfonate using phosphorus oxychloride. To 40.0 g (0.190 mole) of sodium 2-hydroxy-α-toluenesulfonate was added 320 g (2.09 moles) of phosphorus oxychloride. No reaction occurred at room temperature, and the reaction mixture was heated slowly to 125°. Hydrogen chloride fumes started to evolve at 110°, and the mixture was refluxed at 125°.

[13] We recommend avoiding the use of phosphate buffers in the neutral pH range since these buffers catalyze the hydrolysis of 2-hydroxy-5-nitro-α-toluenesulfonic acid sultone.
[14] This synthetic procedure is taken from the Ph.D. thesis of Dr. O. R. Zaborsky, Department of Chemistry, University of Chicago, Chicago, Illinois, 1968.

Heating was continued for 1 hour, the excess $POCl_3$ was removed by distillation, and the cream-colored residue was allowed to cool.

The residue was ground and was transferred slowly into 600 ml of ice water. The white material was left in contact with the water for 4 hours, then suction filtered, thoroughly washed with cold water, and finally air dried. Crystallization from ethanol gave 17.6 g (54% yield) of pure, white crystals, m.p. 86.1–87.1.

2-Hydroxy-α-toluenesulfonic acid sultone is available commercially from Eastman (No. 10333).

Preparation of 2-Hydroxy-5-nitro-α-toluenesulfonic Acid Sultone.[15] 2-Hydroxy-α-toluenesulfonic acid sultone (2.40 g, 14.1 millimoles) was dissolved in 20 ml concentrated sulfuric acid, and the solution was cooled in an ice bath. To this solution 1.07 ml (1.52 g, 16.9 millimoles) of 70.3% nitric acid was added dropwise with stirring over a period of 15 minutes. The resultant solution was pale yellow and was allowed to stand in the ice bath for 10 minutes. Small ice cubes were slowly added until no further precipitate was obtained. The pale yellow precipitate was filtered on a sintered-glass funnel, washed with a small amount of ice-cold water and by suction. Recrystallization from ethanol gave very pale yellow crystals, m.p. 148.5–149.5°. The yield was 2.85 g (94%).

2-Hydroxy-5-nitro-α-toluenesulfonic acid sultone is now available commercially from Eastman (No. 10335). We recommend that the sultone be recrystallized twice from dry ethanol before use.

The Use of Rapidly Reversible Covalently Bound Substrates

Titration of α-Chymotrypsin with p-Nitrophenyl Ester Substrates

As pointed out in the introduction, optimal enzyme titrations should make use of a stoichiometric reaction between the enzyme and a specific substrate. The serine proteases offer an excellent opportunity for such a titration, since the reaction of the enzyme (E) with the substrate (S) comprises the formation of a covalent intermediate, the acyl enzyme (ES'), according to the reaction scheme:

$$E + S \overset{K_s}{\rightleftharpoons} ES \overset{k_2}{\to} ES' \overset{k_3}{\to} E + P_2 \tag{9}$$
$$+$$
$$P_1$$

where P_1 and P_2 are the alcohol and acid portions of the substrate respectively. It is readily apparent from this scheme that a quantitative trans-

[15] This preparation has been described by Dr. K. W. Lo, Ph.D. thesis, Department of Chemistry, University of Chicago, Chicago, Illinois, 1968.

formation of E into ES′ is only possible when $k_3 \ll k_2$ and when the substrate concentration is high enough to transform a reasonable amount of enzyme into the enzyme-substrate complex. It is also obvious that ES′ will exist in the solution only as long as there is an excess of substrate present. Thus both the amount of substrate and the value of k_3 will be important factors in limiting the time span during which the titration has to be performed. For these reasons it would seem that an ideal titrant would have $k_3 = 0$; i.e., instead of being a substrate, it would be an irreversible inhibitor. Unfortunately, it would be then impossible to ascertain whether active enzyme molecules are the only ones reacting with the titrant. Therefore, titrants with a small but measurable k_3 are preferred.

The reaction of titration is quantitated by measuring the change in some physical property of one of the participating molecular species. The spectrophotometric determination of a change in the visible or ultraviolet absorption spectrum is ideally suited for this purpose because of its accuracy and sensitivity, and is therefore presently preferred to any other method. The absorption spectra of E, ES, and ES′ are rather unsuitable for titrations with rapidly reversible covalently bound substrates because of the experimental difficulties involved in the determination of their molar absorptivity. Measurement of the disappearance of the substrate is also suboptimal, since the requirement of excess substrate means that one has to measure a small change of disappearing substrate against a large background of excess substrate, thereby decreasing the accuracy of the method. The measurement of the formation of P_1 seems to be most satisfactory, especially when there is a spectral region where P_1 alone absorbs. As early as 1954, p-nitrophenol was proposed as a P_1 of choice for titrations of α-chymotrypsin.[16] In its ionized form ($pK_a = 7.04$), p-nitrophenol has a high absorbance in the visible region of the spectrum ($\epsilon_{400} = 18,300$)[17] and, therefore, with a good spectrophotometer its concentration can be measured in a $10^{-6}\,M$ solution with less than 2% error. On the other hand, p-nitrophenyl esters have a negligibly small ϵ_{400}. In solutions of pH < 6, protonated p-nitrophenol has to be measured at wavelengths where the absorbance of the ester is not negligible. In this case the choice of the wavelength will be dictated by the optimal value of $(\epsilon_{phenol} - \epsilon_{ester})/\epsilon_{ester}$, which occurs at the neighborhood of $\lambda = 340$ nm for many p-nitrophenyl esters. Finally, p-nitrophenol is an excellent leaving group and its esters have high values of k_2/k_3 with chymotrypsin, while the esters are not too unstable in aqueous solutions at neutral pH values at room temperature.

[16] B. S. Hartley and B. A. Kilby, *Biochem. J.* **56**, 288 (1954).
[17] F. J. Kézdy and M. L. Bender, *Biochemistry* **1**, 1097 (1962).

When a great excess of a p-nitrophenyl ester reacts with the enzyme according to Eq. (9), the concentration of P_1 as a function of time can be expressed by an equation of the form.[18-20]

$$P_1 - P_{1_0} = At + \pi(1 - e^{-bt}) \qquad (10)$$

where

$$A = (k_{cat}E_0S_0)/(S_0 + K_m)$$
$$\pi = E_0[k_2/(k_2 + k_3)]^2/[1 + K_m/S_0]^2$$
$$b = [(k_2 + k_3)S_0 + k_3K_s]/(S_0 + K_s)$$

$$k_{cat} = k_2k_3/(k_2 + k_3)$$
$$K_m = K_sk_3/(k_2 + k_3) \qquad (11)$$

Equation (10) states that for values of $t > 5/b$ the production of P_1 will be linear with time, i.e.,

$$P_1 - P_{1_0} = At + \pi \qquad (12)$$

The slope of this straight line is of course equal to the steady-state reaction rate ($=V$), from which $k_{cat}E_0$ ($=V_{max}$) and K_m can be calculated by routine procedures. Extrapolation of the line to $t = 0$ yields the value of π as the intercept on the P_1 axis. Thus, the determination of π is a very simple experimental procedure: After addition of the enzyme to the substrate solution one measures the steady-state production of P_1 for a short period of time. When the data are plotted as P_1 vs. t, a straight line should be obtained which intersects the P_1 axis at π. This graphically extrapolated value of π is directly proportional to the active site concentration, according to Eq. (11).

In the simplest possible case, when $k_2 \gg k_3$ and $S_0 \gg K_m$, π is equal to E_0. If these simplifying conditions are not fulfilled, then determination of π at several initial substrate concentrations may be necessary. Since Eq. (11) can be transformed into the form

$$1/\sqrt{\pi} = (k_2 + k_3)/k_3 \sqrt{E_0} + (K_s/\sqrt{E_0})(1/S_0) \qquad (13)$$

a plot of $1/\sqrt{\pi}$ vs. $1/S_0$ will yield $(k_2 + k_3)/k_3\sqrt{E_0}$ as the intercept on the axis of the ordinates. If $k_2 \gg k_3$, then E_0 is determined directly from this plot.

K_m is readily measured from turnover kinetics and thus a numerical

[18] H. Gutfreund and J. M. Sturtevant, *Biochem. J.* **63**, 656 (1956).
[19] M. L. Bender, F. J. Kézdy, and F. C. Wedler, *J. Chem. Educ.* **44**, 84 (1967).
[20] For the more complicated case where the formation of ES from E and S cannot be treated as a rapid equilibrium, the reader is referred to the complete steady-state derivation by L. Ouellet and J. A. Stewart, *Can. J. Chem.* **37**, 737 (1959).

correction of π also would lead to the correct value of active site concentrations:

$$E_0 = \pi(1 + K_m/S_0)^2 \tag{14}$$

It is much more difficult to assess the value of k_3/k_2. For the limited case of nonspecific substrates, when k_3 is small, isolation of ES′ is possible by rapid gel filtration of the reaction mixture. Then the kinetics of decomposition of ES′ will yield k_3, and hence from the definition of k_{cat} we can calculate $k_3/k_2 = (k_3/k_{cat}) - 1$. But in general, detailed kinetic study of both the steady state and the presteady state will be necessary in order to determine k_2 and k_3 individually before k_3/k_2 can be calculated.

Experimental Methods for Titrations with Nitrophenyl Esters

In the following discussion a few representative titrations of α-chymotrypsin by p-nitrophenyl esters will be described. These procedures were chosen because of their simplicity and the wide range of experimental conditions and specificity of the titrants.

Materials

p-Nitrophenyl acetate was synthesized by adding a stoichiometric amount of acetyl chloride in chloroform to a dioxane suspension of sodium p-nitrophenolate. The sodium chloride formed is removed by filtration and the solvent evaporated. The ester is recrystallized from chloroform–hexane, m.p. 80.0°.

p-Nitrophenyl trimethylacetate was synthesized similarly; recrystallization from ethanol yielded colorless crystals, m.p. 95°. p-Nitrophenyl N-benzyloxycarbonyl-L-tyrosinate was a product of Mann Laboratories, Inc. p-Nitrophenyl N-acetyl-DL-tryptophanate was purchased from Cyclo Chemical Co. The commercially available esters were used without further purification. Substrate stock solutions were prepared by dissolving an appropriate amount of substrate in acetonitrile.

α-Chymotrypsin (Worthington Biochemical Corp., Freehold, New Jersey) crystallized three times, was dissolved in 0.02 M acetate buffer, pH 4.0, or 0.05 M citrate buffer, pH 3.2, and stored on ice.

Titration Procedure

Three milliliters of the appropriate buffer is pipetted into a 1-cm pathlength cuvette and placed into the thermostatted cell holder of a recording spectrophotometer. Then, 50 μl of the substrate stock solution is added on the flattened tip of a glass rod, and the solution is thoroughly mixed. The absorbancy of the solution is recorded for a short

TABLE III

TITRATION OF α-CHYMOTRYPSIN BY p-NITROPHENYL ESTERS

Titrant, p-nitrophenyl ester of	pH	Buffer	Wavelength (nm)	S₀ Stock solution (mg/ml)	Optimal E_conc. in reaction mixture (mg/ml)	Δε (AU/M)	f	Accuracy (%)	Ref.[a]
N-Benzyloxycarbonyl-L-tyrosine	2.2	0.05 M Citrate	340	0.54	0.05	6014	1.00	±2	21
N-Acetyl-DL-tryptophan	3.2	0.05 M	340	4.55	0.05	6360	1.05	±3	21
Acetic acid	7.8	0.067 M Phosphate	400	11.3	0.3	15,600	1.05	±2	22
Trimethylacetic acid	8.2	0.03 M Tris-HCl	400	0.78	0.3	18,000	1.02	±2	19

[a] Reference numbers refer to footnotes cited in the text.

period of time, in order to allow graphical extrapolation of the absorbance to the time when the enzyme will be added. This extrapolated value will be designated as A_s.

At time zero an aliquot (20–100 μl) of the enzyme stock solution is added (V_e) to the reaction mixture and the time recorded. The solution is rapidly mixed, and recording of the absorbance of the solution is started at a known time t after addition of the enzyme at an appropriate wavelength. When a sufficient portion of the linear steady-state phase has been recorded, the value of π is measured by extrapolating graphically the absorbance vs. time curve to $t = 0$. This yields A_0. The concentration of active sites in the enzyme stock solutions can then be calculated from the following equation:

$$E_0^{\text{stock}} = \frac{3.05 + V_e}{V_e} \times f \times \frac{A_0 - A_s(3.05/3.05 \times V_e)}{\Delta\epsilon} \qquad (15)$$

The corrective factor f is necessary because with some of the substrates the condition $S_0 \gg K_m$ is not fully satisfied. The value of f is calculated from the known value of K_m:

$$f = (1 + K_m/S_0)^2 \qquad (16)$$

Experimental conditions and the accuracy of the measurements on a Cary 14 or 15 recording spectrophotometer are reported for four representative p-nitrophenyl esters in Table III.

Discussion

Titration of α-chymotrypsin by the four substrates described yields values of concentrations of active sites which are in excellent agreement with each other and also with results obtained by other methods, such as titration by *trans*-cinnamoyl imidazole. For all these substrates it has been ascertained by kinetic studies that Eq. (11) is indeed the correct description of the phenomenon and that the value of k_2/k_3 is much larger than 100. Therefore, the correction due to the factor $(1 + k_3/k_2)^2$ is negligible. Because of the high sensitivity and specificity of the substrate, titration of α-chymotrypsin by p-nitrophenyl N-benzyloxycarbonyl-L-tyrosinate is recommended whenever a recording spectrophotometer is available. When the absorbances are measured manually, p-nitrophenyl acetate will be a more appropriate titrant because of its much slower turnover rate.

---

[21] F. J. Kézdy, G. E. Clement, and M. L. Bender, *J. Am. Chem. Soc.* **86,** 3690 (1964).
[22] M. L. Bender *et al., J. Am. Chem. Soc.* **88,** 5890 (1966).

Limitations of the Method

When *p*-nitrophenyl CBZ-tyrosinate is used as a titrant, the pH of the enzyme stock solution should be smaller than 3.5 or greater than 6.0 in order to avoid artifacts due to the slow dedimerization of the chymotrypsin dimer which occurs in the pH range of 4 to 5. Finally, it cannot be over emphasized that titration with even the most specific substrate *will not* distinguish between closely related enzymes, such as α-, β-, and δ-chymotrypsin, or even chymotrypsin and trypsin. The evaluation of the concentration of individual species in such mixtures requires a lengthy kinetic analysis of the data and is far more time consuming than a routine titration. In the case of enzyme mixtures an approach based on the use of selective inhibitors can be used as a general technique to differentiating the active site concentrations of closely related enzyme species.[23]

Conclusions

Historically, titration procedures for the determination of active site concentrations were first developed for chymotrypsin. Since then, in the past decade a great many papers have been devoted to finding newer and better methods and titrants not only for chymotrypsin but also for a host of other enzymes. It is the hope of the authors that the time is not far when enzyme concentrations will be routinely expressed in units of active site concentrations, and when the use of arbitrary enzyme units will be considered as an oddity of the past.

[23] L. Lorand and F. V. Condit, *Biochemistry* **4**, 265 (1965).

[2] Titration of Trypsin, Plasmin, and Thrombin with *p*-Nitrophenyl *p'*-Guanidinobenzoate HCl[1]

By Theodore Chase, Jr., and Elliott Shaw

The theory and desirability of titration methods for proteolytic enzymes has been described.[1a, 2] In the case of serine proteinases, advantage is taken of the fact that in the normal catalytic process an acyl enzyme is formed as an intermediate (Fig. 1). Titrants have been de-

[1] Research carried out at Brookhaven National Laboratory under the auspices of the U.S. Atomic Energy Commission.
[1a] F. J. Kézdy and E. T. Kaiser, see this volume [1].
[2] M. L. Bender, M. L. Begue-Canton, R. L. Blakeley, L. J. Brubacher, J. Feder, C. R. Gunter, F. J. Kézdy, J. V. Kilheffer, Jr., T. H. Marshall, C. G. Miller, R. W. Roeske, and J. K. Stoops, *J. Am. Chem. Soc.* **88**, 5890 (1966).

$$E + S \rightleftharpoons E \cdot S \xrightarrow{k_2} \text{Acyl enzyme} \xrightarrow{k_3} \text{Acid} + E$$
$$\text{Complex}$$

FIG. 1

veloped which also utilize acyl enzyme formation. However, in these cases, deacylation is very slow and the amount of alcohol released is just equivalent to the active enzyme, providing the reagent has enough affinity to convert all active enzyme to acyl enzyme. These ideal conditions are rarely met in practice. The use of a light-absorbing or chromogenic alcohol is the analytical basis of the method.

A number of titrants for trypsin have been suggested: p-nitrophenyl N^α-benzyloxycarbonyl-L-lysinate HCl[2,3] (also used with thrombin[4]), N^α-methyl-N^α-(p-toluenesulfonyl)-L-lysine β-naphthyl ester,[5a] p-nitrophenyl-N^2-acetyl-N^1-benzylcarbazate,[5b] p-nitrophenyl p'-guanidinobenzoate HCl (p-NPGB)[6,7] and p-nitrophenyl-p'-amidinobenzoate HCl.[8] Of these, only the last two afford the high sensitivity of p-nitrophenyl esters usable above the pK_a of the product p-nitrophenol (molar extinction coefficient of p-nitrophenoxide 18,300 at 402 nm,[2] as against 6000 for the un-ionized compound[2]). p-NPGB is preferable for the much slower rate of hydrolysis of the resultant acyl enzyme[7] as well as the greater ease of synthesis of the compound. In addition, because of its high affinity, determinations are not dependent on reagent concentration when an adequate amount is used as described below. Its use has been extended to the titration of the proteolytic enzymes of the blood clotting system, thrombin[7,9] and plasmin (Fig. 2).[7]

Reagents

Buffer. Sodium Veronal, 0.1 M, pH 8.3; for the titration of trypsin or human thrombin the buffer should also be 0.02 M in $CaCl_2$. Sodium Veronal (sodium barbital, sodium diethylbarbiturate) is dissolved to a concentration of slightly over 0.1 M (20.6 g/liter), titrated carefully to pH 8.3 with HCl (the free acid is sparingly

[3] M. L. Bender, J. V. Kilheffer, Jr., and R. W. Roeske, *Biochem. Biophys. Res. Commun.* 19, 161 (1965); M. L. Bender, F. J. Kézdy, and J. Feder, *J. Am. Chem. Soc.* 87, 4953 (1965).

[4] F. J. Kézdy, L. Lorand, and K. D. Miller, *Biochemistry* 4, 2302 (1965).

[5] D. T. Elmore and J. J. Smyth, *Biochem. J.* 107, (a)97, (b)103 (1968).

[6] T. Chase, Jr. and E. Shaw, *Biochem. Biophys. Res. Commun.* 29, 508 (1967).

[7] T. Chase, Jr. and E. Shaw, *Biochemistry* 8, 2212 (1969).

[8] K. Tanizawa, S. Ishii, and Y. Kanaoka, *Biochem. Biophys. Res. Commun.* 32, 893 (1968).

[9] J. R. Baird and D. T. Elmore, *Federation European Biochem. Soc. Letters* 1, 843 (1968).

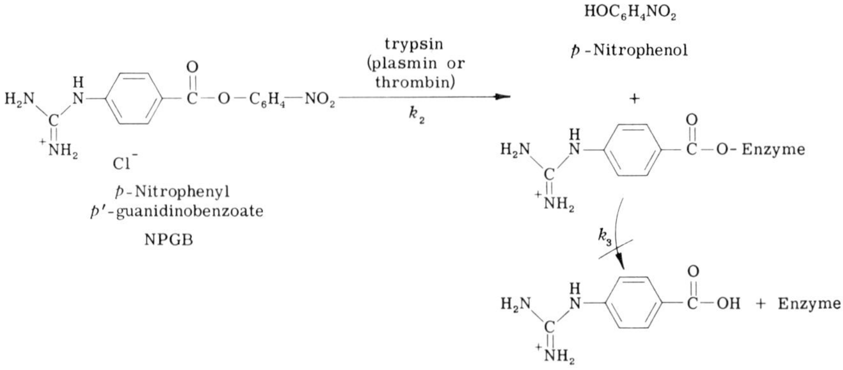

Fig. 2

soluble, and will precipitate if the pH is allowed to drop to 8.0) and diluted to volume. It should be stored at 4°.

p-NPGB. The solid reagent is dissolved in dimethylformamide to a concentration of 0.05 M (16.8 mg/ml), diluted with 4 volumes of acetonitrile (final concentration of 0.01 M), and stored at 4°; it is stable for weeks in this solution (more stable than in dimethylformamide alone, as originally described[6]), and may be used until the solution is visibly yellow.

Synthesis of p-NPGB

p-Guanidinobenzoic Acid Hydrochloride.[10] p-Aminobenzoic acid, 27.4 g (0.2 mole), and 22.8 g NH_4SCN (0.3 mole) are dissolved in 185 ml $H_2O + 20$ ml concentrated HCl. The solution is evaporated to dryness (about 4 hours) in an evaporating dish on a steam bath in the hood. The residue is dissolved in excess NaOH and the product, p-thioureidobenzoic acid, precipitated by addition of excess HCl and chilling. The collected product is dried overnight; yield, 90%.

The dried p-thioureidobenzoic acid is suspended in 200 ml absolute ethanol with 45.6 g CH_3I (about 1.8 equivalents). The suspension is refluxed for $1\frac{1}{4}$ hour in the hood, chilled, and some ether added to complete precipitation. Yield of dried p-(S-methylisothioureido)benzoic acid hydroiodide is 36.9 g (61%).

This material is refluxed with 75 ml concentrated NH_4OH *in the hood* for 3 hours; some precipitation of the zwitterion of p-guanidinobenzoic acid occurs. After cooling, the pH is adjusted to about 6 to ensure full precipitation of the zwitterion. The precipitate is filtered off and dis-

[10] H. C. Beyerman and J. S. Bontekoe, *Rec. Trav. Chem. Pays-Bas* **72**, 643 (1953).

solved in hot 1 N HCl for crystallization; it should be recrystallized at least once to ensure removal of iodide. Yield of dried p-guanidinobenzoic acid hydrochloride is 13.5 g (58%; 31.5% from p-aminobenzoic acid); m.p. 268°.[10]

p-NPGB. p-Guanidinobenzoic acid HCl (1.075 g, 5 millimoles) is dissolved in 7.5 ml dimethylformamide, and dicyclohexylcarbodiimide (1.083 g, 5 millimoles) is dissolved in 7.5 ml pyridine or tetrahydrofuran. The solutions are mixed and p-nitrophenol (0.725 g, 5.2 millimoles) is added. After standing overnight at room temperature the precipitated dicyclohexylurea is filtered off and the solvents removed in vacuo. The oily residue is taken up in 25 ml 0.1 N HCl and the slurry extracted repeatedly with 25-ml aliquots of ethyl acetate, with separation of the phases by centrifugation. Extraction of the residual p-nitrophenol is monitored by shaking the ethyl acetate extract with a few milliliters of weak base (Tris buffer, pH 8); when the Tris extract is only faintly yellow the extraction is complete. The slurry of p-NPGB in HCl is then dissolved in 300 ml water-saturated t-amyl alcohol and any residue filtered off; the t-amyl alcohol solution, after extraction with two 10-ml aliquots of 0.1 N HCl to remove p-guanidinobenzoic acid HCl, is evaporated to dryness below 40°. The residue is dissolved in minimum volume of boiling glacial acetic acid and crystallizes out on cooling (a little ether may be added to the cool solution to prevent freezing at 4°); yield, 68–72%, giving 98–99% of the expected amount of p-nitrophenoxide on hydrolysis in 0.1 N NaOH. The product shows some discoloration and decomposition at about 260°, but does not melt up to 300° except when heated rapidly.

Titration

The following instructions are for use of 1-ml cells and a double-beam recording spectrophotometer, in which nonenzymatic breakdown of p-NPGB can be automatically subtracted by adding an equal amount of the compound to the reference cuvette. A single-beam recording spectrophotometer can be used and the nonenzymatic hydrolysis subtracted manually; larger cells can be used, with suitable multiplication of the volumes of buffer, enzyme, and p-NPGB.

Veronal buffer (1.0 M) is pipetted into the reference cuvette, and enzyme sample and buffer are pipetted into the sample cuvette to a volume of 1.0 ml. (If the enzyme is at low concentration in Tris or other nucleophilic buffer, it may be desirable to include an equal amount of buffer in the reference cuvette, so that nonenzymatic hydrolysis of p-NPGB catalyzed by the buffer will be the same in both cuvettes.) The cuvettes are placed in the spectrophotometer and the instrument (set at

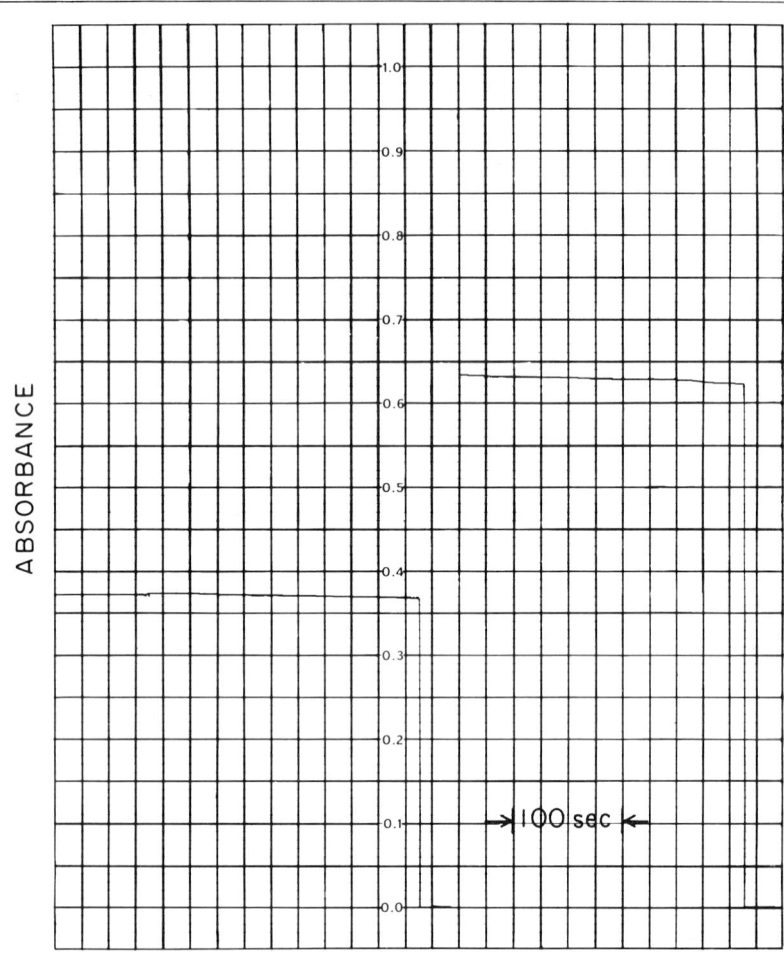

FIG. 3. Titration of (left) 0.05 ml and (right) 0.08 ml of a solution of β-trypsin. The procedure is described in the text. Extrapolation of the absorbance to time zero, which is the chart line just to the right of the initial rise in absorbance, corresponds to enzyme concentrations of $2.22 \times 10^{-5} M$ and $3.76 \times 10^{-5} M$, respectively. The stock solution is thus $4.58 \pm 0.13 \times 10^{-4} M$.

402 or 410 nm) zeroed, using the 0–1 slide wire for expected enzyme concentration (in the cuvette) 0.7–$5 \times 10^{-5} M$, 0–0.1 slide wire for enzyme concentration 0.1–$0.7 \times 10^{-5} M$; the chart is run to a point 10 seconds beyond a chart line. The cuvettes are then removed from the instrument; 5 μl p-NPGB solution is added to the reference cuvette. The solutions are mixed by inversion and the cuvette replaced in the instrument. As quickly

as possible, 5 μl of p-NPGB solution are added to the sample cuvette, the solutions mixed and the cuvette replaced, and the instrument turned on. The optical density increased in an immediate burst, then slowly (see footnotes 2, 6, and 7 for illustrations). The response of trypsin is shown in Fig. 3. The instrument is allowed to run until a straight line of post-burst nitrophenoxide production is established. This line is extrapolated back to the chart line 10 seconds before the burst (to allow for postburst nitrophenoxide production during the mixing and insertion into the instrument, which takes about 10 seconds). The concentration of active enzyme in the cuvette is then calculated from the increase in optical density thus determined and the molar extinction coefficient of p-nitro-phenol at this pH and wavelength.

Considerations

A pH of 8.3 was selected as the pH of titration in a compromise between two considerations: the instability of both trypsin and p-NPGB at higher pH, and the rapid change of the operational extinction coefficient at lower pH (pK_a of p-nitrophenol, 7.2).[11] At pH 8.3 p-nitrophenol is approximately 92% ionized; not only is the sensitivity of the titration near maximal, but only small errors are introduced by small changes of pH (± 0.2 units) caused by the buffering effect of a buffered sample. At a lower pH not only is the sensitivity of the titration less because the product is present partially in the un-ionized form, but the pH must be controlled very tightly so that the correct operational extinction coefficient is used in calculation of the amount of p-nitrophenol released in the burst.[11] If the enzyme sample is buffered at a pH other than 8.3 and the volume or buffer concentration of the sample is large, it may be desirable to check the pH of the enzyme–Veronal mixture and if necessary correct the pH to 8.3 or determine the extinction coefficient of p-nitrophenol at this pH. The authors find that up to 0.5 ml trypsin in $10^{-3}\,M$ HCl 0.02 M CaCl$_2$ can be titrated in a total volume of 1.0 ml without a significant pH drop below 8.3.

Veronal was selected as the preferred buffer for the titration because the rate of nonenzymatic hydrolysis of p-NPGB is lower in it (k_0 in 0.1 M Na-Veronal, pH 8.3, 3.3×10^{-5} sec^{-1}) than in other buffers of useful capacity at this pH (k_0 in Tris-Cl, 0.1 M, pH 8.3, 18.3×10^{-5} sec^{-1};

---

[11] At pH values below 6 the now completely protonated product p-nitrophenol can be measured without severe pH effects, although the extinction coefficient is lower ($\epsilon_{340} = 6000$) and the acylation slower. This situation has been utilized (J. F. Hruska, J. H. Law, and F. J. Kézdy, *Biochem. Biophys. Res. Commun.* **36**, 272 (1969)] to demonstrate the heterogeneity of trypsin preparations by observation of several different rate constants of acylation by p-NPGB.

in 0.1 M K-borate, pH 8.6, 18.0×10^{-5} sec^{-1}) ; although the nonenzymatic hydrolysis is blanked out by adding p-NPGB to the reference cuvette as well as to the sample cuvette, any inequality of such addition will result in an erroneous rate of postburst nitrophenol production and possible error in the extrapolation to the time of addition of the reagent, and this error will be magnified if the rate of nonenzymatic hydrolysis is high; also, in the above procedure the reagent is not added to the reference and sample cuvettes at the same instant, and the amount of p-nitrophenol produced in the reference cuvette before the reagent is added to the sample cuvette will be subtracted from the true burst in the sample cuvette, so it must be kept small.

After the immediate burst of p-nitrophenol produced by the stoichiometric reaction of enzyme active sites with p-NPGB, a slow further production of p-nitrophenol is observed, due either to hydrolysis of the acyl enzyme and reaction of the freed active site with a second molecule of p-NPGB ("turnover") or to nonspecific catalysis of p-NPGB hydrolysis by other nucleophilic groups of the enzyme protein or buffer ("nonspecific hydrolysis"). For discrimination between these, see footnote 7. The turnover rate is an inescapable property of the enzyme; but if nonspecific hydrolysis is rapid enough to make uncertain the determination of the burst by extrapolation to zero time, as is the case with many preparations of human plasmin, it may be minimized by using low concentrations of p-NPGB ($10^{-5} M$) and plasmin ($< 10^{-5} M$) and the 0–0.1 scale. The problem is not significant with trypsin (where both types of postburst nitrophenol production are discernible but small[7]) or either human or bovine thrombin (where only turnover has been detected).[7]

The theory of titration[1,2] shows that enzyme concentration is equal to the observed burst of p-nitrophenol π only if $k_2 \gg k_3$ and the titrant concentration is much greater than the apparent Michaelis constant. If these conditions are not fulfilled, as is the case with many titrants,[2] the size of the burst will be affected by the titrant concentration, and must be determined at several concentrations and extrapolated to infinite titrant concentration (plotting $1/\sqrt{\pi}$ vs. $1/[S]^{1,2}$) to determine the true concentration of active enzyme. Obviously, this consumes time, reagent, and enzyme; but the observation that no effect of p-NPGB concentration on burst size is observed down to $[p\text{-NPGB}] = 2 \times 10^{-6} M$ with trypsin, while with plasmin and thrombin the effect is observed only when $[p\text{-NPGB}] < 5 \times 10^{-6} M$, justifies titrating at only a single concentration ($5 \times 10^{-5} M$, or even $10^{-5} M$). For determination of the apparent k_m and relevant kinetic constants see footnote 7.

A reading of 410 nm was selected and maintained as the wavelength of observation of p-nitrophenol production for reasons of convenience in this laboratory; the absorbance maximum is actually 402 nm. In any case, it is desirable to determine the extinction coefficient with recrystallized p-nitrophenol in the instrument to be used; in our Cary 15 ϵ_{410} at pH 8.3 = 16,595, so that complete hydrolysis of $5 \times 10^{-5} M$ p-NPGB would lead to an optical density of 0.830.

Accuracy and Specificity

The accuracy and reproducibility of the method are probably limited principally by the accuracy of pipetting the enzyme solutions; if considerable care is taken in pipetting 0.100 or 0.200 ml enzyme, with ordinary pipettes rather than micropipettes, reproducibility of $\pm 2\%$ can be obtained (using the 0–1 slide wire), although $\pm 5\%$ is more usual; with the 0–0.1 slide wire accuracy is usually limited by instrument noise. Errors due to free p-nitrophenol in the p-NPGB solution and rapid nonenzymatic hydrolysis in the cuvette are in the double-beam instrument dependent only on the variation in volume of p-NPGB solution added.

p-NPGB has been used successfully for the titration of bovine trypsin, an insect trypsin,[12] bovine and human thrombin, and human plasmin; it presumably can be used for the titration of other closely related serine proteases, although not necessarily with all such, as shown by the observation that it is a normal substrate, not a titrant, for the C'1a esterase of guinea pig complement (David H. Bing, personal communication). With α-chymotrypsin p-NPGB gives a burst followed by rapid turnover; it is therefore not suited for discrimination between trypsin and chymotrypsin (or between plasmin and thrombin). p-NPGB does not react with subtilisin or papain.

---

[12] J. F. Hruska, J. H. Law, and F. J. Kézdy, *Biochem. Biophys. Res. Commun.* **36,** 272 (1969).

Section II

Individual Proteolytic Enzymes

A. The Serine Proteases
Articles 3 through 13

B. The Cysteine Proteases
Articles 14 through 18

C. Cathepsins
Article 19

D. The Acidic Proteases
Articles 20 through 30

E. Proteases Specific for Releasing COOH-Terminal Amino Acids
Articles 31 and 32

F. Proteases Specific for Releasing NH$_2$-Terminal Amino Acids
Articles 33 through 38

G. Some Bacterial and Mold Proteases
Articles 39 through 43

H. Collagenases
Articles 44 and 45

I. Other Proteases
Articles 46 through 53

J. Peptidases
Articles 54 through 58

[3] Serine Proteases

By K. A. WALSH and P. E. WILCOX

Introduction

Characterization of Serine Proteases

The serine proteases comprise a large group of enzymes which is distinguished by the reactivity of a serine residue in the active site.[1] A common test for these enzymes is the inhibition of their hydrolase activity by the reaction of this serine residue with diisopropylphosphorofluoridate (DFP).

Although the natural substrates of the serine proteases are proteins or fragments of proteins, most studies of these enzymes have been based on their hydrolase activity toward synthetic peptides, peptide amides, and esters. All members of the group may be classified as endopeptidases since it is generally observed that the cleavage of terminal peptide bonds is inhibited by the charge on the amino or carboxyl group of a terminal residue. The serine proteases also exhibit strong esterolytic activity toward esters analogous to the specific peptide substrates. This property is useful for kinetic assay methods to be described later in this section. In addition, hydrolytic activity extends to a variety of acyl compounds in which the acyl moiety is bonded to a good leaving group, such as *p*-nitrophenol or imidazole. Depending upon the nature of the acyl group and the pH of the solution, rapid acylation of the enzyme may be followed by very slow hydrolytic deacylation. Reactions such as these form the basis for measurements of functional normality of the serine proteases by methods described in Section I (see this volume, F. J. Kézdy and E. T. Kaiser [1]).

Comparisons of relative rates of catalysis using a variety of synthetic substrates have led to the definition of the specificities of representative members of the class. These range from the narrow specificity of the trypsins, which is directed toward the bond on the carboxyl side of arginine and lysine, to the rather broad specificity of the subtilisins which will attack bonds between a wide variety of amino acids. It should be noted, however, that specificity toward natural protein substrates may be influenced not only by the side chain next to the peptide bond, but also by neighboring residues in the sequence and the surrounding structure of the protein.

---

[1] B. S. Hartley, *Ann. Rev. Biochem.* **29,** 45 (1960).

More recently a quite different method has been utilized to further define their specificity, namely, irreversible inhibition by reaction with synthetic chloromethyl ketones which are analogs of peptide substrates.[2] The general structure may be written: X—NH—CH(R)—CO—CH₂Cl. Specificity is in large part determined by the side chain R; and X may be a blocking group, such as p-toluenesulfonyl or benzyloxycarbonyl, or may be one or more amino acid residues in a peptide chain blocked at the N-terminal.[3] The reaction of a specific chloromethyl ketone with the corresponding enzyme results in the stoichiometric alkylation of a histidine residue in the active site. These reactions provide a discriminating method for selectively blocking the action of one serine protease in the presence of another.

General Distribution and Occurrence of Zymogens

Serine proteases are widely distributed in nature. Historically their characteristic properties were first recognized among the digestive enzymes originating in the pancreas of mammals. More recently, examples of such digestive enzymes have been found throughout the vertebrate subphylum, in fact, in every vertebrate species investigated.[4,5] Furthermore, they are not restricted to digestive functions in the intestinal tract. Thrombin and plasmin, enzymes involved in the formation and dissolution of blood clots, show these characteristic properties, and serine proteases have also been detected in mammalian mast cells.[6]

The discovery of DFP-sensitive proteases extends to invertebrate phyla such as the Coelenterata (sea anemone, *Metridium senile*[7]), Echinodermata (starfish, *Evasterias trochelii*[8]), and the Arthropoda (hornet, *Vespa orientalis*[9]). The isolation of proteases from bacteria, such as the subtilisins from *Bacillus subtilus* and related strains, is well known. These subtilisins and enzymes isolated from quite different bacteria, *Streptomyces griseus*[10] and *Sorangium* sp.,[11] have also been characterized as serine proteases.

[2] E. Shaw, Vol. XI [80], p. 677.
[3] J. C. Powers and P. E. Wilcox, *J. Am. Chem. Soc.* in press, 1970.
[4] E. N. Zendzian and E. A. Barnard, *Arch. Biochem. Biophys.* **122,** 699 (1967).
[5] H. Neurath, K. A. Walsh, and W. P. Winter, *Science* **158,** 1638 (1967); L. B. Smillie, A. Furka, N. Nagabhushan, K. J. Stevenson, and C. O. Parkes, *Nature* **218,** 343 (1968).
[6] Z. Darzynkiewicz and E. A. Barnard, *Nature* **213,** 1198 (1967), and references contained therein.
[7] D. Gibson and G. H. Dixon, *Nature* **222,** 753 (1969).
[8] W. P. Winter and H. Neurath, *Federation Proc.* **27,** 492 (1968).
[9] H. H. Sonneborn, G. Pfleiderer, and J. Ishay, *Hoppe-Seylers Z. Physiol. Chem.* **350,** 389 (1969).
[10] S. Wahlby, *Biochim. Biophys. Acta* **151,** 394 (1968); *ibid.* **185,** 178 (1969).
[11] D. R. Whitaker and C. Roy, *Can. J. Biochem.* **45,** 911 (1967).

In vertebrates with a distinct pancreas, the proteases are present in the gland and in the exocrine juice as inactive precursors, or zymogens, which must be activated before they can be identified by enzymatic assays. All of the known zymogens of the serine proteases are converted into active enzymes by the action of trypsin.[12] Usually the cleavage of one specific peptide bond is sufficient for this activation. Trypsin itself arises from the activation of pancreatic trypsinogen by an autolytic mechanism; *in vivo* this process is initiated by an intestinal enzyme, enterokinase.

In teleost fishes, which do not have a distinct pancreas, serine proteases have been found in the pyloric ceca, but only in the active form.[4] It is likely that the detection and isolation of the corresponding zymogens have been prevented by technical difficulties. One elasmobranch, the spiny Pacific dogfish (*Squalus acanthias*) contains both a discrete pancreas and zymogens.[13]

The proteases involved with blood clots, thrombin and plasmin, also originate as zymogens and circulate as such in the blood. Although the conversion of these zymogens, prothrombin and plasminogen, to active enzymes may be accomplished with trypsin, activations are brought about *in vivo* by complex systems of blood and tissue enzymes.

Procedures for the preparation of gram quantities of the various serine proteases have generally taken advantage of rich and plentiful source materials such as fresh pancreases, acetone powders prepared from them, pancreatic juices, or large-scale bacterial fermentation broths. The classical preparations of bovine trypsin, chymotrypsins A and B, and the respective zymogens from fresh pancreases have been described in Volume II, [2] and [3].

In recent times the isolation and purification of enzymes have been greatly facilitated by ion-exchange chromatography. This is notably demonstrated by the methods presented in the following sections. Not only are these chromatographic methods applicable to the large-scale preparations which are given in detail; they are also useful for the isolation and characterization of proteases in smaller amounts of source material, whether it be some animal tissue, small organism, or bacterial culture.

Variants of Serine Proteases

Specific examination of the amino acid sequence of a variety of serine proteases has revealed common structural features indicating that they belong to a "family" of homologous proteins.[5] Each homologous structure within this group has a specific biological function, apparently

[12] H. Neurath, *Federation Proc.* **23**, 1 (1964).
[13] J. W. Prahl and H. Neurath, *Biochemistry* **5**, 2131 (1966).

acquired during a process of divergent molecular evolution from a common ancestral prototype. For example, specific serine proteases with common structural features are known which catalyze the hydrolysis of polypeptides for the varied biological purposes of digestion, blood clotting, clot lysis, sensing pain, and chemically opening insect cocoons. Some of these adaptations of an ancestral hydrolytic function have apparently occurred by selection during the evolution of species (e.g., trypsin in diverse animals). These may involve major changes in such features as the isoelectric point without alteration of the characteristic function of the protein. In contrast, in many other cases, several distinctly different but homologous enzymes occur within one species (e.g., trypsin, chymotrypsin, elastase, plasmin, and thrombin) and must have resulted from a gene duplication process preceding divergent molecular evolution.[14]

It is clear that more than one family of serine proteases exist since the subtilisins[15] bear no homologous relationship to the family of serine proteases discussed above, but are functionally serine proteases. Both the subtilisin family and the trypsin–chymotrypsin family of serine proteases operate by analogous mechanisms involving specific seryl and histidyl residues, but the amino acid sequences surrounding the serine and the histidine of one family bear no structural resemblance to those of the other family.

The occurrence of several separable serine proteases in one animal species, each catalyzing the hydrolysis of the same test substrate, can lead to confusion regarding their specific identification. This multiplicity may simply reflect an overlapping of specificity of distinctly different enzymes. Alternatively, it may denote the occurrence of several allotypes within a given population (e.g., in dogfish trypsin[16]) or it may indicate that active, but partly degraded, enzyme derivatives have resulted from autolytic or proteolytic degradation.

Autolyzed derivatives of enzymes have been characterized for both chymotrypsin and trypsin and almost surely exist for most serine proteases. In the case of bovine trypsin, chain cleavage without loss of function has been noted at either Arg_{105}-Val_{106}[17] or Lys_{131}-Ser_{132},[18] whereas inactive products are obtained by bond cleavage at Lys_{49}-Ser_{50}[18]

[14] G. H. Dixon, *Essays Biochem.* **2**, 147 (1966).

[15] E. L. Smith, F. S. Markland, C. B. Kasper, R. J. DeLange, M. Landon, and W. H. Evans, *J. Biol. Chem.* **241**, 5974 (1966).

[16] H. Neurath, R. A. Bradshaw, and R. Arnon, *Proc. Intern. Symp. Structure-Function Relationships of Proteolytic Enzymes, Munksgaard, Copenhagen,* in press, 1969.

[17] S. Maroux and P. Desnuelle, *Biochim. Biophys. Acta* **181**, 59 (1969).

[18] D. D. Schroeder and E. Shaw, *J. Biol. Chem.* **243**, 2943 (1968).

or Lys$_{176}$-Asp$_{177}$.[19] The latter autolysis product, "pseudotrypsin," although devoid of activity against BAEE, still possesses the characteristics of a serine protease as indicated by stoichiometric reaction with DFP. Chain breaks have also been identified in chymotrypsinogen prior to the specific chain break inducing activity. These inactive precursors are called neochymotrypsinogens. In both the neochymotrypsinogens and the degraded enzymes, the nomenclature has been the arbitrary one of assigning Greek letters to a characterized product. It is recommended that the specific locus of chain cleavage be clearly stated in the identification of these serine protease variants, e.g., α-trypsin, which has been activated by the removal of residues 1–6 from trypsinogen, and in which Lys$_{176}$-Asp$_{177}$ has autolyzed, could be identified as α-trypsin (7–176, 177–229).

General Methods of Assay

Principle

Enzymatic activity measurements generally fall into three categories, depending upon the experimental objectives:

1. Determination of the amount of a particular enzyme in an impure mixture, for example, a tissue extract.
2. Determination of the amount of activity remaining in a relatively pure enzyme sample after a given chemical or physical treatment.
3. Determination of relative activity toward various specific substrates in order to define the specificity of a given enzyme preparation.

For the assay of a serine protease, any one of these objectives is usually met by using a simple synthetic substrate in which only one bond is susceptible to enzymatic hydrolysis. For such systems the kinetic parameters may be readily defined. The preferred substrates are esters of L-α-amino acids blocked at the N-terminal. Ester substrates have advantages over the corresponding amides in that K_m values are lower and catalytic rate constants are higher. Furthermore, the release of at least one equivalent of protons from an ester permits the use of sensitive potentiometric methods. The general equation is:

$$\underset{\text{X—NH—CH—CO—OR}' + H_2O}{\overset{R}{|}} \rightarrow \underset{\text{X—NH—CH—COO}^- + \text{HOR}' + H^+}{\overset{R}{|}}$$

in which X is a blocking group such as acetyl, benzoyl, tosyl, or benzyl-

[19] R. L. Smith and E. Shaw, *J. Biol. Chem.* **244**, 4704 (1969).

oxycarbonyl; R is a side chain which meets the specificity requirements of the enzyme; and R′ may be a methyl, ethyl, or p-nitrophenyl group.

Potentiometric methods measure the rate of production of protons and are most conveniently accomplished in a recording pH-stat. The pH is usually set at about 8 where the serine proteases show maximal activity.

Spectrophotometric methods depend upon differences in absorption between the products on the right side of the above equation and the substrate on the left. The rate of change in absorption at a selected wavelength is proportional to the rate of hydrolysis and is most conveniently determined on a recording spectrophotometer with a constant speed chart. Substrates which contain aromatic groups absorbing below 300 nm, e.g., the chymotrypsin substrate N-benzoyl-L-tyrosine ethyl ester, have difference spectra with maxima in the region of 240–260 nm. The hydrolysis of p-nitrophenyl esters results in useful difference spectra above 300 nm. Such substrates have special advantages for applications in which interference from UV-absorbing substances is to be avoided. Catalytic rate constants are of a magnitude such that p-nitrophenyl esters can be used at pH values below 6 under circumstances which make this an advantage.

In recent times the general applicability of esterolytic assays has limited the use of protein substrates to special purposes. For a particular investigation the natural specificity may be of predominant interest, for example, the action of thrombin in blood clotting. In other cases, the natural, specific substrate may have some advantage in identifying a particular serine protease activity, for example, the use of elastin for the estimation of elastase in a mixture containing other proteases. A third type of application for natural substrates is represented by an investigation of tissue extracts which may contain protease activities of unknown or partially defined specificity. The rate of digestion of a denatured protein gives a rough measure of the total protease activity. That portion inhibited by DFP may then be identified as due to serine proteases. A more complete characterization of the enzymes may be obtained by the use of specific inhibitors, such as the chloromethyl ketones.[20]

Studies of the action of a protease on small natural peptides of known sequence have also been useful in obtaining a broad characterization of the enzyme specificity. After proteolytic digestion under controlled conditions, the peptides of the digest are separated and identified. In this way the bonds which are specifically cleaved become known and the relative extent of each cleavage can be estimated. For example, the specificity of bovine chymotrypsin B was characterized by investigating its action on glucagon.

[20] E. N. Zendzian and E. A. Barnard, *Arch. Biochem. Biophys.* **122,** 714 (1967).

For measurement of general proteolytic activity, the widely tested method of Kunitz is recommended. A description of this method, which uses casein as substrate and a spectrophotometric procedure, may be found in Vol. II [3].

In the sections immediately below, detailed examples are given of a potentiometric assay and a spectrophotometric assay, each using an ester substrate. Although these examples are designed for bovine trypsin and chymotrypsin A with substrates suitable for their assay, the methods are applicable in principle to the other serine proteases. Furthermore, these general methods have been adapted to a wide selection of substrates for various enzymes. Short descriptions with references to particular assay systems will be found in the sections on individual enzymes.

Examples of the Potentiometric Method

Assay for trypsin using α-N-benzoyl-L-arginine ethyl ester:

Reagents

Buffer. 0.10 M KCl, 0.05 M CaCl$_2$, 0.01 M Tris at pH 7.75 with HCl
Substrate solution. Dissolve 343 mg of α-N-benzoyl-L-arginine ethyl ester hydrochloride (BAEE) in a final volume of 100 ml of buffer. The final solution contains 0.01 M BAEE and can be used for a period of 2 days providing it is stored at 0–4°.
Enzyme. A standard solution of trypsin. It is prepared in 0.01 M HCl at 0° at a concentration of approximately 10 mg per milliliter and dialyzed overnight at 4° against 0.001 M HCl. Any traces of insoluble material are removed by centrifugation. The precise concentration is established by measuring the absorbancy at 280 nm after 1:10 dilution with 0.001 M HCl ($E_{1\,cm}^{1\%} = 15.4$).

Procedure

Assays are performed in a pH-stat at a constant temperature (26°) maintained in a thermostatted 4-ml vessel. The design of the vessel allows for continuous (magnetic) stirring at a rate sufficient to avoid local concentrations of titrant in the vicinity of the electrodes. With the Radiometer pH-stat, model TTT1c, a convenient electrode is the combined glass/calomel electrode, GK2302c. The syringe is filled with standardized 0.1 M NaOH and the titrator set to deliver with a maximum response to small pH increments (on the Radiometer pH-stat this corresponds to a "proportional band" setting of 0.1). The chart drive is set at about ½ inch per minute.

Place 2.0–3.0 ml of substrate solution in the pH-stat vessel (the exact volume is not critical) and allow the solution to reach thermal equilibrium. Set the titrator to raise the pH to 7.80. When this is accomplished add an aliquot (up to 100 μl) of enzyme solution, containing 0.001 to 0.1 mg of trypsin, directly into the substrate solution. In this particular assay the rate of acid production is constant throughout more than 90% of the hydrolysis since the K_m for BAEE is less than $10^{-4}\,M$ and benzoylarginine is not a strong inhibitor of the system. The slope of the linear relationship between NaOH consumed and the time of reaction gives a direct measure of micromoles of substrate consumed per minute.

Definition of Unit and Specific Activity

One unit of activity catalyzes the hydrolysis of 1 micromole of substrate per minute, thus the slope of the straight-line progress curve for the reaction (expressed in micromoles/minute) is equal to the number of units of the enzyme in the aliquot analyzed. The specific activity is expressed as units per milligram of protein. The specific activity of the best preparations of trypsin examined is 70 units per milligram.

Examples of Spectrophotometric Method

1. Assay for chymotrypsin A$_\alpha$ using N-benzoyl-L-tyrosine ethyl ester.[21]

Reagents

Substrate solution. N-Benzoyl-L-tyrosine ethyl ester (BTEE), 0.001 M in 50% (w/w) aqueous methanol. This substrate is readily available from several commercial sources. Dissolve 15.7 mg of BTEE in 30 ml of methanol, spectral grade, and make up to 50 ml with water. Solution should be freshly prepared each day.

Buffer. Tris-HCl, 0.10 M, pH 7.8, containing CaCl$_2$, 0.10 M

Enzyme. Prepare a standard solution of chymotrypsin A$_\alpha$ in 0.001 M HCl at a concentration of approximately 1 mg per milliliter. Determine exact concentration by absorbancy measurement at 282 nm (E$^{1\%}$ = 20.0). Just before use dilute this stock solution 1:50 with 0.001 M HCl.

Procedure

Assays are performed in 10-mm quartz cuvettes of 3.5 ml capacity. The cuvettes are held at 30.0° in a thermostatted compartment. Preferably the spectrophotometer has a recording mechanism with a constant speed chart drive, for example, the Cary 15. Prepare the reference solution by mixing 1.5 ml of substrate solution and 1.5 ml of buffer in a cuvette. Place 1.5 ml of substrate solution and 1.4 ml of buffer in the assay cuvette. Allow 5 minutes for the solutions to come to thermal

[21] B. C. W. Hummel, Can. J. Biochem. Physiol. 37, 1393 (1959).

equilibrium in the cell compartment and zero the instrument at 256 nm, the wavelength of the maximum in the difference spectrum. At zero time add 100 μl of the diluted enzyme solution to the assay cuvette, mix thoroughly for 5 seconds and record absorbancy difference for a period of about 5 minutes. The rate of change of absorbancy should be linear up to a reading of 0.12. For accurate measurements, a 0.0 to 0.1 range setting of the recording spectrophotometer should be used.

Samples of unknown activity should be adjusted in concentration so that a 100-μl aliquot contains an amount of enzyme corresponding to about 1 to 3 μg of chymotrypsin A_α.

Definition of Unit and Specific Activity

Activity is calculated from the slope of the linear portion of the reaction curve. One unit is equal to the hydrolysis of 1 micromole of substrate per minute. The value of $\Delta\epsilon$ (change in molar extinction coefficient) for complete hydrolysis of BTEE is 964. Specific activity is expressed as units per milligram of protein. The specific activities of various chymotrypsin A_α preparations vary, largely because of the presence of 10 to 15% inert protein. Based on the determination of the functional normality of several preparations, the author has obtained an average specific activity of 48 units per milligram for fully active enzyme.

2. Assay for chymotrypsin A_α using N-benzyloxycarbonyl-L-tyrosine p-nitrophenyl ester.[22]

Reagents

Substrate. N-Benzyloxycarbonyl-L-tyrosine p-nitrophenyl ester (ZTNE) may be obtained from a few commercial sources. Synthesis in the laboratory may be carried out from more readily obtained N-benzyloxycarbonyl-L-tyrosine. The following general procedure may be used to prepare substrates from other blocked amino acids.

Dissolve Z-L-tyrosine (1.58 g, 5 millimoles) in 15 ml of anhydrous dioxane containing 1.2 ml of tributylamine. Cool to 5° and maintain anhydrous conditions with stirring during the addition of a solution of ethyl chloroformate (0.65 ml in 5 ml of dioxane). After 30 minutes, add nitrophenol (0.70 g, 5 millimoles) and stir for 1 hour. Remove the solvent under vacuum in a rotary evaporator. The residual oil is diluted with chloroform and the solution is washed with cold 1 N HCl, saturated sodium bicar-

[22] C. J. Martin, J. Golubow, and A. E. Axelrod, *J. Biol. Chem.* **243**, 294 (1959).

bonate, 1 N HCl, and water. Dry the solution over sodium sulfate and remove the solvent under vacuum as before. Add methanol to the oil and induce crystallization of the product. Recrystallize the product from hot chloroform to obtain white material melting at 158–159° (uncorr.). Yield, approximately 900 mg, 55%. MW = 436. $[\alpha]_D = -16.3°$ (1.0% in acetone).

Substrate solution. ZTNE, $1.2 \times 10^{-3} M$ in dioxane, spectral grade. Dissolve 5.23 mg of substrate in 10 ml of solvent.

Methanolic buffer. The complete buffer contains Tris-HCl, 0.03 M, pH 8.0; methanol, 12% (v/v); and $CaCl_2$, 0.1 M.

Enzyme. A standard enzyme solution is prepared as described in Example 1 above. However, the dilution to be made just before calibration of the assay is 1:2500 with 0.12 M $CaCl_2$. Great care must be taken in handling very dilute solutions of the enzyme to prevent loss on the glass surfaces. Calcium ion tends to prevent losses.

Procedure

The instrumentation is the same as that described in Example 1 above. Because the substrate undergoes spontaneous hydrolysis at pH 8, it is necessary to determine the rate of this background reaction and subtract it from the assay rate. Add 2.8 ml of buffer to a cuvette plus 0.10 ml of 0.12 M $CaCl_2$. Equilibrate the cuvette at 30° for 5 minutes in the spectrophotometer and zero the instrument at 400 mμ against water as reference. At zero time add 100 μl of the substrate solution, mix thoroughly for 5 seconds, and record the progress of the reaction for 3 minutes. From the initial slope obtain the rate in units of absorbancy change per minute.

The assay is performed by the addition of 100 μl of enzyme solution to 2.80 ml of buffer in a cuvette at 30°, followed 1 minute later by the addition of 100 μl of substrate solution. Since the value of K_m for this enzymatic reaction is $3.2 \times 10^{-5} M$ and the substrate concentration is $4.0 \times 10^{-5} M$, the reaction curve is not linear. Estimate the value of the initial slope and obtain the corrected value by subtraction of the rate of spontaneous hydrolysis.

Definition of Specific Activity

In general, investigators using this method have expressed activities in terms of specific molar rate constants. p-Nitrophenol has a molar extinction coefficient of about 18,700 under assay conditions. Using this value, the specific rate constant at maximum velocity was reported in the original work to be 545 sec^{-1} (recalculated for enzyme MW = 25,000).

The rate constant at a substrate concentration of $4 \times 10^{-5}\,M$ should therefore be close to 300 sec^{-1}, not far from the value observed in the author's laboratory. However, the difficulties of mixed kinetics and blank corrections are such that all assays should be calibrated with known concentrations of enzyme, using a standard solution of measured functional normality when this is advisable.

Comment

The method described above is the most sensitive assay for chymotrypsin that has been developed in detail. As little as 5 mμg can be measured with fair precision. Reproducibility depends greatly upon precise technique. The substrate (ZTNE) is not specific for chymotrypsin. It is known that trypsin and papain both are effective catalysts for the hydrolysis. However, the method provides a useful model for the assay of a serine protease, the identity of which is not in doubt. The method has been found to be particularly useful in studies of inhibitors under conditions which make it necessary to measure very low levels of activity.

The substrate, ZTNE, may also be used for assay at pH 5 rather than at pH 8, the point of optimal enzyme activity. At the lower pH spontaneous hydrolysis is negligible relative to enzymatic hydrolysis, and no blank correction is needed. Simplification of procedure is accompanied by loss of sensitivity due to lower enzyme activity and to a lower extinction coefficient for the product, p-nitrophenol. In one investigation[23] the following conditions were used: acetate buffer, $0.2\,M$, pH 5.0; 0.65% acetonitrile (v/v); substrate, $1.5 \times 10^{-5}\,M$; temperature, 25°; wavelength of measurement, 340 nm (maximum in difference spectrum at pH 5, $\Delta\epsilon = 6014$). The value of k_{cat} is given as 1.17 sec^{-1} and K_m (app.) as $3 \times 10^{-5}\,M$.

[23] F. J. Kézdy, A. Thomson, and M. L. Bender, *J. Am. Chem. Soc.* **89**, 1004 (1967).

[4] Trypsinogens and Trypsins of Various Species

By K. A. WALSH

The proteolytic enzyme trypsin and its inactive precursor, trypsinogen, were first obtained in crystalline form from bovine pancreatic tissue by Northrop and Kunitz.[1] Trypsinogen is transformed into trypsin as the

[1] J. H. Northrop, M. Kunitz, and R. M. Herriott, "Crystalline Enzymes," 2nd Ed. Columbia Univ. Press, New York, 1948.

result of the cleavage of a single peptide bond (Lys_6-Ile_7) near the N-terminus of the zymogen,[2] and the appearance of activity is accompanied by conformational changes.[3] The activation process is catalyzed by a variety of enzymes including enterokinase,[4] mold proteases,[5] and trypsin itself. The latter autocatalytic process is accelerated by calcium ions which bind to the N-terminal region of the zymogen and promote the specific bond cleavage.[6,7] A striking feature of the enzyme is the narrow specificity of its action, which is almost exclusively directed toward L-lysyl and L-argininyl bonds of polypeptides. Biologically, trypsin serves as the activator of all the other zymogens of pancreatic tissue. Thus, the control of the activation of trypsinogen has broad consequences in terms of the formation of the endopeptidase and exopeptidase components of pancreatic juice.

Methods of Assay of Trypsin

The substrates of trypsin can be described by the general formula, R—CO—X, where the narrow specificity of trypsin is determined by the acyl moiety (R—CO—) and the type of bond cleaved (i.e., peptide, amide, or ester) is defined by the nature of X. Many assay procedures have been described using each type of catalysis. In addition, by the appropriate choice of R and X, the deacylation step of hydrolysis can be slowed so that the initial release of X is a measure of the available active sites. Such compounds serve as active site titrants[8-11] which define the true concentration of functional active sites.

Esterase Activity

The most widely used assays for trypsin follow the esterase activity toward benzoyl-L-arginine ethyl ester (BAEE)[12,13] or p-toluenesulfonyl-L-arginine methyl ester (TAME)[14] by potentiometric[12] or spectrophoto-

[2] E. W. Davie and H. Neurath, *J. Biol. Chem.* **212**, 515 (1955); P. Desnuelle and C. Fabre, *Biochim. Biophys. Acta* **18**, 49 (1955).
[3] H. Neurath and G. H. Dixon, *Federation Proc.* **16**, 791 (1957).
[4] M. Kunitz, *J. Gen. Physiol.* **22**, 429 (1939).
[5] cf. T. Hofmann, this volume [25].
[6] T. M. Radhakrishnan, K. A. Walsh, and H. Neurath, *Biochemistry* **8**, 4020 (1969).
[7] J. P. Abita, M. Delaage, M. Lazdunski, and J. Savrda, *European J. Biochem.* **8**, 314 (1969).
[8] M. L. Bender and F. J. Kézdy, *J. Am. Chem. Soc.* **87**, 4953 (1965).
[9] T. Chase, Jr. and E. Shaw, *Biochem. Biophys. Res. Commun.* **29**, 508 (1967).
[10] D. T. Elmore and J. J. Smyth, *Biochem. J.* **107**, 97, 103 (1968).
[11] R. J. Vaughan and F. H. Westheimer, *Anal. Biochem.* **29**, 305 (1969).
[12] See K. A. Walsh and P. E. Wilcox, this volume [3].
[13] G. W. Schwert and Y. Takenaka, *Biochim. Biophys. Acta* **16**, 570 (1955).
[14] B. C. W. Hummel, *Can. J. Biochem. Physiol.* **37**, 1393 (1959).

metric procedures. Potentiometric measurement of the specific esterase activity toward BAEE under standardized conditions is described in this volume (p. 37). An alternative spectrophotometric method using TAME[14] has the dual advantage over BAEE of greater sensitivity and of greater selectivity (chymotrypsin does not hydrolyze TAME), and is described below.

A buffer solution [0.04 M Tris (hydroxymethyl aminomethane), 0.01 M CaCl$_2$, pH 8.1] is first prepared by dissolving 1.47 g of CaCl$_2$·2 H$_2$O in 200 ml of 0.2 M Tris, adjusting the pH to 8.1 with HCl, and diluting to 1 liter. Then 19.7 mg of TAME·HCl is added to 50 ml of buffer to yield a substrate concentration of $1.04 \times 10^{-3} M$.

The assay procedure employed is exactly analogous to that described for the assay of chymotrypsin A$_\alpha$ with N-benzoyl-L-tyrosine ethyl ester in this volume (p. 38). Three milliliters of substrate solution are measured into each cuvette of the spectrophotometer. Allow 3 minutes for temperature equilibration to be reached in the compartment at 30°. Balance the absorbance of the two cuvettes at 247 nm. Add 0.1 ml of water to the reference cuvette and mix well. At zero time add 0.1 ml of enzyme solution (containing up to 0.45 μg of trypsin) to the sample cuvette and mix well. The rate of increase of absorbance should be linear up to an absorbance of 0.24 (65% hydrolysis) and this rate is directly proportional to the concentration of the enzyme.

One unit of TAME activity is defined as the amount of trypsin catalyzing the hydrolysis of 1 micromole of TAME per minute. The absorbance change accompanying hydrolysis of 1 micromole of TAME per milliliter of assay solution is 0.409 cm^{-1} mM^{-1}. Trypsin, at a concentration of 0.1 μg/ml in the assay cuvette results in an absorbance change of 0.0101 absorbance units (min^{-1}·cm^{-1}). Thus, 1.0 mg of trypsin is equal to 247 TAME units.

Proteolytic Activity Toward Casein ᴗ

A method of assaying trypsin by its proteolytic activity toward casein has been described in detail by Laskowski.[15]

Amidase Activity

Amidase activity can be determined with benzoyl-L-argininamide by the manual method of Schwert *et al.*[16] as outlined in Vol. II [2] or by its adaptation to the automatic analytical procedure described by Lenard

---

[15] M. Laskowski, in Vol. II [2].

[16] G. W. Schwert, H. Neurath, S. Kaufman, and J. E. Snoke, *J. Biol. Chem.* **172**, 221 (1948).

et al.[17] Alternatively, the colorimetric procedure of Erlanger *et al.*[18] using benzoyl-DL-arginine *p*-nitroanilide provides a sensitive amidase assay.

Active Site Titration with p-Nitrophenyl, p'-Guanidinobenzoate (NPGB)[9]

This method is based on the rapid and stoichiometric *p*-guanidino-benzoylation of trypsin by NPGB as measured by the extent of liberation of *p*-nitrophenol. An aliquot of a trypsin solution is diluted with $0.1 M$ Veronal buffer, pH 8.3, containing $0.02 M$ $CaCl_2$ to give 0.99 ml containing 0.05–2.4 mg of trypsin (2-$100 \times 10^{-6} M$ in trypsin). This is placed in a 1-ml cuvette (1-cm light path) in a double-beam recording spectrophotometer and the instrument balanced at 410 nm against a reference cuvette containing buffer alone. Ten microliters of a $0.01 M$ solution of NPGB in dimethylformamide are added to the reference cuvette and the contents thoroughly mixed. This procedure is repeated with another 10-μl aliquot of the NPGB solution in the sample cuvette at zero time. The optical density is followed for 3–5 minutes and a very slow liberation of *p*-nitrophenol extrapolated back to zero time to give a measure of the initial burst of *p*-nitrophenol. The increase of absorbance above baseline during the initial burst multiplied by 6.025×10^{-5} yields the molarity of liberated *p*-nitrophenol (and therefore the molarity of active trypsin). Although this method is less sensitive than the rate assays described above, it has the inherent advantage of providing an absolute standardization of the concentration of active enzyme and avoiding the uncertain variables involved in rate assays.

Purification of Trypsinogen

The classical purification of bovine trypsinogen by Kunitz and Northrop[19] involves acid extraction of fresh pancreas, ammonium sulfate fractionation, and removal of α-chymotrypsinogen as a crystalline by-product. This is described in detail in Volume II [2]. Subsequent purification and crystallization of trypsinogen is described in detail in Volume II [3] and involves treatment of the mother liquor and washings from the crystallization of chymotrypsinogen with acidic 70% saturated ammonium sulfate to form a precipitate of crude trypsinogen. Two alternate methods of crystallization of trypsinogen are then possible.

[17] J. Lenard, S. L. Johnson, R. W. Hyman, and G. P. Hess, *Anal. Biochem.* **11**, 30 (1965).
[18] B. F. Erlanger, N. Kokowsky, and W. Cohen, *Arch. Biochem. Biophys.* **95**, 271 (1961).
[19] M. Kunitz and J. H. Northrop, *J. Gen. Physiol.* **19**, 991 (1936).

Crystallization Procedure A

A method is outlined in detail by Laskowski in Volume II [3] involving $MgSO_4$ at pH 8.0 and yields short triangular prisms of trypsinogen containing approximately 50% $MgSO_4$ by weight. It should be noted that autocatalytic activation of trypsinogen is avoided during this procedure because pancreatic trypsin inhibitor is present during the crystallization. Attempts to recrystallize trypsinogen by this procedure result in autocatalytic activation and loss of the zymogen, although recrystallization has been accomplished by Tietze[20] using $0.01\ M$ diisopropylphosphofluoridate (DFP) at pH 9 as described in Volume II [3].

Crystallization Procedure B

An alternate method of crystallization in ethanol–water mixtures at 4° is described by Balls.[21] In this procedure, 100 g of the crude trypsinogen precipitate formed in acidic 70% saturated ammonium sulfate (above) are dissolved in 300 ml of water and 105 ml of 95% ethanol. The pH is adjusted to 7.5 with 5 N NaOH. After standing for 30 minutes, the solution is filtered through Celite on a large sintered glass funnel. The volume of the filtrate is measured and to each 100 ml are added 14 ml of 95% ethanol. The filtrate is seeded (if possible) with trypsinogen crystals and stored at 4°. After about 3 days the crystals are collected on hardened filter paper, washed with 100 ml of cold 33% ethanol, then (very rapidly) with ethanol, 1:1 ethanol–ether and finally with ether. The crystals can be dried in air and *in vacuo* at 55° over anhydrous calcium sulfate. The yield (about 11 g from 100 g of "crude trypsinogen precipitate") can be increased by 20% by leaving the alcoholic filtrate (not including the washings) at 4° for several more days. Further recrystallizations must be carried out at low pH since pancreatic trypsin inhibitor is no longer present. This is accomplished by suspending 1 g of dried crystals in 100 ml of 25% ethanol at 4°. The pH is adjusted to 3.5 with 1 N HCl and stirred occasionally for 20 minutes. Insoluble material is removed by centrifugation and discarded. Fourteen milliliters of 95% ethanol is added to raise the concentration of ethanol to 33%. This is followed by the addition of 1 ml of saturated ammonium sulfate per 100 ml of solution and the pH adjusted to 7.5 with 2 N NaOH. Traces of precipitate are rapidly removed by centrifugation, the supernatant is seeded with crystals of trypsinogen and left for 24 hours at 4°. Crystals can be collected by filtration on hardened paper and washed with cold 33% ethanol. This recrystallization procedure can then be repeated if desired.

[20] F. Tietze, *J. Biol. Chem.* **204**, 1 (1953).
[21] A. K. Balls, *Proc. Natl. Acad. Sci. U.S.* **53**, 392 (1965).

Crystalline trypsinogen obtained by either procedure A or procedure B still shows minor heterogeneity during ion-exchange chromatography. A highly purified product can be obtained by chromatography on SE-Sephadex at pH 4.2.

Chromatography of Trypsinogen[22]

SE-Sephadex is prepared by swelling 15 g of C-50 grade in three changes of 2 liters each of chromatography buffer (0.20 M NaCl, 0.05 M acetic acid at pH 4.20 with NaOH at room temperature) for 1 day. Do not stir the suspension with a magnetic mixer during any of the preparatory steps as the gel beads are fragile and crush easily, resulting in poor flow rates. After swelling the SE-Sephadex, allow a slurry of the ion-exchanger to settle in a graduate cylinder for 30 to 45 minutes. Remove the slower sedimenting particles with an aspirator. Repeat until all the fine particles are removed as judged by a sharply sedimenting boundary. Failure to do this results in poor flow characteristics. Check that the pH of the suspension is 4.20. Suspend the ion-exchanger in about 1 liter of buffer and pour the slurry into a 50-cm chromatography column (2.5 cm I.D.) with a nylon net support. Allow the gel slurry to settle under gravity for about 5 minutes, then, with the outlet open, allow the remainder of the slurry to settle with no more than 30 cm of water pressure between the top of the buffer in the column and the end of the outlet tube. Allow the remainder of the slurry to settle. Continue to add slurry until the gel bed is 40 cm long. Place the column at 4° and connect to the outlet tube a peristaltic pump adjusted to deliver about 25 ml/hour. (Note: Under no circumstances should the buffer be pumped into the top of the column since it leads to packing of the gel bed and obstruction of flow.) Ideal flow behavior is obtained when the buffer reservoir and pump are at the same height. Allow the column to equilibrate with flowing buffer for 24 hours at 4°, at which time the pH and the ionic strength of the effluent at room temperature should be identical with those of the chromatography buffer.

Dissolve 200 mg of trypsinogen crystals obtained by procedure A or 100 mg obtained by procedure B in 7 ml of cold 10^{-2} M HCl at 0°. The solution should contain about 14 mg of trypsinogen per milliliter in either case. Dialyze overnight against 4 liters of 10^{-3} M HCl at 4° to remove the salts occluded in the crystals. Remove the trypsinogen solution from the dialysis sac, add dropwise with stirring 1.0 ml of 2 M NaCl in 0.5 M acetic acid (at pH 4.20 with NaOH) and dilute to 10.0 ml with water. Remove traces of insoluble material by centrifugation and adjust the pH to exactly 4.20.

[22] N. Robinson, M. Sanders, and K. A. Walsh, unpublished experiments.

This trypsinogen solution is applied to the top of the gel bed by opening the column, removing the buffer by aspiration, and adding the solution by pipette without stirring up the top of the packed gel bed. After this sample flows into the gel bed, it is washed in with several 2-ml washes of buffer and the column top is connected to the chromatography buffer which is pumped at 25 ml per hour into a fraction collector. Five-milliliter fractions are collected and examined at 280 nm for protein.

Trypsinogen begins to elute as a symmetrical peak in about 12 hours. It appears after traces of contaminants and before contaminating trypsin. The contents of tubes containing the trypsinogen are pooled, the pH adjusted to 3.0, and the solution dialyzed for 24 hours at 4° against two changes of 4 liters each of $10^{-3}\,M$ HCl. The trypsinogen solution can then be lyophilized and stored at $-20°$.

Trypsinogen Purification from Other Species

Trypsinogens have been prepared from rat,[23] goat,[24] sheep,[25] pig,[26] horse,[27] and dogfish[28] pancreas as well as from bovine pancreas. The procedures involve ion-exchange chromatography techniques and yield in all cases, material sufficiently pure for characterization of their chemical and physical properties. Several of these fractionation procedures have been summarized by Marchis-Mouren.[28a]

Properties of Trypsinogen

Bovine trypsinogen prepared as described above contains less than 0.2% contaminating trypsin. It has 229 amino acid residues and a molecular weight of 23,991.[29] It is stable at low pH and can be stored either as a lyophilized powder or as a frozen solution in $10^{-3}\,M$ HCl. The amino acid sequence has been described and is given in Table I and Fig. 2. Activation of the zymogen to form trypsin involves the specific cleavage of the Lys_6-Ile_7 bond near the N-terminus.[2] The physical properties of trypsinogen are summarized together with those of trypsin in Table V.

[23] A. Vandermeers and J. Christophe, *Biochim. Biophys. Acta* **188**, 101 (1969).

[24] S. Bricteux-Grégoire, R. Schyns, and M. Florkin, *Arch. Intern. Physiol. Biochim.* **76**, 571 (1968).

[25] S. Bricteux-Grégoire, R. Schyns, and M. Florkin, *Biochim. Biophys. Acta* **127**, 277 (1966).

[26] M. Charles, M. Rovery, A. Guidoni, and P. Desnuelle, *Biochim. Biophys. Acta* **69**, 115 (1963).

[27] C. I. Harris and T. Hofmann, *Biochim. J.* **114**, 82P (1969).

[28] R. Haynes, W. P. Winter, R. Tye, and H. Neurath, personal communication, 1968.

[28a] G. Marchis-Mouren, *Bull. Soc. Chim. Biol.* **47**, 2207 (1965).

[29] K. A. Walsh and H. Neurath, *Proc. Natl. Acad. Sci. U.S.* **52**, 884 (1964).

Autocatalytic Activation of Bovine Trypsinogen[30]

One hundred milligrams of trypsinogen is dissolved in 4 ml of $10^{-2} M$ HCl and dialyzed against $10^{-3} M$ HCl at $4°$ to remove any salts. The absorbance at 280 nm is measured (with an aliquot diluted 1:20 with $10^{-3} M$ HCl) to insure that the trypsinogen concentration is in the range 10–20 mg/ml ($E_{1cm}^{1\%} = 15.0$). The volume is measured, the solution chilled to $0°$ and an equal quantity of cold $0.2 M$ Tris buffer, pH 8.1, containing $0.1 M$ $CaCl_2$ is added. The addition of trypsin to a concentration of about 0.5 mg/ml (1:20 weight ratio of trypsin:trypsinogen) initiates the autocatalytic activation. The progress of trypsinogen activation can be readily followed using 5–25 μl aliquots in the BAEE assay procedure.[12] Maximum activity is seen in about 90 minutes and this is followed by a slow loss in activity, presumably the result of autolysis. To express the results of an absolute basis, the concentration of trypsin added to initiate the activation must be subtracted. The best preparations of trypsinogen give more than 95% of the stoichiometric yield of trypsin.

Nature of the Activation Process

During the activation of trypsinogens from all species examined, a highly charged N-terminal peptide with the sequence X-Asp_4-Lys is

TABLE I
N-TERMINAL SEQUENCES OF VARIOUS TRYPSINOGENS

Species	N-Terminal sequence
Cow[a]	Val-Asp-Asp-Asp-Asp-Lys-
Goat[b]	Val-Asp-Asp-Asp-Asp-Lys-
Sheep[c]	{ Val-Asp-Asp-Asp-Asp-Lys- / Phe-Pro-Val-Asp-Asp-Asp-Asp-Lys-
Pig[d]	Phe-Pro-Thr-Asp-Asp-Asp-Asp-Lys-
Horse[e]	Ser-Ser-Thr-Asp-Asp-Asp-Asp-Lys-
Dogfish[f]	Ala-Pro-Asp-Asp-Asp-Asp-Lys

[a] E. W. Davie and H. Neurath, *J. Biol. Chem.* **212**, 515 (1955).

[b] S. Bricteux-Grégoire, R. Schyns and M. Florkin, *Arch. Intern. Physiol. Biochim.* **76**, 571 (1968).

[c] S. Bricteux-Grégoire, R. Schyns, and M. Florkin, *Biochim. Biophys. Acta* **127**, 277 (1966).

[d] M. Charles, M. Rovery, A. Guidoni, and P. Desnuelle, *Biochim. Biophys. Acta* **69**, 115 (1963).

[e] C. I. Harris and T. Hofmann, *Biochem. J.* **114**, 82P (1969).

[f] Quoted by H. Neurath, R. Arnon, and R. A. Bradshaw, *Proc. Symp. Structure-Function Relationships of Proteolytic Enzymes, Munksgaard, Copenhagen*, in press, 1969.

[30] J. F. Pechère and H. Neurath, *J. Biol. Chem.* **229**, 389 (1957).

cleaved from the N-terminus to expose the N-terminal isoleucyl residue of trypsin.[2, 24-27, 31] The results of these studies are summarized in Table I. Calcium ions promote the activation reaction by binding to the N-terminal hexapeptide of bovine trypsinogen and accelerating the specific bond cleavage process.[6, 7] A conformational change accompanies the appearance of activity[3] and it is not known whether this reflects an essential step in the activation or an incidental step accompanying the activation. By analogy with the activation of chymotrypsinogen,[32] it is generally anticipated that the homologous Ile-Val-Gly sequence at the N-terminus of trypsin may form a salt link with Asp_{182} as an integral part of a refolding process resulting in the assembly of components of the active site.

Purification of Trypsin

Bovine trypsin is readily prepared from trypsinogen by autocatalytic activation and crystallization procedures.[19] The yield was improved by McDonald and Kunitz,[33] who showed that the inclusion of calcium ions in the activation mixture minimized a side reaction leading to the formation of "inert protein." The starting material for trypsin purification is the "crude trypsinogen" obtained as a by-product of the chymotrypsinogen preparation[19] as described in detail by Laskowski in Volume II [2]. A detailed description of the procedure for crystallizing and recrystallizing trypsin is outlined by Laskowski in Volume II [3]. The crystalline product can be stored at 4° as a dry powder. The maximum specific activity observed, expressed in units of BAEE activity (this volume, p. 39) is 70 units per milligram.

Preparations of crystalline bovine trypsin have been found to contain variable quantities of at least five molecular species, apparently resulting from partial autolysis (Table II). Functionally, these species may be completely active, partially active, or inert. Several chromatographic methods have been described to fractionate this mixture and purify the enzyme species (Table III). Of these, the method of Maroux, Rovery, and Desnuelle[34] effectively removes chymotrypsin and inert proteins from trypsin and is described in detail in Volume XI [25]. Higher resolution of the individual components of partially autolyzed trypsin is given by the equilibrium chromatographic procedure of Schroeder and

---

[31] Quoted in H. Neurath, R. Arnon, and R. A. Bradshaw, *Proc. Symp. Structure-Function Relationships of Proteolytic Enzymes, Munksgaard, Copenhagen*, in press, 1969.

[32] P. B. Sigler, D. M. Blow, B. W. Matthews, and R. Henderson, *J. Mol. Biol.* **35**, 143 (1968).

[33] M. R. McDonald and M. Kunitz, *J. Gen. Physiol.* **29**, 155 (1946).

[34] S. Maroux, M. Rovery, and P. Desnuelle, *Biochim. Biophys. Acta* **56**, 202 (1962).

TABLE II

NATURE OF AUTOLYSIS PRODUCTS OF TRYPSIN

Internal bond cleaved	Nomenclature assigned	Active toward[a]	Inhibited by[a]	Unreactive with[a]	Reference
None	β-Trypsin	All substrates	All inhibitors		b
Lys_{49}-Ser_{50}	—	—	—	TLCK	c
Arg_{105}-Val_{106}	—	BAEE	TLCK DFP		c
Lys_{131}-Ser_{132}	α-Trypsin	BANA TAME	NPGB DFP TLCK	—	b
		BAEE	TLCK		c
Lys_{176}-Asp_{177} and Lys_{131}-Ser_{132}	Pseudotrypsin	BANA[d] BAEE[d]	DFP	TLCK	e

[a] Abbreviations: BANA, N-α-benzoyl-DL-arginine p-nitroanilide; BAEE, N-benzoyl-L-arginine ethyl ester; TAME, p-toluenesulfonyl-L-arginine methyl ester; TLCK, 1-chloro-3-tosylamido-7-amino-2 heptanone; DFP, diisopropylphosphofluoridate; NPGB, p-nitrophenyl, p'-guanidinobenzoate.

[b] D. D. Schroeder and E. Shaw, *J. Biol. Chem.* **243**, 2943 (1968).

[c] S. Maroux and P. Desnuelle, *Biochim. Biophys. Acta* **181**, 59 (1969).

[d] The activity is greatly diminished from that of β-trypsin.

[e] R. L. Smith and E. Shaw, *J. Biol. Chem.* **244**, 4704 (1969).

TABLE III

CHROMATOGRAPHIC SYSTEMS FOR THE PURIFICATION OF BOVINE TRYPSIN

Ion-exchanger	pH	Component preventing autolysis during chromatography	Reference
Carboxymethyl cellulose	3.2	Low pH	a
IRC-50	6.1	8 M urea	b
Carboxymethyl cellulose	6.0	Low pH and calcium	c
Carboxymethyl cellulose	5.5	Low pH and calcium	d
SE-Sephadex	2.6	Low pH	e
SE-Sephadex	7.1	Benzamidine	f
SE-Sephadex	3.0	Low pH	g

[a] I. E. Liener, *Arch. Biochem. Biophys.* **88**, 216 (1960).

[b] R. D. Cole and J. M. Kinkade, *J. Biol. Chem.* **236**, 2443 (1961).

[c] S. Maroux, M. Rovery, and P. Desnuelle, *Biochim. Biophys. Acta* **56**, 202 (1962); Vol. XI [25].

[d] S. Maroux and P. Desnuelle, *Biochim. Biophys. Acta* **181**, 59 (1969).

[e] S. Papiannou and I. E. Liener, *J. Chromatog.* **32**, 746 (1968).

[f] D. D. Schroeder and E. Shaw, *J. Biol. Chem.* **243**, 2943 (1968).

[g] J. G. Beeley and H. Neurath, *Biochemistry* **7**, 1239 (1968).

Shaw[35] which separates α-trypsin from β-trypsin (see Table II for definition of these species).

Chromatography of Trypsin[35]

A cation-exchange column (2.2×85 cm) of SE-Sephadex (C-50 beaded) is prepared and equilibrated with chromatography buffer containing $0.1\,M$ Tris chloride, $0.02\,M$ $CaCl_2$, $10^{-3}\,M$ benzamidine at pH 7.1. Precautions to be observed during equilibration and packing of the column are described above in the section outlining the chromatography of trypsinogen. The flow rate is adjusted to 15–25 ml per hour at 4° with as little hydrostatic pressure head as possible and 9–10 ml fractions are collected. Since the chromatographic procedure is very long (5–7 days), autolysis after chromatography is excluded by acidifying the collected fractions. This can be accomplished either by providing each fraction collector tube with 0.4 ml of $1.25\,M$ potassium formate, pH 2.9, and a mixing device or by pumping this pH 2.9 buffer into the effluent stream via a T-tube at the rate of about 1 ml per hour.

The sample of trypsin is prepared by dissolving 250 mg in 25 ml of the chromatography buffer. (If the trypsin preparation contains occluded salts, they should be removed by dialysis against 4 liters of chromatography buffer at 4°). Any insoluble material should be removed by centrifugation at about 2000 g at 4°. The clear supernatant is then applied to the column and the chromatogram developed with 3 liters of buffer. The effluent is monitored at 280 nm (measured against a blank of chromatography buffer since benzamidine absorbs light at 280 nm).

A peak of inert protein appears during the first day of chromatography, α-trypsin during the fourth day, and β-trypsin during the fifth day (at 20 ml per hour). Pooled fractions of α-trypsin or β-trypsin can be concentrated to about 100 ml each by adsorption at 4° on a small SE-Sephadex C-50 column (4.4×4 cm) previously equilibrated with $0.05\,M$ Tris-formate, pH 3.0, and elution (at about 1 liter per hour) with $0.5\,M$ Tris-chloride, pH 7.6, containing $0.02\,M$ $CaCl_2$, $10^{-3}\,M$ benzamidine. The concentrated trypsin fractions are then brought to pH 3 with $2\,M$ HCl, dialyzed against $10^{-3}\,M$ HCl, and lyophilized.

Trypsin Purification from Other Species

Trypsin and trypsinlike enzymes have been prepared from 14 animal species and microbiological organisms largely by chromatographic procedures. The amino acid compositions of this related group of enzymes

[35] D. D. Schroeder and E. Shaw, *J. Biol. Chem.* **243**, 2943 (1968).

TABLE IV
Amino Acid Compositions of Various Trypsins

	Cow[a]	Pig[b]	Sheep[c]	Turkey[d]	Dogfish[e]	Star-fish[f]	Horse[g]	Human[h]	Light chain[i] of human plasmin	Cocoonase[j]	Strepto-myces griseus[k]	Shrimp[l]	Sea pansy[m]	Horshoe crab[n]	Rat 1[o,p]	Rat 2[o,p]
Lys	14	10	12-13	7	6.3	9	7	11	14	13	6.5	5	16	13	9.8	12.2
His	3	4	3	3	7.6	4	4	3	8	4	1.0	5	5	4	5.0	5.3
Arg	2	4	3-4	2-3	6.5	4	7	6	14	6	8.4	3	9	9	3.2	4.1
Asp	22	18-24	20-24	15-16	27.9	31	24	21	15	26	18.1	30	27	22	28.7	25.9
Thr	10	10-11	10-15	11	7.3	15	9	10	14	16	16.4	10	13	9	8.4	10.8
Ser	33	24-25	26-32	24	16.5	18	28-30	24	18	23	14.2	24	23	15	15.1	22.0
Glu	14	17	14-16	12-13	15.5	20	19-20	21	22	15	17.4	24	21	28	19.6	16.7
Pro	9	10	9	8-9	11.3	13	11	9	15	9	8.0	11	14	11	15.2	11.6
Gly	25	25-26	19-26	19-22	27.8	28	22	20	21	22	28.4	28	25	32	22.4	22.8
Ala	14	15-16	16-17	11-13	17.0	16	17	13	12	16	26.0	16	13	13	16.5	14.1
Half-cys	12	12	11-12	8	11.9	8	10	8	9	4	6.1	8	12	6	9.8	11.3
Val	17	16	17	14-15	17.4	19	16	16	22	20	17.8	18	14	13	20.2	20.3
Met	2	2	2	1-2	8.1	2	1	1	2	1-2	2.7	2	2	1	2.0	1.7
Ile	15	15	10-14	10	13.5	11	15	12	8	12	8.0	14	11	11	13.4	10.3
Leu	14	16	14-15	13-14	14.6	13	16	12	20	12	11.0	10	13	12	17.4	17.5
Tyr	10	8	6-11	10	11.5	8	11	7	6	9	8.2	10	6	6	6.8	8.0
Phe	3	4	3-5	1-2	1.2	5	5	4	8	5	5.7	6	6	6	4.8	5.2
Trp	4	4-6	5	5	5.4	5	4	3	5	3	ND[q]	3	ND[q]	ND[q]	3.9	4.8
Total	223	214-226	200-240	174-187	227.3	229	226-229	201	232	216-217	203.9	226	230	211	222.2	224.6

[a] K. A. Walsh and H. Neurath, *Proc. Natl. Acad. Sci. U.S.* **52**, 884 (1964).

[b] M. Charles, M. Rovery, A. Guidoni, and P. Desnuelle, *Biochim. Biophys. Acta* **69**, 115 (1963); J. Travis and I. E. Liener, *J. Biol. Chem.* **240**, 1962 (1965).

[c] R. Schyns, S. Bricteux-Grégoire, and M. Florkin, *Biochim. Biophys. Acta* **175**, 97 (1969); J. Travis, *Biochim. Biophys. Res. Commun.* **30**, 730 (1968).

[d] C. A. Ryan, J. J. Clary, and Y. Tomimatsu, *Arch. Biochem. Biophys.* **110**, 175 (1965); T. Kishida and I. E. Liener, *Arch. Biochem. Biophys.* **126**, 111 (1968).

[e] R. W. Tye, personal communication.

[f] W. P. Winter and H. Neurath, unpublished data, 1969.

[g] T. Hofmann, personal communication.

[h] J. Travis and R. C. Roberts, *Biochemistry* **8**, 2884 (1969).

[i] W. R. Groskoff, B. Hsieh, L. Summaria, and K. C. Robbins, *J. Biol. Chem.* **244**, 359 (1969).

[j] F. C. Kafatos, A. M. Tartakoff, and J. H. Law, *J. Biol. Chem.* **242**, 1477 (1967).

[k] L. Jurasek, D. Fackre, and L. B. Smillie, *Biochim. Biophys. Res. Commun.* **37**, 99 (1969).

[l] B. J. Gates and J. Travis, *Biochemistry* **8**, 4483 (1969).

[m] M. H. Coan and J. Travis, *Comp. Biochem. Physiol.* **32**, 127 (1970).

[n] M. R. Moore and J. Travis, personal communication.

[o] A. Vandermeers and J. Christophe, *Biochim. Biophys. Acta* **188**, 101 (1969).

[p] Trypsinogen

[q] ND = Not determined.

are summarized in Table IV. Porcine trypsin was obtained as a crystalline product[36, 37] and has been characterized in greatest detail.

Properties of Trypsin

The chemical, physical, and enzymatic properties of trypsin have been reviewed by Desnuelle[38] in 1960 and by Cunningham[39] in 1965. Only certain selected features of the enzyme and its zymogen will be mentioned here as a guide to the general characteristics of trypsin.

Stability of Trypsin

Trypsin is stable at pH 3 at low temperatures where it can be stored for weeks without loss of activity. It can be reversibly heat-denatured[40] and the effect of pH on the temperature-induced reversible denaturation of both bovine trypsinogen and bovine trypsin has been studied by Lazdunski and Delaage.[41] Their state diagrams (Fig. 1) reveal that four different forms of the enzyme can be distinguished at low temperatures and reversible transformations between these forms are controlled by pH. Below pH 8 an increase in temperature results in reversible denaturation, whereas above pH 8 elevated temperatures induce irreversible denaturation. A similar state diagram is seen for bovine trypsinogen. Porcine trypsin is more stable than bovine trypsin to thermal denaturation.[42]

Trypsin loses enzymatic activity reversibly in high concentrations of denaturants such as $8\,M$ urea. However, Harris has shown that the enzyme retains about 50% activity toward TAME in $4\,M$ urea[43] where there is apparently an equilibrium mixture of native and denatured enzymes. In $4\,M$ urea the enzyme activity is irreversibly lost as active enzyme attacks its denatured counterpart. Delaage and Lazdunski have examined the susceptibility of the different forms of trypsin (Fig. 1) to urea denaturation and have shown that substrates or competitive inhibitors can protect trypsin from denaturation.[44]

The addition of calcium ions to trypsin solutions retards autolysis.

---

[36] J. Travis and I. E. Liener, *J. Biol. Chem.* **240**, 1962 (1965).

[37] P. J. VanMelle, S. H. Lewis, E. G. Samsa, and R. J. Westfall, *Enzymologia* **26**, 133 (1963).

[38] P. Desnuelle, *in* "The Enzymes," (P. D. Boyer, H. Lardy, and K. Myrbäck, eds.), Vol. 4, p. 119. Academic Press, New York, 1960.

[39] L. Cunningham, *in* "Comprehensive Biochemistry," (M. Florkin and E. H. Stotz, eds.) Vol. 16, p. 85. Elsevier Press, Amsterdam, 1965.

[40] M. L. Anson and A. E. Mirsky, *J. Gen. Physiol.* **17**, 393 (1934).

[41] M. Lazdunski and M. Delaage, *Biochim. Biophys. Acta* **140**, 417 (1967).

[42] M. Lazdunski and M. Delaage, *Biochim. Biophys. Acta* **105**, 541 (1965).

[43] J. I. Harris, *Nature* **177**, 471 (1956).

[44] M. Delaage and M. Lazdunski, *European J. Biochem.* **4**, 378 (1968).

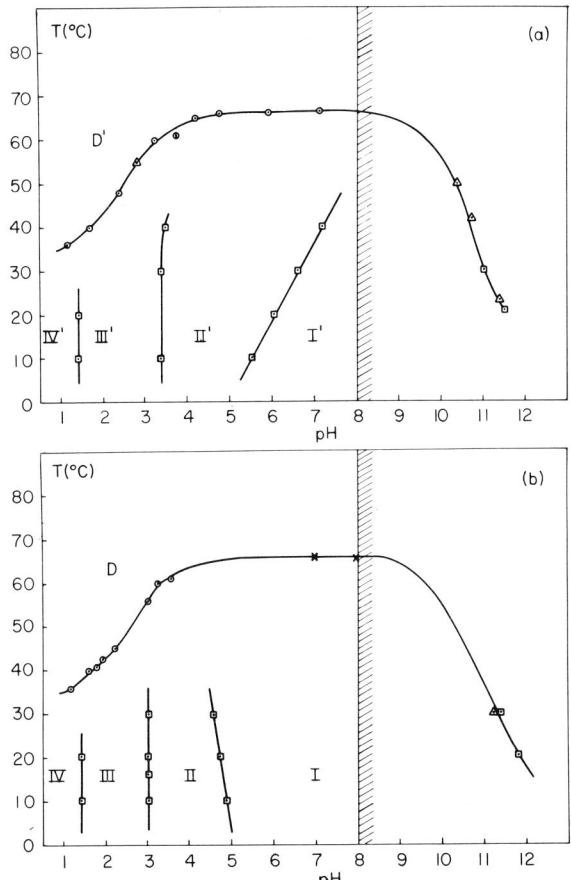

FIG. 1. State diagrams of bovine trypsinogen and trypsin.[41] The figures are separated into two regions by a hatched line. I', II', III', D' for trypsinogen, and I, II, III, D for trypsin represent the different molecular forms in equilibrium at acidic pH's. (⊙) Spectrophotometric studies at constant pH (the inflection points of the $\Delta\epsilon_{293\,nm}$-temperature curves). (□) Spectrophotometric studies at constant temperature (the inflection points of $\Delta\epsilon_{293\,nm}$, $\Delta\epsilon_{288\,nm}$, or $\Delta\epsilon_{300\,nm}$-pH plots). (△) Spectropolarimetric studies. (×) Activity measurements. The curves represent, in fact, isoconcentration lines of two adjacent forms. Only one alkaline transition is represented in the diagrams at alkaline pH. It is mainly due to the reversible unmasking of buried tyrosines. This unmasking is rapidly followed by an irreversible inactivation due to attack of hydroxyl ions on the disulfide bridges. Reproduced from *Biochim. Biophys. Acta* **140**, 422 (1967).

This stabilizing effect is accompanied by a conformational change in the trypsin molecule, apparently induced by calcium and resulting in a more compact structure.[42] The protective effect of calcium ions is much more

```
Horse^h        ILE
S. Griseus^g
Turkey^f       ILE                                                    SER-MET-GLY-CYS-
Porcine^c,d    ILE                                          ASX-SER-GLY-TYR-HIS-PHE-CYS-
Dogfish^e      (ILE,VAL,GLY,GLY,TYR,GLX,CYS,PRO,LYS)ASX(THR,VAL,PRO,TYR,HIS,GLY,ARG)  ASN-SER-GLY-SER-HIS-PHE-CYS-
Bovine^a,b     ILE-VAL-GLY-GLY-TYR-THR-CYS-GLY-ALA-ASN-THR-VAL-PRO-TYR-GLN-VAL-SER-LEU-ASN-SER-GLY-TYR-HIS-PHE-CYS-
                                                                      ASX-VAL-GLY-TYR(HIS,PHE,CYS,
               7           10                          20

S. Griseus^g   GLY-GLY-ALA-LEU                 THR-ALA-ALA-HIS-CYS-VAL(ASN,ASN,GLY,GLY,SER,SER)
Turkey^f       GLY-GLX-SER-LEU                     ALA-ALA-HIS-CYS-TYR-LYS
Porcine^c      GLY-GLY-SER-LEU                     ALA-ALA-HIS-CYS-TYR-LYS
Dogfish^e      GLY,GLY,SER)LEU,ILE-HIS-GLX-GLX                                  ILE-GLN-VAL-ARG,LEU-
Bovine^a,b     GLY-GLY-SER-LEU-ILE-ASN-SER-GLN-TRP-VAL-VAL-SER-ALA-ALA-HIS-CYS-TYR-LYS-SER-GLY-ILE-GLN-VAL-ARG-LEU-
                                       40                                        50

Turkey^f                                                                  ALA-LEU-THR-HIS-PRO-ASX-
Porcine^c      GLY-GLX ASX-ASX-ILE-VAL-LEU                                ILE-ILE-THR-HIS-PRO-ASN-
Dogfish^e      GLY-GLX-HIS-ASX-ILE-SER-ALA-ASX(GLX,GLY,ASX,THR,GLX)TYR-ILE-ASP-SER-SER-MET(VAL,ILE)ARG,HIS-PRO-ASX-
Bovine^a       GLY-GLU-ASP-ASN-ILE-ASN-VAL-VAL-GLU-GLY-ASP-GLU-GLN-PHE-ILE-SER-ALA-SER-LYS-SER-ILE-VAL-HIS-PRO-SER-
Bovine^b       GLN         60                     ASN-GLN  70                                        80

Turkey^f       TYR
Porcine^c      PHE
Dogfish^e      TYR(SER,ASX,LEU,GLY,TYR,ASX=ASN,ASP=ILE,MET,LEU,ILE)LYS(LEU,LYS)PRO-ALA-ALA-SER-LEU-ASX-ARG-ASP-VAL-
Bovine^a       TYR-ASN(PRO,ASN,THR,LEU)ASN-ASN-ASP-ILE-MET-LEU-ILE-LYS-LEU-LYS-SER-ALA-ALA-SER-LEU-ASN-SER-ARG-VAL-
Bovine^b       -SER-ASN-THR-LEU-                 90                                100

Dogfish^e      ASN-LEU-ILE-SER-LEU-PRO-THR-GLY-CYS-ALA-TYR-ALA-MET(GLY,GLX)CYS,LEU-ILE-SER-GLY-TRP-GLY(ASX,THR,ASX,
Bovine^a,b     ALA-SER-ILE-SER-LEU-PRO-THR-SER-CYS-ALA-SER-ALA-GLY-THR-GLN-CYS-LEU-ILE-SER-GLY-TRP-GLY-ASN-THR-LYS-
                                      110                                   120                                130

S. Griseus^g                                                              ALA-ALA-CYS-ARG-SER-
Dogfish^e      GLY)MET-ALA-VAL-SER-   -GLY-ASP-GLN-LEU-GLN-CYS-LEU-ASP-ALA-PRO-VAL(ALA,ASP=LEU,GLU,SER)CYS-LYS,GLY-
Bovine^a       SER-SER-GLY-THR-SER-TYR-PRO-ASP-VAL-LEU-LYS-CYS-LEU-LYS-ALA-PRO-ILE-LEU-SER-ASP-SER-SER-CYS-LYS-SER-
Bovine^b                              140                            150 ASN

S. Griseus^g   ALA-TYR-GLY-ASN-GLU            GLU-ILE-CYS-ALA-GLY-TYR+ASP-THR-GLY-GLY-VAL-ASP-THR-CYS-GLN-GLY-
Porcine^c                                                                ASN-SER-CYS-GLN-GLY-
Dogfish^e      ALA-TYR-PRO-GLY(ASN,ILE,THR,MET)ASN-MET-MET-CYS-VAL^Z GLY-TYR^Z MET^Z GLU-GLY-GLY-LYS-ASP-SER-CYS-(GLN,GLY,
Bovine^a       ALA-TYR-PRO-GLY-GLN-ILE-THR-SER-ASN-MET-PHE-CYS-ALA-GLY-TYR-LEU-GLU-GLY-GLY-LYS-ASP-SER-CYS-GLN-GLY-
Bovine^b                              160                     170           ASN             180

S. Griseus^g   ASP-SER-GLY-GLY-PRO-MET-PHE                           TYR-GLY-CYS-ALA-ARG-PRO-
Porcine^c      ASP-SER-GLY-GLY-PRO-VAL-VAL-CYS-GLY-GLN-GLN-LEU
Dogfish^e      ASP,SER,GLY,GLY,PRO,VAL)VAL-CYS
Bovine^a,b     ASP-SER-GLY-GLY-PRO-VAL-VAL-CYS-SER-GLY-LYS-LEU-GLN-GLY-ILE-VAL-SER-TRP-GLY-SER-GLY-CYS-ALA-GLN-LYS-
                                      190                                  200

Horse^h                                                              (GLN,THR,ILE,ALA)ALA-ASN
S. Griseus^g   GLY-TYR-PRO-GLY-VAL
Turkey^f                                                                  SER-ASN
Porcine^d                                                            THR(ILE,GLN)ALA-ASN
Dogfish^e                                                            THR-ILE-ALA(SER,ALA^Z)
Bovine^a,b     ASN-LYS-PRO-GLY-VAL-TYR-THR-LYS-VAL-CYS-ASN-TYR-VAL-SER-TRP-ILE-LYS-GLN-THR-ILE-ALA-SER-ASN
                                      210                                  220
```

Fig. 2. Partial and complete amino acid sequences of various trypsins. The residue numbering is that for bovine trypsinogen. The structure of bovine trypsinogen was elucidated by two independent groups[a,b] which differ[b] only where indicated. Disulfides of bovine trypsin are paired[i] as follows: 13–143; 31–47; 115–216; 122–189; 154–168; 179–203.

a. Bovine—K. A. Walsh and H. Neurath, *Proc. Natl. Acad. Sci. U.S.* **52,** 884 (1964).

b. Bovine—O. Mikes, V. Holeysovsky, V. Tomasek, and F. Sorm, *Biochem. Biophys. Res. Commun.* **24,** 346 (1966).

c. Porcine—J. Travis and I. E. Liener, *J. Biol. Chem.* **240,** 1967 (1965); R. A. Smith and I. E. Liener, *J. Biol. Chem.* **242,** 4033 (1967).

d. Porcine—M. Charles, M. Rovery, A. Guidoni, and P. Desnuelle, *Biochim. Biophys. Acta* **69,** 115 (1963).

e. Dogfish—R. A. Bradshaw, H. Neurath, K. A. Walsh, R. W. Tye, and W. P. Winter, quoted by H. Neurath, R. A. Bradshaw, and R. Arnon in *Proc. Symp. Structure-Function Relationships of Proteolytic Enzymes, Munksgaard, Copenhagen,* in press, 1969.

f. Turkey—T. Kishida and I. E. Liener, *Arch. Biochem. Biophys.* **126,** 111 (1968).

pronounced with bovine and ovine trypsin than with porcine trypsin.[45] Green and Neurath have reviewed the protective effect of other cations on bovine trypsin.[45a]

Self-digestion of trypsin can be minimized by chemical modification of the ε-amino groups of lysyl residues. Labouesse and Gervais[46] report that ε-N-acetylated trypsin does not lose activity at neutral or slightly alkaline pH during several days of storage at room temperature.

Purity of Trypsin

The variants, intrinsic in partially autolyzed preparations of trypsin, have already been discussed (see Table II). It should be emphasized that the specificity of action and catalytic parameters of these variants have not been examined in detail.

An important feature of trypsin is its narrow specificity of action, which renders it suitable for specific cleavage at lysyl or arginyl residues in amino acid sequence analysis. Any contamination with chymotrypsin obviously reduces its effectiveness in this role. A method of removing traces of chymotrypsin by chromatography is described in detail by Rovery in Volume XI [25]. An alternate method of selectively inactivating the chymotryptic contaminant with tosylphenylalanine chloromethyl ketone (TPCK)[47] is described by Carpenter in Volume XI [26]. The product of the latter treatment, "TPCK-trypsin," retains very low activity toward chymotryptic substrates, but this may be an inherent property of the trypsin molecule.[48]

Structure of Trypsin and Trypsinogen from Various Species

Amino acid compositions and partial amino acid sequences of various trypsins and trypsinlike enzymes are summarized in Table IV and Fig. 2.

---

[45] F. F. Buck, A. J. Vithayathil, M. Bier, and F. F. Nord, *Arch. Biochem. Biophys.* **97,** 417 (1962).
[45a] N. M. Green and H. Neurath, *in* "The Enzymes," (H. Neurath and K. Bailey, eds.), Vol. II,B, p. 1057. Academic Press, New York, 1954.

---

g. *Streptomyces griseus*—L. Jurasek, D. Fackre, and L. B. Smillie, *Biochem. Biophys. Res. Commun.* **37,** 99 (1969).
h. Horse—C. I. Harris and T. Hofmann, *Biochem. J.* **114,** 82P (1969).
i. D. Kauffman, *J. Mol. Biol.* **12,** 929 (1965).
z. May be an alternate sequence in an allotypic variant [see (e)].
* His is inserted at this point in porcine trypsin.
** Corrected sequence reported by K. A. Walsh, L. L. Houston, and R. A. Kenner, in *Proc. Symp. Structure-Function Relationships of Proteolytic Enzymes, Munksgaard, Copenhagen,* in press, 1969.
+ Pro is inserted at this point in *S. griseus* trypsinlike enzyme.

An examination of the partial sequence data reveals the high degree of homology among the structures so far examined. The similarity of amino acid compositions provides a basis for the prediction that most, if not all, of the trypsins from diverse species will possess homologous structures. Thus, it is probable that they have arisen from a common ancestor by a process of divergent molecular evolution.[49]

The recent finding of virtual identity in the three-dimensional conformation of two homologous enzymes, bovine chymotrypsin and porcine elastase,[50] provides a valid basis for attempts to predict the three-dimensional structure of trypsin. Several such hypothetical structures have been described[32, 51, 52] and, at the least, provide a guide to the probable three-dimensional structure of trypsin in advance of its detailed determination by X-ray crystallographic techniques.

Physical Properties of Trypsins and Trypsinogens

Various physical parameters of trypsin and trypsinogen are summarized in Table V. Both the molecular weight data in this table and the amino acid composition data in Table IV indicate that the molecular weight of trypsin is approximately 24,000 regardless of the species examined. The net charge of trypsin varies widely, depending upon the species examined. Although isoionic or isoelectric points have been determined in a few cases only, it has been shown that rat,[23] dogfish,[31] and lungfish[31] trypsins are anionic at neutral pH, whereas bovine and porcine trypsins are cationic (Table V). Thus, the net charge is not an important determinant of specific tryptic function.

The physical forms of trypsin and trypsinogen at various temperatures and pH values are summarized in Fig. 1 and discussed in detail by Lazdunski and Delaage.[41]

Inhibitors of Trypsin

Three categories of inhibitors of trypsin have been studied: (a) protein inhibitors; (b) competitive inhibitors; and (c) active site titrants.

[46] J. Labouesse and M. Gervais, *European J. Biochem.* **2**, 215 (1967).

[47] L-(1-Tosylamido-2-phenylethyl) chloromethyl ketone as designed by G. Schoellman and E. Shaw, *Biochemistry* **2**, 252 (1963).

[48] T. Inagami and J. M. Sturtevant, *J. Biol. Chem.* **235**, 1019 (1960).

[49] H. Neurath, K. A. Walsh, and W. P. Winter, *Science* **158**, 1638 (1967).

[50] D. M. Shotton and H. C. Watson, *Proc. Roy. Soc. (London) Ser. B,* in press, 1969.

[51] K. A. Walsh, L. L. Houston, and R. A. Kenner, *Proc. Symp. Structure-Function Relationships of Proteolytic Enzymes, Munksgaard, Copenhagen,* in press, 1969.

[52] B. Keil, V. Dlouhá, V. Holeyšovsky, and F. Šorm, *Collection Czech. Chem. Commun.* **33**, 2307 (1968).

TABLE V

SOME PHYSICAL PARAMETERS OF VARIOUS TRYPSINOGENS AND TRYPSINS

	Molecular weight	$s_{20,w}$	Isoionic point	f/f_0
Bovine trypsinogen	$23,700^a$	2.48^a	9.3^b	1.15^a
Bovine trypsinogen	$23,991^c$			
Bovine trypsinogen	$24,200^{d,e}$			
Bovine trypsinogen	$24,000^{e,f}$			
Bovine trypsinogen	$25,700^{e,g}$	2.7^e		
Bovine trypsinogen	$22,100^{d,h}$			
Bovine trypsin	$24,000^i$	2.50^i	10.1^b	1.2^i
Porcine trypsinogen	$25,327^j$		7.5^j	
Porcine trypsin	$23,400^{f,k}$	2.77^k	$10.8^{k,l}$	
Equine trypsinogen	$25,600^m$			
Human trypsin	$22,900^n$	2.68^o		
Dogfish trypsinogen	$24,520^p$			

	Optical rotatory dispersion parameters^q		
	$\lambda_c(nm)$	a_0	b_0^r
Bovine trypsin, Form I^s	249 ± 3	$-213 \pm 8°$	$-143 \pm 15°$
Bovine trypsin, Form III	237 ± 2		
Bovine trypsinogen, Form I'	240 ± 3	$-280 \pm 14°$	$-143 \pm 18°$
Bovine trypsinogen, Form III'	236 ± 2		

[a] Sedimentation velocity and diffusion. F. Tietze, *J. Biol. Chem.* **204**, 1 (1953).

[b] N. M. Green and H. Neurath, *in* "The Proteins," (H. Neurath and K. Bailey, eds.) Vol. IIB, p. 1057, Academic Press, New York, 1954.

[c] Sequence analysis, K. A. Walsh and H. Neurath, *Proc. Natl. Acad. Sci. U.S.* **52**, 884 (1964).

[d] Light scattering.

[e] C. M. Kay, L. B. Smillie, and F. A. Hilderman, *J. Biol. Chem.* **236**, 118 (1961).

[f] Approach to sedimentation equilibrium.

[g] Sedimentation—viscosity.

[h] Reported by A. K. Balls, *Proc. Natl. Acad. Sci. U.S.* **53**, 392 (1965).

[i] Diisopropylphosphoryl trypsin. L. W. Cunningham, Jr., F. Tietze, N. M. Green, and H. Neurath, *Discussions Faraday Soc.* **13**, 58 (1953).

[j] M. Charles, M. Rovery, A. Guidoni, and P. Desnuelle, *Biochim. Biophys. Acta* **69**, 115 (1963).

[k] J. Travis and I. E. Liener, *J. Biol. Chem.* **240**, 1962 (1965).

[l] Isoelectric point.

[m] C. I. Harris and T. Hofmann, *Biochem. J.* **114**, 82P (1969).

[n] J. Travis and R. C. Roberts, *Biochemistry* **8**, 2884 (1969).

[o] Sedimentation velocity experiments at a concentration of 2.5 mg/ml.

[p] R. W. Tye, personal communication.

[q] M. Lazdunski and M. Delaage, *Biochim. Biophys. Acta* **140**, 417 (1967).

[r] $\lambda_0 = 212$ nm.

[s] The various physical states of trypsin and trypsinogen are defined in Fig. 1.

These have provided both a variety of reagents for terminating the action of the enzyme and considerable insight into the nature of the catalytic mechanism. The protein inhibitors are discussed separately in this volume [66, 67] and provide stoichiometric inhibition by binding with characteristically large association constants. Studies with a variety of competitive inhibitors have correlated the inhibitory capacity both with the hydrophobic nature of the compound and with the specific location of a positive charge.[53-55] The characteristics of some of these inhibitors are summarized in Table VI.

Active site titrants (Table VI) provide reagents for both irreversibly inhibiting the enzyme and estimating the concentrations of active sites. In addition, active site titrants have led to direct proof of the involvement of His_{46} and Ser_{183} in the active center of bovine trypsin.[39, 56]

Specificity of Trypsin

The specificity of trypsin is directed toward lysyl and arginyl residues[57] in natural and synthetic substrates, thus exhibiting the most restricted specificity of action of the known endopeptidases. The length of the side chain containing the positive charge is apparently an important determinant of this specificity.[58, 59] Some kinetic constants for trypsins with selected substrates are summarized in Table VII. Theories of the mechanism of action of the enzyme have been reviewed by Cunningham,[39] Bender and Kaiser,[60] and Bender and Kézdy.[61] In general, they draw attention to the apparent analogy with the mechanism of chymotryptic action, the intermediate acylation of the enzyme and the specific involvement of Ser_{183} and His_{46} in a combination of general acid, general base, and nucleophilic catalysis. More recently it has been suggested that aspartic $acid_{90}$ accentuates the nucleophilic character of Ser_{183}. Although the specificities of action of trypsin and chymotrypsin are quite different,[57] analogy in their catalytic mechanisms is striking and is apparently dictated by homology in primary structure. It is significant in this connection that the greatest similarity in the two primary struc-

[53] M. Mares-Guia and E. Shaw, *J. Biol. Chem.* **240**, 1579 (1965).

[54] M. Mares-Guia, *Arch. Biochem. Biophys.* **127**, 317 (1968).

[54a] T. Inagami and S. S. York, *Biochemistry* **7**, 4045 (1968).

[55] J. D. Geratz, *Arch. Biochem. Biophys.* **118**, 90 (1967).

[56] E. Shaw, Vol. XI [80].

[57] H. Neurath and G. W. Schwert, *Chem. Rev.* **46**, 69 (1950).

[58] J. B. Baird, E. F. Curragh, and D. T. Elmore, *Biochem. J.* **96**, 733 (1965).

[59] N. J. Baines, J. B. Baird, and D. T. Elmore, *Biochem. J.* **90**, 470 (1964).

[60] M. L. Bender and E. T. Kaiser, *J. Am. Chem. Soc.* **84**, 2556 (1962).

[61] M. L. Bender and F. J. Kézdy, *Ann. Rev. Biochem.* **34**, 49 (1965).

TABLE VI
Some Inhibitors of Bovine Trypsin

	K_I (M)	pH	Temp. (°C)	Ref.
I. Competitive inhibitors[a]				
p-Aminobenzamidine	8.25×10^{-6}	8.15	15.0	b
Benzamidine	1.66×10^{-5}	8.15	15.0	c
β-Naphthamidine	1.46×10^{-5}	8.15	15.0	c
m-Toluamidine	2.27×10^{-5}	8.15	15.0	c
p-Toluamidine	2.65×10^{-5}	8.15	15.0	c
Cyclohexylcarboxamidine	4.27×10^{-4}	8.15	15.0	b
Phenylguanidine	7.25×10^{-5}	8.15	15.0	b
Phenylguanidine	1.4×10^{-4}	7.0	10.0	d
α-N-Benzoyl-L-arginine	5.8×10^{-3}	8.0	25	e
1-Propyl-guanidine	5.3×10^{-4}	8.0	25.0	f
n-Butylamine	1.7×10^{-3}	6.6	25	g
Benzylamine	6.0×10^{-4}	6.6	25	g
Thionine	2.5×10^{-5}	7.6	24	h
Proflavin	4.0×10^{-5}	7.3	25	i
II. Active site titrants				
Diisopropylphosphofluoridate (DFP)				j
1-Chloro-3-tosylamido-7-amino heptanone (TLCK)				k
p-Nitrophenyl-p'-guanidinobenzoate				l
Ethyl-p-guanidinobenzoate				m
α-N-Methyl-α-N-toluene-p-sulfonyl-L-lysine β-naphthyl ester				n
p-Nitrophenyl N^2-acetyl-N^1-benzylcarbazate				o
p-Nitrophenyl α-N-benzyloxycarbonyl-L-lysinate				p
p-Nitrophenyl ethyl diazomalonate				q

[a] The values of K_I reported have been obtained with undefined mixtures of β-trypsin and its autolysis products. The various species of active trypsin may possess different binding affinities for the inhibitors.

[b] M. Mares-Guia and E. Shaw, *J. Biol. Chem.* **240**, 1579 (1965).

[c] M. Mares-Guia, *Arch. Biochem. Biophys.* **127**, 317 (1968).

[d] Equilibrium dialysis by B. M. Sanborn and W. P. Bryan, *Biochemistry* **7**, 3624 (1968).

[e] J. J. Bechet, M. C. Gardiennet, and J. Yon, *Biochim. Biophys. Acta* **122**, 101 (1966).

[f] T. Inagami and S. S. York, *Biochemistry* **7**, 4045 (1968).

[g] T. Inagami, *J. Biol. Chem.* **239**, 787 (1964).

[h] A. N. Glazer, *J. Biol. Chem.* **242**, 3326 (1967).

[i] S. A. Bernhard and H. Gutfreund, *Proc. Natl. Acad. Sci. U.S.* **53**, 1238 (1965).

[j] E. F. Jansen, M. D. F. Nutting, R. Jang, and A. K. Balls, *J. Biol. Chem.* **179**, 189 (1949); G. H. Dixon, D. L. Kauffman, and H. Neurath, *J. Biol. Chem.* **233**, 1373 (1958).

[k] E. Shaw, M. Mares-Guia, and W. Cohen, *Biochemistry* **4**, 2219 (1965); E. Shaw and S. Springhorn, *Biochem. Biophys. Res. Commun.* **27**, 391 (1967).

[l] T. Chase, Jr., and E. Shaw, *Biochem. Biophys. Res. Commun.* **29**, 508 (1967).

[m] M. Mares-Guia and E. Shaw, *J. Biol. Chem.* **242**, 5782 (1967).

[n] D. T. Elmore and J. J. Smyth, *Biochem. J.* **107**, 97 (1968).

[o] D. T. Elmore and J. J. Smyth, *Biochem. J.* **107**, 103 (1968).

[p] M. L. Bender, F. J. Kézdy, and J. Feder, *J. Am. Chem. Soc.* **87**, 4953 (1965).

[q] R. J. Vaughan and F. H. Westheimer, *Anal. Biochem.* **29**, 305 (1969).

TABLE VII
KINETIC CONSTANTS FOR SOME SYNTHETIC SUBSTRATES OF BOVINE TRYPSIN[a]

Substrate	k_m (app.) (mM)	k_{cat} sec^{-1}	pH	Temp. (°C)	Ref.
Amides					
Benzoyl-L-arginine p-nitroanilide	0.939	0.611	8.15	15	b
L-Lysyl-p-nitroanilide	0.364	0.003	8.65	15	b
Benzoyl-L-argininamide	3.1	1.35	8.35	35	c
N-α-Benzyloxycarbonyl-L-arginine p-toluidide	1.4	0.69	8.0	25	d
Esters					
Benzoyl-L-arginine ethyl ester	0.0043	14.6	8.0	25	e
Tosyl-L-arginine methyl ester	0.0125	100.0	8.0	25	f
α-N-Acetyl S-(β-aminoethyl) cysteine ethyl ester	1	18.7	8.0	25	g
α-N-Tosyl-S-(β-aminoethyl) cysteine ethyl ester	0.369	55.6	8.0	25	h
α-N-Acetyl lysine ethyl ester	0.28	80.0	8.0	25	g
Ethyl p-guanidinophenylacetate	0.0115	0.047	8.0	25	i
Neutral substrates					
N-Acetyl-L-tyrosine ethyl ester	42.0	14.5	8.0	25	j
N-Acetyl-glycine ethyl ester	800.0	0.012	7.0	25	k
N-α-Benzoyl-L-citrulline methyl ester	41.0	0.14	7.0	25	l
N-Benzoyl-L-heptyline methyl ester	0.1	9.4	7.0	25	l, m

[a] It should be noted that the kinetic constants have been obtained with undefined mixtures of β-trypsin and its autolysis products. Since α-trypsin and β-trypsin have been shown[35] to differ in their activity toward either N-α-benzoyl-DL-arginine p-nitroanilide or tosyl-L-arginine methyl ester, the constants should be regarded only as a guide to the probable value associated with a particular species of enzyme.

[b] B. F. Erlanger, N. Kokowski, and W. Cohen, Arch. Biochem. Biophys. **95**, 271 (1961).

[c] J. Chevallier and J. Yon, Biochim. Biophys. Acta **112**, 116 (1966).

[d] In 5% dimethylformamide, T. Inagami, J. Biochem. **66**, 277 (1969).

[e] N. J. Baines, J. B. Baird, and D. T. Elmore, Biochem. J. **90**, 470 (1964).

[f] C. G. Trowbridge, A. Krehbiel, and M. Laskowski, Jr., Biochemistry **2**, 843 (1963). Note that the kinetics are complicated by excess substrate activation in the concentration range 1–100 mM.

[g] M. Gorecki and Y. Shalitin, Biochem. Biophys. Res. Commun. **29**, 189 (1967).

[h] D. T. Elmore, D. V. Roberts, and J. J. Smyth, Biochem. J. **102**, 728 (1967).

[i] M. Mares-Guia, E. Shaw, and W. Cohen, J. Biol. Chem. **242**, 5777 (1967).

[j] T. Inagami and J. M. Sturtevant, J. Biol. Chem. **235**, 1019 (1960).

[k] T. Inagami and H. Mitsuda, J. Biol. Chem. **239**, 1388 (1964).

[l] B. M. Sanborn and G. E. Hein, Biochemistry **7**, 3616 (1968).

[m] In 18% acetonitrile.

tures is found in those portions of the two molecules specifically implicated in the catalytic mechanisms.[29, 62]

The action of trypsin toward protein substrates can be redirected by selectively blocking lysyl residues[63] or arginyl residues[64] of the substrate. Alternatively, cystinyl or cysteinyl residues can be converted to the lysyl homolog, S-(β-aminoethyl) cysteine, as detailed by Cole in Volume XI [33]. Thus, by choosing the appropriate modification procedures it is possible to use trypsin to hydrolyze selectively lysyl, cysteinyl, or arginyl residues.

It has recently become clear that the accepted picture of the specificity of trypsin toward lysyl and arginyl residues is incomplete. Trowbridge et al.[65] found that high concentrations of tosyl-L-arginine methyl ester deviated from Michaelis-Menten kinetics. Their observation of activation by excess substrate was explained in terms of a ternary complex involving an auxiliary binding site. Sanborn and Hein[66] have examined the nature and significance of such a ternary complex and found that even neutral compounds can bind in an auxiliary site, resulting in cooperative interaction with the active site. They have indicated that this could be important in explaining anomalies in polypeptide hydrolyses by trypsin[67, 68] since the general environment of a given bond could satisfy secondary specificity considerations. Some synthetic neutral substrates are hydrolyzed by trypsin[66, 69] (Table VII) and apparently bind to the enzyme in a manner which overlaps the primary binding site. It is not clear what relationship exists between the auxiliary site for binding of additional cationic substrate and this neutral substrate binding site.

[62] D. M. Blow, J. J. Birktoft, and B. S. Hartley, *Nature* **221**, 337 (1969).
[63] R. F. Goldberger and C. B. Anfinsen, *Biochemistry* **1**, 401 (1962). See discussion by Goldberger in Vol. XI [34].
[64] K. Toi, E. Bynum, E. Norris, and H. A. Itano, *J. Biol. Chem.* **240**, PC3455 (1965).
[65] C. G. Trowbridge, A. Krehbiel, and M. Laskowski, Jr., *Biochemistry* **2**, 843 (1963).
[66] B. M. Sanborn and G. E. Hein, *Biochemistry* **7**, 3616 (1968).
[67] H. Bachmayer, K. T. Yasunobu, J. L. Peel, and S. Mayhew, *J. Biol. Chem.* **243**, 1022 (1968).
[68] B. V. Plapp, M. A. Raftery, and R. D. Cole, *J. Biol. Chem.* **242**, 265 (1967).
[69] W. Cohen and P. Pétra, *Biochemistry* **6**, 1047 (1967).

[5] Chymotrypsinogens—Chymotrypsins

By P. E. WILCOX

A. Bovine Chymotrypsinogen A and Chymotrypsins A

$$\text{Chymotrypsinogen A} \xrightarrow{\text{T}} \text{ChT A}_\pi \xrightarrow{\text{ChT}} \text{ChT A}_\delta \xrightarrow{\text{ChT}} \text{ChT A}_\gamma$$

$$\text{ChT A}_\gamma \xrightarrow[\text{slow}]{\text{pH 3-4}} \text{ChT A}_\alpha$$

$$\text{Chymotrypsinogen A} \xrightarrow{\text{ChT}} \text{neochymotrypsinogens} \xrightarrow{\text{T}} \text{ChT A}_\alpha$$

$$\text{ChT A}_\alpha \xrightarrow[\text{slow}]{\text{pH 7-8}} \text{ChT A}_\gamma$$

Chymotrypsinogen A is converted to active enzyme at pH 7.6 by tryptic cleavage of the bond between Arg_{15} and Ile_{16} (cf. sequence in Table II). A subsequent autolytic action of chymotrypsin removes the dipeptide, $\text{Ser}_{14}\text{-Arg}_{15}$, to produce chymotrypsin A_δ . Cleavage of two more bonds with the deletion of $\text{Thr}_{247}\text{-Asn}_{248}$ leads to chymotrypsin A_γ. This sequence is the normal course of events under conditions of rapid activation (trypsin-to-zymogen ratio of 1:30, 5°). The γ-form of the enzyme may be crystallized at pH 5.6 from ammonium sulfate solution. When the protein in the activation mixture is precipitated at pH 3 with ammonium sulfate and crystallized at pH 4, the γ-form of the enzyme is transformed into monoclinic crystals of chymotrypsin A_α.[1] The primary structures of the α- and γ-forms of the enzyme do not differ, but they do differ in conformation and physical properties, though not to a large degree.[2]

Under slow activation conditions (trypsin-to-zymogen ratio of 1:10,000, 5°) there is ample opportunity for chymotrypsin to attack the zymogen and produce neochymotrypsinogens. The principal N-terminal residues that appear in the neochymotrypsinogens are Thr, Ala, and Ser; therefore it is probable that these arise by cleavages of the same bonds that occur in the transformation of A_π to A_δ and A_δ to A_γ. The neo-chymotrypsinogens are activated by trypsin to the various forms of the

[1] It has been proposed in the past that chymotrypsin A_α was derived directly from A_δ during activation. However, no enzyme with properties distinctive for A_α, in particular the ability to dimerize at pH 4.5, can be found in rapid activation mixtures even after they have been allowed to stand at pH 7.6 and 5° for 48 hours. (T. Horbett and D. C. Teller, unpublished data.)

[2] H. T. Wright, J. Kraut, and P. E. Wilcox, *J. Mol. Biol.* **37**, 363 (1968); B. W. Matthews, G. H. Cohen, E. W. Silverton, H. Braxton, and D. R. Davies, *J. Mol. Biol.* **36**, 179 (1968).

enzyme. In a slow activation mixture at the end of 48 hours a large portion of the protein has the properties of chymotrypsin A$_\alpha$, although as much as 25% does not.

Chymotrypsinogen A and chymotrypsin A$_\alpha$ are available in large quantities from commercial sources. Zymogen which has been recrystallized several times is usually homogeneous. Samples of enzyme are more or less heterogeneous, depending upon the care of preparation and whether or not they have been subjected recently to purification procedures. The commercial proteins are prepared from bovine pancreas by the classical procedure of Kunitz. Briefly, the zymogen is first isolated from acid extracts of the tissue and purified by recrystallization from ammonium sulfate solution. The pure zymogen is treated with bovine trypsin under slow activation conditions and the enzyme is isolated by crystallization at about pH 4 from ammonium sulfate solution. Salt is removed by dialysis against 10^{-3} M HCl and the product lyophilized.

In the Kunitz procedure, chymotrypsin A$_\gamma$ is obtained by holding a solution of A$_\alpha$ at pH 8 and 5° for 3 weeks and crystallizing the enzyme at pH 5.6 from ammonium sulfate solution. Treatment of the zymogen under conditions of rapid activation for only 60 minutes results in an activation mixture which contains chymotrypsin A$_\delta$ as the principal component. This form of the enzyme cannot be isolated in a pure crystalline state. Dialyzed and lyophilized preparations contain not only A$_\delta$ but also forms of the enzyme with Thr and Ala as additional N-terminal residues, and Tyr as C-terminal. Neither chymotrypsins A$_\delta$ or A$_\gamma$, as they are available commercially, have been thoroughly characterized.

Details of the procedures outlined above have been given by M. Laskowski in Vol. II [2].

Chymotrypsinogen A

Purity

Zymogen which has been purified by manifold recrystallization, as many as five or seven times, appears homogeneous by physical criteria, including sedimentation analysis and free-boundary electrophoresis at pH 4.97.[3,4] Impurities unrelated to the zymogen are absent, but recrystallization is not entirely effective in removing small amounts of chymotrypsin and neochymotrypsinogens, which are sometimes generated during early stages of the preparation. These impurities account for the faint

[3] P. E. Wilcox, J. Kraut, R. D. Wade, and H. Neurath, *Biochim. Biophys. Acta* **24,** 72 (1957).
[4] W. J. Dreyer, R. D. Wade, and H. Neurath, *Arch. Biochem. Biophys.* **59, 145** (1955).

band that is observed in disc gel electrophoresis at pH 4.8 in addition to the sharp, heavy band of the zymogen itself. The faint band has the lower mobility, as is characteristic for the active enzymes.

Test for Chymotrypsin. The most sensitive test is the spectrophotometric assay using N-benzyloxycarbonyl-L-tyrosine p-nitrophenyl ester as substrate (see this volume [3]). The use of 0.1 mg of zymogen in this assay permits the measurement of active enzyme in the range of 0.02% of the weight of total protein. Some commercial samples of zymogen have been found to contain more than 100 times this amount.

Test for End Groups. Impurities may be detected by the Stark method of N-terminal analysis (Vol. XI [11]). Pure samples of zymogen give only traces of hydantoins by this method, less than 0.02 mole per mole of protein.[5] N-Terminal threonine is usually found in the highest amount. Care must be taken to obtain proper correction factors from the blank analyses. *WARNING:* If appreciable amounts of chymotrypsin are present in the zymogen, proteolysis may occur during the treatment with cyanate even in the presence of urea. In that event, the analysis can no longer be a true measure of the inhomogeneity of the zymogen. The difficulty can be avoided by the addition of diisopropylphosphofluoridate (DFP) or phenylmethanesulfonyl fluoride (100 moles per mole of protein) to the solution just before it is to be treated with cyanate at pH 8.

Purification by Chromatography[6]

Principle. Chromatography on high-capacity carboxymethyl cellulose (CM-cellulose) at pH 6.2, using a linear salt gradient of low slope, yields a highly purified chymotrypsinogen. Impurities having the chromatographic properties of active enzyme and showing N-terminal threonine and alanine are reduced to less than 2%. The zymogen is treated with inhibitor before chromatography in order to prevent any enzyme action on the column or in the effluent.

This procedure may also be used in the preparation of inhibited derivatives of chymotrypsin A_π or A_δ (see below) and for the purification of A_γ without the addition of inhibitor. Chymotrypsin A_α does not chromatograph well in this system.

Reagents

Initial eluant. Potassium phosphate buffer, 0.05 M in total phosphate, pH 6.2

Final eluant. The above buffer containing KCl, 0.10 M

CM-cellulose. A high capacity adsorbent is required for satisfactory

[5] G. R. Stark and G. D. Smyth, *J. Biol. Chem.* **238**, 214 (1963).
[6] E. Surbeck and P. E. Wilcox, unpublished data.

resolution of protein components. Whatman CM 52, micro-granular, is recommended. The adsorbent is equilibrated with initial buffer before use.

Inhibitor. DFP, 0.10 M in isopropanol; *or* phenylmethanesulfonyl fluoride, 0.10 M in dioxane. Only a few tenths of a milliliter are needed for each run.

Acid. Acetic acid, 1.0 M; HCl, 10^{-3} M

Procedure. All chromatographic operations are carried out at 5°. Prepare a column of CM-cellulose (2.0 × 70 cm) and wash the column with 2 liters of initial buffer. A satisfactory method for packing the column is given in Whatman Data Manual 2000. The column is fed by a two-reservoir system devised to deliver a linear gradient of buffer solution at a constant rate between 20 and 50 ml per hour. Initially each reservoir holds 3.0 liters of buffer.

Dissolve 500 mg of salt-free protein in about 20 ml of initial buffer. Add 200 μl of one of the inhibitor solutions and let the sample stand for 1 hour. Apply the sample to the column and elute the protein with a gradient running from the initial to the final concentrations of the buffers over the total volume of 6.0 liters. Fractions of about 15 ml are collected, and the effluent is monitored spectrophotometrically at 282 nm. Impurities emerge in the breakthrough fractions and in two or three small peaks that run before the main peak. The main peak appears at a K^+ concentration of about 0.12 M and should be nearly Gaussian in shape. Peak fractions are pooled, the pH is reduced to 4.0 with 1.0 M acetic acid, and the solution dialyzed against 10^{-3} M HCl. The purified protein may be recovered by lyophilization. A concentrated solution is best prepared by ultrafiltration in the Amincon Diaflo apparatus.

Chemical and Physical Characteristics

Amino Acid Composition. Amino acid analyses of bovine chymotrypsinogen A are compared in Table I with the composition that is indicated by the determined sequence. Early analyses gave low values for glutamic acid, probably because of some loss of this amino acid as the pyrrolidone. All analyses give somewhat low values for lysine for reasons that are not clear, although the high serine content may be implicated.

Sequence. Bovine chymotrypsinogens A and B are compared in Table II.

Physical Properties. Physical parameters are collected in Table III. Chymotrypsinogen A is stable in solution at room temperature between pH 3.5 and 10.5; over this range it shows no change in properties that are

TABLE I
Amino Acid Compositions of Chymotrypsinogens Based on Analytical Data[a] and on Sequence Determinations

Amino acid	Bovine A Analyses	Bovine A Analyses	Bovine A Sequence	Bovine B Analyses	Bovine B Analyses	Bovine B Sequence	Porcine A Analysis	Porcine B Analysis	Porcine C Analysis	Dogfish Analysis
Lysine	13.2[b]	13.4[c]	14[d]	10.6[e]	10.9[f]	11[g]	11.4[h]	5.7[i]	7.2[i]	10.1[j]
Histidine	1.9	1.9	2	2.0	2.1	2	2.3	2.8	6.1	4.0
Arginine	4.0	3.9	4	5.1	5.0	5	5.6	7.7	8.9	6.9
Aspartic acid	21.8	22.3	23	19.5	19.8	20	21.0	20.0	25.2	20.4
Threonine	23.0	22.0	23	20.3	22.2	23	19.8	17.4	16.7	14.4
Serine	30.1	29.1	28	20.6	21.7	22	23.9	28.3	22.5	21.4
Glutamic acid	14.2	14.9	15	18.5	19.4	18	16.6	14.9	26.4	12.9
Proline	8.7	9.1	9	12.7	15.1	13	15.8	13.7	13.5	13.7
Glycine	23.3	22.2	23	22.4	23.3	23	21.8	22.0	26.6	21.4
Alanine	21.7	22.0	22	22.0	23.0	23	21.9	21.9	14.9	22.9
Valine	22.4	22.0	23	23.5	23.9	25	24.9	24.6	23.0	22.7
Methionine	1.9	1.8	2	3.8	3.8	4	1.9	2.0	1.0	3.5
Isoleucine	9.9	9.8	10	8.5	8.3	9	10.6	11.2	13.8	9.8
Leucine	18.8	19.2	19	18.6	18.7	19	19.4	16.6	21.7	11.2
Tyrosine	4.1	4.2	4	3.2	3.3	3	4.8	4.2	6.3	5.6
Phenylalanine	6.5	6.2	6	6.8	6.8	7	5.9	7.9	4.2	3.0

Half-cystine	10.0	9.8	10	9.5	9.6	10	10.5	9.6	10.3	8.4
Tryptophan	7.0	6.8	8	6.4	8.4	8	8.4	12.7	12.1	11.7
Amide	24.1	—	23	15.6	18	16	17	—	—	20.1
Mol. wt. $\times 10^{-3}$	25.1	24.9	25.6	25.0	25.6	25.7	26.0	26.0	29.0	24.5
No. of residues	243	241	245	236	245	245	246	243	260	224

[a] Each analysis is based on a series of hydrolyzates prepared with different times of heating. Serine and threonine values have been corrected for destruction by extrapolations to zero time of hydrolysis. Values for amino acids released slowly, such as valine and isoleucine, have been taken from hydrolyzates of maximum time. Tryptophan is determined spectrophotometrically and half-cystine is determined as cysteic acid after performic acid oxidation.

[b] P. E. Wilcox, E. Cohen, and W. Tan, J. Biol. Chem. 228, 999 (1957).

[c] Z. Zmrhal, Collection Czech. Chem. Commun. 27, 2662 (1962).

[d] B. S. Hartley, Nature 201, 1284 (1964). Corrections appear in B. S. Hartley and D. L. Kauffman, Biochem. J. 101, 229 (1966); O. Mikeš, V. Holeyšovský, V. Tomášek, and F. Šorm, Biochem. Biophys. Res. Commun. 24, 346 (1966); and D. M. Blow, J. J. Birktoft, and B. S. Hartley, Nature 221, 337 (1969).

[e] L. B. Smillie, A. G. Enenkel, and Cyril M. Kay, J. Biol. Chem. 241, 2097 (1966).

[f] O. Guy, D. Gratecos, M. Rovery, and P. Desnuelle, Biochim. Biophys. Acta 115, 402 (1966).

[g] L. B. Smillie, A. Furka, N. Nagabhushan, K. J. Stevenson, and C. O. Parkes, Nature 218, 343 (1968).

[h] M. Charles, D. Gratecos, M. Rovery, and P. Desnuelle, Biochim. Biophys. Acta 140, 395 (1967).

[i] D. Gratecos, O. Guy, M. Rovery, and P. Desnuelle, Biochim. Biophys. Acta 175, 82 (1969).

[j] J. W. Prahl and H. Neurath, Biochemistry 5, 2131 (1966).

TABLE II

AMINO ACID SEQUENCES OF BOVINE CHYMOTRYPSINOGEN A AND CHYMOTRYPSINOGEN B[a]

	1	2	3	4	5	6	7	8	9	10	11	12	13	14	15	16	17	18	19	20
A	Cys	Gly	Val	Pro	Ala	Ile	Gln	Pro	Val	Leu	Ser	Gly	Leu	Ser	Arg	Ile	Val	Asn	Gly	Glu
B														*Ala*						

	21	22	23	24	25	26	27	28	29	30	31	32	33	34	35	36	37	38	39	40
A	Glu	Ala	Val	Pro	Gly	Ser	Trp	Pro	Trp	Gln	Ser	Ser	Leu	Gln	Asp	Lys	Thr	Gly	Phe	His
B	*Asp*															*Ser*				

	41	42	43	44	45	46	47	48	49	50	51	52	53	54	55	56	57	58	59	60
A	Phe	Cys	Gly	Gly	Ser	Leu	Ile	Asn	Glu	Asn	Val	Val	Val	Thr	Ala	Ala	His	Cys	Gly	Val
B		*Gly*						*Ser*		*Asp*										

	61	62	63	64	65	66	67	68	69	70	71	72	73	74	75	76	77	78	79	80
A	Thr	Thr	Ser	Asp	Val	Val	Val	Ala	Gly	Glu	Phe	Asp	Gln	Gly	Ser	Ser	Ser	Glu	Lys	Ile
B															*Leu*	*Glu*	*Thr*		*Asp*	*Thr*

	81	82	83	84	85	86	87	88	89	90	91	92	93	94	95	96	97	98	99	100
A	Gln	Lys	Leu	Lys	Ile	Ala	Lys	Val	Phe	Lys	Asn	Ser	Lys	Tyr	Asn	Ser	Leu	Thr	Ile	Asn
B		*Val*				*Gly*						*Pro*		*Phe*	*Ser*	*Ile*			*Val*	*Arg*

	101	102	103	104	105	106	107	108	109	110	111	112	113	114	115	116	117	118	119	120
A	Asn	Asp	Ile	Thr	Leu	Leu	Lys	Leu	Ser	Thr	Ala	Ser	Ser	Phe	Ser	Gln	Thr	Val	Ser	Ala
B									*Ala*		*Pro*	*Gln*				*Glu*				

The sequence of chymotrypsinogen A is given in a continuous line. Every fifth residue number is indicated above the line. Replacements which convert the A sequence into the sequence of chymotrypsinogen B are given below the line in italics.

Line (residues 121–140)

				125					130					135					140
A	Val	Cys	Leu	Pro	Ser	Ala	Ser	Asp	Asp	Phe	Ala	Ala	Gly	Thr	Thr	Cys	Val	Thr	Thr	Gly
B											*Pro*				*Leu*		*Ala*			

Line (residues 141–160)

				145					150					155					160
A	Trp	Gly	Leu	Thr	Arg	Tyr	Thr	Asn	Ala	Asn	Thr	Pro	Asp	Arg	Leu	Gln	Gln	Ala	Ser	Leu
B			*Lys*		*Lys*			*Asp*	*Asn*				*Glu*					*Leu*	*Thr*	

Line (residues 161–180)

				165					170					175					180
A	Pro	Leu	Leu	Ser	Asn	Thr	Asn	Cys	Lys	Lys	Tyr	Trp	Gly	Thr	Lys	Ile	Lys	Asp	Ala	Met
B		*Ile*	*Val*			*Ser*				*Arg*						*Val*	*Thr*			

Line (residues 181–200)

				185					190					195					200
A	Ile	Cys	Ala	Gly	Ala	Ser	Gly	Val	Ser	Ser	Cys	Met	Gly	Asp	Ser	Gly	Gly	Pro	Leu	Val
B																				

Line (residues 201–220)

				205					210					215					220
A	Cys	Lys	Lys	Asn	Gly	Ala	Trp	Thr	Leu	Val	Gly	Ile	Val	Ser	Trp	Gly	Ser	Ser	Thr	Cys
B		*Gln*																		

Line (residues 221–240)

				225					230					235					240
A	Ser	Thr	Ser	Thr	Pro	Gly	Val	Tyr	Ala	Arg	Val	Thr	Ala	Leu	Val	Asn	Trp	Val	Gln	Gln
B						*Ala*									*Met*					*Glu*

Line (residues 241–245)

				245
A	Thr	Leu	Ala	Ala	Asn
B					*Glu*

[a] L. B. Smillie, A. Furka, N. Nagabhushan, K. J. Stevenson, and C. O. Parkes, *Nature* **218**, 343 (1968). See also footnote *d* to Table I. The sequence of chymotrypsinogen A is given in a continuous line. Every fifth residue number is indicated above the line. Replacements which convert the A sequence into the sequence of chymotrypsinogen B are given below the line in italics. Disulfide bridges in chymotrypsinogens A and B are at identical positions as follows: I–IV (residues 1–122), II–III (residues 42–58), V–IX (residues 136–201), VI–VII (residues 168–182), and VII–X (residues 191–220).

TABLE III

PHYSICAL PROPERTIES OF BOVINE CHYMOTRYPSINOGENS A AND B[a]

Parameter	Chymotrypsinogen	
	A	B
$\bar{v}$ (ml/g)	0.733	0.721
$[\eta]$ (ml/g)	2.18	2.26
$s^0_{20,w}$		
pH 3.0	2.49 S	2.58 S
pH 5.0		2.60
pH 7.5	2.58	
$D_{20,w} \times 10^7$ (cm² sec⁻¹) $c \to 0$		
pH 3.0	9.009	
pH 7.5	9.480	
Molecular weight (daltons)		
Light scattering	26,100	25,000
Sedimentation rate—diffusion	24,200	
Approach to sedimentation equil.		24,700
Osmotic pressure		24,000
Amino acid analysis	25,081	24,940
X-ray diffraction	25,000	
Sequence	25,635	25,717
Optical rotatory dispersion		
λ_c (nm)	231.8	231.1
a_0	$-436°$	$-465°$
b_0	$-86°$	$-97°$
$[m']_{231.5}$	$-3300°$	$-3640°$
Absorbancy coefficients		
$\epsilon \times 10^{-4}$ (282 nm)	5.13	4.81
$E^{1\%}_{1\,cm}$ (282 nm)	20.0	18.7

[a] Values for the parameters in this table have been taken from L. B. Smillie, A. G. Enenkel, and C. M. Kay, *J. Biol. Chem.* **241**, 2097 (1966). Additional references to the source of some of the parameters will be found in that paper.

sensitive to conformation. The sedimentation constant is independent of pH in this range at ionic strengths above 0.1; the protein does not tend to dimerize in the acid range as is the case for chymotrypsin A_α. At pH 2 the protein assumes a different conformational state. This process is reversible in dilute solutions, but at concentrations above 2 mg/ml and in the presence of salts such as sodium sulfate, aggregation of the protein prevents return to the native conformation. The reversible transition between the two conformational states is linked to the ionization of a single group with pK = 3.0.[7]

Binding of Ca^{2+}.[7] Chymotrypsinogen binds one Ca^{2+} strongly and

[7] M. Delaage, J. P. Abita, and M. Lazdunski, *European J. Biochem.* **5**, 285 (1968).

thereby becomes stabilized against denaturation. The association constant for the ion at pH 8.4, ionic strength 0.4, and 20° is 1.0×10^3. This constant was derived from the effect of Ca^{2+} concentration on the first-order rate constant for denaturation in urea solution. Denaturation was followed by measuring the change of absorbance at 298 nm, the wavelength at which a maximum occurs in the difference spectrum for the reaction.

The presence of Ca^{2+} has no effect on the rate of activation of the zymogen by trypsin.

Chymotrypsins A

Assay Methods

Kinetic methods are generally suitable for a variety of experimental purposes in studies of the chymotrypsins. However, the specific activity of chymotrypsin changes during activation and further autolytic modification. Some of this change is due to inherent differences among the different species of chymotrypsin, some is due to inhibition by products of autolysis, and some to the formation of inert protein fragments. Therefore, it is advisable for some purposes to determine the functional normality of the sample of enzyme and to calibrate the kinetic assay on the basis of the concentration of active molecules. (See this volume [1].)

Potentiometric Method

Procedure. The principle and procedure generally applicable to the serine proteases may be found in an earlier chapter (K. A. Walsh and P. E. Wilcox [3]). For chymotrypsin the substrate of choice is N-acetyl-L-tyrosine ethyl ester (ATEE) at an initial concentration of 0.01 M. The assay solution should always contain 0.10 M CaCl$_2$. Methanol, ethanol, or acetonitrile at a concentration of 5% may be used to increase the solubility of the substrate without affecting the assay significantly. The optimal enzyme concentration in the assay solution is 1.0 μg/ml when 0.5 N base is used as titrant in the pH-stat. The optimal concentration can be reduced to 0.2 μg/ml by using 0.1 N base, but with some loss of precision. Total protein concentration in stock solutions of chymotrypsin is determined spectrophotometrically at 282 nm. The assumption is made that the molecular extinction coefficient for all species is the same as that for zymogen ($\epsilon = 5.00 \times 10^4$, $E_{1cm}^{1\%} = 20.5$ for A$_\alpha$). The value of K_m for the enzymatic reaction is 0.0007 M; therefore the observed velocity is about 93% of V_{max}. For practical purposes the rate of hydrolysis is constant until 25% of the ATEE is hydrolyzed.

Specific Activity. The specific rate constant (k_{cat}) for the hydrolysis of ATEE by chymotrypsin A$_\alpha$ has been reported to be **193 sec^{-1} at 25°**

and 295 sec^{-1} at 31° (pH 8.6, 1.8% acetonitrile).[8] These constants were calculated on the basis of the normality of active enzyme. This determination is not far from an earlier value reported by Cunningham and Brown,[9] namely 170 sec^{-1} (pH 8.0, 25°, 0.1 M Ca^{2+}; recalculated for $E_{282}^{1\%} = 20.5$), and the agreement is exact if the enzyme used in the earlier work contained 10% inert protein. The amount of inert protein in commercial samples of A$_\alpha$ is generally found to run between 10 and 15%. Expressed in standard units of enzyme activity, a rate constant of 193 sec^{-1} corresponds to 464 micromoles per minute per milligram of enzyme.

When samples of enzyme are taken from a rapid activation mixture, the specific activity measured by the ATEE assay is considerably higher than the value for chymotrypsin A$_\alpha$. Chymotrypsin A$_\delta$ preparations show an activity of about 530 units per milligram ($k_{cat} = 220$ sec^{-1}) in assays at 25°. In the early phase of rapid activation, the peak activity may reach over 600 units/mg, reflecting the long-known fact that chymotrypsin A$_\pi$ is the most active of all forms of the enzyme.

Spectrophotometric Method

The application of the spectrophotometric method to the assay of chymotrypsin has been presented as two examples in a preceding chapter (this volume [3]). The definitions of unit and specific activity for assays using N-benzoyl-L-tyrosine ethyl ester as substrate are also given.

Differences in the specific activity between samples of chymotrypsin A$_\alpha$ and A$_\delta$ have been found in spectrophotometric assays, as well as in the potentiometric measurements noted above. Using the substrate BTEE, the activity of a freshly prepared solution of A$_\delta$ was 130% of the activity of a commercial sample of A$_\alpha$, 59 units per milligram compared to 45 units.[10] It has also been reported that the amide substrate, N-acetyl-L-tryptophan amide, is more effectively hydrolyzed by A$_\delta$ than by A$_\alpha$. The respective values of K_m (app.) are 0.002 M and 0.004 M, and the values for k_{cat} are 0.07 sec^{-1} and 0.05 sec^{-1}.[11]

Comments

The various forms of chymotrypsin A are basic proteins and therefore tend to be adsorbed on glass surfaces as monolayers. Contact of dilute solutions with glass surfaces should be kept to a minimum, otherwise large errors are introduced into the measurements and precision is poor.

[8] M. L. Bender, F. J. Kézdy, and C. R. Gunter, *J. Am. Chem. Soc.* **86**, 3714 (1964).
[9] L. W. Cunningham and C. S. Brown, *J. Biol. Chem.* **221**, 287 (1956).
[10] D. D. Miller and D. C. Teller, unpublished data.
[11] A. Hilmoe, B. C. Park, and G. P. Hess, *J. Biol. Chem.* **242**, 919 (1967).

A high concentration of Ca^{2+}, at least $0.1\,M$, appears to help overcome this problem. Some investigators have suggested the use of poor substrates for assay purposes, for example, acyl-L-valine esters[12] or fatty acid esters of o-hydroxybenzoic acid,[13] thereby permitting the use of higher concentrations of enzyme. Advantages of such methods are lessened by the reduced sensitivity of the assay and by self-aggregation of the enzyme which complicates the kinetic behavior of the system. Another approach has been to add serum albumin as a protective agent.[14]

A procedure for measuring the amidase activity of chymotrypsin, using the substrate N-acetyl-L-tryptophan amide has been recently reported.[11] The rate of the reaction is obtained from the increase in free ammonia, which is determined by means of a Technicon Autoanalyzer.

PURIFICATION OF CHYMOTRYPSIN A_α

Preparations of active chymotrypsin have never been found to be homogeneous, as demonstrated by free-boundary and disc gel electrophoresis and by end-group analysis (see Table IV). Crystalline chymotrypsin A_α contains variable amounts of inert protein (sometimes as much as 15%) and low molecular weight material. Recrystallization is not effective in lowering the level of impurities. In fact, heterogeneity may be increased because autolysis proceeds even at low pH, although slowly, and some impurities appear to be concentrated by cocrystallization. It has been reported that ninhydrin-positive material, probably autolysis products, are bound to the enzyme and interfere with some of its kinetic properties.[15] These low molecular weight impurities can be removed by gel filtration through Sephadex G-25.

Gel Filtration Through Sephadex G-25.[15] All operations are carried out at $4°$. A column of Sephadex G-25 (fine), 4.0×45 cm, is equilibrated with $10^{-3}\,M$ HCl. About 300 mg of chymotrypsin A_α are dissolved in 15 ml of $10^{-3}\,M$ HCl and the pH is adjusted to 3.0. The sample is applied to the column and eluted with $10^{-3}\,M$ HCl at a flow rate of 20 ml per hour. Fractions of 5 ml each are collected and monitored by spectrophotometry at 282 nm. The enzyme appears in about five tubes as a sharp peak. Tests of these fractions show a low ninhydrin color reaction. Emerging after the main peak is a peak of small-sized material giving a strong ninhydrin reaction. The tubes containing the high enzyme concentrations are pooled, but the last one or two tubes on the trailing edge

---

[12] T. H. Applewhite, H. Waite, and C. Niemann, *J. Am. Chem. Soc.* **80**, 1465 (1958).
[13] B. H. J. Hofstee, *Biochim. Biophys. Acta* **32**, 182 (1959).
[14] B. H. J. Hofstee, *J. Am. Chem. Soc.* **82**, 5166 (1960).
[15] A. Yapel, M. Han, R. Lumry, A. Rosenberg, and D. F. Shiao, *J. Am. Chem. Soc.* **88**, 2573 (1966).

TABLE IV

N-Terminal Analyses of Bovine Chymotrypsin Preparations[a]

	Asp	Thr	Ser	Glu	Gly	Ala	Val	Ile	Leu
Rapid activation mixture, trypsin-free, mostly A_δ	0.10	0.25	0.08	t	0.14	0.05	t	1.00	t
Rapid activation mixture held at pH 3, 48 hours, 25°	0.13	1.00	0.26	0.07	0.26	0.10	0.09	1.00	0.04
Rapid activation mixture held at pH 7.5, 3 hours, 25°	0.22	0.56	0.49	0.11	0.24	0.35	0.01	1.00	t
PMS-Chymotrypsin A_δ, purified by chromatography[b]	0.05	0.10	0.04	t	0.08	0.05	t	1.00	t
Chymotrypsin A_α, first crystals from slow activation mixture	0.14	0.24	0.32	0.08	0.18	0.85	0.09	1.00	0.03

[a] Results are expressed in moles of N-terminal residue per mole of N-terminal isoleucine. Trace amounts are indicated by t. Unpublished experiments of D. D. Miller and D. C. Teller.

[b] Results of unpublished work by E. Surbeck and P. E. Wilcox.

are discarded. The enzyme remains free of low molecular weight autolysis products for a day or two at low pH, but purity slowly deteriorates.

PREPARATION OF TRYPSIN-FREE CHYMOTRYPSIN A_δ AND THE PHENYLMETHANESULFONYL DERIVATIVE

Rapid activation of chymotrypsinogen A at 5° is complete and results in a mixture of chymotrypsins A_π and A_δ before substantial amounts of A_γ or neochymotrypsinogens have formed. Such a mixture is probably the most homogeneous sample of the enzyme that is available. However, neither A_π or A_δ can be isolated as pure, stable products because autolysis inevitably converts them to other forms. The predominant product formed at pH 3 is the one derived from A_δ in which the Tyr_{146}-Thr_{147} bond has been cleaved. In the cold this intermediate is relatively stable (see Table IV). In addition to the above complications, enzyme prepared by rapid activation is ordinarily contaminated with 2 or 3% trypsin and its degradation products.

The method described below is designed to yield a preparation that is predominantly chymotrypsin A_δ.[16] The enzyme can be isolated in a pure crystalline state by forming the inhibited derivative, resolving the π- and δ-forms by chromatography, and crystallizing the products from ammonium sulfate solution. By using acetyl-trypsin for the activation, the activating enzyme is eliminated in the chromatographic step.

Materials

Chymotrypsinogen A, 5 × recrystallized
Acetyl-trypsin, Worthington Biochemical Corp. or Mann Research Laboratories, Inc.
$CaCl_2$, 1.0 M
NaOH, 0.5 M
HCl, 1.0 M
Phenylmethanesulfonyl fluoride (PMSF), CalBiochem Corp. Dissolve 21 mg of inhibitor in 6.0 ml of purified dioxane to give a 0.04 M solution.
Sephadex G-25, coarse, Pharmacia
CM-cellulose, Whatman CM 52, microgranular
Phosphate buffers, 0.05 M and 0.08 M, pH 6.2

Procedure

Activation. Weigh 1.1 g of chymotrypsinogen into a 100-ml reaction vessel of a pH-stat. The vessel is jacketed and cooled to 5°. Add 50 ml

---

[16] E. Surbeck and P. E. Wilcox, unpublished data.

of cold, distilled water and 0.5 ml of 1.0 M CaCl$_2$. Dissolve the protein with a magnetic stirring bar and measure absorbance at 282 nm in order to determine protein concentration. Weigh 40 mg of acetyl-trypsin into a small flask, add 2 drops of the CaCl$_2$ solution, and dissolve the protein in 5 ml of water. The concentration, as determined by absorbance ($E_{280}^{1\%} = 15.6$), should be 7 mg/ml. Adjust the pH of the zymogen solution to 7.0 with 0.5 M NaOH and activate the pH-stat to maintain the pH at 7.50. Add the acetyl-trypsin solution and record the base uptake. Activation begins rapidly and continues for about 1 hour. Enzyme assays may be used to follow the activation. Activity should reach a maximum in about 60 minutes and remain high at 90 minutes, above 550 units per milligram in the ATEE assay at 25°.

Isolation of Trypsin-free Chymotrypsin $A_δ$. After 90 minutes of activation, the solution is adjusted to pH 4 by the addition of 1.0 M HCl. Approximately 20 ml of preswollen CM-cellulose that has been equilibrated with 0.05 M phosphate buffer are added to the preparation and the mixture is stirred at 5° for 20 minutes. The adsorbent is sedimented at 18,000 rpm in a preparative centrifuge and the supernatant solution is discarded. Wash the adsorbent twice with 30 ml of cold 0.05 M phosphate buffer, stirring the mixture for 5 minutes during each wash. Elute the enzyme by washing the adsorbent with 30 ml of cold 0.08 M phosphate buffer, pH 6.2. Approximately 70% of the enzyme is recovered in four of these washings. The eluants are immediately pooled and the pH adjusted to 3.0 with 1 M HCl. The protein concentration is about 5 mg/ml. Analyses for N-terminal residues in a preparation of this kind may be found in Table IV.

Preparation of Phenylmethanesulfonyl Chymotrypsin $A_δ$. This preparation is an alternative to the isolation of the active enzyme. After 90 minutes of activation the enzyme is inhibited by the addition of 1.0 ml of the PMSF solution. The base uptake at pH 7.5 is recorded. After 30 minutes one more milliliter of inhibitor solution is added. After an additional 30 minutes the inhibition should be complete and the addition of a third milliliter of inhibitor should produce no increase in the rate of base uptake. Remove a sample of the protein and test for chymotryptic activity by one of the assay methods. The activity should be less than 0.02%. Lower the pH to 4.0 by the addition of 1 M HCl. Prepare a column of Sephadex G-25 that has a void volume of 150 ml and has been equilibrated with 0.05 M phosphate buffer, pH 6.2. Pass the solution of the protein through the column, eluting with the same buffer. Collect the protein peak using absorbancy measurements at 282 nm to monitor the concentration. The recovery of protein is about 80%. The pooled solution is divided into two equal portions and stored frozen until the preparation is chromatographed.

The derivative of A_δ is isolated by chromatography on CM-cellulose according to the procedure that has been given for the purification of chymotrypsinogen A (see pp. 66–67). After preparation of the column, one portion of the inhibited enzyme containing about 400 mg of protein is pumped directly onto the adsorbent, and the gradient elution is begun. A small breakthrough peak contains the acetyl-trypsin, followed later by small amounts of impurities. At a K^+ concentration of about 0.09 M a large peak of PMS-A_δ emerges. This is followed by a smaller peak of PMS-A_π which is well separated from the main product. Fractions containing the large peak are pooled, the pH is adjusted to 4.0 with 1 M HCl, and the solution is concentrated to 15 ml by ultrafiltration in the Amincon Model 400 apparatus. The derivative is crystallized at 25° by adding an equal volume of saturated ammonium sulfate solution and adjusting the pH to 5.6. The protein separates as well-formed tetragonal bipyramids in a yield of about 200 mg. Analysis for N-terminal residues in a preparation of PMS-chymotrypsin A_δ is given in Table IV.

The above procedure may be modified to yield the derivative of chymotrypsin A_π. The conversion of A_π to A_δ is inhibited by the addition of 0.2 M β-phenylpropionate to the zymogen solution before the addition of acetyl-trypsin. At the completion of the activation, most of the β-phenylpropionate is precipitated by lowering the pH to 3. The solution is clarified by centrifugation, 2.0 ml of the phenylmethanesulfonyl fluoride solution are added, the pH is raised to 7.5, and the procedure is continued as described above. Upon chromatography, the peak of PMS-A_π is now the larger of the two main peaks. This peak may then be collected and the protein crystallized. However, in this case it is necessary to have phenylmethanesulfonyl fluoride present at all stages of the procedure to prevent the conversion of the derivative to the δ-form.

PHYSICAL AND CHEMICAL PROPERTIES OF CHYMOTRYPSINS A

Stability. The chymotrypsins A are most stable at pH 3, but even at this pH autolysis proceeds, although very slowly (see Table IV). As the pH is lowered below 3, the enzyme undergoes a reversible conformational change which is similar to that shown by the zymogen. At pH values above 10, the enzyme also undergoes a conformational change. This latter transition is associated with the loss in activity in the alkaline region. The enzyme no longer binds substrates and resembles the zymogen in several of its physical properties.[17] Chymotrypsin A_α becomes denatured

[17] J. McConn, G. D. Fasman, and G. P. Hess, *J. Mol. Biol.* **39**, 551 (1969).

in concentrations of urea above 3.5 M, but even in 8 M urea some activity persists and autolysis is extensive at pH values above 5.

Effect of Ca²⁺ and Other Metal Ions. One Ca^{2+} is bound to chymotrypsin and stabilizes the molecule, just as the binding of Ca^{2+} has been shown to stabilize the zymogen (see p. 72). At a concentration of 0.10 M, Ca^{2+} protects chymotrypsin A_δ and A_γ from denaturation in the presence of urea and greatly slows the autolysis and loss of activity of chymotrypsin A_α at pH 8.[18]

There have been several reports that Ca^{2+} specifically activates chymotrypsin A_α and also A_γ. It is not clear whether this effect is directly related to the enzymatic mechanism or is actually the result of stabilization of the enzyme against autolysis and denaturation, perhaps on glass surfaces. Definitive experiments have not been reported.

Heavy metal ions such as Cu^{2+}, Hg^{2+}, and Ag^+ inhibit chymotrypsin activity.

Physical Properties. The structure of chymotrypsin A_α has been deduced by X-ray crystallography.[19] Physical properties of the enzyme have been studied in great detail, particularly in regard to the influence of pH. A complete exposition is beyond the scope of this volume. Data and references to the details of optical rotatory dispersion and circular dichroism may be found in a report of a recent investigation.[17] The sedimentation behavior of chymotrypsin A_α is markedly dependent on pH, ionic strength, and concentration. At pH values between 3 and 5 a solution of the enzyme exists in a monomer–dimer equilibrium with maximum association occurring at pH 4.3. At 5° the association constant is about 1.5×10^4 moles per liter, but drops to low values at pH 2 and pH 5.5.[20] The molecular weight of the monomer is indistinguishable from that of the zymogen. Chymotrypsins A_δ and A_γ do not show this association behavior. At pH values above 6.5, the chymotrypsins aggregate to high molecular weight oligomers. This aggregation becomes more pronounced as the ionic strength is reduced.

The absorption spectra of the various forms of the enzyme are essentially identical, but differ from that of the zymogen to a small degree. During activation a difference spectrum develops with a peak at 285 nm.[18] The change at 282 nm is so small that the extinction coefficient at this wavelength is not affected significantly.

[18] C. H. Chervenka, *Biochim. Biophys. Acta* **31**, 85 (1959); *ibid. J. Biol. Chem.* **237**, 2105 (1962).

[19] P. B. Sigler, D. M. Blow, B. W. Matthews, and R. Henderson, *J. Mol. Biol.* **35**, 143 (1968).

[20] T. Horbett and D. C. Teller, unpublished data.

B. Bovine Chymotrypsinogen B and Chymotrypsins B

$$\text{Chymotrypsinogen B} \xrightarrow{\text{T}} \text{ChT B}_\pi \xrightarrow{\text{ChT B}} \text{ChT B}_\alpha$$

$$\text{Chymotrypsinogen B} \xrightarrow{\text{ChT B}} \text{neochymotrypsinogens} \xrightarrow{\text{T}} \text{ChT B}_\alpha$$

The activation of chymotrypsinogen B by trypsin is very similar to the activation of chymotrypsinogen A. Under conditions of rapid activation (trypsin-to-zymogen ratio of 1:40, pH 8.0, 0°) the bond between Arg_{15} and Ile_{16} is cleaved to give chymotrypsin B_π (cf. sequence in Table II). The maximum velocity (k_{cat}) for this enzymatic reaction is 1.6 sec^{-1} for chymotrypsinogen B compared with 0.18 sec^{-1} for chymotrypsinogen A, and the values of K_m are 0.57 mM and 1.1 mM, respectively.[21] The activations of the two zymogens differ in that the autolytic conversion of B_π to a form comparable to chymotrypsin A_δ does not occur. In the absence of inhibitors or stabilizers, the autolysis of B_π does proceed, however, with the cleavage of bonds in the region of residues 146 to 150. The modified enzyme has a reduced specific activity and is probably chymotrypsin B_α in large part (see below).

Under conditions of slow activation (trypsin-to-zymogen ratio of 1:2,000, pH 7.8, 0°) neochymotrypsinogens B are formed by the action of chymotrypsin B on the zymogen. The most rapid cleavage occurs at Leu_{149}-Lys_{150}. Cleavages also at Tyr_{146}-Asn_{147} and Asn_{147}-Ala_{148} result in the deletion of the peptides, Asn-Ala-Leu and Ala-Leu.[22] In the course of the activation, the neochymotrypsinogens are activated by trypsin to the corresponding enzymes. Chymotrypsin B_α may be defined as that form of the enzyme with Tyr_{146} in a C-terminal position and Lys_{150} N-terminal. It appears that B_α and the other autolyzed forms will cocrystallize with B_π; there is no evidence for the existence of conformers of the enzyme that are related in the way that chymotrypsin A_γ is related to A_α.

The conversion of chymotrypsin B_π to B_α can be inhibited by the presence of 0.1 M β-phenylpropionate or 0.01 M indole. The presence of 0.05 M Ca^{2+} also inhibits the autolysis by stabilizing the structure of the enzyme.

The isolation of chymotrypsinogen B in a pure state is more difficult than is the case for chymotrypsinogen A because (1) activation by traces of trypsin occurs more rapidly, (2) the B zymogen is found in the bovine pancreas in an amount of only about one-fourth that of the A

---

[21] J. P. Abita, M. Delaage, M. Lazdunski, and J. Savrda, *European J. Biochem.* **8**, 314 (1969).

[22] O. Guy, M. Rovery, and P. Desnuelle, *Biochim. Biophys. Acta* **124**, 402 (1966).

zymogen, (3) the B zymogen does not crystallize as readily as the A zymogen, and (4) the pancreatic deoxyribonuclease is precipitated by ammonium sulfate along with the B zymogen and shows similar chromatographic behavior.

The method of isolation of chymotrypsinogen B presented in Volume II [2] begins with an acid extract of bovine pancreas. The protein precipitated between 0.2 and 0.4 saturated ammonium sulfate is purified by repeated fractionation from salt solutions and is finally crystallized at pH 5.5 by dialysis against 0.01 M acetate buffer. The preparation contains appreciable impurities of active enzyme, neochymotrypsinogens, and deoxyribonuclease which are not removed by recrystallization. The first two of these impurities are reduced but not eliminated by inhibiting enzyme activity with DFP during the isolation and purification. Commercial samples of chymotrypsinogen B have been usually isolated by this general method, without the benefit of DFP.

Alternative procedures have been devised using chromatography on CM-cellulose and CM-Sephadex under acidic conditions which suppress tryptic activation. In one procedure, preliminary fractionation of the extract by ammonium sulfate precipitation at pH 3.0 is followed by gradient elution of the protein from a CM-cellulose column, using citrate buffer at pH 4.37 containing 0.005 M DFP.[23] The product is 90 to 95% activatable, has a very low content of neochymotrypsinogens, and free enzyme activity is less than 0.03%. In the chromatogram, however, the peak of deoxyribonuclease is not well resolved from the zymogen and recovery of pure zymogen is necessarily reduced. The procedure described below was designed specifically to give good yields of chymotrypsinogen B which are free of the nuclease. The method uses chromatography on CM-cellulose with a stepwise gradient to effect a preliminary separation of proteins, followed by chromatography on CM-Sephadex to yield the pure zymogen.

Chymotrypsinogen B

Assay Method

After tryptic activation, the amount of chymotrypsin produced from a given amount of sample is determined by a standard method (see this volume, [3]). For assay of zymogen at various steps in the isolation, it is advisable to use high levels of trypsin in the activation because pancreatic trypsin inhibitor may be present.

[23] A. G. Enenkel and L. B. Smillie, *Biochemistry* **2**, 1445 (1963).
[24] O. Guy, D. Gratecos, M. Rovery, and P. Desnuelle, *Biochim. Biophys. Acta* **115**, 404 (1966).

Procedure.[24] An aliquot containing about 100 μg of zymogen is made up to 5 ml with 0.2 M phosphate buffer, pH 7.6. About 100 μg of bovine trypsin are added and the mixture is incubated for 2 hours at 35°. If the zymogen is in the final stage of purification (CM-cellulose chromatography), only 10 μg of trypsin are used. An aliquot of the activation mixture containing about 10 μg of chymotrypsin is then assayed by the potentiometric method using 0.01 M ATEE as substrate, 3% methanol, pH 7.9, and 25°.

Calculation. The amount of chymotrypsinogen is calculated from the units of activity found by assay and the specific activity of the enzyme that is formed from the zymogen under the given conditions of activation. In the above procedure, 1 mg of chymotrypsinogen B gives about 410 units of enzyme. (One unit is equivalent to the hydrolysis of 1.0 micromole of ATEE per minute.) Under the same conditions the yield of chymotrypsinogen A is about 320 units. Standardization of this assay by using samples of the pure zymogens would be generally advisable.

Protein concentrations are determined spectrophotometrically at 282 nm. For chymotrypsinogen B, $E_{1\,cm}^{1\%} = 18.4$, and for chymotrypsinogen A, $E_{1\,cm}^{1\%} = 20.0$.

Estimation of Deoxyribonuclease

The amount of deoxyribonuclease in samples of chymotrypsinogen B may be estimated by the method given in Volume II [63]. A modification of the spectrophotometric method of Kunitz, which makes use of the hyperchromic effect that is associated with the hydrolysis of DNA, is given here as a convenient method for monitoring the effluent in the chromatography of the zymogen.

Procedure.[24] A 10-mm cuvette of 1.5 ml capacity is used in a recording spectrophotometer equipped with a thermostatted cell compartment. One milliliter of buffer (Tris-HCl, 0.005 M; NaCl, 0.04 M; MgSO$_4$, 0.01 M; pH 7.6) is equilibrated in the cuvette at 37°. Into this solution is mixed 100 μl of a stock solution of DNA (Fluka or Worthington Biochemical Corp., 400 μg/ml). At zero time a sample of 10 μl containing about 0.1 μg of deoxyribonuclease is added with mixing. The reference cuvette contains the same buffered substrate solution without enzyme. The increase in absorbance at 260 nm is recorded for about 5 minutes. One unit of enzyme activity is equal to that quantity of enzyme which produces an increase of one absorbancy unit per minute. Crystalline deoxyribonuclease (Worthington) has a specific activity of about 730 units per milligram. The concentration of the standard solution of the enzyme may be determined by absorbance measurements at 280 nm

($E_{1cm}^{1\%} = 11.5$). Samples of the effluent from chromatographic columns contain citrate buffer and must be diluted 1:10 with water before an aliquot is used for assay. Otherwise the activities of the cations in the enzymatic reaction are unbalanced.

Purification Procedure[24]

Extraction of Bovine Pancreas. Beef pancreases are removed from the animals immediately after slaughter and are rapidly frozen. All operations in the laboratory are carried out at 3–5°. The glands are thawed in $0.25\,N\ H_2SO_4$, fat and connective tissue are removed, and the tissue is minced with a meat chopper in the cold room. The mince is promptly mixed with cold $0.25\,N\ H_2SO_4$ (2 liters per kilogram of fresh tissue). After standing overnight, the acid extract is separated from solids by filtration through gauze and the residue is washed with an additional 400 ml of $0.25\,N$ acid for each kilogram of fresh tissue. The combined extracts contain about 1.5 g of chymotrypsinogen B and 6 g of chymotrypsinogen A for each kilogram of fresh tissue.

Ammonium Sulfate Fractionation. Add solid ammonium sulfate to the extract to bring the concentration to 0.2 saturation (110 g per liter of extract). Clarify the solution by centrifugation at 30,000 g, using for example the No. 1600 rotor of the Spinco Model L centrifuge. The clear liquid is removed from the centrifuge tubes, care being taken to leave behind the floating pellicle of lipid. Add ammonium sulfate to bring the concentration to 0.5 saturation (170 g per liter). Under these conditions most of the chymotrypsinogen B is precipitated, about 1.0 g per kilogram of pancreas. The precipitate contains very little trypsinogen, but about one-fifth is chymotrypsinogen A and a considerable amount of deoxyribonuclease is present. The precipitate is dissolved in water, salt is removed by dialysis against $10^{-3}\,M$ HCl, and the protein is recovered by lyophilization.

Fractionation on CM-Cellulose. A column of CM-cellulose is prepared (3×9 cm) and equilibrated with initial buffer ($0.05\,M$ citrate buffer, pH 4.20). The CM-cellulose should have a capacity of 0.5–0.6 meq of acid per gram of dry weight.[25] About 3.0 g of the dry powder obtained from the salt fractionation is dissolved in initial buffer and introduced onto the column. The column is washed with initial buffer until the absorbance at 280 nm falls to a low value (effluent volume about 250 ml). A small breakthrough peak containing inactive protein is discarded. The

[25] It has been found that high capacity adsorbent (0.9 to 1.0 meq/g) does not give satisfactory results in this procedure because chymotrypsinogen B is not eluted even by 0.2 M citrate buffer. (See footnote 23.) BioRad Cellex CM with a capacity of 0.6 to 0.7 meq/g is recommended.

eluant is changed to 0.05 M citrate buffer, pH 4.60, and fractions of about 15 ml are collected. Chymotrypsinogen B is eluted in a volume of about 350 ml. In the leading edge and in the main portion of the peak, the protein should have a specific activity of 300–330 units per milligram, but the specific activity falls to lower levels on the trailing side of the peak. The principal contaminant throughout is deoxyribonuclease. All fractions with a specific activity over 300 units per milligram are pooled, the solution is dialyzed against 10^{-3} M HCl, and the protein recovered in a yield of about 900 mg by lyophilization. Chymotrypsinogen A, which remains on the column, is eluted in a third step with 0.05 M citrate buffer, pH 4.6, containing 0.1 M NaCl.

Chromatography on CM-Sephadex. A column of CM-Sephadex (Pharmacia) is prepared (2.5 $\times$ 50 cm) and equilibrated with initial buffer (0.05 M citrate buffer, pH 4.60). The 900 mg of zymogen obtained by CM-cellulose fractionation is dissolved in initial buffer and the solution is pumped onto the column. A linear gradient of buffer is delivered to the column from a two-reservoir system. The first reservoir contains 470 ml of initial buffer and the second reservoir contains an equal volume of 0.20 M citrate buffer, pH 4.60. Fractions of about 18 ml are collected at a flow rate of 70 ml per hour. The first peak begins to emerge at an effluent volume of about 520 ml and reaches a maximum near 600 ml, correlated to a buffer concentration of 0.15 M. The main portion of this peak contains pure zymogen with a specific activity of 380–410 units per milligram after activation at 35°. Trailing fractions have lower specific activity. Emerging after the main peak and well separated from it is a peak of deoxyribonuclease. The absorbance of this peak is rather low because of a low extinction coefficient. The position of the nuclease may be determined by enzymatic assay.

All fractions which are free of nuclease and which have specific activity greater than 380 are pooled, dialyzed against 10^{-3} M HCl, and lyophilized. The yield of purified chymotrypsinogen B is about 370 mg from each kilogram of fresh pancreas. The overall yield based on the amount of zymogen in the acid extract is 24%.

Purity. Chymotrypsinogen B prepared by the above method was found to have less than 0.5% impurity as active chymotrypsin and deoxyribonuclease. Presumably the use of DFP in the preparation would reduce the activity further. The preparation appeared homogeneous in free-boundary electrophoresis and equilibrium chromatography on CM-Sephadex with 0.12 M citrate buffer, pH 4.6.[24] No end-group analysis has been reported for such a preparation. In the case of a preparation obtained by gradient elution from CM-cellulose, using a buffer containing DFP, C-terminal analysis (carboxypeptidase A method) gave a trace of tyrosine, 0.07–0.09 mole of leucine, 0.06–0.08 mole of valine, 0.03–0.06 mole of alanine, and 0.0–0.06 mole of phenylalanine.[23] These analyses indicate the presence of about 10 to 20% neochymotrypsinogen, but the presence of smaller amounts of more extensively autolyzed protein is not ruled out.

Chemical and Physical Characterization

Amino Acid Composition. An amino acid analysis of chymotrypsinogen B is compared in Table I (p. 68) with the composition that is indicated by the determined primary structure.

Sequence. Table II (p. 70) presents a comparison of the sequences of bovine chymotrypsinogens A and B. The extensive homology is evident.

Physical Properties. Some physical properties of the zymogen are presented in Table III (p. 72) along with corresponding values for chymotrypsinogen A.

Chymotrypsinogen B is stable in solution from pH 4.0 to 10.5 at temperatures below 37°. As the pH is lowered below 4.0, the protein undergoes a transition to a different conformation as indicated by measurements of absorbance at 298 nm and thermal melting. The transition is reversible at concentrations of 2.5 mg/ml and is linked to the ionization of a single group with $pK = 3.1$.[7]

Binding of Ca^{2+}. The protein possesses one binding site of high affinity for Ca^{2+}, entirely analogous to bovine chymotrypsinogen A (see p. 72). The association constant is 1.3×10^3 at pH 8.4, ionic strength 0.4, and 20°.[7] The binding of Ca^{2+} has a stabilizing effect on the protein, providing some protection against denaturation by heat and by urea. However, this effect is not as pronounced as it is for the A zymogen. The binding of Ca^{2+} does not affect the rate of activation by trypsin, but it has a large effect on the specific activity of the enzyme that is eventually formed because autolyzed species have lower activity.

Chymotrypsins B

The crystallization and physical characterization of chymotrypsin B_π has not been reported. A commercial sample of crystallized chymotrypsin B proved to have a large amount of N-terminal lysine.[26] It had been prepared by slow activation (Vol. II [2]) and probably consisted of a large amount of B_α along with some B_π. The preparation of homogeneous, crystalline B_α has not been reported and it is very likely that homogeneous preparations would be very difficult to achieve, just as is the case for chymotrypsin A_α.

---

[26] C. O. Parkes and L. B. Smillie, *Biochim. Biophys. Acta* **113**, 629 (1966).

Activation of Chymotrypsinogen B[27]

The following conditions have been recommended as optimal for the preparation of chymotrypsin B_π of highest specific activity and least heterogeneity: Zymogen concentration, 10 mg/ml; trypsin concentration, 0.25 mg/ml; buffer, 0.1 M Tris-HCl at pH 8.0; 0.05 M CaCl$_2$; 0.01 M indole; temperature, 0°. These conditions are designed to give maximum activation in 10 minutes without excessive amounts of trypsin, although the final product will contain 2.5% trypsin and its degradation products. Autolysis of the chymotrypsin is kept to a minimum by the presence of indole and calcium, and by the use of low temperature.

Specific Activity. The maximum specific activity of preparations of B_π (see above) is about 515 units per milligram in the potentiometric assay using ATEE as substrate at 25°.[27] At pH 8.0 and 0° the specific activity slowly falls to 480 in 24 hours. Crystalline enzyme prepared by slow activation of chymotrypsinogen B has a specific activity of about 340 units per milligram. Determination of the functional normality of this sample by the use of cinnamoyl imidazole indicated that it contained about 18% inert protein. The preparation appeared to be mainly a mixture of one-third B_π and two-thirds B_α.

Activation of Neochymotrypsinogen B[22]

A preparation consisting largely of chymotrypsin B_α is obtained by tryptic activation of neochymotrypsinogen B rich in N-terminal lysine. The latter protein is prepared by incubating chymotrypsinogen B with chymotrypsin B_π in a ratio of 10:1 at pH 7.8 and 25° for 1 hour. In the ATEE potentiometric assay system (3% methanol, 25°, pH 7.9) the following parameters for the active enzyme were determined: V_{max} was 250 micromoles per minute per milligram of enzyme (compared with 570 for B_π) and K_m was 2.5 mM (compared with 1.1 mM for B_π).

Stability. The relative stability of chymotrypsins B and the zymogen have been determined in solutions of urea at different values of pH.[7] Chymotrypsin B_π is the most stable with chymotrypsin B_α, the zymogen, and neochymotrypsinogen B following in order of decreasing stability.

Specificity. Chymotrypsin B shares with chymotrypsin A a specificity directed mainly toward aromatic side chains in peptides and analogous compounds. However, differences between the two enzymes have been detected by comparing their action on a natural polypeptide, glucagon,[28]

---

[27] C. O. Parkes and L. B. Smillie, *Can. J. Biochem.* **45**, 1459 (1967). Specific rate constants given in this paper have been recalculated to give specific activity in standard units.
[28] A. G. Enenkel and L. B. Smillie, *Biochemistry* **2**, 1449 (1963).

and on the isloated A-chains of bovine chymotrypsins A and B, and porcine A.[29] Differences appear in the rates by which the enzymes cleave the peptide bond on the carbonyl side of leucine. The rates are sensitive to the primary sequence of residues in the vicinity of the leucine residue. In general, chymotrypsin B appears to attack leucine in more diverse structures than does chymotrypsin A.

C. Porcine Chymotrypsinogen A

A cationic chymotrypsinogen is found in porcine pancreas that is very similar to bovine chymotrypsinogen A. The porcine zymogen accounts for 70% of the potential activity as determined by the ATEE assay method. It is readily isolated in a pure state from acid extracts of pig pancreases by a single chromatographic step. The only other proteins in the acid extract are trypsinogen and a small amount of an anionic protein. The anionic chymotrypsinogens B and C are not found in the extract.

Assay Method[24]

The specific activity of the zymogen is determined after tryptic activation at 35° by the method used to monitor the chromatography of bovine chymotrypsinogen B (see p. 83). Activity is expressed in standard units of the potentiometric assay, using ATEE as substrate. Concentration of zymogen is determined spectrophotometrically at 280 nm ($E_{1cm}^{1\%} = 18.0$).

Purification Procedure[30]

Extraction. Pancreases are removed from pigs immediately after slaughter, frozen with dry ice, and transported to the laboratory. All operations are then carried out at 5°. The glands are extracted with acid by the procedure described for bovine chymotrypsinogen B, using 2 liters of $0.10\,M$ H_2SO_4 per kilogram of fresh tissue. Ammonium sulfate (5 g per liter of combined extracts) is added and the liquid is slowly stirred for 24 hours. The extract is clarified by centrifugation for 45 minutes at 30,000 g (No. 1600 rotor in a Spinco Model L centrifuge). The clear liquid is separated from the sedimented pellet and the fatty pellicle that floats on top.

Salt Fractionation. Add 240 g of ammonium sulfate for each liter of extract. Collect the precipitate by filtration on a Büchner funnel. Dissolve the precipitate in 8 times its weight of water and add 0.9 times its

[29] D. Gratecos and M. Rovery, *Biochim. Biophys. Acta* **140**, 410 (1967).
[30] M. Charles, D. Gratecos, M. Rovery, and P. Desnuelle, *Biochim. Biophys. Acta* **140**, 395 (1967).

weight of ammonium sulfate. The precipitate is removed by centrifugation for 45 minutes at 25,000 g and discarded. Add 110 g of ammonium sulfate to each liter of the supernatant solution, and collect the precipitate by filtration on a Büchner funnel. Dissolve the precipitate in water, remove salt by dialysis against 10^{-3} M HCl, and recover the protein by lyophilization.

Chromatography. Dissolve 1.0 g of the dry protein in 20 ml of initial buffer (0.05 M sodium citrate buffer, pH 4.15). Prepare a column of CM-cellulose, 3.0 × 15 cm (BioRad, Cellex CM, capacity 0.64 meq/g), and equilibrate the adsorbent with initial buffer. Apply the sample to the column and wash the adsorbent with initial buffer at a rate of 140 ml per hour until a small breakthrough peak of anionic protein has been washed out. Change the eluant to 0.05 M citrate buffer, pH 4.35, and collect fractions of about 10 ml each. A single peak of zymogen is eluted between an effluent volume of about 200 and 400 ml.[31] The potential specific activity of the zymogen in each fraction is determined. Those fractions with specific activity of 400 units per milligram or more are pooled, the solution is dialyzed against 10^{-3} M HCl, and lyophilized. The overall yield is about 30%, 1.5 g of zymogen for each kilogram of pancreas.

Properties[30]

Purity. The method described above yields a protein that appears homogeneous in polyacrylamide gel electrophoresis at pH 5.0, in equilibrium chromatography on CM-Sephadex using 0.12 M citrate buffer (pH 4.6), and upon sedimentation in the ultracentrifuge at pH 3.6.

Amino Acid Composition. The amino acid analysis of porcine chymotrypsinogen A may be found in Table I (p. 68). The composition resembles that of bovine chymotrypsinogens A and B more than it resembles the composition of porcine chymotrypsinogen C.

Sequence. The sequence of the first 15 residues at the N-terminus of the zymogen, the A-chain, is compared with the A-chains of other zymogens in Table V. The C-terminal residue of the protein is asparagine, and 16 residues around the two histidine residues of the molecule are arranged in the same sequence that occurs in bovine zymogens A and B.

Physical Properties. The molecular weight determined by gel filtration on Sephadex G-100 is 24,600 daltons, and the value obtained by the

[31] The elution volume of the proteins and the optimal pH of the citrate buffer may differ for different batches of CM-cellulose, especially if the capacity differs very much from 0.64 meq/g. A trial run may be necessary in order to determine the best conditions.

TABLE V

COMPARISON OF THE SEQUENCES OF AMINO ACIDS IN THE

A-CHAINS OF THREE CHYMOTRYPSINS

	1	2	3	4	5	6	7	8	9	10	11	12	13	14	15
Bovine A	Cys	Gly	Val	Pro	Ala	Ile	Gln	Pro	Val	Leu	Ser	Gly	Leu	Ser	Arg
Bovine B	Cys	Gly	Val	Pro	Ala	Ile	Gln	Pro	Val	Leu	Ser	Gly	Leu	Ala	Arg
Porcine A	Cys	Gly	Val	Pro	Ala	Ile	Pro	Pro	Val	Leu	Ser	Gly	Leu	Ser	Arg

ultracentrifuge (Archibald method) is $24,300 \pm 1050$. The isoelectric point of the porcine zymogen is at pH 7.2, notably lower than the isoelectric point of the corresponding bovine zymogen. The value of the extinction coefficient, $E_{1\text{cm}}^{1\%}$, at 280 nm is 18.0.

Activation. Upon activation with bovine trypsin, the porcine zymogen is converted into chymotrypsin A_π by cleavage of the bond between Arg_{15} and Ile_{16}. Autolytic cleavage of the A-chain does not occur, and therefore there is no enzyme of the δ-form produced during activation.

D. Porcine Chymotrypsinogen B, Chymotrypsinogen C, and Chymotrypsin C

The porcine pancreas contains three different zymogens that have been isolated as relatively pure proteins, each of which can be activated with trypsin to give an enzyme with strong esterase activity toward ATEE. Porcine chymotrypsinogen A is the subject of the preceding section. It is cationic and possesses other properties that are similar to those of bovine chymotrypsinogen A, including stability at pH 3. Porcine chymotrypsinogen B, although similar to the bovine B zymogen in isoelectric point, molecular weight, and composition, differs in that it is denatured at low pH and cannot be isolated by acid extraction of porcine pancreas. Acid lability is also a property of the third zymogen, chymotrypsinogen C. This anionic protein possesses properties that are closer to those of a subunit of bovine procarboxypeptidase, fraction II (see this volume, P. H. Pétra [31]), than to the properties of either the A or B type zymogens. Most notably, the specificity of the enzyme derived by activation is distinctive. Chymotrypsin C and also the enzyme from fraction II have been found to readily attack bonds on the carbonyl side of leucine in a large variety of simple substrates and larger polypeptides.

By working in the pH region of 6 to 8, the three porcine zymogens can be extracted from acetone powders prepared from pancreases. They can be separated from each other by chromatography of the extract on DEAE-cellulose. The two anionic zymogens are not readily separated, however, from procarboxypeptidases A and B, which are prominent components in the chromatograms. Nevertheless, by applying gel filtration through Sephadex G-100 to selected fractions, the isolation of chymotrypsinogen C was achieved.[32a] The method also leads to the isolation of procarboxypeptidase A.

Acetone powders prepared from autolyzed porcine pancreas give extracts which contain proteases in the activated state. Salt fractionation and chromatography on DEAE-cellulose at pH 8 may be used to isolate porcine chymotrypsin C. This preparation possesses enzymatic properties that are the same as those of the enzyme obtained by activation of the purified zymogen.

In 1962 Gjessing and Hartnett isolated and crystallized a hydrolase of porcine pancreas which they described as an "esteroproteolytic enzyme" in recognition of its marked activity toward valeryl salicylate and poly-L-glutamic acid. Recent evidence strongly suggests that this enzyme was a form of chymotrypsin C.[32a, 32b] A procedure was developed by McConnell and Gjessing for the isolation of the zymogen from which the esteroproteolytic enzyme could be derived.[32b] Extracts of the acetone powder prepared from porcine pancreas were chromatographed on DEAE-cellulose using a pH gradient from 6.5 to 4.8. This system was found to be effective in separating procarboxypeptidase and an *anionic* trypsinogen from the zymogen which may be presumed to be chymotrypsinogen C. Activation during the isolation was prevented by the use of DFP and soybean trypsin inhibitor (STI). A second chromatography at pH 7.2 was found to be necessary in order to remove the last traces of the anionic trypsinogen, which otherwise would autoactivate and bring about the activation of the chymotrypsinogen. Such an activation occurred readily under the conditions for crystallization at pH 5, since at this pH the inhibition of trypsin by STI was incomplete. STI was finally removed by adsorption on a column of hydroxyapatite. The yield of three-times crystallized zymogen was 25 to 50 mg from 15 g of acetone powder.

The pancreatic juice is simpler in composition than neutral, aqueous extracts of the acetone powder prepared from the glands. The juice has proved to be a good starting material for the isolation of the two anionic zymogens. The number of steps in the isolation procedure are fewer and the yields are higher than is the case when the starting material is the acetone powder. After collection, the juice can be stocked in a stable form as a lyophilized powder. Chymotrypsinogens B and C each occur to the extent of about 4% of the total proteins in this powder. The

[32a] J. E. Folk and E. W. Schirmer, *J. Biol. Chem.* **240**, 181 (1965).
[32b] B. McConnell and E. C. Gjessing, *J. Biol. Chem.* **241**, 573 (1966); see also E. C. Gjessing and J. C. Hartnett, *J. Biol. Chem.* **237**, 2201 (1962).

isolation begins with chromatography on DEAE-cellulose at pH 6.0. Each chymotrypsinogen is then separated from procarboxypeptidases by gel filtration through Sephadex G-100. In this procedure advantage is taken of the tendency of chymotrypsinogen B to self-aggregate at low ionic strengths and to behave as monomer at high ionic strengths. Thus gel filtration under two different sets of conditions provides a very effective method for purification.

Chymotrypsinogen B

Assay Method[7,33]

The method that is used for the assay of bovine chymotrypsinogen B is applicable to the porcine zymogens (see p. 82). For crude samples, 200–400 μg of protein in 1.5 ml of 0.1 M Tris-HCl buffer, pH 8.0, are incubated at 25° for 30 minutes with 400 μg of bovine trypsin (Worthington). For more purified fractions, the amounts of protein are 100 μg of sample and 40 μg of trypsin. Activation is followed by the potentiometric assay using ATEE as substrate. Concentrations of zymogen are determined by absorbancy measurements at 280 nm, $E_{1\,cm}^{1\%} = 23.8$. The specific activity of the purest samples of porcine chymotrypsinogen B have been found to be about 420 units per milligram.

Purification Procedure[33]

Pancreatic Juice. The pancreatic juice of young pigs, each weighing about 70 kg, is collected by cannulation of the Wirsung duct. The juice is frozen and lyophilized. It is stable in storage at −15°. The average protein content of the dry powder is about 25%.

Chromatography on DEAE-Cellulose. All operations are carried out at 5°. A portion of the lyophilized pancreatic juice containing about 4 g of protein is dissolved in 30 ml of cold water and 150 mg of soybean trypsin inhibitor (Worthington) are added. The solution is dialyzed overnight against 20 liters of Tris-acetate buffer, pH 6.0, 10 mM in Tris. Prepare a column of DEAE-cellulose (Whatman DE II, capacity 1.0 meq/g), 3.5 × 40 cm, and equilibrate the adsorbent with the 10 mM Tris buffer. The sample is pumped onto the column, cationic proteins are washed from the adsorbent with about 600 ml of initial buffer, and then a linear gradient of NaCl is delivered to the column. The gradient is generated from 2 liters of initial buffer and 2 liters of the same buffer containing 0.4 M NaCl. Fractions of about 18 ml each are collected and

[33] D. Gratecos, O. Guy, M. Rovery, and P. Desnuelle, Biochim. Biophys. Acta 175, 82 (1969).

assayed for chymotrypsinogen. A peak of activity due to the B zymogen emerges at a NaCl concentration of 0.20 M. A second peak of activity arising from chymotrypsinogen C emerges at about 0.27 M NaCl. Fractions containing the B zymogen are pooled. The specific activity of this solution is about 100 units per milligram. The fraction also contains procarboxypeptidases A and B. The protein is concentrated by precipitation at 0.8 saturation of ammonium sulfate. The precipitate is dissolved in a small volume of water and salt is removed by dialysis against water.

The fractions containing chymotrypsinogen C may be treated in a similar fashion. The collected protein then serves as material for the isolation of the purified C zymogen (see pp. 94–95).

Gel Filtration at Low Ionic Strength. Prepare a column of Sephadex G-100 (Pharmacia) with a void volume of 500 ml (3.0 × 190 cm), and equilibrate the gel with sodium citrate buffer, pH 6.0, 30 mM in citrate. Charge the column with 700 mg of protein that has been collected in the preceding step from chromatographic fractions containing the B zymogen. Elute the proteins from the column with the 30 mM buffer at a flow rate of 50 ml per hour. Fractions of about 15 ml each are collected and are assayed for chymotrypsinogen activity. Determine protein concentration by absorbancy measurements at 280 nm.

Chymotrypsinogen B emerges in the first protein peak, well separated from procarboxypeptidase B, but mixed with procarboxypeptidase A. Fractions containing the chymotrypsinogen are pooled, concentrated by precipitation from 0.8 saturated ammonium sulfate solution as before, and dialyzed first against water and then against Tris-acetate buffer, pH 6.0, 0.1 M in Tris and 0.4 M in NaCl. The total amount of protein in the fraction is 400 to 500 mg.

Gel Filtration at High Ionic Strength. Reequilibrate the Sephadex column with Tris-acetate buffer, pH 6.0, 0.1 M in Tris and 0.4 M in NaCl. Introduce the sample of protein from the preceding step into the column and elute with the high ionic strength buffer. Monitor the effluent fractions for total protein by absorbancy measurements at 280 nm and for chymotrypsinogen by the assay using ATEE. Two peaks of protein appear in the chromatogram. The first consists of procarboxypeptidase A. Chymotrypsinogen B is now retarded and appears in the second peak, well separated from the first. A small amount of procarboxypeptidase B may be found between the two main peaks.[34] Determine the specific activity of the chymotrypsinogen fractions and pool all of those which have activities over 380 units per milligram. The protein may be con-

---

[34] As a precaution fractions may be assayed for procarboxypeptidase A and B. Methods will be found in this volume (see P. H. Pétra [31]).

centrated by ultrafiltration in an Amincon Diaflo apparatus. The yield of chymotrypsinogen B is about 60 mg from 15 g of lyophilized juice. The preparation may be stored frozen at $-15°$ in a stable state.

Properties

Purity. The preparation of chymotrypsinogen B described above appears homogeneous in disc gel electrophoresis at pH 8.6, and also in starch gel electrophoresis.

Molecular Weight. Gel filtration through Sephadex G-100 at an ionic strength of 0.5 indicates a molecular weight of about 26,000 daltons. This value is consistent with the amino acid composition.

Amino Acid Composition. An amino acid analysis of porcine chymotrypsinogen B is presented in Table I. In composition, this zymogen resembles bovine chymotrypsinogens A and B more than it resembles porcine chymotrypsinogen C. The protein has been shown to have N-terminal half-cystine and C-terminal asparagine. The A-chain, isolated after activation and performic acid oxidation, was found to have the same composition as that of the A-chain of porcine chymotrypsinogen A.

Activation. Tryptic activation occurs by the cleavage of the bond between Arg_{15} and Ile_{16}, as is the case for other chymotrypsinogens of the A and B type.

Porcine Chymotrypsinogen C

Assay Method

Potentiometric Method Using ATEE.[7] The method that is applied to the assay of chymotrypsinogen B, referred to above and used to follow the isolation of zymogens from pancreatic juice, may also be applied to the assay of chymotrypsinogen C. However, under the given conditions of activation, the highest specific activity is found to be about 180 units per milligram, compared to around 400 units per milligram for the two other porcine zymogens.

Spectrophotometric Assay Using BLEE.[32] Chymotrypsin C shows high catalytic activity toward substrates based on L-leucine, whereas the activity of chymotrypsins of the A and B type toward these substrates is much lower. Therefore an assay using N-benzoyl-L-leucine ethyl ester (BLEE) has certain advantages, especially when one wishes to distinguish chymotrypsinogen C from zymogens of the A or B type. The spectrophotometric method was presented earlier in this volume (K. A. Walsh and P. E. Wilcox [3]). In the present application the following conditions are used: BLEE concentration, $2 \times 10^{-3} M$; 0.04 M Tris-HCl buffer, pH 7.8; 30% methanol (v/v); 25°. The value of $\Delta\epsilon$ at

254 nm for the hydrolytic reaction is 840. Units of activity are expressed as micromoles of BLEE hydrolyzed per minute. BLEE, m.p. 67°–68°, may be obtained from commercial sources or may be prepared by the general method of Ottesen and Spector.[35]

Activation of the zymogen is performed under the following conditions: concentration of sample protein, 0.1–1.0 mg/ml; concentration of trypsin, 0.10 mg/ml; buffer, 0.025 M Tris-HCl, pH 8.0; temperature, 25°. Full activation is acheived during an incubation time of 10 minutes with no loss of activity for as long as 3 hours. The purest samples of chymotrypsinogen C have been found to have a specific activity of 3.00 ± 0.12 units per milligram when assayed by this method.

Purification Procedure[33]

In the chromatography of pancreatic juice on DEAE-cellulose, chymotrypsinogen C is identified as a peak of activity that emerges near 0.27 M NaCl in the elution gradient (see pp. 92–93). Fractions in this peak also contain some chymotrypsinogen B and procarboxypeptidases A and B, as well as other proteins of undetermined nature. The pooled fractions are concentrated by salt precipitation. The protein is dialyzed against water and subjected to gel filtration and chromatography on CM-Sephadex as described below.

Gel Filtration at Low Ionic Strength. A volume of solution containing 900 mg of protein is filtered through a column of Sephadex G-100 (Pharmacia) 3 × 190 cm, equilibrated with sodium citrate buffer, pH 6.0, 30 mM in citrate, at a flow rate of 50 ml per hour. Fractions of about 15 ml each are collected, protein concentrations are determined spectrophotometrically, and every other tube is assayed for chymotrypsinogen activity. The first large peak to emerge contains chymotrypsinogen B mixed with procarboxypeptidases A and B, whereas the second, retarded peak contains chymotrypsinogen C. A small peak of procarboxypeptidase B emerges between the two peaks of chymotrypsinogens.[34] All fractions containing the C zymogen at concentrations higher than 20 units per milliliter (assay method using ATEE) are pooled. The protein is precipitated from 0.8 saturated ammonium sulfate solution as before, dissolved in a little water, and the solution is dialyzed against water in order to remove the salt.

Chromatography on CM-Sephadex. Prepare a column of CM-Sephadex C-50 (Pharmacia) 2.0 × 50 cm, and equilibrate it with Tris-acetate buffer, pH 6.0, 0.05 M in Tris. The protein is pumped onto the column and eluted with a linear gradient of Tris-acetate buffer. The gradient is

[35] M. Ottesen and A. Spector, *Compt. Rend. Trav. Lab. Carlsberg,* **32,** 63 (1960).

generated by mixing 400 ml of 0.3 M buffer, pH 6.0, into an equal volume of the initial buffer. Fractions of about 10 ml each are collected and monitored for total protein and zymogen activity. A large amount of protein is eluted just after the breakthrough volume of the column. This peak is followed by small amounts of impurities and finally by a symmetrical peak of the C zymogen near a concentration of 0.20 M in the buffer gradient. All fractions containing zymogen with a specific activity greater than 150 units per milligram in the ATEE assay are pooled. The solution is concentrated by ultrafiltration in an Amincon Diaflo apparatus. The overall yield from the chromatography on DEAE-cellulose approaches 40%. About 60–70 mg of chymotrypsinogen C are obtained from 15 g of lyophilized pancreatic juice.

Properties

Purity. The preparation of chymotrypsinogen C described above yields a protein that appears homogeneous in disc gel electrophoresis at pH 8.6 and in starch gel electrophoresis at pH 8.2.

Molecular Weight. The molecular weight of porcine chymotrypsinogen C has not been established with certainty. The application of the method of gel filtration through Sephadex G-100 leads to an estimate of 23,000 daltons.[33] In the ultracentrifuge (25°, 0.05 M Tris-HCl buffer, 0.1 M NaCl, pH 7.0) the sedimentation behavior indicates a concentration-dependent association-dissociation equilibrium. Using an $s_{20, w}$ value of 2.78 S and a $D_{20, w}$ value of 7.70×10^{-7} cm^2 sec^{-1} at a protein concentration of 7.3 mg/ml, a molecular weight of 31,800 daltons was calculated.[32] By use of the method of sedimentation equilibrium at 0.25 to 0.4 mg per milliliter in 0.3 M Tris-acetate buffer, pH 6.0, a value of 28,000 ± 1000 daltons was found by extrapolation to zero concentration. The Archibald method gave a value of 30,800 with no evidence of association in the 0.3 M buffer.[33] The best fit of the molecular weight to the data on amino acid composition gave a value of 28,200. These measurements led to the conclusion that the molecular weight of chymotrypsinogen C is close to 29,000 daltons, appreciably higher than the weight of the bovine and porcine zymogens of A and B types.

Amino Acid Composition. An amino acid analysis of porcine chymotrypsinogen C is presented in Table I. It has been pointed out that the composition rather closely resembles that of bovine fraction II, an isolated subunit of procarboxypeptidase A.[32] (See this volume [31].) The N-terminal residue has been identified as half-cystine.

Activation. When chymotrypsinogen C is activated by either bovine or porcine trypsin, a new N-terminal residue appears. This residue is valine, and not isoleucine which is found in the A and B enzymes. The composition of the A-chain isolated from the C enzyme indicates that it contains only 13 residues.

Activation under different conditions gives preparations of enzyme that have identical N-terminal residues but apparently different C-terminal residues. The possibility of autolysis of the enzyme during activation is indicated.

Porcine Chymotrypsin C

Assay Method

The enzyme may be assayed by either the potentiometric or the spectrophotometric method as they are described for the assay of the zymogen on pages 94–95. However, the specific activities of characterized species of the enzyme have not been established. The specific activity of enzyme prepared by activation of the purified zymogen appears to be lower than that of the enzyme isolated from autolyzed pancreas, 3.00 ± 0.12 compared to 3.95 ± 0.15 units per milligram in the spectrophotometric assay using BLEE.[32]

Purification Procedure[32]

Acetone Powder of Autolyzed Pancreas. Frozen whole swine pancreas glands are cut into slices approximately 5 mm thick and the tissue is spread on enameled trays to a depth of 2 to 4 cm. The trays are covered with heavy paper and the tissue allowed to autolyze for 16 hours at room temperature. The entire autolyzate is extracted with 4 volumes of acetone at room temperature for 1 minute in a Waring blendor at high speed and filtered rapidly with suction. The residue is reextracted successively in the blender twice with 4 volumes of acetone, once with 4 volumes of acetone:ethyl ether (1:1 by volume), and once with 4 volumes of ethyl ether. The powder is spread out to dry at room temperature and is stored at −20°.

Extraction and Salt Fractionation. Prepare an extract by gently stirring 100 g of acetone powder with 1 liter of distilled water at room temperature for 30 minutes. The suspension is centrifuged for 30 minutes at 15,000 g and the clear supernatant solution is cooled to 2°. All further operations are carried out at this temperature. Adjust the pH of the extract to 7.2 with 1 N NaOH and add solid ammonium sulfate slowly with stirring, all the while maintaining the pH at 7.2. A total of 243 g of salt per liter of extract are added to bring the solution to 0.4 saturation. Allow the suspension to stand 30 minutes and then remove the precipitate by centrifugation at 15,000 g for 30 minutes. Solid ammonium sulfate is added to the clear solution to 0.6 saturation (132 g per liter), while the pH is again held at 7.2. After the suspension has stood for 30 minutes, the protein precipitate is collected by centrifugation at 15,000 g. The precipitate is dissolved in 50 ml of water and the solution is dialyzed against distilled water overnight.

Chromatography on DEAE-Cellulose. Prepare a column of DEAE-cellulose (capacity 0.9 meq/g) 3.5 ×20 cm and equilibrate the adsorbent with 0.005 M Tris-acetate buffer, pH 8.0. After the protein solution from the preceding step has been clarified by centrifugation, it is pumped onto the column. The chromatogram is developed with a linear gradient generated by mixing 750 ml of 0.005 M Tris-acetate buffer, pH 8.0, containing 0.45 M NaCl into an equal volume of initial buffer. Fractions of 10 ml each are collected, the protein concentration is monitored at 280 nm and the chymotrypsin activity is assayed in each tube after an effluent volume of 750 ml has been reached. A large breakthrough peak of protein is followed by several retarded peaks and finally by a trailing peak of chymotrypsin. Fractions showing high activity are pooled, the protein is concentrated by salt precipitation from 0.65 saturated ammonium sulfate solution, and the partially purified enzyme is dissolved in 5–10 ml of water. The solution is dialyzed against distilled water overnight. The enzyme is rechromatographed on a column of DEAE-cellulose, 3.5 × 10 cm, that has been equilibrated with 0.005 M Tris-acetate buffer, pH 6.0. The protein is eluted with a gradient of NaCl generated from 400 ml of buffer containing 0.45 M NaCl and 400 ml of initial buffer. After the elution of small amounts of impurities, the enzyme emerges as a single peak. The fractions containing enzyme are combined, the protein is precipitated at 0.65 M saturation of ammonium sulfate, and the suspension is stored at 4°. The yield of twice-chromatographed chymotrypsin C is 100–150 mg from 100 g of acetone powder.

Properties[32]

Stability. Chymotrypsin C may be stored at 4° and pH 6 as a suspension in 0.65 saturated ammonium sulfate solution, as a salt-free solution, or as a lyophilized powder for at least 4 months without loss of activity. The enzyme becomes denatured at pH values below 3. At pH 2.6 and 0° the enzyme loses half of its activity in 20 minutes.

Molecular Weight. The behavior of the enzyme in the ultracentrifuge indicates that an association–dissociation equilibrium exists at pH 7.0, ionic strength 0.15. Using a $s_{20,w}$ value of 2.65 S obtained by extrapolation to zero concentration and a value for $D_{20,w}$ of 10.1×10^{-7} cm^2 sec^{-1} for a 0.45% solution, an estimate of 23,000 daltons was calculated for the molecular weight. A partial specific volume of 0.726 cm^3/g was calculated from the amino acid analysis.

As noted above, the "esteroproteolytic enzyme" isolated by Gjessing and Hartnett was undoubtedly a form of chymotrypsin C. The molecular weight of this enzyme has been determined by the sedimentation equilibrium method of Yphantis and compared with the molecular weight of the zymogen (chymotrypsinogen C) determined under the same conditions. A value of about 32,000 daltons was found for both proteins, indicating that autolytic degradation was minimal during the activation of the zymogen.[32b] However, some uncertainty was introduced by the existence of an association-dissociation equilibrium for both proteins under the conditions of sedimentation.

Sequence. N-Terminal analysis by the DNP technique gave 0.9 equivalents of DNP-valine and 0.6 equivalents of DNP-cysteic acid after oxidation and acid hydrolysis. C-Terminal residues have not been identified. The sequence of the amino acids around the reactive histidine residue in the active site have been investigated. The following sequence is postulated:

Ala-Ala-His-Cys-Ile-Asn-Ser-Gly-Thr-Ser-Arg-Thr

The sequence of the first four residues is identical to the corresponding sequence in bovine chymotrypsin A and B, trypsin, and elastase. The succeeding eight residues are different.

Specificity. Compared with chymotrypsins of the A and B type, porcine chymotrypsin C hydrolyzes L-leucyl bonds to other amino acids much more readily. For example, Ser-His-Leu-Val-Glu is split into Ser-His-Leu and Val-Glu, whereas this peptide is not attacked at all by bovine chymotrypsin A or B, and only very slightly by elastase. The value of k_{cat} for the esterolytic activity of chymotrypsin C against benzoyl-L-leucine ethyl ester is 25.4 sec^{-1}, whereas k_{cat} is 0.07 sec^{-1} and 0.11 sec^{-1} for chymotrypsins A and B, respectively. On the other hand, k_{cat} for the activity of chymotrypsin C on benzoyl-L-tyrosine ethyl ester is comparable to the rate constants for the A and B enzymes. However, K_m is higher and the C enzyme does hydrolyze this substrate rather more slowly under the usual assay conditions.

An investigation of the action of chymotrypsin C on various polypeptides obtained from natural sources showed that the enzyme cleaves tyrosyl, phenylalanyl, methionyl, tryptophanyl, leucyl, glutaminyl, and asparaginyl bonds in these substrates.[36]

The distinctive specificity of chymotrypsin C is also found in reactions with chloromethyl ketone inhibitors. *N*-Tosyl-L-phenylalanine chloromethyl ketone, which reacts rapidly and specifically with chymotrypsin A, reacts only very slowly with porcine chymotrypsin C. On the

[36] T. Tobita and J. E. Folk, *Biochim. Biophys. Acta* **147**, 15 (1967).

other hand, N-tosyl-L-leucine chloromethyl ketone reacts rapidly with chymotrypsin C and more slowly with chymotrypsin A.[37]

The specificity of porcine chymotrypsin C is one additional characteristic which indicates a close relationship of this enzyme to bovine fraction II than to chymotrypsins of the A and B type.

E. Dogfish Chymotrypsinogen A and Chymotrypsin A

Among the pancreatic zymogens of the dogfish (*Squalus acanthias*) are found both cationic and anionic chymotrypsinogens. The isolation of the cationic zymogen is described below. The method begins with extraction of an acetone powder of the glands, followed by salt fractionation and chromatographic purification of chymotrypsinogen A on DEAE-Sephadex and CM-cellulose. After activation with trypsin, the enzyme can also be isolated by chromatography.

Chymotrypsinogen A[38]

Assay Method

The method of activation that is described for bovine chymotrypsinogen B (pp. 82–83) may also be applied to the dogfish zymogen. The chymotryptic activity that is generated may be determined by the potentiometric method using ATEE as substrate (see this volume [3]).

Purification Procedure

Preparation of Acetone Powder. The pancreases from freshly collected, living Pacific spiny dogfish are removed by excision and immediately frozen with dry ice. The tissue is allowed to thaw overnight in the cold room at 4°. Stripped of vessels, ducts, and connective tissue, the glands are washed with cold acetone. The tissue is then homogenized in a Waring blendor for approximately 30 seconds. The homogenized material is suspended in cold acetone (1 liter for each 200 g of fresh tissue) and the mixture is stirred for 4 hours. The suspension is filtered on a Büchner funnel and the precipitate is resuspended in an equal volume of cold acetone. This washing procedure is twice repeated. The powder is suspended in a 1:1 mixture of acetone and ethyl ether and finally in ethyl ether, both at room temperature. After the last filtration, the powder is dried in a desiccator under vacuum. The powder is sealed in polyethylene bottles and stored at —20°. The yield is about 1.6 g of acetone powder for each 10 g of fresh tissue. The powder is stable at —20° for at least 18 months.

[37] J. E. Folk and P. W. Cole, *J. Biol. Chem.* **240**, 193 (1965).
[38] J. W. Prahl and H. Neurath, *Biochemistry* **5**, 2131 (1966).

Extraction and Ammonium Sulfate Fractionation. All operations are carried out at 4°. A 100-g portion of acetone powder is stirred with 1.0 liter of cold distilled water. Foaming is suppressed by a few drops of *n*-octyl alcohol. After 4 hours insoluble material is removed by centrifugation at 10,000 rpm for 20 minutes in a Spinco Model L preparative centrifuge. The clear supernatant solution should have a very low activity in the ATEE assay.

The pH of the clear extract is adjusted to 7.5 by addition of $1 N$ NaOH. DEP ($1 M$ in isopropanol) is added to give a final concentration of $1 \times 10^{-3} M$. After 1 hour, solid ammonium sulfate is added to give a final concentration of 0.7 saturation ($3.0 M$) and the pH is readjusted to 7.5. After 4 hours the precipitate is collected by centrifugation at 10,000 rpm for 30 minutes in the Model L.

Chromatography on DEAE-Sephadex. The crude zymogen precipitate is dissolved in water and dialyzed against $0.005 M$ potassium phosphate buffer, pH 7.0. Prepare a column of DEAE-Sephadex A-50 (Pharmacia) 4.5×100 cm, and equilibrate the adsorbent with the same phosphate buffer. Pump the sample of protein onto the column and continue elution with the initial buffer until the cationic proteins have been eluted in the breakthrough peak. This peak represents 15 to 20% of the total absorbance at 280 nm. Fractions containing the cationic proteins are pooled and the solution lyophilized.

The proteins remaining on the column may be eluted by a linear gradient of sodium succinate buffer, pH 6.0, running from 0 to $0.3 M$ in succinate. The first major peak in the chromatogram contains the anionic chymotrypsinogen and the second contains the anionic trypsinogen, both as impure components.

Chromatography on CM-Cellulose. Prepare a column of CM-cellulose (capacity, 0.9 meq/g) 1.4×40 cm, and equilibrate the adsorbent with $0.005 M$ sodium succinate buffer, pH 5.0. The cationic protein is dissolved in 100 ml of $0.1 M$ succinate buffer, pH 5, and the solution is dialyzed against 12 liters of the initial buffer. The protein solution is pumped onto the column and the adsorbent is washed free of unadsorbed protein with initial buffer. A linear gradient is developed by mixing 300 ml of the $0.005 M$ succinate buffer containing $0.20 M$ NaCl into an equal volume of the same initial buffer. Fractions of 10 ml each are collected, monitored for absorbance at 280 nm, and assayed for chymotrypsinogen. A single peak emerges at a NaCl concentration of about $0.08 M$; it is preceded by a small peak and followed by a shoulder. Fractions of specific activity greater than 400 units per milligram are pooled and the solution is dialyzed against the $0.005 M$ succinate buffer. The zymogen is then rechromatographed by the procedure described above. Fractions over 500 units per milligram in specific activity are pooled, the solution dialyzed against water, and the protein recovered by lyophilization.

Properties

Amino Acid Composition. The composition of dogfish chymotrypsinogen A may be found in Table I. The absorbancy coefficient, $E_{1cm}^{1\%}$ at 280 nm, based on the anhydrous residue weights obtained from the analysis, is 21.4.

Molecular Weight. The minimum molecular weight based on the amino acid analysis is $24,568 \pm 724$ daltons. At pH 8.0 the zymogen shows a behavior in the ultracentrifuge that indicates an association–dissociation equilibrium involving oligomers. At pH 3.0, ionic strength 0.1, 20°, the protein behaves as a monomer, $s_{20,w} = 2.73$ S. Sedimentation equilibrium under these conditions gives a molecular weight of 25,400 daltons.

Electrophoresis. At least 95% of the material prepared as described above migrates as a single peak in free-boundary electrophoresis. In monovalent buffers the isoelectric point is at pH 8.7. In phosphate buffers it falls to 6.5, indicating phosphate binding.

Activation. Dogfish zymogen may be activated with bovine trypsin. In Tris-HCl buffer, pH 8.0, at 0° the activation is complete in 90 minutes when the ratio of trypsin to zymogen is 1:10 by weight. β-Phenyl-propionate and Ca^{2+} have no effect on the course of the activation. There is no evidence of the existence of an intermediate which has higher specific activity than the final product. Under conditions of activation, activity decreased only 5% from the maximum value over a period of 24 hours.

N-Terminal analysis by the DNP technique has indicated that the only N-terminal residue in the zymogen is half-cystine. Oxidation and acid hydrolysis of the DNP-protein gave 0.5 equivalents of DNP-cysteic acid (uncorrected). N-Terminal and C-terminal analysis of activation mixtures show that activation is correlated with the cleavage of an Arg-Val bond. A search for peptides that might be split out by autolysis has given negative results. Thus there is no evidence that an enzyme of the δ-form, comparable to bovine chymotrypsin A_δ, is produced during activation of the dogfish zymogen.

Chymotrypsin A[38]

Preparation and Purification

Because a high trypsin-to-zymogen ratio is advantageous for activation, succinylated trypsin is used in the preparation. This anionic form of trypsin is then eliminated in the chromatographic purification of the chymotrypsin. Succinyl-trypsin is prepared by dissolving 300 mg of salt-free trypsin (Worthington) in 8 ml of 50% saturated ammonium acetate solution and treating the enzyme at 0° with 0.2 mg of succinic anhydride, added slowly over 1 hour. The reaction mixture is dialyzed overnight against 0.005 M succinate buffer, pH 5.0. The solution is poured over a small column of CM-cellulose that has been equilibrated with the same buffer and the breakthrough fraction is collected and lyophilized.

Activation is carried out at 0° in a 0.1 M phosphate buffer, pH 8.0, containing 0.1 M β-phenylpropionate. The zymogen concentration is 5 mg/ml, and the succinyl-trypsin is at 0.17 mg/ml. Samples are taken during the course of the activation and are assayed for chymotrypsin. When a plateau of constant maximum activity is reached, add soybean trypsin inhibitor (Worthington) in an amount that is two times the weight of succinyl-trypsin used for activation. Adjust the pH to 5.0 and dialyze the solution against 0.005 M succinate buffer, pH 5.0. The protein is then chromatographed on a CM-cellulose column using a linear gradient of NaCl, as described above for the purification of the dogfish zymogen. A small breakthrough peak of succinyl-trypsin is followed by a single peak of chymotrypsin A which emerges at a NaCl concentration of about 0.08 M. Fractions with specific activity over 580 units per milligram are pooled, the solution is dialyzed against water, and the protein lyophilized.

Enzyme Activity

In the potentiometric assay using ATEE, the specific activity of the dogfish chymotrypsin A is 580–600 units per milligram. This level approaches the activity of bovine chymotrypsin A_π. The dogfish and bovine enzymes are similar in pH optimum, specificity, and inhibition by DFP, PMSF, indole, and β-phenylpropionate. There is some indication that the dogfish enzyme is somewhat broader in its attack on polypeptides such as glucagon, in this respect lying between bovine chymotrypsins A and B.

F. Chymotrypsins of Other Species

Ovine Chymotrypsin A[39]

Method of Preparation

The preparation of the crude zymogen is carried out by the procedure used for bovine zymogen (see Vol. II [2]). The protein is dissolved in 0.1 M borate buffer, pH 8.0, containing 0.1 M CaCl$_2$ and the zymogens

[39] A. Koide, T. Kataoka, and Y. Matsuoka, J. Biochem. (Tokyo) 65, 475 (1969).

are allowed to activate spontaneously at 5°. After dialysis against $10^{-3} M$ HCl and lyophilization, 6 g of crude enzyme is obtained from 1 kg of fresh tissue.

The mixture of enzymes may be separated into fractions by chromatography on CM-cellulose using a stepwise gradient of citrate buffer. The first fraction eluted with 0.02 M citrate, pH 4.0, consists of crude anionic chymotrypsin. The next fraction, eluted by 0.03 M citrate, pH 5.0, contains chymotrypsin A, whereas the last fraction eluted with 0.05 M citrate, pH 5.0, shows activity corresponding to trypsin. The chymotrypsin A fraction may be purified by rechromatography under the same conditions.

Properties

The purified enzyme appears to be homogeneous in disc gel electrophoresis at pH 4.0 and gives a single, symmetrical peak in the ultracentrifuge. The molecular weight determined by the Archibald procedure is 24,000 daltons. The isoelectric point of the enzyme is 8.0 under conditions that give a value of 8.4 for bovine chymotrypsin A_α. The specific activity of the ovine enzyme appears to be close to that of bovine A_α, and the ovine enzyme is inhibited by DFP and N-tosyl-L-phenylalanine chloromethyl ketone.

Fin Whale Chymotrypsin B[40]

Method of Preparation

Fin whale pancreas is extracted with acid by the usual procedure (see Vol. II [2]), and the protein precipitated between 0.2 and 0.4 saturation of ammonium sulfate. The zymogens in this fraction are activated at pH 8.0 and 5°. After dialysis at pH 3, the protein is lyophilized and 300 mg are dissolved in 0.02 M citrate buffer, pH 4.0. The enzymes are separated into two fractions by chromatography on CM-cellulose using a stepwise gradient of citrate buffer. The major peak eluted by 0.02 M buffer, pH 6.0, contains an anionic chymotrypsin B, whereas the major peak eluted by 0.25 M buffer, pH 6.0, contains both a cationic chymotrypsin and a cationic trypsin.

The fin whale chymotrypsin B can be purified by repeated chromatography on CM-Sephadex.[41] Such a preparation appears homogeneous in disc gel electrophoresis, gel filtration, and sedimentation in the ultracentrifuge. A value of $s_{20,w} = 1.62$ S has been obtained and a molecular weight of 17,000 determined by the Archibald technique. These values

[40] Y. Matsuoka and A. Koide, Arch. Biochem. Biophys. 114, 422 (1966).
[41] Y. Matsuoka, personal communication.

are notably smaller than the corresponding values for the bovine and porcine chymotrypsins B. In other respects the fin whale enzyme appears to be similar to the other vertebrate enzymes.

Chicken Chymotrypsin[42]

Method of Preparation

The principal protease extracted from chicken (*Gallus domesticus*) pancreas by acid is a cationic chymotrypsinogen. About 600 g of frozen pancreases are thawed in 1200 ml of ice-cold 0.25 N H_2SO_4. The glands are passed through a meat grinder and returned to the acid. The whole suspension is homogenized for 1 minute in a Waring blendor. The extract is separated by filtration through gauze and the residue is reextracted with 600 ml of acid. Solid ammonium sulfate is added to 0.2 saturation, the precipitate is removed, and the pH of the solution is adjusted to 7.0. The ammonium sulfate concentration is raised to 0.4 saturation, and the precipitate is collected, dissolved in water, and reprecipitated from 0.4 saturated ammonium sulfate solution. The precipitate is dialyzed against water at 5°, the solution clarified, and the protein lyophilized.

The protein is dissolved in 0.005 M Tris-HCl buffer, pH 8, to give a 1% solution. A column of DEAE-cellulose, 2.5 × 10 cm, is equilibrated with the same buffer and the solution is passed through the adsorbent. The cationic protein is not retarded, but large amounts of other proteins are adsorbed. The breakthrough peak is collected, the solution adjusted to pH 3 and dialyzed against 10^{-3} HCl. During these operations the chymotrypsinogen becomes partially activated.

The lyophilized preparation is dissolved in water and activation is completed at pH 7.0 and 25° by the addition of bovine trypsin, 1 mg for each 500 mg of protein. The enzyme may be purified by chromatography on a column of Amberlite CG 50. The enzyme is eluted by a pH gradient of 0.2 M phosphate buffer running from pH 5.6 to 6.9. The enzyme emerges at a pH of about 6.0.[43]

Properties[44]

Amino Acid Composition. The composition of chicken chymotrypsin is similar to the composition of bovine chymotrypsin A. The most notable

[42] C. A. Ryan, *Arch. Biochem. Biophys.* **110,** 169 (1965).

[43] A more efficient chromatographic purification would probably be possible using CM-cellulose or CM-Sephadex as recently described for several mammalian chymotrypsins.

[44] C. A. Ryan, J. J. Clary, and Y. Tomimatsu, *Arch. Biochem. Biophys.* **110,** 175 (1965).

differences are the presence of 5 histidine residues, 8 tyrosine residues, only 6 lysine residues, and 1 methionine residue in the avian enzyme.

Molecular Weight. A molecular weight determined by light scattering was found to be near 20,000. The minimum molecular weight based on the amino acid analysis is 26,300 daltons.

Activity. The chicken enzyme resembles bovine chymotrypsin A in pH optima with several substrates, inhibition with DFP, and inhibition by several natural protein inhibitors.

Salmon Chymotrypsin[45]

Method of Preparation

The ceca of freshly collected chinook salmon (*Oncorhynchus tshawytscha*) are removed by excision at sea and immediately frozen with dry ice. About 100 g of the frozen tissue is homogenized in a Waring blendor containing 500 ml of water and 2.2 g of $CaCl_2$. After 1 minute, 50 ml of CCl_4 is blended into the homogenate. The pH is adjusted to 8.0 and after standing 1 hour the mixture is centrifuged at 2400 rpm. The decanted supernatant solution is mixed with 2 volumes of acetone at 5° in order to precipitate the protein. The precipitate is recovered by centrifugation, dissolved in 0.01 M Tris-HCl buffer, pH 8.8, containing 0.01 M NaCl and 0.005 M $CaCl_2$, and the solution is dialyzed against 2.5 liters of the same buffer at 5°.

Chromatography on DEAE-cellulose at pH 8.8 using a gradient of NaCl from 0.01 M to 0.35 M resolves the protein into four fractions. The unretarded proteases emerge in the breakthrough fraction. Chromatography of this material on CM-cellulose at pH 5.0 using a linear gradient of NaCl gives one peak of activity which is associated with a protein similar to bovine chymotrypsin A_α. A second, complex peak emerges after this chymotrypsin and also shows activity against ATEE, but the peak also contains a component active against trypsin substrates. Gradient elution of the anionic proteins that are adsorbed on the DEAE-cellulose column brings out two peaks with activity against ATEE and a third, double peak with activity toward TAME, a specific substrate for trypsins. The three different proteases with chymotryptic activity behave as single components on rechromatography. The anionic enzymes have isoelectric points near pH 5 and in chromatographic behavior resemble mammalian chymotrypsin B. They also fall into this classification according to their reaction with soybean trypsin inhibitor.

[45] C. B. Croston, *Arch. Biochem. Biophys.* **112**, 219 (1965).

Hornet Protease[46]

The larvae of hornets possess an enzyme in the midgut that has properties similar to those of vertebrate chymotrypsin A.

Method of Preparation

About 250 mg of freeze-dried midgut of male hornet larvae (*Vespa orientalis* F.) are homogenized with 3 ml of 0.01 M Tris-HCl buffer, pH 8.0. The clear extract obtained by centrifugation contains 11 mg of protein per milliliter and shows a specific activity against ATEE of 57.4 units per milligram (spectrophotometric assay). No trypsinlike activity is found in this extract. The enzyme is purified by passing the solution over a column of DEAE-Sephadex A-50 (Pharmacia) and eluting the protein with the 0.01 M Tris-HCl buffer. About 70% of the activity appears in the breakthrough peak. No other type of protease activity is eluted from the column by a gradient of NaCl. The protein is further purified by gel filtration through Sephadex G-75. Activity toward ATEE appears in a symmetrical peak in the effluent solution. In terms of specific activity (950 units per milligram of protein) this fraction represents an enrichment of 16-fold over the activity in the homogenate.

Properties

The molecular weight of the hornet enzyme, determined by gel filtration through Sephadex G-75, is 12,500 daltons, one-half of the weight of vertebrate chymotrypsins A or B. In electrophoresis on cellulose acetate strips at pH 8.6, the purified enzyme shows only a minor impurity. It has a much lower mobility than bovine chymotrypsin A_α. On a molecular basis, the activity of the hornet enzyme is 5-fold greater than the activity of the bovine enzyme. The two enzymes show the same specificity in cleavage of peptide bonds in oxidized B-chain of bovine insulin. The hornet enzyme is inhibited by phenylmethanesulfonyl fluoride and N-tosyl-L-phenylalanine chloromethyl ketone, but not by N_α-tosyl-L-lysine chloromethyl ketone. These are characteristic properties of chymotrypsins.

Sea Anemone Protease A[47]

The gastric filaments of the sea anemone (*Metridium senile*) contain zymogen granules similar to the pancreatic granules of vertebrates. Fresh homogenates of the filaments show little esterase activity toward ATEE,

[46] H.-H. Sonneborn, G. Pfleiderer, and J. Ishay, *Hoppe-Seylers Z. Physiol. Chem.* **350**, 389 (1969).

[47] D. Gibson and G. H. Dixon, *Nature* **222, 753** (1969).

but on incubation spontaneous activation takes place with autocatalytic kinetics and generation of the esterase activity. The evidence points to the presence of an autoactivated trypsinlike enzyme that in turn activates one or more chymotrypsinogens. The system would then resemble the one found in vertebrates. Three cationic proteases with esterase activity toward ATEE may be isolated from homogenates of *Metridium* filaments by chromatographic methods.

Method of Preparation

Sea anemones are collected by scuba divers and the gastric filaments are dissected and quickly frozen with dry ice. The tissue is homogenized with an equal volume of 0.005 M Tris-HCl buffer, pH 8.0, and the homogenate is allowed to stand at 0° until fully activated (3–5 hours). Acetone powder prepared from this material and stored at −20° serves as a stable source of the enzyme.

The enzyme is isolated by the following steps: (1) extraction with 2 M LiCl at pH 8.0, (2) acetone fractionation at −15°, (3) chromatography on DEAE-Sephadex at pH 8.6, (4) chromatography of the breakthrough peak from the preceding step, using a column of Bio Rex 70 (BioRad Laboratories) at pH 6.0 in the presence of 5 mM indole, and (5) rechromatography under the conditions of the preceding step. Three peaks of activity appear in the final chromatogram when the fractions are assayed by the potentiometric method using ATEE. The major component, which is the first to be eluted, is given the name, *Metridium* protease A.

Properties

The known properties of the protease A indicate that it is related to the vertebrate chymotrypsins. The enzyme is inhibited by DFP. A partial acid hydrolyzate of the enzyme labeled with radioactive DFP contains a series of labeled peptides that correspond to those obtained from serine proteases having the sequence Asp-Ser-Gly around the active serine residue. Protease A is readily inhibited by N-tosyl-L-phenylalanine chloromethyl ketone, a specific inhibitor of vertebrate chymotrypsin A, but it does not react with N_α-tosyl-L-lysine chloromethyl ketone, a specific inhibitor of trypsins. In substrate specificity, protease A resembles bovine chymotrypsin A. The value of k_{cat} for the hydrolysis of ATEE is 3 times larger than the value for the bovine enzyme, whereas k_{cat} for the hydrolysis of N-acetyl-L-tryptophan ethyl ester is 6 times smaller. In its action on glucagon, the *Metridium* enzyme cleaves phenylalanyl, tyrosyl, and leucyl bonds. A trypsinlike impurity also results in the cleavage of arginyl bonds. The tryptophanyl bond is not attacked, and in this respect protease A differs from vertebrate chymotrypsin A.

[6] Chymotrypsin C (Porcine Pancreas)

By J. E. FOLK

$$R'R''CHCOR''' \xrightarrow{H_2O} R'R''CHCOO^- + {}^+H_3NR \text{ or } HOR$$

where R′ is (usually) an acylamido group; R″ is the side chain of a variety of neutral amino acids; and R‴ is —NHR or —OR.

Assay Method[1]

Principle. The method is based upon measurement of change in ultraviolet absorbancy that occurs as a result of enzymatic hydrolysis of N-benzoyl-L-leucine ethyl ester to N-benzoyl-L-leucine and ethanol.

Reagents

Tris(hydroxymethyl) aminomethane(Tris)-HCl buffer, pH 7.8, 0.08 M in Tris
N-Benzoyl-L-leucine ethyl ester,[1] 0.01 M in absolute methanol
Absolute methanol
Enzyme. Dissolve enzyme in cold water and dilute with cold water to obtain a solution containing 1 to 10 units/ml.

Procedure. The incubation mixture contains 1.5 ml of buffer, 0.6 ml of substrate, 0.3 ml of methanol, 0.55 ml of water, and 0.05 ml of enzyme. The change in absorbancy is measured in a 0.5-cm quartz cuvette at 254 nm and 25°. The hydrolysis of 1 micromole of substrate causes an increase in absorbancy of 0.222.

Definition of Unit and Specific Activity. One unit of enzyme activity is defined as that amount of enzyme that catalyzes the hydrolysis of 1 micromole of substrate per minute in the above assay. Specific activity is the number of units per milligram of protein; protein concentration is determined by the trichloroacetic acid turbidimetric procedure[2] in early stages of purification and by the use of an $E_{278}^{1\%}$ of 25.0[1] in the chromatography steps.

Purification Procedure[1]

Step 1. Extraction of Acetone Powder. One hundred grams of an acetone powder of autolyzed swine pancreas, prepared as described in this volume for the isolation of carboxypeptidase B [32] is extracted with 1 liter of distilled water at room temperature by gently stirring for

[1] J. E. Folk and E. W. Schirmer, *J. Biol. Chem.* **240**, 181 (1965).
[2] T. Bücher, *Biochim. Biophys. Acta* **1**, 292 (1947).

30–45 minutes. The extract is obtained as a clear supernatant fluid after centrifugation for 30 minutes at 15,000 g.

Step 2. Ammonium Sulfate Fractionation. The extract is cooled to 2°–5° and adjusted to pH 7 to 7.2 with 1 N NaOH. All further operations are performed in this temperature range. Solid $(NH_4)_2SO_4$ (243 g/liter) is added gradually with stirring while the pH is maintained between 7 and 7.2. The precipitate is removed by centrifugation for 15 minutes at 15,000 g.[3] Solid $(NH_4)_2SO_4$ (132 g/liter) is added to the supernatant fluid while the pH is maintained as before. The suspension is stirred for 30 minutes and the precipitate is collected by centrifugation for 30 minutes at 15,000 g. This precipitate is dissolved in about 50 ml of distilled water and is dialyzed against distilled water for 18 hours.

Step 3. First DEAE-Cellulose Chromatography. The dialyzed protein solution is applied to a column, 3.5 × 20 cm, of DEAE-cellulose which has previously been equilibrated with 5 mM Tris-HCl, pH 8.0. The protein solution is washed into the column with about 20 ml of the same buffer and the chromatogram is developed with a linear gradient of NaCl from 0 to 0.45 M NaCl in 1500 ml of 5 mM Tris, pH 8.0. Fractions of 10 ml are collected at a flow rate of 6–8 ml/minute. The chymotrypsin C emerges from the column between 0.3 and 0.4 M NaCl. The fractions containing the enzyme are combined and treated with solid $(NH_4)_2SO_4$ (43 g/100 ml). The precipitate obtained upon centrifuging this suspension is dissolved in 5–10 ml of distilled water. This solution of partially purified enzyme is dialyzed against distilled water for 18 hours.

Step 4. Second DEAE-Cellulose Chromatography. The enzyme solution obtained above is applied to a column, 3.5 × 10 cm, of DEAE-cellulose which has previously been equilibrated with 5 mM Tris-acetate, pH 6.0. The protein solution is washed into the column with 20 ml of the same buffer and the chromatogram is developed with a linear gradient of NaCl from 0 to 0.45 M in 750 ml of Tris-acetate, pH 6.0. Fractions of 5 ml are collected at a flow rate of 4–5 ml/minute. The enzyme emerges from the column between 0.3 and 0.4 M NaCl. The fractions containing the enzyme are combined and treated with $(NH_4)_2SO_4$ (43 g/100 ml). The precipitate of purified enzyme is dissolved in a minimum volume of distilled water and dialyzed against several changes of distilled water. In order to obtain the purest preparations of chymotrypsin C, the dialyzed enzyme may be rechromatographed as outlined in this step. The salt-free dialyzed enzyme solution is lyophilized and stored at 5°.

[3] The precipitate obtained upon 0.4 saturation with $(NH_4)_2SO_4$ may be conveniently used for the purification of porine carboxypeptidase A according to J. E. Folk and E. Schirmer, *J. Biol. Chem.* **238**, 3884 (1963).

PURIFICATION OF CHYMOTRYPSIN C FROM ACETONE
POWDER OF AUTOLYZED PORCINE PANCREAS[a]

Step	Protein (mg)	Units	Specific activity (units/mg protein)	Yield (%)
1. Acetone powder extract	12,000	4000	0.33	100
2. (NH₄)₂SO₄ fractionation	1,600	2360	1.5	59
3. First DEAE-cellulose chromatography, pH 8.0	333	1000	3	25
4. Second DEAE-cellulose chromatography, pH 6.0	150	600	4	15

[a] 100 g of acetone powder is used.

The results of a typical purification are summarized in the table. This procedure has been repeated many times with similar results. The large losses in benzoyl-L-leucine ethyl ester hydrolyzing activity in the first DEAE-cellulose chromatography step may be explained by the fact that this ester is not a specific substrate for chymotrypsin C. Chymotrypsin A, which also hydrolyzes the ester, does not chromatograph on DEAE-cellulose under the present conditions and appears in the forepeak of the first chromatogram.

Properties

Stability. The dry purified enzyme powder has been kept frozen for periods up to 4 years without detectable loss in activity. Solutions of the enzyme in water or Tris buffer at pH 6–8.5 (5–20 mg/ml), under toluene vapors to prevent bacterial contamination, may be kept stored for several weeks without loss in activity. The purified enzyme in solution loses activity rapidly at acid pH. At pH 3 and 0° a 12% loss in activity occurs in 20 minutes.

Purity and Physicochemical Properties. The enzyme prepared by this procedure contains no protein impurities detectable by the usual physical procedures, including disc gel electrophoresis. There is, however, a small amount of contaminating carboxypeptidase A activity present in all preparations. This activity may be eliminated in digestion experiments by addition of 5–10 mM 1,10-phenanthroline, an effective inhibitor of the carboxypeptidases.[4] 1,10-Phenanthroline does not inhibit chymotrypsin C.[1,4]

A preliminary estimate of molecular weight of **23,800** has been made

[4] J. E. Folk and P. W. Cole, *J. Biol. Chem.* **240**, 193 (1965).

on the basis of sedimentation and diffusion ($s_{20,w}^0$ 2.67 and $D_{20,w}$ 10.1 $\times$ 10^{-7} cm^2 sec^{-1}). The sedimentation and diffusion behavior of the enzyme is indicative of a concentration-dependent, association–dissociation equilibrium.

Chymotrypsin C shows the following amino acid composition: Asp_{22} Thr_{14} Ser_{20} Glu_{21} Pro_{12} Gly_{25} Ala_{12} ½ Cys_7 Val_{19} Met_1 Ile_{12} Leu_{19} Tyr_6 Phe_4 Lys_7 His_5 Arg_7 Trp_8 Amide N_{14}.[1] The sequence of amino acids around the essential serine has been determined as Gly-Asp-Ser-Gly[5]; that around an essential histidine as Ala-Ala-His-Cys-Ile-Asp-Ser-Gly-Thr-Ser-Arg-Thr.[6]

Specificity. The purified enzyme catalyzes the hydrolysis of peptide bonds that contain the carboxyl group of a variety of α-amino acids including tyrosine, phenylalanine, tryptophan, methionine, leucine, glutamine, and asparagine.[4] It has been suggested that the broad specificity of chymotrypsin C toward peptide substrates is directed by α-amino acids, the side chains of which contain at the β-carbon atom an uncharged carbon-containing monosubstitution.[4] The enzyme also catalyzes the esterolysis of a variety of N-substituted amino acid esters.[1]

Activators and Inhibitors. The enzyme is active over a broad range of pH from about 5 to 9. It displays optimum activity in Tris or NH_4 HCO_3 buffers at pH 7.8–8. It is readily inactivated by diisopropyl phosphorofluoridate,[1] by diphenylcarbamyl chloride,[1] and by tosyl-L-leucine chloromethyl ketone.[6]

Distribution. The enzyme has been isolated only from swine pancreas glands. It exists in fresh pancreas tissue as an inactive zymogen. The zymogen has been purified and partially characterized.[1] The enzyme is similar to, if not identical with, a crystalline "esteroproteolytic" enzyme of porcine pancreas described by Gjessing and Hartnett.[7] A similar enzymatic activity of bovine pancreatic gland has been described.[8]

[5] T. Tobita and J. E. Folk, unpublished data.
[6] T. Tobita and J. E. Folk, *Biochim. Biophys. Acta* **147**, 15 (1967).
[7] E. C. Gjessing and J. C. Hartnett, *J. Biol. Chem.* **237**, 2201 (1962).
[8] J. R. Brown, M. Yamasaki, and H. Neurath, *Biochemistry* **2**, 877 (1963).

[7] Elastase

By DAVID M. SHOTTON

Assay Methods

Assays Using Elastin as a Specific Natural Substrate

Principle. Porcine pancreatic elastase is an endopeptidase which will digest a wide variety of protein substrates.[1] It is unique among the principal pancreatic endopeptidases, however, in its ability to digest elastin, the elastic fibrous protein of connective tissue.[2] This property forms the basis for many assay methods which involve the measurement of the amount of insoluble elastin solubilized by elastase digestion in a certain time. These have been reviewed by Mandl in an earlier volume in this series[3] and elsewhere,[4] and include the gravimetric determination of the amount of undigested elastin remaining after incubation with elastase,[1,5] and the measurement of the amount of peptide material released into solution by increases in optical density at 280 nm,[6] Folin phenol color,[7] biuret color,[8,9] or refractive index,[10] or by decreases in turbidity.[7] However, the most convenient assays are those which involve the colorimetric determination of the amount of dye released into solution by elastase digestion of dyed elastin substrates. Congo red-elastin,[11] azoelastin,[12] and orcein-elastin[13] have been used for this purpose, of which the first and last are the most commonly employed. The Congo red-elastin method of Naughton and Sanger[11] is a logical development of the use of Congo red dyed fibrin to measure general proteolytic activity,[14] while the orcein-elastin method of Sachar *et al.*[13] has the advantage that orcein is a stain specific for elastin, so that traces of collagen or other

[1] U. J. Lewis, D. E. Williams, and N. G. Brink, *J. Biol. Chem.* **222,** 705 (1956).

[2] S. M. Partridge and H. F. Davis, *Biochem. J.* **61,** 21 (1955).

[3] I. Mandl, Vol. V, p. 665.

[4] I. Mandl, *Advan. Enzymol.* **23,** 163 (1961).

[5] J. Balò and I. Banga, *Biochem. J.* **46,** 384 (1950).

[6] N. H. Grant and K. C. Robbins, *Arch. Biochem. Biophys.* **66,** 396 (1957).

[7] I. Banga, J. Balò, and M. Horvath, *Biochem. J.* **71,** 544 (1959).

[8] D. A. Hall, *Biochem. J.* **59,** 459 (1955).

[9] H. Cohen, H. Megel, and W. Kleinberg, *Proc. Soc. Exptl. Biol. Med.* **97,** 8 (1958).

[10] D. A. Hall and J. W. Czerkawski, *Biochem. J.* **73,** 356 (1959).

[11] M. A. Naughton and F. Sanger, *Biochem. J.* **78,** 156 (1961).

[12] L. Robert and P. Samuel, *Experientia* **13,** 167 (1957).

[13] L. A. Sachar, K. K. Winter, N. Sicher, and S. Frankel, *Proc. Soc. Exptl. Biol. Med.* **90,** 323 (1955).

[14] N. E. Roaf, *Biochem. J.* **3,** 188 (1908).

protein contaminants in the substrate do not affect the validity of the assay. They are essentially similar in principle, and procedures described for one substrate can be used with the other if desired.

Because of the peculiar cross-linked structure of elastin, the rate of release of peptides into solution is not linear with respect to the rate of hydrolysis of peptide bonds in the substrate by elastase. In consequence none of the above-mentioned assays give linear progress curves (see Fig. 1). The orcein-elastin method of Sachar et al.[13] involves the measurement of the amount of digestion after an arbitrarily defined interval of 20 minutes. Consequent inaccuracies arise because the elastin in different assays will be at various different stages of breakdown when the digestions are stopped (see Fig. 1). Calibration curves obtained by this method are far from linear.

In an effort to avoid these difficulties, modifications[15] have been made to the Congo red-elastin method of Naughton and Sanger,[11] whereby the time taken to reach a defined stage of digestion, namely 50% solubilization of the substrate, is measured. This time bears a close inverse ratio to the amount of elastase present.

This modified Congo red-elastin method[11,15] is described in detail below. A similar method has been reported by Gertler and Hofmann.[16]

Reagents

Substrate. Purified powdered bovine or equine ligamentum nuchae elastin, free from collagen and other contaminants,[2] is stained by stirring overnight in a saturated aqueous solution of Congo red. It is then washed well with water until the washings are colorless, and dried by successive washings of acetone and ether.[11] The Congo red-elastin is finely ground in a mortar, sieved through a 120-mesh sieve, and suspended in 0.02 M sodium borate buffer,[17] pH 8.8, at a concentration of 1 mg/ml (Elastin, and also orcein-elastin may be obtained commercially from Sigma Chemical Co., St. Louis, Mo.).

Enzyme. Aqueous solutions of elastase, suitably diluted with 0.02 M sodium borate buffer, pH 8.8, immediately prior to the assay.

Procedure. Seven milliliters of Congo red-elastin suspension are placed in an assay tube, centrifuged, and the blank optical density of the supernatant measured against distilled water. At zero time 1 ml of

---

[15] D. M. Shotton, Ph.D. dissertation, University of Cambridge. England, 1969.

[16] A. Gertler and T. Hofmann, J. Biol. Chem. **242**, 2522 (1967).

[17] Here and throughout this article, the molarity given describes the concentration of the cationic component of the buffer.

elastase solution is added to the tube, which is stoppered and shaken vigorously for about 5 seconds. Elastase binds strongly to the insoluble elastin substrate, so that continuous shaking is not required to produce intimate contact of enzyme and substrate.[11] Digestion is allowed to proceed at room temperature to completion, and at regular intervals, usually every 5 minutes, the contents of the assay tube are mixed, centrifuged for 30 seconds, and the optical density of the supernatant is measured. This can be done by decanting and measuring the optical density at 495 nm in a spectrophotometer, but excellent results are also obtained by the simpler procedure of measuring the optical density directly in the assay tube using a simple colorimeter with a suitable filter.[18] After each measurement the tube is shaken and the reaction is allowed to proceed.

From the smooth sigmoidal progress curve obtained for the assay by plotting the increase in optical density against time, the time taken to achieve the release of 50% of the Congo red into solution is determined. A plot of such times obtained from a series of calibration assays against the reciprocal of the concentration of pure elastase[19] present gives a calibration curve which is almost linear, from which the amount of active elastase present in the solutions being assayed can be determined. The precise shape and gradient of the calibration curve will vary slightly with variations in the particle size and concentration of the Congo-red elastin suspension used, and in room temperature. A separate calibration curve is therefore plotted for each batch of assays. Figures 1 and 2 show the progress curves and calibration curve of a typical series of calibration assays. The report of Gertler and Hofmann[16] that the calibration curve is linear is an approximation which holds true over small ranges of enzyme concentration.

This method has been found to give accurate and reproducible results. Under the conditions of the assay, 50% digestion is achieved in about 15 minutes by 0.1 mg of elastase. The extinction coefficient $E_{495 nm}$ for a 1 mg/ml solution of digested Congo red-elastin is about 1.0.[11]

Definition of Unit and Specific Activity. In most of the elastin-digesting assays, one elastase unit is defined as the amount of enzyme which digests 1 mg of elastin under the assay conditions within an

---

[18] For this purpose an EEL Portable Colorimeter (Evans Electroselerium Ltd., Halstead, Essex, England) was found suitable. Good quality rimless thin-walled glass test tubes, $4 \times \frac{5}{8}$ in., were used as assay tubes, as they conveniently fitted both bench centrifuge and colorimeter, while possessing good optical transmission properties. They could be centrifuged without breakage at the moderate speeds required to sediment the Congo red-elastin.

[19] See p. 117.

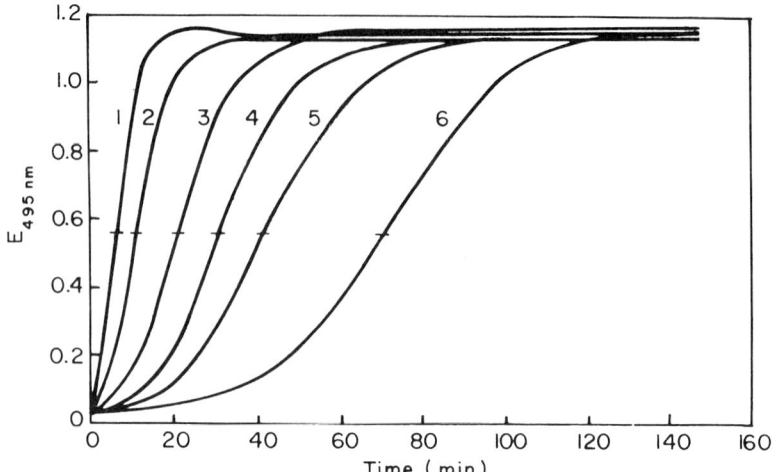

FIG. 1. Digestion progress curves showing the rate of release of Congo red into solution during the digestion of Congo red-elastin by six different amounts of elastase ($1 = 0.500$ mg, $2 = 0.167$ mg, $3 = 0.067$ mg, $4 = 0.040$ mg, $5 = 0.027$ mg, and $6 = 0.013$ mg). The point at which 50% digestion occurred is marked on each curve. A control assay containing no elastase, not illustrated here, showed no increase in optical density.

arbitrarily defined time interval. While any one assay under standard conditions yields fairly reproducible results, the conditions vary so widely from one method to another that direct comparisons of the results obtained from, and the elastase units defined by different methods are

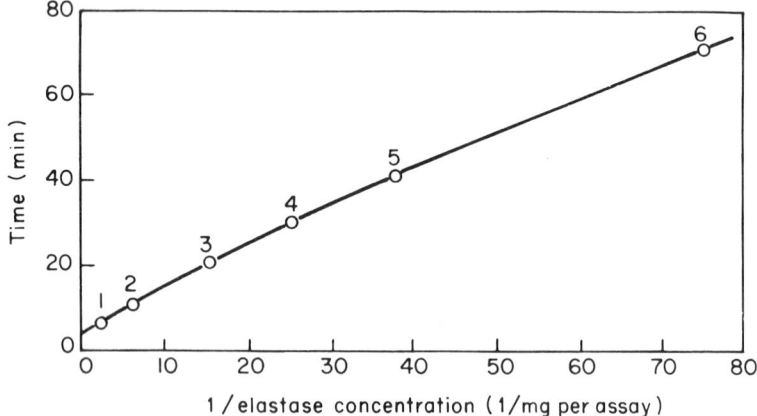

FIG. 2. Congo red-elastin assay calibration curve obtained from the results of the six assays shown in Fig. 1, by plotting the reciprocal of the elastase concentration in each assay against the time taken to achieve 50% digestion of the Congo red-elastin.

generally only of qualitative significance. The various units proposed, and the problems involved in their comparison, are reviewed by Mandl.[3]

Because of the nature of the Congo red-elastin assay, no unit of activity is defined. Instead, elastase activity is expressed in terms of specific activity relative to that of pure elastase which is, by definition, unity. This is possible because of improved purification techniques[19] which yield elastase suitable for use as an absolute standard.

The activity of elastase toward synthetic substrates can be defined much more precisely, and is discussed on pages 118, 119, and 135.

Application of Assay to Crude Preparations. Elastase is the only major pancreatic enzyme which can digest elastin (but see p. 120), so the method described above can be used to assay for the presence of elastase in crude preparations. Several factors complicate the quantitative application of this assay to crude preparations, and should be borne in mind when assaying early steps in the purification of elastase. If the assayed material contains substantial amounts of proteins other than elastase, these may act as alternative substrates, reducing the amount of elastase available to digest the Congo red-elastin and leading to a spuriously low assay result. Conversely, it has been found that the rate of digestion of elastin is substantially increased by the presence of trypsin or chymotrypsin,[20] which are presumably able to hydrolyze certain peptide bonds in the partially digested substrate after initial elastase attack, although they cannot themselves initiate digestion. These enzymes are present in large quantities in the initial extract of the elastase preparation. Furthermore, proelastase, the inactive precursor of elastase which is present in large quantities in the pancreas,[21] is also extracted by the purification procedure, and is present in the early steps (see p. 120).[20] It is rapidly activated by contaminating amounts of trypsin under the conditions of the assay, giving rise to an increase in the assay result. For these reasons the results of Congo red-elastin assays on crude elastase preparations, although very useful, are not of precise quantitative significance (see Table I). Some of these factors also apply for assays of elastase activity using the specific synthetic substrates described below.

Assays Using Specific Synthetic Substrates

Principle. Elastase hydrolyzes peptide bonds on the C-terminal side of amino acids bearing uncharged nonaromatic side chains,[11] in contrast to trypsin and the chymotrypsins which show substrate specificity for

---

[20] A. Gertler and Y. Birk, *European J. Biochem.* **12,** 170 (1970).
[21] H. D. Moon and B. C. McIvor, *J. Immunol.* **85,** 78 (1960).

basic and aromatic amino acid side chains, respectively. This specificity difference is the basis for an assay using N-benzoyl-L-alanine methyl ester as a specific substrate, which has been recently reported by Kaplan and Dugas,[22] and Geneste and Bender.[23] In principle it is identical to the use of N-benzoyl-L-arginine ethyl ester (BAEE) and N-acetyl-L-tyrosine ethyl ester (ATEE) to assay specifically for tryptic and chymotryptic activities, and represents a substantial increase in convenience over the Congo red-elastin assay described on page 113. However, it should be noted that this substrate is also hydrolyzed slowly by trypsin and chymotrypsin, at about 1 and 2% of the rate at which it is hydrolyzed by elastase itself[15, 25] (an effect very similar to that observed for the hydrolysis of BAEE by chymotrypsin and of ATEE by trypsin), so that the assay is not as specific as the elastin-digesting assays. Furthermore, it should be realized that this assay measures esterase activity, while the Congo red-elastin assay measures normal peptidase activity, there being clear evidence[22] that these two activities can vary independently with changes in conditions (see p. 135). The rate of hydrolysis of N-benzoyl-L-alanine methyl ester may be conveniently followed either spectrophotometrically[22] or by titration in a pH-stat.[23] A method[15] based on that reported by Kaplan and Dugas[22] is given below.

Reagents

> Substrate. Stock solution of 10 mM N-benzoyl-L-alanine methyl ester, prepared by the method of Hein and Nieman,[25] in 0.1 M KCl, 0.01 M Tris-HCl buffer, pH 8.0 (storable frozen).
> Enzyme. Aqueous solutions of elastase.

Procedure. Six milliliters of substrate stock solution are placed in a jacketed incubation chamber of a pH-stat controlled at 25°, and stirred magnetically under nitrogen. At zero time about 0.1 mg of elastase is added, and the pH maintained at 8.0 by titration with 0.1 N NaOH. The initial rate of hydrolysis of N-benzoyl-L-alanine methyl ester, as recorded by the rate of titration, is directly proportional to the concentration of elastase present. The assay can thus be directly calibrated using a sample of pure elastase.[19]

Definition of Unit and Specific Activity. One unit of elastase is defined as that amount which hydrolyzes 1 micromole of substrate per minute under the conditions described. The specific activity of pure

---

[22] H. Kaplan and H. Dugas, *Biochem. Biophys. Res. Commun.* **34,** 681 (1969).
[23] P. Geneste and M. L. Bender, *Proc. Natl. Acad. Sci. U.S.* **64,** 683 (1969).
[24] A. C. Reimer, personal communication, 1969.
[25] G. E. Hein and C. Niemann, *J. Am. Chem. Soc.* **84,** 4487 (1962).

elastase[19] against N-benzoyl-L-alanine methyl ester is 13.4 units per milligram of enzyme.[15] Various other kinetic parameters determined for the elastase-catalyzed hydrolysis of this substrate[22, 23] are given on page 135. It should be noted that the rate of hydrolysis varies with changes in substrate, salt, and Tris concentrations as well as in enzyme concentration.[15]

N-Carbobenzoxy-L-alanine p-nitrophenyl ester has also been shown to be a good synthetic substrate[23] (see p. 135).

Assays Using Nonspecific Synthetic Substrates

Principle. A number of synthetic substrates exist which can be hydrolyzed by all serine proteinases and by other hydrolytic enzymes. Esters of p-nitrophenol are substrates whose rate of hydrolysis can be readily followed spectrophotometrically by measurement of the rate of release of p-nitrophenol into solution, and these have been studied by Bender and his collaborators.[23, 26, 27] They have shown[26] that p-nitrophenol diethyl phosphate (Paraoxon), which had previously been shown to react with chymotrypsin and trypsin, irreversibly acylating the active center serine residue, leading to complete inhibition of the enzymes and the liberation of 1 mole of p-nitrophenol per mole of enzyme,[28, 29] also inhibits elastase in a similar fashion. They have proposed this reagent for use in the determination of elastase concentration by titration of the number of active sites.[26] However, it should be noted that this method is capable of giving spuriously low results. Two molecules of p-nitrophenyl diethyl phosphate can react in solution to give 2 molecules of p-nitrophenol, plus 1 molecule of tetraethylpyrophosphate. This latter compound is very highly reactive toward the active center serine residue of serine proteinases, more so than diisopropylphosphofluoridate (DFP) and much more so than p-nitrophenyl diethyl phosphate itself, so that even minute traces of it in the stock p-nitrophenyl diethyl phosphate solution will lead, in conditions of substrate excess, to a rapid and unobserved acylation of a substantial quantity of the enzyme, reducing the amount free to react with p-nitrophenyl diethyl phosphate, and thus reducing the amount of p-nitrophenol release observed.[30] Bender and Marshall have also studied[27] the hydrolysis of other esters of p-nitro-

[26] M. L. Bender, M. L. Begué-Cantón, R. L. Blakeley, L. J. Brubacher, J. Feder, C. R. Gunter, F. J. Kézdy, J. V. Killheffer, Jr., T. H. Marshall, C. G. Miller, R. W. Roeske, and J. K. Stoops, *J. Am. Chem. Soc.* **88**, 5890 (1966).
[27] M. L. Bender and T. H. Marshall, *J. Am. Chem. Soc.* **90**, 201 (1968).
[28] B. S. Hartley and B. A. Kilby, *Biochem. J.* **50**, 672 (1952).
[29] B. A. Kilby and G. Youatt, *Biochem. J.* **57**, 303 (1954).
[30] B. S. Hartley, personal communication, 1969.

phenol, including N-carbobenzoxy glycine p-nitrophenyl ester, which reversibly acylate elastase and are turned over by the enzyme in the normal way, and have analyzed the kinetics of the hydrolysis of p-nitrophenyl trimethyl acetate in detail (see p. 135). Wasi and Hofmann[31] have used N-carbobenzoxy glycine p-nitrophenyl ester as a substrate to measure elastase activity of purified elastase, illustrating the fact that nonspecific substrates can be used to assay for elastase activity provided that the purity of the enzyme, and the absence of other enzymes capable of attacking such substrates, has been previously determined by other means.

For experimental details of these methods the reader is referred to the original papers.[26, 27, 31]

Purification

Discussion

Elastase is present in all mammals investigated, including man,[4] but porcine pancreatic elastase has been the enzyme which has received the most attention, chiefly because it can be easily prepared in good yield from commercially available starting materials. All the preparative methods devised start with some form of acetone powder of porcine pancreas,[1, 8, 32] or fresh pancreas itself.[20, 33] Trypsin 1-300 (Nutritional Biochemicals Corp., Cleveland, Ohio, and elsewhere) and Pancreatin (Merck and Co., Inc., Rahway, N.J. and elsewhere) are two such commercial preparations of acetone powder of porcine pancreas which have been used as starting materials, of which the former is the richer source.[1] These powders are stable, and maintain their activity for years without refrigeration.

Commonly the first step in the preparation is to extract the acetone powder with 0.1 M sodium acetate buffer, pH 4.5[1, 8, 33, 34] or with alcohol solvents containing 1–2% salt,[32] although alkaline buffers have been used.[34] A variety of materials including elastin,[2, 35] alumina gel,[10, 35] and various ion-exchange materials[32, 36-39] have been used with some success to absorb elastase from the crude extract, but the most common puri-

[31] S. Wasi and T. Hofmann, Biochem. J. 106, 926 (1968).
[32] D. Bagdy and I. Banga, Acta Physiol. Acad. Sci. Hung. 11, 371 (1957).
[33] I. Banga, Acta Physiol. Acad. Sci. Hung. 3, 317 (1952).
[34] D. A. Hall and J. E. Gardiner, Biochem. J. 59, 465 (1955).
[35] D. A. Hall, Arch. Biochem. Biophys. 67, 366 (1957).
[36] N. H. Grant and K. C. Robbins, Chem. Abstr. 53, 15171 (1959).
[37] D. Badgy and I. Banga, Experientia 14, 64 (1958).
[38] J. Thomas and S. M. Partridge, Biochem. J. 74, 600 (1960).
[39] F. Lamy, C. P. Craig, and S. Tauber, J. Biol. Chem. 236, 86 (1961).

fication procedures applied to this initial extract involve a fractional
ammonium sulfate precipitation[1,8,33,34] and often an ethanol fractionation,[1,33] followed by dialysis against water to yield a euglobin precipitate.[1,33-35] This precipitate is a mixture of two major components, one
of which is elastolytic, and at least three nonelastolytic minor components, one of which is a highly active insulinase.[1,11,40-42] Lewis et al.[1]
showed that this euglobulin precipitate could be recrystallized to yield
a purer but still inhomogeneous material which contains elastase (50–
80%), a second nonelastolytic and more acidic major component,[1] and
possibly one nonelastolytic minor component.[41] This material is commercially available (from Worthington Biochemical Corp., Freehold,
N.J.; British Drug Houses Chemicals Ltd., Poole, England; and other
sources) under the misleading name of Crystalline Elastase. Lewis et al.[1]
found that the "crystalline elastase" resembled the euglobulin precipitate
by being totally insoluble in water, and only partially soluble in buffers
between pH 5 and pH 9. The two major components of "crystalline
elastase" could be separated by electrophoresis at various pH values,
which enabled the preparation of elastase essentially free from contaminants. They showed that electrophoretically purified elastase was
very soluble in distilled water and dilute salt solutions, and was homogeneous by electrophoresis at six different pH values and by ultracentrifugation and diffusion studies. It showed an increase in specific activity
of 20% over "crystalline elastase," and although measurements of the
degree of contamination by trypsin and chymotrypsin were not made,
this preparation was elastase of quite high purity. It has been commercially available from Worthington under the name of Purified
Elastase. The acidic nonelastolytic second major component of "crystalline elastase" exhibited the solubility behavior of a euglobulin even after
electrophoretic separation from elastase,[1] but has not been further
characterized.

Uriel and Avrameas[43] fractionated all the hydrolytic enzymes from
autolyzed porcine pancreas by ammonium sulfate precipitation followed
by chromatography on DEAE-Sephadex. They isolated two elastolytic
enzymes which were chromatographically, electrophoretically, immunochemically, and catalytically distinct. One enzyme, corresponding to the
elastolytic euglobulin of Lewis et al.,[1] described above, was isolated in
good yield. It showed a high activity toward Congo red-elastin which

[40] U. J. Lewis and E. H. Thiele, J. Am. Chem. Soc. 79, 755 (1957).
[41] U. J. Lewis, D. E. Williams, and N. G. Brink, J. Biol. Chem. 234, 2304 (1959).
[42] J. S. Baumstark, W. A. Bardawil, A. J. Sbarra, and N. Hayes, Biochim. Biophys. Acta 77, 679 (1963).
[43] J. Uriel and S. Avrameas, Biochemistry 4, 1740 (1965).

was not stimulated by cysteine, and negligible activity toward ATEE. Much smaller quantities of the second elastolytic enzyme were isolated (about 5% of the euglobulin yield). This enzyme showed a low activity against Congo red-elastin (about 7% that of the euglobulin) which was increased by 45% in the presence of $2.5 \times 10^{-2} M$ cysteine, and showed appreciable activity toward ATEE (about 60 times that of the euglobulin). Like the euglobulin, it was totally inhibited by $10^{-3} M$ DFP. It has not been properly named, nor further characterized.

Naughton and Sanger[11] isolated elastase from the euglobulin precipitate and from "crystalline elastase," both prepared by the method of Lewis et al.,[1] by chromatography on carboxymethyl cellulose (CM-cellulose) at pH 4.5. They obtained purified elastase which was homogeneous by chromatography and ultracentrifugation and contained low chymotryptic and negligible tryptic activity, and observed an almost 2-fold increase in specific activity over "crystalline elastase."

Baumstark et al.,[42] following the fractionation studies of Lewis and Thiele,[40] purified elastase by absorbing the contaminating proteins of the euglobulin precipitate onto diethylaminoethyl cellulose (DEAE-cellulose) at pH 8.9, and observed a 2-fold increase in specific activity over the euglobulin precipitate. They reported their product to be homogeneous by ultracentrifugation and chromatography on CM-cellulose, and to show no tryptic or chymotryptic activity.

Smillie and Hartley[44] improved the method of Naughton and Sanger[11] by first separating elastase from the contaminants present in "crystalline elastase" by a batchwise fractionation procedure on DEAE-Sephadex at pH 8.8, essentially similar to that of Baumstark et al.,[42] before chromatographing it on CM-cellulose at pH 4.5. They found the DEAE-Sephadex fractionation led to a 3-fold increase in specific activity over "crystalline elastase." Further work[15] has shown that while this method gives reasonably pure elastase in good yield (about 2.5 g per 500 g Trypsin 1-300), slight tryptic and chymotryptic impurities can always be detected by acrylamide gel electrophoresis in 8 M urea at high loadings, using the pH 4.3 method of Reisfield et al.[45] to avoid autolysis, and by assaying for tryptic and chymotryptic activities using BAEE and ATEE, which are not hydrolyzed by pure elastase (see Table I and p. 132). Ling and Anwar[46] have reported an essentially similar method.

Gertler and Hofmann[16] were able to prepare elastase of higher purity, although in lower yield, by slight modifications to this general method. They first dissolved and reprecipitated the euglobulin precipitate three

[44] L. B. Smillie and B. S. Hartley, Biochem. J. **101**, 232 (1966).
[45] R. A. Reisfield, U. J. Lewis, and D. E. Williams, Nature **195**, 281 (1962).
[46] V. Ling and R. A. Anwar, Biochem. Biophys. Res. Commun. **24**, 593 (1966).

times to free it from trypsin contaminants, and chromatographed the product on DEAE-cellulose at pH 9.0 by the method of Baumstark *et al.*,[42] rather than using a batchwise fractionation. They finally chromatographed the elastase eluted from the DEAE-cellulose column on CM-cellulose at pH 5.0 by the method of Naughton and Sanger.[11] Their final product showed about 0.5% tryptic activity and only traces of chymotryptic activity, and apart from these contaminants, no other proteins were detected on acrylamide gel electrophoresis at high loadings. This indicated a purity of greater than 99%, with a yield of 300–350 mg per 200 g starting material, although a later paper from the same laboratory[31] claimed far less purity for another batch of elastase prepared by the same method.

Chromatographically purified elastase is now commercially available from Worthington under the name of Purified Elastase.

It has been recently shown that elastase can be easily crystallized from dilute salt solutions[47] in which all contaminating proteins are fully soluble. This has led to a simple and highly selective procedure[15] for obtaining very pure elastase in good yield from the DEAE-Sephadex filtrate of Smillie and Hartley[44] discussed above. Once crystallized elastase thus obtained sediments as a single sharp symmetrical peak in the ultracentrifuge and has a purity greater than 99.5%, as estimated by ATEE and BAEE assays, which show the degree of chymotryptic and tryptic contamination to be less than 0.1 and 0.2%, respectively, and by acrylamide gel electrophoresis[45] at high loadings, which reveals no minor components. This once crystallized material can be recrystallized with almost quantitative recovery (83%) to yield about 2.7 g of twice crystallized elastase per 500 g. Trypsin 1-300, the purity of which is greater than 99.9%, showing less than 0.04% chymotryptic and less than 0.01% tryptic contamination[15] (see Table I), and less than 0.001% contamination by carboxypeptidase, leucine amino peptidase, ribonuclease, and deoxyribonuclease.[24]

The elastase obtained by this method (commercially available from Whatman Biochemicals Ltd., Springfield Mill, Maidstone, Kent, England) should not be confused with the impure and misnamed "crystalline elastase" preparation of Lewis *et al.*[1] discussed above, which is commercially available from other sources. Details of the method used for its preparation are set out below.

Gertler[20] has recently shown that proelastase also can be prepared from the euglobulin precipitate of Lewis *et al.*[1] The euglobulin is prepared by the usual method, from fresh frozen pancreas, except for the presence

---

[47] D. M. Shotton, B. S. Hartley, N. Camerman, T. Hofmann, S. C. Nyburg, and L. Rao, *J. Mol. Biol.* **32**, 155 (1968).

of added soybean trypsin inhibitor, and then proelastase is absorbed from the redissolved euglobulin precipitate onto elastin at pH 8.5 in the presence of soybean trypsin inhibitor, eluted at pH 3.6 and then chromatographed on CM-cellulose at pH 4.5. This finding raises very interesting questions about the nature of the material normally extracted and worked up during the elastase preparation. Unfortunately no workers have yet investigated the ratios of the zymogen to the active enzyme in the initial extract and at different stages of the preparation, nor the conditions most favorable for the tryptic activation of proelastase into elastase while limiting autolytic breakdown of the activated enzyme, in order to achieve maximal yield of elastase. However, the method outlined below gives very satisfactory results.

Procedure for the Preparation of Elastase

All stages are conducted at $+2°$. AnalaR grade reagents are used throughout.

STAGE ONE

Preparation of the euglobulin precipitate by the method of Lewis et al.[1]

Acetate Extraction. Five hundred grams of Trypsin 1-300 (Nutritional Biochemicals Corp., Cleveland, Ohio) or similar material is stirred with 2500 ml of 0.1 M sodium acetate buffer, pH 4.5, for 3 hours, and then centrifuged for 1 hour at 2000 rpm. The supernatant is decanted and saved, and the residue is reextracted with a further 2000 ml of the acetate buffer, and spun, as before. The two supernatants are pooled and the insoluble residue is discarded.

Ammonium Sulfate Precipitation. The combined extract is brought to 45% saturation with ammonium sulfate (262 g of solid ammonium sulfate per liter of extract) and after stirring for 30 minutes the precipitate is removed by centrifugation for 30 minutes at 12,000 rpm. The precipitate is washed three times with 1000-ml portions of the acetate buffer which has been brought to 45% saturation with solid ammonium sulfate. An electric blender is useful for resuspending the fine precipitate during each washing. After each spin the supernatant is discarded.

Preparation of the Euglobulin Precipitate. The washed ammonium sulfate precipitate is dissolved is 2000 ml of 0.05 M Na_2CO_3-HCl buffer, pH 8.8 and dialyzed against four changes of 20 liters of distilled water for 48 hours to remove all salt. The resulting euglobulin precipitate is centrifuged off and washed twice with 1000 ml of distilled water. The supernatant and washings are discarded.

STAGE TWO

Fractionation of the euglobulin precipitate on DEAE-Sephadex by the method of Smillie and Hartley.[44]

The washed euglobulin precipitate is suspended in 2000 ml of 0.02 M Tris base and dissolved by adjusting to pH 10.4 with 1 M NaOH. The pH is then brought to pH 9.4 with 1 M HCl and the solution stirred with about 1500 ml settled bed volume of DEAE-Sephadex A-50 (60 g dry weight) which has been previously equilibrated with 0.02 M Tris-HCl, pH 8.8. After 4 hours the suspension is filtered under gentle vacuum and the DEAE-Sephadex washed with a further 1000 ml of 0.02 M Tris-HCl, pH 8.8. The combined filtrate is brought to pH 5.0 with glacial acetic acid, dialyzed overnight against three changes of 20 liters of 1 mM acetic acid, and freeze dried.

STAGE THREE

Purification of elastase from the DEAE-Sephadex filtrate by crystallization.[15,47]

First Crystallization. The salt-free, freeze-dried DEAE-Sephadex filtrate is dissolved in 0.01 M sodium acetate buffer, pH 5.0, to give a final protein concentration of about 25 mg/ml. Sodium sulfate, 1 M, is slowly added, with stirring, to a final concentration of 0.1 M. Elastase crystals form immediately, and after 24 hours are filtered on a fine-porosity sintered glass funnel and washed twice with 20-ml portions of 1.2 M sodium sulfate, buffered at pH 5.0 with 0.01 M sodium acetate. The filtrate and washings are discarded and the crystals are suspended in 1000 ml of distilled water, redissolved completely by dialysis overnight against three changes of 20 liters of 1 mM acetic acid, and freeze dried.

Recrystallization. A repetition of the above procedure yields twice crystallized elastase, which is again redissolved by dialysis against 1 mM acetic acid to remove all salt before being finally freeze dried.

Table I gives a summary of the purification achieved in a typical preparation at each step of this procedure.[15]

Properties

Stability

Elastase is readily soluble in water and dilute salt solutions at concentrations up to 50 mg/ml between pH 4 and pH 10.5, and within this pH range at 2° elastase solutions are stable for prolonged periods below pH 6.0, and reasonably stable at higher pH's. It is, however, a powerful proteolytic enzyme and will rapidly autolyze to a mixture of peptides

TABLE I

SUMMARY OF THE PURIFICATION OF ELASTASE

Purification stage	Total protein (extinction units)[a]	Total Congo red-elastolytic activity[b]	Total N-benzoyl-L-alanine methyl esterase activity[b]	% Purity (from N-benzoyl-L-alanine methyl esterase assays)[c]	% Contamination by trypsin (from BAEE assays)[c]	% Contamination by chymotrypsin (from ATEE assays)[c]
Initial sodium acetate pH 4.5 extract[d]	621,000	50,900	17,400	2.8	11.3	6.1
45% Ammonium sulfate precipitate	133,300	17,100	16,000	12.0	10.7	21.2
Euglobulin precipitate	48,900	12,610	10,280	21.0	3.2	15.9
Dialyzed DEAE-Sephadex filtrate	12,500	13,570	8,740	69.9	7.9	15.2
Once crystallized elastase	6,500	6,360	6,440	99.1	0.17	0.09
Twice crystallized elastase	5,400	5,400	5,400	100	0.006	0.03
Trypsin	—	—	—	1.1	100	1.5
α-Chymotrypsin	—	—	—	2.0	3.0	100

[a] Protein content was estimated by determining $E_{280 nm}$. A protein solution having $E_{1 cm., 280 nm} = 1.0$ is defined as possessing one extinction unit per milliliter.

[b] Total activities were obtained by multiplying the activity relative to twice crystallized elastase by the total extinction units at that stage, and are equivalent to that number of extinction units of pure elastase. It will be noted that the apparent elastolytic activity exceeds the apparent esterase activity during the early stages of the preparation. This is the result of factors discussed on p. 118, especially the presence of large amounts of trypsin and chymotrypsin.

[c] The percent purity and contamination were determined by measuring the activities per extinction unit toward the synthetic substrates relative to those of pure enzymes. Against N-benzoyl-L-alanine methyl ester the activity of twice crystallized elastase (the final product), against BAEE that of dialyzed Worthington twice crystallized salt-free trypsin (batch TRSF 6188), and against ATEE that of dialyzed Worthington α-chymotrypsin, crystallized three times (batch CDI 6108-9), were taken to be 100%. The observed activities of these enzymes against the substrates for which they are not specific is shown at the foot of the table.

[d] Starting material 500 g Trypsin 1-300 (Nutritional Biochemicals Corp., Cleveland, Ohio).

if incubated at room temperature at or near its optimum pH of 8.8. At pH 5.0 the proteolytic activity is very slight and solutions at this pH can be used at room temperature with little autolysis.[15]

Gertler and Hofmann,[16] and Wasi and Hofmann[31] have shown that below pH 4.0 elastase undergoes a reversible pH-dependent conformational change, pK_a 3.55, which apparently involves the cooperative protonation of three carboxyl groups. In the acid form the N-terminal amino group is accessible to attack by nitrous acid, which inactivates the enzyme, while at pH's above 4.2 it is completely inaccessible to the reagent, being buried in the interior of the molecule.[48] Brief titration of elastase down to pH 2.6 is fully reversible, but prolonged incubation at acid pH's leads to irreversible denaturation with loss of enzymatic activity, previously noted by Lewis et al.[1] In this behavior elastase differs somewhat from the closely related enzymes trypsin and chymotrypsin, the reactivity of the amino-terminal group of trypsin to nitrous acid at all pH's being equal to that of elastase at pH 3.0.[49]

Kaplan, Stevenson, and Hartley[50] have shown that the N-terminal amino group of elastase remains buried at alkaline pH values up to pH 10.5, as judged by its low reactivity toward acetic anhydride, although its pK_a in the native enzyme is 9.7. Similarly, no correlation of change in optical rotation, which might reveal a conformational change, with the state of ionization of this group could be demonstrated.[22] This is in marked contrast to the situation in chymotrypsin, where a reversible conformational change leading to loss of enzymatic activity occurs on deprotonation of the corresponding N-terminal amino group, pK_a 8.3.[51] This deprotonation breaks an internal ion pair between the amino group and the side chain carboxyl group of an aspartic acid residue adjacent to the active center serine, disturbing the conformation at the active center.[52] As the N-terminal amino group of elastase is involved in an identical ion pair,[48] one is forced to the conclusion that there may be other less obvious factors stabilizing the active conformation of elastase, absent in chymotrypsin, which are strong enough to maintain activity at pH's where this ion pair does not exist. At pH's above 10.5 the reactivity of the N-terminal amino group of elastase toward acetic anhydride increases, indicating that it is becoming unburied.[50]

[48] D. M. Shotton and H. C. Watson, Nature 225, 811 (1970).
[49] S. T. Scrimager and T. Hofmann, J. Biol. Chem. 242, 2528 (1967).
[50] H. Kaplan, K. J. Stevenson, and B. S. Hartley, Federation Proc. 28, 533 (1969).
[51] H. L. Oppenheimer, B. Labouesse, and G. P. Hess, J. Biol. Chem. 241, 2720 (1966).
[52] P. B. Sigler, D. M. Blow, B. W. Matthews, and R. Henderson, J. Mol. Biol. 35, 143 (1968).

Hofmann[53] has studied the inactivation of elastase by reaction with 2,4,6-trinitrobenzene sulfonic acid (TNBS), and has found that this inactivation is strongly pH dependent, showing a single noncooperative ionization with a pK_a of 10.3. No correlation was observed between the degree of inactivation and the degree of reaction of TNBS with either the N-terminal amino group or the three lysine residues of elastase, but Hofmann nevertheless thinks that this alkaline pH effect is not simply due to nonspecific unfolding, but is controlled by one specific group, possibly a tyrosine, which reacts with TNBS. Similarly, decreases of activity when assayed at high pH have been observed in kinetic studies (see p. 135), but Lewis et al.[1] found that elastase could be incubated at pH's up to 12.0 for 24 hours at 5°, and still show full activity when assayed for elastolytic activity at pH 8.8, indicating that high pH-induced conformational changes up to this pH are fully reversible.

Crystals of elastase are stable indefinitely in 1.2 M sodium sulfate, buffered at pH 5.0 with 0.01 M sodium acetate, at room temperature, and show full activity when redissolved and assayed.[15] Freeze-dried elastase prepared after dialysis against 1 mM acetic acid is stable indefinitely at −10°, with no loss of proteolytic activity.[15]

Physical Properties

The elastase molecule is a single polypeptide chain of 240 amino acid residues, containing 4 disulfide bridges. The amino acid composition is shown in Table II, while Table III shows the complete amino acid

TABLE II
AMINO ACID COMPOSITION OF ELASTASE

Amino acid	Number of residues	Amino acid	Number of residues
Lysine	3	Glycine	25
Histidine	6	Alanine	17
Arginine	12	Cysteine	8
Aspartic acid	6	Valine	27
Asparagine	18	Methionine	2
Threonine	19	Isoleucine	10
Serine	22	Leucine	18
Glutamic acid	4	Tyrosine	11
Glutamine	15	Phenylalanine	3
Proline	7	Tryptophan	7
		Total: 240 residues	

[53] L. Rao and T. Hofmann, *Can. J. Biochem.*, in press, 1970.

sequence.[54] The complete three-dimensional structure of the molecule has recently been determined by X-ray crystallographic studies on the crystalline enzyme.[48, 55, 56] The polypeptide chain is folded upon itself in a complicated fashion into two distinct halves to form a compact globular molecule of dimensions 55 Å × 40 Å × 38 Å, as measured along the principal crystallographic axes. There are only two small regions of α-helix in the molecule, the majority of the chain tending to be fully extended, and often looping back on itself to run antiparallel with other regions of the chain in an irregular antiparallel pleated-sheet configuration, stabilized by hydrogen bonds. Extensive amino acid sequence homologies exist between elastase and other mammalian serine proteinases, especially in those regions which are involved in catalysis and which form the hydrophobic molecular interior (see Table III).[54] The overall conformation of the elastase main chain, and the orientation of the side chains, especially in internal regions, are very similar to those of α-chymotrypsin, determined by Blow and his collaborators,[52, 57, 58] "insertions" and "deletions" in the sequences all occurring on the molecular surface, usually at the ends of external loops.[48] A comparison of these two structures has led to plausible explanations for the common catalytic mechanism of these enzymes, and their differing substrate specificities (see p. 132).[48,56]

The molecular weight of elastase, calculated from the amino acid composition, is 25,900.[54] Lewis et al.[1] reported 25,000 from ultracentrifuge studies, Naughton and Sanger[11] calculated it to be 28,500 from specific activity studies of ^{32}P-diisopropylphosphoryl elastase (DIP elastase), and Gertler and Hofmann[16] suggested 24,860 from amino acid analyses.

Both electrophoretically purified elastase[1] and elastase purified by crystallization[15] sediment as a single sharp symmetrical peak in the ultracentrifuge, with a sedimentation constant $s_{20,w}$ of 2.6 S. The observed[1] diffusion constant is 9.5×10^{-7} cm^2 sec^{-1}, and the observed[1] partial specific volume is 0.73 cm^3 g^{-1}, which agrees well with the theoretical value of 0.726 cm^3 g^{-1} calculated from the amino acid composition.[15, 59] The reported[1] frictional ratio f/f_0 of 1.2 is in good agreement

[54] D. M. Shotton and B. S. Hartley, Nature 225, 802 (1970).
[55] H. C. Watson, D. M. Shotton, J. M. Cox, and H. Muirhead, Nature 225, 806 (1970).
[56] D. M. Shotton and H. C. Watson, Phil. Trans. Roy. Soc. B (London), 257, 111 (1970).
[57] B. W. Matthews, P. B. Sigler, R. Henderson, and D. M. Blow, Nature 214, 652 (1967).
[58] J. J. Birktoft, D. M. Blow, R. Henderson, and T. A. Steitz, Phil. Trans. Roy. Soc. B (London), 257, 67 (1970).
[59] E. J. Cohn and J. T. Edrall, "Proteins, Amino Acids and Peptides," p. 370. Reinhold, New York, 1943.

TABLE III

The Amino Acid Sequence of Porcine Elastase^a

```
                                                 10                              20                                    30
VAL-VAL-Gly-GLY-Thr-Glu-Ala-Gln-Arg-Asn-SER-Trp-PRO-Ser-GLN-ILE-SER-LEU-GLN-Tyr-Arg-Ser-Gly-Ser-SER-Trp-Ala-HIS-Thr-CYS-
16  17  18  19  20  21 (22) 23  24  25  26 (27)(28)(29)(30)(31)(32)(33) 34  35  36 36A 36B 36C 37  38  39 (40) 41 (42)

                                                 40                              50                                    60
-GLY-THR-LEU-ILE-Arg-Gln-ASN-TRP-VAL-Met-THR-ALA-ALA-HIS-CYS-Val-Asp-Arg-Glu-Leu-Thr-Phe-Arg-Val-Val-GLY-GLU-GLU-His-
43 (44)(45)(46)(47) 48  49  50 (51)(52)(53)(54)(55) 56  57 (58) 59  60  61  62  63  64  65 65A (66)(67)(68) 69  70  71

                                                 70                              80                                    90
-ASN-Leu-Asn-Gln-Asn-Asn-Gly-Thr-Glu-GLN-Tyr-Val-Gly-Val-Gln-LYS-Ile-Val-His-Pro-Tyr-TRP-Asn-Thr-Asp-Asp-Val-Ala-Ala-
72  73  74  75  76  77  78  79  80  81  82  83  84  85  86  87  88  89  90  91  92  93  94  95  96  97  98  99 99A 99B

                                                 100                             110                                   120
-Gly-Tyr-ASP-ILE-Ala-LEU-LEU-ARG-LEU-Ala-Gln-Ser-Val-Thr-Leu-Asn-Ser-Tyr-VAL-Gln-Leu-Gly-Val-LEU-PRO-Arg-Ala-Gly-Thr-Ile-
100 101(102)(103)(104)(105)(106)107(108)109 110 111(112)113 114 115 116 117(118)119 120(121)122 123(124)125 126 127 128 129

                                                 130                             140                                   150
-Leu-Ala-Asn-Ser-Pro-CYS-Tyr-Ile-THR-GLY-TRP-GLY-Leu-THR-ARG-Thr-Asn-Gly-Gln-Leu-Ala-Gln-Thr-LEU-Gln-Gln-Ala-Tyr-Leu-
130 131 132 133 134 135(136)137(138)(139)(140)(141)(142)143 144 145 146 147 148 149 150 151 152 153 154(155)156 157(158)159(160)

                                                 160                             170                                   180
-PRO-Thr-Val-Asp-Tyr-Ala-Ile-CYS-Ser-Ser-Ser-Ser-Tyr-TRP-Gly-Ser-Thr-VAL-Lys-Asn-SER-MET-Val-CYS-ALA-GLY-Gly-Asn-GLY-Val-
161(162)163 164 165 166 167(168)169 170 171A 170B 171 172 173 174 175(176)177 178 179(180)(181)(182)(183)(184)185 186 187 188

                                                 190                             200                                   210
-Arg-Ser-Gly-CYS-Gln-GLY-ASP-SER-GLY-GLY-PRO-Leu-His-CYS-Leu-Val-Asn-Gly-Gln-Tyr-Ala-Val-His-GLY-VAL-Thr-SER-PHE-Val-SER-
188A 189 190 191 192(193)(194)195 196(197)(198)(199)(200)(201)202 203 204 205 206 207 208(209)(210)(211)(212)(213)(214)(215)216 217

                                                 220                             230                                   240
-Arg-Leu-Gly-CYS-Asn-Val-Thr-Arg-Lys-PRO-Thr-VAL-PHE-Thr-ARG-VAL-Ser-Ala-Tyr-Ile-Ser-TRP-ILE-Asn-Asn-Val-ILE-ALA-Ser-ASN
217A 218 219 220 221 221A 222 223 224 225 226 227(228)(229)230(231)232 233(234)235 236(237)(238)239 240 241 242 243 244 245
```

Disulfide bridges: 42–58, 136–201, 168–182, 191–220

^{a} The bovine chymotrypsinogen-A numbering is given below the elastase sequence. "Insertions" are numbered 36A, 36B, etc. Residues which are identical or chemically similar to corresponding residues in bovine trypsin, chymotrypsin-A and chymotrypsin-B are shown in capitals. (Chemical similarity: Arg = Lys, Asp = Glu, Asn = Gln, Asp = Asn, Glu = Gln, Ser = Thr, Val = Ile, Ile = Leu, and Tyr = Trp = Phe.) Residues whose side chains are internal, judged to be inaccessible to water molecules in the molecular models of both elastase and α-chymotrypsin, are indicated by circles around the residue numbers.

with the observed molecular dimensions obtained from the crystallographic studies.[48] The isoelectric point is pH 9.5 ± 0.5.[1] The molar and specific extinction coefficients for pure elastase are as follows[15]: in $0.1 M$ sodium hydroxide, $\epsilon_{1\,cm,\,280\,nm} = 5.74 \times 10^4$ and $E_{1\,cm,\,280\,nm}^{1\%} = 22.2$; in $0.05 M$ sodium acetate, pH 5.0 $\epsilon_{1\,cm,\,280\,nm} = 5.23 \times 10^4$ and $E_{1\,cm,\,280\,nm}^{1\%} = 20.2$; and in $0.1 M$ sodium hydroxide, following thorough digestion with pepsin, $\epsilon_{1\,cm,\,280\,nm} = 6.11 \times 10^4$ and $E_{1\,cm,\,280\,nm}^{1\%} = 23.6$. Previous reports in undefined solvents[1, 11, 22, 31] agree reasonably with these values, except that of Lewis *et al.*,[1] which is rather low.

Activators and Inactivators

Elastase is produced as a result of tryptic activation of the zymogen, proelastase, which is secreted into the duodenum by the pancreatic acinar tissue.[6, 20, 21, 60] During activation a small peptide of undetermined sequence is removed from the N-terminal end of proelastase by tryptic cleavage,[20] creating a new N-terminal amino group and enabling the molecule to adopt its active configuration, with this new N-terminal valine residue buried in the interior of the molecule, forming an ion pair with the side chain carboxyl group of the aspartic acid adjacent to the active center serine residue.[48] This activation process closely resembles those of trypsinogen and chymotrypsinogen.[52]

Elastase contains no prosthetic groups or metal ions, and is not subject to any allosteric activatory or inhibitory control. Its enzymatic activity results solely from the specific three-dimensional conformation which its single polypeptide chain adopts. Activity is consequently lost by denaturation and conformational changes, as discussed on page 125. In common with other serine proteinases, it is totally and irreversibly inhibited by small amounts of DFP, which specifically and stoichiometrically alkylates the active center serine residue.[11] Similar covalent inhibition at the active center serine has been achieved with various sulfonyl fluorides,[15, 55, 56] and with *p*-dinitrophenol diethyl phosphate (see p. 119).[26]

Kaplan *et al.*[61] have shown that elastase is not inhibited by the substrate analogs tosyl-L-alanine chloromethyl ketone, tosyl-L-alanine bromoethyl ketone, and tosyl-L-valine chloromethyl ketone, thus differing from chymotrypsin which is readily inhibited by its corresponding substrate analog tosyl-L-phenylalanine chloromethyl ketone (TPCK),[62] which reacts covalently with the active center histidine residue.[63]

[60] N. H. Grant and K. C. Robbins, *Proc. Soc. Exptl. Biol. Med.* **90**, 264 (1955).
[61] H. Kaplan, V. B. Symonds, H. Dugas, and D. R. Whitaker, *Can. J. Biochem.* in press, 1970.
[62] G. Schoellmann and E. Shaw, *Biochemistry* **2**, 252 (1963).

Lewis *et al.*[1] showed the elastolytic activity of elastase to be inhibited almost 50% by $10^{-5}\,M$ copper sulfate, and 50% by $7 \times 10^{-2}\,M$ sodium chloride. Similar effects were observed with sodium cyanide, ammonium sulfate, and potassium chloride, but $10^{-3}\,M$ solutions of zinc, manganese, cobalt, magnesium, and calcium chlorides, and $10^{-2}\,M$ sodium cyanide and cysteine were found to be without effect. Thomas and Partridge[38] noticed that cysteine reduced salt inhibition at high ionic strengths, but discounted the possibility of elastase being a thiol enzyme, as it was not affected by iodoacetic acid. Lamy *et al.*[39] found that EDTA, calcium, and magnesium at $10^{-3}\,M$ had no effect.

In contrast, it has been found that the rate of elastase-catalyzed hydrolysis of N-benzoyl-L-alanine methyl ester, while being inhibited by about 14% by $0.1\,M$ sodium chloride, is greatly stimulated by sodium sulfate and Tris, $1.2\,M$ sodium sulfate leading to more than a 3-fold increase in rate, while $0.01\,M$ Tris increases the rate by about 25%, assayed at pH 8.0, 25° in $0.1\,M$ potassium chloride.[15] Tris has been shown to increase the deacylation rate of trimethylacetyl elastase, the acyl enzyme intermediate of the hydrolysis of trimethylacetyl p-nitrophenyl ester (see p. 135).[27]

Various serum and intestinal nondialyzable inhibitors of elastase have been reported, but none of these have been well characterized and their biological function is not understood. Attempts to correlate the amounts of elastase and elastase inhibitors with the onset of various diseases, especially arteriosclerosis, have unfortunately yielded ambiguous but essentially negative results.

The effects of these nondialyzable inhibitors and various salts on the activity of elastase, and the relationships between elastase and arteriosclerosis, are reviewed more fully by Mandl.[4] Soybean trypsin inhibitor does not inhibit elastase.[15, 20]

Enzymatic Activity and Substrate Specificity

Much of the early literature concerning the nature of elastase action is complicated by the beliefs of some workers that elastase should digest only elastin, and not attack other proteins,[3, 4] or that elastase possesses mucolytic[64] or even lipolytic[65] activities. Much of this confusion is due to the use of impure preparations of the enzyme, which also invalidates many of the studies on the detailed proteolytic substrate specificity of elastase.

---

[63] L. B. Smillie and B. S. Hartley, *Meeting Fed. European Biochem. Soc., 1st, London* Abstract A-30. Academic Press, New York, 1964.

[64] J. W. Czerkawski, *Abstr. Intern. Congr. Biochem. 4th Vienna,* p. 43 (1958).

[65] A. I. Lansing, T. B. Rosenthal, M. Alex, and E. W. Dempsey, *Anat. Record* **114**, 555 (1952).

Partridge and Davis[2] were the first to show that elastolysis was accompanied by the creation of new free amino groups, and concluded that elastolysis was truly proteolytic, while Lewis et al.[1] demonstrated that elastase could digest many other proteins besides elastin. Naughton and Sanger[11] showed that the substrate specificity of elastase was rather broad, directed toward peptide bonds involving the carbonyl groups of amino acids bearing uncharged nonaromatic side chains, such as alanine, valine, leucine, isoleucine, glycine, and serine. The minor cleavages after tyrosine and phenylalanine reported by them[11] have been shown to be due to chymotryptic contaminants.[65a] This specificity nicely complements those of the other major endopeptidases active in the duodenum, namely trypsin and the chymotrypsins. Naughton et al.[66] showed that elastase had the same active center sequence, -Asp-Ser-Gly-, as trypsin and the chymotrypsins, suggesting it was a related enzyme.

Since then, the determination of the complete amino acid sequence[54, 67] and three-dimensional structure[48, 55, 56] of elastase have confirmed that elastase is indeed highly homologous with the other serine proteinases. Kinetic studies (see p. 135) have increased our knowledge of the substrate specificity of elastase, and have shown the enzyme to function by a catalytic mechanism similar to trypsin and the chymotrypsins. A detailed comparison[48] of the molecular models of elastase and α-chymotrypsin[52] indicates that the proposals made by Blow et al.[68] for the molecular mechanism of α-chymotryptic activity may also be true for elastase, as the functionally important groups have the same conformation in both enzymes. In this, the unusually high nucleophilicity of the active center serine residue, Ser_{195},[69] is explained by the transfer of charge to it from the carboxylate group of an aspartic acid residue, Asp_{102}, which is buried in a hydrophobic environment in the molecular interior. This relay of charge from the buried aspartic acid to the surface serine residue is mediated by the active center histidine residue, His_{57}, the aromatic imidazole ring of which lies between them, hydrogen bonded to both, forming a resonating "charge-relay system." Peptide bond cleavage is brought about by the nucleophilic attack of the active center serine γ-oxygen onto the carbonyl carbon atom of the peptide bond, resulting

[65a] A. Sampath Narayanan and R. A. Anwar, Biochem. J. 114, 11 (1969).
[66] M. A. Naughton, F. Sanger, B. S. Hartley, and D. C. Shaw, Biochem. J. 77, 149 (1960).
[67] J. R. Brown, D. L. Kauffman, and B. S. Hartley, Biochem. J. 103, 497 (1967).
[68] D. M. Blow, J. J. Birktoft, and B. S. Hartley, Nature 221, 337 (1969).
[69] To facilitate the comparison between elastase and the other serine proteinase, and to emphasize the homologies between them, the numbering of the amino acid residues in the chymotrypsinogen-A sequence has been used to describe the positions of the amino acid residues in these related enzymes.[54]

in bond cleavage and the release of the amide half of the substrate, and the transitory formation of an acyl-enzyme intermediate in which the charge-relay system is disrupted, leaving Asp_{102} fully charged in the hydrophobic molecular interior. For normal substrate deacylation follows almost instantaneously by a reversal of the electron shifts which occur during acylation, reestablishing the reasonance system. This proposed mechanism, believed to be common to all serine proteinases, is described more fully by Blow et al.[68]

The differing substrate specificity observed between elastase and the other serine proteinases is explained by structural differences in that region of the molecule identified as the substrate binding site of α-chymotrypsin by Birktoft et al.[58] The aromatic side chains of chymotrypsin substrates lie in a deep hydrophobic pocket adjacent to the active center. In the amino acid sequences of trypsin and thrombin a serine residue, Ser_{189}, which lies in the bottom of this hydrophobic pocket in α-chymotrypsin and elastase, is replaced by an aspartic acid residue.[70] This explains why these two enzymes specifically bind basic amino acid side chains, which can enter the pocket and neutralize the charge of this buried aspartic acid residue. In the elastase molecule the mouth of this pocket is partially occluded by the side chain of residue 216, a valine, which is absent in the other enzymes, where residue 216 is a glycine. Similarly the interior of the pocket is partially filled by the side chain of Thr_{226}, again a glycine residue in the other enzymes. Bulky aromatic side chains are thus sterically hindered from entering the substrate binding site of elastase, explaining the specificity of elastase for smaller nonaromatic, uncharged side chains.[11] These observations also explain why elastase exhibits complete lack of activity toward ATEE and BAEE, while α-chymotrypsin and trypsin can hydrolyze N-benzoyl-L-alanine methyl ester and similar substrates to a limited extent (see pp. 117, 119, and Table I).

Hofmann[53] has investigated the inhibition of the elastase-catalyzed hydrolysis of N-carbobenzoxy glycine p-nitrophenyl ester by various peptide derivatives bearing aliphatic side chains, toward which elastase shows some specificity.[11] He has found that while the dipeptides L-Ala-L-Ala, L-Val-L-Val, and L-Leu-L-Val, and also N-formyl-L-Ala-L-Ala, N-formyl-L-Leu-L-Val, and L-Ala-L-Ala-NH$_2$ have no inhibitory effects, the tripeptide L-Ala-L-Ala-L-Ala, as well as N-carbobenzoxy- and N-acetyl- derivatives of L-Ala-L-Ala, L-Val-L-Val, L-Leu-L-Val, L-Val-L-Leu, L-Leu-L-Leu, L-Val-L-Ile and L-Ile-L-Val, and also N-acetyl-L-Ala-L-Ala-L-Ala all showed some inhibitory effects, with K_I's between

[70] B. S. Hartley, Phil. Trans. Roy. Soc. B (London), **257**, 77 (1970).

10^{-3} M and 10^{-2} M, confirming the specificity of elastase for this type of amino acid, mentioned above.

This wide substrate specificity of elastase for nonaromatic uncharged side chains explains its ability to digest elastin, a highly cross-linked fibrous protein rich in aliphatic side chains and very poor in charged side chains.[2] This ability is by no means peculiar to elastase, for although trypsin and the chymotrypsins have such a narrow substrate specificity that they are unable to attack sufficient peptide bonds to disrupt the netlike structure of elastin, many other enzymes exist which can do so (see p. 139).

The observation that proelastase binds well to elastin[6, 20] is in accord with the deductions made from the three-dimensional structures of elastase[48] and α-chymotrypsin[52] that activation of the zymogens to form the native enzymes does not involve any major conformational change in the vicinity of the substrate binding site, which is already formed and unobstructed in the zymogens.

In the crystal lattice the active center of crystalline elastase is not obstructed by intermolecular contacts,[48] and the crystalline enzyme has been shown to possess high catalytic activity.[15]

Kinetic Properties

As discussed on page 113, the kinetics of the elastase digestion of elastin is very complex, because of the cross-linked nature of the substrate, and because digestion must be preceded by adsorption of elastase onto the substrate.[12] Various pH optima for elastolysis have been reported,[1, 5, 6, 13, 34] differing with the buffers used, and varying between pH 7.4 and 10.3, although it is likely that some of the higher values are artifacts of inadequate buffering which enabled the actual pH to drop to near the real optimum during the course of the reaction. In 0.1 M carbonate buffer Lewis et al.[1] found the optimum pH to be 8.8 and it is at this pH that elastin-digesting assays are normally conducted. Absolute activities were noted to vary with the buffer used, an effect probably due to different degrees of salt inhibition (see p. 132). In all cases the pH-activity curve was bell shaped.

Kaplan and Dugas[22] and Geneste and Bender[23] have studied the kinetics of hydrolysis of N-benzoyl-L-alanine methyl ester, an excellent specific ester substrate for elastase, and of N-3-(2-furylacryloyl)-L-alanine methyl ester, which is also a good substrate. In both cases maximal activity is observed at all pH's between 8 and 10. At lower pH's the activity drops steadily to almost nothing at pH 5.0. This pH-dependent inactivation involves a single ionizing group with a pK_a variously reported as 6.5,[22] 6.85[23] or for another substrate (see below) 6.7.

This is taken to be that of a histidine imidazole side chain by both laboratories but may be the pK_a of the whole charge-relay system involving Ser_{195}, His_{57}, and Asp_{102} (see p. 133).[68] At higher pH's activity also drops, although measurements of it are complicated by nonenzymatic spontaneous alkaline hydrolysis of the substrate, and by slight enzyme denaturation.[23] The approximate pK_a for this alkaline inactivation is 11.4,[23] which is clearly not that of the N-terminal amino group (pK_a 9.7[50]), but may be related to the conformational changes induced by high pH discussed on p. 135. Kaplan and Dugas[22] estimated the K_m for the reaction with N-benzoyl-L-alanine methyl ester in 0.1 M potassium chloride at pH 8.0 and 25°C to be 19.3 mM, and K_{cat} to be 12.3 sec^{-1}. K_{cat} but not K_m decreased with decreases in pH below 8.0 under the conditions of the assay.[61] The esterase activity, against the same substrate, of fully acetylated elastase, in which the N-terminal amino group and the three lysines are blocked, was identical with that of the native enzyme over the pH range of 5 to 10, proving that the state of ionization of the N-terminal amino group has no effect on this reaction. However, acylation reduced elastolytic activity by approximately 50%, although no change in specificity was observed on the carboxymethylated B-chain of insulin.[22]

Bender et $al.$[26] and Geneste and Bender[23] investigated the rates of hydrolysis of various p-nitrophenyl esters by elastase, both of carboxylic acids, which can be considered as nonspecific acylating agents, and which are hydrolyzed very slowly, and of amino acids, which resemble more closely the natural substrate, and which elastase might be expected to hydrolyze more rapidly, with some selective specificity. Estimations of the purity of their elastase preparations by titrating the active sites with diethyl p-nitrophenyl phosphate indicated that none of the preparations used were better than 85% pure (but see p. 119), and one of the preparations was contaminated by an elastomucase which showed substantial activity toward N-carbobenzoxy-L-tyrosine p-nitrophenyl ester, a chymotrypsin substrate.[23, 26, 27] Nevertheless significant differences in the rates of hydrolysis of these various substrates were observed. N-carbobenzoxy-L-alanine p-nitrophenyl ester was shown to be an excellent ester substrate for elastase, while esters of other amino acids bearing aliphatic side chains were less good, in the order Ala > Leu > Gly > Norleu > Val > Ile. The L-alanine substrate gave the highest value of K_{cat}/K_m yet observed for elastase, from which it was possible to determine an approximate value for K_m of 0.6 mM and for K_{cat} of 110 sec^{-1}. Table IV summarizes the results of both groups of workers[22, 23, 26, 61] on these synthetic substrates. Because of the low solubility of most of these substrates, individual of K_m and K_{cat} have not been determined, except in the two

TABLE IV

ELASTASE ESTERASE ACTIVITY AGAINST SYNTHETIC SUBSTRATES[a]

Substrate	K_{cat}/K_m ($M^{-1}sec^{-1}$)	K_{cat} (sec^{-1})	K_m (mM)	$[E]_0$ (μM)	$[S]_0$ (mM)	pH	Reference
N-Benzoyl-L-alanine methyl ester	638	12.3[b]	19.3[b]	0.051	1.26–5.06	8.0–10.0	22, 61
N-Benzoyl-L-alanine methyl ester	600			8.0–10.0	0.3–0.5	8.0–10.0	23
N-Benzoyl-D-alanine methyl ester	2.7			0.382	4.83	8.0	61
N-Benzoyl-glycine methyl ester	2.9			1.04	1.95–5.05	8.0	61
N-Acetyl-L-alanine methyl ester	63.6			0.423	1.84–4.86	8.0	61
N-Furylacryloyl-L-alanine methyl ester	132			4.0–5.0	0.08–0.12	8.0	23
N-Carbobenzoxy-L-alanine p-nitrophenyl ester	185,000	110.0[b]	0.6[b]	c	c	7.90	23
N-Carbobenzoxy-L-leucine p-nitrophenyl ester	30,400			0.1–0.5	0.005	7.79	26[e]
N-Carbobenzoxy-glycine p-nitrophenyl ester	16,300			c	c	7.82	23
N-Carbobenzoxy-L-norleucine p-nitrophenyl ester	5,800			c	c	7.82	23
N-Carbobenzoxy-L-valine p-nitrophenyl ester	2,200			c	c	7.82	23
N-Carbobenzoxy-L-isoleucine p-nitrophenyl ester	280			c	c	7.86	23
p-Nitrophenyl formate	1,690	0.0224	0.0133[d]	0.5–2.6	0.001–0.02	7.69	26[e]
p-Nitrophenyl isobutyrate	1,620	0.0340	0.0162[d]	0.7–2.6	0.007–0.15	7.68	26[e]
p-Nitrophenyl acetate	410	0.0210	0.0512[d]	7.0	0.16	7.43	26[e]
p-Nitrophenyl trimethylacetate	123	0.0018	0.0143[d]	5.0	0.055	7.33	26[e]

[a] First-order kinetics observed for all substrates except the last four, which exhibit Michaelis–Menten kinetics. All assays conducted at 25°. Solvents: footnotes 22 and 61—0.1 M KCl; footnotes 23—phosphate buffer, $I = 0.1$; footnote 26—phosphate buffer, $I = 0.05$.

[b] Approximate values.

[c] Not quoted, but probably similar to the values given for N-carbobenzoxy-L-amino acid p-nitrophenyl esters in footnote 26.

[d] K_m here represents the ratio of rate constants k_3/k_{II} (see Eq. 1 in text), rather than a binding constant.

[e] Results for two elastase preparations are quoted in footnote 26. Those for the preparation showing high activity toward N-carbobenzoxy-L-tyrosine p-nitrophenyl ester are omitted from this table.

cases already mentioned and for the carboxylic acid p-nitrophenyl esters, where the hydrolysis rates are so slow that near-saturating amounts of substrate can be employed, and where K_m represents the ratio of rate constants k_3/k_{II} (see Eq. 1 below) rather than a binding constant.

Bender and Marshall[27] have studied in detail the kinetics of hydrolysis of p-nitrophenyl trimethyl acetate, the slowest hydrolyzed ester of Table IV. They have found the reaction to be a two-step process, consisting of a fast initial presteady-state acylation of elastase, with the liberation of a burst of p-nitrophenol, followed by a slower linear turnover, in which deacylation is the rate-limiting step. The initial rate of the presteady-state reaction is proportional to the initial substrate concentration and independent of the enzyme concentration, while the magnitude of the burst is proportional to the enzyme concentration. The steady-state reaction can be described thus:

$$ \mathrm{E} + \mathrm{S} \xrightarrow{k_{II}} \mathrm{ES'} + \mathrm{P_1} \xrightarrow{k_3} \mathrm{E} + \mathrm{P_2} \tag{1} $$

where k_{II}, the second-order acylation rate constant, shows a sigmoid pH-rate profile with an inflection point at pH 6.7, and possibly a steady value between pH 8 and 10.9, while the rate constant $k_3 (= K_{cat})$ of the rate-limiting deacylation step shows a bell shaped pH-rate profile with a lower pK_a of 6.7 and a higher pK_a of between 10.5 and 11.0 (high rates of spontaneous alkaline hydrolysis of the substrate in this upper pH range make it difficult to determine the pH dependency more precisely). The pH optimum for this rate-limiting deacylation step is about pH 8.8. In these respects this reaction closely resembles the elastase-catalyzed hydrolysis of N-benzoyl- and N-furylacryloyl-L-alanine methyl ester, discussed above. It will be noted that the above reaction scheme differs from the scheme proposed[71] for the chymotryptic hydrolysis of p-nitrophenyl acetate

$$ \mathrm{E} + \mathrm{S} \underset{k_{-1}}{\overset{k_1}{\rightleftharpoons}} \mathrm{ES} \xrightarrow{k_2} \mathrm{ES'} + \mathrm{P_1} \xrightarrow{k_3} \mathrm{E} + \mathrm{P_2} \tag{2} $$

in that the preacylation (Michaelis) complex ES has not been observed. This state of affairs (Eq. 1) should not be regarded as representing a true bimolecular acylation reaction, but rather as a special case of the more general mechanism described by Eq. (2), which applies when K_s, the equilibrium constant of the Michaelis complex, is much greater than $[S]_0$, as is the case for this substrate of very low solubility, so that the effective concentration of ES is very small. K_s for this substrate was therefore estimated to be greater than 0.5 mM.[27] Evidence in support of the acyl enzyme intermediate, trimethylacetyl elastase, is given by the findings that the deacylation rates of o-nitrophenyl trimethylacetate and

[71] H. Gutfreund and J. M. Sturtevant, *Biochem. J.* **63,** 656 (1956).

p-nitrophenyl trimethylacetate are the same, suggesting a common intermediate, while their basic hydrolysis rates and pre-steady-state acylation rates differ markedly, and that Tris (hydroxymethyl) aminomethane and methanol accelerate the enzymatic hydrolysis rate, an effect consistent with an acceleration of a rate-determining deacylation, such as has been noted for indole with chymotrypsin.[72] For further details the reader is referred to the original paper.[27] It should be noted that the observable accumulation of the acyl intermediate discussed above is due to the use of a nonideal substrate containing a good initial leaving group, p-nitrophenol, resulting in the rapid acylation of elastase by a very poor acyl leaving group, trimethylacetate, illustrating the fact that substrate specificity is expressed in deacylation as well as acylation. For ideal substrates the acyl enzyme intermediate, if indeed it exists as a distinct entity, should be regarded as a fleeting one.

It is thus concluded from these kinetic studies that the catalytic mechanism of elastase closely resembles those of chymotrypsin and trypsin, in agreement with the conclusions drawn from the structural studies discussed on page 132.

Distribution

Lewis et al.[1] first isolated the porcine enzyme, as described above and established its digestive function, but Banga[33] had previously isolated a crystalline preparation showing elastolytic activity from bovine pancreas, and Balò and Banga[5] had studied elastase levels in humans. Lewis et al.[1] reported human pancreas to contain between 0.3 and 6.2 units of elastase per gram, and bovine pancreas about 5 times as much, while quoting the elastase activity of porcine Trypsin 1-300 as 360 units per gram and that of electrophoretically purified porcine pancreatic elastase as 130,000 units per gram. Marrama et al.[73] reported the elastase levels in man, ox, and horse to be roughly equal, while that in rat was about 5 times greater, and those in pig and chicken not much less. Elastase activity has been detected in the pancreatic juices of cats and chickens[1] and dogs,[74] and Moon and McIvor[21] have shown immunological antigenic differences to exist between the enzymes from pig, human, and guinea pig. However, the amount of the enzyme in carnivores has not been reported. Lansing et al.[75] detected an elastase in the teleost fish *Lophius piscatorius*. None of these enzymes have been studied in detail, besides porcine

[72] E. Awad, in B. S. Hartley, *Ann. Rev. Biochem.* **29**, 45 (1960).
[73] P. Marrama, C. Ferrari, R. Lapiccirella, and U. Parisoli. *Ital. J. Biochem.* **8**, 280 (1959).
[74] E. Kokas, I. Foldes, and I. Banga, *Acta Physiol. Acad. Sci. Hung.* **2**, 333 (1951).
[75] A. I. Lansing, T. B. Rosenthal, and M. Alex, *Proc. Soc. Exptl. Biol. Med.* **84**, 689 (1953).

elastase discussed above, and nothing is known about the degree of similarity between them. There are no substantiated reports of vertebrate pancreas lacking elastolytic enzymes. Uriel and Avrameas[43] have detected small quantities of a second, quite distinct, elastolytic enzyme in pig pancreas, in addition to elastase (see p. 120).

Many elastolytic enzymes from plants and microorganisms have been reported[3,38,76,78] which will not be reviewed here. However, the α-lytic protease of *Sorangium* sp. is worthy of special note. It has been studied in detail, and shown to be a serine proteinase exhibiting substantial amino acid sequence homologies with porcine elastase around the catalytically important residues.[76,77] Its kinetic and catalytic properties are almost identical with those of porcine elastase,[78,79] suggesting that its active center conformation will prove to be very similar, despite differences elsewhere in the molecule.

[76] L. B. Smillie and D. R. Whitaker, *J. Am. Chem. Soc.* **89**, 3350 (1967).
[77] L. B. Smillie and D. R. Whitaker, personal communication.
[78] H. Kaplan and D. R. Whitaker, *J. Am. Chem. Soc.* **89**, 3352 (1967).
[79] H. Kaplan and D. R. Whitaker, *Can. J. Biochem.* **47**, 305 (1969).

[8] Horse Prothrombin

By KENT D. MILLER

Preparation

Principle

Horse prothrombin is unique because of its ease of purification and crystallization which may be due to molecular characteristics that distinguish it from the zymogen of other species. Highly purified horse prothrombin was previously obtained by two relatively specific absorptions of the protein on barium citrate gels followed by an isoelectric precipitation, and either chromatography or crystallization.[1] The following procedures involve one absorption of the plasma protein on barium citrate, a second absorption on DEAE-cellulose, and a final chromatography on DEAE-Sephadex. The modified procedures provide consistent 40–60% yields of the zymogen. They are designed to eliminate traces of other coagulation factors having affinities for barium-based gels. They also avoid formation of the modified prothrombin which sometimes develops on precipitation (i.e., ammonium sulfate precipitation) of horse prothrombin before its impurities are removed.

[1] K. D. Miller and A. W. Phlean, *Biochem. Biophys. Res. Comm.* **27**, 505 (1967).

Reagents and Materials

Citrated Horse Plasma. Horse blood (17 parts) drawn from the external jugular vein is added with gentle agitation to one part of 0.165 M sodium citrate. This is just sufficient citrate to prevent clotting of horse blood. The low citrate concentration is utilized to reduce the amount of barium citrate subsequently generated and consequently to reduce the amount of impurities absorbed on the gel along with the prothrombin. Twice this amount of citrate is usually required to anticoagulate other species of blood or when inordinate amounts of tissue juice are introduced during blood collection.

The citrated horse blood is stored at 2°–5°. The high sedimentation rate of horse red cells permits plasma removal after 4 hours' settling without centrifugation. This property renders horse prothrombin particularly amenable to large-scale production. Large volumes (i.e., greater than 10 liters) are usually stored overnight in the cold, the clear plasma then being removed by siphoning. Residual blood cells can be removed from the plasma by centrifugation (1500 g; 15 minutes; 2°–5°) or by continuous flow in a Sharples No. 1 Super Centrifuge or its equivalent). However, presence of a few cells does not affect the prothrombin preparation.

Amberlite IRF-97 (Rohm and Haas, Philadelphia), after conversion to the acid cycle by stirring in 1.0 N HCl, is washed thoroughly with water, then changed to the sodium cycle in 2.0 N NaOH. Fines are removed by decanting 4–5 aqueous suspensions, each after 30 minutes settling. The resin is then thoroughly washed with water on a Büchner funnel and dried. This carboxylic cation-exchanger, employed for the dissolution of the barium citrate gels containing absorbed prothrombin, is selected because of its high affinity for thrombin and thrombinlike proteins which might contaminate the prothrombin. DEAE-cellulose (Whatman DE-52) and DEAE-Sephadex (Pharmacia Fine Chemicals, Inc.) are both cycled according to the manufacturers' directions. The DEAE-cellulose is equilibrated in 0.05 N acetate buffer, pH 6.0, which is 0.15 N with respect to NaCl. The DEAE-Sephadex is equilibrated in 0.05 N acetate buffer, pH 6.0, which is 0.25 N with respect to NaCl.

Reagent-grade barium chloride should contain as little contamination by heavy metals as possible. Visking casings (Visking Co., Chicago) are soaked at least 15 minutes in hot 0.1 M Na$_2$EDTA and rinsed with distilled water before use.

Preparative Procedure

Barium Citrate Absorption and Elution. The citrated horse plasma is cooled to 2°–5° and one-ninth volume of cold 1.0 M BaCl$_2$ is added dropwise with stirring over a period of 10–15 minutes. Rapid absorption of prothrombin to the barium citrate gel thus generated permits immediate centrifugation (1200 g; 10 minutes; 2°–5°: or Sharples No. 1 Super Centrifuge or equivalent with a clarifier bowl containing liquid discharge ports in the neck). The sedimented gel containing the prothrombin is then gently blended (Waring blendor or equivalent) in a volume of cold 0.1 M BaCl$_2$ approximately 10 times that of the packed gel. After centrifugation (1200 g; 5 minutes; 2°–5°) the washed gel is gently blended in a volume of cold water equal to one-half the original plasma volume. The suspension is then stirred while dry Amberlite IRF(Na$^+$) is added in increments. The amount of resin necessary for degradation of the barium citrate varies with the Na$^+$ equivalent of the resin. Usually 3–10 g of resin per 100 ml of suspension are adequate. Degradation is usually complete in 10 minutes. The simple but important test for complete barium citrate dissolution is clearing of the supernatant after standing 1–2 minutes without stirring or agitation. The pH of the clear supernatant should be in the range 7.0–9.0. The resin is removed from the eluate by filtration on fluted Schleicher and Schuell No. 520-B-½ papers, or their equivalent.

DEAE-Cellulose Absorption and Elution. The pH of the cold eluate is reduced to 6.0 by dropwise addition with stirring of 0.2 N HCl. Pre-equilibrated DEAE-cellulose paste (6.0–7.0 g per 100 ml eluate) is added and the suspension stirred for 15 minutes at 2°–5°. For best results the resin containing the absorbed prothrombin is recovered by filtration in a chromatographic tube of appropriate size (height:diameter approximately 7:1) fitted with a coarse sintered glass or similar fast-flowing disk. The resin is washed while in the tube with 0.15 N NaCl until the OD$_{280}$ is less than 0.100. The prothrombin is then eluted from the column with 0.50 N NaCl. In order to keep the elution volume low, only the effluents containing prothrombin activity are collected. Overall yields at this point should approximate 60% of the original plasma prothrombin.

The DEAE-cellulose absorption and elution should be completed during the first day of the preparation. The low pH of 6.0 assures stability of the zymogen while on the resin.

DEAE-Sephadex Chromatography. The above eluate is diluted with an equal volume of cold 0.05 M acetate buffer, pH 6.0, and then applied to a DEAE-Sephadex column of appropriate size. This final chromatographic separation of the zymogen from trace contaminants requires a relatively flat salt gradient. For example, 100–130-mg aliquots of horse prothrombin, each obtained from 1 liter of plasma, were separated from contaminants visible on disc electrophoresis by chromatography at 4° on 2.5 × 30 cm columns employing linear salt gradients generated from 1000 ml of 0.80 N NaCl–0.05 N NaOAc, pH 6.0, added to 1000 ml of the equilibration buffer (0.25 N NaCl–0.05 N NaOAc, pH 6.0). It is advisable to monitor the chromatographic fractions by disc electrophoresis[2] since an impurity is eluted from such columns just before the prothrombin.

Alternative Procedure and Crystallization. Alternatively, the first barium citrate eluate can again be made 0.1 M with respect to $BaCl_2$. The barium citrate gel which is thus formed and which contains the absorbed prothrombin is recovered by centrifugation (1200 g; 5 minutes; 2°–5°. It is suspended in one-half the original plasma volume of cold water and decomposed by stirring with IRF-97(Na+) (1.5–5 g per 100 ml suspension). After removal of the resin by filtration through Schleicher and Schuell No. 520-B-½ papers, the pH of the eluate is adjusted to 4.7. After 1 hour at 2°–5° the precipitated prothrombin is recovered by centrifugation (3000 g; 20 minutes; 2°–5°). It is then suspended in the desired volume of cold water and dissolved by addition of sufficient 0.1 N NaOH or dry IRF-97(Na+) to raise the pH to 6.0–6.5.

Crystallization of the above isoelectrically precipitated horse prothrombin or the active material from DEAE-Sephadex chromatography, or from any other ion-exchange or Sephadex chromatographic effluents[1] is accomplished after exhaustive dialysis against distilled water followed by lyophilization. The dry powders are suspended in cold (−10°C) chloroform–ethanol (2:1) and stirred with a glass rod. A "sheen" due to needle crystals appears on stirring for 5–15 minutes. An equal volume of cold (−10°C) ethanol is added and the crystals are harvested by centrifugation. The residual chloroform is removed by three extractions with cold (−10°C) ethanol, and the crystals are dried *in vacuo.* Recrystallization is accomplished by partial lyophilization of a 1% solution of the prothrombin with reduction to 1/5 to 1/10 its original volume. The lyophilization flask is then placed at 2°–5° before the remaining frozen material thaws completely and seed crystals are added. Further crystallization occurs overnight.

The crystallization procedure denatures impurities present in the unchromatographed preparations, resulting in higher specific activities of the soluble prothrombin.

Properties of Horse Prothrombin

Stability

Solutions of horse prothrombin stored at pH 6.0–6.5 and at −20° are stable for at least 18 months, as are lyophilized preparations. No detectable thrombin was formed in solutions stored 10 days at 5° and at pH 7.7 in the presence of bacteriostatic agents.

[2] B. J. Davis, *Ann. N.Y. Acad. Sci.* **121**, 404 (1964).

Horse prothrombin fractions which contain impurities on disc electrophoresis, such as those of the early preparative steps, sometimes generate a second prothrombin, particularly when precipitation methods are applied. The first protocol outlined above avoids such precipitation steps. This second prothrombin is slowly activated by tissue thromboplastic systems compared to the native zymogen. It can be detected and separated from the native zymogen by chromatography on long (190–200 cm) columns of Sephadex G-200 (63–88 μ) equilibrated and developed with 0.1 M borate buffer, pH 8.0. No such derivative has been noted in horse prothrombins precipitated by salting-out or other procedures after removal of the impurities detectable by disc electrophoresis.

Purity and Physical Properties

The preparations eluted from DEAE-cellulose contain 5–10% impurities migrating close to the prothrombin on disc electrophoresis at pH 8.9 although the zymogen may be ultracentrifugally and immunochemically homogeneous. Proper elution during DEAE-Sephadex chromatography separates the tenacious contaminants. However, disc electrophoresis of the active chromatographic fractions is advisable for detection of possible overlapping impurities, particularly on the leading edge of the prothrombin peak.

Several preparations tested showed no detectable coagulation factors VII, IX, and X activities which in other species of plasma also absorb to barium citrate.

Physical and Chemical Properties

$D^0_{20,w}$	4.4 (immunodiffusion method[3])
$D^0_{20,w}$	4.5 (free diffusion)
$s^0_{20,w}$	6.67
MW	145,000 (Svedberg equation; $D^0_{20,w} - 4.4$; $\bar{V} - 0.74$)
MW	130,000 (method of Andrews[4])
%N	15.0
N-Terminal amino acids	2.04 moles Ala per 130,000 g protein
Disk electrophoresis	
pH 8.9, 8 M urea	1 polypeptide band
pH 8.9, 8 M urea, 0.015 M mercaptoethanol	2 polypeptide bands
Specific activity	750 NIH units per mg protein; 1300 "Iowa" units per mg protein

[3] A. C. Allison and J. J. Humphrey, *Nature* **183**, 1590 (1959).
[4] J. Andrews, *Biochem. J.* **96**, 595 (1965).

The striking feature of horse prothrombin is its molecular weight— nearly twice that reported for the bovine and human zymogens.[5-9] The nature and reason for the difference is not clear. Dimer formation during purification is probably not the cause since the bovine and human zymogens prepared by identical or similar methods are of the lower molecular weight. Also, the molecular weight of rat prothrombin, prepared by similar methods, is reportedly nearly 25% greater than that of the bovine and human proteins.[10] The major molecular differences may be responsible for the ease of purification of the horse prothrombin and its unusual crystallization.

[5] F. Lamy and D. F. Waugh, *J. Biol. Chem.* **203**, 489 (1953).
[6] C. R. Harmison, R. H. Landaburu, and W. H. Seegers, *J. Biol. Chem.* **236**, 1963 (1963).
[7] G. H. Tishkoff, L. C. Williams, and D. M. Brown, *J. Biol. Chem.* **243**, 4151 (1968).
[8] G. F. Lanchantin, J. A. Friedmann, J. DeGroot, and J. W. Mehl, *J. Biol. Chem.* **238**, 238 (1963).
[9] G. F. Lanchantin, J. A. Friedmann, and D. W. Hart, *J. Biol. Chem.* **240**, 3276 (1965).
[10] L. F. Li and R. E. Olson, *J. Biol. Chem.* **242**, 5611 (1967).

[9] Thrombin Assay[1]

By D. Joe Baughman

Thrombin occurs in the blood of all vertebrates as the inactive precursor, prothrombin, which can be converted to thrombin by both *in vivo* and *in vitro* mechanisms.[2,3] As an enzyme, thrombin exhibits distinct specificities for proteolysis of large substrates and for esterolytic hydrolysis of small substrates. The physiologically important thrombin substrates are components of the blood-clotting system; thrombin activates and inactivates many different clotting factors and can initiate aggregation of blood platelets. The substrate most widely studied has been the protein, fibrinogen, which can polymerize into a gel or clot after partial proteolysis by thrombin.

Assays for the specific proteolytic activity of thrombin using current techniques differ in several respects from the assays of most other

[1] The details of the assay procedure described in this chapter were derived from work previously published. D. F. Waugh, D. J. Baughman, and C. Juvkam-Wold, *Thromb. Diath. Haemorrhag.* **20**, 497 (1968). D. J. Baughman, D. F. Waugh, and C. Juvkam-Wold, *Thromb. Diath. Haemorrhag.* **20**, 477 (1968).
[2] R. Biggs and R. G. MacFarlane, "Human Blood Coagulation and Its Disorders," 3rd Ed. F. A. Davis, Philadelphia, 1962.
[3] W. H. Seegers, "Blood Clotting Enzymology." Academic Press, New York, 1967.

enzymes. These differences occur because the proteolytic or clotting activity of thrombin is not uniquely related to its small substrate activity.[3,4] For this reason the assay techniques are dependent on the poorly understood chemical and physical events associated with clotting in which the protein, fibrinogen, is used as substrate. Thrombin assays, then, require specific interactions of two proteins, neither of which can be standardized independently. One consequence has been the somewhat empirical development of assay procedures which are easy to perform but difficult to reproduce.

An important property of any assay system is its reproducibility. Reproducibility can be described in terms of three types of variations: (1) variations between replicates on the same solutions, (2) variations between equivalent solutions on different days, and (3) variations introduced when data are converted into units comparable between laboratories. For enzyme assays in which the substrate is available in a reproducible state, all variations are identical to variations between replicates. For thrombin assays, however, all three types of variations are important since each arises from different sources. As a result the procedures section not only describes techniques important for minimizing variations between replicates but also describes techniques which are important for minimizing day-to-day variations and variations which result from interlaboratory standards.

The effects of these variations will depend on the detailed method by which the assays are performed. By using special equipment and time-consuming methods it is possible to distinguish 1% differences between thrombin solutions on different days.[1] Many experiments do not require this discrimination. The methods described here do not require special equipment and therefore will have less discrimination. If maximum discrimination is desired, the original papers should be consulted.

Most of the specific data summarized in this chapter were obtained on bovine materials. Except where noted, however, the interpretations are believed to be species independent. For this reason few distinctions between species are made.

Assay Method

Principles

The chemical basis of the thrombin assay is the sequence of reactions which occur when thrombin and fibrinogen are combined.[5] These reactions can be summarized as follows:

[4] E. Jorpes, V. Torbjörn, B. Öhman, and B. Blombäck, *J. Pharm. Pharmacol.* **10**, 561 (1958).

(1) Activation: fibrinogen $\xrightarrow{\text{thrombin}}$ monomer fibrin (f_1)

(2) Linear polymerization: $nf_1 \rightarrow f_n$

(3) Side-to-side aggregation and network formation $f_n \rightarrow$ fibrin clot

Activation (1) is the partial proteolysis of fibrinogen during which thrombin releases four small peptides producing monomer fibrin. Once formed, monomer fibrin rapidly polymerizes into linear polymers (2). These polymers continue to grow in both length and diameter. Eventually large fibers are produced which suddenly cross link into a space-filling network, converting the solution into a gel called the clot (3). The chemical events continue until all fibrinogen is converted into fibrin and appears in the network. In the usual clotting assay, thrombin is $\sim 10^{-8} M$, fibrinogen $\sim 10^{-6} M$, gelation occurs in ~ 15 seconds and chemical equilibrium is attained in ~ 2–5 minutes.

An important property of the system is the sensitivity of the polymerization reactions to low concentrations of many different types of chemical agents such as protamine, calcium, acetone, or cyanide.

Dependent on the chemical events of clotting are important physical events which are usually used for studying the clotting system. The description, however, will be confined to those physical events which are visible in a stirred clotting system since these are the changes essential for thrombin assay. For a thrombin–fibrinogen solution which will clot in about 30 seconds, the clotting solution remains transparent for about 24 seconds after mixing. At that time opalescence and then turbidity become apparent with the latter continuously increasing. Suddenly one or several lumps of compacted fibrin will appear. At the same time, the turbidity of the surrounding solution is decreased. If stirring is continued, the events will be repeated. If stirring is stopped, the solution will gel. Since the first appearance of compacted fibrin is preceded by turbidity, appears instantaneous in occurence, and is highly reproducible, this event is used as the clotting end point for thrombin assays. For clotting times between 15 and 60 seconds the end point occurs when approximately 50% of the fibrinogen is activated. Although the end point occurs a few seconds before gelation, for thrombin assay it is not critical.

The parameter used to measure thrombin concentration is called the clotting time, τ. The clotting time is defined as the interval between combining thrombin with fibrinogen and the clotting end point. Clearly, clotting times depend on both thrombin-dependent activation reactions

[5] H. A. Scheraga and M. Laskowski, Jr., in "Advances in Protein Chemistry," Vol. XII (C. B. Anfinsen, Jr., M. L. Anson, K. Bailey, and J. T. Edsall, eds.) p. 1. Academic Press, New York, 1960.

and fibrin-dependent polymerization reactions. When all polymerization variables are constant, the clotting times increase as thrombin concentrations decrease. Empirical linear relations describing the thrombin-clotting time relationship have been described for two different sets of axes,[2,6] ($\log \tau : \log T$) and ($\tau : 1/T$) where T is the thrombin concentration.

The current understanding of thrombin, fibrinogen, and the polymerization reactions is so poorly developed that the clotting properties of thrombin or fibrinogen cannot be measured independently. One of these components then must be used as a clotting standard. By comparison with fibrinogen the purity, stability, and reproducibility of many thrombin preparations clearly indicate that thrombin is the better standard. Not only must different preparations of fibrinogen be calibrated but for many purposes fibrinogen will need to be calibrated every day that assays are performed. For many assay situations a secondary thrombin standard is required. Thus, clotting times are used to calibrate fibrinogen with a standardized thrombin. Unknown thrombin solutions are measured in terms of dilutions of a standard thrombin with a calibrated fibrinogen using clotting times. Often the standardized thrombin is a secondary standard obtained by clotting time calibrations with a primary standard. Since the fibrin polymerization reactions are sensitive to low concentrations of different chemical agents, this assay procedure is valid only if the standard and unknown thrombin are in the same chemical environment. Either protocols must be designed to create this condition, or differences between the chemical environments of thrombin must be shown to be insignificant.

Since thrombin is the standard for clotting assays, it is important that thrombin dilutions and transfers be accurately performed with a minimum of variations. At the thrombin concentrations of most assays, however, significant amounts of thrombin are lost by adsorption at the vessel–solution and air–solution interfaces.[1] Thus, particular attention to the thrombin dilution techniques is necessary to minimize the effect of these variations.

Clotting Conditions. One set of clotting conditions which gives highly reproducible assays is: pH = 7.0, I = 0.15 (0.10 sodium chloride and 0.05 phosphate buffer) ; temperature = 29°; fibrinogen = 1.05 mg clottable protein per milliliter; thrombin = 2.0–0.5 NIH units per milliliter (giving clotting times of 15–60 seconds) and total volume = 0.3 ml. Most of these conditions can be varied within the following limits:[7] pH = 6.5

[6] D. F. Waugh, D. J. Baughman, and K. D. Miller, *in* "The Enzymes," Vol. 4, 2nd Ed. (P. D. Boyer, H. Lardy, and K. Myrbäck, eds.), p. 215. Academic Press, New York, 1960.

[7] W. H. Seegers and H. P. Smith, *Am. J. Physiol.* **137**, 348 (1942).

to 8.5; I = 0.1 to 0.2; temperature = 10° to 40° and total volume = 0.2 to 1.5 ml. If assays must be performed in the presence of calcium, imidazole buffer can be exchanged for phosphate buffer.

Preparing Buffers. The phosphate salts for buffers of clotting solutions should be dissolved at room temperature to reduce day-to-day variation, particularly with respect to fibrinogen. For convenience these salts can be dissolved overnight in deaerated water. The effects of other salts in this respect are not known.

Preparation of Stock-Fibrinogen (Stock-F). Fibrinogen or Stock-F is prepared from Cohn's Bovine fraction I, which can be obtained from commercial sources. Fraction I should be stored at low temperature (below —30°) and opened only when ready to use after equilibration at room temperature. The purification procedure follows closely that recommended by Laki[8] and yields ~60 days of assay at 30 clotting times per day. The procedure usually takes 5 days.

Four 5-g aliquots of fraction I are weighed. Each aliquot is dissolved with gentle stirring at room temperature in 250 ml of 0.1 M phosphate buffer, pH = 6.4. Complete dissolution occurs in 3–4 hours. After dissolution 250 ml of water are added to each aliquot and the solutions are stored at 4° overnight. The next day the precipitated cold-insoluble globulins are removed by centrifugation for 30 minutes at 2500 rpm and discarded. The supernatant is warmed to room temperature. The fibrinogen is precipitated from the supernatant by adding sufficient saturated ammonium sulfate to produce 23% saturation in each container. After centrifugation for 30 minutes at 2500 rpm, the precipitated fibrinogen is collected and dissolved in 240 ml of 0.3 M potassium chloride. The solution is dialyzed against ~8 liters of 0.3 M potassium chloride for 2 nights with at least two changes of dialysis fluid.

After dialysis the fibrinogen solutions are centrifuged to remove particulate matter. Following centrifugation the fibrinogen solutions are combined and an aliquot removed for assay. The remainder of the fibrinogen is frozen in small aliquots of 2.5–3.0 ml and stored frozen at —95°. These frozen aliquots are called Stock-F and have remained stable for at least a year.[1] Although aliquots have been stored as high as 0° for a few weeks without detectable loss, storage at —25° for a long time will introduce significant deterioration. Presumably this deterioration results from repeated freezing and thawing since the eutectic point of 0.3 M KCl is —22°.

Fibrinogen Assay. An aliquot of Stock-F should be removed for assay before freezing. The fibrinogen is diluted with a clotting buffer to

---

[8] K. Laki, *Arch. Biochem. Biophys.* **32**, 317 (1951).

an absorbance of slightly less than 1.0 at 280 nm for a pathlength of 1 cm. Multiple aliquots are clotted with 1 NIH units per milliliter of thrombin. The total absorbance at 280 nm clotted by thrombin is corrected for scatter at 320 nm, and is converted to milligrams protein/ml[9] by multiplying by the factor 0.64. From the dilution factor required to produce that fibrinogen concentration added to clotting tubes (1.58 mg clottable protein per milliliter), a fibrinogen buffer is determined which will maintain constant ionic strength, including the 0.3 M KCl introduced from the Stock-F. Any Stock-F which does not have 95–96% clottable absorbance or contain ~70% of the initial absorbance in fraction I, should not be used.

This fibrinogen assay measures the total amount of protein which appears in the clot. Such an assay is quite unsatisfactory for quantitatively predicting the clotting time response of fibrinogen preparations. A series of Stock-F's diluted to the same clottable protein appeared to have a range of ±25% of the average clotting time for the same thrombin solution. Despite its unreliability, this parameter is still the one used to define the clotting conditions.

Preparation of τ-Tubes. On the days clotting times are to be measured, one or more tubes of Stock-F are thawed at 29° and diluted to 1.58 mg clottable protein per milliliter with a buffer which will give after dilution pH = 7.0 and I = 0.15. Although fibrinogen can be warmed for thawing, it will denature if heated much above 40°. Nonuniformity of thawing is a good practical test for detecting fibrinogen deterioration. Stock-F should be transferred into its diluent with a pipette since using a syringe would increase variations between tubes of Stock-F.

Aliquots of 0.2 ml of diluted fibrinogen are transferred into 12 × 75 mm Pyrex glass tubes, called τ-tubes. The τ-tubes are covered and placed in a water bath at 29° for 2 hours before use in order to eliminate fibrinogen changes which occur during this time. The order of filling the τ-tubes is haphazardly destroyed when placing tubes in the bath.

τ-Tubes of glass rather than translucent plastic are used since currently available plastics introduce larger variations in the assay. The glass tubes must be washed with chromic acid to remove all adsorbed thrombin and the chromate ion must be thoroughly removed by soaking in water at least 24 hours.

Dilution of Thrombin. For quantitatively reproducing dilutions of thrombin solutions, all dilutions should be made in plastic or siliconized glass containers whose walls were previously saturated with thrombin. Thus, two identical dilutions are made. The first dilution is made at least

[9] R. W. Hartley, Jr. and D. F. Waugh, *J. Am. Chem. Soc.* **82,** 978 (1960).

60 minutes before the thrombin is required or, more conveniently, the night before. After thrombin adsorption has been completed, the container is emptied, drained for a minute or so, and the second dilution is performed. Thrombin should always be added to the diluent with the tip of the pipette just below the surface. If serial dilutions are avoided and single dilutions of 1/100 or more are made, thrombin losses are less than 1% and quantitatively reproducible.

It is often convenient to have a large source of dilute thrombin which can be frequently sampled without change. Bottom drain bottles have often been used for this purpose; these consist of a polyethylene bottle with a small plastic tube of Teflon or polyethylene inserted near the bottom. Thrombin dilutions are made as described above into buffers containing $5 \times 10^{-4} M$ KCN for bacteriostasis. Aliquots are drained into small plastic containers whose walls were previously adsorbed with thrombin. The contents of the drain bottle remain unchanged upon sampling, and the losses, less than 1%, occurring in the individual aliquots are quite reproducible. This type of bottle eliminates losses which arise from repeatedly piercing the thrombin film that forms at air–solution interfaces.

Clotting-Time Measurement. All clotting times on any one day should be measured together as quickly as convenient. The order of thrombin assays should be randomized with respect to other variables being studied.

For any one assay a τ-tube is uncovered and 0.1 ml of thrombin withdrawn into a clean serological or blow-out pipette. At the same instant that thrombin is added to the fibrinogen solution, the timer is started. The thrombin is blown into the fibrinogen sufficiently rapidly to produce mixing and the tube is swirled to ensure mixing. The τ-tube is returned to the water bath and occasionally examined for the first appearance of turbidity. Once turbidity appears, the tube is removed from the bath and continuously swirled or tilted. At the first sign of a visibly compacted fibrin mass, the timer is stopped. Before examining the timer, the observer decides whether to repeat the measurement or accept it. If the measurement is accepted, it is recorded. Once confidence and experience are gained, few if any repeat measurements are necessary. Several practice clotting times should be performed before measurements are begun. The best range of clotting times for reproducibility is between 20 and 40 seconds.

The coefficient of variation for this technique is 2.5%. Many other methods which include automatization at different stages have significantly smaller random variations. It is believed, however, that 2.5% is quite satisfactory for many different types of experiments, particularly if they are not designed to study the kinetics of clotting. For improved techniques the original papers[1] should be consulted.

Standardization of Fibrinogen. Day-to-day assay variations will mainly arise from variations between tubes of Stock-F which result from the freezing, thawing, and diluting procedures or between different preparations of Stock-F. The coefficients of variation are 1.7% for the former (it can increase to 5% if phosphate salts are warmed during dissolution) and 12% for the latter. One procedure for reducing these variations is to remove an aliquot from a reservoir of standardized thrombin solution each day, to dilute it and to measure its average clotting time with several determinations. These assays are randomized with respect to thrombin unknowns to be measured that day. By this procedure thrombin solutions are always measured in terms of the standardized thrombin. Then, the day-to-day variations are reduced to the variations of thrombin dilutions, the number of assays for each thrombin solution, and the variations of the individual assays.

There is an alternative procedure for experiments consecutively performed over a maximum of 5 days. On the first day standardized thrombin is diluted with buffer containing $5 \times 10^{-4} M$ KCN. Sufficient thrombin is diluted to last the 5 days and is stored at room temperature (23°). This solution is used as a standard each day. The clotting time of such solutions increase $1.44 \pm 0.27\%$ of the clotting time per day.[10] After 5 days a new dilution should be made.

Standardized Thrombin. With current definitions of primary thrombin standards, most experimental protocols will require secondary thrombin standards to be available. For a secondary standard a large reservoir of stable concentrated thrombin should be prepared, either Stock-T or Sephadex-T are satisfactory. On one day an equal number of dilutions of this reservoir and of the primary standard are made. From preliminary trials the clotting times of each reservoir dilution should be nearly identical to those of each primary standard dilution. All clotting times should lie between 20 and 60 seconds. Multiple clotting times of each dilution are measured in a randomized sequence. Both sets of dilution averages should be plotted with a linear least squares regression on either $(\tau : 1/T)$ axes or $(\log \tau : \log T)$ axes. On the $(\tau : 1/T)$, where τ is the ordinate, both curves should pass through the same ordinate, and the concentration of the reservoir is obtainable from the ratio of the slopes. On the $(\log \tau : \log T)$ axes, where $\log \tau$ is the ordinate, both curves should have the same slope and the concentration of the reservoir is determined

---

[10] In each case the number following the $\pm$ is the standard deviation of the individual measurement which gave rise to the average preceding the $\pm$.

by the difference in the intercepts. The greater the number of dilutions and the greater the number of assays, the more certain is the concentration of the secondary thrombin standard.

Units. The units in which thrombin is described indicate the primary standard of the assay. Currently, two units are being used.

The National Institutes of Health maintain a thrombin standard defined in terms of the NIH unit.[11] Some of this standard is readily available by a written request to the National Institutes of Health, Biologic Controls, Bethesda, Maryland. The activity in terms of NIH units per milligram is given on each vial. The material is quite stable in solution at room temperature. The standard is designed to be used primarily for calibrating a secondary standard rather than for calibration during individual experiments. One NIH unit of thrombin in a volume of 23.5 ml has been measured to produce 0.37 micromole of N-terminal glycine in 20 minutes at pH $= 9.0$, $I = 0.2$ and temperature $= 25°$ when fibrinogen $= 0.3\%$.[12]

There are, however, several disadvantages to the NIH standard. Not all of the material in NIH thrombin is soluble and the recommended dissolving procedure has a coefficient of variation $= 2.5\%$. Since the vial-to-vial variations and lot-to-lot variations of NIH thrombin are not known, the secondary standards of different laboratories can easily differ by more than 10% and possibly as high as 20% or more.

Many laboratories report thrombin activity in Iowa units, which appear to have two definitions, the original definition and the working definition. The original unit was defined as the amount of thrombin required to give a 15-second clotting time with a standardized fibrinogen.[7] The working definition equates 1 unit of thrombin with 1 unit of prothrombin in which pooled bovine plasma contains a standard amount of prothrombin.[13] Although these definitions suggest that each laboratory could maintain its own primary standard, it appears that different laboratories may have different Iowa units.[14] A further disadvantage to this unit is the use of prothrombin activation reagents which introduce greater variations between standardizations than those standardizations which use thrombin directly.

An improved standard for thrombin assays is desirable for further

[11] "Minimum Requirements of Dried Thrombin," Second Revision, Division of Biologics Control, National Institutes of Health, Bethesda, Maryland, 1946.
[12] S. Magnusson, *Thromb. Diath. Haemorrhag.* **4,** 167 (1960).
[13] A. G. Ware and W. H. Seegers, *Am. J. Clin. Pathol.* **19,** 471 (1949).
[14] Compare W. H. Seegers, "Prothrombin," p. 436. Harvard University Press, Cambridge, Massachusetts with K. D. Miller and W. H. Copeland, *Exptl. Mol. Pathol.* **4,** 431 (1965).

kinetic studies of thrombin and the clotting process. A recent report suggests that such a standard may be available.[15]

Preparation[16]

Column Preparations. Two ion-exchange columns are prepared; one 5×7 cm (diameter $\times$ height) containing DEAE-cellulose ($\sim$0.9 meq/g) and the other 2.5×7 cm containing cellulose phosphate ($\sim$0.8 meq/g). After washing both columns with acid and base twice, the columns are equilibrated with phosphate buffer, pH = 7.0 and I = 0.1. Then the output of the DEAE column is connected to the input of the cellulose phosphate column. The flow rate is adjusted to $\sim$7 ml per minute.

Sephadex G-100 is packed in a column 0.9 cm in diameter to a height of 100 cm. The column is equilibrated with sodium chloride pH = 7.0 and I = 1.0 for 4 days before use. Flow rate is $\sim$3.5 ml per hour.

Ion-Exchange Separation. Parke-Davis Thrombin Topical, a bovine thrombin which is commercially available, is used as starting material; $\sim$800 mg or 20,000 NIH units are dissolved in 320 ml of equilibrating buffer and applied to the DEAE column. The dilute thrombin is followed with 400 ml of the same buffer. Although the thrombin activity in Parke-Davis thrombin is slightly retarded on the DEAE column, essentially all the thrombin passes through the DEAE-cellulose and is adsorbed onto the cellulose phosphate. Since Parke-Davis Thrombin is unstable in dilute solutions, it should not be dissolved until just before addition to the DEAE column.

After applying the 400 ml of equilibrating buffer, the columns are separated. Then 250 ml of phosphate buffer pH = 7.0 and I = 0.15 are passed through the cellulose phosphate to remove further impurities. Thrombin is eluted with buffer at I = 1.0 (I = 0.1 phosphate buffer pH = 7.0 and I = 0.9 NaCl). Practically all of the thrombin, 91.1 $\pm$ 8.4%, appears in 50 ml after discarding the first 30 ml. Approximately 70% of the thrombin is in the first 1.2 ml and 85% in the first 12 ml. The thrombin at this step is called Stock-T. On the average Stock-T contains 24 $\pm$ 10% nonthrombin impurities.

Sephadex Chromatography. Final purification, if desired, is achieved by collecting the most concentrated 1.0 ml of Stock-T and applying it to the Sephadex G-100 column. Thrombin appears in an 8-ml volume which peaks at 42.5 ml of effluent. This peak contains $\sim$5% nonthrombin material under the leading edge and the remaining 0.6 of the peak has constant specific activity. The constant specific activity thrombin is

[15] R. D. Rosenberg and D. F. Waugh, *Federation Proc.* **28**, 321 (1969).
[16] The preparative technique essentially follows that of D. J. Baughman and D. F. Waugh, *J. Biol. Chem.* **242**, 5252 (1967).

called Sephadex Thrombin. All thrombins can be stored in concentrated solution at 4° with ionic strength greater than 0.5.

Properties

Stability. Solutions of Stock-T are quite stable.[16] At a temperature of 4° and I = 1.0 in the presence of ~I = 0.1 buffer at pH = 7.0, concentrated solutions of Stock-T have been stored for a year without detectable losses. At room temperature in dilute solution at pH = 7.0, I = 0.1, and $5 \times 10^{-4} M$ KCN (for bacteriostasis) thrombin solutions reproducibly denature at the rate of $1.44 \pm 0.27\%$ of the clotting time per day.

Purity. Stock-T contains $24 \pm 10\%$ nonthrombin impurities. After gel filtration of Stock-T, the last 0.6 of the thrombin peak has been shown to have six electrophoretic bands of different specific activity.[17] When thrombin is prepared from fresh bovine plasma of a single cow and purified as described above, only two or three electrophoretic bands are observed.[15] Within 1% all preparations appear to have had the same specific activity independent of the electrophoretic pattern.

Physical Properties. The molecular weight of thrombin is generally agreed to be near 34,000.[3] Diffusion constant[18] $= 8.76 \times 10^{-7}$ cm^2 sec^{-1} and $s_{20,w}^0 = 3.76 \times 10^{-13}$ sec.

Although not satisfactorily explained, N-terminal amino acid analysis of bovine thrombin has consistently given 1 mole of isoleucine and 0.5 moles of threonine with a sequence of Ile-Val-Glu-Gly.[19] Human thrombin gives only isoleucine as the N-terminal amino acid with a sequence of Ile-Val-Gly-Gly.[19] From radioactive diisopropylphosphofluoridate (DFP) binding studies, part of the active site of thrombin is believed to have the sequence Gly-Asp-Ser-Gly-Glu-Ala[20]; a sequence similar to trypsin and chymotrypsin.

For thrombin concentrations ($\sim 10^{-7} M$) used in the reservoirs of most assay procedures, significant amounts of thrombin adsorb to both the vessel–solution (V/S) and the air–solution (A/S) interfaces.[1] For nonpolar surfaces such as polyethelene or siliconized glass, the V/S loss is near 4% of the total thrombin, can be quite variable between surfaces, is highly reproducible for the same surface, and appears to be reversible. Under these conditions surfaces become saturated in about 60 minutes. For polar surfaces such as glass, the losses may be 5 to 10 times greater.

[17] R. D. Rosenberg and D. F. Waugh, *Federation Proc.* **27**, 628 (1968).
[18] C. R. Harmison, R. H. Landaburu, and W. H. Seegers, *J. Biol. Chem.* **236**, 1693 (1961).
[19] S. Magnusson, *Arkiv Kemi* **24**, 375 (1965).
[20] J. A. Gladner and K. Laki, *J. Am. Chem. Soc.* **80**, 1263 (1958).

For the A/S interfaces the adsorption occurs in about 6 minutes and removes ~1% of the total thrombin at these concentrations. However, the loss is irreversible and can become quite large if the air–surface to solution–volume ratio is large or the air surface is repeatedly broken.

Activators. Thrombin is produced from its blood plasma precursor, prothrombin, by a variety of agents.[2,3] Biological activation can occur by at least two known mechanisms, intrinsic and extrinsic activation. Both methods depend on factors commonly present in plasma but the extrinsic system replaces some of these with tissue extracts. Nonbiologic activation can occur in the presence of 25% citrate, 1%-protamine sulfate, papain, or trypsin. Trypsin and papain activation, however, will also result in the destruction of thrombin activity.

Inactivators. There are several naturally occurring specific thrombin inactivators.[2,3] Blood plasma contains several of these, the most important of which is antithrombin III. Antithrombin III inhibits thrombin activity by forming an irreversible complex with thrombin. The rate of thrombin removal in plasma is markedly increased in the presence of heparin and a plasma component called heparin cofactor. There remains some confusion whether heparin cofactor and antithrombin III are the same or different substances. Heparin, itself, will combine reversibly with thrombin at low concentrations and reduce thrombin activity. Hirudin, a protein produced in the salivary glands of leeches, is a specific and rapid thrombin inactivator whose inactivation is not reversible.[21] DFP will react specifically with the serine residues in thrombin and destroy enzymatic activity, both proteolytic and esterolytic.[20]

Specificity. One of the unusual properties of thrombin is its specificity in hydrolyzing peptide bonds in proteins and small peptides. Compared to the many fragments trypsin produces from fibrinogen,[22] thrombin hydrolyzes only four arginyl-glycine bonds.[3] Conversely, thrombin does not hydrolyze the arginyl-glycine bond in insulin[23] or in those small peptides which have been examined.[24] Secretin,[25] chymotrypsinogen,[26] and trypsinogen[27] are three proteins, not important in the blood clotting system, which have recently been shown to be susceptible to thrombin proteolysis. Most other nonclotting proteins are unaffected by thrombin. The clotting factors which have appeared to be substrates of thrombin

[21] F. Markwardt, "Blutgeunnungshemmende Wirkstoffe ans blutsaugenden Tieren." Fischer, Jena, 1963.
[22] S. Iwanaga, A. Henschen, and B. Blombäck, *Acta Chem. Scand.* **20**, 1183 (1966).
[23] K. Bailey, F. R. Bettleheim, L. Lorand, and W. R. Middlebrook, *Nature* **167**, 233 (1951).
[24] L. Lorand and E. P. Yudkin, *Biochem. Biophy. Acta* **25**, 437 (1957).
[25] V. Mutt, S. Magnusson, F. Jorpes, and E. Dahl, *Biochemistry* **4**, 2358 (1965).
[26] A. Engel and B. Alexander, *Biochemistry* **5**, 3590 (1966).
[27] A. Engel, B. Alexander, and L. Pechet, *Biochemistry* **5**, 1543 (1966).

are factors V and VII, fibrinogen, profibrinolysin, and profibrinase.[2,3] Of the many small synthetic esters which thrombin can hydrolyze,[3] p-toluenesulfonyl-L arginine methyl ester (TAME) was discovered first and appears to be hydrolyzed more rapidly than most others.[28]

Kinetics. The hydrolysis of most small synthetic substrates by thrombin appears to involve the formation of an acyl enzyme intermediate,[29] since its decomposition becomes rate limiting at pH 5.0.

Most kinetic studies of thrombin proteolysis use fibrinogen as substrate and their interpretation is complicated by the polymerization reactions. Studies on the increase of N-terminal glycine in fibrin and on the decrease of N-terminal glutamic acid in fibrinogen imply that most of the two peptides, fibrinopeptides A, are released prior to two other peptides, fibrinopeptides B, and before gelation.[30] From the studies of proteolysis after gelation has occurred, it has been concluded that only one thrombin–fibrinogen contact is rate limiting for the production of monomer fibrin.[31, 32]

[28] S. Sherry and W. Troll, *J. Biol. Chem.* **208**, 95 (1954).
[29] F. J. Kézdy, L. Lorand, and K. D. Miller, *Biochemistry* **4**, 2302 (1965).
[30] B. Blombäck, *Arkiv Kemi* **12**, 321 (1958).
[31] D. F. Waugh and B. J. Livingstone, *J. Phys. Colloid Chem.* **55**, 1206 (1951).
[32] S. Ehrenpreis and H. A. Scheraga, *Arch. Biochem. Biophys.* **79**, 27 (1959).

[9a] Bovine Prothrombin and Thrombin

By STAFFAN MAGNUSSON

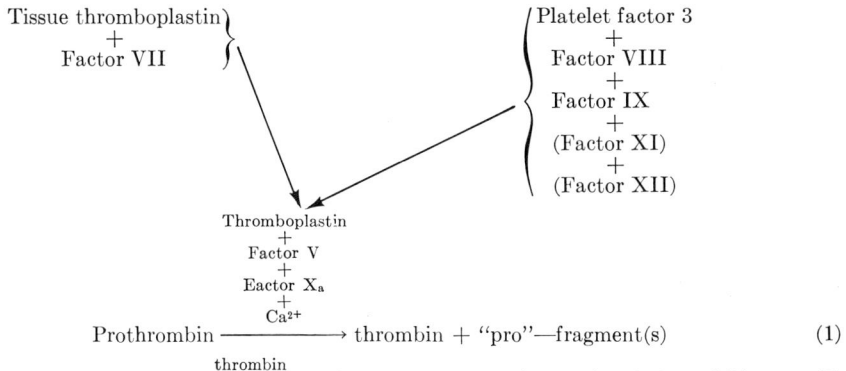

$$\text{Prothrombin} \xrightarrow{} \text{thrombin} + \text{``pro''—fragment(s)} \qquad (1)$$

$$\text{Fibrinogen} \xrightarrow{\text{thrombin}} \text{fibrin monomer} + \text{fibrinopeptide(s) A (and B)} \qquad (2)$$

$$\text{Fibrin monomer} \xrightarrow{\text{polymerization}} \text{fibrin} \qquad (3)$$

For recent review articles on the blood coagulation mechanism, see footnotes 1, 2, and 2a.

Bovine Prothrombin

Assay Methods

Principle

Prothrombin is a plasma protein and by definition the zymogen of thrombin. Therefore it can be properly assayed only in a system where prothrombin is first activated to thrombin and the activation mixture subsequently assayed for thrombin activity. Thrombin activity is defined as fibrinogen clotting activity. Immunological precipitin assays[3-5] have been developed which are at least as sensitive[3a] as the clotting assay. However, in plasma from vitamin K-deficient patients, where little or no prothrombin is found by clotting assays, the "specific" immunoprecipitation assay detects nearly as much "prothrombin" as in normal plasma.[4, 6] There are essentially two different clotting assays, namely the one-stage method[7, 8] where prothrombin activation and fibrinogen clotting take place together in the same reaction mixture, and the two-stage method[9, 10] where samples from the prothrombin activation mixture are assayed for thrombin activity against time. The one-stage method is far simpler to perform and has found wide-spread application in clinical practice. Activities found with this method do not depend solely on prothrombin concentration. The method, therefore, has to be used with caution. The two-stage method is the method of choice for assaying the products of purification procedures.

---

[1] M. P. Esnouf and R. F. Macfarlane, *Advan. Enzymol.* **30**, 255 (1968).

[2] W. H. Seegers, *Ann. Rev. Physiol.* **31**, 269 (1969).

[2a] C. R. Harmison and E. F. Mammen, *in* "Blood Clotting Enzymology" (W. H. Seegers, ed.), p. 23. Academic Press, New York, 1967.

[3] S. Magnusson, unpublished results.

[3a] Using antiserum raised in rabbits against chromatographed bovine prothrombin (see p. 167), washing the immunoprecipitate and then dissolving it in dilute phosphoric acid and determining the absorbancy at 210 nm in a 0.2 ml cuvette. the author obtained a straight line standard curve (plotting absorbancy against prothrombin concentration) in the range 40 to 400 nanograms of prothrombin.

[4] F. Josso, J. M. Lavergne, M. Gouault, O. Prou-Wartelle, and J. P. Soulier, *Thromb. Diath. Haemorrhag.* **20**, 88 (1968).

[5] P. O. Ganrot and J. E. Niléhn, *Scand. J. Clin. Lab. Invest.* **21**, 238 (1968).

[6] J. E. Niléhn and P. O. Ganrot, *Scand. J. Lab. Clin. Invest.* **22**, 17 (1968).

[7] A. J. Quick, *J. Biol. Chem.* **109**, P73 (1935).

[8] P. A. Owren and K. Aas, *Scand. J. Clin. Lab. Invest.* **3**, 201 (1951).

[9] E. D. Warner, K. M. Brinkhous, and H. P. Smith, *Am. J. Physiol.* **114**, 667 (1936).

[10] A. G. Ware and W. H. Seegers, *Am. J. Clin. Pathol.* **19**, 471 (1949).

One-Stage Method[8]

Reagents

Thromboplastin from bovine brain tissue.[11,12] The brains from 5 animals, slaughtered the same day, are carefully freed from membranes and blood vessels. The cerebral hemispheres (about 1 kg) are cut into pieces of about 1 cm³, thoroughly rinsed in tap water to get rid of most of the blood, then homogenized in a Turmix blender (full speed, 60 seconds) with an equal volume of warm (40–42°) 0.15 M NaCl at 37° with stirring for 30 minutes and centrifuged (15–20°, 1250 g, 25 minutes). The whitish supernatant suspension, comprising roughly half the total volume is sucked off and diluted with 0.1 volume of 0.15 M Tris-hydrochloride buffer pH 7.2. The activity of this suspension is assayed by mixing triplicate samples of 0.2 ml suspension and 0.2 ml normal bovine oxalated plasma, incubating for 3 minutes at 37°, and then measuring the time from the addition of 0.2 ml 30 mM $CaCl_2$ to the formation of a clot. If the clotting time in this one-stage test is shorter than 13–14 seconds the batch of thromboplastin is diluted further with a mixture of 0.1 volume 0.15 M Tris-hydrochloride pH 7.2 and 0.9 volume 0.15 M NaCl to obtain the desired clotting time. The batch is then divided into portions of suitable volume which are stored frozen at —20°, at which temperature the preparation is stable for at least 6 months. Individual portions are only thawed once.

Normal bovine citrated plasma. Bovine blood collected at slaughter is allowed to run into a solution of trisodium citrate (one part 3.8% $Na_3C_6H_5O_7 \times 2\ H_2O$ to 9 parts of blood) with thorough stirring. The plasma is sucked off after centrifugation of the blood at 2500 rpm for 20–25 minutes, and then recentrifuged at 2500 rpm for 10–15 minutes. "Normal" bovine citrated plasma is obtained by pooling plasma from 7–8 animals. The pool is divided into small portions which are stored frozen at —20°, and only thawed once.

Barium sulfate-adsorbed bovine oxalated plasma.[13] This reagent is a source of Factor V and fibrinogen and is called "P.P-reagent." Bovine blood is collected as above into a solution of potassium oxalate (100 ml 2.5% $K_2C_2O_4 \times H_2O$ to 900 ml blood). The plasma obtained after two centrifugations (same as above) is

---

[11] P. A. Owren, *Acta Med. Scand. Suppl.* **194** (1947).

[12] S. Magnusson, *Arkiv Kemi* **23**, 285 (1965).

[13] H. Brodthagen, *Scand. J. Clin. Lab. Invest.* **5**, 376 (1953).

mixed with a suspension (made by vigorous Turmix blending) of 20 g BaSO₄ (A. R. Baker or Malinckrodt) in 200 ml 0.15 M NaCl. Sufficient stirring to keep the BaSO₄ in suspension is maintained for 60 minutes at room temperature. (Do not use Turmix at this stage. It would denature the fibrinogen.) The suspension is then centrifuged at 2500 rpm for 20–25 minutes and the sediment discarded. The resulting P.P-reagent is assayed in triplicate for prothrombin-proconvertin-Factor X activity by first preincubating 0.2 ml P.P-reagent and 0.2 ml thromboplastin at 37° for 3 minutes. Then 0.2 ml 30 mM CaCl₂ is added; a clotting time at 37° of more than 10 minutes is satisfactory. The criterion that this reagent has sufficient Factor V activity and fibrinogen is that 0.2 ml normal bovine citrated plasma (1:10), 0.2 ml P.P-reagent, and 0.2 ml thromboplastin when mixed and pre-incubated for 3 minutes at 37° should give a clotting time of 30–32 seconds on addition of 0.2 ml 30 mM CaCl₂ at 37°. The reagent is divided into portions which can be stored frozen at −20° for up to 12 months.

NaCl, 0.15 M

CaCl₂, 30 mM

Tris-hydrochloride buffer, 0.15 M, pH 7.2 (Tris-hydroxymethyl aminomethane, 0.15 M)

Procedure. A series of dilutions for a standard curve is prepared from a stock dilution of 1 part "normal" bovine citrate plasma in 9 parts 0.15 M NaCl. This stock dilution is stable for only 2 hours in an ice-water bath. A series of dilutions in 0.15 M NaCl (100, 90, 80, 70, 60, 50, 40, 30, 20, 15, 10, 5, 3, 2, 1%) is prepared and the average clotting time for triplicate samples of each dilution is plotted against concentration on a log–log scale. A straight line is obtained, at least from 100 to 10%, with clotting times from 30 seconds to 10–12 minutes. The clotting system is as follows: 0.2 ml normal plasma dilution or unknown sample, 0.2 ml P.P-reagent and 0.2 ml thromboplastin which are mixed and preincubated at 37° for 3 minutes. Then 0.2 ml 30 mM CaCl₂ is added and the clotting time recorded. Samples are always tested in triplicate and at two different concentrations. Activities are expressed as prothrombin-proconvertin activity, 100% being the activity of normal plasma.

Two-Stage Method[9, 10, 12]

Reagents

Normal bovine citrated plasma. See p. 159.

Thrombin solution. This solution is used for defibrination of test samples that contain fibrinogen. Commercially available crude thrombin such as Thrombin Topical of Parke Davis & Co., or Topostasin, Hoffmann-La Roche, Basel, Switzerland can be used for this purpose. A solution of 100 NIH-U/ml in 0.15 M NaCl.

Thromboplastin. Bovine brain thromboplastin suspension (see p. 159) is suitably diluted (32–64 times) with 0.15 M NaCl.

"Reaction mixture." A mixture of 2.0 ml 30 mM CaCl$_2$, 3.0 ml suitably diluted thromboplastin and 1.0 ml Tris-hydrochloride buffer, pH 7.2 (0.3 M Tris).

P.P-reagent. See pp. 159–160.

Bovine serum. Bovine blood is collected without the addition of anticoagulant citrate or oxalate, allowed to clot and then incubated at room temperature for 3–5 hours. After centrifugation for 20–25 minutes at 2500 rpm the supernatant serum is collected and stored frozen at −20°. Its prothrombin content by the two-stage method is less than 1% of that of normal bovine citrated plasma.

"P.P-NaCl" solution. A dilution of 0.2–0.4 ml freshly thawed P.P-reagent in 15.0 ml ice-cold 0.15 M NaCl is prepared. This solution is unstable and is stored in ice-water and used within one hour.

"P.P-serum-NaCl" solution. The P.P-reagent supplies the two-stage system with Factor V. In cases where this is not sufficient to obtain activation of some purified prothrombin preparations,[14] the "P.P-serum-NaCl" solution, which provides Factors V, VII, and X is used instead of "P.P-NaCl" solution. It is a mixture of 0.2–0.4 ml freshly thawed P.P-reagent, 0.2–0.4 ml bovine serum and 15.0 ml 0.15 M NaCl. This solution is also unstable, is stored in ice-water and used within 1 hour.

NaCl, 0.15 M

CaCl$_2$, 30 mM

Tris-hydrochloride buffer, pH 7.2 (0.3 M Tris)

Fibrinogen solution. All glassware to be used in the preparation and handling of fibrinogen solutions *must* first be washed in 2 M NaOH to ensure that no trace of thrombin activity is introduced. A stock solution of 1.4–1.6% bovine fibrinogen Fraction I-4[15] in 0.3 M NaCl is used. If fibrinogen of this standard of purity is not available, Fraction I-2[15] or Fraction I-0[15] can also be used. Many commercially available preparations of Fraction I form a clot or a precipitate without the addition of thrombin when left in solution at room temperature for only a few hours. The stock

[14] S. Magnusson, *Arkiv Kemi* **24**, 217 (1965).
[15] B. Blombäck and M. Blombäck, *Arkiv Kemi* **10**, 415 (1956).

solution of Fraction I-4 is stored in suitable portions at $-20°$. It is stable for at least 12 months. For daily use a 0.40–0.45% fibrinogen solution is prepared as follows: 2.0 ml Tris-hydrochloride buffer pH 7.2 (0.3 M Tris), 4.0 ml deionized water, and 2.0 ml stock 1.4–1.6% fibrinogen solution are mixed. This solution kept at room temperature gives a stable clotting time for one day and quite often for a second day if stored at $+4°$ overnight. When prepared aseptically (sterile-filtered) it shows signs of partial clotting only after 5–7 days at room temperature.

Standard thrombin. There is no internationally accepted thrombin standard. The NIH-standard thrombin (Lot B-5) has a specific activity of 21.7 NIH-U/mg dry weight. For the time being it is accepted as the ultimate standard by most workers in this field. This standard is not completely soluble in 0.15 M NaCl, but attempting to dissolve it with stirring for about 10–15 minutes gives a suspension with constant fibrinogen-clotting activity. The author finds it very useful to have an intermediary local standard available for everyday use. Preparations of the type described in the section on purification of thrombin (pp. 173–174) (nonchromatographed, with a specific activity of 250–300 NIH-U/mg dry weight) are satisfactory for this purpose if stored at $-20°$ in freeze-dried condition. A method for careful comparison of the two standards has been described.[16]

Procedure. DETERMINATION OF PROTHROMBIN IN PLASMA. The plasma is defibrinated by mixing 0.4 ml 0.15 M NaCl and 0.1 ml thrombin solution with 0.5 ml test plasma. A glass rod with a ground conical tip is left in the tube. A clot is formed in a few seconds at room temperature and will adhere to the glass rod. The fibrin threads of the clot are collected after 1–2 minutes by slowly winding the glass rod, while pressing its tip gently against the wall of the tube to extrude the clot liquid. This defibrinated plasma is then left for at least 10 minutes to allow antithrombin in the plasma to neutralize the added thrombin. A dilution of 0.3 ml defibrinated plasma with 2.7 ml P.P-NaCl solution is prepared. The prothrombin in 1.0 ml of this diluted, defibrinated plasma is then activated by incubating with 3.0 ml reaction mixture at $37°$. The formation of thrombin is followed by withdrawing 0.2 ml samples from this 4.0 ml incubation mixture at half-minute intervals (or as frequently as possible), adding them to 0.2 ml fibrinogen solution at $37°$ and measuring the clotting time. The highest thrombin activity is found after incubation for 3–5 minutes. The average of the two shortest clotting

[16] S. Magnusson, *Thromb. Diath. Haemorrhag.* **4,** 169 (1960).

times obtained is used in calculating the thrombin concentration from a standard curve. This thrombin concentration, expressed as NIH-U/ml is also taken to be the prothrombin concentration in the sample. Normal bovine citrated plasma has a prothrombin activity of about 155 NIH-U/ml with this technique.

CALIBRATION OF THE STANDARD CURVE. To avoid adsorption of thrombin all glassware used in the handling of thrombin solutions has to be siliconized.[17] A suitable amount of standard thrombin is dissolved and a series of dilutions made in 0.15 M NaCl; 0.3 ml of each of these is further diluted with 2.7 ml P.P-NaCl solution and 1.0 ml of each dilution is mixed with 3.0 ml "reaction mixture" and tested on fibrinogen solution at zero incubation time. The test system is as follows: 0.2 ml sample is added to 0.2 ml fibrinogen solution and the clotting time recorded to the nearest 0.2 of a second. Triplicate samples from each thrombin dilution are tested and the average clotting time for each dilution plotted vs. its thrombin concentration constitutes the standard curve. This is a straight line between 10 seconds and 50–60 seconds. The standard curve is checked with each new batch of thromboplastin, P.P-reagent, and fibrinogen.

DETERMINATION OF PROTHROMBIN ACTIVITY IN PURIFIED PREPARATIONS. One mg of the purified preparation to be tested is dissolved in 1.00 ml 0.15 M NaCl, diluted suitably with 0.15 M NaCl (generally 1:20–1:100) and tested in the manner described for defibrinated plasma. The result is calculated as NIH-U/mg dry weight of material.

Purification Procedure[12]

Principle

Three properties of prothrombin seem to be related to the function of vitamin K: (1) its adsorption from deionized or oxalated plasma onto barium sulfate[4,5] (and perhaps other inorganic adsorbents such as barium citrate, magnesium hydroxide, and calcium phosphate), (2) its activation to thrombin in the one- and two-stage test systems,[5] and (3) its change of electrophoretic mobility when calcium lactate is included in the electrophoresis buffer.[5] The purification methods in the literature all depend on an adsorption step at some stage of the procedure.[12,18-25] The

[17] L. B. Jaques, E. Fidlar, E. T. Feldsted, and A. G. MacDonald, *Can. Med. Assoc. J.* **55**, 26 (1946).

[18] W. H. Seegers, *Record Chem. Progr. Kresge-Hooker Sci. Lib.* **13**, 143 (1952).

[19] R. Goldstein, A. LeBolloc'h, B. Alexander, and E. Zonderman, *J. Biol. Chem.* **234**, 2857 (1959).

[20] M. L. Lewis and A. G. Ware, *Proc. Soc. Exptl. Biol. Med.* **84**, 636 (1953).

[21] K. D. Miller, *J. Biol. Chem.* **231**, 987 (1958).

present author[26] has only used three of the published methods.[12, 19, 20] The preferred procedure[12] is modified to increase yield from Seegers[18] and Miller.[21] It includes isoelectric precipitation of euglobulin at low ionic strength (Mellanby[27]), extraction of prothrombin from the precipitate, adsorption on magnesium hydroxide and subsequent elution with carbon dioxide at 2–4 atmospheres (Fuchs[28]), ammonium sulfate fractionation, isoelectric fractionation, and ion-exchange equilibrium chromatography.[21]

Procedure

Collection of Blood and Separation of Plasma. Blood is obtained within 1–2 minutes after slaughter (or preferably from live animals if that can be arranged) from neck-vein cuts. For each batch 9.0–9.3 liters of blood from each of either 4 or 8 animals (cows, bulls, or oxen; *not* calves or heifers which are too small to give 9 liters quickly enough to avoid partial clotting during collection) is collected with thorough mixing into a separate 10-liter polyethylene bottle containing 700 ml of an anticoagulant solution of low ionic strength (1.85% potassium oxalate $K_2C_2O_4 \times 2\ H_2O$ and 0.50% oxalic acid $H_2C_2O_4 \times 2\ H_2O$ in deionized water). Within 1–2½ hours from the time of blood collection the separation of plasma from blood cells is completed at room temperature in a separator (e.g., Type BP 15 KRE DeLaval Stockholm, Sweden). The separator must be rinsed with $0.15\ M$ NaCl between consecutive blood bottles to avoid agglutination due to different blood groups in individual cattle. All subsequent operations are performed at a temperature of 0 to $+3°$.

Precipitation of Euglobulins. Pooled plasma (10 or 20 liters) is diluted with 15–16 plasma volumes of distilled water and adjusted to pH 5.05–5.15 by the addition of 1% acetic acid. The precipitate formed (euglobulin) is allowed to settle for at least 4 hours, and generally overnight. After most of the supernatant solution has been siphoned off, the precipitate is separated from the remainder (about 15–40 liters) by centrifugation for 10 minutes at $1250\ g$. (The author has been using two refrigerated centrifuges Model SR-3, International Equipment Co., Boston, Mass., modified to handle 8 liters each for this purpose.)

[22] H. Kowarzyk, W. Mejbaum-Katzenellenbogen, Z. Kowarzykowa, and B. Czerwinska-Kossobudzka, *Bull. Acad. Polon. Sci., Ser. Biol.* **12**, 441 (1964).

[23] H. C. Moore, S. E. Lux, O. P. Malhotra, S. Bakerman, and J. R. Carter, *Biochim. Biophys. Acta* **111**, 174 (1965).

[24] G. H. Tishkoff, L. C. Williams, and D. M. Brown, *J. Biol. Chem.* **243**, 4151 (1968).

[25] J. S. Ingwall and H. A. Scheraga, *Biochemistry* **8**, 1860 (1969).

[26] S. Magnusson, *Arkiv Kemi* **23**, 271 (1965).

[27] J. Mellanby, *Proc. Roy. Soc. B,* **107**, 221 (1930).

[28] H. J. Fuchs, *Biochem. Z.* **222**, 470 (1930).

Extraction from Euglobulin Precipitate. The precipitate is then suspended in a Turmix blender ($\frac{1}{2}$ speed, 15 seconds) in 0.3 plasma volumes of a potassium oxalate–sodium chloride solution (0.075% $K_2C_2O_4 \times H_2O$ and 0.85% NaCl w/v). The suspension is then adjusted to pH 6.4–6.7 by slowly adding 0.1 M NaOH with mixing and extracted for 30 minutes using a vibrating mixer (Vibro-Mischer, Chemap A. G. Zürich, Switzerland). The extract is collected by centrifugation (5 minutes, 1250 g). The sediment is discarded.

Adsorption. Then prothrombin is adsorbed from the extract with 39 ml of an aqueous magnesium hydroxide suspension per liter of plasma for 30 minutes. The stirring has to be sufficiently vigorous to keep the magnesium hydroxide in suspension. The aqueous magnesium hydroxide suspension has been prepared in advance by slowly adding 5 liters of concentrated ammonium hydroxide (specific gravity 0.91) to 20 liters of a 20% solution of magnesium chloride, washing the magnesium hydroxide precipitate several times with distilled water, and dialyzing it against distilled water until it gives a negative silver nitrate/nitric acid test for chloride ions. After adsorption the precipitate is collected by centrifugation (20 minutes, 1250 g) and washed twice in a Turmix blender ($\frac{1}{2}$ speed, 15–30 seconds) with 40 ml 0.15 M NaCl per liter of plasma, each time collecting the precipitate by centrifugation (20 minutes, 1250 g). The magnesium hydroxide precipitate carrying the adsorbed prothrombin is then suspended a third time in the same volume of 0.15 M NaCl and subjected to a carbon dioxide atmosphere of 3–4 kp/cm² for 3–4 hours in an open glass flask enclosed in a metal container, placed on a rocking device to ensure continuous mixing. This treatment dissolves the magnesium hydroxide eluting the prothrombin. The pressure is released slowly—over 1–2 hours at least—to avoid foaming with consequent protein denaturation. Quite often a precipitate of magnesium carbonate forms in the eluate. This crystalline precipitate has a dusky gray color, distinct from the milky white opaque of the magnesium hydroxide. The elution time has to be adjusted depending on the shape of the glass flask and the efficiency of stirring during elution, so that one is sure to dissolve all the magnesium hydroxide. The eluate (about 400 ml from 10 liters of plasma or about 800 ml from 20 liters of plasma) contains some floating fibrinous precipitate (probably denatured fibrinogen) and is either filtered through silk cloth or gauze, or cleared by centrifugation (10 minutes, 1250 g).

Fractionation with Ammonium Sulfate. This is carried out in two steps at $\frac{1}{2}$ and $\frac{2}{3}$ saturation, respectively. The filtered eluate containing the prothrombin is maintained at 0 to $+1°$ in a constant temperature alcohol bath. One volume (about 400 ml or about 800 ml) of saturated ammonium sulfate solution at room temperature [760 g $(NH_4)_2SO_4$ in

1000 ml deionized water] is ten added dropwise at a controlled rate so that the temperature of the prothrombin solution can be maintained at 0 to +1° by slightly decreasing (one or two degrees) the temperature of the cooling bath. After centrifugation (20 minutes, 1250 g) the precipitate is discarded. An additional 1 volume of the saturated ammonium sulfate solution is added to the supernatant. The precipitation mixture is then allowed to stand for 30 minutes at −2°, which allows salt crystals, but not the protein to settle. The supernatant solution containing the protein precipitate in fine suspension is decanted from the salt crystal sediment and the protein precipitate collected by centrifugation (45–60 minutes, 1250 g). It is important to inspect the supernatant after this centrifugation to see that it is clear. If not, another centrifugation for 45–60 minutes is recommended with a better balanced centrifuge load. The precipitate is then dissolved in 3–5 ml deionized, distilled water per liter of plasma, and dialyzed in 20 mm Cellophane tubing (Kalle A. G. Wiesbaden, Germany) against deionized, distilled water for 20 hours, the water being exchanged every 20 minutes during the first 6 hours. This gives the prothrombin solution a specific resistance of at least 1750 ohms (Philoscope Type GM 4140, Philips).

Isoelectric Fractionation. The dialyzed solution is adjusted to pH 5.4 with 0.1 M HCl and the supernatant clarified by centrifugation (20 minutes, angle head at 3000 rpm). The precipitate is discarded and the supernatant adjusted to pH 4.6 with 0.1 M HCl. The resulting prothrombin precipitate is collected by centrifugation (5 minutes, angle head at 3000 rpm) and dissolved in 3–5 ml deionized, distilled water per liter of plasma with addition of 0.1 M NaOH to pH 6.7–7.0. The solution can be stored frozen at −20° or freeze dried. About 400–600 mg of freeze-dried white prothrombin material is obtained from 10 liters of plasma. The specific activity is 1250–1450 NIH-U/mg dry weight, corresponding to a yield of 32–56% and a purification of about 700 times from plasma.

Ion-Exchange Equilibrium Chromatography.[21] This step is optional and may be useful if the specific activity after the isoelectric fractionation step is not near 1450 NIH-U/mg. The *trans*-α-glucosylase activity described by Miller[21] which separates from prothrombin in this chromatography step is not adsorbed on magnesium hydroxide to any appreciable extent from 0.15 M NaCl which is the concentration used in the procedure described here.

RESIN. The carboxylic acid (polymethacrylic acid) cation-exchange resin Amberlite IRC-50 (XE-64) is used. It is first purified by removing "fines" washing with water, then washed with acetone, cycled through the Na⁺—form and the acid form, and finally equilibrated with the chromatography buffer. A detailed description of this procedure has been given by Hirs.[29] Alternatively, the chromatographic grade CG-50, type II (200–400 mesh), can be used in which case the removal of fines as well as water and acetone washes can be dispensed with.

PROCEDURE. The chromatography is carried out at a temperature of 0 to $+2°$. A column (100×2 cm diameter) of the resin is equilibrated with a phosphate buffer of pH 6.00–6.10, made by mixing 0.1 M Na_2HPO_4 and 0.1 M NaH_2PO_4 to the desired pH. Up to 300 mg of the prothrombin preparation to be purified is dissolved in 2.0–2.5 ml 0.1 M sodium ortho-phosphate buffer pH 5.6, applied to the column and washed in with an extra 2.0 ml of the same buffer. The column is developed with the equilibrating buffer (flow rate 5–10 ml/hour) collecting fractions of 1–2 ml. One peak (absorbancy at 280 nm) at 2.0 hold-up volumes, or two peaks, at 1.6–1.7 and at 2.0 hold-up volumes, appear depending on whether or not the preparation contained the transglucosylase protein. The fractions that contain prothrombin (emerging at 2.0 hold-up volumes) are pooled and dialyzed for 36–48 hours against several changes of deionized water. The solution is then either frozen or freeze dried. This type of preparation has a specific activity of 1400–1450 NIH-U/mg dry weight.

Properties

Stability

Freeze-dried preparations can be stored for a couple of years at $-20°$, losing about 10–20% of their two-stage activity eventually. When stored at room temperature they quickly lose two-stage activity (up to 30% in 2–3 weeks) but can still be activated almost quantitatively in 25% sodium citrate after many years. Aqueous solutions of prothrombin (5–10 mg/ml) can be stored frozen at $-20°$ with little loss of activity for several months.[30] Incubation at 95–98° (in a boiling water bath) for 5 minutes completely destroys the one-stage activity but leaves at least 97% of the two-stage activity intact.[3]

Purity and Physical Properties

The criteria of constant solubility[31] and homogeneity in ultracentrifugal analysis,[12, 23–25, 32] immunodiffusion,[21] starch gel electrophoresis,[14] moving boundary electrophoresis,[23] and disc electrophoresis[25] have indi-

[29] C. H. W. Hirs, Vol. I, p. 113.
[30] W. H. Seegers, personal communication, 1969.
[31] W. H. Seegers, E. C. Loomis, and J. M. Vandenbelt, Arch. Biochem. 6, 85 (1945).
[32] F. Lamy and D. F. Waugh, J. Biol. Chem. 203, 489 (1953). Thromb. Diath. Haemorrhag. 2, 188 (1958). Physiol. Rev. 34, 722 (1954).

cated that the respective prothrombin preparations studied are at least very nearly pure. Analysis[3] by disc electrophoresis and immunoelectrophoresis of preparations obtained by the present author with the method described has indicated the presence of small amounts of contaminating material (probably less than 2–3% of the total amount of protein) in most batches of prothrombin. It has recently been observed[5] that if calcium ions are included in the buffer the electrophoretic mobility of prothrombin is altered, from that of an α_1 to that of an α_2–β_1 plasma protein. The sedimentation constant has been given as $s_{25,w}^0 = 4.89$ S,[32] $s_{20,w}^0 = 5.22$ S,[33] $s_{20,w}^0 = 4.80$ S,[25] the diffusion constant ($D_{20,w}^0$) as 6.24×10^{-7} cm^2 sec^{-1} (footnote 32) and as 5.17×10^{-7} cm^2 sec^{-1} (footnote 24) and partial specific volume as $\bar{v} = 0.70$ ml g^{-1} (footnote 32). The isoelectric point has been given as pH 4.2[34] and pH 4.1.[24] The molecular weight has been calculated from physical data to be 68,000,[32] 68,500,[33] 70,500 $\pm$ 2,800,[24] and 74,000 $\pm$ 4,100.[25] Quantitative determination of N-terminal alanine[26] indicates a maximal molecular weight of 70,000. Evidence that prothrombin can be dissociated into dimers has been obtained neither by gel filtration in the presence or absence of 8 M urea of unmodified prothrombin; reduced, carboxymethylated prothrombin; reduced, carboxymethylated, maleylated prothrombin,[35] nor by ultracentrifugal analysis of reduced prothrombin in 6 M guanidine HCl.[24, 25] Optical rotatory dispersion data indicate little or no change in conformation with temperature (6°, room temperature, and 75°)[25] or when prothrombin is dissolved in 25% sodium citrate.[36] A value of −35 for b_0 indicates low helix content in native prothrombin.[25]

Amino Acid Analysis

Six different analyses of bovine prothrombin[25, 26, 35, 37–39] are included in Table I and are expressed as grams of amino acid residue per 100 g prothrombin. All analyses were performed by the standard Moore-Stein technique except one[26] (paper chromatography of phenylthiohydantoins of amino acids).

[33] C. R. Harmison, R. H. Landaburu, and W. H. Seegers, *J. Biol. Chem.* **236**, 1693 (1961).

[34] W. H. Seegers, R. I. McClaughry, and J. L. Fahey, *Blood* **5**, 421 (1950).

[35] S. Magnusson and B. S. Hartley, unpublished results.

[36] M. MacAulay, S. Bakerman, H. Moore, and J. R. Carter, *Thromb. Diath. Haemorrhag.* **11**, 289 (1964).

[37] G. F. Lanchantin, D. W. Hart, J. A. Friedmann, N. V. Saavedra, and J. W. Mehl, *J. Biol. Chem.* **243**, 5479 (1968).

[38] K. Laki, D. R. Kominz, P. Symons, L. Lorand, and W. H. Seegers, *Arch. Biochem. Biophys.* **49**, 276 (1954).

[39] W. H. Seegers, E. Marciniak, R. K. Kipfer, and K. Yasunaga, *Arch. Biochem. Biophys.* **121**, 372 (1967).

TABLE I
COMPOSITION OF BOVINE PROTHROMBIN

Residue[g]	a	b	c	d	e	f
Alanine	3.07	3.26	3.17	3.29	3.65	3.36
Arginine	7.87	7.34	8.29	7.25	8.71	7.25
Aspartic acid + asparagine	8.52	8.66	8.22	9.02	10.52	8.80
Half-cystine	2.73	2.67[h]	1.59[h]	2.40	2.77	2.57
Glutamic acid + glutamine	10.80	11.28	11.78	12.37	14.59	11.98
Glycine	3.86	3.41	3.69	3.89	4.39	3.58
Histidine	1.47	1.63	1.98	1.99	1.80	2.05
Isoleucine	2.39	2.71		2.96	3.63	2.91
Leucine	6.88	6.56	9.75[k]	6.74	6.67	6.69
Lysine	5.20	5.02	6.46	4.84	5.88	4.77
Methionine	1.28	1.71	1.02	1.15	0.80	1.36[i]
Phenylalanine	3.98	4.21	4.07	4.06	4.39	4.39
Proline	3.14	4.66	4.69	4.51	5.54	5.24
Serine	5.21	5.13	3.19	4.55	4.83	5.01
Threonine	4.47	3.95	2.82	3.96	4.33	4.20
Tryptophan		3.01		2.97	4.42	
Tyrosine	3.98	3.95	4.65	4.50	3.94	3.88
Valine	4.11	4.13	5.10	4.89	5.14	4.95
Total	78.96	83.29	80.47	85.34	96.00	88.48
Hexose			3.6–5.6			
Hexoseamine			2.0			
NANA[j]			4.2			

[a] G. F. Lanchantin, D. W. Hart, J. A. Friedmann, N. V. Saavedra, and J. W. Mehl, J. Biol. Chem. **243**, 5479 (1968).
[b] K. Laki, D. R. Kominz, P. Symons, L. Lorand, and W. H. Seegers, Arch. Biochem. Biophys. **49**, 276 (1954).
[c] S. Magnusson, Arkiv Kemi **23**, 271 (1965).
[d] W. H. Seegers, E. Marciniak, R. K. Kipfer, and K. Yasunaga, Arch. Biochem. Biophys. **121**, 372 (1967).
[e] J. S. Ingwall and H. A. Scheraga, Biochemistry **8**, 1860 (1969).
[f] S. Magnusson and B. S. Hartley, unpublished results.
[g] Amino acid or monosaccharide residues in grams per 100 g prothrombin.
[h] Half-cystine determined as cysteic acid.
[i] Methionine determined as methionine sulfone.
[j] N-Acetyl neuraminic acid.
[k] Isoleucine + leucine.

Crystallization

Crystalline material containing 21.5% protein, 32% barium, and 8.9% citrate has been described.[24]

Activators and Inactivators

The activation to thrombin involves limited proteolysis of the prothrombin molecule. It seems at present that the proteolytic enzyme normally catalyzing this cleavage is Factor X_a. This activation is greatly facilitated by the presence of Factor V, phospholipids, and calcium chloride. Prothrombin preparations of the type described can also be activated in 25% sodium citrate solution[34] or in the presence of poly-L-lysine.[40] Activation in both 25% sodium citrate and with "biological" activators has been shown to depend on a serine proteinase. DFP does not inhibit the "biological" activators (thromboplastin + serum),[41] but it does inhibit the activation as well as the limited proteolysis of prothrombin in both types of activation system.[26] Therefore, the necessary serine proteinase is formed from the prothrombin preparation itself. It seems unlikely that this serine proteinase is thrombin. Whether it is derived from a small amount of contaminating protein or from the N-terminal half of the prothrombin molecule has yet to be determined. In either case, it seems most probable that this serine proteinase is activated by some means other than 'imited proteolysis. Activation of prothrombin can be prevented, e.g., by adding salts (oxalate, citrate, fluoride) to blood, by exchanging the calcium ions with sodium ions, or by adding heparin.

Distribution

Evidence obtained with specific fluorescent antibodies shows that prothrombin is synthesized in the hepatic cells of the liver.[42] The catabolic half-life of prothrombin that had been labeled with ^{131}I and then injected intravenously was found to be 2.8 days.[43] It was also deduced from those experiments that about 64% of the total body pool of prothrombin is contained in the blood plasma. The concentration in bovine plasma is about 70–100 mg per liter.

Bovine Thrombin (EC 3.4.4.13)

Assay Method[16, 44]

Principle

Because thrombin is defined as the fibrinogen-clotting activity formed in blood, specific assay methods must be based on this activity. The clotting of fibrinogen is essentially a two-step process. The first step is a limited proteolysis of fibrinogen. It produces fibrin monomer and fibrinopeptides. This is followed by polymerization of fibrin monomer

[40] K. D. Miller, J. Biol. Chem. 235, PC 63 (1960).
[41] K. D. Miller and H. van Vunakis, J. Biol. Chem. 223, 227 (1956).
[42] M. I. Barnhart, Am. J. Physiol. 199, 360 (1960).
[43] S. S. Shapiro and J. Martinez, J. Clin. Invest. 48, 1292 (1969).
[44] W. H. Seegers and H. P. Smith, Am. J. Physiol. 137, 348 (1942).

to produce the fibrin clot. Thrombin catalyzes only the proteolytic step. An assay that depends only on this proteolytic step would be as specific as a clotting assay and have certain advantages. Such a method has been developed.[16,45] It measures the increase in N-terminal glycine as the fibrinopeptidyl-fibrin monomer bonds in fibrinogen are split by thrombin. The method has been used for comparing standard preparations,[16] but it is too cumbersome for routine use. Thrombin also has esterase activity, particularly against arginyl[46] and lysyl[47] substrates with a blocked α-amino group. Several esterase assay methods have been developed.[46,48-52] Furthermore, attempts have been made[52-55] to find a synthetic substrate analog that can be used as an active site titrant for thrombin. This has to be a compound that acylates thrombin at a fast rate producing an acyl-enzyme that is sufficiently stable that the "initial burst" molarity of the other reaction product can be accurately measured and taken as the operational molarity of thrombin. The best such compound available at the moment appears to be p-nitrophenyl p'-guanidinobenzoate.[56] This compound reacts not only with trypsin[56] but also with thrombin[54,55] and with plasmin.[55] With all esterase assays so far described one has to bear in mind that they are not specific for thrombin. In blood plasma alone there are at least half a dozen different zymogens with potential arginine esterase activity.

Reagents

> Fibrinogen solution. See two-stage method for prothrombin assay (pp. 161–162).
>
> Thrombin solutions. A suitable amount of thrombin (sample or standard) is weighed out and dissolved (in siliconized[17] glassware or plastic vessels) in ice-cold 0.15 M NaCl. Dilutions are made in ice-cold 0.15 M NaCl and tested as soon as possible after dilution.
>
> NaCl, 0.15 M

[45] E. Jorpes, T. Vrethammar, B. Öhman, and B. Blombäck, *J. Pharmacy (London)* **10**, 561 (1958).

[46] S. Sherry and W. Troll, *J. Biol. Chem.* **208**, 95 (1954).

[47] D. T. Elmore and E. F. Curragh, *Biochem. J.* **86**, 9P (1963).

[48] S. Ehrenpreis and H. A. Scheraga, *J. Biol. Chem.* **227**, 1043 (1957).

[49] R. H. Landaburu and W. H. Seegers, *Proc. Soc. Exptl. Biol. Med.* **94**, 708 (1957).

[50] C. J. Martin, J. Golubow, and A. E. Axelrod, *J. Biol. Chem.* **234**, 1718 (1959).

[51] E. F. Curragh and D. T. Elmore, *Biochem. J.* **93**, 163 (1964).

[52] F. J. Kézdy, L. Lorand, and K. D. Miller, *Biochemistry* **4**, 2302 (1965).

[53] D. T. Elmore and J. J. Smyth, *Biochem. J.* **107**, 97, 103 (1968).

[54] J. B. Baird and D. T. Elmore, *Fed. European Biochem. Soc. Letters* **1**, 343 (1968).

[55] T. Chase, Jr. and E. Shaw, *Biochemistry* **8**, 2212 (1969).

[56] T. Chase, Jr. and E. Shaw, *Biochem. Biophys. Res. Commun.* **29**, 508 (1967).

Procedure

To 0.2 ml thrombin dilution is added 0.2 ml fibrinogen solution, which has been prewarmed in a water bath at 37° in a nonsiliconized test tube (9 × 60 mm). The time from the addition of thrombin to the appearance of a visible clot is determined to the nearest 0.2 second. Triplicate determinations are performed on each dilution. The clotting times obtained range from 10 to 50 seconds. A standard curve is obtained by plotting the average clotting times for each of a series of dilutions of standard thrombin against thrombin activity on a log–log scale. This plot gives a straight line from 10 seconds to 50–60 seconds. The activity of an unknown thrombin sample is determined by triplicate testing of at least two dilutions of different concentration.

Definition of Unit and Specific Activity

Activities are expressed as NIH-U/mg dry weight. The unit is defined by the NIH-standard (see two-stage assay for prothrombin, p. 162). The Iowa-unit that is used in many laboratories is not always clearly related to the NIH-standard. Its activity has been variously given as 0.60 NIH-units,[12] 0.62 NIH-units,[56a] 0.80 NIH-units,[57] and 0.56 NIH-units.[58] Under defined conditions of fibrinogen digestion one NIH-unit of thrombin was found to make 0.4 μmole glycine appear in N-terminal position.[16]

Purification Procedure[59]

Principle

The procedure is the same as that for prothrombin through the steps of euglobulin precipitation, extraction of prothrombin, adsorption on magnesium hydroxide, and elution with carbon dioxide. This produces prothrombin material in good yield which is essentially free from plasma antithrombin activity and can therefore be safely activated to thrombin that is then purified from the activation mixture. Alternative procedures are available,[60, 61] but the author has no personal experience with them.

Procedure

The magnesium hydroxide eluate (see preparation procedure for prothrombin, p. 165) from 20 liters of plasma (about 800 ml) is cleared by

[56a] K. D. Miller, personal communication, 1964.
[57] W. H. Seegers, "Prothrombin." Harvard University Press, Cambridge, Massachusetts, 1962.
[58] K. D. Miller and W. H. Copeland, *Exptl. Mol. Pathol.* 4, 431 (1965).
[59] S. Magnusson, *Arkiv Kemi* 24, 349 (1965).

centrifugation (10 minutes, 1250 g) and dialyzed against deionized water overnight. The pH during and after dialysis is 7.5–7.9. After addition of 1/30 volume (about 27 ml) of 3 M sodium acetate the pH is lowered to 7.2–7.5 by the addition of 1.0 M acetic acid. During all operations described the temperature is maintained at 0–4°.

Activation. At this stage the prothrombin preparation still contains sufficient activities of Factors V, VII, and X to permit relatively rapid activation to thrombin on addition of bovine brain thromboplastin (see one-stage assay for prothrombin, p. 159) and calcium chloride. Small-scale pilot activations are set up to choose the optimum thromboplastin concentration and activation time for each batch. Thus small-scale activation mixtures of 2.0 ml prothrombin solution with 0.2, 0.4, or 0.8 ml thromboplastin suspension and 0.2 ml 0.18 M $CaCl_2$ are incubated at 37° in a water bath. Aliquots of 0.2 ml are withdrawn from each activation mixture at 5-minute intervals and tested for thrombin activity. Maximal thrombin activity is usually found after activation for 20–40 minutes. The optimal conditions found in this pilot test are then used for full-scale activation of the entire batch. This is carried out with constant stirring at 37°. The prothrombin solution, the thromboplastin suspension, and the calcium chloride solution have each been prewarmed to 37° before mixing. After activation the mixture is cooled during a centrifugation (10 minutes, 0°, 1250 g) that separates most of the thromboplastin material.

Fractionation with Ethanol. The supernatant is cooled to 0–1°. Then 135 ml 53.3% (v/v) aqueous ethanol at −5° is added per 1000 ml supernatant. During the addition stirring is maintained and the temperature lowered to −4°. A precipitate forms which is separated by centrifugation (10 minutes, −4°, 1250 g) and discarded. The supernatant is brought to pH 6.0–5.5 with 1.0 M acetic acid. Then the second ethanol precipitate, containing the thrombin, is obtained by adding 1 volume of 65% (v/v) aqueous ethanol at −10° to 1 volume of the supernatant. During the addition and for a further 30 minutes, stirring is maintained and the temperature is lowered to −6°. The precipitate is then collected by centrifugation (30 minutes, −6°, 1250 g), dissolved in 90–120 ml ice-cold deionized water and dialyzed overnight against 10 liters of deionized water at 0–2°. The dialyzed solution is clarified by centrifugation (20 minutes, 0°, 2,500 g, angle-head) and freeze dried. The yield of greyish white to white freeze-dried material is about 900–

[60] W. H. Seegers, W. G. Levine, and R. S. Shepard, *Can. J. Biochem. Physiol.* **36**, 603 (1958).

[61] D. J. Baughman and D. F. Waugh, *J. Biol. Chem.* **242**, 5252 (1967).

1400 mg with a specific activity of 240–290 NIH-U/mg dry weight. This corresponds to a yield from plasma prothrombin of 7–13%. This type of material is quite stable and about 10–15% pure.

Chromatography on IRC-50.[59,62] The resin is pretreated as described under preparation of prothrombin (see p. 166) and equilibrated batchwise with 0.05 M sodium orthophosphate buffer pH 7.0. Two different chromatography systems can be used.

1. A column (60 × 2.0–2.2 cm diameter) is packed and equilibrated at pH 7.0 ± 0.1. About 300 mg crude bovine thrombin (see p. 161) is dissolved in 2–6 ml of the 0.05 M phosphate buffer pH 7.0 and applied to the column. Elution is performed with 0.3 M sodium orthophosphate buffer pH 8.0 at an initial flow rate of 1 drop per 11–15 seconds. Fractions of 7–15 ml are collected (every 20 minutes) and their absorbancy at 280 nm used to locate the protein peaks. Two well separated peaks are obtained (Fig. 1). The first is the breakthrough, containing 85–90% of the protein and lacking thrombin activity. The second is the thrombin peak emerging after 700–800 ml of effluent usually in a volume of 30–

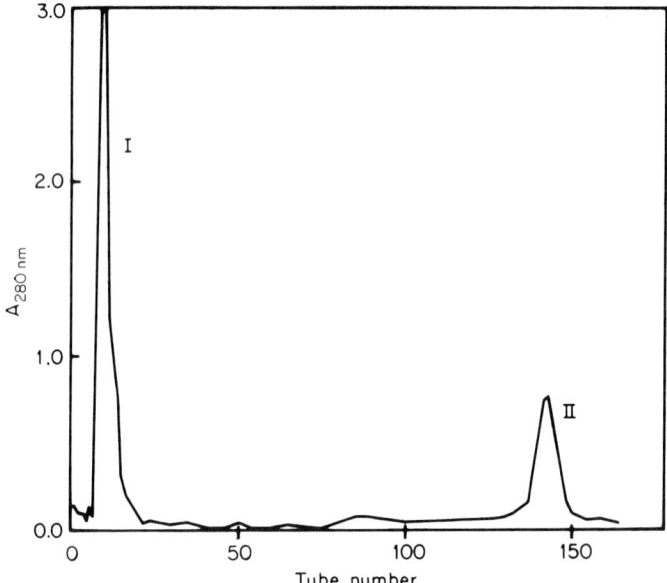

Fig. 1. Chromatography (p. 174) of thrombin (pp. 173–174) on a column (60 × 2.2 cm diameter) of IRC-50, XE-64. Equilibration buffer: Sodium orthophosphate, 0.05 M, pH 7.0. Elution buffer: Sodium orthophosphate, 0.3 M, pH 8.0. Peak I: No thrombin, weak tosyl-L-arginine methyl esterase activity. Peak II (tubes 138–149): Thrombin (which also has tosyl-L-arginine methyl esterase activity).

[62] P. S. Rasmussen, *Biochim. Biophys. Acta* **16**, 157 (1955).

60 ml. Although the separation in this system is excellent, a major disadvantage is the 2–3 days required to elute the thrombin, since the column must be operated at room temperature because of the low solubility of the pH 8.0-buffer.

2. To avoid the long columns, the following slightly modified system with short columns $(20 \times 2.0$–2.2 cm diameter) is used. Instead of starting elution with the 0.3 M buffer of pH 8.0 immediately after applying the solution of crude thrombin, elution is started with the 0.05 M buffer of pH 7.0. This buffer is continued until the absorbancy of the effluent at 280 nm has decreased to 0.05 or less, which takes about 3–5 hours. Then elution with the 0.3 M buffer of pH 8.0 is started at a flow

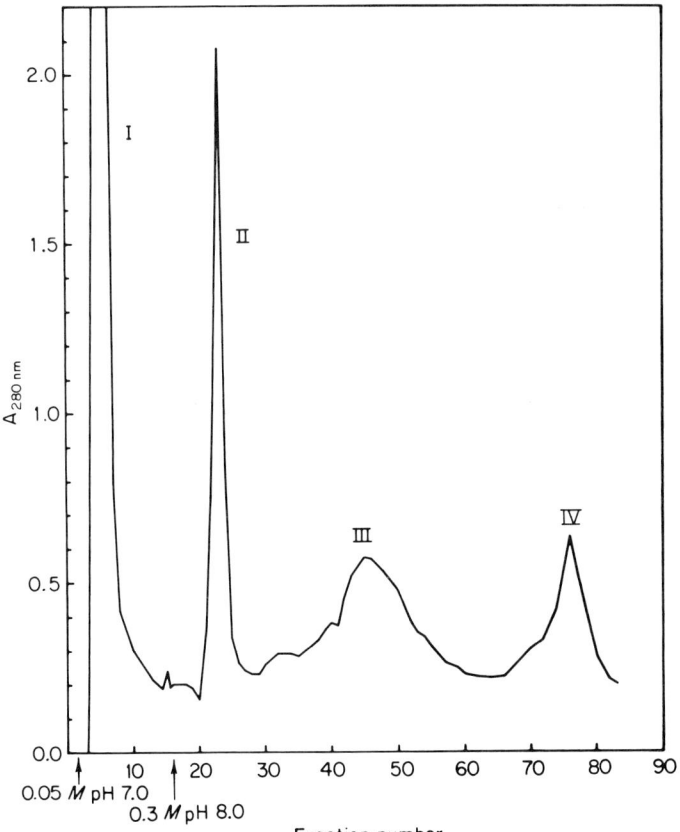

Fig. 2. Chromatography (p. 175) of thrombin (pp. 173–175) on a column $(20 \times 2.0$ cm diameter) of IRC-50, XE-64. Equilibration buffer: Same as Fig. 1. Elution buffers: Indicated in the figure. Stepwise change in fraction 15. Peak I: Weak tosyl-L-arginine methyl esterase. Peak III (fractions 36–60): Thrombin. Peak IV (fractions 70–82): Second thrombin peak.

rate of 1 drop per 15 seconds. Peaks I and III have tosyl-L-arginine methyl esterase activity, peak III has thrombin activity. With several batches of crude thrombin an additional thrombin peak (peak IV) has been obtained (see Fig. 2).[63] The short column system has the advantage that the thrombin can be eluted overnight. The separation is good, but some care has to be exercised in pooling peak III, so that contamination with peak II material is avoided.

Recovering Thrombin from the Column Effluent. This step has turned out to be rather tricky. In the hands of the present author the following procedure has been entirely reproducible, though, and is therefore described in some detail. All tubes from the thrombin peak(s) (III, or III and IV) having an absorbancy at 280 nm greater than 0.25 are pooled (III and IV separately). The solution is then cooled to 0°, 0.70 g solid ammonium sulfate per milliliter is added. The mixture is stirred for 1 hour and then left overnight at −4°. The precipitate is collected by centrifugation (10 minutes, −4°, 25,000 g) in 25 ml polypropylene tubes, dissolved, and transferred to one tube only, using 10 ml deionized water at 0°. Ten milliliters of acetone (analytical reagent, Merck), precooled to −50° to −55°, is added rapidly with stirring. (The precooling of the acetone, if done by submersion in dry ice–alcohol, should be done in a closed vessel. This avoids dissolving a lot of CO_2 that will otherwise bubble away with most of the precious thrombin on mixing the two solutions.) The resulting thrombin precipitate is separated by centrifugation (15–30 seconds, −5 to −8°, 15,000 g) and recovered either as a sediment or as a cake floating between the upper and lower liquid phases of the system. The thrombin precipitate is then redissolved and reprecipitated twice from 50% acetone (only one liquid phase is obtained) and finally dissolved in 20 ml deionized water at 0° and freeze dried. During the entire acetone precipitation procedure the utmost care must be taken to preserve the low temperature (−5 to −8°). The pH is maintained at 6.0–6.5 by adding a few drops of 0.1 M NaOH each time the thrombin is redissolved. Unless this is done pH tends to drop to values of 3–4 with heavy loss of thrombin activity and solubility. The whole process of acetone precipitation is done as quickly as possible taking about 20–30 minutes from the first addition of acetone to the start of freeze drying. In a series of 21 preparations each using about 1200 mg of crude thrombin as starting material and running four columns simultaneously the average yield of thrombin was 74.7 ± 6.5 (S.D.) mg with a specific activity of 2098 ± 134 NIH-U/mg dry weight. This corresponds to a yield of about 48%.[59]

---

[63] B. Magnusson, S. Magnusson, and K. D. Miller, unpublished results, 1966.

Properties

Stability

The nonchromatographed preparations are stable for up to 6 years at —15 to —20° in the freeze-dried state in a dessicator at reduced pressure over anhydrous calcium chloride. Two of the preparations have been used as internal thrombin standards by the author. Standardizations in 1959 and 1964 of BT-1 against two different samples of the NIH standard thrombin (Lot B-3, 21.7 NIH-U/mg) gave 290 (range 261–324)[16] and 273 (range 262–285)[59] NIH-U/mg, respectively. This type of preparation was sterile filtered through cellulose nitrate membranes (Membranfilter No. 6, Sartorius-Werke, A. G., Göttingen, Germany) and freeze dried without change of specific activity. The chromatographed preparations are not as stable in the freeze-dried state. Most preparations are found to maintain at least 50–60% of their activity for a period of 18 months, but a few have been found to completely lose activity and solubility on storage for unknown reasons.

Purity

The chromatographed preparations described above have been found to be homogeneous on ultracentrifugation in phosphate buffer at pH 7.15, $\Gamma/2$ 0.15[59] and on starch gel electrophoresis in four different buffer systems in the pH range 6.1 to 9.0, even in the presence of $7 M$ urea (as judged by staining with Amido Black 10 B),[59] as well as on disc electrophoresis extending the pH range down to 5.1.[3] Thrombin from peak IV[63] was found to be identical to thrombin from peak III by specific activity, amino acid composition, and immunodiffusion on Ouchterlony plates (against antipeak III and antipeak IV sera raised in rabbits). Their disc electrophoretic mobility is so similar that only when the two are run together in the same gel column does one detect a slight difference.[63] Slight caseinolytic activity is found in chromatographed thrombin at pH 7.2 and 50–300 NIH-U of thrombin per milliliter. The increase of this activity on addition of urokinase is so small that the preparation can be used as "plasminogen-free" thrombin in test systems for fibrinolysis.[64]

Physical Properties

On starch gel and paper electrophoresis the mobility of thrombin is similar to that of γ-globulins.[59] The isoelectric point has been determined by free-boundary electrophoresis to be pH 5.3 and 5.75 at ionic strength

---

[64] This analysis was performed by Dr. Kurt Bergström.

0.1 and 0.2, respectively,[65] and by paper electrophoresis to be pH 5.6.[66] The sedimentation and diffusion constants have been determined,[33] and found to be $s_{20,w}^0 = 3.76$ S, $D_{20,w}^0 = 8.70 \times 10^{-7}$ and 8.82×10^{-7} cm^2 sec^{-1} (calculated average for the two methods used gives 8.76×10^{-7} cm^2 sec^{-1}), respectively. The partial specific volume was found[33] to be 0.69 ml g^{-1}. These data were used to calculate a molecular weight of 33,700. The N-terminal amino acids isoleucine and threonine[59] have been quantitatively estimated as 1.0 mole of isoleucine and 0.5–0.6 mole of threonine per 30,000 grams[59] and 1 mole of each per 25,000–34,000 grams.[67] Carboxypeptidase B releases 0.85 mole of C-terminal arginine per 30,000 grams of thrombin.[68] I am not aware of any optical rotatory dispersion data on thrombin. The amino acid composition of bovine thrombin of high specific activity[35,67,69,70] and of undefined specific activity[71,72] has been determined. The results of all these analyses except one[72] are included in Table II and expressed as grams of amino acid residue per 100 grams of protein. Bovine thrombin also contains 1.7% glucosamine,[73] 1.7–1.8% sialic acid,[74] 0.61% galactose and 0.95% mannose,[75] a total of about 5.1% carbohydrate. The amino acid sequence is not completely known. The A-chain of 49 residues has the sequence Thr-Ser-Glu-Asn-His-Phe-Glu-Pro-Phe-Phe-Asn-Glu-Lys-Thr-Phe-Gly-Ala-Gly-Glu-Ala-Asp-Cys-Gly-Leu-Arg-Pro-Leu-Phe-Glu-Lys-Lys-Gln-Val-Glx-Asx-Glx-Thr-Gln-Lys-Glu-Leu-Phe-Glu-Ser-Tyr-Ile-Glu-Gly-Arg.[76] It is disulfide bound to the B-chain of about 265 residues. The preliminary sequence of the B-chain[35] is *Ile*-Val-Glu-Gly-Gln-Asp-Ala-Glu-Val-Gly-Leu-Ser-Pro-Trp-Gln-Val-Met-Leu-Phe-Arg-Lys-Ser-Pro-Gln-Glu-Leu-Leu-Cys-Gly-Ala-Ser-Leu-Ile-Ser-Asp-Arg-Trp-Val-Leu-Thr-Ala-Ala-*His*-Cys-Leu-Leu-Tyr-Pro-(Trp,Pro,Asx,Lys,AsnCBH,Phe)-Thr-

[65] W. H. Seegers, C. R. Harmison, N. Ivanovic, and D. L. Heene, *Thromb. Diath. Haemorrhag.* **15**, 343 (1966).
[66] W. G. Levine and O. W. Neuhaus, *Proc. Soc. Exptl. Biol. Med.* **101**, 64 (1959).
[67] W. H. Seegers, L. McCoy, R. K. Kipfer, and G. Murano, *Arch. Biochem. Biophys.* **128**, 194 (1968).
[68] S. Magnusson and B. Steele, *Arkiv Kemi* **24**, 359 (1965).
[69] K. D. Miller, R. K. Brown, G. Casillas, and W. H. Seegers, *Thromb. Diath. Haemorrhag.* **3**, 362 (1959).
[70] S. Magnusson and T. Hofmann, *Can. J. Biochem.* **48**, 432 (1970).
[71] E. E. Schrier, C. A. Broomfield, and H. A. Scheraga, *Arch. Biochem. Biophys.* **99** (Suppl. 1), 309 (1962).
[72] K. Laki and J. A. Gladner, *Physiol. Rev.* **44**, 127 (1964).
[73] S. Magnusson, J. R. Brown, and B. S. Hartley, unpublished results, 1964.
[74] B. Steele and S. Magnusson, unpublished results, 1964.
[75] Gas chromatographic analysis by Dr. R. J. Winzler (1967) on material supplied by the author (p. 176).
[76] S. Magnusson, E. Merler, J. Wootton, and B. S. Hartley, unpublished results (1967).

TABLE II
Composition of Bovine Thrombin

Residue[g]	a	b	c	d	e	f
Alanine	2.51	2.63	3.05	2.68	3.00	3.03
Arginine	8.83	8.89	10.24	9.01	9.34	9.61
Aspartic acid + asparagine	8.10	8.07	10.12	8.17	11.34	8.72
Half-cystine	2.20	2.67 (2.21)[h]	1.67	2.04[h]	2.15	2.90
Glutamic acid + glutamine	10.86	11.58	12.86	11.71	13.17	12.23
Glycine	3.62	4.11	4.37	3.88	4.21	4.05
Histidine	2.18	2.38	2.74	2.36	2.41	2.60
Isoleucine	3.45	4.25	4.52	4.61	4.38	4.28
Leucine	7.70	8.80	9.38	8.11	9.55	8.57
Lysine	7.93	7.93	9.05	7.30	8.11	8.49
Methionine	1.61	1.80 (1.57)[i]	1.83	1.74	1.85	1.86
Phenylalanine	4.22	4.43	5.49	4.81	5.18	5.57
Proline	3.89	3.88	5.51	4.63	5.47	4.73
Serine	3.51	3.55	3.29	3.48	4.29	4.12
Threonine	3.52	3.57	3.77	3.54	4.62	3.83
Tryptophan	3.24	2.54	5.95	3.67	3.93	5.29
Tyrosine	4.83	5.77	5.16	4.44	5.74	5.41
Valine	4.55	5.29	5.74	6.26	5.58	5.16
Total	87.83	98.77	104.74	92.44	104.32	100.45
Galactosamine	1.08					
Glucoseamine			1.68			
Carbohydrate		1.80				

[a] K. D. Miller, R. K. Brown, G. Casillas, and W. H. Seegers, *Thromb. Diath. Haemorrhag.* **3**, 362 (1959).

[b] E. E. Schrier, C. A. Broomfield, and H. A. Scheraga, *Arch. Biochem. Biophys.* **99**, (Suppl. 1) 309 (1962).

[c] S. Magnusson and T. Hofmann, *Can. J. Biochem.* **48**, 432 (1970).

[d] S. Magnusson and B. S. Hartley, unpublished results.

[e] W. H. Seegers, L. McCoy, R. K. Kipfer, and G. Murano, *Arch. Biochem. Biophys.* **128**, 194 (1968). 3.7 S thrombin.

[f] Same as footnote *e*. 3.2 S thrombin.

[g] Amino acid or monosaccharide residues in grams per 100 g protein.

[h] Half-cystine determined as cysteic acid.

[i] Methionine determined as methionine sulfone.

Val-Asx-Asx-Leu-Leu-(Val,Glu,Ser,Val,Arg),(Ile-Gly-Lys)-His-Ser-Arg-
Thr-Arg-Tyr-Glu-Arg-Lys-Val-Glu-Gln-Lys-Ile-Ser-Met-Leu-Asp-Lys-Ile-
Tyr-His-Pro-Ile-Arg-Tyr-Asn-Trp-Lys-Glu-Asn-Leu-Asp-Arg-*Asp*-Ile-
Ala-Leu-Leu-Lys,(Ala-Ser-Thr-Arg),(Thr-Thr-Ser-Val-Ala-Glu-Val-Gln-
Pro-Ser-Val-Leu),(Gln-Val-Val-Asn-Leu-Pro-Leu),(Val-Glu-Arg-Pro-Val-
Cys-Lys),(Val-Ala-Ile-Trp),(Lys-Gly-Arg-Val-Thr-Gly-Trp-Gly-Asn-Arg),

(Leu-Leu-His-Ala-Gly-Phe-Lys),Gln-Thr-Ala-Ala-Lys-Leu-Lys-Arg-Pro-
Ile-Glu-Leu-Ser-Asp-Tyr-Ile-His-(Cys,Pro,Val,Leu,Pro,Asx)-Lys-Arg-Ile-
Arg-Ile-Thr-Asp-Asn-Met-Phe-Cys-Ala-Gly-Tyr-Lys-Pro-Gly-Glu-Gly-
Lys-Arg-Gly-*Asp*-Ala-Cys-Glu-Gly-*Asp*-*Ser*-Gly-Gly-Pro-Phe-Val-Met-
Lys-Ser-Pro-Tyr-Asn-Asn-Arg-Trp-Tyr-Gln-Met-Gly-Ile-Val-Ser-Trp-
Gly-Glu-Gly-Cys-Asp-Arg-Asn-Gly-Lys-Tyr-*Gly*-Phe-Tyr-Thr-His-Val-
Trp-(Arg-Leu-Lys)-Lys-Trp-Ile-Gln-Lys-Val-Ile-Asp-Arg-Leu-Gly-Ser.[35]
The italic residues are homologous to Ile-16, His-57, Asp-102, Ser-189
(Asp-189 in thrombin), Asp-194, Ser-195, Gly-216, and Gly-226 in α-chy-
motrypsin, respectively.[35] CBH stands for carbohydrate.

Activators and Inactivators

Thrombin is a serine proteinase and diisopropyl fluorophosphate
(DFP) and other compounds reacting with the active serine inhibit both
esterase and clotting activities.[41,77] Hirudin[78] is a specific polypeptide
inhibitor of thrombin, which is produced in the salivary glands of leeches.
It forms a stoichiometric complex with thrombin and also with diisopropyl phosphoryl-thrombin. Soybean trypsin inhibitor does not inhibit
thrombin.[46] Treatment of thrombin with acetic anhydride is said to
produce material with 4% of the initial clotting activity and 160% of
the initial tosyl-L-arginine methyl esterase activity at about 49%
acetylation of amino groups.[79] Plasma antithrombin is usually described
as a specific protein in the α-globulin fraction, which destroys thrombin
enzymatically. It may be identical to heparin-cofactor which in the
presence of heparin inactivates both the clotting and the esterase activities of thrombin.[80] Heparin alone also inhibits the proteolytic activity
of thrombin on fibrinogen.[81]

Specificity

The specific substrate for thrombin is fibrinogen.[82] The reaction
produces two acidic fibrinopeptides (in some species also one or more
variants of either peptide), A and B, per fibrinogen monomer. In both
cases the bond split by thrombin is an Arg-Gly bond. In all fibrinopeptide
A sequences but one there is a Phe residue nine residues before the
C-terminal Arg.

[77] J. A. Gladner and K. Laki, *Arch. Biochem. Biophys.* **62**, 501 (1956).
[78] Markwardt, this volume [69].
[79] R. H. Landaburu and W. H. Seegers, *Can. J. Biochem. Physiol.* **38**, 613 (1960).
[80] See e.g., F. C. Monkhouse, *in* "Blood Clotting Enzymology" (W. H. Seegers, ed.),
   p. 323. Academic Press, New York, 1967.
[81] B. Blombäck, *Arkiv Kemi* **12**, 321 (1958).
[82] For a review see B. Blombäck, *in* "Blood Clotting Enzymology" (W. H. Seegers,
   ed.), p. 143. Academic Press, New York, 1967.

Two other reactions involved in hemostasis are thrombin-dependent. The aggregation of blood platelets has been found to require the presence of fibrinogen in the platelets.[83] Clot retraction is believed to depend on the contraction of the contractile protein in blood platelets which is called thrombosthenin. The myosinlike component of thrombosthenin, thrombosthenin M, has recently been shown to be a substrate for thrombin.[84] In this case also C-terminal Arg is the product of thrombin action. Circumstantial evidence indicates that clot retraction is actually triggered by thrombin. Some plasma clotting factors, namely Factor XIII (plasma transglutaminase), Factor V (proaccelerin), and Factor VIII (antihemophilic factor) are considered by some authors to be activated by thrombin but sufficient chemical evidence is not available to settle these points.

The only "nonspecific" polypeptide or protein substrates known are secretin[85] and cholecystokinin-pancreozymin.[86] In secretin with 27 amino acid residues the Arg-14 to Asp-15 bond is split. There is a phenylalanine residue nine residues before the arginyl bond that is split by thrombin, as is the case in the fibrinopeptides A. Neither fragment is active. In cholecystokinin-pancreozymin, which has 33 amino acid residues, the Arg-6 to Val-7 bond is split and the resulting C-terminal 27 residue fragment (thrombocholecystokinin) has full biological activity on intravenous injection.

Kinetic Properties

Because of the several steps involved in the conversion of fibrinogen to fibrin kinetic analyses based on clotting times are difficult to interpret. Fibrinopeptide A is split off by thrombin at a faster rate than fibrinopeptide B.[81,87,88] Since reptilase, a thrombinlike proteinase from the venom of the snake *Bothrops jararaca*, causes the clotting of fibrinogen,[89,90] but splits off only the A-peptide,[91] it appears that the B-peptide does not have to be split off before polymerization of fibrin monomer can

[83] K. M. Brinkhous, M. S. Read, N. F. Rodman, and R. G. Mason, *J. Lab. Clin. Med.* **73**, 1000 (1969).
[84] I. Cohen, Z. Bohak, A. deVries, and E. Katchalski, *European J. Biochem.* **10**, 388 (1969).
[85] V. Mutt, S. Magnusson, J. E. Jorpes, and E. Dahl, *Biochemistry* **4**, 2358 (1965).
[86] V. Mutt and J. E. Jorpes, *European J. Biochem.* **6**, 156 (1968).
[87] F. R. Bettelheim, *Biochim. Biophys. Acta* **19**, 121 (1956).
[88] B. Blombäck and A. Vestermark, *Arkiv Kemi* **12**, 173 (1958).
[89] B. Blombäck, M. Blombäck, and I. M. Nilsson, *Thromb. Diath. Haemorrhag.* **1**, 76 (1957).
[90] H. W. Hohnen, *Z. Ges. Exptl. Med.* **128**, 427 (1957).
[91] B. Blombäck and I. Yamashima, *Arkiv Kemi* **12**, 299 (1958).

start. In spite of these obvious difficulties attempts have been made to study the kinetics of the thrombin–fibrinogen interaction. Waugh and associates,[92-95] measuring fibrin yield, found that the overall reaction appears to be first order in thrombin but departs from first order in fibrinogen. Shinowara,[96] analyzing clotting times, concluded that the reaction is second order in thrombin, first order in fibrinogen up to $2.3 \times 10^{-6}\,M$, zero order between $2.3 \times 10^{-6}\,M$ and $9.7 \times 10^{-6}\,M$, and modified zero order at higher concentration. He arrived at a Michaelis constant of $4.38 \times 10^{-7}\,M$.

p-Toluenesulfonyl-L-arginine methyl ester was the first synthetic substrate found for thrombin.[46] The K_m was reported as $4 \times 10^{-3}\,M$.[46] It was later found[51] that p-toluenesulfonyl-L-lysine methyl ester is at least as good a substrate. The hydrolysis of these esters is zero order in substrate and first order in thrombin. α-N-Toluene-p-sulfonyl-L-ornithine is a moderately good substrate[51] but α-N-toluene-p-sulfonyl-L-homo-arginine is an inhibitor.[51] The fact that different α-N-benzoyl-L-arginine esters are hydrolyzed at the same rate, whereas different α-N-toluene-p-sulfonyl-L-arginine esters are hydrolyzed at different rates by thrombin[51] indicates that the deacylation step is rate-determining in the former case whereas the acylation rate also contributes significantly to $k_{3(\mathrm{app})}$ in the latter case.[51] Increasing ionic strength inhibits the hydrolysis of tosyl esters but not of benzoyl esters and, therefore, presumably the acylation step, but not the deacylation step.[51] Sodium cholate and glycocholate were found to have the opposite effect, accelerating acylation and decelerating deacylation.[51] The pH dependence of $k_{3(\mathrm{app})}$ of tosyl-arginine esters indicates that hydrolysis depends on the dissociation of a group with pK(app) of about 6.6.[51,52] The $K_{m(\mathrm{app})}$ at pH 8.4 and 25° for α-N-benzoyl-L-arginine methyl ester and for α-N-toluene-p-sulfonyl-L-arginine methyl ester of $16.1 \times 10^{-6}\,M$ and $32.1 \times 10^{-6}\,M$, respectively[51] are much lower than previously found values.[46] This was attributed to enzyme activation by the substrate. Seegers *et al.*[67] claim to have split off and separated a **75** residue fragment, which was not recovered, from thrombin by rechromatography of the original 3.7 S thrombin. These authors give the K_m at pH 8.0 and 22° on α-N-toluene-p-sulfonyl-L-arginine methyl ester as $2.97 \times 10^{-4}\,M$ and $9.5 \times 10^{-5}\,M$ for the 3.7 S and the 3.2 S thrombins, respectively. Some p-nitrophenyl esters of carbobenzoxy-L-amino acids have been found to be good sub-

---

[92] D. F. Waugh and B. J. Livingstone, *Science* **113**, 121 (1951).
[93] D. F. Waugh and B. J. Livingstone, *J. Phys. Colloid Chem.* **55**, 1206 (1951).
[94] D. F. Waugh and M. J. Patch, *J. Phys. Chem.* **57**, 377 (1953).
[95] D. F. Waugh, *Advan. Protein Chem.* **9**, 325 (1954).
[96] G. Y. Shinowara, *Biochim. Biophys. Acta* **113**, 359 (1966).

strates.[97] The K_s at pH 8.0, 30° for carbobenzoxy-L-tyrosine-p-nitro-
phenyl ester ($\Gamma/2$ 0.3, 11.7% MeOH) and for α-N-toluene-p-sulfonyl-
L-arginine methyl ester ($\Gamma/2$ 0.001–0.02) were $8.6 \times 10^{-5}\,M$ and $110 \times$
$10^{-5}\,M$, respectively.[97] Lorand *et al.*[98] found the relative rate constants
for the p-nitrophenyl esters of acetate, carbobenzoxy-L-tyrosinate, carbo-
benzoxy-L-phenylalaninate, and for α-N-toulene-p-sulfonyl-L-arginine
methyl ester (at 30°, pH 8.0, $\Gamma/2$ 0.34 in aqueous 3.3% Me$_2$CO, 13.3%
iso-PrOH) to be 1, 2500, 1500, and 5000, respectively. Kézdy *et al.*[52]
attempted to use the rapid formation and relatively slow hydrolysis of
the acyl thrombins formed with the p-nitrophenyl esters of carbobenzoxy-
L-tyrosinate and -lysinate to determine the operational molarity of
thrombin solutions. They also determined k_{cat} (0.22, 0.72, and 16.5 sec^{-1})
and K_m (7.2, 21, and $56 \times 10^{-6}\,M$) at 25° for human thrombin on the
p-nitrophenyl esters of carbobenzoxy-L-tyrosinate and -lysinate at pH
5.02 and on α-N-benzoxyl-L-arginine ethyl ester at pH 8.75, respectively.
p-Nitrophenyl-p'-guanidinobenzoate (NPGB) which has been developed
as a titrant for trypsin[56] reacts rapidly with thrombin[54] giving a burst
of p-nitrophenol independent of substrate concentration from 16.6–82.0
μM and proportional to thrombin concentration in the range 0.8 to 7.6
μM. Both clotting and esterase activities are inhibited. The first order
rate constant was found to be 4.05 sec^{-1} at pH 8.3, 25°. 7-Amino-1-chloro-
3-toluene-p-sulfonamide-2-butanone (TLCK) was found to react with
thrombin[54] leading to complete inhibition of clotting and esterase activities. First-order kinetics (k about 4.4×10^{-3} sec^{-1}) was observed for
75% of the reaction at 20°, pH 8.3. In the reaction with NPGB slower
acylation and faster deacylation was found[55] for thrombin than for
trypsin and plasmin. The kinetic constants in Table III were determined[55]

TABLE III
Kinetic Constants for Thrombin Reaction[a,b]

	EGB	p-NPGB	m-NPGB
k_2	1.19×10^{-3} sec^{-1}	0.13 sec^{-1}	0.12 sec^{-1}
K_s	$7.3 \times 10^{-3}\,M$	$3.95 \times 10^{-6}\,M$	$1.64 \times 10^{-5}\,M$
k_3		98×10^{-5} sec^{-1}	5.48×10^{-3} sec^{-1}
K_m			$16.7 \times 10^{-6}\,M$

[a] T. Chase Jr. and E. Shaw, *Biochemistry* **8**, 2212 (1969).
[b] Ethyl p-guanidinobenzoate (EGB), p-nitrophenyl-p'-guanidinobenzoate (p-NPGB),
p-nitrophenyl-m'-guanidinobenzoate (m-NPGB).

[97] C. J. Martin, J. Golubow, and A. E. Axelrod, *J. Biol. Chem.* **234**, 1718 (1959).
[98] L. Lorand, W. T. Brannen, Jr., and N. G. Rule, *Arch. Biochem. Biophys.* **96**, 147
(1962).

for the reaction of bovine thrombin with ethyl-*p*-guanidinobenzoate (EGB), *p*-nitrophenyl-*p'*-guanidinobenzoate (*p*-NPGB) and *p*-nitrophenyl-*m'*-guanidinobenzoate (*m*-NPGB). A series of derivatives of benzylamine and benzamidine were found[99] to be more efficient inhibitors of trypsin and plasmin than of thrombin with one exception, namely 4-chlorobenzylamine. The most efficient thrombin inhibitor in this series was found to be 4-amidinophenylpyruvic acid ($K_i = 2 \times 10^{-8} M$).

[99] F. Markwardt, H. Landmann, and P. Walsmann, *European J. Biochem.* **6**, 502 (1968).

[10] Human Plasminogen and Plasmin

By KENNETH C. ROBBINS and LOUIS SUMMARIA

Assay Method

Principle. The method is based on the enzymatic hydrolysis of casein for a fixed period of time followed by termination of the reaction with trichloroacetic acid. After removal of the undigested casein by filtration, the tyrosine- and tryptophan-containing peptides are determined by measuring the absorbance of the solution at 280 nm. Enzymatic activity is related to the liberation of trichloroacetic acid-soluble peptides containing tyrosine.

Reagents

Phosphate buffer, 0.067 *M*, pH 7.4

Tris, 0.05 *M*; lysine, 0.02 *M*; NaCl, 0.10 *M*; EDTA (disodium salt), 0.001 *M*, buffer, pH 9.0

Casein solution. A 5% suspension of casein (Devitaminized-Sheffield Chemical Co., or Hammarsten quality) is prepared in 0.067 *M* phosphate buffer, pH 7.4. Then 2 ml of 1 *N* NaOH is added to each 100 ml of suspension and the mixture is stirred for 1 hour at room temperature. The casein solution is adjusted to pH 7.4 with 1 *N* NaOH and to 4% concentration with the pH 7.4 phosphate buffer. Sodium azide is added to a final concentration of 0.01%. The insoluble residue is removed by centrifugation. The casein solution is stored at 4° and should not be used after 2 weeks.

Plasminogen activator. Streptokinase, approximately 5000 Christensen units per milligram (Varidase, Lederle Laboratories, Pearl

River, N.Y.) is dissolved in 0.067 M phosphate buffer, pH 7.4, to a concentration of 2500 units/ml and can be stored at −30°.

Zymogen or enzyme. The zymogen, or enzyme, is diluted from concentrated solutions in the pH 9.0 Tris-lysine-NaCl-EDTA buffer, to a suitable concentration which will allow 0.01 to 0.4 ml to be pipetted for assay.

15% Trichloroacetic acid

Procedure. The following solutions are pipetted, in order, in a 20 × 150 mm Pyrex test tube at 4° to a total volume of 4.0 ml: (1) 2.0 ml of 4% casein solution, (2) 1.4 ml (for zymogen assay) or 1.6 ml (for enzyme assay) of 0.067 M phosphate buffer, pH 7.4, (3) 0 to 0.4 ml of the pH 9.0 Tris-lysine-NaCl-EDTA buffer, (4) 0 to 0.4 ml of ice-cold zymogen or enzyme, (5) 500 units of streptokinase in 0.2 ml (for zymogen assay). The tube is mixed well, transferred to a 37° water bath and incubated for 30 minutes. The tube is then placed into a 4° water bath and 6.0 ml of 15% trichloroacetic acid is added and the mixture is shaken. After 18 hours at 4°, the tube is allowed to warm up to room temperature and the precipitate is removed by filtration through Whatman No. 50 filter paper (9.0 cm). Refiltration through the same filter paper will remove fine particles. The absorbance of the solution is read at 280 nm against a blank containing casein, buffers, and trichloroacetic acid only. All samples are run in duplicate with appropriate blanks. This method is valid and reproducible with enzyme concentrations of approximately 0.5 to 1.5 casein units which give absorbance readings between approximately 0.03 and 0.22 (linear portion of curve) (Fig. 1). Alpha casein (National Dairy) is a better substrate than either beta or kappa casein, Sheffield casein, or Hammarsten-quality casein[1] (see Fig. 1). The linear portion of the curve is between enzyme concentrations of approximately 0.5 to 2.5 casein units which give absorbance readings of approximately 0.05 and 0.55. Since good quality alpha casein is not readily available, the more complex mixture of caseins has been used as the substrate.

Definition of Unit and Specific Activity. The plasmin casein unit is defined as the amount of enzyme which will liberate 450 μg of trichloroacetic acid-soluble tyrosine in 1 hour under the conditions of the assay.[2] The calculation is made as follows[3]:

[1] K. C. Robbins and L. Summaria, *Immunochemistry* **3**, 29, (1966).
[2] L. F. Remmert and P. P. Cohen, *J. Biol. Chem.* **181**, 431 (1949).
[3] Assay Factors: Factor 2 converts 30 minutes to 1 hour; factor 10 converts absorbance per milliliter to 10 ml total volume; factor 147 converts absorbance per milliliter to micrograms tyrosine per milliliter. The absorbance of a solution containing 147 μg tyrosine per milliliter in 9% trichloroacetic acid, at 280 nm, is 1.00.

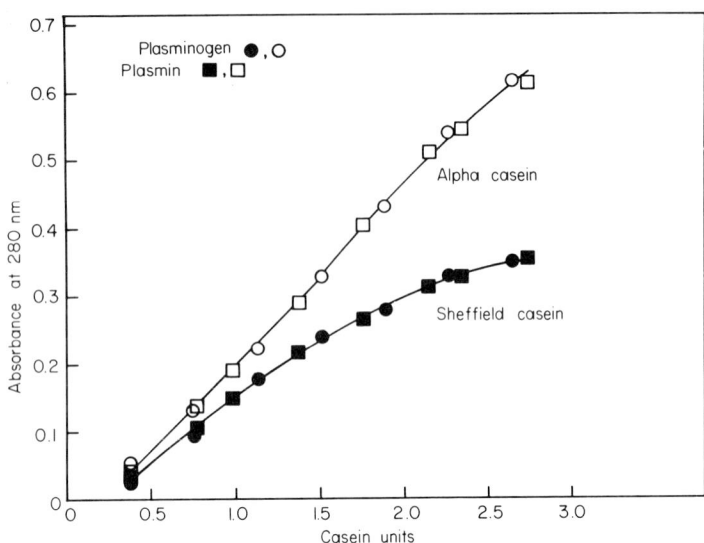

Fig. 1. Typical assay curve for the hydrolysis of casein (Sheffield) (●, ■) and alpha casein (National Dairy) (○, □) by human plasminogen (with streptokinase) (○, ●) and plasmin (□, ■). The alpha casein was prepared by the Warner procedure [R. C. Warner, *J. Am. Chem. Soc.* **66**, 1725 (1944)].

$$1 \text{ casein unit/ml} = \frac{\text{absorbance}}{\text{sample volume (ml)}} \times 2 \times 10 \times 147 \times \frac{1}{450}$$

or

$$1 \text{ casein unit/ml} = \frac{\text{absorbance}}{\text{sample volume (ml)}} \times 6.52$$

The plasmin or plasminogen specific activity is defined in casein units per milligram of protein. The protein concentration of the solution is determined by measuring the absorbance of the solution at 280 nm in 0.1 N NaOH and converting absorbance to protein concentration using $E_{1 cm}^{1\%} = 17.0$[4] for pure zymogen or enzyme. The plasminogen specific activity is actually "potential" casein units per milligram of protein. Plasminogen preparations always contain 3 to 5% plasmin activity which is included in the plasminogen assay value.[4,5]

Purification Procedure

Preparation of Human Plasminogen

The most satisfactory purification procedure for human plasminogen requires the use of human plasma fraction $III_{2,3}$ paste.[4-6] The purity attainable by the procedure described below is dependent on the use of this starting material. Human plasma fraction III or other cruder plasma fractions are not suitable starting materials. In the purification procedure given below, all steps are carried out at 2°–4° unless otherwise indicated, and all reagents used are at the same temperature.

Step 1. Preparation of Extract. A 10% suspension (w/v) of frozen human plasma fraction $III_{2,3}$ paste is prepared in a 0.05 M Tris-0.02 M lysine-0.1 M NaCl-0.001 EDTA (disodium salt) buffer, pH 9.0. Mechanical stirring at a moderate speed for a period of 4–16 hours is necessary to completely disperse the frozen paste. When dispersed, the suspension is opalescent and uniform. The suspension is clarified by centrifugation at approximately 3000 g for 1 hour. The precipitate is discarded. A volume of 2–3 liters, or more, can be prepared at any given time with great ease.

Step 2. Isoelectric Precipitate. The extract is diluted with 9 volumes of ice-cold distilled water and the pH is adjusted to 6.1 $\pm$ 0.1 by adding 1 M KH_2PO_4 (volume varies with fraction $III_{2,3}$ lots). The suspension is allowed to stand for at least 18 hours. The precipitate is removed by centrifugation at 3000 g for 30 minutes and suspended in the pH 9.0 Tris-lysine-NaCl-EDTA buffer, in a volume equal to approximately ⅓ of the original extract volume. The suspension is stirred for 1 to 2 hours for complete solubilization. This solution can be stored for months at −30° before use. In this step, 30–40% of the plasminogen is soluble and is discarded. This soluble plasminogen may be a different form of the zymogen.

Step 3. Chromatography on DEAE-Sephadex. SMALL SCALE. A sample of approximately 6000 absorbance units of protein[7] in a volume of 50 to 100 ml is passed through a column (8 $\times$ 43 cm) of DEAE-Sephadex A-50, equilibrated, and eluted with the pH 9.0 Tris-lysine-NaCl-EDTA buffer. Fraction volumes of 20 ml are collected at a flow rate of approximately 5 ml per minute. The zymogen activity and absorbance of each fraction are determined and the specific activity is calculated. A typical chromatogram is shown in Fig. 2. Approximately 5% of the absorbance units and 90–100% of the plasminogen casein units in the applied sample are unadsorbed to the DEAE-Sephadex column. After samples are taken

---

[4] K. C. Robbins, L. Summaria, D. Elwyn, and G. H. Barlow, *J. Biol. Chem.* **240**, 541 (1965).

[5] K. C. Robbins and L. Summaria, *J. Biol. Chem.* **238**, 952 (1963).

[6] Human plasma fraction $III_{2,3}$ was prepared from fresh plasma and obtained from Cutter Laboratories, Berkeley, Calif. One-thousand liters of plasma will give approximately 5 kg of fraction $III_{2,3}$ paste (approximately 50% protein). The paste is stored at −30° until used.

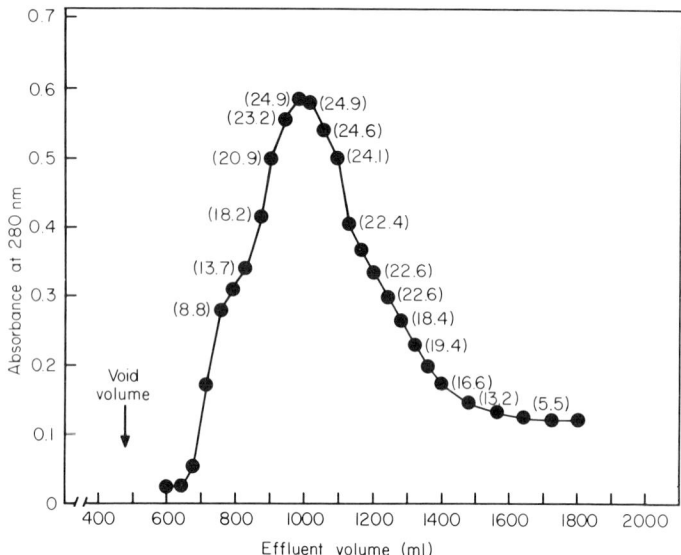

Fig. 2. Chromatography of crude human plasminogen on DEAE-Sephadex. The specific activity of each fraction is shown in parentheses.

for activity measurements and the protein concentration is determined,[7] the zymogen is precipitated in each tube by adding 3.1 g of $(NH_4)_2SO_4$ per 10 ml of solution. After activity measurements are completed, a pool is made of all fractions (as precipitates) containing specific activities over 18 casein units per milligram of protein. The precipitate is removed by centrifugation and dissolved in the pH 9.0 Tris-lysine-NaCl-EDTA buffer in a concentration of 5% or less of the volume of the fraction. The final pool is clarified by centrifugation and is stored at −30° until used for further purification. The specific activity of the final pool may vary between 18 and 22 casein units per milligram of protein.

LARGE SCALE. A sample of approximately 60,000 absorbance units of protein in 500–1000 ml is passed through a column (15 × 100 cm) of DEAE-Sephadex A-50 equilibrated and eluted with the pH 9.0 Tris-lysine-NaCl-EDTA buffer. Fraction volumes of 800 ml are collected at

[7] An absorbance unit of protein is equal to an absorbance reading of 1.00 per milliliter at 280 nm in the pH 9.0 Tris-lysine-NaCl-EDTA buffer. At step 3, before chromatography, the absorbance unit of protein is a relative unit and may be equal to 0.6 to 0.7 mg of protein. After chromatography the absorbance unit of protein is equal to approximately 0.6 mg of protein. The absorbance readings of highly purified plasminogen, at 280 nm in either 0.1 N NaOH or in the pH 9.0 Tris-lysine-NaCl-EDTA buffer are similar.

a flow rate of 10–15 ml per minute. The fractions are handled and pooled as described for the small-scale procedure.

Step 4. Gel Filtration Through Sephadex G-150. A sample of approximately 1000 absorbance units of protein[7] in a volume of 30–50 ml is gel filtered through a column (4.8 × 90 cm) of Sephadex G-150 equilibrated and eluted with the pH 9.0 Tris-lysine-NaCl-EDTA buffer. With small volumes of material, 10–15 ml, a shorter column (4.8 × 25 cm) can be used. Fraction volumes of 10 ml are collected at a flow rate of 0.5–1.0 ml per minute. The zymogen activity and absorbance of each fraction is determined and the specific activity calculated. A typical elution pattern is shown in Fig. 3. After samples are taken for activity measurements and the protein concentration is determined,[7] the zymogen is precipitated in each tube by adding 3.1 g of $(NH_4)_2SO_4$ per 10 ml of solution. After activity measurements are completed, a pool is made of all fractions (as precipitates) containing specific activities over 22 casein units per milligram of protein. The precipitate is removed by centrifugation and dissolved in the pH 9.0 Tris-lysine-NaCl-EDTA buffer in a concentra-

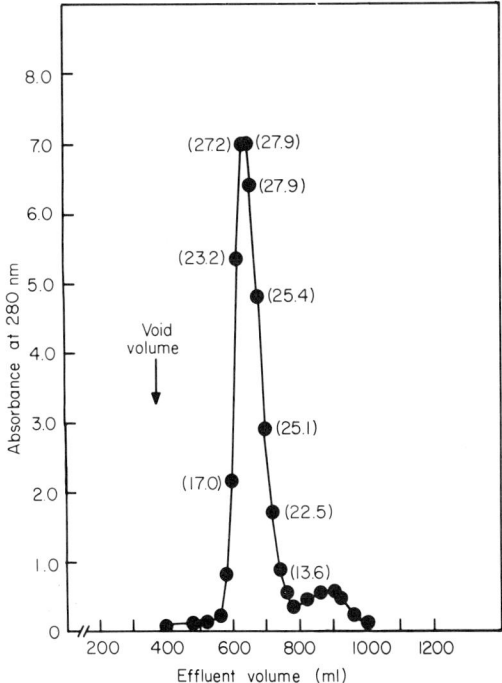

Fig. 3. Gel filtration of purified human plasminogen through Sephadex G-150. The specific activity of each fraction is shown in parentheses.

tion of about $\frac{1}{6}$, or less, of the volume of the fraction. The final pool is clarified by centrifugation and is stored at $-20°$ to $-30°$. The specific activity of the final pool may vary between 22 and 26 casein units per milligram of protein. More careful selection of fractions before pooling will give preparations with specific activities of 25 to 28 casein units per milligram of protein. The results of the fractionation procedure are summarized in Table I.

Preparation of Human Plasmin

Activation of Human Plasminogen with Either Urokinase or Streptokinase at pH 9.0.[4,8] Three milliliters of human plasminogen solution in the pH 9.0 Tris-lysine-NaCl-EDTA buffer, containing 15 to 20 mg of protein per milliliter, are adjusted to 25% glycerol by adding 1 ml of 99.5% glycerol.[9] One-tenth milliliter of urokinase solution,[10] or streptokinase solution,[11] is added at a concentration of 1 CTA unit of urokinase[12] or 3 Christensen units of streptokinase[13] per casein unit of plasminogen. The ratio of activator to plasminogen in this system is 0.25 μg of pure urokinase or 0.75 μg of pure streptokinase to 1.0 mg of plasminogen. The molar ratio of plasminogen[14] to pure activator is approximately 2700 for urokinase and 800 for streptokinase.

The activation mixture is allowed to stand at $25°$ for approximately 20 hours. The solution is cooled to $0°$, diluted 10-fold with ice-cold distilled water, adjusted to pH 6.2 with 1 M KH_2PO_4 and the enzyme is precipitated by adding 3.1 g of $(NH_4)_2SO_4$ per 10 ml of solution. The suspension is allowed to stand for 2 hours; the precipitate is removed by centrifugation at 3000 g for 30 minutes and dissolved in a volume of ice-cold pH 9.0 Tris-lysine-NaCl-EDTA buffer equal to the volume of the starting plasminogen solution. The plasmin solution is clarified by centrifugation and stored in the frozen state at $-20°$ to $-30°$. The

[8] L. Summaria, B. Hsieh, and K. C. Robbins, *J. Biol. Chem.* **242**, 4279 (1967).

[9] Glycerol, 99.5%, synthetic, U.S.P., Shell.

[10] The human urokinase preparation used was obtained from Abbott Laboratories, North Chicago, Ill. It had a specific activity of approximately 41,000 CTA units per milligram and was approximately 40% urokinase.[27] The molecular weight of urokinase is approximately 54,000.[27]

[11] The streptokinase preparation used was Varidase (Lederle Laboratories). It had a specific activity of approximately 5000 units per milligram and was approximately 5% streptokinase.[22] The molecular weight of streptokinase is approximately 48,000.[22]

[12] A. J. Johnson, D. L. Kline, and N. Alkjaersig, *Thromb. Diath. Haemorrhag.* **21**, 259 (1969).

[13] L. R. Christensen, *J. Clin. Invest.* **28**, 163 (1949).

[14] Assuming pure human plasminogen to have a specific activity of approximately 30 casein units per milligram of protein. The molecular weight of human plasminogen is 81,000.[15]

TABLE I
SUMMARY OF PURIFICATION PROCEDURE

Step	Protein (absorbance units/kg III$_{2,3}$ paste)	Protein (mg/kg III$_{2,3}$ paste)	Activity (casein units/kg III$_{2,3}$ paste)	Specific activity (casein units/absorbance unit)	Specific activity (casein units/mg protein)	Yield of zymogen (%)	Purification factor[c] (-fold)
1. Extract[a]	400,000 ± 100,000		190,000 ± 40,000	0.48		100	12
2. Isoelectric precipitate	290,000 ± 75,000		150,000 ± 30,000	0.52		79	13
3. Chromatography on DEAE-Sephadex[b]	7,400 ± 900	4,350 ± 500	86,000 ± 15,000	11.8 ± 1.1	20 ± 2	45	300
4. Gel filtration through Sephadex G-150[b]	5,500 ± 300	3,230 ± 200	71,000 ± 2,000	14.1 ± 1.2	24 ± 2	37	350

[a] Clarified extract.
[b] Selected pools.
[c] Related to normal plasma. The specific activity of normal plasma is approximately 0.07 casein units per milligram protein [S. F. Rabiner, I. D. Goldfine, A. Hart, L. Summaria, and K. C. Robbins, J. Lab. Clin. Med. **74**, 265 (1969)].

plasmin specific activity will depend upon the specific activity of the starting plasminogen solution. Preparations with specific activities of 25 to 28 casein units per milligram of protein can be readily obtained.

Properties

Stability. The zymogen is stored in concentrated solution in the pH 9.0 Tris-lysine-NaCl-EDTA buffer at $-20°$ to $-30°$. It is stable under these conditions for years. The enzyme is most stable in concentrated solution in the pH 9.0 Tris-lysine-NaCl-EDTA buffer containing 25% glycerol at $-20°$ to $-30°$. It is usually stored under these conditions after activation of the zymogen in the glycerol-buffer solution. Freezing and thawing results in loss of enzymatic activity. After removal of the glycerol, the enzyme is unstable at neutral pH's. Both the zymogen and enzyme are very stable at pH's below 4.0. For most physical measurements, it is preferred to use the zymogen and enzyme solutions at acid pH's.

Purity and Physical Properties. The zymogen and enzyme behave as homogeneous monomeric proteins in the ultracentrifuge. At the microgram level, by sedimentation equilibrium analysis using photoelectric scanning optics, the molecular weight of plasminogen was determined to be $81,000 \pm 2900$ and the molecule weight of plasmin was determined to be $75,400 \pm 2800$.[15] The partial specific volumes, $\bar{V}$, determined by falling drop and $H_2O–D_2O$ sedimentation equilibrium methods[15] and calculated from amino acid composition data,[4] for plasminogen were 0.715, 0.714, and 0.720, respectively, and for plasmin were 0.715, 0.709, and 0.717, respectively. For the calculation of molecular weights of zymogen and enzyme, a single average $\bar{V}$ value was used. The $S_{20,w}^0$ values of plasminogen and plasmin, using photoelectric scanning optics at the microgram level, were determined to be 4.2 S and 3.9 S, respectively.[15] The frictional coefficients, f/f_0, of zymogen and enzyme were calculated from sedimentation and diffusion data and were found to be 1.8 and 1.5, respectively.[4]

In moving-boundary electrophoresis, at pH 2.9, the zymogen and enzyme preparations showed major components representing approximately 90% of the total with mobilities (as average descending values) of 9.40 and 7.63×10^{-5} cm² per volt per second, respectively.[4] The single minor component in both zymogen and enzyme preparations appeared adjacent to the major peak but with a slower mobility. This component may represent both a plasmin contaminant in the plasminogen preparation and a different form (inert) of the enzyme in the plasmin prepara-

[15] G. H. Barlow, L. Summaria, and K. C. Robbins, *J. Biol. Chem.* **244**, 1138 (1969).

tion.[16] Electrophoretic analysis of the zymogen and enzyme on cellulose acetate membranes at pH 8.6 showed that they were homogeneous.[17] The mobilities of the zymogen and enzyme in this system were different and the proteins were readily distinguishable. Both proteins were located in the γ-globulin region of the electrophoretogram. In vertical starch gel electrophoresis, at pH 3.1 in 8 M urea, the zymogen and enzyme were found to be homogeneous and to have similar mobilities.[4,16]

The determination of the amino and carboxyl terminal amino acid residues of the zymogen and enzyme (see p. 195) also reflect the purity and homogeneity of the preparations.

Purity and Immunochemical Properties. Specific rabbit precipitating antibodies have been prepared to plasminogen and plasmin.[1,4,5] The zymogen-enzyme pair were found to be immunochemically identical in gel diffusion. The preparations appear to be homogeneous on both gel diffusion and immunoelectrophoretic analyses, with antiplasminogen and antiplasmin sera. In immunoelectrophoretic systems, at pH 8.6, in agarose, the zymogen- and enzyme-antibody complexes were found in the beta region. Therefore, we have classified these proteins as β-globulins and not γ-globulins as found in cellulose acetate electrophoresis. The mobilities of the zymogen and enzyme, at pH 8.6 in agarose, and in cellulose acetate, were different, with the zymogen being more positively charged.

Amino Acid Composition and Chemical Properties. The amino acid composition of plasminogen and plasmin is shown in Table II.[4] The number of residues and amino acid composition of the zymogen and enzyme per mole of protein (calculated per 79,000 g of protein) are very similar. These proteins contain less than 2% carbohydrate. The percentage of nitrogen calculated from amino acid composition for the zymogen and enzyme is 17.1 ± 0.1%. The extinction coefficient, $E_{1cm}^{1\%}$, for both proteins is 17.0 ± 0.3.

Mechanism of Activation of Human Plasminogen to Plasmin. The Peptide Chains of Plasmin.[8,16,18,19] The activation of plasminogen to plasmin in a 25% glycerol buffer at pH 9.0 by low molar ratios of urokinase or streptokinase (see p. 190) was shown to involve the cleavage of a single arginyl-valine bond. The activation of the zymogen by trypsin

[16] K. C. Robbins, L. Summaria, B. Hsieh, and R. J. Shah, *J. Biol. Chem.* **242**, 2333 (1967).

[17] L. Summaria, B. Hsieh, W. R. Groskopf, and K. C. Robbins, *Proc. Soc. Exptl. Biol. Med.* **130**, 737 (1969).

[18] L. Summaria, B. Hsieh, W. R. Groskopf, K. C. Robbins, and G. H. Barlow, *J. Biol. Chem.* **242**, 5046 (1967).

[19] W. R. Groskopf, B. Hsieh, L. Summaria, and K. C. Robbins, *J. Biol. Chem.* **244**, 359 (1969).

TABLE II
Amino Acid Composition of Human Plasminogen and Plasmin[a]

	Residues per 79,000 g of protein		
Amino acid	Plasminogen	Urokinase-activated plasmin	Streptokinase-activated plasmin
Lysine	41	42	40
Histidine	21	22	22
Arginine	37	38	38
Aspartic acid	71	71	71
Threonine	54	58	55
Serine	44	48	42
Glutamic acid	69	70	68
Proline	65	67	68
Glycine	56	56	58
Alanine	32	31	32
Half-cystine	38	41	41
Valine	43	40	42
Methionine	8.5	7.9	8.3
Isoleucine	19	16	18
Leucine	40	37	39
Tyrosine	29	28	29
Phenylalanine	17	15	18
Tryptophan	20	19	19
Amide NH$_3$	(79)	(72)	(82)
Total	705	707	708

[a] K. C. Robbins, L. Summaria, D. Elwyn, and G. H. Barlow, J. Biol. Chem. 240, 541 (1965).

and pig heart activator also involves the cleavage of the same arginyl-valine bond but small amounts of additional peptide bonds are split by these activators.

The cleavage of a single disulfide bond in both plasminogen and plasmin by reduction with 2-mercaptoethanol, in 8 M urea at pH 9.0, resulted in complete loss of proteolytic activity (on casein). The cleavage of this bond resulted in the formation of two major electrophoretic chains from plasmin, but not from plasminogen. Highly purified S-carboxymethyl chain derivatives of plasmin can be prepared by either gel filtration through Sephadex G-200, or by dialysis. The apparent molecular weight of the S-carboxymethyl light chain derivative was determined to be 25,700 ± 200 by sedimentation equilibrium analysis using photoelectric scanning optics. The sedimentation coefficient of this derivative was determined to be 1.4 S. The apparent molecular weight of the S-carboxymethyl heavy chain derivative was determined to be approximately 48,800.[19a] The sedimentation coefficient of this derivative was determined to be 2.8 S.[19a] The amino acid composition of the S-carboxymethyl light chain derivatives obtained from both urokinase-activated and streptokinase-activated plasmin are similar (Table III) confirming the specificity of the activation of plasminogen to plasmin by these two activators. The amino acid composition of the light chain derivative prepared by a dialysis procedure shows some differences (Table III).

Amino and Carboxyl Terminal Amino Acid Residues.[8,16,18] The amino terminal amino acid residue of plasminogen and the heavy chain of plasmin was found to be lysine and the amino terminal amino acid residue of the light chain of plasmin was found to be valine. The carboxyl

TABLE III

AMINO ACID COMPOSITION OF SCM-LIGHT CHAIN DERIVATIVE OF HUMAN PLASMIN[a]

| | Residues per 25,700 g of protein | | |
| | Urokinase-activated | Streptokinase-activated | Urokinase-activated |
Amino acid	Chromatographically purified	Chromatographically purified	Purified by precipitation method
Lysine	10.3	12.9	14.3
Histidine	6.5	6.0	8.1
Arginine	11.2	12.6	14.2
Aspartic acid	19.4	19.8	14.6
Threonine	16.2	16.0	14.3
Serine	18.7	15.6	17.8
Glutamic acid	25.2	24.3	21.7
Proline	19.6	19.6	15.0
Glycine	21.9	21.2	20.6
Alanine	13.5	13.0	12.0
Half-cystine	10.4	11.8	8.7
Valine	21.5	19.1	21.6
Methionine	1.1	2.5	1.9
Isoleucine	9.2	8.1	8.2
Leucine	19.8	18.6	19.6
Tyrosine	5.3	6.5	6.2
Phenylalanine	7.6	7.0	7.7
Tryptophan			4.7
Amide nitrogen			(20)
Total			231.2

[a] L. Summaria, B. Hsieh, W. R. Groskopf, K. C. Robbins, and G. H. Barlow, *J. Biol. Chem.* **242**, 5046 (1967); W. R. Groskopf, B. Hsieh, L. Summaria, and K. C. Robbins, *J. Biol. Chem.* **244**, 359 (1969).

[19a] See footnote 48.

terminal amino acid residue of plasminogen and the light chain of plasmin was found to be asparagine and the carboxyl terminal amino acid residue of the heavy chain was found to be arginine.

Active Center.[18-20] A single DPF[21]-sensitive serine residue and a single TLCK[21]-sensitive histidine residue has been found in the active center of plasmin. Both of these residues are located on the light chain of plasmin. The partial amino acid sequence of a tryptic peptide containing the active serine residue is

Val-Glx-(Ser-Thr, Glx)-Leu-(Gly, Ala)-His-Leu-Ala-
Cys-Asn-(Thr, Gly, Gly)-Ser-Cys-Gln-Gly-Asp-Ser-
(diisopropyl phosphoryl-^{32}P)-Gly-Gly-Pro-Leu-Val-
Cys-Phe-Glu-Lys-Asp-Lys-Tyr

Activators and Inhibitors. Human plasminogen can be readily activated to plasmin by a variety of bacterial and animal substances. The highly purified and well-characterized activators are streptokinase,[22-26] urokinase,[27, 28] trypsin, pig heart activator,[29] rabbit kidney activator,[30] and staphylokinase.[31] The mechanism of activation of human plasminogen by streptokinase, urokinase, trypsin, and pig heart activator is specific and proceeds primarily through the cleavage of a single arginine-valine bond.[8, 16] Activation with staphylokinase may be similar.[32] Urokinase,[33] trypsin,[34] and pig heart activator[29] are capable of hydrolyzing a variety of alpha-amino substituted arginine and lysine esters. Streptokinase will activate DFP- or TLCK-treated plasminogen very rapidly and will also activate plasminogen in 10^{-2} *M* DFP but at a much slower

[20] W. R. Groskopf, L. Summaria, and K. C. Robbins, *J. Biol. Chem.* **244**, 3590 (1969).
[21] The abbreviations used are: DFP, diisopropylphosphorofluoridate; TLCK, L-1 chloro-3-tosylamido-7-amino-2-heptanone.
[22] E. C. De Renzo, P. K. Siiteri, B. L. Hutchings, and P. H. Bell, *J. Biol. Chem.* **242**, 533 (1967).
[23] F. B. Taylor and J. Botts, *Biochemistry* **7**, 232 (1968).
[24] L. Mesrobeanu, N. Mitrica, F. Mihalcu, and D. Movileanu, *Z. Immunitaetsforsch. Allerg.* **136**, 187 (1968).
[25] T. Bilinski, T. Loch, and K. Zakrzewsi, *Acta Biochim. Polon.* **15**, 123 (1968).
[26] T. Loch, T. Bilinski, and K. Zakrewski, *Acta Biochim. Polon.* **15**, 129 (1968).
[27] W. F. White, G. H. Barlow, and M. M. Mozen, *Biochemistry* **5**, 2160 (1966).
[28] A. Lesuk, L. Terminiello, and J. H. Traver, *Science* **147**, 880 (1965).
[29] F. Bachman, A. P. Fletcher, N. Alkjaersig, and S. Sherry, *Biochemistry* **3**, 1578 (1964).
[30] S. Y. Ali and L. Evans, *Biochem. J.* **107**, 293 (1968).
[31] E. Soru and M. Sternberg, *Enzymologia* **25**, 231 (1963).
[32] E. Soru, *Rev. Roum. Biochim.* **5**, 17 (1968).
[33] N. O. Kjeldgaard and J. Ploug, *Biochim. Biophys. Acta* **24**, 283 (1957).
[34] N. M. Green and H. Neurath, *in* "The Proteins" (H. Neurath and K. Bailey, eds.), Vol. 2, Part B, p. 1057. Academic Press, New York, 1954.

rate.[17] Plasmin is therefore not essential for streptokinase activation of the zymogen.

A number of well-characterized natural human plasmin inhibitors have been described. Two homogeneous human plasma protein inhibitors have been prepared, α-1 globulin with a molecular weight of 47,000[35] and an α-2 globulin with a molecular weight of 845,000.[36] Other mammalian tissues contain plasmin inhibitors.[37] Crystalline bovine pancreatic trypsin inhibitor (Kunitz)[38] which has a molecular weight of approximately 6500 is an excellent inhibitor.[1,39] Other bovine inhibitors have been prepared from other tissues, e.g., lung and parotid, which may be similar to the pancreatic inhibitor.[37] Inhibitors have also been prepared from a variety of plants.[37] Highly purified lima bean trypsin inhibitor with a molecular weight of approximately 9000 and crystalline soybean trypsin inhibitor with a molecular weight of approximately 21,000 are excellent human plasmin inhibitors.[1,37,39-41] A glycoprotein has been isolated from kidney bean with a molecular weight of about 10,000 which forms a 1:1 enzyme–inhibitor complex with human plasmin at low concentrations.[42] Specific antibodies to plasminogen and plasmin are excellent inhibitors.[1]

Synthetic inhibitors of human plasmin have been reported. In high concentrations, above $10^{-1} M$, ϵ-amino caproic acid (EACA) is an excellent noncompetitive inhibitor.[43] Amino methyl cyclohexane carboxylic acid (AMCHA) is also an excellent inhibitor.[44] Derivatives of L-lysine and EACA[45] and ω-guanidino acid esters[46] have been prepared which will inhibit human plasmin. The active site reagents DFP and TLCK will irreversibly inhibit human plasmin.[18,19]

Equimolar Human Plasmin-Streptokinase Activator Complex (Bovine Plasminogen Activator). A homogeneous equimolar human plasmin-

[35] A. Rimon, Y. Shamash, and B. Shapiro, *J. Biol. Chem.* **241**, 5102 (1966).
[36] H. G. Schwick, N. Heimburger, and H. Haupt, *Z. Ges. Inn. Med. Ihre Grenzgebiete* **21**, 193 (1966).
[37] R. Vogel, I. Trautschold, and E. Werle, "Natural Proteinase Inhibitors." Academic Press, New York, 1968.
[38] M. Kunitz and J. H. Northrop, *J. Gen. Physiol.* **19**, 991 (1936).
[39] R. E. Maxwell, V. Lewandowski, and V. S. Nickel, *Life Sci.* **4**, 45 (1964).
[40] L. B. Nanninga and M. M. Guest, *Arch. Biochem. Biophys.* **108**, 542 (1964).
[41] R. E. Feeney, G. E. Means, and J. C. Bigler, *J. Biol. Chem.* **244**, 1957 (1969).
[42] A. Pusztai, *European J. Biochem.* **5**, 252 (1969).
[43] M. Iwamoto, Y. Abiko, and M. Shimizu, *J. Biochem. (Tokyo)* **64**, 759 (1968).
[44] S. Okomoto, S. Oshiba, H. Mihara, and U. Okomoto, *Ann. N.Y. Acad. Sci.* **146**, 414 (1968).
[45] M. Muramatu, T. Onishi, S. Makino, S. Fujii, and Y. Yamamura, *J. Biochem. (Tokyo)* **57**, 450 (1965).
[46] M. Muramatu and S. Fujii. *J. Biochem. (Tokyo)* **64**, 807 (1968).

streptokinase activator complex has been prepared and isolated from a mixture of the two components.[47] The sedimentation coefficient of the activator complex in EACA was found to be approximately 5 S and the apparent molecular weight by sedimentation equilibrium analysis in EACA was determined to be approximately 123,900.[48] In lysine buffers at high concentrations, the activator complex was found to have a sedimentation coefficient of approximately 11 S with an apparent molecular weight of approximately 300,000, indicating perhaps dimer formation in this buffer.[47] At low concentrations, in lysine buffers, the molecular weight was determined to be approximately 138,000. Complex formation can occur with the zymogen, or with DFP-treated zymogen or enzyme.

The activator complex will activate both bovine and human plasminogens[17,47] and will digest both casein[47] and human fibrin.[48] The activator complex contains a single DFP-sensitive residue in the active center which is probably the active site serine of the plasmin moiety.[49] Complete loss of both caseinolytic and bovine plasminogen activator activities accompanies DFP treatment of the complex.[49] The active plasmin moiety of the activator complex appears to be similar to native plasmin but the streptokinase moiety has been altered, probably degraded, but is still bound to the plasmin moiety.[50]

Specificity. Human plasmin is similar in specificity to trypsin in that it catalyzes the hydrolysis of α-amino substituted lysine and arginine esters.[51] It also, like trypsin, cleaves the Arg_{22}-Gly and Lys_{29}-Ala peptide bonds of the β-chain of oxidized bovine insulin.[52] In addition, the enzyme will cleave only lysine and arginine peptide bonds in human S-sulfo fibrinogen.[53,54] Only 50% of the available arginine and lysine peptide bonds of S-sulfo fibrinogen are cleaved while trypsin cleaves 80% of them.[55] In the amino terminal part of the human fibrinogen A chain containing the first 50 residues (cyanogen bromide fragment), plasmin cleaves one lysine peptide bond whereas trypsin cleaves all six available lysine and arginine bonds.[53,54]

[47] C.-M. Ling, L. Summaria, and K. C. Robbins, *J. Biol. Chem.* **242**, 1419 (1967).
[48] L. Summaria, G. H. Barlow, and K. C. Robbins, unpublished results.
[49] L. Summaria, C.-M. Ling, W. R. Groskopf, and K. C. Robbins, *J. Biol. Chem.* **243**, 144 (1968).
[50] L. Summaria, W. R. Groskopf, and K. C. Robbins, *Federation Proc.* **28**, 322 (1969).
[51] S. Sherry, N. Alkjaersig, and A. P. Fletcher, *Thromb. Diath. Haemorrhag.* **16**, 18 (1966).
[52] W. R. Groskopf, B. Hsieh, L. Summaria, and K. C. Robbins, *Biochim. Biophys. Acta* **168**, 376 (1968).
[53] B. Blombäck, M. Blombäck, B. Hessel, and S. Iwanaga, *Nature* **215**, 1445 (1967).
[54] S. Iwanaga, P. Wallén, N. J. Gröndahl, A. Henschen, and B. Blombäck, *Biochim. Biophys. Acta* **147**, 606 (1967).
[55] P. Wallén and S. Iwanaga, *Biochim. Biophys. Acta* **154**, 414 (1968).

Kinetic Properties. The Michaelis constant, K_m, with p-tosyl-L-arginine methyl ester (TAME) as the substrate, was determined to be 0.021 M[4] and 0.012 M.[51] With bovine fibrinogen, the K_m was determined to be approximately $3 \times 10^{-5}\,M$.[56,57] The velocity constant, V, with TAME, was found to be 1700 ± 200 moles per minute per mole of enzyme[4] (approximately 0.7 micromoles per minute per casein unit).

Distribution. Plasminogen is a zymogen found in the blood plasma of mammals. Man, monkey, and cat plasminogens can be activated with low concentrations of streptokinase whereas dog and rabbit plasminogens are activated with high concentrations of streptokinase; cow, sheep, pig, mouse, and rat plasminogens are not activated with streptokinase.[58] Urokinase and the equimolar streptokinase-human plasmin complex (streptokinase-human globulin mixtures) show no species specificity and activate all species indicated above.[58] Highly purified bovine plasminogen[59] and dog plasminogen[60] have been prepared. Partially purified plasminogens have been prepared from cat, monkey, rabbit, sheep, pig, mouse, and rat plasmas.[58]

The concentration of plasminogen in normal human plasma determined by a radioimmunoassay method was found to be 206 ± 36 μg per milliliter for adults (male and female) and 90 ± 24 μg per milliliter for infants, ages 0–10 days.[61] Using a single radial immunodiffusion method, the human plasma plasminogen concentration was found to be 170 μg/ml (range 150 to 200 μg/ml).[62] With a fibrinogenolytic assay, the concentration of plasminogen in human plasma determined to be approximately 170 μg/ml.[57]

[56] A. F. Bickford, F. B. Taylor, Jr., and R. Sheena, *Biochim. Biophys. Acta* **92**, 328 (1964).
[57] L. B. Nanninga and M. M. Guest, *Thromb. Diath. Haemorrhag.* **19**, 492 (1968).
[58] R. J. Wulf and E. T. Mertz, *Can. J. Biochem.* **47**, 927 (1969).
[59] J. Y. S. Chan and E. T. Mertz, *Can. J. Biochem.* **44**, 475 (1966).
[60] P. J. Heberlein and M. I. Barnhart, *Biochim. Biophys. Acta* **168**, 195 (1968).
[61] S. F. Rabiner, I. D. Goldfine, A. Hart, L. Summaria, and K .C. Robbins, *J. Lab. Clin. Med.* **74**, 265 (1969).
[62] H. G. Schwick, *Z. Anal. Chem.* **243**, 424 (1968).

[11] The Subtilisins

By Martin Ottesen and Ib Svendsen

Nomenclature

The subtilisins are serine proteases originating from strains of *Bacillus subtilis* or related bacteria. The nomenclature of these enzymes has been changed several times and this has caused considerable confusion in the literature. Originally, the term subtilisin was suggested as the name of the plakalbumin-forming enzyme produced by a *B. subtilis* strain,[1] but this name is now also commonly used for the alkaline protease isolated by Hagihara[2] from *B. subtilis* strain N', and for a bacterial protease isolated at the Novo Pharmaceutical Co., Copenhagen.[3] According to the nomenclature suggested by the Enzyme Commission,[4] these enzymes all belong to the category EC 3.4.4.16 with the recommended trivial name subtilopeptidase A.

In the present treatment, the nomenclature suggested by E. L. Smith[5] for this group of enzymes will be followed: the original subtilisin will be called *subtilisin Carlsberg*, the enzyme isolated by Hagihara and co-workers is called *subtilisin BPN'*, and the enzyme from the Novo Company is called *subtilisin Novo*. Some of the different names which have been used for this group of enzymes are listed in Table I.

The presently available evidence indicates the amino acid sequence

TABLE I
Some Synonyms Used for the Various Subtilisins

Enzyme	Synonyms	Manufacturer[a]
Subtilisin Carlsberg	Subtilisin Subtilisin A Subtilopeptidase A Alcalase Novo	Novo Pharmaceutical Co. Copenhagen, Denmark
Subtilisin BPN'	Nagarse proteinase Nagarse BPN' Subtilopeptidase C	Nagase & Co., Ltd., Ohama, Amagasaki, Japan
Subtilisin Novo	Bacterial proteinase Novo Subtilisin B Subtilopeptidase B	Novo Pharmaceutical Co. Copenhagen, Denmark

[a] Sigma Chemical Co., St. Louis, Mo., lists a bacterial protease type VII, which according to amino acid composition and enzymatic specificity corresponds to subtilisin BPN' or Novo, and a bacterial protease type VIII which corresponds to subtilisin Carlsberg.

Mann Research Laboratories, New York, lists a crystalline subtilisin which corresponds to subtilisin BPN' (I. Svendsen, unpublished observations).

[1] A. V. Güntelberg and M. Ottesen, *Compt. Rend. Trav. Lab. Carlsberg* 29, 36 (1954).
[2] B. Hagihara, *Ann. Rept. Sci. Works. Fac. Sci. Osaka Univ.* 2, 35 (1954).
[3] M. Ottesen and A. Spector, *Compt. Rend. Trav. Lab. Carlsberg* 32, 63 (1960).
[4] Report of the Commission on Enzymes of the International Union of Biochemistry. Macmillan (Pergamon), New York, 1961.
[5] R. J. DeLange and E. L. Smith, *J. Biol. Chem.* 243, 2134 (1968).

of subtilisin Novo to be completely identical with the sequence of subtilisin BPN' (see Fig. 1). However, until this has been definitely proven
we will as far as possible specify the subtilisin type used to obtain the
reported data.

Several other serine proteases of bacillary origin have been reported
in the literature. Among these the alkaline protease isolated by Tsuru
et al.[6] from B. subtilis var. amylosacchariticus, mentioned on page 215,
belongs to the subtilisin group. Another subtilisin-like protease was
isolated by Rappaport et al.[7,8] from a transformable strain of B. subtilis.
This protease has an unusually high molecular weight around 166,000
and it consists probably of several subunits. Studying the formation of
proteases in a transformable strain of B. subtilis, Boyer and Carlton[9]
obtained two enzymes which apparently differ from the well-known
subtilisins in amino acid composition. Keay and Moser[10] purified alkaline
proteases from different species and strains of bacillus, and classified
them according to amino acid composition, enzymatic activity, and immunological properties. They found only two types of proteases: the
enzymes from B. pumilis and B. licheniformis were similar to subtilisin
Carlsberg, while the enzymes from B. subtilis var. amylosacchariticus
and B. subtilis NRRL B3411 were related to subtilisin BPN' or Novo.

In addition to the serine proteases, many strains of B. subtilis produce
neutral proteases, but these enzymes are metalloenzymes, and they are
apparently not related to the subtilisin group of enzymes.

The subtilisins have found an important industrial application as
additives to household detergents used in washing machines, and some
of these enzymes are now prepared on very large scale.

Assay Methods

Esterolytic Activity

Principle. The most convenient method for the assay of esterolytic
activity is the potentiometric titration at constant pH of the hydrogen
ions liberated during the hydrolysis of ester bonds. A variety of ester
substrates are hydrolyzed by the subtilisins, but toluenesulfonyl arginine
methyl ester (TAME) is a preferable choice since this compound is
easily soluble in water.

[6] D. Tsuru, H. Kira, T. Yamamoto, and J. Fukumoto, *Agr. Biol. Chem.* **30,** 1261
(1966).
[7] H. P. Rappaport, W. S. Riggsby, and D. A. Holden, *J. Biol. Chem.* **240,** 78 (1965).
[8] W. S. Riggsby and H. P. Rappaport, *J. Biol. Chem.* **240,** 87 (1965).
[9] H. W. Boyer and B. C. Carlton, *Arch. Biochim. Biophys.* **128,** 442 (1968).
[10] L. Keay and P. W. Moser, *Biochem. Biophys. Res. Commun.* **34,** 600 (1969).

Reagents

TAME (0.025 M) in H_2O. Stable when kept in the refrigerator.
KCl (1 M)
Enzyme Solutions. Subtilisin Carlsberg (approx. 1 mg/ml) or subtilisin BPN' (approx. 10 mg/ml) freshly prepared in a 10^{-2} M phosphate buffer, pH 7–8.

Procedure. To a pH-stat vessel thermostatted at 30° are added 1 ml of KCl solution and 1–9 ml of TAME solution. Water is added to bring the total volume to 10 ml. After equilibrating 5 minutes, the end point of the pH-stat is set to pH 8.0 and the pH is adjusted to this value with 0.1 N NaOH in the burette. The reaction is initiated by the addition of 50 μl of the enzyme solution, and the reaction is allowed to proceed until the recorder has plotted sufficient of the curve to permit the determination of the initial slope. At a TAME concentration of 0.0225 M the initial rate of H^+-release should be 44 μeq/minute/mg subtilisin Novo per milliliter and 580 μeq/minute/mg subtilisin Carlsberg per milliliter. The Michaelis-Menten kinetics is followed and from Lineweaver-Burk plots the following parameters should be obtained for pure subtilisin Carlsberg:

$$K_m = 3.7 \times 10^{-2}\ M, V_{max} = 117\ \text{sec}^{-1}$$

and for pure subtilisin Novo:

$$K_m = 3.0 \times 10^{-2}\ M, V_{max} = 7\ \text{sec}^{-1}$$

Proteolytic Activity. The proteolytic activity of commercial preparations of subtilisins is frequently expressed in Anson units[11] using urea-denatured hemoglobin as substrate. The degradation of casein to trichloroacetic acid-soluble products is also a convenient method to determine the activity of subtilisin preparations,[12] and in the Carlsberg Laboratory the splitting of the very basic protein substrate clupein has been used as a routine method.[3] The degradation was followed in the pH-stat at pH 8.0 and 30°. As a quick method to determine the approximate activity of culture beers, the milk-clotting test can be used.[13]

Isolation Procedures

Subtilisin Carlsberg was obtained from the nutrient medium in which a strain of *B. subtilis* had been grown. The bacterial strain used in the initial studies of the plakalbumin formation did not produce a very active

[11] J. H. Northrop, M. Kunitz, and R. M. Herriott, "Crystalline Enzymes." Columbia University Press, New York, 1948.
[12] B. Hagihara, H. Matsubara, M. Nakai, and K. Okunuki, *J. Biochem.* (*Tokyo*) **45**, 185 (1958).
[13] A. V. Güntelberg, *Compt. Rend. Trav. Lab. Carlsberg* **29**, 27 (1954).

enzyme preparation. However, when the bacteria were spread on an agar plate, colonies could be found which produced a higher amount of enzyme. After a long series of spreading and selections, a strain was obtained which produced a high amount of enzyme.[13] This bacteria was a motile, gram-positive rod, forming a thick, deeply wrinkled pellicle on glucose broth. The culture had lost its ability to form spores, and the strain was not stable. It could be preserved only as long as the repeated spreadings and selections were carried out.

The culture medium contained, besides glucose and salts, casein which had been digested to peptone with the bacterial enzyme. Vigorous aeration and use of an antifoam agent was required, and glucose was added during the growth. From the best bacterial colonies it was possible to obtain an enzyme yield corresponding to more than 2 g of crystalline enzyme per liter culture liquid. After removal of the bacteria by centrifugation and the antifoam agent by carbon tetrachloride extraction, the enzyme was precipitated by addition of 250 g Na_2SO_4 per liter plus some Hyflo Supercel to facilitate filtration. The salt-containing filter cake was dried in a desiccator.

The enzyme was isolated from the crude powder after dissolution in water and addition of $CaCl_2$ to make a solution of 0.1 M in Ca^{2+}.[1] Addition of 1 volume of acetone at room temperature precipitated a dark, mucinous impurity which was removed by centrifugation. Addition of acetone to 75% by volume precipitated the enzyme. The precipitate was collected by centrifugation and most of the acetone was removed by a stream of air before the precipitate was dissolved in water. To induce crystallization, the pH should be decreased to the lowest range of the stability range, i.e., pH 5.5. To avoid overshooting, this adjustment was done with an acetate buffer at pH 4.9. Five grams of Na_2SO_4 were added per 100 ml and the solution left at 30° with mechanical stirring. More Na_2SO_4 was slowly added, and if required, some seed crystals. Within 10 to 30 hours at a Na_2SO_4 concentration of 12 g/100 ml, practically all the enzyme had crystallized out. The redissolution, which was very slow, was best performed in water at 0° with slow mechanical stirring. Recrystallizations were performed in the same manner, but due to autolysis the purity was only slightly increased, while considerable loss of activity occurred. After dialysis against distilled water, a stable product could be obtained by lyophilization, although the autolysis during the dialysis decreased the specific activity by 10–30%. Not all enzyme preparations could be crystallized easily, and it was observed that it was important to harvest the bacterial culture at a relatively early stage, i.e., before the maximal yield was obtained and before autolysis products of the bacteria had seriously contaminated the culture liquid.

Subtilisin BPN' was isolated by Hagihara *et al.*[2,12] from the culture liquid of a bacterial strain, *B. subtilis* N', which had also been carefully selected. The identity of this strain as a *B. subtilis* was later questioned by Welker and Campbell,[14] who found that it should be classified as a species of *B. amyloliquefaciens*.

The bacteria were grown in submerged culture, and the crude enzyme was precipitated with ammonium sulfate. To purify the enzyme, the precipitate was dissolved in water, filtered, and dialyzed against $0.05 M$ sodium phosphate buffer, pH 6.5.[12] The dialyzed solution was adsorbed on a column of a sulfonic acid type of resin, Duolite C-10, and after washing with water and acetone, the enzyme was eluted with a phosphate buffer containing $0.5 M$ NaCl. The enzyme-rich fractions of the eluate were pooled, and ammonium sulfate added to 40 g/100 ml. The enzyme-containing precipitate was collected by filtration after addition of Hyflo Supercel. Most of the ammonium sulfate was washed out from the precipitate with 75% acetone at 40°, and the residue was then suspended in $0.02 M$ sodium acetate. After removal of the Supercel by filtration, and after washing with the sodium acetate solution, the enzyme was precipitated from the combined filtrate and washings by addition of 2.5 volume of acetone at 5–10°. The precipitate was collected by centrifugation, and acetone was removed with a stream of air until the precipitate had almost dissolved. Any insoluble material was removed, and to the concentrated enzyme solution, containing more than 10% of enzyme, acetone was added dropwise until turbidity. Crystallization was allowed to start at 15°, the solution was then cooled to 10° for 2 hours, and finally left at 4–7° for 2 to 3 days. During this time large amounts of needle-like crystals were formed. After washing with 55% acetone and dissolution in $0.02 M$ sodium acetate, recrystallization could be performed as described.

Commercial Subtilisin Preparations. Since the bacterial strains used to produce the subtilisins are not easily available, most investigators are using commercially prepared subtilisin preparations. As far as we are informed, these products are produced according to methods similar to those just described, and they are normally available as lyophilized powders prepared from dialyzed solutions of the crystalline material. These enzymes consistently contain a relatively large percentage of peptide material, sometimes up to 40% peptides, which probably has originated from the autolytic degradation occurring during the dialysis.

The purity of an enzyme preparation can easily be estimated from a determination of the enzymatic activity, or with somewhat greater pre-

---

[14] N. E. Welker and L. L. Campbell, *J. Bacteriol.* **94,** 1124 (1967).

cision from a determination of the operational normality, for instance with *N-trans*-cinnamoyl imidazole titration according to Bender *et al.* (see this volume [1]).

For many applications the presence of the peptide impurities is without significance. However, in cases where the peptide impurities are troublesome, they can be removed simply by a gel filtration on Sephadex G-25 or G-50.

Example. One hundred milligrams of subtilisin are dissolved in 2 ml of 0.01 M phosphate buffer, pH 7.0, and in the cold room applied to the top of a 2 $\times$ 25 cm column of Sephadex G-25 equilibrated with the same buffer. This buffer is also used for the elution with a flow rate around 30 ml/hour, and 3-ml fractions are collected. The ultraviolet absorption is used to locate the enzyme-containing fractions. The purified enzyme should be kept at low temperature and used without delay since the peptide impurities will reappear rapidly. After 0.5–1 hour at room temperature or 1–2 hours at 10° the peptide content is again around 5%.[15]

Chemical and Physical Properties

Primary Structures

The following amino acid compositions have now been established for subtilisin BPN' or Novo: Lys_{11} His_6 Arg_2 Asp_{11} Asn_{17} Thr_{13} Ser_{37} Glu_4 Gln_{11} Pro_{14} Gly_{33} Ala_{37} Val_{30} Met_5 Ile_{13} Leu_{15} Tyr_{10} Phe_3 Trp_3, and for subtilisin Carlsberg: Lys_9 His_5 Arg_4 Asp_9 Asn_{19} Thr_{19} Ser_{32} Glu_5 Gln_7 Pro_9 Gly_{35} Ala_{41} Val_{31} Met_5 Ile_{10} Leu_{16} Tyr_{13} Phe_4 Trp_1 corresponding to molecular weights of 27,532 and 27,287, respectively.[5, 16]

The subtilisins consist of single peptide chains, they are devoid of cysteine and cystine, and no phosphorus, carbohydrate, or metals have been found in these molecules.

The amino acid sequences of subtilisin BPN' and Carlsberg as determined by E. L. Smith and co-workers[16, 17] are listed in Fig. 1.

The peptides produced by tryptic digestion of subtilisin Novo have been investigated by peptide mapping[18] and by chromatographic separation and analysis.[19] From these findings the Novo enzyme appears to be identical in all respects to subtilisin BPN', although minor differences

---

[15] M. Ottesen, unpublished observations.

[16] E. L. Smith, R. J. DeLange, W. H. Evans, M. Landon, and F. S. Markland, *J. Biol. Chem.* **243**, 2184 (1968).

[17] E. L. Smith, F. S. Markland, C. B. Kasper, R. J. DeLange, M. Landon, and W. H. Evans, *J. Biol. Chem.* **241**, 5974 (1966).

[18] J. A. Hunt and M. Ottesen, *Biochim. Biophys. Acta* **48**, 411 (1961).

[19] S. A. Olaitin, R. J. DeLange, and E. L. Smith, *J. Biol. Chem.* **243**, 5296 (1968).

Ala-Gln-Thr-Val-Pro-Tyr-Gly-Ile-Pro-Leu-Ile-Lys-Ala-Asp-Lys-Val-Gln-Ala-Gln-Gly-
Ala-Gln-Ser-Val-Pro-Tyr-Gly-Val-Ser-Gln-Ile-Lys-Ala-Pro-Ala-Leu-His-Ser-Gln-Gly-
10   20

Phe-Lys-Gly-Ala-Asn-Val-Lys-Val-Ala-Val-Leu-Asp-Thr-Gly-Ile-Gln-Ala-Ser-His-Pro-
Tyr-Thr-Gly-Ser-Asn-Val-Lys-Val-Ala-Val-Ile-Asp-Ser-Gly-Ile-Asp-Ser-Ser-His-Pro-
30   40

Asp-Leu-Asn-Val-Val-Gly-Gly-Ala-Ser-Phe-Val-Ala-Gly-Glu-Ala-■-Tyr-Asn-Thr-Asp-
Asp-Leu-Lys-Val-Ala-Gly-Gly-Ala-Ser-Met-Val-Pro-Ser-Glu-Thr-Pro-Asn-Phe-Gln-Asp-
50   60

Gly-Asn-Gly-His-Gly-Thr-His-Val-Ala-Gly-Thr-Val-Ala-Ala-Leu-Asp-Asn-Thr-Thr-Gly-
Asp-Asn-Ser-His-Gly-Thr-His-Val-Ala-Gly-Thr-Val-Ala-Ala-Leu-Asn-Asn-Ser-Ile-Gly-
70   80

Val-Leu-Gly-Val-Ala-Pro-Ser-Val-Ser-Leu-Tyr-Ala-Val-Lys-Val-Leu-Asn-Ser-Ser-Gly-
Val-Leu-Gly-Val-Ala-Pro-Ser-Ser-Ala-Leu-Tyr-Ala-Val-Lys-Val-Leu-Gly-Asp-Ala-Gly-
90   100

Ser-Gly-Ser-Tyr-Ser-Gly-Ile-Val-Ser-Gly-Ile-Glu-Trp-Ala-Thr-Thr-Asn-Gly-Met-Asp-
Ser-Gly-Gln-Tyr-Ser-Trp-Ile-Ile-Asn-Gly-Ile-Glu-Trp-Ala-Ile-Ala-Asn-Asn-Met-Asp-
110   120

Val-Ile-Asn-Met-Ser-Leu-Gly-Gly-Ala-Ser-Gly-Ser-Thr-Ala-Met-Lys-Gln-Ala-Val-Asp-
Val-Ile-Asn-Met-Ser-Leu-Gly-Gly-Pro-Ser-Gly-Ser-Ala-Ala-Leu-Lys-Ala-Ala-Val-Asp-
130   140

Asn-Ala-Tyr-Ala-Arg-Gly-Val-Val-Val-Val-Ala-Ala-Ala-Gly-Asn-Ser-Gly-Asn-Ser-Gly-
Lys-Ala-Val-Ala-Ser-Gly-Val-Val-Val-Val-Ala-Ala-Ala-Gly-Asn-Glu-Gly-Ser-Thr-Gly-
150   160

Ser-Thr-Asn-Thr-Ile-Gly-Tyr-Pro-Ala-Lys-Tyr-Asp-Ser-Val-Ile-Ala-Val-Gly-Ala-Val-
Ser-Ser-Ser-Thr-Val-Gly-Tyr-Pro-Gly-Lys-Tyr-Pro-Ser-Val-Ile-Ala-Val-Gly-Ala-Val-
170   180

Asp-Ser-Asn-Ser-Asn-Arg-Ala-Ser-Phe-Ser-Ser-Val-Gly-Ala-Glu-Leu-Glu-Val-Met-Ala-
Asp-Ser-Ser-Asn-Gln-Arg-Ala-Ser-Phe-Ser-Ser-Val-Gly-Pro-Glu-Leu-Asp-Val-Met-Ala-
190   200

Pro-Gly-Ala-Gly-Val-Tyr-Ser-Thr-Tyr-Pro-Thr-Asn-Thr-Tyr-Ala-Thr-Leu-Asn-Gly-Thr-
Pro-Gly-Val-Ser-Ile-Gln-Ser-Thr-Leu-Pro-Gly-Asn-Lys-Tyr-Gly-Ala-Tyr-Asn-Gly-Thr-
210   220

Ser-Met-Ala-Ser-Pro-His-Val-Ala-Gly-Ala-Ala-Ala-Leu-Ile-Leu-Ser-Lys-His-Pro-Asn-
Ser-Met-Ala-Ser-Pro-His-Val-Ala-Gly-Ala-Ala-Ala-Leu-Ile-Leu-Ser-Lys-His-Pro-Asn-
230   240

Leu-Ser-Ala-Ser-Gln-Val-Arg-Asn-Arg-Leu-Ser-Ser-Thr-Ala-Thr-Tyr-Leu-Gly-Ser-Ser-
Trp-Thr-Asn-Thr-Gln-Val-Arg-Ser-Ser-Leu-Gln-Asn-Thr-Thr-Thr-Lys-Leu-Gly-Asp-Ser-
250   260

Phe-Tyr-Tyr-Gly-Lys-Gly-Leu-Ile-Asn-Val-Glu-Ala-Ala-Ala-Gln-
Phe-Tyr-Tyr-Gly-Lys-Gly-Leu-Ile-Asn-Val-Gln-Ala-Ala-Ala-Gln
270   275

have not yet been completely excluded. As in the mammalian serine proteases, a serine residue (No. 221) and a histidine residue (No. 64) are involved in the enzymatic activity, but apart from this, the subtilisins show hardly any resemblances to the sequence of the mammalian enzymes. The two subtilisins differ from one another in 84 positions of the 275 residues in the peptide chain, and in a deletion at position 56 in the Carlsberg enzyme. It is characteristic for these enzymes that they show many repetitions in the sequence, and it has been suggested that they have evolved by a process of duplication of parts of the sequences resulting in the extension of a shorter peptide chain to the present longer ones.[16]

Secondary and Tertiary Structure

The difficulties connected with the removal of peptide impurities from subtilisin preparations has complicated the determination of physico-chemical constants for these substances, and in some cases it has been preferred to make the investigations with the inactivated diisopropyl-phosphofluoridate (DFP) derivatives of the enzymes. The data listed in Tables II and III should therefore be taken with some reservation. Sedimentation diffusion and viscosity data indicate the subtilisins as compact, almost spherical molecules, and the optical rotation data

TABLE II
PHYSICOCHEMICAL CONSTANTS FOR SUBTILISINS

	Subtilisin Carlsberg	Subtilisin BPN'	Subtilisin Novo
Isoelectric points	9.4^a	7.8^b	9.1^a
$S_{20,w}^0$ (Svedberg units)	2.85^a	2.72^b	2.81^a
$D_{20,w}$ (cm²/second)	—	9.04×10^{-7}	—
Intrinsic viscosity^c	3.2	—	3.6
$E_{280 nm}^{1 mg/ml}$	$0.96^{d,e}$	1.17^b	1.17^d

^{a} M. Ottesen and A. Spector, Compt. Rend. Trav. Lab. Carlsberg 32, 63 (1960).
^{b} H. Matsubara, C. B. Kasper, D. M. Brown, and E. L. Smith, J. Biol. Chem. 240, 1125 (1965). The DFP enzyme was used.
^{c} M. Ottesen, unpublished data.
^{d} J. T. Johansen, unpublished data.
^{e} The previously published value of 0.86 was apparently too low due to contamination by salt remaining from the dialysis.

FIG. 1. The amino acid sequences of subtilisin Carlsberg (upper line) and subtilisin BPN' (lower line). Amino acid residues in identical positions are underlined in order to stress the homology. ■ denotes a deletion in position 56 of subtilisin Carlsberg. [From E. L. Smith, R. J. DeLange, H. Evans, M. Landon, and F. S. Markland, J. Biol. Chem. 243, 2184 (1968).]

TABLE III

Optical Rotation of the Subtilisins

	$[\alpha]_{589}^{10°}$	$[\alpha]_{578}^{20°}$	$[\alpha]_{546}^{20°}$	$[\alpha]_{436}^{20°}$	$[\alpha]_{364}^{20°}$	From Drude plot in nm	b_0 From Moffitt plot
Subtilisin Carlsberg[a]	—	−53°	−62°	−118°	−209°	—	appr. −140°
Subtilisin BPN′[b]	−63	—	—	—	—	246	—
Subtilisin Novo[a]	—	−77°	−89°	−163°	−283°	—	appr. −125°

[a] M. Ottesen, unpublished values based on peptide-free subtilisins.

[b] I. Fuke, H. Matsubara, and K. Okunuki, *J. Biochem.* (*Tokyo*) **46**, 1513 (1959).

(Table III) plotted according to the Moffitt equation indicate a relatively low helix content about 20%. There are some discrepancies in the values for the isoelectric points (Table II), although the determinations are based on free electrophoresis using buffers of similar nature. Furthermore, the experimentally determined values are considerably higher than the isoelectric points calculated from the amino acid composition, i.e., about pH 6–7, suggesting specific binding of cations, or the specific binding of cationic peptides arising from autodigestion.

Our understanding of the subtilisin structure has recently been improved greatly through the X-ray crystallographic investigations of Wright *et al.*[20] who determined the structure of subtilisin BPN′ at a resolution of 2.5 Å.

As expected, the molecule is close to spherical, with a diameter around 42 Å, and the core of the molecule is composed almost entirely of amino acid residues with hydrophobic side chains. The molecule contains approximately 30% of somewhat distorted α-helix distributed in 8 segments, ranging in length from 6 to 16 residues. Four of the longer helical stretches are approximately parallel to one another and they have the same N-terminal to C-terminal sense. Five extended peptide chain segments form a twisted parallel-chain pleated sheet.

The active center serine-221 is located in a shallow groove in the enzyme surface. Its OH group is in a position consistent with the formation of a hydrogen bond to the neighboring histidine-64. On the other side of histidine-64 a carboxylate group is located within hydrogen-bonding distance (aspartic acid-32), and the steric arrangement of these three important residues is closely related to the corresponding arrangement at the active sites of the pancreatic proteases.[21]

The differences between the Carlsberg enzyme and subtilisin BPN′

[20] C. S. Wright, R. A. Alden, and J. Kraut, *Nature* **221**, 235 (1969).

[21] D. M. Blow, J. J. Birktoft, and B. S. Hartley, *Nature* **221**, 337 (1969).

practically all occur in the exterior chain segments, so that the residues determining the interior spatial conformation of subtilisin BPN′ are found in identical positions in subtilisin Carlsberg. On this basis Wright et al.[20] suggested that the two subtilisins must have very similar three-dimensional structures, and this is strongly supported by the similarities in physicochemical properties.

The subtilisin molecules exhibit rather interesting stability properties. They are not completely stable at any pH value, although a solution containing 10–20% autolysis peptides can be kept with constant activity for many hours, around neutral pH and at low temperature. At low pH values, i.e., around pH 1–2, the subtilisins are denatured, while at an intermediary pH around 4, where both active and denatured subtilisin are present simultaneously, a rapid autolysis will take place. Actually, a slow adjustment of the pH of a subtilisin solution from neutral pH to pH 2 will probably not result in the formation of a solution of denatured protein, but in a complex mixture of autolysis peptides. At high pH values, above pH 11, there is also a destabilization of the subtilisin structure. From studies of the succinylated subtilisin Novo, Gounaris and Ottesen[22] could demonstrate that the net charge had no influence on the acid denaturation, and they suggested that the acid instability could be due to protonation of masked histidine residues in the interior, hydrophobic parts of the structure. This prediction agrees well with the model of Wright et al.[20] who found that 4 of the 6 histidines were either inside the structure or in surface crevices.

The heat denaturation of subtilisin BPN′ was studied by Fuke[23] who found it to be rapid above 55°, while the denaturation of the DFP-inhibited enzyme is somewhat slower, again indicating that autodigestion plays a significant role at the stability limit of this enzyme.

The subtilisins are relatively resistant to denaturation by urea and guanidine solutions, and even in solution of $8\,M$ urea enzymatic activity is retained for some time,[15,24] while $4\,M$ urea has practically no effect on the enzymes.

From the location of the majority of the polar amino acid side chains at the surface of the subtilisin BPN′, it would be expected that they were available for chemical modification procedures. However, this is not always the case. The three tryptophans are all located at the surface of the molecular model, but they are unavailable to reaction with Koshland's reagent.[25] The 10 tyrosine residues are also located at the surface,

[22] A. D. Gounaris and M. Ottesen, Compt. Rend. Trav. Lab. Carlsberg 35, 37 (1965).
[23] I. Fuke, J. Biochem. (Tokyo) 53, 304 (1963).
[24] I. Fuke, H. Matsubara, and K. Okunuki, J. Biochem. (Tokyo) 46, 1513 (1959).
[25] K. E. Neet, A. Nanci, and D. E. Koshland, Jr., J. Biol. Chem. 243, 6392 (1968).

but in a spectrophotometric titration[26] only about 6 of these titrate with normal pK values, and they also show very different reactivities toward acylating, nitrating, and iodinating reagents. Among the lysines, one is partially buried while those remaining are free to react with acylating reagents.[22] Although some of the histidines are located at the surface, they all behave toward iodoacetate and similar reagents as if they were completely masked. More work is clearly required before the influence of the various local factors on the chemical reactivity of surface groups in a protein molecule is fully understood.

Activators and Inhibitors

Activators. The subtilisins have no requirement for cofactors.

Irreversible Inhibition. Subtilisin Carlsberg and BPN′/Novo are inhibited by diisopropylphosphofluoridate to less than 1% of the original activity[27,28] through reaction with the essential seryl residue 221.[17,29] Subtilisin Novo[30,31] and subtilisin Carlsberg[32] are also inhibited strongly by phenylmethanesulfonyl fluoride (PMSF) which also reacts with serine-221. The rate of this inactivation is between 3 and 4 times as fast with the Carlsberg enzyme as with the Novo enzyme.[33] With PMSF the enzymes are inhibited to less than 0.1% residual activity. The DFP and the PMSF-inhibited enzymes have been crystallized by methods similar to those described on pages 203–204.[20,27,28] The inactivation of the subtilisins by two other sulfonyl halides, dansyl chloride and 4,4′-biphenyl-enedisulfonyl chloride,[33] is much slower than the reaction with PMSF, but is not investigated in detail.

Although the subtilisins can split bonds which are attacked by trypsin and chymotrypsin, they do not react with TPCK or TLCK which are potent inhibitors of these enzymes. However, Shaw and Ruscica[34] found that the related compound benzyloxycarbonyl phenylalanine bromo-methyl ketone (ZPBK) slowly inactivates the subtilisins by reaction with an essential histidyl residue in the catalytic center, now identified as histidine-64.[35]

[26] F. S. Markland, *J. Biol. Chem.* **244**, 694 (1969).
[27] M. Ottesen and C. G. Schellman, *Compt. Rend. Trav. Lab. Carlsberg* **30**, 157 (1957).
[28] H. Matsubara and S. Nishimura, *J. Biochem. (Tokyo)* **45**, 503 (1958).
[29] F. Sanger and D. C. Shaw, *Nature* **187**, 872 (1960).
[30] L. Polgar and M. L. Bender, *J. Am. Chem. Soc.* **88**, 3153 (1966).
[31] K. E. Neet and D. E. Koshland, Jr., *Proc. Natl. Acad. Sci. U.S.* **56**, 1606 (1966).
[32] I. Svendsen, unpublished observations.
[33] A. O. Barel and A. N. Glazer, *J. Biol. Chem.* **243**, 1344 (1968).
[34] E. Shaw and J. Ruscica, *J. Biol. Chem.* **243**, 6312 (1968).

TABLE IV
SOME INHIBITORS OF THE SUBTILISINS

Inhibitor	Inhibitor conc. (M)	pH	$K_1(M)$ Carlsberg	Novo	BPN'
Indole[a]	0.00625	7.5	0.05	0.05	0.03
Phenol[a]	0.05	7.4	0.10	0.11	0.10
Hydrocinnamate[a]	0.025	8.0	0.14	0.30	0.34
4-(4'-Aminophenylazo) phenylarsonic acid[b]	5.67×10^{-5}	6.0	22×10^{-5}	5.7×10^{-5}	—

[a] A. N. Glazer, *J. Biol. Chem.* **242**, 433 (1967).
[b] A. N. Glazer, *Proc. Natl. Acad. Sci. U.S.* **59**, 996 (1968).

Reversible Inhibition. Indole, phenol, and hydrocinnamate act as competitive inhibitors of the subtilisins[36] (see Table IV). Other inhibitors are acetone, acetonitrile, cyclohexanone, cyclohexanol, and aliphatic alcohols.[32] Alcohols also act as acceptors in transesterification reactions.[37]

4-(4'-Aminophenylazo)phenylarsonic acid was found by Glazer[38] to be reversible, inhibiting both subtilisin Carlsberg and Novo (Table IV). The subtilisins are also inhibited by other phenylarsonates such as p-nitrophenylarsonate, p-tolylarsonate, p-arsanilate, and phenylarsonate.[39] Inhibition by these compounds is time dependent and dependent on an ionizable group on the enzyme with a pK_a of 7.17 (subtilisin Novo), presumably the active histidine.[39]

Matsubara and Nishimura[28] found subtilisin BPN' to be strongly inhibited by extracts from potatoes and broad beans, and Ryan[40] observed that the chymotrypsin inhibitor I from potatoes also inhibited subtilisin Carlsberg. The activity toward tyrosine ethyl ester could be completely abolished, while the milk-clotting capacity and the activity toward hemoglobin were inhibited less strongly. An inhibitor from egg white originally described by Matsushima[41] inhibits both subtilisin Carlsberg[42] and subtilisin BPN'.[43]

[35] F. S. Markland, E. Shaw, and E. L. Smith, *Proc. Natl. Acad. Sci. U.S.* **61**, 1440 (1968).
[36] A. N. Glazer, *J. Biol. Chem.* **242**, 433 (1967).
[37] A. N. Glazer, *J. Biol. Chem.* **241**, 635 (1966).
[38] A. N. Glazer, *Proc. Natl. Acad. Sci. U.S.* **59**, 996 (1968).
[39] A. N. Glazer, *J. Biol. Chem.* **243**, 3693 (1968).
[40] C. A. Ryan, *Biochemistry* **5**, 1592 (1966).
[41] K. Matsushima, *Science* **127**, 1178 (1958).
[42] Y. Tomimatsu, J. J. Clary, and J. J. Bartulovich, *Arch. Biochem. Biophys.* **115**, 536 (1966).
[43] M. B. Rhodes, N. Bennett, and R. E. Feeney, *J. Biol. Chem.* **235**, 1686 (1960).

Specificity

The subtilisins are capable of hydrolyzing both peptide bonds and ester bonds, and they are also effective in transesterification reactions[37] and transpeptidation reactions.[33]

Esterase Activity. Subtilisin Carlsberg hydrolyzes even the methyl esters of simple aliphatic carboxylic acids[1,44] and tripropionin, tributyrin, triacetin are also split by the subtilisins.[45] The broad specificity is also seen from the variety of esters of amino acids the subtilisins are able to degrade. As indicated by the kinetic parameters collected in Table V, the subtilisins have a preference for esters of aromatic amino acids, but care should be taken when the constants of different substrates are compared. Organic solvents like dioxane, acetonitrile, methanol, and ethanol are often added to the medium in order to facilitate the dissolution of sparingly soluble substrates, and these compounds act, as mentioned, as inhibitors of the subtilisins.

Peptidase Activity. A variety of bonds were split in the oxidized B-chain of insulin when subtilisin Carlsberg was allowed to act on this substrate for a long period of time.[46] However, when the initial stages of the degradation of the B-chain were investigated,[47,48] it was demonstrated that the bond Leu_{15}-Tyr_{16} was split much more rapidly than any other bond in the B-chain, both by subtilisin Carlsberg and subtilisin Novo. The great susceptibility of this bond to hydrolysis by the subtilisins is probably dictated by the sequence of amino acid residues on both sides of the bond. These residues are supposed to function in the attachment of the B-chain to the "secondary binding sites"[49] of the enzymes. The active site of subtilisin BPN' has at least an extension of 18 Å, and can be divided into at least 5 "subsites,"[50] which bind the various amino acid residues, as shown from the splitting of peptides of varying chain length and amino acid composition.

pH Optimum. When synthetic ester substrates are used, the activity of the subtilisins increases with pH in a manner indicating that the activity depends on a group with a pK of around 7, presumably histidine-64.[37] The pH optimum of the enzymes toward protein substrates

[44] J. Graae, *Acta Chem. Scand.* **8,** 356 (1954).

[45] G. Sierra, *Can. J. Microbiol.* **10,** 926 (1964).

[46] H. Tuppy, *Monatsh. Chem.* **84,** 996 (1953).

[47] J. T. Johansen, M. Ottesen, I. Svendsen, and G. Wybrandt, *Compt. Rend. Trav. Lab. Carlsberg* **36,** 365 (1968).

[48] K. Morihara and H. Tsuzuki, *Arch. Biochem. Biophys.* **129,** 620 (1969).

[49] I. Schechter and A. Berger, *Biochemistry* **5,** 3371 (1966).

[50] K. Morihara, T. Oka, and H. Tsuzuki, *Biochem. Biophys. Res. Commun.* **35,** 210 (1969).

TABLE V

KINETIC PARAMETERS FOR ESTER HYDROLYSIS BY SUBTILISINS

	Subtilisin Novo		Subtilisin Carlsberg		Subtilisin BPN'		
	K_M	V_{max}	K_M	V_{max}	K_M	V_{max}	Ref.
N-Acetyl tyrosine methyl ester	0.09	1560	0.07	1930			a
N-Acetyl tyrosine ethyl ester	0.07	731	0.09	1316			a
N-Acetyl phenylala- nine methyl ester	0.06	415	0.03	765			a
N-Acetyl tryptophan methyl ester	0.09	415	0.05	820			a
N-Acetyl valine ester	0.28	28	0.19	23			a
Toluenesulfonyl argi- nine methyl ester	0.03	7	0.04	117			b
Toluenesulfonyl argi- nine methyl ester	0.07	15.5	0.04	68.6			c
Benzoyl arginine ethyl ester	0.01	4.6	0.007	16			b
Benzoyl arginine ethyl ester	0.007	3.9	0.007	16.1			c
N-Acetyl alanine methyl ester					0.12	72	d
N-Acetyl leucine methyl ester					0.066	57.5	d
N-Acetyl phenylala- nine ethyl ester					0.017	30.6	d
N-Acetyl tyrosine ethyl ester					0.022	383.3	d
N-Acetyl tryptophan ethyl ester					0.024	35.8	d
N-Acetyl lysine methyl ester					0.091	47.4	d
Benzoyl arginine ethyl ester					0.010	3.1	d

[a] pH 8.0, 37°, 8% dioxane. A. O. Barel and A. N. Glazer, *J. Biol. Chem.* **243,** 1344 (1968).

[b] pH 8.0, 30°. I. Svendsen, *Compt. Rend. Trav. Lab. Carlsberg* **36,** 347 (1968).

[c] pH 8.0, 37°. E. L. Smith, F. S. Markland, and A. N. Glazer, *in* "Structure-Function Relationships of Proteolytic Enzymes" (P. Desnuelle, H. Neurath, and M. Ottesen, eds.) p. 160. Munksgaard, Copenhagen, 1970.

[d] pH 7.5, 30°, 4% ethanol. K. Morihara and H. Tsuzuki, *Arch. Biochem. Biophys.* **129,** 620 (1969).

varies and it depends on the time of incubation, the incubation temperature, and the substrate as well. With subtilisin Carlsberg the pH optimum was reported to be between 10 and 11 when casein was used as the substrate.[1] The pH optimum for the hydrolysis of denatured hemoglobin by subtilisin Novo was between 7 and 8,[51] in the same range as the gelatin splitting by BPN'.[52]

Chemical Modifications

Succinylation of 10 of the 11 lysyl residues in subtilisin Novo did not lead to any change in activity of the enzyme,[22] and in subtilisin Carlsberg 8 out of 9 lysyl residues could similarly be succinylated. Although succinylation led to changes in the activity of the Carlsberg enzyme, these changes were not due to the modification of the lysyl residues, but probably to the modification of certain seryl- and/or threonyl residues.[53] Carbamylation of subtilisin Novo[54] and Carlsberg led to partial inactivation of the enzymes, but reactivation was readily achieved by hydroxylamine.

Nitration and iodination of the tyrosyl residues of both subtilisin Carlsberg and Novo[55] changed the enzymatic properties of the enzymes. The rate of hydrolysis of small ester substrates was practically unaffected, while the splitting of the large peptide substrates clupein and gelatine was markedly increased at low substrate concentrations. These findings were explained by assuming that the lowering of the pK's of certain tyrosyl residues by nitration or iodination causes these residues to be charged at pH 8 and thus creates new "secondary binding sites."[56] The introduction of negative charges on the hydroxyamino acid residues by succinylation or glutarylation of the subtilisins influenced their enzymatic properties in a similar manner.[53]

Both Neet and Koshland[31] and Polgar and Bender[30] found that inactivation of subtilisin Novo with phenylmethanesulfonyl fluoride, followed by treatment of the product with thiolacetate, transformed the enzyme into thiol-subtilisin in which the active serine-221 was transformed into a cystein residue. Thiol-subtilisin was able to hydrolyze p-nitrophenyl acetate, N-*trans*-cinnamoyl imidazole, and p-nitrophenyl N-benzyloxycarbonyl glycinate, but it was inactive toward N-acetyl-

[51] "Novo Enzymes." Pamphlet published by Novo Industries, Copenhagen, Denmark.
[52] H. Matsubara, B. Hagihara, M. Nakai, T. Komaki, T. Yonetani, and K. Okunuki, *J. Biochem.* (*Tokyo*) **45**, 251 (1958).
[53] J. T. Johansen, *Compt. Rend. Trav. Lab. Carlsberg* **37**, 145 (1970).
[54] I. Svendsen, *Compt. Rend. Trav. Lab. Carlsberg* **36**, 235 (1967).
[55] I. Svendsen, *Compt. Rend. Trav. Lab. Carlsberg* **36**, 347 (1968).
[56] J. T. Johansen, M. Ottesen, and I. Svendsen, *Biochim. Biophys. Acta* **139**, 211 (1967).

L-tyrosine ethyl ester, N-acetyl-L-tryptophan methyl ester, and protein substrates.

Subtilisin Amylosaccariticus

An alkaline protease from *B. subtilis* var. *amylosaccariticus* has been isolated by Tsuru *et al.*[6,57] in crystalline form. This enzyme has now been designated subtilisin amylosaccariticus.[58] Great similarity is observed to BPN′ in enzymatic activity and physicochemical properties.[58] However, subtilisin amylosaccariticus is much less soluble than the two better-known subtilisins in the pH range 6 to 9, and V_{max} is independent of pH from 6 to 8 when TAME or ATEE are used as substrates, suggesting that no histidines are involved in the active site of this enzyme.[58] A redetermination of the amino acid composition gave the following values[58]: Lys_8 His_6 Arg_4 Asp_{27} Thr_{17} Ser_{44} Glu_{16} Pro_{13} Gly_{35} Ala_{32} Val_{27} Met_4 Ile_{16} Leu_{16} Tyr_{12} Phe_3 Trp_3, and sequence studies are now being carried out.[59]

[57] D. Tsuru, H. Kira, T. Yamamoto, and J. Fukumoto, *Agr. Biol. Chem.* **31**, 330 (1967).
[58] E. L. Smith, F. S. Markland, and A. N. Glazer, *in* "Structure–Function Relationships of Proteolytic Enzymes" (P. Desnuelle, H. Neurath, and M. Ottesen, eds.) p. 160. Munksgaard, Copenhagen, 1970.
[59] E. L. Smith, personal communication.

[12] Thiolsubtilisin (An Artificial Enzyme)

By MANFRED PHILIPP, LASZLO POLGAR, and MYRON L. BENDER

Thiolsubtilisin is a hydrolase that is active toward *p*-nitrophenyl esters and imidazole amides.[1] It is synthesized from natural subtilisin by chemical substitution of a thiol group for a hydroxyl group on the active site serine. In contrast to natural subtilisin, it is inactive toward other ester and amide substrates,[2] even though the rest of the primary structure and, according to crystal structure work, all of the tertiary structure is maintained.[3]

This chapter will deal with the preparation of thiolsubtilisin BPN′.

[1] L. Polgar and M. L. Bender, *J. Am. Chem. Soc.* **88**, 3153 (1966).
[2] K. E. Neet and D. E. Koshland, Jr., *Proc. Natl. Acad. Sci. U.S.* **56**, 1606 (1966).
[3] J. Kraut, R. A. Alden, C. S. Wright, and F. C. Westall, Abstract No. 63, Division of Biological Chemistry, 158th National Meeting, American Chemical Society, New York, 1969.

A similar preparation of thiolsubtilisin Carlsberg is given in an article by L. Polgar.[3a]

Assay Method[4]

Principle. The method is based on the fact that thiolsubtilisin catalyzes the hydrolysis of *p*-nitrophenyl acetate (PNPA). The hydrolysis is followed in the visible region, and, under the conditions used, the enzymatic reaction rate for the greater part of the reaction is constant and directly proportional to the enzyme concentration.

Confirmation that the activity measured is due to thiolsubtilisin is done using *p*-mercuribenzoate (PMB), which inhibits thiolsubtilisin, but not the enzyme from which it was made.

Reagents

> Phosphate buffer, 0.01 M, pH 6.5
> Phosphate buffer, 0.10 M, pH 6.5
> Phosphate buffer, 0.05 M, pH 6.5
> Phosphate buffer, 0.10 M, pH 7.5
> *p*-Nitrophenyl acetate (PNPA) ($3 \times 10^{-3} M$) in acetonitrile
> Novo or nagarse subtilisin (EC 3.4.4.16)
> Potassium thioacetate
> Phenylmethanesulfonyl fluoride (PMSF), 8.8 mg/ml in dioxane
> *p*-Chloromercuribenzoic acid (PMB) (1.15 mg/ml) in pH 7.5 buffer

The phosphate for the buffers is analytical reagent grade, used without further purification. The *p*-nitrophenyl acetate is an Aldrich product, recrystallized four times from absolute ethanol and stored in a vacuum dessicator. Potassium thioacetate is an Eastman product, which was purified by dissolving the salt in water and filtering off the colored insoluble impurities with the aid of carbon black. The colorless solution was concentrated in a rotary evaporator until white crystals appeared. These were washed with ether and stored under vacuum in the dark. Phenylmethanesulfonyl fluoride was a Calbiochem product and was used without further purification. Dioxane was reagent grade, and the acetonitrile was Mallinckrodt "Nanograde" material.

Assay Procedure. Three milliliters of a 0.1 M phosphate, pH 7.5, buffer are pipetted into a 10 mm glass or silica cuvette positioned in a spectrophotometer recording absorbance at 400 nm. To this is added 0.100 ml of the *p*-nitrophenyl acetate solution, and the liquids stirred. The temperature of the buffers will be $25.0 \pm 0.2°$. After observing the spontaneous

[3a] L. Polgar, *Acta Biochim. Biophys. Hung.* **3**, 397 (1968).
[4] L. Polgar and M. L. Bender, *Biochemistry* **6**, 610 (1967).

hydrolysis rate for a time sufficient to obtain an initial rate for it, 0.100 ml of a thiolsubtilisin solution is added, and the mixture stirred again. In this procedure it is important that in the initial stirring, little if any of the solution be lost, as any loss will slightly increase the actual enzyme concentration, when it is added later, and will increase it in a nonreproducible manner. If the enzyme is reasonably pure, a solution which has an absorbance at 278 nm of about 5.0 (or contains about 4 mg/ml) will give a reaction rate that is easily followed.

This procedure is then repeated, but now 0.100 ml of the PMB solution is added together with the PNPA solution, and again the resulting spontaneous and enzymatic rate measured. If purified thiolsubtilisin is being used, the enzymatic rate now should be negligible. In all cases it is advantageous to reverse the order of addition to the buffer solution, once the spontaneous hydrolysis rate is known.

It has been shown that, under the conditions used here, the steady-state velocity observed in M/sec is equal to 6.15×10^{-3} sec^{-1} (footnote 4) times the molar enzyme concentration. This was determined using the concentration of SH-active sites indicated by a PMB titration and was corrected for any non-PMB inhibitable activity.

When the assays are performed, it will be noted that before the steady-state reaction is established, a faster first-order pre-steady-state reaction, representing acylation of the enzyme, is observed. The rate to be used is the zero-order rate following this. This can be seen in the linear portion of Fig. 1.

This rate is measured in absorbance units per second and can be converted to M sec^{-1} using the $\Delta\epsilon_{400}$ of p-nitrophenol at pH 7.5, 1.36×10^4 abs. M^{-1}.[5]

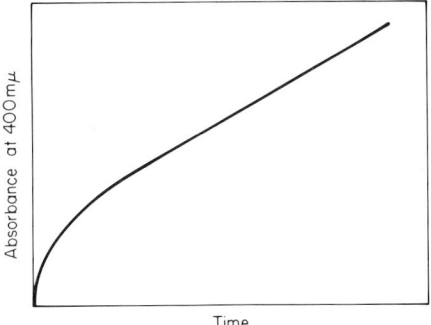

FIG. 1. Successive first-order and zero-order components of p-nitrophenylacetate hydrolysis by thiolsubtilisin.

[5] F. J. Kézdy and M. L. Bender, *Biochemistry* **1**, 1097 (1962).

Purification Procedure

Preparation of Purified PMS-Subtilisin[6]

Two hundred milligrams of subtilisin are dissolved in 8 ml 0.10 M phosphate buffer at pH 7.0; 0.2 ml of an 8.8 mg/ml solution of phenyl-methanesulfonyl fluoride (PMSF) in dioxane are then added. After 10 minutes, the solution is dialyzed against 0.01 M, pH 6.5 phosphate buffer overnight, and then passed through a 0.9 × 15 cm carboxymethyl cellulose column, applying a linear concentration gradient, starting with 0.01 M phosphate buffer and ending with 0.1 M phosphate, both at pH 6.5. The second, and larger, of the two peaks appearing at 280 nm is isolated and can be lyophilized.

Preparation of Purified Thiolsubtilisin[6]

For this one can use the CM-cellulose effluent containing PMS-sub-tilisin directly, or take the lyophilized material, and dissolve it in 0.1 M pH 6.5 phosphate buffer to a concentration of 100 mg/4 ml.

To this is added purified potassium thioacetate to a final concentra-tion of 100 mg/ml. The pH is then adjusted to 5.25 with dilute HCl and the solution stored at room temperature for 72 hours. It is then cen-trifuged to remove any precipitate, and passed through a Sephadex G-25 (or Biogel P-10) column equilibrated with 0.5 M pH 6.5 phosphate. The effluent solution containing the breakthrough fraction is then stored at 0° under nitrogen. Final separation of any residual subtilisin and PMS-subtilisin is done on the CM-cellulose column using the same buffer system as before. Thiolsubtilisin appears in a 280 nm peak at low ionic strength. This solution is stored again, at 0° under nitrogen and may be lyophilized.

It may be noted that a shortened preparation[4] is feasible, by which the appropriate amount of potassium thioacetate (to make 100 mg/ml at pH 5.25) is added directly after treatment of subtilisin with PMSF. Following this, after 72 hours, the solution is centrifuged, gel filtered on the phosphate equilibrated G-25 (or P-10) column, and lyophilized. Such a preparation appears to be sufficiently pure for some purposes, such as investigation of PNPA and cinnamoyl imidazole kinetics.

In some preparations of thiolsubtilisin using the shortened prepara-tion procedure, it was found that the concentration of thiol groups titrable with PMB exceeded the molar enzyme concentration as determined by the 278 nm protein absorbance. In these cases, the enzyme was again gel filtered and the thiol content redetermined.[6a] This was done using the

---

[6] L. Polgar and M. L. Bender, *Biochemistry* **8**, 136 (1969).

[6a] Neet *et al.*[8] note that 3 moles of thioacetate per mole of enzyme were bound to both native and thiolsubtilisin, but found that this did not affect any activity.

absorbance change at 250 nm of 7.6×10^3 M^{-1} following the method of Boyer.[7]

This actual cysteine content may also be determined by amino acid analysis. A convenient procedure is given in the paper by Polgar and Bender.[4]

Kinetic Properties

SUBSTRATES

p-Nitrophenyl acetate[4]

k_{cat} (limit)	$9.3 \pm 0.7 \times 10^{-3}$ sec^{-1}
pK of k_{cat}	7.15 ± 0.15
k_{cat}/K_m (limit)	570 ± 60 M^{-1} sec^{-1}

Cinnamoyl imidazole[4]

k_3 (limit)	$4.3 \pm 0.3 \times 10^{-4}$ sec^{-1}
pK of k_3	7.50 ± 0.15
k_3 (limit) H_2O/k_3 (limit) D_2O	2.86
k_4 (cinnamoylation of glycinamide at pH 9.30)	4.0×10^{-2} M^{-1} sec^{-1}

INHIBITORS

L-N-Acetyltyrosine ethyl ester[8, 8a]

K_I	2.9×10^{-2} M

D-N-Acetyltyrosine ethyl ester[8, 8a]

K_I	1.1×10^{-2} M

L-N-Acetyltryptophan ethyl ester[8]

K_I	$1.7 \pm 0.8 \times 10^{-2}$ M

Iodoacetamide[8]

k (second-order inactivation rate)	
pH 7–8	3.4×10^2 M^{-1} min^{-1}
pH 11	8.6×10^2 M^{-1} min^{-1}

Properties

Specificity. The enzyme is active toward PNPA, N-benzyloxycar-bonyl(CBZ)glycine p-nitrophenyl ester, cinnamoyl imidazole,[4, 6] CBZ-

[7] P. D. Boyer, *J. Am. Chem. Soc.* **76**, 4331 (1954).
[8] K. E. Neet, A. Naci, and D. E. Koshland, Jr., *J. Biol. Chem.* **243**, 6392 (1968).
[8a] Fluorometric

tyrosine *p*-nitrophenyl ester, and acetyl phenylalanine *p*-nitrophenyl esters.[8] Preliminary work also indicates activity toward several other CBZ amino acid *p*-nitrophenyl esters. A specific substrate of subtilisin, *N*-acetyltyrosine ethyl ester, is bound nearly as strongly to thiolsubtilisin as to the native enzyme, but is not hydrolyzed.[8] Of the activity toward PNPA 99–100%, and of the activity toward *N*-CBZ glycine *p*-nitrophenyl ester 60–70%, is inhibitable by *p*-mercuribenzoate.[6] *p*-Mercuribenzoate reacts specifically with the active site cysteine (none other being present in the molecule).

In addition, the enzyme acts as an acyl group transferase. In the presence of 0.02 *M* glycinamide in pH 9.3 carbonate, kinetic data indicate that only 25% of the reacting cinnamoyl moiety in *S*-cinnamoyl subtilisin is released as cinnamic acid, the rest presumably reacting to give *N*-cinnamoyl glycinamide.[4]

Inhibitors. Thiolsubtilisin is subject to all the SH inhibitors tried, such as iodoacetamide,[8] PMB,[1] *N*-ethylmaleimide,[6,8] and 5,5'-dithiobis-(2-nitrobenzoic acid).[8] Several specific ethyl esters also bind to the protein, and, in the case of both isomers of *N*-acetyl tyrosine ethyl ester, they enhance the fluorescence emission of the enzyme in the same way in which that of the native enzyme is enhanced.[8]

Stability. The enzyme appears to be stable under all conditions investigated, although the stability of the enzyme has not been thoroughly investigated. However, no rapid inactivation has yet been seen under any of the conditions under which it has been used and stored. A slow inactivation occurs when in solution, even under nitrogen at 0°, that is presumably due to reaction with residual oxygen. Since the enzyme does not have the capability to hydrolyze amide bonds, no autolysis is observed.

Mechanism of Action

For PNPA and cinnamoyl imidazole, it appears certain that there is an *S*-acyl intermediate in the course of catalysis. In the case of cinnamoyl imidazole, the kinetic constants are such that the acyl enzyme can easily be isolated and its physical properties investigated.

The presence of a pK of 7.1–7.5 in the k_{cat} of PNPA and the k_3 of *S*-cinnamoyl subtilisin indicates the participation of the active site histidine in the catalysis. The deuterium isotope effect toward cinnamoyl imidazole hydrolysis indicates that the histidine imidazole acts as a general base in deacylation.

Recent crystal structure data show that this imidazole is in a good position to do so.[3]

The experimental K_m of PNPA hydrolysis decreases markedly (toward better binding) below pH 7.[4] This has not been explained, since the K_m of the native enzyme shows no pH effect toward specific[9] or non-specific[4] substrates. It may be that the improvement in binding at low pH is caused by a change in the k_2/k_3 ratio, since $K_m = (k_3/(k_2 + k_3))$ K_s[10] where k_2 is the acylation rate constant, k_3 the deacylation rate constant, and K_s the true binding constant for the substrate. The existence of a lower K_m for PNPA in thiolsubtilisin even at high pH over that in the native enzyme is most probably due to such kinetic effects, since the nonkinetic dissociation constants of N-acetyltyrosine ethyl ester and acetyltryptophan ethyl ester change only slightly, and toward poorer binding.[8] In addition, physical methods, determination of the crystal structure, ultracentrifugation[8] and fluorescence studies show that the tertiary structure, including the active site structure, remains unchanged after the OH to SH replacement.

This decrease in the experimental K_m also, in the steady-state catalysis of PNPA hydrolysis, counteracts the low k_{cat}, so that the specificity constant k_{cat}/K_m of thiolsubtilisin toward PNPA is about the same as in the subtilisin system.[4]

[9] A. N. Glazer, *J. Biol. Chem.* **242**, 433 (1967).
[10] B. Zerner and M. L. Bender, *J. Am. Chem. Soc.* **86**, 3669 (1964).

[13] Cocoonases

By JEROME F. HRUSKA and JOHN H. LAW

Cocoonases are a group of proteinases produced by silkworm moths for the purpose of softening the end of the cocoon to permit the escape of the adult animal.[1-4] The cocoon is woven by the caterpillar in such a way that the end designed for exit consists of silk fiber loops cemented together with sericin, an amorphous protein.[1,2] As the pupa develops into an adult, epidermal cells in the maxillary galeae (mouth parts) secrete a concentrated solution of a virtually homogeneous proteolytic enzyme. The droplets dry to a partially crystalline deposit which may be harvested with fine forceps.[3]

Silkworm pupae may sometimes be obtained from dealers (for example, Jim Oberfoell, Bowman, N.D.) or they may be reared in the

[1] F. Duspiva, *Z. Naturforsch.* **5**, 273 (1950).
[2] F. C. Kafatos and C. M. Williams, *Science* **146**, 538 (1964).
[3] F. C. Kafatos, A. M. Tartakoff, and J. H. Law, *J. Biol. Chem.* **242**, 1477 (1967).
[4] F. C. Kafatos, J. H. Law, and A. M. Tartakoff, *J. Biol. Chem.* **242**, 1488 (1967).

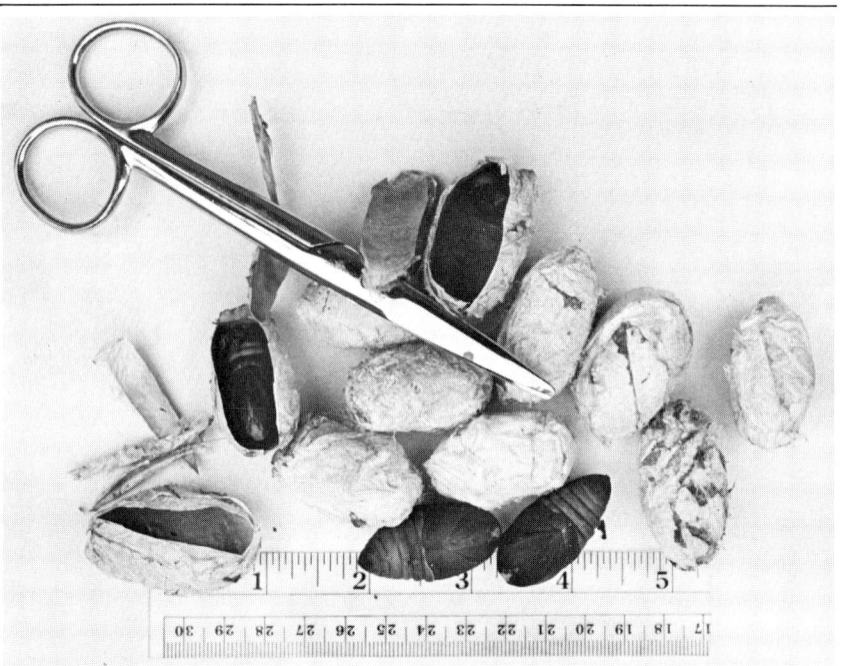

FIG. 1. Cocoons and pupae of *A. polyphemus;* Cocoons are cut open with blunt-ended scissors. Pupae at the top are male, left, (note larger antennae) and female, right.

laboratory. A most useful and enlightening book on the subject has recently appeared.[5] Artificial diets have been devised that permit rearing silkworm larvae at any season.[6,7] The domestic silkworm, *Bombyx mori*, is very easily reared on mulberry leaves. The diapausing resting eggs of this species can be kept in·the cold for long periods and are readily available from dealers (for example, Turtox Biological Supplies, Chicago, Ill.). The giant silkworms of the *Antheraea* group are more difficult to rear, but they are more convenient to use because the pupae diapause and can be kept in the cold for long periods until needed. Furthermore, they are larger and have much more enzyme than *B. mori*.

Cocoonase from *Antheraea polyphemus*

The enzyme is harvested from adult months just before they emerge from their pupal skins. Correct timing is critical because tardiness results in enzyme loss by dilution with a physiological buffer that the insect secretes to dissolve the enzyme for its normal mode of use.

[5] P. Villiard, "Moths and How to Rear Them." Funk and Wagnalls, New York, 1969.
[6] L. Riddiford, *Science* **157,** 1451 (1967).
[7] L. Riddiford, *Science* **160,** 1461 (1968).

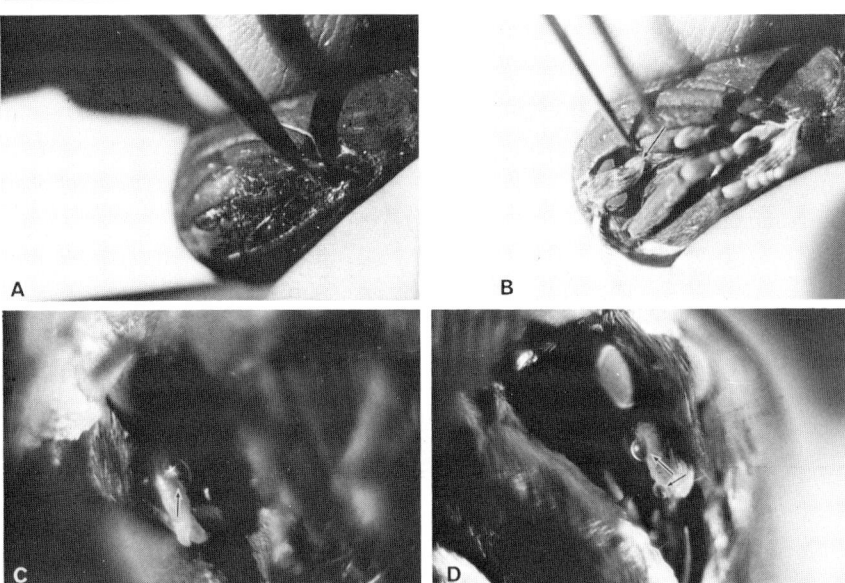

Fig. 2. Cocoonase harvest from *A. polyphemus*. (A) Cuticle is broken with fine forceps. (B) Cuticle is removed from face and one antenna. Arrow points to galeae. (C) and (D) A magnified view of face, the arrows point to droplets of cocoonase solution. The droplets are too wet for harvest and must be dried over desiccant before removal.

Pupae of *A. polyphemus* are in diapause when received in the fall. They must be stored at about 4° for at least 6 weeks before they will terminate diapause. The cocoons can be loosely packed in jars or plastic bags in a cold room or refrigerator. After cold storage, the tough cocoons are opened with a pair of blunt-ended scissors. Care must be taken to avoid injuring the pupa (Fig. 1).

The age of a developing pupa can be quickly determined by feeling the hardness of its cuticle. When pupae are first removed from their cocoons, their cuticles are tough and hard. After being stored for 2 weeks at room temperature with light for 8 hours a day their cuticle becomes softer. Shortly thereafter, when metamorphosis is completed, the cuticle becomes extremely dry and crisp. The crispness is due to reabsorption of molting fluid between the pupal and adult cuticles, and is a sign that the enzyme is ready for harvesting. This stage immediately precedes emergence of the adult.

The enzyme is harvested with the aid of a magnifying glass or a dissecting microscope as shown in the figures. Forceps are used to penetrate and pull away the triangle of cuticle which covers the galeae between the antennae (Fig. 2A). The cuticle covering the antennae often pulls free also. Usually droplets of the enzyme solution can be seen on the galeae (Fig. 2D). The animal is placed in a desiccator over Drierite or other suitable desiccant. Cooling to 4°–10° will delay emergence and keep the adult moth quiescent until the enzyme is harvested. After 2–3 hours the deposits are usually dry and can then be removed with fine forceps. After a small amount of practice the operation becomes simple and efficient.

Cocoonase from *Bombyx mori*

The harvesting operations for *B. mori* are similar to those for *A. polyphemus* except that the galeae are much smaller and the deposits of enzyme cannot be grasped by forceps. The surfaces of the galeae are therefore scraped with a flattened tool. A tip of a pair of fine forceps is an adequate instrument. Care should be taken not to contaminate the enzyme with scales from the animal. However, if this should occur, the scales can be removed manually at a later time, or the enzyme can be dissolved in buffer and filtered before use.

Properties of Cocoonases

When stored in the dry state at 4°, cocoonases are stable indefinitely. Frozen solutions at pH 7–8 are stable for several months. At low pH (below pH 4) solutions lose activity rapidly. Typical preparations of cocoonases from *A. polyphemus* and *B. mori* have the absorption at 280 nm and activity shown in Table I.

Three cocoonases (*A. polyphemus*, *A. pernyi*, and *B. mori*) which have been extensively investigated are all trypsinlike enzymes. They can

TABLE I
UV EXTINCTION COEFFICIENT AND V_{max} OF THE COCOONASES[a]

Cocoonase	Extinction coefficient (E_{280}) $\dfrac{AU \times ml}{mg}$	$V_{max} \times 10^8 \left(\dfrac{moles}{sec \times liter} \right)$
A. polyphemus	1.30	35
B. mori	0.98	65

[a] The concentration of cocoonase in a 0.1 N sodium hydroxide solution may be calculated from its observed absorption at 280 nm using the extinction coefficient given in this table. The maximum velocities were determined according to Michaelis-Menten kinetics from BAEE rate assays with a final enzyme concentration of approximately 7.0×10^{-4} mg/ml. [For a specific discussion of the assay, see F. J. Kézdy, L. Lorand, and K. D. Miller, *Biochemistry* **4**, 2302 (1965).] Our experience with the cocoonases has shown that *A. polyphemus* contains 70% active material by weight and *B. mori* contains 60%.

be assayed with typical trypsin substrates, for example, N-benzoyl-L-arginine ethyl ester (BAEE)[3] or they can be titrated with p-nitrophenyl-p-guanidinobenzoate (NPGB).[8] *B. mori* and *A. pernyi* cocoonases give single sharp bands on acrylamide gel at pH 4, while *A. polyphemus* cocoonase gives one intense band and one very weak band at high concentration under the same conditions.[3,9] A number of other chromatographic and electrophoretic methods have failed to give any indication of more than traces of contaminating proteins in cocoonase preparation.[3] Small peptides are sometimes present, however.

The enzymes are inhibited by diisopropylphosphofluoridate (DFP) and 1-chloro-3-tosylamido-7-amino-2-heptanone (TLCK).[4,9] The enzymes from *A. polyphemus* and *A. pernyi* are inhibited by soybean trypsin inhibitor[4,10] and the *A. polyphemus* enzyme interacts with the inhibitor at the same reactive site as does bovine trypsin.[10]

TABLE II
AMINO ACID COMPOSITION OF COCOONASES

	A. pernyi cocoonase[a]	*A. polyphemus* cocoonase[b]
Lys	13	12
His	4	4
Arg	6	9
Gly	22	26
Ala	16	16
Val	19	19
Ile	11	14
Leu	12	15
Thr	16	15
Ser	23	18
Pro	9	10
Met	1–2	2
Half-Cys	4	6
Asp + Asn	26	23
Glu + Gln	15	14
Tyr	9	9
Phe	5	4
Trp	3	2

[a] From F. C. Kafatos, A. M. Tartakoff, and J. H. Law, *J. Biol. Chem.* **242**, 1477 (1967).

[b] From J. F. Hruska and J. H. Law, unpublished.

[8] J. F. Hruska, J. H. Law, and F. J. Kézdy, *Biochem. Biophys. Res. Commun.* **36**, 272 (1969).

[9] J. F. Hruska and J. H. Law, unpublished.

[10] H. F. Hixson, Jr., and M. Laskowski, Jr., *Biochemistry* **9**, 166 (1970).

Molecular weight estimates from Sephadex chromatography, ultra-centrifugation, and amino acid analysis give values of about 24,000 for the enzymes of *A. pernyi* and *A. polyphemus*, and about 20,000 for *B. mori* cocoonase.[9] Amino acid compositions are shown in Table II.

[14] Papain

By RUTH ARNON

Papain is one of the sulfhydryl proteases isolated from the latex of the green fruit of *Carica papaya*. It was first isolated in the crystalline form by Balls and co-workers,[1] and later by Kimmel and Smith[2] from dried latex. Since then the enzyme has been extensively studied, and its properties were reviewed by Kimmel and Smith in 1960.[3]

Assay Methods

Papain activity may be determined either by measuring the rate of digestion of proteins, or by following the rate of hydrolysis of small molecular weight synthetic substrates such as esters or amide derivatives of amino acids. Both types of substrates are widely used, and therefore the various assays will be described in detail. Since in its native state papain exhibits very low activity, *all* assays are carried out in the presence of activators, comprising cysteine ($0.005 M$) and EDTA ($0.002 M$).

Assay of Proteolytic Activity

Principle. The method is based upon the estimation of the amount of the small molecular weight digestion products (trichloroacetic acid-soluble material) formed from proteins in the presence of the enzyme. Either urea-denatured hemoglobin or casein may serve as substrates. Only the procedure for the assay with casein, based upon the method of Kunitz[4] as adapted for papain,[5] is given below.

Reagents

Tris-HCl buffer, $0.05 M$, pH 8.0
Activating Agents. Freshly prepared solution of $0.05 M$ cysteine +

[1] A. K. Balls, H. Lineweaver, and R. R. Thompson, *Science* **86**, 379 (1937).
[2] J. R. Kimmel and E. L. Smith, *J. Biol. Chem.* **207**, 515 (1954).
[3] E. L. Smith and J. R. Kimmel, *in* "The Enzymes" (P. D. Boyer, H. Lardy, and K. Myrbäck, eds.), p. 133. Academic Press, New York, 1960.
[4] M. Kunitz, *J. Gen. Physiol.* **30**, 291 (1947).
[5] R. Arnon and E. Shapira, *Biochemistry* **6**, 3942 (1967).

0.02 M ethylenediaminetetraacetic acid (EDTA), adjusted to pH 8.0 with NaOH.

Casein. One gram of casein (Hammersten quality) is suspended in 100 ml of the Tris buffer, and the suspension is heated for 15 minutes in boiling water to bring about complete dissolution. This solution, designated as 1% casein, is stable for at least 1 week at 4°.

Enzyme. Crystalline papain is dissolved in distilled water to a concentration of approximately 0.05 mg/ml [$E_{1cm}^{1\%} = 25.0^6$].

5% Trichloroacetic acid (TCA)

Procedure. The casein solution is brought to 37° prior to the assay. Papain solution in amounts of 0.1 to 0.8 ml is pipetted into 20 ml tubes. The activating agent solution (0.2 ml) is added to each tube, and the volume is adjusted to 1 ml with Tris buffer. One milliliter aliquots of the casein solution are added to each tube in intervals of 1 minute. After mixing and incubation at 37° for 10 minutes, the reaction is stopped, and the residual large molecular weight material is precipitated by the addition of 3 ml TCA solution. The precipitates formed are filtered after standing 1 hour or longer at 25°. The concentration of the split products in the supernatant solution is determined by measuring the absorbancy of the solutions at 280 nm. The readings are corrected for blank solutions in which no papain was present, and are plotted vs. protein concentration. This calibration curve is not linear, but unknown papain solutions may be quantitated on its basis.

Definitions of Units and Specific Activity. One unit of enzyme activity is defined as the activity which gives rise, under the conditions described, to an increase of one unit of absorbancy at 280 nm per minute digestion. For this calculation the initial rate (the straight-line tangent to the first part of the curve) is used. The specific activity is the number of units of activity per milligram of protein.

Assay of Esterase Activity

Principle. The method is based on the cleavage of an ester bond in a small molecular weight substrate, and the assay of either the free carboxylic group or the liberated alcohol. The first approach can be exemplified by the use of benzoyl arginine ethyl ester (BAEE) as substrate, and titration of the formed carboxylic group in the pH-stat, and the second one by using *p*-nitrophenyl carbobenzoxy tyrosine (CTNP) as substrate and following the rate of liberation of the colored *p*-nitrophenol in a spectrophotometer.[7] Whereas the second method is very

[6] A. N. Glazer and E. L. Smith, *J. Biol. Chem.* **236**, 2948 (1961).

sensitive and is suitable for kinetic studies, the first one is the most widely used assay and will be described in detail below.

Titration of the Rate of Hydrolysis of BAEE[8]

Reagents

Substrate. 0.08 M benzoyl-L-arginine ethyl ester hydrochloride (BAEE)

Activating Agents. Freshly prepared solution of 0.05 M cysteine and 0.02 M EDTA, adjusted to pH 6.2 with NaOH.

NaCl, 3 M

Enzyme. Crystalline papain is dissolved in H_2O to a concentration of approximately 0.1 mg/ml.

NaOH, 0.1 M, standardized

Procedure. The reaction is carried out in the pH-stat, at 25° or 37°. The reaction mixture consists of 7 ml of the substrate solution, 1 ml of the activators, and 1 ml of 3 M NaCl. One milliliter of the enzyme solution is added and the reaction mixture is immediately adjusted to pH 6.2 by addition of 0.1 M NaOH as a titrant. The rate of the hydrolysis, which is linear under these conditions, is followed by the rate of consumption of alkali. (The final concentration of the substrate [0.05 M] is high enough to insure saturation of the enzyme, so that the kinetics approach zero order).

Units and Specific Activity. One unit of enzyme activity is defined as the amount which will hydrolyze 1 micromole of BAEE per minute at 25°. The specific activity is expressed as the number of units of enzyme per milligram of protein.

Assay of Amidase Activity

Principle. Similarly to the assay of esterase activity, this method is based upon the cleavage of an amide bond in a small molecular weight synthetic substrate. The activity may be followed either by the titration of the free carboxylic group or by the determination of the amine. An example for the first case is the hydrolysis of benzoyl-L-arginine amide (BAA), which is followed in the pH-stat similarly to the hydrolysis of BAEE, as described above. The difference between the two is that the rate of hydrolysis of BAA is first-order[2] and the specific activity (C_1)

---

[7] M. L. Bender, M. L. Begue-Cantón, R. L. Blakely, L. J. Brubacher, J. Feder, C. R. Gunter, F. J. Kézdy, J. V. Killheffer, Jr., T. H. Marshall, C. G. Miller, R. W. Roeske, and J. K. Stoops, *J. Am. Chem. Soc.* **88**, 5890 (1966).

[8] E. L. Smith and M. J. Parker, *J. Biol. Chem.* **233**, 1387 (1958).

is expressed as K (first-order rate constant calculated in decimal logarithms) per mg/ml of enzyme protein.

The estimation of the free amine is exemplified here by the spectrophotometric determination of the p-nitroaniline formed during the hydrolysis of benzoyl arginine-p-nitroanilide (BAPA). The assay is based on the method of Erlanger et al.[9] as adapted for papain.[10]

Reagents

Substrate. 43.5 mg of benzoyl-DL-arginine-p-nitroanilide hydrochloride are dissolved in 1 ml of dimethyl-sulfoxide and the volume is adjusted to 100 ml with 0.05 M Tris buffer, pH 7.5, containing 0.005 M cysteine and 0.002 M EDTA. (Care must be taken that *all* the BAPA is dissolved prior to the addition of the buffer.) The solution should be kept at a temperature above 25°.

Enzyme. Papain is dissolved in H_2O to a concentration of approximately 0.1 mg/ml.

Acetic acid. A 30% solution.

Procedure. The papain solution, in amounts of 0.1 to 1 ml is pipetted into 15-ml tubes, and the volume in each tube is adjusted with distilled water to 1 ml. Five milliliters of the substrate solution are added to each tube at 1-minute intervals. After 25 minutes of incubation at 25° the reaction is terminated by the addition of 1 ml of 30% acetic acid. The quantity of the liberated p-nitroaniline is estimated spectrophotometrically at 410 nm. Control tubes containing the reagents in the absence of enzyme show no self-hydrolysis.

Units and Specific Activity. The amount of substrate hydrolyzed by the enzyme can be calculated according to the molar extinction of p-nitroaniline at 410 nm ($E = 8800$); one unit of BAPA activity is the amount of enzyme which will hydrolyze 1 micromole of substrate per minute under the above conditions. The specific activity is expressed as number of units per milligram protein.

Comparison and Evaluation of the Various Assay Methods

As already mentioned, the most widely used method for papain assay is the hydrolysis of BAEE. This is also the assay used for standardization of commercial preparations. However, the other methods are often used for completion of the data. For example, in the study of papain derivatives, which may differ from papain in their relative activities toward

[9] B. F. Erlanger, N. Kokowsky, and W. Cohen, *Arch. Biochem. Biophys.* **95**, 271 (1961).

[10] R. Arnon, *Immunochemistry* **2**, 107 (1965).

different substrates, or in the study of specific inhibitors, which may exert different extents of inhibition depending on the nature and size of the substrate. The activity units are in all cases arbitrary units, but whereas with regard to the small molecular weight substrates they refer to molar amounts of hydrolyzed substrate, the use of a protein substrate complicates the process by offering many potential cleavage sites. The corresponding definition of activity is, therefore, even more arbitrary, and rarely used; the casein assay is of value mainly for comparison purposes.

Purification Procedure

Isolation and Crystallization[2]

Papain is prepared either from fresh papaya latex or from the commercially available dried latex. The method is based on fractional precipitation with ammonium sulfate and sodium chloride.

Step 1. Preparation of Crude Aqueous Extract. Dried papaya latex (180 g) is mixed with 100 g of Celite and 150 g of washed sand, and ground thoroughly at room temperature in a mortar with 200–300 ml of 0.04 M cysteine (6.3 g of cysteine hydrochloride dissolved in 1000 ml of 0.054 M NaOH; the excess of alkali is required in order to bring the aqueous extract to pH 5.7). Alternatively, the dried latex may be extracted in a Waring blendor. The suspension is allowed to settle and the supernatant is decanted. The grinding and extraction is repeated with another portion of 300 ml cysteine solution; the mortar is then washed with cysteine solution to adjust the volume of the extract to 1 liter. The resulting suspension is then filtered in the cold on an 18-cm Büchner funnel with mild suction through a 0.5 cm layer of Hyflo Super-Cel on Whatman No. 1 filter paper. The time required for filtration depends on the coarseness of the dried latex used, and the process may be very slow. The filtrate (fraction 1) is opalescent and greenish yellow, and should be close to pH 5.7. The remainder of the procedure is carried out in the cold, unless specified otherwise.

Step 2. Removal of Material Insoluble at pH 9. Fraction 1 is adjusted to pH 9 by slow addition, with stirring, of approximately 110 ml of 1 M NaOH. The resultant gray precipitate is removed either by filtration or by centrifugation at 2600 rpm for 1 hour. The supernatant (fraction 2) should be clear. The precipitate which contains denatured protein is discarded.

Step 3. Ammonium Sulfate Fractionation. Fraction 2 is brought to 40% saturation with ammonium sulfate, and after 1–2 hours the suspension is centrifuged at 2500 rpm. The supernatant is discarded (or may be saved for preparation of chymopapain) and the precipitate (fraction 3) is washed once with 400–500 ml of 40% saturated ammonium sulfate.

Step 4. Sodium Chloride Fractionation. Fraction 3 is dissolved in 600 ml of $0.02\,M$ cysteine (pH 7–7.5), and the papain is precipitated by slow addition of 60 g of solid NaCl. After 1 hour the precipitate (fraction 4) is collected by centrifugation for 1 hour at 2500 rpm and the supernatant is discarded.

Step 5. Crystallization. Fraction 4 is suspended in 400 ml of $0.002\,M$ cysteine at pH 6.5 at room temperature. The pH of the suspension is readjusted to pH 6.5 following the addition of the protein. The suspension is allowed to stand at room temperature for about 30 minutes, at which time it develops a crystalline sheen, and is then maintained at 4° overnight. The crystals (fraction 5) are collected by centrifugation in the cold at 2500 rpm for 4–5 hours. The supernatant is discarded.

Step 6. Recrystallization.[11] Fraction 5 is dissolved in the minimal amount of distilled water (protein concentration about 1%) at room temperature, and saturated NaCl solution (10 ml per 300 ml of protein solution) is added very slowly with stirring. When about 75% of the solution has been added, papain will begin to crystallize at room temperature. The suspension is maintained at 4° overnight, and the crystals are collected as described above. The specific activity may be slightly increased by an additional recrystallization according to the same procedure.

Alternatively, papain may be recrystallized by solution in 70% methanol followed by salting-out with lithium sulfate or lithium chloride. The results of a typical purification are summarized in Table I. The

TABLE I
PURIFICATION OF PAPAIN[a]

Fraction no.	Purification step	Protein (mg)	Specific activity[b] C_1	Total activity (units)
1	Crude extract	45	0.19	1300
2	After removal of inert material	43	0.23	1500
3	$(NH_4)_2SO_4$ fractionation	13	0.30	580
4	NaCl precipitation	7		
5	Crystallization	2.1	0.84	270

[a] According to Kimmel and Smith (text footnote 2).
[b] Activity was assayed with benzoyl-L-arginine amide as substrate at 39°, pH 5.5 in the presence of $0.005\,M$ cysteine and $0.001\,M$ EDTA. The specific activity was expressed as the first-order rate constant per milligram of protein. However, the specific activity is not a good index of the extent of purification in this case, since there are other proteolytic enzymes in the crude papaya latex.

[11] A. K. Balls and H. Lineweaver, *J. Biol. Chem.* **130**, 669 (1939).

yields and the specific activity of the crystallized papain may differ from one batch of dried papaya latex to the other.

Further Purification by Conversion to Mercuripapain and Activation.[12]

A suspension of 1 g crystallized papain in pH 4.5 acetate buffer is made 0.03 M in cysteine, and after about 30 minutes is centrifuged at 10,000 g for 2 minutes at 2°. The sedimented papain is washed with 2 ml of H_2O, centrifuged, and dissolved in 14 ml of 0.01 M $HgCl_2$, giving a total volume of 20 ml. Sufficient 95% ethanol is added to give a total volume of 30 ml and the solution is centrifuged to remove a small amount of undissolved material. The supernatant is made up to 40 ml with 95% ethanol and allowed to crystallize at 2°. Over a period of 1 week 95% ethanol is added in increments to bring the final concentration up to 70%. The resultant crystals of mercuripapain are soluble in water, and such a solution can be stored for several months at 2° without noticeable loss of activity. It is activated to papain by gently shaking with 2 volumes of a toluene solution of 4-methylbenzenethiol (10 mg/ml) for 5 minutes (the mercury is apparently complexed by the thiol and extracted into the toluene layer). The aqueous layer is gel filtered through a freshly prepared Sephadex G-25 column previously equilibrated with 0.05 M acetate buffer, pH 5.2, $\mu = 0.038$. The fractions containing the papain are pooled and stored at 2° in a flask flushed with nitrogen.

There are two advantages in employing the above procedure: (1) small amounts of impurities are removed; and (2) for storage of papain for long periods of time, it is preferable to keep it as the inactive mercury derivative, which is more stable, and does not lose activity. The activation step is carried out on a small sample at a time, for use within a relatively short period.

Methods of Crystallization for X-Ray Studies[13]

Papain crystals sufficiently large for X-ray analysis can be grown from solutions in 66% methanol. Four crystal forms are obtained in that way, denoted A, B, C, and D, of which form C is the dominant. Since in 66% methanol an appreciable loss in papain activity is observed, crystallization was carried out also from a medium of 15% dimethylsulfoxide in which the papain is active. The best conditions for crystallization in this medium were the following: Papain solution, 3% protein in 15% dimethylsulfoxide, was dialyzed (in an inverted tube which had been sealed with a dialysis membrane) against 15% dimethylsulfoxide

[12] L. J. Brubacher and M. L. Bender, *J. Am. Chem. Soc.* **88**, 5871 (1966).
[13] J. Drenth, W. G. J. Hol, J. W. E. Wisser, and L. A. Æ. Sluyterman, *J. Mol. Biol.* **34**, 369 (1968).

containing 0.2 M ammonium sulfate. The dialysis was carried out for 3 weeks at room temperature, followed by storage in the cold for several months. The crystals obtained in this way were of adequate quality for X-ray studies and appeared to be identical to the B crystals obtained from methanol.

Purity

The enzyme prepared by the above procedure for isolation and crystallization (pp. 231–232) contains small amounts of impurities. Although Finkle and Smith[14] obtained a single symmetrical peak upon chromatography on Amberlite IRC-50, starch-gel electrophoresis demonstrated the presence of such impurities[15] and when the enzyme was chromatographed on CM-cellulose (Fig. 1) a slight peak of inactive material preceded the main peak. The shape of the main peak itself indicated the presence of a tailing additional compound, which had, however, specific activity identical to that of the main component. The presence of impurity in crystalline papain was also demonstrated by sensitive immunological techniques. In both immunodiffusion and immunoelectrophoresis,[10] papain gave more than one precipitation band with the respective antiserum.

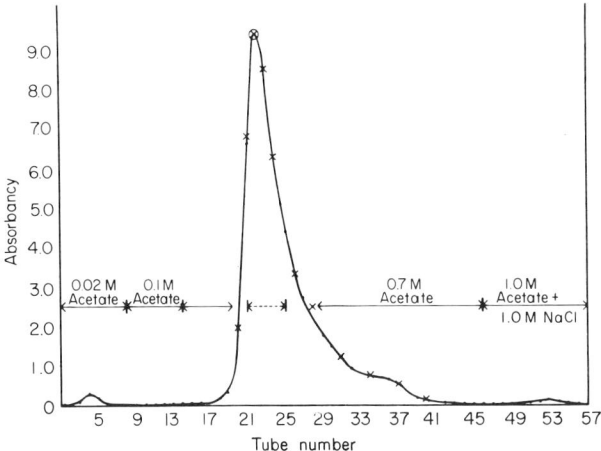

FIG. 1. Chromatography of papain on CM-cellulose. Points indicate absorption at 280 nm, crosses indicate activity on benzoyl arginine amide. In fraction 22 (encircled point) the relative activity is arbitrarily equated to ultraviolet absorption. Reproduced from *Biochim. Biophys. Acta* **85**, 305 (1964) by permission of the copyright owner, Elsevier, Amsterdam.

[14] B. J. Finkle and E. L. Smith, *J. Biol. Chem.* **230**, 669 (1958).
[15] L. A. Æ. Sluyterman, *Biochim. Biophys. Acta* **85**, 305 (1964).

Only specifically purified antibodies gave a single precipitin band with the enzyme.[5]

Another method for estimation of the purity of the enzymes is by titrating the active site[16] and relating this titration, which had been shown to be a stoichiometric reaction,[17] to a spectrophotometric determination of protein concentration. Using this method, Bender et al.[7] demonstrated only 58–75% purities for different crystalline papain preparations. The impurities in this case were not characterized at all and could have been the result of partial inactivation or autodegradation of originally active enzyme.

Properties

Stability

The crystalline enzyme shows a high degree of stability. As crystals in suspension in sodium chloride solution, it can be kept at 4° for months without detectable loss in activity. In solution, the mercury derivative may be kept for months without losing potential activity, whereas the active enzyme loses 1–2% of its activity per day, probably due to autolysis and/or oxidation. However, the more striking property of papain is its stability when exposed to high temperatures and organic solvents and reagents which cause denaturation of other enzymes. It has been known for many years[18] that papain powder resists dry heat at 100° for 3 hours, and even in solution it shows a remarkable temperature stability; the latter, however, is pH dependent[19]—the enzyme is unstable under acidic conditions. Papain is unaffected to a remarkable degree by denaturing agents which are known to cause conformational changes in proteins. Thus no change in optical rotatory dispersion was observed in 70% methanol[13] and no change in the viscosity in 50% methanol[20]; any decrease in activity in this medium was attributed, therefore, to interference with hydrophobic interactions between enzyme and substrates and not to conformational changes in protein. Consequently, the enzyme may be crystallized in 70% methanol to yield fully active papain. In solutions of dimethylsulfoxide containing up to 20% of the organic solvent, no decrease in activity has been noticed. Similarly, papain retains its activity in $8 M$ urea solutions[19-21] and does not undergo any

[16] G. Lowe and A. Williams, *Proc. Chem. Soc.* p. 140 (1964).
[17] B. S. Hartley and B. A. Kilby, *Biochem. J.* **56**, 288 (1954).
[18] K. Hwang and A. C. Ivy, *Ann. N.Y. Acad. Sci.* **54**, 161 (1951).
[19] H. Lineweaver and S. Schwimmer, *Enzymologia* **10**, 81 (1941).
[20] L. A. Æ. Sluyterman, *Biochim. Biophys. Acta* **139**, 418 (1967).
[21] E. Shapira and R. Arnon, *J. Biol. Chem.* **244**, 1026 (1969).

conformational changes, as inferred by optical rotation measurements.[22] On the other hand, exposure to stronger denaturing agents such as 10% trichloroacetic acid or $6 M$ guanidine hydrochloride causes irreversible changes in papain which are expressed both in a drastic change in specific optical rotation[22] and in loss of activity.[21, 22] Similarly, under acidic conditions, at pH values below 2.8 the enzyme suffers drastic decrease of activity.

Activators and Inactivators

Activation. Papain is a sulfhydryl enzyme—namely, it requires a free sulfhydryl group for its catalytic activity. In the native crystalline protein, prepared by the method of Kimmel and Smith[2] described above, this thiol group appears to be blocked, mainly in the form of mixed disulfide with half-cysteine,[23, 24] and may exist in part at the oxidation level of a sulfonic acid[23]; in that state it exhibits extremely low proteolytic activity and a correspondingly low thiol titer.[14, 22] Activation is achieved by mild reducing agents such as cysteine, sulfide, sulfite, as well as by cyanide.

Several studies have been concerned with the mechanism of this activation. In an earlier work,[25] the activation by cyanide was reported to be different from that effected by reducing agents, in that it involved an aldehyde group as a part of the active site. This assumption proved to be wrong. By employing thiol compounds[24, 26] as well as cyanides,[27] the activation was shown to consist solely of the removal of half-cysteine residue from the enzyme, with concomitant liberation of a free thiol group on the enzyme, and was found to be associated with little, if any, conformational change.[26]

Optimum activation was found to occur upon simultaneous application of a thiol compound like cysteine or thioglycolate and a heavy metal-binding agent like EDTA,[2] or by the addition of BAL (2,3-dimercaptopropanol), a compound which combines the functions of both a thiol compound and a metal binder.[28] The standard activation conditions which are used in activity assays of papain require a medium containing $0.005 M$ cysteine and $0.001–0.002 M$ EDTA.

Inactivation. Papain is reversibly inactivated in the presence of air

[22] R. L. Hill, H. C. Schwartz, and E. L. Smith, *J. Biol. Chem.* **234**, 572 (1959).
[23] A. N. Glazer and E. L. Smith, *J. Biol. Chem.* **240**, 201 (1965).
[24] L. A. Æ. Sluyterman, *Biochim. Biophys. Acta* **139**, 430 (1967).
[25] T. Masuda, *J. Biochem.* (*Japan*) **46**, 1489 (1959).
[26] A. O. Barel and A. N. Glazer, *J. Biol. Chem.* **244**, 268 (1969).
[27] I. B. Klein and J. B. Kirsch, *Biochem. Biophys. Res. Commun.* **34**, 575 (1969).
[28] A. Stockel and E. L. Smith, *J. Biol. Chem.* **227**, 1 (1967).

and a low concentration of cysteine. This inactivated enzyme is indistinguishable from the native crystalline papain (indicating that in the native state the enzyme indeed exists as a mixed anhydride with half-cysteine) and can be reactivated in the presence of higher cysteine concentrations.

Heavy metal ions such as Cd^{2+}, Zn^{2+}, Fe^{2+}, Cu^{2+}, Hg^{2+}, and Pb^{2+} are inhibitory for papain.[24] The inactive complexes can be totally reactivated by addition of both cysteine and EDTA. The readily reversible formation of a stable inactive complex with mercury has already been mentioned as a useful step in the purification of the enzyme.[2,12]

All sulfhydryl reagents act as papain inhibitors. Thus p-chloromercuribenzoate forms a stable complex with the enzyme and can serve for titration of the free —SH group.[14] Iodoacetic acid or iodoacetamide also react with the free sulfhydryl group of papain, causing thereby irreversible inactivation.[21,29]

The reaction of papain with the chloromethyl ketones of phenylalanine and lysine (TPCK and TLCK) was also found to bring about total loss of activity.[30] In this case, the reagents act specifically on the active sulfhydryl group of the enzyme rather than on the imidazole group of particular histidyl residues as they do in the case of trypsin and chymotrypsin, and thus the inactivation of papain is a stoichiometric reaction.

Aldehyde reagents, such as phenylhydrazine, hydroxylamine, etc., have also been reported to inactivate papain,[25] but their effect seems to be dependent on the route of the initial activation of the enzyme—only cyanide-activated papain was inhibited. The actual role of an aldehyde group in the active site of the enzyme is, therefore, unclear. The presence of papain inhibitors in horse serum has also been described in the literature,[31] but no detailed information about these inhibitors is available.

Besides the inhibitors described so far, which are mostly group-specific, there are several structural inhibitors which are specific for papain. Carbobenzoxy-L-glutamic acid has been described[2] as an inhibitor in the region of pH 3.9 to 4.5, causing a noncompetitive or a mixed-type inhibition. A series of papain inhibitors, all containing phenylalanine as a second residue from the C-terminal, have been reported recently.[32] Since these are strong competitive inhibitors, presumably because they occupy part of the active site, they have been used for the mapping of the active site of the enzyme,[33] a subject which will be discussed in more detail in a later section.

[29] J. R. Kimmel, H. J. Rogers, and E. L. Smith, *J. Biol. Chem.* **240**, 266 (1965).
[30] J. R. Whitaker and J. Perez-Villaseñor, *Arch. Biochem. Biophys.* **124**, 70 (1968).
[31] J. Pochon, *Compt. Rend. Soc. Biol.* **138**, 211 (1944).
[32] I. Schechter and A. Berger, *Biochem. Biophys. Res. Commun.* **32**, 898 (1968).

Physical Properties

Various physical properties of papain are listed in Table II. The electrophoretic behavior was studied in the Tiselius apparatus[34] and served for the determination of the isoelectric point. The sedimentation constants were determined in the ultracentrifuge and calculated for infinite dilution; these studies[34] were carried out in the presence of cysteine and EDTA, since in their absence papain showed a tendency to aggregate at pH values 5–7. The diffusion constant was determined independently in three different laboratories[11, 34, 35] with excellent agree-

TABLE II
PHYSICAL PROPERTIES OF PAPAIN

Property	Value	Reference
Isoelectric point	pH 8.75	34
Sedimentation constant, $S_{20,w}$	2.42 ± 0.04 S	34
Diffusion constant, $D_{20,w}$	$10.27 \pm 0.13 \times 10^{-7}$ cm^2 sec^{-1}	34
Molecular weight (by sedimentation diffusion)	20,700	34
Molecular weight (approach to equilibrium)	20,900	3
Extinction coefficient, $E_{1\,cm}^{1\%}$	25.0	6
K_m (BAEE)	0.023 M	8, 50
Optical rotation (pH 5.7, 25°), $[\alpha]_D$	$-66.7°$	22
Parameters of Moffit equation		
a_0	378 ± 5	20
b_0	-135	
Calculated α-helix content	17%	13
Region of Cotton effect	290 nm	13

ment. The molecular weight was calculated from these sedimentation-diffusion constants and also by the approach to the equilibrium method.[3] The various values are listed in Table II. The actual molecular weight of the molecule, according to recent findings,[36] has been reported as 23,000. This value was calculated from the amino acid composition of the molecule,[37] as corrected on the basis of data obtained from X-ray analyses.

[33] A. Berger and I. Schechter, *Phil. Trans. Royal Soc. (London)* B **257**, 249 (1970).
[34] E. L. Smith and J. R. Kimmel, *J. Biol. Chem.* **207**, 533 (1954).
[35] J. Close, Étude des Enzymes de la Papayotine. Thèse de doctorat de la Faculté des Sciences, Université de Liége, 1951.
[36] J. Drenth, J. N. Jansonius, R. Koekoek, H. M. Swen, and B. G. Wolthers, *Nature* **218**, 929 (1968).
[37] A. Light, R. Frater, J. R. Kimmel, and E. L. Smith, *Proc. Natl. Acad. Sci. U.S.* **52**, 1276 (1964).

Spectroscopic studies and fluorescence measurements have been employed to gain information concerning the secondary structure of papain. The spectroscopic studies[6] indicated that out of the 17 tyrosine residues, 11 to 12 residues ionize normally and reversibly at pH values below 12, whereas the remaining phenolic groups ionize slowly and irreversibly at higher pH. These data suggest that a portion of the secondary structure of papain is highly inaccessible to hydroxyl ions. Studies of the fluorescence emission spectrum of papain,[26] a property attributable to the tryptophan residues in the molecule, and the investigation of the pH dependence of the fluorescence intensity[38] which showed a marked change in the range of pH 5 to 8.5, both suggested the existence of a well-defined histidine–tryptophan complex in papain. As will be described later, this finding was corroborated by studies from other directions as well.

The papain molecule consists of a single polypeptide chain. Its amino acid sequence was first reported by Light and collaborators,[37] who described both the tentative linear sequence of the 198 amino acids and the assignment of the positions of the disulfide bridges in the molecule, as well as of the active sulfhydryl group.

The exact primary structure of the papain molecule has been derived with the help of X-ray analysis. These studies[36] indicated that in order to fit the data of the electron-density map, a peptide had to be inserted in the tentative sequence suggested by Light et al.[37] With the determination of the sequence of this peptide, and with the insertion of several other minor corrections, both the primary sequence and the three-dimensional structure of the 212 residues in the chain are completely known and are given in Figs. 2 and 3.

The structural conformation is stabilized by three disulfide bridges—their complete rupture results in the destruction of the protein as indicated by the loss of biological activity, catalytic as well as immunological.[21] However, not all three disulfide bonds are essential for the performance of biological functions. Specific cleavage of only one disulfide bond, achieved by carrying out the reduction of the molecule in 8 M urea, leads to the formation of a partially reduced enzyme derivative which has been shown to be catalytically active. Furthermore, it is capable of cross-reacting with 60% of the antibodies to native papain, thus suggesting that the remaining two disulfide bridges are sufficient for maintaining the required conformation of the molecule.[21]

A direct approach to the determination of the three-dimensional structure of papain was achieved by the elucidation of its crystal structure by X-ray study.[36, 39] It has revealed that the conformation of the

[38] M. Shinitzky and R. Goldman, European J. Biochem. 3, 139 (1967).
[39] J. Drenth, J. N. Jansonius, and B. G. Wolthers, J. Mol. Biol. 24, 443 (1967).

chain is irregular, with the exception of one short segment of beta structure and four short α-helical segments. The total α-helix content of the molecule amounts to about 20%. This value is in excellent agreement with the value of about 17% α-helix content found by analysis of optical rotatory dispersion data.[13, 20] The shape of the molecule is spheroidal, with dimensions of about $36 \times 48 \times 36$ Å, and the main chain is folded into two distinct parts which are divided by a cleft.

The active site, according to the X-ray studies, lies at the surface of this dividing cleft, the active sulfhydryl group of cysteine residue 25 is on its left, and histidine residue 158 (106 according to the numbering by Light *et al.*[37]) on its right,[36] with a distance of 4 Å between the sulfur and the imidazole ring. This is in agreement with the evidence for the presence of histidine in the active site of papain[40, 41] suggested by chemical studies accomplished by the use of a bifunctional reagent. This reagent, 1,3-dibromoacetone, binds preferentially to the active sulfhydryl and can subsequently bind to another nucleophile within the range of 5 Å. By isolation of the corresponding peptide it was proven that His_{158} (106 according to the numbering of Light *et al.*[37]) is indeed involved in the active site.

An important feature of the active site is its size. In order to "measure" it, Schechter and Berger[32, 33, 42] used a series of substrates and inhibitors large enough to interact with the furthermost parts of the binding site of papain, and with their aid they established that the active site is large enough to accommodate and combine with seven amino acid residues. Furthermore, with the help of both the wire model[36] and the space-filling model of part of the molecule, they predicted the manner in which such a heptapeptide, covering all seven subsites of the combining site, would bind to papain.[33] It is of interest to note that in this model, the spatial proximity of tryptophyl residue-128 (176 according to the numbering of Light *et al.*[37]) to both Cys_{25} and His_{106} (158) is readily observed, corroborating the evidence for the existence of a histidine–tryptophan complex,[38] mentioned earlier.

Biological Properties

Enzymatic Specificity. Papain is noted for its wide specificity, regarding both proteins and small molecular weight substrates. Studies with synthetic peptides and peptide derivatives have contributed significantly to the knowledge of the hydrolytic specificity of papain, and

[40] S. S. Hussain and G. Lowe, *Biochem. J.* **108**, 855 (1968).
[41] S. S. Hussain and G. Lowe, *Biochem. J.* **108**, 861 (1968).
[42] I. Schechter and A. Berger, *Biochem. Biophys. Res. Commun.* **27**, 157 (1967).

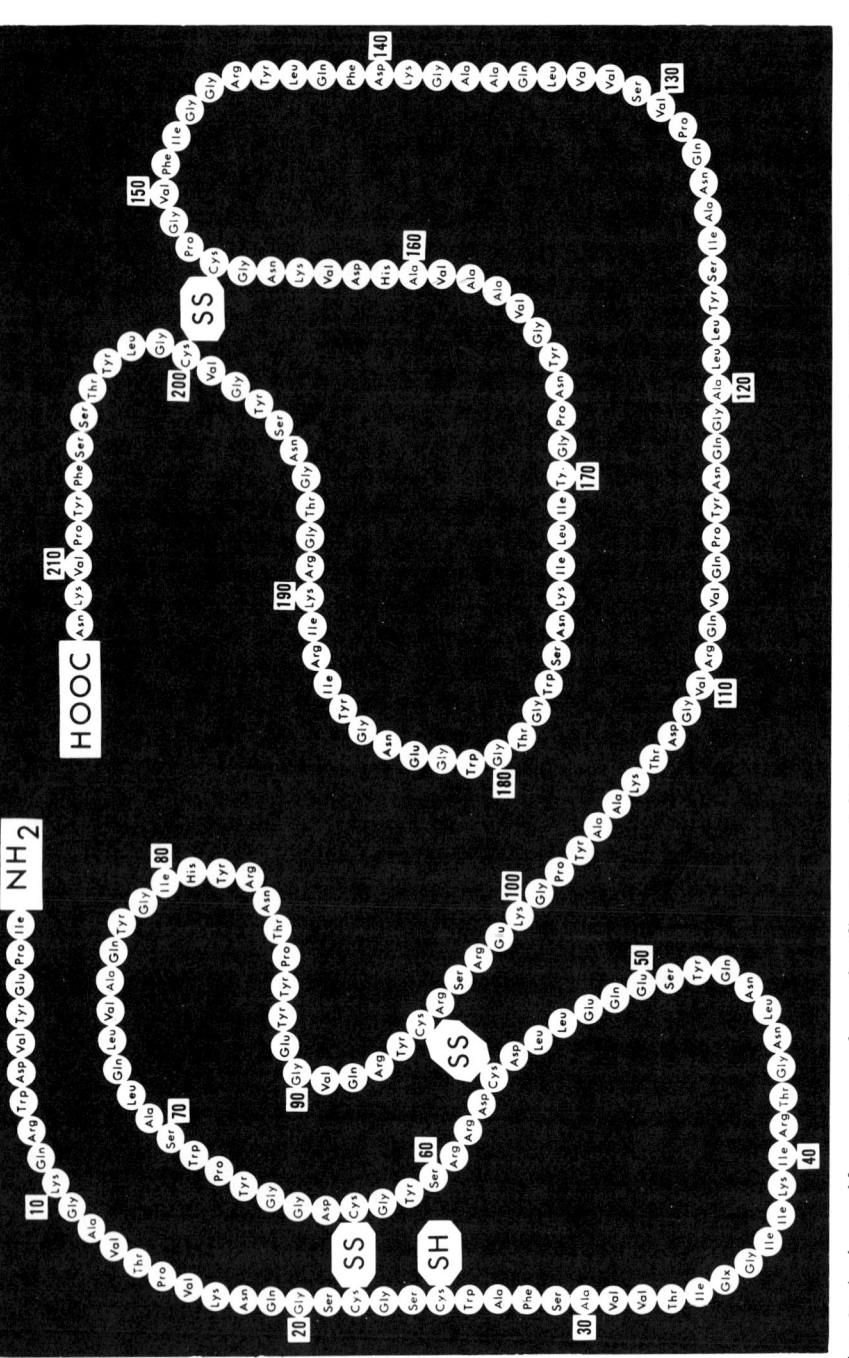

Fig. 2. Amino acid sequence of papain. Courtesy of Dr. J. Drenth, and with permission from Macmillan (Journals), London.

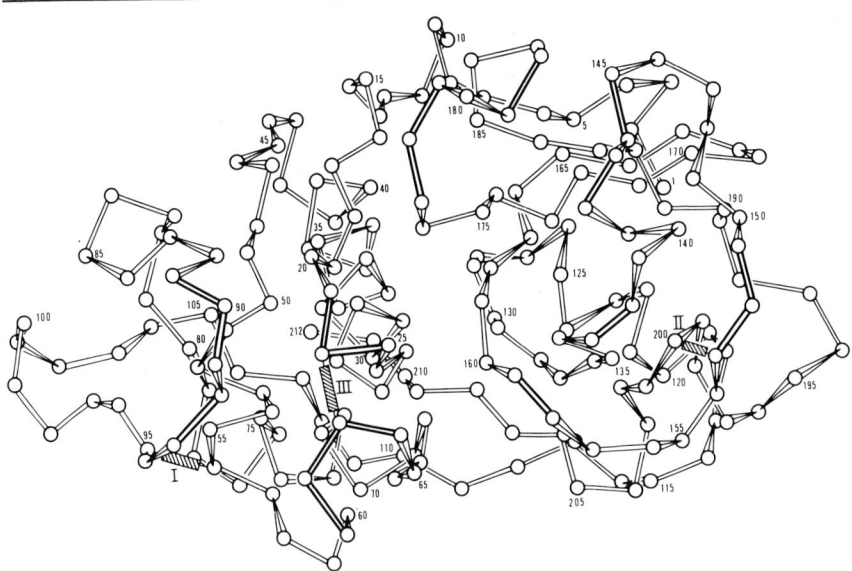

FIG. 3. Perspective drawing of the main chain conformation of papain. The circles represent the α-C atoms for the 212 residues. Courtesy of Dr. J. Drenth, and with permission from Macmillan (Journals), London.

were reviewed extensively by Smith and Kimmel.[3, 43] These studies demonstrated that indeed most peptide bonds are hydrolyzed to some extent by papain, although the relative rates of hydrolysis for different synthetic substrates may differ by three orders of magnitude. The most susceptible peptide bonds were found to be those formed by the carboxyl groups of α-amino substituted lysine and arginine[44]; benzoyl-L-arginine amide was found to be the most sensitive substrate.[2] A new kind of specificity of papain was recently defined—the bond next-but-one to the carboxyl group of phenylalanine[32]; this phenylalanine specificity was found to dominate the pattern of hydrolysis. In addition to the hydrolysis of peptide bonds, papain is also very effective as an esterase or thiol-esterase, and also possesses transferase activity.[3] Regarding the latter specificity, papain is capable of catalyzing not only transamidation and transpeptidation, but transesterification reactions as well.[45]

In its degradation of proteins papain exerts, in agreement with the results of studies with small peptides, a very broad specificity and will degrade most protein substrates more extensively than trypsin, pepsin,

[43] J. R. Kimmel and E. L. Smith, Advan. Enzymol. 19, 267 (1957).
[44] R. L. Hill, Advan. Protein Chem. 20, 37 (1965).
[45] A. N. Glazer, J. Biol. Chem. 241, 3811 (1966).

or chymotrypsin, in many cases giving rise to free amino acids.[44] This is one way of stating that papain shows little specificity and will cleave the peptide bond adjacent to most amino acid residues. It is noteworthy, however, that the extent of hydrolysis of bonds of similar type may vary considerably and the susceptibility of a particular bond is influenced partly by the ease with which other bonds in its immediate vicinity are hydrolyzed, as was shown by Konigsberg and Hill[46, 47] in their studies on the sequence of human hemoglobin. In these studies it was demonstrated that it is the wide specificity of papain which makes it an ideal choice in sequence studies for degradation of peptides from tryptic or chymotryptic digests into smaller peptides which are more amenable to sequence analysis.

Worthy of special mention is a particular reaction of papain, which offered a breakthrough in the study of the structure of antibodies, namely its degradative effect on immunoglobulins. Papain, and as shown at a later stage for other proteases as well, was found to cause limited hydrolysis of native immunoglobulins, to yield biologically active fragments separable by chromatography.[48] These pioneering studies, which were followed by an extremely extensive research, have been of the utmost importance in the elucidation of the structure of immunoglobulins and constitute a major contribution to the field of immunochemistry.

Immunological Properties. Immunological studies indicate that papain is highly immunogenic in rabbits.[5, 10] The purified specific antibodies react with the crystalline enzyme to yield a single precipitation band in gel diffusion and are capable of inhibiting its catalytic activity. This inhibition was shown to be of the pure noncompetitive type,[49] and to be due mainly to steric hindrance, rather than to a direct interaction of the antibodies with the catalytic site.

Kinetic Properties

Many studies have been performed on the mechanism and kinetics of papain action, and are beyond the scope of the present article. The subject was reviewed in 1960 by Smith and Kimmel,[3] mainly in regard to the hydrolysis of benzoyl arginine amide[28] and benzoyl arginine ethyl ester,[8] and since then the mechanism has been reinvestigated in various laboratories for the action of papain on the same substrates[50] as well as

[46] W. Konigsberg and R. J. Hill, *J. Biol. Chem.* **237**, 3157 (1962).
[47] J. Goldstein, W. Konigsberg, and R. J. Hill, *J. Biol. Chem.* **238**, 2016 (1963).
[48] R. R. Porter, *Nature* **182**, 670 (1958).
[49] E. Shapira and R. Arnon, *Biochemistry* **6**, 3951 (1967).
[50] J. R. Whitaker and M. L. Bender, *J. Am. Chem. Soc.* **87**, 2728 (1965).

on other substrates such as hippuric esters,[51] esters of carbobenzoxy glycine,[52] p-nitrophenyl, benzyl and methyl esters of carbobenzoxy-L-lysine,[53] and α-N-benzoyl citrulline methyl ester,[54] which is closely related to benzoyl arginine methyl ester.

The results of the different studies are consistent with a mechanism of a two-step catalytic process—namely, acylation and deacylation, the rates of which determine the turnover rate constant (K_{cat}) for the various substrates.

Distribution

Papain is present in, and obtained from, the latex of *Carica papaya*. This latex contains several enzymes, and papain represents only a minor part of the total proteolytic activity. Another proteolytic enzyme that has been isolated in crystalline form[55] is chymopapain. A third enzyme which has been isolated from the same latex and characterized is lysozyme.[56]

Although chymopapain will be described and discussed separately in another section of this book, it is pertinent here to compare some of the properties, physical and biological, by which it is related to papain.[57] The papain molecule is smaller than chymopapain—their molecular weights are 23,000 and 27,000, respectively. In their acid stability the two enzymes differ entirely—chymopapain is much more stable, even at pH as low as 2. The high solubility of chymopapain in salt solution is another property by which the two enzymes can be distinguished. On the other hand, chymopapain is a sulfhydryl enzyme like papain, and the two are quite similar to each other in their substrate specificities, although some differences do exist in the relative activities toward different substrates.[57] The two enzymes are also related in their immunological properties,[58] and although detailed antigenic mapping is not yet available, it has been possible, by immunological techniques, to detect common determinants present on both enzymes, and to predict their proximity to the region of the catalytic center. And last but not least, the amino acid sequence near the active sulfhydryl of papain and

[51] G. Lowe and A. Williams, *Biochem. J.* **96**, 199 (1965).
[52] J. F. Kirsch and M. Igelstrom, *Biochemistry* **5**, 783 (1966).
[53] M. L. Bender and L. J. Brubacher, *J. Am. Chem. Soc.* **88**, 5880 (1966).
[54] W. Cohen and P. J. Petra, *Biochemistry* **6**, 1047 (1967).
[55] E. F. Jansen and A. K. Balls, *J. Biol. Chem.* **137**, 459 (1941).
[56] E. L. Smith, J. R. Kimmel, D. M. Brown, and E. O. Thompson, *J. Biol. Chem.* **215**, 67 (1955).
[57] M. Ebata and K. T. Yasunobu, *J. Biol. Chem.* **237**, 1086 (1962).
[58] R. Arnon and E. Shapira, *Biochemistry* **7**, 4196 (1968).

chymopapain was compared[59] and found to be quite different. This does not exclude, however, similarity in the active sites of the two enzymes, as a result of an appropriate spatial conformation. More detailed studies of these and other related enzymes will shed light on this point.

Acknowledgments

The author is grateful to Dr. J. Drenth for generously providing prints of Fig. 2 and Fig. 3, and to Dr. L. A. Æ. Sluyterman for his permission to use Fig. 1.

I also wish to express my gratitude to Professor Hans Neurath for his encouragement in the preparation of this chapter, during my stay in his laboratory at the University of Washington, Seattle.

[59] J. N. Tsunoda and K. T. Yasunobu, J. Biol. Chem. **241**, 4610 (1966).

[15] Chymopapain B

By Donald K. Kunimitsu and Kerry T. Yasunobu

Assay Method

Principle. Although several assay methods exist for determining the activity of proteolytic enzymes using either protein or synthetic substrates,[1-5] we have generally employed the casein digestion method of Kunitz[6] because it has proved to be the most convenient, particularly where a large number of routine determinations is required. Moreover, the addition of certain reducing agents (some of these activators are enumerated below), which is required to effect maximal activity of chymopapain, presents no difficulty with this assay method but could conceivably complicate many of the other assays, especially the colorimetric ones. Of course, the possibility of complication introduced by these reducing compounds could be circumvented in these latter type assays by appropriate modifications of the published procedure (see, for example, the papers by Whitaker,[7] Inagami and Murachi,[8] and Smith, et al.[9]).

[1] Vol. XI, [1], [2], [3], [5], [7], [8].
[2] N. C. Davis and E. L. Smith, *Methods Biochem. Anal.* **2**, 215 (1955).
[3] E. L. Smith and M. J. Parker, *J. Biol. Chem.* **233**, 1387 (1958).
[4] A. Reidel and E. Wunsch, *Z. Physiol. Chem.* **316**, 61 (1959).
[5] B. F. Erlanger, N. Kokowsky, and W. Cohen, *Arch. Biochem. Biophys.* **95**, 271 (1961).
[6] M. Kunitz, *J. Gen. Physiol.* **30**, 291 (1947).
[7] J. R. Whitaker, *Nature* **189**, 662 (1961).
[8] T. Inagami and T. Murachi, *Biochemistry* **2**, 1439 (1963).
[9] E. L. Smith, V. J. Chavre, and M. J. Parker, *J. Biol. Chem.* **230**, 283 (1958).

The Kunitz method is discussed in detail elsewhere,[1,6] but the assay method as routinely employed by the authors will be described here as it does involve slight modifications incorporated specifically for assaying the activity of —SH proteases, such as chymopapain.

Reagents

Casein (1%). Dissolve 1 g of casein (Hammersten quality, reprecipitated with glacial acetic acid and repeatedly washed with acetone and ether and dried) in 100 ml of 0.10 M phosphate buffer, pH 7.2, by heating in boiling water for 15 minutes. This solution should be stable for about a week if suitably refrigerated.

Enzyme Activator Solution (prepared fresh as needed). 0.10 M cysteine in 0.10 M phosphate buffer, pH 7.2, containing 0.01 M ethylenediaminetetraacetic acid (EDTA).

Enzyme. 0.05–0.20 mg/ml. (The concentration of purified chymopapain B can be determined spectrophotometrically using $E_{1cm, 280}^{1\%} = 18.4$ at pH 7.0.)

Procedure. For most of the assays carried out in this study, activity measurements were made at 35°. The activation of chymopapain, when required, is carried out by adding an equal volume of the enzyme activator solution to a suitable volume of enzyme solution and incubating the resulting mixture at 35° for at least 5 minutes. Subsequently, various aliquots of this activated enzyme solution (up to a volume of 1 ml) are introduced into a series of test tubes and the volume adjusted to 1 ml with the activator solution. The reaction is initiated by pipetting 1 ml of casein solution into the various tubes at 30-second intervals. The digestion is allowed to proceed for exactly 10 minutes and then terminated by the addition of 3 ml of 5% trichloroacetic acid (TCA). The mixture is allowed to stand at room temperature for at least 30 minutes and then filtered or centrifuged. The absorbance of the clear supernatant is measured at 280 nm in a spectrophotometer. The readings are corrected for the values of the blanks, such blanks being prepared by first mixing 1 ml of casein solution with 3 ml of TCA solution and then adding 1 ml of the highest concentration of enzyme used or 1 ml of the activator solution. The corrections for blanks for the intermediate concentrations of chymopapain are calculated by interpolation.

Definition of Unit and Specific Activity. A unit of enzyme activity is defined as the amount of enzyme required to cause a net change of 0.001 absorbance unit. Specific activity is expressed as the total enzyme units per minute per microgram enzyme. This value is determined from the initial slope of the plot of enzyme units vs. enzyme concentration (μg/ml of reaction mixture), divided by 10 (digestion time in minutes).

Purification Procedure

For the purpose of simplifying the discussion to follow, we have provisionally designated the enzyme crystallized by Jansen and Balls[10] as chymopapain, the enzyme purified by Ebata and Yasunobu[11] as chymopapain A, and the protease described by Kunimitsu and Yasunobu[12, 13] as chymopapain B. It can be noted here, also, that Cayle and Lopez-Ramos[14, 15] have independently described the fractionation of chymopapain into two fractions (also designated chymopapain A and chymopapain B) utilizing Sephadex chromatography and acetone fractionation. According to these workers, these fractions can be distinguished electrophoretically, with chymopapain B being the more basic protein.[14, 15] Coincidentally, our provisional designation, based on chromatographic behavior on Amberlite XE-64 (and other properties as well), seems consistent with their observations.

The procedure of Jansen and Balls[10] was initially used to determine whether or not homogeneous preparations of chymopapain could be obtained from commercially prepared (dried) papaya latex. This procedure failed to yield homogeneous preparations of the enzyme.[11-15] The following purification procedure was finally adopted in our laboratory with the isolation of chymopapain B as the principal goal.[12, 13]

Step 1. Extraction and Ammonium Sulfate Fractionation. All steps are carried out at 4°, unless otherwise indicated. "Crude Standardized Papain" (500 g, Paul Lewis Labs, Milwaukee, Wis.) is dissolved in deionized water (1:3, w/v) and stirred for 1 hour. The preparation is then filtered and the insoluble material discarded. The filtrate is next brought to 0.45 saturation by the addition of solid ammonium sulfate (277 g/liter of solution) and stirred for 1 hour or longer. The mixture is then centrifuged at 20,000 g. The precipitate can be kept for the isolation of papain.[11] The supernatant solution containing chymopapain is brought to pH 2 by the gradual addition of 1 N HCl and the small amount of precipitate formed centrifuged off. The clear supernatant is then brought to 0.65 saturation with solid ammonium sulfate (134 g/liter of solution), stirred for about an hour, and centrifuged. The supernatant fluid is discarded and the precipitate dissolved in a small amount of 0.02 M acetate buffer, pH 5.0, containing EDTA (0.001 M) and dialyzed overnight against the same buffer.

[10] E. F. Jansen and A. K. Balls, *J. Biol. Chem.* **137**, 459 (1941).

[11] M. Ebata and K. T. Yasunobu, *J. Biol. Chem.* **237**, 1086 (1962).

[12] D. K. Kunimitsu, Ph.D. thesis, University of Hawaii, Honolulu, 1964.

[13] D. K. Kunimitsu and K. T. Yasunobu, *Biochem. Biophys. Acta* **139**, 405 (1967).

[14] T. Cayle and B. Lopez-Ramos, *Abstr. 140th Meeting Am. Chem. Soc.* p. 19c (1961).

[15] T. Cayle, personal communication.

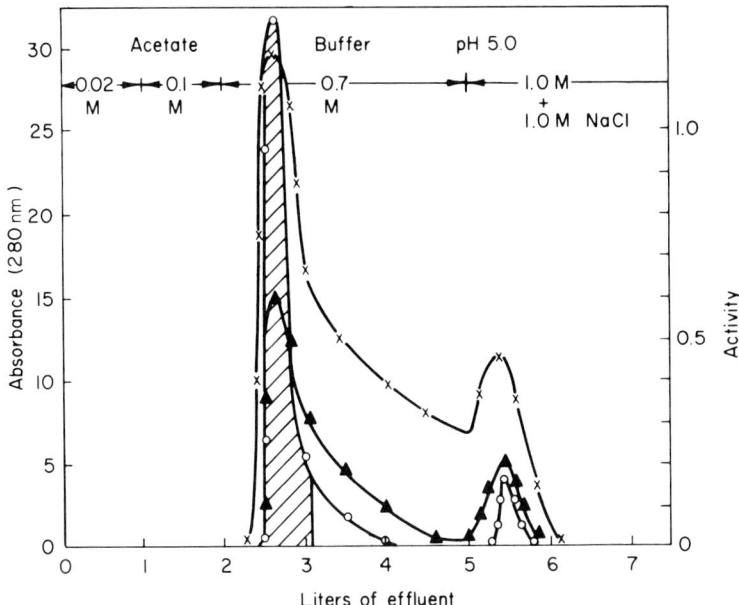

FIG. 1. Chromatography of 0.65 saturated ammonium sulfate precipitate on CM-cellulose. In the experiment, 8.67 g of protein was introduced to a 3.5 × 39 cm column and stepwise elution with 1 liter of 0.02 M acetate buffer, pH 5.0; 1 liter of 0.1 M acetate buffer, pH 5.0; 3 liters of 0.7 M acetate buffer, pH 5.0; and 1.0 M acetate buffer containing 1 M NaCl was performed. (○) Represents protein, (▲) casein digestion units, and (×) casein digestion units after activation with 0.05 M cysteine. The diagonal hatched area represents the fraction which was pooled for further purification.

Step 2. Carboxymethyl Cellulose Column Fractionation. The dialyzed solution is further purified on a carboxymethyl cellulose column. The solution from step 1 is applied to the cellulose column (ca. 20 mg protein/ml resin), which has been equilibrated with the same buffer. Stepwise elution with 0.02 M, 0.1 M, and 0.70 M acetate buffers, pH 5 is carried out and the chymopapain-rich fractions which are eluted with the 0.70 M buffer pooled and dialyzed exhaustively, first against deionized water containing 0.001 M EDTA and then against 0.25 M phosphate buffer, pH 5.90. A typical chromatogram is shown in Fig. 1.

Step 3. Amberlite XE-64 Column Chromatography. The chymopapain fraction obtained from step 2 is applied to a column of Amberlite XE-64 equilibrated with 0.25 M phosphate buffer, pH 5.90 (ca. 15 mg protein/ml resin). Elution is started with the same buffer until the concentration of protein being eluted drops to a very low level and at least two chromatographically distinguishable components are eluted (see Fig. 2).

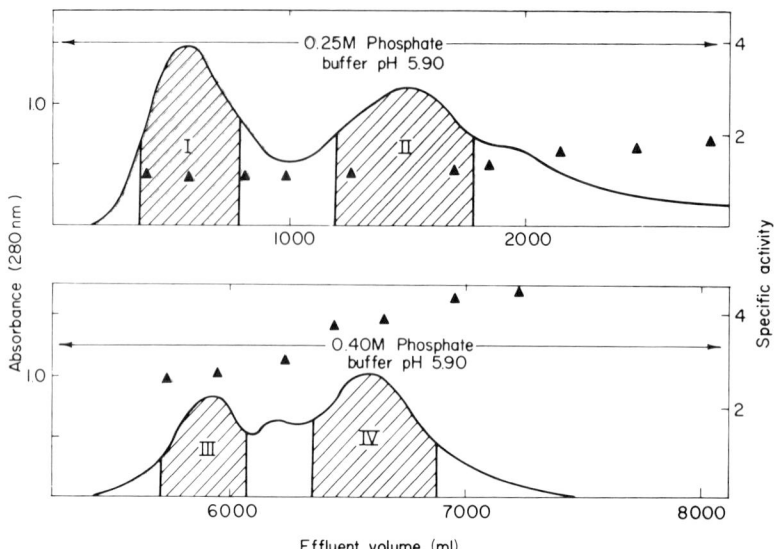

Fig. 2. Chromatography of the 0.7 M acetate buffer fraction on XE-64. In the experiment, 1.74 g of protein were chromatographed on a 2.0×25 cm column. Stepwise elution with 3 liters of 0.25 M phosphate buffer, pH 5.9; 2360 ml of 0.3 M phosphate buffer, pH 5.9 (not shown in figure); and finally 2.5 liters of 0.4 M phosphate buffer, pH 5.9. Diagonally hatched areas were pooled and peak 4 contained chymopapain B in the purest form. The solid line represents protein while the triangles represent specific activity using the casein assay.

Thereafter, the eluant is changed to 0.40 M phosphate buffer, pH 5.90, and the elution continued until an additional fraction or two is eluted. These latter fractions are pooled separately as chymopapain B (peaks III and IV) and dialyzed against 0.02 M acetate buffer, pH 5.0, containing 0.001 M EDTA. The presence of chymopapain B in these fractions should be confirmed by demonstrating that tyrosine is indeed the NH$_2$-terminus of the protein present in these fractions. Moreover, end group analyses should indicate the relative homogeneity of the chymopapain B fractions (generally 90% or greater at this stage of purification). The dilute enzyme solutions can be concentrated by introducing them to a small column of carboxymethyl cellulose and eluting the enzyme with 0.02 M acetate buffer, pH 5.0, containing 3.0 M NaCl.

Step 4. Crystallization. If an attempt at the crystallization of this protease is desired, then the crystallization should begin with a highly concentrated solution. Sodium chloride is gradually added to the solution whose pH is adjusted to 2.0. The crystallization, when successful, should take several days' time, but subsequent recrystallizations become

TABLE I
PURIFICATION OF CHYMOPAPAIN B

Procedure	Volume (ml)	Protein (mg/ml)	Specific activity	Total protein (mg)	Total activity	Yield (%)
1. Crude extract	1785	33.51	2.80	59,820	167,500	100.0
2. 0.45 Saturated	2080	24.28	2.50	50,510	126,280	75.4
3. 0.65 Saturated precipitate	1240	25.50	3.08	31,620	97,390	58.1
4. 0.70 M CMC-eluate	961	12.43	2.70	15,610	42,150	25.2
5. XE-64[a] fraction	1060	2.14	3.70	2,270	8,399	5.0
6. Once crystallized chymopapain B	74	21.42	3.40	1,590	5,406	3.2

[a] Fraction IV of Fig. 2.

routine if the seeding technique is utilized. In our hands, crystallization has met with fair success, giving rise to needle-shaped crystals. However, from the standpoint of the degree of purification to be achieved, it has been shown that crystallization is not required in obtaining homogeneous preparations of chymopapain B.

A typical protocol, summarizing data on the specific activity of the various fractions and the yield obtained from 500 g of commercially dried papaya latex, is presented in Table I.

It is important to point out that starting material used was "Crude Standardized Papain," obtained from the Paul Lewis Labs, Inc. Different starting materials obtained from other companies may give different results. However, by analyzing all fractions for various properties of chymopapain (e.g., specific activity, N-terminus), the present procedure should yield purified chymopapain.

Properties

Homogeneity of the Enzyme. Chymopapain B is homogeneous as determined by sedimentation, free-boundary electrophoresis in several different buffers in the pH range 4 to 10.5, NH$_2$-terminal analyses by dinitrophenylation and by rechromatographic studies.[12, 13]

Physicochemical Properties.[12] Some of the physicochemical properties of the enzyme are summarized in Table II.

The amino acid composition of chymopapain B is summarized in Table III.

Specificity. Preliminary experiments with simple substrates indicate that chymopapain resembles papain in its ability to hydrolyze a wide variety of peptide and amino acid derivatives, with the derivatives of

TABLE II
Some Physicochemical Properties of Chymopapain A and B

Property	Chymopapain A	Chymopapain B
1. Molecular weight (Archibald method)	$36,400 \pm 1,500$	$34,500 \pm 1,500$
2. Sedimentation coefficient, $s_{20,w}$	2.71	2.82
3. Partial specific volume	0.721	0.728
4. NH$_2$-terminal residue	Glutamic acid	Tyrosine
5. $E^{1\%}_{1\,cm}$ at 280 nm	18.7	18.4
6. Nitrogen content	15.7%	15.8%
7. Specific activity (Activated) (casein digestion units)	0.9–1.07	3.4
8. Crystalline form (from sodium chloride)	Rods	Needles
9. Cysteine residues/mole protein	1–2	1–2
10. dn/dc at 436 nm	0.193	0.193
11. Isoelectric point	10.1	10.4

TABLE III
Amino Acid Composition of Chymopapain B

Amino acid	GM AA Residue/100 g protein	Calculated no. of residues/35,200[a] g protein	Assumed no. of residues	No. of residues[b] (Cayle and Lopez-Ramos)
Lysine	9.73	26.7	27	27
His	1.86	4.8	5	4
Arg	4.48	10.1	10	10
NH$_3$	1.42	29.4	29	30
Asp	9.17	28.7	29	23
Thr	4.88	17.0	17	17
Ser	5.97	24.1	24	22
Glu	11.06	30.1	30	25
Pro	3.90	14.1	14	14
Gly	6.64	40.9	41	39
Ala	4.09	20.2	20	19
½-Cys[c]	3.29	11.2	11	9
Val	7.60	27.0	27	21
Met[c]	0.40	1.1	1	1
Ile	4.27	13.3	13	9
Leu	5.05	15.7	16	15
Tyr	9.38	20.2	20	18
Phe	2.87	6.9	7	7
Trp[d]	3.10	5.9	6	4
Total	99.09		318	284

[a] Minimum molecular weight based on amino acid analyses.

[b] 16-hour hydrolyzate.

[c] Performic acid oxidized values.

[d] Estimated by the method of T. L. Goodwin and R. A. Morton, *Biochem. J.* **40**, 628 (1946).

arginine being the most susceptible to hydrolysis.[11,12] However, the rate of cleavage of the various bonds is slower when compared to the action of papain.

Other studies designed to elucidate the specificity of chymopapain include the action of this enzyme on the A- and B-chains of performic-acid oxidized insulin. Ebata and Takahashi have shown that chymo-papain preferentially hydrolyzes those peptide bonds containing acidic (glutamic, aspartic, cysteic acid) residues as well as aromatic (tyrosine, phenylalanine) residues in the performic-acid oxidized B-chain of insulin.[16] Cayle notes that the enzyme preferentially hydrolyzes the glutamyl bonds as shown by studies with A- and B-chains of insulin, as well as with several glutamyl peptides of varying lengths.[15]

pH Optimum. The pH optimum of casein digestion by chymopapain appears to be quite broad (pH 7–9) and exhibits a dual optimum (7.2, 9.5), a result which probably reflects the heterogeneity of the casein.[13] With hemoglobin as substrate, the optimum is 7.0 when assayed by Anson's method,[1] and with benzoyl-L-argininamide, the optimum is a plateau between pH 5.5 and 8.

Cayle has made the interesting observation that chymopapain ex-hibits two pH optima when assayed with hemoglobin, which are dependent upon exposure of the enzyme to urea. In the absence of urea, the pH optima is 4.0 on acid-denatured hemoglobin but shifts to 7.0 when the enzyme is assayed with urea-denatured hemoglobin.[15,16]

Activation and Inhibition. Chymopapain, like papain, ficin, and bromelain, has the properties of a sulfhydryl protease. It exhibits maxi-mal activity in the presence of various reducing compounds (e.g., cysteine, 2,3-dimercaptopropanol, thioglycolic acid, glutathione) supplemented by a metal-chelator such as EDTA, and is inhibited by various reagents which show a pronounced reactivity toward the free sulfhydryl group (e.g., PCMB, iodoacetate, iodosobenzoate, H_2O_2, NEM, and heavy metals such as Hg^{2+}, Ag^+, Cu^+, Zn^+). Studies with chymopapain indicate that probably two sulfhydryl groups are present in the enzyme manifesting maximal activity and that one of them is preferentially labeled by N-(4-dimethyl-amino-3,5-dinitrophenyl) maleimide with concomitant loss of enzymatic activity. The sequence around this cysteine residue has been elucidated.[17]

Stability. Chymopapain is very stable when kept at slightly acidic pH's. In fact, one of the earliest characteristics noted was the remarkable stability of this enzyme at a pH as low as 2 when kept at temperatures

[16] M. Ebata and Y. Takahashi, *Biochem. Biophys. Acta* **118**, 198 (1966).
[17] J. N. Tsunoda and K. T. Yasunobu, *J. Biol. Chem.* **241**, 4610 (1966).

below 10°.[10] It will be noted that this stability is required when the enzyme is subjected to salt fractionation, as it is soluble in saturated NaCl above pH 3.0. Complete precipitation of this enzyme requires that the pH of the solution be adjusted to approximately 2.

[16] Streptococcal Proteinase

By S. D. ELLIOTT and TEH-YUNG LIU

Assay Methods

Three different methods have been used for the assay of streptococcal proteinase: (1) measurement of its hydrolytic activity; (2) estimation of its serological reactivity; (3) for highly purified material, measurement of its absorbancy at 280 nm.

Measurement of Proteolytic Activity. The milk-clotting action of streptococcal proteinase affords a quick means of assay reproducible over a narrow range of enzyme concentration.[1] Alternatively, assay with protein substrates such as casein can be performed by the method of Kunitz[2] modified by Liu *et al.*,[3] as described here. The reactions are carried out at 40° and pH 7.6 in the presence of $1 \times 10^{-5} M$ EDTA and $5 \times 10^{-6} M$ mercaptoethanol or dithiothreitol. The concentration of casein (Hammarsten) in the reaction mixture is 0.05% (w/v) and the amount of enzyme used is 5–10 μg/ml. After 20 minutes' digestion, enzyme action is stopped by the addition of an equal volume of 5% trichloroacetic acid, the mixture centrifuged or filtered, and the supernatant or filtrate examined in a spectrophotometer at 280 nm. Appropriate controls should be run. The proteolytic unit, U, is the activity that gives rise to an increase of one unit of absorbance at 280 nm per minute of digestion.

Peptidase Activity. Assay with a peptide substrate such as N^a-Z-Lys-Phe[4] or N-Z-Phe-Tyr[5] is carried out in 0.2 *M* N-ethylmorpholine acetate buffer containing 0.2 *M* NaCl at 37° in the presence of $1 \times 10^{-5} M$ EDTA and $5 \times 10^{-6} M$ dithiothreitol. The concentration of the peptide

[1] S. D. Elliott and V. P. Dole, *J. Exptl. Med.* **85**, 305 (1947).

[2] M. Kunitz, *J. Gen. Physiol.* **30**, 291 (1947).

[3] T. Y. Liu, N. P. Neumann, S. D. Elliott, S. Moore, and W. H. Stein, *J. Biol. Chem.* **238**, 251 (1963).

[4] T. Y. Liu, N. Nomura, E. K. Jonsson, and B. G. Wallace, *J. Biol. Chem.* **244**, 5745 (1969).

[5] B. I. Gerwin, W. H. Stein, and S. Moore, *J. Biol. Chem.* **241**, 3331 (1966).

in the reaction mixture is $4 \times 10^{-3}\,M$ and the amount of enzyme used is 0.2–0.4 mg/ml. At time intervals of 0, 2, 4, 8, 12, 16, and 20 minutes, 100-μl aliquots of the sample are withdrawn and the amino acids released from the peptide are quantitatively measured either by the amino acid analyzer[4] or by the ninhydrin procedure.[5] For the synthetic substrate N^{a}-Z-Lys-Phe, the turnover rate of the purified enzyme is 2.25 ± 0.2 mole of substrate hydrolyzed per mole of enzyme per second.

Esterase Activity. Assay with an ester substrate such as N^{a}-Z-Lys-ONp (Cyclo Chemical Co.) may be followed spectrophotometrically at 340 nm[4] with the Cary Model 14 or Model 15 CM recording spectrophotometer. The reactions are performed in $0.2\,M$ Na-acetate buffer at pH 5.5 and 25° in the presence of $1 \times 10^{-5}\,M$ EDTA and $5 \times 10^{-6}\,M$ dithiothreitol. The concentration of substrate in the reaction mixture is $2 \times 10^{-4}\,M$ and the amount of enzyme used is 20 to 40 μg/ml. Substrate is stored in solution in acetonitrile and the final concentration of acetonitrile in the assay mixture is 1.6%. The turnover number of the purified enzyme under the experimental condition specified is 140 ± 15 moles of substrate hydrolyzed per mole of enzyme per second.

Estimation of Serological Reactivity

Both the enzyme and the inactive zymogen from which it is derived are antigenic in rabbits.[6] Zymogen and proteinase have distinct immunological specificities, and precipitation reactions with the appropriate antisera afford a rapid means of assessing the concentrations of either protein. Serological tests, like the test for milk-clotting activity, are of particular value in the rapid assay of crude culture fluids.

Measurement of Absorbance at 280 nm

For highly purified preparations such as those which result from chromatographic treatment of crystalline material, the enzyme concentration can be estimated from the amount of protein in solution as determined by its absorbance at 280 nm. An absorbance of 1.64 (1 cm; 280 nm) corresponds to 1 mg proteinase per milliliter. The corresponding specific absorbancy of the zymogen is 1.37.[3]

Preparation and Purification

General Considerations

In broth cultures of group A streptococci the proteinase first appears in the form of an inactive zymogen or precursor. The zymogen accumulates extracellularly during active multiplication of the microorganisms

[6] S. D. Elliott, *J. Exptl. Med.* **92**, 201 (1950).

and reaches its maximum concentration at the end of the logarithmic phase of growth when, from some strains (e.g., B220 of Dr. Rebecca Lancefield's collection at The Rockefeller University), yields up to about 150 mg/liter may be expected. From filtrates of such cultures the zymogen can readily be separated in crystalline form by treatment with $(NH_4)_2SO_4$ and thereafter it can be converted to the proteinase by digestion with trypsin.[6,7] In streptococcal cultures of sufficient reducing potential, zymogen is converted in the absence of trypsin by an autocatalytic reaction but, although the enzyme formed in this way crystallizes in the presence of $(NH_4)_2SO_4$, preparations of this kind have been found to be less homogeneous than those obtained from crystalline zymogen by the action of trypsin.

Preparation of the Culture Medium

To facilitate purification of zymogen, the streptococci utilized for its production are grown in a medium all constituents of which are dialyzable. When a highly productive strain is grown under appropriate conditions in such a medium, zymogen protein constitutes more than 90% of the nondialyzable material present in filtrates of fully grown cultures. The medium here described is a modification of that recommended by Dole[8] and consists of combined dialyzates of beef heart infusion and peptone to which are added salts and dextrose. The recipe is for 20 liters, a volume easy to handle and sufficient to yield 2–3 g of crystalline zymogen.

Preparation of Dialyzate of Beef Heart Infusion for 20 Liters of Broth. After stripping off fat, 10 lb of fresh beef heart are first ground and then infused overnight in 2 liters of tap water at 4°. The infusion is heated to 85° for 30 minutes, cooled, and filtered through paper (Eaton and Dikeman No. 195). The filtrate is concentrated approximately 10-fold in a rotary evaporator and then dialyzed at 4° against three changes of distilled water, each for 12 hours. Thick-walled, seamless, cellulose casings of 24/32 inch inflated diameter (Visking Corp.) are used in dialysis. The three dialyzates are combined, concentrated *in vacuo* to approximately 400 ml, and stored, frozen. One lot of concentrate is needed for 20 liters of complete medium.

Preparation of Peptone Dialyzate for 20 Liters of Broth. One pound (454 g) Phanstiehl (R.I.) peptone is dissolved in 2 liters of distilled water, and 20 g charcoal (decolorizing carbon, General Chemical Co., New York) is then added. The preparation is brought to 80° after which

[7] T. Y. Liu and S. D. Elliott, *J. Biol. Chem.* **240**, 1138 (1965).
[8] V. P. Dole, *Proc. Soc. Exptl. Biol. Med.* **63**, 122 (1946).

the suspension is filtered through Filter Cel. The filtrate is dialyzed overnight against 14 liters (7 volumes) of distilled water at 4°, care being taken to allow for at least a 20% increase of volume of the fluid in the dialysis sacs. The dialyzate is reabsorbed with charcoal (1% w/v) as before and then brought to a volume of 20 liters by the addition of concentrated beef heart dialyzate (1 lot of approximately 400 ml, see above) and distilled water.

Final Stage in Preparation of the Complete Medium. To the mixture of meat infusion and peptone dialyzates (20 liters) are added NaCl, NaHCO$_3$ and dextrose to final concentrations of 0.1, 0.2, and 2.5% respectively. Finally, the pH level is adjusted to 7.8 by the addition of 5 N NaOH; approximately 50 ml are usually required for 20 liters of broth. The complete medium is sterilized by filtration through Coors P3 filter candles and should be used immediately.

Production of Zymogen in Broth Cultures of Streptococci

One liter of dialyzate broth is inoculated with 10 ml of an 18-hour culture of a strain of group A streptococci potent in the production of proteinase. The dialyzate broth culture is incubated at 37° for approximately 10 hours and then added to 20 liters of dialyzate medium preheated to 37°. After inoculation, incubation is continued at 37° and the reaction maintained at a pH level between 6.5 and 7.0 by the frequent addition of sterile sodium bicarbonate solution (10% w/v). Under these conditions, zymogen usually reaches its maximum concentration in the culture supernatant after about 8 hours at which time glycolysis becomes much reduced.

Assay of Zymogen in Culture Supernatant

The concentration of zymogen in the culture fluid may be roughly estimated by either of the two rapid methods, serological precipitation reaction[6] or milk-clotting activity.[1] The most convenient method makes use of precipitation reactions between the culture supernatant and zymogen-specific rabbit antiserum. Alternatively, the culture fluid may be treated with trypsin, thereby converting zymogen to enzyme which, after reduction, can be estimated by its milk-clotting activity. Small samples are taken from the dialyzate broth culture at intervals during incubation and tested by one or other of these methods. Although imprecise, they give a rapid estimate of zymogen concentration so that the culture can be harvested as soon as the yield is satisfactory. Incubation for too long may result in the loss of zymogen through autocatalytic conversion to the active enzyme.

Separation and Crystallization of Zymogen from Broth Cultures

When zymogen in the culture fluid has reached the desired concentration, the supernatant is separated from the cocci by centrifugation. (Collected with sterile precautions, the cocci may be stored at −70° and used for further zymogen production by resuspending them in dialyzate medium and incubating under the appropriate conditions).[6]

To the culture supernatant is added $(NH_4)_2SO_4$ to a concentration of 560 g/liter (0.8 saturation). The resulting precipitate contains all the zymogen and it is collected by filtration through coarse paper using Filter Cel and negative pressure. The mixture of precipitate and Filter Cel is immediately resuspended in chilled $(NH_4)_2SO_4$ (0.4 saturation) in a volume approximately 0.01 that of the original culture. Under these conditions the zymogen protein dissolves in the cold $(NH_4)_2SO_4$ solution provided that the final concentration of $(NH_4)_2SO_4$ does not rise above 0.5 saturation through the addition of excessive salt entrained in the mixture of protein and Filter Cel. The insoluble residue of extraneous streptococcal metabolites and Filter Cel is removed by filtration at 4°. The filtrate usually contains zymogen protein in a concentration of approximately 1.0% w/v but it is advisable to confirm this either by tests for serological reactivity or by absorbance at 280 mμ. If present in a concentration of approximately 1.0%, zymogen will crystallize from solution when dialyzed at 37° against half-saturated $(NH_4)_2SO_4$ at pH 8[6]; at 4° it usually forms an amorphous precipitate.

Chromatographic Procedure for Purification of Crystalline Zymogen

Crystalline zymogen can be further purified by gel filtration on Sephadex G-75 followed by ion-exchange chromatography on sulfoethyl-Sephadex.[7] Twice crystallized zymogen is essentially homogeneous as shown by chromatographic and electrophoretic procedures.

Conversion of Zymogen to Enzyme by Trypsin

Proteolytic cleavage of the zymogen to the enzyme can be effected by the action of trypsin, subtilisin, or the streptococcal proteinase itself.[7]

The crystalline zymogen, after being spun down from the suspension in $(NH_4)_2SO_4$ solution, is dissolved in 0.04 M NaCl containing either 0.1% disodium tetrathionate or 0.1% iodoacetamide for the gel filtration. After gel filtration, the zymogen solution (100 mg in 10–15 ml) is adjusted to pH 6.7 ± 0.1 with 1 M sodium phosphate buffer. The final concentration of phosphate is 0.1 M. Trypsin (twice crystallized) is added in an amount equal to 0.5% of the weight of the zymogen. The mixture is left at 25° and after 4 hours and 8 hours further aliquots of trypsin are added. At the end of 18–20 hours the transformation is usually complete as judged by immunological test. The reaction mixture is then transferred to a Sephadex G-75 column for gel filtration followed by ion-exchange chromatography on SE-Sephadex.

Transformation of zymogen to enzyme can also be effected by subtilisin. The conditions are the same as those used for transformation by trypsin; the transformation is complete in 1 hour.

Activated streptococcal proteinase, freed of sulfhydryl reagent by gel filtration or dialysis under nitrogen can also be used to activate zymogen. When 1 mg of this activated proteinase is added to 100 mg of the zymogen under the conditions used with trypsin, the transformation is complete in 2 hours at 37° under nitrogen.

Of the three methods of effecting the transformation of zymogen to enzyme, the best method for the preparation of a homogeneous sample of enzyme in high yield (88 to 98%) is by the action of trypsin.

Properties

Stability. After removal of $(NH_4)_2SO_4$ with Sephadex G-25, solutions of both the enzyme and the zymogen are stable for several days at 4°.[3] There was no loss of potential enzymatic activity or change in chromatographic behavior when solutions or crystal suspensions of zymogen in half-saturated $(NH_4)_2SO_4$ were stored at −20° for 4 years. If, however, the zymogen or the enzyme was kept as a crystalline suspension in half-saturated ammonium sulfate at room temperature, there was a gradual modification to yield inactive protein. Activated proteinase free from sulfhydryl reagents can be stored under nitrogen in the presence of 1×10^{-5} M EDTA without loss of more than 3–5% of activity per week.[7]

Purity and Physical Properties. The twice crystallized zymogen is essentially homogeneous in several chromatographic systems (CMC, SE-Sephadex, and IRC-50).[3] The purity of the zymogen preparation was further established by amino terminal residue analysis and ultracentrifugal studies. Thus it was shown that the crystalline zymogen produced in the culture medium has a molecular weight of 44,000, and has an aspartic acid residue in the NH_2-terminal position.

The enzyme produced by the action of trypsin on the zymogen was shown to be homogeneous by gel filtration on Sephadex 75 or 100 and by ion-exchange chromatography on SE-Sephadex.[7] Analytical data also provide evidence of the homogeneity of the enzyme obtained by tryptic action. For example, a special property of the streptococcal proteinase is the presence of 1 half-cystine residue per molecule; the trypsin-produced enzyme is pure by this analytical criterion. It is also homogeneous by end group analysis; 0.9 to 1.0 residue of glutamic acid was found as the sole aminoterminal residue by the cyanate method. The molecular weight determined by ultracentrifugation was 32,000 ± 800, which agrees with the minimum molecular weight calculated from the amino acid composition.

Activators and Inactivators. Streptococcal proteinase is a sulfhydryl enzyme like papain and ficin.[9] After conversion from zymogen to enzyme, reduction is essential for catalytic activity as a proteinase, peptidase,[4, 5] or esterase.[4] Reduction may be brought about by thiol compounds such as thioethanol or thioglycolic acid, or by cyanide or bisulfide. Once the enzyme is activated by a reducing agent, the presence of the activator is no longer necessary for the catalytic activity of the enzyme. Thus, these activators cannot be functioning as coenzymes.

The reduced enzyme is reversibly inactivated by disodium tetra-thionate. The inactive *S*-sulfenylsulfonate form of the enzyme upon exposure to a sulfhydryl reducing agent will regenerate full enzymatic activity. If, however, the *S*-sulfenylsulfonate derivative of the enzyme is exposed to α-*N*-bromoacetylarginine methyl ester, 1 molecule of the reagent is incorporated per molecule of protein. The modified protein possessed no enzymatic activity when the sulfhydryl group was regenerated by reduction.[10]

The reduced enzyme is irreversibly inactivated by iodoacetic acid,[9] iodoacetamide, and *N*-ethylmaleimide (NEM)[3] with the irreversible modification of the single sulfhydryl group of the enzyme.

Specificity. Mycek, Elliott, and Fruton[11] observed that streptococcal proteinase hydrolyzes typical trypsin substrates such as benzoylarginine amide, benzoyl lysinamide, benzoyl histidinamide, and benzoyloxycarbonylisoglutamine. Liu *et al.*[3] noted that Z-Gly-Phe, a typical substrate for carboxypeptidase, was also hydrolyzed by the enzyme. However, the rate of hydrolysis of these substrates was low; for example, streptococcal proteinase is only about 1% as effective as trypsin in hydrolyzing benzoyl arginine amide.

With the reduced, carboxymethylated phenylalanyl chain of insulin as substrate, Gerwin *et al.*[5] investigated the specificity of streptococcal proteinase and identified one of the bonds most rapidly hydrolyzed as Phe-Tyr linkage in the -Phe-Phe-Tyr sequence. With these results as a lead, a series of synthetic peptides was tested. The results of these experiments showed that the chief requirement for hydrolysis is the presence of a bulky side chain on the amino acid adjacent, on the amino terminal

[9] S. D. Elliott, *J. Exptl. Med.* **81**, 573 (1945).
[10] T. Y. Liu, *J. Biol. Chem.* **242**, 4029 (1967).
[11] M. Mycek, S. D. Elliott, and J. S. Fruton, *J. Biol. Chem.* **197**, 637 (1952).

side, to the residue contributing the carbonyl to the susceptible peptide bond.

A spectrophotometric method was developed by Liu et al.[4] to demonstrate that streptococcal proteinase is an efficient esterase in the cleavage of phenyl and p-nitrophenyl esters of the general type A-X-Y, where A = benzyloxycarbonyl, X = lysine, norleucine or glutamic acid, and Y = phenyl or p-nitrophenyl. Tosylnorleucylphenylalanine and tosyllysine phenyl ester were found to be resistant. It was shown that streptococcal proteinase exhibits a preference for X, the acyl component of the sensitive bond, in the decreasing order Lys > Nle > Glu. This sequence is not the same as for peptide bond cleavage in A-X-Y where Y = phenylalanine; the most susceptible peptide was the one in which X = Nle; and the least susceptible one was the one where X = Lys. Evidence of the exacting specificity of streptococcal proteinase with respect to A is provided by the observation that when A is changed from Z to Tos- as in Tos-Nle-Phe and Tos-Lys-Phe the peptide bonds were resistant to proteolytic cleavage[4] and that when A is lacking altogether, or when it is small, hydrolysis is impaired.[5]

Kinetic Properties. The effect of pH on the streptococcal proteinase-catalyzed hydrolysis of N^a-Z-LysONp has been examined over a range between pH 4 and 6.9 by Liu et al.[4] The results of these studies reveal that K_{cat} and K_m values are fairly constant above 5.0 for this substrate. The ratio of log K_{cat}/K_m for the enzyme was shown to depend on an ionizable group with a pK of 4.8. From the values of K_m and V_{max} for the streptococcal proteinase-catalyzed hydrolysis of Z-Phe-Tyr over the pH range 5 to 10, Gerwin et al.[5] have inferred a pK_a of 6.4 for a catalytically important group in the free enzyme. This group was suggested to be an unprotonated histidine residue at the active site of the enzyme. Direct chemical evidence for the involvement of a histidine residue in the active site of the enzyme was demonstratd by Liu.[10] The apparent discrepancy of 1.6 pK units (4.8 vs. 6.4) between the results of Liu et al.[4] and Gerwin et al.[5] on the pK of a catalytic entity in the streptococcal proteinase-catalyzed hydrolysis of peptide and ester substrates emphasizes the necessity of having independent chemical evidence for assigning a particular functional group at the active site of an enzyme. On the basis of the kinetic data alone, one would hesitate to assign to an imidazole group a p$K = 4.8$. However, on the basis of the chemical evidence cited,[10] an imidazole group with p$K_a = 4.8$ may be present in the active site of streptococcal proteinase for the hydrolysis of Z-Lys-ONp. It is not possible to rule out the alternative that this group might actually be a carboxyl group.

Amino Acid Composition and the Sequence of Amino Acid Residues

AMINO ACID COMPOSITION OF THE STREPTOCOCCAL ZYMOGEN AND ENZYME

Amino acids[a]	Zymogen (residues/molecule)		Enzyme (residues/molecule)	
	Found	Expressed to nearest integer	Found	Expressed to nearest integer
Aspartic acid	51.5	52	39.1	39
Glutamic acid	38.4	38	28.3	28
Glycine	44.4	44	36.6	37
Alanine	31.4	31	21.8	22
Valine	24.8	25	20.1	20
Leucine	21.0	21	17.0	17
Isoleucine	21.7	22	12.2	12
Serine[b]	36.9	37	26.3	26
Threonine[b]	16.4	16	11.1	11
Half-cystine	1.03[c] 0.94[d]	1	1.02[c]	1
Methionine	6.85[e] 6.45[f]	7	5.02[e]	5
Proline	14.0	14	14.1	14
Phenylalanine	16.4	16	12.0	12
Tyrosine[a]	22.9	23	18.8	19
Tryptophan	4.4[g] 4.6[h]	5		—
Histidine	7.6	8	8.1	8
Lysine	29.4	29	17.0	17
Arginine	10.9	11	8.7	9
Amide NH_2	45.5	46		

[a] Determined by ion-exchange chromatography. The results are expressed as the calculated number of residues per molecule, based on leucine = 21 for the zymogen and leucine = 17 for the enzyme. The finding of one half-cystine residue per molecule serves to place the minimal molecular weights calculated from the chemical analyses at 44,347 for the zymogen and 32,000 for the enzyme, which are compatible with the ultracentrifugally determined value of 44,000 ± 500 for the zymogen and 32,000 ± 800 for the enzyme. [D. H. Spackman, W. H. Skin, and S. Moore, *Anal. Chem.* **30**, 1190 (1958).]

[b] The results for the labile amino acids were corrected approximately for decomposition during 22 hours of acid hydrolysis. The factors applied were: serine, 0.90; threonine, 0.95; tyrosine, 0.95. [A. M. Crestfield, S. Moore, and W. H. Stein, *J. Biol. Chem.* **238**, 622 (1968).]

[c] Measured as cysteic acid after performic acid oxidation.

[d] Determined as S-carboxymethyl cysteine after iodoacetate treatment of the reduced zymogen.

[e] Determined as methionine sulfone after performic acid oxidation. [S. Moore, *J. Biol. Chem.* **238**, 235 (1963).]

[f] Determined as methionine on a hydrolyzate of the unoxidized protein.

[g] Determined spectrophotometrically. [T. W. Goodwin and R. A. Morton, *Biochem. J.* **40**, 628 (1946).]

[h] Determined chromatographically after hydrolysis by alkali. [A. Drèze, *Bull. Soc. Chim. Biol.* **42**, 407 (1960).]

Around the Sulfhydryl Group at the Active Site of Streptococcal Proteinase. The amino acid compositions of the zymogen and the enzyme produced by the action of trypsin on the zymogen are listed in the table.

Val-Lys-Pro-Gly-Glu-Gln-Ser-Phe-Val-Gly-Gln-Ala-Ala-Thr-Gly-His-Cys(Cm)-Val-Ala-
Thr-Ala-Thr-Ala-Gln-Ile-Met-Lys

Fig. 1. The sequence of amino acid residues around the sulfhydryl group at the active site of streptococcal proteinase.

The sulfhydryl group at the active site of streptococcal proteinase was labeled with iodoacetic-1-^{14}C acid and the labeled protein was hydrolyzed with trypsin. A single ^{14}C-containing peptide of 27 amino acid residues was isolated and the sequence of amino acid residues in this peptide (shown in Fig. 1) was determined by Edman degradation and enzymatic hydrolysis with carboxypeptidase, chymotrypsin, pepsin, and papain. One feature of this sequence is the presence of a histidine residue adjacent to the essential sulfhydryl residue.[12]

[12] T. Y. Liu, W. H. Stein, S. Moore, and S. D. Elliott, *J. Biol. Chem.* **240**, 1143 (1965).

[17] Ficin

By Irvin E. Liener and Bernard Friedenson

Assay Method

Principle. The activity of ficin (systemic number, EC 3.4.4.12) may be determined by measuring the rate of hydrolysis of protein or synthetic substrates. Procedures involving the use of both of these two types of substrates will be described.

Reagents

Casein Method
  Sodium phosphate buffer, 0.1 M, pH 7.0, containing 0.007 M mercaptoethanol and EDTA 0.001 M.
  For other reagents, see Vol. II, p. 33.
Synthetic substrate, p-nitrophenyl carbobenzoxyglycinate (p-NPZG).
  Substrate solution. p-NPZG (41.2 mg) dissolved in 3.0 ml of acetonitrile (Eastman spectro grade).
  Sodium phosphate buffer, 0.02 M, containing 0.005 M KCN, 0.001 M EDTA, and 20% acetonitrile, pH adjusted to 6.8.

Enzyme solution. A solution of ficin containing 2.5–250 μg/ml prepared by suitable dilution of a stock enzyme solution with the 0.02 M phosphate buffer described above.

Procedure with Casein as Substrate. The use of casein as a substrate for measuring ficin activity is based on the procedure originally described by Kunitz[1] for the determination of tryptic activity. The details of this method have already been presented (see Vol. II, p. 33). For ficin, the method has been somewhat modified by Englund et al.[2] to include an activator (mercaptoethanol) and EDTA. A solution containing 0–8 μg of ficin is diluted to 1.0 ml with 0.1 M phosphate buffer, pH 7.0, containing mercaptoethanol and EDTA. The enzyme solution is mixed with an equal volume of 1% casein made up in the same buffer solution. Following incubation at 37° the remainder of the procedure is essentially the same as described previously and involves measurement of the absorbance of the trichloroacetic acid filtrate at 280 nm. If blank corrections are carefully made, this assay procedure can be a rapid and fairly accurate method for following the purification of ficin from the crude latex and for verifying the validity of assays with synthetic substrates.

Procedure with Synthetic Substrate. Synthetic substrates which have been used for measuring the amidase activity of ficin include hippurylamide[3] and N-benzoyl-D,L-arginine p-nitroanilide.[2] Esterase activity has been measured on such substrates as N-benzoyl-D,L-arginine ethyl ester,[3,4] methyl hippurate,[3,5] and the p-nitrophenyl esters of hippuric acid[5] and carbobenzoxyglycine.[6,7] Use of the latter substrate forms the basis of an assay procedure which has proved convenient and reliable in our laboratory.

Three milliliters of enzyme solution are pipetted into a 1-cm cuvette which is placed into the cell compartment of a spectrophotometer in which thermospacers are used to maintain the temperature at 25°. After an equilibration period of about 10 minutes, a 50-μl aliquot of p-NPZG solution is carefully pipetted onto the foot of a small footed stirring rod which is rapidly raised and lowered several times inside the cuvette in order to mix the enzyme and substrate. The rate of change of the absorbance at 400 nm is carefully recorded. All readings are made against an identical solution from which the enzyme has been omitted.

---

[1] M. Kunitz, *J. Gen. Physiol.* **30**, 291 (1947).
[2] P. T. Englund, T. P. King, and L. C. Craig, *Biochemistry* **7**, 163 (1968).
[3] B. R. Hammond and H. Gutfreund, *Biochem. J.* **72**, 349 (1959).
[4] S. A. Bernhard and H. Gutfreund, *Biochem. J.* **63**, 61 (1956).
[5] G. Lowe and A. Williams, *Biochem. J.* **96**, 199 (1965).
[6] J. F. Kirsch and M. Igelström, *Biochemistry* **5**, 783 (1966).
[7] K. Fossum and J. R. Whitaker, *Arch. Biochem. Biophys.* **125**, 367 (1968).

This assay is most conveniently carried out in an automatic recording spectrophotometer (Gilford model 2000). The increase in absorbance at 400 nm should be linear with time. A standard curve relating increase in absorbance per unit time (Δ absorbance per minute) to enzyme concentration obtained under the conditions described here is shown in Fig. 1. Because of the lack of specificity inherent in nitrophenyl assays (for example, sulfhydryl compounds catalyze the hydrolysis of p-NPZG[6,8]) this assay procedure should be used only for fairly pure enzyme preparations.

Definition of Unit and Specific Activity. With casein as the substrate, the activity may be expressed in terms of activity units by constructing a standard activity curve as described for trypsin by Kunitz[1] (see Fig. 1, Vol. II, p. 33). When p-NPZG is used as the substrate according to the procedure described here, one unit of activity is arbitrarily defined as that amount of ficin which will cause an increase of 1 absorbance unit per minute at 400 nm. Specific activity may then be expressed as units per milligram of protein.

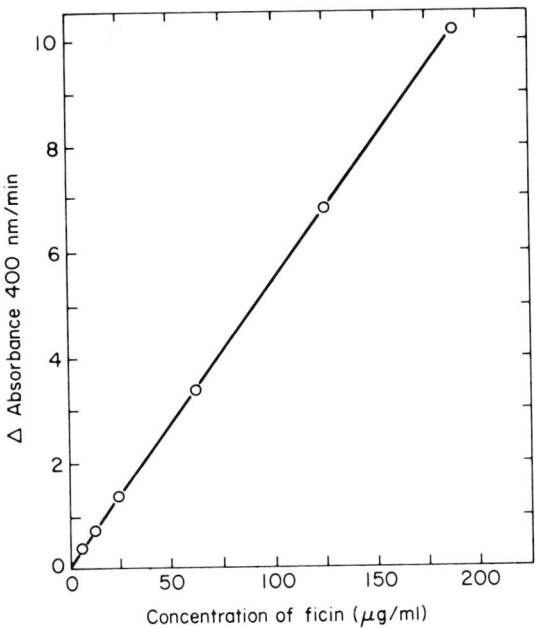

FIG. 1. Standard assay curve for ficin with p-nitrophenyl carbobenzoxyglycinate as the substrate. Reprinted from *Biochemistry* **7**, 163 (1968). Copyright 1968 by the American Chemical Society. Reprinted by permission of the copyright owner.

[8] B. Friedenson, E. Oosterom, and I. E. Liener, unpublished observations.

Purification Procedure

Ficin was first crystallized from fresh fig latex by Walti,[9] and, although simple, this method is not always reproducible, perhaps because of the variability of the starting material. More recent studies have centered on attempts to purify proteolytically active components from dried fig latex, which is inexpensive and commercially available. In general, most procedures involve preliminary purification by means of precipitation with salt, followed by ion-exchange chromatography,[2, 10] or, in some instances, by gel filtration[11] or curtain electrophoresis.[12] Since several active components are invariably observed, the designation of a particular component as "ficin" has been quite arbitrary and often inconsistent from one research group to another.

A number of procedures for the preliminary purification of the active enzyme by salt precipitation have been described.[3, 11, 13, 14] The procedure given here is taken from Englund et al.[2] and has the advantage that the enzyme is stabilized as an inactive derivative during the course of the purification procedure. About 20 g of crude latex powder are suspended in 200 ml of 0.01 M sodium tetrathionate containing 0.001 M EDTA. Tetrathionate is used to reversibly inactivate ficin via conversion of the sulfhydryl group to a sulfenyl thiosulfate $(SSSO_3)^-$ and thus to minimize losses due to oxidation and autolysis. The pH of the suspension is brought to neutrality with 1.0 M NaOH, and the mixture is stirred for 3 hours at room temperature. After centrifuging off the insoluble residue, 17.7 g NaCl are added to each 100 ml of the supernatant at 4°. After standing for several hours at 4°, the precipitate is collected by centrifugation and suspended in 100 ml of pH 7.1 buffer containing 0.005 M sodium phosphate, 0.001 M sodium tetrathionate, and 0.001 M EDTA. The preparation is then dialyzed overnight against the same buffer and may then be used directly for chromatography (see below). Restoration of enzymatic activity can be accomplished at any stage of the purification process by adding a 5000-fold molar excess of mercaptoethanol or a 300-fold molar excess of dithioerythritol.

Subsequent resolution of the proteolytically active components of

[9] A. Walti, J. Am. Chem. Soc. 60, 493 (1938).
[10] V. C. Sgarbieri, S. M. Gupte, D. E. Kramer, and J. R. Whitaker, J. Biol. Chem. 239, 2170 (1964).
[11] G. Porcelli, J. Chromatog. 28, 44 (1968).
[12] R. A. Messing and W. P. Van Ness, Enzymologia 23, 373 (1961).
[13] R. M. Metrione, R. B. Johnston, and R. Seng, Arch. Biochem. Biophys. 122, 137 (1967).
[14] G. B. Marini-Bettolo, P. U. Angelatti, M. L. Salvi, L. Tentori, and G. Vivaldi, Gazz. Chim. Ital. 93, 1239 (1963).

the salt-purified fraction has been generally accomplished by chromatography on CM-cellulose. Three representative chromatograms and the conditions under which they were obtained are presented in Fig. 2 and Table I. All of the chromatograms were obtained with preparations of ficin from *Ficus glabrata* latex. Although the validity of comparing chromatographic data obtained in different laboratories under different conditions may be questioned, an attempt has neverthless been made in Fig. 2 to indicate the possible relationship of the various active components observed in three different laboratories. There would appear to be at least three major active components (arbitrarily designated as fractions I, II, and III in Fig. 2) and several minor enzymatically active components in the latex of *F. glabrata*. One of these (referred to as component "G" by Sgarbieri *et al.*,[10] see Fig. 2B) is apparently the major component of ficin crystallized according to the procedure of Walti.[9] As many as ten active components have been reported to be present in the latex from *F. carica*.[16] That these multiple components are probably not active products of autolysis is indicated by the fact that identical chromatograms are obtained from fresh latex and from latex which had been incubated at 35° for 6 hours.[10] Since ultracentrifugation, electrophoresis, and chromatography on CM-cellulose at pH 4.4 all fail to resolve this mixture of enzymes,[3,17] these active components must be very similar with respect to their physical and chemical properties. All of the active components appear to be sulfhydryl proteases since they are all activated by reducing agents and inactivated by reagents which combine with sulfhydryl groups.[15,16] Radioautographs of peptide maps derived from radioactive *S*-carboxymethyl derivatives of the three major components shown in Fig. 2C show little difference, indicating that the amino acid sequences around the reactive SH group are probably identical.[17] Whether the difference in the chromatographic behavior of the various active components of "ficin" can be accounted for by minor differences in size, amino acid composition, carbohydrate content, or charge remains to be elucidated.

Properties

Stability. At 50° for 2 hours ficin (described by the author[18] as crystalline without further details regarding source) showed maximum stability in a broad region between pH 4.5 and 9.5 except for a minimum in a narrow zone around pH 8.[18] This minimum was attributed to the

[15] D. E. Kramer and J. R. Whitaker, *J. Biol. Chem.* **239**, 2178 (1964).
[16] I. E. Liener, *Biochem. Biophys. Acta* **53**, 332 (1961).
[17] B. Friedenson and I. E. Liener, unpublished observations.
[18] W. Cohen, *Nature* **182**, 659 (1958).

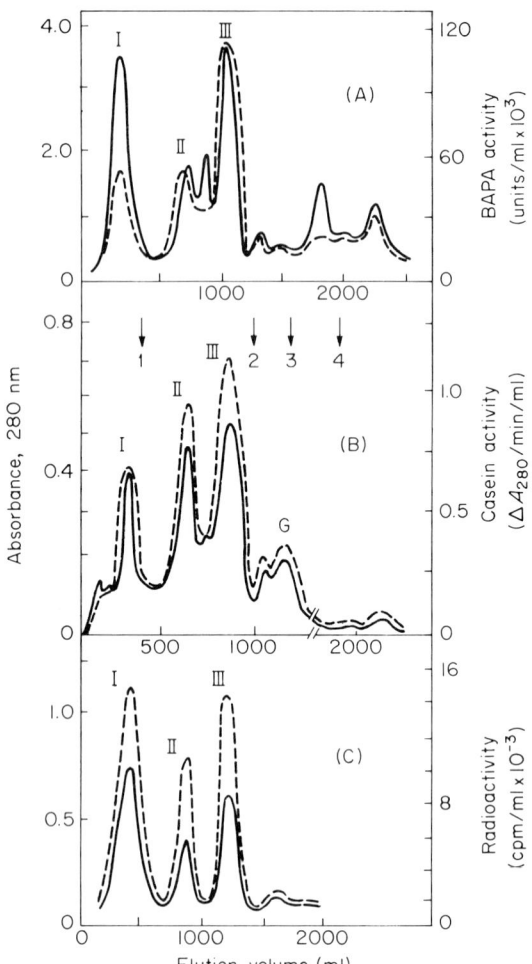

Fig. 2. Chromatography of preparations of ficin on CM-cellulose as performed in several different laboratories. (A) Taken from Englund et al.,[2] with permission of the American Chemical Society; (B) taken from Sgarbieri et al.,[10] with permission of J. Biol. Chem.; (C) Friedenson and Liener.[15] Conditions of chromatography are described in Table I. Solid line denotes protein as measured by absorbance at 280 nm Dotted line denotes activity on benzoyl-D,L-arginine-p-nitroanilide (BAPA) in A, casein in B, and radioactivity in C.

oxidation of an essential sulfhydryl group since, on raising the concentration of cysteine from 0.01 M to 0.025 M, this decline in activity was no longer observed, and the true point of maximum stability was pH 7.5. A study of the stability of 10 active components from the latex of

TABLE I

CONDITIONS USED FOR OBTAINING THE CHROMATOGRAMS SHOWN IN FIGURE 2

Chromatogram	Column size (cm)	Flow rate (ml/min)	Temperature (°C)	Preparation used	Elution schedule
A	4 × 20	2.5	25	Salt-purified and tetrathionate inactivated as described in text.	Sodium phosphate buffer, 0.005 M, pH 7.1, containing 0.001 M EDTA and 0.001 M tetrathionate until peak I was eluted. Followed by linear gradient with 0.185 M sodium phosphate buffer, pH 6.9, containing 0.001 M EDTA and 0.001 M tetrathionate. Total volume of gradient, 3 liters.
B	2 × 45	1.5	4	Active enzyme, salt-purified by the procedure of Hammond and Gutfreund.[3] Subsequently absorbed to small quantity of CM-cellulose in 0.005 M citrate–0.01 M sodium phosphate, pH 4.9, washed, and transferred to top of column.	Stepwise elution with 0.01 M phosphate buffer, pH 7, containing increasing of NaCl in the following order (indicated by arrows in Fig. 2B): (1) 0.04 M, (2) 0.1 M, (3) 0.12 M, (4) 0.20 M.
C	2 × 22	0.8–1.0	4	Same as 2A except tetrathionate and EDTA were omitted, and free SH groups were labeled with ICH$_2$-^{14}COOH.	Same as 2A except that linear gradient was produced with 0.06 M sodium phosphate buffer, pH 7.0, as the upper limit. Total volume of gradient, 2.4 liters.

F. carica var. *Kadota* revealed that 5 components retained at least 75% of their activity after 1 hour at pH 7 and 55°, whereas the remaining 5 components were much less stable under the same conditions.[16]

Solutions of ficin have been reported to lose activity after prolonged periods of storage in the frozen state, presumably due to partial oxidation of an inactive form of the enzyme.[3,16] Lyophilization of solutions of ficin in water or volatile salts causes almost complete inactivation.[2,16,19] In fact, mere removal of salt by dialysis is sufficient to cause almost complete and irreversible inactivation of enzymes from *F. carica* var. *Kadota*.[19] This loss in activity could be prevented by lyophilization from 0.01 *M* phosphate buffer, pH 7, containing 0.1 *M* NaCl.

Effect of pH on Activity. Curves relating to the effect of pH on the initial rate of hydrolysis of benzoyl-L-arginine ethyl ester, benzoyl-L-argininamide, hippuryl amide, and hippuryl methyl ester are virtually superimposable and show the optimum pH to be 6.5.[3] Three of the isolated active components from the latex of *F. carica* var. *Kadota* had a pH optimum of 6.6 on benzoyl-L-argininamide, whereas 4 other components displayed a higher optimum of pH 7.2.[15]

Purity. If pure, the individual active components separated by column chromatography should show a single component when examined by one or more of the following techniques: (1) chromatography on CM-cellulose under the conditions employed for the initial resolution,[2,10] or under conditions somewhat different from those employed for the original purification[2]; (2) chromatography on Bio-Rex 70 at pH 6.1[2]; (3) moving-boundary electrophoresis[10]; or (4) zone electrophoresis on strips of cellulose acetate at pH 7.[10]

Physical Properties. A summary of the known physical properties of ficin is presented in Table II, and in Table III are shown amino acid data reported by three different laboratories. These data must be accepted with the reservation that the homogeneity of the enzyme preparation may have been in doubt, that variations may exist with respect to the particular active component which was isolated, and that preparations from different varieties of fig may not be identical.[10]

Activators and Inhibitors. The proteolytic enzymes in fig latex exhibit enhanced activity in the presence of metal chelating agents such as EDTA,[20] particularly when employed in conjunction with one of a variety of reducing agents such as cyanide,[18,21] cysteine,[4,18,21] 1,2-dimercaptopropanol,[4,16] or mercaptoethanol.[2] The extent of activation of crystalline ficin by either cysteine or 1,2-dimercaptopropanol was constant

[19] D. E. Kramer, Ph.D. thesis, University of California, Davis, California (1964).
[20] J. R. Whitaker, *Food Res.* **22**, 483 (1957).
[21] T. Winnick, W. H. Cone, and D. M. Greenberg, *J. Biol. Chem.* **153**, 465 (1944).

TABLE II

PHYSICAL PROPERTIES OF FICIN

Source	Method of purification	Sedimentation coefficient ($s^0_{20,w}$)	Molecular weight	Isoelectric point	$\dfrac{A_{280}}{A_{260}}$	$E^{1\%}_{280}$	λ_e (nm)	Reference
F. glabrata	Salt precipitation followed by chromatography on CM-cellulose, pH 7		25,500 ± 750[a] 23,800 ± 700[b]		1.95	21.0	267	Englund et al.[c]
?	Cystalline	2.66 S						Cohen[d]
F. glabrata	Salt precipitation only	2.56 S	26,000[e]	9.0				Bernhard and Gutfreund[f] Liener[g]
F. glabrata	Salt precipitation followed by chromatography on CM-cellulose, pH 4.4	2.61 S						
F. carica var. Kadota	Individual active components isolated directly from latex by chromatography on CM-cellulose, pH 7			>9.6	1.88	20.2		Kramer and Whitaker[h]
F. anthelmintica	Salt precipitation only		26,500[a]					Marini-Bettolo et al.[i]

[a] Based on ultracentrifugal data.

[b] Based on amino acid analysis.

[c] P. T. Englund, T. P. King, and L. C. Craig, Biochemistry 6, 164 (1968). These data pertain to fraction III only (see Fig. 2A).

[d] W. Cohen, Nature 182, 659 (1958).

[e] Based on osmotic pressure measurements.

[f] S. A. Bernhard and H. Gutfreund, Biochem. J. 63, 61 (1956).

[g] I. E. Liener, Biochim. Biophys. Acta 53, 332 (1961).

[h] D. E. Kramer and J. R. Whitaker, J. Biol. Chem. 239, 2178 (1964). Each of the ten active components resolved by chromatography had approximately the values shown here.

[i] G. M. Marini-Bettolo et al., Gazz. Chim. Ital. 93, 1239 (1963).

TABLE III

AMINO ACID COMPOSITION OF FICIN REPORTED BY VARIOUS INVESTIGATORS[a]

Amino acid	Marini-Bettolo et al.[b,c]	Metrione[c,d]	Gould[c,e]	Englund[f,g]
Lysine	7	9	9	5
Histidine	2	2	2	1
Ammonia	—	—	52	25
Arginine	8	7	10	10
Aspartic acid	20	21	21	17
Threonine	10	8	10	8
Serine	13	10	16	14
Glutamic acid	22	23	25	25
Proline	11	12	13	11
Glycine	28	30	32	28
Alanine	19	21	21	20
Half-cystine	7	4	9	8
Valine	15	15	19	18
Methionine	3	3	4	5
Isoleucine	7	10	10	7
Leucine	14	17	17	15
Tyrosine	11	14	15	15
Phenylalanine	6	5	6	5
Tryptophan	3	—	9	6

[a] Number of residues per mole of protein.
[b] G. B. Marini-Bettolo et al., Gazz. Chim. Ital. **93,** 1239 (1963).
[c] Purified by salt fractionation only.
[d] R. M. Metrione, Ph.D. thesis, University of Nebraska, Lincoln, Nebraska (1963).
[e] N. R. Gould, Ph.D. thesis, University of Minnesota, St. Paul, Minnesota (1964).
[f] P. T. Englund et al., Biochemistry **6,** 164 (1968).
[g] Fraction III obtained by chromatography on CM-cellulose (see Fig. 2A).

between $0.005\ M$ and $0.05\ M$.[18] On the other hand, maximum activation of ficin was produced by $0.075\ M$ of HCN, but this level of activity was below that observed in the presence of cysteine.

Because of the direct participation of a sulfhydryl group in its catalytic mechanism,[22] ficin may be inactivated by stoichiometric levels of such thiol reagents as mercuric chloride,[16] N-ethylmaleimide,[16] iodoacetic acid,[16, 23] and chloroacetamide or iodoacetamide.[24] It is also this essential cysteine residue which reacts with those inhibitors which have been found to inactivate animal proteinases, namely diisopropylphosphofluoridate (DFP) and the chloromethyl ketone derivatives of N-tosyl-

[22] G. Lowe and A. Williams, Biochem. J. **96,** 189 (1965).
[23] R. C. Wong and I. E. Liener, Biochem. Biophys. Res. Commun. **17,** 470 (1964).
[24] M. R. Hollaway, A. P. Mathias, and B. R. Rabin, Biochim. Biophys. Acta **92,** 111 (1964).

L-phenylalanine and N-tosyl-L-lysine.[25] The inhibition by DFP, how-ever, has been attributed to the presence of an unknown impurity in commercial lots of DFP.[26] 1,3-Dibromoacetone has been used to in-activate ficin by combining with its essential cysteine residue as well as with a histidine residue which is estimated to be 5 Å removed from it. The latter is also considered by Husain and Lowe[27] to be an essential component of the active site. The inhibition of ficin by a protein com-ponent of egg white has been recently reported[7] presumably involving the formation of an inactive complex which has a dissociation constant of about $1 \times 10^{-9} M$.

Sequence of Amino Acids Around the Active Site. Little is known about the primary structure of ficin except for a limited sequence of amino acids in the immediate vicinity of the reactive sulfhydryl group. This sequence, along with the sequences reported for other thiol plant proteases, is presented in Table IV. The homology revealed by these sequences is indeed striking and may reflect a common mechanism of catalysis. Although the sequence shown here was determined on a salt-purified preparation of ficin which was subsequently shown to consist of several active components,[10] radioautography of peptides derived from each of these components, in which the active SH group had been labeled with radioactive iodoacetate, proved them to be identical.[17]

TABLE IV

SEQUENCE OF AMINO ACIDS IN THE VICINITY OF THE ACTIVE SULFHYDRYL
GROUP OF FICIN COMPARED WITH PAPAIN AND BROMELAIN[a]

Enzyme	Sequence
Ficin[b,c]	(Val, Ser)-Pro- Ile -Arg- Gln-Gln-Gly- Gln- Cys-Gly-Ser-Cys*
Papain[d]	Val-Thr -Pro- Val- Lys-Asn-Gln-Gly- Ser- Cys-Gly-Ser-Cys*
Bromelain[e]	Asn-Gln- Asp-Pro- Cys-Gly- Ala- Cys*

[a] Active cysteine residue denoted in sequence by an asterisk. Homologous sequences enclosed in solid lines.

[b] R. C. Wong and I. E. Liener, *Biochim. Biophys. Res. Commun.* **17,** 470 (1964).

[c] Amino acids in parentheses based on amino acid composition only. (B. Friedenson and I. E. Liener, unpublished data.)

[d] A. Light, R. Frater, J. R. Kimmel, and E. L. Smith, *Proc. Natl. Acad. Sci. U.S.* **52,** 1276 (1964).

[e] L.-P. Chao and I. E. Liener, *Biochem. Biophys. Res. Commun.* **27,** 100 (1967); S. S. Husain and G. Lowe, *Chem. Commun.* p. 1387 (1968).

[25] M. Stein and I. E. Liener, *Biochem. Biophys. Res. Commun.* **26,** 376 (1967).
[26] N. R. Gould and I. E. Liener, *Biochemistry* **4,** 90 (1965).
[27] S. S. Husain and G. Lowe, *Biochem. J.* **110,** 53 (1968).

Specificity. Ficin catalyzes the hydrolysis of peptide bonds of a wide variety of natural substrates, including gelatin,[28, 29] collagen,[30, 31] milk proteins,[32, 33] hemoglobin,[31] elastin,[34] soy proteins,[33] fibrin and fibrinogen,[31, 35] and even living *Ascaris*.[28] Ficin exhibits broad specificity with respect to the hydrolysis of peptide, ester, and amide bonds of synthetic substrates. Employing the general formula, R'—NH—CHR—CO—X, cleavage of the —CO—X bond has been demonstrated when R represents the side chain from glycine,[7] serine,[36] threonine,[36] methionine,[36] lysine,[8] arginine,[3, 36] citrulline,[8] leucine,[36] isoasparagine,[36] and tyrosine.[37] Substrates in which R is derived from alanine, phenylalanine, or tryptophan can function as either donors or acceptors in transpeptidation reactions catalyzed by ficin.[37] The X component of the substrate may be derived not only from amino acids but also from alcohols,[4] ammonia,[3] mercaptans,[13] aniline,[2] or *p*-nitrophenol.[7] A benzoyl or carbobenzoxy group has been generally employed as the R' substituent of the synthetic substrate.

Again it must be borne in mind that most of the studies regarding the specificity of ficin have been carried out with preparations of doubtful purity, and the possibility that the various active components may exhibit differences in specificity cannot be ruled out. Englund *et al.*[2] have examined the specificity of one of the active components of ficin by its action on the oxidized B-chain of insulin. Seven points of cleavage were definitely established, and two more splits were labeled as "possible." Although no clear-cut pattern of specificity was obvious, there did seem to be a preference for the hydrolysis of peptide bonds involving aromatic amino acids. The failure to detect splits of bonds containing lysine and arginine was unexpected since such bonds have been generally regarded as most favorable for ficin-catalyzed hydrolysis.[38]

Distribution. It has been reported[39] that proteolytic activity could be

[28] B. H. Robbins, *J. Biol. Chem.* **87**, 251 (1930).
[29] J. R. Whitaker, *Food Res.* **22**, 468 (1957).
[30] J. S. Hinrichs and J. R. Whitaker, *J. Food Sci.* **27**, 250 (1962).
[31] S. Sherry, W. Troll, and E. D. Rosenblum, *Proc. Soc. Exptl. Biol. Med.* **87**, 125 (1954).
[32] J. R. Whitaker, *Food Res.* **23**, 364 (1958).
[33] J. R. Whitaker, *Food Technol.* **13**, 86 (1959).
[34] E. Yatco-Manzo and J. R. Whitaker, *Arch. Biochem. Biophys.* **97**, 122 (1962).
[35] T. Astrup and N. Alkjaersig, *Nature* **169**, 314 (1952).
[36] C. A. Dekker, S. P. Taylor, Jr., and J. S. Fruton, *J. Biol. Chem.* **180**, 155 (1949).
[37] F. Janssen, M. Winitz, and S. W. Fox, *J. Am. Chem. Soc.* **75**, 704 (1953).
[38] E. L. Smith and J. R. Kimmel, *in* "The Enzymes" (P. D. Boyer, H. Lardy, and K. Myrbäck, eds.), Vol. 4, p. 168. Academic Press, New York, 1960.
[39] K. P. Hajnal, M. M. Racz, L. V. Vigyazo, E. H. Volgyesi, and M. Lak, *Elelmiszertudomany* **1**, 67 (1967); cited in *Chem. Abstr.* **68**, 93485d (1968).

detected only in the latex of *F. carica* and was absent in the leaves, stem, and fruit portions of the plant.

Studies of the ficin content of different species of the genus *Ficus* have revealed significant differences in proteolytic activity.[40,41] Of 46 species of *Ficus* examined, only 13 exhibited appreciable proteolytic activity; the latex of *F. stenocarpa* had the highest specific activity, followed closely by the latices of *F. carica* and *F. glabrata*.[41] A total of as many as 26 chromatographically distinct active components were detected in 6 species of *Ficus*.

Variations in activity have also been observed in different varieties of the same species of *Ficus*. Thus, in an examination of 25 varieties of *F. carica*, Whitaker[32] observed a 2-fold variation in the casein-digesting activity of the latex. The number and relative amounts of the active components of *F. carica* and *F. glabrata* differed among varieties within the same species.[15,41] For example, in the latices of 9 varieties of *F. carica*, a total of at least 16 chromatographically distinct components were detected, the Kadota variety alone showing 10 distinct active components.

The complexity of this problem is further aggravated by the fact that there are more than 1800 named species of *Ficus*, a number of subspecies, and at least 700 varieties of *F. carica* alone.[10] The importance of defining the source of material and the preparation and properties of the purified components cannot be emphasized too strongly.

[40] B. H. Robbins and P. D. Lamson, *J. Biol. Chem.* **106**, 725 (1934).
[41] D. C. Williams, V. C. Sgarbieri, and J. R. Whitaker, *Plant Physiol.* **43**, 1083 (1968).

[18] Bromelain Enzymes

By TAKASHI MURACHI

Nomenclature

Bromelain enzymes are found in tissues of plant family Bromeliaceae of which pineapple, *Ananas comosus* (L), is the best known. The proteolytic enzyme found in the juice of pineapple stem is called stem bromelain and the enzyme in the fruit was first described under the name of bromelin[1] and is now called fruit bromelain.[2] Systemic number EC 3.4.3.24 is given to bromelain.

[1] R. H. Chittenden, *Trans. Connecticut Acad. Sci.* **8**, 281 (1892).
[2] R. M. Heinicke and W. A. Gortner, *Econ. Botany* **11**, 225 (1957).

Stem Bromelain

Assay Method

Principle. Like most proteolytic enzymes, stem bromelain catalyzes the hydrolysis of protein substrates as well as synthetic substrates. Casein is most commonly used as a protein substrate and the assay is made by measuring increase in the amount of trichloroacetic acid-soluble peptides as a function of time. Various modifications of the Kunitz method[3,4] are used. The esterase activity of stem bromelain may be assayed by following the hydrolysis of α-N-benzoyl-L-arginine ethyl ester (BAEE) in a pH-stat.[5] Ammonia liberated is determined by a combination of the microdiffusion technique and the indophenol method with sodium nitroprusside as catalyst.[6]

Assay for Caseinolytic Activity[4]

Reagents

> Casein, 1.2 g/dl. Dissolve 3 g of Hammarsten grade casein in 250 ml of 0.03 M potassium phosphate buffer, pH 7.5, by heating in a boiling water bath for 15 minutes. The final pH is 7.2.
> L-Cysteine, 0.15 M, freshly prepared.
> Trichloroacetic acid solution. Dissolve 9 g of trichloroacetic acid, 15 g of sodium acetate, and add 19.5 ml of glacial acetic acid in water to make a final volume of 500 ml.
> Enzyme. The concentration of the enzyme solution should be more than 20 μg/ml.

Procedure. Place 5.0 ml of casein solution and 0.2 ml of cysteine in a 1.8 × 18 cm test tube and add water and enzyme to give a final volume of 6.0 ml. After incubation at 35° for 10 minutes, add 5.0 ml of trichloro-acetic acid solution rapidly, shake the mixture well and then allow it to stand for 30 minutes at 35°. Remove the coagulated protein either by filtration or by centrifugation. Read the absorbance of the filtrate or the supernatant solution at 275 nm against the blank. Inclusion of cysteine in the blank run is necessary, since cysteine causes a significant increase in absorbance at 275 nm of the casein filtrate. The optimal range for the enzyme is 40 to 80 μg enzyme protein per tube.

Definition of a Unit and Specific Activity. A proteinase unit is defined

[3] M. Kunitz, *J. Gen. Physiol.* **30**, 291 (1947).
[4] B. Hagihara, H. Matsubara, M. Nakai, and K. Okunuki, *J. Biochem.* (*Tokyo*) **45**, 185 (1958).
[5] T. Inagami and T. Murachi, *Biochemistry* **2**, 1439 (1963).
[6] A. L. Chaney and E. P. Marbach, *Clin. Chem.* **8**, 130 (1962).

as the amount of enzyme that gives an increase in absorbance at 275 nm equivalent to 1 μg of tyrosine per minute at pH 7.2 and at 35°.[4] Specific activity is expressed as units of enzyme per milligram of protein.

Assay for Esterolytic Activity[5]

   Reagents

   α-*N*-Benzoyl-L-arginine ethyl ester (BAEE), 0.1 M, adjusted to pH 6.0
   L-Cysteine, 0.05 M, freshly prepared
   KCl, 1.0 M
   NaOH, 0.1 N
   Enzyme. The concentration of the enzyme solution should be more than 1 mg/ml.

   Procedure. The reaction is carried out at 25°. A Radiometer model SBR2/SBU1/TTT1 Autotitrator (pH-stat) is used. A syringe-type buret of 1-ml capacity is filled with 0.1 N NaOH. Place 5.0 ml of BAEE, 1.0 ml of L-cysteine, and 1.0 ml of KCl in a reaction vessel of 30-ml capacity. Adjust the pH to 6.0, and initiate the reaction by adding water and enzyme to give a final volume of 10.0 ml. Follow the hydrolysis of the ester substrate by recording the uptake of alkali for a period of 10 to 20 minutes during which the rate of the hydrolysis is practically constant.

   Definition of a Unit and Specific Activity. An esterase unit is defined as the amount of enzyme that catalyzes the hydrolysis of 1 micromole of BAEE per minute at pH 6.0 and 25°. Specific activity is expressed as units of enzyme per milligram of protein.

Assay for Amidase Activity[5]

   Reagents

   α-*N*-Benzoyl-L-arginine amide (BAA), 0.05 M, adjusted to pH 6.0
   L-Cysteine, 0.05 M, freshly prepared
   KCl, 0.1 M
   Potassium phosphate buffer, 0.15 M, pH 6.0
   Enzyme. The concentration of the enzyme solution should be more than 2 mg/ml.

   Procedure. Incubation is carried out at 25°. Place 0.5 ml of BAA, 0.5 ml of L-cysteine, 0.5 ml of KCl, and 1.0 ml of buffer in a 1.5 × 15 cm test tube. Add water and enzyme to give a final volume of 5.0 ml. Withdraw 0.5-ml aliquots from the reaction mixture at 10-minute intervals and determine the amount of ammonia liberated in each aliquot. The rate of the reaction is practically constant for the initial 40 minutes.

Definition of a Unit and Specific Activity. An amidase unit is defined as the amount of enzyme that catalyzes the hydrolysis of 1 micromole of BAA per minute at pH 6.0 and at 25°. Specific activity is expressed as units of enzyme per milligram of protein.

Other Methods of Assay

Hemoglobin may be used as a substrate for proteinase activity.[7] Ninhydrin reaction can be used for determining ammonia liberated from BAA.[8] α-N-Benzoyl-DL-arginine p-nitroanilide is an alternative amide substrate.[9] 2-Mercaptoethanol at a final concentration of 0.005 M can be used for L-cysteine as an activator.[9] Bromelain can be assayed also by measuring its milk-clotting activity.[10]

Purification Procedure[11]

Starting Material. Commercial bromelain may be obtained from the Dole Company, Honolulu, Hawaii. The product is acetone-dried powder of the juice of pineapple stem.[2]

Step 1. Extraction. Ten grams of commercial bromelain is suspended in 100 ml of 0.05 M potassium phosphate buffer, pH 6.1, at room temperature for 30 minutes. The suspension is centrifuged at 5000 rpm for 20 minutes and the precipitate is discarded.

Step 2. Treatment with Anion-Exchange Resin. This and all of the following steps of the procedure are carried out at cold room temperature. The supernatant fluid (fraction 1) is applied to a 4.5 × 19 cm column of Duolite A-2 (coarse), which has been equilibrated overnight with 0.05 M potassium phosphate buffer, pH 6.1. The initial 100 ml of the effluent is discarded and the following 400-ml effluent (fraction 2) is collected by a continuous flow of the same buffer. A part of the colored material is removed by this treatment.

Step 3. Treatment with Cation-Exchange Resin. Fraction 2 is applied to a 4.5 × 31.5 cm column of Amberlite CG-50, Type 1 (100–200 mesh), which has been equilibrated with 0.2 M potassium phosphate buffer, pH 6.1. The column is washed with 200 ml of the same buffer and then with 5 liters of 0.05 M potassium phosphate buffer, pH 6.1, overnight.

[7] T. Murachi and H. Neurath, *J. Biol. Chem.* **235**, 99 (1964).
[8] S. Moore and W. H. Stein, *J. Biol. Chem.* **211**, 907 (1954).
[9] S. Ota, S. Moore, and W. H. Stein, *Biochemistry* **3**, 180 (1964).
[10] M. El-Gharbawi and J. R. Whitaker, *Biochemistry* **2**, 476 (1963).
[11] T. Murachi, M. Yasui, and Y. Yasuda, *Biochemistry* **3**, 48 (1964).

The effluent is discarded. The enzyme adsorbed is eluted by washing the column with 800 ml of 0.2 M potassium monohydrogen phosphate containing 1 M KCl. The initial 250 ml of the effluent is discarded and the following 250-ml eluate (fraction 3) is collected.

Step 4. Gel Filtration. To fraction 3 is added 100 g of ammonium sulfate. After 1 hour the precipitate formed is collected by centrifugation at 8000 rpm for 20 minutes and is dissolved in 50 ml of 0.05 M sodium acetate buffer, pH 5.2 (fraction 4). Fraction 4 is applied to a 4.5 × 31.5 cm column of Sephadex G-100 (140–400 mesh) which has been washed with 0.05 M sodium acetate buffer, pH 5.2. The column is washed with the same buffer and the initial 320-ml effluent is discarded. The following 150 ml effluent (fraction 5) is collected.

Step 6. Ammonium Sulfate Fractionation. To fraction 5 is added 38.5 g of ammonium sulfate to make 0.42 saturation. After 1 hour the moderately turbid mixture is centrifuged and to the supernatant fluid is added 8.8 g of ammonium sulfate to bring the suspension to 0.50 saturation. Bromelain thus precipitated is collected by centrifugation at 8000 rpm for 20 minutes, dissolved in 30 ml of water, and dialyzed against four changes of 5 liters of water for 48 hours (fraction 6).

The whole procedure of purification up to fraction 6 can be completed within 5 days, including 2 days for final dialysis. Table I summarizes the purification procedure. The gel filtration is essential to remove the contaminant carbohydrates present in fraction 4. Care should be given not to collect earlier fractions of the effluent which contain carbohydrates of larger molecular size.[11] The purified preparations are kept frozen and retain full activity for over a year.

Alternate Purification Procedure[9]

Two hundred milligrams of commercial bromelain preparation is dissolved in 4.2 ml of cold 0.02 M sodium citrate buffer, pH 5.5. After insoluble material has been removed by centrifugation at 30,000 g for 15 minutes, the supernatant solution (4.0 ml) is applied to a 2 × 75 cm column of Sephadex G-75 (medium particle size) which has been equilibrated at 25° with 0.02 M sodium citrate buffer, pH 5.5, saturated with phenylmercuric acetate (less than 10^{-4} M). The column is washed with the same buffer at a flow rate of 10 ml per hour at room temperature. Several effluent fractions with high proteolytic activity are pooled and added directly to a 2 × 35 cm column of sulfoethyl-Sephadex (C-25, fine mesh size, 2.0 meq/g) equilibrated with 0.3 M sodium citrate buffer, pH 6.0, containing 5 × 10^{-4} M phenylmercuric acetate. The column is washed with the same buffer at room temperature and the fractions that have high proteolytic activity are pooled. The main component isolated

TABLE I

PURIFICATION OF STEM BROMELAIN[a]

Step	Volume (ml)	Total protein (g)	Carbohydrates (g/g protein)	Proteinase activity		Esterase activity	
				Units/g protein (10^6)	Total units (10^6)	Units/g protein (10^3)	Total units (10^3)
Commercial product		Total solid, 10					
1. Extraction	101	5.12	0.657	1.15	5.90	0.224	1.15
2. Duolite A-2	400	4.04	0.758	1.23	4.98	0.226	0.913
3. Amberlite CG-50	250	2.72	0.0255	1.42	3.86	0.232	0.631
4. $(NH_4)_2SO_4$ pptn.	50	1.75	0.0283	1.50	2.62	0.252	0.441
5. Gel filtration	150	1.18	0.0210	1.62	1.92	0.250	0.295
6. $(NH_4)_2SO_4$ fractionation	30	0.871	0.0205	1.64	1.44	0.252	0.219

[a] T. Murachi, M. Yasui, and Y. Yasuda, *Biochemistry* **3**, 48 (1964). The specific activities shown here are one-thousand times as high as those described in the original publication which contained computational errors.

accounts for 42% of the total protein and 60% of the caseinolytic activity of the original crude stem bromelain.

Properties

Unless otherwise noted, description of properties will be made below for the step 6 preparation obtained by the procedure of Murachi *et al.*[11]

Stability. The enzyme retains full activity against casein when kept at 5° for 24 hours over a range of pH from 4 to 10.[5] The enzyme is stable in 25% (v/v) methanol at 25° for 20 minutes, while it loses 33% caseinolytic activity in 20% (v/v) ethanol at 37° for 20 minutes.[12] A 50% loss of the activity is caused by heating the enzyme solution at 55° for 20 minutes at pH 6.10.[10] Lyophilization causes 27% decrease in the activity.[11]

Purity. The purified enzyme is homogeneous by ultracentrifugal sedimentation, free-boundary electrophoresis, and diffusion analyses.[11] Chromatography on Amberlite CG-50, CM-Sephadex, DEAE-Sephadex, or SE-Sephadex yields a single symmetrical peak.[9, 11] The homogeneity of the enzyme is verified by disc electrophoresis on acrylamide gel.[13] The chromatographically purified enzyme, which migrates as a single band upon electrophoresis on cellulose acetate,[9] contains small but significant amounts of extraneous end groups in addition to the

TABLE II
PHYSICAL PROPERTIES OF STEM BROMELAIN

Sedimentation constant, $s_{20,w}^0$	2.73 S
Diffusion constant, $D_{20,w}^0$	$7.77 \times 10^{-7} \text{ cm}^2 \text{ sec}^{-1}$
Partial specific volume, $\bar{V}$	0.743 ml/g
Intrinsic viscosity, $[\eta]$	0.039 dl/g
Frictional ratio, f/f_0	1.26
Isoelectric point, pI	9.55
Absorbancy, $A_{1\,cm}^{1\%}$ at 280 nm	20.1[a]
ORD parameters, λ_c	241 nm
$-a_0$	190°
$-b_0$	78
Molar ellipticity, $[\theta]_{222}$	-4200[b]
Molecular weight	$33,200$,[c] $32,100$,[d] $33,500$[e]

[a] Revised value [T. Murachi, T. Inagami, and M. Yasui, *Biochemistry* **4**, 2815 (1965)].

[b] T. Sakai, K. Ikeda, K. Hamaguchi, and T. Murachi, *Biochemistry* **9**, 1939 (1970).

[c] By sedimentation-diffusion.

[d] From sedimentation constant and intrinsic viscosity.

[e] By Archibald method.

[12] T. Murachi, unpublished observations.

[13] L. P. Chao and I. E. Liener, *Biochem. Biophys. Res. Commun.* **27**, 100 (1967).

terminal valine.[9] In view of this and other facts, particularly the observations made by Whitaker and associates[10, 14] that stem bromelain can be fractionationated further into 5 proteolytically active components, each having a different amino acid composition, strict homogeneity of the "step 6 preparation" can hardly be claimed at the present moment.

TABLE III

AMINO ACID COMPOSITION OF STEM AND FRUIT BROMELAINS

	Stem bromelain[a]				Fruit bromelain[b]	
Amino acid	1	2	3	4	Green fruit	Ripe fruit
Lysine	20	23	12	20.2	7.8	8.3
Histidine	1	2	1	1.34	1.4	1.3
Arginine	10	12	6	10.2	8.6	9.1
Aspartic acid	27	29	16	28.9	29.8	29.8
Threonine	12	14	8	12.4	13.5	13.8
Serine	24	28	16	24.9	32.2	32.0
Glutamic acid	20	23	12	23.0	23.2	23.4
Proline	13	14	8	13.1	11.6	12.0
Glycine	29	35	19	30.5	32.6	32.2
Alanine	30	35	20	32.4	23.8	24.4
Half-cystine	11	10	5	7.4	10.0	10.0
Valine	19	22	12	20.7	19.8	20.1
Methionine	4	5	2	5.0	6.0	5.8
Isoleucine	20	21	12	18.4	16.4	16.2
Leucine	9	10	5	9.0	10.0	10.0
Tyrosine	19	21	11	18.2	22.4	22.2
Phenylalanine	9	9	5	8.0	7.6	8.0
Tryptophan	8	8	5	—	5.6	—
Total	(285)	(321)	(179)			
Ammonia (amide)	25	42	19	—	43.0	43.4
Glucosamine	2[c]	6	4	—	<0.2	<0.2
Carbohydrate (%)	2.1	1.46	2.0	—	3.2	3.3

[a] Sources of values are as follows. Column 1: T. Murachi, *Biochemistry* **3**, 932 (1964). Number of residues per mole of MW 33,000. Column 2: S. Ota, S. Moore, and W. H. Stein, *Biochemistry* **3**, 180 (1964). Number of residues per mole of MW 35,730. Column 3: G. Feinstein and J. R. Whitaker, *Biochemistry* **3**, 1050 (1964). For component II, taken methionine as two residues per molecule. Column 4: S. S. Husain and G. Lowe, *Biochem. J.* **110**, 53 (1968). Taken leucine as nine residues per molecule.

[b] S. Ota, S. Moore, and W. H. Stein, *Biochemistry* **3**, 180 (1964). Mole ratios with leucine set as 10.

[c] Revised value [N. Takahashi, Y. Yasuda, M. Kuzuya, and T. Murachi, *J. Biochem. (Tokyo)* **66**, 659 (1969)].

[14] G. Feinstein and J. R. Whitaker, *Biochemistry* **3**, 1050 (1964).

Physical Properties.[11] Physical constants of stem bromelain protein are listed in Table II. The data suggest that the enzyme is a basic protein with a molecular weight of approximately 33,000. It is more basic and about 1.5 times larger in size if compared with papain. From values of $-b_0$ and molar ellipticity at 222 nm, the content of α-helix in stem bromelain can be calculated to be approximately 10%. A time-dependent and finally irreversible conformational change occurs at pH values higher than 10.3.[15, 15a]

Chemical Properties. In Table III are shown the amino acid compositions of stem bromelain reported by different investigators.[14, 16, 17] The principal amino terminal residue is valine[9] and the carboxyl terminal is glycine.[18] The enzyme is a glycoprotein having one oligosaccharide moiety per molecule which is covalently linked to the peptide chain.[19, 19a] The proposed structure of undecaglycopeptide isolated from the pepsin digest of stem bromelain is [20-22a]:

$$
\begin{array}{c}
\text{L-Fuc} \qquad\quad \text{D-Xyl} \\
\Big| \;(\alpha 1 \to 6 \text{ or } 2) \;\; \Big|\; (\beta) \\
\text{D-Man} \!\!-\!\!-\!\!- \text{D-Man} \!\!-\!\!-\!\!- \text{D-Man} \!\!-\!\!-\!\!- \text{D-GlcNAc} \\
(\alpha 1 \to 2) \qquad (\alpha 1 \to 2 \text{ or } 6) \qquad (\alpha) \qquad\quad \Big| \;(\beta 1 \to 3 \text{ or } 4) \\
\text{D-GlcNAc} \\
\Big| \;(\beta 1 \to \beta \text{NH}_2\text{-N}) \\
\text{Ala-Arg-Val-Pro-Arg-Asn-Asn-Glu-Ser-Ser-Met}
\end{array}
$$

Stem bromelain has one reactive sulfhydryl group per molecule as determined by titration with *p*-chloromercuribenzoate. This sulfhydryl group is essential for catalytic activity.[23] The reported amino acid sequences of the active site are:

Cys-Gly-Ala-*Cys*-Trp[13]
Asn-Gln-Asp-Pro-Cys-Gly-Ala-*Cys*-Trp[24]

[15] A. Tachibana and T. Murachi, *Biochemistry* **5**, 2756 (1966).
[15a] T. Murachi and M. Yamazaki, *Biochemistry* **9**, 1935 (1970).
[16] T. Murachi, *Biochemistry* **3**, 932 (1964).
[17] S. S. Husain and G. Lowe, *Biochem. J.* **110**, 53 (1968).
[18] S. Ota, *Seikagaku* **37**, 433 (1965) (Abstract, in Japanese).
[19] T. Murachi, A. Suzuki, and N. Takahashi, *Biochemistry* **6**, 3730 (1967).
[19a] J. Scocca and Y. C. Lee, *J. Biol. Chem.* **244**, 4852 (1969).
[20] K. Kito and T. Murachi, *J. Chromatog.* **44**, 205 (1969).
[21] N. Takahashi, Y. Yasuda, M. Kuzuya, and T. Murachi, *J. Biochem. (Tokyo)* **66**, 659 (1969).
[22] T. Murachi and N. Takahashi, *in* "Structure-Function Relationships of Proteolytic Enzymes" (P. Desnuelle, H. Neurath, and M. Ottesen, eds.), p. 298, Munksgaard, Copenhagen, 1970.
[22a] Y. Yasuda, N. Takahashi, and T. Murachi, *Biochemistry* **9**, 25 (1970).
[23] T. Murachi and M. Yasui, *Biochemistry* **4**, 2275 (1965).

where Cys indicates the reactive cysteinyl residue.

The reagent 1,3-dibromoacetone cross links a cysteinyl and a histidyl residue within the same enzyme molecule. The amino acid sequence around the latter residue is described[24]:

His-Ala-Val-Thr-Ala-Ile-Gly-Tyr

The result is interpreted as showing the presence of the imidazole group of a histidyl residue within 5 Å of the reactive sulfhydryl group.

Activators and Inhibitors. The purified enzyme obtained by the routine procedure[11] shows 60–70% activity as assayed by casein hydrolysis in the absence of activators.[23] The enzyme can be fully activated in the presence of 0.005 M cysteine, 2-mercaptoethanol, or dithiothreitol. KCN is less effective.[7] Stem bromelain is reversibly inhibited by inorganic mercuric ion,[7] organic mercurials,[9, 23] and tetrathionate.[22] Irreversible inhibition occurs when stem bromelain is reacted with N-ethylmaleimide, N-(4-dimethyl-3,5-dinitrophenyl)maleimide (DDPM),[25] monoiodoacetic acid,[13] and 1,3-dibromoacetone.[24] These reagents alkylate the essential sulfhydryl group of the enzyme protein. Chloromethyl ketone derivative of N-tosyl-L-phenylalanine (TPCK) and 1-chloro-3-tosylamido-7-amino-2-heptanone (TLCK) also alkylate the SH group, resulting in inactivation of the enzyme.[22, 26] The second-order rate constant of the reaction between the enzyme and TPCK or TLCK is 2.3 or 11.9 1 mole^{-1} sec^{-1}, respectively, as determined at pH 7.0 and 30°.[22] Diisopropylphosphorofluoridate (DFP) does not inhibit stem bromelain but it alkylphosphorylates the enzyme-protein at pH 8.2 without inhibition of proteinase activity.[23] The alkylphosphorylation occurs at the phenolic hydroxyl groups of tyrosyl residues.[27] The diisopropylphosphorylated enzyme shows an alteration in specificity toward synthetic substrates.[28]

Specificity and Kinetic Properties. The specificity can be described as being broad, since a variety of synthetic substrates are hydrolyzed by stem bromelain as shown in Table IV. Basic amino acyl residues are preferred, but the preference is less strict than in the case of papain. The fact that in bromelain catalysis k_{cat} for BAEE is 140 times as large as k_{cat} for BAA, is in sharp contrast to the finding that with papain or ficin k_{cat} values are almost identical for the ester and for the corresponding amide substrates, suggesting some important difference in the

[24] S. S. Husain and G. Lowe, *Chem. Commun.* p. 1387 (1968).
[25] T. Murachi, T. Miyake, and M. Mizuno, *Abstr. Intern. Congr. Biochem. 7th August 1967, Tokyo,* p. 765.
[26] T. Murachi and K. Kato, *J. Biochem. (Tokyo)* **62,** 627 (1967).
[27] T. Murachi, T. Inagami, and M. Yasui, *Biochemistry* **4,** 2815 (1965).
[28] T. Murachi, *Abstr. Intern. Congr. Biochem. 6th July 1964, New York,* p. 324.

TABLE IV
SUBSTRATE SPECIFICITY AND KINETIC PARAMETERS OF STEM BROMELAIN

| Substrate | Rate of hydrolysis of 0.01 M substrate | | $K_{m(app)}$ (M) | k_{cat} (sec^{-1}) |
	$\times 10^3$ (sec^{-1})	(%)		
Benzoyl-L-arginine ethyl ester (BAEE)	29.4	100	0.17	0.50
Benzoyl-L-arginine amide (BAA)	3.1	10.5	0.0012	0.0035
Benzoyl-L-arginine methyl ester	91.7	312	0.032	0.11
Tosyl-L-arginine methyl ester	14.8	50.3		
Tosyl-L-lysine methyl ester	3.96^a	13.4^a	0.084^a	0.035^a
L-Lysine ethyl ester	2.81	9.56		
L-Histidine ethyl ester	<0.1	—		
Glycine ethyl ester	5.40	18.4		
Benzoylglycine ethyl ester	11.9	40.5	0.21^a	0.36^a
Benzoylglycine amide	0.28	0.96		
Acetylglycine ethyl ester	<0.2	—	33^a	0.55^a
Benzoyl-DL-alanine ethyl ester	3.96	13.5		
L-Leucine ethyl ester	5.48	18.6		
L-Phenylalanine ethyl ester	8.60	29.3		
L-Tyrosine ethyl ester	6.39	21.7		

[a] T. Murachi, unpublished data.

mechanism of catalysis.[5] Stem bromelain rapidly cleaves glucagon at either Arg(18)-Ala(19) or Ala(19)-Gln(20) bond, while it leaves intact Lys(12)-Tyr(13) and Arg(17)-Arg(18) bonds.[7] The B-chain of oxidized insulin is a relatively poor substrate for bromelain.[7] Bradykinin is rapidly and almost exclusively cleaved between Phe(5) and Ser(6).[22, 29]

Fruit Bromelain

Assay Method

The assay procedures and reagents are the same as those for stem bromelain except that a smaller amount of fruit enzyme is needed for assaying the hydrolyses of casein, α-N-benzoyl-L-arginine amide, and α-N-benzoyl-DL-arginine p-nitroanilide.[9] There is no report of the assay for the esterolytic activity of fruit bromelain.

Purification Procedure[9]

Acetone Powder of Fruit Juice. The fresh fruit, green or ripe, is freed of leaves and epithelium and the juice is obtained by pressing with a

[29] T. Murachi and T. Miyake, *Physiol. Chem. Phys.* **2**, 97 (1970).

hydraulic press. The juice (pH 3.2–3.5) is cooled to 0°–4° and 1 volume of cold acetone is added. The precipitate is discarded. The enzyme is precipitated by the addition of two more volumes of cold acetone and the precipitate is collected by centrifugation and dried under reduced pressure. The dried product is ground to a powder in a mortar.

DEAE-Cellulose Chromatography. Two hundred milligrams of the acetone powder is added to 10 ml of cold 0.02 M sodium citrate buffer, pH 6.0. After centrifugation at 30,000 g for 15 minutes, 9 ml of the clear supernatant solution is applied to a 2 $\times$ 20 cm column of DEAE-cellulose (0.96 meq/g) which has been equilibrated at 25° with 0.02 M sodium acetate buffer, pH 6.0, containing 5 $\times$ 10^{-4} M phenylmercuric acetate. The buffer change elutes the adsorbed enzyme. Several fractions that have high proteolytic activity are obtained and are pooled.

Yields of the products are summarized in Table V.

TABLE V
PURIFICATION OF FRUIT BROMELAIN

Step	Yield
Acetone powder	3.3–3.7 g per liter of juice
DEAE chromatography	
Green fruit	43% of total protein; 88% of activity
Ripe fruit	32% of total protein; 87% of activity

Properties[9]

Unlike stem bromelain, the fruit enzyme is an acidic protein. The enzyme is apparently homogeneous as judged by rechromatography and electrophoresis on cellulose acetate. The principal NH$_2$-terminal residue is alanine (0.9 residue per 3 $\times$ 10^4 g protein), but additional NH$_2$-terminal residues are noted (valine 0.3, serine 0.2, and glycine 0.1). The amino acid compositions of the enzymes from green and ripe fruits are shown in Table III. The fruit enzyme also contains carbohydrate that cannot be removed by the purification procedures used thus far. Fruit bromelain is much more active against BAA than the stem enzyme. The enzyme catalyzes synthesis of acylamino acid anilides.[30] Fruit bromelain is inhibited by mercurials and the activity is restored by cysteine.

---

[30] S. Ota, T. Fu, and R. Hirohata, *J. Biochem. (Tokyo)* **49**, 532 (1961).

[19] Cathepsins

By Mary J. Mycek

Introduction

The term, cathepsin ($\chi\alpha\theta\acute{\epsilon}\psi\iota\nu$ = to digest), was proposed by Willstätter and Bamann[1] for acidic proteinase activity found in aqueous extracts of a variety of animal tissues. The multiplicity of proteolytic activities associated with the extracts, labeled cathepsin, was recognized largely because of their action on synthetic peptide substrates of known structure, and led to the formulation of a classification of the enzymes in 1941,[2] which was revised in 1952.[3] At that time three major cathepsins derived from beef spleen were recognized. On the basis of their specificity toward N-benzylcarbonyl-α-L-glutamyl-L-tyrosine, benzoyl-L-argininamide, and glycyl-L-phenylalaninamide, these were assigned the names, cathepsin A, B, and C, respectively. Since that time, two additional cathepsins have been reported, cathepsin D[4, 5] and E.[6] The catalytic activity of these latter two enzymes has not been defined in terms of synthetic substrates since their action appears to be primarily directed against proteins. In addition, a catheptic carboxypeptidase has been isolated and partially purified.[7] The specificity of this enzyme is directed primarily against N-benzylcarbonyl-α-L-glutamyl-L-phenylalanine.

Of the cathepsins, the one most extensively studied and purified is cathepsin C.[3, 8-10] As a result of this work, Fruton has suggested that it be renamed dipeptidyl transferase[9] in order to more accurately define its action. It might be expected, therefore, that further purification and study of the other cathepsins will eventually lead to a revised nomenclature which will more meaningfully describe the nature of the reaction catalyzed by the enzyme. At the present time, cathepsin, followed by the appropriate letter, remains the term applied to those intracellular enzymes derived from animal tissue extracts which act on proteins or on synthetic substrates for well-defined proteinases at acidic pH.[3] These

[1] R. Willstätter and E. Bamann, *Hoppe-Seylers Z. Physiol. Chem.* **180**, 127 (1929).
[2] J. S. Fruton, G. W. Irving, Jr., and M. Bergmann, *J. Biol. Chem.* **141**, 763 (1941).
[3] H. H. Tallan, M. E. Jones, and J. S. Fruton, *J. Biol. Chem.* **194**, 793 (1952).
[4] E. M. Press, R. R. Porter, and J. Cebra, *Biochem. J.* **74**, 501 (1960).
[5] C. Lapresle and T. Webb, *Biochem. J.* **76**, 538 (1960).
[6] C. Lapresle and T. Webb, *Biochem. J.* **84**, 455 (1962).
[7] L. M. Greenbaum and R. Sherman, *J. Biol. Chem.* **237**, 1082 (1962).
[8] J. S. Fruton and M. J. Mycek, *Arch. Biochem. Biophys.* **65**, 11 (1956).
[9] R. M. Metrione, A. G. Neves, and J. S. Fruton, *Biochemistry* **5**, 1597 (1966).
[10] I. M. Voynick and J. S. Fruton, *Biochemistry* **7**, 40 (1968).

enzymes have been largely located in the lysosomal fraction within the cell[11] and thereby differ from proteinases which are elaborated by cells (e.g., trypsin, chymotrypsin). A review of this topic appeared in 1960.[12]

Assay Methods

A fair number of catheptic activities have been recognized and the number of methods used to detect them have been multiple. In general, these methods are the same as those employed in identification and assay of other proteinases and peptidases. Rather than repeat the assay procedures for each separate cathepsin, it was considered to be more convenient to group them according to the type of reaction being monitored. Thus, in this section, the following assays will be described in general terms: proteinase, amidase, esterase, hydrolase, and transferase. The appropriate substrates, buffers, and activators that pertain to the particular activity being measured will be noted.

Proteinase Assay

Principle. This method, based on one described by Anson,[13] has found wide application in the detection and estimation of proteolytic activity. It depends on the appearance of the free amino acids which are released in the course of the digestive process. The presence of these components can be conveniently measured in the acid filtrate of the terminated incubation mixture by (1) direct spectrophotometric determination at 280 nm,[3] or (2) reaction with biuret reagent and estimation of the colored product at 555 nm,[5] or (3) formation of a blue product with the Folin-Ciocalteau phenol reagent[13] and colorimetric determination at 500 nm, or (4) reaction with ninhydrin.[7]

Cathepsins A and B exert their optimal proteolytic effects at pH values of 4–5, while cathepsins D and E are most active in the region of pH 2.5–3.5. Highly purified preparations of cathepsin C (dipeptidyl transferase) are devoid of proteinase activity. For the desired assay pH, the nature of the buffer used in the incubation medium is varied. Of the proteolytic cathepsins, only cathepsin B requires a sulfhydryl activator, e.g., cysteine.

Because of its simplicity and time-saving feature, the direct spectrophotometric procedure has been singled out for description here.

---

[11] C. de Duve, B. C. Pressman, R. Gianetto, R. Wattiaux, and F. Appelmans, *Biochem. J.* **60**, 604 (1955).

[12] J. S. Fruton, *in* "The Enzymes" (P. D. Boyer, H. Lardy, and K. Myrbäck, eds.) Vol. 4, p. 233. Academic Press, New York, 1960.

[13] M. L. Anson, *J. Gen. Physiol.* **22**, 79 (1938).

Reagents

Bovine Hemoglobin Substrate Powder. This material which has been denatured by acid is readily procured from a number of commercial sources (e.g., Worthington Biochemical Corp., Freehold, N.J.). A 2.5% (w/v) solution is employed. To retard bacterial growth, 1 ml Merthiolate can be added per 40 ml of solution. The solution should be kept refrigerated until use and it is preferable to make up fresh solutions frequently. Alternative substrates which may be employed are serum albumin (10% w/v), casein (1% w/v), or gelatin (5% w/v).

Buffers. Sodium citrate, 0.04 M with respect to citrate, pH 2.8; or 1.35 M acetic acid, 0.02 M ammonium sulfate, adjusted to pH 3.5; or 0.4 M citric acid, 0.8 M NaOH, adjusted to pH 5.0.

Cysteine·HCl in water, 0.07 M, pH 3.5, or adjusted to pH 5.0 with NaOH.

5% Trichloroacetic acid (TCA)

Enzyme

Procedure. The following reagents are pipetted into test tubes, mixed, and allowed to equilibrate at 38° for 5 minutes: 1 ml buffer, 4 ml hemoglobin solution, 1 ml cysteine (or water). At the end of this time, 1 ml of the enzyme solution is added. Incubation is continued at 38° for 10 minutes after addition of the enzyme. The reaction is stopped by the addition of 9 ml TCA. The contents of the tube are thoroughly mixed and allowed to stand at room temperature for 10 minutes. Filtration through Whatman No. 3 filter paper (9.0 cm) removes the denatured material and the clear filtrate is read spectrophotometrically at 280 nm either directly or after dilution (1:10) with water. (At this step, an aliquot may be further reacted with the Folin-Ciocalteau reagent, or biuret reagent, or ninhydrin.) The experimental tubes are compared to a suitable blank, prepared by adding the TCA to a mixture of buffer, enzyme, and activator, mixing, and then introducing the substrate.

Definition of Units. A unit of catheptic activity has been arbitrarily defined as that amount of enzyme which, in the 7 ml of test solution, causes the optical density of the acid filtrate to increase 1.000 in 10 minutes in excess of the blank reading or as that amount of enzyme which causes an increase of 0.001 unit of extinction/minute of digestion. Specific activity is usually expressed as the units of catheptic activity per milligram protein N, but also can be related to milligrams of enzyme protein.

Amidase Assay

This assay has been found convenient to use for the estimation of cathepsins B and C (dipeptidyl transferase). No amide substrates for the other cathepsins are known at this time.

Principle. Enzymatic attack at a susceptible amide linkage produces ammonia which can be quantitatively measured by a number of procedures. Among those which have been employed are two microdiffusion assays. The ammonia is liberated by alkali and absorbed into a boric acid–indicator solution and then titrated with standard acid in a modification of the Conway method.[14] In another method, described by Seligson and Seligson,[15] the ammonia released by the added alkali is trapped into sulfuric acid and subsequently determined colorimetrically after being reacted with Nessler's reagent. The latter procedure has several advantages over the former; in particular, it is less time consuming.

Reagents

Substrate solution, 0.25 M. In assays of cathepsin B, benzoyl-L-argininamide is the preferred substrate, while for cathepsin C (dipeptidyl transferase), glycyl-L-phenylalaninamide is generally employed.

β-Mercaptoethylamine·HCl, 0.04 M

Citrate buffer, 1 M, pH 5.0

Enzyme

Saturated potassium carbonate solution

Sulfuric acid, 1 N

Nessler's solution diluted 1:10 before use

Procedure. Into a 2-ml volumetric flask are placed 0.4 ml of substrate solution, 0.1 ml β-mercaptoethylamine, and 0.1 ml citrate buffer. After the mixture has been brought to equilibrium in a constant temperature bath at 37° for 5 minutes, a suitable dilution of enzyme is introduced, water is added to volume, and it is then mixed. A zero time sample (0.2 ml) is withdrawn and the tube is replaced into the bath. Samples can be removed at desired time intervals. A blank tube without enzyme is run in parallel. The aliquot (0.2 ml) of the incubation medium is pipetted into a suitable bottle equipped with a glass rod mounted in a stopper (see reference for design; or it can be procured from a commercial source). Saturated potassium carbonate (1 ml) is added so as not to come into contact with the sample to be assayed. A drop of the 1 N sulfuric acid is placed onto the glass rod and the stopper and rod are replaced in the

[14] R. B. Johnston, M. J. Mycek, and J. S. Fruton, *J. Biol. Chem.* **185**, 629 (1950).
[15] D. Seligson and H. Seligson, *J. Lab. Clin. Med.* **38**, 324 (1951).

bottle. The assembly is placed on a rotator and the liberated ammonia is trapped in the sulfuric acid as ammonium sulfate. Time necessary for maximal diffusion can be ascertained by carrying out the procedure on a standard ammonium sulfate solution (0.4716 g/liter); an hour is usually found to be sufficient. At the end of the diffusion process, the rod is removed and the ammonium sulfate washed off with 10 ml of diluted Nessler's solution. The color intensity is determined by reading at 420 nm in a colorimeter against the blank tube.

The amount of ammonia liberated is calculated from a standard curve prepared from a standard ammonium sulfate solution and Nessler's reagent.

Definition of Units. A unit of catheptic amidase activity has been defined as that amount of enzyme which produces a 1% hydrolysis of substrate amide per minute. Specific activity is expressed as units per milligram enzyme protein.

Esterase Assay

As with the amidase activity, the only cathepsins which possess esterolytic activity are B and C (dipeptidyl transferase).

Principle. The availability of an automatic titration assembly and recorder permits a simple, convenient, and precise method[16] for measuring the above catheptic activities by cleavage of ester substrates. During the course of the reaction, acid is formed and results in a drop of the pH of the unbuffered incubation medium. Standardized NaOH is automatically added to a pH value preset to coincide with the optimal pH of the reaction (5–6 in the case of the above enzymes). The amount of alkali delivered per unit time is recorded automatically onto a chart moving at a constant rate of speed; thus, initial reaction rates may be readily calculated from the data obtained.

Reagents

Substrate ester, 0.01 M. Typical substrates are benzoyl-L-arginine ethyl ester for cathepsin B and glycyl-L-phenylalanyl ethyl ester for cathepsin C (dipeptidyl transferase). No substrates of this class have been reported for cathepsins A, D, or E.

β-Mercaptoethylamine·HCl, 4 mM

Sodium chloride, 0.1 M, containing 5×10^{-5} M EDTA

Enzyme, 0.5–1.0 unit/ml

Standardized 0.01 N NaOH

[16] G. I. Tesser, R. J. F. Nivard, and M. Gruber, *Biochem. Biophys. Acta* **89**, 303 (1964).

Procedure. The substrate and activator dissolved in 5 ml of the NaCl–
EDTA solution are placed into the jacketed reaction vessel maintained
at 37° and allowed to equilibrate for 5 minutes. The pH is adjusted to
5.0 and the enzyme is then introduced and washed in with enough water
to bring the total volume to 5 ml. The reaction is allowed to proceed
for about 15–30 minutes and the initial rates calculated from the slopes
of the charted curve. Control determinations of the nonenzymatic breakdown of substrates are also performed and subtracted.

Definition of Units. A unit of catheptic esterase activity is that
amount of enzyme which under the above reaction conditions results in
a hydrolysis of 1% of the substrate per minute. Specific activity is expressed as units per milligram enzyme protein.

Transferase Assay

Cathepsins B and C (dipeptidyl transferase) exhibit their maximal
hydrolytic activities at pH values around 5.0; however, at pH 6.5 and
above, they have been shown to catalyze transfer reactions in which the
terminal amide or ester group is replaced by a nucleophilic compound
(e.g., hydroxylamine, amino acid amide, or peptide). Although this type
of reaction has not been reported to be catalyzed by cathepsins A, D,
and E, it would be surprising if they did not possess it inasmuch as it is
characteristic of a number of proteinases.

Principle. The incorporation of hydroxylamine to form a hydroxamic
acid derivative as a consequence of an enzyme-catalyzed transfer reaction provides a simple method for the estimation of the extent of this
reaction. The presence of the hydroxamic acid is readily detected by
complexing with ferric chloride under acidic conditions to yield a claret-
colored product easily measured spectrophotometrically. This method
is the one originally described by Johnston et al.[14] based on that of
Lipmann and Tuttle.[17]

Reagents

> Substrate amide or ester, 0.25 M, adjusted to pH 6.8 with 0.1 N
> NaOH. Typical substrates for cathepsin B and C (dipeptidyl
> transferase) are benzoyl-L-argininamide or ethyl ester and glycyl-
> L-phenylalaninamide or ethyl ester, respectively.
>
> β-mercaptoethylamine, 0.125 M, freshly prepared from a stock
> solution of the hydrochloride adjusted to pH 6.8 with 0.1 N
> NaOH.
>
> Hydroxylamine, 2 M, prepared fresh from 4 M hydroxylamine·
> HCl by adjusting to pH 6.8 with 4 N NaOH.

[17] F. Lipmann and L. C. Tuttle, *J. Biol. Chem.* **159**, 21 (1945).

20% Trichloroacetic acid
5% $FeCl_3 \cdot 6\ H_2O$ in 0.1 N HCl
Enzyme

Procedure. The following reagents are pipetted into a 2-ml volumetric flask: 0.1 ml of 2 M NH_2OH, 0.1 ml of 0.125 M β-mercaptoethylamine, 0.1 ml of substrate, and 0.1 ml of water. After bringing the contents to equilibrium in a constant temperature bath at 37° for 5 minutes, 0.1 ml of enzyme solution (appropriately diluted) is added. After 10 minutes, addition of 0.5 ml of 20% TCA, followed by 0.5 ml of 5% ferric chloride reagent stops the reaction, the volume is brought to 2 ml with water and the absorbance is read at 510 nm against a control which does not contain enzyme.

The amount of hydroxamic acid produced is calculated from a standard curve based on the color formed with benzoyl-L-hydroxamic acid and ferric chloride reagent.[14]

Definition of Units. Under the conditions of assay described above, the amount of enzyme required to catalyze the formation of 1 mole of hydroxamic acid per minute is considered one enzyme unit of catheptic transferase activity. The specific activity is defined as units per milligram of enzyme protein.

Hydrolase Activity

Principle. The release of substances, such as amino acids, which can react with ninhydrin to yield colored products is the basis for a colorimetric method for the determination of the hydrolytic activity of cathepsins.[7] Inasmuch as the quantitative results depend on the relationship between color yield and concentration and this in turn depends on the particular substance reacting with ninhydrin,[18] it is best to apply the reaction to a system in which the product of the reaction has been identified and then to employ that substance in determining a standard curve.

Reagents

Substrate peptide, 0.01 M, e.g., *N*-benzylcarbonyl-α-L-glutamyl-L-tyrosine
Phosphate buffer, 0.15 M, pH 4.0, or 0.15 M phosphate-acetate buffer, pH 4.0
Cysteine·HCl (not required for cathepsin A), 0.004 M
Enzyme. Cathepsin A or catheptic carboxypeptidase.
Ninhydrin reagent according to Moore and Stein[18]

[18] S. Moore and W. H. Stein, *J. Biol. Chem.* **211**, 907 (1954).

Procedure. In a total final volume of 1.03 ml are placed the substrate made up in the buffer, activator, and after equilibration for several minutes at 38°, the enzyme. Aliquots of either 0.1 or 0.2 ml are withdrawn at zero time and appropriate time intervals and pipetted into ice-cold water, either 0.8 or 0.9 ml, in test tubes kept in an ice-bath. The ninhydrin method of Moore and Stein[18] is then used to develop the color of the reaction product (in this case, tyrosine). The optical density of the experimental tubes is compared to the standard curve obtained from the reaction of tyrosine with ninhydrin under identical conditions.

Definition of Units. A unit of catheptic hydrolase activity is that amount of enzyme which causes 1% hydrolysis per minute of the peptide substrate. Specific activity is expressed as units per milligram of enzyme protein.

Cathepsin A

In the early attempts to resolve the activities associated in the complex of cathepsins, one proteinase was detected which had a pH optimum at about 5.5, was not dependent on sulfhydryl-containing compounds, nor was it inhibited by iodoacetate.[19,20] It was characterized by being able to cleave the pepsin substrate, N-benzylcarbonyl-α-L-glutamyl-L-tyrosine. This enzyme has been partially purified from beef spleen extracts through an ammonium precipitation step. A more recent description[21] of an isolation of cathepsin A from chicken breast muscle has appeared and will be presented below. Although the degree of purification is low, it has been selected because cathepsin A is separated from cathepsin D. As will be seen later, these two enzymes possess a striking similarity in their enzymatic behavior inasmuch as both show a preference for those bonds which are susceptible to pepsin attack.[4,22] Cathepsin A, however, hydrolyzes the synthetic substrate while cathepsin D apparently does not have this capacity.

Purification Procedure

Step 1. Extraction. All the steps are carried out at 4°. Chicken breast muscle, 240 g, is minced and then homogenized in 960 ml of 2% KCl solution in a Waring blendor for two 1.5-minute periods, being careful to chill between periods. (It is convenient to carry out the blending in

[19] J. S. Fruton and M. Bergmann, *J. Biol. Chem.* **130**, 19 (1939).
[20] J. S. Fruton, G. W. Irving, Jr., and M. Bergmann, *J. Biol. Chem.* **138**, 249 (1941).
[21] A. A. Iodice, V. Leong, and I. M. Weinstock, *Arch. Biochem. Biophys.* **117**, 477 (1966).
[22] N. M. Lichenstein and J. S. Fruton, *Proc. Natl. Acad. Sci. U.S.* **46**, 787 (1960).

batches of 50 g of muscle with 200 ml KCl.) The homogenate is allowed to stand overnight and then centrifuged at 13,000 g for 45 minutes.

Step 2. Treatment with Acid. The pH of the extract (715 ml) is adjusted to 3.8 by the dropwise addition of 1 N HCl with stirring. After 10 minutes of additional stirring, the suspension is centrifuged at 13,000 g for 15 minutes. The supernatant fluid obtained is treated with 1 N KOH until the pH reaches 6.8 when the centrifugation is repeated.

Step 3. Heat Treatment. The supernatant fluid (655 ml) is divided into two portions and heated in a 50° water bath with stirring. The temperature of the extract is allowed to reach 46° (about 8 minutes) and is maintained at this temperature for 3 minutes. After cooling, the suspension is centrifuged at 13,000 g for 20 minutes and the residue is discarded.

Step 4. Precipitation with Ammonium Sulfate. The supernatant solution from the above step (615 ml) is brought to 0.60 saturation with ammonium sulfate (240 g) added slowly with stirring. Stirring is continued for another 30 minutes, and the suspension is recentrifuged at 13,000 g for 30 minutes. The residue is suspended in 0.01 M potassium phosphate buffer, pH 6 (final vol. approx. 45 ml). After dialysis overnight against 8 liters of 0.005 M potassium phosphate buffer, pH 6, the insoluble material is removed by centrifugation. The clear extract which remains is about 48 ml.

Step 5. Chromatography on DEAE-Sephadex. DEAE-Sephadex (obtained from Pharmacia Chemicals) is prepared for chromatography in the following manner. Fine particles are removed by repeated suspension of the resin in distilled water followed by decantation. The resin is then washed on a filter alternating with 0.5 N HCl, distilled water, and 0.5 N NaOH until the washing fluid is near neutrality. The dextran is then equilibrated with 0.01 M potassium phosphate buffer, pH 6.

The dialyzed solution from step 4 (43 ml) is placed onto a column of DEAE-Sephadex (2.5 × 25 cm). Elution with 0.01 M potassium phosphate buffer, pH 6, is allowed to proceed until a large protein peak lacking enzymatic activity emerges. Cathepsins D and A are then eluted by a stepwise procedure of increasing the ionic strength of the developing solution. After the absorbance at 280 nm drops to about 0.05 OD, 0.12 M NaCl in the 0.01 M potassium phosphate buffer, pH 6, is passed through the column. Cathepsin D is eluted and the most active fractions are pooled (45 ml), dialyzed overnight against 8 liters of 0.005 M potassium phosphate buffer, pH 6, and concentrated to about 10 ml by lyophilization. The elution is continued until absorbancy at 280 nm drops to about 0.10. The eluting solution is changed to 0.4 M NaCl in 0.01 M K_2HPO_4, pH 8.5. Cathepsin A emerges in very few fractions.

TABLE I
PURIFICATION OF CATHEPSINS A AND D FROM CHICKEN MUSCLE[21]

Fraction	Cathepsin D[a]					Cathepsin A[b]		
	Volume (ml)	Protein (mg/ml)	Activity (micromoles/ ml/hour)	Yield (%)	Specific activity (micromoles/ mg/hour)	Activity (micromoles/ ml/2 hours)	Yield (%)	Specific activity (micromoles/ mg/2 hours)
KCL extract	715	16.80	1.28	100	0.076	4.70	100	0.280
Supernatant solution after acid treatment	665	6.65	1.40	102	0.210	3.72	73	0.560
Supernatant solution after heat treatment	625	4.99	1.40	94	0.280	1.85	34	0.370
0-0.60 Saturated (NH₄)₂SO₄ fraction (dialyzed)	48	13.25	11.84	62	0.893	18.80	27	1.40
DEAE-Sephadex								
Eluates (80–88)D	45	1.31	7.96	39	6.07	—	—	—
Eluates (124–129)A	30	3.78	—	—	—	23.65	21	6.30

[a] D = Assayed at pH 2.8. Specific activity of homogenate: 0.03 micromoles/mg/hour.
[b] A = Assayed with N-benzylcarbonyl-α-L-glutamyl-L-tyrosine: 0.14 micromoles/mg/2 hours.

The pH of these pooled eluates is about 6.5. (See Table I for a summary of the purification of cathepsins A and D.)

Cathepsin D may be further purified by chromatography on SE-Sephadex. For this information, the reference should be consulted.

Properties

Stability. Some uncertainty exists about the stability of cathepsin A. The earlier cruder preparations were characterized as very labile unless cysteine was present.[21] Exposure to 50° for 10 minutes resulted in a loss of one-third of the catalytic activity. However, the preparation of cathepsin A from chicken breast muscle presented above appears to be more stable. Although no data on the stability toward heat are available, the authors indicate that the enzyme can be stored frozen for months with little to no loss of activity.

That the muscle enzyme may represent a more stable species of this enzyme is suggested by the results of dialysis experiments. Dialysis of beef spleen and beef liver cathepsin A against $0.01 M$ Tris-HCl buffer, pH 8, overnight resulted in a loss of about 95% of the enzymatic activity. Under the same conditions, the chicken muscle enzyme lost only about 25% of its activity.

Purity and Physical Properties. In the above isolation, the muscle enzyme exists as a single chromatographic peak when eluted from a DEAE-cellulose column; no other criteria have been reported as to its homogeneity or physical properties.

Activators and Inhibitors. There are no known activators of this enzyme. In fact, it is distinguished from the carboxypeptidase present in cathepsin preparations partly on this basis.[19] Carboxypeptidase and cathepsin A have very similar specificities; however, the carboxypeptidase is activated by sulfhydryl-containing compounds whereas cathepsin A is not.[7, 19, 21]

As might be expected, the substances which interact with sulfhydryl groups to inhibit enzymatic activity also are without effect on cathepsin A. Additional work is required to determine whether ions such as heavy metals can block cathepsin A action.

Specificity. The cathepsin A derived from spleen was characterized on the basis of its optimal hydrolytic activity toward N-benzylcarbonyl-α-L-glutamyl-L-tyrosine near pH 5.6.[19] Because of the action on this substrate, it was called a pepsinlike enzyme. It was also shown to hydrolyze the substrate in which the tyrosyl residue was replaced by a phenylalanyl moiety. More recent work with this enzyme has supported its role as an endopeptidase because rupture of the glutamyl-tyrosine bond of N-benzylcarbonyl-α-L-glutamyl-L-tyrosylglycine was observed.[22]

The muscle enzyme has also been found to have an optimal hydrolytic activity at pH 5.0–5.4; however, N-benzylcarbonyl-α-L-glutamyl-L-phenylalanine is hydrolyzed at twice the rate as the corresponding peptide with tyrosine in the COOH-terminal position.[21] This may be due to a species variation, for both chicken liver and muscle cathepsin gave this result while cathepsin A prepared from rabbit muscle and beef liver was the opposite, i.e., N-benzylcarbonyl-α-L-glutamyl-L-tyrosine was the preferred substrate.

Hemoglobin is not readily hydrolyzed by the chicken muscle enzyme and myoglobin and serum albumin are also relatively resistant to attack.[21] Glucagon, however, is hydrolyzed at an appreciable rate and on the basis of a study with this substrate, it has been suggested that cathepsin A is a carboxypeptidase.[23]

It is most important to determine whether cathepsin A, cathepsin D, and catheptic carboxypeptidase are distinct enzymes. It is very interesting to note that the activity and properties ascribed to cathepsin A do overlap many that are reported for the other two enzymes.[22] Obviously, only a careful examination of purified preparations will resolve this state of confusion.

Kinetic Properties. At this time no information is available concerning the kinetic properties of cathepsin A.

Distribution. As with all the cathepsins, this enzyme is widely distributed in animal tissues. It is present in spleen, muscle, and is apparently the major cathepsin present in rat kidney lysosomes.[24, 24a] It has been reported to be absent in brain.[25]

Cathepsin B

The sulfhydryl-dependent enzyme, cathepsin B, has been partially purified from beef spleen.[26, 26a] The suggestion that further purification studies would reveal newer aspects of the nature of cathepsins finds support in the work of Otto[26a] who recognized the presence of a cathepsin B′ in cathepsin B preparations of bovine spleen. He was able to separate the two activities by subjecting the Amberlite IRC-50 treated fraction (described below) to Sephadex G-75 chromatography.

The procedure described below is that of Greenbaum and Fruton[26] and consists of four steps: (1) the extraction of the ground spleen at

[23] A. A. Iodice, *Arch. Biochem. Biophys.* **121**, 241 (1967).
[24] S. Shibko and A. L. Tappel, *Biochem. J.* **95**, 731 (1965).
[24a] S. Y. Ali and L. Evans, *Biochem. J.* **112**, 427 (1969).
[25] N. Marks and A. Lajtha, *Biochem. J.* **97**, 74 (1965).
[26] L. M. Greenbaum and J. S. Fruton, *J. Biol. Chem.* **226**, 173 (1957).
[26a] K. Otto, *Hoppe-Seylers Z. Physiol. Chem.* **348**, 1449 (1967).

pH 2.6; (2) fractional precipitation with $(NH_4)_2SO_4$; (3) treatment with the cation-exchange resin Amberlite IRC-50-XE64; and (4) fractional precipitation with $HgCl_2$ in ethanol. The resultant enzymatic activity represents a 150- to 200-fold purification over that present in the crude tissue extract.

Purification Procedure

Step 1. Extraction. After being chilled for 1–1.5 hours, three fresh beef spleens were trimmed of the outer fascia and fat and passed once through a meat grinder. To the ground tissue (2120 g), 4240 ml of a cold solution containing 2% NaCl and $0.001\,M$ Versene in $0.015\,N$ HCl was added with stirring. The pH was 5.0 after stirring for 5 minutes. In order to lower the pH to 2.6, solid NaCl (40 g) was added to maintain the concentration at 2%, and cold $1\,N$ HCl (320 ml) was introduced dropwise with stirring over a 1-hour period. Toluene (15 ml) to retard bacterial growth was added and the stirring was continued slowly at 2° for 18 hours. At the end of this time the pH had risen to 3.2. The pH was adjusted to 3.6 by the dropwise addition of cold $1\,N$ NaOH (ca. 125 ml). The crude extract was obtained by squeezing the suspension through a double layer of cheesecloth; an aliquot of this dark brown suspension was withdrawn for enzymatic assay and protein determination. The crude extract was then subjected to centrifugation at 2° and top speed in the International PR-2 centrifuge utilizing the No. 845a rotor and 8 oz plastic bottles. After careful decantation of the supernatant fluid (3800 ml), the residue was discarded.

Step 2. Fractionation with Ammonium Sulfate. To the above solution solid $(NH_4)_2SO_4$ (700 g) was added to 40% saturation with stirring at 2° over a period of 20 minutes. Subsequent to a further 20 minutes of stirring, the suspension was centrifuged as described above. The resulting orange supernatant fluid (3150 ml) was brought to 70% ammonium sulfate saturation by addition of 575 g of the solid salt at 2° with stirring. The precipitate was collected by centrifugation as before and suspended in 40 ml of cold water. The suspension (85 ml) was placed in 36/32 Visking casing and dialyzed at 2° for 24 hours against 6 liters of $0.001\,M$ Versene (pH 4.9) with one change of medium, and then against distilled water for another 24 hours. Centrifugation of the dialyzed suspension for 10 minutes at 4000 g at 2° resulted in a clear brown supernatant fluid which was removed and reserved (140 ml). The residue was washed with 10 ml of cold water and centrifuged. The washings were combined with the supernatant fluid; the pH of the mixture was 4.6. The pH was brought to 5.4 by adding 2 ml of $1\,N$ NaOH. The total volume of this fraction, designated 40–70 SAS, was 151 ml.

Step 3. Resin Treatment. The procedure of Boardman and Partridge[27] was followed in regenerating Amberlite IRC-50-XE64 resin. It was then washed with distilled water until the pH of the wash fluid was about 7.0. Equilibration of the resin with 0.08 M sodium citrate buffer, pH 5.2, required about 20 liters of the buffer. For convenience, 1 ml of resin was defined as that amount which settles to 1.0 ml in 20 minutes when a suspension is placed in a graduated cylinder.

Into each of two 250-ml centrifuge tubes kept at 2° were placed 75 ml of resin and 75 ml of the 40–70 SAS fractions. The mixture was stirred at 2° for 5 minutes and after standing another 5 minutes was centrifuged. The supernatant fluid was removed and the residue was washed with three 20-ml portions of the citrate buffer; the third washing was obtained by filtration through a medium porosity sintered glass funnel. The washings were combined with the supernatant fluid (250 ml total). The solution was divided into four equal parts each of which was treated with 65 ml of fresh resin; the stirring and centrifugation were repeated and the supernatant fluid reserved. Each of the 4 resin samples was washed twice with 20-ml portions of citrate buffer; the washings were added to the supernatant fluid. In order to remove fine particles of resin, the combined solution was filtered through Whatman No. 41 filter paper.

To concentrate the enzyme, the pH of the solution (360 ml) was adjusted to 3.6 by the addition of 1 N H_2SO_4 (34 ml) at 0° with stirring, followed by the addition of 275 g of solid ammonium sulfate to 100% saturation. After being stirred for 30 minutes, the suspension was centrifuged at 4000 g for 15 minutes and the supernatant fluid discarded. Following suspension of the precipitate in about 20 ml of cold water, it was dialyzed at 2° against 6 liters of 0.001 M Versene, pH 5.0, for 24 hours with one change of dialyzing medium, and then for 24 hours against 6 liters of cold distilled water. The pH of the cloudy dialyzed solution was 4.2. It was brought to pH 5.0 by the careful addition of 1 ml of 0.1 N NaOH and centrifuged. The clear supernatant fluid (35 ml) was termed the "resin-treated enzyme" and was assayed for enzymatic activity.

Step 4. Mercuric Chloride–Ethanol Fractionation. The resin-treated fraction (34 ml) was lyophilized and the resulting powder was dissolved in 14 ml of cold 0.001 M $HgCl_2$ to give a solution containing 1% protein. To this solution, 0.45 ml of 0.001 M $HgCl_2$ in 95% ethanol was added dropwise with stirring at −2° (final concentration of ethanol 3%). After standing for 15 minutes, the mixture was centrifuged at −2°. The resulting supernatant fluid (14 ml) was treated with 1.33 ml of the

[27] N. K. Boardman and S. M. Partridge, *Biochem. J.* **59**, 543 (1955).

TABLE II
PARTIAL PURIFICATION OF CATHEPSIN B FROM BEEF SPLEEN[26]

Fraction	Total vol. (ml)	Total protein (mg)	[C.U.]^{BAA} (units)	[C.U.]$^{BAA}_{mg. Pr}$ (sp. act.)	Purification	Recovery (%)
Crude extracts	3800	74,100	2660	0.036		100
40–70 SAS	151	1,163	1590	1.36	38	60
Resin-treated	35	147	595	4.05	112	22
Hg-Ethanol	7.2	33	223	6.80	189	8

$HgCl_2$–ethanol solution at −6° (final concentration of ethanol, 11%). After standing for 15 minutes, the mixture was centrifuged at −6° and the residue was immediately suspended in 8 ml of cold water. Centrifugation sedimented the small amount of insoluble material present. This precipitate was extracted with 1 ml of cold water and the washing was added to the supernatant fluid (total volume, 8.5 ml). Dialysis of 7.2 ml of this solution was carried out at 2° against 6 liters of 0.001 M Versene (pH 4.9) for 5 hours and continued against water for another 5 hours. After centrifugation, 7.0 ml of the supernatant fluid (pH 3.6) was obtained. Careful addition of 0.2 ml of 0.1 N NaOH raised the pH to 5.0. The "Hg-ethanol" fraction was stored in the deep-freeze until use.

Properties

Stability. The purified preparation of cathepsin B described above has not been studied in terms of its stability. However, it had been noted in earlier studies that the activity attributed to cathepsin B loses about 50% of its hydrolytic activity both toward the synthetic substrate and proteins after heating at 52° for 30 minutes.[28]

Dilute solutions of cathepsin B′ in 50 mM acetate buffer (pH 4.5) are fairly stable when stored in the cold. It was noted that the pH range of optimal stability varies according to the buffer employed.[26a]

Purity and Physical Properties. The homogeneity of the preparation outlined remains to be elucidated. It is known to contain as an impurity a catheptic carboxypeptidase,[7] which requires sulfhydryl compounds as activators and exerts its catalytic activity optimally at pH 3.0–4.0. Further work is essential to obtain preparations free of impurities to characterize the enzyme protein definitively.

As previously mentioned, cathepsin B′ has been separated from the preparation of cathepsin B from beef spleen.[26a] On the basis of its behavior on Sephadex columns, a molecular weight of about 25,000 has

[28] H. R. Gutmann and J. S. Fruton, *J. Biol. Chem.* **174**, 851 (1948).

been assigned to B'. Inasmuch as the activity ascribed to cathepsin B precedes that of B' in the column effluent, a higher molecular weight for that enzyme may be inferred.

Activators and Inhibitors. The presence of sulfhydryl compounds is necessary in order that a maximal enzymatic response be elicited for both cathepsin B and B'. Most effective in this regard is β-mercaptoethylamine, cysteine less so, followed by 2,3-dimercaptopropanol, β-mercaptoethanol, and glutathione. Even though cyanide ion can substitute for sulfhydryl compounds in stimulating the hydrolytic activity of other SH-dependent proteinases (e.g., papain) it is without effect on cathepsin B.

In the presence of cysteine, (0.04 M), complete inhibition of enzymatic activity will occur if 0.001 M iodoacetamide is added; p-chloromercuribenzoate (0.0001 M) will also inhibit to a limited extent but N-ethylmaleimide (0.001 M) does not. Chloroquine has been shown to inhibit the cathepsin B activity of rabbit ear cartilage.[28a] Despite the similarity in its catalytic activity to trypsin, compounds such as diisopropylphosphofluoridate (DFP), dinitrophenol, or soybean trypsin inhibitor display no inhibitory action on the spleen-derived enzyme.

An inhibitor of cathepsin B activity has been reported in the cell-soluble fraction of rat liver.[29] However, a more recent communication indicates that no inhibitor of cathepsin B could be found in this tissue.[30]

Specificity. Cathepsin B acts on substrates similar to those attacked by crystalline pancreatic trypsin.[19] The spectrum of activity is less broad, however. For example, although both enzymes will split benzoyl-L-argininamide and ethyl ester as well as the related lysyl compounds, only trypsin will cleave the nonacylated analogs, L-argininamide or ethyl ester and L-lysinamide or ethyl ester. Further differences are apparent from a systematic study comparing trypsin, papain, and cathepsin B.[16] The cathepsin resembles papain more closely in hydrolyzing phthaloyl-glycyl-L-4-thialysine methyl ester at a greater velocity than phthaloyl-glycyl-L-lysine methyl ester.

Like trypsin, cathepsin B is able to activate trypsinogen to form trypsin.[31] On the other hand, cathepsin B treatment of prothrombin does not yield thrombin as contrasted to the activation by trypsin.[31a] The acid pH required for the cathepsin reaction may be a contributing factor.

[28a] S. Y. Ali, L. Evans, E. Stainthorpe, and C. H. Lack, *Biochem. J.* **105**, 549 (1967).

[29] J. T. Finkenstaedt, *Proc. Soc. Exptl. Biol.* **95**, 302 (1957).

[30] J. M. W. Bouma and M. Gruber, *Biochem. Biophys. Acta* **89**, 545 (1964).

[31] L. M. Greenbaum, A. Hirshkowitz, and I. Shoichet, *J. Biol. Chem.* **234**, 2885 (1959).

[31a] M. I. Barnhart, *Am. J. Physiol.* **198**, 899 (1960).

Although both cathepsins B and B' catalyze the hydrolysis of benzoyl-L-argininamide, only B' has significant hydrolytic activity in degrading the enzyme, glucokinase, thereby providing a method for distinguishing these two activities.[26a]

Besides carrying out hydrolase activities, cathepsin B can also function as a transferase although it is less efficient than are cathepsin C (dipeptidyl transferase) and papain.[32, 33] As has been demonstrated for these two enzymes,[34, 35] cathepsin B also exhibits a preference for replacement agents.[32] For example, glycyl-L-leucine can serve as a replacing nucleophilic compound while glycyl-D-leucine cannot. L-Leucylglycine is incorporated into peptide linkage by the transferase type reaction whereas D-leucylglycine or L-leucylglycylglycine are excluded. Thus, cathepsin B shows a specificity not only for the substrate it attacks but also for the agent which is being transferred into the susceptible linkage.

Kinetic Properties. The lack of a homogeneous preparation precludes the possibility of any meaningful kinetic data. Some indication of its affinity for substrates are the published K_m values[16] for the hydrolysis of a number of related compounds; that of phthaloylglycyl-L-lysine methyl ester is $4 \times 10^{-2} M$, for phthaloylglycyl-L-4-thialysine methyl ester, $8 \times 10^{-2} M$, for phthaloylglycyl-DL-4-oxalysine methyl ester, $6 \times 10^{-2} M$, and for phthaloyl-L-alanyl-L-4-oxalysine methyl ester, $1 \times 10^{-1} M$.

A comparison of the K_m values of cathepsins B and B' when benzoyl-L-argininamide is employed as the substrate reveals that the former is about half of that of the latter, e.g., $1.25 \times 10^{-2} M$ and $3.33 \times 10^{-2} M$, respectively.

The activity of a given preparation of cathepsin B was frequently not proportional to the enzyme concentration, giving an apparent increase in specific activity at lower concentrations of test solutions.[26] It was suggested that this observation might be explained by the reported presence of a cathepsin B inhibitor in rat liver extracts (cf. above).

A preparation of cathepsin B obtained from an acetone powder of calf liver lysomal-mitochondrial fraction has appeared recently.[35a] It represents a 200-fold purification over the liver homogenate. A single component is found on preparative gel electrophoresis (phosphate buffer, pH 6–7); but the activity cannot be recovered from the gel. On the basis of gel filtration the molecular weight is estimated to be about 25,000–

[32] S. Fujii and J. S. Fruton, *J. Biol. Chem.* **230**, 1 (1958).
[33] J. Durell and J. S. Fruton, *J. Biol. Chem.* **207**, 487 (1954).
[34] Y. P. Dowmont and J. S. Fruton, *J. Biol. Chem.* **197**, 271 (1952).
[35] M. J. Mycek and J. S. Fruton, *J. Biol. Chem.* **226**, 165 (1957).
[35a] O. Snellman, *Biochim. J.* **114**, 673 (1969).

30,000. The K_m values reported are in the same range as those reported for previous preparations.

Distribution. The enzyme is found in many tissues,[30] being particularly rich in spleen. Moderately high activities are observed in the lysosomal fractions of thymus, lung, liver, kidney, and adrenal. It has also been localized in the mitochondrial fraction of rat lymph nodes. Very low to no activity is present in the lysosomal fractions of heart, testis, brain, and skeletal muscle of the rat. Evidence for presence of cathepsin B in rabbit ear cartilage has also been presented.[28a]

The cathepsin B activity of rat liver was found to be distributed throughout the cell fractions in contradistinction to those of cathepsin A and catheptic carboxypeptidase which were localized in the mitochondria.[36]

The exposure of rats to total body X-radiation results in a significantly higher specific activity for both cathepsins B and C while that of cathepsin D was considerably lower.[30] As a consequence of these observations, the suggestion has been put forth that these enzymes are associated with the phagocytic cells of the spleen and may play an important role in the degradation of phagocytosed material.

Cathepsin C (Dipeptidyl Transferase)

As mentioned above, of all the catheptic activities studied, the one which has undergone the most extensive investigation is cathepsin C. As a consequence of the findings by Planta and Gruber[37, 38] and of studies over the last 17 years in the laboratories of Fruton, it was recognized that the activity referred to as cathepsin C could better be termed dipeptidyl transferase.[9] This designation will be used herein together with the older term. It should be noted that another name has also been proposed for cathepsin C—dipeptidyl amino peptidase I.[39] However, in the author's estimation the latter term, although accurate in describing the hydrolytic properties of the enzyme, has no advantage over the former, which is based on its striking transfer ability. For the sake of clarity, it would be best to settle on one designation and adhere to it. As has already been pointed out, it is to be hoped that further study of the cathepsins will allow for the establishment of a nomenclature which will be more precise in its description of the individual enzymatic activities.

[36] W. Rademaker and J. B. J. Soons, *Biochem. Biophys. Acta* **24**, 451 (1957).
[37] R. J. Planta and M. Gruber, *Biochem. Biophys. Acta* **53**, 443 (1961).
[38] R. J. Planta, J. Gorter, and M. Gruber, *Biochem. Biophys. Acta* **89**, 511 (1964).
[39] J. K. McDonald, B. B. Zeitman, T. J. Reilly, and S. Ellis, *J. Biol. Chem.* **244**, 2693 (1969).

Purification Procedure

The method of purification is that of Metrione *et al.*[9] and uses fresh beef spleen as the source of the enzyme. An earlier preparation can be found in Volume II of this treatise.[39a]

Step 1. Extraction. After the removal of the outer membranes from about 12 lb of fresh beef spleen, the organs were cut into narrow strips (ca. 5 cm wide), placed into a plastic pan, and frozen for 8 hours. Thawing was allowed to take place overnight in the cold room, after which the tissue was passed through a meat grinder twice. The ground material (4667 g) was suspended in twice the volume (9334 ml) of water containing 5.6 g of EDTA. As the mixture was being stirred mechanically, 150 ml of 6 N H_2SO_4 was added to bring the pH to 3.5. Stirring was continued at room temperature for 1 hour. At the end of this time, the pH was readjusted to 3.5 with 10 ml of 6 N H_2SO_4. Autolysis of the suspension took place by incubating at 38° for 22 hours (slow stirring for the first 5 hours). When the solid material had settled, the supernatant fluid was siphoned off. Centrifugation of the residue at 10,000 g resulted in additional supernatant fluid which was combined with that siphoned off. This was designated "acid extract."

Step 2. Precipitation with Ammonium Sulfate. The acid extract was treated with 2960 g of ammonium sulfate to bring it to 40% saturation. Stirring was continued for 2 hours and 40 g of acid-washed Hyflo filter cell was added to facilitate filtration with suction through large sheets of soft filter paper (Schleicher and Schull, No. 589). The addition of 2750 g of ammonium sulfate to the filtrate brought its concentration to 70% saturation. Precipitation was allowed to proceed overnight in the cold room following which the suspension was filtered through hardened filter paper (Schleicher and Schull, No. 576). After being suspended in about 70 ml of water, it was extensively dialyzed against 0.15 M (0.9%) NaCl. This was called the "40–70 AS fraction."

Step 3. Heat Treatment. The above fraction was divided into portions of about 15 ml each and heated at 65° for 40 minutes. After chilling in ice, the precipitate which had formed was removed by centrifugation. The supernatant fluids were pooled and labeled "heated fraction."

Step 4. Chromatography on Sephadex G-200. This step and the two subsequent ones were carried out at room temperature without loss of activity. A portion (50 ml) of the heated fraction was placed onto a Sephadex G-200 column (5 × 73 cm) and eluted with 0.9% NaCl. Four large areas of material absorbing at 280 nm came off the column in an effluent volume of about 1 liter. The enzyme activity was localized in

[39a] G. de la Haba, P. S. Cammarata, and J. S. Fruton, Vol. II, p. 64.

the fractions between 440 and 615 ml. These fractions were combined and concentrated by precipitation with 80% ammonium sulfate (special enzyme grade, Mann Research Laboratories). After 30 minutes, the suspension was centrifuged and the residue obtained was suspended in 5 ml of water. Dialysis against $0.0005\,M$ sodium phosphate buffer (pH 6.8) containing $0.0002\,M$ β-mercaptoethanol resulted in the "G-200 fraction." This operation was repeated until all of the heated fraction had undergone the step.

Step 5. Chromatography on DEAE-Cellulose. A column of DEAE-cellulose (Brown Co., Selectacel reagent grade, 0.83 meq/g) of dimensions 4×35 cm was prepared. Equilibration was carried out with $0.005\,M$ sodium phosphate buffer, pH 6.8, containing $0.002\,M$ β-mercaptoethanol. After a small sample of the G-200 fraction (5.2 ml) had been placed on the column, gradient elution was started by passing $0.25\,M$ phosphate buffer, pH 6.7, containing $0.002\,M$ β-mercaptoethanol into a mixing chamber containing 250 ml of the $0.005\,M$ buffer. As in the previous step, this operation can be conducted at room temperature without loss of activity. After the emergence of a twin-peaked 280 nm absorbing fraction devoid of transferase activity, the enzyme appeared in a rather broad peak (612–822 ml). These fractions were combined and concentrated by precipitation with 80% ammonium sulfate, suspension in water as before, and dialysis against $0.005\,M$ sodium acetate buffer, pH 5.0, containing $0.002\,M$ β-mercaptoethanol. The result was the "DEAE fraction."

Step 6. Chromatography on CM-Cellulose. A column of CM-cellulose (Mann Research Laboratories, Mannex-CM, type 20) of dimensions 2×32 cm was prepared and equilibrated with $0.005\,M$ sodium acetate buffer, pH 5.0, containing $0.002\,M$ β-mercaptoethanol. A portion of the DEAE fraction (4.7 ml) was applied to the column and gradient elution was begun immediately. This was done by passing $0.35\,M$ sodium acetate

TABLE III

PURIFICATION OF BEEF SPLEEN CATHEPSIN C[9,a]

Fraction	Volume (ml)	Total activity (units)	Total protein (mg)	Specific activity (units/mg)
Acid extract	12,180	34,100	120,000	0.28
40–70 AS fraction	194	22,700	12,300	1.85
Heated fraction	178	19,200	6,870	2.79
G-200 fraction	18.5	12,100	666	19.2
DEAE fraction	16.8	6,040	286	21.1
CM fraction	7.2	3,690	142	25.9

[a] Reprinted from *Biochemistry* **5**, 1597 (1966). Copyright (1966) by the American Chemical Society. Reprinted by permission of the copyright owner.

buffer, pH 5.0, containing 0.002 M β-mercaptoethanol into a mixing chamber containing 250 ml of the 0.005 M buffer. After the emergence of a 280 nm absorbing substance early in the effluent, the enzyme activity appears as a sharp peak between 245–272 ml. These fractions were combined and again concentrated by precipitation with 80% ammonium sulfate. After suspension in a small volume of water, it was dialyzed against 0.9% NaCl and stored at 3°. The end material or "CM-fraction" represents a 90-fold purification over the initial acid extract.

See Table III for a summary of the purification of cathepsin C from beef spleen.

Properties

Stability. In crude extracts, the enzymatic activity is stable to acid (pH 3.5) and heat (65°); these properties are important in the purification procedure where they serve to denature other less stable proteins which are present as impurities.

During the chromatographic purification, it was found that the enzyme preparation became cold labile and lost activity upon freezing. This together with its molecular size suggested that it may be an oligomeric protein composed of several subunits.

Purity and Physical Properties. Over a pH range of 5.3–6.4 ($\gamma/2$, 0.1) the "CM fraction" appears to be nearly homogeneous upon ultracentrifugation with an $s_{20,w}$ value of 9.73. Two sedimentation equilibrium experiments resulted in z average molecular weights 213,000 ± 2000 and 200,000 ± 2000. More recent work indicates that 8 sulfhydryl groups are present per unit molecular weight.[9] The partial specific volume was found to be 0.73 ml/g. The molecular weight of the rat liver enzyme is in the same range.[39]

Treatment of the purified enzyme with 0.001 M p-chloromercuribenzoate resulted in the appearance of a new component in the ultracentrifugal sedimentation pattern performed at pH 5.4. In addition to active enzyme which sedimented at 9.5 S, a material with a sedimentation coefficient of 5.5 S was observed, which accounted for about one-fourth of the total protein. The original sedimentation pattern could be restored by dialysis against 400 volumes of 0.001 M β-mercaptoethanol. It remains to be determined whether the 5.5 S component retains enzymatic activity after reaction with mercuribenzoate or whether the reappearance of 9.5 S component after dialysis represents a reconstitution of the less dense material.

All the protein was converted to a 1.9 S component after treatment of the beef spleen enzyme with 1% sodium dodecyl sulfate; this deaggregation could not be reversed under any conditions. Under comparable conditions the rat liver enzyme could also be dissociated into subunits.[39]

The behavior of rat liver cathepsin C during acrylamide gel electrophoresis at three separate pH values indicated that it is a more acidic protein than the beef spleen enzyme.[39]

It has been shown that exposure of the enzyme to $2\,M$ urea causes about a 60% inhibition of overall transferase activity utilizing L-alanyl-L-phenylalaninamide as substrate.[40] This, along with the sedimentation behavior discussed above, would support the possibility of an aggregated protein.

The enzyme preparation does contain carbohydrate material which may or may not be an integral part of the protein. Exposure to neuraminidase did not result in any decrease in the specific activity of the enzyme.

Activators and Inhibitors. The work carried out on the early cruder preparations of beef spleen cathepsin C[8] has been confirmed with the purer preparations now available.[9,10] The addition of sulfhydryl compounds to the assay mixtures results in a marked stimulation of the enzymatic activity. Most potent in this regard are β-mercaptoethylamine, cysteine, 2,3-dimercaptopropanol, and dithiothreitol. Thioglycolic acid and glutathione exhibit much less stimulatory activity. Cyanide which is known to stimulate papain, another sulfhydryl requiring proteinase, also effects a mild stimulation of cathepsin C. It has been suggested[10] that cathepsin C (dipeptidyl transferase) forms an intermediate thiol ester between a reduced sulfhydryl group on the enzyme and the reactive carbonyl group of the dipeptidyl substrate. The activator would presumably play a role in keeping the enzyme sulfhydryl in a reduced state. McDonald *et al.*[39] have reported that Cl⁻ (5–10 mM) stimulated the hydrolysis of glycyl-L-phenylalanine-β-naphthylamide by the rat liver enzyme. These same investigators indicated that Gorter and Gruber have found the same to be true of the beef spleen enzyme.

In a study of inhibitors,[8] it was found that diisopropylphosphofluoridate and 2,4-dinitrophenol had no effect on the enzymatic activity. The lack of inhibition by $1 \times 10^{-5}\,M$ p-chloromercuribenzoate was a surprising finding, especially in view of the complete inhibition effected by $0.001\,M$ iodoacetate which occurs even in the presence of $0.004\,M$ cysteine. Formaldehyde ($0.05\,M$) also serves to completely inhibit the enzyme.

Specificity. Substrates most susceptible to enzymatic attack are dipeptidyl amides or esters, bearing a free α-amino (or α-imino) group in the NH$_2$-terminal position.[8,41] The nature of the NH$_2$-terminal amino acid is not as important in determining potential substrate activity as is the structure of the amino acid bearing the susceptible carbonyl bond.

[40] C. P. Heinrich and J. S. Fruton, *Biochemistry* **7**, 3556 (1968).
[41] N. Izumiya and J. S. Fruton, *J. Biol. Chem.* **218**, 59 (1956).

Most of the compounds found to possess substrate activity have glycyl, L-alanyl, or L-seryl residues as the NH$_2$-terminus. If either leucine or lysine is present in this position, then the substrate activity is lost. The importance of the free NH$_2$ group has been appreciated through studies of compounds in which this group was masked[41] or replaced by a diazoacetyl moiety[10]; in both instances, no hydrolysis of the potentially susceptible bond takes place.

Since the enzyme activates the terminal carbonyl group, it is not surprising to find the structural requirement of the amino acid in that position to be more restricted. Preference is shown for those substrates with hydrophobic side chains in this position, e.g., tryptophan, tyrosine, phenylalanine, or leucine. Glycylglycine ethyl ester is also a substrate. The recent studies of McDonald et al.[39] in which compounds such as glycyl-L-arginine-β-naphthylamide served as substrates for the rat liver enzyme are perplexing and it would seem that a reinvestigation of specificity is in order.

It is of interest that the interior —NH— group is apparently not strictly required for substrate activity for the depsipeptide, glycyl-(β-phenyl)-L-lactic acid methyl ester undergoes cleavage readily by cathepsin C (dipeptidyl transferase).[10] This is in marked contrast to the resistance displayed by glycyl-N-methyl-L-phenylalanine ethyl ester to hydrolysis.[41] The resistance of the latter compound to enzymatic attack may be a consequence of steric hindrance conferred by the N-methyl group, thereby blocking enzyme-substrate interaction.[10]

Hydrolysis of most dipeptidyl derivatives at the susceptible amide or ester linkage predominates when the pH of the enzymatic reaction is near pH 5. At pH values between 7 and 8, replacement of the amide or ester moiety by nucleophilic compounds prevails and oligopeptides are formed.[42,43] However, when a substrate, such as L-prolyl-L-phenylalaninamide, is used, the greatest hydrolytic activity is observed at pH 7.[8] This substrate does not undergo polymerization to form oligopeptides.[32] On the basis of this observation and those which indicate that purified preparations of this enzyme do not attack proteins,[37] it is questionable that cathepsin C (dipeptidyl transferase) should be considered as a cathepsin in the classical sense.

The unique ability of cathepsin C (dipeptidyl transferase) to attack the carbonyl bond of dipeptide amides and esters has been utilized to

---

[42] M. E. Jones, W. R. Hearn, M. Fried, and J. S. Fruton, *J. Biol. Chem.* **195**, 645 (1952).

[43] J. S. Fruton, W. R. Hearn, V. M. Ingram, D. S. Wiggans, and M. Winitz, *J. Biol. Chem.* **204**, 891 (1953).

study sequences of oligopeptides, such as angiotensin II, ACTH, glucagon, and the oxidized B-chain of insulin.[39] Since the presence of an NH_2-terminal lysine or arginine will retard hydrolysis, the method is best applicable to tryptic digests of proteins.

The ability of this enzyme to catalyze replacement reaction was recognized in 1952[42] when it was shown to promote the formation of hydroxamic acid derivatives in the presence of hydroxylamine. Furthermore, in studies of the action of the enzyme on the substrate glycyl-L-phenylalaninamide at pH 7.2, a heavy flocculent precipitate was noted which was found to be a polymer with an average chain length of eight amino acid residues (i.e., a tetramer of the substrate). An extension of this work utilizing glycyl-L-tyrosine amide as substrate suggested that the polymerization proceeds as a single-chain process without the intermediate formation of dimers or trimers, inasmuch as the addition of the dimer glycyl-L-tyrosyl-L-glycyl-L-tyrosyl amide did not accelerate the process.[44] More recent experiments have resulted in the detection of such intermediates.[40, 45] Of the peptides studied for polymerization efficiency, L-alanyl-L-phenylalaninamide stands out.

The hypothesis has been advanced that a suitable dipeptidyl amide reacts with the enzyme to form a complex which can be converted to a dipeptidyl enzyme which in turn can react either with water to yield free dipeptide or with the amino group of a suitable acceptor (e.g., another molecule of substrate) to yield a tetrapeptide amide. This process could then be repeated with the intermediate formation of a tetrapeptidyl enzyme which would act as donor. This presumably would favor the presence of the shorter chain peptides over those higher in the series. However, the data do not support this supposition and consequently, the hypothesis has been revised to suggest that the oligomeric nature of the enzyme may permit a cooperative mechanism of interacting enzyme subunits, whereby dipeptidyl moieties on adjacent subunits react to propagate the chain. This interesting suggestion awaits further clarification.

Not only can oligopeptides arise as a result of the polymerization but also by the incorporation of the α-amino group of a second peptide. As mentioned earlier (cf. cathepsin B), there is also a specificity directed toward this replacement agent, i.e., donor of the nucleophilic moiety.[32, 38, 42, 45] In a systematic study of peptides which can serve as acceptors of the reactive carbonyl,[38] it was found that those with NH_2-terminal valine, leucine, tyrosine, arginine, and histidine can function

[44] J. S. Fruton and M. H. Knappenberger, Biochemistry 1, 674 (1962).
[45] K. K. Nilsson and J. S. Fruton, Biochemistry 3, 1220 (1964).

in this capacity. Those with NH_2-terminal glutamic acid, alanine, serine, and glycine cannot. Glycine and alanine derivatives may or may not function in this capacity. As has been pointed out, glycyl-L-phenylalaninamide and glycyl-L-tyrosinamide as well as alanyl-L-phenylalaninamide can undergo polymerization.

Kinetic Properties. K_m values for a series of dipeptidyl ethyl ester substrates range from $6.3 \pm 0.4 \times 10^{-3} M$ for glycylglycine ethyl ester, while poorer substrates, e.g., glycyl-(β-phenyl)L-lactic acid and sarcosyl-L-phenylalanyl ethyl ester, had K_m values of $1.9 \pm 0.1 \times 10^{-2} M$ and $6.1 \pm 0.4 \times 10^{-2} M$, respectively.[10] These data obtained with the more pure preparation of cathepsin C (dipeptidyl transferase) are in reasonable agreement with those calculated for the dipeptide amides and esters tested with the less pure enzyme.[8]

It would be of some interest to examine the K_m values for the hydrolysis of the amide substrates with the more homogeneous protein, inasmuch as earlier work had shown these data to be higher than those obtained for the corresponding esters.[8] For example, glycyl-L-phenylalaninamide had a K_m of $1.05 \times 10^{-2} M$ while the K_m is $1.3 \times 10^{-3} M$ for the corresponding ethyl ester. Since the V_{max} values are very nearly equal, the deacylation step is probably rate-limiting.[8, 10, 38]

Approximate K_m values for the hydrolysis of the β-naphthylamide derivatives of glycyl-L-arginine, seryl-L-methionine, alanyl-L-alanine, and glycyl-L-phenylalanine by the purified rat liver enzyme ranged from 1×10^{-4} to $1.9 \times 10^{-4} M$; for the corresponding esters and amides the values ranged from $4–16 \times 10^{-4} M$ and $1.8–5.1 \times 10^{-3} M$, respectively.

Distribution. In a study of the lysosomal fractions of various rat tissues,[30] cathepsin C (dipeptidyl transferase) was most active in the liver followed by the spleen. Other organs possessing measurable activity were the kidney, pancreas, lung, and thymus. The result with the kidney lysosomes is open to question, since in another paper, this fraction was reported to lack cathepsin C activity.[24] Brain has also been found not to contain this enzyme.[25, 30]

It has been suggested[38] that cathepsin C may act to degrade peptides liberated as a result of proteolysis.

Cathepsin D

Several methods are available for the isolation of cathepsin D[4, 5, 6, 6a, 21] from various organs, bovine spleen,[4] rabbit spleen,[5, 6] rabbit liver,[6a] and chicken breast muscle.[21]

The method selected for description is that published by Lapresle and Webb[5, 6]; the source of the enzymatic activity is rabbit spleen. A similar purification procedure has been described by Iodice et al.,[21] using chicken breast muscle as the starting material. The following steps are carried out at 4°.

Purification Procedure

Step 1. Extraction. Spleens are removed from exsanguinated rabbits and frozen. The frozen spleens are broken up in a small volume of water and blended for 1 minute. Following the dialysis of the resulting suspension against distilled water for 1 hour with stirring, the mixture is centrifuged at 15,000 g for 1 hour. The supernatant fluid is decanted and the residue is discarded. Addition of NaCl to a final concentration of 0.15 M and HCl to pH 5.0 causes the formation of a precipitate which is removed by centrifugation.

Step 2. Precipitation with Ammonium Sulfate. The clear supernatant fluid is precipitated by the addition of 2 volumes of saturated $(NH_4)_2SO_4$ solution, pH 7. After 30 minutes, the suspension is centrifuged and the supernatant discarded. The precipitate is dialyzed free of $(NH_4)_2SO_4$ after being dissolved in a small volume of water and then is stored frozen. At this stage the specific activity is increased about 3-fold over the aqueous extract.

Step 3. Chromatography on DEAE-Cellulose. The crude preparation from the above step is dialyzed against 0.05 M potassium phosphate buffer, pH 8.0 (overall recovery at this stage is 50% and specific activity 4 times that of the aqueous extract), and passed through a DEAE-cellulose column (3 g of cellulose/g of protein) equilibrated with the same buffer. The enzymatic activity emerges in the reddish breakthrough fraction.

The pooled eluates containing enzymatic activity are dialyzed against 15 mM potassium phosphate buffer, pH 6.75, and passed through a carboxymethyl cellulose column equilibrated with the same buffer (15 g of cellulose/g of protein) which adsorbs the inactive reddish material. The active enzyme passes through the column. The pooled eluates are dialyzed against 5 mM potassium phosphate buffer, pH 8.0, and adsorbed on a column of DEAE-cellulose equilibrated with the same buffer. Following the passage of the inactive breakthrough fraction, the active enzyme is eluted in a small volume by 0.1 M potassium phosphate buffer, pH 7.0.

Step 4. Chromatography on CM-Sephadex. The enzyme solution is again dialyzed against 15 mM potassium phosphate, pH 6.75, and passed through a column of carboxymethyl-Sephadex. The enzyme is retained on the column. After emergence of an inactive material, elution with 15 mM potassium phosphate buffer containing 0.5 M NaCl results in the appearance of cathepsin D with a specific activity of 130 units/mg protein (about a 25-fold purification) in the eluate.

Properties

Stability. Exposure of the rabbit spleen cathepsin D to 60° for 40 minutes results in a loss of all the enzymatic activity[4]; the enzyme from chicken muscle may be more stable at this temperature. It has been noted that storage of a dilute solution of the muscle-derived enzyme at −14° will cause a decrease in activity, but this can be prevented by concentrating the enzyme by lyophilization.

The spleen enzyme is also inactivated by pH conditions below 2.5 which may be due either to denaturation or autolysis. The rabbit liver enzyme can be stored at 4° at pH 3–4 or at pH 8 for days without loss of activity. Furthermore in the presence of 20% (w/v) glycerol and storage in polystyrene containers no appreciable loss of activity is noted for months.[6a]

Purity and Physical Properties. The preparations of this proteinase cannot be considered homogeneous and are contaminated by other cathepsins and possibly other proteins. For example, cathepsin A and E activity has been identified in the preparations of the muscle[21] and rabbit spleen[6] enzymes, respectively. Due to these possible impurities, it remains to be unequivocally proven that the ten forms recognized in the beef spleen preparation[4] as a consequence of their chromatographic and electrophoretic behavior are truly isozymes of cathepsin D. These forms appear to differ only in their charge distribution at pH 5.5 and 8.4. The amino acid, glycine, has been identified as the NH_2-terminal position of all.

The preparation from rabbit liver[6a] emerges as a single peak from Sephadex G100 column. It has been estimated to have a molecular weight of 50,000–52,000 because of its effluent position between bovine serum albumin and egg albumin. However, the homogeneity of this material remains to be proved. This value for molecular weight is in good agreement with that of 58,000 published by Press *et al.*[4]

At pH 8.2, the rabbit spleen enzyme exhibits an electrophoretic mobility in agar gel of -1.7×10^{-5} cm²/V/second.[5]

Activators and Inhibitors. It is very unlikely that sulfhydryl groups play a role in the activity of the enzyme since addition of typical sulfhydryl activators (e.g., cysteine) or inhibitors (e.g., p-chloromercuribenzoate, iodoacetate) do not alter its activity.[4] L-Tosylamido-2-phenylethyl chloromethylketone, ascorbate, and a number of other compounds are also without effect.[6a] The rabbit liver enzyme[6a] is inhibited by 3-phenylpyruvic acid and dithioerythritol.

Earlier work with the rabbit spleen preparation had indicated that enzymatic activity could be enhanced by the presence of Ca^{2+} and Mg^{2+}; however, these ions were later found to have no effect and presumably were required for removal of impurities in the more crude preparations.[5] Fe^{2+} is also without effect.

Specificity. No synthetic substrate of defined chemical composition has been found for cathepsin D as yet. The enzyme is without effect on those compounds ordinarily cleaved by cathepsin A, B, or dipeptidyl transferase (cathepsin C). When the B-chain of insulin serves as a substrate, a specificity similar to, but more limited than that of pepsin is observed.[4]

Cathepsin D acts as a true proteinase. At pH 2.8–3.0, the rabbit and ox spleen and chicken muscle enzymes hydrolyze denatured hemoglobin optimally, while serum albumin undergoes cleavage optimally at pH values close to 4. The pepsinlike activity has been noted not only in the case of the B-chain of insulin, as mentioned above, but also with human serum albumin as substrate. The same immunoelectrophoretic pattern is obtained by human serum albumin degraded either by pepsin or the rabbit spleen cathepsin D.[5] Until studies of substrates of a more defined structure are carried out, the specificity will not be wholly understood.

Kinetic Properties. A more extensively purified enzyme preparation, as well as synthetic substrates, would permit the acquisition of meaningful kinetic data. It has been reported that the cathepsin D activity from chicken muscle is not strictly linear with enzyme concentration but tends to fall off gradually.[21] Whether this is due to the presence of an inhibitor (cf. cathepsin B) or to possible product inhibition remains to be elucidated.

Distribution. Cathepsin D occurs in a variety of tissues. Spleen is a particularly rich source; cathepsin D is apparently the principal protease of ox spleen.[4] It has also been found in ox kidney, rabbit bone marrow,[6] rabbit liver,[6a] and chicken muscle.[21] A similar enzyme has been reported in pituitary,[46] brain,[25, 47] erythrocyte stroma,[48] and thyroid,[49] and rabbit ear cartilage.[24a]

The lysosomal fractions of various rat tissues were assayed for cathepsin D activity.[30] Particularly high in activity were the spleen and adrenal, followed by the kidney and thymus. Lung and liver had about one-third the activity of spleen. In this same study, it was shown that the cathepsin D activity of the spleen decreased as a consequence of exposure of the animal to total body X-radiation.

[46] E. Adams and E. L. Smith, *J. Biol. Chem.* **191**, 651 (1951).
[47] M. W. Kies and S. Schwimmer, *J. Biol. Chem.* **145**, 685 (1942).
[48] W. L. Morrison and H. Neurath, *J. Biol. Chem.* **200**, 39 (1953).
[49] K. Balasubramanian and W. P. Deiss, Jr., *Biochem. Biophys. Acta* **110**, 564 (1965).

The suggestion has been offered that cathepsin A and D act in concert with each other to break down protein substrates.[21] The former enzyme possesses little or no hydrolytic activity toward proteins such as denatured hemoglobin, yet when it is present in an incubation with cathepsin D and denatured hemoglobin, a greater breakdown of the substrate is observed than could be explained by a simple additive effect of both enzymatic activities.

Cathepsin E

Purification Procedure

The source of enzyme activity used in the isolation procedure of Lapresle and Webb[6] is the long bones of exsanguinated rabbits stored in the frozen state.

Step 1. Extraction. The frozen marrow is removed from the broken bones and suspended in a small amount of water by a blender. After dialysis against water for 2 hours, an aqueous extract is obtained by centrifuging the suspension at 55,000 g for 45 minutes. The clear supernatant fluid is siphoned off in such a way as not to disturb either the lipid layer collected at the top or the packed precipitate.

Step 2. Chromatography on DEAE-Cellulose. After dialysis of the aqueous extract against 0.05 M potassium phosphate, pH 8.0, it is passed through a column of DEAE-cellulose prepared and equilibrated with the 0.05 M potassium phosphate buffer, pH 8.0. Under these conditions cathepsin E is absorbed onto the column, while cathepsin D comes through in a breakthrough fraction. Elution by 0.05 M potassium phosphate, pH 8.0, containing 0.5 M NaCl results in the passage of cathepsin E.

Step 3. Elution by pH 5.78 Buffer. The pooled fractions containing cathepsin E are subject to dialysis against 5 mM potassium phosphate, pH 5.78, and subsequently adsorbed on DEAE-cellulose equilibrated with the same buffer. Inactive material appears in the breakthrough fraction. Elution with 5 mM phosphate buffer containing 0.15 M NaCl results in the more inactive material appearing in the eluate. Cathepsin E is washed off the column by 5 mM phosphate buffer containing 0.25 M NaCl.

Step 4. Salt Gradient Elution. Another dialysis and adsorption procedure is performed on the pooled eluates containing cathepsin E obtained from step 3. Dialysis against 5 mM potassium phosphate buffer, pH 5.78, is followed by adsorption on DEAE-cellulose equilibrated with the same buffer. More inactive material is removed by elution with the same buffer containing 0.15 M NaCl, while the enzyme is eluted in a relatively sharp peak by passage of 5 mM phosphate buffer containing 0.5 M NaCl.

Step 5. Gel Filtration Through Sephadex G-75. Filtration of the enzyme-containing fractions through Sephadex G-75 prepared in 0.15 M NaCl causes the retention of inactive material while the enzyme emerges with the solvent front.

See Table IV for a summary of this preparation.

TABLE IV
PREPARATION OF CATHEPSIN E FROM BONE MARROW[6]

Step	Total units	Specific activity
1. Aqueous extract	—	—
2. Elution from DEAE-cellulose by 0.05 M phosphate buffer, pH 8.0	1840	6.5
3. Elution from DEAE-cellulose by 5 mM phosphate, pH 5.78, containing 0.25 M NaCl	1800	57.0
4. Elution from DEAE-cellulose by 5 mM phosphate, pH 5.78, containing 0.5 M NaCl	1750	103.0
5. Gel filtrate on Sephadex G-75	1750	145.0

Properties

Stability. Cathepsin E has not been extensively studied in terms of its specificity toward substrates of a defined chemical composition. This is not a consequence of its stability, for under the conditions of preparation described above, it is a very stable enzyme and is recoverable in very high yield. The enzyme is fairly stable to heat; exposure to 60° for 60 minutes results in an 80% loss in activity while heating to 80° for 10 minutes causes complete inactivation.

Purity and Physical Properties. Inasmuch as the preparations have not achieved a high degree of purity, a paucity of information exists regarding its chemical composition, physical constants (e.g., molecular weight), as well kinetic data. Electrophoresis in agar gel showed a mobility of -7.2×10^{-5} cm^2/V/second.

Activators and Inhibitors. Cysteine does not appear to markedly alter the activity of cathepsin E at its pH optimum of 2.5 although it may stimulate slightly at pH 3.5–4.5. It has been suggested that the enhanced activity caused by cysteine at the higher pH values is due to an interaction of the sulfur-containing amino acid with the protein substrate. Unlike other cathepsins, iodoacetate does not inhibit nor does diisopropylphosphofluoridate.

Specificity. The specificity of cathepsin E remains to be elucidated.

It has no activity toward the substrates attacked by cathepsin A, B, or dipeptidyl transferase (cathepsin C). Human serum albumin is commonly employed as its substrate; the extent of hydrolysis is measured by reacting the neutralized trichloroacetic acid filtrate of the proteinase assay with biuret reagent.

Distribution. Cathepsin E is present in large amounts in polymorphonuclear cells. In smaller amounts it occurs in macrophages, while in lymphocytes only traces are to be found. This distribution probably accounts for the low activity found in the spleen as compared to the bone marrow.[6]

Catheptic Carboxypeptidase

Earlier work[19, 26] with the cathepsins of beef spleen revealed the presence of carboxypeptidase activity which is dependent on sulfhydryl compounds to express its optimal activity at pH values between 3 and 4, and is inhibited by iodoacetate. This carboxypeptidase occurs as a contaminant of cathepsin B.[7] It has not been purified from this preparation, consequently its properties have not been defined.

A study of its action carried out with the Hg-EtOH fraction of cathepsin B indicates that it is distinct from either cathepsin A or the pancreatic carboxypeptidases.[7] Cathepsin A hydrolyzes the same peptide substrates (e.g., *N*-benzylcarbonyl-α-L-glutamyl-L-tyrosine) as carboxypeptidase; however, it does so at a pH near 5.5 and in the absence of sulfhydryl compounds. The requirement for sulfhydryl groups is also not exhibited by the pancreatic carboxypeptidase activities. Zn^{2+} and Co^+ can enhance carboxypeptidase A activity,[50] whereas Zn^{2+} inhibits catheptic carboxypeptidase and Co^+ is without effect.

The preferred substrate for catheptic carboxypeptidase is *N*-benzyl-α-L-glutamyl-L-tyrosine followed in order by *N*-benzylcarbonyl-α-L-glutamyl-L-phenylalanine, *N*-benzylcarbonyl-DL-methionyglycine, and a number of other peptides.

The reported K_m for the hydrolysis of *N*-benzylcarbonyl-α-L-glutamyl-L-tyrosine is 2.2×10^{-3} M. In view of the lack of purity of the enzyme, this is an approximation at best.

This activity awaits more extensive purification and study.

[50] H. Neurath, *in* "The Enzymes" (P. D. Boyer, H. Lardy, and K. Myrbäck, eds.), Vol. 4, p. 11. Academic Press, New York, 1960.

[20] The Porcine Pepsins and Pepsinogens

By A. P. RYLE

Nomenclature

The major acid gastric protease of the pig, pepsin, which has been the subject of an article in this series[1] and of reviews elsewhere[2,3] is referred to by some authors and by the Enzyme Commission[4] as pepsin (EC 3.4.4.1) without an attached letter although other authors refer to the same enzyme as pepsin A. Pepsin A has also been used[5] to designate an unusually highly active enzyme obtained by fractional crystallization. A proposal has been made[6] for the designation of pepsins by letters to indicate their origin and numbers for their electrophoretic mobility, but this article will follow as closely as possible the recommendations of the Enzyme Commission. Thus the main enzyme is simply pepsin, the minor enzymes originally[7] called parapepsins I and II are pepsins B and C, respectively, and the most recently characterized minor enzyme[8] is pepsin D. The zymogens are named correspondingly.

Assay Methods—General Considerations

Two substrates, hemoglobin and acetyl-L-phenylalanyl-L-diiodotyrosine[9] have been used for assay of the minor pepsins and of pepsin. Pepsin and pepsin D are active in both assays, pepsin B only with APD and pepsin C only with Hb. Several other substrates of low molecular weight,[2] some having groups giving enhanced solubility[10-13] have been described

---

[1] R. M. Herriott, Vol. II, p. 3.

[2] F. A. Bovey and S. Yanari, *in* "The Enzymes" (P. D. Boyer, H. Lardy, and K. Myrbäck, eds.) 2nd ed. Vol. 4, p. 63. Academic Press, New York, 1960.

[3] R. M. Herriott, *J. Gen. Physiol.* **45**, 57 (1962).

[4] Report of the Commission on Enzymes of the International Union of Biochemistry, Macmillan (Pergamon), London, 1961.

[5] R. M. Herriott, V. Desreux, and J. H. Northrop, *J. Gen. Physiol.* **24**, 213 (1941).

[6] D. J. Etherington and W. H. Taylor, *Nature* **216**, 279 (1967).

[7] A. P. Ryle and R. R. Porter, *Biochem. J.* **73**, 75 (1959).

[8] D. Lee and A. P. Ryle, *Biochem. J.* **104**, 742 (1967).

[9] The abbreviations used will be: Hemoglobin, Hb; acetyl-L-phenylalanyl-L-diiodotyrosine, APD; diethylaminoethyl, DEAE.

[10] K. Inouye, I. M. Voynick, G. R. Delpierre, and J. S. Fruton, *Biochemistry* **5**, 2473 (1966).

[11] K. Inouye and J. S. Fruton, *Biochemistry* **6**, 1765 (1967).

[12] T. R. Hollands, I. M. Voynick, and J. S. Fruton, *Biochemistry* **8**, 575 (1969).

[13] M. Schlamowitz and R. Trujillo, *Biochem. Biophys. Res. Commun.* **33**, 156 (1968).

for pepsin, and poly-L-glutamate[14] and aromatic sulfite esters[15] are also rapidly hydrolyzed. A very sensitive assay based on peptic inactivation of ribonuclease has been described.[16] These substrates have not been used for routine assay purposes.

In general, assays using small synthetic substrates should be preferred since the substrates are more easily characterized, but when very crude enzyme or zymogen solutions are to be assayed, methods dependent on reaction with ninhydrin give unacceptably high blank values and a protein substrate must be used.

Assay with Hemoglobin as Substrate

Principle. After partial digestion of hemoglobin at pH 1.7, undigested protein is precipitated with trichloroacetic acid. The products soluble in this reagent are estimated spectrophotometrically or by the Folin-Ciocalteu reagent.

Various modifications of the earlier assays[1,17,18] have been described.[7,19] The modifications eliminate a tedious dilution step and require less material. The method is suitable for pepsin, pepsin C, and pepsin D. The corresponding zymogens are activated so rapidly under the conditions of the assay that the method is also suitable for them. Pepsin B and pepsinogen B have almost no activity in the assay. If the enzyme or zymogen is dissolved in a buffered solution, the $0.30\,N$ HCl solution used for acidifying the hemoglobin must be replaced by HCl of such greater concentration as will give a pH of 1.7 in the final digestion mixture (i.e., the additional normality should equal the normality of the cations in the buffer whose counterions are the anions of weak acids).

Reagents

HCl, $0.30\,N$ (or other suitable concentration, see above)

Trichloroacetic acid, 4.0% (w/v)

Dialyzed hemoglobin, 2.5% (w/v). Bovine hemoglobin enzyme substrate powder (Armour Pharmaceutical Co.) is dissolved in water and dialyzed against two lots of distilled water (each about 10 volumes) for 2–4 hours each or overnight in the cold room. The solution is diluted to bring the concentration to 2.5% and

[14] H. Neumann, N. Sharon, and E. Katchalski, *Nature* **195**, 1002 (1962).
[15] T. W. Reid and D. Fahrney, *J. Am. Chem. Soc.* **89**, 3941 (1967).
[16] A. Berger, H. Neumann, and M. Sela, *Biochim. Biophys. Acta* **33**, 249 (1959).
[17] M. L. Anson, *J. Gen. Physiol.* **22**, 79 (1939).
[18] J. H. Northrop, M. Kunitz, and R. M. Herriott, "Crystalline Enzymes" 2nd ed., p. 305. Columbia Univ. Press, New York, 1948.
[19] T. G. Rajagopalan, S. Moore, and W. H. Stein, *J. Biol. Chem.* **241**, 4940 (1966).

filtered through Whatman No. 54 paper. Thiomersal (2.5 mg/100 ml) is added as a preservative and the solution can be kept a few days in the cold.

Procedure. The hemoglobin solution is acidified as required each day by the addition of 0.25 volume of HCl of the appropriate concentration and brought to 37°. This substrate (1.0 ml) is added to 0.2 ml of the enzyme solution (0.05–0.2 milli [PU]^{Hb} of pepsin or pepsin D; 0.05–0.12 milli [PU]^{Hb} of pepsin C) at 37° and after 10 minutes the reaction is stopped by the addition of 5 ml of 4% trichloroacetic acid. Blank reactions are performed when very crude material is being assayed by adding the trichloroacetic acid to the enzyme before the hemoglobin; with even partly purified material the total possible contribution of the enzyme to the blank is negligible and a blank value is obtained by using water or buffer in place of enzyme solution. Blank and test samples are run in duplicate. After 5–10 minutes, the samples are filtered through Whatman No. 3 papers (7.0 cm circles) and the absorbances at 280 nm are read against water.

Units of Activity. The mean blank values are subtracted from the mean test values and the ΔE_{280} obtained is converted into [PU]^{Hb} by

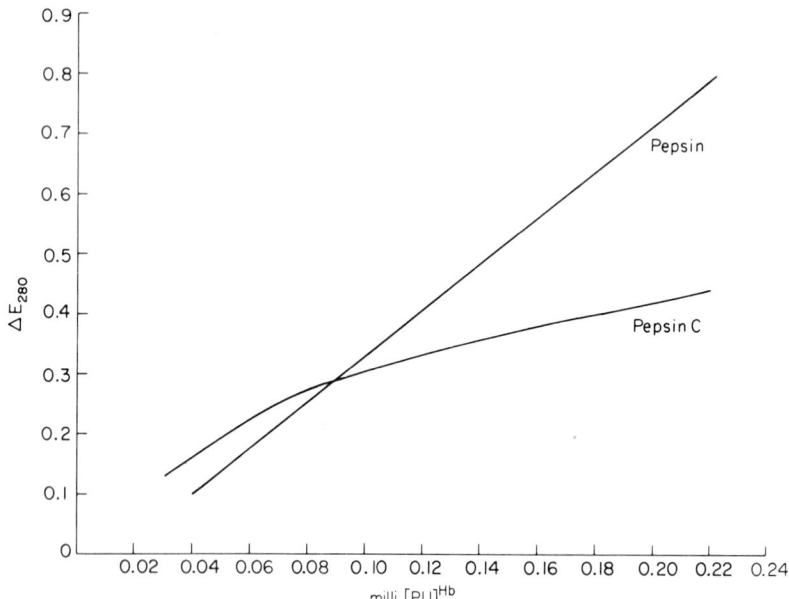

FIG. 1. Standard curves for assay of pepsin and pepsin C with hemoglobin as substrate. See text for details.

reference to a standard curve prepared with pure enzyme. The curve obtained may vary slightly from one batch of hemoglobin to another.

One [PU][Hb] was defined by Anson[17] as the amount of enzyme which liberated material giving the same color as 1 mg of tyrosine with the Folin-Ciocalteu reagent under conditions slightly different from the above. The curve for pepsin (Fig. 1) was prepared by using a solution of crystalline pepsin which had been previously standardized by Anson's original method.[17] The unit for pepsin C, which gives a curve far from linear, is arbitrarily defined so that $\Delta E_{280} = 0.29$ corresponds to 0.09 milli [PU][Hb]. The curves for pepsin and pepsin C coincide at this point so that assays performed with mixtures of the enzymes near this level of total activity will not be seriously in error. For pepsin C the assay becomes unreliable when ΔE_{280} exceeds 0.35 so that assays must be restricted to not more than 0.12 milli [PU][Hb].

Assay with *N*-Acetylphenylalanyl-L-diiodotyrosine as Substrate[20]

Principle. Hydrolysis of the substrate in 0.01 *N* HCl liberates diiodotyrosine which is estimated by its reaction with ninhydrin. The concentration of the acid used in the digestion mixture is adjusted so that the final concentration of HCl, after titration of the anions of weak acids, is 0.01 *N*. For pepsin[21] $K_m = 0.075$ m*M*, $k_0 = 0.2$ sec^{-1} at pH 2 and 37°.

Reagents

> *N*-Acetylphenylalanyl-L-diiodotyrosine, 1 m*M* in 0.01 *N* NaOH. The acetyl peptide (which is available commercially) is dissolved in the calculated volume of 0.1 *N* NaOH and then diluted to the final concentration. The solution is stable for weeks in the cold.
>
> HCl, 0.05 *N* (or other concentration if the enzyme or zymogen is dissolved in a buffered solution)
>
> Ninhydrin reagent.[22] Ninhydrin (20 g) and hydrindantin (6 g) are dissolved in 750 ml of 2-methoxyethanol. To this is added 250 ml of acetate buffer (544 g $CH_3COONa \cdot 3 H_2O$ plus 100 ml of glacial acetic acid per liter). Care is taken to avoid getting bubbles of air into the reagent during its preparation and it is stored under nitrogen in a dark bottle from which it can be dispensed.
>
> Ethanol, 60% (v/v)

Procedure. The procedure adopted is suitable for assay of pepsin, pepsin B, pepsin D, and their zymogens. Pepsin C has very little activity

[20] A. P. Ryle and M. P. Hamilton, *Biochem. J.* **101**, 176 (1966).
[21] W. T. Jackson, M. Schlamowitz, and A. Shaw, *Biochemistry* **4**, 1537 (1965).
[22] S. Moore and W. H. Stein, *J. Biol. Chem.* **211**, 907 (1954).

in the assay. The procedure includes a 10-minute incubation of the enzyme or zymogen in acid before the substrate (or ninhydrin in the blank reactions) is added. This allows the release of peptides which occurs on activation of the zymogens to be completed in both test and blank tubes before the assay proper begins; it has no effect on the measured activity of the enzymes. Test and blank reactions are performed in duplicate.

To 0.5 ml of the enzyme or zymogen solution (0.001–0.006 APD units) at 37° is added 0.25 ml of HCl of appropriate normality. After 10 minutes 0.25 ml of APD solution is added to the test reactions and 1.0 ml of ninhydrin reagent is added to the blanks. After 20 minutes more, 1.0 ml of ninhydrin reagent is added to the test reactions and at any time 0.25 ml of APD solution is added to the blanks. All the tubes are placed in a boiling water bath for exactly 15 minutes and are then cooled in a bath of cold water. The contents of the tubes are diluted with 5 ml of 60% (v/v) ethanol and the tubes are then shaken to mix the solutions. The absorbance of the solutions at 570 nm is read against water and ΔE_{570} (mean test value minus mean blank value) calculated. If ΔE_{570} does not exceed 0.4, the reaction is linear with respect to time and the activity can be expressed in APD units.

Definition of Unit. One APD unit is the quantity of enzyme which liberates 1 micromole of diiodotyrosine per minute under the above conditions. Since E_{mmol} for the color reaction of diiodotyrosine is 22.8, the volume of the solution is 7 ml, and the time of incubation is 20 minutes. ΔE_{570} is converted to APD units by multiplying by 0.0153. Specific activity is expressed as APD units/mg N. In some early work, assays were performed at 35.5° and activities are expressed as rate of liberation of diiodotyrosine at that temperature.

Assay of Mixtures of Enzymes and Zymogens

In the assays with hemoglobin or APD as substrate described above, the zymogens are activated and assayed as the active enzyme. It is possible to assay enzymes in the presence of zymogen by using substrates whose hydrolysis can be followed at pH values of 5 or above at which the activation of the zymogens is extremely slow. Assays based on the clotting of milk for pepsin C[23] and for pepsin[24] and on the liquefaction of gelatin solutions for pepsin B[25] and for pepsin[24] have been described. These laborious assays are most unsuitable for use as routine procedures and they will not be described in detail here.

[23] A. P. Ryle, *Biochem. J.* **75**, 145 (1960).
[24] R. M. Herriott, *J. Gen. Physiol.* **21**, 501 (1938).
[25] A. P. Ryle, *Biochem. J.* **96**, 6 (1965).

Since pepsin is rapidly inactivated at pH 8, pepsin and pepsinogen in a mixture can be determined by assay of the activity at low pH directly (giving the sum of pepsin and pepsinogen) and assay at low pH after alkaline inactivation of the enzyme (giving the potential activity of pepsinogen). The activity of the enzyme is found by difference.[24] Mixtures of pepsin C and pepsinogen C may be assayed similarly but a pH of 8.5 is required for the rapid inactivation of the pepsin C.[26]

Assay of the Pepsins in the Presence of one Another

It should be possible to make use of the differences of stability and of substrate specificity to determine pepsin B, pepsin C, and pepsin plus pepsin D in a mixture. Assay with APD will give the sum of pepsin, pepsin D, and pepsin B. The same assay after treatment of the enzyme solution at pH 6.9, at which pepsin B is stable,[7] will give the activity of pepsin B alone, and assay with hemoglobin will give the sum of pepsin, pepsin D, and pepsin C. This method has not been tested beyond its use in identifying the enzymes and zymogens on preparative chromatograms.

Purification of the Zymogens

The preparation of crystalline porcine pepsinogen has been described[1, 24, 27] and essentially homogeneous crystalline pepsinogen from which a small contaminant can be removed by chromatography[19] is available commercially. The zymogen has also been purified by chromatography on DEAE-cellulose[28] and is obtained as a byproduct in the chromatographic purification of pepsinogens B, C, and D. This method of ion-exchange chromatography gives a demonstrable separation of pepsinogen from pepsinogens B, C, and D; it is not known whether the method of Liener[28] gives such a separation or not. A description is therefore given of the joint purification of pepsinogen and pepsinogens B, C, and D. The procedure is summarized in Table I.

Preparation of the Mucosal Extract.[25] The body (fundic) region of the stomach of pigs (recognizable by the darker brown or red of the mucosa) is collected on ice as soon as possible after slaughter and brought to the laboratory. It is convenient to work with 10–12 stomachs. All subsequent operations are performed in the cold room.

The mucosa is stripped from the muscle, scraped firmly with a microscope slide to remove as much as possible of the adhering mucus,

[26] K. K. Oduro, M.Sc. Thesis, University of Edinburgh, Edinburgh, Scotland, 1967.
[27] R. M. Herriott, *in* "Crystalline Enzymes" (J. H. Northrop, M. Kunitz, and R. M. Herriott, eds.) 2nd ed., p. 259. Columbia Univ. Press, New York, 1948.
[28] I. E. Liener, *Biochim. Biophys. Acta* **37**, 522 (1960).

TABLE I

SUMMARY OF PREPARATION OF ZYMOGENS AND ENZYMES[a]

| | Ion-exchange chromatography | | Exclusion chroma- | Freeze dried | |
	1st	2nd	tography	/E_{280}	/mg N
Pepsinogen B	—	0.20	0.22	0.23	2.22[b]
Pepsin B	0.11	0.21	0.22	—	1.9, 2.1[b]
Pepsinogen C	0.01–0.02	0.03–0.035	0.040	0.040	0.37
Pepsin C	0.018–0.025	0.026–0.030	0.033	0.035	0.41
Pepsinogen D	—	0.02	0.023	—	0.18
Pepsin D	0.02	0.02	0.02	—	0.20
Pepsinogen	—	—	0.020	0.022	0.21
Pepsin	0.02	0.019	0.020	0.020	0.20

[a] The figures show units of (potential) activity/ml/E_{280} and for the freeze-dried material units/mg N. For pepsin(ogen) B APD units, and for the other enzymes and zymogens [PU]^{Hb} are given. For pepsin and pepsinogen 1.0 [PU]^{Hb} corresponds to 6.8 APD units, for pepsin D and pepsinogen D a corresponding ratio of 7.0 has been found. Recovery of 70–80% is usually found on ion-exchange chromatography and close to 100% on exclusion chromatography.

[b] Specific activities of 1.84 and 1.86 were reported earlier for pepsinogen B and pepsin B, respectively [A. P. Ryle, *Biochem. J.* **96**, 6 (1965)]. Other preparations have given these higher figures.

and minced in a Latapie mincer. The mince is stirred for 2 hours or overnight with 4 liters/kg of 0.02 M phosphate buffer, pH 6.9 (0.01 M NaH_2PO_4, 0.01 M Na_2HPO_4), containing 5000 units of penicillin and 50 mg of streptomycin per liter. Filter-Cel (100 g/liter of buffer) and Hyflo Super-Cel (50 g/liter of buffer) are then stirred in and the liquid is separated by filtration or by centrifugation (1000 g, 10 minutes). The extract (which may still be cloudy) contains about 60 potential [PU]^{Hb}/ liter; assays with APD are impossible because of the high blank values.

Chromatographic Separation of the Zymogens.[25] The extract obtained above is stirred for ½ to 1 hour with 0.15 volume of a suspension of DEAE-cellulose[29] (capacity about 1 meq/g) equilibrated with 0.02 M phosphate buffer, pH 6.9 whose concentration is such that the volume of the settled bed of ion-exchanger is about half that of the suspension. The DEAE-cellulose with adsorbed proteins is centrifuged off (1000 g, 10 minute). The potential activity of the supernatant is now 5–10% of that before adsorption of the zymogens. More of the zymogens can be removed by further treatment with DEAE-cellulose, but the yields of the minor pepsinogens are not thereby noticeably increased.

The DEAE-cellulose is then washed three times (1000 g, 10 minutes)

[29] E. A. Peterson and H. A. Sober, *J. Am. Chem. Soc.* **78**, 751 (1956).

with 0.02 M phosphate buffer, pH 6.9, by centrifugation in order to remove mucus which otherwise makes the subsequent chromatography very slow. The DEAE-cellulose with the adsorbed zymogens is resuspended and transferred to the top of a column of DEAE-cellulose (10 × 5 cm diameter) already equilibrated with 0.02 M phosphate buffer, pH 6.9. The total height of the packed bed is about 20 cm. Elution is performed with a rising gradient of sodium chloride concentration obtained by passing 0.02 M phosphate buffer containing 0.45 M NaCl into a closed mixing vessel containing 2 liters of 0.02 M phosphate buffer, pH 6.9, the buffers being saturated with toluene, and 50-ml fractions are collected. A typical elution profile is shown in Fig. 2.

The pepsinogen C is found in the peak emerging after 4 liters. The fractions from about 2.5–3 liters contain pepsinogen B, detected by the

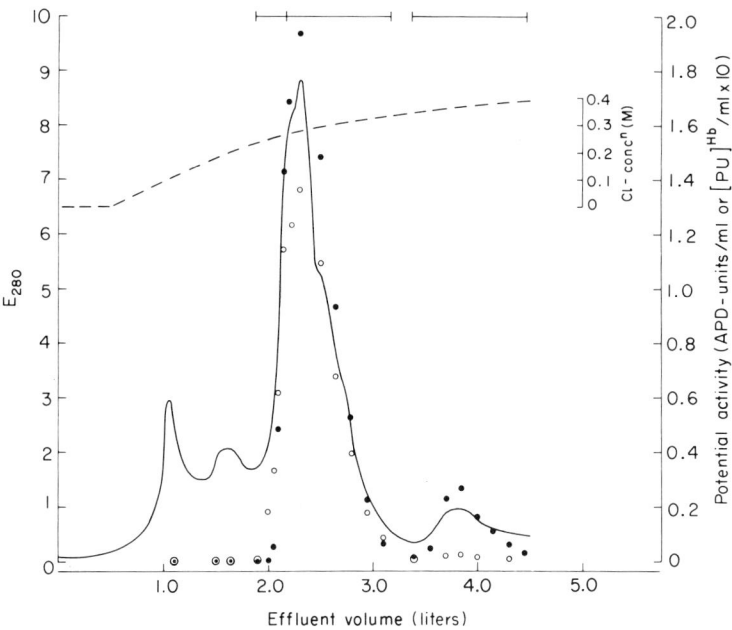

FIG. 2. Chromatography on DEAE-cellulose of zymogens adsorbed from a neutral extract of pig fundic mucosa. The column (20 × 5 cm) was packed in 0.02 M phosphate buffer, pH 6.9 and elution was performed at approx. 50 ml/hour with a rising gradient of sodium chloride concentration in the same buffer. ———, E_{280}; - - - -, chloride concentration; ●, activity against Hb; ○, activity against APD (curves for the activities have not been drawn in). The bars at the top show, from left to right, the fractions pooled for preparation of pepsinogens B and D and pepsinogen, fractions containing mainly pepsinogen, and fractions pooled for preparation of pepsinogen C.

high ratio of potential activity against APD to that against Hb (APD units/[PU]^{Hb} > 7.0) ; they also contain pepsinogen D and pepsinogen. The remainder of the large peak contains almost exclusively pepsinogen. In one chromatogram run on Whatman DEAE-cellulose, DE-52, a similar elution profile was found, but all the peaks emerged about 0.5 liter earlier.

Appropriate fractions are pooled (the rechromatography of pepsinogen B is unsuccessful if too much pepsinogen is present so fractions beyond the top of the peak should not be included in the pool), and the zymogens are concentrated either by dialysis against polyethylene glycol or by ultrafiltration through cellulose (Visking $^{24}/_{32}$ inch tubing) or more rapidly through a Diaflo membrane UM20E (Amicon N.V., 43 Heemskerckstraat, The Hague, Holland). The recovery of potential activity against hemoglobin on such chromatograms is 70–80%. The zymogens are dialyzed against 0.02 M phosphate buffer, pH 6.9 in preparation for rechromatography.

Repeated Ion-Exchange Chromatography of the Zymogens.[20, 25, 30] Rechromatography is performed by applying the dialyzed solutions to columns of DEAE-cellulose (30 × 2.9 cm) equilibrated with 0.02 M phosphate buffer, pH 6.9. Elution is achieved as above with the same (2 liter) size of mixing vessel so that the effective sodium chloride gradient is less steep. For the rechromatography of pepsinogen C, the mixing vessel may initially contain buffer with 0.1 M NaCl without loss of resolution. A typical elution pattern for the fraction containing pepsinogen B, pepsinogen D, and pepsinogen is shown in Fig. 3. Successful separation into three peaks is not achieved if the quantity of pepsinogen greatly exceeds that of the other zymogens.

The chromatograms of pepsinogen C show a single peak of potential activity emerging at 3 liters (or earlier if the initial NaCl concentration is 0.1 M) with small peaks of inactive material.

Suitable fractions are pooled and concentrated by ultrafiltration. Polyethylene glycol cannot be used at this stage because it penetrates the dialysis sac[31] and is not completely removed by the subsequent exclusion chromatography. The recovery of potential activity on rechromatography as above is 70–80%.

Further Purification of Pepsinogen B.[25] In the original report of the isolation of pepsinogen B, material obtained as above was chromatographed on Sephadex G-100 (43.5 × 2.2 cm) in 0.02 M phosphate buffer, pH 6.9, containing 0.45 M NaCl. This failed to give complete separation from inactive material, and the material from the active fractions was

[30] D. Lee and A. P. Ryle, *Biochem. J.* **104**, 735 (1967).
[31] A. P. Ryle, *Nature* **206**, 1256 (1965).

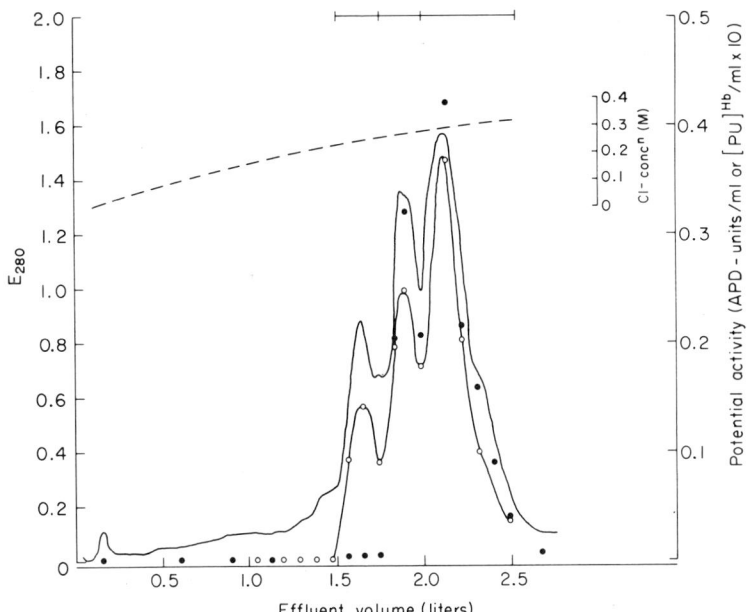

FIG. 3. Rechromatography of pepsinogen and pepsinogens B and D on DEAE-cellulose. The conditions were as in Fig. 2 except that the column was 21 × 2.4 cm and the flow rate 22 ml/hour. Symbols as in Fig. 2 except that the bars at the top show, from left to right, the fractions pooled as pepsinogen B, pepsinogen D, and pepsinogen.

chromatographed by ion-exchange again as described above. The material from this chromatogram gave a single symmetrical peak of absorbance at 280 nm and of potential activity against APD when chromatographed on Sephadex G-200. The low peak of potential activity against hemoglobin which was found very slightly preceded the peaks of absorbance and of potential activity against APD, indicating some residual contamination with pepsinogen or with pepsinogen D. The recovery of potential activity against APD was 100%. This method of purification has not been investigated much further. In some experiments the chromatography on Sephadex G-100 successfully separated pepsinogen B from inactive material. The specific potential activity of pepsinogen B obtained as described above was 1.84 micromoles of diiodotyrosine liberated per minute per mg N at 35.5°.

Exclusion Chromatography of Pepsinogen C.[20] The concentrated material obtained by ion-exchange chromatography as above is chromatographed on Sephadex G-100 which provides a better separation from inactive, ultraviolet-absorbing material than does Sephadex G-75 used originally. The column used is 40×2.2 cm, the buffer is $0.02\ M$ phosphate, pH 6.9, containing $0.1\ M$ NaCl, and 10 ml fractions are collected. The active material (recovery is over 90%) emerges between two peaks of inactive material about midway between the void volume and the elution volume of small molecules. The zymogen is dialyzed against distilled water and freeze dried. The specific potential activity of zymogen so obtained is fairly consistently about 0.37 $[PU]^{Hb}$/mg N.

Exclusion Chromatography of Pepsinogen D.[30] Chromatography on Sephadex G-100 is carried out exactly as for pepsinogen C above with a similar separation from inactive material and recovery over 90%. The pooled material is dialyzed against 0.5 mM NH$_4$OH (exhaustive dialysis against water results in some activation of the zymogen) and freeze dried. The specific potential activity of the zymogen is 0.18 $[PU]^{Hb}$/mg N and 1.75 micromole of diiodotyrosine liberated per minute per milligram N at 35.5°.

Further Purification of Pepsinogen. Unpublished experiments show that pepsinogen can be further purified by repeated ion-exchange chromatography and exclusion chromatography on Sephadex G-100 in the same way as pepsinogens C and D. The material is dialyzed against 0.5 mM NH$_4$OH and freeze dried. The specific potential activity of the zymogen is then 0.21 $[PU]^{Hb}$/mg N (identical with that of crystalline pepsinogen[1]) and 1.4 APD unit/mg N.

Preparation of the Enzymes

The preparation of crystalline pepsin was reported long ago.[1,32,33] Pepsin B was first found in commercial crystalline pepsin, and pepsins B and C were purified from commercial crude pepsin. The heterogeneity of pepsin preparations has been observed many times, most recently by chromatography on hydroxylapatite.[19] The authors of the latter paper describe a procedure for preparing homogeneous pepsin from crystalline pepsinogen; this is given below after the preparations of the minor enzymes which are summarized in Table I (see p. 322).

Pepsins B and C.[7] The operations are all conducted in the cold room. Pepsin B.P. (500 g) is dissolved in 1 liter of 0.5 N H$_2$SO$_4$, the solution is filtered, and the enzymes are precipitated by addition of 1 liter of saturated MgSO$_4$, suspended in 0.1 M acetate buffer, pH 5.6, and dialyzed against the same buffer. (This procedure avoids the use of NaOH[7,32] for re-solution of the enzymes.) Any precipitate is removed, and the supernatant containing 6–8 g of pepsin is run onto a column (35×5 cm)

[32] J. H. Northrop, *J. Gen. Physiol.* **13**, 739 (1930).
[33] J. H. Northrop, *J. Gen. Physiol.* **30**, 177 (1947).

of DEAE-cellulose equilibrated with 0.1 M acetate buffer, pH 4.0. The column is washed with the same buffer until E_{280} of the effluent is less than 1.0 when a rising gradient of NaCl concentration (obtained by passing 0.1 M acetate buffer pH 4.0 containing 0.2 M NaCl into a closed mixing vessel containing 2.6 liters of buffer free of NaCl) is applied. Fractions of 50 ml are collected and pepsin B, pepsin C, and pepsin emerge with maxima at about 3, 4, and 7.5 liters of effluent. Pepsins B and C are recognized by their characteristic differential activities with the substrates hemoglobin and APD.

The pepsin B and pepsin C are pooled separately, concentrated by ultrafiltration, dialyzed against 0.1 M acetate buffer, pH 5.6 (enzyme which may have precipitated redissolves) and further separated from one another by rechromatography at pH 4.0 on DEAE-cellulose columns (40 $\times$ 2.4 cm) with a gradient obtained as above, except that the volume of the mixing vessel is 560 ml. Pepsin B emerges at 0.75 liter and pepsin C at 1.1 liters. The pooled fractions (recovery of enzymatic activity 70–80%) are concentrated by ultrafiltration to about 5 ml and dialyzed against 0.1 M acetate buffer, pH 5.6, in readiness for purification by exclusion chromatography.

Exclusion Chromatography of Pepsin B. In unpublished experiments pepsin B, obtained as above, has been further purified by exclusion chromatography on columns of Sephadex G-75 in 0.1 M acetate buffer, pH 5.6 or in 0.02 M phosphate buffer, pH 6.9, containing 0.45 M sodium chloride. (Any contaminating pepsin or pepsin C was inactivated at the pH of the second experiment.) In both cases the enzyme emerged as a symmetrical peak shortly after a small peak of inactive material, the recovery of activity being close to 100%. The specific activities of the freeze-dried, dialyzed enzymes obtained were 1.86 and 2.11 micromoles of diiodotyrosine liberated per minute per milligram N at 35.5°, respectively.

Exclusion Chromatography of Pepsin C.[20] The material (10–20 [PU][Hb]) obtained by ion-exchange chromatography is applied to a column (40 $\times$ 2.2 cm) of Sephadex G-100 equilibrated with 0.1 M acetate buffer, pH 5.6, and eluted with the same buffer, 10-ml fractions being collected. The enzyme emerges at about 100 ml, before a peak of ultraviolet-absorbing, inactive material. The fractions containing enzymatic activity are pooled, dialyzed against distilled water, and freeze dried. The specific activity of the enzyme obtained is 0.41 [PU][Hb]/mg N.

Pepsin D.[8] The enzyme is not detected in chromatograms of pepsin run at pH 4.0 but has been prepared from commercial crystalline pepsin by ion-exchange chromatography at pH 3.2.

A column (16 × 5 cm) of DEAE-cellulose is prepared by equilibrating the ion-exchanger with 0.1 M citrate buffer, pH 3.2, containing 0.1 M NaCl and then washing it with 0.1 M citrate buffer, pH 3.2, until chloride cannot be detected in the effluent. Successful separation of pepsin D from pepsin is not obtained unless the DEAE-cellulose is pretreated in this way. The enzyme (1.5 g), dissolved in 0.1 M citrate buffer, pH 3.2, is applied to the column and elution is performed with a gradient of sodium chloride in the same buffer obtained by passing buffer containing 0.1 M NaCl into a closed mixing vessel containing 2 liters of buffer free of sodium chloride. Pepsin D emerges at 1 liter and pepsin at 1.5 liters. The fractions containing pepsin D are concentrated to about 5 ml by ultrafiltration and dialyzed against 0.1 M acetate buffer, pH 4.0.

Exclusion Chromatography of Pepsin D.[8] Filtration of the enzyme obtained as above through a column of Sephadex G-75 (19.5 × 4.8 cm), equilibrated with 0.1 M acetate buffer, pH 4.0, gives a peak of enzymatic activity emerging just after the void volume and separated from a peak of inactive material emerging later. The enzyme can be dialyzed against distilled water and freeze dried. The specific activity found is 0.20 [PU][Hb]/mg N and 1.61 micromoles of diiodotyrosine liberated/minute/mg N at 35.5°.

Preparation of Pepsin from Pepsinogen.[19] The method affords pepsin which is homogeneous as tested by a number of methods. A solution of 75 mg of crystalline pepsinogen in glass-distilled water (the solution has a pH of 5–6) is brought to 14° and acidified to pH 2.0 (glass electrode) while being stirred, by the addition of 0.75 ml of a chloroacetic acid–hydrochloric acid mixture (9 ml of 2 N chloroacetic acid and 1 ml of 1 N HCl). The acidification must be done rapidly, in 20–25 seconds. If the preparation of pepsinogen contains much salt, more acid is required and the concentration of the HCl should be increased; alternatively the pepsinogen may be freed of salt by previous passage through a Sephadex G-25 column in water.[34] The solution is maintained at 14° and pH 2.0 for exactly 20 minutes and the pH is then raised to 4.40 by the addition of 0.75 ml of 4 N sodium acetate buffer, pH 5.0 (544 g of $CH_3COONa \cdot 3 H_2O$ + 100 ml of acetic acid made up to 1 liter). The activation mixture is applied within 5 minutes to a column (30 × 2.5 cm) of sulfoethyl Sephadex C-25 equilibrated with 0.40 N sodium acetate buffer pH 4.38 ± 0.03 (40.29 g of $CH_3COONa \cdot 3 H_2O$ plus 29 ml of glacial acetic acid made up to 2 liters). The column is operated at 4° under gravity at about 60 ml/hour. The effluent is collected in 4-ml fractions and those emerging at the void volume and having absorbance in the ultraviolet

---

[34] S. Moore, personal communication, 1968.

are pooled. (The authors monitored the column effluent at 254 nm; measurements at 280 nm would also be suitable). The pooled fractions are brought to pH 5 by addition of $2 N$ sodium acetate, divided into small lots, frozen, and stored at $-20°$. The enzyme can be transferred to water or to buffer of any desired composition by passage through a Sephadex G-25 column.

Properties

Stability. Pepsin is unstable at pH values above 6 but is relatively stable at pH values even as low as pH 1. Pepsinogen is unstable at pH values below 5, being activated to pepsin. The zymogen undergoes reversible denaturation at elevated temperatures at pH 7 or at pH values in the range 8.5–11 at room temperature.[1]

Pepsin B is stable at pH 6.9 and 25° (footnote 7) and pepsin C is much more stable than pepsin under these conditions, having a half-life of 70 minutes.[7] At pH 8.5 pepsin C is rapidly inactivated.[26] Pepsin D[8] is very unstable at pH values above 6. All the zymogens are converted to enzyme at pH values below 5; enzymes and zymogens are stable at pH 5.6.

No study of the alkali-lability of pepsinogen B has been made; pepsinogen C is stable at pH 8.5 and 37° (footnote 26) and pepsinogen D starts to lose activity at pH values above 7.5 at 35.5°, this inactivation being at least partly reversible.[30]

Purity and Physical Properties. Tables II and III list some of the properties of the enzymes and zymogens.

The preparations described above yield enzymes and zymogens which appear chromatographically homogeneous. Both pepsinogen B[25] and pepsin B[7] also appear homogeneous in the ultracentrifuge but on electrophoresis in starch gel at pH 6.9 both are found to contain two components with activity or potential activity against APD, and analysis suggests that the two components of pepsinogen B have different amino-terminal residues.[25]

Pepsin C[7] and pepsinogen C,[23] even without purification by exclusion chromatography, appeared homogeneous in the ultracentrifuge. Staining for protein after electrophoresis in starch gel showed them to be homogeneous, but very slight contamination by pepsin (or pepsin D) and pepsinogen (or pepsinogen D) could be shown by the sensitive method of detecting hemoglobin-digesting activity in the gel. The amino acid compositions[20] (Table IV) of the enzyme and zymogen provide further evidence of essential homogeneity since nearly integral values were found for the amino acids.

Pepsinogen D and pepsin D, examined by electrophoresis in starch

TABLE II
SOME PROPERTIES OF PEPSINOGENS AND PEPSINS

	$S_{20,w}$ (sec $\times 10^{13}$)	$D_{20,w}$ (cm²/sec $\times 10^7$)	Molecular weight	Amino terminus
Pepsinogen B[a]	2.99	7.40	39000	0.8 Met, 0.3 His
Pepsin B[b]	3.26		38600	Ala
Pepsinogen C[c]	3.36		40500, 41400[d]	Ser
Pepsin C[b]	3.32		40700, 36000[d]	Ser and Leu or Ile
Pepsinogen D[e]			35000[f]	Leu
Pepsin D[g]			41000[f]	Ile
Pepsinogen	3.24 ± 0.15[h]		40400 ± 1600[h] 38944[i]	Leu
	3.2–3.35[j]		41000 ± 1000[j]	
	3.6[k]	7.54[k]	42240[k]	
Pepsin	2.88 ± 0.15[h]		32700 ± 1200[h] 34163[i]	Ile
	2.96–3.0[l]		35000,[l] 36212[l]	
	3.25[k]	8.7[k]	32930[k]	

[a] A. P. Ryle, *Biochem. J.* **96**, 6 (1965).
[b] A. P. Ryle and R. R. Porter, *Biochem. J.* **73**, 75 (1959).
[c] A. P. Ryle, *Biochem. J.* **75**, 145 (1960).
[d] A. P. Ryle and M. P. Hamilton, *Biochem. J.* **101**, 176 (1966). Molecular weight from amino acid composition.
[e] D. Lee and A. P. Ryle, *Biochem. J.* **104**, 735 (1967).
[f] From exclusion chromatography.
[g] D. Lee and A. P. Ryle, *Biochem. J.* **104**, 742 (1967).
[h] R. C. Williams and T. G. Rajagopalan, *J. Biol. Chem.* **421**, 4951 (1966).
[i] T. G. Rajagopalan, S. Moore, and W. H. Stein, *J. Biol. Chem.* **241**, 4940 (1966). Molecular weight from amino acid composition.
[j] R. Arnon and G. E. Perlmann, *J. Biol. Chem.* **238**, 653 (1963).
[k] V. N. Orekhovitch, V. O. Shpikiter, and V. I. Petrova, *Dokl. Akad. Nauk. S.S.S.R.* **11**, 401 (1956).
[l] O. O. Blumenfeld and G. E. Perlmann, *J. Gen. Physiol.* **42**, 533 (1959). Molecular weight 36212 from amino acid composition.

TABLE III
SOME OPTICAL PROPERTIES OF PEPSINOGEN AND PEPSIN

	λ_c(nm)	$-[\alpha]_{366}$	E^{278}_{Mol}
Pepsinogen	236[a]	212°[a]	51300 (M = 41000)[b]
Pepsin	216[c]	232°[c]	50990 (M = 35000)[b]

[a] S. Rimon and G. E. Perlmann, *J. Biol. Chem.* **243**, 3566 (1968).
[b] R. Arnon and G. E. Perlmann, *J. Biol. Chem.* **238**, 963 (1963).
[c] G. E. Perlmann, *Proc. Natl. Acad. Sci. U.S.* **45**, 915 (1959).

TABLE IV
Amino Acid Composition of Pepsinogens and Pepsins

	Moles of amino acid per mole of protein			
	Pepsinogen C[a]	Pepsin C[a]	Pepsinogen[b,c]	Pepsin[b,d]
Lys	12	4	10	1
His	2	1	3	1
Arg	7	4	4	2
Cys	6	6	6	6
Asp	30	28	44	40
Thr	25	25	26	25
Ser	35	35	46	43
Glu	46	41	28	26
Pro	20	18	19	16
Gly	35	32	35	34
Ala	23	21	19	16
Val	22	20	23	20
Met	5	4	4	4
Ile	16	14	25	23
Leu	40	34	33	28
Tyr	22	18	17	16
Phe	24	21	15	44
Trp	6	6	6	6
NH_3	34	32	27	27

[a] A. P. Ryle and M. P. Hamilton, *Biochem. J.* **101**, 176 (1966).

[b] T. G. Rajagopalan, S. Moore, and W. H. Stein, *J. Biol. Chem.* **241**, 4940 (1966).

[c] Slightly different analyses are given by R. Arnon and G. E. Perlmann, *J. Biol. Chem.* **238**, 653 (1963) and H. van Vunakis and R. M. Herriott, *Biochim. Biophys. Acta* **32**, 600 (1957).

[d] A slightly different analysis is given by O. O. Blumenfeld and G. E. Perlmann, *J. Gen. Physiol.* **42**, 553 (1959).

gel, give rather broad zones of hemoglobin-digesting activity but show essential freedom from contamination by pepsinogen and pepsin. Analysis shows about 0.3 g atom of phosphorus per mole in both zymogen and enzyme. Treatment with a phosphatase fails to alter the electrophoretic mobility of the enzyme or activated zymogen and the residual phosphate may represent contamination with inorganic phosphate or nucleic acid.

Crystalline pepsinogen and pepsin prepared from it have been shown to be homogeneous by the criteria of chromatography, amino acid analysis and carboxyl terminal analysis,[19] and by their behavior in the ultracentrifuge.[35]

Amino Acid Composition and Sequences. Table IV shows the compositions of pepsin, pepsin C, and their zymogens. No analyses have been

[35] R. C. Williams and T. G. Rajagopalan, *J. Biol. Chem.* **241**, 4951 (1966).

TABLE V

Amino terminus of pepsinogen[a]

Leu-Val-Lys-Val-Pro-Leu-Val-Arg-Lys-Lys-Ser-Leu-Arg-Gln-Asn-Leu-Ile-Lys-Asp-
    Gly-Lys-Leu-Lys-Asp-Phe-Leu-Lys-Thr-His-Lys-His-Asn-Pro-Ala-Ser-Lys-Tyr-
    Phe-Pro-Ala-Glu

Amino terminus of pepsin[b]

Ile-Gly-Asp-Glu-Pro

Tryptophan peptides of pepsin[c]

Val-Phe-Asp-Asn-Leu-Trp-Asp-Gln-Gly
Leu-Trp-Val-Pro-Ser
Val-Glu-Gly(Trp,Gln)
Leu-Asn-Trp-Val-Pro

Cystine peptides of pepsin[d]

Cys-Ser-Ser-Ile-Asp-Glu   Glx-Asx(Asx,Ser)Cys(Thr,Ser,Asp,Glu)
Cys-Ser-Gly-Gly-Cys-Glu
Cys-Ser-Ser-Leu-Ala-Cys-Ser-Asx-His(Glx,Asx)

Cyanogen bromide cleavage peptides of pepsin[e]

$$\overset{\text{P}}{|}$$

Ile-Gly-Asp-Glu. . .Glu-Ala-Thr-Ser-Glu-Glu-Leu. . .(Ile,Thr)Met   (Peptide B-2,
    amino terminus of pepsin, 130–140 residues)
Val-Ile. . .Met   (32 residues)
Asp-Gly-Glu-Thr-Ile. . .Met   (38–39 residues)
Asp-Val-Pro-Thr-Ser-   (continues to C-terminus.)

Peptides from peptide B-2 above[f]

Asp-Gly-Ile-Gly-Leu-Leu

$$\overset{\text{P}}{|}$$

Glu-Ala-Thr-Ser-Glu-Glu-Leu-Ser(Thr,Ile)Tyr
Ser-Val-Tyr
Cys-Ser-Ser-Leu(Cys,Ser$_2$,Ala,Asx$_5$,His,Thr,Glx,Pro)Phe-Phe
Gly-Leu-Ser-Glu-Thr-Glu-Pro-Gly-Ser-Phe
Leu-Tyr
Tyr-Ala-Pro-Phe
Leu-Gly-Gly-Ile(Ser,Asp)Ser-Tyr-Tyr

Carboxyl terminus of pepsin[g]

Asp-Val-Pro-Thr-Ser-Ser-Gly-Glu-Leu-Trp-Ile(Asp,Thr,Ser$_2$,Glu,Pro,Gly,Val,Leu,
    Tyr,Phe)Ile[h]-Leu-Gly-Asp-Val-Phe-Ile[i]-Arg[j]-Gln-Tyr-Tyr-Thr-Val-Phe-Asp-Arg-
    Ala-Asn-Asn-Lys-Val-Gly-Leu-Ala-Pro-Val-Ala

Peptide sequence round an active aspartate residue[k]

Ile-Val-Asp-Thr-Gly-Thr-Ser

---

[a] E. B. Ong and G. E. Perlmann, *J. Biol. Chem.* **243,** 6104 (1968).

[b] V. M. Stepanov, *Abstr. FEBS Meeting 5th, Prague,* no. 1089, (1968).

[c] T. A. A. Dopheide and W. M. Jones, *J. Biol. Chem.* **243,** 3906 (1968).

[d] J. Tang and B. S. Hartley, *Biochem. J.,* in press.

reported for pepsinogen B or pepsin B. The amino acid compositions of pepsinogen D and pepsin D are not distinguishable from those of pepsinogen and pepsin, respectively.[35a] Table V lists the amino acid sequences identified in pepsin and pepsinogen. It has recently been reported that pepsinogen contains about 3 moles of sugar per mole and that the sugar is lost from the protein on activation to pepsin.[36]

Activators and Inactivators. No activators of pepsin have been reported. The protease activity is destroyed to the extent of **70–80%**, and the activity with benzyloxycarbonyl-L-glutamyl-L-tyrosine is destroyed completely, by reaction with *p*-bromophenacyl bromide.[37] The analogous α-diazo-*p*-bromoacetophenone causes complete inactivation, and is also capable of reacting with, and completely inactivating, pepsin previously treated with *p*-bromophenacyl bromide.[38] With both reagents 1 mole of inhibitor is bound per mole of enzyme. The reagents apparently react with different amino acid residues in the enzyme.

With diphenyldiazomethane[39] 50% loss of activity against hemoglobin is achieved with reaction of about 2.4 moles of reagent per mole of enzyme. The reagent also reacts with, and prevents the activation of, pepsinogen, and the inactivation of pepsin is enhanced by the presence of peptide substrates.[10] These facts suggest that inactivation does not occur solely by reaction at the active center of the enzyme.

[35a] A. P. Ryle and F. Falla, unpublished observations, 1970.

[36] H. Neumann, U. Zehavi, and T. D. Tanksley, *Biochem. Biophys. Res. Commun.* **36**, 151 (1969).

[37] B. F. Erlanger, S. M. Vratsanos, N. Wasserman, and A. G. Cooper, *J. Biol. Chem.* **240**, PC 3447 (1965).

[38] B. F. Erlanger, S. M. Vratsanos, N. Wasserman, and A. G. Cooper, *Biochem. Biophys. Res. Commun.* **28**, 203 (1967).

[39] G. R. Delpierre and J. S. Fruton, *Proc. Natl. Acad. Sci. U.S.* **54**, 1161 (1965).

e V. I. Ostoslavskaya, I. B. Pugacheva, E. A. Vakhitova, V. F. Krivtsov, G. L. Muratova, E. D. Levin, and V. M. Stepanov, *Biochemistry (U.S.S.R.)* **33**, 276 (1968),

f E. A. Vakhitova, I. B. Pugacheva, M. M. Amirkhayan, R. Valiulis, L. G. Senyutenkova, B. G. Belen'kii, and V. M. Stepanov, *Izv. Akad. Nauk S.S.S.R.; Ser. Khim.* p. 2415 (1968); *Chem. Abstr.* **70**, 25942s (1969).

g V. Kostka, L. Morávek, I. Kluh, and B. Keil, *Biochim. Biophys. Acta* **175**, 459 (1969).

h From this residue to terminus; see also T. A. A. Dopheide, S. Moore, and W. H. Stein, *J. Biol. Chem.* **242**, 1833 (1967).

i From this residue to the terminus; see also R. A. Matveeva, V. F. Krivtsov, and V. M. Stepanov, *Biochemistry (U.S.S.R.)* **33**, 142 (1968).

j From this residue to the terminus; see also R. N. Perham and G. M. T. Jones, *European J. Biochem.* **2**, 84 (1967).

k R. S. Bayliss, J. R. Knowles, and G. B. Wybrandt, *Biochem. J.* **113**, 377 (1969).

Inhibition has also been found in the presence of cupric ions with a series of aliphatic diazocarbonyl compounds,[40] some having structural resemblance to pepsin substrates. The substrate analog L-1-diazo-4-phenyl - 3 - tosylamidobutanone (tosyl - L - phenylalanyldiazomethane),[41] causes complete inhibition in the presence of cupric ions when 1 mole of inhibitor has reacted per mole of enzyme and does not react with pepsinogen or denatured pepsin. Benzyloxycarbonyl-L-phenylalanyldiazomethane[42] inhibits pepsin similarly, except that cupric ions are not necessary. The reaction of other chromophoric phenylalanine diazoketones has also been investigated.[43]

The diazoacetyl amino acid ester derivatives of norleucine,[44] glycine,[45] and phenylalanine[46] also cause complete inhibition at a 1:1 molar ratio of reaction and the aspartic acid at which the phenylalanine compound reacts[46] may also be the site of reaction of 1-diazo-4-phenyl-butanone-2 (footnote 47 and Table V) and of N-diazoacetyl-N'-2,4-dinitrophenylethylenediamine.[48]

Pepsin C, like pepsin, is inactivated by diazoacetyl-DL-norleucine methyl ester.[49]

The activation peptides of pepsinogen C[26,50] and those of pepsinogen D,[51] like those of pepsinogen,[52] contain material which is inhibitory to the milk-clotting activity of the enzymes.

Specificity. Pepsin has a rather broad specificity which has been examined with numerous synthetic peptides[2,11] and by a survey of the bonds it is known to hydrolyze in natural polypeptides.[53] Peptides containing *p*-nitrophenylalanine,[11] 3,5-dibromotyrosine,[54] or 3,5-dinitrotyro-

[40] L. V. Kozlov, L. M. Ginodman, and V. N. Orekhovich, *Biochemistry (U.S.S.R.)* 32, 839 (1967).
[41] G. R. Delpierre and J. S. Fruton, *Proc. Natl. Acad. Sci. U.S.* 56, 1817 (1966).
[42] E. B. Ong and G. E. Perlmann, *Nature* 215, 1492 (1967).
[43] R. A. Badley and F. W. J. Teale, *Biochem. J.* 108, 15P (1968).
[44] T. G. Rajagopalan, W. H. Stein, and S. Moore, *J. Biol. Chem.* 241, 4295 (1966).
[45] R. L. Lundblad and W. H. Stein, *J. Biol. Chem.* 244, 154 (1969).
[46] R. S. Bayliss, J. R. Knowles, and G. R. Wybrandt, *Biochem. J.* 113, 377 (1969).
[47] K. T. Fry, O-K. Kim, J. Spona, and G. A. Hamilton, *Biochem. Biophys. Res. Commun.* 30, 489 (1968).
[48] V. M. Stepanov and T. I. Vaganova, *Biochem. Biophys. Res. Commun.* 31, 825 (1968).
[49] J. Kay and A. P. Ryle, unpublished observations, 1968.
[50] K. K. Oduro and A. P. Ryle, *Abstr. Intern. Congr. Biochem. 7th Tokyo,* F-306 (1967).
[51] D. Lee, *Can. J. Biochem.* 45, 1002 (1967).
[52] H. van Vunakis and R. M. Herriott, *Biochim. Biophys. Acta* 22, 537 (1960).
[53] J. Tang, *Nature* 199, 1094 (1963).
[54] E. Zeffren and E. T. Kaiser, *J. Am. Chem. Soc.* 88, 3129 (1966).

sine[55] as well as ones containing diiodotyrosine[21] are hydrolyzed and have been used in kinetic investigations. While the general conclusion is that activity is enhanced by the presence of bulky nonpolar sidechains in the amino acids on either side of the sensitive bond, activity is not restricted to such bonds. Poly-L-glutamic acid has been shown[14] to be very rapidly hydrolyzed to oligopeptides; the initial rate of splitting of peptide bonds at pH 2.3 and 35° corresponds to about 100 moles of peptide bond per mole of pepsin per minute when the substrate is about 0.22 millimolar. A free amino group on the acyl moiety of the peptide bond is inhibitory,[2] and a free carboxyl group on the amino moiety is inhibitory when it is dissociated.[10, 56, 57] Curiously, replacement of the terminal phenylalanine methyl ester in the substrate benzyloxycarbonyl-phenylalanylphenylalanine methyl ester by phenylalaninol (replacement of —COOMe by —CH$_2$OH) produces a substance resistant to hydrolysis.[11]

In addition to hydrolyzing peptide bonds, pepsin catalyzes trans-peptidation reactions, shown to be of the amino-transfer type[58, 59] although the product of activation of pepsinogen at pH 3 lacks such activity.[60]

Pepsin has recently been shown to act as an esterase. It catalyzes the hydrolysis of the ester bond in acetyl-L-phenylalanyl-L-(β-phenyl)-lactic acid[61] and the corresponding bond in benzyloxycarbonyl-L-histidyl-p-nitro-L-phenylalanyl-L-(β-phenyl)-lactic acid methyl ester.[11] It also catalyzes the hydrolysis of aromatic sulfite esters.[15]

The specificity of the minor pepsins has not been extensively investigated, but their sites of attack on the B-chain of oxidized insulin have been examined.[7, 8, 62] Pepsin D is indistinguishable by this test from pepsin; pepsins B and C show a slightly more restricted specificity than pepsin. Pepsin B is a highly active gelatinase and a poor milk-clotting enzyme. The most marked difference between the enzymes is that shown by their relative activities against APD and Hb. Pepsin C has been shown to catalyze transpeptidation reactions of the amino-transfer type[63] and although it has little activity against N-acylated di- and tripeptides

[55] J. R. Knowles, H. Sharp, and P. Greenwell, *Biochem. J.* **113**, 343 (1969).
[56] J. L. Denburg, R. Nelson, and M. S. Silver, *J. Am. Chem. Soc.* **90**, 479 (1968).
[57] A. J. Cornish-Bowden and J. R. Knowles, *Biochem. J.* **113**, 353 (1969).
[58] H. Neumann, Y. Levin, A. Berger, and E. Katchalski, *Biochem. J.* **73**, 33 (1959).
[59] J. S. Fruton, S. Fujii, and M. H. Knappenberger, *Proc. Natl. Acad. Sci. U.S.* **47**, 759 (1961).
[60] H. Neumann and N. Sharon, *Biochim. Biophys. Acta* **41**, 370 (1960).
[61] L. A. Lokshina, V. N. Orekhovich, and V. A. Sklyankina, *Nature* **204**, 580 (1964).
[62] A. P. Ryle, J. Leclerc, and F. Falla, *Biochem. J.* **110**, 4P (1968).
[63] A. P. Ryle, *Bull. Soc. Chim. Biol.* **42**, 1223 (1960).

containing peptide bonds which are sensitive in the B-chain of oxidized insulin, longer peptides containing these bonds are hydrolyzed.[62]

Distribution. Examination of the zymogens found in the gastric mucosa of individual pigs[64] has shown that some animals make all four zymogens but some (at the time of slaughter) apparently lacked one or more of the zymogens. Pepsin C is relatively more abundant than the other zymogens in the pyloric mucosa, but the absolute amounts found in the fundus are much greater. All three minor pepsins as well as pepsin have been found in the stomach contents of four pigs examined.[65]

Status of the Minor Pepsins and Pepsinogens. Since pepsins are acid proteases, and since commercial pepsin is obtained after extensive acid treatment of mucosal extracts and is heterogeneous,[19] one must consider the possibility that the minor pepsins are degradation products of pepsin (or vice versa) or are partial activation products of pepsinogen. The finding of a zymogen in neutral mucosal extracts for each of the enzymes found in crude or crystalline pepsin lends considerable weight to the belief that the enzymes and zymogens are not such degradation products. There is little evidence beyond this to establish pepsin B and pepsinogen B as proteins in their own right except that they do not contain phosphate and so cannot be autolytic precursors of pepsin or pepsinogen. If one is to maintain that pepsin B is a degradation product of pepsin (and similarly for the zymogens) one must postulate either loss of the phosphate residue or its excision as a phosphoserine peptide during the isolation of both zymogen and enzyme under their different conditions of pH.

Pepsinogen D and pepsin D have not been found to differ from pepsinogen and pepsin in any property except chromatographic behavior, electrophoretic mobility, a small difference in specific activity, and phosphate content. Pepsin and pepsinogen become electrophoretically indistinguishable from pepsin D and pepsinogen D after treatment with phosphatase. It is therefore suggested[8, 30] that the minor forms may merely be the major ones without the phosphate. It is an open question whether the phosphate is lost during isolation or was never attached during the biosynthesis of the zymogen.

The amino acid analyses of pepsinogen C and pepsin C and those of pepsinogen and pepsin (Table IV) make it impossible that either enzyme should be a degradation product of the other or either zymogen a degradation product of the other (see, for example, the contents of aspartic and glutamic acids).

[64] A. P. Ryle, *in* "The Council for International Organizations of Medical Sciences Symposium The Role of the Gastrointestinal Tract in Protein Metabolism" (H. Munro, ed.), p. 25. Blackwell, Oxford, 1964.

[65] A. P. Ryle, *Biochem. J.* **98**, 485 (1966).

[21] Bovine Pepsinogen and Pepsin

By Beatrice Kassell and Patricia A. Meitner

Introduction

Bovine pepsin was isolated from gastric juice and crystallized by Northrop[1] in 1933. Bovine pepsinogen was isolated by Chow and Kassell[2] and shown to differ considerably in amino acid composition from porcine pepsinogen [20].

Assay Methods

The Hemoglobin Method for Total Potential Peptic Activity[2,3]

Principle. The extent of digestion of denatured hemoglobin substrate is determined by measuring the formation of trichloroacetic acid[4] soluble material as increase in absorbance at 280 nm.

Reagents

Buffer 1. HCl, 0.01 M, containing 0.1 M NaCl (pH 2.0)
Buffer 2. Sodium acetate, 0.1 M, pH 5.3
HCl, 0.3 M
Substrate. Dissolve 1 g of crystalline hemoglobin (Pentex Biochemicals, Kankakee, Ill.) in 32 ml of water; add 8 ml of 0.3 N HCl. Filter through a coarse sintered glass filter. The solution may be kept at 4° for 1 week.
TCA, 5% aqueous solution
Standard Enzyme Solution. Dissolve a few milligrams of crystalline porcine pepsin (Worthington Biochemical Corp., Freehold, N.J.) in 5 ml of the acetate buffer. Measure the absorbance at 280 nm ($A_{280} \times 676 = \mu$g/ml). Prepare several dilutions containing 30–100 μg/ml in the same buffer.
Unknown Pepsinogen or Pepsin Solution. Dilute to about 50 μg/ml with Tris phosphate buffer, 0.05 M, pH 7.

Procedure. Predetermine the amount of 0.3 N HCl required to bring 0.1 ml of each assay solution to pH 2.0. Place polycarbonate centrifuge tubes (16 $\times$ 100 mm) in a 37.6° water bath, add the required amount of 0.3 N HCl plus sufficient buffer 1 to bring the volume to 0.9 ml. Add 0.1

[1] J. H. Northrop, *J. Gen. Physiol.* **16**, 615 (1933).
[2] R. B. Chow and B. Kassell, *J. Biol. Chem.* **243**, 1718 (1968).
[3] M. L. Anson, *J. Gen. Physiol.* **22**, 79 (1938).
[4] Abbreviations used are: TCA, trichloroacetic acid; DEAE, diethylaminoethyl.

ml of the solution to be assayed. With timing, add 1 ml of substrate solution previously equilibrated at 37.6°, mix, and after exactly 5 minutes, terminate the digestion by adding 2 ml of 5% TCA. The timed additions are best made with automatic pipettes. Place the tubes in ice for 10 minutes and centrifuge at 28,000 g for 60 minutes. Use cellulose nitrate tubes, 17 × 95 mm, as adapters for the polycarbonate tubes in the SM-24 rotor if a Sorvall centrifuge is used. Measure the absorbance of the clear supernatant at 280 nm against an appropriate blank. Usually a reagent blank is adequate, but for the dark-colored solutions of the early purification steps, prepare a blank using the same amount of assay solution, but reverse the order of addition of TCA and hemoglobin. Include standard determinations in each set of assays.

Definition of Unit and Specific Activity. Prepare a standard curve with porcine pepsin. In our hands, 5 μg of porcine pepsin shows a net increase of 0.47 in A_{280}. The potential specific activities of bovine zymogen and enzyme are expressed as microgram equivalents of porcine pepsin per A_{280} of the solutions.

The Milk-Clotting Method to Detect Active Pepsin[2,5]

Principle. The hemoglobin procedure does not differentiate between pepsin and pepsinogen. To detect pepsin in the presence of the zymogen, either to determine contamination in the crude preparation or in the study of activation, the amount is measured by the milk-clotting procedure. The pepsinogen is not activated at pH 5.3 in the time required to form the clot. The clotting time varies inversely with the amount of pepsin, but the relationship is not exactly proportional. Therefore choose the amount of unknown to give a clotting time near 30 seconds.

Reagents

Buffer solution 2 for the hemoglobin assay
$CaCl_2$, 0.1 M in water
Stock Milk solution. Dissolve 16.8 g of nonfat dried milk (Carnation) in 100 ml of water. Store at 4° up to 3 days.
Dilute milk solution. 50 ml of stock milk solution, 50 ml of buffer 2, and 10 ml of 0.1 M $CaCl_2$.
Standard Enzyme. Same as for hemoglobin assay (50 μg/ml).
Unknown Solution. Same as for hemoglobin assay.

Procedure. Place a series of glass tubes (15 × 125 mm) in a plastic rack in a 37.6° water bath, and add to each 2 ml of buffer 2 and 0.5 ml of dilute milk solution. Stir the test solution by means of an air-driven

[5] M. J. Seijffers, H. L. Segal, and L. L. Miller, *Am. J. Physiol.* **205**, 1099 (1963).

magnetic stirrer placed in the water bath beneath the tube. Add 0.1 ml of preequilibrated unknown solution to the tube and simultaneously start a stopwatch. To observe the clot formation, siphon the solution slowly through a transparent Tygon tube (1.5 mm inside diameter, with the end 0.5 mm in diameter). Note the time for the first flock to appear in this tube. The milk-clotting time is about 30 seconds for 5 μg of porcine pepsin standard. Up to 0.5 ml of the unknown solution may be assayed by adjusting the volume of buffer 2 accordingly. A solution containing 25 μg of pepsinogen is considered to be free of active pepsin if it does not clot in 2 minutes.

Definition of Unit and Specific Activity. The specific activity is expressed as microgram equivalents of porcine pepsin per A_{280} unit of the bovine pepsinogen.

Purification Procedure for the Pepsinogen[2]

General Precautions. Pepsinogen solutions are very susceptible to accidental activation by acid, to denaturation, and to bacterial contamination. Do not freeze solutions containing NaCl. Always stir slowly without whipping. Store buffers and protein solutions under toluene (1 ml per 4 liters) in the cold. Avoid long storage, particularly in test tubes, during chromatography steps.

Step 1. Collection of the Mucosa.[6] Remove the abomasa or fourth stomach from about eight cows immediately after slaughter, turn inside out, and wash thoroughly with cold water. Transfer to the laboratory on ice.[6a] Cut out the glandular fundic portion and dissect the mucosa from the muscular stomach wall. Keep the tissue cold by means of an ice bag during this procedure. Gently scrape and wash away the slimy layers. At this stage, the preparation may be frozen for storage.

Step 2. Extraction. All subsequent procedures should be carried out at 5°. Grind 5 kg of the mucosa in a meat grinder (fine setting) and extract with 10 liters of 0.1 M sodium phosphate buffer of pH 7.3 for 1 hour. Add filter aid (Solca Floc BW-20, 2 kg) and filter the mixture on three Büchner funnels, 24 cm in diameter. Reextract the filter cakes twice with 10-liter portions of the pH 7.3 buffer.

Step 3. Ammonium Sulfate Fractionation. Add solid ammonium sulfate gradually to the combined filtrates to a final saturation of 23%, maintaining the pH close to pH 7.3 by intermittent addition of 1 M NaOH. Stir to avoid local excess of acidic ammonium sulfate or of NaOH to

---

[6] R. M. Herriott, *J. Gen. Physiol.* **21**, 501 (1938).
[6a] Stomach mucosa, obtained frozen from Pel-Freez Biologicals, Inc., Rogers, Ark., may be used, but these are not free of active pepsin.

prevent activation or denaturation, respectively. Add 2 kg of BW-20 again and filter the mixture as before. Wash the cakes with 5 liters of buffer saturated to 23% with ammonium sulfate and readjusted to pH 7.3. Combine the filtrate and washings and saturate further to 63% by the same procedure. Add 250 g of Perlite 4106 and allow to settle overnight. Discard the supernatant and collect the precipitate by filtration. It may be stored at −20°. The average yield is 1200–1300 g, containing about 1.5 g of pepsinogen. Batches vary in pepsinogen content.

Suspend 1200 g of the ammonium sulfate cake in 1 liter of 0.01 M sodium phosphate buffer of pH 7.0, and dialyze against two changes of 18 liters of the same buffer. (The molarity of the Tris-phosphate buffers is determined by the concentration of Tris; adjust to pH 7.0 ± 0.05 with phosphoric acid.)

Centrifuge the suspension for 3 hours at 13,000 g in a refrigerated centrifuge. Resuspend the solid in an equal volume of buffer and centrifuge again. Combine the supernatants and centrifuge at 36,000 g in an ultracentrifuge for 3 hours. Wash the sediment as before. Just before the batch absorption, filter the supernatant and washings through Schleicher and Schuell No. 588 folded paper. The filtrate should be clear.

Step 4. First Batch Absorption on DEAE-Cellulose. Precycle the DEAE[4]-cellulose with 0.5 M HCl and 0.5 M NaOH,[7] neutralize, and equilibrate with 0.05 M Tris-phosphate buffer of pH 7.0. Remove fines by settling several times in the buffer and then settle overnight. Dilute the crude pepsinogen solution, containing about 80,000 A_{280} units to 7900 ml with 0.05 M Tris-phosphate buffer and mix with 2100 ml of settled DEAE-cellulose (equivalent to about 190 g). Stir the suspension gently mechanically for 1 hour and filter through cloth on a 24-cm Büchner funnel. Suspend the moist cake in fresh buffer and transfer to a chromatographic column (7 × 60 cm). The settled height of the resin should be 30 to 35 cm. Wash the column at a rate of 300–400 ml per hour with 0.05 M Tris-phosphate buffer, about 10 liters, until the absorbance at 280 nm of the effluent is about 0.3. Then elute the pepsinogen with 0.45 M NaCl in the same buffer at the rate of 200 ml per hour. Pool the portion of the peak having a potential specific activity over 40, concentrate to 750 ml by ultrafiltration in a DiaFlo model 400 ultrafiltration cell with reservoir (Amicon Corp., Cambridge, Mass.), using a No. UM-10 membrane. Remove the remaining salt by dialysis against two changes of 0.005 M Tris-phosphate buffer of pH 7.0. The pepsinogen solution now contains 15,000 to 20,000 A_{280} units and about 1 g of pepsinogen.

Step 5. Second Batch Absorption on DEAE-Cellulose. Measure the

---

[7] Whatman Data Manual No. 2000, H. Reeve Angel, Clifton, New Jersey.

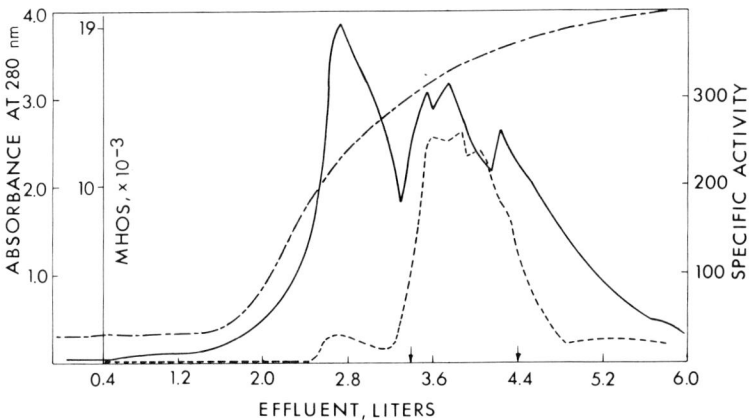

FIG. 1. Gradient elution of pepsinogen after the second batch absorption on DEAE-cellulose (see the text, step 5). The column was 5×50 cm; sample weight, 910 mg. The closed mixing chamber contained 2250 ml of 0.05 M Tris-phosphate buffer of pH 7.0. The reservoir contained 0.45 M NaCl in the same buffer. ———, A_{280}; - - - -, specific activity; — · — · —, conductivity expressed as reciprocal ohms $\times 10^{-3}$.

DEAE-cellulose needed as follows: fill a chromatographic tube (5×60 cm) with cleaned DEAE-cellulose, and wash with 0.05 M Tris-phosphate buffer of pH 7.0 for 1 hour. The final packed height should be 50 cm. Remove this DEAE-cellulose from the tube and repack the column to a height of 15 cm. Mix the remainder of the measured DEAE with the pepsinogen solution and adjust the volume to 5 liters with the 0.05 M buffer. Stir the mixture for 1 hour and filter on a Büchner funnel as in step 4. Resuspend the filter cake in 1200 ml of 0.05 M buffer and pour carefully onto the 15 cm of DEAE-cellulose in the chromatography tube. Elute the column with a gradient[8]: starting buffer, 2250 ml of 0.05 M Tris-phosphate, pH 7.0, in a constant volume bottle; gradient solution, 0.45 M NaCl in the same buffer. Collect 20-ml fractions at the rate of 100 ml per hour. Figure 1 shows a typical pattern. Pool the fractions[9] having potential specific activity greater than 100 and reduce the volume to less than 200 ml, using the ultrafilter. The pepsinogen solution now contains 2000 to 3000 A_{280} units of protein and about 750 mg of pepsinogen.

Step 6. Recycling Gel Filtration on Sephadex G-100. Use a chromatographic tube (5×100 cm) fitted for upward flow (Pharmacia, Piscataway,

[8] R. Arnon and G. E. Perlmann, *J. Biol. Chem.* **238**, 653 (1963).

[9] In our earlier report,[2] only the center portion of the peak was pooled to separate partially the main pepsinogen from the others. This is now done on hydroxylapatite (step 8).

N.J.), with a recycling valve (LKB Instruments, Rockville, Md.) and à pump (Accu-Flo Pump, Beckman-Spinco, Palo Alto, Calif.) with an added pressure shutoff valve. If the recycling device is not available, two separate gel filtrations may be carried out. Fill the column with Sephadex G-100, bead form, after removing fines by settling in 0.05 M Tris-phosphate buffer of pH 7.0. Allow the Sephadex to settle evenly through a layer of buffer while packing and avoid excessive pressure.[10] Pack the column to the maximum height, pump the buffer upward for 24 hours, then adjust the upper fitting to the level of the beads if any settling has occurred.

Apply the sample through the pump in a volume of 100 ml containing not more than 1500 A_{280} units (usually one half of the material of step 5) at a flow rate of 20 ml per hour. Wash it in with three 4-ml portions of buffer. Increase the flow rate to 40 ml per hour, taking care that no back pressure develops in the column. Collect 20-ml fractions. In the initial run, collect the effluent without recycling. From the elution pattern, determine the position of the potentially active peak. In subsequent chromatographies, recycle this peak once. A typical chromatography is shown in Fig. 2. The column may be used many times by washing overnight with the buffer between chromatographies. Pool the fractions in the recycled peak having potential activity greater than 400, and store under toluene until two such chromatographies have been run. The pepsinogen solution now contains about 700 mg of protein. Reduce the volume to 100 ml on the ultrafilter.

Step 7. Chromatography on DEAE-Cellulose. Prepare a column (5 × 50 cm) of DEAE-cellulose equilibrated with 0.05 M Tris-phosphate buffer of pH 7.0. Apply the pepsinogen sample, in a volume of 100 ml, and elute with the same gradient as for step 5 (2250 ml of starting buffer). Almost all the protein is in the main peak with uniform activity across the peak. Pool the fractions having a specific activity greater than 500 and reduce the volume to less than 100 ml. Dialyze against 6 liters of 0.01 M Tris-phosphate buffer of pH 7.0, 6 liters of 0.01 M sodium acetate of pH 7.0 to displace phosphate and Tris, and finally three times against 6 liters of distilled deionized water, carefully adjusted to pH 7.0 and tightly covered to avoid pH change. Filter, lyophilize, and store at −20°. This purified pepsinogen has a specific activity of about 550, and usually contains no active pepsin. The yield is about 500 mg. The preparation at this stage is practically free of extraneous material, but is still a mixture of several pepsinogens, differing only in organic phosphate content, and variable in relative amount. If it is desired to separate the individual pepsinogens,

---

[10] Sephadex Technical data sheet No. 6, Pharmacia Fine Chemicals, Inc., Piscataway, N.J.

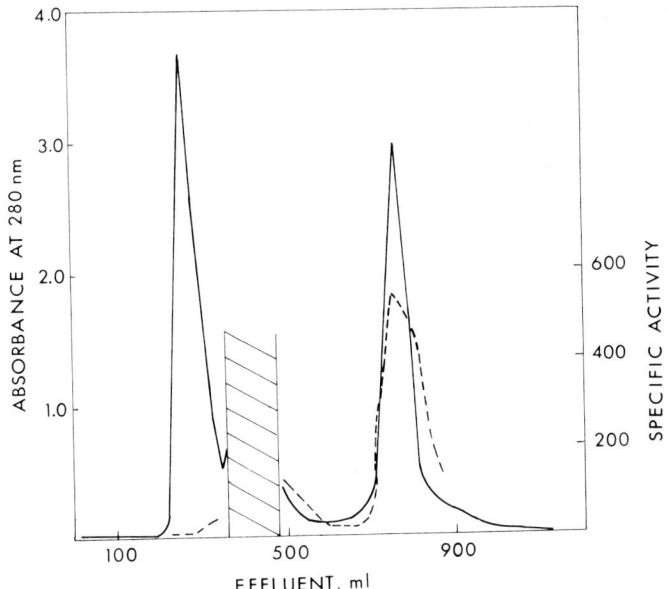

FIG. 2. Recycling gel filtration on Sephadex G-100 (see the text, step 6). The column for this experiment was one half size (3.2 × 100 cm). Sample: 114 mg in 50 ml; buffer: 0.05 M Tris-phosphate, pH 7.0; upward flow rate: 17 ml per hour; void volume of the column: 225 ml. ——— A_{280}, - - - - specific activity.

continue with step 8, which also separates active pepsin if any remains (see Fig. 3).

Step 8. Hydroxylapatite Chromatography of Pepsinogens.[11,12] A small-scale experiment is described. It may be scaled up to a 2-cm column, but with some loss of resolution. Bio-Gel HT (BioRad Laboratories, Richmond, Calif.) may be purchased as a suspension in phosphate buffer. Resuspend the solid by gentle agitation, without stirring.[13] Allow to settle and remove fines. Suspend in 1 volume of starting buffer (0.02 M sodium phosphate, prepared by dilution[11] of 0.1 M sodium phosphate of pH 7.3). Pour into a 0.9 × 25 cm column, closed at the bottom, and allow to settle for 5 minutes before starting the flow. Adjust the rate of flow to 8 ml per hour, and pack the column to a final height of 20 cm. Equilibrate until the pH and conductivity of the effluent equal the starting buffer (at least overnight). Apply a sample of 10–20 mg of pepsinogen in 2 ml of starting

[11] T. G. Rajagopalan, S. Moore, and W. H. Stein, *J. Biol. Chem.* **241**, 4940 (1966).
[12] P. A. Meitner and B. Kassell, manuscript in preparation. The procedure is similar to that used by Rajagopalan *et al.*[11] for porcine pepsinogen.
[13] Follow the manufacturer's directions.

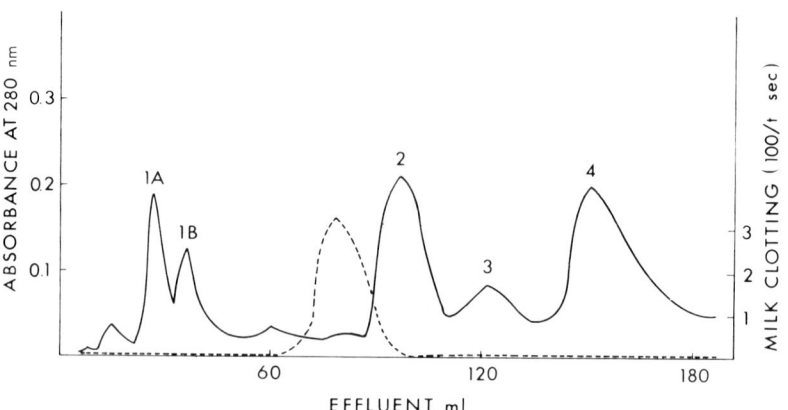

FIG. 3. Gradient elution of the pepsinogens from hydroxylapatite. Column 0.9 × 20 cm; sample 9.65 A_{280} units in 5 ml of starting buffer. Starting buffer: 125 ml of 0.02 M sodium phosphate, pH 7.3. Gradient buffer: 125 ml of 0.1 M sodium phosphate buffer, pH 7.3. The two buffers were in paired interconnected open bottles. ——— A_{280}; - - - - milk clotting activity measured as 100/t sec where t = time of formation of clot by 0.1 ml of effluent.

buffer. Elute with a linear gradient (125 ml of starting buffer and 125 ml of 0.1 M sodium phosphate, pH 7.3, in paired interconnected open bottles). Collect 1.8-ml fractions at the rate of 10 ml per hour. Several peaks separate (Fig. 3), all with equal potential activity, except the small unlettered peaks which are inactive. Pool the individual peaks, concentrate to small volumes by means of a syringe ultrafilter (Amicon), dialyze against water at pH 7.0 as before, and lyophilize.

Preparation of the Pepsin[12]

Activation of Purified Pepsinogen. The same procedure may be used to activate either the mixture of pepsinogens from step 6 or 7 or individual pepsinogens prepared in step 8. The experiment described is the activation of the individual pepsinogens. The procedure is similar to the activation of porcine pepsinogen.[11]

Recycle Bio-Rex 70, minus 400 mesh (BioRad Laboratories) or Amberlite CG-50 (Mallinckrodt Chemical Works, St. Louis, Mo.) with NaOH and HCl, wash with water, and adjust to pH 5.6. Equilibrate and remove fines in 0.2 M sodium acetate buffer of pH 5.6. Prepare a 2 × 40 column, and equilibrate overnight with buffer. Dissolve 100 mg of pepsinogen in 4 ml of cold water in a plastic centrifuge tube and place in an ice-water bath. Insert a microelectrode, a small magnetic flea, and a fine Teflon tube leading from a microburet containing 2.5 M HCl. Place the bath on a magnetic stirrer. With stirring, add the HCl rapidly until the solution is at pH 2.0. Wait 4 minutes, then raise the pH to pH 5.6 ± 0.1 by adding cold 4 M sodium acetate. It is desirable for rapid additions to predetermine on an aliquot the amounts of HCl and sodium acetate required. Centrifuge if necessary, and apply the activation mixture to the column. Elute with 0.2 M sodium acetate buffer of pH 5.6, collecting 20-ml fractions. The pepsin is not retarded by the resin and appears in a single sharp peak. The activation peptides are held to the resin and may be eluted with 0.1 M ammonium hydroxide. Pool the pepsin fractions of highest activity, concentrate on the ultrafilter, and dialyze against water adjusted to pH 5.6.

Hydroxylapatite Chromatography of Pepsins. The chromatography is exactly analogous to step 8, except that a gradient of sodium phosphate buffer from 0.075 M to 0.15 M is used. The eluting solutions are prepared from a stock solution[11] of 0.3 M sodium phosphate buffer of pH 5.7. A series of chromatographies is shown in Fig. 4.

Properties

Stability. There is no loss of activity when pepsinogen solutions are stored at pH 7.0 for 9 days at 5°, or when the lyophilized material is stored for 6 months at −20°. Below pH 6.0, however, the pepsinogen is slowly activated to pepsin. Above pH 7.8, it is denatured.

Purity. PEPSINOGENS. The potential peptic activity is evenly distributed across the peaks in steps 7 and 8; the molar ratios of amino acids present in amounts less than 10 residues per molecule are close to integral values; ultracentrifugation gives a single symmetrical peak. However, disc electrophoresis shows a faint band of impurity for some of the pepsinogens.

PEPSINS. When prepared from purified pepsinogen (step 7) and chromatographed on hydroxylapatite, each pepsin gives a single peak upon rechromatography on hydroxylapatite, with uniform activity across the peak.

Physical Properties. The molecular weight of the pepsinogen determined by ultracentrifugation is 37,500 and by amino acid analyses is 38,943. The absorbance at 280 nm is 1.305 for 1 mg of protein per milliliter. For the pepsin, the corresponding absorbance is 1.48.

Activators and Inactivators. The pepsinogen is activated maximally to pepsin by exposure to pH 2.0 for 4 minutes. The enzyme is inhibited 100% by treatment with diazoacetyl norleucine methyl ester[14] with the concomitant addition of 1 mole of norleucine ester per mole of protein.

[14] T. G. Rajagopalan, W. H. Stein, and S. Moore, *J. Biol. Chem.* **241**, 4295 (1966).

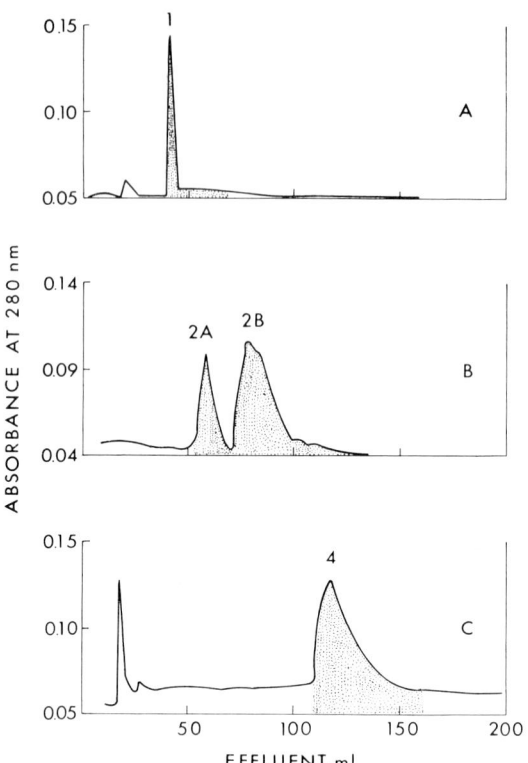

Fig. 4. Gradient elution of the pepsins produced upon activation of separated pepsinogens. The same column (0.9 × 20 cm) was used for each chromatography after exhaustive washing with 0.2 M sodium phosphate buffer, pH 5.7, followed by equilibration with 0.03 M buffer, pH 5.7. Samples were added in 2 ml of 0.03 M buffer. Starting buffer: 150 ml of 0.075 M sodium phosphate buffer, pH 5.7. Gradient buffer: 150 ml of 0.15 M sodium phosphate buffer, pH 5.7. Other conditions as in Fig. 3. Shaded areas indicate active pepsins. A, B, and C are activation products from 2.5 A_{280} units of pepsinogen 1, 5 A_{280} units of pepsinogen 2, and 5 A_{280} units of pepsinogen 4, respectively.

Specificity. All of the bovine pepsins act on hemoglobin substrate (see assay) and on the synthetic substrate acetylphenylalanyl diiodo-tyrosine; the bovine pepsin is about one third as active as porcine pepsin on the synthetic substrate and about two thirds as active on hemoglobin substrate.[15] They have milk-clotting action at pH 5.3.

Amino Acid Composition. The composition, given in the table, shows that bovine pepsinogen resembles porcine pepsinogen[11] in the large

[15] H. M. Lang, S.S.J., and B. Kassell, unpublished results.

Amino Acid Composition of Bovine Pepsinogen[a]

Amino acid	Residues per molecule[b]	Amino acid	Residues per molecule[b]
Lysine	8	Half-cystine	6
Histidine	2	Valine	25
Arginine	6	Methionine	4
Aspartic acid	40	Isoleucine	32
Threonine	27	Leucine	25
Serine	50	Tyrosine	18
Glutamic acid	32	Phenylalanine	15
Proline	15	Tryptophan	6
Glycine	35	Amide groups	37
Alanine	16		
		Total	362 ± 2

[a] R. B. Chow and B. Kassell, *J. Biol. Chem.* **243**, 1718 (1968).
[b] Molecular weight 38,943; total nitrogen 14.56%.

number of acidic and small number of basic residues and in the size of the molecule. However, there are at least 22 amino acid substitutions.

Distribution. The pepsinogen is found only in the fundic mucosa of the abomasum or fourth stomach. The total pepsinogen content of a single mucosa varies from 0.4 to 1 g.

Terminal Groups. The bovine pepsinogens all have amino terminal Ser-Val- and carboxyl terminal -Val-Ala. The bovine pepsins have amino terminal valine and carboxyl terminal alanine.

[22] Chicken Pepsinogen and Chicken Pepsin[1]

By Zvi Bohak

Assay Methods

Chicken pepsin catalyzes the hydrolysis of acid-denatured hemoglobin at pH 1.5–4.0, and the extent of digestion under standard conditions is a measure of the amount of enzyme present. The digestion of hemoglobin is followed either by determining the amount of split products released in a given time[1a, 2] or by determining the rate of hydrolysis of peptide bonds with the aid of a pH-stat.[3]

[1] This investigation was supported by Grant GM-12971 from the National Institutes of Health, U.S. Public Health Service.
[1a] M. L. Anson, *J. Gen. Physiol.* **22**, 79 (1938).
[2] M. L. Anson, *in* "Crystalline Enzymes" (J. H. Northrop, M. Kunitz, and R. M. Herriott, eds.), 2nd ed., p. 305. Columbia Univ. Press, New York, 1948.

Chicken pepsin possesses a marked milk-clotting activity at pH 5.0–6.5, and can therefore be assayed by the procedure of Berridge for the assay of rennin[4] or by following the splitting of casein[5] at pH 6.0. As these assays are carried out under conditions where negligible conversion of chicken pepsinogen to chicken pepsin takes place, they can be employed to assay the enzyme in the presence of the zymogen. The sensitivity of these assays is, however, much smaller than that of the assays based on the hydrolysis of hemoglobin.

The potential peptic activity of chicken pepsinogen is determined by any of the above methods after conversion to pepsin at pH 1.8–2.0.

METHOD I: THE ASSAY OF CHICKEN PEPSIN BY THE SPECTROPHOTOMETRIC DETERMINATION OF TCA-SOLUBLE HYDROLYSIS PRODUCTS OF ACID-DENATURED HEMOGLOBIN

Reagents

Substrate. Dissolve 25 g of substrate grade hemoglobin (Worthington) in 1 liter of distilled water by stirring for 30 minutes. Dialyze the solution (20/32 Visking tubes) for 3–4 days at 4° against 10 liters of distilled water changed twice daily. Divide the solution into 20-ml portions and keep frozen. For the assay thaw one 20-ml portion, add 5 ml 0.3 N HCl, and keep at room temperature for 15–30 minutes. The pH of this solution is 1.8 ± 0.05.

Enzyme. Prepare a solution containing 0.5–1 mg per milliliter of chicken pepsin in 0.03 N HCl or of chicken pepsinogen in 0.005 M phosphate buffer, pH 7.5, and determine its absorbance at 280 nm. Calculate protein concentration taking E 1 mg/ml = 1.46 for chicken pepsin and E 1 mg/ml = 1.26 for chicken pepsinogen. Dilute the solution with 0.03 N HCl to a final concentration of 5–25 μg ml. If the zymogen is assayed, leave this solution for about 5 minutes at room temperature to allow for complete activation.

5% Trichloroacetic acid solution. Use within 1 week of preparation.

Procedure. Pipette 1.25-ml portions of the substrate solutions into three 16 × 150 mm test tubes and keep 10 minutes at 37°. To one tube,

[3] C. F. Jacobsen, J. Léonis, K. Linderstrøm-Lang, and M. Ottesen, *in* "Methods of Biochemical Analysis" (D. Glick, ed.), Vol. IV, p. 171. Wiley (Interscience), New York, 1957.
[4] N. J. Berridge, Vol. II, p. 69.
[5] M. Kunitz, *in* "Crystalline Enzymes" (J. H. Northrop, M. Kunitz, and R. M. Herriott, eds.), 2nd ed., p. 308. Columbia Univ. Press, New York, 1948.

the blank, add 2.5 ml 5% TCA, shake well, then add 250 μl of the enzyme solution. To each of the other tubes add 250 μl enzyme solution, incubate for 10 minutes at 37°, then add 2.5 ml 5% TCA and shake well. Filter the suspension through hardened filter paper (Whatman No. 50 or S & S No. 576), read the absorbance of the filtrate at 280 nm against distilled water, and subtract the blank reading from the readings obtained for the test reaction mixtures. The figure thus obtained (ΔA_{280}) represents the absorbance of the hydrolysis products.

Calculations. (a) Pure chicken pepsin yields $\Delta A_{280} = 0.167$ per μg enzyme taken for assay and pure chicken pepsinogen yields $\Delta A_{280} = 0.140$ per μg zymogen taken for assay. (b) The pepsin unit defined by Anson[2] is the amount of enzyme which produces an increase in the absorbance at 280 nm of TCA-soluble hydrolysis products of 0.001 per minute when the assay is carried out in a reaction mixture whose volume is 4 times that used in the present procedure. Specific activities in Anson's units are calculated from ΔA_{280} values obtained as described above by employing the equation

$$\text{Units/mg} = \frac{\Delta A_{280}}{\mu\text{g enzyme in assay}} \times 25 \times 10^3$$

Notes. The assay is linear in time and in enzyme concentration up to $\Delta A_{280} = 0.8$.

The duplicate test readings should agree within 5%. Blank values are 0.15–0.25, and blank readings higher than 0.3 indicate unsuitable hemoglobin solutions, which often yield completely nonlinear calibration curves.

Turbidity of the TCA-filtrate is often due to the presence of traces of some detergents or to protein adhering to glassware repeatedly used for this assay. Glassware, in particular filter funnels, should be thoroughly rinsed with water immediately after the experiment before washing with detergent, and after repeated use funnels should be cleand by overnight immersion in a mixture of conc. HNO_3 (3 vols) and conc. H_2SO_4 (1 vol).

METHOD II: AUTOMATED ASSAY OF CHICKEN PEPSIN BY DETERMINATION OF DIALYZABLE HYDROLYSIS PRODUCTS OF HEMOGLOBIN, EMPLOYING FOLIN'S REAGENT

This method employing the Technicon Autoanalyzer was originally developed for clinical investigations.[6] It was found very convenient for monitoring peptic activity in column effluents. The method is suitable for the assay of chicken pepsin concentrations of 10–250 μg/ml.

[6] J. Vatier, A. M. Cheret, and S. Bonfils, *in* "Automation in Analytical Chemistry," Vol. II, p. 371. Mediad, New York, 1967.

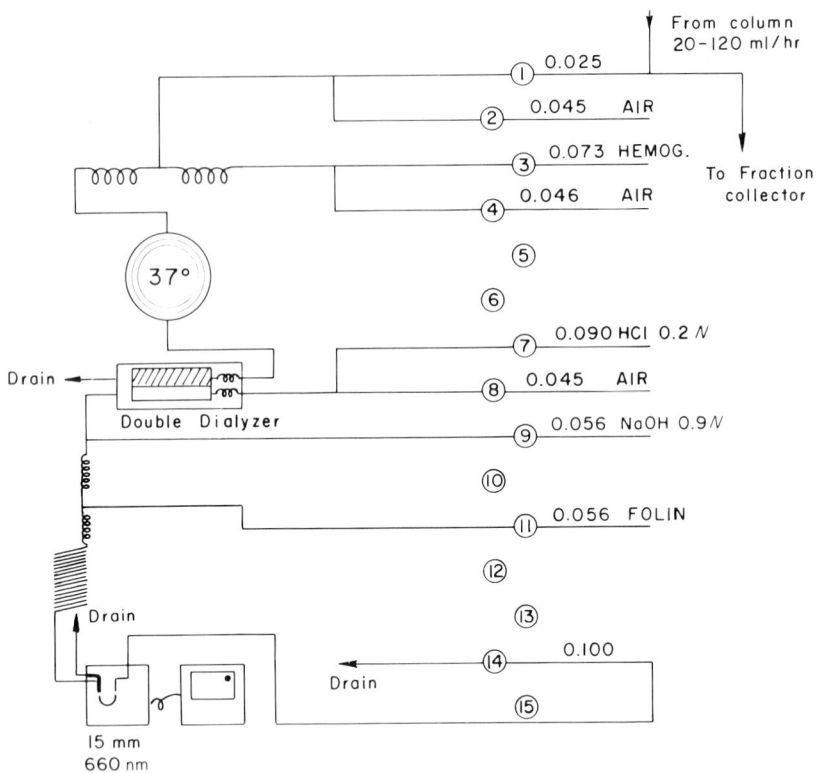

Fig. 1. Autoanalyzer flow diagram for the assay of peptic activity.[6]

Reagents and Apparatus

Hemoglobin. Dissolve 20 g substrate hemoglobin in 400 ml distilled water and add 100 ml 0.3 N HCl. The solution should be used within 2–3 days of preparation and stored at 4° when not in use.

Folin Reagent. Mix 500 ml commercial Folin-Ciocalteu 2 N (Harleco) with 1500 ml distilled water. Store in amber bottles.

HCl, 0.2 N, and NaOH, 0.9 N

Equipment. The manifold arrangement is given in Fig. 1.

METHOD III. pH-STATIC ASSAY OF CHICKEN PEPSIN

Reagents and Apparatus

Substrate. Prepare a 2% solution of acid-denatured hemoglobin as described in Method I. For assay this solution is titrated with 2 N NaOH to pH 3.1–3.2, kept at 0°, and used on the day prepared.

Enzyme. Bring the solution of the enzyme to pH 3.0, and dilute with 0.001 N HCl to a final concentration of 0.5–2.5 mg/ml.

HCl, 0.05 N, standardized.

Equipment. Autotitrator operated as pH-stat (Radiometer) equipped with an 0.5 ml microburet and a 10-ml jacketed titration vessel thermostatted at 37°.

Procedure. Pipette 4 ml substrate solution into the titration vessel, and allow 5–10 minutes for temperature equilibration. Set end point to 3.0 and start titrator. A small acid uptake, required to bring the substrate solution to pH 3.0, is registered and a flat base line is then recorded for at least 1 minute. Add 50 μl of enzyme solution and record acid uptake for 2–3 minutes. Measure the initial slope of the curve and calculate from it proton uptake in micromoles per minute.

Calculations. To obtain the rate of hydrolysis of peptide bonds, divide the rate of proton uptake by 0.83.[6a]

The number of moles of peptide bonds hydrolyzed per mole of pure chicken pepsin is 1.5×10^3 per minute.

Isolation Procedures

Avian pepsinogen is produced and stored in gastric glands located in the forestomach (proventriculus) which is the part of the alimentary canal leading into the stomach. In addition to zymogen granules, these glands produce acid-containing granules and are rich in mucinlike materials[7] To extract the pepsinogen without activation, an alkaline solution must be employed for homogenization of the tissue in order to neutralize the acid. To minimize losses of the zymogen during isolation, the mucinlike materials must be removed from the extract in an early step of purification. In the original method for the isolation of pepsinogen from chicken stomachs,[8] extraction was carried out with 0.1 M NaHCO$_3$, mucopolysaccharides were adsorbed on basic cupric sulfate, and the zymogen was precipitated from the extract between 0.2 and 0.6 saturation with ammonium sulfate. In the procedure[9] described below, mucin-

---

[6a] Calculated from Linderstrøm-Lang's formula[3] for the conversion factor β,

$$\beta = \frac{1}{1 + 10^{pH-pK}}$$

employing the experimentally determined value $pK = 3.7$ for the dissociation of the carboxylic groups liberated in the reaction.

[7] P. D. Sturkie, *in* "Avian Physiology," p. 152. Cornell Univ. Press, Ithaca, New York, 1954.

[8] R. M. Herriott, Q. R. Bartz, and J. H. Northrop, *J. Gen. Physiol.* **21**, 575 (1938).

[9] Z. Bohak, *J. Biol. Chem.* **244**, 4638 (1969).

like materials are rendered insoluble by treating the forestomach homogenate with acetone, and homogenization is therefore carried out with a solution of Tris-acetate buffer which is miscible with acetone. Acetone precipitation is also employed for the isolation of the zymogen from the tissue extract.

Chicken pepsinogen is then purified by chromatography on DEAE-cellulose followed by gel filtration on Sephadex G-100. The pepsinogen emerges from the DEAE-cellulose columns as a single peak, when the elution schedule described below is employed. Multiple peaks apparently containing chicken pepsinogen together with some contaminants can be obtained by employing other elution schedules, and rechromatography on DEAE-cellulose is then required.[9, 10]

Pure chicken pepsin is obtained by activation of the pure zymogen and separation of the products by gel filtration on Sephadex G-100 in an acid solution. Chicken pepsin preparations were also obtained from extracts of chicken forestomachs by acidification to pH 2 and precipitation by the addition of solid NaCl to a final concentration of 250 g per liter.[11] The crude pepsin thus obtained was further purified by gel filtration to yield an enzyme preparation possessing 70–90% of the specific activity of the pure enzyme. Both procedures for the preparation of chicken pepsin are described below.

Preparation of Chicken Pepsinogen

Step 1. Chicken Forestomachs. The forestomachs were removed from fowl immediately after slaughter, sliced open, rinsed, and adhering fat was removed. The forestomachs were kept frozen. A single forestomach weighs 6–8 g.

To determine the total potential activity in the tissue, small pieces were removed from the central part of several forestomachs and homogenized with 10 ml/g of 0.1 M NaHCO$_3$. A 0.1 ml portion of the homogenate was diluted with 5 ml HCl 0.03 M and activity was determined as described in assay Method I. The activity found in 1 g of forestomachs corresponds to 5–12 mg of pure pepsinogen.

Step 2. Homogenization and Acetone Treatment. One hundred grams of forestomachs were partly thawed to a semisolid state, and homogenized in a Sorvall Omnimix with 200 ml of ice-cold 0.4 M Tris-acetate buffer, pH 8.6 (0.4 M Tris solution adjusted to pH 8.6 with 50% acetic acid). Homogenization was carried out for 10 minutes at full speed while cooling the jar in an ice–salt mixture. The slurry was poured into 1 liter of acetone precooled to −15°. The mixture was stirred for 5 minutes and filtered with suction through Whatman No. 3 MM paper moistened with

[10] T. P. Levchuck and V. N. Orekovich, *Biokhimiya* (Engl. Transl.) **28**, 738 (1963).
[11] D. Gabison, Y. Levin, E. Katchalski, and Z. Bohak, unpublished results.

acetone. The filter cake was squeezed and kept on the pump until it started turning brown at the edges.

Step 3. Extraction of Chicken Pepsinogen and Precipitation with Acetone. The wet filter cake was suspended in 400 ml of 0.4 *M* Tris-acetate buffer, pH 8.6, and homogenized with cooling for about 5 minutes. The suspension was then stirred 30 minutes at room temperature, and filtered through a double layer of gauze. The cake was squeezed out by hand and kept at 0°. The red-brown milky filtrate was mixed with 50 g of Celite 535 and centrifuged for 1 hour at the maximum speed (about 11,000 *g*) in the Sorval RC-2 centrifuge employing the large volume head, and the supernatant was carefully decanted. In most preparations a slightly opalescent solution was obtained, whereas in some runs the supernatant had a milky appearance and was mixed with 25 g Celite 535 and recentrifuged for 30 minutes. The precipitates, collected by gauze filtration and by centrifugation, were combined and reextracted with 200 ml 0.4 *M* Tris-acetate buffer, pH 8.6, employing the same procedure as for the first extraction, and the two extracts were combined.

The crude extract was poured into 1.4 liter acetone precooled to −15° and 3 g cellulose powder (Whatman standard grade) were added. The mixture was allowed to stand at 0° for about 30 minutes, until most of the precipitate settled. It was then filtered through a layer 3–5 mm thick of cellulose powder spread on a sintered polyethylene filter (Bel Art, Labpor filter). The solution was transferred slowly to the filter bed starting with the upper layer and filtration was started with light suction which was slowly increased as filtration slowed down. The filter cake was kept on the pump until all the liquid was sucked out.

The filter cake was homogenized for 10 minutes with 100 ml 0.4 *M* Tris-acetate, pH 8.6, buffer. The mixture was stirred for 15 minutes at room temperature then centrifuged in the small head on the Sorval centrifuge at maximum speed (27,000 *g*) for 15 minutes. The supernatant was kept at 0°, and the precipitate was reextracted as above. The supernatants were combined to give about 190 ml of a clear pinkish brown solution. The solution was transferred to 18/32 Visking dialysis tubing and dialyzed against 4 liters of 0.005 *M* borate buffer, pH 7.6 (0.005 *M* boric acid adjusted to pH 7.6 with 4 *N* NaOH), at 4°. The outer buffer was changed after 2–3 hours and then twice daily for 2 days.

Step 4. Chromatography on DEAE-Cellulose. The dialyzed solution was run at a rate of 120 ml/hour, into a 2.5 × 30 cm column of DEAE-cellulose preequilibrated with 0.05 *M* borate buffer, pH 7.6. The column was then eluted at a rate of 120 ml per hour with a solution 0.05 *M* in boric acid and 0.1 *M* in NaCl adjusted to pH 7.6 with 4 *N* NaOH, the absorbance at 280 nm of the effluent was monitored, and the effluent was discarded. After about 1.5 liters of the wash solution passed through the

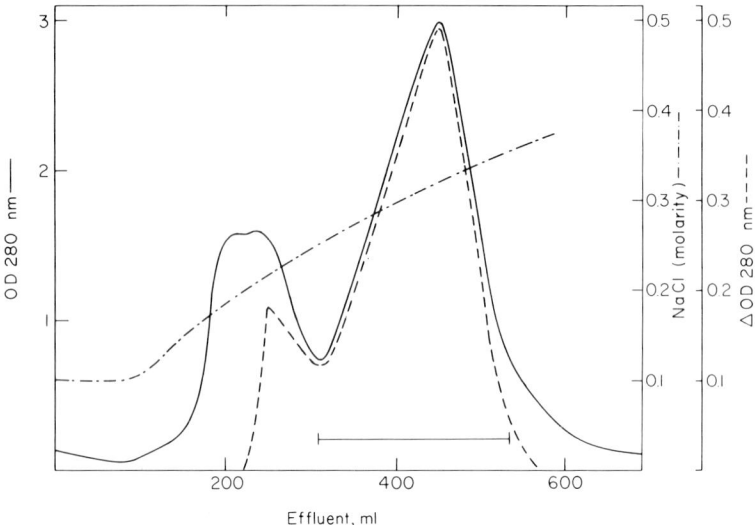

FIG. 2. Chromatography of crude chicken pepsinogen on DEAE-cellulose. The zymogen preparation contained 11.4×10^5 potential peptic units, was chromotographed on a 2.5×20 cm column employing the elution schedule described in step 3. The right ordinate represents ΔA_{280} values obtained when 10 μl aliquots of the effluent were mixed with 1.5 ml of HCl 0.03 N and the activity was determined using 250 μl of this solution employing assay Method I.[9]

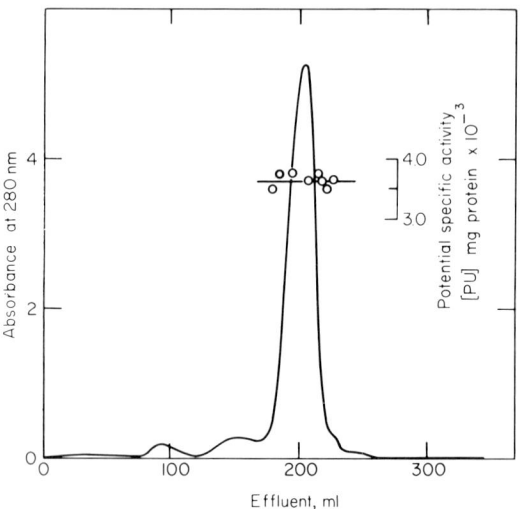

FIG. 3. Gel filtration of chicken pepsinogen on a column (1.8×95 cm) of Sephadex G-100 in 0.04 M ammonium bicarbonate, pH 7.2.[9]

column, the absorbance dropped to 0.1. Chicken pepsinogen was then eluted with a logarithmic gradient formed by running a solution 0.05 M in boric acid and 0.5 M in NaCl adjusted to pH 7.6 with 4 N NaOH into a closed mixing vessel containing 450 ml of the 0.05 M borate, 0.1 M NaCl buffer, pH 7.6. Gradient elution was carried out at 120 ml/hour, the absorbance at 280 nm, NaCl concentration in the effluent was determined, and the effluent was collected in 10-ml fractions. Chicken pepsinogen emerged as a large peak with its center at 0.28 M NaCl (Fig. 2). A small protein peak (or shoulder) containing potential proteolytic activity preceded the main peak and emerged from the column at about 0.15 M NaCl. This material was either added to the next batch to be chromatographed or discarded. The fractions containing the main peak were pooled and desalted in 150–200 ml portions on a 4 × 50 cm column of an ion-retardation resin (Biorad AG11A8). The column was eluted with water at a rate of 200 ml/hour and fractions of 25 ml were collected. The protein peak in the effluent was located by monitoring the absorbance at 280 nm, and the appropriate fractions pooled and lyophilized.

Step 5. Gel Filtration on Sephadex G-100. One half of the lyophilized powder was dissolved in 15 ml of 0.04 M ammonium bicarbonate buffer, pH 7.3, and applied to a 1.8 × 95 cm column of Sephadex G-100 equilibrated with the same buffer. The column was eluted at a rate of 20 ml/hour and the effluent collected in 10-ml fractions. A minor protein peak appeared in the breakthrough volume and the chicken pepsinogen was eluted as a single peak with constant specific activity (Fig. 3). The fractions containing the zymogen were pooled and lyophilized.

See the table for a summary of the purification of chicken pepsinogen.

RECOVERY OF POTENTIAL PEPTIC ACTIVITY IN THE
PURIFICATION OF CHICKEN PEPSINOGEN

Step in isolation procedure	Purification step	Volume (ml)	Total (PU × 10⁵)	Yield (%)	Specific [PU]/mg protein
1	Chicken forestomachs 100 g	—	29.2	100	
2	Crude extract	550	30.0	103	
3	Precipitate obtained by addition of acetone. After dissolution and dialysis.	190	24.7	85	1.1 × 10³
4	After chromatography on DEAE-cellulose.	210	22.5	77	3.1 × 10³
5	After gel filtration on Sephadex G-100.	40	18.0	61	3.5 × 10³

Preparation of Chicken Pepsin

Preparation of Chicken Pepsin from Chicken Pepsinogen. One hundred milligrams of chicken pepsinogen was dissolved in 7 ml 0.005 M phosphate buffer, pH 7.5. The zymogen was activated by acidifying the solution to pH 1.8 with 1 N HCl, and keeping the acid solution for 15 minutes at room temperature. Chicken pepsin was separated from the peptide fragment split off during activation by filtration through a 2 × 83 cm column of Sephadex G-100 (Fig. 4). Elution was carried out with 0.001 N HCl, and the pepsin in the effluent was located by determination of the absorbance at 280 nm and determination of activity on aliquots of the effluent. The pepsin-containing fractions were pooled and lyophilized.

Preparation of Chicken Pepsin from Chicken Forestomachs. One hundred grams of chicken forestomachs was ground in a meat grinder and stirred for 3 hours with 300 ml of a solution containing 30 g NaCl and 7 g NaHCO₃ per liter. The slurry was filtered through two layers of gauze and the filtrate was acidified to pH 2 with 3 N HCl. The suspension was centrifuged for 30 minutes at top speed (11,000 g) in a Sorvall RC-2 centrifuge employing the large head. The clear supernatant was decanted

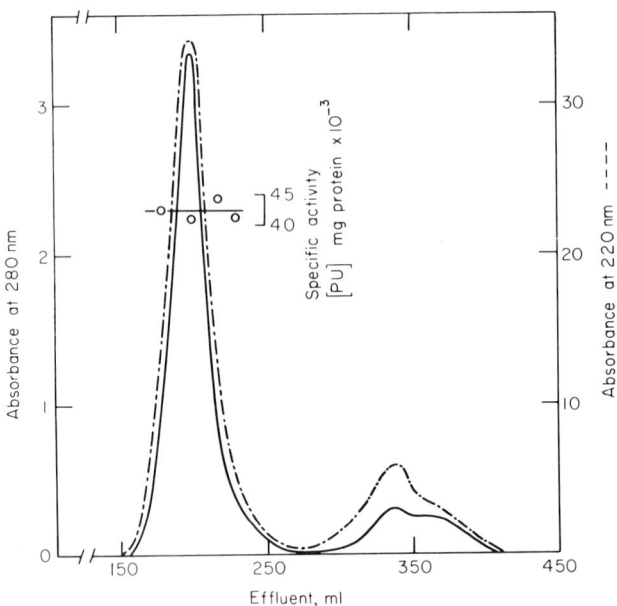

Fig. 4. Separation of chicken pepsin from the peptides formed during the activation of chicken pepsinogen. A Sephadex G-100 column (2 × 83 cm) was used and elution was carried out with HCl 0.001 N.⁹

and 28 g of solid NaCl were added for each 100 ml supernatant. The mixture was stirred for 30 minute at room temperature, and the precipitated crude pepsin was collected by centrifugation at 11,000 g for 15 minutes, and lyophilized. The dry powder thus obtained contained about 10–15% by weight of chicken pepsin and about 50% by weight NaCl. The recovery of activity was 45 ± 5%.

Two grams of this crude pepsin preparation was mixed with 3.4 ml HCl 0.02 N in a Sorvall centrifuge tube and the mixture stirred with a glass rod until all lumps disintegrated and a smooth suspension was obtained. The suspension was centrifuged for 15 minutes at 27,000 g and the clear supernatant discarded. The precipitate was mixed with 10 ml 0.02 N HCl and stirred for 5 minutes. The suspension was centrifuged for 30 minutes at 27,000 g and the supernatant filtered through a Sephadex G-100 column as described in the previous section.

Properties[8–11]

Stability. Chicken pepsinogen is stable at 25° between pH 7 and pH 10.5 and is irreversibly inactivated in more alkaline solutions. It is stable in solution in 0.1 M phosphate buffer, pH 7.6, up to 55° and is inactivated at higher temperatures, the inactivation being partly reversible on cooling to 25°.

Chicken pepsin is stable at 25° between pH 1 and pH 8. In more alkaline solutions it is inactivated, with concomitant autolysis, at a rate which increases with pH and temperature.

Purity and Physical Properties. Chicken pepsinogen and chicken pepsin are homogeneous on disc electrophoresis at pH 7.6. The zymogen sediments in 0.05 M Tris-HCl buffer, pH 7.5, 0.1 M in KCl as a monodisperse material whereas the enzyme yields an asymmetric sedimentation boundary indicating some aggregation. The molecular weight of chicken pepsinogen calculated from its sedimentation and diffusion coefficients ($s_{20,w} = 3.58$; $D_{20} = 8.28 \times 10^{-5}$ cm^2 sec; $\bar{V}$ assumed 0.75), from gel filtration, and from amino acid composition are all nearly 43,000. The molecular weight of chicken pepsin calculated from gel filtration and from amino acid composition is nearly 35,000.

Composition. CHICKEN PEPSINOGEN. Lys_{18}, His_8, Arg_7, Asp_{43}, Thr_{28}, Ser_{39}, Glu_{30}, Pro_{19}, Gly_{32}, Ala_{18}, ½ Cys_7, Val_{26}, Met_9, Ile_{23}, Leu_{30}, Tyr_{24}, Phe_{21}, Trp_5. Amide NH$_3$-32. Glucosamine-2, Hexoses 6–7.

CHICKEN PEPSIN. Lys_8, His_3, Arg_4, Asp_{35}, Thr_{24}, Ser_{35}, Glu_{23}, Pro_{14}, Gly_{27}, Ala_{15}, ½ Cys_7, Val_{21}, Met_9, Ile_{20}, Leu_{20}, Tyr_{20}, Phe_{18}, Trp_5, Amide NH$_3$-20. Glucosamine-2, hexoses 6–7.

Inactivators. Chicken pepsin contains one free sulfhydryl group and is inactivated with mercuric chloride and on oxidation with ferricyanide.

Inactivation appears to be due to the intermolecular linking of sulfhydryl groups, since no activity is lost on reacting the SH groups with N-ethyl-maleimide, p-mercurybenzoate, and 5,5'-dithiobis-(2-nitrobenzoic acid).

p-Bromophenacylbromide and α-diazo, p-bromoacetophenone, reported to react with the carboxyl groups at the active site of swine pepsin,[12] also inactivate chicken pepsin at pH 6.

Reaction of the free amino groups of chicken pepsinogen with trinitro-benzenesulfonate or with benzyl acetamidate, but not with methyl ace-timidate, leads to complete loss of potential peptic activity.

[12] B. F. Erlanger, S. M. Vratsanos, N. Wasserman, and A. G. Cooper, *Biochem. Biophys. Res. Commun.* **28**, 203 (1967).

[23] Chicken Pepsinogens

By Sam T. Donta and Helen Van Vunakis

The purification of the chicken pepsinogens[1] is accomplished by ion-exchange and molecular sieve chromatographic procedures similar to those used for the swine[2] and dogfish[3] systems. Most of the purification of the combined pepsinogens takes place on the first DE-11 cellulose column, with subsequent columns mainly used to achieve a separation of the zymogens from each other. The pepsins were generated from the purified precursors and no attempt was made to isolate the various enzymes from acidified extracts of the mucosae or from gastric juice. The preparation of chicken pepsinogens and pepsin has been reported in other laboratories.[4-6]

Assay procedures for proteolytic activities utilize hemoglobin,[7] milk,[8] and synthetic peptides as substrates. Modifications of the original assay methods have been described.[3]

Purification Procedures

The proventriculus ("stomach equivalent" of the chick) is the source of the chicken pepsinogens and pepsins and is that area of the

[1] S. T. Donta and H. Van Vunakis, *Biochemistry* **9**, 2791 (1970).
[2] A. P. Ryle, this volume [20].
[3] E. Bar-Eli and T. G. Merrett, this volume [24].
[4] R. M. Herriott, Q. R. Bartz, and J. H. Northrop, *J. Gen. Physiol.* **21**, 575 (1938).
[5] T. P. Levchuk and V. N. Orekhovich, *Biokhimiya* **28**, 738 (1963).
[6] Z. Bohak, *J. Biol. Chem.* **244**, 4638 (1969); this volume [22].
[7] M. L. Anson, *J. Gen. Physiol.* **22**, 79 (1938).
[8] R. M. Herriott, *J. Gen. Physiol.* **21**, 501 (1938).

gastrointestinal tract just proximal to the gizzard. The whole stomach is used as the starting material, as the mucosal layer containing the enzymes is too delicate and thin to be dissected away easily. Approximately 40–50 stomachs should be used as a minimum to obtain all three pepsinogens with relative ease; a lesser number decreases the chances of obtaining adequate amounts of pepsinogen C. Phosphate buffers at pH 6.9 are used unless otherwise stated and all operations are carried out in the cold.

The stomachs from freshly killed chickens are immediately placed in 0.02 M phosphate buffer, the fat removed from the outer layer of stomach, the organ opened, and the internal contents washed out with buffer. The stomachs, minced by one coarse and one fine grind with a meat grinder are then placed in 1 liter of buffer and stirred for 1 hour. The tissue is removed by centrifugation at 10,000 rpm for ½ hour. The supernatant is then applied to an ion-exchange column (35 × 6 cm) containing 750 ml of packed DE-11 cellulose equilibrated with 0.02 M phosphate buffer, and the resin is washed with sufficient buffer (2–3 liters) to elute the inactive protein. This wash can be performed quickly and is continued until readings at OD 280 nm indicate that the peak has come off. The effluent should contain no potential proteolytic activity. The column is eluted with a linear gradient of 2 liters 0.02 M phosphate buffer connected to 2 liters of 0.02 M phosphate, 0.8 M NaCl, pH 6.9. The pepsinogens, located by suitable assay procedures, emerge in a pattern similar to the dogfish proteins[3] with pepsinogen C sometimes appearing as a slightly later shoulder (Fig. 1A). Pepsinogen B, the fraction active against N-Cbz-L-Glu-L-Tyr, is pooled and purified further by DE-11 and DE-52 cellulose chromatography using a gradient of 0.02 M phosphate, to 0.02 M phosphate containing 0.18 M NaCl, pH 6.9, followed by Sephadex chromatography.

No attempt to separate the pepsinogens with potential enzyme activity toward hemoglobin is made here and they are pooled together. After dialysis against buffer (0.02 M phosphate), this pooled fraction is reapplied to a smaller DE-11 column (30 × 3 cm), washed with 500 ml of buffer, and eluted with a shallower salt gradient of 2 liters 0.02 M phosphate, to 2 liters 0.02 M phosphate, 0.55 M NaCl, pH 6.9. A separation of two fractions with potential proteolytic activity may be observed, the smaller and later emerging peak being pepsinogen C. If pepsinogen C is not readily separated from the pepsinogen A–D peak on this column, the pepsinogen peaks are pooled, dialyzed against phosphate buffer, then adsorbed onto a DE-52 cellulose column and eluted with a similar salt gradient (Fig. 1B). Even though a separation of zymogens is not always achieved on the second DE-11 cellulose column, this chromatographic step greatly improves the flow properties of the solutions through the

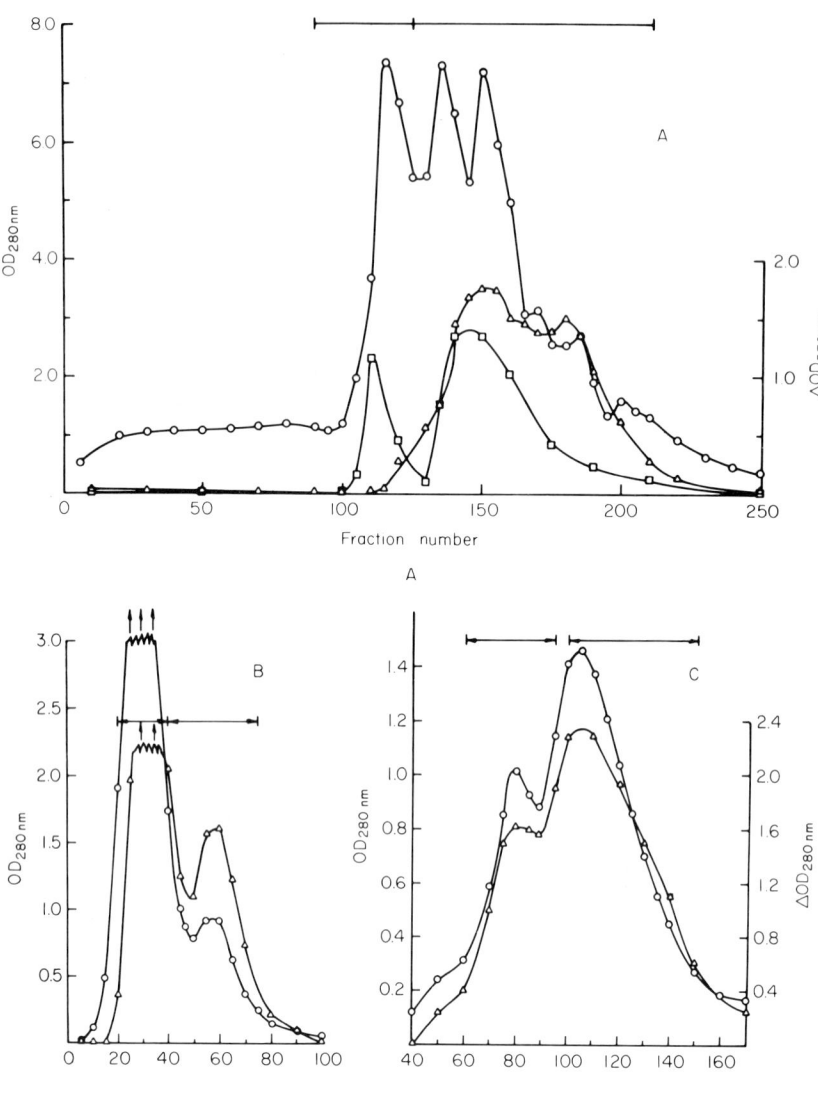

Fig. 1 (A). Chromatography of original crude stomach extract on DE-11 ion-exchange cellulose. The extract from 72 chicken stomachs was applied to a column (35 × 6 cm) and eluted with a gradient of 2.0 liters 0.02 M phosphate buffer, pH 6.9, to 2.0 liters 0.02 M phosphate + 0.9 M NaCl, pH 6.9, with 20-ml fractions being collected. Arrows indicate fractions pooled. (B). Separation of pepsinogen C from main pepsinogen fraction with DE-52 column (8 × 3 cm) and gradient of 2.0 liters, 0.02 M phosphate, pH 6.9, to 2.0 liters, 0.02 M phosphate, 0.6 M NaCl, pH 6.9; 15-ml fractions were collected. (C). Separation of pepsinogens D and A; DE-52 column (8 × 3 cm) eluted with gradient of 1.0 liter, 0.02 M phosphate, pH 6.9 to 1.0

DE-52 cellulose columns used in subsequent steps of the purification procedure (Figs. 1B and 1C). On DE-52 cellulose columns, the major fraction consisting of pepsinogens D and A should separate easily at salt concentrations of 0.15 and 0.20 M NaCl, respectively, while pepsinogen C is eluted at 0.35 M NaCl. Acrylamide electrophoresis should be used to follow the purification, especially the final steps, as sometimes pepsinogen A can be found in the early eluted portion of an apparently homogeneous (protein concentration paralleling potential proteolytic activity) pepsinogen C preparation.

The purified DE-52 pepsinogens are concentrated by ultrafiltration (Amicon Corp. Diaflo—Membrane with 10,000 MW cutoff) and filtered through a Sephadex G-100 column (100 × 1 cm) equilibrated with 0.02 M phosphate, 0.15 M NaCl, pH 6.9 as the final purification step for each pepsinogen. Only with pepsinogen D was any further slight purification noticed. As an approximation, there are 30 mg of total pepsinogen in each stomach, in roughly a ratio of 10:15:5 between D, A, and C. In the chromatographic separations, only the midareas of protein peaks were pooled to obtain homogeneous preparations of the individual pepsinogens; thus the final yields of the purified precursors were between 10–20%. During acrylamide electrophoresis at pH 8.5, the zymogens migrate toward the anode. Pepsinogen C moves well in front of pepsinogens A and D and right behind or coincident with the bromphenol blue dye marker (Fig. 2). Pepsinogen D has a slightly slower mobility than pepsinogen A; this difference is detectable when the zymogens are electrophoresed in a single tube.

The pepsins are derived from their respective purified pepsinogens by first incubating the pepsinogens at 37° and pH 2.0 for 4 minutes (conditions of maximal activation), then dialyzing at 0° against 0.1 N acetate, pH 4.0 for 8 hours followed by dialysis against 0.02 M phosphate, 0.15 M NaCl, pH 6.9, for 8 hours. The activation mixtures are then chromatographed on Sephadex under conditions used for the pepsinogens.

Properties

The amino acid compositions of chicken pepsinogens A, D, and C and the pepsins derived from these purified precursors are given in the table and are based on the molecular weights obtained from high-speed sedimentation to equilibrium determinations. Amide, tryptophan, phosphate, and polysaccharide analyses have not yet been done.

---

liter, 0.02 M PO$_4$ + 0.45 M NaCl, pH 6.9; 10 ml fractions were collected. (○) Optical density at 280 nm; (△) activity against hemoglobin at pH 2 expressed as increase of optical density at 280 nm of digested trichloroacetic acid-soluble material; (□) optical density at 570 nm used to determine hydrolysis of synthetic substrate N-Cbz-L-Glu-L-Tyr by ninhydrin assay.

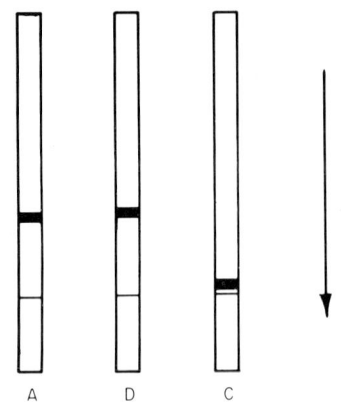

A   D   C

Pepsinogens

FIG. 2. Schematic representation of polyacrylamide gel electrophoresis of the pepsinogens in Tris-HCl buffer system, pH 8.5. Arrow indicates direction of electrophoresis; bottom line in each case represents dye front.

AMINO ACID COMPOSITIONS OF CHICKEN PEPSINOGENS AND PEPSINS

	Pepsinogens			Pepsins		
	A	D	C	A	D	C
Lys	17	15	5	11	10	3
His	7	7	3	5	5	1
Arg	7	7	3	6	6	2
Asp	44	44	41	44	44	35
Thr	28	28	33	31	32	32
Ser	39	40	40	44	46	39
Glu	32	29	48	31	29	43
Pro	19	20	22	19	19	17
Gly	33	34	40	35	34	39
Ala	21	20	22	20	19	18
Half-Cys	(6)	(6)	(6)	(6)	(6)	(6)
Val	27	28	22	27	28	21
Met	10	10	9	10	10	10
Ile	24	24	24	25	26	23
Leu	30	29	30	29	29	28
Tyr	24	27	24	24	24	20
Phe	25	22	27	23	22	24
MW	42,000	42,000	42,000	42,000	42,000	38,500

From this data and that obtained from immunochemical analysis,[9] pepsinogens A and D appear to be closely related to each other and different from pepsinogen C. Pepsins A and D closely resemble their precursors, being smaller by some 15 amino acid residues. Using immuno-

[9] S. T. Donta and H. Van Vunakis, *Biochemistry* **9**, 2798 (1970).

chemical techniques which can detect small conformational differences in macromolecules,[10] pepsins A and D were shown to be very similar to their precursors and to each other. Pepsin C (but not its precursor, pepsinogen C) also reacted with antipepsinogen A.

The pepsinogens were homogeneous by a number of criteria, i.e., superposition of activity and optical density curves in the elution patterns obtained from DEAE-cellulose and Sephadex chromatography, ultracentrifugal analysis, the formation of single bands with the homologous antibodies on immunodiffusion and immunoelectrophoresis and migration of each protein as a single band on disc-gel electrophoresis. The pepsins generated from these purified precursors also meet these criteria for homogeneity except that several specific bands are observed for each pepsin on acrylamide electrophoresis and two bands are observed for pepsin C on immunoelectrophoresis.

Pepsinogen B has been obtained in small yields only and is still heterogeneous.

Stabilities. Chicken pepsins are much more stable than swine pepsin,[1, 4–6, 9] they maintain their activities throughout the purification procedure which is carried out at neutral pH and are stored in phosphate buffer, pH 6.9, either refrigerated or frozen. Pepsinogen preparations are stable if stored in 0.03 M Tris buffer, pH 7.5, either refrigerated or frozen. Preparations frozen in phosphate buffer at pH 6.9 were converted into pepsin'. It is strongly suggested that the pepsinogen samples be assayed routinely by the milk clot assay to assure that conversion has not taken place. Neither the pepsinogens nor the pepsins should be lyophilized since they tend to lose activity in this process.

Specificities. All the enzymes are active at acid pH. The comparative proteolytic activities of swine pepsin and chick pepsins D, A, and C toward hemoglobin are 1.0, 3.4, 2.3, and 1.7, respectively, and 1.0, 0.29, 0.33, and 0.18 in the milk clot assay. The greater reactivity of swine pepsin compared to chick pepsin in the milk clot assay has already been noted.[5] Even at concentrations of 1 mg/ml and incubation times of 1 hour, chick pepsins D, A, and C did not hydrolyze N-Cbz-L-Glu-L-Tyr, either at pH 2 or pH 4, the latter being the pH optimum of the chick peptidase (pepsinogen B). Conversely, on exposure to acid pH, the protein corresponding to pepsinogen B showed no enzymatic activity in the hemoglobin or milk clot assays, even at concentrations (100-fold) and incubation times (10-fold) greater than those used for the other pepsins. Under standard assay conditions, each milligram of peptidase B contains 0.5 units of enzyme activity using N-Cbz-L-Glu-L-Tyr as substrate.

[10] L. Levine and H. Van Vunakis, Vol. XI, p. 928.

[24] Dogfish Pepsinogens

By Estelle Bar-Eli and Terence G. Merrett

Four pepsinogens have been separated from the stomach mucosae of the smooth dogfish, *Mustelus canis*.[1]

In order to retain some semblance of order in an already confused terminology, we have continued the nomenclature suggested by Ryle for swine pepsinogens[2,3] using as our standard of comparison the order of elution of the proteins from DEAE-cellulose columns at pH 7.5, i.e., pepsinogens B, D, A, and C, respectively. Using this terminology, we find pepsinogen B has enzymatic activity toward synthetic substrates (e.g., *N*-carbobenzoxy-L-glutamyl-L-tyrosine) at pH 2 and that the other three, after activation, digest protein substrates. Pepsinogens A and D are similar to each other and differ from pepsinogen C.

Assay Methods

The Estimation of Pepsin Using Hemoglobin as Substrate[4]

The pH at which this test is carried out also converts pepsinogen into pepsin; therefore, this assay gives peptic activity of pepsin and potential peptic activity of pepsinogen. Denatured hemoglobin is digested under standard conditions, the undigested hemoglobin is precipitated with trichloroacetic acid, and the amount of acid-soluble digested products is estimated by reading the absorbance at 280 nm. The original Anson[4] procedure has been modified as follows:

Reagents

Hemoglobin Substrate Powder (Worthington Biochem. Corp.). A 2% solution in $0.06 N$ HCl is exhaustively dialyzed against $0.06 N$ HCl in order to eliminate material that would contribute to the blank.

Trichloroacetic acid, 5% in H_2O

Procedure. One milliliter of $0.06 N$ HCl is added to 0.1 ml of the zymogen or enzyme followed by 5.0 ml of 2% hemoglobin solution and the reaction mixtures allowed to incubate at 37° for 5–15 minutes. The conditions

[1] T. G. Merrett, E. Bar-Eli, and H. Van Vunakis, *Biochemistry* 8, 3696 (1969).
[2] A. P. Ryle, *Biochem. J.* 96, 6 (1965).
[3] A. P. Ryle, this volume [20].
[4] M. L. Anson, *J. Gen. Physiol.* 22, 79 (1938).

convert the zymogens to the active enzymes rapidly and no preincubation in acid is required. The solution is deproteinized by the addition of 5 ml of 5% TCA, incubated a further 10 minutes, filtered, and read at 280 nm. The increase in absorbance at 280 nm of the filtrate was a measure of the enzyme activity, and was referred to a standard curve prepared for purified swine pepsin or pepsinogen. All assays were carried out in duplicate together with blank reactions in which the hemoglobin was added after the addition of TCA.

The Estimation of Pepsin in the Presence of Pepsinogen Using the Milk Clot Assay at pH 5.4

This modification of the original Klim substrate[5] was developed in the laboratory of Dr. R. M. Herriott.

Reagents

Bordens Evaporated Milk (13-oz can)
Acetate buffer, 0.22 M, pH 4.6. 170 ml of the buffer is added to the milk with continuous stirring.

The substrate is stable for approximately 1 week if kept refrigerated.

The Estimation of Peptic Activity with Synthetic Substrates

N-Cbz-L-Glu-L-Tyr was used as substrate and the extent of hydrolysis determined by the ninhydrin reaction.[6] Activity was determined by adding a 0.1 ml sample to a 1.0 ml solution containing 5 micromoles of N-Cbz-L-Glu-L-Tyr in 0.02 M NaCl at pH 2.0. Incubation was at 37° for 15 minutes. All assays were carried out in duplicate, together with duplicate blank reactions in which ninhydrin reagent was added to activated enzyme prior to the addition of substrate. After the color was developed in the usual manner, the mean increment of the optical densities at 570 nm was determined and related to a standard tyrosine calibration curve. One unit of enzyme liberates 1 micromole of tyrosine in 1 minute under standard conditions.

Purification Procedures

All operations were carried out at 4° and at pH 7.5 unless otherwise stated. The protein content of crude mucosal extracts in 0.05 M Tris buffer at pH 7.5 varied in different preparations from 4 to 6% of the initial tissue weight. Approximately 4% of the total protein was represented by enzyme(s) that digested the blocked dipeptide, N-Cbz-L-Glu-

[5] R. M. Herriott, J. Gen. Physiol. 21, 501 (1938).
[6] S. Moore and W. H. Stein, J. Biol. Chem. 211, 907 (1954).

L-Tyr and another 8% by enzymes that readily digested hemoglobin at acid pH. The pepsinogens were extracted from the mucosae by dilute buffer since buffers containing $(NH_4)_2SO_4$ fail to extract all of the pepsinogens.

Extraction. Smooth dogfish, *Mustelus canis,* caught at the Woods Hole Marine Biological Laboratories, were killed by a blow on the head, the stomachs removed, washed, and the mucosae stripped and stored in a deep freeze. In a typical run, the mucosae from 7 dogfish stomachs (210 g, although this weight varied with size of stomach) were partially thawed, and homogenized for 30 seconds with 360 ml of 0.05 M Tris buffer. The homogenate was stirred for 1 hour, passed through gauze, and the filtrate centrifuged at 7500 rpm for 45 minutes. The supernatant was further clarified by ultracentrifugation for 45 minutes at a speed of 35,000 rpm and then dialyzed overnight against 0.03 M Tris. High-speed centrifugation of the extract prior to chromatography was essential in order to maintain satisfactory flow properties with DEAE-cellulose columns.

DEAE Chromatography. The pepsinogens were adsorbed batchwise from the dialyzed solution onto 250 ml of DEAE-cellulose type 11, previously equilibrated with buffer. After stirring for 30 minutes, the DEAE-cellulose was collected by centrifugation, washed with 400 ml of buffer, and added to a column of DEAE-cellulose (24 × 5 cm) and elution was started using the same 0.03 M Tris buffer. Five hundred milliliters of buffer were sufficient to elute a very large peak of inactive protein and a gradient of increasing molarity was applied by connecting a cylindrical bottle with 0.5 M Tris buffer (2000 ml) with a mixing cylindrical chamber containing the starting buffer (2000 ml). Assays using hemoglobin and N-Cbz-L-Glu-L-Tyr as substrates were performed and the peaks pooled accordingly (Fig. 1). The initial purification by the ion-exchanger eliminated a large peak of inactive protein and yielded three smaller peaks which had no enzymatic activity in the milk clot assay at pH 5.4. At acid pH, however, the protein in the first peak hydrolyzed N-Cbz-L-Glu-L-Tyr while the proteins in the second and third peaks were active against hemoglobin.

Pepsinogens D, A, C. After dialysis against 0.03 M Tris, the proenzymes in the second (pepsinogens A and D) and third peaks (pepsinogen C) were purified further by repeating the DEAE-cellulose chromatography twice and it was then that the second peak split into two peaks, both of which were capable of digesting hemoglobin after activation (Figs. 2 and 3).

Prior to gel filtration on Sephadex G-100, it was necessary to concentrate the individual pepsinogens contained in a volume of approximately 400 ml by adsorbing them onto a small DEAE-cellulose column

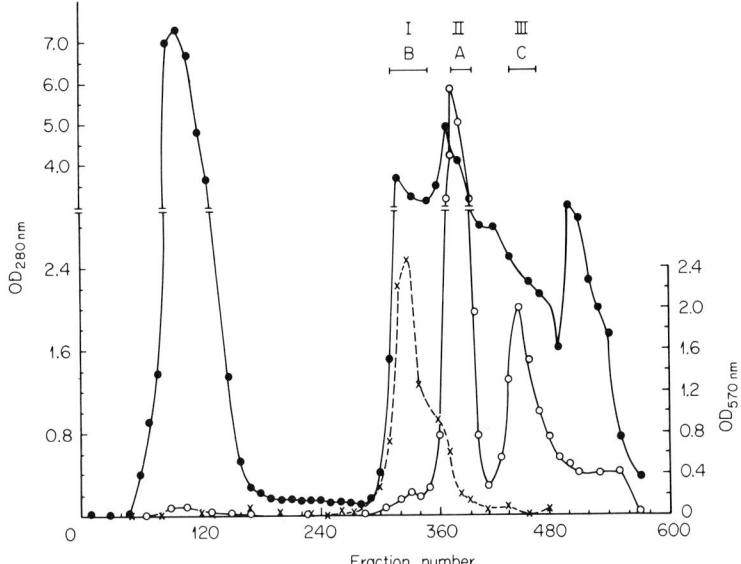

FIG. 1. Chromatography of crude dogfish pepsinogens extracted from 7 stomachs on a DEAE-cellulose column in 0.03 M Tris, pH 7.5. Tris gradient from 0.03–0.5 M applied at tube No. 240. Flow rate was 40 ml/hour and fractions of 14 ml were collected. The bars show the fractions pooled for further purification. (●——●) OD 280 nm; (○——○) activity against hemoglobin expressed as increase of OD at 280 nm of digested TCA-soluble material; (×——×) ninhydrin assay used to follow digestion of N-Cbz-L-Glu-L-Tyr (OD 570 nm).

(30 × 3 cm) in 0.03 M Tris, and eluting with 0.6 M Tris buffer. The proenzymes were further concentrated about 10-fold, to a volume of 5 ml by ultrafiltration; each was applied to a Sephadex column by the sucrose density layering method and eluted with 0.1 M Tris buffer. Enzyme assays were performed and the potentially active fractions pooled, concentrated by ultrafiltration, and filtered once again through the Sephadex column.

The proenzymes were stored frozen in 0.05 M–0.20 M Tris buffer at pH 7.0–7.5. Compared to purified swine pepsinogen, the potential enzymatic activities for pepsinogen A, D, and C were 3.8, 5.3, and 3.0, respectively, in the hemoglobin assay. In the milk clot assay, the enzymatic activities of pepsinogens A, D, and C after conversion into pepsin were all about 5% that of purified swine pepsin.

The pepsins were generated from purified pepsinogens D, A, or C by incubating the precursor at a concentration of approximately 1 mg/ml at 37°, pH 2.0, for 3 minutes. The temperature was brought down to

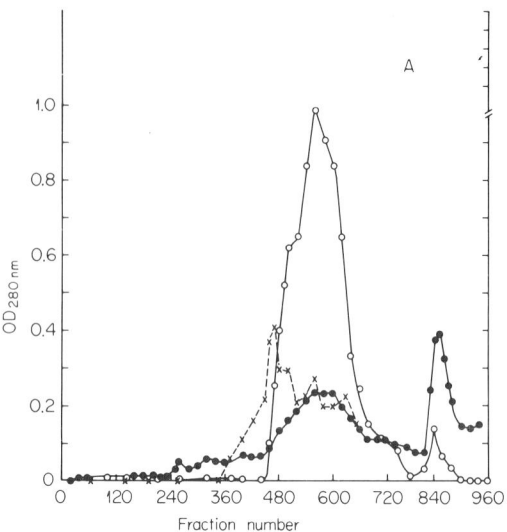

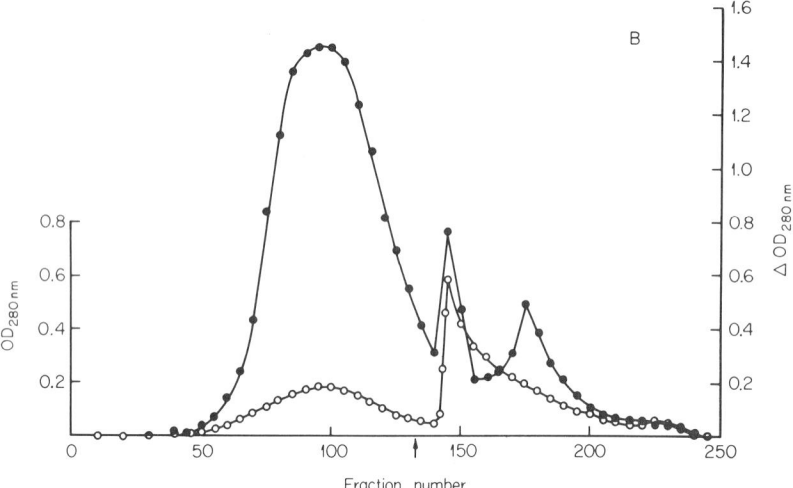

FIG. 2. (A) Rechromatography of pepsinogen A on a DEAE-cellulose column (5 × 36 cm) in 0.03 M Tris, pH 7.5. The elution was carried out with a linear gradient of 0.03–0.25 M Tris applied at tube No. 1 followed by a gradient of 0.25–0.5 M Tris applied at tube No. 420. Fraction volume was 10 ml. (B) Rechromatography of the leading edge of pepsinogen A peak from DEAE 2nd (Fig. 2A), on a DEAE-cellulose column (3 × 36 cm) in 0.03 M Tris, pH 7.5. The elution was carried out with a linear gradient of 0.03–0.25 M Tris followed by a gradient of 0.25–0.5 M Tris. The arrow indicates the application of the second gradient. The first peak represents pepsinogen D, followed by pepsinogen A and pepsinogen C. Fraction volume was 10 ml. (●——●) OD 280 nm; (○——○) activity against hemoglobin.

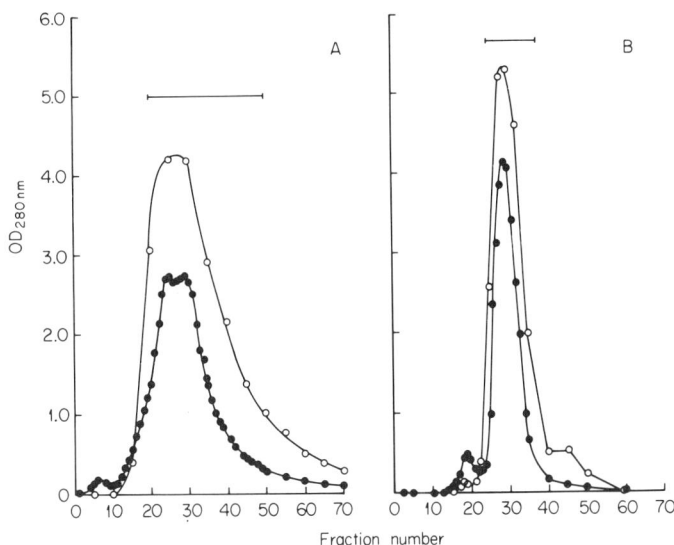

Fig. 3(A). Rechromatography of pepsinogen C on a DEAE-cellulose column (3 × 36 cm) in 0.3 M Tris, pH 7.5, Tris gradient 0.3–0.6 M applied at tube No. 1. (B) Second gel filtration of pepsinogen C on Sephadex G-100 (2 × 90 cm) in 0.1 M Tris, pH 7.5. Fraction volume was 3 ml and flow rate 10 ml/hour. (●——●) OD 280 nm; (○——○) activity against hemoglobin.

0°, the pH was adjusted to 4.0, and dialysis was carried out against 0.1 M acetate buffer, pH 4.0, to eliminate small peptides. Passage of the digested mixture through Sephadex G-100 also served to remove small peptides. Under these activating conditions, the pepsinogens were converted completely to pepsin (as estimated by milk clot assay) and loss of the generated pepsins by autocatalytic digestion was not observed. The enzymes were stored in 0.1 M acetate buffer, pH 4.6, in a frozen state.

Pepsinogen B. The peak which was active against Cbz-L-Glu-L-Tyr was pooled and purified by a second passage through a DEAE-cellulose type 11 column followed by gel filtration chromatography. Although a single symmetrical peak coincident with activity was obtained, polyacrylamide gel electrophoresis at pH 8.5 and high-speed sedimentation to equilibrium demonstrated the presence of 2 components. Pepsinogen B was applied to a column of DEAE-cellulose type 52 equilibrated in 0.03 M Tris and eluted with a gradient from 0.03–0.25 M Tris. The fractions with potential activity toward the synthetic substrate were pooled and concentrated by ultrafiltration and then applied to a Sephadex G-100 column (Fig. 4). These steps resulted in a 2-fold increase in

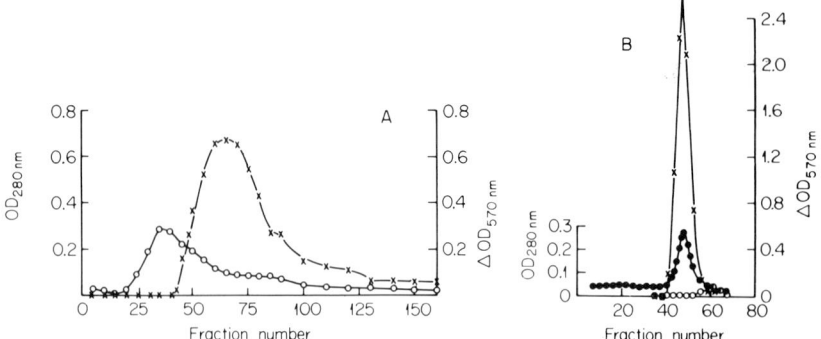

Fɪɢ. 4(A). Chromatography of pepsinogen B on a DEAE-52 cellulose column (2 × 25 cm) in 0.03 M Tris, pH 7.5. Tris gradient 0.03 M–0.25 M, fraction volumes 4.0 ml, and flow rate 16 ml/hour; (B) Gel filtration of fraction numbers 55–95 from DEAE-52 on a Sephadex G-100 column (1.5 × 100 cm) in 0.1 M Tris, pH 7.5. (●——●) OD 280 nm; (○——○) hemoglobin assay—ΔOD 280 after 45 minutes incubation; (×——×) ninhydrin assay was used to follow digestion of N-Cbz ʟ-Glu ʟ-Tyr (ΔOD 570).

specific activity. Disc-gel electrophoresis indicated the presence of a single component. The final yield from 10 stomachs was 4 mg of protein.

Each milligram of protein contained 17 units of activity toward Cbz-ʟ-Glu-ʟ-Tyr at pH 2.0 under standard conditions. The enzyme is active with synthetic substrates only at acid pH. The molecular weight and amino acid compositions are similar before and after acidification. It has been difficult, therefore, to obtain conclusive evidence for a precursor to enzyme conversion since properties which could distinguish the two molecules have not yet been found. Two possibilities exist, the conversion may occur rapidly at acid pH and not involve great changes in structure, or what we are calling pepsinogen B is in reality a peptidase active at acid pH.

Properties

From physical, chemical, and immunochemical data,[1,7] pepsinogen A and D appear to be closely related to each other and different from pepsinogen C. The pepsins generated from the proenzymes also show this similarity, i.e., only pepsin A and D resemble each other. Pepsinogen B appears unrelated to either pepsinogen A and D or pepsinogen C systems.

The molecular weights of pepsinogen D, A, and C were found to be approximately 42,000 and their amino acid compositions are shown in

[7] T. G. Merrett, L. Levine, and H. Van Vunakis, submitted to *Immunochemistry.*

AMINO ACID COMPOSITIONS OF DOGFISH PEPSINOGENS

	Dogfish pepsinogens		
	A	C	D
Lys	14	5	11
His	7	4	7
Arg	14	8	15
Asp	44	44	45
Thr	23	27	24
Ser	43	51	38
Glu	39	41	38
Pro	19	22	17
Gly	40	48	43
Ala	18	18	16
Half-Cys	7	7	7
Val	23	27	25
Met	7	6	7
Ile	22	18	22
Leu	28	24	25
Tyr	19	20	21
Phe	17	17	18
Trp	5	—	—

the table. The amide phosphate and/or sugar contents have not yet been determined nor has any separate determination been made to estimate relative contents of cysteine/cystine. At pH 8.5 on acrylamide gels, the order of their mobilities toward the anode was $C > A > D$ as would be expected from their amino acid compositions and chromatographic properties on DEAE.

The pepsins which are generated from the purified precursors have molecular weights of 34,000 to 36,000 by ultracentrifugal analysis, indicating a loss of fragments with molecular weights of approximately 7000 during the conversion process.

The molecular weight of pepsinogen B was 96,000 from sedimentation–equilibrium runs and 45,000 from molecular sieve chromatography on a calibrated Sephadex G-100 column. It is more basic than pepsinogens D, A, and C.

Stabilities. The dogfish enzymes are more stable than swine pepsin at neutral pH. Lyophilization of both precursors and enzymes should be avoided since it leads to aggregation with loss in potential enzymatic or enzymatic activities.

Specificities. Dogfish pepsins A, D, and C are active on protein substrates at acid pH only and show no activity toward the synthetic substrate, N-Cbz-L-Glu-L-Tyr. Pepsin A has a broader pH optimum than swine pepsin using hemoglobin as a substrate. Differences in specificities exist since dogfish pepsins A, D, and C digest various protein substrates differently from each other and from swine pepsin. Their specificities on protein substrates of known sequence are unknown.

Pepsin B has no activity toward hemoglobin. There is, however, a second pepsinogen B fraction in which the activities toward N-Cbz-L-Glu-L-Tyr and hemoglobin have proven inseparable by the chromatographic methods used thus far. Pepsin B is active only at acid pH and can hydrolyze the N-Cbz derivatives of L-Tyr-L-Tyr, Gly-L-Tyr, L-Glu-L-Phe, Gly-L-Phe, L-Phe-L-Tyr, L-Tyr-L-Phe, and L-Phe-L-Phe in addition to the test substrate N-Cbz-L-Glu-L-Tyr.

[25] Microbial Acid Proteinases

By J. ŠODEK *and* T. HOFMANN

Introduction

The presence of proteolytic enzymes with a pH optimum in the acid pH range (pH 1–5) has been reported in a variety of microorganisms where they occur both intracellularly and extracellularly. The enzymes from Eumycota (true fungi) have been studied most extensively and many have been isolated and purified. Acid proteinases are also found in many protozoa, both as intracellular and extracellular enzymes.[1] The intracellular enzyme from *Tetrahymena pyriformis* has been partially purified.[2] On the other hand, there are only few reports of the occurrence of acid proteinases among the bacteria: Several strains of *Clostridia* (*acetobutylicum* and *butyricum*)[3] and lactobacilli[4] have been shown to produce weak proteinase activity with an acid pH optimum. No information appears to be available on acid proteinases in algae.

The only enzymes which have been fully purified are of fungal origin and are extracellular. The intracellular acid enzyme (proteinase A) from bakers' yeast was first noted by Dernby[5] and described by Lenney[6] and was partially purified by Hata *et al.*[7]

[1] M. Muller, *in* "Chemical Zoology" (G. W. Kidder, ed.), Vol. I, p. 374. Academic Press, New York, 1967.
[2] N. Dickie and I. E. Liener, *Biochim. Biophys. Acta* **64**, 41 (1962).
[3] F. Uchino, K. Miura, and S. Doi, *J. Ferment. Technol.* **46**, 188 (1968).
[4] V. Bottazzi, *Intern. Dairy Congr. Proc. 16th Copenhagen* **2**, 522 (1962).
[5] K. G. Dernby, *Biochem. Z.* **81**, 107 (1917).
[6] J. F. Lenney, *J. Biol. Chem.* **221**, 919 (1956).
[7] T. Hata, R. Hayashi, and E. Doi, *Agr. Biol. Chem.* **31**, 357 (1967).

The following enzymes will be described: (a) the trypsinogen-activating enzyme from *Pencillium janthinellum*[8] (penicillopepsin, previously called peptidase A); (b) penicillocarboxypeptidase from the same organism[9]; (c) "trypsinogen kinase" from *Aspergillus oryzae* (Takadiastase)[10]; (d) *Candida albicans* acid proteinase[11]; (e) *Paecilomyces varioti* acid proteinase[12]; (f) *Rhizopus chinensis* acid proteinase.[13] The trypsinogen-activating acid proteinase from *Aspergillus saitoi*[14] (aspergillopeptidase A) and the milk-clotting enzymes from *Endothia parasitica*[15, 16] and from *Mucor pusillus* (mucor rennin)[17, 18] are described elsewhere in this volume (Articles [26], [29], and [30], respectively).

The acid proteinase from *Penicillium notatum* has been partially purified. It rapidly and extensively digests a variety of different proteins at pH 3.8–4.2.[19]

The acid proteinase from *Trametes sanguinea* has been purified and crystallized[20] but no information is available on its properties.

Two comments on the action of fungal acid proteinases may be made at this point. First, the enzymes described are all endopeptidases with the exception of penicillocarboxypeptidase (p. 385). In spite of an extensive search, no small synthetic substrates have been found for penicillopepsin[20a] (p. 383) and aspergillopeptidase A (see this volume [26]). Although it has been reported that the "trypsinogen-kinase" from *Aspergillus oryzae* (p. 386), the *Paecilomyces varioti*[21] acid proteinase (p. 391), and the *Rhizopus chinensis*, acid proteinase (p. 391) hydrolyze di- and tripeptides, the rates of hydrolysis are very low and require incubation periods up to 24 hours. It is probable that these activities are due to contaminations with penicillocarboxypeptidaselike enzymes and

[8] T. Hofmann and R. Shaw, *Biochim. Biophys. Acta* **92**, 543 (1964).
[9] R. Shaw, *Biochim. Biophys. Acta* **92**, 558 (1964).
[10] K. Nakanishi, *J. Biochem. (Tokyo)* **46**, 1263 (1959).
[11] H. Remold, H. Fasold, and F. Staib, *Biochim. Biophys. Acta* **167**, 399 (1968).
[12] J. Sawada, *Agr. Biol. Chem.* **27**, 677 (1964).
[13] J. Fukumoto, D. Tsuru, and T. Yamamoto, *Agr. Biol. Chem.* **31**, 710 (1967).
[14] F. Yoshida and M. Nagasawa, *Bull. Agr. Chem. Soc. Japan* **20**, 257 (1956).
[15] J. L. Sardinas, *Appl. Microbiol.* **16**, 248 (1968).
[16] K. Hagemeyer, I. Fawwal, and J. R. Whitaker, *J. Dairy Sci.* **51**, 1916 (1968).
[17] J. Yu, S. Iwasaki, G. Tamura, and K. Arima, *Agr. Biol. Chem.* **32**, 1051 (1968).
[18] G. A. Somkuti and F. J. Babel, *J. Bacteriol.* **95**, 1407 (1968).
[19] W. E. Marshall, R. Manion, and J. Porath, *Biochim. Biophys. Acta* **151**, 414 (1968).
[20] K. Tomoda and H. Shimazono, *Agr. Biol. Chem.* **28**, 770 (1964).
[20a] A current study of the action of penicillopepsin on synthetic peptides shows that it will hydrolyze the Phe·Phe bond in CbzGly·Gly·Phe·Phe·4-(propyloxy) pyridine, a recently described pepsin substrate. [G. P. Sachdev and J. S. Fruton, *Biochemistry* **8**, 4231 (1969).]
[21] J. Sawada, *Agr. Biol. Chem.* **28**, 869 (1964).

other enzymes. On the other hand, the hydrolysis of many proteins by the fungal acid endopeptidases is extensive[8] and indicative of a low degree of specificity. The failure to find small synthetic substrates indicates that these enzymes require a yet-unknown minimal length of peptide chain for activity, and unlike pepsin and such alkaline endopeptidases as chymotrypsin, subtilisin, and many others, are unable to act on di- and tripeptides. Second, special mention should be made of the unique ability of several fungal acid proteinases to activate pancreatic trypsinogen from several species[22] at pH 3–4, an activity first discovered by Kunitz[23] for an unidentified strain of *Penicillium*. Since then this activity has been demonstrated in *Aspergillus oryzae*,[10] in *Aspergillus carbonarius*,[24] and in the following penicillia: two strains of *P. janthinellum*, and in *brevi-compactum, expansum, italicum, notatum, spinulosum, stipitatum, vermiculatum*, and *wortmanii* (all reported by Hofmann[24]). It is probable that all strains of aspergilli and penicillia and other ascomycetes produce an acid proteinase with this activity. Recently, this activity has also been detected[25] in the commercially available acid proteinase from *Rhizopus chinensis*, a fungus from the class Phycomycetes. On the other hand, another species of Ascomycetes, *Saccharomyces cerevisiae*,[25] does not produce a trypsinogen-activating enzyme. The acid proteinase from *Mucor pusillus*, a fungus closely related to *Rhizopus*, also lacks this ability.[26]

The mechanism of activation of bovine trypsinogen by the mold enzymes is identical with that of autocatalytic activation and involves the cleavage of the lysine-6 isoleucine-7 bond with the liberation of the hexapeptide valyl(aspartyl)₄lysine.[27-30] In addition, however, breakdown products of the hexapeptide are formed. These differ for the different

[22] Although most of the studies have been carried out with the bovine zymogen it has been demonstrated that equine (Harris and Hofmann, to be published) and dog-fish trypsinogen (Prahl and Hofmann, unpublished observation) are also activated by penicillopepsin.

[23] M. Kunitz, *J. Gen. Physiol.* **21**, 601 (1938).

[24] T. Hofmann, *Pharm. Acta Helv.* **38**, 634 (1963).

[25] J. Šodek and T. Hofmann, to be published.

[26] S. Iwasaki, T. Yasui, G. Tamura, and K. Arima, *Agr. Biol. Chem.* **31**, 1421 (1967).

[27] E. W. Davie and H. Neurath, *J. Biol. Chem.* **212**, 515 (1955).

[28] T. Hofmann, *Bull. Soc. Chim. Biol.* **42**, 1279 (1960).

[28a] Recent experiments (G. Mains and T. Hofmann, unpublished) show that the highly purified enzyme liberates only the hexapeptide, Val·Asp·Lys₄, and does not produce smaller peptides. It is possible that the *Aspergillus* trypsinogen activating enzymes also specifically liberate the hexapeptide. In each case the production of the smaller peptides could then be attributed to the action of contaminant peptidases.

[29] K. Nakanishi, *J. Biochem* (*Tokyo*) **46**, 1553 (1959).

[30] C. Gabeloteau and P. Desnuelle, *Biochim. Biophys. Acta* **42**, 230 (1960).

enzymes (*Penicillium janthinellum*:valyl-aspartyl-aspartic acid and (aspartyl)$_2$lysine,[28, 28a] *Aspergillus oryzae*: valine and (aspartyl)$_3$aspartic acid and lysine,[29] and *Aspergillus saitoi*: valyl-(aspartyl)$_2$-aspartic acid and aspartyl-lysine[30]). The activation of trypsinogen by these enzymes is very specific; bovine chymotrypsinogen is activated by aspergillo-peptidase A at a rate which is only about 2% of that of the trypsinogen activation[30] while porcine proelastase is not activated by penicillopepsin.[31] The trypsinogen-activating property is specific for only one of the several peptidases which many fungi secrete[32] and thus provides a convenient assay for this proteinase which can be used even when other acid and alkaline peptidases are present. This assay is also very sensitive and allows the detection of a few nanograms of the enzyme.

The acid proteinases from *Mucor* species[33, 34] and *Endothia parasitica*[15, 16] are being studied for their ability to clot milk; they are potentially useful for replacing calf rennin in the cheese-making process.

Nomenclature

A few remarks about the nomenclature of the microbial acid proteinases may be appropriate. It is very likely that at least some of the acid proteinases will turn out to be homologous. The enzyme from *Penicillium janthinellum* has many properties in common with porcine pepsin[35] and, like pepsin, has an aspartic acid residue at the active site.[35a] The sum of these properties suggests that this enzyme is homologous to pepsin. As mentioned above, it is one of the trypsinogen-activating enzymes and it is likely that all these enzymes are related to each other. It is therefore imperative for the sake of clarity that there should be uniformity of nomenclature among these enzymes. The enzyme from *Aspergillus saitoi* has been named aspergillopeptidase A (EC 3.4.4.17) by the International Commission on Enzymes.[36] Similarly, the *Penicillium janthinellum* enzyme has been called *Penicillium janthinellum*

---

[31] Y. Birk and A. Gertler, *Abstr. Fed. European Biochem. Soc. 6th Meeting, Madrid,* No. 469 (1969).

[32] The neutral proteinase from *Aspergillus oryzae* has a weak trypsinogen-activation activity at pH 5.5, but not at 3.4.

[33] S. Iwasaki, G. Tamura, and K. Arima, *Agr. Biol. Chem.* **31**, 546 (1967).

[34] G. H. Richardson, J. H. Nelson, R. E. Lubnow, and R. L. Schwarberg, *J. Dairy Sci.* **50**, 1066 (1967).

[35] J. Šodek and T. Hofmann, *J. Biol. Chem.* **243**, 450 (1968).

[35a] The sequence of the active site peptide which contains this aspartic acid has recently been determined.[35b] It is Ile·Ala·Asp·Thr·Gly·Thr·Thr·Leu· and shows extensive homology to the active site peptide of porcine pepsin which has the sequence Ile·Val·Asp·Thr·Gly·Thr·Ser.[35c]

[35b] J. Šodek and T. Hofmann, *Can. J. Biochem.* **48**, 425 (1970).

[35c] R. S. Bayliss, J. R. Knowles, and G. B. Wybrandt, *Biochem. J.* **113**, 377 (1969).

peptidase A.[8] This nomenclature is likely to lead to confusion. Thus, the name aspergillopeptidase B has been given to an alkaline serine proteinase from *Aspergillus oryzae*[37] and the name penicillium peptidase B to an acid carboxypeptidase from *Penicillium janthinellum*[9] (see also p. 384). We should like to suggest here that the giving of letters and trivial names to acid proteinases should be avoided and that for those enzymes where there is sufficient evidence of homology with a well-known enzyme, such as pepsin, the trivial name of the enzyme should be added after the generic name or after the name of the organisms, e.g., penicillopepsin or aspergillopepsin, or if the species is to be identified, *Penicillium janthinellum* pepsin or *Aspergillus oryzae* pepsin. The acid proteinase from *Mucor pusillus* has been named in a similar manner mucor rennin,[38] but this was done solely on the basis of its ability to clot milk. No other evidence for a relationship to rennin is at present available. It is interesting to note that this principle for nomenclature was used by Dernby[5] who called the acid proteinase in a yeast autolyzate "Hefe pepsin."

The nomenclature suggested would be in accordance with the practice in mammalian enzymes and would be based on firm molecular criteria.

Penicillopepsin (formerly Peptidase A) from *Penicillium janthinellum* (C. M. I. 75589 = NRRL 905)

Assay Methods

Principle. The assay for penicillopepsin is based on its ability to activate trypsinogen at pH 3.4. The trypsin formed is determined by one of the standard methods at pH 8.0; for routine work, the most convenient one is that described by Kunitz[39] with casein as substrate. The complete assay given below is based on that of Kunitz[23] with the modifications given by Hofmann and Shaw.[8] The enzyme can also be assayed with bovine serum albumin at pH 2.6.

TRYPSINOGEN ACTIVATION ASSAY

Reagents

Activation step
Sodium citrate buffer at pH 3.4, 0.1 M
KH_2PO_4, 0.1 M

[36] Report Commission on Enzymes, Intern. Union Biochem., p. 114. Macmillan (Pergamon), New York, 1961.
[37] A. R. Subramanian and G. Kalnitsky, *Biochemistry* 3, 1861 (1964).
[38] K. Arima, J. Yu, S. Iwasaki, and G. Tamura, *Appl. Microbiol.* 16, 1727 (1968).
[39] M. Kunitz, *J. Gen. Physiol.* 30, 291 (1947).

HCl, 0.0025 M

Trypsinogen, 10^{-5} M. Commercial trypsinogen, 5.0 mg (containing 50% MgSO$_4$), are dissolved in 10 ml of 0.0025 M HCl. This is stable under refrigeration for 1 day.

Enzyme Dilutions. These are made in 0.1 M KH$_2$PO$_4$.

Trypsin assay

Sodium phosphate buffer at pH 8.0, 0.1 M

Tris [tris-(hydroxymethyl)amino-methane], 0.37 M

8.5% w/v Trichloroacetic acid (TCA)

2% Casein in phosphate buffer at pH 8.0. Suspend 2 g casein (Hammarsten) in 100 ml of the phosphate buffer (pH 8.0). Heat for 15 minutes in a boiling water bath with occasional stirring, cool, and adjust the pH to 8.0. This solution can be kept for many months at −20°.

0.05% Trypsin in 0.0025 M HCl for preparing a standard curve. The concentration of active enzyme is determined by active site titration with carbobenzoxy-L-lysine p-nitrophenyl ester at pH 3.0.[40]

The following solutions are prepared daily:

Solution A—1 volume trypsinogen plus 2 volumes citrate buffer, pH 3.4.

Solution B—1 volume casein plus 1 volume Tris buffer.

A mixture of 3 volumes A, two volumes B, and 1 volume KH$_2$PO$_4$ should give pH 7.9–8.1. If this pH is not obtained, the concentration of the Tris can be adjusted so that the mixture will give this pH, which is essential for the trypsin assay.

Procedure

Preparation of Trypsin Standard Curve. Tris buffer (0.37 M, 20 ml) is mixed with 2% casein (20 ml), citrate buffer (0.1 M, pH 3.4, 20 ml), and 0.1 M KH$_2$PO$_4$ (10 ml). Aliquots of 3.5 ml of this solution are pipetted into ten test tubes and warmed to 36°. At 1-minute intervals, 0.5 ml trypsin standard solutions (containing 6–140 μg trypsin per milliliter) are added. After exactly 10 minutes, undigested casein is precipitated with TCA (2 ml). The tubes are left at 36° for 10 minutes and filtered through Whatman No. 42 (7 cm) filter paper. The extinction values at 280 nm are read against a blank prepared without trypsin. They are plotted as a function of the total amount of trypsin (in nanomoles) in each tube.

[40] M. L. Bender, M. L. Begue-Canton, R. L. Blakeley, L. J. Brubacher, J. Feder, C. R. Gunter, F. J. Kezdy, J. V. Killheffer, T. H. Marshall, C. G. Miller, R. W. Roeske, and J. K. Stoops, *J. Am. Chem. Soc.* 88, 5890 (1966).

Assay

Activation Step. Solution A (1.5 ml) is warmed to 36°. Five-tenths milliliter of enzyme solution in 0.1 M KH_2PO_4 or the growth medium containing the equivalent of 5–20 ng of the pure enzyme is added to start the reaction.

Trypsin Assay. After exactly 10 minutes, solution B (2 ml) is added and after a further 10 minutes the undigested casein is precipitated by 8.5% TCA (2 ml). After 10 minutes of standing at 36°, the precipitate is filtered through Whatman No. 42 (7 cm) filter paper. The extinction values at 280 nm are read against a blank containing the reagents. The number of nanomoles of trypsinogen activated is obtained from the standard curve.

Definition of Unit and Specific Activity. One unit of penicillopepsin activity is defined as the amount of enzyme required to activate 1 micromole trypsinogen per minute at 25°, pH 3.4, and saturating substrate concentration. Routine assays are carried out, however, at 36° with suboptimal substrate concentrations so that a conversion is required to obtain the defined units. The conversion factor is calculated from the known maximum velocities at 25° and 36° (420 and 685 micromoles trypsinogen activated per micromole penicillopepsin per minute) to obtain the temperature correction. The correction for nonsaturating conditions is calculated from the Michaelis-Menten equation using a value for $K_m = 7.6 \times 10^{-6}\ M$. The combined conversion factor = 0.405. Thus the defined units = (micromoles trypsinogen activated at 36°) $\times$ 0.405. The specific activity is taken as the number of units of activity per milligram of enzyme.

ASSAY WITH BOVINE SERUM ALBUMIN

Reagents

1% Bovine serum albumin (fraction IV, B-grade, Calbiochem) in McIlvaine's citric acid-phosphate buffer[41] at pH 2.6
8.5% Trichloroacetic acid (TCA)

Procedure

The solution of bovine serum albumin (1 ml) is warmed to 36°. The enzyme solution (50 μl) containing the equivalent of 1 to 25 μg of the pure enzyme is added to start the reaction. After exactly 10 minutes, undigested albumin is precipitated with TCA (2 ml). Ten minutes later

[41] R. G. Bates, *in* "Handbook of Biochemistry" (H. Sober, ed.), p. J-197. The Chemical Rubber Co., Cleveland, Ohio, 1968.

the precipitate is filtered (Whatman No. 42, 7 cm). The extinction values at 280 nm are read against a blank prepared without enzyme. The enzyme concentration is read off a standard curve. It may be mentioned that the reading for 5 μg of the purest enzyme preparation is 0.200 OD units.

Purification Procedure

Step 1. Cultivation of Penicillium janthinellum. The medium used for growing the organism is that described previously[8] where stationary cultures were used. However, higher enzyme levels are obtained in a shorter time with submerged cultures.

The medium is made up as follows: 7.2 g sucrose, 3.6 g glucose, 1.23 g $MgSO_4 \cdot 7 H_2O$, 13.62 g KH_2PO_4, 2 g KNO_3, 1.1 g $CaCl_2 \cdot 6 H_2O$, and 1 liter of distilled water. The following trace elements are added to 1 liter of medium to provide for better growth[8]: 1.5 mg $ZnSO_4 \cdot 6 H_2O$, 2.8 mg $Fe(NH_4)_2(SO_4)_2 \cdot 6 H_2O$, 0.32 mg $CuSO_4 \cdot 5 H_2O$, 0.14 mg $MnCl_2 \cdot 4 H_2O$, and 0.08 mg $(NH_4)_2MoO_4$. Usually four 1-liter culture flasks, each containing 400 ml of the complete medium, are used to grow an inoculum for a 10-liter cultivation in a Microferm apparatus (New Brunswick Scientific Co., New Brunswick, N.J.). The medium and apparatus are sterilized by autoclaving at 115 psi for 30 minutes. Separately sterilized lactic acid (10% v/v) is then added to the medium (100 ml per liter) to adjust the pH to approximately 3.0. The culture flasks are seeded with spores from an agar slant.

The original cultures are obtainable from the Commonwealth Mycological Institute, Kew, Surrey, U.K. (strain 75589) or from the Northern Utilization Research Branch, U.S.D.A., Peoria, Ill. (strain 905). These two strains are identical. Another strain (CMI 75588 = NRRL 904) produces lower enzyme levels in stationary culture.[24]

Incubation is carried out on a mechanical shaker operating at 180 cycles/minute and at 25°. When a good growth of viable mycelium has been produced (2–3 days), the contents of the flasks are used to inoculate the medium contained in the Microferm. Rapid growth is attained at 28° with continuous stirring at 200–300 rpm. Vigorous and uninterrupted aeration (4–6 liters/minute) is essential for the production of penicillopepsin.

Maximum enzyme production is usually reached after 4 days. The initial addition of lactic acid is usually sufficient to give a pH between 5.0 and 5.6 when maximum enzyme levels are reached. This pH range provides optimal conditions for adsorbing the enzyme on DEAE-cellulose. If the pH rises too quickly, however, further additions of lactic acid may be required as the enzyme is unstable above pH 6.0. At harvest the culture medium is filtered twice through Whatman No. 1 filter paper in preparation for adsorption on DEAE-cellulose.

Step 2. Adsorption on DEAE-Cellulose. The penicillopepsin contained in the culture medium (about 10 liters) is concentrated by adsorption on DEAE-cellulose (Whatman DE-11). Approximately 45 g of resin, sufficient to pack a column (4×28 cm) with a bed volume of about 400 ml is washed successively with $0.5 M$ NaOH, water, $0.5 M$ HCl, and then equilibrated to pH 5.2 with $0.005 M$ acetate buffer. After the column is packed at $4°$, the whole medium is passed through the bed. The enzyme is almost quantitatively adsorbed. It is eluted with a mixture of equal volumes of $0.2 M$ citrate buffer and $0.2 M$ ammonium sulfate to give a final pH of 3.2. Most of the penicillopepsin activity is eluted in 300–400 ml and is thus concentrated approximately 30 times. Further concentration is achieved with Diaflo apparatus using a UM-1 membrane (Amicon Corp., Cambridge, Mass.) until the volume is reduced to approximately 40 ml.

Step 3. Chromatography on Aminoethyl Cellulose (AE-Cellulose). Before adsorption on AE-cellulose, the concentrated enzyme is dialyzed free of ammonium sulfate. At this stage the preparation still contains high cellulase activity and it is advisable to use double dialysis tubing. When a test of the dialysis water with acidified $BaCl_2$ indicates that it is free of SO_4, the material is equilibrated to pH 3.8 with $0.002 M$ citrate buffer. The AE-cellulose (Whatman AE-11, 25 g), after being washed as described for the DEAE-cellulose, is equilibrated with the same citrate buffer at pH 3.8, and packed into a column at $4°$ (2×40 cm). The enzyme is applied and is adsorbed almost completely. Two elution buffers are then used consecutively. The first, $0.1 M$ citrate, pH 5.0, elutes a protein band essentially free of penicillopepsin. When all this has been removed, the second buffer, $0.25 M$ ammonium sulfate, adjusted to pH 3.6 with H_2SO_4 is passed through the column. This brings out a single protein peak containing the total penicillopepsin activity. The enzyme is concentrated again by Diaflo filtration to a volume of approximately 5–10 ml.

Step 4. Chromatography on Phosphocellulose. The enzyme is again dialyzed free of ammonium sulfate and then equilibrated with $0.005 M$ acetate buffer, pH 3.8. Meanwhile, unused phosphocellulose (Whatman P-11, 2.5 g) is directly equilibrated with the same buffer and packed into a column (0.5×20 cm) in the cold. The penicillopepsin passes through the column while the remaining impurities are retarded. Penicillocarboxypeptidase is the most likely contaminant since it chromatographs closely behind penicillopepsin. Its presence can be detected by assaying with carbobenzoxyglutamyl tyrosine. The purification procedure is summarized in Table I.

TABLE I
SUMMARY OF PURIFICATION PROCEDURE OF PENICILLOPEPSIN

Fraction	Volume (ml)	Protein conc. (mg/ml)	Activity (units/ml)	Specific activity (units/mg protein)	Yield (%)
Medium	10,000	0.416	0.1	0.244	100
DEAE-cellulose eluate	153	0.801	5.0	6.25	76
Diaflo I	41	2.90	17.9	6.2	72
AE-cellulose eluate	45	1.72	15.4	8.99	69.4
Diaflo II	7	8.88	79.2	8.9	57
Phosphocellulose eluate	11.0	2.48	43.7	18.0	48

The pure enzyme can be crystallized directly or precipitated with 60% ammonium sulfate and stored under refrigeration. It can also be freeze dried with little loss of activity.

A successful preparation has also been made in cooperation with the Connaught Medical Research Laboratories in a 1500-liter fermentor using essentially the conditions and methods described[35b] above. The yield was 2.6 g.

Properties

Stability. The stability of purified penicillopepsin has been determined under various conditions of pH and temperature.[35b] It is completely stable at pH 4.9 for 70 hours. The enzyme is most stable toward higher temperatures at pH 4.9, retaining full activity after 1 hour at 55°. In the crystalline state at 20°, it retains full activity for several weeks.

Purity. Penicillopepsin from the phosphocellulose column is homogeneous as shown by moving-boundary electrophoresis at pH 4.25, by electrophoresis in acrylamide at pH 8.6, and in starch gel at pH 3.0, 4.5, 6.0, and 8.6.[8] Boundary analysis according to the Fujita equation of a sedimenting peak in the Model E Ultracentrifuge also indicates homogeneity.[8]

Molecular Properties

The enzyme has the following physical constants[8]: molecular weight $= 32,000$; sedimentation constant $s_{20,w} = 3.1 \times 10^{-13}$ sec; diffusion constant $D_{20,w} = 8.4 \ (\pm 0.23) \times 10^{-7}$ cm^2 sec^{-1}; partial specific volume $\bar{V} = 0.718$ (calculated from the amino acid composition); electrophoretic mobility (in 0.22 M acetate, pH 4.25) is $\mu = -4.47 \pm 0.02$ cm$^2 V^{-1}$ sec^{-1};

0.5 mm

FIG. 1. Crystals of penicillopepsin from $(NH_4)_2SO_4$ (0.3 saturated) pH 4.4. Reproduced by permission of the National Research Council of Canada from the Canadian *Journal of Biochemistry* **48**, 430 (1970).

isoelectric point below pH 3.8; molecular extinction at 280 nm is $\epsilon_m = 43,200$; specific extinction $E_{1\,cm}^{1\,mg/ml} = 1.35$.

Amino Acid Composition.[8] Lys_5, His_3, Asp_{36}, Thr_{28}, Ser_{42}, Glu_{28}, Pro_{12}, Gly_{39}, Ala_{23}, Val_{22}, Ile_{12-13}, Leu_{20}, Tyr_{14}, Phe_{19}, Trp_{4-5}, $(NH_2)_{43}$. There is no arginine, half-cystine, or methionine. Alanine is both N-terminal and C-terminal.[35b] An essential aspartic acid residue is at the active site.[35, 35a]

Crystals.[42] The enzyme crystallizes from ammonium sulfate (0.3 saturated) at pH values between 3.5 and 6. Crystals for X-ray are obtained at pH 4.4 and are shown in Fig. 1. The crystals are monoclinic, space group C2. The unit cell dimensions are $a = 97.81$ Å, $b = 46.73$ Å, $c = 65.69$ Å, and $\beta = 115.55°$. The unit cell contains 4 molecules.

Enzymatic Properties

Inhibitors. Penicillium janthinellum pepsin is inhibited competitively by the pepsin and carboxypeptidase A substrates carbobenzoxyglutamyl

---

[42] N. Camerman, T. Hofmann, S. Jones, and S. C. Nyburg, *J. Mol. Biol.* **44**, 569 (1969).

tyrosine (K_i app. $5 \times 10^{-5} M$), carbobenzoxyglutaminyl tyrosine, carbo-benzoxyvalyl tyrosine, and also by carbobenzoxyglutamic acid, but not by carbobenzoxytyrosine and carbobenzoxyphenylalanine.[8]

The pepsin inhibitors diazoacetyl norleucine methyl ester and diazo-acetyl phenylalanine methyl ester in the presence of Cu^{2+} inhibit the enzyme by reacting covalently with an aspartic acid residue.[35, 42a] Another pepsin inhibitor, p-bromophenacylbromide does not inhibit. Millimolar concentrations of diisopropylphosphofluoridate, p-hydroxymercuriben-zenesulfonic acid, iodoacetic acid, ethylenediaminetetraacetic acid, F^{1-}, Fe^{3+}, Cu^{2+}, and Mn^{2+} have no effect on the enzyme.

Specificity. Penicillopepsin shows a specificity similar to pepsin toward the B-chain of S-sulfo insulin. It cleaves about 15% of the bonds in native bovine serum albumin.[8] It does not act significantly on Congo red elastin, polyglutamic acid, polyalanine, and polyglycine; polylysine is slowly attacked, and the copolymer of glutamic acid, tyrosine, and lysine is readily hydrolyzed.[43]

It does not hydrolyze any of some fifty N-substituted and nonsub-stituted di- and tripeptides, amino acid esters, and amino acid amides[8]; among them are the commonly used, commercially available proteinase and peptidase substrates. It is especially notable that it has no effect on carbobenzoxyhistidylphenylalanyl tryptophan ethyl ester,[43] a recently described excellent pepsin substrate.[20a, 44] The enzyme appears to have a high specificity with regard to size of the peptide, but relatively low specificity for the bonds involved.

The most specific substrates are trypsinogens from various species. Bovine trypsinogen is activated rapidly with the specific cleavage of the bond lysine-6-isoleucine-7. Chymotrypsinogen A and proelastase are not activated.

pH Optimum. Optimal trypsinogen activation is obtained at pH 3.4[8]; the pH optimum of hydrolysis of bovine serum albumin is 2.6.[25]

Kinetic Constants. K_m for trypsinogen activation is $(7.6 \pm 2) \times 10^{-6} M$ at $0°$, pH 3.4.[8] The molecular activities (turnover numbers) for trypsinogen activation are listed in Table II. It should be noted that these measure specifically the cleavage of the activating peptide bond.

[42a] Other enzymes inhibited by diazoketones are pepsin, cathepsin D, an enzyme from *Aspergillus awamori* and thyroid acid proteinase. [T. G. Rajagopalan, W. H. Stein, and S. Moore, *J. Biol. Chem.* **241**, 4925 (1966); V. M. Stepanov, V. N. Orekhovich, L. S. Lobareva, and T. I. Mzhel'skaya, *Biokhimiya* **34**, 209 (1969); V. M. Stepanov, L. S. Lobareva, N. F. Frolova, and L. I. Oreshenko, *Izv. Akad. Nauk SSSR, Ser. Khim.* **12**, 2840 (1968); G. D. Smith, M. A. Murray, L. W. Nichol, and V. M. Trikojus, *Biochim. Biophys. Acta* **171**, 288 (1969); respectively.]

[43] G. Mains and T. Hofmann, unpublished observations.

[44] K. Inouye, I. M. Voynick, G. R. Delpierre, and J. S. Fruton, *Biochemistry* **5**, 2473 (1966).

TABLE II
MOLECULAR ACTIVITIES OF PENICILLOPEPSIN[a]

Temperature (°C)	Trypsinogen activated (micromoles/minute/micromole enzyme)
0	64
10	105
15	180
20	260
25	420
30	540
35	660

[a] Conditions: 0.01 M sodium citrate, pH 3.4; trypsinogen $= 10^{-4}\,M$.

Penicillocarboxypeptidase, (formerly Peptidase B)[45] from *Penicillium janthinellum*

This enzyme has been described by Shaw.[9] It is obtained in low yield as a byproduct at the last stage of the purification of penicillopepsin described in the preceding section.

Assay Method

Principle. The enzyme hydrolyzes carbobenzoxyglutamyl tyrosine and similar compounds. The hydrolysis can be followed by a ninhydrin procedure.[46]

Reagents

Substrate. Carbobenzoxy-L-glutamyl-L-tyrosine $(2 \times 10^{-3}\,M)$ in 0.02 M sodium acetate, pH 4.7.

Sodium acetate buffer, 4 M, pH 5.3, containing $2 \times 10^{-4}\,M$ NaCN. (Prepared fresh daily from 4.08 M Na-acetate and $10^{-2}\,M$ NaCN.)

3% Ninhydrin in Methyl Cellosolve

Isopropanol–water, 1:1

Procedure. Equal volumes (0.5 ml) of the enzyme in distilled water or very dilute buffers and the substrate are incubated at 35° for 10 minutes. Ninhydrin (0.25 ml) and cyanide–acetate buffer (0.25 ml) are added. The mixture is heated to 100° for 20 minutes. After adding 2.5 ml isopropanol–water the extinction is read at 570 nm.

[45] In order to avoid confusion with the enzyme aspergillopeptidase B which is a DFP-inhibited alkaline endopeptidase and because of its enzymatic properties, which are those of a carboxypeptidase, the term peptidase B will be dropped.

[46] H. Rosen, *Arch. Biochem. Biophys.* **60**, 10 (1957).

Definition of Units. One unit is defined as the amount of enzyme which liberates 1 micromole tyrosine per minute under the standard assay conditions.

Purification Procedure

Penicillocarboxypeptidase is eluted from the phosphocellulose column (last stage of penicillopepsin) by sodium acetate (0.01 M, pH 4.0), as a single peak in about 6% yield as measured from the second purification stage (chromatography on DEAE-cellulose, see Table I).

Properties

Penicillocarboxypeptidase is homogeneous as judged by moving-boundary and paper electrophoresis.[9] The optimum for hydrolysis of carbobenzoxyglutamyl tyrosine is pH 4.7. K_m is $5 \times 10^{-4} M$.

Inhibitors. The enzyme is inhibited by Fe^{3+}, to a small extent by Ba^{2+} and Cd^{2+}, but not by Ca^{2+}, Zn^{2+}, and Mg^{2+} (all $5 \times 10^{-3} M$). Fatty acids, most di- and tricarboxylic acids and divalent inorganic anions show inhibition at a concentration of $5 \times 10^{-2} M$. The inhibition by aliphatic monocarboxylic acids increases with increasing chain lengths. It is about 40% with butyric acid ($10^{-2} M$) and approaches 100% with heptanoic and octanoic acid ($10^{-2} M$). It is not affected by thiol reagents ($10^{-3} M$).

Specificity. The enzyme is a carboxypeptidase with limited specificity as shown in Table III. Most notable is the absence of activity on a substrate with C-terminal arginine and on substrates with glycine as the residue next to the C-terminal. It does not act on bovine serum albumin.[8]

TABLE III
ACTION OF PENICILLOCARBOXYPEPTIDASE ON PEPTIDES[a]

Cbz-Glu-Tyr	100	Leu-Tyr	17
Cbz-Glu-Phe	105	Gly-Phe	0
Cbz-Tyr-Glu	72	Gly-Leu	0
Cbz-Glu(NH₂)-Tyr	55	Ala-Gly	0
Cbz-Val-Tyr	48	Leu-Gly-Gly	0
Cbz-Tyr-Glu(OMe)[b]	33	Gly-Tyr	0
Tosyl-Lys-Gly	17	Glutathione	0
Cbz-Phe-Ser	12	Glu-Asp	0
Tosyl-Arg-Gly	8	Gly-His	0
Cbz-Gly-Phe	0	Gly-Asp(NH₂)	0
Cbz-Gly-Glu	0		
ClAc-Gly-Leu	0		
Cbz-Leu-Arg	0		

[a] Values are relative to that for Cbz·Glu·Tyr(= 100).
[b] γ-Methyl ester.

Trypsinogen-Kinase from *Aspergillus oryzae* (Takadiastase)[10]

Assay Method

The assay of this enzyme using trypsinogen activation is carried out as described for penicillopepsin (p. 376). As an alternative, casein digestion at pH 3.0 can be used.

Purification Procedure

The enzyme is prepared from commercially available Takadiastase by extraction with water (1500 ml per 30 g powder). After adjustment of the extract to pH 3.5, the enzyme is adsorbed on a Duolite CS 101 column (column volume 300 ml) at pH 3.5. The enzyme is eluted with $0.5 M$ sodium acetate (355 ml) and precipitated with $(NH_4)_2SO_4$. The fraction between 0.6 and 0.8 saturation contains the enzyme. After dialysis, the enzyme is precipitated with ethanol between 50 and 60%.

Properties

The enzyme gives essentially one peak on paper electrophoresis at pH 5.5 and 7.4. It hydrolyzes casein and, in contrast to penicillopepsin, a number of small peptides with a low degree of specificity[47]: carbobenzoxyglutamyl tyrosine, carbobenzoxyleucyl arginine, glutathione, L-leucyl-L-arginine, benzoyl-L-arginyl-L-leucine, and L-phenylalanyl-L-arginine. However, the turnover numbers for these substrates are very low and incubation times of 24 hours are used to measure their hydrolysis. As mentioned in the introduction, this activity could be due to contamination by an enzyme like penicillocarboxypeptidase (p. 385).

Similar Enzymes

Bergkvist[48] has partially purified an acid proteinase ("proteinase III") from a strain of *Aspergillus oryzae*. Two types of acid proteinases were isolated from *Aspergillus niger* var. *macrosporus*.[49] It is not known whether these enzymes activate trypsinogen. Aspergillopeptidase A from *Aspergillus saitoi* is described elsewhere in this volume (article [26]).

Candida albicans Acid Proteinase

The enzyme is purified from the culture medium of *Candida albicans* strain C.B.S. 2730.[11]

[47] K. Nakanishi, *J. Biochem.* (*Tokyo*) **47**, 16 (1960).
[48] R. Bergkvist, *Acta Chem. Scand.* **17**, 1541 (1963).
[49] Y. Koaze, H. Goi, K. Ezawa, Y. Yamada, and T. Hara, *Agr. Biol. Chem.* **28**, 216 (1964).

Assay Method

Principle. The acid proteinase activity is determined by digestion of bovine serum albumin at pH 3.2. The extent of hydrolysis is derived from the amount of TCA-soluble material remaining after digestion.

Reagents

1% w/v Bovine serum albumin in 0.05 M sodium citrate buffer, pH 3.2

5% w/v Trichloroacetic acid

Procedure. Enzyme samples (0.5 ml) are incubated at 37° with 2.0 ml of the 1% bovine serum albumin substrate (pH 3.2) for 60 minutes. The reaction is terminated by the addition of 5% TCA (5 ml) and the precipitated albumin removed by centrifugation. Two milliliters of the supernatant is taken to determine the amount of TCA-soluble products by the method of Lowry *et al.*[50]

Specific Activity. The unit of proteinase activity has not been defined and thus specific activity is reported in arbitrary units.

Purification Procedure

Step 1. Cultivation. The culture medium used to provide good proteinase production consists of the following; 0.5 g $MgSO_4$, 1 g KH_2PO_4, 20 g glucose, 2 g human or bovine serum albumin, and 1.25 ml Protovita (polyvitamin preparation, Hoffmann-La Roche Co.) in 1 liter of water. The pH is adjusted to 4.0 and the total medium sterilized by filtration. The medium is inoculated to contain 500 viable cells per milliliter and incubated with shaking at 26° for 3–5 days. During this time the pH usually decreases to 3.0–3.2. In order to harvest, the medium containing the enzyme is filtered to remove cellular material.

Step 2. Concentration. The medium containing 600 mg protein per liter is concentrated to a workable volume by pervaporation.[51] The volume is reduced from 1 liter to approximately 40 ml.

Step 3. Gel Filtration. The concentrated protein is applied to a column of Sephadex G-75 (3 × 100 cm) equilibrated to pH 4.0 with 0.05 M sodium citrate buffer. Three major protein peaks are eluted after 380 ml, 450 ml, and 560 ml. The second of these contains the acid proteinase activity. It is pooled (approx. 70 ml) and dialyzed against 0.01 M sodium citrate buffer (pH 6.5).

[50] O. H. Lowry, N. J. Rosebrough, A. L. Farr, and R. J. Randall, *J. Biol. Chem.* **193**, 265 (1951).

[51] This step is probably replaceable by a step using a Diaflo (Amicon Corp.) apparatus.

Step 4. Chromatography on DEAE-Sephadex A-25. The dialyzate is applied to a DEAE-Sephadex A-25 column (0.8 × 10 cm). Elution with the equilibration buffer removes a peak with no activity. After approximately 15 ml, a linear gradient from 0.01 M sodium citrate to 0.2 M sodium citrate using a gradient volume of 100 ml is applied. This elutes the enzyme as two partially separated peaks.

Rechromatography of each of the peaks under the same conditions produces in each case a major protein peak and a minor peak with identical specific activities and pH optima. They appear to be two very similar forms of the enzyme and can be separated by acrylamide gel electrophoresis.

The purification is summarized in Table IV.

TABLE IV
PURIFICATION OF *Candida albicans* ACID PROTEINASE

Fraction	Protein concentration (mg/ml)	Volume (ml)	Specific activity (arbitrary units)
Culture medium	0.6	1000	1
Conc. medium	8.8	39	1.4
Sephadex G-75 eluate	0.93	102	4.1
DEAE-Sephadex eluate	0.27	7.9	140

Properties

Stability. Gradual loss of activity occurs on standing in solution in the cold or when frozen. Activity is rapidly lost on repeated freezing and thawing and lyophilization.

Purity. The two acid proteinases appear homogeneous on acrylamide disc electrophoresis at pH 4.5 and pH 8.0 and give a single peak in the ultracentrifuge.

Molecular Properties. Molecular weight determined by gel filtration (Sephadex G-75 and 100): 42,000; sedimentation constant: $s_{20, w} = 3.5 \times 10^{-13}$ seconds (density gradient) or 3.2×10^{-13} seconds (sedimentation velocity).

Enzymatic Properties. pH optimum: for bovine serum albumin: pH 3.2.

Inhibition. Millimolar concentrations of EDTA, mercaptoethanol, $CaCl_2$, $MgCl_2$, $MnCl_2$, $(NH_4)_2SO_4$, p-chloromercuribenzoate, kallikrein (equal weight to enzyme) do not affect the activity.

The two pepsin inhibitors p-bromophenacylbromide and diazoacetylnorleucine methyl ester were without effect.

Specificity. The *Candida albicans* acid proteinase cleaves 9% of the peptide bonds in bovine serum albumin. Digestion of the B-chains of insulin indicates a preference for peptide bonds involving the carboxyl group of amino acids with hydrophobic side chains. However, there are no indications of any strict specificity.

Paecilomyces varioti Acid Proteinase[12]

Assay Procedure

The enzyme is assayed with a 0.6% casein solution in Sorensen buffer[41] at pH 3.0, 40°. Undigested casein is precipitated with trichloroacetic acid (0.2 M final concentration) and centrifuged. Digested casein in the supernatant is determined with the Folin-Ciocalteau reagent.[52]

Definition of Units. One unit is equal to the amount of enzyme which liberates the equivalent of 1 microgram of tyrosine per minute.

Purification Procedure[12]

The organism, *Paecilomyces varioti* Bainier TRR-220, is grown on wheat bran (10 kg) and rice husks (3 kg) mixed with 5 liters water at 30°. After 60 hours the culture is extracted with 50 liters water. After centrifugation the enzyme is precipitated from the supernatant with 2 volumes ethanol precooled to —25°. The precipitate is dried *in vacuo*. It is extracted with water (1 liter for 200 g) and centrifuged. The residue is washed with 200 ml water. Both supernatants (950 ml) are adjusted to pH 4.0 with conc. HCl; cold ethanol (475 ml) is added slowly so that the temperature is maintained below 0°. Cornstarch (300 g) is stirred in at —5° and removed after 1 hour by centrifugation. The ethanol concentration of the supernatant is increased to 40% and the cornstarch treatment repeated. (This treatment serves to remove amylase.) After removal of the starch, the ethanol concentration is raised to 65%. The precipitate is centrifuged off and dissolved in 0.1 M sodium acetate buffer, pH 3.0 (250 ml). Insoluble material is centrifuged off and discarded. The superntant (240 ml) is adjusted to pH 4.0; calcium acetate (1 M, 80 ml) is added and a precipitate removed. The supernatant is dialyzed against running water, centrifuged, and brought to 40% saturation with ammonium sulfate at pH 5.3. After removing a precipitate, the ammonium sulfate concentration is increased to 85% saturation and the pH adjusted to 3.5. The precipitate is centrifuged off and dissolved in 0.005 M KH$_2$PO$_4$ (pH 5.0, 50 ml). Ammonium sulfate is removed on a Sephadex G-25 column (3 × 70 cm) equilibrated with 0.005 M KH$_2$PO$_4$.

[52] M. L. Anson, *J. Gen. Physiol.* **20**, 561 (1937); *ibid.*, **22**, 79 (1938).

The fraction containing the enzyme is adjusted to pH 3.0, a small amount of precipitate is removed, and after adjusting to pH 4.0 cold acetone is added to 60%. The precipitate is removed, dissolved in 0.005 M KH_2PO_4 at pH 3.5 to give a 2% protein concentration and centrifuged to clarify. After adjusting to pH 4.0 the enzyme is crystallized by adding acetone to 50%. On recrystallization 960 mg of the proteinase crystals are obtained.

The purification is summarized in Table V.

TABLE V
SUMMARY OF PURIFICATION OF *Paecilomyces* PROTEINASE

Fraction	Specific activity (units/mg protein)	Yield (%)
Crude extract	378	100
First starch treatment	670	88.5
Second starch treatment	810	84.0
Ethanol precipitate	1100	82.4
85% Saturated $(NH_4)_2SO_4$ precipitate	1460	51.9
60% Acetone precipitate	1600	46.2
First crystals	1650	19.2
Second crystals	1940	14.3

Properties

Stability. The enzyme is stable between pH 3 and 6.5. It is most stable to elevated temperatures at pH 5.0.

Purity. The crystalline enzyme gives a single sedimenting boundary in the ultracentrifuge and single bands on continuous paper electrophoresis at pH 7.35 and 8.25.

Molecular Properties.[53] Molecular weight 37,300; sedimentation constant at 15°, $S_{15}^{\circ} = 2.75 \times 10^{-13}$ seconds; isoelectric point 3.8; amino acid composition: Lys_{28}, His_7, Arg_3, Asp_{45}, Thr_{27}, Ser_{39}, Glu_{23}, Pro_{20}, Gly_{33}, Ala_{26}, $CySH_1$, Val_{18}, Met_1, Ile_{15}, Leu_{21}, Tyr_{16}, Phe_{14}, Trp_3. There is evidence for the presence of one SH group. Leucine or isoleucine is N-terminal, alanine C-terminal.

Enzymatic Properties. pH optimum: This is 3.0 for the digestion of casein, ovalbumin, and fibrin; between 3.0 and 4.5 for hemoglobin[54]; 3.5 for carbobenzoxy-L-glutamyl-L-tyrosine and 5.5 for benzoyl-L-arginine amide.

Inhibitors. The enzyme is inhibited by $5 \times 10^{-3} M$ Hg^{2+} but not by other metal ions. It is partially inhibited by p-hydroxymercury benzoic acid and iodoacetic acid, and by bacitracin, tetracycline, and nitrofuryl-acrylamide. The effect of many other compounds on the activity is also

[53] J. Sawada, *Agr. Biol. Chem.* **30**, 393 (1966).
[54] J. Sawada, *Agr. Biol. Chem.* **28**, 348 (1964).

given by Sawada.[54] Many divalent metal ions increase the heat stability of the enzyme.

Specificity. The enzyme readily digests casein and hydrolyzes some peptides, notably, carbobenzoxy-L-glutamyl-L-tyrosine (I) and benzoyl-L-arginine amide (II). Approximate turnover numbers calculated from Sawada's paper[21] are 0.05 to 0.1 micromoles substrate per minute per micromole enzyme. As with the *Aspergillus oryzae* enzyme (this volume [26]) it is possible that the hydrolysis of these small substrates is due to contaminating enzymes.

Kinetic Constants. K_m for (I) $= 4.1 \times 10^{-3} M$; for (II) $= 6.6 \times 10^{-3} M$.

Acid Proteinase from *Rhizopus chinensis*

The acid proteinase has been purified by Fukumoto *et al.*[13] from the culture medium of *Rhizopus chinensis* Saito. It is also available commercially as a $3 \times$ crystallized preparation from Miles Laboratories, Elkhart, Ind. and from Seikagaku Kogyo Ltd., Tokyo, Japan.

Assay Method

Principle. The activity of the enzyme is determined using casein at pH 3.1 as a substrate.[13, 55] Alternatively, the enzyme may be assayed by the trypsinogen-kinase assay described on page 376, by digestion of bovine serum albumin (p. 378) or by a milk-clotting assay.

Reagents

> 0.6% Casein in $5 \times 10^{-2} M$ lactic acid, pH 3.1
> Trichloroacetic acid (TCA) precipitant: $0.22 M$ TCA–$0.33 M$ acetic acid–$0.11 M$ sodium acetate

Procedure. The enzyme solution (1 ml) is added to the casein (5 ml) at 30°. After 10 minutes, 1 ml TCA precipitant is added. The precipitate is filtered off after 30 minutes at 30° and the extinction at 275 nm of the filtrate measured.

Definition of Units. One unit is defined as the amount of enzyme which liberates extinction units equivalent to 1 μg tyrosine per milliliter per minute. Specific activity is defined as the enzyme units per milligram protein.

Purification Procedure

Method of Fukumoto et al.[13] *Rhizopus chinensis* Saito is cultured by the Koji method. The culture medium, which consists of 1 kg of wheat bran (Senkan) and 100 g of cornstarch in 1 liter of tap water, is sterilized

---

[55] D. Tsuru, T. Yamamoto, and J. Fukumoto, *Agr. Biol. Chem.* **30**, 651 (1966).

at 120° for 30 minutes. The mixture is spread on a pan, inoculated with spores, and left under a cover at 25° for 70 hours.

The medium containing the enzyme is extracted with 8 liters of deionized water, pH 5.8, with occasional stirring for 1.5 hours at 30°. Insoluble material is removed with a filter press and benzalkonium chloride added to give a concentration of 0.2% at pH 5.2. After standing overnight, a precipitate forms which is removed by filtration through a layer of Celite. The filtrate is passed through a column of Asmit 173 N (4×70 cm) equilibrated to pH 5.2 with $10^{-2} M$ acetate buffer to remove colored impurities. The total effluent, including 500 ml equilibration buffer, is collected.

Ammonium sulfate is added to the combined effluent to 70% saturation. After standing for 5 hours, the precipitate is collected by centrifugation and dissolved in $10^{-2} M$ acetate buffer (500 ml), pH 5.4. The presence of cellulase in this solution prevents dialysis. Sulfate ions are therefore precipitated with barium acetate and after adjusting the pH to 5.4, the barium sulfate precipitate is centrifuged off. The supernatant is diluted with deionized water until the concentration of ammonium acetate is $5 \times 10^{-2} M$.

The proteinase is next adsorbed on a Duolite A-2 column (4×70 cm), equilibrated with ammonium acetate at pH 5.4. The cellulase, glucoamylase, and some hemicellulases pass through. One liter of equilibration buffer is used to wash the column before the proteinase is eluted with $0.3 N$ acetic acid (1.5 liters). The proteinase emerges when the pH drops below 4.2.

Ammonium sulfate is added to 70% saturation. The precipitate formed overnight is centrifuged off and dissolved in $10^{-2} M$ acetate buffer (30 ml), pH 4.5 and applied to a Sephadex G-100 column (2.8×80 cm) equilibrated with the same buffer. The column is operated at a flow rate of 30 ml/hour and the enzymatic fraction (150 ml) collected. The desalted material is adsorbed on a CM-cellulose column (2.3×70 cm) equilibrated with $10^{-2} M$ acetate buffer, pH 4.5. After washing the column with 200 ml of the equilibration buffer, the acid proteinase is eluted with a linear sodium chloride gradient (0 to $0.4 M$, prepared from 1 liter buffer and 1 liter buffer–$0.4 M$ sodium chloride) at a flow rate of 30 ml/hr. The proteinase activity emerges as a single peak at $0.2 M$ sodium chloride.

The combined fractions are brought to 75% saturation with ammonium sulfate; the precipitate is collected by centrifugation and dissolved in $10^{-2} M$ acetate buffer, pH 4.2 (170 ml). The acid proteinase can be crystallized at this stage by dialyzing against the acetate buffer containing $2 \times 10^{-3} M$ calcium acetate and stirring for 24 hours. Acetone is added until the material becomes slightly cloudy. After leaving at 0°, needle crystals appear in 10 minutes and additional drops of acetone will crystallize out most of the remaining enzyme within 24 hours.

The crystals are collected by centrifugation, washed twice with small quantities of $2 \times 10^{-3} M$ calcium chloride solution, pH 5.0, and then dissolved by the dropwise addition of 0.2 N ammonium hydroxide, bringing the pH to 7.0. Any insoluble material is removed by centrifugation and the supernatant adjusted to pH 4.8–5.4 with 0.2 N acetic acid before dialyzing against $10^{-3} M$ calcium chloride solution, pH 5.2. Pillar-shaped crystals form 1 or 2 days after dialysis. Dropwise addition of acetone to a final concentration of 20% completes the crystallization.

The purification is summarized in Table VI.

TABLE VI

PURIFICATION OF *Rhizopus chinensis* ACID PROTEINASE

Fraction	Volume (ml)	Total activity $\times 10^{-4}$ (units)	Specific activity (units/mg)	Yield (%)
Crude extract	8000	344	—	100
Benzalkonium chloride treatment	8000	320	50	93
Asmit 173-N eluate[a]	8500	290	90	84.4
Barium acetate treatment	6000	270	230	78.5
Duolite A-2 eluate[a]	900	130	615	37.8
Sephadex G-100 eluate	150	115	1140	33.4
CM-cellulose eluate[a]	170	89	1840	25.9
Acetate buffer dialyzate	35	70	2140	20.4
Acetone crystals	—	65	2200	18.9
Recrystallization	—	62	2200	18.0
Lyophilized product	—	60	2200	17.4

[a] After concentration by ammonium sulfate precipitation.

Purification of Commercial Enzyme.[56] The acid proteinase is available as a 3 × crystallized product from Miles Laboratories. It is prepared in a manner similar to that described in the preceding paragraphs.[57] The preparation shows three major components on both acrylamide disc electrophoresis at pH 9.0 (Fig. 2; A,0) and analytical isoelectric focusing[58,59] (Fig. 2; B,0). It can be purified by isoelectric focusing on a preparative scale by the method of Svensson.[60]

[56] J. Šodek and T. Hofmann, to be published.
[57] D. Tsuru, personal communication.
[58] H. Svensson, *Acta Chem. Scand.* 15, 325 (1961).
[59] A. Dale, and A. L. Latner, *Lancet,* p. 847 (1968).
[60] H. Svensson, *Arch. Biochem. Biophys. Suppl.* I, 132 (1962).

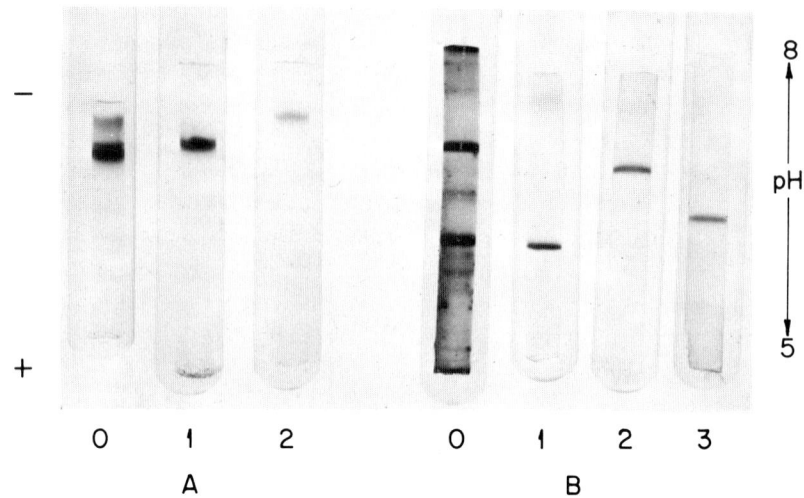

FIG. 2. Electrophoresis of commercial *Rhizopus chinensis* acid proteinase on acrylamide gel, pH 9.0 (A) and by analytical electrofocusing (B). (0) Starting material (3 × crystallized preparation) (1–3) Peaks I, II, and III, respectively, after purification by preparative electrofocusing.

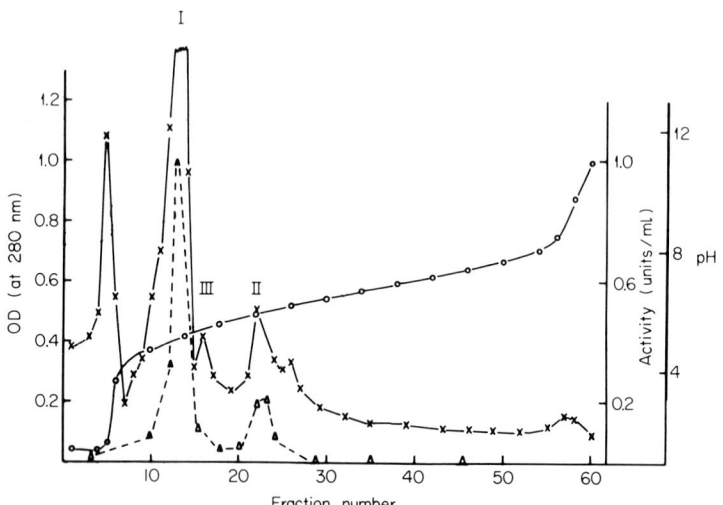

FIG. 3. Electrofocusing of commercial *Rhizopus chinensis* acid proteinase on a preparative LKB electrofocusing column 8101 (110 ml). Fractions 1.5 ml. (○——○) pH, (×——×) OD at 280 nm, (△——△) trypsinogen-activating activity.

The preparative method is carried out on an LKB electrofocusing column (No. 8101) using 2% carrier ampholyte (pH range 5–8) in a sucrose gradient. The freeze-dried Miles preparation is dissolved in distilled water (approximately 20 mg in 2 ml of water) and dialyzed against distilled water overnight. It is added to those sucrose solutions that will form the midregion of the gradient to prevent the protein from being exposed to either anode or cathode buffer solution. A constant voltage (500 V) is applied over a period of 24 hours. During this time the current passing through the column drops to zero as the pH gradient is formed and the proteins focus at their respective isoelectric points. The sucrose gradient is eluted in 1.5 ml fractions; the pH, optical density at 280 nm, and the activity are determined (for a typical run see Fig. 3).

One peak is eluted from the acidic region of the gradient; two major peaks with isoelectric point 5.2 (peak I) and isoelectric point 6.0 (peak II) and a minor peak (III, isoelectric point = 5.45) are eluted subsequently. Peak I and peak II have comparable specific activities using both bovine serum albumin and trypsinogen as substrates. Peak III and the other, unlabeled peaks appear to be inactive in both assays.

Stability.[13] *Rhizopus chinensis* acid proteinase is stable for 24 hours between pH 2.8 and 6.5 at 30°. At pH 3.1, the enzyme retains full activity after 15 minutes of exposure to temperatures up to 60°.

Purity. Fukumoto Preparation. Ultracentrifugation and behavior on elution from sulfoethyl Sephadex suggest homogeneity of the enzyme.[13] On disc electrophoresis, isoelectric focusing, and moving-boundary electrophoresis, a minor component accounting for less than 5% of the total protein is occasionally observed.[57]

Purified Commercial Preparation. Peak I and peak II appear as a single band on acrylamide disc electrophoresis (Fig. 2; A, 1 and 2) and on analytical isoelectric focusing (Fig. 2; B, 1 and 2).

Molecular Properties

The enzyme has the following physical constants[61]: molecular weight = 35,000; sedimentation constant, $S_{20,w} = 2.83 \times 10^{-13}$ seconds; partial specific volume, $\bar{V} = 0.73$ (calculated from the amino acid composition); isoelectric point, 5.2; specific extinction, $E_{1\,cm}^{1\,mg/ml} = 1.29$. The optical rotary dispersion parameters are $[\alpha]_D = -33$, $a_o = -200$, $b_o = 0$, $\lambda_c = 208$ nm and $[m']$ 227 nm $= -2.100$, suggesting the absence of α-helix structure in the molecule.

Peak I from the commercial preparation is identical in molecular weight, partial specific volume, isoelectric point, and specific extinction.

Amino Acid Composition. The amino acid composition of the *Rhizopus*

TABLE VII

AMINO ACID COMPOSITIONS OF *Rhizopus chinensis* ACID PROTEINASE
FUKUMOTO AND PURIFIED "MILES" PREPARATIONS

	Rhizopus chinensis[a] residues[c]	Miles P I[b] residues	P II[b] residues
Lysine	12	13	14
Histidine	0	0	0
Arginine	9	9	8
Aspartic acid	43	43	40
Threonine	32	30	28
Serine	26	27	30
Glutamic acid	21	21	21.5
Proline	14	13	14
Glycine	39	46	46
Alanine	23	22	22
Valine	19	20	21
Methionine	3	3	1.4
Isoleucine	21	21	20
Leucine	23	21	21
Tyrosine	14	15	12
Phenylalanine	16	17	16
Tryptophan	5	d	d
Half-cystine	4	4	d
Amide NH$_2$	32	d	d

[a] D. Tsuru, A. Hattori, H. Tsuji, and J. Fukumoto, *J. Biochem. (Tokyo)*, **67**, 415 (1970).
[b] J. Šodek and T. Hofmann, to be published.
[c] Number of residues per molecule (MW = 35,000).
[d] Not determined.

acid proteinase[61] is listed in Table VII. Also listed for comparison are the composition for Miles peak I and peak II.[56] A noticeable feature in all three is the absence of histidine. In spite of the clear electrophoretic difference between peaks I and II, there are only minor differences in composition between them. Peak I is probably identical to the Fukumoto enzyme. Alanine is the N-terminal amino acid.

Enzymatic Properties

Inhibition. Millimolar concentrations of EDTA, *o*-phenanthroline, CuSO$_4$, HgCl$_2$, FeSO$_4$, Pb-acetate, monoiodoacetic acid, *p*-chloromercuribenzoate, and ascorbic acid have no appreciable affect on the activity.[13] Ferric sulfate, 2.0 mM, inhibits approximately 75%.[13]

The pepsin inhibitor, diazoacetyl norleucine methyl ester in the

[61] D. Tsuru, A. Hattori, H. Tsuji, and J. Fukumoto, *J. Biochem. (Tokyo)*, **67**, 415 (1970).

presence of Cu^{2+}, completely inhibits activity of the Miles peak I and peak II when assayed with serum albumin or with trypsinogen as substrate.[56] In each case, 1.3 and 2.0 residues, respectively, of norleucine per molecule are incorporated.

Specificity. The digestion of the B-chain of oxidized beef insulin by *Rhizopus chinensis* acid proteinase is similar to that obtained by pepsin.[62] During the digestion of casein about 15% of the peptide bonds are broken.

The enzyme acts on polyglutamic acid and polylysine.[62] The following di- and tripeptides are hydrolyzed[62] (after incubation for 18 hours at 37°): lysyl-tyrosyl-glutamic acid, carbobenzoxyglutamyl-tyrosine, carbo-benzoxyglutamyl-phenylalanine, carbobenzoxyphenylalanyl-tyrosine, car-bobenzoxy-glycyl-phenylalanine, glycyl-leucyl-tyrosine, glycyl-glutamyl-tyrosine, leucyl-glycyl-phenylalanine, glutamyl-glycyl-phenylalanine, leucyl-tyrosine, glutamyl-tyrosine, and leucyl-phenylalanine. Except for the first substrate, where both peptide bonds are cleaved and transpepti-dation occurs, a C-terminal aromatic amino acid is liberated in all cases. Other di- and tripeptides lacking the C-terminal aromatic amino acid are not hydrolyzed.[62] The enzyme also has milk-clotting activity.

pH Optimum. The pH optimum of casein digestion is between 2.9 and 3.3.[13]

Similar Enzymes. A milk-clotting acid proteinase from *Rhizopus oligosporus* (NRRL-3271) has been partially purified.[63]

Acknowledgments

The authors are most grateful to Dr. Tsuru for making available unpublished observations. Experimental work by the authors reported here was supported by the Medical Research Council (Canada) Grants MT-1982 and MA-2438. J. S. is the holder of an M.R.C. Studentship.

[62] D. Tsuru, A. Hattori, H. Tsuji, T. Yamamoto, and J. Fukumoto, *Agr. Biol. Chem.* **33**, 1419 (1969).
[63] H. L. Wang, D. I. Ruttle, and C. W. Hesseltine, *Can. J. Microbiol.* **15**, 99 (1969).

[26] Purification and Mode of Assay for Acid Proteinase of *Aspergillus saitoi*

By Eiji Ichishima

Assay Method

Principle. Methods developed by Anson, Kunitz, and Hagihara[1] are modified for estimating the extracellular acid proteinase of *Aspergillus saitoi* (EC 3.4.4.17, aspergillopeptidase A). A convenient procedure is based on the determination of split products, soluble in 0.4 M trichloroacetic acid, of a standard protein. The optimal pH for milk casein digestion is in the range of 2.5 to 3.0.

Reagents

HCl-CH$_3$COONa buffer, 0.1 M, pH 2.7

Two percent milk casein solution at pH 2.7 is prepared by suspending 2.0 g milk casein (Hammarsten, E. Merck Co.) in about 10 ml of water. This suspension is allowed to stand for about 10 to 15 minutes at 30°. The mixture is prepared by kneading in a volume of about 60 ml of water. To the milk casein suspension is added 1.5 ml 1.0 N HCl and the mixture is dissolved by stirring on a boiling water bath. The solution is then adjusted to pH 2.7, made up to 100 ml, and stored in the refrigerator.

Trichloroacetic acid, 0.4 M

Na$_2$CO$_3$, 0.4 M

Folin-Ciocalteau reagent.[2] The commercially available Folin-Ciocalteau phenol reagent is diluted 1:5 with distilled H$_2$O before use.

Standard Tyrosine Solution. The tyrosine standard is prepared by dissolving 18.10 mg of pure, dry tyrosine in 1 liter of 0.2 M HCl.[1]

Procedure. One milliliter of a 2% milk casein solution of pH 2.7 is digested for 10 minutes at 30° by 1 ml of a suitably diluted acid proteinase solution. The enzyme solution is diluted with 0.1 M sodium acetate buffer at pH 2.7. The reaction is terminated by the addition of 2 ml of 0.4 M trichloroacetic acid and the mixture is filtered through Toyo filter paper No. 2 or Whatman paper No. 2. The tyrosine and tryptophan in an aliquot of the filtrate are determined by the blue color given with the Folin-Ciocalteau phenol reagent in alkaline solution. The color is compared against a standard tyrosine solution.

One milliliter of the tricholoroacetic acid filtrate is placed in a 25-ml test tube and 5 ml of 0.4 M Na$_2$CO$_3$ and 1 ml of the diluted Folin-Ciocalteau phenol reagent are added while the solution is kept at 30° during addition of the phenol reagent. After standing for 20 to 30 minutes, the color is read at 660 nm against a standard prepared in the same manner. A blank is run with each series of determinations.

Determination of Units and Specific Activity. One unit of the acid

[1] B. Hagihara, *Ann. Rept. Sci. Works, Fac. Sci. Osaka Univ.* **2**, 35 (1954).
[2] O. Folin and V. Ciocalteau, *J. Biol. Chem.* **73**, 627 (1927).

proteinase is defined as the amount of enzyme which yields the color equivalent to 1 micromole of tyrosine per minute in 2 ml of digestion mixture at pH 2.7 and 30°.

The protein nitrogen of the trichloroacetic acid-precipitated fraction of the enzyme solution is determined by the micro-Kjeldahl method. Using a 5-ml beaker, 0.5 ml of 21% trichloroacetic acid is added to 1.0 ml of the enzyme solution. The mixture is allowed to stand for about 30 minutes to obtain complete precipitation of the protein.[1]

The specific activity of acid proteinase per milligram enzyme is expressed in terms of micromoles of tyrosine equivalents produced per minute during casein hydrolysis at pH 2.7 and 30°, as determined by the increase of the absorbancy at 660 nm through application of the Folin-Ciocalteau reagent.

Purification Procedure

Cultivation of Aspergillus saitoi

Cultivation of molds of the genus *Aspergillus* is carried out both in solid culture (Koji-culture) and liquid culture (surface or submerged culture).[3] The first extracellular acid proteinase of *Aspergillus saitoi* was found by Yoshida.[4]

The production of acid proteinase either on solid or in liquid media involves many variable factors: medium components, sterilization of media, size of inoculum, type of antifoam reagents, temperature control, aeration and agitation, and sterility control during the operation.

Organism. The organism used in the preparation of this enzyme is *Aspergillus saitoi* R-3813 (ATCC-14332) of black *Aspergillus*.[5,6] In all the strains of black *Aspergillus* isolated by Sakaguchi, Iizuka, and Yamazaki[5,6] (*A. saitoi, A. usamii, A. inuii, A. aureus, A. awamori,* and *A. nakazawai*), the characteristics of the conidial wall surface are quite different from those of ordinary *A. niger*. In black *Aspergillus* the wall could be either rough in appearance or perfectly smooth, while in *A. niger* it is distinctly echinulated or with colored bars, as mentioned by Thom and Raper.[7] The nature of the conidial wall as well as the size, shape, and color of conidia of *Aspergillus* are the essential features used in the

[3] K. Arima, *in* "Global Impacts of Applied Microbiology" (M. P. Starr, ed.), p. 277. Almgvist and Wikselle Boktryckeri Aktiebolag, Uppsala, Sweden, 1964.
[4] F. Yoshida, *J. Agr. Chem. Soc. Japan* **28**, 66 (1954).
[5] K. Sakaguchi, H. Iizuka, and S. Yamazaki, *J. Agr. Chem. Soc. Japan* **24**, 138 (1951).
[6] H. Iizuka, *J. Gen. Appl. Microbiol. (Tokyo)* **1**, 10 (1955).
[7] C. Thom and K. B. Raper, *in* "A Manual of the Aspergilli," p. 214. Williams & Wilkins, Baltimore, Maryland, 1945.

taxonomy of the genus *Aspergillus*.[5, 8] The electron microscope is an indispensable tool for studying the conidial wall of *Aspergillus*.[6]

The mold *A. saitoi* is maintained on Koji agar slant at 10° and transferred twice per year. Czapeck's medium and the modified Czapeck medium, containing glucose and peptone, can be substituted for the Koji agar slant. *A. saitoi* grows well in a solid or liquid medium, with added wheatbran or/and defatted soybean.

Koji Culture.[4, 9] A wheat bran medium is prepared by kneading 5 g of wheat bran and 3.5 ml of tap water in a 150-ml Erlenmeyer flask and autoclaving at 15 lb pressure for 30 minutes. The production of acid proteinase is increased by about 60% by adding 50 mg NH₄Cl.[9] The sterilized bran is inoculated with spores of *A. saitoi* and is then incubated at 30° for about 60 hours or more. A large number of molds grow satisfactorily on this medium. For the sake of temperature control it is a good procedure to shake and crush germinating or growing mycelium twice a day. Contamination should be avoided.

Submerged Culture.[10, 11] A liquid medium prepared from wheat bran, defatted soybean, and NH₄Cl is used. Wheat bran, 1.8 g, and crushed defatted soybean, 1.2 g, are mixed with 3 ml of tap water containing 0.03 ml conc. HCl, and are then sterilized at 15 lb for 2 hours. The sterilized mixture with or without 1 g of NH₄Cl is prepared in 100 ml tap water in a 500-ml flask and is autoclaved at 15 lb pressure for 2 hours. For the germination of black *Aspergillus*[10, 11] the pH is adjusted to about 5.5. After inoculation with spores, a culture is incubated at 30 to 35° and is shaken continuously at 140 7-cm strokes per minute for 60 hours or more in a reciprocating-type shaker. The production of acid proteinase falls by 30 to 50% if the 1% NH₄Cl is omitted from the medium.[10, 11]

The described fermentation unit accommodates flasks from 500-ml shake types to the 20-liter Waldhof's type[12, 13] jar fermentor.[11] A suspension of rapidly growing cells is used for inoculation at a level of 5%. This suspension is obtained by inoculating 100 ml of inoculum medium into a 500-ml shake flask and incubating for 24 hours on the shaker at 35°. Fermentation is carried out at 35° in a base medium consisting of 0.66% wheat bran, 1.33% defatted soybean, and 1% NH₄Cl at pH 5.5.

[8] K. Sakaguchi and K. Yamada, *J. Agr. Chem. Soc. Japan* **20**, 65, 141 (1944).

[9] E. Ichishima and F. Yoshida, *Agr. Biol. Chem. (Tokyo)* **26**, 547 (1962).

[10] E. Ichishima and F. Yoshida, *Agr. Biol. Chem. (Tokyo)* **26**, 554 (1962).

[11] E. Ichishima, Y. Gomi, T. Watarai, and F. Yoshida, *Agr. Biol. Chem. (Tokyo)* **27**, 302 (1963).

[12] K. Yamada, T. Ito, A. Koyama, and Y. Handa, *J. Agr. Chem. Soc. Japan* **25**, 173 (1951).

[13] K. Yamada, T. Ito, Y. Handa, and A. Koyama, *J. Agr. Chem. Soc. Japan* **26**, 335 (1952).

In all cases, 10 liters of inoculated medium are stirred and aerated in a 20-liter fermentor. The rate of agitation and aeration are 300 to 400 rpm and 2.5 to 5 liters/minute. Sterilized soy sauce oil,[14,15] rapeseed oil, or soybean oil at a concentration of 0.5% is used as a defoaming agent.[11] Maximal production of acid proteinase is reached in 60 to 90 hours.

Isolation and Purification of Acid Proteinase of Aspergillus saitoi

Preparation of Crude Enzyme from Culture Filtrate of Koji or Submerged Cultures. The major part of acid proteinase produced in the Koji culture is extracted with 10 volumes of tap water. The mixture is adjusted to pH 4.0 with 1 N HCl and is allowed to stand in the cold room for 30 to 60 minutes.[16] After adjusting the culture filtrates to pH 4.0, the acid proteinase is precipitated with cold (5°) ethyl alcohol to obtain a final ethanol concentration of 70%. The precipitate is separated by centrifugation and dried *in vacuo* or lyophilized (fraction M_1).[16] The acid proteinase can be separated from the liquor by methyl alcohol, isopropyl alcohol, or acetone, and can be salted out by addition of 53 g solid $(NH_4)_2SO_4$ per each 100 ml of solution.[16,17] The dried material is ground up in a mortar and stored in the cold.

Chromatographic Purification of Acid Proteinase.[18] Chromatographic elution is accomplished at low temperature (e.g., 5°). Six grams of the crude enzyme preparation are added to 10 ml of water, and the mixture is prepared by kneading and is added to about 30 ml of water. The enzyme solution is adjusted to pH 3.5 with 1 N HCl in 60 ml; it is allowed to stand for 1 hour in the refrigerator and is subsequently filtered through Celite. The filtrate containing the acid proteinase is dialyzed against 0.02 M sodium acetate buffer at pH 3.5.

Commercial cation-exchange resin, Duolite CS-101 or Amberlite CG-50, 100 mesh, has been found suitable for chromatographic separation of the acid proteinase from culture filtrates.[18] Commercial Duolite CS-101 or Amberlite CG-50 should be thoroughly washed before use. The dry material is suspended in 1.0 N NaOH. The suspension is filtered and washed with additional 1 N NaOH. The Na form of the resin is filtered and washed with water until no more alkali is removed. This is followed by the addition of sufficient 1 N HCl to make a strongly acid suspension, which is filtered and washed free of acid with water. The filter cake is adjusted to the pH of the starting buffer. The column (3 × 60 cm) is

[14] S. Uchida, *J. Chem. Soc. Japan* **37**, 442 (1934).

[15] T. Kubo, *J. Agr. Chem. Soc. Japan* **23**, 335 (1947).

[16] E. Ichishima, M. Funabashi, and F. Yoshida, *Agr. Biol. Chem.* (*Tokyo*) **27**, 310 (1963).

[17] F. Yoshida and M. Nagasawa, *Bull. Agr. Chem. Soc. Japan* **20**, 252 (1956).

[18] E. Ichishima and F. Yoshida, *Biochim. Biophys. Acta* **99**, 360 (1965).

first equilibrated with 0.02 M sodium acetate buffer at pH 3.5. Fifty milliliters of the dialyzed enzyme filtrate is applied to the column. Elution is performed with 0.15 M sodium acetate buffer at pH 5.2 with increasing concentration of salt and a pH gradient, using a mixing chamber of 1000 ml. The effluent is collected in 10-ml fractions at a flow rate of 30 ml per hour.

The absorbance at 280 nm is used to scan the eluate and to locate the protein-containing fractions. The acid proteinase activity is determined according to the method described. It has been found that the stepwise effluent fraction[19] of the acid proteinase on Duolite CS-101 is contaminated with a cellulase and on testing with free boundary electrophoresis reveals several components.[18] The modified procedure using gradient elution chromatography is more successful. In this chromatography the mobility of the acid proteinase is greater than that of cellulase. The yield of the acid proteinase in this step is more than 70% and the specific activity increased about 4 times with respect to the fraction M_1.

The active fraction is dialyzed against 0.005 M sodium acetate buffer of pH 4.1. The further purification on a DEAE-cellulose column (3×70 cm) yields a further enriched active fraction. Commercial cellulose absorbent, DEAE-cellulose, is prepared and packed into a column according to Peterson and Sober.[20] Elution is performed with 0.01 M sodium acetate buffer, pH 4.1, with increasing concentrations of NaCl, using two 500-ml mixing chambers. The sodium acetate buffer of pH 4.1 in the reservoir, containing 0.1 M NaCl, is tightly connected to the mixing chambers. Ten-milliliter portions of the effluent are collected and the flow rate is about 30 ml/hour. Proteolytic activity is not found in any other fraction except in the last emerging peak. This procedure results in a 2- to 3-fold purification and a 98% yield of the acid proteinase. The active fraction is dialyzed against 0.01 M acetic acid or 0.02 M ammonium formate buffer[21] and is lyophilized. In electrophoresis, the enzyme preparation obtained with DEAE-cellulose at pH 5.0 to 7.0 showed two peaks, the major one containing about 90% of the protein. In contrast, the preparation appeared to be homogeneous on ultracentrifugation.

Repeated purification of 50 mg of the partially purified lyophilized preparation is attempted using a column (2×50 cm) of SE-Sephadex at pH 4.1. The SE-Sephadex is prepared by the method of Peterson and Sober.[20] The salt gradient procedure is the same as that used for DEAE-cellulose chromatography. Ten-milliliter portions of the effluent are

[19] F. Yoshida and M. Nagasawa, *Bull. Agr. Chem. Soc. Japan* **22**, 32 (1958).
[20] E. A. Peterson and H. A. Sober, Vol. V, p. 3.
[21] C. H. W. Hirs, S. Moore, and W. H. Stein, *J. Biol. Chem.* **195**, 669 (1952).

collected at a flow rate of 30 ml/hour. A major enzymatically active component and a minor inactive component are separated by this procedure. This ionic strength gradient at pH 4.1 is more successful for the separation of the contaminated compound than the pH gradient procedure. A symmetrical peak is obtained by chromatography on SE-Sephadex. The active fraction is dialyzed against 0.01 M acetic acid or 0.02 M ammonium formate buffer[21] and is lyophilized. The lyophilized preparation is further purified on commercial Sephadex G-100 with 0.01 M acetic acid or 0.02 M ammonium formate buffer to separate the minor autodigestive products formed. The specific activity is about 150 times that of the initial culture filtrate. The preparation is stored in a desiccator in the refrigerator.

Properties

Stability. There is no appreciable difference in the residual activities of the acid proteinase of *A. saitoi* on Koji or submerged culture filtrates kept for 24 hours at pH 3.0 to 5.0 and at temperatures of 5° to 30°.[16] The acid proteinase of *A. saitoi* in the culture filtrate appears to be most stable at pH 4.0,[16] which strongly indicates the possibility of concentrating the filtrate at pH 4.0.[16]

At pH 5.5 and at 22° to 25°, no autodigestion of aspergillopeptidase A occurs. After 16 hours, the loss of the enzymatic activity is about 5%.[22] At pH 2.7 and 30°, there is a drastic decrease in enzymatic activity of aspergillopeptidase A. It is likely that this decrease in enzymatic activity may be due to autolysis of the proteinase during incubation. The decrease in enzymatic activity of aspergillopeptidase A as a function of time has been observed to parallel the increase of amino nitrogen liberated. These results indicate that presence of the substrate prevents autodigestion to a considerable extent.

The optical rotatory dispersion properties of aspergillopeptidase A in the far-ultraviolet region do not change in the pH range of 2.7 to 5.5.[23] At pH 2.4, the optical rotatory dispersion of aspergillopeptidase A shows a small pH dependence.[23] The conformational change of aspergillopeptidase A starts at pH 2.0.[23]

Denaturation and inactivation of aspergillopeptidase A begins above pH 6.0 to 7.0.[16, 22, 23] At pH 7.0 and 22° the Cotton effect above the 210 nm zone is no longer detected in the optical rotatory dispersion spectrum; it is replaced by a smooth (featureless) dispersion curve.[23] When examined at pH 7.0 and 0°, a strong effect of 2 M NaCl has been observed

[22] E. Ichishima and F. Yoshida, *Biochim. Biophys. Acta* **128**, 130 (1966).
[23] E. Ichishima and F. Yoshida, *Biochim. Biophys. Acta* **147**, 341 (1967).

in the stabilization of the enzymatic activity of aspergillopeptidase A.[24, 25]

The enzymatic activity completely disappears when an enzyme solution of pH 5.5 is maintained at 55° for 20 minutes.[23] The ultraviolet optical rotatory dispersion curves obtained on heat denaturation show a plain dispersion curve similar to that obtained in the presence of weakly alkaline solutions of pH 8.0.[23] Ca^{2+} does not prevent heat inactivation of the enzyme.[26]

The degree of inactivation in the presence of 5 M urea of an apparent pH of 5.5 has been 46% after 30 minutes and 59% after 1 hour's incubation.[22] At low concentration of urea, neither a conformational change nor inactivation of the enzyme occurs. The decrease in the enzymatic activity of aspergillopeptidase A as a function of the urea concentration parallels the behavior of the $-a_o$ value and the behavior of the difference spectra at 294 nm, in the region of tryptophan absorption.[22]

Purity and Physical Properties. Purified aspergillopeptidase A appears to be homogeneous on free boundary electrophoresis over a pH range of 2.0 to 10.0.[18] The sedimentation patterns show aspergillopeptidase A to be monodisperse; the sedimentation constant $s_{20,w}^0$ is 3.33 S.[18]

The isoelectric point is at pH 3.65.[18] The molecular weight of aspergillopeptidase A is in the range of 34,900 to 34,200, calculated with the aid of the Scheraga-Mandelkern formula and the Yphantis procedure, respectively.[27, 28] Using the method of Andrews a molecular weight of 34,500 has been estimated.[29] It is apparent that these three values of the molecular weight are in good agreement.

The conformation of aspergillopeptidase A has been investigated in aqueous solution.[22, 23, 30] The optical rotation, $[\alpha]_D$, is $-35°$. The optical rotatory dispersion constant, λ_c, is 207 nm, and the Moffitt-Yang parameter, b_o, is zero. The $-a_o$ value in the Moffitt-Yang parameter or levorotation of the aspergillopeptidase A molecule increases markedly in the presence of urea, while the value of b_o remains unchanged. In the native state in the ultraviolet optical rotatory dispersion, aspergillopeptidase A has a minimum at 220 nm with $[m'] = -2400$ and a maximum 203 nm with $[m'] = 900$, respectively.[23, 30]

The conformation of aspergillopeptidase A is apparently converted

[24] E. Ichishima and F. Yoshida, *Nature* **215**, 412 (1967).

[25] E. Ichishima and F. Yoshida, *in* "Microbial Enzymes," (K. Tsuda, ed.), p. 186. Tokyo Univ. Press, Tokyo, 1967.

[26] F. Yoshida and M. Nagasawa, *Bull. Agr. Chem. Soc. Japan* **20**, 257 (1956).

[27] E. Ichishima and F. Yoshida, *Nature* **207**, 525 (1965).

[28] E. Ichishima and F. Yoshida, *Biochim. Biophys. Acta* **110**, 155 (1965).

[29] K. Hayashi, D. Fukushima, and K. Mogi, *Agr. Biol. Chem. (Tokyo)* **31**, 1171 (1967).

[30] E. Ichishima and F. Yoshida, *Agr. Biol. Chem. (Tokyo)* **31**, 507 (1967).

by the anionic detergent, sodium lauryl sulfate, into an α-helical conformation, as judged from the optical rotatory dispersion curve in the ultraviolet region.[23,30] The $[m']_{233}$ and $[m']_{198}$ values have been found to be -3400 and 11600, respectively.[23,30] It is concluded that aspergillopeptidase A is devoid of a complete α-helical strand as judged from the visible and ultraviolet optical rotatory dispersion.[22,23,30]

The infrared result indicates that the deuterium-exchanged aspergillopeptidase A exists in the antiparallel β structure.[22] The location of an amide I band at 1632 cm^{-1} has been observed. The spectrum has shown the presence of a weak band around 1685 cm^{-1}.[22]

Aspergillopeptidase A has the following amino acid composition and a total of 283–289 residues: Lys_{11}, His_3, Arg_1, Asp_{34-35}, Thr_{25}, Ser_{42-43}, Glu_{22}, Pro_{10}, Gly_{31-32}, Ala_{20}, $Cys\ Cys_1$, Val_{22}, Met_0, Ile_{11-12}, Leu_{19-20}, Tyr_{17-18}, Phe_{13}, Trp_1, amide-ammonia$_{31-32}$.[28] The molecule of aspergillopeptidase A consists of a single polypeptide chain with serine as the NH$_2$-terminus and with alanine as the COOH-terminal residue.[31]

Activators and Inactivators. No activators are required for this acid proteinase.

Aspergillopeptidase A is inhibited 50% by a 100 M excess of N-acetylimidazole in 0.033 M phosphate buffer containing 2.0 M NaCl, pH 7.0.[25,27,32] The degree of inactivation is 30% after 1 hour and 60% after 5 hours inactivation with a 1000 molar excess of iodine at pH 7.0, 0°.[25,32] The enzymatic activity of aspergillopeptidase A is completely destroyed in 20 minutes if 40 moles of N-bromosuccinimide are present for each molecule of enzyme at 0° and pH 4.1.[25,27,32] Aspergillopeptidase A is inactivated by photooxidation at pH 5.5.[25,27,32] Aspergillopeptidase A is rapidly inactivated by methyl alcohol-HCl,[33] which suggests that the carboxyl group of the enzyme may be esterified.[25,32]

The residual activity of aspergillopeptidase A falls to zero at sodium lauryl sulfate concentrations between 5 and 10^{-3} M at pH 5.5.[23] The conformational transition of aspergillopeptidase A by the detergent has been observed.[23]

Specificity. Aspergillopeptidase A is capable of activating trypsinogen and chymotrypsinogen A at acidic pH range[34] (optimum pH $= 3.5$). This activation of trypsinogen is associated with the usual cleavage of the Lys-Ile bond and liberation of the hexapeptide.

Aspergillopeptidase A does not hydrolyze any of the small synthetic

[31] E. Ichishima and F. Yoshida, *J. Biochem. (Tokyo)* **59**, 183 (1966).
[32] E. Ichishima and F. Yoshida, *2nd Intern. Symp. Gärungsind. Tech. Mikrobiol. Enzymol. Leipzig May, 1968*, pp. 166, 572.
[33] H. Fraenkel-Conrat, Vol. IV, p. 247.
[34] C. Gabeloteau and P. Desnuelle, *Biochim. Biophys. Acta* **42**, 230 (1960).

peptides, amides, and esters.[32] The acid proteinase is inactive in the milk clotting test.[35]

Kinetic Properties. The activation system of trypsinogen follows Michaelis-Menten kinetics, giving $K_{m\ app} = 1.25 \times 10^{-5}\ M$ (at 30°, pH 3.5) and a $V_{max} = 1.5 \times 10^{-2}\ M$/minute/mg aspergillopeptidase A. Initial velocity, v, is measured as μmoles trypsin/minute/mg aspergillopeptidase A. Kinetic properties have also been obtained by Abita *et al.*[36] These two values of the kinetic data are in good agreement.

Distribution. The culture products of *Aspergillus usamii, A. inuii, A. aureus, A. awamori, A. nakazawai, A. niger, A. oryzae, A. sojae,* and other strains of genus *Aspergillus* also catalyze the hydrolysis of proteins at low pH range.[9-11, 17, 37]

In enzymatic properties, *Penicillium janthinellum* peptidase A,[38-40] *Paecillomyces varioti* acid proteinase,[41] *Rhizopus chinensis* proteinase,[35] and *Trametes sanguinea* acid proteinase[42] resemble aspergillopeptidase A.

[35] J. Fukumoto, D. Tsuru, and T. Yamamoto, *Agr. Biol. Chem. (Tokyo)* **31**, 710 (1967).
[36] J. P. Abita, M. Delaage, M. Lazdunski, and J. Savrda, *European J. Biochem.* **8**, 314 (1969).
[37] K. Nakanishi, *J. Biochem. (Tokyo)* **46**, 1263 (1959).
[38] T. Hofmann and R. Shaw, *Biochim. Biophys. Acta* **92**, 543 (1964).
[39] A. Thangamani and T. Hofmann, *Can. J. Biochem.* **44**, 579 (1966).
[40] J. Sŏdek and T. Hofmann, *J. Biol. Chem.* **243**, 450 (1968).
[41] J. Sawada, *Agr. Biol. Chem. (Tokyo)* **27**, 677 (1963).
[42] K. Tomoda and H. Shimazono, *Agr. Biol. Chem. (Tokyo)* **28**, 770 (1964).

[27] Gastricsin and Pepsin

By JORDAN TANG

Assay Methods

Hemoglobin Assay

This assay can be used for both pepsin and gastricsin. The method is essentially that of Anson and Mirsky[1] but modified slightly to give more sensitivity and reproducibility.

Principle. Acid-denatured bovine hemoglobin is used as substrate in this assay. The small peptides, which are produced in the proteolytic digestion, are measured in the filtrate of trichloroacetic acid by means of their absorption at 280 nm.

[1] M. L. Anson and A. E. Mirsky, *J. Gen. Physiol.* **16**, 59 (1932).

Reagents

10% Hemoglobin Solution. This solution is stable for several months when stored at 4°. Therefore, if the assay is carried out routinely, it should be prepared in a larger quantity. Bovine hemoglobin easily forms lumps and is difficult to dissolve. A 50-ml glass tissue grinder (Kontes K-88545) is a useful tool in speeding the solution of the hemoglobin.

HCl solution, 0.3 N

Sodium citrate buffer, 0.1 M, pH 3.1

5% Trichloroacetic acid (TCA)

Procedure. The acidified hemoglobin substrate was freshly prepared by addition of 7 ml of 0.3 N HCl to 20 ml of 10% hemoglobin solution. This solution is diluted with distilled water to 100 ml. The final pH of the solution should be pH 3.1 (adjust to that pH value using small aliquots of NaOH or HCl if the final pH is off by more than 0.1 unit). The sample to be assayed is diluted with water so that each 0.5 ml contains approximately 2 to 15 μg of gastricsin (or 5 to 30 μg of pepsin). All the solutions should be warmed to 37° in the water bath prior to the mixing. To 0.5 ml of enzyme solution, 0.5 ml of 0.1 M sodium citrate buffer, pH 3.1, and 5 ml of acidified hemoglobin solution are added. The solutions are rapidly swirled by a Vortex mixer and placed in a water bath at 37°. After 10 minutes of incubation, the reaction is stopped by the addition of 10 ml of 5% TCA. After filtering through Whatman No. 50 filter paper (d = 7 cm), the optical density of the filtrate at 280 nm is determined with a spectrophotometer. A blank should be run in which the TCA is added before the enzyme solution. The blank usually has an optical density at 280 nm of about 0.2 to 0.3. All determinations and blanks are run in duplicates and averages are used.

Calculation and Cautions. The specific activity which is calculated as ΔOD 280 nm/10 minutes incubation/μg enzyme, is 0.015 for pepsin from human and hog and 0.050 for gastricsin from human. It is important to adjust the enzyme quantities in the assay so that ΔOD 280 nm values are below 1.0. Above that value, the linear relationship between absorbancy and enzyme amount is no longer held.

Determination of Pepsin in a Mixture Containing Gastricsin Using Synthetic Substrate

Principle. The quantity of pepsin in a mixture can be determined by using N-acetyl-L-diiodotyrosine (APDT) as substrate, which is not hydrolyzed by gastricsin.[2] The amount of pepsin is determined by the

[2] L. Chiang, L. Sanchez-Chiang, S. Wolf, and J. Tang, *Proc. Soc. Exptl. Biol. Med.* **122**, 700 (1966).

amount of free diiodotyrosine produced during the incubation, which is analyzed by using the ninhydrin reaction of Rosen.[3]

Reagents

HCl, 0.01 N

N-acetyl-L-phenylalanyl-L-diiodotyrosine (Cyclo Chemical Corp.) 0.002 M, in 0.005 M NaOH (freshly prepared before assay)
Sodium acetate buffer, 4 M, pH 5.3
Sodium cyanide solution, 0.01 M
3% Ninhydrin in ethyleneglycol monomethyl ether
50% Isopropanol in water

Procedure. To 50 μl of enzyme solution, 0.7 ml of 0.01 N HCl and 0.25 ml of APDT solution are added. (For the analysis of pepsin in human gastric juice, 1 ml of sample is mixed with 6.5 ml of 0.01 N HCl. An aliquot of 0.75 ml is taken and mixed with 0.25 ml of APDT.) The mixture is then incubated in a water bath at 37° for 1 hour. The incubation mixture is then analyzed for ninhydrin color using the procedure of Rosen[3] as in the following. The acetate-cyanide buffer is made by the addition of 0.2 ml of sodium cyanide to 10 ml of sodium acetate buffer. To each tube, containing 1 ml incubation mixture, 0.5 ml of acetate-cyanide buffer and 0.5 ml of ninhydrin solution are added. The tubes are heated in a boiling water bath for 15 minutes. At the end of incubation, 5 ml of 50% isopropanol is added, which is followed by vigorous shaking for about 30 seconds. After the tubes are allowed to cool to about room temperature, the optical density at 570 nm is determined in a spectrophotometer.

Calculation and Cautions. The optical density at 570 nm produced from the substrate by 1 μg of human pepsin was found to be 0.0111. However, porcine pepsin hydrolyzes APDT at a somewhat faster rate and gives a higher value.[4] Therefore, when pepsin from a source other than human is used the new specific activity should be determined using pure enzyme. Samples of human gastric juice contain small amounts of ninhydrin-positive substances.[2] A blank run for zero time incubation usually corrects this problem.

Determination of Gastricsin in a Mixture Containing Pepsin Using Hemoglobin and APDT Assays

The hemoglobin assay described above determines the sum of activities of pepsin and gastricsin in a mixture. On the other hand, APDT

[3] H. Rosen, *Arch. Biochem. Biophys.* **67**, 10 (1957).
[4] L. Chiang, L. Sanchez-Chiang, J. Mills, and J. Tang, *J. Biol. Chem.* **242**, 3098 (1967).

assay determines the amount of pepsin in the mixture. The amount of gastricsin in the sample, therefore, can be calculated by using the results of both assays.[2]

$$
\begin{array}{l}
\mu\text{g of} \\
\text{gastricsin in} \\
\text{the incubation} \\
\text{mixture}
\end{array}
=
\dfrac{\begin{array}{l}\text{Total } \Delta\text{OD}_{280\,nm} \\ \text{in hemoglobin assay}\end{array} - \left(\begin{array}{l}\text{specific activity} \\ \text{of pepsin in} \\ \text{hemoglobin assay}\end{array}\right) \times \left(\begin{array}{l}\mu\text{g of pepsin} \\ \text{found in APDT} \\ \text{assay}\end{array}\right)}{\text{specific activity of gastricsin in hemoglobin assay}}
$$

$$
= \dfrac{\Delta\text{OD } 280 \text{ nm in hemog. assay} - (0.015 \times \mu\text{g pepsin APDT assay})}{0.050}
$$

Resolution and Determination of Human Pepsin and Gastricsin from a Mixture Using a Small Column of Amberlite IRC-50 (CG-50)

Principle. Since human pepsin and gastricsin can be easily separated on a column of Amberlite IRC-50 (CG-50), quantitative measurement of the enzymes can be made using this process.[5] The elution of the enzymes from the column is pH dependent. Therefore, the stepwise elutions are used in the recovery of the enzymes and the activity of each is determined using the hemoglobin assay.

Reagents

> Amberlite IRC-50 (CG-50) resin (Mallinckrodt #3341, 200–400 mesh)
> Sodium citrate buffer, 0.2 M, pH 3.0
> Sodium citrate buffer, 0.2 M, pH 4.16
> Sodium citrate buffer, 0.2 M, pH 4.60
> The reagents used in the hemoglobin assay

Procedures. A column of 0.4 × 5.5 cm is prepared and washed with 20 ml of 0.2 M sodium citrate buffer, pH 3.0. A convenient way of preparing the column, especially when a large number is needed, is to use commercially available disposable capillary Pasteur pipettes (such as Fisher No. 13-678-5, 5¾ inches long). The tips of the capillary is cut to about 1 cm in length and a small plug of fine glass wool is inserted into the bottom of the column before the resin is poured. A sample containing 0.1 to 0.3 mg of a mixture of the two enzymes is dissolved in 2 to 5 ml of the same buffer. If the sample has been previously dissolved in another solution, it is necessary to dialyze it in the cold against several changes of citrate buffer, pH 3, before subjecting it to chromatography. The solution is then allowed to pass into the column at an eluate flow rate of approximately 1 ml per 10 minutes. After the enzymes have been absorbed on the resin, an additional 5 ml portion of 0.2 M sodium citrate buffer, pH 3, is passed through the column. Human pepsin is then eluted

[5] J. Tang and K. Tang, *J. Biol. Chem.* **238**, 606 (1963).

with 20 ml of citrate buffer, pH 4.16, and gastricsin is eluted with 20 ml of citrate buffer, pH 4.6. The amount of enzyme in the eluants is then analyzed using the hemoglobin assay as described above.

Cautions. Since the eluants contain considerable buffer capacity, the aliquots taken for analysis in the hemoglobin assay should be as small as possible. The error of this method should be about 5%.

Purification Procedure

PURIFICATION OF HUMAN GASTRICSIN AND PEPSIN

Human gastricsin and pepsin have been purified from gastric juice[6,7] and human gastric mucosa extract.[5] Only human gastricsin has been obtained in crystalline form.[7]

Starting Material. The best source of human gastricsin and pepsin is gastric juice. For a large-scale preparation, convenient collections are made overnight in the surgical ward of the hospital where gastric juices are removed from patients scheduled for stomach operations on the following day. This "overnight juice" usually has a green color, due to the contamination of bile pigments during the collection. In a large hospital, it is not difficult to collect several liters of gastric juice in a week. Each liter of such juice usually contains about 300–500 mg of pepsin and 100–200 mg of gastricsin.

There are two common methods of storing gastric juice. If the volume involved is quite small, it can be filtered through fine glass wood, dialyzed against several changes of distilled water, and lyophilized. Gastricsin and pepsin are stable in lyophilized gastric juice powder at 4° for months. However, the recovery of these enzymes after a storage of a few years can be considerably less. The second method of storage is to pool and filter the gastric juice through glass wool, then store in 2-liter polyethylene bottles at −30°. This method is suitable for handling large quantities of juice.

Fractionation of Human Gastricsin and Pepsin on a Column of Amberlite IRC-50. Two sizes of columns are most commonly used. A column of 4.5 × 30 cm can fractionate up to 3 g of lyophilized gastric juice powder. A large column 8 × 25 cm, could be used to fractionate up to 10 g of lyophilized gastric juice powder, or 3 liters of gastric juice if that is the starting material (see below). A typical fractionation procedure is described in the following:

Three grams of lyophilized gastric juice powder is dissolved in 30 ml

[6] V. Richmond, J. Tang, S. Wolf, R. E. Trucco, and R. Caputto, *Biochim. Biophys. Acta* **29**, 453 (1958).

[7] J. Tang, S. Wolf, R. Caputto, and R. E. Trucco, *J. Biol. Chem.* **234**, 1174 (1959).

of 0.2 M sodium citrate buffer, pH 3.0. The solution flows freely and should not have a jellied appearance. If the latter is the case, more of the pH 3.0 buffer can be used to dilute the starting material. The solution is then centrifuged in a Spinco model L centrifuge at 30,000 rpm for 30 minutes. The supernatant is retained while the mucouslike sedimenting material is discarded.

The resin Amberlite IRC-50 (CG-50) is treated according to the procedure outlined by Hirs et al.[8] and equilibrated with the starting buffer, 0.2 M sodium citrate, pH 3.0. The column chromatography is operated at room temperature. After the buffer level in the column is brought to the same height as that of the resin, the supernatant of the gastric juice is carefully layered on the top of the resin bed and allowed to flow slowly into the resin. The column is then washed with 300 ml of the same pH 3.0 citrate buffer. The flow rate of the column is approximately 1 ml per minute and fractions of 10 ml per tube are collected. The influent buffer is then changed to 0.2 M sodium citrate, pH 3.80. The elution with this buffer is continued until the effluent pH has reached 3.80 and its optical density at 280 nm is less than 0.1. The elution buffer is then changed to 0.2 M sodium citrate buffer, pH 4.2, with which human pepsin is eluted in a sharp peak. The effluent at the top of the human pepsin peak should be pH 4.0. After the effluent pH has reached 4.2, the elution buffer is changed to 0.2 M sodium citrate, pH 4.6. Human gastricsin is eluted from the column by this buffer at an effluent pH of 4.4. A typical chromatographic pattern is shown in Fig. 1. The material under the enzyme peaks is pooled, dialyzed in cold against at least three changes of distilled water, and lyophilized. The enzymes prepared in the fractionation should be free from cross contamination and have purities better than 80%.

If the large-sized column is used, the procedures are the same. However, gastric juice samples can be directly absorbed onto the column. Three liters of glass wool filtered gastric juice are measured into a flask containing 80.6 g of citric acid and 51.9 g of sodium citrate. After complete solution, the gastric juice should have a pH of about 3.0. Solutions of 0.2 M citric acid or 0.2 M sodium citrate can be used to adjust the gastric juice to the starting pH. The starting sample is then allowed to flow slowly through the column. Nearly all the proteolytic activity is retained by the column. The elution and resolution of pepsin and gastricsin are usually not satisfactory after this absorption procedure. Instead, the two enzymes are eluted together with pH 4.60 buffer into a flask. After dialysis and lyophilization, the enzymes are separated

---

[8] C. H. W. Hirs, S. Moore, and W. H. Stein, *J. Biol. Chem.* **200**, 493 (1953).

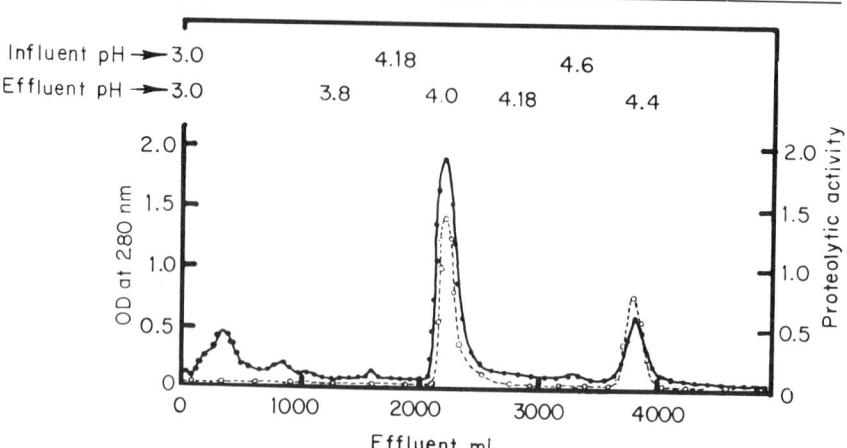

Fig. 1. Chromatography of dialyzed human gastric juice on a column of Amberlite IRC-50 (CG-50). Solid line: protein concentration. Broken line: proteolytic activity.

by repeating the column chromatography using the small size column (4.5 × 30 cm) as described above.

Crystallization of Human Gastricsin. The gastricsin solution obtained from the Amberlite column is dialyzed against distilled water and lyophilized. About 20–30 mg of the powder obtained is dissolved with 2.0 ml of distilled water at 4° and centrifuged at 2000 rpm for 5 minutes. The supernatant solution is poured into a centrifuge tube, put into an ice bath, and 0.26 g of crystalline $(NH_4)_2SO_4$ is added. The protein precipitate formed is centrifuged at 2000 rpm for 10 minutes, and the supernatant solution discarded. The precipitate is dissolved with 2.0 ml of water and the previous precipitation repeated. The precipitate is dissolved with 2 ml of cold 0.2 M sodium acetate buffer (pH 5.0). A clear solution is obtained to which crystalline $(NH_4)_2SO_4$ is added in small portions until the first indication of protein precipitation is observed [approximately 0.2 g of $(NH_4)_2SO_4$ is required]. The tube is then placed in a water bath at 20° and left for 5 minutes. If the turbidity disappears, drops of saturated $(NH_4)_2SO_4$ solution are added until it reappears. The tube is then moved to a 40° water bath, in which, after 5 minutes, the turbidity usually disappears. When it does not, however, the insoluble material is removed by centrifugation. The tube is placed in a beaker containing 2 liters of water at 30° and moved into the 4° cold room where, after 6–8 hours, a white precipitate is formed. The first precipitate, however, is mostly noncrystalline and is removed by centrifugation in the cold room and at a low speed. The supernatant solution is left stand-

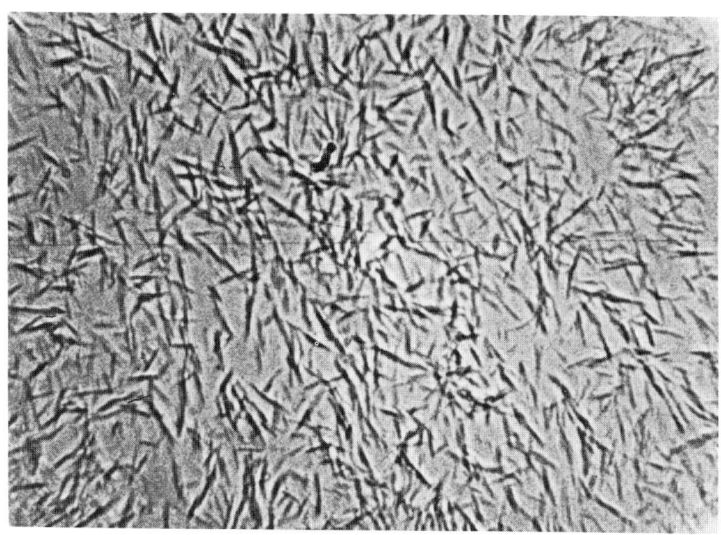

Fig. 2. Crystalline human gastricsin at 480 × magnification.

ing at 4° for a period of 20–30 hours during which crystals are formed. The crystals of gastricsin, shown in Fig. 2, are transparent, and the microscopic observations are best made in dim light. Seeding of crystalline gastricsin reduces the time required for the crystals to appear. After 2–3 days, the first collection of crystals is made by centrifugation at low speed. The supernatant solution still contains enzyme activity, and the formation of more crystals can be obtained by further addition of drops of saturated $(NH_4)_2SO_4$ solution.

Table I shows a protocol of purification procedure. The specific

TABLE I

PURIFICATION AND CRYSTALLIZATION OF GASTRICSIN[a]

	Dry weight (mg)	Specific activity ($\Delta OD_{280\ nm}$/mg protein)	Total activity (%)
Freeze-dried gastric juice	2000	4.6[b]	100
Chromatographic fraction	36	153.6	60
First crystals	7.2	227.5	17.8
Second crystals	6.7	242.5	17.6

[a] The activities were measured according to the method of Anson and Mirsky, *J. Gen. Physiol.* **16,** 59 (1932).

[b] The value for gastricsin given in this table was calculated from the total proteolytic activity recovered from the column after adding the values of both peaks and then assuming that the percent recoveries for pepsin and gastricsin were the same.

activities of the crystalline gastricsin at 25 to 30% higher than that of the lyophilized materials obtained by chromatography, and the yields, related to the same starting material, vary from 20 to 30%.

The shape and the activity of the crystals remain unaltered after storage for a long time (up to 2 years), and this is the best way to preserve the enzyme.

Further Purification of Human Pepsin. Human pepsin obtained from the Amberlite IRC-50 (CG-50) column is usually not free of contamination (see below), especially the material obtained from the fractionation on the large column. Repeating the fractionation on a column of Amberlite IRC-50 is necessary to obtain a pure preparation.

Purification of Human Gastricsin and Pepsin from Gastric Mucosa. Human gastricsin and pepsin can be obtained from the acidified gastric mucosa.[5] About 100 g of human gastric mucosa are minced and homogenized in a Waring blendor with 0.1 M NaHCO$_3$ solution. The operation should be carried out in the cold. The homogenate is centrifuged at 30,000 g for 60 minutes. The supernatant is dialyzed against distilled water which has been adjusted to pH 8 with ammonium hydroxide and lyophilized. The powder of mucosa extract is then dissolved in 0.2 M sodium citrate buffer, pH 3.0, in a concentration of about 100 mg/ml. The solution is allowed to stand in room temperature for 30 minutes and is then transferred to a column of Amberlite IRC-50. The fractionation procedure for gastricsin and pepsin is the same as described above. However, if the starting material is small in quantity, the size of the column should also be reduced.

PURIFICATION OF PORCINE GASTRICSIN

Gastricsin can be isolated from porcine gastric mucosa or commercial crude pepsin preparations.[4]

Starting Material. Whole porcine stomachs can be obtained from slaughterhouses. They are placed in an ice chest immediately after removal from the hogs. The fundus mucosae is separated from each stomach, and the samples which are not processed the same day are frozen and stored. Crude porcine pepsin (1:10,000) can be purchased from commercial companies.

Extraction. The fundus mucosae from the hog stomachs are ground in a chilled meat grinder. The ground tissue (1 to 2 kg, wet weight) is extracted with four times its weight of 0.45 saturated ammonium sulfate (at 25°) in 0.1 M sodium bicarbonate. The suspension is filtered.

Ammonium Sulfate Fractionation. The filtrate is brought to 68% saturation of ammonium sulfate by the addition of 458 g of the salt to each liter of filtrate. The resulting precipitate is recovered by filtration, then dissolved in 0.02 M phosphate buffer, pH 7.0, and dialyzed at 4° against distilled water adjusted to pH 8.0 with KOH. The solution is dialyzed at 4° against several changes of 0.2 M sodium citrate buffer, pH 4.2, for 24 hours in order to activate the zymogens.

Fractionation on Amberlite IRC-50 (CG-50) Column. The enzyme solution obtained by the above procedure is equilibrated with 0.2 M sodium citrate buffer, pH 2.1, by dialysis overnight and then passed into a column (2.5 × 20 cm) of Amberlite IRC-50 resin which has been equilibrated with the same buffer. Porcine pepsin, which accounted for about 95% of the proteolytic activity of the starting material, passes through the column unabsorbed. The column is then washed with 0.2 M sodium citrate buffers of pH 3.8 and 4.2. Proteolytic activity should not appear in the effluent of either of these buffers. After the effluent pH has reached 4.2, porcine gastricsin is eluted from the column by 0.2 M sodium citrate buffer, pH 4.6, as shown in Fig. 3. The fractions containing porcine gastricsin are pooled, dialyzed against distilled water, and lyophilized.

Fractionation of Sephadex G-75 Column. The lyophilized enzyme

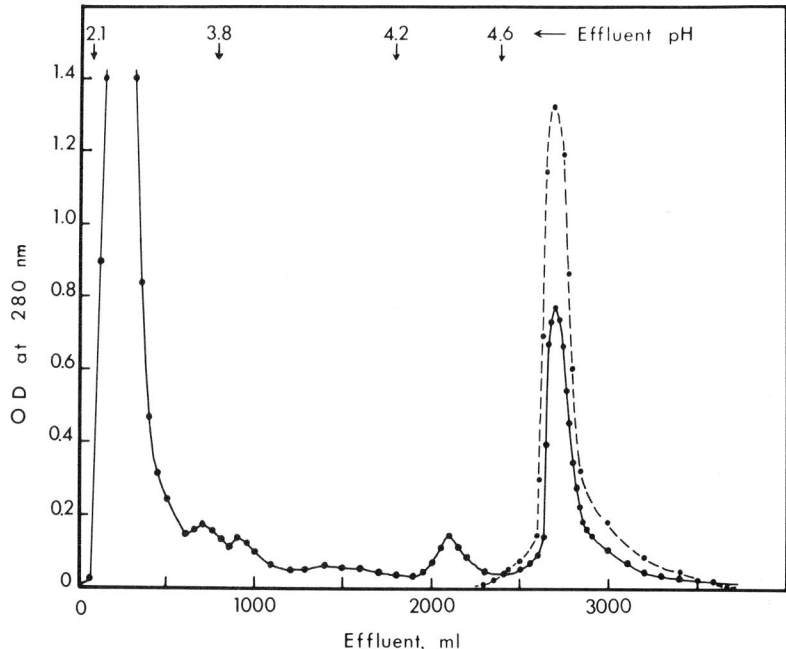

Fig. 3. Chromatography of porcine gastricsin on a column of Amberlite IRC 50 (CG-50). Porcine gastricsin is eluted in a sharp peak at 2700 ml of the effluent.

powder is dissolved in 3 ml of 0.05 M citrate buffer, pH 5.0, and fractionated on a column (2 × 150 cm) of Sephadex G-75. The enzyme is eluted at about 220 ml of effluent with the same buffer. Usually only minor contaminating peaks can be seen in the chromatogram.

Purification of Porcine Gastricsin from Commercial Crude Pepsin Preparations. One hundred grams of commercial crude pepsin (1:10,000) are dissolved in 100 ml of distilled water. The solution is adjusted to pH 2.1 with 3 N HCl. The solution is then dialyzed against 0.2 M sodium citrate buffer, pH 2.1. The solution is then transferred onto a column of Amberlite IRC-50 and fractionated as described in the preceding paragraphs.

Properties

There are considerable similarities in the enzymatic properties of gastricsin and pepsin from human and porcine sources. The properties reviewed in the following paragraphs, therefore, will be made in a comparative manner.

General Enzymatic Properties. Gastricsin as well as pepsin exhibit milk-clotting activity in addition to proteolytic activity. The former can be tested by using the method of Berridge.[9] The specific activity in milk-clotting for gastricsin and pepsin is about the same, however, either one is only about 45% of the specific activity of rennin.[7] The enzymes have somewhat different optimal pH in proteolysis when bovine hemoglobin is the substrate.[6] The optimal pH for gastricsin is 3.0, while for pepsin it is 1.9. The specific proteolytic activity for the enzymes is dependent on the nature of the protein substrate. Using hemoglobin as substrate, the specific proteolytic activity of pepsin is only about 74% of that of gastricsin. However, in the case of egg albumin substrate, pepsin hydrolyzes the substrate more than twice as fast as gastricsin.[10] Their comparative enzymatic properties are shown in Table II.

Stability. Gastricsin as well as pepsin is stable in acidic solutions. However, both lose activity rapidly in neutral or alkaline solutions. The enzymes are rather stable at room temperature. Column chromatography for the fractiontiaon of the enzymes can be carried out at room temperature without significant loss of activity. The lyophilized powder of the enzymes can be stored at 4° indefinitely. However, prolonged storage (over a year) often results in decreasing amounts of soluble enzyme when one attempts to redissolve the powder. The best storage form of gastricsin is its crystal. The suspension of enzyme crystals in saturated

---

[9] N. J. Berridge, Vol. II, p. 69.
[10] J. Tang, J. Mills, L. Chiang, and L. deChaing, *Ann. N.Y. Acad. Sci.* **140**, 688 (1967).

TABLE II

COMPARISON OF SOME PROPERTIES OF GASTRICSIN AND PEPSIN FROM HUMAN AND PIG

Enzyme	Optimal pH[a]	Relative proteolytic activity	Relative milk-clotting activity	Effluent pH[b]	Molecular weight	Sedimentation coefficient $(S_{20,w} \times 10^{13})$	Diffusion coefficient $(D_{20,w} \times 10^7)$	Intrinsic viscosity, $[\eta]$ (dl/g)
Human gastricsin	3.0	100	100	4.4	31,400	3.32	9.6	0.100
Porcine gastricsin	3.0	106	115	4.4	32,500	3.44	9.54	
Human pepsin	1.9	74	103	4.0	34,000	3.14	8.7	0.045
Porcine pepsin	2.0	75	125	4.0	34,200	2.88		

[a] Hemoglobin substrate.
[b] From Amberlite IRC-50 column.

ammonium sulfate can be stored at 4° for several years without significant loss of activity.

The heat stability of gastricsin and pepsin is somewhat dependent on the pH of the solution.[7] Both enzymes are more heat stable at pH 2 than at pH 3.2.

Purity and Physical Properties. The 2✕ crystalline human gastricsin shows a single band in starch gel electrophoresis and a single sedimenting boundary in the schlieren pattern of an ultracentrifuge.[7] It shows a single amino terminal residue, serine.[10] The 2✕ chromatographed human pepsin (see purification above) satisfies the same homogeneity criteria. The human enzymes obtained from a single column chromatography on Amberlite IRC-50, however, are not free of impurities. This is evident since gastricsin increases its specific activity in the first crystallization. Also, human pepsin from the first column chromatography shows additional bands in electrophoresis.[7]

Porcine gastricsin, after chromatographic separation on columns of Amberlite IRC-50 and Sephadex G-75, shows one boundary in ultracentrifugation and a single band in electrophoresis.[4]

The molecular weights for human pepsin and gastricsin are about 34,000 and 31,400, respectively, as determined by the sedimentation–diffusion method, sedimentation equilibrium studies, and amino acid analysis data.[11] The sedimentation coefficient, $s_{20,w}$, are 3.14 S for human pepsin and 3.32 S for gastricsin. The sedimentation coefficient for gastricsin is concentration dependent, which is consistent with the high intrinsic viscosity observed (Table II). The diffusion coefficients $D_{20,w}$ were found to be 8.7×10^{-7} cm^2 sec^{-1} and 9.6×10^{-7} cm^2 sec^{-1} for human pepsin and gastricsin, respectively. The partial specific volume calculated from the amino acid composition is 0.724 ml/g for both enzymes. The intrinsic viscosity as well as the axial ratio for gastricsin are considerably higher than those of pepsin (Table II). Porcine gastricsin appears to have a somewhat larger molecular weight, 32,500, than human gastricsin, 31,400.[4]

Activators and Inactivators. There is no activator known for human gastricsin and pepsin. A number of pepsin inhibitors, diazoacetylglycine ethyl ester and phenylbenzoyldiazomethane[12] and diphenyldiazomethane,[13] also inhibit human gastricsin to various degrees.[14]

Specificity. Both human gastricsin and pepsin show broad specificity in the hydrolysis of peptide bonds in protein and peptide substrates. Additionally, these specificities in two enzymes are qualitatively quite

[11] J. N. Mills and J. Tang, *J. Biol. Chem.* **242**, 3093 (1967).
[12] L. V. Kozlov, L. M. Ginodman, and V. N. Orekhovich, *Biokhimiya* **32**, 1011 (1967).
[13] G. R. Delpierre and J. S. Fruton, *Federation Proc.* **54**, 1161 (1965).
[14] J. Mills, personal communication, 1969.

TABLE III
The Hydrolysis of Synthetic Dipeptides by Human Gastricsin and Pepsin

Substrate	Specific activity[a]	
	Gastricsin	Pepsin
Z^b-L-Tyrosyl-L-alanine	91.0	<2.0
Z-L-Tyrosyl-L-threonine	52.1	<2.0
Z-L-Tyrosyl-L-leucine	27.2	<2.0
Z-L-Tyrosyl-L-serine	19.3	<2.0
Z-L-Tyrosyl-L-valine	8.4	4.2
Z-L-Tyrosyl-L-tyrosine	7.6	3.4
Z-L-Tyrosyl-L-phenylalanine	2.1	21.2
Z-L-Tyrosyl-L-glycine	0	0
Z-L-Seryl-L-tyrosine	0	0
Z-L-Alanyl-L-tyrosine	0	0
Z-L-Phenylalanyl-L-serine	0	0
Z-L-Tryptophanyl-L-alanine	40.0	<2.0
Z-L-Tryptophanyl-L-serine	<2.0	<2.0
Z-L-Tryptophanyl-L-leucine	0	0
Z-L-Glutamyl-L-tyrosine	17.0	34.0
Z-Glycyl-L-phenylalanine	2.0	6.0
Z-L-Glutamyl-L-phenylalanine	1.0	24.0
N-Acetyl-L-phenylalanyl-L-tyrosine	230.0	770.0
N-Acetyl-L-phenylalanyl-L-diiodotyrosine	0	3920.0

[a] Millimicromoles amino acid released per milligram enzyme per hour.
[b] N-Carbobenzoxy group.

similar. However, definite differences in the specificities of gastricsin and pepsin have been demonstrated in both protein and synthetic peptide substrates.[15] Using glucagon and oxidized ribonuclease A as substrates, there are a total of 23 common hydrolytic sites. Two pepsin-specific

TABLE IV
Kinetics of the Hydrolysis of Synthetic Dipeptides by Human Gastricsin at pH 0

Dipeptide substrates	K_m (M)	V_{max} (micromoles/ mg/hour)	K_3 (hours^{-1})
Z-L-Tyrosyl-L-alanine	7.4×10^{-4}	0.40	12.56
Z-L-Tyrosyl-L-serine	4.8×10^{-3}	0.19	5.81
Z-L-Tyrosyl-L-threonine	7.9×10^{-3}	0.13	4.02
Z-L-Tryptophanyl-L-alanine	9.2×10^{-4}	0.39	12.24

[15] W. Y. Huang and J. Tang, *J. Biol. Chem.* **244**, 1085 (1969).

cleavage sites are found in oxidized ribonuclease A, which are Val-Ala, and Phe-Val bonds. Four gastricsin-specific sites are found in the two protein substrates, which are Tyr-Ser and Tyr-Leu bonds in glucagon and Tyr-Ser and Tyr-Gln bonds in ribonuclease. The preferential hydrolysis of gastricsin on Tyr-X bonds is clearly shown in synthetic dipeptide substrates, as shown in Table III. Four carbobenzoxy tyrosyl dipeptides are hydrolyzed exclusively by gastricsin. The second residue in these substrates consists, in the order of preference, of alanine, threonine, leucine, and serine. In addition, carbobenzoxy-tryptophanyl-alanine is also a specific substrate for gastricsin. Gastricsin hydrolyzes rather poorly most of the known synthetic substrates for pepsin (Table III). As described in the assay section, gastricsin does not hydrolyze N-acetyl-L-phenylalanyl-L-diiodotyrosine, which is the best synthetic substrate for pepsin.

Kinetic Properties. Several gastricsin-specific substrates have been measured for their kinetic parameters, K_m, V_{max}, and K_3.[15] As shown in Table IV, the kinetic values are of the same magnitudes as is found

TABLE V

AMINO ACID COMPOSITION OF GASTRICSIN AND PEPSIN FROM HUMAN AND PIG

Amino acid	Porcine gastricsin	Human gastricsin (residues/molecule protein)	Human pepsin (residues/molecule protein)	Porcine pepsin
Lysine	4	0	0	1
Histidine	1	1	1	1
Ammonia	(36)	(44)	(50)	(27)
Arginine	4	3	3	2
Aspartic acid	26	26	40	40
Threonine	23	21	27	25
Serine	32	32	43	43
Glutamic acid	39	39	31	26
Proline	15	17	19	16
Glycine	31	33	35	34
Alanine	19	18	18	16
Half-cystine	6	6	6	6
Valine	19	23	27	20
Methionine	4	5	5	4
Isoleucine	13	13	25	23
Leucine	30	25	22	28
Tyrosine	16	17	15	16
Phenylalanine	19	15	15	14
Tryptophan	3	4	5	6
Total	304	298	337	321

for the peptic hydrolysis of acetyl-L-tyrosyl-L-tyrosine, acetyl-L-tyrosyl-L-phenylalanine, and acetyl-L-phenylalanyl-L-phenylalanine.[16]

Chemical Composition and Partial Structure. The amino acid compositions of gastricsin and pepsin are shown in Table V. The N-terminal residue for human gastricsin is serine and for human pepsin is valine.[10] The C-terminal sequence in two human enzymes is known,[17] as shown in the following:

human gastricsin: -Ile-Arg-Gln-Phe-Tyr-Thr-Val-Phe-Asp-Arg-Ala-
    Asn-Gln-Lys-Asp-Gly-Leu-Ala-Pro-Val-Ala-COOH
human pepsin: -Ile-Leu-Gly-Asp-Val-Phe-Ile-Arg-Gln-Phe-Tyr-Thr-
    Val-Phe-Asp-Arg-Ala-Asn-Asn-Gln-Val-Gly-Leu-Ala-Pro-Val-
    Ala-COOH

From these structures it is clear that human gastricsin and pepsin show a definite homology in their primary structure. The structure from the enzymes is also homologous to that of bovine rennin.[18]

Distribution. Gastricsin is known to be present in human and porcine stomachs. It is quite possible that this enzyme has a wide distribution in higher mammals.

[16] W. T. Jackson, M. Schlamowitz, and A. Shaw, *Biochemistry* 5, 4105 (1966).
[17] W. Y. Huang and J. Tang, *J. Biol. Chem.* 245, 2189 (1970).
[18] B. Foltmann and B. S. Hartley, *Biochem. J.* 104, 1064 (1967).

[28] Prochymosin and Chymosin (Prorennin and Rennin)

By B. FOLTMANN

Chymosin or rennin is the milk-clotting enzyme from the fourth stomach of the calf. The name rennin has been widely used, especially in the English-speaking countries, but unfortunately this name resembles renin (EC 3.4.4.15) very much, and mistakes quite often occur.

Until now I have accepted the general use of the English term rennin, but considerably more papers are published about renin than about rennin; furthermore, it should be emphasized that the name chymosin is 50 years older than rennin.[1,2] Because of these reasons I have arrived at the conclusion that a change in the trivial English nomenclature is advisable. In this chapter, therefore, the name chymosin and pro-

[1] Deschamps, *Journ. Pharm.* 26, 412 (1840).
[2] A. S. Lea and W. L. Dickinson, *J. Physiol.* 11, 307 (1890).

chymosin (EC 3.4.4.3) are used for the milk-clotting enzyme and its zymogen.

The methods of isolation and assay of the enzyme have previously been reviewed by Berridge.[3] The methods described in this chapter have been used in my investigations. Although the principles of the milk-clotting test and crystallization of chymosin are the same as those described by Berridge, the performance differs on several points. The biochemistry of the enzyme and its precursor has recently been reviewed.[4,5]

Assay of Chymosin

Principle. In most cases chymosin is assayed by clotting of cow's milk. The term milk refers in the following discussion to cow's milk. The clotting process consists of two stages. First a limited proteolysis of the stabilizing fraction of the casein (κ-casein) takes place, and second the micelles of Ca-caseinate form larger and larger aggregates until visible particles occur. Different milk samples exhibit great variations in their clotting ability. This is partly overcome by standardizing the test on one batch of spray-dried milk.

Reagents

> A batch of high quality spray-dried skim milk powder (kept in the cold room)
>
> CaCl$_2$-solution, 0.01 M, prepared by dilution from a 1 M stock solution
>
> Sodium phosphate buffer, 0.05 M, pH 6.3 for dilution of the enzyme

The substrate is prepared by reconstitution of 12 g of skim-milk powder in 100 ml 0.01 M CaCl$_2$ with vigorous stirring without foaming for 5 minutes. The reconstituted skim milk will often change its clotting ability upon standing, therefore the substrate is left for 1 hour at room temperature before it is used, and it should not be used later than 3 hours after reconstitution. The enzyme is diluted with the phosphate buffer to an activity corresponding to a clotting time of 4–5 minutes or about 0.005 mg of chymosin per milliliter.

Apparatus. Bifurcated glass tubes with the two legs at right angles to each other with each branch containing 15–18 ml and a water thermostat ($30° \pm 0.1$) with a glass front are used. The observation of the

[3] N. J. Berridge, Vol. II, p. 69.

[4] B. Foltmann, *Compt. Rend. Trav. Lab. Carlsberg* **35**, 143 (1966).

[5] B. Foltmann, *in* "The Milk Proteins" (McKenzie, ed.). Academic Press, New York (1971).

clotting is facilitated if the thermostat has a black background and is illuminated from the side.

For longer series of clotting tests it is convenient to have a rack containing up to 6 vessels immersed in the thermostat. By means of a crank arrangement it should be possible to tilt the rack 100° along its axis.

Procedure. Ten milliliters of reconstituted skim milk is placed in one branch and 1 ml of chymosin solution in the other branch of a bifurcated glass tube. The vessel is incubated in the thermostat, after 15 minutes the contents of the two branches are mixed, and a stopwatch is started. During the reaction the milk is kept flowing slowly from one branch to the other and back again. The moment when the thin film of milk breaks into visible particles is noted as the clotting time.

Unit and Calculation. There is no internationally agreed upon unit for the milk-clotting activity. Berridge[6] defined one unit as the activity which would clot 10 ml of reconstituted skim milk in 100 seconds at 30°. [One rennin unit (RU), in the changed nomenclature corresponds to one chymosin unit (CU). Due to variations between different batches of skim-milk powder, this is not an absolute unit.] The British Standards Institution[7] recommends references of rennet powder and skim-milk powder; such references are supplied from Chr. Hansen Lab. and United Dairies, respectively. Local standards based on chymosin or milk powder may be used as well. The milk-clotting unit used in my previous publications have been based on a standard of crystalline chymosin. This local standard was tested against a batch of good quality skim-milk powder according to Berridge and exhibited an activity of 900 CU/mg N.

Different preparations of crystalline rennin have shown activities from ca. 700 to 1000 CU/mg N or from ca. 100 to ca. 140 CU per milligram dry enzyme.[8]

By coincidence it was found that chromatographically purified chymosin-B had a ratio of ca. 100 between the milk-clotting activity measured in CU per milliliter and the absorbancy at 278 nm (see p. 429). In future work from this laboratory we are going to use chromatographically purified chymosin-B as the primary standard, defining the milk clotting unit as: A chromatographically purified preparation of chymosin-B with an absorbancy at 278 nm of 1.00 containing 100 CU per milliliter.

The chymosin activity is often calculated under the assumption of inverse proportionality between clotting time and amounts of enzyme

[6] N. J. Berridge, *Biochem. J.* **39,** 179 (1945).
[7] British Standards Institution, B.S. 3624:1963.
[8] B. Foltmann, *Acta Chem. Scand.* **13,** 1927 (1959).

used. This is not always true. Holter has introduced a correction for the time lag of the second stage in the milk-clotting process.[9] The following equation is recommended for calculation of chymosin activity:

$$CU = k/(T - t)$$

T is the observed clotting time, t is an empirical correction for the time lag of the secondary coagulation stage, and k is a constant. k and t depend on the conditions of the experiment. The determinations take place either by testing chymosin-B, by definition of a local standard, or by comparison with a sample of known activity.

Note. The milk-clotting test is very sensitive to changes in pH and salt concentration. Care must be taken that the test solution is diluted so that the effect of salt and buffers are negligible. Blank experiments in which diluted test solvent is added together with a known amount of pure chymosin are recommended.

British Standards Institution[7] has adopted a method for the determination of milk-clotting activity which is similar to that described by Berridge.[3] The general proteolytic activity of rennin may be assayed by a technique analogous to that of Anson.[10,11] This method, however, is seldom used for routine assay of chymosin.

Assay of Prochymosin

The potential milk-clotting activity of prochymosin may be assayed after conversion into chymosin, and tested as described above. The activation of prochymosin takes place at pH below 5; the process is much dependent on pH and ionic strength. For an analytical evaluation of potential chymosin activity, a rapid activation at pH 2 is recommended. At room temperature and pH 2 the activation is completed in 1 or 2 hours if the reaction mixture is salt-free. At the same pH and temperature the activation is completed in 10 minutes if the solution contains $0.1 M$ NaCl (see also pp. 430 and 432). In determination of the potential chymosin activity, it is especially important to observe that added acid or salt does not influence the milk-clotting test.

In general the presence of prochymosin in its inactive form will not invalidate a chymosin test. However, by testing small concentrations of chymosin in the presence of high concentrations of prochymosin, large amounts of potential milk-clotting activity are added to the milk and a slight activation during the milk-clotting test cannot be excluded.[4]

[9] H. Holter, *Biochem. Z.* **255**, 160 (1932).
[10] M. L. Anson, *J. Gen. Physiol.* **22**, 79 (1939).
[11] B. Foltmann, *Compt. Rend. Trav. Lab. Carlsberg* **34**, 319 (1964).

Purification of Prochymosin

In all published purifications of prochymosin advantage has been taken by the fact that prochymosin is stable at weakly alkaline pH while chymosin is denatured at pH above 7. The procedure described below is based on my experience with preparations of prochymosin from dried calf stomachs.[12,13]

Step 1. Extraction. Dried calf stomachs are cut, e.g., in an impact mill. Two hundred grams of the cut stomachs are extracted with 3 liters of 2% $NaHCO_3$ for 2 hours at room temperature (mechanical stirring). The tissue is removed by centrifugation.

Step 2. Clarification. With continuous stirring, 100 ml of $1/3\,M$ $Al_2(SO_4)_3$ are slowly added. The precipitate is removed by centrifugation. The supernatant is subjected to a second precipitation with 80 ml $1/3\,M$ $Al_2(SO_4)_3$ followed by addition of 80 ml $1/3\,M$ Na_2HPO_4. The precipitate is removed by centrifugation.

During the addition of $Al_2(SO_4)_3$, CO_2 escapes with vigorous effervescence. The presence of carbon dioxide makes exact determinations of pH difficult. After the first addition of aluminum sulfate pH should be about 7 and after the second about 6. Addition of phosphate should raise the pH to about 7.

The absolute amount of aluminum sulfate and phosphate used might be adjusted according to the starting material of calf stomachs.

Step 3. Precipitation. The clear supernatant from step 2 is precipitated by saturation with NaCl. The suspension is left overnight at room temperature. The precipitate is collected by centrifugation and redissolved in the minimum amount of sodium phosphate buffer ($0.05\,M$, pH 6.3).

Step 4. Reprecipitation takes place by saturation with NaCl. After standing overnight the precipitate is collected by centrifugation.

Step 5. Dialysis. The precipitate is dissolved in 100 ml of $0.05\,M$ sodium phosphate buffer, pH 6.3, and dialyzed against 3×2 liters of $0.02\,M$ sodium phosphate buffer for 2, 6, and 18 hours in the cold room.

Step 6. Chromatography. A column of DEAE-cellulose (3×20 cm) is equilibrated with $0.05\,M$ sodium phosphate, pH 6.0.[13a] The solution of partly saltfree purified prochymosin from step 5 is applied to the column. Most of the nonenzymatic material is washed out with 500 ml

---

[12] B. Foltmann, *Acta Chem. Scand.* **14**, 2247 (1960).

[13] B. Foltmann, *Compt. Rend. Trav. Lab. Carlsberg* **32**, 425 (1962).

[13a] The buffers used for the chromatographic fractionations contain Na_2HPO_4 and NaH_2PO_4 in the ratio 1:9. The decrease in pH with increasing buffer concentration is an effect of the ionic strength.

TABLE I

PURIFICATION OF PROCHYMOSIN

Step	Volume (ml)	PCU/ml	mg N/ml	PCU/mgN	% Recovery
Extraction	2100	224	3.02	74.2	100
Clarification	1650	190	1.65	115	67
Precipitation	210	810	1.77	458	36
Reprecipitation	135	1150	2.32	496	33
Dialysis	185	790			31
Chromatography	340	315			23
Freeze-dried prochymosin[a]	610 mg	120 PCU/mg	156 mgN/g[b]	770	16

[a] Preformed chymosin activity ca. 1 CU/mg.
[b] Determined after drying to constant weight at 100°.

of 0.1 M phosphate buffer. Prochymosin is eluted with 750 ml of 0.2 M phosphate buffer.

Step 7. Concentration. The main part of the prochymosin containing fractions from step 6 is pooled and diluted with a 3-fold volume of distilled water. The prochymosin from this solution is absorbed on a short column of 2.5 g DEAE-cellulose. From this column it is again eluted with 100 ml of 1 M NaCl in 0.01 M phosphate buffer, pH 6.1.

Step 8. Dialysis and Freeze Drying. The concentrated prochymosin solution is dialyzed against 4×2 liters of 0.005 M pyridine for 2, 6, 16, and 24 hours. Finally the preparation is freeze-dried.

See Table I for a summary of this purification.

The procedure described above has been used in my laboratory for seven preparations of prochymosin, all with essentially the same results. Rand and Ernstrom[14] have used a method similar to this while Bundy et al.,[15] starting with an acetone extraction of the gastric mucosa, omitted clarification by precipitation.

By performing the chromatographic fractionation with low prochymosin concentration and using gradient elution with a shallow gradient of phosphate buffer from 0.1 to 0.25 M, I have been able to separate prochymosin in two chromatographic components.[13] These have been named prochymosin-A and -B. After activation the precursors are converted into chymosin-A and -B, respectively. A similar fractionation was not obtained by Bundy et al.[15]

Purification of Chymosin

Chymosin may be prepared from calf stomachs directly[8,16] or from commercial preparations of cheese rennet. The latter starting material is the most easily obtained. In this section a procedure for crystallization of chymosin from cheese rennet powder is described. However, crystalline chymosin is not a homogeneous substance and a method for chromatographic fractionation is outlined.

Crystallization of Chymosin from Cheese Rennet Powder

Step 1. One hundred grams of cheese rennet powder (preferably without pepsin) are dissolved in 500 ml of distilled water. The pH is adjusted to 5.6. The turbid solution is dialyzed in the cold room against 2×5 liters of 0.01 M sodium phosphate buffer pH 6.1 for 6 and 18 hours.

Step 2. Clarification. After dialysis, clarification takes place by addi-

[14] A. G. Rand and C. A. Ernstrom, *J. Dairy Sci.* **47**, 1181 (1964).
[15] H. F. Bundy, N. J. Westberg, B. M. Dummel, and C. A. Becker, *Biochemistry* **3**, 923 (1964).
[16] C. Alais, *Lait* **36**, 26 (1956).

tion of $1/3\,M$ Al$_2$(SO$_4$)$_3$ to a pH of 3.5 (ca. 5 ml). The pH is again raised to 5.5 by addition of $1/2\,M$ Na$_2$HPO$_4$ (ca. 12 ml). The gelatinous precipitate is removed by centrifugation.

Step 3. Precipitation. The clarified liquid is precipitated by saturation with NaCl. After standing overnight at room temperature, the precipitate is collected by centrifugation and redissolved in the minimum amount of $0.01\,M$ phosphate buffer, pH 6.1 (30–50 ml; if necessary pH should be adjusted to 5.5 by slow addition of $0.5\,M$ Na$_2$HPO$_4$).

Step 4. Reprecipitation. Reprecipitation takes place by a second saturation with NaCl. After standing overnight as before, the precipitate is collected by centrifugation and redissolved in the minimum amount of $0.01\,M$ phosphate buffer, pH 6.1 (15–20 ml).

Step 5. Crystallization. The purified solution is left in the cold room. Occasionally scrape at the glass wall until the first crystals are produced. After 3–4 weeks about half of the chymosin may have crystallized. The crystals are collected by centrifugation. Washing of the crystals takes place first with $1\,M$ NaCl and second by ice-cold distilled water. Subsequently the crystals are dissolved in 10–15 ml of $0.01\,M$ phosphate buffer, pH 6.1.

Step 6. Recrystallization. The solution is made $1\,M$ with respect to NaCl and is left in the cold room for a second crystallization. Finally the crystals are washed with $1\,M$ NaCl and stored suspended in $4\,M$ NaCl.

In addition to preparation of chymosin from calf stomachs and from liquid cheese rennet, about 10 preparations have been performed in the author's laboratory according to the method described above. If centrifuges with sufficient capacity are available, it is easy to scale the preparation up with a factor of 10.

Table II summarizes this crystallization.

TABLE II
CRYSTALLIZATION OF CHYMOSIN

Step	Volume (ml)	CU/ml	mgN/ml	CU/mgN	% Recovery
1. Dissolution and dialysis	700	172	0.74	232	100
2. Clarification	620	166	0.69	241	85
3. Precipitation	38	1900	4.80	396	60
4. Reprecipitation	25	2450	4.71	520	51
5. Crystallization	15	2100	2.31	909	26
6. Suspension after recrystallization	10	2160	2.34	923	18

Chromatographic Fractionation of Chymosin

By chromatography of columns of DEAE-cellulose it is possible to resolve crystalline chymosin into three chromatographic components, each with a characteristic ratio between the milk-clotting activity of the solution and its absorbancy at 278 nm,[4, 17] The three components are called chymosin-A, -B, and -C in order of decreasing ratio between the milk-clotting activity and A_{278} nm. Separate zymogens have been isolated for chymosin-A, and -B, while chymosin-C probably is a mixture, partly consisting of degradation products.

A sample of crystalline chymosin corresponding to approximately 750 mg is dissolved by dialysis against 0.01 M phosphate buffer, pH 6.1. A column of DEAE-cellulose (2 cm in diameter, 20 cm high) is equilibrated with 0.2 M sodium phosphate buffer of pH 5.8.

The sample of dialyzed chymosin is applied at the column. Elution takes place with a linear gradient of 1100 ml of 0.2 M phosphate buffer, pH 5.8, and 1100 ml 0.3 M phosphate buffer pH 5.6, flow rate 30 ml per hour. An example of such a fractionation is shown in Fig. 1. The chromatogram is evaluated by the ratio between the milk-clotting activity

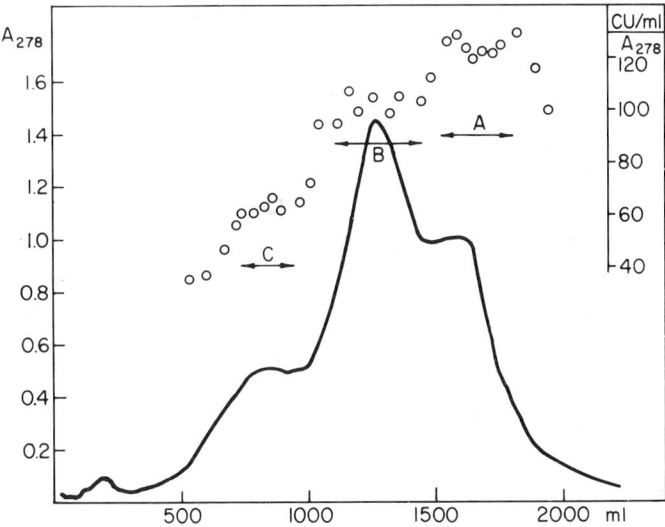

Fig. 1. Chromatographic fractionation of chymosin. Sample: 750 mg of chymosin. Column: DEAE-cellulose 2 cm in diameter 20 cm high. Elution: linear gradient of 1100 ml 0.2 M sodium phosphate buffer, pH 5.8, and 1100 ml 0.3 M phosphate buffer, pH 5.6, flow rate 30 ml per hour. Solid line shows absorbancy at 278 nm. Circles indicate the ratio between milk-clotting activity and A_{278}.

[17] B. Foltmann, *Acta Chem. Scand.* **14**, 2059 (1960).

and the absorbancy at 278 nm. The fractions containing the central part of each peak are pooled as indicated by the double-headed arrows.

Each of the pooled solutions of chromatographically purified chymosin is diluted with a 3-fold volume of distilled water. From these solutions chymosin is absorbed on short columns of DEAE-cellulose equilibrated with 0.1 M phosphate buffer, pH 5.8, and chymosin is again eluted with 1 M NaCl in 0.01 M phosphate buffer, pH 6.0. Most of the chymosin will be eluted with a concentration higher than 2 mg/ml. From such solutions the chromatographically purified components may crystallize again.

Preliminary experiments have shown that it is possible to proceed directly from step 4 of the purification of chymosin to the chromatographic fractionation. In this case salt is removed by dialysis against 0.05 M phosphate buffer, pH 5.9. The dialyzed solution of partly purified chymosin is applied to a column of DEAE-cellulose equilibrated with 0.1 M phosphate buffer, pH 5.8. Nonenzymatic material is washed through the column with 0.1 M phosphate buffer, and subsequently the chymosins are eluted with a linear gradient of sodium phosphate buffer from 0.1 to 0.3 M. Preparations with yields of more than 50 mg of the individual components have not yet been performed.

Properties

Stability and Activation of Prochymosin. Solutions of prochymosin are reasonably stable in the pH range of 5.5 to 9.[4] At pH below 5 prochymosin is converted into chymosin by a limited proteolysis during which a peptide segment is cleaved from the N-terminus. The N-terminal residue of prochymosin is alanine[12,15] while that of chymosin is glycine. Table III[18] shows a balance sheet of the amino acid compositions of chromatographically purified preparations of prochymosin and chymosin and of peptides released during the activation of prochymosin.

The reaction is very dependent on pH, salt concentration, and temperature. At pH 5 and room temperature the activation is completed in 2 or 3 days,[14] while at pH 2, room temperature, and ionic strength of 0.1, the activation is completed in 5–10 minutes.[13]

If the increase of milk-clotting activity is plotted against time, the course of the activation at pH 4.5 to 5 appears as more or less S-shaped curves, indicating that the reaction is at least partly autocatalytic. Experiments carried out at lower pH show a different course of reaction. The lower the pH, the higher is the initial rate of the reaction and the S-shape of the curve disappears. Some results are illustrated in Fig. 2.[13]

[18] B. Foltmann, *Compt. Rend. Trav. Lab. Carlsberg* **34**, 275 (1964).

TABLE III

AMINO ACID COMPOSITION OF PROCHYMOSIN-B, CHYMOSIN-B, AND OF PEPTIDES
SPLIT OFF DURING THE ACTIVATION OF PROCHYMOSIN

	Residues per molecule				Molar ratio of amino acid in activation peptides calculated relative to arginine	Difference between proposed formulas of prochymosin and chymosin
	Prochymosin-B		Chymosin-B			
	Residues per 36200 g	Nearest integer	Residues per 30700 g	Nearest integer		
Lysine	12.97	13	8.14	8	4.91	5
Histidine	5.01	5	4.10	4	1.11	1
Arginine	7.01	7	4.85	5	2.00	2
Aspartic acid	33.37	33	29.99	30	3.40	3
Threonine	20.42	20	18.45	18	2.48	2
Serine	30.56	31	26.68	27	4.44	4
Glutamic acid	36.31	36	28.93	29	7.20	7
Proline	13.98	14	12.02	12	1.96	2
Glycine	28.49	28	24.50	25	4.89	3
Alanine	15.01	15	12.82	13	2.87	2
Half-cystine	5.08	6[a]	4.88	6[a]	[b]	0
Valine	22.80	23	20.90	21	1.99	2
Methionine	6.76	7	6.66	7	0.14	0
Isoleucine	19.04	19	15.16	15	3.42	4
Leucine	25.56	26	18.54	19	6.50	7
Tyrosine	18.34	18	15.38	15	2.79	3
Phenylalanine	16.45	16	13.79	14	2.43	2
Tryptophan	4.05	4	4.19	4	[b]	0
Amide N	34.03	34	30.70	31		

[a] From sequence studies.
[b] Not determined.

I have obtained the highest yield of chymosin by a rapid activation at pH 2, while Rand and Ernstrom prefer a slow activation at pH 5.[14]

For a discussion of the activation kinetics see footnotes 4 and 5.

Stability of Chymosin. Cheese rennet powder and suspensions of chymosin crystals appear to be very stable when stored at temperatures below 5°. Solutions of chymosin have optimum stability at pH between 5.3 and 6.3. At pH around 3.5 and above 7 the enzyme rapidly loses its activity. At pH 2, however, chymosin solutions are relatively stable[8]; if the solutions are saltfree, the stability is reduced by increasing salt concentration.[19]

Solubility of Chymosin. Most determinations of the solubility of chymosin have been carried out in solutions of NaCl. At about pH 5.5

[19] R. Mickelsen and C. A. Ernstrom, *J. Dairy Sci.* **50**, 645 (1967).

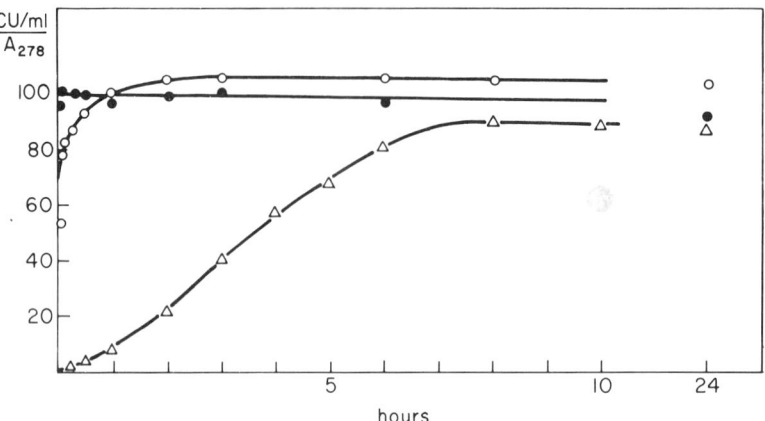

FIG. 2. Activation of prochymosin. Prochymosin concentration ca. 1 mg per ml reaction mixture. Temperature 25°. Experiment I (△): 1 ml prochymosin solution + 1 ml 0.2 M acetate pH 4.7. Experiment II (○): 1 ml prochymosin solution + 1 ml 0.025 N HCl, pH 2.0. Experiment III (●): 1 ml prochymosin solution + 1 ml 0.025 N HCl/0.2 M NaCl, pH 2.0.

crystalline chymosin has a solubility less than 1 mg/ml in solutions containing more than 0.5 M NaCl. Crystalline chymosin prepared in the author's laboratory has been soluble by dialysis against distilled water, while Berridge[6] has reported that his crystals were insoluble in distilled water.

In equilibrium with amorphous precipitates chymosin is very soluble in solutions of 1 M NaCl and pH about 5.5. In 2 M solutions of NaCl the solubility is of the order 20 mg/ml and it is less than 1 mg/ml in saturated solutions of NaCl. In salting out experiments, amorphous precipitates of chymosin are more soluble at 2° than at 25°.

At pH between 5 and 6 chymosin has a strange solubility minimum at ionic strength of 0.1. Thus a transient amorphous precipitate is observed when NaCl is added during the recrystallization.[20]

Chymosin has an isoelectric point of about pH 4.6. At ionic strength of 0.005, chymosin is very insoluble from pH 4.4 to 4.8, the solubility being increased by increasing ionic strength.

Molecular Weights, Amino Acid Compositions, and Primary Structures. The data from ultracentrifugation, diffusion, and gel filtration have been compared with the minimum molecular weights obtained from the amino acid composition.[18, 21] Rounded off to the nearest thousand,

[20] B. Foltmann, *Acta Chem. Scand.* 13, 1936 (1959).

[21] R. Djurtoft, B. Foltmann, and A. Johansen, *Compt. Rend. Trav. Lab. Carlsberg* 34, 287 (1964).

molecular weights of 36,000 and 31,000 are found for prochymosin and chymosin, respectively. The individual fractions of chymosin cannot be distinguished by their molecular weights. (See Table IV.)

On the basis of sedimentation equilibrium studies Cheeseman[22] has recently reported that at pH 3.1 solutions of chymosin contain a monomer of molecular weight about 30,000 while at pH 5.8 to 6.5 a dimer is mainly present (apparent molecular weight about 64,000).

TABLE IV
MOLECULAR WEIGHT OF PROCHYMOSIN AND CHYMOSIN

Mol. wt. (Daltons)	Method	$S_{20,w}$ (S)	$D \cdot 10^{-7}$ ($cm^2\ sec^{-1}$)	$\bar{V}$ ($ml \cdot g^{-1}$)	Ref.
	Prochymosin				
36,500	Sedimentation/diffusion	3.5	ca. 8.5	0.73	a
36,000	Gel filtration				a
36,200	Amino acid composition				b
	Chymosin				
40,000	Sedimentation/diffusion	4	9.5	0.75	c
34,400	Sedimentation/diffusion	2.6	6.8		d
34,000	Sedimentation/diffusion	3.2	8.5	0.73	a
33,000	Gel filtration				a
31,000	Gel filtration				e
34,000	Gel filtration				f
30,700	Amino acid composition				b
30,400	Amino acid composition				f

[a] R. Djurtoft, B. Foltmann, and A. Johansen, *Compt. Rend. Trav. Lab. Carlsberg* **34,** 287 (1964).
[b] B. Foltmann, *Compt. Rend. Trav. Lab. Carlsberg* **34,** 275 (1964).
[c] H. Schwander, P. Zahler, and Hs. Nitschmann, *Helv. Chim. Acta* **35,** 553 (1952).
[d] L. Friedman, *Dissertation Abstr.* **20,** 4510 (1960).
[e] P. Andrews, *Biochem. J.* **91,** 222 (1964).
[f] P. J. de Koning, Thesis, Univ. of Amsterdam, The Netherlands, 1967.

Within the experimental error of the amino acid analyses published so far, it has not been possible to show any significant differences between amino acid compositions of total crystalline chymosin and chromatographically purified chymosin-B. The results of analyses of chromatographically purified preparations are illustrated in Table III, which also shows the amino acid composition of the peptides liberated during the limited proteolysis of the activation process.

Most of the sequence studies carried out until now have been performed with total crystalline chymosin. The results from some major

[22] G. C. Cheeseman, *J. Dairy Res.* **36,** 299 (1969).

TABLE V

AMINO ACID SEQUENCES OF PEPTIDES FROM CHYMOSIN[a,b,c] PARTLY COMPARED WITH PEPSIN[d,e]

N-Terminal sequence: H Gly Glu Val Ala Ser Val Pro Leu Thr AsN Tyr Leu Asp Ser GlN Tyr Phe Gly Lys Ile Tyr

Cystine bridges:

(A)   Asp Ile Asp Cys Asp AsN Leu Glu Gly Ser
      Tyr Thr Ser GlN Asp GlN Gly Phe Cys Thr Ser Gly Phe

(B)   Ala Cys Glu Gly Gly Gly Cys GlN
                  Ser

(C)   Cys Lys Ser AsN Ala Cys Lys AsN His GlN Arg Phe Asp Pro Arg
           Ser   Leu   Ser Asp   AsN GlN   AsN   Asp

C-Terminal sequence:

Try Ile Leu Gly Asp Val Phe Ile Arg Glu Tyr Tyr Ser Val Phe Asp Arg Ala AsN AsN Leu Val Gly Leu Ala Lys Ala Ile OH
                                     GlN   Thr   Lys   Pro Val Ala OH

[a] In the cystine bridges B and C, and in the C-terminal sequence the dashed lined frames show the differences between calf rennin and porcine pepsin. The lower amino acids in these frames indicate the composition of the pepsin peptides (see the text).

[b] B. Foltmann and B. S. Hartley, *Biochem. J.* **104**, 1064 (1967).

[c] B. Foltmann, *Phil. Trans. Roy. Soc. London B* **257**, 147 (1970).

[d] J. Tang and B. S. Hartley, *Biochem. J.* **118**, 611 (1970).

[e] T. A. A. Dopheide, S. Moore, and W. H. Stein, *J. Biol. Chem.* **242**, 1833 (1967).

fragments are illustrated in Table V. It should be emphasized that peptide patterns obtained from chromatographically purified chymosin-B are consistent with the results shown in Table V. To what extent the different chymosins are identical remains to be investigated.

General Proteolytic Activity and Specificity. All the chromatographic fractions of chymosin show general proteolytic activity around pH 3.5. It should be noted, however, that the ratio between the milk-clotting and the proteolytic activity is lower for chymosin-C than it is for chymosin-A, and -B.[11]

The proteolytic specificity has been tested with the B-chain of oxidized insulin as substrate. It has not been possible to demonstrate any differences between the proteolytic specificities of the individual fractions by this method. The specificity has some resemblance to that of pepsin.[23, 24]

Homology with Pepsin. Distribution. In many respects the properties of chymosin are similar to those of pepsin. Both are acidic gastric proteases of similar molecular weights, and they are secreted as precursors which are converted into active enzymes by a limited proteolysis from the N-terminus.

The similarities between chymosin and pepsin probably arise from similarities in primary structures. We are able to compare only relatively short sections of the amino acid sequences of pepsin and chymosin, but three of the peptide fragments illustrated in Table V show a pronounced homology. In the C-terminal sequence, and the sequences around the S-S bridges marked B and C the amino acid residues outside the frames are identical in the two enzymes, the differences are marked in the frames. The amino acid sequences around the third S-S bridge (A) show a faint degree of homology, this is not marked in the table. However, the N-terminal amino acid sequence of chymosin shows no obvious resemblance to any known sequence from pepsin.

The amino acid sequence around the active center of chymosin is not yet known, but after chymotryptic digestion a peptide has been found which is very similar to the active center peptide from pepsin found recently.[25] The homology suggests that the chymosin peptide in fact also represents the active center. The results are shown in Table VI.

The similarities between chymosin and pepsin are pertinent for an evaluation of the distribution of the enzyme. Chymosin has only been isolated from calf, but the scattered information suggests that a proteoly-

[23] J. C. Fish, *Nature* **180**, 345 (1957).
[24] V. Bang-Jensen, B. Foltmann, and W. Rombauts, *Compt. Rend. Trav. Lab. Carlsberg* **34**, 326 (1964).
[25] R. S. Bayliss, J. R. Knowles, and G. Wybrandt, *Biochem. J.* **113**, 377 (1969).

TABLE VI
HOMOLOGY BETWEEN AN ACTIVE CENTER PEPTIDE FROM HOG PEPSIN[a] AND A
CHYMOTRYPTIC FRAGMENT FROM CHYMOSIN[b]

			*						
Pepsin	Ile	Val	Asp	Thr	Gly	Thr	Ser		
Chymosin			Asp	Thr	Gly	Ser	Ser	Asp	Phe

[a] R. S. Bayliss, J. R. Knowles, and G. Wybrandt, *Biochem. J.* **113**, 377 (1969).
[b] B. Foltmann, unpublished results, 1969.
* Indicates the active aspartic acid residue in pepsin.

tic enzyme with a high milk-clotting activity is characteristic for the gastric juice of the young ruminants. As the animal grows up, a change takes place from predominant chymosin to predominant pepsin secretion. The possibility exists that a similar change in the composition of the gastric juice may occur during the ontogenesis of other animals, but it has not been noticed since it is not accompanied with an easily observable change in milk-clotting activity. My opinion is, that had it not been because of cheese-making, chymosin would have been known as calf pepsin with addition of some odd letter.

[29] Protease of *Endothia parasitica*

By JOHN R. WHITAKER

Assay Method

Principle. The renninlike protease produced by the fungus *Endothia parasitica* catalyzes a specific proteolysis leading to clotting of milk as well as a general proteolysis of proteins. Under carefully standardized conditions, the primary phase (proteolysis) is the rate-determining step in milk-clotting and measurement of the clotting time can be used as a very sensitive, quick, and reliable assay for monitoring the purification procedure. A criterion that the enzymatic phase, and not the coagulation phase, is rate-determining is that a plot of clotting time versus reciprocal of enzyme concentration is a straight line with an intercept of zero. When greater precision is required, the assay method based on the initial rate of hydrolysis of casein or hemoglobin is recommended.

MILK-CLOTTING ASSAY

Reagents

Nonfat dry milk powder. Carnation Instant Nonfat Dry Milk powder is excellent. The product should be as fresh as possible and it should be stored at 2°, protected from moisture after opening.

Sodium acetate buffer, 0.67 M, pH 4.60

Procedure.[1,2] Twenty grams of milk powder are dissolved in 50 ml distilled water at room temperature. Care should be taken that all the milk powder is dissolved at this point. Then 30 ml of sodium acetate buffer, pH 4.60, are added slowly from a pipette with continuous stirring and with care to prevent excess foaming. The solution is made to 100 ml in a volumetric flask with water. The pH should be 5.10 ± 0.05. The substrate solution is permitted to stand at room temperature for not less than 45 minutes before use. It should not be stored in a refrigerator or used for more than 1 day. One-milliliter aliquots of substrate are pipetted into the bottom of 13 × 125-mm Pyrex text tubes which are then incubated in a water bath at 35.0° for not less than 5 minutes or more than 15 minutes. An aliquot (preferably 0.05 ml) of the enzyme solution is added rapidly from a 0.10 ml pipette (mark to mark) by placing the tip of the pipette just below the milk surface. A timer, calibrated to 0.1 second, is started immediately, the contents of the tube are mixed thoroughly but gently so as not to cause a bubble, and the tube is rotated slowly around the short axis while held at a 30°–40° angle. The film of milk around the wall of the tube is observed and the end point is the first appearance of a grainy texture. For greater precision, an enzyme concentration should be chosen which gives a clotting time of not less than 50 seconds and not more than 100 seconds. The determination is done in triplicate and the clotting times should agree within 2%. For the greatest precision, a high-intensity light mounted immediately above the water bath and a test tube rack (wire) so positioned in the bath that when the tube rests against one of the openings in the rack and the side of the bath it forms a 30° angle are helpful.

Unit and Specific Activity. One unit is defined as that amount of enzyme which will clot 1 ml of milk in 60 seconds. Specific activity is expressed as units per milligram protein.

Remarks. Lower concentrations of substrate can be used with a proportional decrease in clotting time at a fixed enzyme concentration.[3] Clot formation can be detected even at 2% substrate concentration. However, at lower substrate concentrations the coagulation phase (nonenzymatic) becomes more important and, because of the large number

---

[1] J. R. Whitaker, *Food Technol.* **13**, 86 (1959).

[2] A. K. Balls and S. R. Hoover, *J. Biol. Chem.* **121**, 737 (1937).

[3] M. K. Larson, "Some Chemical and Enzymatic Properties of *Endothia parasitica* Protease." M.S. thesis, Univ. of California, Davis, 1969.

of factors which affect the coagulation phase, the precision of the method is decreased.

Protein Hydrolysis Method

Reagents

1% Casein or hemoglobin. One gram of casein (Hammarsten quality) is dissolved in 25 ml 0.2 M phosphoric acid, 55 ml of distilled water are added, and the pH is adjusted with 1 N NaOH to 2.50 using a pH meter. The solution is diluted to 100 ml with distilled water. For activity determination at pH 6.0, 1 gram casein is dissolved in 25 ml 0.2 M sodium acetate by brief warming in a boiling water bath, the solution is cooled, 55 ml distilled water are added, and the pH is adjusted with 1 N HCl to 6.0. The solution is diluted to 100 ml with water.

The hemoglobin must be denatured before use.[4] Two grams of Hemoglobin Substrate Powder (Mann Research Lab.) are dissolved in 75 ml of 0.06 N HCl, the pH is adjusted to 1.8 with 0.5 N HCl, and the solution is held at room temperature for 20–24 hours. After exhaustive dialysis of the solution against distilled water at 2°, it is diluted to 100 ml with water. The stock solution is stable for several weeks in the refrigerator. For hydrolysis at pH 2.5, 25 ml of 0.2 M phosphoric acid are added to 50 ml of stock hemoglobin solution, the pH is adjusted to 2.5 with 1 N NaOH, and the solution is diluted to 100 ml. For hydrolysis at pH 6.0, 25 ml of 0.2 M acetic acid are used in place of phosphoric acid and the pH is adjusted to 6.0.

5% Trichloroacetic acid (TCA)

Procedure.[5] Fifteen milliliters of substrate are placed in a 20 × 175-mm test tube and equilibrated for 10 minutes at 35.0° in a water bath. An aliquot of enzyme (0.05 ml, equivalent to 0.03 mg purified enzyme) is added, mixed, and 2-ml aliquots are removed at zero time and subsequently at 5-minute intervals into 3 ml of 5% TCA. The contents are mixed thoroughly, allowed to stand at room temperature for 1 hour, centrifuged at 2000 rpm (International Clinical centrifuge) for 20 minutes, and the absorbance of the supernatant liquid read at 280 nm.

Alternatively, 0.10-ml aliquots may be removed at intervals into 0.90 ml 0.1 N HCl and the number of peptide bonds split determined

[4] M. L. Anson, *J. Gen. Physiol.* **22**, 79 (1938).
[5] M. Kunitz, *J. Gen. Physiol.* **30**, 291 (1947).

with the quantitative ninhydrin reagent[6] using 0.1–0.2-ml aliquots of the HCl-diluted sample.

Unit and Specific Activity. One unit is defined as that amount of enzyme which will produce a change of 0.001 in absorbance per minute measured as an initial rate. Specific activity is expressed as units per milligram protein.

Remarks. The progress curve is not linear at either pH 2.5 or 6.0. The nonlinearity is much greater at pH 6.0. Only the initial rates of hydrolysis are linearly related to enzyme concentration.

Purification Procedure

Production of Enzyme.[7] Cultures[8] of the organism are maintained on potato dextrose agar; the organism will grow in a medium containing a source of carbohydrate, nitrogen, and inorganic salts. For the production of protease the following procedure has been successful. The growth medium contains (per liter): 5.0 g N-Z amine B (hydrolyzed casein preparation, Sheffield Chemical Co., Inc.), 2.0 g soya protein digest (Soytone, Difco Lab.), 2.0 g yeast extract, 10.0 g soluble starch, 5.0 g D-mannitol, 15.0 mg $FeSO_4 \cdot 7\ H_2O$, and 1.0 ml of trace elements—Fe [as $Fe(NH_4)_2(SO_4)_2$], 1.0 mg/ml; Zn (as $ZnSO_4$), 1.0 mg/ml; Mn (as $MnSO_4$), 0.50 mg/ml; Cu (as $CuSO_4$), 0.08 mg/ml; Co (as $CoSO_4$), 0.10 mg/ml; B (as H_3BO_3), 0.10 mg/ml. One liter of the medium is placed in a 2.8-liter Fernbach flask and sterilized in an autoclave for 45 minutes at 121°. The organism is washed from the potato dextrose agar slant into the flask under sterile conditions. The flask is then incubated in a rotating shaker at 28° for 96 hours.

The enzyme production medium contains (per liter): 30.0 g soybean meal, 10.0 g Cerelose (glucose), 10.0 skim milk, 3.0 g $NaNO_3$, 0.50 g KH_2PO_4, and 0.25 g $MgSO_4 \cdot 7\ H_2O$. The pH of the medium is about 6.8. The medium is sterilized in 2-liter portions in 4-liter fermentation vessels in an autoclave for 1 hour at 121°. After cooling the enzyme production medium, 100 ml of the growth medium (inoculum) are added to each 2 liters of sterile growth medium. The cultures are agitated at 1700 rpm at 28° while air is introduced at the rate of 0.5 volumes air per volume medium per minute. After 48 hours of incubation, 0.5 ml of the medium will clot 5 ml sweet milk within 3 minutes.

Preparation of Crude Enzyme.[7] The culture medium (2 liters) is

[6] S. Moore and W. H. Stein, *J. Biol. Chem.* **211**, 907 (1954).

[7] J. L. Sardinas, U.S. Pat. 3,275,453, Sept. 27, 1966. (Assigned to Charles Pfizer and Co., New York.)

[8] The organism is available from American Type Culture Collection, Washington, D.C. where it is designated as ATCC 14729.

filtered on a Büchner funnel containing a 0.5-cm layer of Hyflo Super-Cel and the filter cake is washed with 100 ml of deionized water. The filtrate is freeze dried and the freeze-dried solids (clotting specific activity ~2.6 units/mg) are reconstituted in 400 ml deionized water. In the subsequent steps nitrogen is flushed over the container. After cooling the solution to 1°, 2 volumes of acetone, at —25°, are added slowly with constant stirring. The acetone–water layer is decanted from the congealed solid, the excess acetone–water is removed from the solid by pressing with a spatula, and most of the acetone is removed in a rapid stream of nitrogen. Immediately after, the solid is suspended in an equal volume of deionized water and freeze dried. The freeze-dried solids, containing 80–85% of the original activity, are dissolved in water to give a 15% (w/v) solution. The solution is cooled to 1° and acetone (approx. 1.5 vol), at —25°, is added slowly with continual stirring until the precipitate just begins to congeal. The precipitate is allowed to settle and the liquid is removed by decantation. An additional 0.75 volumes of acetone are added slowly to the supernatant liquid. The liquid is removed and the solids are freeze dried as described for the first acetone precipitation. The yield is about 40% of the original activity with a clotting specific activity of about 15 units/mg. Exhaustive dialysis of the solids (dissolved in an equal volume of water) against deionized water increases the specific activity to about 30–32 units/mg. Enzyme of this purity is available from Charles Pfizer and Co., Milwaukee, Wis.

Purification of Enzyme[9]

Further purification of the enzyme can be obtained by $(NH_4)_2SO_4$ precipitation, Sephadex G-100 filtration, and chromatography on DEAE-cellulose. All subsequent steps are carried out at 0° to 2°. Centrifugation is performed at 10,400 g in a Sevall refrigerated centrifuge for 20 minutes. Dialysis is carried out against 0.05 M acetate buffer, pH 4.60.

The dialyzed solution of the crude enzyme (above) is diluted to give a 2.8% solution (protein concentration). Solid $(NH_4)_2SO_4$ is added slowly to the solution (fraction 1, Table I) to 60% saturation (0°). After 1 hour the suspension is centrifuged and the supernatant liquid (fraction 3) is discarded. The precipitate is dissolved in water to give a 3.5% solution and dialyzed for 48 hours (fraction 2). Solid $(NH_4)_2SO_4$ is added to fraction 2 to 40% saturation and the suspension is centrifuged after 1 hour. The precipitate (fraction 4) is discarded. Additional $(NH_4)_2SO_4$ is added to the supernatant liquid to 55% saturation and the suspension centrifuged after 1 hour. The supernatant liquid (fraction 6) is dis-

[9] K. Hagemeyer, I. Fawwal, and J. R. Whitaker, *J. Dairy Sci.* **51**, 1916 (1968).

carded. The precipitate is dissolved in 0.05 M acetate buffer, pH 4.6, to give a 3.5% solution and is dialyzed overnight (fraction 5).

At this stage the preparation still contains a brown pigment which is not removed by Polyclar AT or by charcoal but is readily removed by filtration on Sephadex G-100. Fifty-milliliter aliquots of fraction 5 may be added at one time to a 4.1 × 45-cm column equilibrated with 0.05 M acetate buffer —0.4 M NaCl, pH 4.60. The column is irrigated with the same buffer. All the milk-clotting activity is obtained in the major peak (280 nm absorbance) eluted at 2.5 times the void volume of the column while the pigment is eluted as a separate peak at 4.5 times the void volume. The tubes containing the majority of the enzyme are combined (fraction 7) and precipitated with $(NH_4)_2SO_4$ to 90% saturation. The supernatant liquid (fraction 9) is discarded and the precipitate (very white) is dissolved in 0.05 M acetate buffer, pH 4.6, to give a 3.5% solution.

After dialysis against 0.2 M acetate buffer, pH 4.6, up to 0.25 g of fraction 8 are added to the top of a 2.1 × 45-cm column of DEAE-cellulose equilibrated against the same buffer. DEAE-cellulose is prepared for chromatography by grinding through an 80-mesh sieve of a Wiley Mill, removing of fines, washing with 0.25 M NaCl–0.25 M NaOH, with water until free of Cl⁻, with 0.25 M HCl, with water until free of Cl⁻ and then five times with the starting buffer. The column is poured in one segment by means of an extension tube and, after placing in a cold room, the column is washed with a minimum of 3 void volumes of starting buffer before use. The protein is eluted from the column with a linear gradient. The mixing vessel contains 500 ml of 0.2 M acetate buffer, pH 4.6, and the second vessel contains 500 ml of 0.2 M acetate buffer– 0.6 M NaCl, pH 4.6 (0.6 M acetate buffer, pH 4.6, may also be used). Ten-milliliter fractions are collected at a rate of 3 fractions per hour. Several minor protein components (up to 4) are eluted in rapid succession from the column. The major protein peak, and the only one with milk-clotting activity, begins to come off at fraction 55 and reaches a maximum concentration in fraction 60. The fractions containing the majority of the enzyme are combined, and the enzyme precipitated by addition of $(NH_4)_2SO_4$ to 90% saturation. The precipitate, after recovery by centrifugation, is dissolved in a minimum volume of water and the solution is dialyzed against 0.33 M acetate buffer, pH 4.6 (fraction 10).

Fraction 10 is rechromatographed on a DEAE-cellulose column (2.1 × 45 cm) equilibrated with 0.33 M acetate buffer, pH 4.6. The protein is eluted with the starting buffer. A minor inert protein peak is eluted first, followed by the enzyme which is eluted as a single peak (center of peak is at 2.5 times the void volume of the column). The fractions containing the majority of the activity are combined, and the enzyme is precipitated with $(NH_4)_2SO_4$ to 90% saturation. After 1 hour the precipitate is removed by centrifugation and is dissolved in a minimum volume of water and dialyzed against 0.05 M acetate buffer, pH 4.60 (fraction 11). Our best preparations of enzyme have 12–17% the specific activity in clotting milk of our crystalline rennin preparations.

A typical purification is presented in Table I.

Crystallization. A typical example is given. Cold (0°) isopropanol is added dropwise to 10 ml of a preparation equivalent to fraction 11 (Table I) which contains 33.5 mg/ml protein until the first trace of turbidity is seen. This requires an isopropanol concentration of 28.6%. The solution is allowed to stand at 2°. After 24 hours there is some amorphous precipitate but no crystals. After 48 hours the tube is filled with crystals. After 96 hours, the needle-shaped crystals are collected by centrifugation, washed with a little cold 28.6% isopropanol (isopropanol diluted with 0.05 M acetate buffer, pH 4.60) and dried in a desiccator over P_2O_5 at 2°. The yield is 72% with a specific activity of 125 units/mg.

Crystallization of the enzyme is extremely easy. We have never failed to obtain crystals and even less pure preparations may be crystallized. However, the specific activity is always decreased about 10–12% by the crystallization procedure. For this reason, storage of the enzyme solution at −15° or lyophilization of the enzyme is preferred. When the enzyme solution is equilibrated against 0.01 M acetate buffer, pH 4.6, and frozen at −25° to −30° it may be lyophilized with no loss in activity.

Properties

Stability.[3] The enzyme is maximally stable at pH 3.8 to 4.5. At 50°, only 30% activity is lost in 30 minutes in this pH range. Below pH 2.5 loss in activity is associated with an increase in ninhydrin-reactive groups, fragmentation, and an activation energy of inactivation, E_a, of 14,100 cal/mole. Above pH 6.5 activity is lost rapidly (almost instantaneously above pH 8) and is associated with no increase in ninhydrin-reactive groups, a decrease in solubility, and an E_a of 51,500 cal/mole.

Purity.[9] *Physical and Chemical Properties.* Fraction 11 (Table I) is at least 95% chromatographically pure by starting buffer elution on an analytical DEAE-cellulose column. It migrates as a single band on cellulose acetate electrophoresis at pH 5. The molecular weight has been reported to be 37,500 and 34,000 as determined by Sephadex G-100[9] and Bio-Gel P-100[10] filtration, respectively. The isoelectric point appears

[10] J. L. Sardinas, *Appl. Microbiol.* **16**, 248 (1968).

TABLE I

PURIFICATION OF *E. parasitica* PROTEASE[a]

Fraction	Treatment	Volume (ml)	Total protein[b] (mg)	Total activity[c] (units)	Specific activity (unit/mg)	Recovery (%)	Purification (-fold)
1	Dissolved, dialyzed	624	17,440	568,000	32.6	100	1.00
2	60% (NH₄)₂SO₄, ppt.	460	15,548	525,000	33.8	92.5	1.04
3	60% (NH₄)₂SO₄, supernatant	656	260	1,730	6.65	0.30	0.20
4	40% (NH₄)₂SO₄, ppt.	164	7,740	290,000	37.5	51.1	1.15
5	40–50% (NH₄)₂SO₄, ppt.	173	6,228	370,000	59.5	65.0	1.82
6	40–55% (NH₄)₂SO₄, supernatant	460	302	11,500	38.2	2.0	1.17
7	Sephadex G-100	550	3,630	352,000	97.0	62.0	2.98
8	90% (NH₄)₂SO₄, ppt.	87	2,790	270,000	96.8	47.5	2.97
9	90% (NH₄)₂SO₄, supernatant	655	66	5,190	78.0	9.1	2.39
10	1st DEAE-cellulose	61	880	110,000	125	19.4	3.84
11	2nd DEAE-cellulose	52	468	66,000	141	11.6	4.32

[a] Reproduced in part from text footnote 9 by permission of the American Dairy Science Association.

[b] Protein determined by the biuret method (E. Layne, Vol. III, p. 450).

[c] One unit is defined as that amount of enzyme which will clot 1 ml of substrate in 60 seconds at 35°.

TABLE II

AMINO ACID COMPOSITION OF *E. parasitica* PROTEASE[a]

Amino acid	Determinations[b]	Residues per 37,500	
		Residues	Integer
Aspartic acid	11	25.7 ± 1.2	26
Threonine[c]	12	50.0	50
Serine[c]	12	43.9	44
Glutamic acid	13	14.6 ± 0.8	15
Proline	9	13.2 ± 1.4	13
Glycine	12	37.6 ± 2.7	38
Alanine	12	28.6 ± 1.7	29
Valine	12	22.2 ± 2.4	22
Half-cystine[d]	11	1.62 ± 0.71	2
Methionine	11	0.435 ± 0.180	1
Isoleucine	13	17.4 ± 1.5	17
Leucine	12	18.7 ± 0.9	19
Phenylalanine	13	13.1 ± 1.4	13
Tyrosine	13	$19.4 \pm 1.4 \ (18.5)^e$	19
Ammonia[c]	12	31.1	31
Lysine	12	11.6 ± 1.2	12
Histidine	12	3.05 ± 1.02	3
Arginine	10	1.70 ± 0.35	2
Tryptophan[e]	2	3.18	3

[a] Hydrolysis performed by method of Moore and Stein (Vol. VI, p. 819) and analysis performed on Technicon Auto Analyzer.

[b] Hydrolyses performed for 20, 40, and 70 hours.

[c] Corrected for destruction by extrapolation to zero time.

[d] Determined both as cystine and as cysteic acid.

[e] Determined spectrophotometrically by method of Fraenkel-Conrat (Vol. IV, p. 252).

to be below pH 4.6 since the enzyme is adsorbed strongly to DEAE-cellulose in 0.2 M acetate buffer, pH 4.6.[9] However, Sardinas[10] has reported that his preparation of enzyme behaved as a cation at pH 4.5 on cellulose paper. The amino acid composition of the enzyme is shown in Table II.[11]

Activators and Inhibitors.[9] The enzyme is not inhibited by sodium tetrathionate ($1 \times 10^{-3} M$), $HgCl_2$ ($1 \times 10^{-3} M$), N-ethylmaleimide ($2 \times 10^{-3} M$), iodoacetamide ($2 \times 10^{-3} M$), p-chloromercuribenzoate ($1 \times 10^{-3} M$), cysteine ($1 \times 10^{-2} M$), TPCK[12] ($1 \times 10^{-3} M$), and TLCK[12] ($1 \times 10^{-3} M$).

Specificity. The enzyme does not hydrolyze benzyloxycarbonyl-L-

[11] J. R. Whitaker, unpublished data.

[12] The chloromethyl ketones derived from N-tosyl-L-phenylalanine and N-tosyl-L-lysine, respectively. E. Shaw, Vol. XI, p. 677.

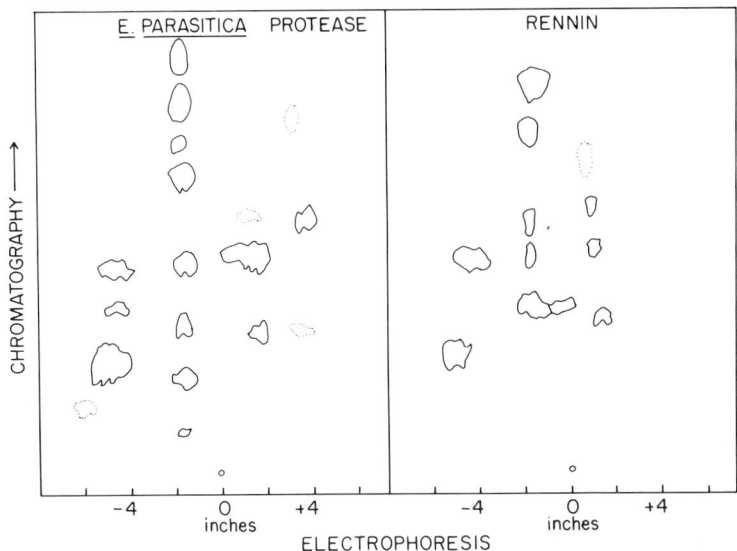

Fig. 1. Peptide maps of *E. parasitica* protease and rennin digests of oxidized B-chain of insulin.[13] The reaction consisted of 5 mg of substrate and 0.01 mg rennin or 0.005 mg *E. parasitica* protease in 1 ml of pyridine acetate buffer, pH 3.6 (0.55 M with respect to total acetate). Incubation was for 20 hours at 35°. Electrophoresis was in pyridine–acetic acid–water (100:4:900, v/v), pH 6.4, at 3000 V for 50 minutes and descending chromatography was in the organic phase of butanol–acetic acid–water (4:1:5, v/v).

glutamyl-L-tyrosine[3] which is a typical substrate for pepsin and rennin. It splits 29.0% of the peptide bonds of casein while pepsin and rennin split 8.0 and 10.2% of the bonds, respectively.[11] *E. parasitica* protease and pepsin produce 10 and 6 peptides, respectively, from κ-casein at pH 3.4.[3] Only one of the peptides produced by the two enzymes is electrophoretically and chromatographically identical. Peptide maps of digests of oxidized B-chain of insulin by *E. parasitica* protease and by rennin are shown in Fig. 1.[13] The more extensive proteolysis produced by *E. parasitica* protease is shown by the 18 peptides produced vs 12 for rennin. Only 9 of the peptides might possibly be identical.

Distribution. The enzyme used here was produced by the strain of organism from the American Type Culture Collection, Washington, D.C. which is designated as ATCC 14729. Other strains of the organism are available from Centralbureau voor Schimmelculture, Baarn, Holland (six strains designated as: CBS Luino, CBS Brescia, CBS Angera III, CBS Luinoxanthostroma, CBS 36, and CBS 54) and from Commonwealth Mycological Institute, Kew, England (strain designated as CMI 59815).

[13] M. K. Larson, unpublished data, 1969.

[30] Milk-Clotting Enzyme from *Mucor pusillus* var. *Lindt*

By Kei Arima, Juhyun Yu, and Shintiro Iwasaki

Introduction

Rennin, the milk-curdling enzyme used in cheese-making is obtained from the stomach contents (rennet) of the unweaned calf. Because of the difficulty in acquiring sufficient quantities of rennet for the cheese industry, other sources for the enzyme have been sought for many years.[1-17] After some 800 strains of microbes were investigated[5,8-23] a fungus isolated from soil and identified as *Mucor pusillus Lindt*[5,8,9] (Table I), was found to produce a suitable milk-curdling enzyme (mucor rennin).

Several types of cheese may be made with mucor rennin without altering any of the conventional manufacturing procedures.[8,9,24-33] Organoleptic tests show that the products are satisfactory (Table II).

[1] I. Y. Velselov, P. Y. Tipograf, and T. A. Pentia, *Prikl. Biokhimiya Mikrobiologiya* **1**, 52 (1965).

[2] T. Tsugo and K. Yamauchi, *Intern. Dairy Congr. 15th London* **2**, 634, 636 (1959).

[3] N. S. Paleva and N. Y. Povova, *Ferment. Spirt. Prom.* **31**, 6 (1965).

[4] P. F. Dyachenck and V. V. Slavayanova, *Intern. Dairy Congr. 16th.* **IV-I**, 349 (1962).

[5] K. Arima, S. Iwasaki, and G. Tamura, *Agr. Biol. Chem.* **31**, 540 (1967).

[6] R. A. Srinivasan *et al., Intern. Dairy Congr. 14th* **B401**, 506 (1962).

[7] J. L. Shimwell and J. E. Evans, Brit. Patent 565,788 (1944).

[8] I. Emanuilof, *Intern. Dairy Congr. 16th* **2**, 200 (1956).

[9] D. Ya. Tipograf *et al., Prikl. Biokhim. Mikrobiol.* **2**, 45 (1966).

[10] A. A. Yulius and O. Kh. Tiryakova, *Prikl. Biokhim. Mikrobiol.* **2**, 670 (1966).

[11] Ilie *et al., Ann. Meeting Agr. Chem. Soc. Japan*, p. 136 (1966).

[12] Z. Puhan, *Intern. Dairy Congr. 17th*, D199 (1966).

[13] J. G. Wahlen, *J. Bacteriol.* **16**, 355 (1928).

[14] K. Morihara, *Agr. Chem. Soc. Japan* **39**, 514 (1965).

[15] G. A. Somukuti and F. J. Babel, *J. Dairy Sci.* **49**, 700 (1966).

[16] J. L. Sardinas and G. Ferry, U.S. Patent 3275453 (1966).

[17] S. G. Knight, *Can. J. Microbiol.* **12**, 420 (1966).

[18] S. Iwasaki, J. Yu, G. Tamura, and K. Arima, *Intern. Congr. Biochem. 7th Tokyo, Japan, August,* **IV**, 758 (1967).

[19] S. Iwasaki, J. Yu, G. Tamura, and K. Arima, *Intern. Ferment. Symp. 3rd, September, 1968.*

[20] S. Iwasaki, G. Tamura, and K. Arima, *Agr. Biol. Chem.* **31**, 546 (1967).

[21] S. Iwasaki, T. Yasui, G. Tamura, and K. Arima, *Agr. Biol. Chem.* **31**, 1421 (1967).

[22] T. Tsugo, U. Yoshino, K. Taniguchi, A. Ozawa, Y. Miki, S. Iwasaki, and K. Arima, *Japan J. Zootech. Sci.* **35**, 229 (1964).

[23] T. Tsugo, K. Taniguchi, U. Yoshino, A. Ozawa, and K. Arima, *Japan J. Zootech. Sci.* **45**, 229 (1964).

TABLE I
LIST OF MICROORGANISMS PRODUCING MILK-CLOTTING ENZYME

	References
Aspergillus candidus	1
Aspergillus oryzae	2
Aspergillus parasiticus	3
Aspergillus terricola	1, 4
Absidia lichtheimi	5
Ascochyta vicae	7
Bacillus brevis	7
Bacillus cereus	6, 5
Bacillus firms	5
Bacillus fusiformis	7
Bacillus mesentericus	8, 9, 10
Bacillus polymyxa	11
Bacillus sphericus	5
Bacillus subtilis	6, 2, 9, 12, 5
Bacillus prodigiosus (*Serratia marcescens*)	13, 14
Bacillus thermoproteolyticus	14
Byssochlamys fulva	17
Colletotrichum atramentarium	15
Colletotrichum lindenruthianun	5
Corynebacterium hoagi	5
Chaetomium brasillieuse	5
Endothia parasilica	16
Monascus anka	5
Mucor mandahuricus	5
Mucor pusillus	5
Mucor spinecenes	5
Mucor rouxii	1
Pseudomonas aeruginosa	14
Pseudomonas myxogenes	2
Pseudomonas schuylkilliensis	5
Rhizopus achlamydosporus	5
Rhizopus batelae	5
Rhizopus candidus	5
Rhizopus chinensis	5
Rhizopus chinensis var. *liquefaciens*	5
Rhizopus chinniary	5
Rhizopus chungkuoensis	5
Rhizopus chungkuoensis isofermentaricus	5
Rhizopus delemar	5
Rhizopus delemar var. *minimus*	5
Rhizopus japanicus	5
Rhizopus nigricans	5
Rhizopus niveus	5
Rhizopus nodosus	5
Rhizopus oryzae	5
Rhizopus peka[11]	5

TABLE I (*Continued*)

	References
Rhizopus pseudokiensis	5
Rhizopus salebrosus	5
Rhizopus thermosus	5
Serratia marcescens	5
Schlerotium oryzae sativa	5
Streptomyces albus	5
Streptomyces chimensis	5
Streptomyces faecalis var. *liquefaciens*	15
Streptomyces griseus	2
Streptomyces grisechromogenus	5
Streptomyces hachijoensis	5
Streptomyces rimosus	5
Streptomyces rubescens	5

The enzyme may be extracted from cultures grown on wheat bran, purified, and crystallized.[34-42]

Assay Method

Principle. Activity determinations are based on the time required to clot a 10% solution of skim-milk powder.

[24] M. Grimberg, *Fette Seifen Anstrichmittel* **67**, 271 (1956).
[25] Jaarverslag, *Rijkszuivelstation-Melle* (*Belgium*), p. 29 (1966).
[26] *Ann. Rept. Dairy Res. Inst.* (*New Zealand*) *38th* p. 16, 1966.
[27] P. S. Robertson et al., *N. Z. J. Dairy Technol.* **1**, 91 (1966).
[28] *Ann. Rept. Division Dairy Research, C. S. I. R. O., Australia* **7**, (1966).
[29] G. H. Richardson et al., *J. Dairy Sci.* **50**, 1066 (1967).
[30] M. E. Schulz et al., *Milchwissenschaft* **22**, 139 (1967).
[31] J. Blaauw, *Stremsei en stremselvervangers, Officieel Orgaan* **59**, 300 (1967).
[32] *Ann. Rept. Dairy Res. Inst.* (*New Zealand*) *39th* p. 24, 1967.
[33] T. Kikuchi, et al., *15th Ann. Meeting Japanese Food. Erg. Assoc. Tokyo. April, 1968* (presented orally).
[34] J. Yu, S. Iwasaki, G. Tamura, and K. Arima, *Agr. Biol. Chem.* **32**, 1051 (1968).
[35] J. Yu, S. Iwasaki, G. Tamura, and K. Arima, *Ann. Meeting Agr. Chem. Soc. Japan* p. 203 (1968).
[36] K. Arima, J. Yu, S. Iwasaki, and G. Tamura, *Appl. Microbiol.* **16**, 1727 (1968).
[37] K. Arima and J. Yu, *3rd Meeting of Corporation of Japan Appl. Enzyme Osaka, December* (*1967*).
[38] J. Yu, G. Tamura, and K. Arima, *Agr. Biol. Chem.* **32**, 1048 (1968).
[39] J. Yu, G. Tamura, and K. Arima, *Biochim. Biophys. Acta* **171**, 138 (1969).
[40] J. Yu, G. Tamura, and K. Arima, *J. Agr. Chem. Soc., Japan* **43**, 60 (1969).
[41] J. Yu, W. Liu, G. Tamura, and K. Arima, *Agr. Biol. Chem.* **32**, 1482 (1968).
[42] J. Yu, H. Ozawa, and K. Arima, *Agr. Biol. Chem.*, in press, 1970.

TABLE II
BIBLIOGRAPHY ON CHEESE-MAKING EXPERIMENTS WITH MUCOR RENNET

Reporter	Ref.	Year	Type of cheese
T. Tsugo *et al.*	23	1964	Gouda, Camembert, Cottage
M. Grimberg	24	1965	
Jaarverslag	25	1966	Gouda
38th Annual Rep. (N. Z.)	26	1966	Cheddar
P. S. Robertson *et al.*	27	1966	Cheddar
Annual Report (Australia)	28	1966	Cheddar
G. H. Richardson *et al.*	29	1967	Brick, Cheddar, Pezza, Parmesan
M. E. Schulz *et al.*	30	1967	Butter, Edam, Tilsit
J. Blaauw	31	1967	Gouda
39th Annual Rep. (N. Z.)	32	1967	Cheddar
T. Kikuchi *et al.*	33	1968	Gouda, Cheddar

Reagents

Skim-milk powder, 10% solution

Hammarsten casein

Calcium chloride solution, 0.01 M

Potassium phosphate buffer, 0.02 M, pH 6.5

Trichloroacetic acid, 0.44 M

Sodium carbonate, 0.55 M

Folin reagent

Crude Enzyme. *Mucor pusillus* is grown on wheat bran for 72 hours at 30°.[20] Wheat bran (2 kg) is mixed well with tapwater (1.4 liters), and autoclaved at 115° for 20 minutes and inoculated with the spore suspension. Inocula are prepared earlier by culturing on wheat bran in 200-ml Erlenmeyer flasks (7 flasks) for 7 days at 30°, and then suspending sporangiospores in sterilized water. The inoculated bran is extracted with tap water (10 liters) for 20 hours at room temperature and filtered. A 6 liter filtrate contains about 800 units/ml of milk-clotting activity (enzyme yield: 2400 units/g of wheat bran).

One liter of the above extract is blended well and precipitated with 3 liters of ethanol at 14°. The precipitate is separated by centrifugation and dried at 5°. (This crude enzyme is produced by Meito Co. Ltd., Japan.)

Procedure

Determination of Milk-Clotting Activity. To 0.5 ml of enzyme solution in a 25 ml test tube, 5 ml of 10% (w/w) solution of skim-milk powder (Snow Milk Products Co. Ltd., Japan) containing 0.01 M calcium chloride is added. (All reagents are preincubated at 35° for at least 10 minutes.) In order to better visualize curd formation, a smaller tube filled with dilute red ink is inserted into the larger test tube. The time elapsing between the mixing of reagents and the first appearance of solid material against the red-dye background is measured.

Determination of Proteolytic Activty.[36] The proteolytic activity was measured by a modification of Anson's method.[4] To 2.5 ml of 0.5 to 1.5% (w/w) casein (Hammarsten casein) solution in 0.02 M potassium phosphate buffer (pH 6.5) 0.5 ml of the enzyme solution is added, and the mixture is incubated at 35° for 10 minutes. Then 2.5 ml of 0.44 M trichloroacetic acid is admixed and the precipitate is removed by filtration. One milliliter of Folin reagent (diluted 3 times) and 2.5 ml of 0.55 M sodium carbonate are added to 1 ml of the filtrate. Color is developed at 35° for 20 minutes; optical density is read at 660 nm.

Definition of Unit and Specific Activity. Using the above conditions of assay, that amount of enzyme which clots the milk solution in 1 minute is defined to contain 400 units of activity.

Specific activity is expressed as the milk-clotting activity per absorbancy units of protein at 280 nm.

Purification and Crystallization[36]

The crude enzyme is dissolved in 5 liters of tap water, and the solution (OD_{280} 1.134 × 10^5 which is 5.2 × 10^7 units) is adjusted to pH 3.5 with 1 N HCl. The enzyme solution is adsorbed onto an Amberlite CG-50 column (10 × 60 cm). The pigment-containing fraction is eluted with 150 liters of 0.05 M sodium acetate buffer (pH 3.5) as shown in Fig. 1. The fraction with milk-clotting activity is eluted with sodium acetate buffer (pH 5.0) in a total volume of 15 liters. The activity is 5.1 × 10^7 units (total OD_{280} 24 × 10^3; specific activity, 2140 units/ OD_{280}). Then the eluate is adjusted to pH 3.5 with 1 N HCl. The eluted solution is again adsorbed on an Amberlite CG-50 column which has been equilibrated with 0.05 M sodium acetate buffer, pH 3.5. With the aid of this rechromatography step, 16.5 liters of enzyme solution of a specific activity of 2820 units (OD_{280} 14 × 10^3) is obtained. Total activity is 4024 × 10^7 units.

The enzyme solution is then adsorbed on a DEAE-Sephadex A-50 column (6 × 24 cm), equilibrated with 0.05 M sodium acetate buffer (pH 5.0). The elution is carried out with a linear salt gradient. The reservoir contained 1 liter of 0.05 M sodium acetate buffer (pH 5.0) and 0.5 M KCl, and the mixing chamber contained 1 liter of the same buffer but without potassium chloride. A typical chromatographic pattern is shown in Fig. 2. The protein peak with milk-clotting activity

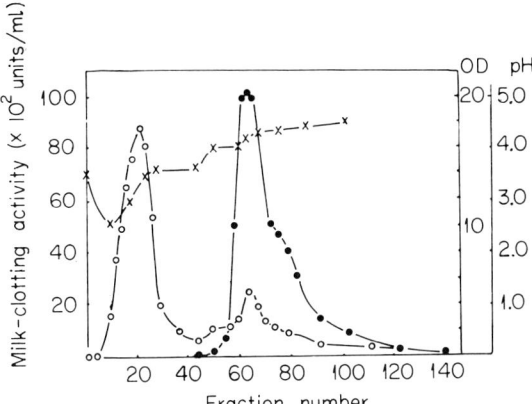

Fig. 1. Column chromatogram of mucor rennin on Amberlite CG-50 (10 × 60 cm). (●) milk-clotting activity, (○) OD 280 nm, (×) pH of the solution. Eluted volume, 300–330 ml, 0.05 M sodium acetate buffer; fractions 1 to 50, pH 3.5; fractions 51 and above, pH 5.0.

is eluted in the 0.30 and 0.45 M KCl concentration range. The active fractions are pooled and the concentration of KCl is lowered to less than 0.2 M by dilution with 0.05 M sodium acetate buffer of pH 5.0. The solution is then rechromatographed as before on DEAE-Sephadex A-50. Following this procedure, 780 ml of the enzyme solution is obtained (Fig. 3).

Ammonium sulfate is added to the enzyme solution to a saturation

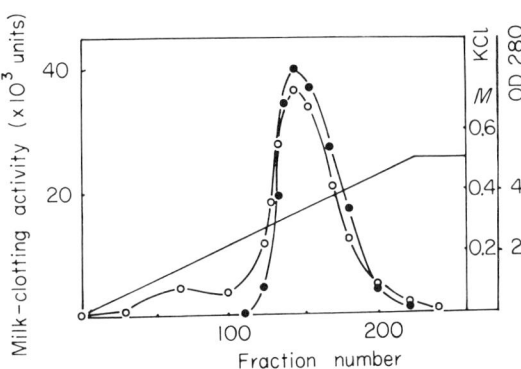

Fig. 2. Chromatographic pattern of mucor rennin on DEAE-Sephadex A-50 (6 × 24 cm). (●) Milk-clotting activity, (○) OD 280 nm; concentration of KCl, solid line. Elution buffer, 0.05 M sodium acetate buffer (pH 5.0); eluted volume, 20 ml.

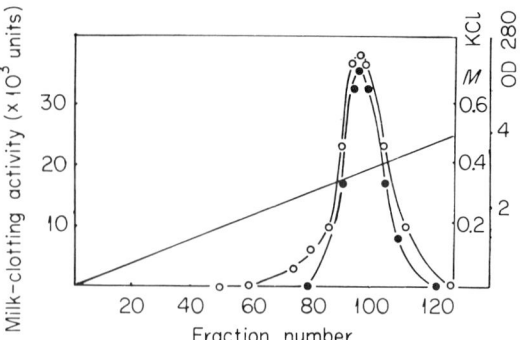

Fig. 3. Chromatographic pattern of mucor rennin on DEAE-Sephadex A-50 (4 × 29 cm). (●) Milk-clotting activity, (○) OD 280 nm; concentration of KCl, solid line. Elution buffer, 0.05 M sodium acetate buffer (pH 5.0); elution volume, 20 ml.

of 70%. After standing at 5° for 10 hours, the precipitate is collected by centrifugation at 10,000 g for 20 minutes at 0°. The sedimented material is dissolved in a minimal amount of 0.05 M sodium acetate (pH 5.0). The solution is then passed through a Sephadex G-100 column, using 0.05 M sodium acetate buffer of pH 5.0 for elution. As seen in Fig. 4, a portion of the pigment fraction is separated by this procedure. Repeated gel filtration on Sephadex G-100 is used to remove remaining pigment residues. The active fractions (20.31 × 10⁶ units) are pooled and ammonium sulfate is admixed slowly to 70% saturation. After 10 hours at 0°, the precipitate is collected by centrifugation and is dissolved in 0.1 M sodium acetate buffer of pH 5.0 to a 2–3% protein concentration.

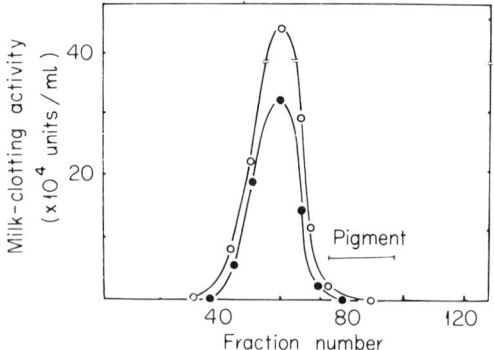

Fig. 4. Column, chromatogram of mucor rennin on Sephadex G-100 (3 × 8 cm). (●) Milk-clotting activity, (○) OD 280 nm. Elution buffer, 0.05 M sodium acetate buffer (pH 5.0).

Saturated ammonium sulfate solution is then added with gentle stirring until a slight turbidity is formed. This solution is clarified by centrifugation. The ammonium sulfate concentration of the supernatant is slowly increased to 50% saturation by adding further increments of a saturated ammonium sulfate solution. Crystallization is allowed to take place during storage at 3°–4° and occurs as the concentration of both ammonium sulfate and enzyme increase by evaporation (ca. 2 months). It can be greatly hastened, however, by seeding the solution with enzyme crystals obtained earlier. With such seeding crystallization starts in about 10–15 hours. Various crystal forms were observed (Fig. 5). Nevertheless, all crystalline forms have the same specific activity.

It is conceivable that the actual shape of the crystal depends on the method employed for crystallization. The mentioned procedure of evaporation over a period of 2 months in the refrigerator produced elliptical crystals.

A crystallization method based on dialysis induces the formation of different crystals. The enzyme solution (2–3% protein) is put into a cellophane bag for dialysis against 0.1 M sodium acetate buffer which is 50% saturated with ammonium sulfate. Saturated solution of ammonium sulfate is then added dropwise to the outside dialyzing

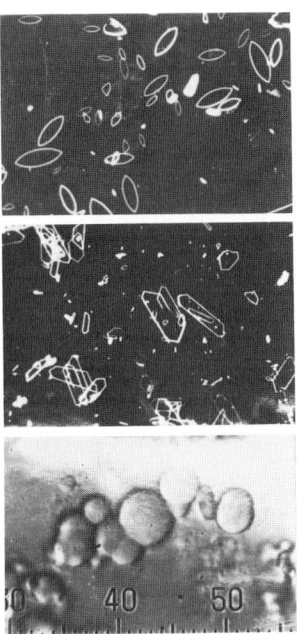

FIG. 5. Different forms of crystalline mucor rennin.

TABLE III
PURIFICATION PROCESS OF MUCOR RENNIN CRYSTAL

Determination	OD 280 nm		Clotting activity		Specific activity	
	OD 280 nm	Yield (%)	$\times 10^3$ Units	Yield (%)	OD 280 nm	Ratio of purification
Crude enzyme	113,400	100	52,000	100	460	1.0
Amberlite CG-50	24,000	21.5	51,000	99.3	2,140	4.7
Amberlite CG-50	14,000	12.4	40,240	77.3	2,870	6.2
DEAE-Sephadex A-50	6,660	5.9	29,667	57.2	4,450	9.1
DEAE-Sephadex A-50	4,220	3.7	25,480	49.0	6,040	13.2
(NH₄)₂SO₄, saturated 0.7						
Sephadex G-100	3,850	3.4	23,320	44.8	6,050	13.3
Sephadex G-100	3,050	3.0	20,310	38.7	6,650	14.4
(NH₄)₂SO₄	2,910	2.6	19,300	37.1	6,670	14.4
Mucor rennin crystal,[a] 2.35 g	2,580	2.6	17,200	33.1	6,670	14.5

[a] The crystals were prepared by the dialysis crystallization method described in the text.

fluid to give a final saturation of 70%. As the concentration of ammonium sulfate inside the cellophane bag slowly rises, crystals of the enzyme form. The crystals may be collected in about a week.

The enzyme crystals are separated by centrifugation at 10,000 g for 10 minutes and are dissolved in distilled water. Following dialysis against water for 24 hours, the solution is freeze dried and yields 2.35 g of enzyme powder (corresponding to 2580 OD_{280} units). The recovered total milk-clotting activity is 172×10^7 units with a specific activity of 6670 units per OD_{280}.

The purification procedure is summarized in Table III.

Properties

Stability.[35,36] As judged by retention of activity at 60° over a 10-minute period, mucor rennin appears to be most stable at pH 5.0. It is recalled that the calf enzyme is stable in a similar pH range.

Purity and Physical Properties.[36] Homogeneity of the crystalline enzyme (in 0.1 M sodium acetate buffer, pH 5.0) was attested both by Tiselius electrophoresis and by ultracentrifugal analysis. Electrophoresis was performed at 90 V and 5 mA. The enzyme migrated as a single symmetrical peak. The ultracentrifugal analysis was carried out at 55,400 rpm at 11.3° for 3 hours and 17 minutes. The sedimentation pattern showed only a single symmetrical peak.

Relation of Milk-Clotting to Proteolytic Activity.[36] Ratio of the milk-clotting activity of mucor-rennin crystals to their proteolytic activ-

TABLE IV

RATIO OF CLOTTING ACTIVITY TO PROTEOLYTIC ACTIVITY OF VARIOUS PROTEASES[36]

Protease	Clotting activity (units/ml)	Proteolytic activity (OD 660 nm)	Ratio (units/OD 660 nm)
Rennin	293	0.04	7350
Mucor rennin crystal	511	0.11	4650
Pfizer microbial rennin	750	0.29	2590
Papain	216	0.59	367
Trypsin	1.6	0.44	4
Molsin	1.3	0.18	7
Ficin	267	0.68	393
Biodiastase	115	0.83	138

ity was compared with those of other proteases. Values for this ratio, expressed as milk-clotting units per the OD_{660}'s obtained in the proteolytic measurements, were as follows. Calf rennin, 7350; mucor rennin, 4650; Pfizer rennin from *Endothia parasitica*, 2590. Molsin (Kikkoman Co. Ltd.), ficin, trypsin, papain, and Biodiastase (Amano Co. Ltd.) gave a ratio of less than 393 (Table IV).

Optimum pH for Proteolytic Activity. Optimum pH is 4.5 for κ-casein and 4.0 for hemoglobin. For Hammarsten casein it is around pH 3.5. Thus the optimum pH of activity of the crystalline mucor rennin enzyme is similar to that of calf rennin.[43-45]

Effect of pH on Milk Coagulation Activity. The crystalline mucor enzyme clots milk most rapidly at pH 5.5. Clot formation at pH 7.0 is greatly delayed. Similar results are also obtained with the crude enzyme.[20,46] Calf rennet activity[46] is likewise pH dependent and is reported to be weaker in alkaline conditions than in acidic conditions.

Physical Properties and Amino Acid Composition of Crystalline Mucor Rennin.[39] SEDIMENTATION, DIFFUSION, AND FRICTIONAL COEFFICIENTS. Though a sedimentation value for the purified enzyme[21] was reported earlier, no correction was made to standard conditions. We now measured an $s_{20,w}^0$ of 2.39 S for the crystalline protein in 0.1 M sodium acetate buffer, pH 5.0.

Density measurements indicate a partial specific volume of 0.74. Using the maximum ordinate method and the maximum ordinate area method,[47] a diffusion $D_{20,w}^0$ of $7.9 \pm 0.3 \times 10^{-7}$ cm^2 sec^{-1} was obtained.

[43] M. L. Anson, *J. Gen. Physiol.* **22**, 79 (1938).
[44] H. Schwander, P. Zahler, and H. Nitschman, *Helv. Chem. Acta* **25**, 553 (1952).
[45] B. Foltmann, *Acta Chem. Scand.* **13**, 1927 (1959).
[46] G. H. Richardson and J. H. Nelson, *J. Dairy Sci.* **50**, 1066 (1967).
[47] S. Mizushima and S. Akabori, *Chem. Protein* **2**, 383, 385 (1954).

Frictional ratio is 1.33.

Calculation of molecular weight according to the Svedberg equation[48] gives a value of about $29,000 \pm 1400$. Using the Yphantis method,[49] the molecular weight of crystalline mucor rennin is 30,600.

Amino acid analysis for crystalline mucor enzyme is given in Table V and is compared with those of rennin and prorennin.[50] The 277–281

TABLE V
AMINO ACID COMPOSITION OF MICROBIAL RENNIN, PRORENNIN, AND RENNIN

Amino acid	Microbial rennin		Rennin[a] MW 31,100	Prorennin[a] MW 36,200
	MW 30,600	Residues		
Half-cys	2.31	2		
Met	3.17	3	7	7
Asp	43.6	44	31	33
Thr	21.2	21	28	21
Ser	22.1	22	27	31
Glu	19.7	20	29	36
Pro	14.1	14	12	14
Gly	33.5	34	25	29
Ala	16.4	16–17	13	15
Val	23.8	24	21	23
Ile	11.8	12	15	19
Leu	14.8	15	19	26
Tyr	12.8	13	15	18
Phe	18.9	19	14	14
His	1.42	1–2	4	5
Lys	11.4	11–12	8	13
Arg	4.1	4	7	5
Trp	2.45	2–3		
	277.52	277–281	263	313
NH$_3$	1.4.3			

[a] Ref. 11.

residues in mucor rennin are distributed as follows: $\tfrac{1}{2}Cys_2$, Met_3, Asp_{44}, Thr_{21}, Ser_{22}, Glu_{20}, Pro_{14}, Gly_{34}, Ala_{16-17}, Val_{24}, Ile_{12}, Leu_{15}, Tyr_{13}, Phe_{19}, His_{1-2}, Lys_{11-12}, Arg_4, and Trp_{2-3}. Elementary analysis gives C, 49.24–49.59%; H, 6.99 and 7.44%; N, 13.86–14.19%; S, 0.57–0.65%. These values agree with the theoretical values obtained from amino acid composition and NH_3.

For purposes of comparison, some data regarding the physical prop-

[48] T. Svedberg and K. O. Pedersen, "The Ultracentrifuge," Clarendon, Oxford, 1940.
[49] D. A. Yphantis, Ann. N.Y. Acad. Sci. 88, 586 (1960).
[50] H. Schwander, P. Zahler, and H. Nitschman, Helv. Chem. Acta 35, 553 (1952).

TABLE VI

Property		Value		
		Mucor rennin	Rennin[a]	Prorennin[a]
Frictional coefficient	f/f_0	1.33	0.98^b	
Sedimentation coefficient	$s_{20,w}^0$	2.39×10^{-13}	3.2×10^{-13} 4×10^{-13b}	3.5×10^{-13}
Diffusion coefficient	$D_{20,w}^0$	7.9×10^{-7}	9×10^{-7} 9.5×10^{-7b}	
Partial specific volume	$\bar{V}$	0.742	0.749^b	
Molecular weight				
Svedberg		29,000	31,100 $40,000^b$	36,200
Yphantis		30,600		
Andrews		32,500		
Amino acid analysis		29,690 30,213 (approx.)		

[a] Ref. 11.
[b] Ref. 21.

erties of rennin are given in Table VI. Foltmann and Hartley[45,51] reported on the amino acid composition of this enzyme. It could also be mentioned that the optimum pH for hydrolysis of both casein[52] and hemoglobin[53] by rennin is pH 3.7.

Altogether, the general properties of mucor rennet[20,46] and the mucor crystalline rennin[34-42] differ only slightly from those of animal rennet and rennin.

Active Site of the Enzyme and the Role of Ca-Ions in Milk Coagulation.[38] Inhibition of the enzyme activity by iodine is pH dependent. Almost complete inhibition (>90°) occurs at pH 6.7, but there is little loss of activity (~10%) if the treatment is carried out in the 3.0–5.0 pH range.

Figure 6 shows the loss of activity of the enzyme caused by photooxidation and also indicates the concomitant changes in the tyrosine, tryptophan, and histidine content. The loss of histidine parallels the inactivation of the enzyme.

Treatment of mucor rennin by *p*-chloromercuribenzoic acid, iodoacetamide, or *N*-ethylmaleimide does not affect the milk-clotting activity.

[51] B. Foltmann and S. Hartley, *Biochem. J.* **104**, 1064 (1967).
[52] E. Schram, S. Moore, and E. J. Bavel, *J. Dairy Sci.* **49**, 700 (1966).
[53] N. J. Berridge, *Biochem. J.* **39**, 179 (1945).

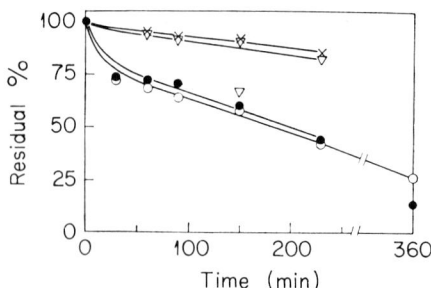

Fig. 6. Relation between activity of mucor rennin and residual amino acid by photooxidation. (●) Histidine, (▽) tyrosine, (✕) tryptophan, (○) residual milk-clotting activity.

The use of diazonium-1H-tetrazole (DHT) for the spectrophotometric determination of histidine residues in protein has been applied to several proteins.[54] When DHT is added to a solution of mucor rennin, optical density at 480 nm increases[55] and the milk-clotting activity is completely lost. One of the two histidine residues of the protein is suggested to be critical for activity.

While Ca ions do not affect the proteolytic activity of mucor rennin, they enhance the apparent activity in the milk-clotting assay (Fig. 7). Calcium ions seem to influence the clotting of casein *per se* rather than affecting the enzyme reaction itself.

Specificity.[42] Among 58 synthetic peptides tested, the following were

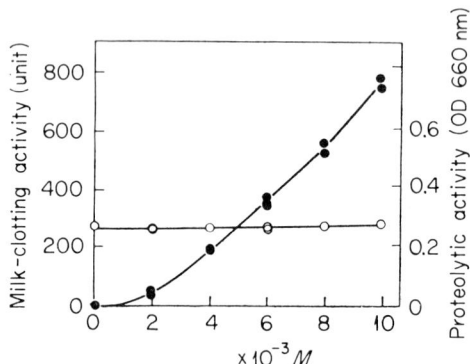

Fig. 7. Effect of CaCl₂ on clotting and proteolytic activity. (●) Final concentration of CaCl₂ milk-clotting activity, (○) proteolytic activity.[38]

[54] H. Horinishi, Y. Hachimori, K. Kurihara, and K. Shibata, *Biochim. Biophys. Acta* **86**, 447 (1964).
[55] J. Yu, G. Tamura, and K. Arima, will be published in *Agr. Biol. Chem.* (1970).

hydrolyzed by mucor rennin: Z-L-Glu-L-Tyr-OH, A-L-Glu-L-Phe-OH, Z-L-Phe-L-Tyr-OH, Z-L-Phe-L-Leu-OH, Z-L-Tyr-L-Leu-OH, L-Leu-Gly-OH, L-Leu-Gly-Gly-OH, and Z-Gly-Pro-L-Leu-Gly-OH. Altogether, the specificity of mucor rennin appears to be similar to that of pepsin and calf rennin but somewhat narrower than that of *Aspergillus saitoi* and *Streptomyces griseus* proteases.

[30a] The Acid Protease of *Mucor miehei*

By Martin Ottesen and W. Rickert

A milk-clotting acid protease which resembles pepsin, rennin, and the protease from *Mucor pusillus*,[1] has recently been isolated from a strain of *Mucor miehei*, CBS number 370.65.[2] The enzyme was obtained from commercially available culture concentrate[1a] by ammonium sulfate fractionation, batchwise adsorption on DEAE-Sephadex, SE-Sephadex, and DEAE-Sephadex chromatography and gel filtration. Although the enzyme has not as yet been crystallized, it appeared homogeneous by paper electrophoresis within the pH range 4.0–8.0, by free moving boundary electrophoresis in the pH range 5.2–3.2, and during ultracentrifugation within the pH range 5.0–8.0. The isoelectric point was approximately 4.2 in acetate buffer, and the sedimentation coefficient, $s_{20,w}^0$, 3.35 Svedberg units. Boundary spreading in the ultracentrifuge indicated a value for the diffusion coefficient, $D_{20,w}^0$, of approximately 7.6×10^{-7} cm^2 sec^{-1}. From this value, and an estimate of 0.72 for the partial specific volume, a molecular weight of around 38,000 was indicated. Optical rotatory dispersion measurements indicated the Moffitt parameter b_0 to be close to zero in agreement with the findings of other acid proteases. The amino acid composition expressed as moles per 100,000 grams of protein was as follows: Asp$_{110-111}$ Thr$_{72-73}$ Ser$_{88-91}$ Glu$_{61-62}$ Pro$_{43-44}$ Gly$_{83-84}$ Ala$_{65-66}$ ½Cys$_{9-10}$ Val$_{63-64}$ Met$_{15-16}$ Ileu$_{46-47}$ Leu$_{49-50}$ Tyr$_{48-49}$ Phe$_{52-54}$ Tryp$_{7-8}$ Lys$_{22-23}$ His$_{4-5}$ Arg$_{15-16}$.

Although no covalently bound phosphorus was detected in the enzyme preparation, the *Mucor miehei* protease did have a 6% carbohydrate content made up of hexosamine and neutral hexoses. Since neither heat denaturation nor exposure of the enzyme to 8 *M* urea diminished the carbohydrate content, it appeared to be covalently bound to the enzyme. The carbohydrate content, the sedimentation coefficient, and the amino

[1] J. Yu, G. Tamura, and K. Arima, *Biochim. Biophys. Acta* **171**, 138 (1969).

[1a] "Rennilase," Novo Industry A/S, Copenhagen, Denmark.

[2] M. Ottesen and W. Rickert, *Compt. Rend. Trav. Lab. Carlsberg* **37**, 301 (1970).

acid composition distinguished the *Mucor miehei* protease from the *Mucor pusillus* protease described by Arima *et al.*[1]

Mucor miehei protease was remarkably stable. After 8 days of incubation at 38° between pH 3.0 and 6.0 more than 90% of the activity was retained, and in 8 M urea at pH 6 no detectable loss in activity occurred after 11 hours incubation. The pH optimum for the enzyme was close to 4, when the B-chain of oxidized insulin was used as substrate. Splittings of the substrate were observed at the bonds Phe(1)–Val(2), Leu(15)–Tyr(16), Tyr(16)–Leu(17), Phe(24)–Phe(25), and Phe(25)–Tyr(26). This indicated that among the bonds cleaved by both pepsin and rennin, only those bonds involving aromatic residues were hydrolyzed by the *Mucor miehei* protease.[3]

[3] W. Rickert, *Compt. Rend. Trav. Lab. Carlsberg* **38**, 1 (1970).

[31] Bovine Procarboxypeptidase and Carboxypeptidase A

By Philip H. Pétra

Bovine Procarboxypeptidase A

Bovine procarboxypeptidase A (PCP-A) occurs in the pancreas as a zymogen from which the active form can be generated by limited proteolysis. The existence of the zymogen *in vivo* was first inferred by Anson[1] in 1935 and was established 20 years later by Keller, Cohen, and Neurath.[2,3] Some of the properties of this early preparation of the zymogen have already been reviewed.[4] In recent years, much work has been directed toward the elucidation of the activation process[5-9]; however, up to the present, the definition of the molecular events describing the generation of carboxypeptidase A activity is incomplete. The reasons for this difficulty can be stated as follows. First, bovine procarboxypeptidase A, unlike trypsinogen and chymotrypsinogen, exists in aggregate forms.

[1] M. L. Anson, *Science* **81**, 467 (1935).

[2] P. J. Keller, E. Cohen, and H. Neurath, *J. Biol. Chem.* **223**, 457 (1956).

[3] P. J. Keller, E. Cohen, and H. Neurath, *J. Biol. Chem.* **230**, 905 (1958).

[4] H. Neurath, Vol. II, p. 77.

[5] H. Neurath, *Proc. Intern. Congr. Biochem. 7th, Tokyo, 1967.*

[6] H. Neurath, R. A. Bradshaw, P. H. Pétra, and K. A. Walsh, *Phil. Trans. Proc. Roy. Soc. (London)* B **257**, 159–176 (1970).

[7] J. H. Freisheim, K. A. Walsh, and H. Neurath, *Biochemistry* **6**, 3020 (1967).

[8] J. H. Freisheim, K. A. Walsh, and H. Neurath, *Biochemistry* **6**, 3010 (1967).

[9] P. Piras and B. L. Vallee, *Biochemistry* **6**, 348 (1967).

One of these is an aggregate of two subunits (PCPA-S5)[10] and the other of three subunits (PCPA-S6).[11,12] Subunit I is common to both forms of the zymogen and is the immediate precursor of the enzyme. So far it has not been possible to isolate subunit I in its native state. Second, the peptide fragment released upon activation contains approximately 60 amino acids as compared to the dipeptide or hexapeptide released during the activation of chymotrypsinogen or trypsinogen. And third, at least three forms of the active enzyme, differing in the structure of their amino terminal region, can be generated from the zymogen. The relative proportion of these three forms depends upon the method of activation. Brown et al.[12] have proposed a mechanism of activation shown in Fig. 1. The process can be divided into two steps involving the tryptic activation of the endopeptidase, subunit II, which then acts in concert with trypsin to convert subunit I into carboxypeptidase. Subunit III of PCPA-S6 has not yet been fully characterized, but preliminary chemical evidence suggests that it may be an inactive form of subunit II.

Purification of Procarboxypeptidase

Step 1. Extraction.[12,13] One hundred grams of pancreatic acetone

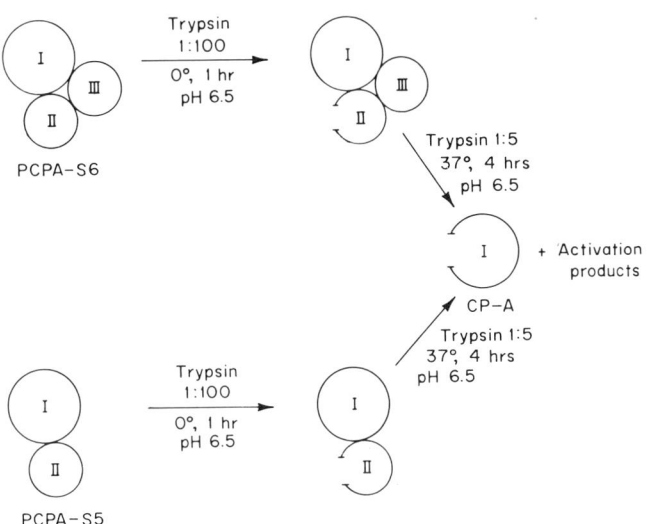

Fig. 1. Schematic diagram describing the mechanism of activation of procarboxypeptidase A-S5 and A-S6, adapted from J. R. Brown et al.[12]

[10] J. R. Brown, M. Yamasaki, and H. Neurath, *Biochemistry* **2**, 877 (1963).

[11] J. R. Brown, D. J. Cox, R. N. Greenshields, K. A. Walsh, M. Yamasaki, and H. Neurath, *Proc. Natl. Acad. Sci. U.S.* **47**, 1554 (1961).

powder[14] (which can be prepared according to published methods[2, 15]) is suspended in 1 liter of cold distilled water containing $10^{-3} M$ diisopropylphosphofluoridate (DFP). (One gram of powder is equivalent to approximately 5 g of ground pancreas.) After extraction for 4–12 hours in the cold room using gentle stirring, the suspension is centrifuged for 20 minutes in the Spinco Model L centrifuge (No. 21 rotor) at 20,000 rpm. The residue is discarded and the supernatant directly chromatographed on DEAE-cellulose.

Step 2. Chromatography of Crude Extract.[13, 15] DEAE-cellulose (Selectacel–DEAE standard type no. 70 purchased from Schleicher and Schuell Co.) is washed with 0.25 M NaCl, 0.5 M NaOH, followed by

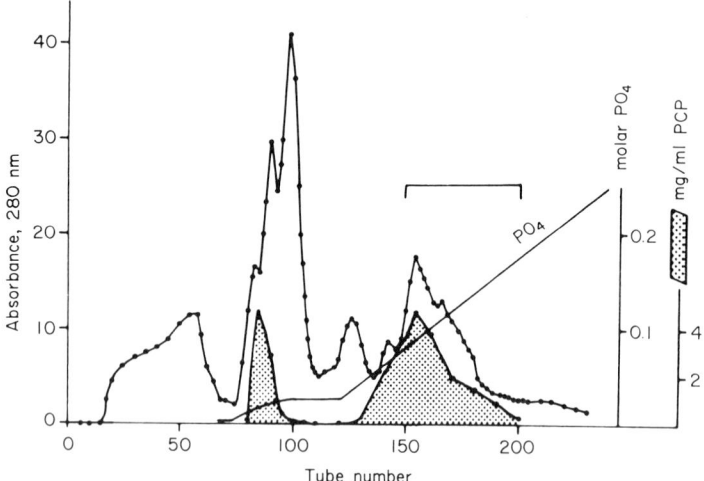

Fig. 2. Chromatography of crude aqueous extracts from pancreatic acetone powder on DEAE-cellulose at pH 8.0. Experimental conditions: column, 3.4 × 60 cm; sample volume, 800–850 ml; flow rate, 100 ml/hour; fraction volume, 20 ml; the linear salt gradient is applied from 2.2 liters of 0.06 M to 2.2 liters of 0.4 M potassium phosphate, pH 8.0 (see text for further details). The shaded areas represent HPLA activity measured after trypsin activation as described in the text. The

[12] J. R. Brown, R. N. Greenshields, M. Yamasaki, and H. Neurath, *Biochemistry* **2**, 867 (1963).

[13] M. Yamasaki, J. R. Brown, D. J. Cox, R. N. Greenshields, R. D. Wade, and H. Neurath, *Biochemistry* **2**, 859 (1963).

[14] The powder must be tested for carboxypeptidase A and trypsin activity before use. Extraction of 5 g of powder with 50 ml of distilled water at 5° should yield a level of HPLA activity less than 10–15% of the total obtained upon full activation with trypsin. There must be no trypsin activity in the extract.

[15] J. R. Brown, Ph.D. dissertation, University of Washington, Seattle, Washington, 1963.

distilled water and equilibrated with $0.005\,M$ potassium phosphate buffer, pH 8.0. A column (3.4×60 cm) is packed under 5 psi at room temperature and transferred to the cold room ($4°$) where it is washed with 1 liter or more of the same buffer. The supernatant (800–850 ml) is pumped onto the column at a flow rate of 100 ml/hour followed by approximately 1 liter of $0.05\,M$ potassium phosphate buffer, pH 8.0, containing $10^{-3}\,M$ DFP. A linear salt gradient is then applied from $0.06\,M$ to $0.4\,M$ potassium phosphate, pH 8.0, as shown in Fig. 2. The column is monitored by optical density at 280 nm and by activity against hippuryl-DL-β-phenyllactic acid (HPLA) by taking 0.5 ml aliquots from every ten tubes and adding trypsin to a concentration of 100 μg/ml (a standard trypsin solution of 20 mg/ml of $0.001\,N$ HCl is used). The solutions are incubated at $37°$ for 2 hours and the standard HPLA assay is performed as described in the carboxypeptidase A section (see below). Under these conditions, the hydrolysis of HPLA by carboxypeptidase A results in the uptake of 200 μeq OH$^-$/minute/mg protein using $E_{280}^{0.1\%} = 1.9$.[2] The expected rate constant obtained after trypsin activation can be calculated as follows:

$$k° = 200 \times (34{,}300/87{,}000) = 79.1 \ \mu\text{eq} \ \ \text{OH}^-/\text{minute/mg protein}$$

Where 34,300 and 87,000 are the molecular weights of carboxypeptidase A (CPA) and PCPA-S6, respectively. This rate constant is used to calculate the total yield of the zymogen described in the right ordinate of Fig. 2. The first peak separating at tube 75 is procarboxypeptidase B.[16-19] The main fraction eluting from tubes 130–200 contains both PCPA-S5 and PCPA-S6.

Step 3. Purification of Procarboxypeptidase A-S5.[12, 15] The fraction obtained from the chromatographic elution shown in Fig. 2 (shaded area, tubes 130–200) is dialyzed against 6 liters of $0.005\,M$ potassium phosphate buffer, pH 6.5, for 24 hours at $5°$ (buffer is changed 3 times) and rechromatographed on DEAE-cellulose at pH 6.5 as shown in Fig. 3.

[16] J. E. Folk, *J. Am. Chem. Soc.* **78**, 3541 (1956).
[17] J. F. Pechère, G. H. Dixon, R. H. Maybury, and H. Neurath, *J. Biol. Chem.* **233**, 1364 (1958).
[18] E. Wintersberger, D. J. Cox, and H. Neurath, *Biochemistry* **1**, 1069 (1962).
[19] D. J. Cox, E. Wintersberger, and H. Neurath, *Biochemistry* **1**, 1082 (1962).

Fig. 2 (*Continued*)

fraction separating at tube 75 is procarboxypeptidase B. Procarboxypeptidase A-S5 and S6 are both eluted in tubes 130 to 200. The total yield of procarboxypeptidase A described on the right ordinate is calculated by using $k° = 79.1$ μeq OH$^-$/minute/mg protein (see text).

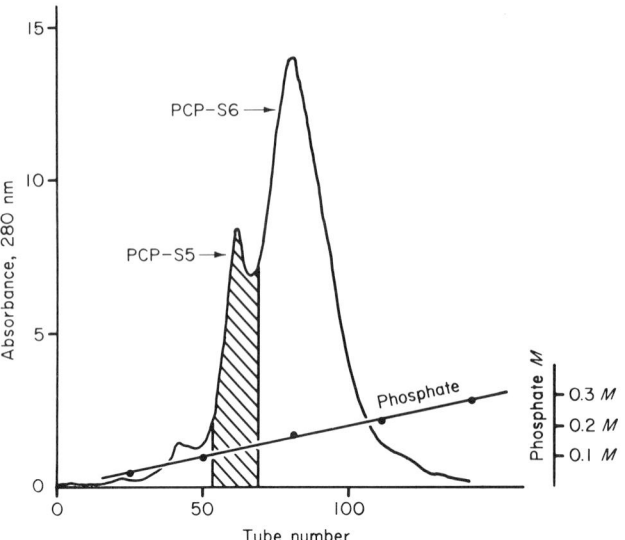

Fig. 3. DEAE-cellulose chromatography at pH 6.5 of the fraction obtained from the elution pattern described in Fig. 2 (pooled tubes No. 130 to 200). Experimental conditions: column, 3.4 × 52 cm; sample volume, 1750 ml (2.5 g protein); flow rate, 100 ml/hour; fraction volume, 25 ml. The salt gradient (same as described in Fig. 2) is started directly after the application of the sample. The fractions corresponding to the hatched portion of the curve were pooled and rechromatographed as shown in Fig. 4.

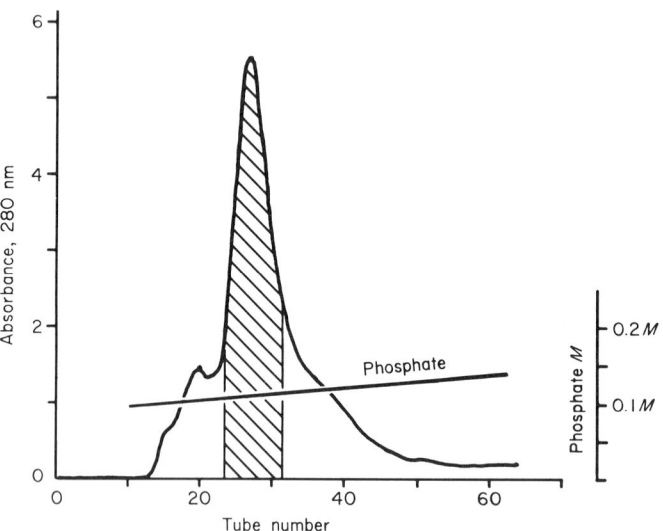

Fig. 4

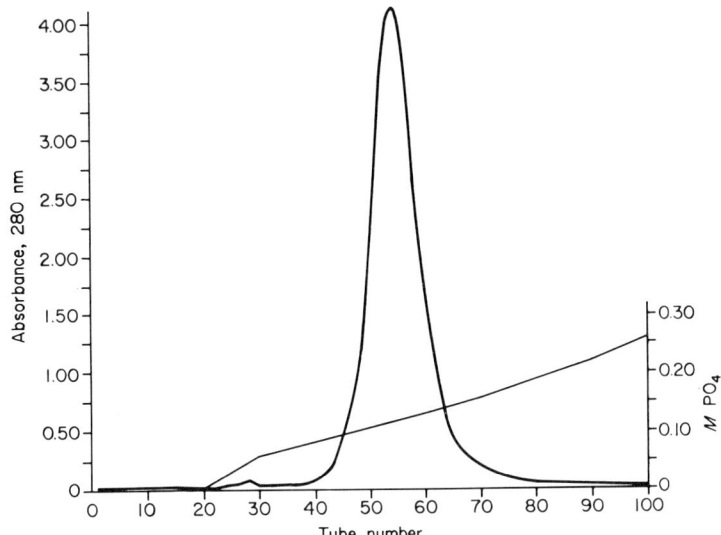

FIG. 5. Fourth chromatography of procarboxypeptidase A-S5 on DEAE-cellulose, pH 6.5, of the pooled fractions corresponding to hatched area in Fig. 4. Experimental conditions: column, 2×30 cm; sample volume, 80 ml (304 mg protein); flow rate, 50 ml/hour; fraction volume, 8 ml. A linear gradient from $0.05\,M$ to $0.3\,M$ potassium phosphate, pH 6.5, is started after 240 ml had passed through the column.

PCPA-S5 separates at the leading shoulder of the main peak which contains PCPA-S6. The fraction represented by the hatched area in Fig. 3 is dialyzed against 6 liters of $0.005\,M$ potassium phosphate containing $10^{-3}\,M$ DFP, pH 8.0, for 24 hours at 5° and is purified further on DEAE-cellulose at pH 8.0. The elution pattern is shown in Fig. 4. A final purification step on DEAE-cellulose at pH 6.5 is needed to yield a chromatographically homogenous preparation as shown in Fig. 5. One hundred grams of acetone powder yields approximately 300 mg of PCPA-S5. Crystallization of the purified zymogen has so far been unsuccessful; however, the material can be stored as a lyophilized powder at −20°.

Step 4. Purification of Procarboxypeptidase A-S6.[13, 15] Ammonium sulfate is added at 0° to the pooled fractions obtained from Fig. 3 (tubes 75 to 100) to 40% saturation. The suspension is centrifuged for 30

FIG. 4. Third chromatography on DEAE-cellulose at pH 8.0 of the pooled fractions corresponding to the hatched area in Fig. 3. Experimental conditions: column, 2×37 cm; sample volume, 350 ml (600 mg protein); flow rate, 50 ml/hour; fraction volume, 8 ml; linear gradient, $0.1\,M$ to $0.15\,M$ potassium phosphate, pH 8.0. The fractions corresponding to the hatched area were pooled and rechromatographed as shown in Fig. 5.

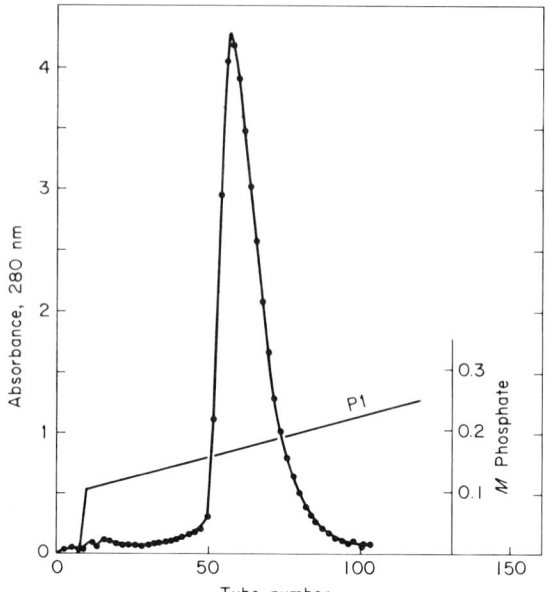

Fig. 6. DEAE-cellulose chromatography of procarboxypeptidase A-S6 at pH 6.5. Experimental conditions: column, 2 × 37; sample, 425 mg lyophilized powder; flow rate, 50 ml/hour; fraction volume, 9.5 ml. The linear gradient is from 0.1 to 0.3 M potassium phosphate.

minutes in the International Centrifuge HR-1 at 8000 rpm (No. 858 rotor). One hundred milliliters of distilled water containing $10^{-3} M$ DFP is added to the precipitate and the solution is stirred at 4° for 1 to 2 hours. The solution is dialyzed against three changes of 6 liters of 0.005 M potassium phosphate, pH 6.5, $10^{-3} M$ DFP (4°), centrifuged, and the supernatant lyophilized. A final chromatography on DEAE-cellulose at pH 6.5 is required to yield a homogeneous preparation as shown in Fig. 6.

An alternative method[13, 15] for the preparation of PCPA-S6 is to pool the fractions represented by the bracket in Fig. 2 (tubes 150–200), removing most of PCPA-S5, and precipitating with ammonium sulfate at a 40% saturation, thereby omitting the chromatography step described in Fig. 3. The precipitate is dissolved and dialyzed as described above, and after centrifugation the solution is directly rechromatographed at pH 8.0 as shown in Fig. 7. The collected fractions are pooled, dialyzed, and the zymogen is stored as a lyophilized product at −20°. Another method for the preparation of PCPA-S6 based on the one discussed above has been published.[20]

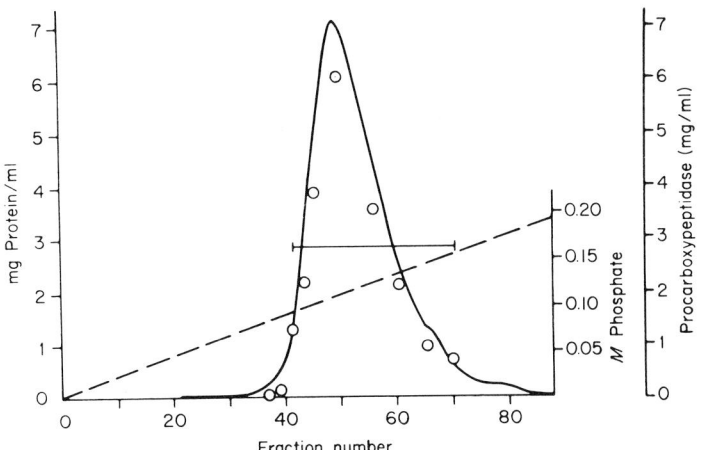

Fig. 7. DEAE-cellulose chromatography of procarboxypeptidase A-S6 at pH 8.0. Experimental conditions: column, 3×20 cm; sample volume, 100 ml; flow rate, 100 ml/hour; fraction volume, 20 ml; linear gradient, $0.005\,M$ to $0.33\,M$ potassium phosphate. The column is equilibrated with $0.005\,M$ potassium phosphate before application of the sample. (———) Optical density at 280 nm; (O—O) activity against HPLA after tryptic activation; (- - - -) calculated gradient.

Properties of Procarboxypeptidase A-S6

Stability and Chemical Properties. Solutions of PCPA-S6 kept at pH 10.5 (20°) or in $8\,M$ urea are irreversibly denatured as shown by the inability to catalyze the hydrolysis of HPLA after tryptic activation.[13] The changes in the structure of the zymogen produced by these treatments are discussed later in this chapter. Solutions of PCPA-S6 (5 mg/ml) kept at 7.5 and 0° are stable provided that $10^{-3}\,M$ DFP is present; if higher protein concentrations are used, DFP should be increased accordingly. PCPA-S6 is soluble in distilled water and stable in solutions of most salts at low or high ionic strength.

PCPA-S6 is a metalloprotein containing 0.84 ± 0.16 g-atom of Zn^{2+} per molecular weight of 87,000.[9,21] The metal can be removed by dialyzing below pH 5.6 or above pH 9.[9] The apparent dissociation constant for $[Zn^{2+}]$ [apoprocarboxypeptidase A-S6] is one order of magnitude higher than $[Zn^{2+}]$ [apocarboxypeptidase A (Allan)], indicating weaker binding of the metal in the zymogen.[9] Small amounts of iron and nickel are also present and these account for a total of 1 g-atom of metal

---

[20] R. J. Peanasky, D. Gratecos, J. Baratti, and M. Rovery, *Biochim. Biophys. Acta* **181**, 82 (1969).
[21] J. H. Freisheim, H. Neurath, and B. L. Vallee, in preparation.

per mole when added to the zinc content. A study of the influence of Zn^{2+} on the activation process indicates that the metal must be bound in PCPA-S6 prior to activation in order to generate carboxypeptidase A activity from the zymogen.[9] The action of trypsin on apoprocarboxy-peptidase A-S6 does not yield an active product even with the subsequent addition of Zn^{2+}, suggesting that certain features of the binding site of carboxypeptidase A exist in the zymogen.[9, 22] On the other hand, activation of succinyl fraction I (see below) in the absence of bound Zn^{2+} results in a product which upon addition of certain metal ions, including Zn^{2+}, has enzymatic activity associated with the various metallocarboxy-peptidases.[7] Furthermore, PCPA-S6 exhibits low esterase activity before tryptic activation,[13] further implying intrinsic catalytic activity; however, contamination of activation products could not be ruled out in that particular investigation. PCPA-S6 can be activated under mild conditions to yield ATEE activity without destroying the trimeric sub-unit structure as shown in Fig. 1.[2, 12] This can be accomplished with 1:100 weight ratio of trypsin/PCPA-S6 at 0° for 1 to 1.5 hours,[23] or with 1:20 molar ratio of trypsin/PCPA-S6 at 0° for 15 minutes.[20] At the end of this period, a 50-fold molar excess of L-(tosylamido-2-phenyl) ethyl chloromethyl ketone (TPCK), or a 200-fold molar excess of (DFP) must be added to inactivate fraction II so as to prevent the generation of carboxypeptidase from subunit I with concomitant dis-aggregation of the trimeric complex. The mild activation is carried out in the presence of $0.1\ M$ β-phenylpropionate or 1,10-phenylanthroline to reversibly inactivate any carboxypeptidase that may be formed.[20]

Purity and Physical Properties. The zymogen isolated by this method shows a high degree of homogeneity which is in accord with the results obtained on earlier preparations.[3] Homogeneity is shown by ultracen-trifugation and moving-boundary electrophoresis with the calculated mobility for the ascending and descending boundary of 4.1 and 4.0 $\times$ 10^{-5} cm^2 V^{-1} sec^{-1} as determined in $0.1\ M$ sodium cacodylate–$0.15\ M$ NaCl, pH 6.7. Table I summarizes the physical properties of PCPA-S6.

Amino and Carboxyl Terminal Analysis.[13] The reaction of 1-fluoro-2,4-dinitrobenzene (FDNB) with three different preparations of PCPA-S6 performed according to Levy[24] indicates that aspartic acid or aspara-gine, lysine, and half-cystine are amino terminal residues. No other DNP-amino acids are found in stoichiometrically significant amounts. The presence of three amino terminal residues agrees with the view that

[22] H. Neurath and J. H. Freisheim, *Federation Proc.* **25,** 408 (1966).
[23] W. D. Behnke and H. Neurath, *Biochemistry* **9,** in press, 1970.
[24] A. L. Levy, "Methods of Biochemical Analysis" (D. Glick, ed.), Vol. II. p. 359. Wiley (Interscience), New York, 1955.

TABLE I
PHYSICAL PROPERTIES OF PROCARBOXYPEPTIDASE A

	PCPA-S5	PCPA-S6
Molecular weight		
Light scattering sedimentation equilibrium	72,500[a]	96,000[b]
		84,100 ± 2,290[c]
Sedimentation and diffusion	—	94,000[b]
		87,000[c]
$\bar{V}$	—	0.720 ± 0.016[c]
$s_{20,w}$	5.0 S[a]	6.12 S[c]
$D_{20,w}(cm^2\ sec^{-1})$	—	6.23 × 10^{-7[b]}
Isoelectric point (0.2 ionic strength)	—	4.5[b]
$E^{1\%}_{280}$	—	19[b]
	—	1.82 × 10^{-5[b]}
Zinc content (g-atom/mole)	—	0.84 ± 0.16[d]

[a] J. R. Brown, M. Yamasaki, and H. Neurath, *Biochemistry* **2**, 877 (1963).
[b] P. J. Keller, E. Cohen, and H. Neurath, *J. Biol. Chem.* **223**, 457 (1956).
[c] M. Yamasaki, J. R. Brown, D. J. Cox, R. N. Greenshields, R. D. Wade, and H. Neurath, *Biochemistry* **2**, 859 (1963).
[d] P. Piras and B. L. Vallee, *Biochemistry* **6**, 348 (1967).

PCPA-S6 is composed of three polypeptide chains held toegther in a trimeric complex by noncovalent bonds.[10, 11] Digestion of PCPA-S6 with carboxypeptidase A results in a release of leucine, tyrosine, valine, and phenylalanine in decreasing order of appearance. However, the identification of carboxyl terminal residues cannot be deduced on the basis of these data since asparagine is the C-terminal residue of carboxypeptidase A,[25] the direct product of activation of subunit I. Therefore, amino acids released during carboxypeptidase A digestion must arise from the carboxyl terminal region of either subunit II or subunit III, or both. The total amino acid composition of PCPA-S6 is given in Table II (see p. 471).

Disaggregation.[12, 15] When PCPA-S6 (5 mg/ml) is dissolved at 0° in 0.28 *M* LiCl, 0.01 *M* DFP, 0.1 *M* glycine buffer, pH 10.5, and allowed to stand in the dark for 48 hours at 20°, complete disaggregation occurs as shown by the change in the sedimentation coefficient from 6 to 3 S. The solution is then dialyzed against 0.005 *M* phosphate buffer, pH 8.0, centrifuged, and chromatographed on DEAE-cellulose as described in Fig. 8. The sharp breakthrough peak contains denatured subunit I (fraction I) which precipitates in the tubes as chromatography proceeds.[26] The second peak corresponds to fraction II and the third peak

[25] E. O. P. Thompson, *Biochem. Biophys. Acta* **10**, 633 (1953).
[26] In some preparations fraction I precipitates during the dialysis step or even during the pH 10.5 treatment.

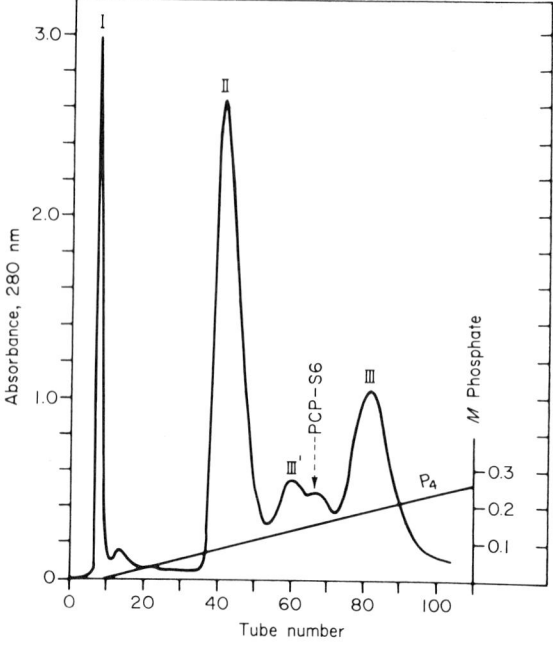

FIG. 8

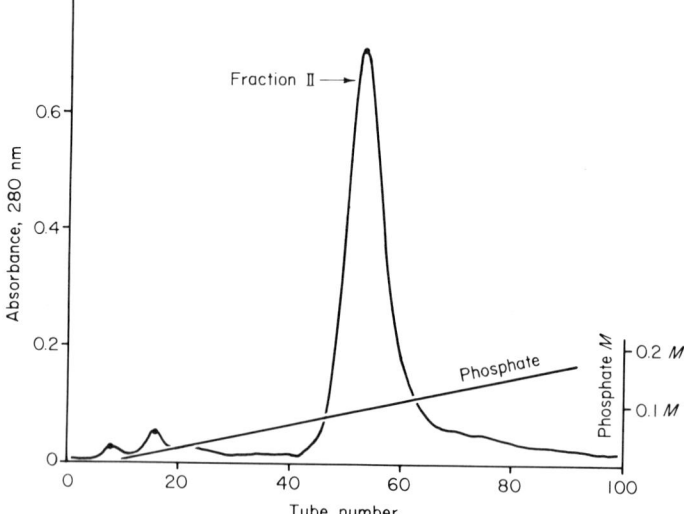

FIG. 9

TABLE II
Amino Acid Composition of Procarboxypeptidase A and Derivatives[a]

Amino acid	PCPA-S6[b]	PCPA-S5[c]	CPA$_a$[d]	Succinyl-fraction I[e]	Fraction II[f]	Fraction III[g]
Asp	77	54	29	33	24–25	22
Thr	49	39	26	25	16	10
Ser	62	43	32	31	14	17
Glu	78	60	25 (26)	40	22	24
Pro	39	25	10	15	11	13
Gly	66	44	23	27	22	20
Ala	53	39	20 (19)	24	15–16	15
Half-Cys	18	10	2	2	8	6
Val	58	40	16 (16)	22	19	19
Met	7	6	3	4	1	2
Ile	45	34	20 (19)	22	13	11
Leu	65	47	23 (24)	30	20–21	17
Tyr	29	24	19	20	7	5
Phe	29	21	16	18	7	7
Tryp	29	19	7	8	13	11
Lys	30	24	15	17	7	8
His	19	14	8	9	5	6
Arg	29	24	11	14	8–9	7

[a] Residues per mole.

[b] M. Yamasaki, J. R. Brown, D. J. Cox, R. N. Greenshields, R. D. Wade, and H. Neurath, *Biochemistry* **2**, 859 (1963).

[c] J. R. Brown, M. Yamasaki, and H. Neurath, *Biochemistry* **2**, 877 (1963).

[d] Based on the amino acid sequence (R. A. Bradshaw, L. H. Ericsson, K. A. Walsh, and H. Neurath, *Proc. Natl. Acad. Sci. U.S.* **63**, 1389 (1969). The values in parentheses represent the linked amino acid replacements [P. H. Pétra, R. A. Bradshaw, K. A. Walsh, and H. Neurath, *Biochemistry* **8**, 2762 (1969)].

[e] J. H. Freisheim, K. A. Walsh, and H. Neurath, *Biochemistry* **6**, 3010 (1967).

[f] Based on molecular weight of 26,000 [D. Gratecos, O. Guy, M. Rovery, and P. Desnuelle, *Biochim. Biophys. Acta* **175**, 82 (1969)].

[g] J. R. Brown, R. N. Greenshields, M. Yamasaki, and H. Neurath, *Biochemistry* **2**, 867 (1963).

Fig. 8. Chromatographic separation of the disaggregation products of procarboxypeptidase A-S6 on DEAE-cellulose at pH 8.0. Experimental conditions: column, 0.9×30 cm; sample volume, 35 ml (176 mg protein); fraction volume, 4.2 ml; pH 8.0 potassium phosphate buffer gradient, 250 ml 0.005 M vs. 250 ml 0.25 M. Peaks I, II, and III contain fractions I, II, and III of procarboxypeptidase A-S6, respectively. Fraction III′ is discussed in the text.

Fig. 9. Chromatography of disaggregated procarboxypeptidase A-S5 on DEAE-cellulose at pH 8.0. Fraction I was removed prior to chromatography by centrifugation. Experimental conditions: column, 0.9×27 cm; sample volume, 6 ml (13 mg protein); fraction volume, 4.6 ml; pH 8.0 potassium phosphate buffer gradient, 250 ml 0.005 M vs. 250 ml 0.25 M.

to fraction III. The existence of fraction III′ will be discussed under Properties of Fraction III (pp. 473, 474). Disaggregation of PCPA-S6 can also take place in urea solutions above 5 M.[12]

PROPERTIES OF FRACTION I AND SUCCINYL FRACTION I. The procedure used to disaggregate PCPA-S6 discussed above yields denatured subunit I (fraction I) which cannot be activated to generate carboxypeptidase A activity. The amino acid composition of this material is similar to that of the active enzyme,[12] suggesting that the denatured product may have undergone limited proteolysis during disaggregation. Therefore, the estimated amino acid composition would not be expected to reflect the true amino acid composition of subunit I. Another method of disaggregation using succinic anhydride produces a chemically modified subunit I (succinyl fraction I) which can be activated with trypsin to yield succinylcarboxypeptidase A.[7] This chemically modified form of carboxypeptidase A has been characterized and shown to be active.[27] The amino acid composition of chromatographically purified succinyl fraction I is given in Table II. Based on the number of residues, it is markedly larger than carboxypeptidase A_α consistent with its proposed role as precursor of the active enzyme. Sedimentation equilibrium yields an approximate molecular weight of 39,800–42,400, assuming $\bar{V} = .073$ ml/g, and 42,106 on the basis of amino acid composition. (This includes a value of 1900 for 19 succinyl groups assumed to be located on 13 lysyl and 6 tyrosyl residues.) Sedimentation analysis gives $s^0_{20,w} = 4.04$ and $E^{1\%}_{280} = 18.3$. Activation of succinyl fraction I with trypsin yields an active product which has the known chemical and enzymatic properties of succinylcarboxypeptidase A and an activation peptide containing 62 amino acid residues.[7]

PROPERTIES OF FRACTION II. Fraction II is one of the products isolated from the chromatographic purification of disaggregated PCPA-S6 or PCPA-S5 (Figs. 8 and 9). It has no carboxypeptidase A activity but, after tryptic activation, catalyzes the hydrolysis of acetyl-L-tyrosine ethyl ester (ATEE), acetyl-L-tryptophan ethyl ester, p-nitrophenyl acetate, casein, and glucagon.[12] Fraction II contains an amino terminal half-cystinyl residue and tryptic activation exposes valine as the newly formed amino terminal residue.[12] Recent data indicate that the amino acid composition of fraction II (based on molecular weight of 26,000) resembles that of porcine chymotrypsinogen C[28] (Table II). Furthermore, the amino acid sequence of the A-chain obtained after

[27] J. F. Riordan and B. L. Vallee, *Biochemistry* **3**, 1768 (1964).
[28] D. Gratecos, O. Guy, M. Rovery, and P. Desnuelle, *Biochim. Biophys. Acta* **175**, 82 (1969).

mild tryptic activation of PCPA-S6 and performic acid oxidation, shows a high degree of homology with the A-chain of porcine chymotrypsinogen C.[20] Fraction II appears homogeneous in the ultracentrifuge; however, recent experiments reveal that it undergoes a concentration-dependent self-association resulting in the formation of aggregates.[29] A previously reported value for the molecular weight[9] (27,000) was estimated at protein concentrations where such polymers would be expected to exist. A redetermination of several parameters gives a molecular weight of 22,900.[30] $s_{20,w}^0 = 2.90 \pm 0.05$,[29] and $E_{280}^{1\%} = 15 \pm 0.5$.[29] Sedimentation equilibrium experiments at low protein concentrations reveal the presence of small amounts of contaminating material.[30] Activated fraction II is inhibited by DFP,[12,23] and L-(1-carbobenzoxyamide-2-phenyl) ethyl chloromethyl ketone (ZPCK).[23] The amino acid sequences of the isolated DI[32]P and ^{14}C-ZPCK peptides are homologous to the corresponding active site regions in the amino acid sequence of α-chymotrypsin, affirming that activated fraction II is a serine proteinase.[23] These "all or none" assays, however, reveal that only 25% of the molecules are active, implying that either the method of activation is inadequate for the generation of maximum endopeptidase activity from PCPA-S6, or that a high degree of denaturation occurs during disaggregation as shown in the case of fraction I (see above).[23] It can be concluded from these data that only 25% of the material isolated as fraction II corresponds to subunit II, the postulated inactive precursor of a chymotrypsinlike endopeptidase which is part of the trimeric complex of PCPA-S6.

PROPERTIES OF FRACTION III. Fraction III is isolated from the chromatographic purification of disaggregated PCPA-S6 (Fig. 8). This fraction is obtained in two forms, III and III'. Fraction III' is believed to be a degradation product of fraction III, since an additional amino terminal threonine can be generated during the conversion of one form into the other. Both fractions appear homogeneous in the ultracentrifuge with $s_{20,w}^0 = 2.9$ at a protein concentration of 0.7%. The amino acid composition of fraction III, shown in Table II, indicates some resemblance to fraction II. The assumed molecular weight of 24,000 is obtained by difference between PCPA-S6 and fractions I and II. The amino terminal analysis yields 0.3 residues of DNP-apartic acid (uncorrected) and 0.2 residue of di-DNP-lysine (uncorrected) per molecular weight of 24,000. No enzymatic activity has been found for either fractions III and III' when tested against HPLA, ATEE, p-nitrophenylacetate, elastin, casein, glucagon, methylbutyrate, DNA, RNA, and uridine-3',5'-phosphate. The apparent lack of enzymatic activity may be

[29] W. D. Behnke, R. D. Wade, and H. Neurath, *Biochemistry* **9**, in press, 1970.
[30] D. C. Teller, *Biochemistry* **9**, in press, 1970.

due to either the inappropriate choice of substrate or to the denaturation during the isolation procedure as in the case of fraction I. Recent data strongly suggest that fraction III is a degradation product of fraction II and exists as an artifact generated by the disaggregation procedure. Inhibition of activated PCPA-S6 by ^{14}C-ZPCK yields 0.69 moles inhibitor/ mole PCPA-S6 as compared to 0.25 moles inhibitor/mole activated fraction II when inhibition is carried out on the isolated fraction.[24] The data imply that 2 moles of fraction II are associated per mole of PCPA-S6 and that during the second stage of activation, producing carboxy-peptidase A activity, a total of 1 mole of fraction II is converted into fraction III, most likely by limited proteolysis. The close similarity in molecular weight, amino acid composition, and diagonal peptide maps[31] between fractions II and III strengthens this hypothesis. It has also been suggested that fractions II and III are derived from an ancestral endopeptidase, and that the latter lost its biological activity through evolutionary changes.[5] Nevertheless, more research is needed to determine the existence and physiological significance of subunit III.

Properties of Procarboxypeptidase A-S5

Stability. Solutions of PCPA-S5 are stable at pH 7.5 in the presence of $10^{-3} M$ DFP. However, when the zymogen is exposed to pH 10.5, or to 8 M urea, irreversible denaturation occurs, resulting in the disaggregation of the dimer into its individual subunits.

Purity and Physical Properties.[10] The zymogen prepared as described above appears homogeneous in the ultracentrifuge. The molecular weight measured by sedimentation equilibrium gives a mean value of 72,500 which is slightly higher than the value calculated from the sum of the molecular weights of succinyl-fraction I (40,000) and fraction II (22,900), i.e., 62,900. Homogeneity is also confirmed by moving-boundary electrophoresis. The calculated mobility of the ascending boundary is 4.4×10^{-5} cm^2 V^{-1} sec^{-1}.

N-Terminal Analysis and Amino Acid Composition.[10] The results (uncorrected) of the FDNB reaction with PCPA-S5 indicate that lysine and half-cystine are the only amino terminal residues. Three different preparations were tested and the data show no significant differences. The amino acid composition of the zymogen is shown in Table II.

Disaggregation.[10] When 35.4 mg of zymogen is dissolved at 0° in 6 ml of 0.28 M LiCl, 0.01 M DFP, 0.1 M glycine buffer, pH 10.5, and allowed to stand in the dark for 48 hours at 20°, complete disaggregation occurs as shown by the change in the sedimentation coefficient from

---

[31] W. D. Behnke, unpublished experiments.

4.4 to **2.15** S. The solution is then dialyzed against 0.005 M potassium phosphate and centrifuged. The precipitate is denatured fraction I based on amino acid composition and lack of enzymatic activity. The supernatant is further purified on DEAE-cellulose as described in Fig. 9. The elution pattern shows the presence of only one peak identified as fraction II on the basis of amino acid composition, N-terminal analysis, and enzymatic properties. When the elution pattern is compared to that obtained for the disaggregation products of PCPA-S6 (see above) it is evident that fraction III is totally absent from PCPA-S5. Furthermore, the amino acid compositions of fraction I-S5 and II-S5 are similar to fractions I-S6 and II-S6, respectively, suggesting that the subunit structure of PCPA-S6 is composed of fractions I, II, and III, while that of PCPA-S5 contains fractions I and II.

Bovine Carboxypeptidase A

A study of bovine carboxypeptidase A was first initiated in 1929 by Walschmidt-Leitz and Purr[32] who found that the hydrolysis of N-acylated peptides having a free carboxyl group was catalyzed by pancreatic extracts. In 1937, Anson crystallized the enzyme from partially autolyzed exudate of freshly collected pancreas glands.[33] Since these early developments, much research has been devoted toward the study of this enzyme and several reviews describing its chemical nature and biological function have been published.[4, 6, 34-38]

Assay Methods

Carboxypeptidase A activity is determined by measuring the rate of hydrolysis of ester or peptide substrates using either titrimetric or spectrophotometric methods.

Assay of Peptidase Activity

Principle. The method is based upon the rate of cleavage of the peptide bond in N-acylated dipeptide substrates. The substrates most

[32] E. Walschmidt-Leitz and A. Purr, *Chem. Ber.* **62B**, 2217 (1929).
[33] M. L. Anson, *J. Gen. Physiol.* **20**, 663 (1937).
[34] E. L. Smith, *Advan. Enzymol.* **12**, 191 (1951).
[35] H. Neurath, "The Enzymes," (P. D. Boyer, H. Lardy, and K. Myrbäck, eds.), Vol. 4, p. 11. Academic Press, New York, 1960.
[36] H. Neurath, R. A. Bradshaw, L. H. Ericsson, D. R. Babin, P. H. Pétra, and K. A. Walsh, *Brookhaven Symp. Biol.* **21**, 1 (1968).
[37] B. L. Vallee and J. F. Riordan, *Brookhaven Symp. Biol.* **21**, 91 (1968).
[38] W. N. Lipscomb, J. A. Hartsuck, G. N. Reeke, F. A. Quiocho, P. H. Bethge, M. L. Ludwig, T. A. Steitz, H. Muirhead, and T. C. Coppola, *Brookhaven Symp. Biol.* **21**, 24 (1968).

widely used are carbobenzoxyglycyl-L-phenylalanine[39,40] (CGP), and hippuryl-L-phenylalanine.[41] The rate of the reaction can be measured either by the colorimetric method of Moore and Stein[42] for the estimation of liberated phenylalanine, or spectrophotometrically by the loss in absorption in the range of 220–236 nm. The former method has already been reviewed,[4] and only the latter using CGP will be described. A spectrophotometric assay method employing hippuryl-L-phenylalanine as substrate has also been published.[43] Any substrate concentration can be used for the standard assay, however, because of the complexities of the kinetics recently reviewed[37] (see below), it is preferable to operate in a region of initial CGP concentration of $8.0 \times 10^{-4} M$ to $2.5 \times 10^{-3} M$ where Michaelis-Menten kinetics prevail.[44,45]

Reagents

Substrate solution. CGP, $1.18 \times 10^{-3} M$ (Fox Chemical Co.); KCl, $0.45 M$; Tris-Cl, $0.05 M$; pH 7.5. This arbitrary choice of substrate concentration is convenient because a first-order rate constant at this concentration has been reported[44] (Table III).

Enzyme. The crystals are dissolevd at $0°$ in $1 M$ NaCl, $5 \times 10^{-3} M$ potassium phosphate, pH 7.5, to a concentration value of 0.4 mg/ml ($E_{278}^{0.1\%} = 1.88$). Carboxypeptidase solutions which have optical densities higher than 1.2 do not obey Beer's law.[46]

Procedure.[44,47] Two cuvettes, one containing 3 ml of substrate solution and the other 3 ml of buffer are placed in a double beam spectrophotometer and allowed to equilibrate at $25°$. The wavelength is selected (223–224 nm) so that the reaction mixture reads 1.8 OD units against the buffer solution. After the wavelength has been selected, the spectrophotometer is zeroed against buffer and the recorder is balanced. Fifty microliters of buffer is added to the blank cuvette and 50 μl of enzyme solution to the reaction cuvette. Recording is started immediately and the reaction is followed to completion in order to convert OD units to units of activity.

[39] K. Hofmann and M. Bergmann, *J. Biol. Chem.* **134**, 225 (1940).

[40] H. Neurath, E. Elkins, and S. Kaufman, *J. Biol. Chem.* **170**, 221 (1947).

[41] J. E. Snoke and H. Neurath, *J. Biol. Chem.* **181**, 789 (1949).

[42] S. Moore and W. H. Stein, *J. Biol. Chem.* **176**, 367 (1948).

[43] J. E. Folk and E. W. Schirmer, *J. Biol. Chem.* **238**, 3884 (1963).

[44] J. R. Whitaker, F. Menger, and M. L. Bender, *Biochemistry* **5**, 386 (1966).

[45] R. C. Davies, J. F. Riordan, D. S. Auld, and B. L. Vallee, *Biochemistry* **7**, 1090 (1968).

[46] W. O. McClure, H. Neurath, and K. A. Walsh, *Biochemistry* **3**, 1897 (1964).

[47] P. H. Pétra and H. Neurath, *Biochemistry* **8**, 5029 (1969).

Units and Specific Activity. One unit of enzyme activity can be defined as that amount which hydrolyzes 1 micromole of substrate per minute. The specific activity is expressed in units per milligram protein. However, since the molecular weight of carboxypeptidase A is known (Table V), the specific activity can also be expressed as a first-order rate contant, k (sec^{-1}) $= V/E_0$, where V is the initial velocity expressed in micromoles of substrate hydrolyzed per second, and E_0 is micromoles of enzyme in the cuvette calculated using $\epsilon_{278} = 6.42 \times 10^4\,M^{-1}$ cm^{-1}.[48] The rate constant obtained for the assay conditions described above is: $k = 1134$ min^{-1} or 18.9 sec^{-1}.[44] The specific peptidase activities of several preparations of carboxypeptidase A are shown in Table III.[47]

Assay of Esterase Activity

Principle. The method is based upon the rate of cleavage of the ester bond present in hippuryl-DL-β-phenyllactic acid (HPLA).[49] The substrate is an ester analog of the usual peptide substrates of carboxypeptidase A. The catalyzed hydrolysis is monitored titrimetrically for the release of protons,[50] or spectrophotometrically by the increase in optical density at 254 nm[44, 46] which occurs when benzoylamino acid derivatives are hydrolyzed.[51] Only the former method will be described because of its usefulness as a simple routine assay.

Reagents

Substrate solution, DL-HPLA, 10^{-2} M (Fox Chemical Co., the sodium salt or the free acid), 5×10^{-3} M sodium Veronal, 4.5×10^{-2} M NaCl, pH 7.5

Enzyme solution. The crystals are dissolved at $0°$ in 1 M NaCl, 5×10^{-3} M potassium phosphate, pH 7.5, to a concentration range of 0.02–0.5 mg/ml enzyme.

Base. 0.1 N NaOH, standardized

Procedure.[50] The assay is carried out in a pH-stat with the syringe containing the standardized base solution. Three milliliters of substrate solution are introduced into the thermostatted vessel and allowed to equilibrate at $25°$ for 15 minutes. The pH is adjusted to 7.5 and after establishing a base line, 10 to 150 μl of enzyme solution are added. The reaction rate is linear for more than 60% of the total hydrolysis.

[48] R. T. Simpson, J. F. Riordan, and B. L. Vallee, *Biochemistry* **2**, 616 (1963).
[49] J. E. Snoke, G. W. Schwert, and H. Neurath, *J. Biol. Chem.* **175**, 7 (1948).
[50] J.-P. Bargetzi, K. S. V. Sampath Kumar, D. J. Cox, K. A. Walsh, and H. Neurath, *Biochemistry* **2**, 1468 (1963).
[51] G. W. Schwert and Y. Takenaka, *Biochim. Biophys. Acta* **16**, 570 (1955).

TABLE III
Specific Activity of Several Preparations and
Species of Bovine Carboxypeptidase A[a,b]

	CGP[c] (sec^{-1})	HPLA[c] (sec^{-1})
CPA (Cox)	19.7	120[d]
CPA (Anson) (Lot C2560)	18.9[e]	100[f]
CPA (Anson) (Lot CoA-7HA)	19.0	122
CPA (Allan)	—	122[g]
CPA$_\alpha^{Val}$	19.4	111
CPA$_\beta^{Val}$	23.6	121
CPA$_\beta^{Leu}$	21.9	115
CPA$_\gamma^{Val}$	20.3	122
CPA$_\gamma^{Leu}$	22.0	119

[a] P. H. Pétra and H. Neurath, *Biochemistry* **8**, 2466 (1969); *ibid.* **8**, 5029 (1969).
[b] These values, except for those referred to other work, represent an average of three determinations using the method described in the assay procedure.
[c] Calculated using molecular weight 34,600.
[d] J.-P. Bargetzi, K. S. V. Sampath Kumar, D. J. Cox, K. A. Walsh, and H. Neurath, *Biochemistry* **2**, 1468 (1963).
[e] J. R. Whitaker, F. Menger, and M. L. Bender, *Biochemistry* **5**, 386 (1966).
[f] M. L. Bender, J. R. Whitaker, and F. Menger, *Proc. Natl. Acad. Sci. U.S.* **53**, 711 (1965).
[g] B. J. Allan, P. J. Keller, and H. Neurath, *Biochemistry* **3**, 40 (1964).

Units and Specific Activity. One unit of esterase activity can be defined as 1 micromole of HPLA hydrolyzed per second, and the specific activity is the pseudo zero-order rate constant, k_0, expressed in units per milligram of protein. Using a molecular weight of 34,600 for carboxypeptidase A, k_0 is equal to 7197 minutes^{-1} or 120 sec^{-1}.[50] Table III lists the specific esterase activities of several preparations of carboxypeptidase A. Experience shows that the specific activity of commercially available preparations varies from lot to lot.

Preparation and Purification Procedures

Bovine carboxypeptidase A is known to exist in at least 8 forms; 4 of these differ from each other in either their chemical or functional properties.[52,53] These are: carboxypeptidase A$_\alpha$, A$_\beta$, A$_\gamma$, A$_\delta$ having as N-terminal amino acids alanine, serine, asparagine, and asparagine, respectively.[54] The enzyme can be prepared by three different methods, all

[52] K. S. V. Sampath Kumar, J. B. Clegg, and K. A. Walsh, *Biochemistry* **3**, 1728 (1964).
[53] T. L. Coombs, Y. Omote, and B. L. Vallee, *Biochemistry* **3**, 653 (1964).
[54] J.-P. Bargetzi, E. O. P. Thompson, K. S. V. Sampath Kumar, K. A. Walsh, and H. Neurath, *J. Biol. Chem.* **239**, 3767 (1964).

yielding heterogeneous crystalline products known as: carboxypeptidase A (Anson)[33] which is the commercially available preparation; carboxypeptidase A (Cox)[55]; and carboxypeptidase A (Allan).[56] Recent developments in the chromatographic purification of the first two preparations have resulted in the isolation of pure crystalline carboxypeptidase A_α, A_β, and A_γ.[47,57] No chromatographic investigation has been done on carboxypeptidase A (Allan), but N-terminal analysis and other studies[53] indicate that the major product is carboxypeptidase A_δ containing small amounts of carboxypeptidase A_α and A_β.[54] In addition to these chemical variations, each form of carboxypeptidase A also exists as two allotypes arising from amino acid replacements due to allelomorphism[58,59]; these are carboxypeptidase A_α^{Val}, A_α^{Leu}, A_β^{Val}, A_β^{Leu}, A_γ^{Val}, and A_γ^{Leu} (for nomenclature see footnotes 57 and 59). The allotypic variants of the delta enzyme have not yet been investigated.

Preparation of Carboxypeptidase A (Anson), A_γ^{Val}, and A_γ^{Leu}

Carboxypeptidase A_γ is the major product occurring in carboxypeptidase A (Anson). The two allotypic variants of the former are obtained by a chromatographic purification of the latter.[57] The preparation of carboxypeptidase A (Anson) has already been reviewed in detail[4] and only a brief account of the procedure will be given in step 1.

Step 1. Carboxypeptidase A (Anson).[4,33] An exudate of pancreatic glands is collected in the cold room and activated at 37° by raising the pH to 7.8. After 1 hour, the pH of the activated juice is adjusted to 4.6 with acetic acid and immediately diluted 10-fold with cold distilled water, keeping the pH at 4.6. After 5 hours at 4°, the euglobin precipitate is collected by centrifugation, suspended in cold distilled water, and adjusted to pH 6 at 4° with $Ba(OH)_2$. The suspension is centrifuged and the precipitate resuspended in water and adjusted to pH 9.5 with $Ba(OH)_2$. After centrifugation, the clear supernatant is adjusted to pH 7.2 with acetic acid and seeded to induce crystallization. Recrystallization is carried out by dissolving the crystals in 10% LiCl and dialyzing against distilled water at 4°.

Step 2. Chromatographic Purification of Carboxypeptidase A (Anson).[57] DEAE-cellulose (DE-52, Reeve Angel Co.) is recycled with 0.5 N HCl and 0.5 N NaOH, and equilibrated in 0.05 M β-phenylpro-

[55] D. J. Cox, F. C. Bovard, J.-P. Bargetzi, K. A. Walsh, and H. Neurath, *Biochemistry* **3**, 44 (1964).

[56] B. J. Allan, P. J. Keller, and H. Neurath, *Biochemistry* **3**, 40 (1964).

[57] P. H. Pétra and H. Neurath, *Biochemistry* **8**, 2466 (1969).

[58] K. A. Walsh, L. H. Ericsson and H. Neurath, *Proc. Natl. Acad. Sci. U.S.* **56**, 1339 (1966).

[59] P. H. Pétra, R. A. Bradshaw, K. A. Walsh, and H. Neurath, *Biochemistry* **8**, 2762 (1969).

pionate (β-pp), 0.04 M LiCl, 0.05 M Tris, pH 7.5 (pH is adjusted with 10 N NaOH at room temperature). A column (2.5 × 100 cm) is packed at 300 ml/hour to a height of 90 cm and transferred to the cold room (4°). A flow adaptor is placed on top of the adsorbent bed, and the column is equilibrated overnight at a flow rate of 150 ml/hour with the pH 7.5 buffer. All subsequent operations are done at 0–4°. Eight hundred milligrams of carboxypeptidase A (Anson) crystals are dissolved in 4.5 ml of 2 M Tris, 0.05 M β-phenylpropionate, 0.04 M LiCl, pH 7.5,[60] and the solution is centrifuged at 3000 rpm for 10 minutes. Thirty-seven milliliters of 0.05 M β-phenylpropionate, 0.04 M LiCl, pH 7.5, are added to the supernatant and dialysis is carried out against 2 liters of the chromatography buffer for 4 hours.[61] The sample is centrifuged for 10 minutes at 3000 rpm and after readjustment of the flow adaptor, the supernatant is applied on the column at a flow rate of 48 ml/hour. Elution is continued at 93 ml/hour and completed in approximately 72 hours. The results are shown in Fig. 10. It has been difficult to separate peaks II and III (Fig. 10) with some batches of DE-52, in which case a longer column with a bed adsorbent height of 110 cm should be used. The flow rate can also be increased to 150 ml/hour to reduce the total amount of time required for the elution.

Step 3. Crystallization of Carboxypeptidase A$_\gamma^{Val}$ and Carboxypeptidase A$_\gamma^{Leu}$. Peaks IV and V (Fig. 10) are pooled separately, concentrated to 10–20 ml at 4° in the Amicon diaflow pressure cell equipped with a UM-1 membrane, and dialyzed against 6 liters of cold 0.005 M Tris-Cl, pH 7.5. Crystals are formed overnight. These are carefully washed with cold 0.005 M Tris-HCl, pH 7.5, several times and recrystallized twice by dissolving in cold 5 M NaCl, 0.05 Tris-Cl, pH 7.5, diluting with 40 ml distilled water and dialyzing against 0.005 M Tris-Cl, 10^{-7} M ZnCl$_2$, pH 7.5, 4°. This step insures the removal of β-phenylpropionate which probably remains bound to the enzyme during the first crystallization. The crystals are carefully washed and stored in 0.005 M Tris-Cl, pH 7.5 under an atmosphere of toluene.

---

[60] Some CPA (Anson) preparations are more insoluble than others, in which case recrystallization is performed as follows: 1 g of crystals is suspended in 50 ml of 5 M NaCl, 0.005 M Tris-Cl, pH 7.5 at 0°. The pH is raised to 9.5 with 1 N NaOH and stirred at 0° for 5 minutes until most of the enzyme is in solution. The pH is immediately brought back to 7.5 with 0.1 N HCl and the solution centrifuged. The supernatant is dialyzed overnight at 4° against 6 liters of 0.005 M Tris-Cl, pH 7.5. The crystals obtained can now be dissolved and chromatographed as described in step 2.

[61] Dialysis for more than 4 hours will produce a supersaturated enzyme solution from which the enzyme will crystallize. If this occurs, the whole process of sample preparation must be repeated on the collected crystals.

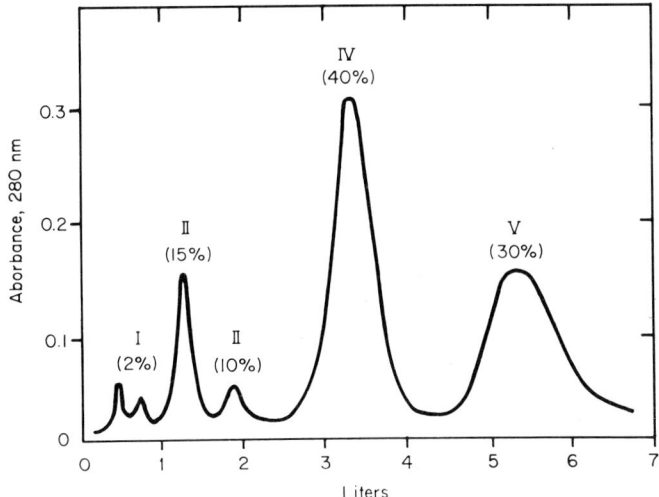

Fig. 10. Chromatographic purification of carboxypeptidase A (Anson). Experimental conditions: column, 2.5×100 cm; sample volume, 41 ml (800 mg protein); fraction volume, 19 ml; flow rate, 93 ml/hour; elution is carried out at $4°$ with $0.05\ M$ β-phenylpropionate, $0.04\ M$ LiCl, $0.05\ M$ Tris, pH 7.5. The fractions obtained from each peak are: peak I: CPA_α^{Val}; peak II: CPA_β^{Val}; peak III: CPA_β^{Leu}; peak IV: CPA_γ^{Val}; peak V: CPA_γ^{Leu}.

Preparation and Crystallization of Carboxypeptidase A_β^{Val} and Carboxypeptidase A_β^{Leu}

The allotypic variants of carboxypeptidase A_β are obtained mainly from the chromatographic purification of carboxypeptidase A (Anson), since large amounts of enzyme can be processed by this method. The fractions corresponding to peaks II and III shown in Fig. 10 are pooled and crystallized according to the method described in step 3 (p. 480). Carboxypeptidase A_β^{Leu} is also obtained from the chromatographic elution of carboxypeptidase A (Cox) (see below, peak 3 in Fig. 11).

Preparation of Carboxypeptidase A (Allan)[56]

The first preparation of carboxypeptidase described by Anson (see above), although satisfactory for the early studies on the enzyme, is not only laborious but also involves steps for which conditions cannot be easily controlled, such as extensive autolysis of the glands and collection of the exudate over a period of several days. Therefore, several modifications have resulted in the development of two new methods of preparation, each yielding carboxypeptidase A (Allan) and A (Cox); the latter being described in the next section. Carboxypeptidase A

(Allan), which has not yet been chromatographically purified, contains mainly carboxypeptidase A$_\delta$ probably with small amounts of carboxypeptidase A$_\alpha$ and A$_\beta$.[54]

Step 1. Extraction of Acetone Powder. Five hundred grams of powder (prepared according to published procedures[2,62]) is extracted with 6 liters of cold distilled water for 12–24 hours in the cold room. The suspension is centrifuged for 1 hour at 15,000 rpm in a Spinco Model L centrifuge with rotor No. 21. The supernatant extract can be kept at 4° under an atmosphere of toluene or can be stored frozen.

Step 2. Activation. The soluble extract isolated as described is warmed to 37° and the pH adjusted to 7.8 with 1 N NaOH. Four hundred milligrams of trypsin is added, and the activation of procarboxypeptidase is followed by removing 10-ml aliquots and assaying against HPLA and CGP. When maximum activity is obtained (corresponding to approximately 1 mg/ml enzyme) the solution is allowed to stand for an additional hour and cooled at 0°.

Step 3. Precipitation. Ammonium sulfate (391 g/liter of extract) is slowly added in portions while maintaining the pH at 7.5 with 1 N NaOH. After the precipitate has settled overnight, the supernatant is carefully removed by siphoning and discarded, and the remaining suspension centrifuged at 20,000 rpm for 1 hour in the Spinco Model L centrifuge using rotor No. 21. The pellet is dissolved in cold distilled water to a concentration of approximately 20 mg/ml estimated at 278 nm and centrifuged at 13,000 rpm (using rotor No. 30) for 30 minutes to remove any turbidity that may be present. The supernatant is dialyzed exhaustively against six changes of 25-fold excess of distilled water. A precipitate is formed during dialysis. The suspension is removed from the dialysis bags and is further processed or can be stored frozen without loss of activity.

Step 4. Isoelectric Precipitation and Extraction. This step must be performed at 0° as quickly as possible to prevent prolonged exposure of the enzyme at low pH. Cold distilled water is added to the suspension to normalize the protein concentration to 15 mg/ml and pH adjusted to 5.5 with 0.1 N acetic acid added dropwise with continuous stirring. After 5 minutes at pH 5.5, the suspension is centrifuged at 15,000 rpm for 30 minutes. The collected precipitate is resuspended in the same volume as used above and the pH adjusted to 6.6 with 0.05 M Ba(OH)$_2$. After 2 hours of gentle stirring, the suspension is centrifuged for 30 minutes at 15,000 rpm, and the precipitate resuspended in 4/5 of the volume used above of distilled water. The suspension can be kept overnight if the pH is adjusted to 7.5 with Ba(OH)$_2$.

[62] E. H. Fischer and E. A. Stein, *Arch. Sci. (Geneva)* **7**, 31 (1954).

Step 5. Crystallization. The pH of the suspension is raised to 10.0 (never higher than 10.4) by the slow addition of cold 0.05 M Ba(OH)$_2$, and this pH is maintained for not more than 3 to 4 hours during which time most of the precipitate dissolves. The turbid solution is centrifuged at 22,500 rpm in rotor No. 40 for 30 minutes. The pH of the clear supernatant containing the enzyme is lowered very slowly by adding 0.1 N acetic acid using efficient stirring until a slight opalescence or color change is detected, indicating the beginning of crystallization. Seeds of carboxypeptidase A (Allan) are added to the solution while mixing for 10 to 15 minutes, and the solution is allowed to stand for 2 to 3 hours. If crystallization does not occur during this time, the process is repeated. After several days, the pH is lowered to 7.5 and the crystals collected by centrifugation. These are usually tan in color and are 60 to 70% pure as determined by enzymatic assay.

Step 6. Recrystallization. The crystals (50 mg/ml) are dissolved at 0° in 1 M NaCl, and the solution centrifuged at 15,000 rpm for 15 minutes. The clear supernatant is dialyzed against 2 liters of each of the following solutions successively for 8 hours :

(1) 0.5 M LiCl–0.02 M sodium Veronal, pH 8.0
(2) 0.2 M LiCl–0.02 M sodium Veronal, pH 8.0
(3) 0.1 M LiCl–0.02 M sodium Veronal, pH 8.0

The crystals are stored in distilled water under an atmosphere of toluene.

Preparation of Carboxypeptidase A (Cox), A_α^{Val}, and A_α^{Leu}

Carboxypeptidase A_α is the major product present in carboxypeptidase A (Cox)[54] from which it is isolated by chromatography.[47] Carboxypeptidase A (Cox) is prepared by activating partially purified procarboxypeptidase A containing both PCPA-S5 and PCPA-S6.[55] Since the enzyme is generated from both dimeric and trimeric forms of the zymogen, the activation can be performed on the mixture.

Step 1. Ammonium Sulfate Precipitation. The solution (1750 ml, tubes 130–200 in Fig. 2) isolated after step 2 in the purification of procarboxypeptidase A (see above)[63] is saturated to 40% ammonium sulfate by adding 283 g/liter over a period of 1 hour at 0°. Gentle stirring is continued for 2 hours and the pH is maintained at 7.2 by adding 4 M NaOH dropwise. The solution is centrifuged at 6000 g for 1 hour and the

---

[63] The starting material is 100 g of acetone powder. The procedure can be scaled up to 150 g of powder by employing the following conditions in the chromatography of Fig. 2: column 4.4 × 50 cm, flow rate 170 ml/hour, linear gradient from 0.005 to 0.33 M phosphate, pH 8.0, with 2 liters per vessel.

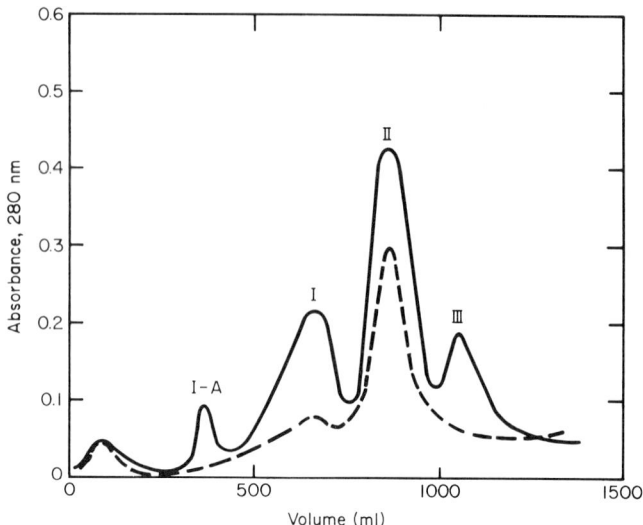

FIG. 11. Chromatographic purification of carboxypeptidase A (Cox). Experimental conditions: column, 1.5×90 cm; sample volume, 40 ml (150 mg protein); fraction volume, 20 ml; flow rate, 50 ml/hour; pH 7.5 complex gradient (see text), temperature 4°. Peak I is CPA_α^{Val}, peak II contains a mixture of CPA_α^{Leu} and CPA_β^{Leu}, and peak III is CPA_β^{Leu}. Peak I-A has full HPLA activity, but remains to be chemically characterized. The dashed line represents the elution pattern of a pure sample of CPA_β^{Val} obtained from peak II of Fig. 10.

precipitate dissolved in less than 150 ml of cold distilled water. The solution is dialyzed at 0° against two 6-liter changes of distilled water.

Step 2. Activation and First Crystallization. The precipitate, which accumulates during dialysis, is removed by centrifugation at 6000 g for $\frac{1}{2}$ hour. The clear yellow supernatant is brought to 0.25 M potassium phosphate, pH 8.0, by adding 1 M potassium phosphate buffer. 250 mg of trypsin (containing 50% magnesium sulfate)[64] are added and the solution is incubated for 16 hours at 37°. A few crystals of thymol are added as a preservative. The solution is centrifuged at 6000 g for $\frac{1}{2}$ hour, and the supernatant dialyzed at 5° against two changes of 1 liter of 0.001 M potassium phosphate buffer, pH 8.0, for 48 hours. Crystals form at the end of the second change.

Step 3. Recrystallization. The crystals are collected by centrifugation and dissolved in 20 ml of 1 M NaCl. The solution is dialyzed against three changes of 2 liters of 0.001 M potassium phosphate, pH 8.0, for 3 days after which the crystals are formed. An additional recrystallization may be needed to obtain an HPLA specific activity of 120 sec^{-1}.

[64] Results are identical when desalted trypsin is used.

Using the new nomenclature,[57] this crystalline preparation is called carboxypeptidase A (Cox).

Step 4. Chromatographic Purification of Carboxypeptidase A (Cox).[47] A column (1.5 × 90 cm) is packed at a flow rate of 120 ml/hour with DEAE-cellulose (DE-52, Reeve Angel Co.) previously equilibrated with 0.05 M β-phenylpropionate, 0.004 M LiCl, 0.05 M Tris, pH 7.5 (the pH is adjusted with 10 N NaOH at room temperature). When the column is filled, a flow adaptor is placed on top of the adsorbent and the column is transferred to the cold room (4°) where it is equilibrated overnight with the same buffer at a flow rate of 50 ml/hour. One hundred fifty milligram of carboxypeptidase A (Cox) crystals are dissolved in 1.0 ml of 0.05 M β-phenylpropionate, 0.004 M LiCl, 2 M Tris, pH 7.5 (adjusted with HCl at room temperature). Thirty-nine milliliters of 0.05 M β-phenylpropionate, 0.004 M LiCl, pH 7.5, are added and the solution centrifuged at 3000 rpm for 10 minutes. The supernatant is pumped onto the column at a flow rate of 50 ml/hour, and after application of the sample, elution is started at the same flow rate using a complex gradient formed with the aid of a 9-chamber Varigrad filled as follows: 200 ml of 0.05 M β-phenylpropionate; 0.004 M LiCl; 0.05 M Tris, pH 7.5 in each of chambers 1, 2, 3, and 5, and 200 ml of 0.122 M β-phenylpropionate; 0.04 M LiCl; 0.05 M Tris; pH 7.5, in each of chambers 4, 6, 7, 8, and 9. The chromatography is completed in approximately 30 hours. The elution pattern is shown in Fig. 11. DEAE-cellulose has a tendency to pack during the chromatography, preventing reuse even after reequilibration with starting buffer. Experience with the method indicates that small errors in LiCl concentration of each buffer, as well as the purity of β-phenylpropionate,[65] have dramatic effects on the reproducibility of the elution pattern.

Step 5. Crystallization of Carboxypeptidase A_α^{Val} and A_α^{Leu}. These are obtained from peaks I and II, respectively (Fig. 11), and crystallized in the same manner as described in step 3 (p. 480). The crystals are also stored in 0.005 M Tris-Cl, pH 7.5, under an atmosphere of toluene in the refrigerator at 4°. Carboxypeptidase A_α^{Leu} (peak II, Fig. 11) contains a 30% contamination of carboxypeptidase A_β^{Val} which cannot be removed by rechromatography or electrophoresis.[47] Consequently, it must be isolated from a single animal homozygous for the Leu trait and subjected to chromatography as described in Fig. 11.[66] The resulting crystals contain pure carboxypeptidase A_α^{Leu}.

---

[65] β-Phenylpropionic acid is purchased from Baker and should be recrystallized from an ethanol–water mixture.
[66] P. H. Pétra, R. W. Tye, R. Sande, and H. Neurath, in preparation.

Physical and Chemical Properties

Most of the physicochemical studies on the native enzyme reported in the literature have been done on carboxypeptidase A (Anson) and A (Allan), the latter being the least chemically characterized preparation of the enzyme. On the other hand, most of the chemical modification and structure determination studies have been performed on carboxypeptidase A (Cox) and A (Anson). Since some of the properties which are important in the routine handling of the enzyme, such as stability, solubility, and purity vary among the three preparations, while others, such as catalytic activity and substrate specificity seem to be similar, it is imperative to define which preparation is used in a particular experiment. The differences and similarities between the properties of carboxypeptidase A (Cox), A (Anson), and A (Allan) are obviously dictated by the relative amounts of carboxypeptidase A_α, A_β, A_γ, or A_δ present in each preparation. These were found to vary in the case of carboxypeptidase A (Cox) and A (Anson).

Solubility. All forms of bovine carboxypeptidase A are insoluble in water. However, upon the addition of salts, the enzyme slowly goes into solution. The various forms of the enzyme differ in their solubility properties and they can be arranged as follows in the order of increasing solubility in salt solutions: $CPA_\alpha > CPA$ (Cox) $> CPA$ (Allan) $> CPA_\beta > CPA$ (Anson) $> CPA_\gamma$.[55,67] No comparative data are available for CPA_δ. There are indications also that the Val enzymes are generally more soluble than the Leu enzymes.[67] For routine work CPA (Anson) can be dissolved with $1 M$ NaCl at $0°$ buffered at pH 7.5.[68] However, when concentrated enzyme solutions are needed (14–19 mg/ml) about 4 hours are required for complete solubilization under these conditions. In order to increase the rate of solubilization, it is advantageous to dissolve the crystals at $0°$ directly in $5 M$ NaCl, $0.005 M$ potassium phosphate (or $0.005 M$ Tris-Cl), pH 7.5, and dilute with buffer to the desired salt concentration. This treatment saves approximately $2\frac{1}{2}$ hours and does not affect the enzymatic activity.[67] The method can also be used with other salts, depending upon the experiment, as long as these do not modify the activity of the enzyme. Sodium β-phenyl-propionate and Veronal (or Tris) buffers solubilize CPA (Anson) in the absence of NaCl or LiCl,[57,69] an important property which was used for the development of chromatographic procedures for the purification

---

[67] Author's unpublished observations.
[68] If the enzyme is to be used as a reagent for sequence analysis of peptides or proteins, treatment at high pH is sufficient to dissolve the enzyme. However, for structure–function studies it is not recommended that such conditions be used.
[69] J. L. Bethune, *Biochemistry* **4**, 2698 (1965).

of the enzyme. Table IV (see p. 488) describes some of the conditions in which the enzyme has been dissolved and can serve as a starting point for the planning of new experiments.

Stability. All three preparations of the enzyme are stable within the pH range of 7.0 to 10.0 at 0°. More extensive studies indicate that CPA (Allan) is stable between pH 4.6 and 11.3 at 0° for 62 hours at concentrations of 0.62 to 1.27 mg/ml in citrate, Tris, glycine, and phosphate buffers containing 1 M NaCl.[70] Rapid denaturation occurs outside this pH range, manifested first by reversible precipitation of the protein at pH 3.5 which becomes irreversible at pH 2.5, and second by complete loss of activity at pH 13.8 with 40% precipitation.[71] CPA (Anson) is stable at pH 7.5, 4°, for 50 hours at concentrations of $8.4 \times 10^{-6} M$ in 0.05 M Tris-Cl, 1 M NaCl; however, after 24 hours at 25° the enzyme loses 1/6, 1/3, and 1/3 of its esterase activity at pH 7.5, 5.5, and 9.0, respectively.[72] CPA (Anson) (1.5 mg/ml) is stable in 4 M urea at 0°, pH 7.5 for 24 hours.[73] It is 50% inactivated in 5 M urea and totally inactivated in 7 M urea under these conditions. The loss of activity in urea is completely reversible up to at least 7 M urea,[37] beyond this concentration denaturation becomes irreversible, probably due to the digestion of the unfolded carboxypeptidase by the remaining active enzyme.[54] Chymotrypsin impurities sometimes present in the commercial preparations do not contribute to the loss of activity in the presence of urea since DFP treatment does not reverse the effect.[73] Carboxypeptidase A (Anson) is irreversibly denatured in 33% ethanol–water mixtures,[54] an insoluble slurry appears which slowly turns into a precipitate after 16 hours at room temperature. Organic solvents have been shown to affect the activity of carboxypeptidase.[74] Studies on the influence of methanol, ethanol, isopropanol, and dioxane on the peptidase activity indicate that methanol lowers k_{cat} with little influence on K_m while the other solvents have marked effects on both kinetic parameters. The activity is also dependent upon the ionic strength,[41, 74] Maximum peptidase activity is obtained above 0.25 M but decreases rapidly below this value. An extensive study on the heat denaturation at 50° indicates that all six forms of carboxypeptidase A can be arranged in the following order of increasing stability[47, 57]: $\text{CPA}_\alpha^{Val} > \text{CPA}_\alpha^{Leu} > \text{CPA}_\beta^{Val} > \text{CPA}_\beta^{Leu} > \text{CPA}_\gamma^{Val} > \text{CPA}_\gamma^{Leu}$.

[70] J. A. Rupley and H. Neurath, *J. Biol. Chem.* **235**, 609 (1960).
[71] J. A. Rupley, Ph.D. dissertation, University of Washington, Seattle, Washington, 1959.
[72] F. W. Carson and E. T. Kaiser, *J. Am. Chem. Soc.* **88**, 1212 (1966).
[73] B. J. Allan, M.A. dissertation, University of Washington, Seattle, Washington, 1957.
[74] R. Lumry and E. L. Smith, *Discussions Faraday Soc.* **20**, 105 (1955).

TABLE IV
SOLUBILITY PROPERTIES OF CARBOXYPEPTIDASE A

Enzyme preparation	Solvent	Temp. (°C)	Approx. solubility limit	References
CPA (Anson)	1 M NaCl, 0.02 M Na Veronal, pH 7.5[a]	4, 22	19 mg/ml[b]	d
CPA (Anson)	2 M NaCl, 0.02 M Na Veronal, pH 7.5[a]	4, 22	28 mg/ml[b]	d
CPA (Anson)	2.5 M NaCl, 0.02 M Na Veronal, pH 7.5[a]	4, 22	45 mg/ml[b]	d
CPA (Anson)	0.1 M β-pp,[c] 0.02 M Na Veronal, pH 7.5	4, 22	17 mg/ml	e
CPA (Anson) and A_γ	1 M Tris-Cl, 0.05 M β-pp,[c] 0.04 M LiCl, pH 7.5	0	100 mg/ml	f
CPA (Anson) and A_γ	2 M Tris-Cl, 0.05 M β-pp,[c] 0.04 M LiCl, pH 7.5	0	200 mg/ml	f
CPA (Allan)	1 M NaCl, 0.02 M Na Veronal, pH 8.0	4	30 mg/ml	g
CPA (Cox)	1 M NaCl, 0.02 M Na Veronal, pH 8.0	4	30 mg/ml	h
CPA (Cox) and A_α	2 M Tris-Cl, 0.05 M β-pp,[c] 0.004 M LiCl, pH 7.5	0	150 mg/ml	i

[a] LiCl can be substituted for NaCl. The enzyme is slightly more soluble in LiCl.
[b] It may be necessary to dialyze overnight against the solvent to obtain these concentrations.
[c] β-Phenylpropionate.
[d] J. L. Bethune, *Biochemistry* **4**, 2691 (1965).
[e] J. L. Bethune, *Biochemistry* **4**, 2698 (1965).
[f] P. H. Pétra and H. Neurath, *Biochemistry* **8**, 2466 (1969).
[g] J. E. Coleman, B. J. Allan, and B. L. Vallee, *Science* **131**, 350 (1960).
[h] D. J. Cox, F. C. Bovard, J.-P. Bargetzi, K. A. Walsh, and H. Neurath, *Biochemistry* **3**, 44 (1964).
[i] P. H. Pétra and H. Neurath, *Biochemistry* **8**, 5029 (1969).

Purity. Carboxypeptidase A (Anson), A (Allan), and A (Cox) are heterogeneous protein preparations as determined by N-terminal analysis,[54] chromatography,[47,57] and disc gel electrophoresis.[47] However, this heterogeneity, at least in the case of CPA (Anson) and A (Cox), is not due to contamination by other proteins, but arises first from the activation of procarboxypeptidase A yielding enzyme species which vary in their N-terminal amino acid sequences, and second from allotypic amino acid replacements. Both of these phenomena are responsible for the existence of at least 8 species of the enzyme. The three enzyme preparations are actually 98 to 100% pure since the heterogeneous products are all carboxypeptidases A. Some preparations of CPA (Anson) contain an insoluble pigmented impurity, some of which can be removed by dissolving the crystals in $5 M$ NaCl and centrifuging at 7000 rpm for 15 minutes[55]; all of it, however, can be removed by chromatography. The isolated species of carboxypeptidases A_α, A_β, and A_γ are all pure, as shown by disc gel electrophoresis and amino terminal analysis.[47,57]

Physical Properties. The physical constants for various forms of carboxypeptidase A are listed in Table V. Electrophoretic studies were performed in the Tiselius apparatus at 1° in $0.2 M$ LiCl using cacodylate, Tris, and glycine buffer, pH 5.4 to 9.3 in the case of CPA (Anson),[75] and in $0.3 M$ LiCl using the same buffers in the range of 5.5 to 10.5 for CPA (Allan).[70] The former preparation gives a single component below pH 8.5, but at pH 9.3 a fast-moving component is detected corresponding to 20% of the amount of enzyme. CPA (Allan) is homogeneous in free-boundary electrophoresis. Sedimentation studies on CPA (Anson) in the pH range of 5.9 to 9.0 using protein concentrations from 0.06 to 0.91% indicate no significant variation in $s_{20,w}^0$ under these conditions.[76] However, certain preparations of CPA (Anson) are contaminated with a faster moving component having $s_{20,w}^0 = 9.0 \times 10^{-3}$ cm sec^{-1} which is insoluble at low ionic strength.[76] Using a corrected $D_{20,w}^0$ of 8.67×10^{-7} cm^2 sec^{-1} (see footnotes 35 and 75) and $\bar{V}$ of 0.732 obtained from amino acid compositions,[77] a molecular weight of 34,300 is calculated.[76] This value is in reasonable agreement with 31,600 calculated from viscosity measurements,[75] and is in excellent agreement with 34,489 obtained from the amino acid sequence of CPA$_\gamma$.[78] Sedimentation studies done on CPA (Allan) in a protein concentration range of 2 to 19 mg/ml at pH 7.0 and 7.5 in $1.0 M$ NaCl–$0.1 M$ Tris-Cl (or $0.02 M$ Veronal) and at pH

[75] F. W. Putnam and H. Neurath, *J. Biol. Chem.* **166**, 603 (1946).
[76] E. L. Smith, D. M. Brown, and H. T. Hanson, *J. Biol. Chem.* **180**, 33 (1949).
[77] E. L. Smith and A. Stockell, *J. Biol. Chem.* **207**, 501 (1954).
[78] R. A. Bradshaw, L. H. Ericsson, K. A. Walsh, and H. Neurath, *Proc. Natl. Acad. Sci. U.S.* **63**, 1389 (1969).

TABLE V

PHYSICAL PROPERTIES OF BOVINE CARBOXYPEPTIDASE A

Constant	CPA (Anson)	Ref.	CPA (Allan)	Ref.	CPA (Cox)	Ref.
Molecular weight						
Diffusion	31,600	a	—	—	—	—
Sedimentation	34,300	b	—	—	—	—
Amino acid composition	34,440	c	—	—	34,600	o
Amino acid sequence	34,489	d,q	—	—	35,268	d,r
Isoelectric point						
0.2 ionic strength	5.95	a	—	—	—	—
0.3 ionic strength	5.7	e	5.60	k	—	—
$E^{1\%}_{278\,nm}$	—	—	19.4	l	18.8	o
$\epsilon_{278}(M^{-1}\,cm^{-1})$	6.42×10^4	f	6.42×10^4	f	6.49×10^4	o
$\epsilon_{222.5}(M^{-1}\,cm^{-1})$	5.27×10^5	g	—	—	—	—
Zinc content	0.95–1.04	h	0.98–1.03	f	—	—
(g-atoms/mole)	0.98–1.03	f	—	—	—	—
Zinc dissociation constant	—	—	4.7×10^{-9}	m	—	—
$S^0_{20,w}(\times 10^{-13}\,cm\,sec^{-1})$, pH 7–8	3.07 ± 0.14	b	3.06 ± 0.04 3.2 ± 0.14	k n	3.2	p
$S^0_{20,w}(\times 10^{-13}\,cm\,sec^{-1})$, pH 4.5	—	—	3.21 ± 0.04	k	—	—
$D^0_{20,w}(cm^2\,sec^{-1})$	8.67×10^{-7} 9.2×10^{-7}	i j	— —	— —	— —	— —
$D^0_{25,w}(cm^2\,sec^{-1})$	9.94×10^{-7}	a	—	—	—	—
$\bar{V}$	0.732	c	—	—	—	—
$[\eta]$ (ml/g)	3.4 ± 0.2	j	—	—	—	—
$[\alpha]_D$	—	—	-18	k	—	—
$[\alpha]^{10}_{546}$	-26 ± 2	j	—	—	—	—
$[\alpha]^{25}_{235}$	-4200 ± 200	j	—	—	—	—
$\lambda_c(nm)$	263 ± 3	j	—	—	—	—
b_0	-146 ± 10	j	—	—	—	—

a F. W. Putnam and H. Neurath, *J. Biol. Chem.* **166**, 603 (1946).

b E. L. Smith, D. M. Brown, and H. T. Hanson, *J. Biol. Chem.* **180**, 33 (1949).

c E. L. Smith and A. Stockell, *J. Biol. Chem.* **207**, 501 (1954).

d R. A. Bradshaw, L. H. Ericsson, K. A. Walsh, and H. Neurath, *Proc. Natl. Acad. Sci. U.S.* **63**, 1389 (1969).

e H. Neurath, "The Enzymes," (P. D. Boyer, H. Lardy, and K. Myrbäck, eds.), Vol. 4, p. 11. Academic Press, New York, 1960.

f R. T. Simpson, J. F. Riordan, and B. L. Vallee, *Biochemistry* **2**, 616 (1963).

g J. R. Whitaker, F. Menger, and M. L. Bender, *Biochemistry* **5**, 386 (1966).

h B. L. Vallee and H. Neurath, *J. Biol. Chem.* **217**, 253 (1955).

i E. L. Smith, *Advan. Enzymol.* **12**, 191 (1951).

j J. L. Bethune, D. D. Ulmer, and B. L. Vallee, *Biochemistry* **6**, 1955 (1967).

k J. A. Rupley and H. Neurath, *J. Biol. Chem.* **235**, 609 (1960).

l B. L. Vallee, J. A. Rupley, T. L. Coombs, and H. Neurath, *J. Biol. Chem.* **235**, 64 (1960).

m J. E. Coleman and B. L. Vallee, *J. Biol. Chem.* **236**, 2244 (1961).

4.5 in $0.96\,M$ NaCl–$0.96\,M$ sodium acetate, indicate homogeneity under these conditions.[70,79] The value of $s^0_{20,w} = 3.06$ (Table V) is uncorrected for adiabatic expansion of the rotor at high speeds.[79] Both CPA (Anson) and A (Allan) polymerize in solutions of $2.5\,M$ NaCl and $1\,M$ NaCl– $0.1\,M$ β-phenylpropionate, pH 7.5, suggesting stabilization through hydrophobic interaction between β-phenylpropionate and individual protein molecules.[69] The sedimentation of CPA (Cox) performed at protein concentrations ranging from 1.3 to 13 mg/ml in $1\,M$ NaCl–$0.01\,M$ Tris-Cl, pH 8.0, indicate homogeneity at all concentrations.[55] The specific optical rotation $[\alpha]^0_D$, is insensitive to pH changes from 6 to 12[70] and has an unusually low value when compared to other proteins, yielding an estimated helical content of about 70%. However, recent measurements of optical rotary dispersion parameters indicate a helical content of about 35 to 40%[80] which agrees quite well with the estimated 30% value obtained from X-ray diffraction measurements.[38]

The Chemical and Crystal Structure. The amino acid sequence of bovine carboxypeptidase A_α has recently been deduced[78] and is shown in Fig. 12. The amino acid composition based on the sequence is shown in Table II. The alpha enzyme is composed of 307 residue joined together in a polypeptide chain having a disulfide bond at positions 138 and 161[38,81] with alanine and asparagine at amino and carboxyl terminal positions, respectively.[54] Positions 179, 228, and 305 have been identified as the location of the three linked amino acid replacements characterizing the two allotypic forms of carboxypeptidase A.[59] One form of the enzyme contains isoleucine, alanine, and valine, respectively at these positions, whereas the other one has valine, glutamic acid, and leucine at the same positions. The model obtained from X-ray diffraction studies,[38] shown in Fig. 13, reveals a compact elipsoid structure with dimensions of $52 \times 44 \times 40$ Å stabilized by long helical structures and

[79] J. L. Bethune, *Biochemistry* **4**, 2691 (1965).
[80] W. N. Lipscomb, J. C. Coppola, J. A. Hartsuck, M. L. Ludwig, M. Muirhead, J. Searl, and T. A. Steitz, *J. Mol. Biol.* **19**, 423 (1966)
[81] K. A. Walsh, L. H. Ericsson, R. A. Bradshaw, and H. Neurath, *Biochemistry* **9**, 219 (1970).

Table footnotes (*Continued*)

[n] J. L. Bethune, *Biochemistry* **4**, 2691 (1965).
[o] J.-P. Bargetzi, K. S. V. Sampath Kumar, D. J. Cox, K. A. Walsh, and H. Neurath, *Biochemistry* **2**, 1468 (1963).
[p] D. J. Cox, F. C. Bovard, J.-P. Bargetzi, K. A. Walsh, and H. Neurath, *Biochemistry* **3**, 44 (1964).
[q] This represents the value for the molecular weight of CPA_γ.
[r] This represents the value for the molecular weight of CPA_α.

```
                    5                    10                   15
Ala-Arg-Ser-Thr-Asn-Thr-Phe-Asn-Tyr-Ala-Thr-Tyr-His-Thr-Leu-
  ↑       ↑                       ↑
  α       β                       γ


                   20                   25                   30
Asp-Glu- Ile -Tyr-Asp-Phe-Met-Asp-Leu-Leu-Val-Ala-Gln-His-Pro-


                   35                   40                   45
Glu-Leu-Val-Ser-Lys-Leu-Gln- Ile -Gly-Arg- Ser- Tyr- Glu-Gly- Arg-


                   50                   55                   60
Pro- Ile -Tyr-Val-Leu-Lys-Phe-Ser-Thr-Gly- Gly-Ser-Asn-Arg-Pro-


                   65                   70                   75
Ala- Ile -Trp- Ile -Asp-Leu-Gly- Ile -His-Ser-Arg-Glu -Trp- Ile -Thr-


                   80                   85                   90
Gln-Ala-Thr-Gly-Val-Trp-Phe-Ala-Lys-Lys-Phe-Thr-Glu-Asn-Tyr-


                   95                  100                  105
Gly-Gln-Asn-Pro-Ser-Phe-Thr-Ala- Ile -Leu-Asp-Ser-Met-Asp- Ile -


                  110                  115                  120
Phe-Leu-Glu- Ile -Val-Thr-Asn-Pro-Asn-Gly-Phe-Ala-Phe-Thr-His-


                  125                  130                  135
Ser-Glu-Asn-Arg-Leu-Trp-Arg-Lys-Thr-Arg-Ser-Val-Thr-Ser-Ser-


                  140                  145                  150
Ser-Leu-Cys-Val-Gly-Val-Asp-Ala-Asn-Arg-Asn-Trp-Asp-Ala-Gly-


                  155                  160                  165
Phe-Gly-Lys-Ala-Gly-Ala-Ser-Ser-Ser-Pro-Cys-Ser-Glu-Thr-Tyr-


                  170                  175            Ile 180
His-Gly-Lys-Tyr-Ala-Asn-Ser-Glu-Val-Glu-Val-Lys-Ser-——-Val-
                                                     Val

                  185                  190                  195
Asp-Phe-Val-Lys-Asn-His-Gly-Asn-Phe-Lys-Ala-Phe-Leu-Ser- Ile -


                  200                  205                  210
His-Ser-Tyr-Ser-Gln-Leu-Leu-Leu-Tyr-Pro-Tyr-Gly-Tyr-Thr-Thr-


                  215                  220                  225
Gln-Ser- Ile -Pro-Asp-Lys-Thr-Glu-Leu-Asn-Gln-Val-Ala-Lys-Ser-


            Ala  230                  235                  240
Ala-Val-——-Ala-Leu-Lys-Ser-Leu-Tyr-Gly-Thr-Ser-Tyr-Lys-Tyr-
            Glu

                  245                  250                  255
Gly-Ser- Ile - Ile -Thr-Thr- Ile -Tyr-Gln-Ala-Ser-Gly-Gly-Ser- Ile -


                  260                  265                  270
Asp-Trp-Ser-Tyr-Asn-Gln-Gly- Ile -Lys-Tyr-Ser-Phe-Thr-Phe-Glu-


                  275                  280                  285
Leu-Arg-Asp-Thr-Gly-Arg-Tyr-Gly-Phe-Leu-Leu-Pro-Ala-Ser-Gln-


                  290                  295                  300
Ile - Ile -Pro -Thr-Ala-Gln-Glu-Thr-Trp-Leu-Gly-Val-Leu-Thr- Ile -


                  305
                  Val
Met-Glu-His-Thr-—— -Asn-Asn
                  Leu
```

Fig. 13. Three-dimensional representation of the polypeptide chains of carboxy-peptidase A derived from X-ray crystallographic studies. From Lipscomb *et al.* *Brookhaven Symp. Biol.* **21**, 24 (1968).

by a twisted pleated sheet located in the center of the molecule made up of 4 pairs of parallel and 3 pairs of antiparallel chains which rotate 120° with respect to the bottom chain. The helical region which makes up about 20% of the structure is composed of four main stretches, all located at the surface and on one side of the molecule. There are: residues 14–29, 72–88, 215–233, and 288–305. The model is described in great detail elsewhere,[38] but nevertheless, it is clear that these general structural features explain the great stability of the enzyme.

Inhibitors and Activators

STRUCTURAL ORGANIC INHIBITORS. A large number of compounds re-versibly inhibit carboxypeptidase A activity.[34, 35, 82, 83] In view of the

Fig. 12. Amino acid sequence of bovine carboxypeptidase A$_a$. From R. A. Bradshaw *et al., Proc. Natl. Acad. Sci. U.S.* **63**, 1389 (1969).

TABLE VI

COMPETITIVE INHIBITORS OF CARBOXYPEPTIDASE A (ANSON)

Inhibitor	$K_I(10^4\ M)$ (Esterase)	Substrate (pH 7.5, 25°)	Ref.	$K_I(10^4\ M)$ (Peptidase)	Substrate (pH 7.5, 25°)	Ref.
3-Phenylpropionic acid	1.52	Hippuryl-L-mandelate	b	0.62	CBZ-glycyl-L-phenylalanine	g
Indole acetic acid	—	—	—	0.78f	CBZ-glycyl-L-tryptophan	h
4-Phenylbutyric acid	13.7	Hippuryl-L-mandelate	b	11.8	CBZ-glycyl-L-phenylalanine	g
Phenylacetic acid	4.1	Hippuryl-L-mandelate	b	3.9	CBZ-glycyl-L-phenylalanine	g
Phenylacetic acid	8.0^a	—	c	—	—	—
p-Nitrophenylacetic acid	—	—	—	25	CBZ-glycyl-L-phenylalanine	g
D-Phenylalanine	—	—	—	20	CBZ-glycyl-L-phenylalanine	g
D-Histidine	—	—	—	200	CBZ-glycyl-L-phenylalanine	g
Carbobenzoxyglycine	160	O-Cinnamoyl-DL-β-phenyllactate	d	—	—	—
L-Mandelic acid	17.6	Acetyl-L-mandelate	e	—	—	—
Glycyl-L-tyrosine	—	—	—	20	CBZ-glycyl-L-phenylalanine	i

[a] Measured by equilibrium dialysis at 0°.

[b] E. T. Kaiser and F. W. Carson, *Biochem. Biophys. Res. Commun.* **18**, 457 (1965).

[c] J. A. Rupley and H. Neurath, *J. Biol. Chem.* **235**, 609 (1960).

[d] S. Awazu, F. W. Carson, P. L. Hall, and E. T. Kaiser, *J. Am. Chem. Soc.* **89**, 3627 (1967).

[e] E. T. Kaiser and F. W. Carson, *J. Am. Chem. Soc.* **86**, 2922 (1964).

[f] This is only an approximate since the substrate concentration used (0.05 M) is in a region of the V-vs-[S] curve where Michaelis-Menten kinetics do not prevail.[91]

[g] E. Elkins-Kaufman and H. Neurath, *J. Biol. Chem.* **178**, 645 (1949).

[h] E. L. Smith, R. Lumry, and W. J. Polglase, *J. Phys. Colloid Chem.* **55**, 125 (1951).

[i] S. Yanari and M. A. Mitz, *J. Am. Chem. Soc.* **79**, 1154 (1957).

kinetic anomalies observed for most substrates (see below), different types of inhibition can arise from either excess substrate, reaction products, inorganic ions, or substrate analogs. With one exception, Table VI lists only the competitive inhibitors for which equilibrium constants have been calculated from kinetic analyses carried out at several substrate or inhibitor[84] concentrations. The common structural requirements for an inhibitor are a free carboxyl group and an aromatic or heterocyclic function.[35, 82] The most potent organic inhibitors which contain these features are β-phenylpropionic acid,[85-87] and indole acetic acid.[87] It is important to state whether an ester or peptide is used as substrate for a particular inhibition study since distinct binding sites for each have been implicated.[88] An interesting example may illustrate this point in the case of *trans* or *cis*-cinnamic acids which show no inhibition in the case of carbobenzyloxyglycyl-L-tryptophan hydrolysis,[87] a peptide substrate, but inhibit the hydrolysis of furylacryloylphenyllactate,[89] an ester substrate.

METAL CHELATORS AND INORGANIC INHIBITORS. APOCARBOXYPEPTIDASE A. Native carboxypeptidase A is a metal enzyme containing 1 g-atom of zinc per mole of protein.[90] The metal is located in the active site as demonstrated by metal exchange studies in the presence of substrates,[91] and by the X-ray model.[38] Zinc can be removed by dialysis against 1,10-phenanthroline at 0°, pH 7.0, resulting in a proportional loss of enzymatic activity.[92] The inactive protein, apocarboxypeptidase A, has the same physical properties as the native enzyme and can be crystallized in the same manner.[70, 91] The addition of zinc to the apoenzyme results in the regeneration of full enzymatic activity. There are no detectable changes in conformation as measured by optical rotatory dispersion during the removal or reassociation of the metal, and the regenerated active

[82] H. Neurath and G. W. Schwert, *Chem. Rev.* **46**, 69 (1950).

[83] J. E. Coleman and B. L. Vallee, *Biochemistry* **3**, 1874 (1964).

[84] The substrate concentration used in these studies was in a region where Michaelis-Menten kinetics are obeyed [E. T. Kaiser and F. W. Carson, *Biochem. Biophys. Res. Commun.* **18**, 457 (1965)].

[85] E. Elkins-Kaufman and H. Neurath, *J. Biol. Chem.* **175**, 893 (1948).

[86] E. Elkins-Kaufman and H. Neurath, *J. Biol. Chem.* **178**, 645 (1949).

[87] E. L. Smith, R. Lumry, and W. J. Polglase, *J. Phys. Colloid Chem.* **55**, 125 (1951).

[88] B. L. Vallee, J. F. Riordan, J. L. Bethune, T. L. Coombs, D. S. Auld, and M. Sokolovsky, *Biochemistry* **7**, 3547 (1968).

[89] W. O. McClure and H. Neurath, *Biochemistry* **5**, 1425 (1966).

[90] B. L. Vallee and H. Neurath, *J. Am. Chem. Soc.* **76**, 5006 (1954).

[91] B. L. Vallee, *Federation Proc.* **23**, 8 (1964).

[92] B. L. Vallee, J. A. Rupley, T. L. Coombs, and H. Neurath, *J. Biol. Chem.* **235**, 64 (1960).

enzyme cannot be distinguished from native carboxypeptidase A.[70,91] Zinc can also be removed by lowering the pH to 5.5, the rate of removal being faster at pH 3.4 where the process is completed after 10 hours.[92] N-Substituted dipeptides, polypeptides, and proteins have been shown to bind to apocarboxypeptidase A (Allan) and retard the association of the metal when it is subsequently added.[93-95] In contrast, the ester HPLA, N-acylamino acids, and small organic inhibitors such as β-phenyl-propionic acid, and phenylacetic acid bind only to the holoenzyme and prevent the dissociation of the metal.[70,83,96] These data suggest the existence of two distinct mechanisms of binding for either peptide or ester substrates. A large number of metallocarboxypeptidases have been prepared by combining apocarboxypeptidase A (Allan) with various metals.[97,98] The resulting metal enzymes differ in their catalytic properties depending upon the nature of the substituted metal.[98,99] For example, [Mn^{2+}], [Co^{2+}], and [Ni^{2+}] carboxypeptidases A hydrolyze both peptides and esters while [Hg^{2+}], [Cd^{2+}], and [Pb^{2+}] carboxypeptidases A show no peptidase activity but higher esterase activity than the native enzyme using the standard assay. On the other hand, [Cu^{2+}] carboxypeptidase A is totally inactive.

A large variety of small molecular weight substances known to be metal chelators inhibit carboxypeptidase A activity[34]; 0.01 M pyrophosphate, 0.01 M oxalate, and 0.01 M cysteine completely inhibit the enzyme after 12–15 hours at neutral pH and 25°. Cyanide, sulfide, and citrate also show inhibitory capacity, whereas fluoride and azide do not alter the activity.

Chemical Modifications. Various studies on the chemical modification of carboxypeptidase have been carried out with the view of identifying functionally important amino acid residues in the active center.[37,100] Acetylation of carboxypeptidase A (Anson) or A (Allan) with acetic anhydride and N-acetylimidazole produces a modified enzyme having a 6- to 7-fold increase in esterase (HPLA) activity with a complete loss of peptidase (CGP) activity using standard assays.[101,102] The changes in

[93] J. E. Coleman and B. L. Vallee, *J. Biol. Chem.* **237**, 3430 (1962).
[94] J.-P. Felber, T. L. Coombs, and B. L. Vallee, *Biochemistry* **1**, 231 (1962).
[95] T. L. Coombs and B. L. Vallee, *Biochemistry* **5**, 3272 (1966).
[96] J. E. Coleman and B. L. Vallee, *Biochemistry* **1**, 1083 (1962).
[97] B. L. Vallee, J. A. Rupley, T. L. Coombs, and H. Neurath, *J. Am. Chem. Soc.* **80**, 4750 (1958).
[98] J. E. Coleman and B. L. Vallee, *J. Biol. Chem.* **235**, 390 (1960).
[99] J. E. Coleman and B. L. Vallee, *J. Biol. Chem.* **236**, 2244 (1961).
[100] B. L. Vallee, J. F. Riordan, and J. E. Coleman, *Proc. Natl. Acad. Sci. U.S.* **49**, 109 (1963).
[101] J. F. Riordan and B. L. Vallee, *Biochemistry* **2**, 1460 (1963).
[102] R. T. Simpson, J. F. Riordan, and B. L. Vallee, *Biochemistry* **2**, 616 (1963).

activity are directly related to the acetylation of two tyrosyl residues as shown by the similarity in the spectral properties of O-acetyltyrosine with the chemically modified enzyme. The modification can be prevented by β-phenylpropionate and reversed by hydroxylamine, resulting in deacetylation with a return to normal esterase and peptidase activity.[102] Acetylation with dicarboxylic acid anhydrides, such as succinic anhydride, gives similar results except that spontaneous deacetylation of the tyrosine residues takes place time dependently without the addition of a nucleophilic reagent.[28] The increase in the rate of deacylation of succinyl-carboxypeptidase is thought to occur through an intramolecular nucleophilic attack by the free carboxyl group on the covalently bound reagent.[28] Acetylation of ϵ-amino groups does not account for the changes in activity, suggesting that amino groups are not important in the catalytic function of the enzyme.[100, 102] Nitration of carboxypeptidase A (Anson) with tetranitromethane (TNM) results in a chemically modified enzyme having 170% esterase and less than 10% peptidase activity.[103] Spectral titration and protection experiments with β-phenylpropionate indicate that modification results in the nitration of one tyrosyl residue. Reaction of carboxypeptidase A (Anson) with an 8-fold molar excess of 5-diazonium-1H-tetrazole (DHT) results in a 200% increase of esterase activity but no change in peptidase activity.[104] Spectroscopic evidence suggests that one tyrosyl residue has been converted into 5-azo-1H-tetrazole tyrosine. When a 30-fold molar excess of DHT to enzyme is used, there is no further increase in esterase activity but complete loss of peptidase activity occurs with additional modification of 1 histidyl and 4 tyrosyl residue.[104] Histidine modification seems to be closely related to the loss of peptidase activity under these conditions. The addition of β-phenylpropionate protects against activity changes and histidine and tyrosine modifications but does not prevent coupling with the ϵ-amino group of lysine residues as demonstrated by amino acid analysis. Lysine modification is not responsible for the observed activity changes since N-succinylazocarboxypeptidase A (Anson) exhibits the same spectral and enzymatic properties as those of azocarboxypeptidase A (Anson).[104] It has been suggested that Tyr_{248} is modified during nitration while mild diazotization (8-fold molar excess DHT) modifies Tyr_{198} since peptidase activity is not destroyed in the latter case and, apart from lysyl residues which are catalytically unimportant, no other residue is modified.[38, 104] A tetrapeptide, isoleucyl-tyrosyl-glutam(in)yl-alanyl, was isolated from iodocarboxypeptidase A (Anson) paired-labeled with ^{125}I-hypoiodite, and ^{131}I-hypoiodite in the presence of β-phenylpro-

[103] J. F. Riordan, M. Sokolovsky, and B. L. Vallee, *Biochemistry* **6**, 358 (1967).
[104] M. Sokolovsky and B. L. Vallee, *Biochemistry* **6**, 700 (1967).

pionate.[105] This peptide contains Tyr$_{248}$ as shown by the X-ray model[106] and the chemical sequence.[78] Iodocarboxypeptidase A (Anson) has similar enzymatic properties as acetyl- and succinylcarboxypeptidase A (Anson), suggesting that the same tyrosyl residue is modified in both cases.[107] Preliminary investigation on arginine modification with diacetyl shows loss of peptidase activity and a 3- to 4-fold increase of esterase activity which can be reversed by prolonged dialysis against borate.[37] Amino acid analysis indicates that one or two arginine residues have been modified. β-Phenylpropionate does not protect against modification and HPLA is hydrolyzed by diacetyl-treated enzyme. Except in the case of iodocarboxypeptidase A, peptides containing the chemically modified residues have not yet been isolated from any of the carboxypeptidase A derivatives just discussed.

Substrate Specificity and Mechanism of Action. Several reviews describing the specificity of carboxypeptidase A have been published.[34, 35] The enzyme catalyzes the hydrolysis of a large variety of small and N-acylated peptide and ester substrates. The only absolute requirement for specificity is the presence of an L-amino acid in the terminal position[108] with a free α-carboxyl group.[39] Optimum activity is obtained when the side chain is an aromatic amino acid although any aromatic group in this position will suffice.[109] As a general rule, when the side chain is aspartic acid, glutamic acid, arginine, lysine, proline, or hydroxyproline, a low degree of catalysis is realized at neutral pH. The influence of the penultimate peptide bond on the rate of hydrolysis has been emphasized[110] in that it must be bound to the enzyme surface strongly enough to permit optimum cleavage of the first peptide, or ester bond. Carboxypeptidase A has also been used extensively as a reagent in the study of C-terminal sequence determinations of proteins and its enzymatic action is therefore not limited to small synthetic substrates. The specificity and experimental use on polypeptides and proteins have been described.[111-113]

[105] O. A. Roholt and D. Pressman, *Proc. Natl. Acad. Sci. U.S.* **58**, 280 (1967).
[106] G. N. Recke, Jr., J. A. Hartsuck, M. L. Ludwig, F. A. Quiocho, T. A. Steitz, and W. N. Lipscomb, *Proc. Natl. Acad. Sci. U.S.* **58**, 2220 (1967).
[107] R. T. Simpson and B. L. Vallee, *Biochemistry* **5**, 1760 (1966).
[108] M. A. Stahman, J. S. Fruton, and M. Bergmann, *J. Biol. Chem.* **145**, 247 (1946).
[109] F. W. Dunn and E. L. Smith, *J. Biol. Chem.* **187**, 385 (1950).
[110] H. T. Hanson and E. L. Smith, *J. Biol. Chem.* **175**, 833 (1948).
[111] J. I. Harris, "Symposium on Peptide Chemistry," Special publication no. 2, p. 71. Chemical Society, London, 1955.
[112] H. Neurath, J. A. Gladner, and E. W. Davie, "The Mechanism of Enzyme Action," p. 50. Johns Hopkins Press, Baltimore, Maryland, 1954.
[113] R. P. Ambler, Vol. XI, pp. 115, 436.

A large number of kinetic studies on the catalyzed hydrolysis of peptide and ester substrates by carboxypeptidase A have been reported.[34-37] Tables VII, VIII, and IX summarize some of the kinetic parameters for substrates which vary in structure. The results indicate that, in general, the relationship between the initial velocity and substrate concentration does not obey classical Michaelis-Menten kinetics. The reported values of k_{cat} found in Tables VII, VIII, and IX were calculated by extrapolating "linear" portions of the velocity-vs-[S] curves. These pseudo first-order rate constants do not describe the turnover velocity of the enzyme as pictured in the Michaelis-Menten model,

TABLE VII

KINETIC PARAMETERS FOR THE PEPTIDASE ACTIVITY OF
CARBOXYPEPTIDASE A (ANSON)

Substrate (25°C, pH 7.5)	Substrate conc. range ($10^2 M$)	Ionic strength (M)	K_m ($10^3 M$)	k_{cat} (sec^{-1})	References
CBZ-glycyl-L-tryptophan	0.1-2	0.5	6.1	90.5	b
CBZ-glycyl-L-tryptophan	—	0.5	7.2	73.4	c
CBZ-glycyl-DL-phenylalanine	0.1-2	0.5	5.0	150	b
CBZ-glycyl-L-phenylalanine	0.2-2	0.5	13.1	189	b
CBZ-glycyl-L-phenylalanine	0.4-5.0	0.5	14	198	d
CBZ-glycyl-L-phenylalanine	0.4-4.0	1	5.9	125[a]	e
CBZ-glycyl-L-phenylalanine	1.3-7.5	0.1	37	186	f
CBZ-glycyl-L-phenylalanine	0.1-0.4	1	1.95	90[a]	e
CBZ-glycyl-L-phenylalanine	0.08-0.4	0.5	5.83	106	d
CBZ-glycyl-L-phenylalanine	—	0.5	16.6	195	c
CBZ-glycyl-L-tyrosine	0.1-2	0.5	12.2	85	b
CBZ-glycyl-L-leucine	0.1-7.5	0.5	31	103	b
CBZ-glycyl-L-leucine	—	0.5	33	27.2	c
N-trans-3-(-3-Indoleacryloyl)-L-phenylalanine	0.03-7.5	1.1	0.05	0.0014	g
Hippuryl-L-phenylalanine	1.3-5.2	0.1	11	182	h
Hippuryl-L-phenylalanine	—	0.5	1.75	130	c
Hippuryl-L-phenylalanine	0.1-5.1	0.5	1.91	118	d
Hippuryl-L-phenylalanine	0.1-1.0	1	2.7	83[a]	e
Hippuryl-L-phenylalanine	0.01-0.04	1	0.81	139[a]	e

[a] Estimated from Figs. 2 and 3 of R. C. Davies et al., Biochemistry 7, 1090 (1968).
[b] R. Lumry and E. L. Smith, Discussions Faraday Soc. 20, 105 (1955).
[c] J. E. Folk and E. W. Schirmer, J. Biol. Chem. 238, 3884 (1963).
[d] J. R. Whitaker, F. Menger, and M. L. Bender, Biochemistry 5, 386 (1966).
[e] R. C. Davies, J. F. Riordan, D. S. Auld, and B. L. Vallee, Biochemistry 7, 1090 (1968).
[f] E. Elkins-Kaufman and H. Neurath, J. Biol. Chem. 175, 893 (1948).
[g] W. O. McClure and H. Neurath, Biochemistry 5, 1425 (1966).
[h] J. E. Snoke and H. Neurath, J. Biol. Chem. 181, 789 (1949).

TABLE VIII

KINETIC PARAMETERS FOR THE ESTERASE ACTIVITY OF
CARBOXYPEPTIDASE A (ANSON)

Substrate (25°, pH 7.5)	Ionic strength (M)	K_m (10^4)	k_{cat} (sec^{-1})	References
Hippuryl-DL-β-phenyllactate	0.1	$\sim$0	138	a
Hippuryl-DL-β-phenyllactate	0.5	0.15	1060	b
Hippuryl-DL-β-phenyllactate	0.07	0.76	476	c
Hippuryl-DL-β-phenyllactate	0.1	0.51	467	d
Hippuryl-L-β-phenyllactate	0.5	0.88	578	e
Acetyl-β-phenyllactate	0.1	130	3.6	a
Chloroacetyl-β-phenyllactate	0.1	$\sim$0	106	a
Bromoacetyl-β-phenyllactate	0.1	16	81	a
O-trans-3-(2-Furylacryloyl)-D-L-β-phenyllactate	0.1	1.32	47	f
Acetyl-L-mandelate	0.5	700	0.72	g
O-Cinnamoyl-DL-β-phenyllactate	0.5	1.55	76.7	h

a J. E. Snoke and H. Neurath, *J. Biol. Chem.* **181**, 789 (1949).

b J. E. Folk and E. W. Schirmer, *J. Biol. Chem.* **238**, 3884 (1963).

c R. C. Davies, J. F. Riordan, D. S. Auld, and B. L. Vallee, *Biochemistry* **7**, 1090 (1968).

d W. O. McClure, H. Neurath, and K. A. Walsh, *Biochemistry* **3**, 1897 (1964).

e M. L. Render, J. R. Whitaker, and F. Menger, *Proc. Natl. Acad. Sci. U.S.* **53**, 711 (1965).

f W. O. McClure and H. Neurath, *Biochemistry* **5**, 1425 (1966).

g E. T. Kaiser and F. W. Carson, *J. Am. Chem. Soc.* **86**, 2922 (1964).

h S. Awazu, F. W. Carson, P. L. Hall, and E. T. Kaiser, *J. Am. Chem. Soc.* **89**, 3627 (1967).

since they vary with substrate concentration. Furthermore, they must always be qualified by describing the concentration range from which they were calculated. This is clearly shown in the catalyzed hydrolysis of CGP at $0.5 \, M$ ionic strength where three different values for k_{cat} can be calculated depending upon the substrate concentration range used. These are: 55 sec^{-1}, 106 sec^{-1}, 198 sec^{-1} for $0.008 \, M$ to $0.098 \, M$, $0.08 \, M$ to $0.4 \, M$, and $0.4 \, M$ to $5.0 \, M$ CGP, respectively (Tables VII and IX). In addition to these apparent substrate activation phenomena, substrate inhibition can also be demonstrated, for example, in the case of CGP above $1 \times 10^{-2} \, M$,[44,45,114] and HPLA above $1 \times 10^{-3} \, M$.[46,89,115] These kinetic anomalies have been interpreted to reflect either an unusual

114 R. Lumry, E. L. Smith, and R. R. Glantz, *J. Am. Chem. Soc.* **73**, 4330 (1951).

115 M. L. Bender, J. R. Whitaker, and F. Menger, *Proc. Natl. Acad. Sci. U.S.* **53**, 711 (1965).

TABLE IX

KINETIC PARAMETERS FOR VARIOUS FORMS OF BOVINE CARBOXYPEPTIDASE[a]

Enzyme	CGP[b]		HPLA[c]	
	K_m ($10^3\ M$)	k_{cat} (sec^{-1})	K_m ($10^4\ M$)	k_{cat} (sec^{-1})
CPA (Anson)	2.0 ± 0.1	55 ± 3	1.3 ± 0.01	466 ± 17
CPA$_\alpha^{Val}$	1.7 ± 0.1	38 ± 3	1.4 ± 0.02	439 ± 35
CPA$_\beta^{Val}$	1.9 ± 0.2	55 ± 5	1.0 ± 0.02	500 ± 33
CPA$_\beta^{Leu}$	1.5 ± 0.2	47 ± 5	2.3 ± 0.03	666 ± 53
CPA$_\gamma^{Val}$	1.8 ± 0.2	48 ± 5	1.4 ± 0.01	476 ± 13
CPA$_\gamma^{Leu}$	2.4 ± 0.2	62 ± 4	0.9 ± 0.004	396 ± 8

[a] Unpublished experiments of Jack Uren, University of Washington, reported in Neurath et al., Brookhaven Symp. Biol. **21**, 1 (1968).

[b] Carbobenzoxyglycyl-L-phenylalanine. Substrate concentration range: 0.74 to $9.8 \times 10^{-4}\ M$, $\mu = 0.5$.

[c] Hippuryl-DL-β-phenyllactic acid. Substrate concentration range: 0.73 to $7.3 \times 10^{-4}\ M$, $\mu = 0.5$.

feature in the structure of the binding site, inadequacies in the structure of the substrate which prevent it from binding in the same manner as natural substrates, or both. There is evidence which supports the second premise in that longer substrates such as benzoylglycylglycyl-L-phenyl-alanine and carbobenzoxyglycylglycyl-L-phenylalanine do not exhibit substrate activation.[116] Several models describing substrate binding based on kinetic evidence have been proposed.[46, 88, 114] The most recent one[88] describes the existence of discrete binding sites productive for either peptide or ester substrates but not for both. An overlap may exist between the esterase and peptidase site depending upon the relative position of the individual substrate molecules involved in the actual binding. Substrate inhibition results if there is an overlap between the productive and unproductive sites[117] and substrate activation occurs in the absence of overlapping. The inclusion of several esterase and peptidase sites is in keeping with earlier proposed kinetic models which postulate their existence.[46, 114] Interpretation of a 2.0 Å difference electron density map between the native and glycyl-L-tyrosine bound enzyme reveals a large amount of information on the mode of binding and the mechanism of action.[38] These studies along with those done on model building of

[116] D. S. Auld, Federation Proc. **27**, 781 (1968).

[117] This is not immediately obvious since the authors point out that "the kinetic consequences of such an inhibition would not be apparent in a V-vs-[S] profile." B. L. Vallee, J. F. Riordan, J. L. Bethune, T. L. Coombs, D. S. Auld, and M. Sokolovsky, Biochemistry **7**, 3547 (1968).

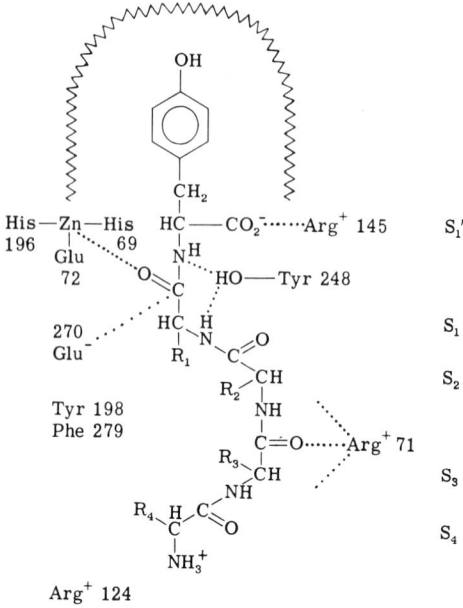

FIG. 14. Schematic drawing of the substrate peptide chain in the active site of carboxypeptidase A obtained from model building and X-ray diffraction data. Lipscomb *et al., Brookhaven Symp. Biol.* 21, 24 (1968). The positions of residues 196 and 279 were established by the chemically determined amino acid sequence. R. A. Bradshaw *et al., Proc. Natl. Acad. Sci. U.S.* 63, 1389 (1969).

longer substrates yield the following conclusions which are schematically described in Fig. 14.[38] The C-terminal side chain of the substrate binds in a dead-end pocket which does not contain any specific binding group but is large enough to accommodate a tryptophan side chain. The free α-carboxyl group in peptide substrates interacts with arginine-145. The —NH— of the hydrolyzable peptide bond interacts with Tyr_{248}. This residue moves 12 Å into proper position when the substrate binds. The carbonyl group of the first peptide bond must be orientated toward zinc in order to have the correct alignment of the groups. Recent experiments on the binding of small and large substrates suggest that the size of the active site extends to 18 Å corresponding to about five amino acid residues.[118] This is substantiated by the X-ray model which indicate that tyrosine-198, phenylalanine-279, arginine-71, and arginine-124 are located close enough to interact with the amino acids in positions S_2, S_3, and S_4 along the substrate peptide chain (Fig. 14). These observations

[118] N. Abramowitz, I. Schechter, and A. Berger, *Biochem. Biophys. Res. Commun.* 29, 862 (1967).

may possibly explain the unusual kinetics obtained by using small substrates which may have strong affinity to those additional binding sites originally postulated by the kinetic model. It also supports recent evidence that the use of longer peptide substrates eliminates these anomalies by returning to normal Michaelis-Menten kinetics.[116]

A mechanism of action for the peptidase activity based on kinetic evidence, chemical modification of the active site, and X-ray diffraction data has recently been proposed,[37, 38] and is briefly described as follows: Starting with the substrate bound in the active site as described above (Fig. 14), tyrosine-248 is specifically located to serve in the role of proton donor to the —NH— group of the peptide bond to be cleaved. At the same time the polarization of the carbonyl group is facilitated by the interaction of the zinc which is a strong electron withdrawing group. Both of these interactions help to reduce the double bond character of the peptide unit and change the carbonyl carbon to a tetrahetral configuration. Interaction at carbon can now take place by either a nucleophilic attack of glutamic acid-270 forming an anhydride intermediate, or attack by water through a general base mechanism generated by glutamic acid-270. Although the data so far obtained make it impossible to distinguish between one of these two mechanisms, the former one seems, nevertheless, to be favored on stereochemical grounds.[38] The pH dependency of CGP activity is a bell-shaped curve with inflection points of pH 6.7 and 8.5.[82, 101] An extensive kinetic analysis on the catalyzed hydrolysis of O-acetyl-L-mandelate by CPA (Anson) also reveals a bell-shaped curve for the pH dependency of k_{cat}/K_m from which a base of $pK_a = 6.9$ and an acid of $pK_a = 7.5$ can be calculated.[72] Assuming only two ionizing groups in the enzyme, the decrease of activity in the alkaline region could be due to the ionization of tyrosine-248, which could no longer serve as a proton donor, and decrease in the acid region would result in the protonation of glutamic acid-270, thereby destroying the nucleophilic character of this residue. Arguments on these proposals, as well as on other possible mechanisms, are discussed in detail elsewhere.[38] It has been proposed that peptides and esters are hydrolyzed by two different mechanisms.[112] Although the data indicate that these are bound differently in the active site of carboxypeptidase A, at present there is too little information available to define a mechanism of action of ester hydrolysis. Certain aspects of this topic have been discussed.[37, 38]

[32] Carboxypeptidase B (Porcine Pancreas)[1]

By J. E. FOLK

$$RCONHCR(R')COO^- \xrightarrow{H_2O} RCOO^- + H_2NCH(R')COO^-$$
$$RCOOCH(R')COO^- \xrightarrow{H_2O} RCOO^- + HOCH(R')COO^-$$

where R' is the side chain of a basic amino acid

Assay Method[1a,2]

Principle. The method is based upon measurement of change in ultra-violet absorbancy that occurs as a result of enzymatic hydrolysis of N-benzoylglycyl-L-arginine to N-benzoylglycine and arginine. The enzyme may also be assayed by colorimetric measurement of arginine formed in the same reaction.[3]

Reagents

Tris(hydroxymethyl) aminomethane(Tris)-HCl buffer, pH 8.0, 0.05 M in Tris

N-Benzoylglycyl-L-arginine,[3] 0.01 M in H_2O

Enzyme, dilute enzyme with water or 0.005 M Tris buffer, pH 8.0, to obtain a solution containing 1.5 to 15 units/ml.

Procedure. The incubation mixture contains 1.5 ml of buffer, 0.3 ml of substrate, 0.05 ml of the enzyme and water to a final volume of 3 ml. The change in absorbancy is measured in a 1-cm quartz cuvette at 254 nm and 25°. The hydrolysis of 1 micromole of substrate causes an increase in absorbancy of 0.12.

Definition of Unit and Specific Activity. One unit of enzyme activity is defined as that amount of enzyme that catalyzes the hydrolysis of 1 micromole of substrate per minute in the above assay. Specific activity is the number of units of activity per milligram of protein; protein concentration is determined by the trichloroacetic acid turbimetric procedure[4] in the early stages of purification and by the use of an $E_{278}^{1\%}$ of 21.4[1] in the chromatography steps.

[1] IUB classification: 3.4.2.2.

[1a] J. E. Folk, K. A. Piez, W. R. Carroll, and J. A. Gladner, *J. Biol. Chem.* **235**, 2272 (1960).

[2] E. C. Wolff, E. W. Schirmer, and J. E. Folk, *J. Biol. Chem.* **237**, 3094 (1962).

[3] J. E. Folk and J. A. Gladner, *J. Biol. Chem.* **231**, 379 (1958).

[4] T. Bücher, *Biochim. Biophys. Acta* **1**, 292 (1947).

Purification Procedure[1]

Preparation of Acetone Powder. Whole swine pancreas glands are cut into slices of approximately 5 mm thickness, spread in enameled trays to a depth of 2 to 4 cm, and allowed to autolyze for 16 to 36 hours at room temperature. The autolyzate is treated with 4 volumes of acetone at room temperature for 1 minute in a Waring blendor at high speed and is filtered rapidly with suction. The residue is reextracted successfully in the blender twice with 4 volumes of acetone, once with 4 volumes of acetone–ether (1:1 by volume), and once with 4 volumes of ether. The powder thus obtained is dried free of ether and stored at 4°. This powder has been stored for several years without loss in enzyme activity.

Step 1. Extraction of Acetone Powder. An extract of this powder is prepared by gently stirring 100 g of the powder with 1 liter of distilled water at room temperature for 30 to 45 minutes and centrifuging this suspension for 30 minutes at 15,000 g. This procedure yields about 980 ml of clear light yellow extract.

Step 2. Ammonium Sulfate Fractionation. The extract is cooled to 2°–5° and adjusted to pH 7–7.2 with 1 N NaOH. All further operations are performed in this temperature range. Solid $(NH_4)_2SO_4$ (209 g/liter) is added gradually with stirring while the pH is maintained between 7 and 7.2. The precipitate is removed by centrifugation for 15 minutes at 15,000 g.[5] Solid $(NH_4)_2SO_4$ (164 g/liter) is added to the supernatant fluid while the pH is maintained as before. The suspension is stirred for 30 minutes and the precipitate is collected by centrifugation for 30 minutes at 15,000 g. This precipitate is dissolved in about 100 ml of 0.05 M Tris, pH 7.25, and is dialyzed for 18 hours against several changes of distilled water.

Step 3. Batch Adsorption and Elution from DEAE-Cellulose. The dialyzed protein solution is made 5 mM in Tris and is adjusted to pH 7.25. Fifty grams of DEAE-cellulose which has been previously equilibrated with 5 mM Tris, pH 7.25, and sucked to damp-dryness on a Büchner funnel is added and the mixture is stirred for 20 minutes. The absorbent containing the enzyme is collected by suction filtration and is washed 4 times on the funnel by stirring for 2–3 minutes with 300-ml portions of 5 mM Tris, pH 7.25. Finally, the enzyme is eluted from the absorbent by stirring successively for 2 to 3 minutes with one 200-ml and two 100-ml portions of 0.1 M NaCl in 5 mM Tris, pH 7.25. To com-

---

[5] The precipitate obtained upon 0.35 saturation with $(NH_4)_2SO_4$ may be conveniently used for the purification of porcine carboxypeptidase A according to J. E. Folk and E. W. Schirmer, *J. Biol. Chem.* **238**, 3884 (1963). The precipitate is dissolved in 150 to 200 ml of distilled water and stored frozen until ready for use in this preparation.

bined eluates is added solid $(NH_4)_2SO_4$ (43 g/100 ml). After equilibration with stirring for 30 minutes the suspension is centrifuged for 30 minutes at 15,000 g. The precipitate is suspended in about 10 ml of 0.05 M Tris, pH 7.25, and is dialyzed for 18 hours against distilled water. Occasionally some protein precipitates during the dialysis. This may be redissolved by adjusting the pH of the suspension to 7.5.

Step 4. First DEAE-Cellulose Chromatography. The dialyzed enzyme solution from step 3 is applied to a column, 3.5×20 cm, of DEAE-cellulose which has previously been equilibrated with 5 mM Tris, pH 7.5. The protein solution is washed into the column with about 20 ml of the same buffer and the chromatogram is developed with a linear gradient of NaCl from 0 to 0.2 M in 1000 ml of 5 mM Tris, pH 7.5. Fractions of 5–7 ml are collected at a flow rate of 5 to 6 ml/minute. The carboxypeptidase B activity emerges from the column between 0.08 and 0.13 M NaCl. Fractions of specific activity 165 and above are combined. Solid $(NH_4)_2SO_4$ (43 g/100 ml) is added with stirring. The precipitate obtained upon centrifuging this suspension is taken up in 5–7 ml of 0.05 M Tris, pH 7.5, and is dialyzed for 18 hours against 5 mM Tris, pH 7.5. Dialysis against distilled water at this stage generally results in crystallization of the enzyme protein.

Step 5. Second DEAE-Cellulose Chromatography. The final chromatography of the enzyme at pH 6 does not increase its specific activity notably. It does, however, remove a significant contamination of carboxypeptidase A activity. The dialyzed enzyme solution from step 4 is adjusted to pH 6.0 with 0.2 M acetic acid and is applied to a column, 2.5×20 cm, of DEAE-cellulose which has previously been equilibrated with 5 mM Tris-acetate, pH 6.0. The enzyme solution is washed into the column with about 20 ml of the same buffer and the chromatogram is developed with a linear gradient of NaCl from 0 to 0.15 M in 500 ml of 5 mM Tris-acetate, pH 6.0, at a flow rate of about 5 ml/minute. The carboxypeptidase B emerges from the column between 0.03 and 0.05 M NaCl while the contaminating carboxypeptidase A is eluted above 0.1 M NaCl. The fractions containing carboxypeptidase B are combined and the enzyme is precipitated with solid $(NH_4)_2SO_4$ (43 g/100 ml). The precipitate is taken up in 3–5 ml of 0.05 M Tris, pH 7.5, and is dialyzed for 18 hours against several changes of distilled water or 5 mM Tris, pH 7.5. The crystalline suspension or solution of dialyzed enzyme is stored frozen at $-10°$. Frozen solutions of the enzyme occasionally form heavy amorphous precipitates upon thawing. Adjusting the pH of these suspensions to 7.5–8.0 with dilute NaOH gives clear fully active enzyme solutions.

The results of a typical purification are summarized in the table. The procedure has been repeated many times with similar results.

PURIFICATION OF CARBOXYPEPTIDASE B FROM ACETONE POWDER[a]
OF AUTOLYZED PORCINE PANCREAS

Step	Protein (mg)	Units	Specific activity (units/mg protein)	Yield (%)
1. Acetone powder extract	12,000	84,000	7	100
2. $(NH_4)_2SO_4$ fractionation	1,850	50,000	27	60
3. DEAE-cellulose, batch treatment	420	42,000	100	50
4. First DEAE-cellulose chromatography, pH 7.5	130	22,000	170	26
5. Second DEAE-cellulose chromatography, pH 6.0	115	20,000	175	24

[a] One hundred grams of acetone powder are used.

Properties

Stability. The purified enzyme has been kept frozen for periods up to 1 year without detectable loss in activity. No loss in activity is observed after several thawings and refreezings. Substantial losses (25–45%) in activity occur upon lyophilization of either solutions or crystalline suspensions of the enzyme.

Purity and Physicochemical Properties. The enzyme prepared by this procedure contains no protein impurities detectable by the usual physical procedures. There are, however, small amounts of contaminating enzymatic activities in the preparations. Treatment of solutions with diisopropylphosphorofluoridate in a manner similar to that outlined for carboxypeptidase A (see footnote 6) eliminates the trace amounts of tryptic and chymotryptic activities. Even the purest preparations contain traces of carboxypeptidase A activity. There is some question as to whether this carboxypeptidase A activity is an inherent property of carboxypeptidase B or contamination in the preparations (for further discussion see footnote 6).

A molecular weight of 34,300 ± 600 has been calculated, based on sedimentation and diffusion ($S_{20,w}^0$ 3.25 and $D_{20,w}^0$ 8.16 × 10^{-7} cm^2 sec^{-1}) and using a pycnometric determined partial specific volume of 0.720 ± 0.005 cm^3/g.

Carboxypeptidase B shows the following amino acid composition: Asp_{32} Thr_{30} Ser_{18} Glu_{25} Pro_{13} Gly_{23} Ala_{25} ½ Cys_8 Val_{11} Met_5 Ile_{17} Leu_{23} Tyr_{20} Phe_{12} Lys_{18} His_6 Arg_{10} Trp_9 Amide N_{28}.[1]

Specificity. The purified enzyme catalyzes the hydrolysis of peptide bonds of the carboxyl terminal basic amino acids, lysine,[3,7] arginine,[3,7]

[6] J. T. Potts, Jr., Vol. XI [76b].
[7] J. E. Folk, *J. Am. Chem. Soc.* **78**, 3541 (1956).

ornithine,[3] homoarginine,[3] and S-(β-aminoethyl)cysteine.[8] It displays an esterase activity toward carboxyl terminal argininic acid (α-hydroxy-γ-guanidino-n-valeric acid).[2,9]

Activators and Inhibitors. The optimal activity is observed in Tris, NH$_4$HCO$_3$, or N-ethyl morpholine acetate buffers between pH 7 and 9. Significant decreases in activity are found in citrate and phosphate buffers below pH 7.5. Basic amino acids as well as a number of basic amino acid derivatives and analogs inhibit the enzyme in a competitive manner.[2] Certain metal-chelating agents, such as 1,10-phenanthroline and 2,2′-dipyridyl, are effective inhibitors of the enzyme.[1] This is in accord with reports that carboxypeptidase B contains 1 g-atom per molecule of zinc and that this metal atom is essential for enzymatic activity.[1,10]

Distribution. The enzyme has been found in extracts of autolyzed pancreas tissue of several mammals, including pig, cow, and rat. It exists in fresh pancreas tissue as an inactive zymogen. A carboxypeptidase B-like enzyme has been found in blood plasma.[11]

[8] F. Tietze, J. A. Gladner, and J. E. Folk, *Biochim. Biophys. Acta* **26**, 659 (1957).
[9] J. E. Folk and J. A. Gladner, *Biochim. Biophys. Acta* **33**, 570 (1959).
[10] J. E. Folk and J. A. Gladner, *Biochim. Biophys. Acta* **48**, 139 (1961).
[11] E. G. Erdös, H. Y. T. Yang, L. L. Tague, and N. Manning, *Biochem. Pharmacol.* **16**, 1287 (1967).

[33] Leucine Aminopeptidase from Swine Kidney

By S. RALPH HIMMELHOCH

Assay Method

The presence of several peptidases in crude extracts of swine kidney makes the use of the specific substrate, leucine amide, mandatory even though available methods for measuring the products formed are more cumbersome than those for colorogenic substrates such as leucine-p-nitroanilide. The reaction catalyzed by the enzyme with this substrate is:

$$\text{Leucine amide} \xrightarrow{\text{Leucine Aminopeptidase}} \text{leucine} + NH_3$$

The most convenient assay procedure for large numbers of samples was first described by Himmelhoch and Peterson.[1] Although previously described methods for assay of the porcine enzyme, including that cited, employ preactivation with manganese to enhance activity, the demon-

[1] S. R. Himmelhoch and E. A. Peterson, *Biochemistry* **7**, 2085 (1968).

stration that native leucine aminopeptidase is a zinc metalloenzyme eliminates the need for this cumbersome and uncertain step.[2]

Reagents

Assay mixture. 0.05 M leucine amide·HCl in 0.04 M Tris–HCl buffer, pH 8.6 (0.04 M Tris–0.01 M HCl).

Michl's buffer. 10 ml of pyridine and 100 ml of glacial acetic acid diluted to 2 liters with water (pH 3.6).

Ninhydrin solution. 0.5 g of ninhydrin dissolved in 950 ml of acetone and diluted to 1000 ml with water.

Procedure. A suitable aliquot of enzyme is added at zero time to 0.1 ml of the assay mixture equilibrated at 40°. The reaction is stopped after the desired interval (30 seconds to 20 minutes) with an equal volume of 10% trichloroacetic acid. Aliquots (10 μl) are spotted on dry Whatman 3 MM paper which is then wet with Michl's buffer. Up to 100 samples can be accommodated at 2-cm intervals along horizontal lines drawn 3 inches apart on a sheet of 46 $\times$ 57 cm paper.

Electrophoresis of the samples is carried out in any standard high voltage electrophoresis apparatus for 10 minutes at 35 V/cm and 25°. The paper is dried in an oven at 60° for 30 minutes, dipped in ninhydrin solution, and the color developed for 20 minutes at 60°. Active samples are identified by the presence of a spot near the origin (leucine). The blue spots corresponding to leucine and leucine amide are eluted with 5 ml of 70% ethanol, and the absorbance of the resulting solutions at 570 nm is determined. The fraction of the substrate hydrolyzed $= (L - B)/[A + (L - B)]$, where L is the absorbance at 570 nm of the eluted leucine spot, A the absorbance at 570 nm of the corresponding leucine amide spot, and B the absorbance at 570 nm of the leucine spot of a sample from which the enzyme has been omitted. The blank correction for leucine is necessitated by the presence of small amounts of leucine in commercially available preparations of leucinamide.

Definition of Unit and Specific Activity. Under these conditions, the reaction follows zero-order kinetics. An enzyme unit is defined as the quantity of enzyme necessary to produce 1% hydrolysis per minute. The purest preparations obtained have had specific activities of about 2900 units/mg of Lowry protein.

Purification of Leucine Amino Peptidase

The following method of purification is recommended for those seeking preparations of the enzyme suitable for peptide or protein sequencing experiments in which endopeptidases contaminating the preparation can

[2] S. R. Himmelhoch, *Arch. Biochem. Biophys.* **134**, 597 (1969).

completely invalidate the results.[3] Its success depends upon the use of kidneys from an inbred strain of hog,[3a] since physical heterogeneity observed in leucine aminopeptidase extracted from lots of mongrel hog kidneys prevents successful chromatographic purification of the enzyme.[1] The accompanying table gives typical results obtained with each step described below.

Preparation of Homogenate. Four kilograms of inbred Yorkshire hog kidney[4] which has been stored at −70° from the time of slaughter is thawed at 4° overnight, dissected free of medulla and connective tissue, and homogenized in 12 liters of $0.04 M$ Tris–$0.005 M$ succinic acid–$0.001 M$ MgCl₂–$0.25 M$ sucrose with a Waring blendor for 30 seconds. To the continuously stirred homogenate is added 600 ml of 10% (w/v) hexadecyltrimethyl ammonium bromide (Technical, Eastman Organic Chemical Co., no. T 5650, "CETAB"). The detergent-treated homogenate is centrifuged at 2300 rpm and 4° in the 4-liter swinging bucket head of the International PR-2 centrifuge. The supernatant solution ("CETAB–supernatant solution") contains the renal leucine aminopeptidase. It is to be noted that the quantity of detergent added should be in proportion to the quantity of kidney employed in making the homogenate. When excessive detergent is added, the supernatant solution is brown rather than pink in color. If this occurs, the enzyme has precipitated and is inactivated and the preparation must be discarded.

Exchange Filtration of the CETAB–Supernatant Solution

An 8.5×25-cm bed of DEAE-cellulose (about 25 g of Whatman DE-23)[5] is equilibrated with $0.04 M$ Tris–$0.005 M$ succinic acid–$0.001 M$ MgCl₂, pH 8.6. (For a detailed description of the proper methods for preparing such columns see Vol. V.) The CETAB–supernatant solution is passed at 2.4 liters/hour through the DEAE-cellulose column which is then washed with 2 liters of starting buffer. If skewing of the colored bands begins to appear during application, owing to the accumulation of fatty material in a dense layer on the upper surface of the adsorbent bed, the upper portion of the adsorbent should be stirred with a rod and allowed to settle again before application is continued. The enzyme, which is adsorbed to the column, is eluted sharply with 2 liters of $0.04 M$

---

[3] R. Frater, A. Light, and E. L. Smith, *J. Biol. Chem.* **240**, 253 (1965).

[3a] Spotted Poland China and Yorkshire strains of hog kidneys have also been successfully employed as starting material. Although other inbred strains may be suitable, no practical experience with them has been obtained.

[4] Available from the Arborgast and Bastion Meat Packing Co., Allentown, Pa.

[5] The capacity of all adsorbents should be tested by the methods described in S. R. Himmelhoch and E. A. Peterson, *Anal. Biochem.* **17**, 383 (1966).

PURIFICATION OF LEUCINE AMINOPEPTIDASE FROM 4 kg OF YORKSHIRE SWINE KIDNEY

Fraction description	Volume (ml/fraction)	Protein[a] (mg/fraction)	Enzyme (units/fraction)[b]	Specific activity (units/mg)	Cumulative yield (%)
CETAB–supernatant solution	10,000	106,000	168,000	2.9	100
Ion-exchange filtrate	660	10,000	168,000	36	100
Ammonium sulfate precipitate	120	2,500	133,000	107	80
Sephadex fraction	240	200	140,000	1460	84
Fractions from gradient chromatography	300	30	87,000	2900	50

[a] Determined according to O. H. Lowry, N. J. Rosebrough, A. L. Farr, and R. J. Randall, *J. Biol. Chem.* **193**, 265 (1951).
[b] See section titled Definition of Unit and Specific Activity.

Tris–0.005 M succinic acid–0.001 M MgCl₂ buffer, pH 8.6 containing 0.2 M NaCl. The effluent is collected in 200-ml fractions. The active fractions are located by assay and pooled (ion-exchange filtrate).

Ammonium Sulfate Precipitation. To the 660 ml of active fractions at 4° is added 207 g of (NH₄)₂SO₄ with constant stirring. The solution is allowed to stand for 10 minutes and then centrifuged at 10,000 g for 10 minutes. To the supernatant solution, containing 90% of the original activity, is added an additional 149 g of (NH₂)SO₄, after which the solution is allowed to stand for 10 minutes and centrifuged at 10,000 g for 30 minutes. The second precipitate is dissolved in 100 ml of buffer and contains approximately 80% of the original leucine aminopeptidase activity.

Sephadex G-200 Chromatography. This step is most conveniently performed utilizing a 5–6 liter bed of Sephadex G-200 in a column such as the model K100/150 available from Pharmacia, Rahway, N.J. Without specially designed columns, it is better to divide the sample into aliquots and chromatograph it on smaller beds in view of the difficult mechanical aspects of chromatography on Sephadex G-200. If a 2.5 × 135 cm column is to be used, the redissolved ammonium sulfate precipitate is divided into 2–60-ml fractions which are chromatographed at about 15 ml/hour on a bed equilibrated with 0.01 M Tris–0.0025 M HCl–0.001 M MgCl₂, pH 8.6, collecting 120 6-ml fractions. The enzyme activity emerges well ahead of the main peak of absorbance at 280 nm. Its position is identified by assay and the active fractions are pooled. Evidence has indicated that the active fractions from Sephadex chromatography do not contain detectable endopeptidase activity even though their specific activity is only 50% of that attained in the purest samples obtained to date. Thus it would be worthwhile for those with a known application in mind to test the enzyme at this level of purity. If it is not satisfactory at this stage, the final step may be appended.

Gradient Chromatography. The active fractions from Sephadex chromatography are applied directly to a 0.9 × 26 cm bed of DEAE-cellulose (Whatman DE-23) equilibrated with 0.04 M Tris–0.02 M succinic acid–0.001 M MgCl₂ (pH 7.4) at 66 ml/hour and 4°. The column is then washed with 80 ml of starting buffer and eluted with an 800-ml linear gradient to 0.05 M succinic acid–0.11 M Tris–0.001 M MgCl₂ (pH 7.2). Although several peaks of enzyme activity are partially resolved in this gradient, it has been the general practice of those using the enzyme for sequencing purposes to pool all of the active fractions prior to concentration. The contaminants removed by this step are still bound to the column at the completion of the gradient.

Concentration of the Purified Enzyme. The enzyme may be concentrated in good yield either by lyophilization from 0.01 M ammonium acetate or by adsorption to and step elution from DEAE-cellulose. For adsorption to the concentrating column, the pooled fractions are adjusted to pH 8.6 with 1 M Tris and then diluted to 3/2 of their initial volume. The diluted and adjusted fractions are passed through a 2-g bed of DEAE-cellulose (Whatman DE-32) equilibrated with 0.04 M Tris–0.005 M succinic acid–0.001 M MgCl$_2$, pH 8.6. The enzyme is eluted with a step of 0.4 M NaCl in the same buffer. It can be recovered in 90% yield in less than 2 ml.

Known Properties of the Enzyme

Leucine aminopeptidase from porcine kidney is an exopeptidase of broad specificity with a pH optimum in the region from pH 7.8–9.0. Previous detailed studies of its specificity[6] cannot be accepted without reservation, since the preparations used in these earlier studies were later shown to be contaminated with other proteolytic enzymes.[3] The enzyme prepared by the present method shows the same general preferences nonetheless. It will release asparagin and glutamine from peptides, but lysyl and arginyl residues have proved refractory to hydrolysis. In one instance,[7] it has been shown to release proline from the NH$_2$-terminal position of a peptide chain.

The enzyme has a molecular weight of approximately 300,000[1] and, when isolated by the present procedure under metal-free conditions, contains 4–6 gram atoms of zinc per mole of protein. The zinc is at or near the active site of the enzyme and is necessary for activity. It can be partially replaced with manganese, with a resultant 2-fold increase in activity with respect to leucine amide.[2]

Although extensive data have been accumulated indicating that renal endopeptidases are effectively removed by this purification procedure,[1] past experience with other procedures[3] dictates caution in the interpretation of sequence data obtained with the enzyme. Such data should always be corroborated by control experiments demonstrating that no internal degradation of the peptide under investigation by endopeptidase contaminants has occurred.

Another leucine aminopeptidase of very similar properties has been isolated from bovine lens by Hanson and his co-workers.[8]

[6] E. L. Smith, D. H. Spackman, and W. J. Polglase, *J. Biol. Chem.* **199**, 801 (1952).
[7] J. Potts and L. Deftos, personal communication.
[8] H. Hanson, D. Glasser, and H. Kirschke, *Z. Physiol. Chem.* **340**, 107 (1965).

[34] Particle-Bound Aminopeptidase from Pig Kidney

By G. Pfleiderer

Assay Method

The simplest method of assay is measuring the hydrolysis rates of *p*-nitroanilides of amino acids, especially alanine or leucine, by the procedure described by Tuppy and co-workers.[1] In neutral or light alkaline medium the splitting of *p*-nitroanilide effects an absorption increase, which can be estimated in a filter photometer at 405 nm ($\epsilon_{405\,nm} = 9620\,M$).

$$R-CH(NH_2)-C-ONH(C_6H_4-NO_2) \xrightarrow{H_2O} R-CH(NH_2)-COOH + H_2N-C_6H_4-NO_2$$

Reagents

> Phosphate buffer, 0.06 *M*, pH 7 [do not use Tris buffer because Tris(hydroxymethyl)aminomethane is a competitive inhibitor]. Stock solution of 1.66 × 10^{-2} *M* L-leucine-*p*-nitroanilide or L-alanine-*p*-nitroanilide. The diluted enzyme is added to 0.06 *M* phosphate buffer pH 7.0 containing 1.66 × 10^{-3} *M* L-leucine-*p*-nitroanilide or L-alanine-*p*-nitroanilide. Total test volume is 2.0 ml.

The absorption increase at 405 nm was measured at 37°. One enzyme unit is defined as the amount of enzyme which produces 1 micromole *p*-nitroanilide per minute at 37°. The specific activity is defined as the enzyme units per milligram of protein. The latter was estimated by the biuret method.

Purification Procedure[2]

Pig kidney (2900 g) was minced and homogenized in 0.1 *M* Tris-HCl buffer, pH 7.3 (9 liters), and stirred at 2° for 30 minutes. After centrifugation at 3000 *g* for 15 minutes the supernatant was adjusted to pH 5.0 with acetic acid, and centrifuged at 3000 *g* for 20 minutes. The sediment was suspended in 0.1 *M* Tris-HCl buffer, pH 7.3, and diluted to approximately 6 liters. Two liters of toluene were slowly added and the suspension was homogenized for 30 minutes at a temperature of 38° to 40° and then stirred overnight at 2°. The suspension was centrifuged at 12,000 *g* for 1 hour. Four layers were visible. The topmost layer

---

[1] H. Tuppy, W. Wiesbauer, and E. Wintersberger, *Z. Physiol. Chem.* **329**, 278 (1962).

[2] D. Wachsmuth, I. Fritze, and G. Pfleiderer, *Biochemistry* **5**, 169 (1966).

was toluene and the second layer was solid (white to light yellow in color). Beneath this was an aqueous layer and on the bottom of the tube a very small quantity of sediment. Using a filter pump, the toluene and the aqueous layers were easily removed, whereupon the second solid layer could be scraped from the tube. This second layer contained the enzyme and was suspended in $0.01\,M$ Tris-HCl buffer, pH 7.3 (end volume 900 ml). Solid trypsin (20 mg/100 g of kidney) was added, the suspension was incubated at 37° for 1 hour, and then cooled to 4°. Ammonium sulfate was added to give a final concentration of 20% saturation and the suspension was then centrifuged at 12,000 g. The aqueous supernatant, containing 50–65% of the total enzymatic activity, was saturated to approximately 80% with solid ammonium sulfate and centrifuged. The sediment was dissolved in 30 ml of $0.06\,M$ phosphate buffer, pH 7.3, and the fraction which precipitated between 60 and 80% saturation with ammonium sulfate was further purified. The enzyme solution (total volume = 30 ml) was desalted on a Sephadex G-50 column (3×70 cm) in $0.02\,M$ phosphate buffer, pH 7.3, and then directly adsorbed at room temperature on a column DEAE-Sephadex A-50 coarse (1.5×30–40 cm, equilibrated with $0.02\,M$ phosphate buffer, pH 7.3). The activity was eluted with an NaCl gradient which increased from 0 to $0.3\,M$. The aminopeptidase activity was eluted at approximately 0.15–$0.2\,M$ NaCl. The enzyme was repeatedly precipitated between 65 and 80% saturation of ammonium sulfate in the cold until no further increase of the specific activity could be achieved (Table I).

By gel filtration[3] over a column of Sephadex G-75 coarse equilibrated with $0.1\,M$ ammonium bicarbonate buffer, pH 7, the specific activity with L-leucine-p-nitroanilide as substrate could be increased to 66.

A modified isolation procedure was described by H. Hanson and co-workers.[4]

Since the properties of the enzymes obtained by both methods seem to be nearly identical, the isolation procedure should be briefly described.

Isolation of Aminopeptidase from Pig Kidney[4]

The kidney was homogenized with the 14-fold excess of ice-cold $0.25\,M$ saccharose solution and pressed through a Perlon net. The suspension was centrifuged at 600 g for 15 minutes. After centrifugation for 120 minutes at 105,000 g the sediment of mitochondria and microsomes was obtained. This sediment was washed several times with saccharose solution. Finally, the remaining sediment was suspended in

---

[3] U. Femfert, Doctoral thesis, Ruhr-University, Bochum, Germany, 1969.
[4] H. Hanson, H. J. Hütter, H. G. Mannsfeldt, K. Kretschmer, and C. Sohr, *Hoppe-Seylers Z. Physiol. Chem.* **348**, 680 (1967).

TABLE I
Isolation of Aminopeptidase from Pig Kidney

Isolation step	Total activity (units micromoles/min)	Specific activity against leucine p-nitroanilide	Purification factor
2900 g kidney homogenized, extracted, and centrifuged. Supernatant (suspension)	26,200	0.177	1
Sediment from the pH precipitation resuspended in 0.1 M Tris, pH 7.3	22,000	0.33	1.9
After toluene and trypsin treatment. Fraction precipitating between 20 and 80% saturation of ammonium sulfate	14,200	3.3	19
Reprecipitation with ammonium sulfate between 60 and 80% saturation	14,000	20.7 3 bands in IEA[a]	120
Eluate from DEAE-Sephadex at 0.2 M NaCl	8,100	31 2 bands in IEA[a]	175
Repeated reprecipitation with ammonium sulfate	5,000	37 1 band in IEA[a]	210

[a] IEA = immunoelectrophoretical analysis.

twice its volume of 0.25 M saccharose solution containing 1% Triton X-100, and homogenized at 0°. After half an hour incubation at 4° and centrifugation at 15,000 g for 20 minutes the prepared sediment was washed one or two times with 0.25 M Triton-saccharose solution. After centrifugation the supernatants were combined and incubated at 4° for 10 to 12 days. Then the sediment was removed by centrifugation and discarded. The supernatant was adjusted to pH 6.5 with 0.1 M sodium hydroxide and saturated with crystalline ammonium sulfate to 40%. The precipitate not containing enzyme activity was collected and decanted. The supernatant was dialyzed against 0.005 M Tris-HCl buffer pH 8 to 4° for 48 hours. The dialyzed solution was concentrated 10-fold by ultrafiltration using the LKB Stockholm apparatus. By continuous-flow paper-electrophoresis in 0.02 M Tris-HCl buffer, pH 8, a further purification of the enzyme could be obtained.

Properties

The pure enzyme could be lyophilized without loss of activity and also concentrated in solution by vacuum evaporation if the salt concentration remained small. Incubation at 65° did not cause a loss of activity. At 70° 50% of enzyme activity was lost in 110 minutes. At room temperature and neutral pH the enzyme is stable for more than 3 hours, pH values smaller than 3 and higher than 11.5 cause fast denaturation. Urea concentrations of 6 M or smaller did not denature the aminopeptidase during 100 hours of incubation at room temperature. A concentration of 0.5 M guanidine causes irreversible denaturation. This was, however, observed after adding alcohols. With increasing chain length a smaller concentration was necessary to cause this denaturation of the amino peptidase. Thus three times as much methanol as 1-butanol was required for 50% denaturation in 1 hour.

Purity and Physical Properties

The enzyme was chromatographed on DEAE-Sephadex at pH 7.2 in phosphate buffer. Only one peak was eluted which contained both the ultraviolet-absorbing material (253 nm) and the enzymatic activity. The same result was obtained using a Sephadex G-100 column (1.5 m long). The enzyme with the highest purity represents immunochemically one fraction with only a single band in the immunoelectrophoretic analysis against rabbit antiserum. The enzymatic activity was demonstrated in this band by using enzyme-specific staining procedures. Some preparations contain small impurities of carboxypeptidases, especially carboxypeptidase B. Since aminopeptidase is only slightly inhibited by excess of chelating agents, this fact could be used to inactivate the contaminating exopeptidase completely and thus eliminate its interfering effect. In addition carboxypeptidase impurities could be destroyed by heating.

The molecular weight of aminopeptidase was determined to be 280,000 by gel filtration through Sephadex G-200.[5] The UV absorption shows a characteristic protein spectrum. The quotient E_{280}/E_{260} is 1.8.

Activators and Inactivators. Heavy Metal Ions

Although the classical leucine aminopeptidase[6] is maximally activated by 0.001 M cobalt and magnesium ions, the described aminopeptidase is not significantly affected by heavy metal ions; however, contradicting data were published in this field. We[7] found only a small activation of the purified aminopeptidase with different bivalent ions. Hanson and co-workers describe no effect of manganese, magnesium, or cobalt ions up to a concentration of $1 \times 10^{-5} M$ in 0.05 M triethanolamine-HCl buffer pH 7 (substrate L-leucine-β-naphthylamide). Cobalt ions at concentrations of 10^{-3} to $10^{-2} M$ cause twofold activation of the en-

[5] F. Auricchio and T. B. Bruni, Biochem. Z. 340, 321 (1964).
[6] E. L. Smith, Vol. II, p. 88.
[7] G. Pfleiderer, P. G. Celliers, M. Stanulovic, E. D. Wachsmuth, H. Determann, and G. Braunitzer, Biochem. Z. 340, 552 (1964).

zyme. At higher concentrations a strong decrease of the activity was observed. Femfert[3] found with a purified aminopeptidase in HEPES buffer pH 7 [4.77 g N-2 hydroxyethylpiperazine-N-ethane-2-sulfonic acid (Calbiochem) dissolved with water to 100 ml] after a preincubation of 1 hour at 25° no effect of Co^{2+} up to 10^{-4} M and for higher concentration a strong inhibition, which could be reversed by 24-hour dialysis against the buffer alone. He could demonstrate that the activation effect depends on the purity of the aminopeptidase. Partially purified preparations could be activated by 10^{-3} to 10^{-2} M cobalt ions.

Chelating Agents

The necessity of heavy metal groups for enzymatic activity could be demonstrated using chelating agents. Amounts of o-phenanthroline of 10^{-5} M or higher cause an inhibition of the enzyme. This inhibition can be reversed by dialysis against H$_2$O. Aminopeptidase which was dialyzed for 24 hours against 10^{-2} M phenanthroline had a residual activity of 2%. It could be restored to nearly 100% after incubation with cobalt but

TABLE II

INHIBITION OF AMINOPEPTIDASE HYDROLYSIS OF
LEUCINE p-NITROANILIDE BY AMINO ACIDS

	Amino acid (K_I/K_m)		Amino acid (K_I/K_m)
L-Norleucine	5.0	L-Lysine	90
L-Phenylalanine	7.0	L-Glutamic acid	230
L-Leucine	12.3	L-Serine	230
L-Isoleucine	14	D-Leucine	360
L-α-Aminobutyric acid	15	L-Aspartic acid	390
L-Valine	30	D-Alanine	770
L-Arginine	36	Glycine	1000
L-Glutamine	40	L-Proline	1800
L-Histidine	46	L-Asparagine	3100
L-Alanine	63	Sarcosine	3400

not with cadmium-, zinc-, manganese-, or molybdenum ions. Similar results were obtained with α,α'-dipyridyl. In contrast, no inhibition was observed after adding diisopropylphosphofluoridate, phenylmethylsulfonylfluoride, or 10^{-2} M iodoacetamide or bromoacetamide.

Inhibition by Amino Acids

In addition an inhibition of the hydrolysis of leucine-p-nitroanilide by aliphatic acids and amines was found (Table II). Aliphatic amines cause a competitive inhibition which increases linearly with the length of the aliphatic chain.

The straight-chain fatty acids also inhibit the hydrolysis of the artificial substrate uncompetitively. The inhibition does not increase linearly with the length of the aliphatic chain and the acids were some eight times more weakly bound than the corresponding amines.

Although α, β, γ, and ϵ amino acid amides were not hydrolyzed by our aminopeptidase, their free acids were good inhibitors of the enzymatic hydrolysis of leucine-p-nitroanilide.

Specificity

The pure enzyme hydrolyzes only derivatives of L-α-amino acids. The amide group of L-asparagine, L-glutamine, or β-alanine amide was not split even after a long incubation. In contrast to the leucine amino-peptidase isolated by E. L. Smith,[6] the particle-bound aminopeptidase splits with a nearly identical velocity L-amino acid amides and L-amino acid-p-nitroanilides. The alanine derivatives show the highest turnover number. The relative rates of hydrolysis of different amino acid-p-nitro-anilides are alanine 100, phenylalanine 83, leucine 71, glycine 22. Decreasing rates were described by Hanson and co-workers[4] for the β-naphthyl-amides of alanine, phenylalamine, tyrosine, and leucine. The splitting of proline peptide bonds could not clearly be demonstrated. After a long incubation time proline-containing peptides appear to be slowly hydrolyzed. This effect is strongly decreased after incubation at 65°. Therefore, it seems that contaminating prolinase or prolidase impurities cause this effect. D-Amino acid peptide bonds were not attacked.[8] From the peptide H-Tyr-Leu-Gly-D-Glu-Phe-OH only tyrosine and leucine were liberated. The enzyme stops at γ peptide bonds.

The enzyme appears to be very useful in the study of peptide sequences, because it hydrolyzes alanine or glycine peptide bonds much faster than the leucine aminopeptidase. It should be possible to analyze peptide sequences by quantitatively measuring the amount of liberated amino acids after different incubation times.

Amino acid residues sensitive against chemical hydrolysis remain unchanged after enzymatic hydrolysis by aminopeptidase (glutamine, asparagine, tryptophan). The enzyme is also useful in identifying a chemical modified amino acid residue in a protein. For example, it is possible to isolate from the incubation mixture S-carboxymethyl methionine, monoazotyrosine, and so on using an enzyme mixture (pronase or subtilisin and aminopeptidase).

[8] G. Pfleiderer and P. G. Celliers, *Biochem. Z.* **339**, 186 (1963).

Kinetic Properties

The K_m values of the different *p*-nitroanilides are as follows: glycine 1.75 m*M*, phenylalanine 3.2, alanine 0.6, leucine 0.24 m*M* (measured at pH 7 in 0.06 *M* phosphate buffer). The K_m values for amino acid amides are generally higher. The K_m decreases with increasing hydrophobic character of the side chain.

Distribution

In mammalian cells, especially in kidney or liver, there exist at least two different aminopeptidases, one cytoplasmic and one particle bound. They differ strongly in their substrate specificity and their activation by heavy metals. In addition, they are immunologically different. The cytoplasmic enzyme appears to be identical with leucine aminopeptidase and also with the crystalline peptidase of bovine lens.[9]

TABLE III
AMINO ACID COMPOSITION OF PARTICLE-BOUND AMINOPEPTIDASE

Amino acid	Residues per mole	Amino acid	Residues per mole
Aspartic acid	340	Isoleucine	145
Threonine	197	Leucine	296
Serine	192	Tyrosine	104
Glutamic acid	364	Phenylalanine	144
Proline	197	NH₃	280
Glycine	127	Lysine	144
Alanine	210	Histidine	60
Valine	165	Arginine	100
Methionine	40	Cysteine	10

Functional Amino Acids

The amino acid composition of the enzyme is shown in Table III. Kinetic measurements suggest that some of the tyrosine and histidine side chains, with apparent p*K* values of 7.2, 8.7, and 9.7, are involved in catalysis.

At low pH H⁺ is a competitive inhibitor of the ammonium group of the substrate.

Diazonium-1-*H*-tetrazole, a substance reacting rapidly with the imidazole side chain of histidine residues in proteins, causes a strong decrease of the enzymatic activity. The inhibition correlates to increasing

[9] H. Hanson, D. Glatter, and H. Hirschke, *Hoppe-Seylers Z. Physiol. Chem.* **340**, 107 (1965).

absorbance at 478 nm, due to the formation of histidine-bisazo-1-H-tetrazole. Total inhibition corresponds to the reaction of five histidine residues.[10] Wachsmuth described the inactivation of aminopeptidase by iodine, which produces mono and diiodo tyrosine.[11] Femfert[12] found with tetranitromethane—a high specific reagent for tyrosine—a decrease of enzymatic activity due to formation of nitrotyrosine. The nitrated enzyme is partially active even in a high excess of the reagent.

[10] G. Pfleiderer and U. Femfert, *FEBS Letters* **4**, 265 (1969).
[11] E. D. Wachsmuth, *Biochem. Z.* **346**, 446 (1967).
[12] U. Femfert and G. Pfleiderer, *FEBS Letters* **4**, 262 (1969).

[35] Aminopeptidase-P

By A. YARON and A. BERGER

Assay Method

Principle. Aminopeptidase-P is an exopeptidase cleaving the bond between any N-terminal amino acid residue and a following proline residue.[1] The assay is based on the measurement by the acid ninhydrin colorimetric method[2,3] of proline released from the substrate poly-L-proline.

Since the enzyme is an exopeptidase, the rate of appearance of proline is proportional to the concentration of polymer molecules, c.

$$c = \frac{\text{mg of polymer}}{\text{ml} \times \text{average molecular weight}} \tag{1}$$

The amount of poly-L-proline necessary to give a solution of known concentration depends therefore on the number average molecular weight of the sample employed. This is determined for a given batch by end group titration.

Reagents

Poly-L-proline.[4,5] Commercial poly-L-proline of molecular weight of 5000 to 10,000 is purified and freed from low molecular weight

[1] A. Yaron and D. Mlynar, *Biochem. Biophys. Res. Commun.* **32**, 658 (1968).
[2] W. Troll and J. Lindsley, *J. Biol. Chem.* **215**, 655 (1955).
[3] S. Sarid, A. Berger, and E. Katchalski, *J. Biol. Chem.* **234**, 1740 (1959).
[4] A. Berger, J. Kurtz, and E. Katchalski, *J. Am. Chem. Soc.* **76**, 5552 (1954).
[5] E. Katchalski, M. Sela, H. I. Silman, and A. Berger, *in* "The Proteins" Vol. II. (H. Neurath, ed.), p. 405. Academic Press, New York, 1964.

fractions by gel filtration on Sephadex G-25. This procedure is necessary in order to reduce blank values and remove heavy metal impurities.

A 2.7 × 90 cm column of Sephadex G-25 was washed with 0.1 M EDTA (1 liter) followed by double distilled water (2 liters). The sample (500 mg) was applied in 10 ml and eluted under gravity with water (flow rate, 48 ml/hour). The effluent was monitored at 240 nm and 168 ml from the first appearance of absorption were collected. Later fractions showing positive acid ninhydrin reaction were rejected. The solution was lyophilized and dried *in vacuo* over concentrated sulfuric acid at room temperature. The yield was 360 mg.

The number average molecular weight is determined by potentiometric titration of the amino end groups ($pK_a = 9.0$) on a 150-mg sample in 3 ml of 0.1 M NaCl using a 0.1 M NaOH titrant delivered from an Agla microburet. Since the end point is not easily detected, it is convenient to evaluate the number of amino groups from the buffering capacity (i.e., the slope of the titration curve) at pH 9.0:

$$\text{moles of NH in sample} = \frac{d(\text{moles of base})}{d\text{pH}} \quad (2)$$

Poly-L-proline stock solution ($1.66 \times 10^{-3} M$) is prepared by dissolving the polymer in ice-cold water (poly-L-proline dissolves well in cold water and precipitates on heating). If turbid, the solution is cleared by centrifugation.

Veronal buffer, 0.05 M, pH 8.6. The buffer is passed through a Chelex-100 (analytical grade chelating resin, mesh 50–100, Biorad Lab., Richmond, Calif.) column in Na⁺ form, preequilibrated with the buffer, and stored at 4°.

Sodium citrate, 0.4 M. The trisodium salt dihydrate is passed through a Chelex-100 column in Na⁺ form, preequilibrated with the sodium citrate solution, and stored at 4°.

Manganous chloride, 0.1 M. A 1 M solution of MnCl₂ in water (500 ml) is passed through a 25-ml Chelex-100 column in Na⁺ form. The first 100 ml of the effluent are discarded. The manganese concentration is determined by complexometric titration with Versene, using Eriochrome black as the indicator. The solution is diluted to 0.1 M with double distilled (metal-free) water.

Sodium hydroxide, 0.1 M. 0.1 M sodium hydroxide is passed through a Chelex-100 column in Na⁺ form.

EDTA (sodium salt), 0.01 M

Ninhydrin reagent. Ninhydrin (3 g) is dissolved in a mixture of glacial acetic acid (60 ml) and phosphoric acid (6 M, 40 ml) by warming at 70°.

Glacial acetic acid

Enzyme. The enzyme preparation (in 0.05 M acetate buffer pH 6.5, or 0.1 M phosphate buffer, pH 7.0, about 15 units per milliliter) can be stored at 4° or frozen for several months.

Mn-citrate reagent. This reagent is prepared fresh before the assay by mixing 2.5 ml 0.05 M Veronal buffer, 0.5 ml of 0.4 M sodium citrate, 0.5 ml of 0.1 M manganous chloride, and 0.5 ml of 0.1 M sodium hydroxide.

Procedure

To a test tube is added 0.025 ml of poly-L-proline stock solution, 10 μl of 0.01 M Versene, 0.75 ml of Veronal buffer, and 0.2 ml of Mn-citrate reagent. The solution is placed into a 40° water bath, and 15 μl of enzyme solution (suitably diluted to produce 5–20 μg proline in 30 minutes) is added. After 30 minutes, the reaction is stopped by adding 2.5 ml of ninhydrin reagent and the solution cooled with tap water. At this stage samples can be stored for several hours. For determination of proline formed during the hydrolysis, 2.5 ml of glacial acetic acid is added and the solution heated at 100° for 30 minutes. After cooling, the intensity of the red color is measured in a Klett-Summerson photoelectric colorimeter with a filter No. 52. The amount of proline formed is calculated from a calibration curve constructed with known amounts of proline.

Definition of Unit and Specific Activity. One unit of activity is defined as the amount of enzyme which produces 1 mg proline per milliliter of incubation solution under the conditions of the above-described assay.

Since the low concentration of poly-L-proline in the assay solution does not correspond to the saturation region of the substrate concentration curve, the unit defined depends on the poly-L-proline concentration. The assay is also very sensitive to the purity of reagents which thus affects the unit as defined.

The specific activity is expressed in units per milligram of protein.

It should be noted that enzyme activity is quite proportional to enzyme concentration in the range from zero to about 20 milliunits per milliliter. At higher concentration deviation from linearity is pronounced, due to association (see below).

Assay of Enzyme in Tissue Extracts. Aminopeptidase-P is not the only enzyme-producing proline from poly-L-proline. Proliniminopeptidase[3,6] is also known to cause similar hydrolysis. The differentiation

---

[6] S. Sarid, A. Berger, and E. Katchalski, *J. Biol. Chem.* **237**, 2207 (1962).

between the two enzymes is based on their different specificity. The sequential polymers poly-(Pro·Gly·Pro) and poly-(Gly·Pro·Pro) can be used for the differentiation, the former being readily hydrolyzed by proliniminopeptidase, the latter by aminopeptidase-P. Tissue extracts when dialyzed against buffers contain very little material reacting to the acid ninhydrin test and can therefore be easily assayed by the described procedure. Bradykinin may be used as an alternative substrate for aminopeptidase-P. In this case 1 mole of arginine and 1 mole of proline are released.

Purification Procedure

Because the enzyme is very sensitive to heavy metal contaminants, all buffers and salt solutions are passed through columns of Chelex-100 (analytical grade chelating resin, mesh 50–100, Biorad Lab., Richmond, Calif.) preequilibrated with the reagent. Metal-free water is used throughout.

The enzyme is isolated from *Escherichia coli* (wild type) and purified 850 times. The enzyme is unstable to lyophilization, but can be kept frozen for several months without loss of activity. The procedure is described for a 1300-g batch of wet cells and is performed at 4°.

Preparation of Cell-Free Extract. Mass cultures of *Escherichia coli* strain B are grown in lots of 14 liters with aeration at 37° on yeast extract (1.0%), K_2HPO_4 (2.18%), KH_2PO_4 (1.70%), and glucose (1.0%). The cells are harvested after 5 hours in a Sharples ultracentrifuge (4 g wet cells per liter). The wet cells are kept frozen at −20° until a sufficient amount is collected by repeated growing of bacteria.

The frozen mass is dispersed at 4° by occasional shaking during 2 days in a 0.9% potassium chloride solution (4.6 ml/g).

The cells are then ruptured by passing the cold (4°) suspension through a Manton-Gaulin homogenizer (Model 15 M Everett, Mass.) at 4500 psi. (Changing the pressure to 3000 psi or to 6000 psi resulted in a specific activity lower by 17%. Repeating the homogenization did not increase the yield.)

Cell debris is removed by centrifugation for 30 minutes in a Sorvall Superspeed RC-2 automatic centrifuge with the GSA-rotor at 23,000 *g* at 0°–2°. The extract can be stored frozen at −20° for several weeks without loss of activity.

Heat Precipitation. The cell-free extract is warmed to room temperature (24°) with tapwater and heat precipitated by passing the extract under gravity through a spiral glass tubing of 0.50 cm internal diameter and 125 ml capacity immersed in a water bath kept at 53°.

The extract is passed at a flow rate of 200 ml/minute and emerges at a temperature of 50°. Fractions of 400 ml are collected in polyethylene 0.5-liter bottles preheated to 50°. Each fraction is then kept for 15 minutes in a 50° water bath, cooled in an ice bath, and stored at 4°.

In preliminary experiments it was shown that the enzyme activity at 50° was constant for 20 minutes. The specific activity after 5, 10, 16, and 20 minutes was 0.084, 0.086, 0.094, and 0.094, respectively. At 60° a sample of 1095 units/ml contained after 3, 7, 11, and 15 minutes 860, 685, 520, and 470 units/ml, respectively.

Ammonium Sulfate Precipitation. A saturated ammonium sulfate solution is neutralized to pH 7.0 by adding solid potassium carbonate (approx. 1 g per 1 liter). The solution is filtered if necessary and passed at a 700 ml/hour flow rate through a Chelex-100 column in NH_4^+-form $(3.0 \times 45$ cm) equilibrated with the saturated ammonium sulfate solution. To the metal-free solution was added sodium citrate $(Na_3C_6H_5O_7 \cdot 2$ H_2O, 0.59 g/liter) and the solution was stored at 4°, where the excess of ammonium sulfate crystallizes out and the supernatant is used for the fractionation procedure.

Saturated ammonium sulfate solution is added at 2° to the extract containing the precipitated proteins (787 ml ammonium sulfate solution per each liter of the extract). The clear supernatant is obtained by centrifugation and mixed with another portion of saturated ammonium sulfate (600 ml ammonium sulfate solution per each liter of the supernatant). The precipitate formed is isolated by centrifugation and dissolved in 0.05 M sodium acetate, pH 5.6. (For volumes used to dissolve the precipitate, see Table I.) The solution obtained loses 12% of its activity in 10 days when kept frozen at −20°, or 23% when kept at 4° for the same period.

In preliminary experiments it was found that reversing of the two purification steps, namely precipitating with ammonium sulfate first and heat precipitating thereafter, resulted in appreciable loss of activity. No advantage in terms of yield or purification was gained by removing the denatured proteins prior to the precipitation by ammonium sulfate.

Acetone Precipitation. Cold acetone (850 ml) is added during 20 minutes at −3° with stirring to the solution from the previous purification step (1000 ml). The formed precipitate is isolated with the aid of a Sorvall Superspeed RC-2 automatic refrigerated centrifuge (11,000 rpm), rotor GSA. The precipitate is well dispersed in 0.05 M sodium acetate, pH 5.6 (1000 ml) at 0° and extracted for 30 minutes. The undissolved material is removed by centrifugation and the supernatant is frozen and stored at −20°. No loss of activity was observed after 3 months of storage.

TABLE I

PURIFICATION OF AMINOPEPTIDASE-P[a]

Procedure	Volume (ml)	Activity[b] (units/ml)	Total units	Protein[c] (mg/ml)	Specific activity (units/mg)	Yield (%)	Purification factor
Cell-free extract	6,300	2.3	14,500	16.9	0.136	100	1
Heat precipitation	6,300	2.23	14,100	11.3	0.198	96.7	1.45
Ammonium sulfate fractionation	1,000	10.5	10,500	22.6	0.465	72	3.4
Acetone fractionation	1,000	9.5	9,500	3.78	2.52	65	18.4
Ca-phosphate gel	2,000	4.03	8,070	0.23	17.3	56	127
DEAE-cellulose	29	197	5,720	4.62	42.5	39.3	313
Gel filtration	107	42.7	4,560	0.37	116	31.4	855

[a] Figures are given for 1300 g frozen wet cells of *E. coli* B.
[b] The activity was determined as described in the text. One unit of activity is defined as the amount of enzyme which produces 1 mg proline per milliliter of incubation solution under the conditions of the above-described assay.
[c] Protein concentration was determined by the method of Lowry.[9] Bovine serum albumin was used as standard.

Adsorption on Calcium Phosphate. A calcium phosphate gel suspension in water (22 mg/ml) is prepared according to Keilin and Hartree.[7] The suspension (30 ml) is added to the enzyme solution obtained in the previous step (1000 ml) and stirred at 4° for 10 minutes. The solid is removed by centrifugation and another portion of the gel suspension (210 ml) is added to the supernatant and mixed for 15 minutes. Both portions of the gel are resuspended in a solution of 1 M KCl in 0.05 M sodium acetate, pH 5.6 (1000 ml), mixed for 15 minutes and the supernatant discarded after centrifugation. The gel sediment is then extracted with two 1-liter portions of 0.05 M sodium phosphate buffer, pH 6.0 (1000 ml), for 25 minutes. The extract can be stored at 4°.

Fractionation on DEAE-Cellulose. The solution from the previous step (2000 ml) is applied under gravity at a flow rate of 72 ml hour to a DEAE-cellulose column (1.7 × 43 cm) which was previously equilibrated with 0.05 M sodium acetate buffer, pH 5.6, containing 0.002 M sodium citrate. The column is washed with the same buffer (125 ml) followed by a solution of 0.05 M KCl in 0.1 M sodium acetate, pH 5.6, containing 0.002 M sodium citrate (500 ml). The adsorbed activity is finally eluted with 0.4 M ammonium sulfate in 0.1 M sodium acetate, pH 5.6, containing 0.002 M sodium citrate at a 30 ml/hour flow rate. The activity started to appear after 60 ml and emerged in a 29-ml volume.

Gel Filtration on Sephadex G-200. The solution obtained in the previous step (29 ml) is applied to a Sephadex G-200 (particle size 40–120 μ, Pharmacia, Uppsala, Sweden) column (3.8 × 191 cm) equilibrated with 0.1 M sodium acetate, pH 5.6, containing 0.002 M sodium citrate. The same buffer is used for elution at a 10 ml/hour flow rate. The active fraction starts to appear soon after the void volume and emerges in an approximately 150-ml volume. Fractions of about 15 ml are collected and checked for specific activity. Most of the enzyme (80%) is eluted in fractions having constant specific activity (116 units per milligram). This solution can be kept frozen for more than a year.

Properties

Stability. The enzyme is stable when kept either in 0.1 M sodium acetate buffer, pH 5.6, containing 2 × 10^{-3} M sodium citrate or in 0.1 M phosphate buffer, pH 7.0, containing 0.05 M EDTA. It loses activity on lyophilization. However, the enzyme can be conveniently concentrated by pressure filtration through a Diaflo ultrafiltration membrane UM-2, (solute cutoff 1000 MW, Amicon, Lexington, Mass.). Alternatively, the enzyme can be applied to a DEAE-cellulose column and eluted at high

---

[7] D. Keilin and E. F. Hartree, *Proc. Roy. Soc. (London) Ser. B* **124**, 397 (1938).

ionic strengths. In a representative run, an enzyme solution obtained from the last purification step (46 ml) was applied to a 10 ml DEAE-cellulose column equilibrated with 0.05 M sodium acetate buffer, pH 5.6, 2 $\times$ $10^{-3} M$ in sodium citrate and eluted with 0.4 M ammonium sulfate in 0.1 M sodium acetate, pH 5.6, 2 $\times 10^{-3} M$ in sodium citrate. The activity (75%) emerged in a 2-ml volume. The salt can be removed by dialysis against the acetate or phosphate buffer containing EDTA. Membranes should be washed with EDTA solution.

Purity and Physical Properties. The protein composition of the enzyme preparations obtained after the different purification steps was followed by polyacrylamide gel electrophoresis.[8] After the last purification step (factor of purification: 860) a single band was observed.[1]

Absorption Spectrum. The enzyme displayed a typical protein absorption peak at 278 nm and a 1.8 ratio of absorbance at 278 and 260 nm, respectively. The extinction was $OD_{280} = 1.03$ (1 cm) at a concentration of 1 mg/ml (determined from the Kjeldahl nitrogen using a factor of 6.25). The color yield in the Lowry method[9] is the same as for bovine serum albumin.

Sedimentation and Diffusion Coefficients. These coefficients of the pure enzyme were measured in 0.407 M ammonium sulfate, 0.1 M in Na-acetate, pH 5.6, containing 0.002 M sodium citrate. The enzyme sedimented as a single symmetric boundary in the ultracentrifuge. From the sedimentation coefficient $s^0_{20,w} = 9.1$ S, obtained after extrapolation to zero concentration, the diffusion coefficient $D^0_{20,w} = 4.4 \times 10^{-7}$ cm²/sec, and an assumed partial specific volume of 0.74, a molecular weight of 205,000 was calculated. A sedimentation equilibrium run with a 0.15% solution of the enzyme in the same solvent evaluated by the Yphantis midpoint method yielded a molecular weight of 230,000.

Amino Acid Analysis. The amino acid composition of aminopeptidase-P is recorded in Table II. The protein obtained after the last gel filtration purification step was hydrolyzed with 6 N HCl in evacuated sealed tubes at 100° for 24, 48, and 72 hours. Amino acid analysis was performed with an automatic amino acid analyzer. The values for serine, threonine, and tyrosine were corrected for decomposition during hydrolysis by extrapolating to zero time of hydrolysis. Also, the amide nitrogen was calculated from the ammonia content of the 24, 48, and 72-hour hydrolyzates after extrapolation to zero time. Valine and isoleucine were incompletely released at 24 hours, and for that reason their values were calculated from the 48- and 72-hour hydrolyzates. No cystine was found

[8] B. J. Davis, *Ann. N.Y. Acad. Sci.* **121**, 404 (1964).
[9] O. H. Lowry, N. J. Rosebrough, A. L. Farr, and R. J. Randall, *J. Biol. Chem.* **193**, 265 (1951).

TABLE II
AMINO ACID COMPOSITION OF AMINOPEPTIDASE-P

Amino acid	Micromoles amino acid per 100 micromoles N
Asp	5.45
Thr	2.83
Ser	3.20
Glu	8.60
Pro	1.93
Gly	4.76
Ala	5.12
Half-cys	—
Val	4.52
Met	1.26
Ile	3.90
Leu	5.54
Tyr	1.88
Phe	2.15
Trp	0.47
Lys	2.10
His	2.08
Ammonia	5.80
Arg	4.50
Cys-SO$_3$H	0.50[a]
Cysteine	0.60

[a] Determination of both cystine plus cysteine as cysteic acid was performed by performic oxidation and analysis according to Moore.[10]

[b] Determination of —SH groups was performed by the Ellman method[11] in presence of 1% sodium dodecylsulfate. The number of cysteic acid residues per 100 micromoles N was calculated from the average of two determinations of cysteic acid, and the average molar ratio of alanine, leucine, glutamic acid, and aspartic acid to cysteic acid was used to express the number of cysteic acid micromoles per 100 micromoles of nitrogen.

by the amino acid analysis of the hydrolyzates. Cysteic acid was quantitatively determined after performic acid oxidation according to Moore.[10] Sulfhydryl content in the intact protein was determined by the Ellman procedure[11] in presence of 1.0% sodium dodecylsulfate and 10^{-4} M EDTA. Tryptophan was determined spectrophotometrically.[12] The amino acid analysis accounts for 87% of the total nitrogen.

pH Profile. The dependence of activity on pH determined in 0.05 M Veronal buffer at three Mn^{2+} concentrations is given in Fig. 1. The bell-shaped curve shows a maximum at pH 8.6.

[10] S. Moore, *J. Biol. Chem.* **238**, 235 (1963).
[11] G. L. Ellman, *Arch. Biochem. Biophys.* **82**, 70 (1959).
[12] T. W. Goodwin and R. A. Morton, *Biochem. J.* **40**, 628 (1946).

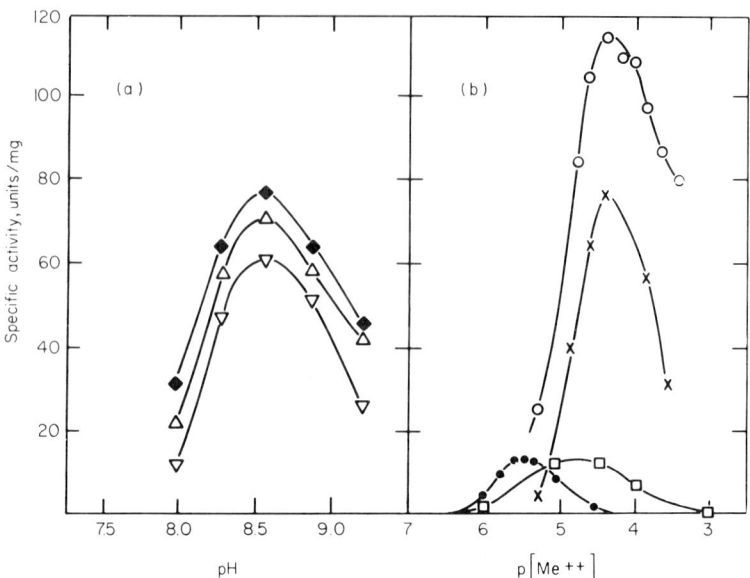

Fig. 1. (a) The dependence of specific activity on pH in $0.05\,M$ Veronal buffer. Mn^{2+} concentrations: ($\blacklozenge$) $3.7 \times 10^{-5}\,M$, ($\triangle$) 5.0×10^{-5}, ($\triangledown$) 2.5×10^{-5}. (b) The dependence of specific activity on metal ion concentration. Metal added as ($\times$) $MnCl_2$, ($\bigcirc$) Mn^{2+} citrate metallobuffer system, ($\bullet$) $CoCl_2$, ($\square$) $CdCl_2$.[13]

Metal Ion Requirement. Exhaustive dialysis of the enzyme against buffers containing EDTA leads to a completely inactive preparation. Activation can be achieved by removing the EDTA by dialysis and adding Mn^{2+}, Co^{2+}, Cd^{2+}, or Ni^{2+}.[13] The dependence of specific activity on the negative log of the concentration of the metal ions is given in Fig. 1. Also for the metals bell-shaped curves are obtained. The highest activity was achieved with Mn^{2+}, although Co^{2+} activates at a ten times lower concentration. A quite sharp optimum in activity occurs at a Mn^{2+} concentration of $3.7 \times 10^{-5}\,M$. This maximum was observed at all pH values at which activity could be tested (7.0–9.2). The pH of highest activity was always 8.6.

The required concentration of free Mn^{2+} in the assay solution is reached by using $MnCl_2$–sodium citrate as a metallobuffer.[14] Incorporation of citrate in the assay solution was found also to protect the enzyme against the inhibitory effects of traces of heavy metals.

[13] R. Granoth, Ph.D. thesis, Feinberg Graduate School, The Weizmann Institute of Science, Rehovot, Israel, to be published.
[14] J. Raaflaub, *in* "Methods of Biochemical Analysis" (D. Glick, ed.), Vol. 3, p. 301. Wiley (Interscience), New York, 1956.

Specificity. Aminopeptidase-P acts on polypeptides, cleaving the peptide bonds formed between the carboxyl of an N-terminal amino acid residue and the secondary amine of a proline residue: H-A$-\!\!^\downarrow$-Pro-B-C . . .-OH. The following low molecular weight peptides are hydrolyzed: L-prolyl$-\!\!^\downarrow$-L-prolyl-L-alanine, glycyl$-\!\!^\downarrow$-L-prolyl-glycine, glycyl-L-proline, L-valyl-L-proline, L-alanyl-L-proline, and L-prolyl-L-proline. The dipeptide glycyl-L-proline is digested considerably slower than higher peptides and the same applies to L-valyl-L-proline and L-alanyl-L-proline. Bradykinin (Arg·Pro·Pro·Gly·Phe·Ser·Pro·Phe·Arg) is digested rapidly (enzyme concentration 4.4 μg/ml), one arginine being released per one bradykinin molecule in less than 5 minutes and the following proline residue is released within 1 hour (80% in 30 minutes). No additional splits were observed after 24 hours. Reduced and carboxymethylated papain, which has the N-terminal sequence Ile·Pro·Glu·Tyr·Val . . . , was digested rapidly. At an enzyme concentration of 1.3 μg/ml, 1 mole of isoleucine was liberated within 1 hour (substrate 10^{-4} M). No other amino acids were released. The terminal amino group seems to be essential, as dinitrophenyl poly-L-proline is resistant to hydrolysis, in contrast to the rapidly hydrolyzed poly-L-proline. Hydroxyproline cannot be substituted for proline in the substrates of aminopeptidase-P, since poly-L-hydroxyproline and L-prolyl-L-hydroxyprolyl-L-alanine are not digested. Substrates of leucine aminopeptidase such as L-leucyl-L-serine are not digested, neither are peptides with an N-terminal L-proline residue when not followed by another L-prolyl, i.e., L-prolylglycine, L-prolyl-L-phenylalanyl-L-lysine, or salmine which are substrates of proline iminopeptidase,[3,6] are not hydrolyzed and were found to inhibit competitively the hydrolysis of poly-L-proline by aminopeptidase.

Because of the specificity described above, aminopeptidase-P may be expected to find useful application in studies of protein structure mainly in proline-rich proteins. It is pertinent to note that leucineaminopeptidase is capable of removing amino acid residues from a polypeptide until the residue preceding proline. Aminopeptidase-P is capable of cleaving this residue, revealing an N-terminal proline, which in turn can be removed by proline iminopeptidase.

Kinetic Parameters. DEPENDENCE OF SPECIFIC ACTIVITY ON ENZYME CONCENTRATION. It was observed that enzyme activity does not increase linearly with enzyme concentration (see Fig. 2). Since in gel filtration experiments it was observed[13] that the apparent molecular weight of the enzyme decreases from about 200,000 at 25 μg/ml (in the sample applied) to about 100,000 at 1.0 μg/ml, it was tentatively assumed that only the 100,000 MW species (M) is enzymatically active, whereas the 200,000 MW species (D) is inactive. Assuming the dissociation equilibrium

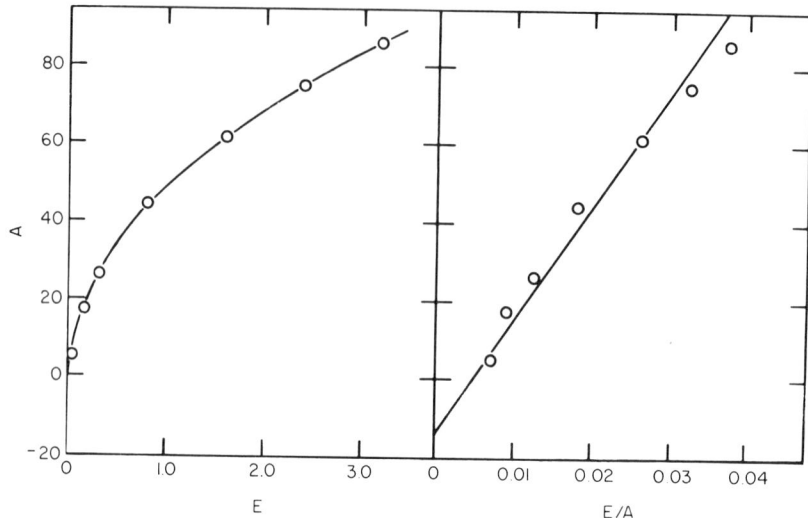

Fig. 2. The dependence of activity A (in enzyme units), on enzyme concentration E (in μg/ml). From the intercept on the E/A coordinate of the A vs. E/A plot, a maximum specific activity of 230 units per milligram protein is obtained. The dissociation constant calculated from the two intercepts is $K = 0.1$ μg/ml or $10^{-9} M$ (based on a molecular weight of 10^5).

$$D \rightleftharpoons 2 M \tag{3}$$

where

$$K = M^2/D$$

a plot of activity vs. the reciprocal specific activity should be linear.

If E is the total enzyme concentration (as monomer), it follows that $K = 2 M^2/(E - M)$. Taking the activity as $A = \alpha M$, where α is a constant depending on the definition of the unit of activity, then $A = (\alpha^2 KE/2A) - (\alpha K/2)$, E/A being the reciprocal specific activity.

The intercept (I') on the reciprocal specific activity coordinate gives the maximum specific activity and the product $2II'$ (where I is the intercept on the activity coordinate) gives the dissociation constant. This relationship was indeed observed (see Fig. 2).[13]

MICHAELIS-MENTEN PARAMETERS. The parameters $\bar{K}_m$ ($=1/K_m$) and k_{cat}, obtained from Lineweaver-Burk plots using initial rates, are given in Table III. The plots were linear in the substrate concentration ranges investigated (10^{-5} to $10^{-3} M$ for poly-L-proline and 10^{-3} to $10^{-2} M$ for L-prolyl-L-prolyl-L-alanine).

The very pronounced substrate specificity for L-proline indicates the presence of a specific binding site for a proline residue in the active site

TABLE III
Kinetic Parameters

Substrate or competitive inhibitor	$\bar{K}_m$ (M^{-1})	$\bar{K}_i{}^a$ (M^{-1})	$k_3{}^b$ (sec^{-1})
Poly-L-proline (MW 6000)[c]	11300	—	105
H·Pro·Pro·Ala·OH[d]	870	—	1210
H·Pro·Phe·Lys·OH[e]	—	8500	—
H·Pro·Phe·OH[e]	—	13300	—
H·Pro·Ala·OH[e]	—	1000	—
H·Ala·Phe·Lys·OH[e]	—	<50[f]	—

[a] Obtained from Lineweaver-Burk plots. The composition of the reaction mixture was that of the standard assay; substrate Pro · Pro · Ala 10^{-3} to 10^{-4} M.
[b] Assumed MW 100,000.
[c] Enzyme concentration 0.3 μg/ml.
[d] Enzyme concentration 0.06 μg/ml.
[e] Enzyme concentration 0.057 μg/ml.
[f] No inhibition at 2.3 × 10^{-3} M peptide.

of the enzyme. Product inhibition by peptides with N-terminal proline could therefore be expected. Indeed, it was found that L-prolyl-L-phenylalanine is a strong competitive inhibitor. By comparison of the reciprocal inhibition constants $\bar{K}_i$ $(= 1/K_i)$ in Table III, it can be seen that most of the binding energy is contributed by the proline residue. When the residue next to proline is phenylalanine, about ten times stronger inhibition is observed than in the case of L-prolyl-L-alanine. This indicates a considerable contribution of the aromatic residue to the binding energy of the inhibitor.

Occurrence. Occurrence in the microsome fraction of swine kidney of an aminopeptidase having a specificity similar to the bacterial aminopeptidase-P was reported by Nordwig and Dehm.[15]

A substrate specificity similar to that of aminopeptidase-P was repeatedly claimed[16-19] for the dipeptidase prolidase.[20] However, pure prolidase was shown to digest specifically dipeptides with a C-terminal proline or hydroxyproline, whereas the tripeptide Gly·Pro·Gly was not digested.[20] In contrast, aminopeptidase-P digests Gly·Pro·Gly (and bradykinin) much more readily than the dipeptide Gly·Pro.

The observations that prolidase can to some extent cleave glycine

[15] A. Nordwig and P. Dehm, *Biochim. Biophys. Acta* **160**, 293 (1968).
[16] A. Light and J. Greenberg, *J. Biol. Chem.* **240**, 258 (1965).
[17] R. L. Hill and W. R. Schmidt, *J. Biol. Chem.* **237**, 389 (1962).
[18] R. Frater, A. Light, and E. L. Smith, *J. Biol. Chem.* **240**, 253 (1965).
[19] C. Nolan and E. Smith, *J. Biol. Chem.* **237**, 453 (1962).
[20] N. C. Davis and E. L. Smith, *J. Biol. Chem.* **224**, 261 (1957).

from Gly·Pro·Leu,[17] isoleucine from denatured papain,[18] glutamine from GluNH$_2$·Pro·Ser·Val·Val·Leu,[16] and lysine from a peptide Lys·Pro·Arg·Glu . . . ,[19] seem to indicate the presence of aminopeptidase-P in the prolidase preparations used, all of which were of low degree of purity.

[36] Brain Aminopeptidase Hydrolyzing Leucylglycylglycine and Similar Substrates

By NEVILLE MARKS and ABEL LAJTHA

$$R_1—R_2\cdots R_n \longrightarrow R_1 + R_2\cdots R_n$$

Intracellular α-aminopeptide amino acid hydrolases (EC 3.4.1) are widely distributed in animal tissues but have been studied in brain only recently.[1-5]

Peptide hydrolases as such are not clearly defined in the literature, but the purified brain aminopeptidase described below appears to be distinct from the enzymes that hydrolyze L-leucyl peptides (leucine aminopeptidase, EC 3.4.1.2.), aminoacyloligopeptides (aminopeptidase, EC 3.4.1.2), and amino acyldipeptides (aminotripeptidase, EC 3.4.1.2).[5] All tissues appear to contain a complex mixture of peptide hydrolases of varying specificity. Brain aminopeptidase is a stable enzyme extracted from the soluble supernatant of whole pig brain and defined by its chromatographic profile on DEAE-cellulose, its specificity against tripeptides such as LLL[5a] and LGG, and by its unique cleavage of some alanyl oligopeptides.

Assay Methods

Two different methods are described: (1) a microassay procedure for monitoring enzyme purification and the rapid screening of materials for potential substrate activity; and (2) a quantitative procedure for the

[1] N. Marks, R. K. Datta, and A. Lajtha, J. Biol. Chem. 243, 2882 (1968).
[2] R. K. Datta, N. Marks, and A. Lajtha, Federation Proc. 17, 302 (1967).
[3] N. Marks, Intern. Rev. Neurobiol. 11, 57 (1968).
[4] A. S. Brecher and R. E. Sobel, Biochem. J. 105, 641 (1967).
[5] N. Marks in "Handbook of Neurochemistry" (A. Lajtha, ed.), Vol. 3, p. 133. Plenum Press, New York, 1970.
[5a] Abbreviations used: amino acids in peptide linkage are according to the single letter code of IUPAC-IUB [Biochem. J. 113, 1 (1969)]. A, alanyl; L, leucyl; G, glycyl; Y, tyrosine. —βNA denotes the naphthylamide moiety of the arylamide substrate. All amino acids unless defined otherwise are of the L-configuration.

isolation of the degradation products themselves for purposes of determining the stoichiometry of peptide breakdown.

Principle. The chief disadvantage of the standard quantitative ninhydrin procedure is the relatively high background of the substrate compared to the liberated products. This method can be modified to reduce E (the molar extinction coefficient) for di- and tripeptides (the substrates) compared to the liberated amino acids. A comparison of E for peptides used in this report based on the modified procedures of Yemm and Cocking[6] and Matheson and Tattrie[7] is given in Table I.

TABLE I
COMPARISON OF E VALUES WITH DIFFERENT NINHYDRIN REAGENTS[a]

| | E Values | |
Substrate	Ninhydrin reagent (modified)	Ninhydrin hydrindantin reagent
Leu-	1.95	2.15
Ala-	1.79	2.30
Gly-	2.01	2.17
Leu-Gly	0.35	2.47
Leu-Gly-Gly	0.38	2.05
Ala-Ala	0.35	4.10
Gly-Gly	1.39	2.04
Gly₃	1.13	1.91
Gly-Leu	1.44	2.18

[a] Data derived from Matheson and Tattrie.[7]

Stoichiometric studies of peptide breakdown require the isolation of degradation products. Paper chromatographic methods have been suggested[1,8] but the losses after elution and the unpredictability of color yield are adverse factors. The method of choice is the use of the standard amino acid analyzer as described below.

Microassay Procedure

Reagents

Citrate buffer, 0.2 M, pH 5.0 (21.0 g $C_6H_8O_7 \cdot H_2O$ dissolved in 200 ml water + 1 N NaOH diluted to 500 ml)

NaCN, 1 mM (4.9 mg NaCN in 100 ml), KCN can be substituted. Store in a glass-stoppered bottle. Stable 6 months at 4°.

[6] E. W. Yemm and E. C. Cocking, *Analyst* **80**, 207, (1955).
[7] A. T. Matheson and B. L. Tattrie, *Can. J. Biochem.* **42**, 95 (1964).
[8] S. Ellis and J. M. Nuenke, *J. Biol. Chem.* **242**, 4623 (1967).

Ethanol 50% v/v

Solution A. Dilute one part of NaCN solution with 12.5 parts of methyl Cellosolve.

Solution B. Ninhydrin 1% w/v in methyl Cellosolve

Reagent mixture. (Prepare fresh for daily use.) Mix 1 part of A with 5 parts of B. Although solutions A and B are stable 1 month or longer at room temperature, the best reproducibility was found when all solutions (except the NaCN) were prepared fresh.

Procedure. Enzyme (10–40 μg protein, depending on purification) is assayed in 1 ml containing 10 micromoles Tris-HCl buffer, pH 7.6, and 0.25 micromoles of peptide substrate. Assays are performed in duplicate and the control experiment is fixed at zero time with 0.5 ml of citrate buffer at 0°. After incubation the reaction is stopped by the addition of 0.5 ml citrate buffer, and color is developed by the addition of 1.2 ml of the reagent mixture, heating for 7.5 minutes in a boiling water bath, and cooling immediately in ice. Samples are read after dilution with 50% ethanol; the degree of dilution is varied according to the color yield. Standards are conveniently prepared with 50 and 100 nanomoles of L-glutamic acid; with 2 ml 50% ethanol as diluent. The OD at 570 nm was 0.26 and 0.51, respectively, and with 5 ml diluent it was 0.15 and 0.295 (Spectronic 20, Bausch and Lomb; tubes 1.2 cm internal diameter).

Autoanalyzer

Two separate methods can be employed: (1) the standard long-column procedure of Hamilton[9] but with the use of modified buffers; or (2) the accelerated short-column procedure of Catravas.[10]

The first procedure employs a resin height of about 120 × 0.6 cm of Aminex A-6 (Biorad) or chromobeads (type-B, Technicon). With alteration of the three standard buffers[9] to a pH of (1) 2.95, (2) 3.9, (3) 5.35, the peaks obtained for typical peptides used in the study of brain aminopeptidase are depicted in Fig. 1. Alteration of the buffers gave a particularly good separation of alanyl oligopeptides compared to the standard buffers which did not give a good separation of Ala and Ala₃. Because of the good separation of alanyl and leucyl peptides, it is possible to run pairs of experiments simultaneously on one column: AAA and GGL, AAA and LLL, AAAA and LLL, Ala₆ and LL, etc. The chief disadvantage of the long column with Aminex A-6 was the low flow rates caused by the

---

[9] P. B. Hamilton *in* "Techniques of Amino Acid Analysis" (D. I. Schmidt, ed.), p. 1. Technicon Instruments, Chertsey, England.

[10] G. N. Catravas *in* "Technicon Symposia 1965" (L. T. Skeggs, ed.). Mediad, New York (1966).

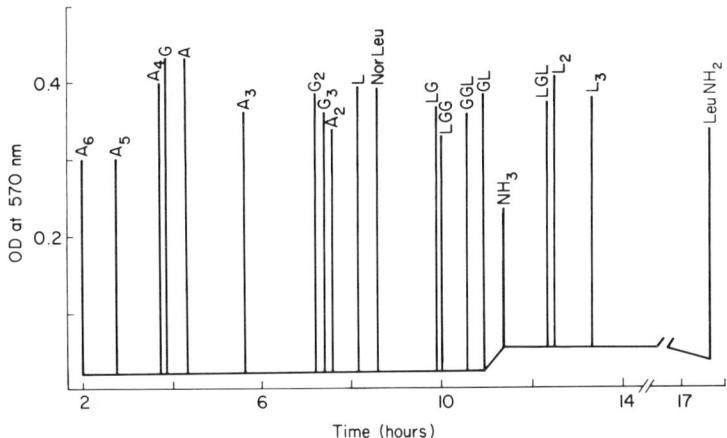

F<small>IG</small>. 1. Appearance of peptides on the Autoanalyzer. Column 120 × 0.6 cm. Chromobeads (type-B, Technicon) or Aminex A-6 (Biorad). Buffers employed in the Varigrad were of pH's 2.95, 3.9, and 5.35. Norleucine, 0.1 micromole, was used as the internal standard.

accumulation of enzyme protein. This necessitated regeneration of the column at frequent intervals (passage of 0.2 N NaOH at 90° facilitated the removal of protein). Protein could be precipitated after incubation by the normal procedures with 0.1 M hydrochloric, perchloric, or sulfosalicylic acid, but in addition to the increased complexity of handling, low pH carried the danger of substrate precipitation. It is an advantage in stoichiometric studies to measure the substrate remaining as well as the products formed by degradation.

An alternative, rapid procedure is the accelerated short-column (75 × 0.6 cm) procedure of Catravas.[10] At flow rates of 1 ml per minute the average run is reduced from 20 to 6 hours, with peptides appearing in almost the same order as with the long-column method. Owing to the closeness of peaks, only one experiment can be run at one time. Short columns with Aminex A-4 or A-5 are subject to less clogging by protein and in the case of Aminex A-4 have a lower back pressure. Generally 0.05 micromoles of product can be detected (expressed as norleucine equivalent), or with scale expansion as little as 1–3 μg of product. At the extreme limits of sensitivity it is essential to run adequate controls to exclude extraneous peaks due to contamination from the incubation ingredients or the enzyme itself. Most commercial substrates tested for purity by paper chromatography showed trace contamination on the analyzer. Tris itself gave a small but reproducible peak ahead of the norleucine peak with the modified buffers suggested for the long-column procedure.

Enzyme Purification (Hog Brain)

Step 1. Hog brain obtained from fresh slaughterhouse material is extracted with 10 volumes of 0.32 M sucrose and centrifuged at 40,000 g to remove crude debris, nuclei, and mitochondria. The supernatant is centrifuged at 100,000 g in a high-capacity SW 27 Spinco rotor to remove microsomes (Table II).

TABLE II
PURIFICATION OF AMINOPEPTIDASE FROM HOG BRAIN

Step	Volume	Total protein	Total activity[a]	Specific activity[b]
1. Supernatant	335	1273	4050	3.2
2. (NH₄)₂SO₄ 43% w/v	330	343	2075	6.0
3. (NH₄)₂SO₄ 60% w/v	12.5	162	1850	11.4
4. Dialysis	12.5	155	1625	10.5
5. DEAE-cellulose	50	11	820	75.0
6. DEAE-Sephadex	30	4	700	175.0
7. CM-cellulose	60	0.5	350	700.0

[a] Activity is expressed as nanomoles of glutamic acid equivalents as determined by the microassay procedure in 30 minutes.

[b] Nanomoles glutamic acid equivalents per milligram protein.

Step 2. Postmicrosomal supernatant is adjusted to pH 5.5 with acetic acid, added with stirring to 43% w/v of enzyme grade (NH₄)₂SO₄, let stand at 0° for 1 hour and spun at 50,000 g for 45 minutes (Sorvall rotor, SS-34 at 17,500 rpm.).

Step 3. Step 2 is repeated with 60% w/v (NH₄)₂SO₄.

Step 4. The precipitate from step 3 is dissolved in 12.5 ml buffer containing 2 mM T.E.S. (N-tris (hydroxymethyl) methyl-2-amino-ethane sulfonic acid; Calbiochem, Calif.) and dialyzed overnight against distilled water containing the same buffer.

Step 5. The enzyme is placed on a DEAE-cellulose (Cellex-D, 0.8 meq/g, Biorad, Richmond, Calif.) column 20 × 2.5 cm previously equilibrated with Tris-HCl buffer 40 mM, pH 7.6, a further 25 ml Tris buffer is added, and the column is eluted with different concentrations of NaCl dissolved in the same buffer: (1) 100 ml of 0.0125 M NaCl, (2) 0.055 M NaCl, (3) 0.11 M NaCl, (4) 100 ml 0.2 M NaCl (Fig. 2).

Step 6. The eluate (0.055 M NaCl) is placed on a short column of DEAE-Sephadex (A-50, coarse, 3.5 meq/g, Pharmacia Inc., N.J.) equilibrated with Tris-HCl buffer 40 mM, pH 7.6, and eluted with 0.2 M NaCl to remove aminopeptidase. Preliminary experiments have shown that aryl-amidase is tightly bound by the absorbant and only eluted by 0.5 M NaCl

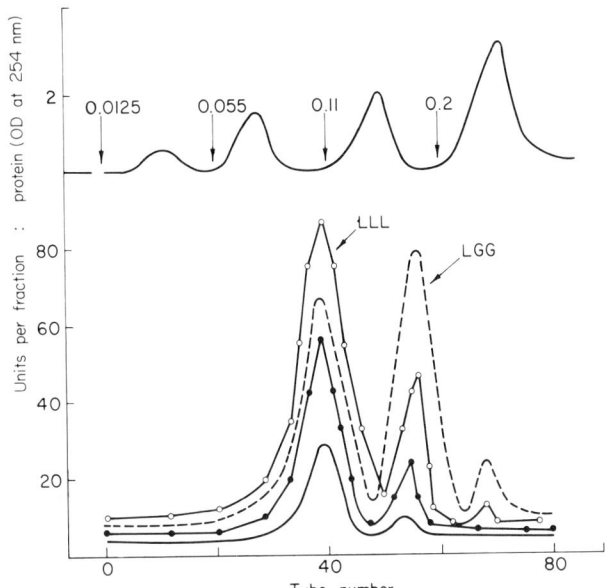

F<small>IG</small>. 2. Purification of hog brain extracts after passage through (a) DEAE-cellulose and (b) DEAE-Sephadex. In (a) the peaks represent the protein eluted by NaCl as monitored by a continuous flow cell (254 nm). The most active peak (0.055 M) was further purified as shown in (b); the first peak was eluted with 0.2 M NaCl and the second with 0.5 M. Specificity with tripeptide substrates is compared with LL (●). Protein is indicated by a solid line.

(Fig. 2). Preliminary separation of aminopeptidase and arylamidase on DEAE-Sephadex results in losses caused by low flow rates and by high salt concentrations for the subsequent purification procedures.

Step 7. Aminopeptidase is further purified by passage through CM-cellulose (Cellex-CM, 0.47 meq/g, Biorad) equilibrated with 50 ml acetate buffer, pH 5.5. Enzyme is loosely bound and can be eluted with 30 ml of acetate buffer; Cl and other anions can be removed by passage through Sephadex G-100 (Pharmacia, N.J.).

Properties

Homogeneity and Molecular Weight. The best preparations gave a single band on disc gel electrophoresis when performed under standard conditions. Molecular weight was 86,000 when estimated by gel exclusion chromatography; the exclusion volume for purified aminopeptidase placed on a 50 × 2.4 cm column of Sephadex G-100 was 78.3 ml Tris buffer (40 mM, pH 7.6) compared to 107.2 ml for soybean trypsin inhibitor (MW 21,500), 87.6 for bovine serum albumin (MW 68,000),

TABLE III
AMINO ACID COMPOSITION[a]

Residue	Micromoles of amino acids/100 residues	Amino acid (residues/mole enzyme)[b]
Half-cys	1.41	2
Asp	7.73	13
Thr	3.86	6
Ser	13.05	21
Glu	15.14	25
Pro	7.41	12
Gly	9.36	15
Ala	9.29	15
Val	4.80	8
Ile	2.82	5
Leu	5.33	9
Tyr	0.83	14
Phe	2.61	4
Lys	7.10	12
His	1.88	3
Arg	5.11	9

[a] One milligram enzyme protein hydrolyzed 20 hours at 110° *in vacuo*. Values are without correction for losses of Ser and Thr; Trp, Glu, NH₃ and Asp·NH₂ were not determined.

[b] Based on a MW of 86,000.

and 60.9 ml for γ-globulin (MW 160,000). The partial amino acid composition is shown in Table III; values for tryptophan and amide content were not determined.

Stability. Preparations stored at a concentration of 0.2–0.4 mg protein/ml in 40 mM Tris buffer or in sucrose media, 0.32 M, showed no appreciable loss of activity after 1 year of storage at −20°. The enzyme resisted freezing and thawing but became inactivated upon lyophilization. One advantage of aging was the inactivation of other contaminating peptidase activities. The enzyme lost activity when heated for 10 minutes at 60°.

Specificity. Information on specificity can be determined by two separate procedures: (1) rapid screening of substrates by the quantitative microassay procedure (Table IV); and (2) isolation and determination of the end products by quantitative chromatographic methods (i.e., amino acid analyzer) (Table V).

Microassay. The first method is suitable for the calculation of the rate constants of enzyme velocity, K_m, pH, temperature sensitivity, etc. For accurate quantitative comparison of different substrates it is necessary to know the individual ninhydrin color yields of all products—

a difficult task in a mixture of peptides and amino acids. All tissues are likely to contain a spectrum of different peptide hydrolases (EC 3.4.1.) and it is essential to carefully map the peptide specificities of the different chromatographic peaks obtained from DEAE-Sephadex columns. The activity present in different peaks clearly demonstrates the presence of a family of related aminopeptidases. Fractions from peak I for 1 year at $-20°$ in $0.25 M$ sucrose showed the highest activity with LLL (Table IV). In contrast, enzyme in the second peak (Fig. 2) split LLL at a rate less than 40% compared to that with LGG. Peak I in both fresh and aged preparations was inactive in the hydrolysis of monoacyl and dipeptidyl arylamides (Leu-, Arg, Arg-Arg, Lys-Ala-, Ala-Ala-βNA, etc.). These are considered to represent a different class of peptide hydrolases[3,5] (arylamidases), and are present in fresh preparations of peak II.[1] Storage led to a complete loss of arylamidase activity within 3 months although crude extracts retained activity for a longer period. The effect of storage on the specificity of aminopeptidases appears to be quite complex and some change in the relative ratios of LLL and LGG hydrolysis occurred in the two peaks on storage.

Autoanalyzer. Analysis of the products formed from LGG showed a 1:1 release of Leu and GG (Table V). This result can be interpreted as an N-terminal cleavage and is consistent with the microassay procedure, which demonstrated an absence of glycyl–glycine dipeptidase (EC 3.4.3.1). Earlier studies have shown absence of activity with N-protected dipeptides, thus excluding the presence of carboxypeptidases. The results for both the microassay and Autoanalyzer procedures showed a complete cleavage of LLL within 90 minutes to form free Leu. In

TABLE IV

SPECTRUM OF AMINOPEPTIDASES PRESENT IN HOG BRAIN SUPERNATANT EXTRACTS
ELUTED FROM DEAE-SEPHADEX

	% Activity[a]	
	Peak I	Peak II
LLL	*100*	*100*
LGG	40	250
LG	37	50
GYG	23	50
LL	80	20

[a] Activity is expressed as a percentage of the values for LLL which are italicized. The peaks refer to those illustrated in Fig. 2. Peptidase activities were determined by the quantitative microassay procedure on samples stored in $0.25 M$ sucrose for 1 year at $-20°$C.

TABLE V
SPECIFICITY AS DETERMINED BY THE AUTOANALYZER

	Time of incubation	% Split	Products[b]					
			L	G	LL	LG	GL	GG
Substrate[a]								
Leu·NH₂	90	27	+	−	−	−	−	−
LG	90	22	1	1	−	−	−	−
LGL	90	64	1	−	−	−	1	−
GGL	90	0	−	−	−	−	−	−
LGG	90	72	1	−	−	−	1	−
LLL	30	60	3	−	1	−	−	−
LLL	90	90	+	−	−	−	−	−
GGG	90	20	−	−	−	−	−	1
Alanyl peptides[a]			A₁	A₂	A₃	A₄	A₅	
A₂	60	5	+	−	−	−	−	
A₃	5	23	1	1	−	−	−	
A₄	60	23	3	1	−	−	−	
A₅	90	10	+[c]	+	+	+	−	
A₆	90	19	3	1.5	−	1	0[c]	
	180	46	5	2	−	1	0[c]	

[a] Abbreviations: L, leucine; G, glycine; A, alanine.

[b] When more than one product is obtained, the approximate ratio to each other is indicated.

[c] Concentration of the products too low to provide an estimate of ratios. Peptidase activity was determined utilizing the Autoanalyzer as described in the text.

a typical Autoanalyzer experiment, 225 nanomoles in 1 ml with 20 μg of enzyme gave at 90 minutes 680 nanomoles of product equivalent to a 100% cleavage of the two available peptide bonds.

With alanyl peptides containing two to six residues, a complex picture emerged. Substrates forming intermediates that are themselves hydrolyzed do not lend themselves easily to stoichiometric analysis. A typical example illustrates the method of utilizing Autoanalyzer techniques for ascertaining the mechanism of peptide breakdown. AAA incubated under standard conditions (225 nanomoles substrate, 20 μg enzyme in 1 ml, etc.) gave the following products after 5 minutes, expressed in terms of 1000 nanomoles starting material: 765 nanomoles of AAA (i.e., 23.5% degraded), 220 nanomoles of AA, and 260 nanomoles of free Ala. If it was assumed from the results summarized in Table V that AA was split 5% (max), then the quantity of Ala to be expected is 235 + 11.7 (i.e., 258.4 nanomoles) leaving 235 − 11.7, or 223.3 of AA. These results agree within the limits of experimental error (5%) with the values experimentally determined. Similar calculations for Ala₄-Ala₆

require additional assumptions concerning the rate constants for a series
of competing reactions (since several intermediate peptides are formed).
At a first approximation, the values obtained do not accord with a
simple N-terminal cleavage as the only mechanism. The accumulation
of intermediates suggests the release of dipeptide units in addition to
the removal of a single amino acid. Since the purified brain aminopeptidase reported here is inactive in the degradation of polypeptides such
as glucagon, and was unaffected by cysteine or other sulfhydryl compounds, and showed no metal or anion dependency, it would appear to
differ from other enzymes recently reported to remove N-terminal dipeptides.[11] It is also noticeable that brain aminopeptidase did not split
monoacyl-, and dipeptide arylamides, and was unaffected by arylamidase inhibitors such as puromycin and its analogs.[1,12]

In summary, the present aminopeptidase can be defined as the hydrolase activity present in supernatant fractions that is precipitated by
43–60% $(NH_4)_2SO_4$, is eluted from DEAE by $0.055\,N$ NaCl, hydrolyzes
LLL at a greater rate than LGG (in aged preparations), and shows
a unique specificity with higher oligopeptides of alanine.

Activators and Inhibitors. Exhaustive dialysis against chelating
reagents such as EDTA was without effect on brain aminopeptidase
activity. The lack of effect on addition of Mg^{2+}, Ca^{2+}, Mn^{2+}, and Co^{2+}
differentiates this enzyme from leucineaminopeptidase, and the metal-
activated dipeptidases. Heavy metals such as Cu^{2+} inhibited enzyme
activity by 90–100% at 0.1 mM concentration, and Hg^{2+} inhibited
activity at 0.1 mM by 50%. Addition of sulfhydryl compounds such as
cysteine, β-mercaptoethanol, and dithiothreitol caused a small but variable increase of 6–20% at 0.1 mM and unlike brain arylamidases did
not affect the stability of the enzyme. One or more sulfhydryl groups
may be involved at the catalytic site since the enzyme was markedly
inhibited by pCMB and o-iodosobenzoate. As a further differentiation
of brain aminopeptidase from arylamidase capable of splitting oligopeptides, this enzyme was unaffected by the addition of puromycin.[1,12]

pH Optima and K_m. The pH optima was between 7.3–7.6 for most
substrates. K_m for LGG was $6 \times 10^{-4}\,M$.

Acknowledgments

These studies were supported in part by U.S.P.H.S. Grant No. NB-03226. We
would also like to thank Dr. Amos Neidle for assistance with the Autoanalyzer and
the technical help of Paula Mela and Steven Bodansky.

[11] J. K. McDonald, B. J. Zeitman, T. J. Reilly, and S. Ellis, *J. Biol. Chem.* **244**,
2693 (1969).
[12] N. Marks and A. Lajtha *in* "Protein Metabolism of the Nervous System" (A.
Lajtha, ed.), p. 39. Plenum Press, New York, 1969.

[37a] Thermophilic Aminopeptidases: AP I
from *Bacillus stearothermophilus*

By G. RONCARI and H. ZUBER

Various strains of *Bacillus stearothermophilus* contain three types of thermostable (thermophilic) aminopeptidases—AP I, AP II, and AP III.[1] Depending on the optimum temperature for their growth, various strains of *Bacillus stearothermophilus* produce structurally different AP I (II and III), which vary in specificity and thermostability. The three aminopeptidases split all amino acids from the amino end of polypeptides; like leucine aminopeptidase,[2] they preferentially hydrolyze peptides containing leucine, valine, etc., and also those with aromatic amino acid residues. However, if proline occupies a penultimate position, then the ultimate amino acid residue will not be liberated. Peptides with cysteine, cystine, and cysteic acid have not been tested. This article describes the details of the membrane-bound AP I from *Bacillus stearothermophilus*, which is the most thermostable of the three aminopeptidases.

Assay Methods

Principle. Three methods can be used for determining the activity of AP I: (A) hydrolysis of leucine-*p*-nitroanilide and spectrophotometric determination of *p*-nitroaniline at 410 nm; (B) hydrolysis of Gly-Leu-Tyr (or other tripeptides) and determination of the liberated glycine as ninhydrin–copper complex[3] or ninhydrin–cadmium complex[4] by the chromatographic method; or (C) hydrolysis of Leu-Gly (or other dipeptides) and determination of the liberated leucine and glycine with an amino acid analyzer. With crude extracts method A is recommended.

Method A

Reagents

CoCl₂, 0.001 M, in Tris-HCl buffer pH 7.2, 0.05 M
Leucine-*p*-nitroanilide, 0.001 M
Enzyme, in Tris-HCl buffer pH 7.2, 0.001 M CoCl₂

[1] G. Roncari and H. Zuber, *Intern. J. Protein. Res.* 1, 45 (1969).
[2] See this volume [33] and [34].
[3] F. G. Fischer and H. Dörfel, *Biochem. Z.* 324, 544 (1953).
[4] G. N. Atfield and J. O. R. Morris, *Biochem. J.* 81, 606 (1961).

Procedure. Leucine-*p*-nitroanilide can be hydrolyzed either by adding the enzyme solution (1–20 μl) to 3 ml leucine-*p*-nitroanilide solution, followed by incubation at 65°, and measurement at 410 nm in the spectrophotometer (1-cm cuvette), or directly by adding the enzyme solution to a 1-cm cuvette containing 1.5 ml leucine-*p*-nitroanilide solution, placed in a spectrophotometer with a cell holder heated at 65°. The blank contains no enzyme. Calibration curve: 1–15 μg/ml *p*-nitroaniline yields an absorption of 0.06–1.50.

Method B

Reagents

$CoCl_2$, 0.001 M, in Tris-HCl buffer, pH 7.2, 0.05 M
Gly-Leu-Tyr, 0.045 M
NaOH, 0.1 N
Chromatography solvent: *n*-butanol:acetic acid:water = 4:1:5
Color reagents as described[3,4]

Procedure. Five-tenths milliliter of the Gly-Leu-Tyr solution is mixed with the appropriate amount of enzyme solution and incubated at 65°. If necessary, the pH is kept constant by adding 0.1 N NaOH (glass electrode). Aliquots of 40 μl of the incubating mixture are chromatographed together with 10–20 μl glycine (comparison solution) chromatography paper: Wt 1). Glycine is determined as described.[3,4]

Method C

Reagents

$CoCl_2$, 0.001 M, in Tris-HCl buffer, pH 7.2, 0.05 M
Leu-Gly, 0.03 M
Sodium citrate buffer, pH 2.2 (amino acid analyzer)

Procedure. One milliliter Leu-Gly solution is incubated with 1–10 μl enzyme solution at 65°. From this solution 10 μl is taken and added to 3 ml sodium citrate buffer, pH 2.0. Aliquots of 1.0 ml are analyzed in the amino acid analyzer and the quantities of leucine and glycine are determined.

Definition of Unit and Specific Activity. One unit enzyme is defined as the amount of enzyme that hydrolyses 1 micromole substrate (methods A, B, and C) per minute under the above conditions. Specific activity is expressed as units per milligram of protein. The protein concentration in the enzyme preparations upon isolation was determined either with

TABLE I

PURIFICATION OF AMINOPEPTIDASE I[a] FROM *Bacillus stearothermophilus*

Step	Fraction	Volume[b] (ml)	Total protein[b,c] (mg)	Units/ml ($\times 10^{-3}$)[d]	Total[b] units	Yield in activity (%)	Specific activity (units/mg protein) ($\times 10^{-3}$)
	Crude extract	6,000	25,980	1,622	9,732	—	374
1.	Sephadex G-150	3,280	5,444	78[e]	256	100	47
2.	Heating	3,280	2,690	96	315	123	117
3.	DEAE-Sephadex						
	Fr. 46–52 (I)	459	87	119	55	21	626
	Fr. 53–60 (II)	525	215	371	195	76	904
	Fr. 61–66 (III)	393	106	127	50	19	470
4.	Sephadex G-200	656	105	290	190	74	1,812
5.	Prep. electrophoresis (Ultrafiltration)	1.3	49[f]	134,000	175	68	7,100[f]

[a] Starting from 800 g bacteria (net weight).

[b] Calculated for total extract (6000 ml from 800 g bacteria).

[c] Calculated by reference to Vol. III, p. 454 (260/280 nm).

[d] 1 Unit: Hydrolysis of 1 μM Leu-p-nitroanilide min^{-1} (65°).

[e] Marked decrease in activity as a result of separation of the more active AP II on Sephadex G-150.

[f] The exact amount of protein (total protein) was calculated from the amino acid analysis (as a basis for the specific activity). This value also holds good for step 4, as the enzyme protein was already pure at this stage (gel electrophoresis). Determination of the amount of protein by method C yields a doubled value, presumably because of the content of nucleic acids.

a method described by Lowry[5] or with one described by Layne.[6] For the pure enzyme, in the determination of activity the values obtained from amino acid analysis (sum of the amino acid residues) were used as a basis for calculating enzyme concentration (Table I).

Purification Procedure

Preparation of Cells. Bacillus stearothermophilus (NCIB 8924) cells are grown on the liquid medium described by Hartmann.[7] The pH of the medium following sterilization (120°, 15 minutes) was 7.1. The following other liquid media could be employed: (1) Brain heart infusion, BBL; or (2) Lab Lemco medium: 3 g Lab Lemco beef extract, 2 g Bacto trypton (Lab Lemco), 2 g glucose, 3 g K_2HPO_4, 1 g KH_2PO_4 per liter, pH 7.1. *Bacillus stearothermophilus* is kept in stock cultures both on brain heart infusion agar slants (BBL) and on CTA medium (BBL) at 5°. For large batches the bacteria are inoculated from CTA medium on to 10 ml liquid medium and up to larger volumes (1 liter, 10 liters, 100 liters, etc.) by repeated inoculations of 10 vol. % (18 hours, 55°).

The large volume is incubated for 15–24 hours, at 55°; 10 liters per minute are aerated. At the end of incubation the cells are centrifuged off with the disc bowl self-cleaning centrifuge (type: α-Laval BRPX-207) or with the Sharples centrifuge. Under the conditions described, the yield is about 1 g bacteria (wet weight) per liter of the medium. Only the AP I in the bacterial cell is worked up, since the medium is discarded.

Disintegration of Cells. The bacterial cells are homogenized at 5° by a recycling process in a sand mill containing Ottawa sand [treatment of 1 kg bacteria per 6000 ml, 0.05 M Tris-HCl buffer, pH 7.2 (0.001 M $CoCl_2$) 2 × 15 hours]. The bacterial residue and the sand are removed by centrifugation with the Sharples centrifuge (17,000 rpm). The homogenate is concentrated in a rotary evaporator (10–20°) or with the "Diaflo" ultrafiltration cell Model 401 (Amicon Corp., Lexington, Mass., membrane: UM-20 E) at room temperature to about 2 liters per 1000 g bacteria homogenized. An alternative procedure for disintegrating small amounts of bacteria is by sonification. One gram bacterial cells (wet weight) is suspended in 10 ml Tris-HCl buffer, pH 7.2. The mixture is treated in a sonic vibrator (Ultra Disintegrator, Schoeller USIG 750 LF, Frankfurt, Germany, 20 kc/sec, 60 W/cm²)

[5] O. H. Lowry, N. J. Rosebrough, A. L. Farr, and R. J. Randall, *J. Biol. Chem.* **193**, 265 (1951).

[6] E. Layne, Vol. III, p. 454.

[7] P. A. Hartmann, R. Wellerson, and P. A. Tetrault, *Appl. Microbiol.* **3**, 11 (1955).

for 3 minutes. The temperature is kept below 10°. The total protein concentration in the crude extract is about 0.03 to 0.04 g per gram bacteria.

Step 1. Fractionation on Sephadex G-150. The concentrated bacterial homogenate is fractionated on a Sephadex G-150 column (10×120 cm, volume: 10 liters) in 500-ml portions. The column is equilibrated with $0.05\,M$ Tris-HCl buffer, pH 7.2 ($0.001\,M$ CoCl₂) at room temperature and is then eluted with the same buffer. The enzyme solutions and the buffer solutions are pumped through the column from below (pumping rate: 250 ml per hour, 50 fractions of 150 to 200 ml each, the run is performed at room temperature). The aminopeptidase activity is in the following fractions: AP I, fractions 7–13 (high molecular weight fraction); AP II and AP III, fractions 15–26 (low molecular weight fractions). The more thermolabile AP II and III were separated at this stage.

Step 2. Heat Treatment. The solutions of AP I from the Sephadex G-150 column, divided into portions of 500 ml are heated to 80° for 30 minutes under stirring. They are left to stand overnight at room temperature and the precipitate is then centrifuged off for 1 hour at 40,000 rpm in a Sorvall centrifuge, Rotor S-34.

Step 3. Chromatography on DEAE-Sephadex. The 500-ml portions obtained from step 2 are chromatographed on a DEAE-Sephadex A-50 column (2.5×40 cm). The DEAE-Sephadex is pretreated as per instructions (Pharmacia, Uppsala, Sweden) with $0.5\,N$ NaOH and $0.5\,N$ HCl and rinsed with $0.05\,M$ Tris-HCl buffer, pH 7.2 ($0.001\,M$ CoCl₂). The enzyme solution is loaded onto the column and the latter is washed with Tris-HCl buffer until the UV-absorption has the same value as before application of the enzyme solution (UVICORD, LKB, 3 mm cell). The column is developed with 1000 ml of a linear sodium NaCl gradient (500 ml $0.5\,M$ NaCl in buffer solution is added to 500 ml $0.1\,M$ NaCl solution in buffer). Pumping rate is 45.6 ml per hour, 10 ml fractions. AP I active fractions are fractions 46–66. The following fractions are pooled: fractions 46–52 (I, 70 ml); fractions 53–60 (II, main fraction, 65% of total activity eluted from the DEAE column, 80 ml); fractions 61–66 (III, 60 ml).

Step 4. Fractionation on Sephadex G-200. The main fraction II (fractions 53–60) is concentrated to 3.5 ml by ultrafiltration ("Diaflo" ultrafiltration cell, Amicon Corp. Model 401 and 10, membrane UM-20 E). The enzyme concentrate is placed on a Sephadex G-200 column (3×92 cm) which has been equilibrated with Tris-HCl buffer, pH 7.2 ($0.001\,M$ CoCl₂). The AP I is fractionated by recycling (2 cycles: pumping rate: 27.6 ml per hour, upward flow) on Sephadex G-200. The enzyme is eluted from the column after 22 hours (bleeding in the second cycle, 10-ml fractions). The activity is in fractions 1–10.

Step 5. Preparative Polyacrylamide Gel Electrophoresis. The AP I after step 4 is uniform and in a pure state in polyacrylamide gel electrophoresis with respect to protein content. Some nucleic acids which are still present can be completely separated by preparative polyacrylamide gel electrophoresis. The electrophoresis is carried out with the Poly-Prep[8] apparatus manufactured by Buchler Instruments, N.J. A 5.63% gel (80 ml) is prepared and used for separation purposes. The gel and the buffer solution are prepared as described, using the following reagents:

Preparation of the gel
    Gel buffer: 24 ml 1 N HCl, 18.15 g Tris, 0.25 ml N,N,N',N'-tetramethylethylenediamine (TEMED)
    Polymerization: 30 g acrylamide and 0.8 g bisacrylamide per 100 ml H_2O, 0.35% ammonium persulfate
Buffer
    Upper reservoir buffer: 6.32 g Tris, 3.94 g glycine per 1000 ml
    Lower elution buffer: 12 g Tris, 50 ml 1 N HCl
    Membrane holder buffer: 4× concentration of the lower buffer

The "concentrating gel" is omitted and the aminopeptidase is layered directly on the top of the separating gel in 0.05 M Tris-HCl buffer, pH 7.2, containing 3% sucrose. After separation the AP I is eluted at an elution volume between 925 and 1070 ml (current 50 mA; pumping rate: 87 ml per hour, location of the protein UV-absorption [UVICORD, LKB, 3 mm cell]; collection of 10-ml fractions in a fraction collector). The active fractions are pooled and the solution is concentrated in the Amicon "Diaflo" ultrafiltration cell (see above).

Properties

Purity and Physical Properties. The purified enzyme appears to be homogeneous in the ultracentrifuge and in gel electrophoresis.[9] From the sedimentation constant ($s_{20,w}^0 = 16.5 \times 10^{-13}$ seconds) and the diffusion coefficient ($D_{20,w}^0 = 3.9 \times 10^{-7}$ cm^2 sec^{-1}) the molecular weight was calculated to be 395,000 ± 30,000 assuming a partial specific volume of 0.739 ml g^{-1} (calculated from the amino acid composition with the aid of partial specific volumes of amino acid residues as calculated by Cohn and Edsall.[10] The molecular weight based on equilibrium centrifugation ($\bar{M}_{app}$) is 400,000 ± 45,000. In 8 M urea and 0.01 M EDTA at pH 5.6 AP I dissociates into subunits ($s_{20,w}^0 = 1.95 ± 0.1 \times 10^{-13}$ seconds,

[8] T. Jovin, A. Chrambach, and M. A. Naughton, *Anal. Biochem.* 9, 351 (1964); B. J. Davis, *Ann. N.Y. Acad. Sci.* 121, 404 (1964).

$D^0_{20,w} = 5.0 \pm 0.25 \times 10^{-7}$ cm² sec⁻¹, $M_{S,D} = 36,500 \pm 4000$). On the basis of the amino acid analysis (protein concentration) and the molecular weight, the extinction coefficient was found to be 10.2 and the molar extinction coefficient $E_{280} = 4.1 \times 10^5$ (.05 M Tris-HCl buffer, pH 7.2, 0.001 M CoCl₂).

Stability. The purified, native AP I is stable in solution (0.05 M Tris-HCl buffer, pH 7.2, 0.001 M CoCl₂) for several months if kept in a refrigerator. The enzyme is stable for 15 hours at 80°. The activity increases by 20 to 30% within 30 minutes (increase in V_{max}, K_m is constant). In cobalt-free buffer the thermostability of AP I is markedly reduced (1 hour, 80°: 35% residual activity). The apoenzyme is also very thermolabile. In 8 M urea (at pH 7.2) AP I is not denatured (in gel electrophoresis the activity is only slightly, if at all, reduced in the hydrolysis of peptides). After treatment with 0.3% sodium dodecylsulfate, the activity is reduced by 8%.

Enzyme Structure. See Table II for the amino acid composition of *B. stearothermophilus.* AP I is a metal enzyme (complex), Co²⁺ ions conferring the highest activity upon activation (see Stability) and upon recombination from apoenzyme with Co²⁺ ions. The Co²⁺ ions are strongly bound at pH 8.1; the metal enzyme complex is fairly stable up to an EDTA concentration of 0.01 M (only a slight decrease in activity occurs). At a pH of less than 6.0 the metal enzyme is converted into a stable, inactive apoenzyme following dialysis against 0.01 M EDTA.

Specificity. The specificity of AP I is similar to that of other known aminopeptidases (broad substrate specificity). It releases not only neutral but also acid and basic amino acids, including proline, from the amino end. In contrast to leucine aminopeptidase[2] from pig kidneys or eye lenses, it splits leucine-p-nitroanilide relatively more slowly than the peptides [specific activity: (a) leucine-p-nitroanilide hydrolysis, 6.6 units mg⁻¹; (b) hydrolysis of Leu-Gly, 400 units mg⁻¹; hydrolysis of Gly-Leu-Tyr (splitting of Gly), 900 units mg⁻¹]. Besides rapidly hydrolyzing aliphatic amino acid residues from the amino end, it also quickly splits off aromatic amino acid residues.

Kinetic Properties. The pH range of maximum activity for the substrate Gly-Leu-Tyr is between 9.2 and 9.4 and for the substrate leucine-p-nitroanilide between 7.5 and 8.0. The optimum temperature for the hydrolysis of Leu-Gly (pH 7.2) is 90° (10 minutes) (activation energy 16,300 cal mole⁻¹). At pH 7.2 the K_m was found to be 95 mM for Leu-Gly and 8 mM for leucine-p-nitroanilide (65°). Assuming that AP I has a

[9] P. Moser, G. Roncari, and H. Zuber, *Intern. J. Protein Res.,* in press (1970).
[10] E. J. Cohn and J. T. Edsall, "Proteins, Amino Acids and Peptides," p. 375. Reinhold, New York, 1943.

TABLE II

AMINO ACID COMPOSITION OF AP I OF *Bacillus stearothermophilus*

Amino acid	Amino acid residues/ml hydrolyzate[a] (μg/ml)	Amino acid residues/100 g protein (g)	Minimal molecular weight	Calculated no. of residues/molecule (MW = 400,000)	No. of residues to nearest integer	Calculated molecular weight
Lysine	18.60	6.55	1,954	204.79	205	400,590
Histidine	8.22	2.89	4,740	84.37	84	398,202
Arginine	19.81	6.98	2,235	178.97	179	400,047
Amide-NH$_3$	3.31[b]	1.17	—	275	275	—
Aspartic acid	28.29	9.97	1,153	346.80	347	400,229
Threonine	16.80[c]	5.92	1,706	234.46	234	399,204
Serine	7.29[c]	2.57	3,385	118.16	118	399,453
Glutamic acid	37.54	13.23	975	410.25	410	399,750
Proline	11.83	4.17	2,326	171.96	172	400,089
Glycine	15.39	5.42	1,051	380.37	380	399,668
Alanine	18.67	6.58	1,079	370.71	371	400,309
Valine	23.36	8.23	1,203	332.52	332	399,362
Isoleucine	19.72	6.95	1,626	246.03	246	399,946
Leucine	23.39	8.24	1,371	291.69	292	400,419
Tyrosine	8.46[c]	2.98	5,503	72.68	73	401,740
Phenylalanine	12.49	4.40	3,341	119.72	120	400,908
Methionine	4.79	1.69	7,751	51.60	52	403,072
Tryptophan	8.86	2.79	6,666	60.00	60[d]	399,960
Total	283.56	100.75			3,675	

[a] Average or extrapolated value (all the values, except those of serine, threonine, tyrosine, and ammonia have been calculated from the average of the 24-, 48-, and 72-hour analyses, since no significant trend seems to exist in the rest of the data).

[b] Average value from zero time extrapolation (6 N HCl) and from hydrolysis with 2 N HCl [A. C. Chibnall, L. J. Mangan, and M. W. Rees, *Biochem. J.* **68**, 111 (1958)].

[c] Extrapolated to zero time.

[d] Average value (rounded off value): Tryptophan was determined by two spectrophotometric methods. Method (1): G. H. Beaven and E. R. Holiday, *Adv. Protein Chem.* **7**, 319 (1952); method (2): calculation from E$_{280}$ and from the tyrosine value (73 Tyr/mole).

specific activity of 900 units mg⁻¹ (Gly-Leu-Tyr) and that the molecular
weight of the functional unit is 400,000, it may be calculated that the
molecular activity (catalytic center activity) is approximately 360,000.

Acknowledgment

The authors wish to thank the *International Journal of Protein Research* for
permitting us to quote from an original paper published in that journal.

[37b] Thermophilic Aminopeptidases: AP I
from *Talaromyces duponti*[1]

By R. CHAPUIS and H. ZUBER

Three types of aminopeptidases (AP I, II, and III) are also present
in thermophilic fungi such as *Talaromyces duponti*, *Talaromyces emer-
sonii*, *Talaromyces auranticus*, *Humicula lanigunosa*, *Mucor mihei*, and
Mucor pusillus. The membrane-bound AP I, which is the most stable
of the three types, displays, in line with the lower optimum temperature
for fungal growth, less thermostability than AP I from *Bacillus stearothermophilus* (see Stability). The purification and characteristics of AP I
from *Talaromyces duponti* are described below.

Assay Method

The assay procedures and reagents used were the same as for AP I
from *Bacillus stearothermophilus* (pp. 544–552), except that 0.0001 M
CoCl₂ was present in the buffer.

Purification Procedures

Preparation of Cells. The fungus *Talaromyces duponti* can be cultivated on the liquid medium 5% Mycophil Broth (BBL). It is then incubated for 72 hours at 55° under stirring (450 rpm) at 1 atm excess
pressure; 1 liter of air per liter of medium is blown through per minute.
The mycelium is filtered off and the medium discarded. The yield of
mycelium is approximately 14 g per liter of medium. *Talaromyces duponti*
is stored in stock cultures on 5% Mycophil Broth Agar (BBL) or as
lyophilized spores (from phopshate buffer pH 7) at 5°. From these stock
cultures the following amounts are reinoculated successively on 5%

[1] This material was taken from the thesis of R. Chapuis, Eidgen. Technische Hochschule, Zürich. See also, *Proc. Intern. Symp. Structure-Function Relationship of
Proteolytic Enzymes* (P. Desnuelle, H. Neurath, and M. Ottesen, eds.), Munksgaard, Copenhagen, 1969.

Mycophil medium: 100 ml in a shaking flask, a further 100 ml in a shaking flask, 5 liters in a fermentor, 30 liters in a fermentor, etc.

Disintegration of Cells. The fungal mycelium is homogenized in an Atomix homogenizer (MSE) in 250 g portions at 10,000 rpm (1 g mycelium per 2.5 ml 0.05 M Tris-HCl buffer, pH 7.2 [0.001 M CoCl$_2$] 2°–8°). More effective disintegration (yield 1.5–3 times higher) is obtained with the Schoeller Ultra Disintegrator (3 minutes, 4°, 1 g mycelium per 20 ml buffer), or with the "Bronwill MSK" homogenizer (Bronwill Scientific, Rochester, N.Y., 5 g mycelium per 30 ml buffer, mixed with 37 ml glass beads, 2 × 15 seconds at maximum speed, 4°–15°). The disintegrated mycelium is centrifuged off with a Spinco L–2 (rotor 19) at 19,000 rpm. The clear mycelium extract is concentrated to 0.6 ml per gram homogenized mycelium in a "Diaflo" ultrafiltration cell (Model 401, membrane: UM-20 E).

Step 1. Fractionation on Sephadex G-150. The concentrated homogenate is placed on a Sephadex G-150 column (10 × 90 cm, volume: 7.2 liters) in portions of 500 ml each and fractionated. The column is first equilibrated with 0.05 M Tris-HCl buffer, pH 7.2 (0.001 M CoCl$_2$); elution is performed at low temperatures with the same buffer which is pumped through the column from below (pumping rate: 122 ml per hour, 25-ml fractions are removed). The high molecular weight AP I is contained in fractions 100–140 (1000 ml) and the low molecular weight AP II and III in fractions 160–220. AP II and III are separated at this stage.

Step 2. Chromatography on DEAE-Sephadex (I). The combined fractions 100–140 from step 1 (1000 ml) are slowly run into a DEAE-Sephadex A-50 column (2.5 × 30 cm). The column is first equilibrated, as described in connection with *Bacillus stearothermophilus* (p. 544). The column loaded with the enzyme is washed with Tris-HCl buffer, pH 7.2, until the absorption of the buffer is the same as before addition of the enzyme solution. AP I is then eluted with an NaCl gradient (500 ml 0.5 M NaCl in 500 ml buffer is added to 500 ml buffer). The pumping rate is 41 ml per hour, 17 ml fractions are collected. The aminopeptidase is in fractions 23–38 (0.15–0.23 M NaCl in the buffer).

Step 3. Chromatography on DEAE-Sephadex (II). The combined fractions 23–38 from step 2 are dialyzed against Tris-HCl buffer, pH 7.2 (0.001 M CoCl$_2$). Chromatography is performed under the same conditions as in step 2. The NaCl gradient, however, differs from the one in step 2, inasmuch as 500 ml of 0.3 M NaCl in 500 ml buffer is run into 500 ml buffer. The AP I activity is in fractions 23–49 (fractions 23–29 contain 20% of the activity, fractions 30–38 contain 40%, and fractions 39–40 contain 30%).

Step 4. Preparative Polyacrylamide Gel Electrophoresis. Fractions 23–29, 30–38, and 39–49 are combined in each case and separated in three distinct runs with the aid of preparative polyacrylamide gel electrophoresis, as described in step 5 of the procedure used for *Bacillus stearothermophilus* (p. 544). For this purpose, the three portions are each dialyzed against 0.05 M Tris-HCl buffer and then concentrated to 20 ml in a "Diaflo" ultrafiltration cell. Electrophoresis conditions: 50 mA, 200–300 V, 4°. Elution rate (elution buffer): 87 ml per hour. The AP I activity is in fractions 51–73 (230 ml, 10-ml fractions).

Properties

Purity and Physical Properties. Purified AP I from *Talaromyces duponti* seems to behave uniformly in polyacrylamide gel electrophoresis. Its molecular weight is approximately 400,000, as determined by gel filtration on Sepharose 6 B (Pharmacia Uppsala), using AP I from *Bacillus stearothermophilus* as a reference protein.

Stability. The aminopeptidase is stable in solution (0.05 M Tris-HCl buffer, pH 7.2, 0.001 M CoCl$_2$) when stored in a refrigerator and also at 55° (8 hours). At 55° activation occurs within 120 minutes (buffer, 0.0001 M CoCl$_2$, increase of 120% in activity), whereas at 65° under the same conditions the enzyme is activated by 158% within 60 minutes. At 65°, however, activity decreases again markedly after 4 hours (denaturation, 50% residual activity).

Enzyme Structure. AP I from *Talaromyces duponti* is a metal enzyme, Co^{2+} ions conferring the highest activity upon activation and also upon

PURIFICATION OF AMINOPEPTIDASE I FROM *Talaromyces duponti*[a]

Step	Fraction	Volume (ml)	Units/ml[b] ($\times 10^{-3}$)	Total units	Activity yield (%)
	Crude extract	500	—	—	—
1.	Sephadex G-150	1025	154	846	100
2.	DEAE-Sephadex A-50 (I)	272	590	780	92
3.	DEAE-Sephadex A-50 (II)				
	Fr. 23–29	122	253	156	20
	Fr. 30–38	157	392	312	40
	Fr. 39–49	192	244	234	30
	Fr. 23–49 (total)			702	90
4.	Prep. electrophoresis				
	Fr. 30–38	230	410	308	98

[a] Starting from 860 g *T. duponti* (wet weight).
[b] 1 Unit: Hydrolysis of 1 μM leucine *p*-nitroanilide min^{-1} (65°).

recombination from the apoenzyme. The enzyme is fully active in 0.001 M EDTA, but is completely inhibited in 0.005 M EDTA. At pH 5.6 and 0.01 M EDTA the inactive apoenzyme is formed.

Specificity and Kinetic Properties. Specific activity (pH 7.2, 0.0001 M $CoCl_2$, 65°) is 4.1 mg^{-1} units in the hydrolysis of leucine-p-nitroanilide, and 300 units mg^{-1} in the hydrolysis of Gly-Leu-Tyr (splitting of glycine). The K_m value is 5 mM (pH 7.2, 65°) in the hydrolysis of leucine-p-nitro-anilide, and the optimum pH 6.9, measured at 65°.

[38] Pyrrolidonecarboxylyl Peptidase

By R. F. DOOLITTLE

Pyrrolidonecarboxylyl peptidase is an enzyme which hydrolytically removes pyrrolidonecarboxylyl (pyroglutamyl) residues from the amino-terminals of peptides and proteins[1]:

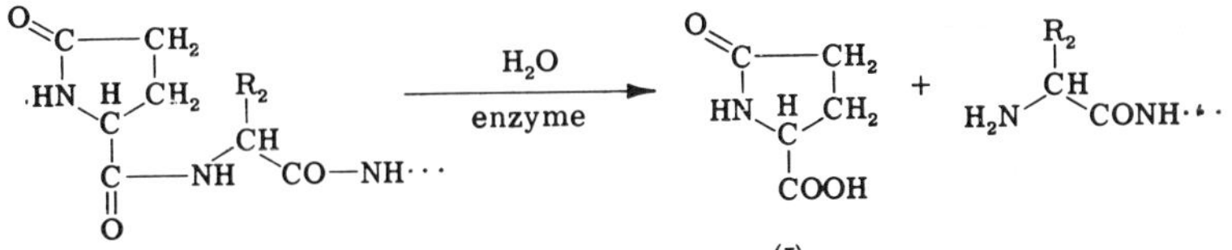

(I)

In earlier publications the enzyme was called pyrrolidonyl peptidase,[2,3] a name which does not precisely convey the nature of the reaction catalyzed. Although the enzyme was first discovered in a strain of *Pseudomonas fluorescens*, the same activity has subsequently been demonstrated in rat liver[4] as well as in a strain of *Bacillus subtilis*.[3] This article is concerned only with studies on the enzyme isolated from the original pseudomonad strain. The emphasis is on application of the enzyme to protein structural studies rather than on a full characterization of the enzyme itself.

The subject of PCA-terminating peptides and proteins has been

---

[1] In this article the abbreviation PCA will be used for both free pyrrolidonecar-boxylic acid (I) and the residue in peptide linkage. In the past we have used the designation Pyr for the latter.

[2] R. F. Doolittle and R. W. Armentrout, *Abstr. Intern. Congr. Biochem* 7th p. 608 (1967).

[3] R. F. Doolittle and R. W. Armentrout, *Biochemistry* 7, 516 (1968).

[4] R. W. Armentrout and R. F. Doolittle, *Arch. Biochem. Biophys.* 132, 80 (1969).

exhaustively reviewed in an earlier volume of this series.[5] It bears mentioning, however, that a large number of naturally occurring peptides and proteins have now been found to have PCA as their amino terminal residue, a situation presumed to arise from the cyclization of terminal glutaminyl or glutamyl residues. PCA-peptides can also arise artifactually during the isolation of peptides after proteolysis, glutamine-terminating peptides being especially liable to cyclize. The difficulties and frustrations of determining the structures of peptides with blocked amino terminals[6]—many of which apparently were of the PCA type—led the author to search for an enzyme which could *open* terminal pyrrolidone rings. A fluorescent pseudomonad was subsequently isolated from the soil which grew on PCA as its sole source of carbon and nitrogen. Disappointingly, crude extracts did not exhibit any hydrolytic activity toward the amide linkage of free PCA when tested *in vitro*. When these crude extracts were tested on a pyrrolidonecarboxylyl dipeptide, L-pyrrolidonecarboxylyl L-alanine (PCA-Ala), however, free PCA and alanine were produced. This result was entirely unexpected since no other peptidase or protease has been reported which will hydrolyze pyrrolidonecarboxylyl dipeptides, and in our own laboratory ten available proteases and peptidases had no detectable activity toward the same dipeptide. Even if completely nonspecific, then, the enzyme would have been useful for identifying PCA in peptides. As it turned out, the enzyme is highly specific and selectively removes PCA from the amino terminals of peptides and proteins. It has already proved very useful in the structural elucidation of a number of polypeptides.

Assay Method

The standard assay used in our laboratory has depended on the enzyme-catalyzed hydrolysis of L-pyrrolidonecarboxylyl L-alanine (PCA-Ala). The progress of the reaction is followed by monitoring the exposure of the alanine α-amino groups by use of the ninhydrin method.

A stock solution of PCA-Ala (MW 200) containing 25 micromoles/ml (5.0 mg/ml) is prepared in glass-distilled water. Ordinarily, 50 μl of buffered (pH 7.3) enzyme solution is added to 25 μl of the stock substrate solution, the final PCA-Ala concentration being $8.3 \times 10^{-3} M$. Digestions are carried out at 30° in stoppered tubes. At appropriate times 1.0 ml absolute ethanol is added to the mixture to poison the enzyme and precipitate the bulk of the protein. The stoppered tubes are placed in the cold for at least 15 minutes to aid precipitation and then centrifuged in a table-top centrifuge. The supernatants are decanted into

[5] B. Blombäck, Vol. XI, pp. 398–411.
[6] B. Blombäck and R. F. Doolittle, *Acta Chem. Scand.* **17**, 1816 (1963).

clean tubes and 0.5 ml aliquots removed for ninhydrin assay. In the latter case we have followed the procedure of the Rockefeller group rather literally, adding 0.5 ml of the ninhydrin reagent, capping loosely, and placing in a boiling water bath for exactly 15 minutes. After cooling, the solutions are diluted with 5 ml of 50% isopropanol and the absorbance read at 570 nm. All of these operations, including the original digestions, can be carried out in disposable flint glass tubes (13 × 100 mm). These tubes fit well into a Bausch & Lomb Spectronic 20 and are matched well enough to assure adequate precision for most routine assay situations. The ninhydrin reagent is made up according to Hirs et al.,[7] 2.0 g of ninhydrin and 0.3 g of hydrindantin being dissolved in 75 ml ethylene glycol monomethylether (methyl Cellosolve) followed by the addition of 25 ml of 4 M sodium acetate buffer, pH 5.5. Under these conditions a leucine standard of 100 nanomoles corresponds to about 0.40 absorbance units. If all the PCA-Ala in a digestion tube is hydrolyzed, a nearly full-scale deflection of 1.2 absorbance units above background is obtained. In crude extracts of the enzyme there is frequently a significant exposure of amino groups which is independent of the presence of added substrate and presumably results from endogenous proteolysis. As a control, then, a separate set of incubation mixtures is set up without substrate (PCA-Ala) but containing instead an equivalent amount of pyrrolidonecarboxylic acid (PCA).

A unit of enzyme activity is defined as the amount of enzyme which releases one millimicromole of alanine per minute under the conditions described above, the pH being maintained at 7.3 with phosphate buffers. We have found it convenient to calculate relative specific activities using the absorbance at 280 nm (A_{280}) of the enzyme solutions as an index of protein concentration.

Synthesis of Pyrrolidonecarboxylyl Dipeptides

L-Pyrrolidonecarboxylyl dipeptides are readily synthesized by the dicyclohexyl carbodiimide (DCCI) method starting with L-pyrrolidone-carboxylic acid (PCA) and the t-butyl ester derivative of the desired L-amino acid.[8] The t-butyl esters of amino acids can be prepared[9] by suspending 1–5 g of the appropriate amino acid in a dioxane–conc. sulfuric acid mixture in the proportions 1 g amino acid : 10 ml dioxane : 1 ml H_2SO_4. After the addition of 10 ml (or proportionately more) of isobutylene at dry ice/isopropanol temperatures, the mixtures are stirred at room temperature (in glass bottles with rubber stoppers wired on)

[7] C. H. W. Hirs, S. Moore, and W. H. Stein, J. Biol. Chem. 219, 623 (1956).
[8] J. A. Uliana and R. F. Doolittle, Arch. Biochem. Biophys. 131, 561 (1969).
[9] R. W. Roeske, Chem. Ind. (London) p. 1121 (1959).

for 24–48 hours until all the material has gone into solution and reacted. The more hydrophobic amino acids are more readily derivatized by this method than some of the more polar amino acids. The latter derivatives are more easily prepared starting with the N-carbobenzoxy (N-CBZ) amino acids, the carbobenzoxy groups being removed by hydrogenation after the formation of the t-butyl ester.

The first step in the synthesis of the dipeptide is accomplished by dissolving equivalent amounts of the L-amino acid t-butyl ester and L-pyrrolidonecarboxylic acid (PCA) in dimethylformamide (DMF). After cooling the mixture, one and a half equivalents of dicyclohexyl carbodiimide (DCCI), also dissolved in DMF, are added and the mixture stirred in the cold for 24–48 hours. At the end of that time a few drops of glacial acetic acid are added to convert any unreacted DCCI to dicyclohexylurea. The DMF is removed on a rotary flash evaporator and the crude residue dissolved in ethylacetate, any insoluble material being removed by filtration. After washing with a 5% sodium bicarbonate solution, the ethylacetate is evaporated and the residue dissolved in ether; insoluble material is discarded. We have generally had difficulty crystallizing the intermediate dipeptide t-butyl esters, and usually find it expedient to move directly to the deblocking step. After the evaporation of the ether solvent, the t-butyl group is removed by treatment with trifluoroacetic acid (TFA) at room temperature in a matter of minutes. The TFA can be blown off with a stream of nitrogen. The dipeptides are readily crystallized and recrystallized from mixtures of ethanol and ether.

L-Pyrrolidonecarboxylyl L-alanine (m.p. 206°) has been the substrate of choice for most enzyme purification and characterization work attempted so far. Commercial pyrrolidonecarboxylyl dipeptides are available from the Cyclo Chemical Corp., Los Angeles. We have tested a batch of their L-pyrrolidonecarboxylyl L-alanine and found it to be as good a substrate for the enzyme as the dipeptide we synthesized ourselves.

Culture of *Pseudomonas* Strain

Slants of the original pseudomonad strain have been sent to numerous laboratories around the world, several of which are now carrying the strain in culture. The strain can also be obtained from the American Type Culture Collection, Rockville, Md. (No. 25289). It has been our custom to use a recipe for slants which contains PCA as the sole source of carbon and nitrogen. Care must be taken in maintaining the strain, however, since Stanier has subsequently found that 166 of 175 strains of *Pseudomonas* in his collection will grow on PCA, but when one of these was chosen at random and sent to us, it failed to exhibit pyrrolidonecarboxylyl peptidase activity.

We have used two different liquid media in tandem for large-scale culturing of the organism. First, slants of the strain are rinsed into a minimal medium containing 0.5% PCA and the usual salts (but not ammonium salts) and trace metals. When this inoculum is grown up, it is transferred to a large volume (5 liters) of a glucose-citrate minimal medium (including ammonium salts). When the latter culture is grown up it can be harvested by centrifugation or used as an inoculum for a 100-liter batch in a fermentor, the glucose–citrate medium being used in the latter case also. The substitution of the glucose–citrate–ammonium medium for the PCA type is primarily a matter of economy, although growth is somewhat more vigorous on the glucose type. Each of the steps involves 24–48 hours' growth at 30°. The recipes used for the preparation of these culture media are summarized in Table I.

The bacteria are readily harvested after a Fermacell (fermentor) run using a Sharples-type centrifuge. Our average yield from 100 liters of culture is about 500 g wet weight. The bacteria are then freeze dried, the dry weight yield ranging from 80 to 120 g. After pulverization the material is stored at −20° until needed. The dried cells readily withstand a few days at room temperature for mailing purposes. Cells prepared according to this procedure can be purchased from the Cyclo Chemical Corp.

TABLE I

LIQUID CULTURE MEDIA FOR GROWING THE *Pseudomonas* STRAIN USED AS A SOURCE OF PYRROLIDONECARBOXYLYL PEPTIDASE

Trace metal solution (g)		PCA-minimal medium (g)		Glucose-minimal medium (g)	
CuSO₄·5 H₂O	1.02	KH₂PO₄	2.38	KH₂PO₄	2.00
FeSO₄·7 H₂O	1.76	K₂HPO₄·3 H₂O	5.66	K₂HPO₄·3 H₂O	9.20
MnCl₂·4 H₂O	1.26	MgSO₄·7 H₂O	1.00	MgSO₄·7 H₂O	0.10
ZnSO₄·7 H₂O	0.24			(NH₄)₂SO₄	1.00
				Na₃ Citrate·2 H₂O	0.60
Dissolve in H₂O, bring to 1 liter. Store frozen.		Dissolve in ca. 800 ml H₂O. Add 6.25 ml trace metals solution, and then bring volume to 1 liter. Autoclave. Add Millipore filtered stock PCA solution and neutralize with sterile 5% NaOH solution. For slants, 2.1 g Difco Noble agar is dissolved in 100 ml bulk salt-trace metal solution (above) before autoclaving, and the sterile PCA and NaOH added later before solidification occurs.		Dissolve in ca. 800 ml H₂O. Add 6.25 ml trace metals solution, and then bring volume to 1 liter. Autoclave. Add sterile stock glucose solution.	

Preparation of the Bacterial Enzyme

The bacterial enzyme can be prepared either from bacterial pellets frozen directly after harvesting or from pulverized freeze-dried suspensions. The latter material is especially useful when shipment at room temperature is necessary. In either case, cell disruption is readily accomplished by sonication. The cells are suspended in a 0.05 M phosphate buffer, pH 7.3, containing 0.01 M mercaptoethanol and 0.001 M EDTA. This buffer, which is used throughout most of the early purification steps, should also contain 0.1 M 2-pyrrolidone as an enzyme stabilizer.[4] The actual sonication conditions will vary with the type of sonifier employed; we use a glass "flow-around" vessel packed in ice and submit the suspension to a series of intermittent blasts with a Branson sonifier. Care must be taken to maintain the temperature of the suspension in the range 0° to 10°.

After sonication the material is centrifuged at 39,000 g at 0° for 30–60 minutes. The supernatant liquid containing the enzyme activity is carefully decanted, and if necessary, it should be centrifuged a second time. The A_{280} of the supernatant should be at least 50; if it is higher, it can be diluted with the phosphate buffer described above.

Nucleic acids may be precipitated by adding 1 volume of 1% protamine sulfate to 6 volumes of cold supernatant containing the enzyme. Although this step is not absolutely necessary and was not used in our early preparations, the removal of nucleic acids does seem to facilitate the subsequent gel filtration step of Sephadex G-200. After centrifugation (15,000 g for 20 minutes), the pellet is discarded and the supernatant used immediately for ammonium sulfate fractionation.[4]

A saturated ammonium sulfate solution is added directly, with stirring, to the supernatant fluid from the protamine sulfate step until 42% saturation is achieved. After standing at 0° for 1 hour, the preparation is centrifuged (39,000 g for 30 minutes). The supernatant fluid is carefully decanted and discarded, the walls of the tubes being wiped dry with tissues. The pellets are dissolved in a minimal volume of phosphate buffer and diluted to an A_{280} of approximately 250. The material is stored frozen in 2-ml portions.

The next step in the purification entails gel filtration on Sephadex G-200. In our early work it was necessary that this step be carried out swiftly to prevent activity losses during the chromatography run. For this reason we employed moderate size columns (2.5 × 30–40 cm), preferring to run several of these rather than a single larger column with a longer run time. The discovery that 2-pyrrolidone added to the buffer stabilizes the enzyme obviates the need for very fast runs, but we still find it convenient to adhere to our original scheme. The enzyme activity comes

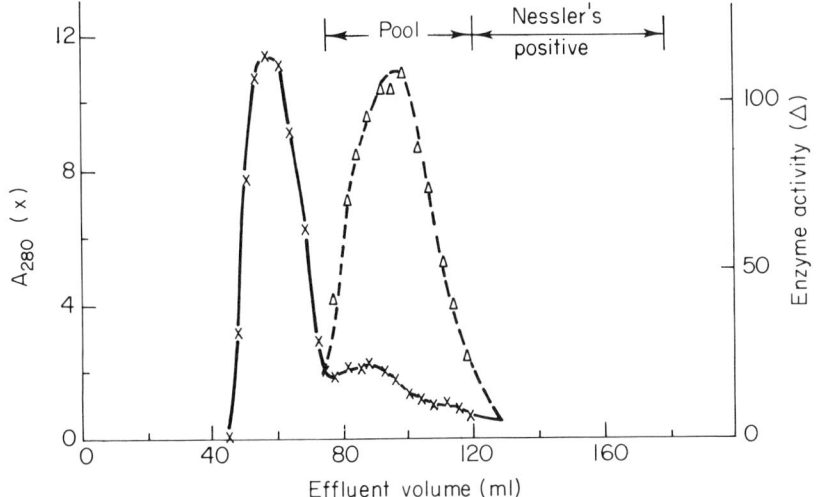

FIG. 1. Gel filtration on Sephadex G-200 (2.5×32 cm) of 42% ammonium sulfate cut.

behind the main protein peak on G-200 (Fig. 1). In fact, the procedure is so reproducible and the separation so dependable that we often omit assaying individual tubes, pooling the entire region from the end of the main A_{280} peak through to the beginning of the Nessler's positive material (ammonium sulfate). After removal of aliquots for assay and adsorbance measurement, the enzyme activity is precipitated from the pool by the addition of solid ammonium sulfate (0.45 g/ml). After at least 1 hour at 0°, the material is centrifuged and the pellets stored frozen. If no further purification is to be undertaken, the slurried pool (usually about 60 ml per column) is divided into several portions before centrifuging and the subpellets frozen separately (Table II). Enzyme preparations at this stage are generally quite suitable for digestion of PCA-peptides (see below), having a specific activity 20- to 40-fold that of the original crude extract (sonicate supernatant).

Further purification of the enzyme is readily accomplished by ion-exchange chromatography on DEAE-Sephadex (A-25). The pooled pellets from four or five G-200 columns (above) are dissolved in a minimal volume (ca. 5 ml) of $0.05\ M$ phosphate buffer, pH 8.0, and desalted either by rapid dialysis or gel filtration on Sephadex G-25. The enzyme activity is readily eluted at pH 8.0 ($0.005\ M$ phosphate) using a 0–3% NaCl gradient. The active tubes, which come early in the gradient, are pooled and solid ammonium sulfate added (0.6 g/ml) to precipitate the enzyme. Once again, if no further purification steps are

TABLE II
Suggested Isolation Scheme for Pyrrolidonecarboxylyl Peptidase

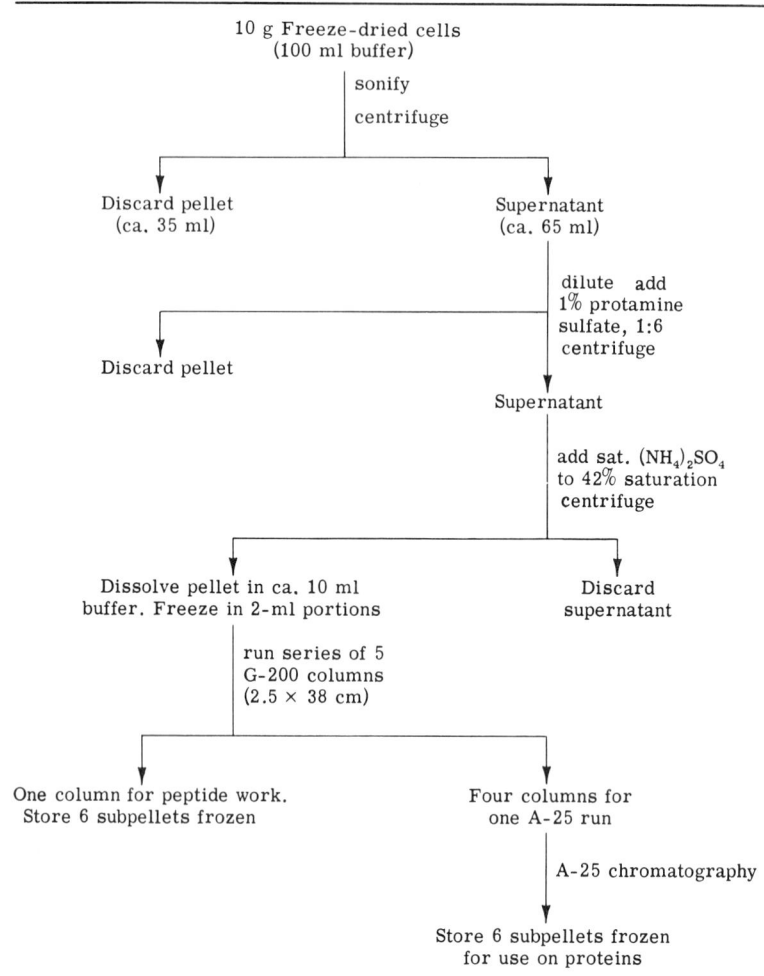

planned, it is convenient to divide the slurry into a half-dozen portions before centrifuging. The pellet or subpellets are stored frozen at −20°. Enzyme preparations at this stage of purification have been employed successfully in the removal of PCA residues from a large molecular weight protein without detecting any other proteolysis.[4] The material ranges from 60- to 100-fold purified over the crude bacterial extract.

Additional purification is not necessary if the enzyme is only to be used as a reagent in protein structure work. In fact, because the enzyme activity tends to become less stable with increasing purification, it is usually undesirable to carry the purification further. On the other hand, complete characterization of the enzyme necessitates purer material, and a Buchler Polyprep gel electrophoresis can be usefully employed at this stage. This approach takes advantage of the great mobility of the enzyme in acrylamide gels in Tris-borate buffer systems at pH 8.4. Suitable stabilizing agents must be added to the system; in particular, thioglycolate is substituted for mercaptoethanol. The active pools from these electrophoresis runs exhibit two bands when examined on analytical gel electrophoresis and are about 200-fold purified relative to starting crude extracts.[4]

Digestion of PCA-Peptides and PCA-Proteins

PCA-Peptides. Peptides suspected of having terminal PCA are usually studied with "G-200 enzyme." The designation "peptide" here includes any material which has a significant mobility on paper electrophoresis at pH 2.0, since this system offers a very convenient method of scanning a preliminary digest on a small amount of unknown material. At pH 2.0, peptides which lose a terminal PCA residue have their cathodic mobility significantly increased by the exposure of the (penultimate) α-amino group. Ninhydrin-negative peptides also become ninhydrin-positive, of course.

Typically, a frozen G-200 subpellet is thawed and dissolved in 1.0 ml 0.05 M phosphate buffer containing 0.01 M mercaptoethanol and 0.001 M EDTA. The material is then dialyzed against the same buffer to remove the residual ammonium sulfate and 2-pyrrolidone (the stabilizer is also an inhibitor). Usually three successive 1-hour dialyses against 250 ml buffer each are sufficient. Alternatively, the subpellet can be dissolved in 0.5 ml of the same buffer and passed over a small (1.0 $\times$ 10 cm) Sephadex G-25 column to remove the ammonium sulfate and the 2-pyrrolidone. The A_{280} of the final enzyme solutions will vary from preparation to preparation but is generally in the range of 2.0 to 4.0. The optimum concentrations of enzyme and substrate will vary with the particular peptide under study, and it is useful to make a preliminary examination of the reaction progress on a small amount of the peptide before attempting to prepare a large amount for subsequent sequential degradation or other follow-up characterization. To this end, about 0.1 micromoles of peptide is dissolved in 100 μl of the enzyme solution and incubated at 30°. Aliquots (20 μl each) are removed at 0, 2, 4, 8, and 20 hours and applied directly to electrophoresis paper to stop the reaction. The papers are then electrophoresed at pH 2.0 (8% acetic acid, 2% formic acid). We are pleased with our results on an LKB low voltage apparatus and use 300 V for 3 hours. On the other hand, any rig will suffice and we have used 2000 V for an hour with the same buffer in a Savant water-cooled plate apparatus.

If the original peptide contains lysine, ninhydrin staining can be used to follow the progress of the reaction. If the peptide contains arginine, the Yamada-Itano stain[10] is recommended because of its great sensitivity and the subsequent saving of material during the initial scan. If the peptide contains neither lysine nor arginine, then the chlorine gas method[11] is useful. One cannot depend only on the appearance of ninhydrin-positive material and still know when all the material has been digested.

Besides the change in mobility of the substrate peptide, successful removal of terminal PCA residues can be observed by following the appearance of the PCA itself. At pH 4.1 PCA has a mobility (anodic) twice that of free aspartic acid. We use a 0.1 M pyridine–acetic acid buffer, pH 4.1, 300 V, 2 hours and locate the PCA with the chlorine gas method.[11]

If the digestion appears satisfactory and complete in the pilot study, another subpellet is thawed, dialyzed, and proportionately more enzyme solution used to digest a substantial amount of the peptide for an appropriate length of time. The new species, without the terminal PCA, can be recovered by preparative electrophoresis or ion-exchange chromatography on Dowex 50-X2. Alternatively, the digestion can be terminated by the addition of pyridine in order that stepwise degradation using the phenylisothiocyanate method may be undertaken directly.

PCA-Proteins. Removal of terminal PCA moieties from proteins has been less well studied. In general we have favored sustained digestions with A-25 preparations of the enzyme. Removal of the PCA can be determined by various quantitative end group procedures directed toward the newly exposed penultimate amino acid. Enzyme incubations of 0, 8, and 24 hours at 30° are recommended. The digestions are stopped by the addition of the appropriate organic solvent, depending on which amino terminal procedure will be attempted (phenylisothiocyanate, isocyanate, fluorodinitrobenzene, dansyl, etc.). If availability of material permits, control incubations of the enzyme solution alone and protein alone should be run simultaneously. In our hands, "A-25 enzyme" solutions of $A_{280} = 1.0$–3.0 do not reveal significant quantities of endogenous end groups when incubated as controls. The same preparations readily remove the terminal PCA group from the β-chain of native bovine fibrinogen, however, exposing the penultimate phenylalanine residue. No other amino terminals are exposed during the digestion.[4]

[10] S. Yamada and H. A. Itano, *Biochim. Biophys. Acta* **130**, 538 (1966).
[11] F. Reindel and W. Hoppe, *Chem. Ber.* **87**, 1103 (1954).

Although the terminal PCA in native bovine fibrinogen is readily removed by pyrrolidonecarboxylyl peptidase, other proteins may present more difficult situations. It is generally accepted that the fibrinopeptide portions of fibrinogen molecules are on the surface of the protein. It might have been anticipated, then, that the terminal PCA would be accessible. In other proteins, however, the terminal PCA may be tucked more inside, and it may be necessary to unfold the target protein by suitable means before digestion with the enzyme.

Properties of Bacterial Pyrrolidonecarboxylyl Peptidase

Stability. Pyrrolidonecarboxylyl peptidase is a sulfhydryl-dependent enzyme and is easily poisoned by iodoacetamide and organomercurials. It is necessary to keep a reducing agent present in solutions employed for isolation steps as well as in the final mixture. The active sulfhydryl group can be reversibly blocked by reaction with tetrathionate under conditions similar to those used to block other sulfhydryl proteases.[12] Subsequent short-term incubations with mercaptoethanol restore full activity.

The bacterial enzyme is remarkably stabilized by the presence of 0.1 M 2-pyrrolidone (II).

$$
\begin{array}{c}
O \!\! = \!\! C \!\!-\!\! CH_2 \\
| \qquad | \\
HN \quad CH_2 \\
\diagdown C \diagup \\
H_2
\end{array}
$$

(II)

The substance, which was selected as a potential substrate analog, behaves as a noncompetitive, fully reversible inhibitor of the enzyme.[4] It must be dialyzed away before the enzyme is used for digestion purposes.

Generally speaking, the purer the preparation of pyrrolidonecarboxylyl peptidase, the more liable it is to decay during storage. Hence, "G-200 enzyme," frozen as pellets after precipitation with ammonium sulfate, is stable at −20° for several months. "A-25 enzyme" stored the same way loses about half its activity in 6 weeks.[4] Purer preparartions decay even more rapidly.[13]

Physicochemical Properties. Not very much physicochemical characterization has been accomplished on the enzyme because of the limitations on the amounts of highly purified material. Only those charac-

[12] P. T. Englund, T. P. King, L. C. Craig, and A. Walti, *Biochemistry* **7**, 163 (1968).
[13] R. W. Armentrout, Ph.D. thesis, University of California, San Diego, 1969.

teristics which can be ascertained on impure preparations are known with any reliability. The molecular radius,[14] as determined by gel filtration on G-200, is slightly larger than that of bovine serum albumin and corresponds to a molecular weight of 70–80,000. Armentrout has found that the comparable enzyme activity isolated from rat liver is retarded to a significantly greater degree on G-200 and apparently has a molecular weight of only half the bacterial enzyme.[13]

Bacterial pyrrolidonecarboxylyl peptidase has a substantial negative charge at physiological pH. Its mobility on 5% polyacrylamide gels at pH 8.4 is comparable to that of other moderately acidic proteins such as bovine serum albumin.[4]

Specificity. A consideration of specificity for pyrrolidonecarboxylyl

TABLE III

TABULATION OF AMINO ACIDS PENULTIMATE TO TERMINAL PYRROLIDONECARBOXYLIC
ACID RESIDUES AND RESULTS OF ENZYME ACTION

Penultimate amino acid	PCA removed by enzyme	Ref.
Alanine	Yes	a
Arginine	—	
Asparagine	—	
Aspartic acid	Yes	b
Cysteine	—	
Glutamine	—	
Glutamic acid	—	
Glycine	Yes	c
Histidine	Yes	a, c
Isoleucine	Yes	d
Leucine	Yes	d
Lysine	No	e
Methionine	—	
Phenylalanine	Yes	a, d
Proline	No	d
Serine	—	
Threonine	Yes	f
Tryptophan	—	
Tyrosine	Yes	d
Valine	Yes	a

[a] R. F. Doolittle and R. W. Armentrout, *Biochemistry* **7**, 516 (1968).

[b] H. Itano, personal communication, 1969.

[c] R. L. Jackson and C. H. W. Hirs, *J. Biol. Chem.* **245**, 624 (1970).

[d] J. A. Uliana and R. F. Doolittle, *Arch. Biochem. Biophys.* **131**, 561 (1969).

[e] L. J. Greene, personal communication, 1969.

[f] E. Appella, R. G. Maye, S. Dubiski, and R. A. Reisfeld, *Proc. Natl. Acad. Sci. U.S.* **60**, 975 (1968).

[14] G. K. Ackers, *Biochemistry* **3**, 723 (1964).

peptidase has two aspects. On the one hand, we want to know if it will attack all or most PCA-peptides, no matter what the nature of the nature of the adjacent amino acid. Second, we have to know if the enzyme is specific for the peptide bond linking PCA to other amino acids or whether this activity is only part of a much broader range of specificity.

With regard to the first aspect, the enzyme has successfully removed PCA from 10 different penultimate amino acids to date (Table III). It does not cleave L-pyrrolidonecarboxylyl L-proline (PCA-Pro) peptide bonds, and a failure has been reported in a situation thought to involve lysine as the penultimate amino acid.[15] The remaining 8 naturally occurring amino acids have not yet been tested.

The rate at which the enzyme hydrolyzes different PCA-dipeptides varies considerably. Of all the substrates tried so far, L-pyrrolidone-carboxylyl L-alanine (PCA-Ala) is the most readily attacked (Table IV). The second most easily hydrolyzed of the PCA dipeptides is PCA-Ile, which is attacked at 50% the rate of PCA-Ala. Next come PCA-Val, PCA-Leu, and PCA-Phe, all of which are hydrolyzed at about 20% the rate of PCA-Ala. Crude extracts exhibit the same order of relative activity, negating any notion that we have inadvertently purified one enzyme of a family by using PCA-Ala as our test substrate during the isolation procedure.

TABLE IV

RELATIVE INITIAL ACTIVITY OF PYRROLIDONECARBOXYLYL PEPTIDASE
DIRECTED TOWARD A VARIETY OF PCA-DIPEPTIDES[a]

PCA-dipeptide[b]	Percent of activity found against PCA-Ala		
	Expt. I	Expt. II	Expt. III
PCA-Ala	100	100	100
PCA-Ile	57	50	49
PCA-Val	20	22	24
PCA-Phe	NR[c]	14	14
PCA-Leu	17	19	19
PCA-Tyr	5	9	10
PCA-Pro	0	NR	0

[a] Adapted from J. A. Uliana and R. F. Doolittle, Arch. Biochem. Biophys. 131, 561 (1969).

[b] In all cases the final substrate concentration was approximately 8×10^{-3}, except in the case of PCA-Pro where it was $5 \times 10^{-3} M$.

[c] NR = not run in that particular experiment.

[15] L. J. Greene, personal communication, March, 1969.

The K_m for PCA-Ala is 2×10^{-3} (footnote 4); values for K_m have not yet been determined for most of the other dipeptides in the series. The enzyme is specific for L- isomers and does not hydrolyze D-PCA-L-Ala or L-PCA-D-Ala. Furthermore, the presence of either of these D-isomers does not significantly influence the rate of hydrolysis of L-PCA-L-Ala.

With regard to the strictness of specificity toward PCA-peptide linkages, it is perhaps premature to state rigidly that this is the only bond which the enzyme hydrolyzes. We know that "G-200 enzyme" preparations (20- to 40-fold purified) can be clean enough that no other peptide bonds are detectably cleaved in bovine fibrinopeptide B (21 residues), and A-25 preparations (60 to 100-fold purified) do not cleave other peptide bonds in native bovine fibrinogen (MW = 330,000). On the other hand, it is possible that the enzyme may cleave other terminal residues which lack α-amino groups. The *Pseudomonas* enzyme does not remove acetyl groups from N-acetyl tyrosine ethyl ester, but it has not been tested on N-acetylated or N-formylated peptides. One unnatural situation which may be attacked by the enzyme is that in which a carboxyamidecysteinyl residue exists at the amino terminal as a result of enzymatic or other cleavage after routine reduction and alkylation of a protein. We have demonstrated that at neutral pH carboxyamidecysteine cyclizes at a rate comparable to that of glutamine, forming a 6-membered ring equivalent to the 5-membered ring of PCA. We are currently synthesizing cyclized L-carboxyamidecysteinyl L-alanine in order to investigate this possibility.

Distribution and Function

The enzyme discussed in this article occurs in a strain of *Pseudomonas fluorescens* isolated by the author from a soil enrichment culture using L-pyrrolidonecarboxylic acid (PCA) as the sole source of carbon and nitrogen. The original intention was to search for an enzyme which would open the pyrrolidone ring by hydrolyzing the internal amide linkage. The discovery of an enzyme which actually removes the entire PCA residue from PCA-peptides was an unanticipated but highly desirable event, even though there is no apparent connection between the possession of pyrrolidonecarboxylyl peptidase activity and the method of isolation. It bears repeating, however, that the three other strains of *Pseudomonas* which have been tested do not have pyrrolidonecarboxylyl peptidase activity. *Escherichia coli* does not possess the enzyme either, but crude extracts of *Bacillus subtilis* (strain 23 WT) exhibit a similar ability to hydrolyze PCA-Ala. Recently the same activity has been reported in a number of other bacterial species, but absent in several others.[16] At this time nothing definitive is known about the function of the enzyme. The presence of substantial pyrrolidonecarboxylyl peptidase activity in rat liver further confuses the present picture, but suggests some tantalizing, albeit premature, notions with regard to protein biosynthesis.[4]

In summary, an enzyme preparation is available which can be used to advantage in structural studies on PCA-peptides and PCA-proteins. The enzyme not only allows the possibility of positively identifying the nature of many blocked α-amino groups, but also permits stepwise degradation techniques to be employed on peptides which were formerly invulnerable to such methods.

Acknowledgments

I would like to express my appreciation to R. W. Armentrout who undertook the final purification of pyrrolidonecarboxylyl peptidase as a part of his Ph.D. thesis, and to R. Chen, J. A. Uliana, and M. Weinstein, each of whom contributed to one or another aspect of the enzyme characterization. I am also grateful to Mrs. Louise Schmidt and Mr. Francis Lau for preparing vast quantities of bacteria and semipurified enzyme, much of which was shipped to other investigators throughout the world.

[16] A. Szewczuk and M. Mulczyk, *European J. Biochem.* **8**, 63 (1969).

[39] *Bacillus subtilis* Neutral Protease

By KERRY T. YASUNOBU and JAMES McCONN

Assay Method

Principle. The enzyme may be assayed using the casein digestion method or by the use of the synthetic substrate hippuryl-L-leucinamide.

Reagents for the Casein Digestion Method

1% Casein in 0.1 N NaCl adjusted to pH 7.3 with 1 N NaOH
Trichloroacetic acid, 0.1 M–NaOAc, 0.22 M–HOAc, 0.33 M

Procedure. The method of Kunitz[1] as modified by Hagihara[2] is used. A 1% solution of casein (Hammersten quality, reprecipitated with glacial acetic acid and repeatedly washed with acetone and ether) is dissolved in 0.1 N NaCl adjusted to pH 7.3 with NaOH. The reaction mixture

[1] M. Kunitz, *J. Gen. Physiol.* **30**, 291 (1947).
[2] B. Hagihara, H. Matsubara, M. Nakai, and K. Okunuki, *J. Biochem.* (*Tokyo*) **45**, 185 (1958).

contains 5 ml of the 1% casein solution and 1 ml of enzyme. Digestion is allowed to proceed for 10 minutes at 30°. After the addition of the 5 ml of the trichloroacetic acid solution and vigorous shaking, the mixture is allowed to stand at room temperature for 30 minutes and then filtered. The absorbancy is measured at 275 nm in a Beckman DU spectrophotometer.

Definition of Unit and Specific Activity. A unit of activity $[PU]_{275\,nm}^{casein}$ is defined as the enzymatic activity which gives the extinction at 275 nm equivalent to 1 μg of tyrosine liberated in 1 minute. The enzyme blank is prepared by the addition of 5 ml of the precipitating agent to 5 ml of the casein solution before the addition of 1 ml of enzyme. Specific activity is expressed as $[PU]_{275\,nm}^{casein}$/mg enzyme. The value is linear up to a total of 8 μg of pure enzyme.

Reagents for the Hippuryl-L-leucinamide Assay

  Hippuryl-L-leucinamide, 0.03 M in 0.1 M cacodylate buffer, pH 7.0 containing 0.004 M Ca(OAc)$_2$
  Ninhydrin reagent-(20 g ninhydrin, 3 g hydrindantin, 250 ml of 4.0 M acetate buffer, pH 5.5, and 750 ml of methyl Cellosolve)
  Acetate buffer, 4.0 M, pH 5.5

Procedure. For the assay, 2 ml of the 0.01 M hippuryl-L-leucinamide in 0.1 M cacodylate buffer, pH 7.0, containing 0.004 M Ca(OAc)$_2$ and 0.1 ml of the enzyme solution containing 4–10 μg of enzyme are added to several tubes. At 0, 3, 6, 9, 12, and 15 minutes, 0.2-ml aliquots are taken and added to 0.4 ml of the ninhydrin reagent and 0.2 ml of 4.0 M acetate buffer, pH 5.5. The resulting mixture is shaken and placed in a boiling water bath for 15 minutes, cooled to room temperature and 10 ml of 50% ethanol is added. The absorbancy of this solution is measured at 550 nm in a spectrophotometer. The blank is obtained from the zero-time sample. From a plot of the absorbancy reading (which can be converted to concentration terms by the standard curve obtained with leucinamide) vs. time, the rate of the reaction (expressed as absorbancy change per unit of time, moles per liter leucinamide formed per unit time, or percent hydrolysis per unit of time) is calculated. The assay yields linear results up to 10 μg of pure enzyme.

Purification of Enzyme

A partially purified form of the enzyme which is obtained from bacteria grown on solid media can be obtained from the Pacific Enzyme Laboratories, Honolulu, Hawaii. A similar preparation obtained from submerged culture techniques can be obtained from Monsanto Co., St. Louis, Mo.

The purification procedure to be described is modified from an earlier publication report[3] and the material from the Pacific Enzyme Laboratories was used. For the experiment, 30 g of dry powder was dissolved in 400 ml of 0.001 M Tris-maleate buffer which contains 0.002 M calcium acetate, the final pH of which was 6.4. The mixture is stirred gently for 30 minutes. The mixture was then centrifuged for 15 minutes at 18,000 g. DEAE-cellulose (equilibrated with starting buffer) is added so that the resin-to-protein ratio is 8:1 and the mixture is stirred for 30 minutes. After filtration through a large sintered glass funnel, acetone is added to 66% (final v/v) while the mixture is cooled gradually to −15°. The precipitate obtained by centrifugation at 8000 g for 10 minutes is dissolved in 150 ml of 0.001 M Tris-maleate buffer containing 0.002 M calcium acetate with a final pH of 6.4. The solution is then applied to a CM-cellulose column (3 × 45 cm) which is equilibrated with the starting buffer. After all of the enzyme is absorbed on the column, it is washed with 1.0–1.5 liters of the starting buffer. The enzyme is then eluted by 0.033 M NaCl. A second peak is eluted with 0.1 M NaCl but it is an impurity. A typical chromatogram is shown in Fig. 1. The results of the purification procedure are summarized in Table I. The elution of the enzyme can also be monitored by analyzing the eluant for zinc.

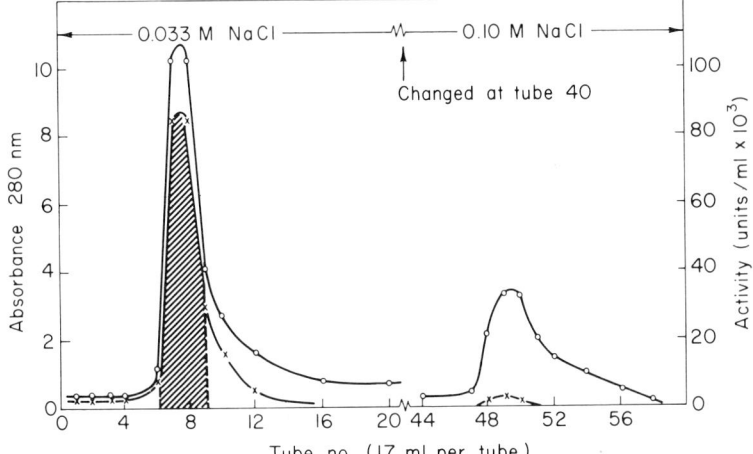

FIG. 1. Chromatography of the acetone precipitate on CM-cellulose. A CM-cellulose column (3 × 45 cm) was charged with 15.5 g of protein containing 2.4 × 10⁷ units. The flow rate was 40 ml per hour and fractions of 17 ml were collected. Stepwise elution was used after the column had been washed with 1.5 l of the starting buffer (0.001 M Tris-maleate + 0.002 M calcium acetate) pH 6.4. At the points indicated, 0.033 M and 0.1 M NaCl were used as eluants. The fraction shown by diagonal lines was collected, (○) and (×) represents absorbance at 280 nm and protease activity (units per milliliter, respectively).

[3] J. D. McConn, D. Tsuru, and K. T. Yasunobu, *J. Biol. Chem.* **239**, 3706 (1964).

TABLE I

DATA ON THE PURIFICATION OF THE *B. subtilis* NEUTRAL PROTEASE[a]

Procedure	Volume (ml)	Activity[a] (units/ml)	Protein (mg/ml)	Zinc (μg/ml)	Specific activity	Total units	Total protein	Yield (%)	μg Zn/mg protein	Purification
Original solution	400	12.9×10^4	118[b]	23.5	1090	51.4×10^6	47.2	100	0.20	—
DEAE-cellulose filtrate	880	3.35×10^4	21.3	7.65	1572	29.5×10^6	18.7	57.4	0.36	1.44
Acetone precipitate	150	2.3×10^5	55	30.1	4347	35.5×10^6	8.25	69.0	0.55	4.00
CM-cellulose 0.03 M eluant	475	2.61×10^4	2.03	4.67	11346	12.4×10^6	1.09	24.0	2.33	20.4

[a] The casein digestion assay was used to assay the enzyme.

[b] The actual weight of the starting material was 30 g. The protein values were determined using the $E_{1\ cm}^{1\%}$ at 280 nm of 13.6 which is the value for the pure enzyme.

Properties

Purity of Enzyme. Since the enzyme is quite prone to autolysis in the absence of calcium ions, all the purity studies were carried out in the presence of calcium ion. The purified enzyme was rechromatographed on CM-cellulose and a single symmetrical peak was observed when the eluate was analyzed for zinc, protein, or enzymatic activity.[4] A single symmetrical peak was observed in the analytical ultracentrifuge and likewise during the free-boundary electrophoresis experiments at several pH's. NH_2-Terminal amino acid and COOH-terminal amino acid analyses also point to the fact that the enzyme is very highly purified.[5]

Metalloenzyme Derivatives.[6] It is possible to prepare Cu, Hg, Pb, and Cd derivatives containing 1 mole of metal by exchanging these metals into the site containing zinc. In the case of the mercury derivative, it is possible to prepare a derivative containing 2 moles of Hg per mole of enzyme. The ability of the metalloenzyme derivatives to hydrolyze casein differed, but the ability of the derivatives to hydrolyze hippuryl-L-leucinamide is not greatly different from the zinc enzyme. It is possible that the metal is required for conformational stability of the enzyme.

Specificity. The enzyme hydrolyzes peptide bonds in which the —NHCHR— group is occupied by hydrophobic amino acid residues alanine, valine, isoleucine, norleucine, leucine, or phenylalanine.[7-10] An enhancement in the rate of hydrolysis was obtained by the presence of an alanyl, seryl, threonyl, or histidyl group as the amino acid whose carboxyl group formed the bond.[7] The enzyme also hydrolyzes peptide bonds in which proline contributed the carboxyl group.[9] In addition, there are other peptide bonds not following the specificity pattern discussed above, but by keeping the hydrolysis time short, it may be possible to prevent hydrolysis of these secondary specificity sites. The specificity pattern is very similar to that observed with thermolysin but differs slightly.[11]

Inhibitor Specificity. The enzyme is not inhibited by sulfhydryl reagents, diisopropylphosphofluoridate, TLCK or TPCK.[4] The enzyme is inhibited by metal chelating agents such as EDTA, *o*-phenanthroline, dithiozone, sodium cyanide, and sodium diethyldithiocarbamate as well

[4] J. D. McConn, Ph.D. thesis, University of Hawaii, Honolulu, 1965.
[5] D. Tsuru, J. D. McConn, and K. T. Yasunobu, *J. Biol. Chem.* **240**, 2415 (1965).
[6] J. D. McConn, D. Tsuru, and K. T. Yasunobu, *Arch. Biochem. Biophys.* **120**, 479 (1967).
[7] J. Feder, *Biochemistry* **6**, 2088 (1967).
[8] A. M. Benson and K. T. Yasunobu, *Arch. Biochem. Biophys.* **126**, 653 (1968).
[9] K. Morihara, T. Oka, and H. Tsuzuki, *Arch. Biochem. Biophys.* **132**, 489 (1969).
[10] K. Morihara, T. Oka, and H. Tsuzuki, *Arch. Biochem. Biophys.* **135**, 311 (1969).
[11] H. Matsubara, *Biochem. Biophys. Res. Commun.* **24**, 427 (1966).

TABLE II

SOME PHYSICOCHEMICAL PROPERTIES OF THE *B. subtilis* NEUTRAL PROTEASE

Property	Value
1. Molecular weight	$44,700 \pm 800$
2. $s_{20,w}^0$	3.24 ± 0.5 S
3. Partial specific volume	0.746 ± 002 ml g^{-1}
4. $E_{1\,cm}^{1\%}$ at 280 nm	13.6 ± 0.2
5. dn/dc at 546 nm	0.182 ± 0.003
6. Isoelectric point	8.95 ± 0.06
7. Nitrogen content	$15.9 \pm 0.2\%$
8. NH$_2$-terminal residue	Alanine
9. COOH terminal sequence	-Ala-Leu-COOH
10. Zinc content	1 g-atom Zn/mole enzyme
11. Specific activity	13,600 casein digestion units

TABLE III

AMINO ACID COMPOSITION OF *B. subtilis* NEUTRAL PROTEASE

Amino acid	Residues per mole[a]
Aspartic acid	61
Threonine	37
Serine	42
Proline	14
Glutamic acid	33
Glycine	37
Alanine	35
Valine	24
Half-cystine	0
Methionine	5
Isoleucine	18
Leucine	26
Tyrosine	29
Phenylalanine	14
Lysine	20
Histidine	7
Arginine	10
Ammonia	(35)
Tryptophan	4
Total	416

[a] Based on a molecular weight of 44,700. Average value from the 18, 24, 48, and 72-hour hydrolyzates. Serine and threonine values obtained by extrapolation to zero hydrolysis time. Maximum values were taken for valine, methionine, and isoluecine. Amide content determined by the Conway microdiffusion method.

as phosphate ions.[3] In addition, the enzyme is inhibited by high concentrations of many metal ions such as Cu^{2+}, Ni^{2+}, Hg^{2+}, Pb^{2+}, Cd^{2+}, Fe^{3+}, and Fe^{2+} in the $1 \times 10^{-4} M$ range.[6] Ethylenediaminetetraacetic acid is a potent inhibitor of the enzyme because it removes the zinc from the enzyme.[9]

Stability of Enzyme.[3] The presence of calcium and other metallic ions greatly stabilizes the enzyme from autolysis. In the presence of calcium ions, the enzyme is stable from pH 5.5–10 for several hours. Longer periods lead to some inactivation of the enzyme. The enzyme is stable up to 50° for 15 minutes at pH 7.4 in the presence of calcium ions. The enzyme can be stored in 60% acetone at −15° for long periods of time.

pH and Temperature Optimum.[3] When casein is used as the substrate, the pH optimum lies between 6.5 and 7.5. The temperature optimum occurs at about 57° in the presence of calcium ions.

Relative Activity.[3] The enzyme is one of the most active proteinases in digesting casein. The comparative specific activities for the *B. subtilis* neutral proteinase, papain, chymotrypsin, trypsin, and chymopapain with casein as a substrate were 13,600, 2550, 1813, 1653, and 850, respectively. K_m and k_{cat} were $0.0471 M$ and 20.0 sec^{-1} when the synthetic substrate hippuryl-L-leucinamide was hydrolyzed at 30° and at pH 7.0.

Physicochemical Properties.[5] Some of the properties of the enzyme are summarized in Table II. In addition, the amino acid composition of the enzyme is summarized in Table III.

Distribution

An enzyme with very similar properties has also been isolated from *Bacillus amyloliquifaciens*.[12] In addition, thermolysin[13, 14] has many properties, including substrate specificity, which are similar to the neutral protease.

[12] D. Tsuru, T. Yamamoto, and J. Fukumoto, *Agr. Biol. Chem. (Tokyo)* **30**, 651 (1966).
[13] S. Endo, *J. Ferment. Technol.* **40**, 346 (1962).
[14] H. Matsubara, *Biochem. Biophys. Res. Commun.* **24**, 427 (1966).

[40] Extracellular Proteinase from *Penicillium notatum*[1]

By MAKONNEN BELEW and JERKER PORATH

Assay Method

Principle. This assay[2] makes use of the general procedure for proteolytic enzymes using casein as substrate, with slight modifications.[1]

Reagents

Tris-acetate buffer, 0.05 M, pH 8.0, containing 0.02% (w/v) NaN_3
Trichloroacetic acid, 5%
Casein according to Hammarsten

The substrate solution is prepared by suspending 0.5 g of casein in 100 ml of the Tris-acetate buffer, heating the suspension in a boiling water bath for 10 minutes and finally adjusting the volume to 100 ml with buffer. Stirring of the suspension may be necessary to dissolve all the casein. The solution is cooled to room temperature before use or can be stored for about a week if kept stoppered in a refrigerator at +4°.

Procedure. To 2 ml of the substrate solution that has been preincubated at 37° for 5 minutes is added a suitable amount of enzyme preparation (maximum 100 μl). The mixture is incubated for exactly 10 minutes and then 3 ml of 5% trichloroacetic acid is added. The blank is prepared by first precipitating 2 ml of the substrate solution with 5 ml of the trichloroacetic acid solution and then adding the same amount of enzyme as used for the sample. The tubes are kept for about 30 minutes at room temperature and filtered. Occasionally, the filtrates may be cloudy, in which case they are filtered again. The absorbance of the filtrates is measured at 280 nm in a 1-cm cuvette using distilled water as reference. The enzyme activity is evaluated by using the difference in absorbance between the sample and blank.

Definition of Unit and Specific Activity. To follow up the purification procedure, it is not necessary (in this case) to have a generally acceptable unit for proteolytic activity. However, for purposes of comparison and convenience, an arbitrary unit is employed. One unit is defined as that

[1] The method outlined here has been described by us in *European J. Biochem.* 6, 425 (1968).
[2] W. Rick, *in* "Methods of Enzymatic Analysis" (H. Bergmeyer, ed.), p. 811. Academic Press, New York, 1963.

amount of enzyme which, after being incubated with casein under the conditions defined above, liberates hydrolysis products that give an absorbance of 0.50 at 280 nm in 10 minutes at 37°. The specific activity is calculated by dividing the units of proteolytic activity obtained by the absorbance value at 280 nm of the enzyme solution used.

Purification Procedure

The initial stages of the purification of this enzyme from a commercially available crude extract[3] are identical to steps 1 and 2 of the report by Pettersson and Porath in Vol. VIII (p. 603). The method of purification described below has proved to be easily reproducible in the authors' laboratory over a period of more than 1 year. All operations are carried out at 0° to +4° unless otherwise specified and all buffers contained 0.02% (w/v) NaN$_3$ as a bacteriostatic agent.

Step 1. Ammonium Sulfate Precipitation. The active material eluted with 0.1 M pyridine acetate buffer, pH 5.0, from DEAE-Sephadex A-25 chromatography (for details refer to Vol. VIII, p. 603) is combined. Solid (NH$_4$)$_2$SO$_4$ is added to 80% saturation, and the flocculent precipitate formed is kept at +4° for 4 hours or overnight. It is then centrifuged in a refrigerated centrifuge (0°) for 15 minutes at 10,000 g. The supernatant is discarded and the precipitate dissolved in the smallest amount of 0.1 M sodium acetate buffer, pH 4.5. The resulting solution is stored at +4° or deep-frozen at −15° until used. In a deep-frozen state, the preparation is stable for more than 6 months.

Step 2. Gel Filtration on Sephadex G-75. The following procedure is employed to prepare the gel bed.[4] The dry Sephadex gel is allowed to swell overnight in a large excess of 0.1 M sodium acetate buffer, pH 4.5. After removing finer particles by decantation, the suspension is deaerated and packed in a 3.5 × 95 cm column of the Recychrome type.[5] For large-scale operations, a large column (7.5 × 115 cm) is used which gives an identical separation pattern as the smaller one.

The proteinase extract (10 ml) is applied to the column by means of a peristaltic pump. Elution is carried out by pumping the buffer upward through the column at a flow rate of 20 ml/hour. Fractions of 6 ml are collected and the distribution of material analyzed spectrophotometrically at 280 nm. Activity for each fraction is also determined. Four distinct peaks appear and activity is localized solely to peak III which is eluted after about 450 ml of buffer has passed through the

[3] The tannin precipitated culture fluid from *P. notatum* is provided as a dry powder by Astra AB, Södertälje, Sweden.

[4] J. Porath, *Biochim. Biophys. Acta* **39**, 193 (1960).

[5] J. Porath and H. Bennich, *Arch. Biochem. Biophys.* Suppl. 1, 152 (1962).

column. The activity curve does not coincide exactly with the protein peak, indicating heterogeneity. The active fractions are pooled (about 130 ml) and precipitated with $(NH_4)_2SO_4$ (80% saturation). After 4 hours or overnight, the precipitate is centrifuged (as in step 1). The sediment is dissolved in a small amount of buffer (or distilled water) and kept at $+4°$ or deep-frozen at $-15°$ until used.

Step 3. Ion-Exchange Chromatography. The ion exchanger is prepared as follows: About 30 g of SE-Sephadex C-50 is allowed to swell in a large excess of distilled water overnight and fine particles removed by decantation. The swollen gel is transferred to a Büchner funnel and 2 liters of 0.5 M HCl passed quickly through the gel, followed by washing with about 2 liters of distilled water. It was then washed with 2 liters of 0.5 M NaOH, and again washed with about 2 liters of distilled water. Finally, the gel is continuously washed with about 10 to 15 liters of 0.1 M sodium acetate buffer, pH 4.5, with continuous stirring until the pH and conductivity of the washings and equilibrating buffer match exactly. The equilibrated ion exchanger is then packed into a column (3.5 × 30 cm) of the type described above. A larger column (7 × 30 cm) can be used for large-scale preparations since the separation pattern obtained is comparable to that obtained using the smaller column. If necessary, additional buffer is pumped through the column until the pH and conductivity of the effluent buffer matches exactly with that used for elution.

The active material obtained from step 2 is desalted on a column of Sephadex G-25 equilibrated with 0.1 M sodium acetate buffer, pH 4.5. Of the desalted sample, 20 ml is applied to the ion-exchange column. Since the gel shrinks with increase in the ionic strength of the eluant buffer, elution is done by pumping the buffer downward at a flow rate of 18 ml/hour. Fractions of about 6 ml are collected and after passage of 400 ml of buffer, the adsorbed material is displaced by using 500 ml of 0.3 M sodium acetate buffer, pH 4.5.

The distribution of material and activity in the fractions is determined. The active peak appears after passage of a total of about 500 ml of buffer. At this stage, the protein and activity curves correspond well. The fractions containing the proteinase are pooled (about 30 ml) and precipitated with $(NH_4)_2SO_4$ up to 80% saturation. The sediment obtained after centrifugation, as in step 1, is dissolved in a small amount of distilled water. The solution is desalted on a 1 × 50 cm column of Sephadex G-25 equilibrated with 0.025 M Tris-acetate buffer, pH 8.0. The fractions containing the active material are pooled and concentrated to a volume of about 3 ml in a collodium bag subjected to mild pressure on the outside. Care must be taken that the pressure is not very low since the activity will diminish considerably.

Step 4. Column Zone Electrophoresis. Cellulose powder[6] was used as a support and anticonvection media. It was packed in a glass column (2 × 50 cm) and in a manner similar to that described elsewhere.[7] The buffer used is 0.05 M Tris-acetate, pH 8.0. The homogeneity of the packed bed of cellulose is tested by passing a solution of a marker dye, dinitrophenyl-ethanolamine, through the column.

The buffer above the bed is removed. The sample (3 ml), which is equilibrated with 0.025 M Tris-acetate buffer, pH 8.0, so as to get a narrow starting zone,[8] is carefully layered on top of the bed. Electrophoresis is carried out for about 60 hours (anode direction downward) with an applied voltage of 800 V and with a current of 20 to 22 mA. The temperature of the cooling water circulating through the external jacket is +8°. After electrophoresis, the fractionated material is eluted with fresh buffer at a flow rate of 10 ml/hour and 2.5 ml fractions are collected. For each fraction the distribution of material and enzyme activity are determined. The enzyme activity is located in the major peak eluted between fractions 37–48. At this stage, the protein and activity peaks coincide exactly. The other peaks eluted before the proteinase peak are inactive and are comparatively low in concentration. A summary of the purification procedure is outlined in the table.

Properties

Stability. The enzyme is stable in the pH range of 4 to 10, but is inactivated below pH 3. It is not markedly affected by solutions of 7 M urea or 5 M guanidium chloride in the pH range of 5 to 8. The enzyme solution slowly loses its activity on storage at +4°, but at pH values around 5 it is stable for at least 2 weeks in the presence of $(NH_4)_2SO_4$. Lyophilization causes a marked decrease, or complete loss in its activity. It can be kept at room temperature for 48 hours without any detectable loss in activity.

Purity and Physical Properties. The purified enzyme did not contain protein impurities as tested by the following criteria of purity:

1. It showed only one band after electrophoresis in analytical polyacrylamide gels[9] and on agarose slides[10] at pH 4.5 and 8.0. The same result is obtained after electrophoresis in starch gels using 7 M urea[11]

---

[6] Obtained from Grycksbo Pappersbruk AB, Grycksbo, Sweden.

[7] J. Porath and S. Hjertén, *in* "Methods of Biochemical Analysis" (D. Glick, ed.), Vol. IX, p. 194. Wiley (Interscience), New York, 1962.

[8] H. Haglund and A. Tiselius, *Acta Chem. Scand.* 4, 957 (1950).

[9] S. Hjertén, S. Jerstedt, and A. Tiselius, *Anal. Biochem.* 11, 219 (1965).

[10] S. Hjertén, *Biochim. Biophys. Acta* 53, 514 (1961).

[11] G. M. Edelman and M. D. Poulik, *J. Exptl. Med.* 113, 861 (1961).

SUMMARY OF PURIFICATION PROCEDURE

Step	Material	Volume (ml)	Total absorbance (A_{280} units)	Total activity (units)	Specific activity (units/A_{280} units)	Yield (%)
1.	Ammonium sulfate precipitate (80%)	1.5	142.15	2640	18.6	100
2.	Gel filtration on Sephadex G-75	4.0	62.0	1752	28.3	66.4
3.	Ion-exchange chromatography on SE-Sephadex C-50	2.0	28.4	1053	36.9	39.9
4.	Zone electrophoresis	18.0	19.3	885	45.8	33.5

and at pH 8.0, but at pH 3.5 and under the same conditions, four bands are obtained.

2. It was monodisperse in the ultracentrifuge. Crystals are obtained when the enzyme is concentrated at pH 4.5 and at $+4°$.

3. It showed only one peak when run in an isoelectric focusing apparatus.[12]

The enzyme has a sedimentation constant of 1.6 and a molecular weight close to 20,000 which is in approximate agreement with that calculated from its amino acid composition. It is acidic with an isoelectric point of 4.92 as determined by the isoelectric focusing apparatus. The enzyme contains no carbohydrate.

pH Optimum. Using denatured hemoglobin as substrate, the pH optimum lies in the range of pH 7.5 to 9.5.

Activators and Inhibitors. Trypsin inhibitors from egg white and soy bean had no inhibitory effect. Other chemicals such as EDTA, Ca^{2+}, glutathione, or hydroxylamine did not activate or inhibit it. However, a $10\,M$ excess of diisoprophylphosphofluoridate (DFP) irreversibly inhibits the enzyme.

Specificity. The exact bond specificity of the enzyme is not known. However, it attacks different kinds of protein substrates such as denatured hemoglobin, albumin, and conjugated proteins such as ceruloplasmin and glycoproteins. It also attacks synthetic substrates, e.g., benzoylarginine ethyl ester. It hydrolyzes the oxidized B-chain of insulin to about 7 components not yet identified. Such results indicate that the enzyme is probably not specific in its action. A similar conclusion has been reached by Marshall *et al.*[13] when investigating the properties of a proteinase from a different strain of the same organism.

[12] O. Vesterberg and H. Svensson, *Acta Chem. Scand.* **4**, 957 (1960).
[13] W. E. Marshall, R. Manion, and J. Porath, *Biochim. Biophys. Acta* **151**, 414 (1968).

[41] Alkaline Proteinases from *Aspergillus*

By YASUSHI NAKAGAWA

It has been reported that *Aspergillus oryzae* produces three types of proteolytic enzymes with acid, neutral, and alkaline pH-activity optima, respectively.[1] Ichishima describes a procedure for the isolation of the acid proteinase from *A. saitoi.*[2] The following section deals with the isolation and purification of alkaline proteinase.

Recently, reports have been published on the isolation, purification, and characterization of alkaline proteinases from *A. flavus*,[3] *A. oryzae*,[4-6] *A. sojae*,[7] and *A. sydowi*.[8] These isolation and purification methods are similar to each other and two are shown here as examples. These enzymes have pH optima in the alkaline range,[9] but there are some differences in the inhibitory effects of metal ions, substrates specificity, and amino acid compositions. Further studies will be required to identify these alkaline proteinases.

Assay Method

Principle. The proteolytic activity is determined by measuring the increase in optical density at 280 nm of a trichloroacetic acid filtrate based on liberation of tyrosine and tryptophan by the enzymatic hydrolysis of hemoglobin or casein.[10]

In an alternate procedure, the trichloroacetic acid filtrate is allowed to react with the phenol reagent (Folin-Ciocalteau reagent) and the color developed is measured at 660 or 750 nm.

Reagents

> Hemoglobin solution: Prepared by a modification of the method by Anson.[11] "Hemoglobin substrate powder" (Worthington Biochemical Corp., Freehold, N.J.); 2 g is dissolved in 100 ml of 0.1 *M* sodium phosphate buffer, pH 7.5, and filtered through glass wool.
> Casein solution: 0.5% solution of casein (Hammarsten)[12]

---

[1] B. Hagihara, *in* "The Enzymes" (Boyer, Lardy, and Myrbäck, eds.) Vol. 4, p. 193. Academic Press, New York, 1960.

[2] E. Ichishima, this volume, [26].

[3] J. Turková, O. Mikeš, K. Gančev, and B. Boublik, *Biochim. Biophys. Acta* **178**, 100 (1969).

[4] R. Bergkvist, *Acta Chem. Scand.* **17**, 1521 (1963).

[5] A. R. Subramanian and G. Kalnitsky, *Biochemistry* **3**, 1861 (1964).

[6] A. Nordwig and W. F. Jahn, *European J. Biochem.* **3**, 519 (1968).

[7] K. Hayashi, D. Fukushima, and K. Mogi, *Agr. Biol. Chem. (Tokyo)* **31**, 1237 (1967).

[8] G. Danno and S. Yoshimura, *Agr. Biol. Chem. (Tokyo)* **31**, 1151 (1967).

[9] A pH-activity optimum occasionally changes depending on a substrate and a medium used (i.e., hemoglobin, synthetic peptides, or a type of buffer solution) or during the process of purification.[6]

[10] The use of casein as substrate is limited since the isoelectric point of casein is pH 4.6.

[11] M. L. Anson, *J. Gen. Physiol.* **22**, 79 (1938); J. H. Northrop, M. Kunitz, and R. M. Herriott, "Crystalline Enzymes," p. 303. Columbia University Press, New York, 1946.

[12] M. Laskowsky, Vol. II [2], [3].

Trichloroacetic acid solution: 5% (w/v) aqueous solution
Buffer solution: 0.1 M sodium phosphate solution, pH 7.5[13]

Procedure. A 2-ml aliquot of hemoglobin solution is added to a test tube (14 × 2 cm) and equilibrated in a 37° bath for 5 minutes. To the hemoglobin solution is added 0.5 ml of a properly diluted enzyme solution and the mixture is incubated at 37° for 10 minutes. At the end of the incubation period the reaction is terminated by addition of 4 ml of 5% trichloroacetic acid. After standing for 5 minutes the suspension is filtered through thick filter paper [e.g., Whatman No. 3 or S & S (Schleicher & Schuell) No. 576] or is centrifuged. The optical density at 280 nm of the clear solution is measured in a silica cuvette with 1-cm pathlength. A zero time incubation blank is prepared by reversing the sequence of the addition of the enzyme solution and the 5% trichloroacetic acid. For the blank solution 0.5 ml of water is substituted for the enzyme solution.

In the second procedure an aliquot of trichloroacetic acid filtrate is allowed to react with the phenol reagent[14] and the color developed is measured at 660 or 750 nm.[15] In this case the absorption must not exceed 0.8–0.9.[6] An 0.1 unit absorption is equivalent to 0.035 μeq of tyrosine/minute.[6]

The protein concentration is determined by the micro-Kjeldahl method or by absorption at 280 nm.

Definition of a Unit. The hemoglobin activity is conventionally expressed as a one-unit increase in the optical density at 280 nm/mg of enzyme per 10 minutes of incubation. This value can be converted to express the results in micromoles of tyrosine[6, 11] or leucine.[16]

Purification Procedure

Production of Enzymes[17]

An aqueous solution containing 1–1.5% soybean meal or peanut meal, 2–4% carbohydrate, and magnesium and phosphorus salts is adjusted

---

[13] G. Gomori, Vol. I, [16].

[14] The phenol reagent is commercially available from Fisher Scientific Co.

[15] Several modified procedures of this method were reported. O. H. Lowry, N. J. Rosebrough, A. L. Farr, and R. J. Randall, *J. Biol. Chem.* **193**, 265 (1951); W. Rick, *in* "Methods of Enzymatic Analysis" (H. U. Bergmeyer, ed.), p. 808. Academic Press, New York, 1963; C. E. McDonald and L. L. Chen, *Anal. Biochem.* **10**, 175 (1965).

[16] T. G. Rajagopalan, S. Moore, and W. H. Stein, *J. Biol. Chem.* **241**, 4940 (1966).

[17] "Fungal protease" prepared from *A. oryzae* is commercially available from Miles Chemical Co., Inc., Clifton, N.J.[5]

to pH 6.5 to 7.0,[4] or the following synthetic medium is also used[8,18]: glucose, 3%; milk casein, 0.5%; yeast extract, 0.1%; NH_4NO_3, 0.2%; K_2HPO_4, 0.1%; KH_2PO_4, 0.05%; $MgSO_4 \cdot 7\ H_2O$, 0.05%; $FeSO_4 \cdot 7\ H_2O$, 1 mg/100 ml; and $ZnSO_4 \cdot 7\ H_2O$, 1 mg/100 ml. One hundred milliliters of this solution is placed in a 500-ml flask and 1 g of sterilized $CaCO_3$ is added before inoculation.

After sterilization and cooling, the medium is inoculated with spores of *A. oryzae* (or other species) and incubated by shaking at 35° for 4 days. At the end of the incubation, the broth is filtered through four layers of gauze, followed by filtration through a 1-inch layer of Celite. The filtrate is used for the isolation of alkaline proteinase.

Isolation of the Enzyme

METHOD OF BERGKVIST[4]

The flow sheet shows the isolation steps.

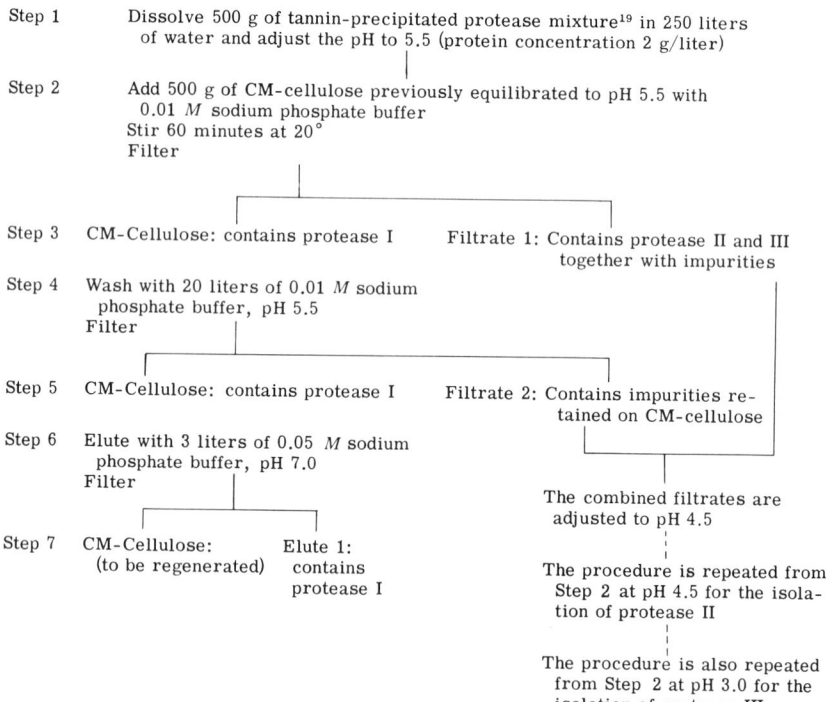

A Separation Method by Bergkvist[4]

Step 1   Dissolve 500 g of tannin-precipitated protease mixture[19] in 250 liters of water and adjust the pH to 5.5 (protein concentration 2 g/liter)

Step 2   Add 500 g of CM-cellulose previously equilibrated to pH 5.5 with 0.01 M sodium phosphate buffer
Stir 60 minutes at 20°
Filter

Step 3   CM-Cellulose: contains protease I   Filtrate 1: Contains protease II and III together with impurities

Step 4   Wash with 20 liters of 0.01 M sodium phosphate buffer, pH 5.5
Filter

Step 5   CM-Cellulose: contains protease I   Filtrate 2: Contains impurities retained on CM-cellulose

Step 6   Elute with 3 liters of 0.05 M sodium phosphate buffer, pH 7.0
Filter

Step 7   CM-Cellulose: Elute 1:
         (to be regenerated)   contains
                                protease I

The combined filtrates are adjusted to pH 4.5

The procedure is repeated from Step 2 at pH 4.5 for the isolation of protease II

The procedure is also repeated from Step 2 at pH 3.0 for the isolation of protease III

---

[18] A slightly different synthetic medium is used by W. G. Crewther and F. G. Lennox, *Australian J. Biol. Sci.* **6**, 410 (1953).

Finally each enzyme fraction is purified on a DEAE-cellulose column
(3 × 30 cm). Protease I is eluted from the column with an 0.01 *M*
sodium phosphate buffer, pH 6.0. Protease II is isolated from the impure
proteins by washing the column with 0.01 *M* sodium phosphate, pH 6.0,
and then 0.10 *M* sodium phosphate buffer, pH 6.0. Protease III is
purified by washing the column with 0.01 *M* and 0.1 *M* sodium phosphate
buffers, pH 6.0. It is eluted with 0.50 *M* phosphate buffers, pH 6.0.

METHODS OF SUBRAMANIAN AND KALNITSKY[5]

Step 1. Five hundred grams of "fungal protease"[17] is dissolved in
2 liters of ice-cold water with gentle stirring. Centrifuge the solution and
wash the precipitate twice with 200 ml of cold water. The combined
supernatant and washings are dialyzed against cold water for 24 hours.

Step 2. Three hundred grams of anion-exchange cellulose (ECTEOLA,
0.48 meq/g, Bio-Rad Laboratories, Richmond, Calif.) is added and stirred
for 20–30 minutes, and is subsequently filtered through a Büchner funnel
with slow suction to prevent foaming. The cellulose is washed twice
with 200 ml of water. To remove most of the pigments this step is
repeated using 200 g of ECTEOLA-cellulose (ECTEOLA-cellulose:protein = 3:1 was found to be the best).

Step 3. The volume of the filtrate is measured and ammonium sulfate is added to reach 75% saturation.[20] (It is not necessary to adjust
the pH of the solution.) After addition of ammonium sulfate, the solution
is kept at 0° for 6–10 hours, and is subsequently centrifuged. Ammonium sulfate is added to the supernatant to bring the ammonium sulfate
concentration to 85% saturation. The suspension is centrifuged and
the precipitated protein is dissolved in distilled water. The enzyme solution is then dialyzed against water until no sulfate can be detected
and is lyophilized.

Step 4. The enzyme is dissolved in 0.01 *M* sodium phosphate, pH 6.5
(protein concentration is adjusted to 1.5 g in 25 ml). The solution is
chromatographed on an Amberlite CG-50 column (40 × 6.5 cm). The
enzyme is eluted in a cold room with the same buffer at a flow rate of
180 ml/hour. The elution pattern is shown in Fig. 1.

[19] The enzymes are precipitated from the broth by the slow addition of 10%
tannin solution to make the final concentration of 3.0 g of tannin per 100 ml of
the filtrate at pH 5.5. After 2 hours the precipitated protein is recovered by
centrifugation. The tannin is removed by washing several times with acetone, and
repeating the centrifugation. The precipitate is dried in vacuum. This step retains
80–100% of the original proteinase in the culture filtrate.[4]

[20] A. A. Green and W. L. Hughes, Vol. I [10]. The required amount of ammonium
sulfate is also estimated by the formula given by E. A. Noltmann, C. G. Gubler,
and S. A. Kuby, *J. Biol. Chem.* **236**, 1225 (1961).

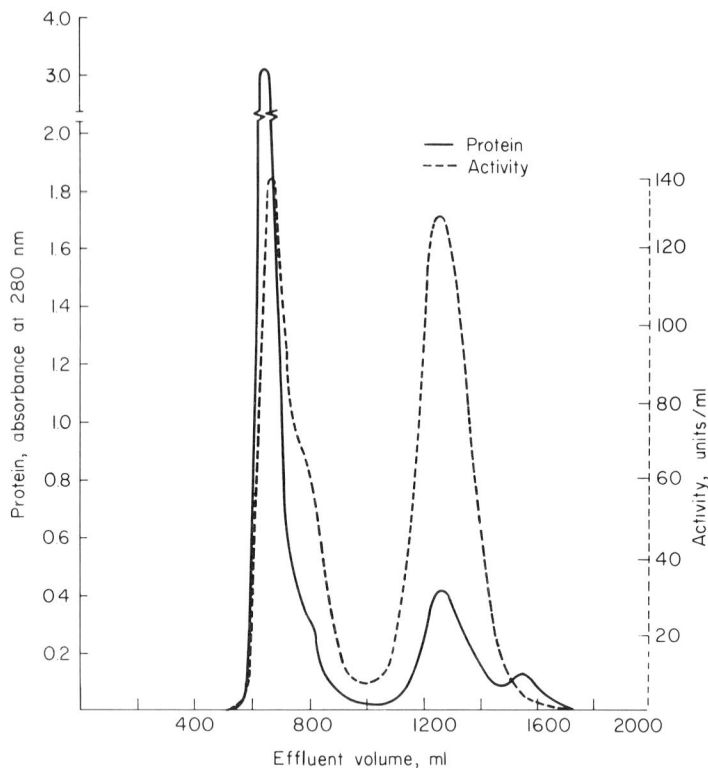

FIG. 1. Chromatography of aspergillopeptidase B on Amberlite CG-50, pH 6.5. A 40 × 6.5 cm preparative column was prepared in 0.01 *M* sodium phosphate buffer, pH 6.5. The enzyme was eluted with the same buffer at the rate of 180 ml/hour in the cold room. Reprinted from *Biochemistry* **3**, 1861 (1964). Copyright (1961) by the American Chemical Society. Reprinted by permission of the copyright owner.

Step 5. The Amberlite CG-50 column chromatography (30 × 0.8 cm) is repeated with 0.01 *M* sodium phosphate, pH 7.5 at a flow rate of 6.6 ml/hour. The chromatogram is shown in Fig. 2. The enzyme eluant is dialyzed against cold water for 24–36 hours and is lyophilized.

See Table I for a summary of this purification.

Properties

Stability. The alkaline proteinase (protein concentration of 0.002%) of *A. sydowi* is completely inactivated after incubation in 0.01 *M* sodium phosphate, pH 7.5, at 55° for 10 minutes.[8] The enzyme is most stable at pH 7.0, 35°.

Alkaline proteinase from *A. sojae* has a pH activity optimum at 9–10.[21]

[21] K. Hayashi, D. Fukushima, and K. Mogi, *Agr. Biol. Chem.* (*Tokyo*) **31**, 642 (1967).

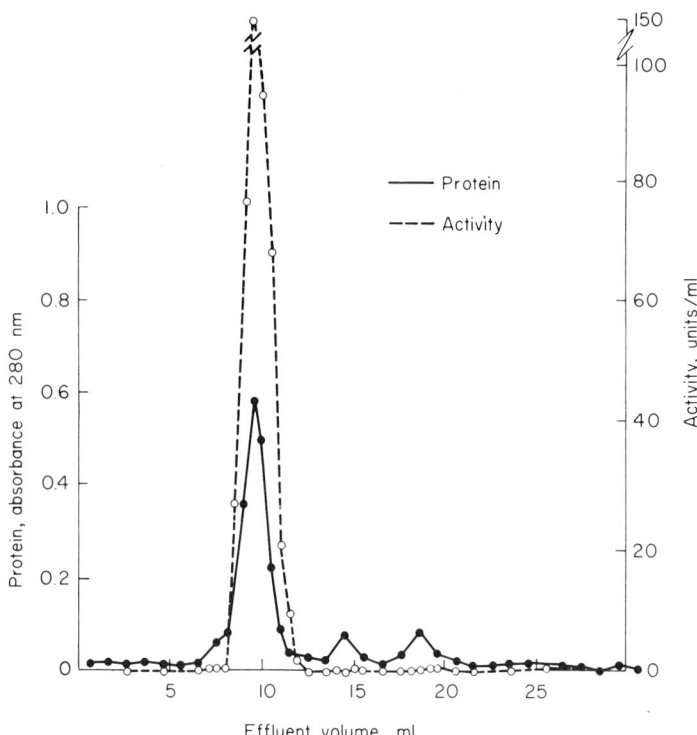

Fig. 2. Chromatography of aspergillopeptidase B on Amberlite CG-50, pH 7.5. A 30×0.8 cm column was prepared in $0.01\ M$ sodium phosphate buffer, pH 7.5. Proteinase was eluted with the same buffer at the rate of 6.6 ml/hour. Activity recovered, 96%; protein recovered, 96%. Reprinted from *Biochemistry* 3, 1861 (1964). Copyright (1961) by the American Chemical Society. Reprinted by permission of the copyright owner.

TABLE I

PURIFICATION OF ASPERGILLOPEPTIDASE B[5]

	Weight (g)	Total activity (units $\times 10^4$)	Purity (units/ $OD_{280\ nm}$)	Yield (%)	Purification
None	500	203	5.6	100	1
Step 1. Dialysis	150	199	8.0	98	1.5
Step 2. Anion cellulose	55	171	39	84	7.0
Step 3. AmmSO₄, 75–85%	6.5	75	106	37	19
Step 4. Amberlite $0.01\ M$ phosphate, pH 6.5	1.1	30	295	15	53
Step 5. Amberlite $0.01\ M$ phosphate, pH 7.5	0.9	28	310	14	55

Bergkvist reported that Protease I is stable between pH 5 and pH 8.5, Protease II has a broader stability range from pH 4.5 to pH 10.5, and Protease III is stable between pH 3 and pH 6.[22] Three enzymes are completely inactivated in less than 2 minutes at 60°, but lose their activities slowly at 40°.

Purity and Physical Properties. The purity of alkaline proteinase (aspergillopeptidase B) prepared by the method of Subramanian and Kalnitsky was examined by moving-boundary and paper electrophoresis, sedimentation velocity ultracentrifugation, and chromatography. These methods showed the preparation was homogeneous.[5] These techniques were also used to demonstrate the purity of alkaline proteinase from *A. sojae.*[7]

Polyacrylamide gel electrophoresis of the enzyme from *A. sydowi* showed a single band at pH 6.0, 7.6, and 8.6, respectively.[8]

The homogeneity of the alkaline proteinase from *A. flavus* was confirmed by disc electrophoresis in Tris-glycine buffer and by immunoelectrophoresis.[3]

The molecular weights reported for alkaline proteinases are as follows:

Alkaline proteinase from *A. flavus*[3]	18,000
Aspergillopeptidase B (*A. oryzae*)[23]	18,000
Aspergillopeptidase C (*A. oryzae*)[5]	19,650
Alkaline proteinase from *A. sojae*[24]	25,500

The optical rotatory dispersion of alkaline proteinase from *A. sojae* was measured at pH 7.0[24]:

$$[\alpha]_D^{20} = -16.5°$$
$$\lambda_c = 256 \text{ nm}$$
$$a_0 = -78$$
$$b_0 = -74$$
$$[M']_{200} = 11,700$$
$$[M']_{233} = -2,400$$

From the b_0 value the α-helix content was estimated as 12%. The apparent isoelectric pH of this enzyme is 5.1.[24]

Both alkaline proteinases from *A. flavus*[3] and *A. oryzae* (aspergillopeptidase B)[23] had an $E_{1cm}^{1\%}$ 9.04 and 9.0 at 280 nm and pH 5.0, respectively.

[22] R. Bergkvist, *Acta Chem. Scand.* **17**, 1541 (1963).
[23] A. R. Subramanian and G. Kalnitsky, *Biochemistry* **3**, 1868 (1964).
[24] K. Hayashi, D. Fukushima, and K. Mogi, *Agr. Biol. Chem. (Tokyo)* **31**, 1171 (1967).

TABLE II
AMINO ACID COMPOSITIONS OF ALKALINE PROTEINASES

Amino acid	Aspergillo-peptidase B (*A. oryzae*)[23]	Aspergillo-peptidase C (*A. oryzae*)[6]	Alkaline proteinase (*A. sojae*)[24]	Alkaline proteinase (*A. flavus*)[3]
Lysine	11–12	12	14	11
Histidine	4	4	5	3–4
Arginine	2	3	3	2
Aspartic acid	21	21–22	31	21
Threonine	11	13	18	11
Serine	19	23	28	20
Glutamic acid	12	13	19	12–13
Proline	4	5–6	6	4–5
Glycine	19	21	27	20
Alanine	23	23	32	23
Half-cystine	0	0	2	0
Valine	15	16	18	15
Methionine	0	1	2	1
Isoleucine	9–10	10–11	14	9–10
Leucine	9	10	14	9
Tyrosine	5	4–5	8	5
Phenylalanine	5	6	7	5
Tryptophan	2	2	2	2
Amide-NH_3	15	—	20	17
Total	171–173	187–191	250	173–177
N-terminal	Glycine	—	—	Glycine
C-terminal	Alanine	—	—	Alanine
Carbohydrate	1–2% as mannose	—	No sugar	1 mole as ribose per mole of protein

Chemical Properties. The amino acid compositions, N and C termini, and carbohydrate components are summarized in Table II.

The active site sequence of alkaline proteinase has been studied using ^{32}P-diisopropylphosphoryl derivatives. Thus the sequence for the alkaline proteinase from *A. flavus* is -Gly-Thr-Ser*-Met-Ala-,[25] and for that of *A. oryzae* it is -Thr-Ser*-Met-Ala-.[26]

Activator and Inhibitor. An activator for alkaline proteinase from *Aspergilli* has not been reported. However, the collagenase activity of aspergillopeptidase C increases with Ca^{2+}.[6] Also Ca^{2+} at concentration of 10^{-4} to $10^{-3} M$ protects a proteolytic activity of the enzyme from *A. sydowi* at pH 7.1 to 7.8.[27]

[25] O. Mikeš, J. Turková, N. Bao Toan, and F. Šorm, *Biochim. Biophys. Acta* **178**, 112 (1969).

[26] F. Sanger, *Proc. Chem. Soc.* p. 76 (1963).

[27] G. Danno and S. Yoshimura, *Agr. Biol. Chem. (Tokyo)* **31**, 1159 (1967).

Alkaline proteinases from *A. oryzae*, *A. sydowi*, and *A. flavus* have no free sulfhydryl groups and, therefore, are "serine enzymes."[28] They are not inhibited by sulfhydryl reagents, *p*-chloromercuribenzoate, cysteine, or KCN. However, they are inhibited by diisopropylphosphofluoridate, and *N*-bromosuccinimide. Nordwig and Jahn[6] showed the inhibition of aspergillopeptidase C by phenylmethanesulfonyl fluoride[29] and diphenylcarbamyl chloride[30] (inhibitors for serine-195 of *α*-chymotrypsin).

The potato inhibitor[31] is reported to be an inhibitor of alkaline proteinase from *A. sydowi*.[8] EDTA does not inhibit the enzymes which have been examined. (The collagenase activity of aspergillopeptidase C is decreased 40% by the addition of 1 m*M* of EDTA.[6])

The differences between these alkaline proteinases are observed in the inhibition of their activity by divalent metal ions. The results are summarized in Table III.

TABLE III
EFFECTS OF METAL IONS ON ALKALINE PROTEINASE ACTIVITY[a]

Metal ion	Alkaline proteinase (*A. sydowi*)[8] ($10^{-3}\,M$)	Aspergillopeptidase B (*A. oryzae*)[b]	Aspergillopeptidase C (*A. oryzae*)[6] ($10^{-2}\,M$)	Protease		
				I	II	III
					(*A. oryzae*)[22] ($10^{-2}\,M$)	
Ca^{2+}	95%		99%	95%	41%	80%
Ni^{2+}	84			54	17	66
Cu^{2+}	96	−	48	3	8	12
Zn^{2+}	83	−	22	10	7	3
Hg^{2+}	60	+	65			

[a] The activity is expressed in relative activity as percent of control.
[b] From A. R. Subramanian, S. Spadari, and G. Kalnitsky, *Federation Proc.* **24**, 593 (1965).

The proteinases prepared by Bergkvist are somewhat different from other alkaline proteinases. The caseinolytic activity of *Aspergillus* Proteases I and III were only slightly inhibited by ascorbic acid; however, Protease II was inhibited. EDTA and L-cysteine were inhibitors for Protease II, but these had no effect in Proteases I and III. Although

[28] M. L. Bender and F. J. Kézdy, *Ann. Rev. Biochem.* **34**, 49 (1965).
[29] A. M. Gold and D. Fahrney, *Biochemistry* **3**, 783 (1964).
[30] B. F. Erlanger and W. Cohen, *J. Am. Chem. Soc.* **85**, 348 (1963).
[31] K. Matushima, *J. Agr. Chem. Soc. Japan* **29**, 883 (1955); *Chem. Abstr.* **51**, 8838a (1957).

Proteases I and III were inhibited by sodium laurylsulfonate, Protease II was not. However, laurylamine inhibited all these proteinases.[22]

Source

As already mentioned, *A. oryzae, A. sojae, A. flavus,* and *A. sydowi* have been used as sources of mold alkaline proteinases.

In addition to these *Aspergillus* species, *A. fumigatus,*[32] *Penicilium cyaneo-fulvum,*[33] *Alternaria tenuissima,*[34] and *Gliocladium roseum*[35] produce alkaline proteinases.

Acknowledgments

The author wishes to acknowledge the kindness of Dr. G. Kalnitsky who made available the use of his figures. He also wishes to thank Dr. R. Bergkvist for giving permission to use his data.

[32] A. G. Jönsson and S. M. Martin, *Agr. Biol. Chem. (Tokyo)* **28,** 734 (1964).
[33] K. Singh and S. M. Martin, *Can. J. Biochem. Physiol.* **38,** 969 (1960).
[34] A. G. Jönsson and S. M. Martin, *Agr. Biol. Chem. (Tokyo)* **29,** 787 (1965).
[35] T. Kishida and S. Yoshimura, *Agr. Biol. Chem. (Tokyo)* **30,** 1183 (1966).

[42] Myxobacter AL-1 Protease

By RICHARD L. JACKSON and GARY R. MATSUEDA

In 1965 Ensign and Wolfe[1] reported the isolation of an organism which was capable of lysing cell walls of *Arthrobacter crystallopoietes.* From the culture media of the organism, called *Myxobacter* strain AL-1, a protease was purified which was able to hydrolyze the interglycan peptides in the cell walls of *A. crystallopoietes*[2] and *Staphylococcus aureus.*[3] For convenience the enzyme is called the AL-1 protease.

Assay Method

Due to an apparent substrate size requirement[4] (cf. Substrate Specificity, p. 598), a small soluble substrate has not yet been found for the AL-1 protease. Therefore, one must rely upon general methods of proteolytic assays.

[1] J. C. Ensign and R. S. Wolfe, *J. Bacteriol.* **90,** 395 (1965).
[2] T. A. Krulwich, J. C. Ensign, D. J. Tipper, and J. L. Strominger, *J. Bacteriol.* **92,** 734 (1967).
[3] J. M. Ghuysen and J. L. Strominger, *Biochemistry* **2,** 1110 (1963).
[4] R. L. Jackson and R. S. Wolfe, *J. Biol. Chem.* **243,** 879 (1968).

Cell Wall Lytic Assay: Turbidimetric[5]

At time zero an aliquot of suspended cell walls of *A. crystallopoietes* is added to a buffered enzyme solution. As the peptides linking the carbohydrate chains together are hydrolyzed, the reaction mixture clears.

Reagents

Tris-HCl, 0.05 M, pH 9.0, containing 10^{-3} M EDTA

A. crystallopoietes cell wall suspension with a turbidity of 2.0 at 660 nm. The cell walls are prepared by sonication of a wet cell mass. After centrifugation at 13,000 g, collection is accomplished by gently washing the white flocculent upper layer of the pellet into another centrifuge tube. The enriched cell wall fraction is resuspended in distilled water and recentrifuged. The last traces of intact cells disappear after collection and centrifugation have been repeated about ten times.

Procedure. To a culture test tube (13 $\times$ 100 mm), 0.2 ml of buffer solution and an aliquot of enzyme solution are added. The volume is then adjusted to 1.20 ml with distilled water. After 0.3 ml of the cell wall suspension is added, the timer is started as the turbidity at 660 nm is read. The blank, prepared with 0.2 ml buffer and 1.0 ml of water, is similarly treated. After a 15-minute incubation at 38° the turbidity of the reaction mixture is measured at 660 nm relative to the blank.

Specific Activity. A unit of enzyme activity is arbitrarily defined as that amount of enzyme required to decrease the OD_{660} by 0.001 absorbance units. One milligram of pure AL-1 protease will give approximately 175,000 units under the described conditions.

Azocoll Assay: Colorimetric[5]

Azocoll is collagen complexed with an azo dye to provide an insoluble substrate. Proteases digesting this insoluble substrate release color to the solution. The reaction is terminated by simply filtering off the substrate. The absorbancy of the filtrate measured at 580 nm is an index of proteolytic activity.

Reagents

Azocoll (Calbiochem) used without purification
Tris-HCl, 0.05 M, pH 9.0, containing 10^{-4} M EDTA

Procedure. The reaction is conveniently carried out in a 25-ml Erlenmeyer flask containing 1.0 ml buffer solution and 1.5 ml of enzyme

---

[5] J. C. Ensign and R. S. Wolfe, *J. Bacteriol.* **90**, 395 (1965).

solution. The reaction is started by adding 10 mg of Azocoll. After shaking for exactly 10 minutes at 38°, the reaction mixture is filtered through a sintered glass filter. The absorbancy of the filtrate is then measured at 580 nm relative to a blank to which no enzyme had been added.

Specific Activity. A unit of activity is arbitrarily defined as that amount of enzyme required to increase the OD_{580} by 0.001 absorbance units. One milligram of AL-1 protease will give approximately 130,000 units under the described conditions.

Purification

The *Myxobacter* strain AL-1 is mass cultured using standard aseptic techniques.[4] The organism is transferred from a solid stock culture to 300 ml of sterile 1% yeast extract (Difco) in a 500-ml Erlenmeyer flask to produce an actively growing culture after shaking for 24 hours at 30°. The culture is then transferred to a 3-liter Erlenmeyer flask containing 1 liter of sterile 1% yeast extract, and shaking at 30° is continued for 12 hours. Finally 15 liters of sterile 1% yeast extract in a 20-liter carboy is inoculated with the 1-liter culture while vigorously aerating the medium through a sterile glass-wool plug to two sterile Nalgene gas dispersion tubes at a rate of approximately 12 liters/minute. After 1 ml of sterile polypropylene glycol has been added as antifoam agent, sterile aeration is continued for 30–36 hours at 30°. The cells are subsequently harvested with a 10-liter Sharples centrifuge, and the effluent liquor containing the AL-1 protease is collected in the rinsed culture carboys and stored at 4°.

To the cooled growth liquor, about 60 ml of 3 M HCl is added dropwise until pH 6.7 is attained on a pH meter. Meanwhile, 350 ml of 10% (w/v) $ZnCl_2$ is heated on a steam bath to facilitate solubilization and is then filtered while hot. As this filtrate is added with stirring to the acidified growth liquor, certain proteins including the AL-1 protease begin to precipitate. Precipitation is completed with a 30-minute cooling period at 4°. The zinc precipitate is collected using the 500-ml Sharples centrifuge. A typical purification yields 27 g of zinc precipitate per 14 liters of growth liquor. At this point the zinc precipitate may be conveniently frozen at −20° until a sufficient supply has accumulated.

In the next step, 30 g of thawed zinc precipitate is first homogenized with 900 ml of 0.05 M Tris-HCl–$10^{-3} M$ EDTA, pH 9.0, with short bursts in a Waring Blendor. During a 30-minute slow stirring period, more of the contaminating proteins are solubilized. The remaining zinc precipitate should be then easily collected by centrifugation at 10,000 g. In the following crucial and sometimes temperamental citrate extraction step, the AL-1 protease is solubilized. First the collected zinc precipitate is roughly divided into thirds. Approximately one-third of the Tris-extracted zinc precipitate is returned to the blender for another quick homogenization with 900 ml of $0.15\,M$ Na citrate, pH 5.0. The homogenate is transferred to a beaker for 5 minutes of slow stirring at $0°$, after which the remaining chunks of zinc precipitate are allowed to settle for 5 minutes. To the decanted amber solution, solid $(NH_4)_2SO_4$ is added with gentle stirring at $0°$ to 70% saturation. After 5 minutes, the $(NH_4)_2SO_4$ precipitate which contains the AL-1 protease is collected at 9000 g. The other two portions of Tris-extracted zinc precipitate are similarly treated, and the ammonium sulfate precipitates from each of the three portions combined. The AL-1 protease is resolubilized by treating the combined ammonium sulfate precipitates with 250 ml of $0.025\,M$ Tris-HCl, pH 7.5, and stirring for 1 hour at $0°$. After centrifugation, the amber supernatant is dialyzed for 2 hours against 14 liters of 0.025 Tris-HCl, pH 7.5, in preparation for carboxymethyl cellulose batch chromatography.

The dialyzed solution is divided among three CM-cellulose columns $(3.5 \times 1.5$ cm) each equilibrated with $0.025\,M$ Tris-HCl, pH 7.5. After sample application, the columns are washed with the same buffer until the effluent OD_{280} is less than 0.1. The AL-1 protease-containing fraction is then eluted with $0.05\,M$ glycine-NaOH–$0.3\,M$ NaCl, pH 10.0. The eluants from the three columns are combined and the protein precipitated with solid $(NH_4)_2SO_4$ at 90% saturation. After centrifugation the ammonium sulfate precipitate is resuspended in 25 ml of $0.025\,M$ Tris-HCl, pH 7.5, and stirred until dissolved at $0°$. The solution is quantitatively transferred to a dialysis tubing and dialyzed for 2 hours against 10 liters of $0.025\,M$ Tris-HCl, pH 7.0, in preparation for the final column chromatography. Alternatively, the solution may be exhaustively dialyzed against distilled water and lyophilized for convenient storage. At this point, the slightly brown preparation has been purified 65 times over the growth liquor, and has a specific activity approaching 160,000.

To obtain a homogeneous enzyme preparation, gradient column chromatography is necessary. Lyophilized enzyme obtained from 15 liters of growth liquor is dissolved in 10 ml of $0.025\,M$ Tris-HCl, pH 7.0, and dialyzed against the same buffer for 4 hours. The dialyzed sample is quantitatively applied on an equilibrated low-capacity carboxymethyl Biogel-30 (Bio-Rad) column $(2.0 \times 20$ cm) and washed into the resin with the starting buffer, $0.025\,M$ Tris-HCl, pH 7.0. Immediately a pH gradient is started from $0.025\,M$ Tris-HCl, pH 7.0, to $0.025\,M$ glycine-NaOH, pH 10.0 using 400 ml of each buffer. When the pH gradient is complete, a linear salt gradient is immediately started from $0.025\,M$ glycine-NaOH,

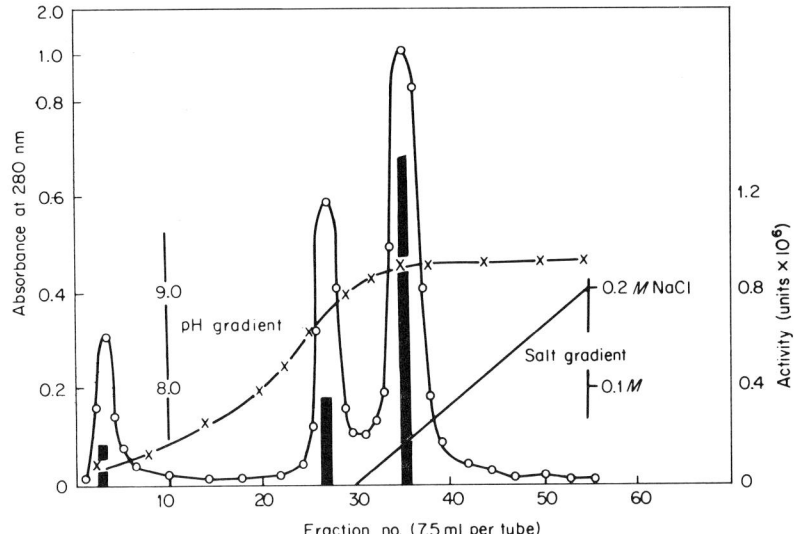

Fig. 1. CM-Biogel column chromatography of the CM-cellulose batch eluted AL-1 protease. For the experiment, 1.87×10^6 units of enzyme in 10 ml was applied to the column (1.5×11 cm) equilibrated with $0.025\,M$ Tris-HCl, pH 7.0. A pH gradient ($\times$) from $0.025\,M$ Tris-HCl, pH 7.0 to $0.025\,M$ glycine-NaOH, pH 10.0 was immediately started. Upon completion of the pH gradient, a salt gradient (——) was quickly started from $0.025\,M$ glycine-NaOH, pH 10.0 to $0.025\,M$ glycine-NaOH–$0.2\,M$ NaCl, pH 10.0. In this step, 100 ml of each buffer was used. Cell wall lytic activity (vertical solid bars) was determined with aliquots from each tube under the peaks of the protein profile ($\bigcirc$).

pH 10.0, to $0.025\,M$ glycine-NaOH–$0.2\,M$ NaCl, pH 10.0 using 400 ml of each buffer. For a typical run, 5 hours is required to complete each gradient.

The fraction corresponding to the last peak in Fig. 1 is now free from amber proteins and exhibits proteolytic as well as cell wall lytic activity. The pooled enzyme is dialyzed against distilled water until the dialyzate gives a negative chloride test with $0.03\,M$ AgNO$_3$. The dialyzed enzyme is then lyophilized and stored at $-20°$ over a desiccant.

Table I summarizes yield data for a typical purification which takes 3 working days. Four days are required for the mass culture and harvest to produce the growth liquor.

Purity and Physical Properties

Three criteria have been used to judge the homogeneity of enzyme preparations[4]: (1) electrophoresis on cellulose acetate strips at pH 7; (2) Ouchterlony double diffusion against rabbit antibody; and (3) poly-

TABLE I

SUMMARY OF PURIFICATION OF AL-1 PROTEASE

Stage of purification	Volume (ml)	Units/ml[a]	Total units	Total (mg)[b]	Specific activity	Yield (%)
Growth liquor	14,000	625	8.7×10^6	3,640	2,400	100
Tris-EDTA extraction of Zinc precipitate[c]	950	30	28,500	70	400	—
Tris-solubilized $(NH_4)_2SO_4$ precipitate	250	23,900	6×10^6	167	35,600	69
CM-cellulose column effluent and wash[c]	350	3,000	1×10^6	154	6,800	—
Eluant from CM-cellulose column	250	20,000	5×10^6	31	160,000	57
Eluant from CM-Biogel column	96	38,000	3.6×10^6	20	175,000	42

[a] Cell wall lytic assay.
[b] Protein determination method of Lowry et al.[e]
[c] Discarded fractions.
[e] O. H. Lowry, N. J. Rosebrough, A. L. Farr, and R. J. Randall, J. Biol. Chem. 193, 265 (1951).

TABLE II
PHYSICAL CHEMICAL PROPERTIES OF THE AL-1 PROTEASE[4,7]

Property	Value
$E_{1\ cm}^{1\%}$ at 280 nm	15.8
$s_{20,w}^{0}$	2.40 S
$D_{20,w}^{0}$	14×10^{-7} cm^2 sec^{-1}
$[\eta]$, intrinsic viscosity	1.66 ± 0.1 ml g^{-1}
(f/f_0), frictional coefficient	1.0
MW, amino acid composition	13,407
MW, sedimentation diffusion	14,340
Hexose	0.88 mole/mole protein
Metal content	Insignificant; Cu and Zn 0.27 moles/mole protein
N-terminal amino acid of carboxymethylated protein	Undetected by DNFB, DNS-Cl
C-terminal amino acid	Valine by hydrazinolysis

acrylamide disc electrophoresis at pH 6.6. Disc electrophoresis is routinely used to check the purity of peak tubes from gradient column chromatography.

Table II lists the physicochemical properties of the AL-1 protease. Table III contains amino acid composition data for the AL-1 protease.

Enzymatic Properties

pH Optimum. A sharp pH optimum at 9.0 is observed with both the cell wall lytic and Azocoll assays.[5]

Stability.[5] At 4° in 0.02 M Tris-HCl, pH 9.0, the AL-1 protease is stable for at least 8 hours and retains 93% of its cell wall lytic activity after 48 hours at 4°. At 35° under identical conditions, 87% of the original activity remains after 8 hours. From 40° to 50° enzymatic activity is slowly lost, such that 30% of the original activity remains after 8 hours at 50°.

From pH 6.5 to pH 9.5, the enzyme exhibits maximum stability when incubated for 1 hour at 45° in 0.02 M buffers ranging from pH 3.0 to pH 11.0. At pH 3, less than 5% of the initial cell wall lytic activity remains.

Inactivators.[5] Studies with inhibitors have not revealed any outstandingly potent enzyme inactivators. Heavy metal ions at approximately $10^{-3}\ M$ concentrations quench activity, but this is not surprising as zinc chloride at $10^{-2}\ M$ efficiently precipitates the enzyme in the purification scheme.

What is perhaps most revealing is the ineffectiveness of such compounds as *p*-chloromercuribenzoate, iodoacetic acid, diisopropylphospho-

TABLE III
AMINO ACID COMPOSITION OF THE AL-1 PROTEASE[4]

Amino acid	Assumed no. of residues	
Lysine	2	
Histidine	6	
Arginine	3	
Cys-SO$_3$[a]	2	
CM-Cys[b]	2	
Aspartic acid	16	
Threonine	11	
Serine	17	
Glutamic acid	7	
Proline	5	
Glycine[c]	17	
Alanine	10	
Valine	3	
Methionine	2	
Isoleucine	3	
Leucine	7	
Tyrosine	8	
Phenylalanine	4	
Tryptophan[d]	3	
Total	128	MW 13,500
Amides[e]	15	
Cysteine as CM-Cys[f]	0	

[a] Determined with performic oxidized protein.
[b] Determined with carboxymethylated protein.
[c] Determined with trichloroacetic-acid-precipitated protein.[7]
[d] Method of Spies and Chambers.[8]
[e] Method of Johnson[9] using amino acid analyzer.
[f] Determined by alkylation of unreduced protein with iodoacetic acid in 8 M urea and by method of Boyer.[10]

fluoridate (DFP) and dinitrofluorobenzene in the 10^{-4} M range and metal chelators like citrate and EDTA in the 10^{-2} M range.

Specificity. Experiments to date have not been able to establish any well-defined specificity. However, the data indicate an affinity toward peptide bonds formed by at least one hydrophobic amino acid in a structure at least as large as a tetrapeptide. A case in point: The Ala-Leu bond in the B-chain of insulin is hydrolyzed; but CBZ-Ala-Leu-NH$_2$

[7] G. R. Matsueda, unpublished data.
[8] J. R. Spies and D. C. Chambers, *Anal. Chem.* **21**, 1249 (1949).
[9] M. J. Johnson, *J. Biol. Chem.* **137**, 575 (1941).
[10] P. D. Boyer, *J. Am. Chem. Soc.* **76**, 4331 (1954).

is not.[4] Furthermore, pentaglycine is readily hydrolyzed; tetraglycine is hydrolyzed with difficulty, and tri- and diglycine are not attacked at all.[4] Such a size requirement would be useful in amino acid sequence work where large peptides must be degraded, preferably to tetra- and pentapeptides.

Distribution

Inhibition studies[5] suggest that the AL-1 protease mechanism of action is unlike any proposed by Hartley.[11] Present data eliminate mechanism of action proposed for the sulfhydryl proteinases, the acid proteinases, the metal proteinases, and the serine proteinases if sensitivity to DFP is a criterion. In this respect, the AL-1 protease is similar to the zinc containing β-lytic protease from *Sorangium*.[12]

[11] B. S. Hartley, *Ann. Rev. Biochem.* **29**, 45 (1960).
[12] D. R. Whitaker, L. Jurasek, and C. Ray, *Biochem. Biophys. Res. Commun.* **24**, 173 (1966).

[43] The α-Lytic Protease of a Myxobacterium

By D. R. WHITAKER

Introduction

The bacterium, "Myxobacter 495," which produces this enzyme was isolated in Ottawa during a study of Myxobacterales from local soil.[1,2] Its taxonomy is still under investigation,[3] but it has been reported to be a species of *Sorangium*.[2] Culture filtrates of the organism have strong bacteriolytic activity toward a number of other soil bacteria[2] and lyse soil nematodes.[1] Two extracellular, basic proteases, which have been designated the "α-lytic protease" and the "β-lytic protease" are responsible for most of the bacteriolytic activity.[4] The two enzymes can be readily distinguished: the β-enzyme is not a serine protease and, for example, on electrophoresis at pH 8 in cellulose acetate it migrates just behind egg-white lysozyme whereas the α-enzyme migrates slightly ahead of egg-white lysozyme. The organism produces several other proteases as well and some, like the α-enzyme, are serine proteases but they are

[1] H. Katznelson, D. C. Gillespie, and F. D. Cook, *Can. J. Microbiol.* **10**, 699 (1964).
[2] D. C. Gillespie and F. D. Cook, *Can. J. Microbiol.* **11**, 109 (1965).
[3] F. D. Cook, personal communication.
[4] D. R. Whitaker, *Can. J. Biochem.* **43**, 1935 (1965).

much less basic and have different sequences around their reactive serine residue.

The unique feature of the α-enzyme is that it is a bacterial enzyme with amino acid sequences around its reactive serine residue[5] and around its single histidine residue[6] which classes it as a homolog of the pancreatic serine proteases. Its enzymatic properties closely match those of porcine elastase.

Assay Methods

Three assay methods are listed below. The first two are not specific for the α-enzyme but are useful routine assay methods for production and purification work. The third method gives a precise measure of esterase activity.

Proteolytic Activity Toward Casein

Five milliliters of a 0.2% solution of casein (Hammarsten quality) in 0.025 M Tris buffer of pH 8.5 is mixed at 25° with 20–150 μg of enzyme in 1 ml of the same buffer. The reaction is stopped after 10 minutes by the addition of 10 ml of 5% trichloroacetic acid. The subsequent analysis with Folin-Ciocalteau reagent is exactly as described by Herriott.[7]

The color yield per 100 μg of α-enzyme is equivalent to a production of 3.4 micromoles of tyrosine per 20 mg of casein. At least five other proteases in the culture filtrate have caseinase activity.

Bacteriolytic Activity Toward Arthrobacter globiformis Cells[8]

Cells for this assay are obtained from 24-hour culture of *Arthrobacter globiformis* (Canada Department of Agriculture, Microbiology Research Institute, Strain No. 616) in a basal salts medium[9] supplemented by 1% sucrose and 0.2% casamino acids. The culture solution is centrifuged and the sediment of packed cells is thoroughly mixed with 11 volumes of acetone at −15°. The cells are left to settle at room temperature, washed with ether, and freed of solvent *in vacuo*. The dried cells are mixed with distilled water to give a thick suspension, and the treatment with acetone and ether is repeated.

For assays of bacteriolytic activity, 5.0 ml of a 0.03% suspension of *A. globiformis* cells in 0.025 M Tris–0.004 M KCl–HCl buffer of pH 9.0

[5] D. R. Whitaker and C. Roy, *Can. J. Biochem.* **45**, 911 (1967).
[6] L. B. Smillie and D. R. Whitaker, *J. Am. Chem. Soc.* **89**, 3350 (1967).
[7] R. Herriott, Vol. II, p. 3.
[8] D. R. Whitaker, F. D. Cook, and D. C. Gillespie, *Can. J. Biochem.* **43**, 1927 (1965).
[9] The salt composition is the same as that of the medium for Myxobacter 495 with the addition of CaCl₂ (0.2 g/liter) and NaCl (0.1 g/liter). [A. G. Lockhead and M. O. Burton, *Can. J. Microbiol.* **1**, 319 (1955).]

is mixed at 25° with 0.5–6 μg of α-enzyme in 200 μl of the same buffer. The absorbance per centimeter at 660 nm is measured at 5-minute intervals. The rate of change of the reciprocal of the absorbance, i.e., dA^{-1}/dt, is a linear function of the enzyme concentration.

The value of dA^{-1}/dt per microgram of enzyme varies with the cell preparation; a representative value is 0.01 reciprocal absorbance units per minute per microgram. The β-enzyme is the only other protease of Myxobacter 495 which seriously interferes with the assay.[4]

Esterase Activity Toward N-Benzoyl-L-alanine Methyl Ester[10]

The substrate can be prepared by the following modification of the method of Hein and Niemann.[11] Seventeen milliliters of methanol is cooled to −10° in an ice–salt bath and stirred continuously while 4.4 ml of thionyl chloride is added dropwise. Five grams of L-alanine is added in small portions. The temperature is slowly raised to 40° and maintained at 40° for 2 hours. During this period, the alanine dissolves completely. The methanol is then removed in vacuo. The residue of crude ester hydrochloride is dissolved in 20 ml of water. The solution is brought to pH 8.0 in a pH-stat by titration with 1 N NaOH and maintained at pH 8.0 while 6.6 ml of benzoyl chloride is slowly added. The N-benzoyl ester separates from the solution as a milky white liquid and is crystallized from high-boiling petroleum ether. Yield: 5.4 g (47%). Melting point: 58°–59°; $[\alpha]_D^{25°} = +30.8°$.

Esterase activity is measured in a pH-stat maintained at pH 8.0 with 0.02 N NaOH. Suitable reaction conditions are: Solvent, 0.10 M KCl; initial substrate concentration ($[S]_0$), $5 \times 10^{-3} M$; enzyme concentration ($[E]$), from 10^{-8} to $10^{-7} M$, i.e., 0.02–2 μg/ml. The rate of hydrolysis is directly proportional to $[S]_0$ and $[E]$. The second-order velocity constant,[12] i.e., moles of alkali consumed per liter per second/ $[E]_0 \cdot [S]_0$, is equal to 720 M^{-1} second^{-1}.

Production and Purification

Production

The production of enzyme in a 130-liter fermentor has been described[9] but substantial amounts of enzyme can be obtained with much simpler

[10] H. Kaplan and D. R. Whitaker, J. Am. Chem. Soc. 89, 3352 (1967); ibid., Can. J. Biochem. 47, 305 (1969).

[11] G. E. Hein and C. Niemann, J. Am. Chem. Soc. 84, 4487 (1962).

[12] As the overall kinetics are Michaelis-Menten kinetics, the second-order velocity constant is equal to k_{cat}/K_m where k_{cat} is the catalytic rate coefficient, i.e., $V_{max}/[E]$, and K_m is the Michaelis-Menten constant.

equipment.[13] The final stage of the procedure described below requires a rotary shaker capable of holding four 40-liter carboys, but it can be scaled down to the capacity of any rotary shaker.

Culture on Agar Slopes. Slopes of 0.2% Difco tryptone-1% agar are inoculated and incubated at 25° for 48 hours. The slopes should be used immediately if they are to provide the inocula for the first stage of shake cultures, but can be stored for several days at 4° if they are merely to provide the inocula for other slopes. If prolonged storage is required, cells from the slopes can be dispersed in sterile 15% glycerol and stored at −40°.[14]

Shake Cultures. The composition of the culture medium is as follows:

Casamino acids[15]	10	g/liter
Glucose	1	g/liter
K_2HPO_4	1	g/liter
KNO_3	0.5	g/liter
$MgSO_4 \cdot 7 H_2O$	0.2	g/liter
$FeCl_3 \cdot 6 H_2O$	0.01	g/liter
$ZnSO_4$	0.001	g/liter

The glucose is autoclaved separately; the solution containing the other components is adjusted to pH 7.0–7.1 with NaOH before it is autoclaved. It is immaterial whether technical grade or "vitamin-free" casamino acids are used, but if a "salt-free" grade is used, sodium chloride (3.5 g/liter) should be added to the medium.

The organism is grown at 25° in three stages of shake culture. The first two stages are in flasks with cotton-wool plugs. The third stage is in 40-liter carboys plugged by rubber stoppers pierced by air-inlet and air-outlet tubes, the former extending through the stopper to within a few inches of the surface of the medium. The air-inlet and outlet tubes are connected to air filters (glass bulbs packed with glass wool) which are autoclaved *in situ;* before entering the inlet filter, the air is scrubbed in a carboy containing 0.2% sulfuric acid and then freed of entrained acid in a spray trap. The rotary shaker for the third stage is set to describe a 1-inch circle at about 150 rpm.

The first stage of shake culture—24-hour growth in 500-ml flasks containing 75 ml of medium—provides the inoculum for the second stage—24-hour growth in 10-liter flasks containing 2 liters of medium.

[13] D. R. Whitaker, *Can. J. Biochem.* **45**, 991 (1967).
[14] M. T. Clement, *Can. J. Microbiol.* **10**, 613 (1964).
[15] Casamino acid is a neutralized, acid-hydrolyzate of casein.

The second stage provides the inoculum for two 40-liter carboys, each containing 11 liters of medium. Four such carboys are shaken for 48 hours with continuous aeration at approximately 1 liter of air per minute per carboy.

The medium is virtually exhausted of nutrients by the end of the third stage and growth is in stationary phase. The absorbance at 660 nm per centimeter of culture solution is 4–5.

Cell Removal. The solution is freed of cells by passage through a Sharples supercentrifuge. This operation should not be delayed unduly or the cells may autolyze.

The absorbance of the cell-free solution depends to some extent on the casamino acid preparation used in the medium; typical values for solutions from a "salt-free" casamino acid medium are: $A_{280\,nm} = 3.8$–4.1 per centimeter, $A_{260\,nm} = 3.6$–3.7 per centimeter. The pH is 8.1–8.5.

Purification

The procedures of this laboratory were designed to isolate the β-enzyme as well as the α-enzyme and might be simplified if only the latter were required. The two enzymes are displaced selectively from the cation-exchange resin, Amberlite CG-50. An earlier procedure[4] used a concentration gradient of buffer to effect this separation. It is a more informative procedure than the stepwise displacement described below but the latter is simpler and requires less attention.[13]

The Amberlite IR-45 and IR-120 resins used in the initial treatment of the filtrate are analytical grade, 16–50 mesh resins and require no special treatment, but the Amberlite CG-50 resin is 400–600 mesh and must be completely cleared of fines before it can be used in columns. Fines are removed in this laboratory by stirring the resin in a large volume of solution (40 liter per pound of resin) and leaving it to settle for various periods before the supernatant solution is discarded. A typical sequence of settling times is: 5 hours, 2½ hours, overnight, 2 hours, and 1½ hours in 0.5% NaOH–0.25% sodium citrate in tapwater; overnight in 2.5% acetic acid (in distilled water); 2 hours in 0.5% NaOH, overnight in water; 1½ hours in 1% HCl; 8 hours in 0.5% sodium citrate–0.05% Versene and overnight in 0.5% NaOH. The resin particles from this operation had a diameter of 29 ± 2 μ. The resin is equilibrated with a given buffer, e.g., acetic acid–0.10 N NaOH of pH 4.95 by dispersing the resin in a large volume of 0.10 N NaOH, adjusting the pH of the dispersion to 4.95 with acetic acid, redispersing the resin in a large volume of buffer, and readjusting the pH, if necessary, with acetic acid. A suitable support for the resin can be made from layers of acid-washed beads of graded diameter (e.g., 40 μ, 100 μ, and 500 μ Ballotini) above a small disk of porous polyethylene in the outlet tube of the column.

Step 1. Mixed-Bed Ion Exchange. Five hundred milliliters (settled volume) of Amberlite IR-45 (acetate) and 400 ml of Amberlite IR-120 (NH_4^+) are stirred in the cell-free culture solution and the dispersion is left to cool in a cold room (2°–5°). All subsequent steps are in the cold room. The dispersion is filtered through a Büchner funnel containing 100 ml of each resin on a pad of nylon cloth. The resin is drained and washed with a liter of water.

Step 2. Adsorption on Amberlite CG-50 at pH 5.0. The filtrate from step 1 is brought to pH 4.95 with 20% acetic acid (roughly 1 liter). Eight hundred milliliters (settled volume) of Amberlite CG-50, equilibrated with sodium hydroxide-acetic acid buffer which is 0.10 M with respect to sodium and of pH 5.00, is added in small portions with vigorous stirring and then left to settle overnight. The supernatant solution is removed and the resin is dispersed in 7 liters of the above-mentioned acetate buffer and again left to settle. Finally it is washed once more with 4 liters of acetate buffer.

Step 3. Titration and Washing of the Resin at pH 6.25. If several batches of shake culture are being prepared, it is convenient to combine two sets of resin from step 2 at this point. The following procedure assumes that this has been done. The resins from step 2 are dispersed in 3 liters of cold sodium hydroxide-citric acid buffer which is 0.033 M with respect to citrate and of pH 6.25. The dispersion is titrated to pH 6.4 by the addition, with stirring, of 0.5 M sodium hydroxide (approximately 1800 ml) through a capillary tube. The dispersion is added to a large column (12 cm diameter) which contains 1 liter of the above-mentioned citrate buffer above a 2-cm bed of resin which has been equilibrated with similar buffer. The column is left to drain until there is no supernatant solution above the settled resin.

Step 4. Stepwise Displacement of the α- and β-enzyme. The subsequent inputs to the column are:

(a) 4.2 liters of sodium hydroxide–citric acid buffer which is 0.160 M with respect to sodium and of pH 5.88

(b) 1 liter of sodium hydroxide–citric acid buffer which is 0.210 M with respect to sodium and of pH 6.00

(c) 4.5 liters of sodium hydroxide–citric acid buffer which is 0.270 M with respect to sodium and of pH 6.20

The last 3 liters of effluent from (a) plus the first 500 ml from (b) contain most of the β-enzyme with minor contamination by less basic proteases. The last 500 ml of effluent from (b) contain small and roughly equal amounts of both the α- and β-enzyme. The effluent from (c) contains practically all the α-enzyme with slight contamination by β-enzyme.

Step 5. Refractionation. The enzyme in the 4.6-liter eluate from buffer (c) is refractionated by the same procedure as was used for the initial fractionation. The solution is brought to pH 5.0 with acetic acid and mixed with 500 ml (settled volume) of Amberlite CG-50 in acetate buffer. The resin is washed with acetate buffer, dispersed in 0.33 M citrate buffer, and titrated to pH 6.4 with 0.5 M NaOH. The dispersion is added to a column containing two layers of resin: a lower layer (2 cm) of resin which had been equilibrated with 0.033 M citrate buffer of pH 6.25 and an upper layer (7 cm) of resin which had been equilibrated similarly but, before addition to the column, had been titrated to pH 6.4 with 0.5 M sodium hydroxide. The column is left to drain as before and the contaminating β-enzyme is displaced with 1 liter of the 0.160 M citrate buffer. The α-enzyme is then displaced with 6 liters of 0.270 M citrate. The effluent from this buffer is collected in 50–100 ml portions at the start and in 100–200-ml portions toward the end; effluent with an absorbance less than one-third the maximum absorbance at 280 nm is discarded; the remainder for the next step.

Step 6. Precipitation and Dialysis. The solution from step 5 is adjusted to pH 8 with dilute ammonia, brought to roughly 60% saturation with "enzyme grade" ammonium sulfate (455 g/liter of solution), and left in the cold room for 48 hours. The precipitated enzyme settles cleanly and most of the supernatant solution can be siphoned off; the remainder is removed by high-speed centrifugation. The sediment is dissolved in 0.1 M KCl and recentrifuged (30 minutes at about 20,000 g) to remove trace amounts of insoluble material. The solution is dialyzed in 3- to 4-foot lengths of ¼-inch dialysis tubing which are suspended in glass columns in which a slow downward flow of 0.1 M KCl is maintained for 24 hours, and then followed by a flow of glass-distilled water for 48 hours.

The resulting salt-free solution is freeze-dried. A typical yield is about 2 g.

Crystallization.[16] Crystals suitable for X-ray diffraction studies can be obtained by dialysis in capillary tubes such as those described by Zeppezauer *et al.*[17] The capillary, loaded with a 1% solution of enzyme in glass-distilled water, is immersed in 1.7 M ammonium sulfate which

---

[16] L. B. Smillie, personal communication.
[17] M. Zeppezauer, H. Eklund, and E. S. Zeppezauer, *Arch. Biochem. Biophys.* **126**, 561 (1968).

TABLE I

AMINO ACID SEQUENCE OF α-LYTIC PROTEASE[a]

	1	2	3	4	5	6	7	8	9	10	11	12	13	14	15	16	17	18	19	20
1	Ala	Asn	Ile	Val	Gly	Gly	Ile	Glu	Tyr	Ser	Ile	Asn	Asn	Ala	Ser	Leu	Cys	Ser	Val	Gly
21	Phe	Ser	Val	Thr	Arg	Gly	Ala	Thr	Lys	Gly	Phe	Val	Thr	Ala	Gly	*His*	Cys	Gly	Thr	Val
41	Asn	Ala	Thr	Ala	Arg	Ile	Gly	Gly	Ala	Val	Val	Gly	Thr	Phe	Ala	Ala	Arg	Val	Phe	Pro
61	Gly	Asn	Asp	Arg	Ala	Trp	Val	Ser	Leu	Thr	Ser	Ala	Gln	Thr	Leu	Leu	Pro	Arg	Val	Ala
81	Asn	Gly	Ser	Ser	Phe	Val	Thr	Val	Arg	Gly	Ser	Thr	Glu	Ala	Ala	Val	Gly	Ala	Ala	Val
101	Cys	Arg	Ser	Gly	Arg	Thr	Thr	Gly	Tyr	Gln	Cys	Gly	Thr	Ile	Thr	Ala	Lys	Asn	Val	Thr
121	Ala	Asn	Tyr	Ala	Glu	Gly	Ala	Val	Arg	Gly	Leu	Thr	Gln	Gly	Asn	Ala	Cys	Met	Gly	Arg
141	Gly	Asp	*Ser*	Gly	Gly	Ser	Trp	Ile	Thr	Ser	Ala	Gly	Gln	Ala	Gln	Gly	Val	Met	Ser	Gly
161	Gly	Asn	Val	Gln	Ser	Asn	Gly	Asn	Asn	Cys	Gly	Ile	Pro	Ala	Ser	Gln	Arg	Ser	Ser	Leu
181	Phe	Glu	Arg	Leu	Gln	Pro	Ile	Leu	Ser	Gln	Tyr	Gly	Leu	Ser	Leu	Val	Thr	Gly		

Disulfide bridges: 17–37; 102–112; 138–171

[a] M. O. J. Olson, N. Nagabushman, M. Dzwiniel, L. B. Smillie, and D. R. Whitaker, manuscript in preparation.

has been adjusted to pH 7.35 with ammonia. The deposition of crystals commences in about 10 days.

Amino Acid Sequence and Related Properties

The amino acid sequence of the enzyme is in Table I, the amino acid composition in Table II. The following points are noteworthy:

The sequence around the active serine residue, Ser_{144}, i.e., Gly·Asp· Ser·Gly·Gly, is identical with the active serine sequences of the pancreato-peptidases.

The sequence around the single histidine residue, His_{36}, i.e., Val·Thr· Ala·Gly·His·Cys·Gly is almost identical with the sequence around the catalytically functional histidine residue of chymotrypsin, i.e., Val·Thr· Ala·Ala·His·Cys·Gly.

Only one aspartic acid residue, Asp_{63}, is available to function in a charge-relay system such as that which Blow et al. have postulated for chymotrypsin.[18]

The molecular weight of the enzyme was estimated to be 19,100–19,300 from the rate of approach to sedimentation equilibrium and 19,100 from the elution volume for Sephadex G-75.[19] The molecular weight calculated from the sequence in Table I is 19,778.

It is apparent from the amino acid composition that the isoelectric point of the enzyme will tend to be very high. In Tris buffer of pH 9, the mobility is still strongly positive.

TABLE II

AMINO ACID COMPOSITION OF α-LYTIC PROTEASE

Amino acid	No. of residues[a]		Amino acid	No. of residues[a]	
Histidine	1	(1.06 ± 0.09)	Glycine	32	(32.2 ± 0.82)
Lysine	2	(2.03 ± 0.03)	Alanine	24	(23.89 ± 0.31)
Arginine	12	(12.08 ± 0.12)	Valine	19	(18.63 ± 0.22)
Aspartic acid	2 }	(15.75 ± 0.41)	Isoleucine	8	(8.00 ± 0.09)
Asparagine	13 }		Leucine	10	(9.95 ± 0.11)
Glutamic acid	4 }	(13.07 ± 0.13)	Phenylalanine	6	(5.96 ± 0.24)
Glutamine	9 }		Tyrosine	4	(3.95 ± 0.06)
Serine	20	(20.85 ± 1.22)	Tryptophan	2	2–4
Threonine	18	(16.47 ± 0.91)	Methionine	2	1.84 ± 0.04
Proline	4	(4.94 ± 0.31)	Half-cystine	6	5.95
			Amide groups	22	19.5

[a] The values in parenthesis are the estimates from the original amino acid analyses. [L. Jurášek and D. R. Whitaker, Can. J. Biochem. 45, 917 (1967).] The estimate for proline and the higher value for tryptophan are known to be overestimates.

[18] D. M. Blow, J. J. Birktuft, and B. S. Hartley, Nature 221, 337 (1969).
[19] L. Jurašek and D. R. Whitaker, Can. J. Biochem. 43, 1955 (1965).

Properties

Stability. Freeze-dried preparations can be stored at $-10°$ for at least 4 years without a detectable loss in activity. Frozen solutions of the enzyme are also stable although repeated freezing and thawing leads to losses in activity. The enzyme is subject to autodigestion if it is dissolved in concentrated urea at a pH above pH 5 and, if a chemical modification entails treatment with concentrated urea, the enzyme's initial exposure should be at a pH below 5.

Purity. Enzyme prepared by the foregoing procedures satisfies the following criteria. It is homogeneous (a) on electrophoresis in cellulose acetate in a variety of buffers ranging in pH from 3.0 to 9.3, including buffers containing 7 M urea,[19] (b) on electrophoresis in starch gel at pH 8.0,[19] (c) on electrophoresis at pH 4.3 in polyacrylamide gels prepared from 7.5% and from 15% solutions of polyacrylamide, with and without 8 M urea,[20] (d) on sedimentation in the ultracentrifuge in buffers of pH 5.0 and 8.0,[19] and (e) on chromatography in columns of hydroxylapatite and of Sephadex G-75.[19] A single N-terminal residue is detectable with 1-fluoro-2,4-dinitrobenzene,[10] no anomalous peptides were detected in the course of the determination of the sequence and the amino acid composition assigned by the sequence is in good agreement with the composition deduced from the original amino acid analyses of the enzyme (Table II).

Physical Properties. The enzyme is very soluble in distilled water. The sedimentation coefficient, $s_{20,w}$, is approximately 2.2 Svedberg units.[19] The ultraviolet absorption spectrum has no unusual features; the absorbance at 280 nm/cm of solution can be related to enzyme concentration by the equation,

$$[E] = 5.16 \times 10^{-5} \times A_{280}^{1cm}$$

where $[E]$ is the enzyme concentration in moles/liter.[19] The optical rotatory dispersion spectrum at wavelengths above 220 nm has one major Cotton effect with its trough at 230–232 nm and small Cotton effects with troughs at 290–291 nm and at 283–285 nm.[21] Its most notable feature is the low magnitude of the rotations, for example, $[M']_{230}$, the reduced mean residue rotation at the trough of the main Cotton effect, is approximately $-1650°$ cm²/decimole (cf. poly-L-lysine, where $[M']_{233} = -1800°$ when the content of α-helix is 0% and $[M']_{233} = -16,000°$ when it is 100%.) The ORD spectrum is not pH-dependent from pH 5 to 10.

[20] G. M. Paterson, personal communication. The buffer system used was that described by R. A. Reisfeld, O. J. Lewis, and D. E. Williams, *Nature* **195**, 281 (1962).

[21] G. M. Paterson and D. R. Whitaker, *Can. J. Biochem.* **47**, 317 (1969).

Crystallographic Properties.[22] The following data refer to crystals obtained by the procedure of L. B. Smillie.

Crystal system	trigonal
Crystal habit	hexagonal prism and dipyramid elongated along the C axis
Cell dimensions	$a = b = 66.5$ Å, $C = 80.2$ Å
Space group	P3$_1$21 or P3$_2$12

Activation and Inactivation. The enzyme does not require any specific activator. It is inactivated rapidly by diisopropylphosphorofluoridate (DFP) and isopropyl methylphosphonofluoridate (sarin), e.g., treatment of $2.4 \times 10^{-4} M$ enzyme at pH 7.7 with $6 \times 10^{-4} M$ DFP at 25° gives a 95% inhibition in 10 minutes.[5] Inhibition is accompanied by irreversible esterification of Ser$_{143}$. By analogy with the inhibition of chymotrypsin by chloromethyl ketones derived from phenylalanine and the inhibition of trypsin and subtilisin by chloro- or bromomethyl ketones derived from lysine, the α-enzyme's specificity would suggest that chloro- or bromomethyl ketones derived from neutral, aliphatic amino acids might be effective inhibitors. However, tests with chloro- and bromomethyl ketones derived from N-tosyl glycine and N-tosyl valine have been uniformly negative—even after prolonged incubation (18 hours) with up to a 100-fold excess of reagent.[10, 23] Tests on porcine elastase[24] were also negative.

Specificity. The bacteriolytic activity of the α-enzyme is derived from its ability to cleave the peptides which cross link chains of amino sugars in the cell walls of susceptible bacteria.[25] It is not a specific property of the α-enzyme; for example, porcine elastase can lyse *Arthrobacter globiformis* cells, its specific activity being about half that of the α-enzyme.[23] Conversely, the α-enzyme has appreciable activity toward elastin, its specific activity in assays with orcein-impregnated elastin[26] being from 30–50% of that of porcine elastase.[23]

Table III lists the amino acids on either side of linkages which are

[22] M. N. G. James and L. B. Smillie, *Nature* **224**, 694 (1970).
[23] H. Kaplan, V. B. Symonds, H. Dugas, and D. R. Whitaker, *Can. J. Biochem.* **48**, in press.
[24] References to porcine elastase refer to crystalline elastase prepared from "Trypsin 1-300" by Smillie and Hartley's modification of the method of Lewis *et al.* [L. B. Smillie and B. S. Hartley, *Biochem. J.* **101**, 232 (1966).]
[25] C. S. Tsai, D. R. Whitaker, L. Jurášek, and D. C. Gillespie, *Can. J. Biochem.* **43**, 1971 (1965).
[26] L. A. Sachar, K. K. Winter, N. Sichler, and S. Frankel, *Proc. Soc. Exptl. Biol. Med.* **90**, 323 (1955).

TABLE III

AMINO ACID RESIDUES (—RCO— AND —NHR) AT PEPTIDE LINKAGES
CLEAVED BY α-LYTIC PROTEASE[a]

—R·CO	—·NH· R—	
	Group 1	Group 2
	In A-Chain of Performate-Oxidized Bovine Insulin[b]:	
Ala		Ser_9
Ser	Val_{10}, Leu_{13}	
Val	Cys_{11} (SO_3H)	Glu_4
	In B-Chain of Performate-Oxidized Bovine Insulin[b]:	
Ala	Leu_{15}	
Ser		His_{10}
Val	Glu_{13}, $Cys_{19}(SO_3H)$	
Gly		Ser_9
Leu		Tyr_{16}
	In S-Aminoethyl α-Lytic Protease[c]	
Ala	Ala_{95}, Asn_2, Asn_{81}, Asn_{122}, Ser_{175}, Val_{96}, Arg_{45}	Val_{128}, Arg_{57}, Gln_{73}, Gln_{155}, Glu_{26}, Gly_{35}, Lys_{117}, Trp_{66}, Phe_{21},
Gly	Asn_{62}, Ala_{98}, Ala_{127}, Gly_6, Gly_{48}, Thr_{113}, Leu_{131}	Phe_{31}
Val	Asn_{41}, Arg_{89}, Val_{51}, Thr_{197}	Ala_{81}, $Cys_{101}(—(CH_2)_2NH_2)$, Gly_{20}, Gly_{52}, Gly_{97}
Ser	Asn_{166}, Gln_{190}, Leu_{69}, Leu_{195}	Leu_{180}, Ala_{72}, Ala_{151}, Gly_{104}, Gly_{160}
Thr	Ala_{34}, Ala_{116}, Gln_{133}, Val_{88}	Ala_{44}, Ala_{121}, Arg_{25}, Lys_{29}, Phe_{54}, Ser_{71}, Ser_{150}, Leu_{75}
Ile	Leu_{188}	
Leu		Phe_{181}
Met		Ser_{159}

[a] The experimental conditions for these digestions were as follows. (1) For digestion of the A- and B-chains of insulin[b]: $[S]_0 = 0.8$ mM, $[E] = 0.02$ mM, solvent: 0.025 M ammonium carbonate at 25°. Peptides were isolated by high-voltage electrophoresis. Cleavages are listed in Group 1 if the relevant peptides were major components of 10-minute digests and in Group 2 if the peptides were major components of 60 minutes but not of 10-minute digests.

(2) For digestion of S-aminoethyl α-lytic protease[c]: $[S]_0 = 0.5$ mM, $[E] = 0.02$ mM, solvent: 0.001 M ammonium chloride maintained at pH 8.0 and 25°, reaction period: 60 minutes. The peptides were isolated by extensive column chromatography and high-voltage electrophoresis. Cleavages are listed in Group 1 if the relevant peptides were isolated in yields exceeding 6% and in Group 2 if the yields were less than 6% but greater than 0.6%.

[b] D. R. Whitaker, C. Roy, C. S. Tasi, and L. Jurášek, *Can. J. Biochem.* **43**, 1961 (1965).

[c] M. O. J. Olson and L. B. Smillie, personal communication.

known to be cleaved during digestions of three polypeptides by the α-enzyme.

A comparison of the data for the insulin chains with that of Naughton and Sanger[27] for digestions by porcine elastase indicates that the α-enzyme is essentially elastaselike in specificity but it cleaves the chains more selectively than elastase. For example, the fast cleavages at Val_{10} and at Cys_{11} of the A-chain and at Glu_{13} and Leu_{15} of the B-chain are alternative, not consecutive, fast cleavages—if one linkage is cleaved, the other is cleaved much more slowly. The third substrate provides a much more thorough test of specificity. Clearly, linkages cleaved by the α-enzyme are linkages which involve the carbonyl group of neutral, aliphatic amino acids.

TABLE IV

ESTERASE ACTIVITY OF α-LYTIC PROTEASE AND PORCINE ELASTASE TOWARD METHYL ESTERS OF N-SUBSTITUTED AMINO ACIDS AT pH 8.0 AND 25.0° [23]

Acyl group of methyl ester	α-Lytic Protease			Porcine elastase		
	K_m (mM)	k_{cat} (sec^{-1})	k_{cat}/K_m (M^{-1} sec^{-1})	K_m (mM)	k_{cat} (sec^{-1})	k_{cat}/K_m (M^{-1} sec^{-1})
Benzoyl glycine			8.1			2.9
Benzoyl L-leucine			28			21
Benzoyl L-valine			230			19
Benzoyl L-alanine	35	23	720	20	17	640
R-oxycarbonyl-L-alanine						
R=phenyl	38	6.4	170	13	3.9	300
R=methyl	32	1.4	43	13	0.44	33
R=ethyl	26	10	390	22	0.99	46
R=n-propyl	20	13	650	21	3.5	160
R=n-butyl	23	9.4	410	21	5.5	260
R=isobutyl	26	11	430	36	7.8	220
R=t-butyl	8.8	11	1200	41	4.7	120

The same specificity is shown in hydrolyses of amino acid esters. Substrates for trypsin such as N-benzoyl arginine methyl ester and substrates for chymotrypsin such as N-acetyl tyrosine methyl ester and N-acetyl-tryptophan methyl ester are not hydrolyzed by the α-enzyme; neither is N-benzoyl-D-alanine methyl ester, the D-isomer of one of the best substrates of the enzyme.[10] As shown in Table IV, the enzyme is sensitive to the nature of the N-substituent; hydrolyses by the α-enzyme and porcine elastase show quite similar trends in the values for k_{cat} but

[27] M. A. Naughton and F. Sanger, *Biochem. J.* **78**, 156 (1961).

opposing trends are evident in the values for K_m. (See Note Added in Proof, p. 613.)

Kinetic Properties.[10] The more general kinetic properties of the enzyme between pH 4 and 10 can be summarized as follows:

(1) the Michaelis-Menten constant, K_m, is independent of pH

(2) the catalytic rate coefficient, k_{cat}, is dependent on an ionization with a pK of 6.7 in H_2O and of 7.3 in H_2^2O

(3) $\Delta(\log k_{cat}/K_m)/\Delta$pH for this ionization is equal to 1.0

(4) k_{cat}/K_m is reduced by 50% when H_2O is replaced by H_2O

(5) when *p*-nitrophenyl trimethyl acetate is hydrolyzed by the enzyme, the initial burst of *p*-nitrophenol is proportional to the enzyme concentration; the magnitude of the proportionality factor and the rate of attainment of a steady state are consistent with the condition, $k_2 > k_3$ where k_2 is the rate of acylation of the enzyme and k_3 is the rate of deacylation.

Property (1) is held in common with elastase[28] but not with chymotrypsin and trypsin. In agreement with the indication that substrate binding is uninfluenced by ionizations of α-amino or ϵ-amino groups, the α-enzyme and elastase retain their esterase activity when these groups are completely acetylated by acetic anhydride.[29] As previously mentioned, the optical rotatory dispersion spectra of the enzyme also shows no evidence of pH-dependent changes in the range of pH under consideration. Properties (2)–(5) are held in common with all the pancreatopeptidases; the first three are consistent with general basic catalysis by a single, unprotonated histidine residue; the last property may merely indicate that esters of *p*-nitrophenol and its derivatives are powerful acylating agents for all these enzymes.

The catalytic processes of the α-enzyme are quite efficient as evident from the values of k_{cat} in Table IV, and from the rate at which a steady state is attained when *p*-nitrophenyl trimethyl acetate is the substrate. For example, under reaction conditions which give a transient phase of roughly 400 seconds for chymotrypsin and 150 seconds for elastase, the transient phase for the α-enzyme is no more than about 40 seconds.[10]

Distribution. At present, Myxobacter 495 is the only bacterium known to produce an enzyme with such a combined functional and structural relationship to the pancreatopeptides. However the trypsinlike component of "Pronase," a protease preparation from *Streptomyces griseus*,

[28] H. Kaplan and H. Dugas, *Biochem. Biophys. Res. Commun.* **34**, 681 (1969).

[29] The enzymes were acetylated at 0° in a pH-stat, maintained at pH 6.8 while the acetic anhydride was slowly added, and then dialyzed. The acetylated enzymes must not be freeze dried.

has been shown recently to be a homolog of trypsin.[30] Like the α-enzyme, it has only one histidine residue.

Note Added in Proof

Narayanan and Anwar have recently demonstrated that porcine elastase is more selective than was previously thought. When thoroughly purified, it cleaves the B chain of insulin rapidly at two linkages—between the Ala_{14} Leu_{15} linkage and Val_{18} and cysteic acid$_{19}$. [A. S. Narayanan and R. A. Anwar, *Biochem. J.* **114**, 11 (1969).] The evidence in Table IV of a very high specificity for alanine supports their findings.

[30] L. Jurášek, D. Fackre, and L. B. Smillie, *Biochem. Biophys. Res. Commun.* **37**, 99 (1969).

[44] Collagenases[1]

By SAM SEIFTER and ELVIN HARPER

Native collagen + H_2O → peptides and/or larger fragments

Introduction

By definition, collagenases are enzymes that catalyze the hydrolytic cleavage of undenatured collagen. This is a broad definition permitting inclusion of enzymes that make either few or multiple scissions. The newer knowledge of the collagen molecule[1a] introduces novel considerations into a discussion of collagenases.

The basic collagen unit, the tropocollagen molecule, consists of three chains (α1, α2, and a second α1 or an α3), each of approximately 95,000 molecular weight. These are intertwined into the characteristic triple-stranded collagen helix. The structure is stabilized by hydrogen bonding, pyrrolidine residue stabilization, and occurrence of interchain covalent cross links. Among the latter, in certain vertebrate collagens, is an aldol condensation product of two residues of α-aminoadipic acid δ-semialdehyde. In certain invertebrate collagens, such as the collagen of the cuticle of *Ascaris*, disulfide bonds occur linking smaller peptide lengths into a larger structure of approximately 900,000 molecular weight.

Accordingly, depending on their mode of attack, collagenases con-

[1] This chapter was written while the authors were under the auspices of grants from the United States Public Health Service, AM 3172 and AM 3564. One of us (EH) is a Research Fellow of The Arthritis Foundation.
[1a] S. Seifter and P. M. Gallop, *in* "The Proteins," 2nd Ed. (H. Neurath, ed.), Vol. IV, p. 293. Academic Press, New York, 1966.

ceivably may cause proteolytic scissions in a single α-chain without necessarily causing collapse of the whole collagen structure. Alternatively, a collagenase could cleave simultaneously across three chains in a lateral fashion to yield segments of the triple-stranded structure retaining features of the collagen fiber observable in electron photomicrographs. At the other end of the range of action of collagenases is a group of enzymes, originally described for microorganisms but now also being detected in animal tissues, that cleave undenatured collagen more extensively, yielding small molecular weight peptides as well as certain larger ones.

The newer knowledge of the structure of collagen also explains the limited action of certain common proteases on this protein. The terminal portions of α-chains have characteristics differing from those of the main chain portions, i.e., they are almost globular and contain little if any of the specific collagen helix. Trypsin, chymotrypsin, pepsin, and pronase, among other common proteases, can promote a limited digestion of these ends without disrupting the main collagen structure. In this sense they could be classified as collagenases, but one prefers to state that they are proteases with a limited action on undenatured collagen. The true collagenases would then be required, by definition, to attack the characteristic helical regions of the collagen molecule.

Thus several enzymes are known that can attack undenatured collagens, and a range of specificities is encountered. On the other hand, almost all proteases can attack denatured collagens (gelatins), i.e., collagens in which the native triple-stranded structure has been melted out. The occurrence of proline and hydroxyproline regularly throughout the amino acid sequence of α-chains in gelatin may impose some restriction on the action of certain proteases, but these do not seem to be too severe. The actions of several exopeptidases, such as carboxypeptidase and amino acid peptidases, would be expected to be limited by the large content of imino acids in the substrate.

The specificities of some of the collagenases have been defined by the nature of terminal and penultimate amino acids appearing in peptides after cleavage of the collagen molecule. In the case of clostridial collagenase, and perhaps in the case of certain recently described animal collagenases, specificity has also been determined by action on synthetic peptides of known amino acid sequence. The synthetic peptides often resemble sequences occurring in α-chains of collagen, but the action of an enzyme on such peptides should not be used as a sole criterion for collagenases. The *sine qua non* for a collagenase must remain its capacity for attack on undenatured collagen, including attack on the portion of the collagen structure characterized by the polyproline-type of helix.

Another problem in definition of a collagenase arises from the nature of the collagen used for study. Although collagens throughout the animal kingdom have certain features in common, there are widely disparate aspects of their composition. Thus a single collagenase could act with different rates on different species of collagen, and in certain cases with a modified specificity.

A few words should be said about the inferred functions of collagenases. Most of the described collagenases are found as secreted extracellular enzymes. In higher organisms, some collagenases purportedly are lysosomal in origin, although again their action on tissue collagen may be extracellular. Generally speaking, the microorganisms producing collagenases are host-invasive, and presumably the enzymes contribute to pathogenecity by allowing the organisms to penetrate a connective tissue barrier. In this group of organisms are the Clostridia, *M. tuberculosis*, and certain fungi. In higher organisms, collagenases appear to be elaborated by specific cells involved in repair and remodeling processes; thus we have the classic example of a collagenase involved in resorption and remodeling of the back skin and tail of the tadpole. The elaboration of collagenases, whether by microbes as a mechanism of invasion or by higher organisms undergoing remodeling, generally is accompanied by elaboration of other enzymes capable of acting on different connective tissue components or membrane structures. Thus collagenases may occur together with elastases, specific peptidases, hyaluronidases, and phospholipases.

The collagenases have proved to be of considerable use to biochemists and biologists in general as investigative tools. In the field of collagen research, the highly specific clostridial collagenase has proved of inestimable value in mapping out the repeating nature of amino acid sequences in α-chains. It has been used to localize the aldehyde and carbohydrate functional groups of the tropocollagen molecule. The tissue collagenases have provided limited fragments of collagen amenable to sequence studies in connection with the cyanogen bromide cleavage technique. These fragments have helped to define the nature of collagen dissolution in repair and remodeling processes. Another important use of collagenases has been for the preparation of single cell types; for example, the employment of bacterial collagenase to liberate β-cells of the pancreatic islet tissue has aided in proving the elaboration of a proinsulin.

In accord with the definition of collagenases used here, we have suggested that definitive assays ultimately must be based on the use of undenatured collagen as a substrate. Generally, physical criteria for collagenolytic activity must be used; these represent either dissolution

TABLE I

COLLAGENASES AND RELATED ENZYMES

Enzyme	Source	Substrates and assays	Selected properties	References
Collagenase A (Clostridiopeptidase A: EC 3.4.19)	C. histolyticum	Collagen; gelatin; synthetic peptides, e.g., Z-Gly-Pro-Gly-Gly-Pro-Ala-OH. Assay with collagen is viscometric; all other assays can be with ninhydrin.	MW, 105,000; Ca dependent; Zn probably intrinsic; cleavage of substrate gives peptides with NH_2-terminal Gly.	(4–11)
Collagenase B	C. histolyticum	Same as Collagenase A	Except for MW of 57,400, properties similar to A above.	(7–11)
"Pseudocollagenase"	C. histolyticum	Gelatin, not collagen substrate; also attacks Z-Gly-Pro-Gly-Gly-Pro-Ala·OH.		(12)
Collagenase (two separable forms)	M. tuberculosis	Collagen, assayed by viscosity drop; Z-Gly-Pro-Leu-Gly-Pro-OH, by ninhydrin. Ca used in assay.	MW of one fraction, 77,000.	(13)
Collagenase[a] (two separable forms)	P. aeruginosa	Not tested against collagen, but active against Z-Gly-Pro-Gly-Gly-Pro-Ala·OH to give same products as does clostridial collagenase. Ca not used in assay. Acts on azocoll.		(14)
Collagenase	Bacteroides melaninogenicus	Collagen, by drop in viscosity	Particulate. Inhibited by EDTA and H_2O_2. Activated by cysteine.	(14a, 14b)

Collagenase	Streptomyces madurae	Collagen (drop in viscosity); azocoll; Z-Gly-Pro-Gly-Gly-Pro-Ala·OH and PZ-Pro-Leu-Gly-Pro-D-Arg·OH by ninhydrin assay	MW, 35,000; EDTA inhibits, reversed by Ca; irreversibly inhibited by cysteine and urea; inhibited by metal sequestration. No –SH groups present. Specificity like clostridial enzyme.	(15)
Collagenase	Trichophyton schoenleinii	Collagen (drop in viscosity); azocoll	MW, 20,000; pH optimum, 6.5; inhibited by EDTA, but not reversed by Ca or Mg; inhibited by cysteine. Not an —SH enzyme.	(16)
Collagenase (Aspergillopeptidase)	Aspergillus oryzae	Denatured hemoglobin; synthetic peptides; N-acetyl tyrosine ethyl ester; gelatin; native collagen (by drop in optical rotation)	MW, 20,000; inhibited by diisopropylphosphorofluoridate; not activated by metal ions; pH optimum, 9–10; is not so specific as clostridial collagenase, and much less active against native collagen.	(17)
Collagenase	Tadpole, Rana catesbiana	Collagen (by drop in viscosity); gelatin; synthetic peptides	Optimum pH, 8–9; inhibited by EDTA, reversed by Ca; inhibited by cysteine. Inhibited by serum.	(18-22)
Collagenase[a]	Complete tissue culture of mouse fibroblasts and of HeLa cells. Direct extracts made.	Assayed by splitting of yellow synthetic peptide, PZ-Pro-Leu-Gly-Pro-D-Arg·OH, which is split at -Leu-Gly- bond to give a yellow peptide extractable into organic solvents.		(23-25)

(Continued)

TABLE I (*Continued*)

Enzyme	Source	Substrates and assays	Selected properties	References
Collagenase	Human granulocyte, tissue culture, direct extraction.	Collagen, by drop in viscosity; ^{14}C-peptide release from gels.	Optimum pH, 7.6; inhibited by EDTA and by cysteine; not inhibited by serum.	(26)
Collagenase	Human skin tissue culture.	Collagen, by drop in viscosity; gel inhibition assay; ^{14}C-peptide release from gels.	Optimum pH, 7–8; inhibited by EDTA, cysteine and by human serum. Gives TC^A and TC^B fragments (75% and 25% respectively).	(27)
Collagenases (detected)	Tissue cultures of bones from man, goats and rats.	Collagen, by drop in viscosity; gel inhibition assay.	Optimum pH, 7–9; requires Ca; inhibited by EDTA and partially by cysteine.	(28)
Collagenase	Tissue culture of human rheumatoid synovium.	Collagen, by drop in viscosity; ^{14}C-peptide release from gels.	Optimum pH, 7.6; inhibited by EDTA and by human serum. Gives characteristic SLS fragments.	(29, 30)
Collagenases (miscellaneous)	Resorbing postpartum uterus (rat); diseased human skin; edges of healing wounds; growing rat bone; inflamed human gingiva; regenerating newt limbs.	Collagen, by drop in viscosity; gel inhibition.		(31–36)
Collagenase	Hepatopancreas of crab, *Uca pugilator*.	Collagen, by drop in viscosity.	Directly extracted. "Serine protease." Inhibited by DFP. Not inhibited by EDTA or cysteine.	(37)

[a] Presumptive collagenases not yet confirmed by action against native collagen.

[b] References cited in table:

[4] S. Seifter and P. M. Gallop, Vol. V, p. 659.

[5] A. Nordwig, *Leder* **13**, 10 (1962).

[6] S. Seifter, P. M. Gallop, L. Klein, and E. Meilman, *J. Biol. Chem.* **234**, 285 (1959).

[7] N. H. Grant and H. E. Alburn, *Arch. Biochem. Biophys.* **82**, 245 (1959).

[8] I. Mandl, S. Keller, and J. Manahan, *Biochemistry* **3**, 1737 (1964).

[9] E. Harper, S. Seifter, and V. Hospelhorn, *Biochem. Biophys. Res. Commun.* **18**, 627 (1965).

[10] E. Yoshida and H. Noda, *Biochim. Biophys. Acta* **105**, 562 (1965).

[11] E. Harper, Ph.D. dissertation, Albert Einstein College of Medicine, New York, 1966.

[12] W. M. Mitchell, *Biochim. Biophys. Acta* **159**, 554 (1968).

[13] S. Takahashi, *J. Biochem. (Japan)* **61**, 258 (1967).

[14] G. Schoellmann and E. Fisher, Jr., *Biochim. Biophys. Acta* **122**, 557 (1966).

[14a] R. G. Gibbons and J. B. MacDonald, *J. Bacteriol.* **81**, 614 (1961).

[14b] F. Hausmann and E. Kaufman, *Biochim. Biophys. Acta* **194**, 612 (1969).

[15] J. W. Rippon, *Biochim. Biophys. Acta* **159**, 147 (1968).

[16] J. W. Rippon, *J. Bacteriol.* **95**, 43 (1968).

[17] A. Nordwig and W. F. John, *European J. Biochem.* **3**, 519 (1968).

[18] C. M. Lapiere and J. Gross, *in* "Mechanisms of Hard Tissue Destruction." Publication No. 75, p. 663. Am. Assoc. Advance. Sci., Washington, D.C., 1963.

[19] Y. Nagai, C. M. Lapiere, and J. Gross, *Abstr. 6th Intern. Congr. Biochem.* **II**, p. 170 (1964).

[20] Y. Nagai, C. M. Lapiere, and J. Gross, *Biochemistry* **5**, 3123 (1966).

[21] A. H. Kang, Y. Nagai, K. A. Piez, and J. Gross, *Biochemistry* **5**, 509 (1966).

[22] J. Gross, *in* "Aging of Connective Tissue," Thule International Symposium, p. 1. Nordiska Bokhandelns Förlag, Stockholm, 1969.

[23] L. Strauch and H. Vencelj, *Z. Physiol. Chem.* **348**, 465 (1967).

[24] L. Strauch, *Umschau* **69**, 310 (1969).

[25] E. Wünsch and H.-G. Heidrich, *Z. Physiol. Chem.* **333**, 149 (1963).

[26] G. S. Lazarus, J. R. Daniels, R. S. Brown, H. A. Bladen, and H. M. Fullmer, *J. Clin. Invest.* **47**, 2622 (1968).

[27] A. Z. Eisen, J. J. Jeffrey, and J. Gross, *Biochim. Biophys. Acta* **151**, 637 (1968).

[28] H. M. Fullmer and G. Lazarus, *Israel J. Med. Sci.* **3**, 758 (1967).

[29] J. M. Evanson, J. J. Jeffrey, and S. M. Krane, *Science* **158**, 499 (1967).

[30] E. D. Harris, Jr., and S. M. Krane, *Abstr. Symp. Biochem. Physiol. Connective Tissue*, p. 124. Geoindustria, Prague, 1969.

[31] J. J. Jeffrey and J. Gross, *Federation Proc.* **26**, 670 (1967).

[32] A. Z. Eisen, *J. Clin. Invest.* **46**, 1052 (1967).

[33] H. C. Grillo and J. Gross, *Develop. Biol.* **15**, 300 (1967).

[34] D. G. Walker, C. M. Lapiere, and J. Gross, *Biochem. Biophys. Res. Commun.* **15**, 397 (1964).

[35] H. M. Fullmer and W. Gibson, *Nature* **209**, 728 (1966).

[36] H. C. Grillo, C. M. Lapiere, M. Dresden, and J. Gross, *Develop. Biol.* **17**, 571 (1968).

[37] A. Z. Eisen and J. J. Jeffrey, *Biochim. Biophys. Acta* **191**, 517 (1969).

of insoluble collagen or a change in one of the physical properties of collagen in solution. Dissolution can be quantified by examining the supernatant fluid either with a biuret-phenol reagent or by ninhydrin reactivity. Action on collagen in solution can be measured most conveniently by a decrease in viscosity as the enzyme digests the substrate, although change in optical rotatory activity could also be used.

A word of caution is required regarding assay of collagenases in the impure form.[2,3] Tissues elaborating a collagenase generally also contain peptidases that carry out the further degradation of peptides produced by the primary collagenolytic action. These peptidases ultimately should prove to have specificities for amino acid sequences found in α-chains, and their presence may be detected by action on synthetic peptides with sequences of this nature. Thus, in course of the preparation of the tadpole collagenase, one can separate a peptidase that cannot act on collagen but can cleave the synthetic peptides, Z-Gly-Pro-Gly-Gly-Pro-Ala-OH and PZ-Pro-Leu-Gly-Pro-D-Arg-OH, both characterized by sequences similar to those found in collagen.[2]

Table I lists a group of enzymes that may qualify as collagenases according to the criteria discussed above. Only a few of these have been isolated in fairly purified form, and the others have been detected by one or another of the assay procedures used for collagenases. The only collagenase available commercially, and indeed the only one prepared thus far in quantity, is the collagenase of *Clostridium histolyticum*.

In the following sections we describe: (1) a general method employed for screening and preparation of collagenases in a variety of animal tissues, namely a surviving cell tissue culture method; (2) the preparation and properties of purified tadpole collagenase obtained by use of the tissue culture method; and (3) the preparation and properties of the collagenases of *Clostridium histolyticum*.

Surviving Cell Tissue Culture Method

This method was introduced by Gross and Lapiere[38] in 1962. Direct extraction procedures for collagenases, although applicable in a few cases involving animal tissues,[26] frequently fail because the enzyme is either absent or present in minute quantities unless its synthesis is induced or stimulated. Extraction also may be difficult because of the tight binding of collagenases to insoluble matrices of collagenous material.

In the tissue culture method, the medium employed contains balanced salts, glucose, and antibiotics to limit bacterial contamination. When a

[2] E. Harper and J. Gross, *Biochim. Biophys. Acta* **198**, 286 (1970).

[3] H.-G. Heidrich, D. Prokopová, and K. Hannig, *Z. Physiol. Chem.* **350**, 1430 (1969).

[4-37] See Table I, p. 619.

[38] J. Gross and C. M. Lapiere, *Proc. Natl. Acad. Sci. U.S.* **48**, 1014 (1962).

tissue capable of producing a collagenase is maintained in this medium, the enzyme is detectable on the third day of incubation. The enzyme probably digests endogenous collagen in the tissue and is then leached into the medium.

The method in this or somewhat modified form has been used to demonstrate the production or preparation of specific collagenases in a number of tissues, including bullfrog tadpole tail fin, back skin and gills,[20] newt regenerating limbs,[37] goat bone,[28] rabbit and guinea pig wound tissues,[33] human bone,[28] human synovium,[29,39] cholesteotoma,[40] postpartum uterus,[31] and skin.[27] The general method is described below in its specific application for the preparation of bullfrog tadpole collagenase.[20]

Preparation of Collagenase from Bullfrog Tadpoles

Bullfrog tadpoles (*Rana catesbiana*), ranging in length from 3 to 5 inches, are exposed to a mixture of penicillin (300,000 units), streptomycin (0.25 mg), and chloramphenicol (100 mg) per liter of aquarium water for 24 hours prior to removal of back skin and tail fins under sterile conditions. These tissues are then cut into thin strips about 1 mm wide, spread upon filter paper disks floating in 10 ml amphibian Tyrode's solution in 105 mm petri dishes, and cultivated at 37° in a moist atmosphere containing 5% CO_2–95% O_2. The tissues from four animals fill one dish; the tail fins and back skins are kept separate. The culture medium has the following composition: NaCl, 8 g; KCl, 200 mg; $CaCl_2$, 200 mg; NaH_2PO_4, 50 mg; $NaHCO_3$, 1 g; glucose, 2 g; penicillin, 230,000 units; streptomycin, 200 mg; and water, 1500 ml. At intervals of 24 hours the fluid in each dish is assayed for collagenolytic activity and examined for bacterial contamination by use of a phase microscope and by bacterial culture. Contaminated cultures are discarded. Enzyme activity usually reaches a peak after 3 days of incubation, and the medium at this time is processed for preparation of the enzyme.

The medium is centrifuged at 25,000 g for 15 minutes at 2°. The medium is collected and replenished for each 24 hours subsequent to the first collection. The centrifuged medium is dialyzed against 0.01 M Tris buffer containing 0.001 M $CaCl_2$, pH 7.4. The material is then lyophilized.

Crude lyophilized powder (1 g) is dissolved in 50 ml of Tris buffer, 0.05 M, pH 7.4, containing 0.005 M $CaCl_2$ (temperature about 2°). The solution is centrifuged in the cold at 10,000 g for 10 minutes and the residue discarded. To the clear amber supernatant solution in an ice

[39] J. M. Evanson, J. J. Jeffrey, and S. M. Krane, *J. Clin. Invest.* **47**, 2639 (1968).
[40] M. Abramson, *Ann. Otol. Rhinol. Laryngol.* **78**, 112 (1969).

bath, solid ammonium sulfate is added to 20% saturation, and the preparation is allowed to stand for 30 minutes. Any visible precipitate is removed by centrifugation. More ammonium sulfate is added to 50% saturation and the suspension allowed to stand for 1 hour before removing the precipitate in the centrifuge at 10,000 g for 10 minutes. The supernatant solution is discarded. The precipitate is dissolved in 5–10 ml of Tris-CaCl$_2$ buffer, dialyzed against another Tris buffer, 0.01 M, pH 7.4, containing 0.2 M NaCl and 0.005 M CaCl$_2$. The material is then centrifuged to remove any undissolved residue.

Gel filtration is then carried out on this solution. This is accomplished in a water-jacketed column (2°) of Sephadex G-200 or 8% Agarose. The column dimensions are 1.5 × 65 cm. Tris buffer, 0.01 M, pH 7.4, containing 0.2 M NaCl and 0.005 M CaCl$_2$, is used for elution. The effluent is monitored at 230 and 280 nm, respectively. Collagenolytic activity is determined for each tube, and the active contents are pooled.

The final purification steps require the use of the gel slab electrophoresis method of Raymond[41] using the buffer and gel systems of Davis[42] and Ornstein,[43] but substituting riboflavin (0.0005%) for ammonium persulfate as described by Brewer[44] and by Fantes and Furminger.[45] The enzyme solution obtained after the gel filtration step is layered on the gel slab and run at 4° and 10 mA. The slab is then cut into slices at the left and right ends and in the center to provide guide strips for staining. The removed strips are stained with amido black, washed and then returned to their positions in the gel slab to permit marking out of the protein bands. The protein bands from the unstained portions are removed and eluted with Tris-CaCl$_2$ buffer of pH 7.4. The materials are then assayed for collagenolytic activity.

The enzyme has also been purified using DEAE-cellulose columns,[20] but appears to be less stable under these conditions. The enzyme appears unstable to manipulations such as lyophilization and, in solution, to repeated freezing and thawing.

Assay Systems for Collagenases Prepared in Surviving Cell Cultures

At least three types of assay systems have been described. These include one type, not elaborated in this discussion, called the "change of opacity assay".[20] In this assay, collagen is acted on by the collagenase and extent of activity determined by the subsequent failure of the collagen, under defined conditions, to form fibrils.

[41] S. Raymond, *Clin. Chem.* **8**, 455 (1962).
[42] B. J. Davis, *Ann. N.Y. Acad. Sci.* **121**, 404 (1964).
[43] L. Ornstein, *Ann. N.Y. Acad. Sci.* **121**, 321 (1964).
[44] J. M. Brewer, *Science* **156**, 256 (1967).
[45] K. H. Fantes and I. G. S. Furminger, *Nature* **215**, 750 (1967).

Viscosity Assay. This is a variation of the viscosity method introduced by Gallop *et al.*[46] and described in detail by Seifter and Gallop in another volume of this series.[4] The modification employed usually utilizes guinea pig skin collagen (acid-extracted) dissolved to a concentration of 0.15% in buffer, pH 7.4, containing 0.05 M Tris and 0.4 M NaCl. The reaction is conducted at 27°, and the drop in specific viscosity with time is recorded. The assay does not provide a measure of units of enzyme activity. If a measure of specific units is required, considerations described by Seifter and Gallop[4] must be employed.

Assay Based on Release of ^{14}C-*Glycine-Containing Peptides.* This assay, designed by Gross and his colleagues,[18, 20] is rapid and extremely useful, but requires use of ^{14}C-labeled collagen and counting equipment. The substrate is prepared from the skins of guinea pigs previously injected with ^{14}C-glycine. The collagen is isolated from the skins by standard procedures of extraction with acetic acid.[38] Gels prepared from 200 μl of collagen solution containing about 400 μg of collagen (800–1000 cpm) in 1 ml plastic centrifuge tubes are incubated for 12 hours at 37° in a water bath. The tubes must be scrupulously clean so that the resultant gels will be equally opaque. The gels are disrupted with a steel needle, diluted with 200 μl of 0.1 M Tris buffer, pH 7.4, containing 0.001 M $CaCl_2$, and then reincubated for 1 hour at 37°. The enzyme solutions to be tested are then added, and the tubes gently agitated and reincubated for various times. The reaction is stopped by addition to each tube of 100 μl of 0.6 M EDTA; incubation is then continued for 1 hour. The tubes are centrifuged at room temperature for 15 minutes at 20,000 g to separate the undissolved collagen. An aliquot of the solution is assayed for radioactivity in a scintillation or gas flow counter. The activity is expressed as counts per minute released in the reaction mixture minus the counts per minute in the supernatant fluid of control gels.

A solution assay with ^{14}C-collagen can also be employed.[20] The preparation of the assay mixture is identical with the fibril method, except that preincubation at 37° to form the gel is omitted. The enzyme is added to the collagen solution and incubation is carried out at 33° in solution phase rather than in gel phase. At the end of the prescribed periods of time, phosphotungstic acid is added to a final concentration of 0.01 N and HCl to a final concentration of 0.5 N. The unreacted collagen is precipitated and products of enzymatic digestion remain in solution. The material is centrifuged at 15,000 g for 5 minutes, and the supernatant fluid is counted for radioactivity. Again, enzymatic activity is expressed in terms of counts per minute released in the reaction mixture minus the counts per minute in the supernatant fluids of control gels.

---

[46] P. M. Gallop, S. Seifter, and E. Meilman, *J. Biol. Chem.* **227**, 891 (1957).

Properties of Tadpole Collagenase

The purified enzyme has an optimum activity between pH 8 and 9.[20] As prepared,[20] the enzyme contains a small amount of caseinolytic activity, presumably an impurity, but curiously inhibited in a parallel manner by conditions causing inhibition of collagenolytic activity. Both, for instance, are inactivated by heating for 10 minutes at 50° and 60°, and both are irreversibly destroyed at pH 3.5. Both activities are inhibited by 0.002 M EDTA, the inhibition being reversible by dialysis against solutions of $CaCl_2$. Cysteine, 0.005 M, reversibly inhibits both activities. Diisopropylphosphorofluoridate in concentrations as high as $10^{-2} M$ has no effect on either collagenolytic or caseinolytic activity.[20] In all of these properties the tadpole enzyme fully resembles the collagenase of *Clostridium histolyticum*, discussed below.

Specificity of Tadpole Collagenase. Unlike the collagenase of *Clostridium histolyticum* which degrades collagen or gelatin to a great number of small peptides, the tadpole enzyme cleaves the collagen molecule across the three strands at one point in each strand.[21] Thus, native rat skin collagen is cleaved into two unequal pieces, both retaining the triple helical collagen structure.[21] The larger piece, called TC^A (denoting a portion of the tropocollagen molecule containing the end designated as A by electron microscopists), represents 75% of the molecule, while the smaller piece, TC^B (arising from the B end of the tropocollagen molecule as designated by electron microscopists), represents 25% of the original molecule. The two pieces of the cleaved collagen molecule can be separated or, alternatively, the mixture can be denatured by mild heat and fractionated to yield corresponding pieces of cleaved α-chains, named according to the chain from which they derive and the end of the molecule (A or B). Thus, the smaller pieces, designated $\alpha 1^B$ and $\alpha 2^B$, are each 24,000 molecular weight, and the larger pieces, $\alpha 1^A$ and $\alpha 2^A$, are each 71,000 molecular weight. The A ends contain the original N-terminals of the α-chains. Figure 1 demonstrates exquisitely the action of tadpole collagenase on collagen as viewed by electron microscopy. The technique of using enzymes to cleave collagen, then reconstituting the pieces as in this instance, and finally aligning electron photomicrographs as shown here, is proving of inestimable aid in locating the site of enzymatic action.

Nagai et al.[19] reported that the action of the tadpole collagenase on native collagen liberated only C-terminal glycine and N-terminal leucine and isoleucine. On denatured collagen (gelatin), they found more extensive cleavage and liberation of N-terminal leucine, isoleucine, valine, alanine, and phenylalanine; glycine was the only C-terminal amino acid

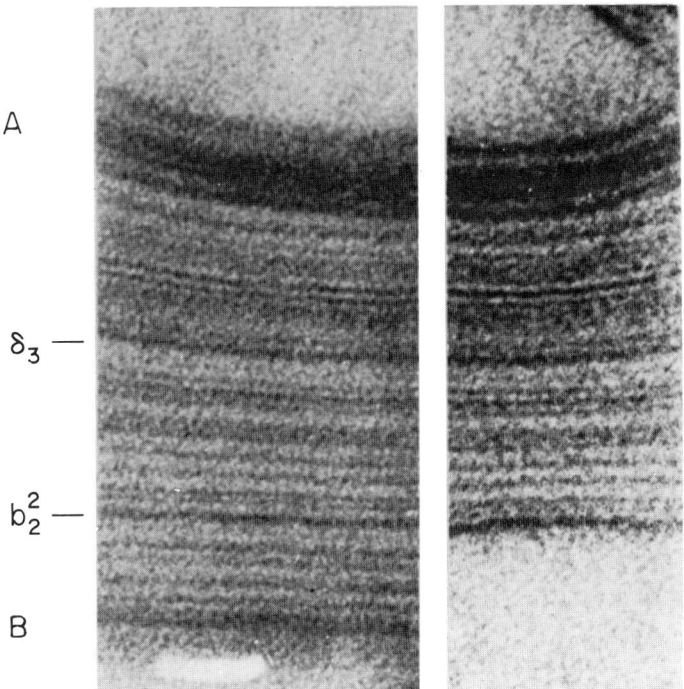

A

δ_3 —

b_2^2 —

B

Fig. 1. Electron micrograph of reconstituted SLS (segmented long spacing) fibers of collagen before and after the action of tadpole collagenase. The photograph to the left shows the alignment of complete tropocollagen molecules with A and B ends labeled. The photograph to the right shows the TC^A fragment obtained after action of the enzyme, and demonstrates that the tropocollagen molecule is cleaved to yield a segment from the A end comprising about 75% of the molecule. Another fragment, not shown here, is obtained and comprises the remaining 25% from the B end (TC^B). (Courtesy of Jerome Gross.)

found in this instance also. Certain tripeptides with a Gly-Leu-X sequence were found to be weakly cleaved at the Gly-Leu linkage. Synthetic peptides known to be cleaved by clostridial collagenase were not cleaved by the tadpole enzyme.[2,19] However, the exact specificity of the tadpole collagenase is yet to be determined since it is not definite that all peptidases are excluded from the preparation.

Collagenases of Clostridium histolyticum

This enzyme has been described in an earlier volume of this series.[4] In this article we shall present newer methods for the purification of collagenases A and B, and newer information concerning their properties and mode of action. We shall not include a description of the viscosity assay fundamental for the study of collagenases here since the method as presented in the earlier volume[4] can be used without modification. Nor shall we report the method of preparation of the collagen substrate since that also is given in another volume of this series.[47]

Culture of *Clostridium histolyticum*

In our laboratory we have maintained and used a stock culture of *Clostridium histolyticum* 47Q5 obtained from Dr. George Warren of Wyeth Laboratories. The bacteria were grown in the medium of Warren and Gray[48] with modifications introduced by Takahashi and Seifter[49] as described below. Other strains of the organism have been used for commercial preparations of collagenase.

Takahashi and Seifter[49] have found that the sodium thioglycolate used as a reducing agent in the culture medium can be replaced altogether or in part by other reducing agents or conditions. Thus, an atmosphere of nitrogen, an atmosphere of hydrogen generated with sodium borohydride, or inclusion of sodium bisulfite can promote growth of the organisms and production of collagenase. Since the organism generates its own reducing medium after a lag period, one may even omit reducing substances other than the glucose of the medium. For best results, however, the medium of Warren and Gray[48] modified to include 0.5% sodium bisulfite and 0.5% oxidized sodium thioglycolate, should be used. The use of this medium yields collagenases of uniformly high specific activity and considerably diminished amounts of general, so-called "nonspecific," proteases. This latter point is of great importance, since the most difficult problem in purification of bacterial collagenase is separation from other proteases in the culture medium.

An inoculum is prepared in 500 ml of medium, and after 24 hours it is added to 3 liters of fresh medium. After incubation for 48 hours at 24°–25°, maximum yield of enzyme is obtained.

Preparation and Purification of Collagenases A and B

The following procedure, taken from the doctoral dissertation of Harper,[11] can be used for 8–10 liters of culture medium after growth of the organisms. The bacterial filtrate or centrifugal supernatant fluid after 48 hours growth is treated with solid ammonium sulfate to 90% saturation. The mixture is allowed to stand overnight at 4° and then either filtered through Celite 545 filter aid at room temperature or centrifuged at 13,200 g for 20 minutes at 10°. The resulting precipitate is dialyzed against successive changes of water at 4° until free of am-

[47] P. M. Gallop and S. Seifter, Vol. VI [196].

[48] G. H. Warren and J. Gray, *Nature* **192**, 755 (1961).

[49] S. Takahashi and S. Seifter, *Biochim. Biophys. Acta*, in press.

monium sulfate, lyophilized, and then dissolved in distilled water. Alternatively, the precipitate is not dialyzed but dissolved in water and clarified by centrifugation. In either case the resulting aqueous solution is then placed on a column (2.5 × 85 cm) of Sephadex G-50 (fine). The column is eluted with water at room temperature and 5-ml fractions are collected at the rate of 60 ml/hour. Elution is monitored by measurement for protein at 280 nm and for enzymatic activity. The screening assay consists of incubation with 2 ml of 2% gelatin in water in 13 × 100 mm test tubes for 15 minutes at 37°, and observing the degree of gelation subsequent to immersion in an ice bath for 5 minutes. If enzyme is present, proteolysis occurs and gelation does not take place.

Fractions containing enzymatic activity by this assay are pooled, lyophilized, and the pool assayed viscometrically. The protein is dissolved in 0.06 M sodium-phosphate buffer, pH 7.4, and the solution placed on a DEAE-cellulose column of 1.5 × 17 cm. Elution is then conducted using the same phosphate buffer followed by 0.6 M acetate buffer of pH 5.6. Fractions (2 ml) are collected at the rate of 60 ml/hour. Two enzymatically active protein fractions, labeled respectively collagenase A and collagenase B, are obtained. These fractions are dialyzed against distilled water at 4°, lyophilized, and again assayed for enzymatic activity.

Table II summarizes a typical purification by this method. The preparation is free of contaminating activities against casein, denatured hemoglobin, benzoyl arginine amide, and several glycosides.* Collagenase A, prepared by this method from a culture medium containing sodium thioglycolate but no sodium bisulfite, often is obtained in specific activity of 600 viscometric units[9] per milligram of protein, and collagenase B with specific activity of 300 to 400 units per milligram protein. Takahashi and Seifter,[49] starting with a culture filtrate from a medium containing both sodium bisulfite and oxidized sodium thioglycolate, were consistently able to obtain collagenase A with a specific activity of 1100 viscometric units per milligram of protein. The modified procedure used by them, starting with 10 liters of growth medium, is as follows.

The bacterial filtrate or centrifugate obtained after 48 hours' growth is treated with solid ammonium sulfate to 90% saturation. The mixture is allowed to stand overnight at 4° and then filtered through Celite 545 filter aid. The filtrate is discarded, and the cake is extracted 3 times with water to give a total extract of about 50 ml. The extract is then subjected to pressure dialysis at 4° using an Amicon membrane to exclude material of 50,000 molecular weight. The concentrate (about 30 ml) is placed on

---

* Note added in proof. A personal communication from Dr. M. Bernfield indicates that the preparation is active against chondroitin sulfate.

TABLE II
SUMMARY OF PURIFICATION PROCEDURE

	Total protein (mg)[a]	Units per mg protein	Total units	Initial yield (% of units)
Crude (supernatant fluid)	834	28	23,400	100
90% saturation with $(NH_4)_2SO_4$ (precipitate)	650	30	19,500	83
Sephadex G-50 (eluate)	125	98	12,250	52
DEAE-cellulose (eluate)				
Fraction A (elution with phosphate)	12	525	6,300	27
Fraction B (elution with acetate)	10	286	2,860	12

[a] Protein determined using method of Lowry et al. (See text footnote 61.)

a column of Sephadex G-100 (3 × 100 cm), and eluted with 0.005 M Tris buffer, pH 7.5, containing 0.005 M $CaCl_2$. Elution is at room temperature, and 5-ml fractions are collected. Fractions are monitored for protein content and screened for enzymatic activity as described above in the first method of preparation. The pattern of elution from the Sephadex column is characterized by four major protein peaks, the second of which contains most of the collagenolytic activity. The contents of the tubes comprising this peak are pooled and concentrated by pressure dialysis using an Amicon 50,000 membrane. The concentrate (approximately 30 ml) is placed on a DEAE-cellulose column (1.5 × 17 cm) and eluted with a gradient of Tris buffer, pH 7.5, constantly 0.005 M with respect to $CaCl_2$; the Tris concentration varies from 0.01 to 0.2 M. Collagenase A elutes in the region of 0.1 M Tris, and the contents of the peak in this region are combined. The pooled material is then concentrated by pressure dialysis, again using the 50,000 Amicon membrane. The concentrate is then lyophilized, although it has been kept active in solution at 2° for 1 month. Collagenase A, prepared in this manner, is free of contaminating "nonspecific" proteases, of amidase, and of certain glycosidases. It resembles fully the collagenase A prepared by the first method, except that it consistently has the higher specific activity of 1100 viscometric units per milligram of protein. Collagenase B can also be obtained from the DEAE-cellulose column, but its characteristics after this kind of separation have not yet been studied.

Physical Properties of Collagenases A and B

Prepared by the method described by Harper et al., collagenase A was shown by sedimentation equilibrium analysis to have a molecular weight of 105,000 and collagenase B to have a molecular weight of 57,400.[50] Thus, collagenase A is similar in molecular weight to preparations described by Seifter *et al.*,[6] Mandl *et al.*,[8] Yoshida and Noda,[10] Strauch and Grassmann,[51] and Levidkova *et al.*[52] Yoshida and Noda[10] described a second clostridial collagenase, perhaps identical with collagenase B, to which they assigned a molecular weight of 79,000.

Seifter *et al.*, using another method of preparation, had obtained a collagenase, most probably collagenase A, with a sedimentation constant of $s_{20,w}^{0} = 5.4\,\mathrm{S}$ and a diffusion constant of $D_{20,w}^{0} = 4.3 \times 10^{-7}\ \mathrm{cm}^2/$ second. An isoelectric point of 8.6 has been reported by Nordwig.[5] Using another preparation, Levdikova *et al.*[52] reported a diffusion constant of $5 \times 10^{-7}\ \mathrm{cm}^2/$second.

Possibility of Subunits of Collagenase A

Levdikova *et al.*[52] suggested the possibility that clostridial collagenase is composed of subunits. They presented evidence to show that a basic structural unit of the enzyme is an inactive "monomer" of molecular weight of 25,000. Thus, in their hands, a collagenase preparation of 100,000 molecular weight could be dissociated into 4 molecules of 25,000 molecular weight. This dissociation was effected by exposure of the enzyme to either pH 3 or 12, and perhaps by the use of chelating agents.

Harper *et al.*[9] pursued the idea of subunits of collagenases A, suggesting that indeed collagenase B could be an active dimer (MW 57,400) of the 25,000 molecular weight "monomer" of Levdikova *et al.* Evidence for the relationship between collagenases A and B is similarity of amino acid composition (see below), immunological cross reactivities of the purified enzymes with antisera prepared to either enzyme, molecular weight relationship between A and B, and, under certain conditions, the binding of a specific inhibitor (cysteine) in the molecular ratios of 2:1 with collagenase A and 1:1 with collagenase B.[11,53] Arguments against a subunit relationship of B to A are primarily against the notion of identical subunits. These are that the amino acid compositions are similar but not identical, and that significant differences exist between the number of charged amino acid residues in collagenases B and A. However, it must be noted that neither enzyme has yet been prepared in crystalline form, and reproducibility of preparations used

[50] E. Harper and S. Seifter, *J. Biol. Chem.*, in press.

[51] L. Strauch and W. Grassmann, *Z. Physiol. Chem.* **344**, 140 (1966).

[52] G. A. Levdikova, N. Orekhovich, N. I. Solov'eva, and V. O. Shpikiter, *Dokl. Akad. Nauk S.S.R.* **153**, 725 (1963); *English Transl.* **153**, 1429 (1964).

[53] E. Harper and S. Seifter, *Federation Proc.* **24**, 359 (1965).

for amino acid analysis has not always been of high order. The likelihood is great that the collagenases have undergone some degree of degradation or deamidation by the potent "nonspecific" proteases and amidases in the growth medium, leading to difficulty both in obtaining crystals and obtaining preparations without some degree of difference in composition. Thus the problem of subunit relationships awaits further standardization of the B as well as the A enzyme preparation.

Amino Acid Compositions of Collagenases

Table III shows the composition of collagenases A and B prepared by the first method presented here. Similar results were reported by Harper et al.[9] Generally the two enzymes show few differences in composition, although the differences may prove to be significant. Their contents of glycine, tyrosine, phenylalanine, lysine, and histidine are almost identical. Among the preparations analyzed, differences in serine contents were notable but cannot be evaluated at present since hydrolysis was done only at one time. A large difference in proline contents between collagenases A and B was noted, and as yet has not been rationalized on the parent-subunit hypothesis.

Of special significance for many studies with the collagenases is the absence, first noted by Grant and Alburn,[7] of sulfur-containing amino

TABLE III

AMINO ACID COMPOSITIONS OF COLLAGENASES A AND B[a]

	Collagenase A	Collagenase B
Aspartic acid	159	170
Threonine	54	60
Serine	27	32
Glutamic acid	85	93
Proline	56	40
Glycine	96	96
Alanine	73	68
Valine	66	60
Isoleucine	56	61
Leucine	73	74
Tyrosine	50	48
Phenylalanine	50	49
Lysine	103	103
Histidine	16	18
Arginine	37	29
Tryptophan	7	6
Amide	67	not done

[a] Residues per 1,000 total residues; uncorrected for losses during hydrolysis.

acids. Thus inactivation of the enzymes by reagents containing —SH groups cannot be attributed to disulfide exchange or reduction.

Functions of Ca²⁺ Ions in Collagenolytic Action

Ca^{2+} ions are required for both the binding of the enzyme to the collagen substrate and for full catalytic activity. Magnesium ions cannot substitute for calcium ions.[46] Thus agents such as EDTA inhibit collagenolytic activity in the first instance by binding calcium[46]; this is true for both collagenase A and B.[9] Many studies indicate that calcium ions are required to keep the enzyme in a conformation necessary for its catalytic function. Thus Takahashi and Seifter[54] showed that collagenase A can be photoinactivated in the presence of methylene blue, and that enzymatic activity is fully lost after destruction of 4 of the total of 16 residues of histidine; calcium ions, but not magnesium ions, were shown to offer strong protection against this inactivation. Furthermore, as will be discussed under inhibitors, the inactivation of collagenase A by cysteine proceeds differently in the presence or absence of calcium ions, being in the one case reversible by dialysis and in the other case seemingly irreversible. Also, as will be discussed, there is much evidence that an intrinsic metal component, most likely a zinc atom, is present at the active site of collagenase. Taken with the facts already presented, it would seem that calcium ions make the enzyme protein conform around the putative zinc atom so that the latter is chelated with two or three amino acid residues in a manner resembling that of zinc in carboxypeptidase A. If calcium ions are not present in sufficient amount, the protein conformation will be different, and perhaps only one or two of the chelating positions of the zinc atom are occupied by amino acid residues of the enzyme; this would leave two or three positions around the zinc atom able to bind other substances. Presumably one or two of these positions ordinarily binds to the substrate. One may understand, then, that if a multifunctional inhibitor such as cysteine is brought into the environment in absence of calcium ions, its binding would be more firm, because a different species of enzyme inhibitor should have been formed. Thus, inhibition would be less effective in the presence of calcium ions, a matter that appears to be experimental fact.

Inhibition by Cysteine

Maschmann[55] originally described inhibition of collagenase by cysteine, and Seifter et al.[6] and Harper and Seifter[53] studied details of the inhibition. In addition to its inhibition by cysteine, clostridial collagenase

---

[54] S. Takahashi and S. Seifter, *Federation Proc.* **28**, 407 (1969); *Biochim. Biophys. Acta* **198** (1970) in press.

[55] E. Maschmann, *Biochem. Z.* **295**, 391 (1938); *ibid.* **300**, 89 (1938).

is inhibited by reduced glutathione,[56] by 2,3-dimercaptopropanol, and by dithiothreitol. It is poorly inhibited by —SH containing substances that do not contain a second functional group capable of chelating strongly.[6] Seifter *et al.*[6] suggested that cysteine and similar reagents inhibit by their properties of chelation, and that they were acting on a probable zinc component in the enzyme. At pH 7.4, 0.01 M cysteine is required for inhibition. At pH 9.0, only 0.001 M cysteine is required, showing that more of the chelating species is present at the higher pH. One can infer, from the pH data, that —SH and —NH$_2$ groups of cysteine are most important for chelation to the proposed metal component of the enzyme. It has also been found[53] that under optimum conditions approximately two molecules of cysteine chelate with one of collagenase A, and that approximately one molecule of cysteine chelates with one molecule of collagenase B. In either case the inactivated enzyme can be recycled through a column of DEAE-cellulose, and the modified enzyme, with bound cysteine, isolated. The enzyme in this condition can no longer bind to the substrate at 0°, as can active collagenase, showing that the cysteine covers the binding site (probably the zinc atom). On the other hand, titration of the cysteine-enzyme with Elman's reagent or with p-hydroxymercuribenzoate does not show an available —SH group, demonstrating that the cysteine is bound through that group to the enzyme.

Recently, Seifter *et al.*[56a] have performed the following experiment offering the strongest evidence yet that cysteine binds to zinc in collagenase. Collagenase A was prepared from a clostridial culture medium in which ^{65}Zn was incorporated. The purified enzyme contained radioactivity and could be followed in the course of treatment with chelating agents. The ^{65}Zn-collagenase was first treated with 2,3-dimercaptopropanol and the solution was then passed through a Sephadex column that resulted in a ^{65}Zn-free protein, enzymatically inactive, and ^{65}Zn-thiol complex. The zinc-free, inactive collagenase was then treated with ^{35}S-cysteine and this solution then passed through a column of Sephadex. The elution pattern showed recovery of nonradioactive, enzymatically inactive collagenase and of all the cysteine added. Thus, prior removal of zinc from collagenase rendered it incapable, subsequently, to bind with cysteine.

Inhibition by Histidine, Imidazole, and Related Compounds

Harper and Seifter[53] have demonstrated that 0.1 M to 0.05 M histidine at pH 7.4 can inhibit either collagenase A or B. Imidazole also inhibits, but less effectively. This is also true of imidazole derivatives

[56] A. A. Tytell and K. Hewson, *Proc. Soc. Exptl. Biol. Med.* **74**, 555 (1950).

[56a] S. Seifter, S. Takahashi, and E. Harper, *Biochim. Biophys. Acta* **198** (1970), in press.

containing fewer possible chelating groups than does histidine. Inhibition by histidine is reversible by dialysis against water or by passage through suitable columns. Most importantly, inhibition by histidine appears to be competitive with a collagen substrate, so that in a viscometric assay in which histidine concentration is kept constant but collagen concentration varied, one gets typical competitive inhibition kinetics. It would seem that histidine and the substrate compete for the intrinsic metal atom of the enzyme.

The degree of inhibition of the collagenases and the ease of reversibility of inhibition follow closely the stability constants known for the various inhibitors in combination with zinc as chelates.

Other Inhibitors of Collagenase

Both collagenases A[6] and B are inhibited by a number of metal-chelating agents, including o-phenanthroline, α,α-bipyridyl, and 8-hydroxyquinoline. These are generally effective in concentrations ranging from 0.01 to 0.001 M. Their action would seem to be on the proposed metal component other than calcium. Diisopropylphosphorofluoridate does not inhibit collagenase,[6] eliminating this enzyme from the class of "serine proteases."

Effects of Cobalt and Manganese

Yagisawa et al.[57] had suggested some degree of activation of collagenase by cobalt. Takahashi and Seifter[49,54] have now studied the effects of metals in detail with the following conclusions. First, if cobalt ions are incorporated into the culture medium, enzyme with specific activity 1.8 times higher than that obtained in the usual culture medium can be prepared. Second, if collagenase A, prepared from the usual culture medium, is subjected to equilibrium dialysis in the presence of both calcium ions and cobalt ions, the activity of the enzyme is increased 1.8 times. Significant, but less striking results are obtained with manganese and iron ions in place of cobalt ions (Co^{2+}, Mn^{2+}, Fe^{2+}). Thus, as in the case of other zinc-containing enzymes, collagenase activity is improved by substitution of cobalt or manganese; in this case, however, calcium ions exert a strong directing effect on the replacement of the metal atom.

Isolation of Peptides from the Active Center

Harper and Seifter[58] and Takahashi and Seifter[49] have been using a double-label technique to isolate portions of the collagenase molecule,

[57] S. Yagisawa, F. Morita, Y. Nagai, H. Noda, and Y. Ogura, J. Biochem. (Japan) 58, 407 (1965).

[58] E. Harper and S. Seifter, Abstr. Am. Chem. Soc. 152nd Meeting Sept. 1966 p. C-39.

presumably at its active center. ^{65}Zn-collagenase A is obtained by purification of growth medium in which this isotope has been present. The purified enzyme is then reacted with ^{35}S-cysteine, and from the mixture an inactive enzyme isolated that contains both ^{65}Zn and ^{35}S-cysteine. This is digested with trypsin and chymotrypsin, and labeled peptide fragments are isolated by column and paper electrophoretic techniques. Several relatively small peptides have been isolated which contain both radioactive labels; they are characterized by inclusion of residues of lysine and glutamic acid and, in one case, also of histidine. Thus it would seem that the zinc atom is chelated to one or more basic amino acid residues and perhaps a glutamic acid residue. More work has yet to be performed to show the direct linkage of the metal atom to one or more of these residues.

Binding of Collagenases to Collagen at Low Temperature

Collagenase has been purified partially by taking advantage of its capacity to bind firmly at 0° to a suspended collagen substrate.[46] The binding is dependent on presence of calcium ions and is inhibited by EDTA. In the presence of calcium and suspended or insoluble collagen, the enzyme is adsorbed, and can be isolated as a complex which, on raising of the temperature, becomes the kinetic complex and exhibits dissolution of the collagen. Using ^{14}C-labeled collagenase A, Takahashi and Seifter[49] have been able to isolate enzyme-substrate complexes at 0° which appear to be saturated when the ratio is 17 collagen molecules to 1 enzyme molecule. This is the lowest figure obtained in many experiments and is achieved by reacting enzyme and substrate in presence of the reversible inhibitor, histidine, and then dialyzing against $0.002\,M$ phosphate buffer of pH 8. The dialysis allows collagen fibrils to form in the presence of enzyme.

Specificity of Collagenases A and B

Both enzymes act against native and denatured collagens to yield seemingly identical peptide mixtures. Both act against synthetic peptide substrates designed for screening collagenolytic activities, Z-Gly-Pro-Gly-Gly-Pro-Ala-OH[59] and PZ-Pro-Leu-Gly-Pro-D-Arg-OH.[25] In general, but perhaps in not all cases, the specificity resides in a sequence, common in the collagen molecule, of -P-X-Gly-P-Y-, enzyme action occurring between -X-Gly- residues to yield a peptide with C-terminal X and always peptides with N-terminal glycine as reviewed elsewhere.[4, 5]

---

[59] E. Wünsch, Z. Physiol. Chem. **332**, 295 (1963).

P (either pro or hypro) seems to be required in the 1 and 4 positions of the defined sequence. It is of interest that the collagen of the earthworm cuticle, which contrary to most collagens, contains more hypro (17%) than pro (about 1%), is also digested by clostridial collagenase, albeit to a different end point.[60]

One might conjecture that the requirement for pyrrolidine residues in positions 1 and 4 of a susceptible sequence is related to a positioning of the metal atom of the enzyme between them.

To sum up the status of specificity, one could say that the general sequence required has been defined but experiments with different types of collagens and with synthetic peptides demonstrate that this specificity is not so absolute that minor escapes are not encountered. A full discussion of specificity is given by Seifter and Harper.[61]

Kinetics of Collagenase Action

This topic was briefly considered in the chapter on collagenases in the previous volume of this series.[4] Yagisawa et al.[57] have considered the kinetics of action of collagenase on synthetic substrates. Seifter and Harper[61] have reviewed in detail the kinetics of action on collagen and gelatin substrates.

[60] D. Fujimoto, *Biochim. Biophys. Acta* **154**, 183 (1968).
[61] S. Seifter and E. Harper, *in* "The Enzymes" (P. Boyer, H. Lardy, and K. Myrbäck, eds.), Vol. IV. Academic Press, New York, 1971.

[45] Clostripain

By WILLIAM M. MITCHELL and WILLIAM F. HARRINGTON

Introduction

Clostripain (clostridiopeptidase B, EC 3.4.4.20) is a sulfhydryl proteolytic enzyme from the culture filtrate of *Clostridium histolyticum*. The enzyme possesses esterase, amidase, and protease activity with a highly limited specificity directed primarily at the carboxyl linkage of arginine. Although the remarkable specificity of clostripain was not reported until 1953 by Ogle and Tytell,[1] the existence of this protease

[1] J. D. Ogle and A. A. Tytell, *Arch. Biochem. Biophys.* **42**, 327 (1953).

has been known for many years.[2-11] Subsequent studies on the purification and properties of the enzyme from bacterial culture have been reported by Labouesse et al.[12-15] Despite the unusual behavior and possible significance of such a highly specific enzyme in peptide sequence work, the difficulties inherent in handling potentially pathogenic bacterial cultures has inhibited more intensive investigation of clostripain. Recently, clostripain has been purified to apparent homogeneity from a crude commercial collagenase preparation, allowing a more readily available source of the enzyme.[16] The following discussion outlines in detail the methods for purification from the commercial source, a simple and accurate assay procedure, and a summary of the known properties of clostripain.

Standard Assay

A sensitive and accurate assay of clostripain activity is obtained by measurement of the initial hydrolytic rate against benzoylarginine ethyl ester (BAE). Since the esterolysis of BAE results in an increased ultraviolet absorption by the product, benzoylarginine, a convenient means of following the initial rate is by monitoring the change in optical density at 253 nm with time.[17] Although manual spectrophotometric determinations may be utilized, much greater accuracy and speed is realized by the utilization of an automatic recording spectrophotometer. Standard reaction conditions consist of sodium phosphate buffer, pH 7.8, $\Gamma/2 = 0.05$, BAE, $2.5 \times 10^{-4} M$; and dithiothreitol, $2.5 \times 10^{-3} M$ at 25°. If an automatic recording spectrophotometer is used, a recorder full range of 0.2 optical density units is most appropriate. A straight edge is used to isolate the initial zero-order kinetic element of the reaction[18]; a molar

[2] W. Kochalaty, L. Weil, and L. Smith, Biochem. J. 32, 1685 (1938).

[3] E. Maschmann, Biochem. Z. 295, 391 (1938).

[4] W. E. Van Heyningen, Biochem. J. 34, 1540 (1940).

[5] R. C. Bard and L. S. McClung, J. Bacteriol. 56, 665 (1948).

[6] W. Kochalaty and L. E. Krejci, Arch. Biochem. 18, 1 (1948).

[7] C. L. Oakley and G. H. Warrack, J. Gen. Microbiol. 4, 365 (1950).

[8] I. H. Lepow, S. Katz, J. Pansky, and L. Pillemer, J. Immunol. 69, 435 (1953).

[9] R. DeBellis, I. Mandl, J. D. MacLennan, and E. L. Howes, Nature 174, 1191 (1954).

[10] J. D. MacLennan, I. Mandl, and E. L. Howes, J. Gen. Microbiol. 18, 1 (1958).

[11] A. Nordwig and L. Strauch, Hoppe-Seylers Z. Physiol. Chem. 330, 145 (1963).

[12] R. Monier, G. Litwack, M. Somlo, and B. Labouesse, Biochim. Biophys. Acta 18, 71 (1955).

[13] B. Labouesse and P. Gros, Bull. Soc. Chim. Biol. 42, 543 (1960).

[14] P. Gros and B. Labouesse, Bull. Soc. Chim. Biol. 42, 559 (1960).

[15] B. Labouesse, Bull. Soc. Chim. Biol. 42, 1293 (1960).

[16] W. M. Mitchell and W. F. Harrington, J. Biol. Chem. 243, 4683 (1968).

[17] G. W. Schwert and Y. Takenaka, Biochim. Biophys. Acta 16, 570 (1955).

absorptivity difference of $1150 \, M^{-1} \, cm^{-1}$ is used to express the results as micromoles of BAE hydrolyzed per minute at $25°$.[19] Within the enzyme concentrations applicable for the spectrophotometric assay, the initial rates of hydrolysis are proportional to concentration.

Purification Procedure

Although clostripain may be isolated directly from the culture filtrate[20] of *C. histolyticum*, a more convenient method[16] is afforded through hydroxylapatite and Sephadex column chromatography by utilization of the commercially available Worthington collagenase. Yields vary greatly, depending on the lot. However, unusually high yields have been obtained recently from a "high protease" collagenase preparation (Worthington Lot No. 95). All procedures are carried out at room temperature except for the pressure concentration and column effluents which are collected

---

[18] If the enzyme is in the oxidized state prior to assay, a 30–60 second lag will generally be observed.

[19] J. R. Whitaker and M. L. Bender, *J. Am. Chem. Soc.* **87**, 2728 (1965).

[20] Method of Labouesse and Gros (footnote 13): Bacteria are grown anaerobically at $27°$ for 25–28 hours at pH 7.4 (initial) in 10 liters of medium consisting per liter of: peptone, 60 g; KH_2PO_4, 1.9 g; $Na_2HPO_4 \cdot 12 \, H_2O$, 22 g; $FeSO_4 \cdot 7 \, H_2O$, 0.6 g; $MgSO_4$, 12 mg; $MnCl_2$, 6 mg. The cultures are chilled and bacteria removed by centrifugation for 15 minutes at 3500 g. Clostripain is precipitated from solution by the addition of solid ammonium sulfate to 70% saturation at 0–5°. The precipitate is suspended in 500 ml of 40% ammonium sulfate, pH 7.2, plus 20 g Hyflosupercel (Johns-Manville, New York). The suspension is filtered through a Büchner funnel and extracted with 40% ammonium sulfate until no clostripain activity is obtained. The pooled active solution (approximately 2 liters) is brought to 70% saturation and clostripain again precipitated. This precipitate is dissolved by dialysis against water and subsequently incubated in 50 mM cysteine and 10 mM $CaCl_2$ at 35° for 4–6 hours. This incubation is terminated by increasing the temperature to 60° for 5 minutes. The precipitate is removed by centrifugation, with the major portion of clostripain activity remaining in the supernatant. The latter is dialyzed against pH 7.2, 25 mM Veronal buffer containing 10 mM $CaCl_2$ and passed through a DEAE-cellulose column (15×2.5 cm) equilibrated with the same buffer. This step retains the culture pigment on the column while allowing the enzyme to pass through the column. The column wash is passed over a second identical DEAE column, dialyzed against water, and concentrated approximately 10-fold in a rotary evaporator. The concentrated solution is brought to a 1 mM final concentration of phosphate buffer, pH 7.0, and passed through a hydroxylapatite column (10×2 cm) which is equilibrated with the same buffer. Stepwise buffer molarity increments in 1 mM, 10 mM, and 100 mM are made every 150 ml with clostripain emerging on the final 100 mM step increase. This procedure realizes an approximately 50% yield with a 500-fold purification. The reported purification profile may be artificially high since clostripain is easily oxidized and the large purification realized by the cysteine incubation step may be secondary to clostripain reduction and concomitant activation.

in a refrigerated fraction collector and monitored at OD_{280}. A 1- or 2-g sample of the crude collagenase is solubilized, giving a 2% solution with $0.10\,M$ sodium phosphate buffer, pH 6.7. The resulting solution is clarified by low-speed centrifugation in a clinical centrifuge and applied to a hydroxylapatite column, 2.5×12 cm. After several column volumes of buffer are passed through the hydroxylapatite gel, the crude enzyme preparation is pumped onto the surface of the column at 60 ml per hour. This method of application prevents cracking of the gel, provided that the protein concentration is not greater than 2%. Following emergence of the initial peak, and after the optical density ($\lambda = 280$ nm) has decreased to about 0.1 absorbance unit, the concentration of the eluting buffer is increased to $0.20\,M$ sodium phosphate (pH 6.7) in order to elute the desired column-bound components. The pooled fractions are concentrated under N_2 in an Amicon pressure cell (Amicon Corp., Lexington, Mass.) using an XM-50 filter at 5° and chromatographed on a column, 3.5×125 cm, of Sephadex G-75 in $0.05\,M$ sodium phosphate, pH 6.7, at a flow rate of 50 ml per hour. Two distinct, well-separated peaks emerge. The first peak is eluted in the column void volume (i.e., at the position of the blue dextran marker) and contains various collagenolytic enzymes. The second peak consists of clostripain. Although the clostripain fraction at this point will generally be of high purity, some lots of Worthington collagenase contain a pigment which is not completely resolved by the above procedure and which yields enzyme preparations of low specific activity (i.e., 60 micromoles BAE hydrolyzed per minute per milligram enzyme). This situation may be circumvented by the following procedure. The pooled clostripain fraction is dialyzed for approximately 12 hours at 5° against 0.1 M, pH 7.8, sodium phosphate buffer containing $10^{-3}\,M$ dithiothreitol and applied to a second hydroxylapatite column as previously described. After initial loading, a linear gradient (0.10 to 0.30 M sodium phosphate, pH 6.7) in a total volume of 800 ml is developed. A single peak is resolved which is pigment free and of high specific activity.[21]

Clostripain prepared in the above manner exhibits a high degree of purity[22] as judged by sedimentation velocity and sedimentation equilibrium ultracentrifugation, immunodiffusion, immunoelectrophoresis, and polyacrylamide gel electrophoresis. Erroneous conclusions of heterogenity may arise in polyacrylamide gel electrophoresis and gradient-developed hydroxylapatite columns if precautions are not taken to insure that the enzyme is fully reduced.[21, 22a]

[21] Failure to fully reduce clostripain results in double peaks of identical although diminished specific activity.

[22] Although purified clostripain appears highly homogenous with respect to physical

Properties of the Enzyme

pH Optimum and Ion Effects

The pH optimum for hydrolysis of L-arginine methyl ester in phosphate buffer is about 7.2 whereas maximum activity against α-benzoylarginine ethyl ester occurs in the range pH 7.4–7.8 in Tris-HCl buffer. Calcium is unique among metal ions in enhancing the esterase activity (50% increase in initial rate of cleavage at $10^{-2}\,M$) and also acts to depress thermal inactivation. The work of Labouesse and Gros indicates that binding of calcium ion is an essential prerequisite of activity[13] whereas significant inhibition of esterase activity is observed in the presence of other cations such as Co^{2+}, Cu^{2+}, Cd^{2+}, Na^+, and K^+.[16] Conversely, EDTA completely inhibits enzymatic activity at concentrations as low as 10 μM. Citrate, borate, Veronal, and Tris anions partially inhibit esterase activity.

Sulfhydryl Requirement. The enzymatic activity of clostripain is rapidly lost on incubation in the presence of oxidizing agents such as hydrogen peroxide, but activity may be regenerated completely by reduction with dithiothreitol.[16] Labouesse and Gros[13] have investigated the effect of cysteine concentration on the esterase activity of the enzyme and report that one molecule of cysteine is required per mole of enzyme for activation; their studies also demonstrate inactivation of the enzyme by *p*-chloromercuribenzoate (PCMB) with complete loss of activity occurring at a 1:1 molar ratio of enzyme to inhibitor.

Physical Properties. The physical properties of clostripain are listed in Table I. The molecular weight of 50,000 g/mole is large for a protease,[16] but is about half the value reported by Labouesse and Gros[13] from sedimentation-diffusion data. Since the higher value was obtained in a low salt environment, this discrepancy may be the result of dimerization in the low-salt medium. This interpretation is consistent with the measured sedimentation coefficients: $s_{20,w}^0 = 4.43$ S (high salt)[16] and $s_{20,w}^0 = 6.70$ S (low salt).[13] Hydrodynamic evidence and electron microscopy reveal the enzyme to be a globular protein of approximately 33–35 Å radius.[23] Clostripain possesses an acid isoelectric point[24] and a low α-helix content.[16] The amino acid content[16] per thousand residues is

structure by all analytical procedures employed to date, recent studies (W. H. Porter and W. M. Mitchell) with affinity labeling (tosyllysine chloromethyl ketone) of the active site of clostripain suggest that only a fraction of the molecules are fully active.

[22a] W. M. Mitchell, *Biochem. Biophys. Acta* **147**, 171 (1967).

[23] W. M. Mitchell and J. J. Schmidt, *Biochim. Biophys. Acta* **175**, 207 (1969).

[24] W. M. Mitchell, *Biochim. Biophys. Acta* **178**, 194 (1969).

TABLE I
Physical Properties of Clostripain

Molecular weight[a]	50,000
Sedimentation coefficient $(s_{20,w}^{0})^{a}$	4.43 S
Partial specific volume, $\bar{V}$, calculated	0.72 ml/g
Intrinsic viscosity, $[\eta]$	0.053 dl/g
Particle shape	$\sim$spherical
Particle radius (from hydrodynamic measurements and electron microscopy)	$\sim$33–35 Å
Isoelectric point (8°)	4.8–4.9
Optical rotatory dispersion (4.5°)	
Cotton effects minimum	233 mμ
b_0 (Moffitt)	$-19.7°$
λ_c (one term Drude)	229°
% α-Helicity, estimated	30%
Helix-coil transition temperature	$\sim$50°

[a] See text.

as follows: Asp_{148}, Thr_{49}, Ser_{74}, Glu_{99}, Pro_{30}, Gly_{82}, Ala_{51}, Val_{47}, Isl_{43}, Leu_{77}, Tyr_{35}, Phe_{39}, Lys_{91}, His_{24}, Arg_{24}, Trp_{19}, $\frac{1}{2}$-Cys_{42}, Met_{27}.

Substrate Specificity

Synthetic Esters and Amides. The relative rate of hydrolysis of a number of simple substrates is presented in Table II. Among the synthetic esters and amides which have been examined, arginine-containing substrates are much more rapidly hydrolyzed than identically substituted lysine compounds. The cleavage rate of L-lysine methyl ester is anomalously low and the amides of L-phenylalanine, L-leucine, glycine, and L-proline as well as the peptides L-leucylglycine, glycylglycine, and L-leucylglycine are not attacked. No demonstrable activity can be detected against the standard chymotryptic substrates N-acetyl-L-tyrosine ethyl ester and N-benzoyl-L-tyrosine ethyl ester. Thus the enzyme appears to have a specificity similar to that of trypsin in that cleavage of the simple substrates occurs at the carboxyl terminal bonds of arginine and lysine but a striking preference is shown for cleavage at arginine residues.

Synthetic Homopolymers. Polyarginine is completely resistant to hydrolysis by clostripain although the homopolymer is digested to di- and tripeptides by trypsin.[16] This phenomenon may be explained by the observed isoelectric points of clostripain and trypsin (4.9 and 11, respectively).[24] At the pH of digestion (pH 8), polyarginine is positively charged while trypsin carries a formal positive charge and clostripain a negative charge. Thus at the pH optimum of hydrolysis, a strong electrostatic interaction of the charged homopolymer with clostripain may

TABLE II

COMPARISON OF ESTERASE AND AMIDASE ACTIVITIES FOUND IN
C. histolyticum CULTURE FILTRATES

Esters or amides	I[a,b]	II[b]	III[b]	IV	V
Arginine					
N-Benzoyl-L-arginine ethyl ester		+		+	+
N-Benzoyl DL arginine naphthyl amide	+			+	
N-Benzoyl-L-arginine amide	+	+	+	+	+
N-Benzoyl-L-arginine methyl ester	+				
p-Toluene sulfonyl-L-arginine methyl ester	+			+	+
L-Arginine methyl ester		+	+		+
D-Arginine methyl ester	−				
N-Benzoyl-L-arginine benzyl ester		+			
N-Benzoyl-L-arginine isopropyl ester		+			
Lysine					
p-Toluene-sulfonyl-L-lysine methyl ester				−	(±)[c]
L-Lysine ethyl ester	−				
L-Lysine methyl ester		(±)[d]			−
Others					
N-Acetyl-L-tyrosine ethyl ester				−	−
N-Benzoyl-L-tyrosine ethyl ester				−	
D,L-Phenylalanine methyl ester		−			
N-Benzoyl-D,L-phenylalanine ethyl ester		−			
N-Benzoyl-D,L-phenylalaninamide		−			
L-Leucine methyl ester		−			
L-Leucinamide		−			
Benzoylglycinamide		−			
Glycylglycine ethyl ester		−			
L-Proline benzyl ester		−			
L-Prolinamide		−			
Benzoylglycine phenylalaninamide	−				
Benzoylglycine leucinamide	−				

[a] I = Nordwig and Strauch[11]; II = Ogle and Tytell[1]; III = DeBellis et al.[9]; IV = Mitchell and Harrington,[16] V = Gros and Labouesse.[14]

[b] Assays on relatively crude preparations.

[c] Conversion at 8% the rate of the arginine homolog.

[d] Conversion was found with this substrate. However all other hydrolyses were reported virtually complete so that initial rates and relative magnitudes of activity cannot be compared.

prevent hydrolysis. The like charge of polyarginine and trypsin under normal hydrolytic conditions prevents association and thus allows cleavage to occur.

The specificity regarding arginine versus lysine R groups is also found in the synthetic homopolymers. Although neither polyarginine nor polylysine are hydrolyzed by clostripain, both are effective inhibitors of the esterase activity against BAE. However, polyarginine functions as an inhibitor some two orders of magnitude more efficiently than polylysine, a phenomenon consistent with a preferential binding of the arginine moiety to the active site of the enzyme.[16]

Heteropolymers. The preference of clostripain for an arginine moiety in synthetic esters, amides, and homopolymers is also present in naturally occurring polypeptides and proteins where hydrolysis occurs most rapidly at the carboxyl terminal peptide bond of arginine. Although cleavage also occurs at lysine residues, hydrolytic rates are generally one to two orders of magnitude less. Moreover, this unusually specific protease is distinguished by its ability to hydrolyze the arginylproline peptide bond,[25] a bond usually resistant to proteolysis.

Conclusion

Clostripain is a unique enzyme with regard to its specificity. This property may be of advantage in protein sequence work where large initial peptide fragments are desired as well as in active site studies regarding the origin of its arginine specificity. The stringent sulfhydryl conditions may prove of great interest with regard to the active site chemistry of the enzyme as well as in evolutionary comparisons with the plant sulfhydryl proteases.

[25] W. M. Mitchell, *Science* **162**, 374 (1968).

[46] Purification and Assay of Thermolysin

By HIROSHI MATSUBARA

Assay Method

Principle. Two procedures will be described—casein digestion and hydrolysis of synthetic substrates. The first method is based on that of Kunitz[1] which involves digestion of casein by thermolysin under defined conditions. The digestion mixture is treated with trichloroacetic acid (TCA) to remove undigested casein. The extent of digestion is measured by the amount of unprecipitated products which contain tyrosine and tryptophan, and absorb at about 280 nm.[2] The second method uses synthetic peptides, such as carbobenzoxy(Cbz)-glycyl-L-phenylalanin-amide and Cbz-L-threonyl-L-leucinamide, as the substrates. The extent

[1] M. Kunitz, *J. Gen. Physiol.* **30**, 291 (1947).
[2] The reader may refer to the methods described previously: M. Laskowski, Vol. II, p. 8; B. Hagihara, H. Matsubara, M. Nakai, and K. Okunuki, *J. Biochem.* **45**, 185 (1958); N. C. Davis and E. L. Smith, *Methods Biochem. Anal.* **2**, 215 (1955).

of hydrolysis is measured by a colorimetric ninhydrin method based on that of Moore and Stein[3] which estimates the number of α-amino groups liberated after hydrolysis of the glycylphenylalanine or threonylleucine bond.[4]

The enzyme units are calculated for each case from a standard curve.

Reagents

Casein Digestion—Method A
  Tris-HCl buffer, 0.01 M, pH 8.0, with 0.01 M CaCl$_2$
  NaOH, 0.2 N
  CH$_3$COOH, 0.2 N
  2% Casein Solution. Two grams of casein (Hammarsten) are dissolved in 90 ml of 0.1 M Tris with the aid of 0.2 N NaOH; the pH of the solution is then adjusted to 8.0 by addition of 0.2 N CH$_3$COOH with stirring. The total volume of the solution is adjusted to 100 ml with water. The solution may be stored at −10° for at least several weeks.
  Trichloracetic acid (TCA), 1.2 M

Hydrolysis of Synthetic Substrates—Method B
  Tris-HCl buffer as described above.
  CH$_3$COOH, 0.1 N
  Substrates (5 × 10^{-3} M). Cbz-Gly-L-Phe-amide (Mann Research Lab.) (7.1 mg) or Cbz-L-Thr-L-Leu-amide (Cyclo Chemical) (7.3 mg) are dissolved in 0.2 ml of ethanol by warming in a bath at 60° for 1–2 minutes. To the solution is added 3.8 ml of 0.01 M Tris-HCl buffer equilibrated to 40°. The final solution is kept in a bath at 40°.
  Ninhydrin solution. This solution is made according to Moore.[5]
  Other preparations[6,7] give similar results.
  50% Ethanol solution

Procedure

Casein Digestion. The casein solution is incubated in a water bath at 35° for at least 5 minutes. Thermolysin solutions,[8] appropriately diluted with 0.01 M Tris-HCl buffer, pH 8.0, are pipetted into test tubes

---

[3] S. Moore and W. H. Stein, *J. Biol. Chem.* **176**, 367 (1948).
[4] Refer also to the methods described previously: H. Neurath, Vol. II, p. 77; N. C. Davis and E. L. Smith, *Methods Biochem. Anal.* **2**, 215 (1955).
[5] S. Moore, *J. Biol. Chem.* **243**, 6281 (1968).
[6] S. Moore and W. H. Stein, *J. Biol. Chem.* **211**, 907 (1954).
[7] D. H. Spackman, W. H. Stein, and S. Moore, *Anal. Chem.* **30**, 1190 (1958).
[8] To make a stock solution of thermolysin, the following procedure is convenient. About 10 mg of crystals are suspended in 4 ml or 0.01 M Tris-HCl buffer containing 0.01 M CaCl$_2$, pH 8, in an ice bath and dissolved with 0.2 N NaOH (0.5–1.0 ml). The

and incubated at 35°. One milliliter of casein solution is pipetted into tubes with enzyme at proper intervals, e.g., at every 30 seconds. The solutions are mixed well and the reaction is carried out for 10 minutes, at the end of which 2 ml of 1.2 M TCA is added and, after shaking, the tubes are kept in the bath for about 30 minutes. The precipitate is filtered off through Whatman No. 1 filter paper, 7 cm, and the absorbance of the filtrate is measured at 280 nm against the blank. The blank is prepared by first mixing the casein solution with TCA and then adding the enzyme solution to the casein-TCA mixture. The standard curve for thermolysin is illustrated in Fig. 1. Two curves are shown for different temperatures.

Hydrolysis of Synthetic Substrates. The synthetic peptide solution (0.5 ml) is incubated with the enzyme solution[8] (0.5 ml) in a test tube at 40°, for 10 minutes. At the end of the reaction, 0.5 ml of 0.1 N CH₃COOH is added. Next 1 ml of ninhydrin solution is added and the mixture is placed in a boiling water bath for 15 minutes. It may be

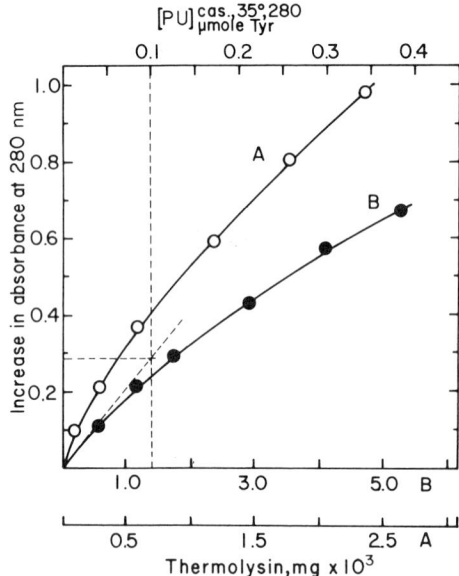

Fig. 1. Standard curve for thermolysin activity using casein substrate. (A) 60°, (B) 35°.

pH of the solution is immediately adjusted to 8.0 with 0.2 N CH₃COOH (1.0–1.5 ml) and the solution is kept at −10°. The activity of thermolysin remains nearly constant for over 1 month. The solution with an absorbance of 5×10^{-3} at 277 nm is suitable for the assay by method (A), and 1×10^{-2} for method (B).

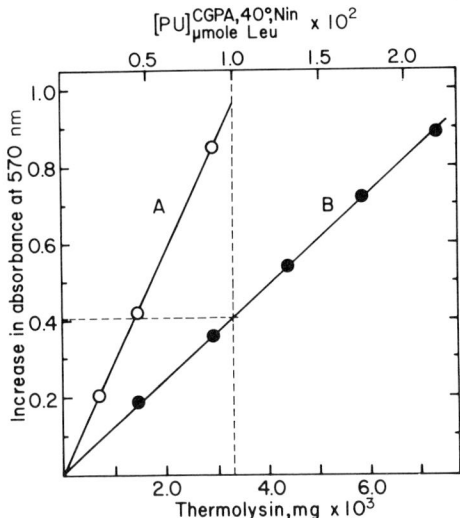

Fig. 2. Standard curve for thermolysin activity using synthetic substrates. (A) Cbz-L-Thr-L-Leu-amide, (B) Cbz-Gly-L-Phe-amide.

convenient to put a small glass funnel or a marble on the top of the tube to lessen evaporation of water from the reaction mixture. The tube is cooled in an ice bath and the solution is diluted with 2 ml of 50% ethanol solution. The absorbance of ninhydrin color at 570 nm is measured against the blank. The blank is made by first mixing substrate with acetic acid and then adding the enzyme solution to the mixture. The standard curves are shown in Fig. 2.

Definition of Unit and Specific Activity

Casein Digestion. One unit of thermolysin activity is defined as that liberating material which has absorbance at 280 nm in the casein digest equivalent to 1 micromole of tyrosine in 1 minute at 35°. The unit is corrected by a standard curve and represented by $[PU]_{\mu mole\ Tyr}^{cas.,35°,280}$. Specific activity is expressed as the unit activity per milligram of protein.[9] It is calculated to be 2060 for pure thermolysin.

Hydrolysis of Synthetic Substrates. One unit of thermolysin activity is defined as liberating the α-amino groups from Cbz-Gly-L-Phe-amide equivalent to 1 micromole of leucine in 1 minute at 40° assayed by the ninhydrin method and represented by $[PU]_{\mu mole,\ Leu}^{CGPA,\ 40°nin}$. Specific activity is expressed as the unit activity per milligram of protein[9] and calculated to be 3.0 for pure thermolysin.

[9] Y. Ohta, Y. Ogura, and A. Wada, *J. Biol. Chem.* **241**, 5919 (1966). The present author recalculated $E_{277\ nm}^{1\%}$ to be 18.3 from the data reported.

Purification Procedure

Production. The following brief procedure for the enzyme production is that described by Endo[10, 11] for a semimanufacturing scale. About 40 liters of culture medium containing 3% wheat flour, 1% lactose, 1.5% soy bean cake, and 0.3% casein (pH 6–7) are sterilized at 120° for 30 minutes in a 70-liter tank. After cooling to 55°, about 100 ml of a culture of *Bacillus thermoproteolyticus* Rokko[12] are added and incubated at 55° for 13 hours with stirring (400 rpm) and aeration (30 liters per minute). Thermolysin is excreted from the cells. The presence of zinc ion in the medium stimulates enzyme production.[13]

Purification of Thermolysin. The purification procedure is essentially that described by Endo.[10] As crystalline material is now commercially available,[14] only recrystallization is required for further purification.

AMMONIUM SULFATE FRACTIONATION. The culture solution (10 liters) is filtered through a filter press with the aid of radiolite (or Celite) and the enzyme in the filtrate is precipitated with solid ammonium sulfate (45% saturation at pH 7.0). The precipitate is dissolved in $0.02 M$ Ca-acetate (1.5 liters) and the solution is brought to 30% saturation with solid ammonium sulfate. After centrifugation, the supernatant solution is brought to 45% saturation with solid ammonium sulfate (pH 7.0). The precipitate is dissolved in $0.02 M$ Ca-acetate (950 ml) at pH 7.0 and the solution is dialyzed overnight against $0.01 M$ Ca-acetate, pH 7.0 at 5°.

ACETONE FRACTIONATION AND CRYSTALLIZATION. To the dialyzed solution (1030 ml) chilled acetone (824 ml) is gradually added and the viscous precipitate formed is removed by centrifugation. The enzyme in the supernatant solution is precipitated by the addition of acetone (750 ml). The precipitate is collected by centrifugation and dissolved in $0.02 M$ Ca-acetate (200 ml). Crystals appear within 30 minutes and crystallization is completed by standing at 5° overnight. Total recovery of enzyme activity is about 25% (approximately 1.5 g).

RECRYSTALLIZATION. The crystals (or commercial preparation) (about 1 g) are suspended in 40 ml of cold $0.02 M$ Ca-acetate, pH 7.0, in an ice bath, and the pH of the suspension is adjusted to 11.5[15] by dropwise addition of cold $0.2 N$ NaOH (4–5 ml). The crystals dissolve completely and if any precipitate appears, the solution is quickly centrifuged (3000 g,

---

[10] S. Endo, *J. Ferment. Technol.* **40**, 346 (1962) (in Japanese).

[11] S. Endo and Y. Noguchi, Japanese Patent, 5447, 1964.

[12] A new species isolated from soil. Not registered.

[13] S. Endo, personal communication.

[14] Calbiochem., Los Angeles; grade B (1968).

[15] pH test papers (e.g., Hydrion pH papers) are convenient.

2 minutes). The pH of the clear solution is then adjusted to 8.5–9.0 with cold 0.2 N CH_3COOH added dropwise. Any precipitate formed is centrifuged off[16] and the clear solution is adjusted to pH 7.0–7.2 with 0.2 N CH_3COOH. The crystals appear and grow on standing in a refrigerator (5°) overnight. They grow very rapidly when the solution is kept at room temperature (25°). The crystals are collected by centrifugation and suspended in 0.02 M Ca-acetate (about 30 ml). Recrystallization is repeated as described above three or four times. The recovery of thermolysin after four recrystallizations is about 200 mg. The suspension may be lyophilized.

Properties

Stability. The lyophilized enzyme is very stable for months if stored in a refrigerator. The enzyme solution can be kept for weeks in a frozen state without marked loss of activity. At pH 6.5 to 8.5, full activity is observed in the presence of 0.002 M Ca-acetate even after treatment at 60° for 1 hour.[9, 10, 17] The half-life of thermolysin activity at 80° is about 1 hour.[10, 17] Without calcium ion, thermolysin is very unstable.[10] The enzyme is stable in 8 M urea, 20% ethanol or methanol, and 0.12% cetyl trimethylammonium bromide with 100% activity at room temperature, but these reagents promote thermal denaturation of the enzyme.[18] Slight inactivation, about 5–10%, occurs in n-butanol or n-propanol.[18]

Physical and Chemical Properties. Free boundary electrophoresis,[9, 10] sedimentation,[9] Sephadex gel filtration,[9] polyacrylamide gel electrophoresis,[19] and amino terminal analysis[19] showed that the crystalline preparation of thermolysin was homogeneous. Ohta *et al.*[9] determined the molecular weight of the enzyme to be 37,500 by a sedimentation equilibrium method using an assumed partial specific volume of 0.73 cm^3/g; the sedimentation constant, $s_{20,w}^0 = 3.54$–3.63×10^{-3} seconds; and the molar extinction coefficient at 280 nm is 66,300.

The amino acid composition[9] was reported to be Asp_{43}, Thr_{23}, Ser_{23}, Glu_{20}, Pro_8, Gly_{36}, Ala_{28}, $\frac{1}{2}$-Cys_0, Val_{24}, Met_2, Ile_{18}, Leu_{17}, Tyr_{29}, Phe_{10}, Lys_{12}, His_9, Arg_{10}, Trp_5, $(NH_3)_{38}$. Another composition was also reported on the basis of a higher molecular weight.[17] The amino terminal sequence is Ile-Thr-Gly-Thr-.[19]

---

[16] The precipitate as well as mother liquor at every crystallization step may contain thermolysin and be saved for further purification, e.g., acetone or ammonium sulfate fractionation.

[17] H. Matsubara, *in* "Molecular Mechanism of Temperature Adaptation" (C. L. Prosser, ed.), p. 283. American Association for the Advancement of Science, Washington, D.C., 1967.

[18] Y. Ohta, *J. Biol. Chem.* **241**, 509 (1966).

[19] H. Matsubara, unpublished results.

Inhibitors. EDTA and 1,10-phenanthroline were found to be strong inhibitors.[10,20] The enzyme activity is lost completely with 0.005 M EDTA at 40° in 3 minutes. Oxalate, citrate, and phosphate also have inhibitory effects. Mercuric chloride and silver nitrate in concentrations of 0.005 M cause complete inactivation.[10] No inhibition was observed with diisopropylphosphofluoridate (DFP),[17,20] TPCK (L-tosylamido-2-phenyl) ethyl chloromethyl ketone,[19] dibromoacetophenone,[19] diphenylcarbamyl chloride,[19] potato inhibitor,[10,20] soybean trypsin inhibitor,[20] cysteine,[20] or sodium cyanide.[20]

Specificity. A broad survey of the substrate specificity of thermolysin was carried out by hydrolyzing proteins,[17,20-24] and it was concluded that in general thermolysin hydrolyzes peptide bonds involving the amino groups of hydrophobic amino acid residues with bulky side chains,[22] e.g., isoleucine, leucine, valine, phenylalanine, methionine, and alanine. However, it is not strictly limited to these amino acid residues. The amino sites of tyrosine, glycine, threonine, and serine residues were also found to be susceptible in some cases, although to a minor extent. These latter cases were especially observed when a longer reaction time or a higher enzyme-substrate ratio was employed to hydrolyze proteins or peptides. There is so far no clear example of the hydrolysis of tryptophan site. The table summarizes the peptide bonds hydrolyzed by thermolysin in various proteins and peptides. Besides these common cases, several unusual hydrolyses were also observed at the bonds, Tyr-Glu,[19] Thr-His,[23] Thr-Asn,[24] Ser-CySO₃H,[24] Glu-CySO₃H,[24] Gly-His,[24] Arg-His,[25] His-Asn,[26] Tyr-Asn,[27] and Glu-Thr.[28] However, it is not clear at the present time whether these hydrolyses were caused by thermolysin or by contaminating enzymes.

Several studies were conducted with synthetic substrates[20,29-35] and

[20] K. Morihara and T. Tsuzuki, *Biochim. Biophys. Acta* **118**, 215 (1966).
[21] H. Matsubara, A. Singer, R. M. Sasaki, and T. H. Jukes, *Biochem. Biophys. Res. Commun.* **21**, 242 (1965).
[22] H. Matsubara, R. M. Sasaki, A. Singer, and T. H. Jukes, *Arch. Biochem. Biophys.* **115**, 324 (1966).
[23] H. Matsubara and R. M. Sasaki, *J. Biol. Chem.* **243**, 1732 (1968).
[24] R. P. Ambler and R. J. Meadway, *Biochem. J.* **108**, 893 (1968).
[25] R. J. Delange, R. M. Fambrough, E. L. Smith, and J. Bonner, *J. Biol. Chem.* **244**, 319 (1969).
[26] A. Tsugita, M. Kobayashi, T. Kajihava, and B. Hagihara, *J. Biochem.* **64**, 727 (1968).
[27] D. M. Blow, J. J. Birktoft, and B. S. Hartley, *Nature* **221**, 337 (1969).
[28] K. Dus, K. Sletten, and M. D. Kamen, *J. Biol. Chem.* **243**, 5507 (1968).
[29] Y. Ohta and Y. Ogura, *J. Biochem.* **58**, 607 (1965).
[30] H. Matsubara, *Biochem. Biophys. Res. Commun.* **24**, 427 (1966).

AMINO ACID RESIDUES INVOLVED IN THERMOLYSIN HYDROLYSIS[a-p]

$R_2{}^q$:	R_1
Ile	Lys,Arg,Cys(Cm),[r]Asp,Glu,Asn,Gln,Thr,Ser,Pro,Gly,Ala,Val,Met, Ile,Leu,Tyr,Phe
Leu	Lys,His,Asp,Glu,Asn,Gln,Thr,Ser,Gly,Ala,Val,Ile,Leu,Tyr,Phe,Trp
Val	Lys,Arg,Cys(Cm),Asp,Glu,Asn,Gln,Thr,Ser,Pro,Gly,Leu,Tyr,Phe, Trp,Lys(Me)
Phe	Lys,Arg,Asp,Glu,Gln,Thr,Ser,Gly,Ala,Met,Leu,Phe
Ala	Lys,Arg,CySO₃H,Asp,Asn,Thr,Ser,Ala,Tyr,*Phe*[s]
Met	Lys,Asp,Gly,Ala,Leu,Tyr,*Gln*
Tyr	Arg,Gln,Tyr,Phe,*Lys,Ala*
Thr	Lys,His,Arg,Ser,*Tyr*
Ser	Arg,Ser,Ala,Tyr,*Phe*
Gly	Thr

[a] H. Matsubara, *in* "Molecular Mechanism of Temperature Adaptation," (C. L. Prosser, ed.), p. 283. American Association for the Advancement of Science, Washington, D.C., 1967.

[b] K. Morihara and T. Tsuzuki, *Biochim. Biophys. Acta* **118**, 215 (1966).

[c] H. Matsubara, A. Singer, R. M. Sasaki, and T. H. Jukes, *Biochem. Biophys. Res. Commun.* **21**, 242 (1965).

[d] H. Matsubara, R. M. Sasaki, A. Singer, and T. H. Jukes, *Arch. Biochem. Biophys.* **115**, 324 (1966).

[e] H. Matsubara and R. M. Sasaki, *J. Biol. Chem.* **243**, 1732 (1968).

[f] R. P. Ambler and R. J. Meadway, *Biochem. J.* **108**, 893 (1968).

[g] K. Sugeno and H. Matsubara, *Biochem. Biophys. Res. Commun.* **32**, 951 (1968).

[h] A. Benson and K. T. Yasunobu, *J. Biol. Chem.* **244**, 955 (1969).

[i] J. N. Tsunoda, K. T. Yasunobu, and H. R. Whiteley, *J. Biol. Chem.* **243**, 6262 (1968).

[j] R. J. Delange, D. M. Fambrough, E. L. Smith, and J. Bonner, *J. Biol. Chem.* **244**, 319 (1969).

[k] T. C. Vanaman, S. J. Wakil, and R. L. Hill, *J. Biol. Chem.* **243**, 6409, 6420 (1968).

[l] T. Ando and K. Suzuki, *Biochim. Biophys. Acta* **140**, 375 (1967).

[m] A. Tsugita, M. Kobayashi, T. Kajihara, and B. Hagihara, *J. Biochem.* **64**, 727 (1968).

[n] D. M. Blow, J. J. Birktoft, and B. S. Hartley, *Nature* **221**, 337 (1969).

[o] W. Rombauts and H. Fraenkel-Conrat, *Biochemistry* **7**, 3334 (1968).

[p] K. Dus, K. Sletten, and M. D. Kamen, *J. Biol. Chem.* **243**, 5507 (1968).

[q] R_1 and R_2 represent the amino acid residues involved in thermolysin hydrolysis of a peptide bond, $NH_2 \ldots R_1 - R_2 \ldots COOH$.

[r] Cys(Cm), S-carboxymethylcysteine; Lys(Me), ε-N-methyllysin.

[s] Italics represents minor hydrolysis.

[31] H. Matsubara, A. Singer, and R. M. Sasaki, *Biochem. Biophys. Res. Commun.* **34**, 719 (1969).

[32] K. Morihara and M. Ebata, *J. Biochem.* **61**, 149 (1967).

[33] K. Morihara, *Biochem. Biophys. Res. Commun.* **26**, 656 (1967).

[34] K. Morihara, H. Tsuzuki, and T. Oka, *Arch. Biochem. Biophys.* **123**, 572 (1968).

[35] K. Morihara and T. Oka, *Biochem. Biophys. Res. Commun.* **30**, 625 (1968).

supported the general conclusion obtained with natural substrates. Thermolysin requires L-configuration at the sensitive residue.[32] Complete absence of a free amino group in this area is essential. Slightly less essential is the absence of free carboxyl groups from the immediate vicinity of the sensitive peptide bond.[30, 32] Thus thermolysin is an endopeptidase. However, ω-amino or carboxyl groups alter the activity only slightly.[24] The appearance of specificity is affected not only by the sensitive residue but also by the nature of at least five residues in its neighborhood.[35] This might lead to unexpected cleavages in proteins. Thermolysin does not have amidase and esterase activity,[17, 20] but it has a strong elastase activity.[20]

Thermolysin does not cleave the peptide bond at the amino site of a hydrophobic amino acid residue which has a proline residue at the carboxyl site,[17, 24] regardless of the presence or absence of the second residue attached to the carboxyl group of proline.[31] The specificity of thermolysin was shown to be the same at extremely different temperatures.[17, 24]

The unique specificity of thermolysin led to its broad and effective use in studies of protein sequences.[23]

Kinetic Properties. The Michaelis-Menten constant (K_m) and maximum velocity (V_{max}) determined at $0.1 \times 10^{-6} M$ enzyme, pH 8.0, and 40°, are $19.2 \times 10^{-3} M$ and $161.3 \times 10^{-6} M$ per minute per milligram protein N, respectively, for Cbz-Gly-L-Pro-L-Leu-Gly-L-Pro; $26 \times 10^{-3} M$ and $178.7 \times 10^{-6} M$ per minute per milligram protein N for Cbz-Gly-L-Leu-amide.[32] The maximum velocity of native and heat-treated (80°, 1 hour) enzyme was the same, but the Michaelis constant is different: heat-treated enzyme showed a larger value than that of native enzyme, suggesting that no large alteration appeared to occur in its conformation.[9]

Distribution. The enzymes having a similar specificity are found in various organisms: *B. subtilis* neutral proteinase,[36, 37] *Pseudomona aeruginosa* elastase,[38] *Staphylococcus griseus* neutral proteinase,[34] *Aspergillus oryzae* neutral proteinases,[34] snake venom proteinase,[39] hog thyroid proteinase,[40] and *Bacillus megaterium* proteinase.[41]

[36] J. D. McConn, D. Tsuru, and K. T. Yasunobu, *J. Biol. Chem.* **239**, 3706 (1964).
[37] J. Feder and C. Lewis, Jr., *Biochem. Biophys. Res. Commun.* **28**, 318 (1967).
[38] K. Morihara and H. Tsuzuki, *Arch. Biochem. Biophys.* **114**, 158 (1966).
[39] G. Pfliederer and A. Krauss, *Biochem. Z.* **342**, 85 (1965).
[40] L. F. Kress, R. J. Peanasky, and H. M. Klitgaard, *Biochim. Biophys. Acta* **113**, 375 (1966).
[41] J. Millet and R. Acher, *Biochim. Biophys. Acta* **151**, 302 (1968).

Note Added in Proof

Readers should also refer to a spectrophotometric assay procedure which was originally applied to the *Baccillus subtilis* neutral protease [J. Feder, *Biochem. Biophys. Res. Commun.* **32**, 326 (1968)]. An example of the hydrolysis at a tryptophan residue by thermolysin was observed during the sequence study of carboxypeptidase A [R. A. Bradshaw, *Biochemistry* **8**, 3871 (1969)].

[47] Pronase

By YOSHIKO NARAHASHI

Pronase is a mixture of several proteolytic enzymes, including endopeptidases and exopeptidases which are produced by a strain of *Streptomyces griseus* K-1.[1-6] Those that have been ascertained are neutral and alkaline proteinases, aminopeptidases, and carboxypeptidase.[7] Methods for the determination of activity of these enzymes will be described below.

Assay Method

DETERMINATION OF PROTEINASE ACTIVITY BY THE CASEIN DIGESTION METHOD

Principle. The casein-275 nm method, based on the method of Kunitz[8] and modified by Hagihara *et al.*[9] is used with further modification for the determination of both neutral and alkaline proteinases. This method determines the activity producing trichloroacetic acid-soluble product from casein using extinction at 275 nm. For the activity of both neutral and alkaline proteinases, pronase solution is directly used without pretreatment which involves inactivation of only neutral proteinases by EDTA. The distribution ratio of neutral and alkaline proteinases entails

---

[1] Y. Narahashi and M. Yanagita, *Sci. Papers Inst. Phys. Chem. Res. (Tokyo)* **59**, 44 (1965).

[2] Y. Narahashi and M. Yanagita, *J. Biochem. (Tokyo)* **62**, 633 (1967).

[3] A. Hiramatsu and T. Ouchi, *J. Biochem. (Tokyo)* **54**, 462 (1963).

[4] M. Nomoto, Y. Narahashi, T. Ouchi, and A. Hiramatsu, *Abstr. Intern. Congr. Biochem. 6th New York* **4**, 123 (1964).

[5] S. Wählby, *Biochim. Biophys. Acta* **151**, 394 (1968).

[6] M. Trop and Y. Birk, *Biochem. J.* **109**, 475 (1968).

[7] Y. Narahashi, K. Shibuya, and M. Yanagita, *J. Biochem. (Tokyo)* **64**, 427 (1968).

[8] M. Kunitz, *J. Gen. Physiol.* **30**, 291 (1947).

[9] B. Hagihara, *Ann. Rept. Sci. Works Osaka Univ.* **2**, 35 (1954).

two determinations, one with pronase dissolved in a buffer, which gives the sum of neutral and alkaline proteinases, the other with the pronase solution dialyzed against 0.05 M ethylenediaminetetraacetic acid disodium salt (EDTA), which measures the alkaline proteinases. The difference in the two values is the neutral proteinases.

Reagents

> Substrate, 2% casein. The substrate solution is made by suspending 2 g of casein (Hammarsten, "Merck" Ltd.) in 80 ml of 0.1 M sodium borate. The solution is heated for 10 minutes in boiling water and its pH is adjusted to 7.5 with 0.5 N hydrochloric acid. The volume of the solution is adjusted to 100 ml with distilled water. This casein solution should be stored in a freezer. Prior to the assay, the frozen casein is dissolved in a water bath at 40°.
>
> Trichloroacetic acid (TCA), 0.11 M, containing 0.22 M sodium acetate and 0.33 M acetic acid.
>
> Enzyme solution containing 4 to 40 μg of pronase per milliliter in 0.1 M sodium borate–0.05 M hydrochloric acid buffer containing 5 mM calcium chloride, pH 7.5. (For the determination of neutral and alkaline proteinases.)
>
> Pronase solution dialyzed against 0.05 M EDTA. The concentration of pronase before dialysis is about 0.001–0.010%. (For the determination for alkaline proteinases.)

Procedure. To 2 ml of 2% casein in a test tube, equilibrated at 40°, is added 2 ml of a suitably diluted enzyme solution, also equilibrated at 40°. Four milliliters of 0.11 M TCA solution is added to the reaction mixture exactly 10 minutes after addition of the enzyme. The resulting suspension is kept at 40° for 20 minutes and is then filtered through Toyo No.-5B filter paper. The filter paper used should not contain materials giving absorption at 275 nm. The optical density of the filtrate is read at 275 nm. The reading is corrected for the value of blank which is mixed with enzyme solution and TCA solution before the addition of substrate.

Definition of Activity Unit. One unit of caseinolytic activity is defined as the quantity of an enzyme giving an absorbancy equivalent to 1 μg tyrosine per minute under the conditions described. The unit is determined by the standard curve shown in Fig. 1, and represented by $[PU]_{\mu g\,Tyr}^{Cas.275}$.

According to this method, pronase-P, also known as pronase-research grade, shows 1,250,000 $[PU]_{\mu g\,Tyr}^{Cas.275}$ per gram. Several types of product

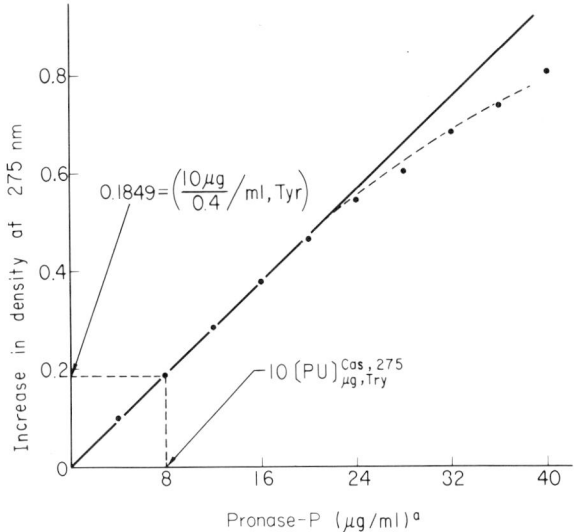

Fig. 1. Standard activity curve of pronase-P by casein-275 nm method. Indicated concentrations of pronase represent the concentrations of enzyme solution used for the assay, not those of enzyme in the reaction mixture.

such as pronase-E, -research grade, -AS, -AF, and -D are commercially available and graded by the Kaken Chemical Co., Ltd., Tokyo, Japan. For instance, pronase-research grade has been shown to have 45,000 PUK/g. PUK is an abbreviation for proteinase activity unit of Kaken with casein as substrate at pH 7.4. One unit of PUK has been defined as the activity of enzyme which liberates, per minute at 40°, a digestion product equivalent to 25 μg of tyrosine.

Determination of the Esterase Activity of Proteinases by a Titrimetric Method

The esterase activity of pronase results from three alkaline proteinases; one of them hydrolyzes benzoyl-L-arginine ethyl ester (BAEE) and the other two enzymes hydrolyze actyl-L-tyrosine ethyl ester (ATEE), and, therefore, this method is the most useful when alkaline proteinases are isolated from pronase. Wählby[5] measured the activity of alkaline proteinases which hydrolyze acetyl-L-tyrosine ethyl ester, using p-nitrophenyl acetate as a substrate, by means of the spectrophotometric method of Huggins and Lapides.[10]

Principle. The potentiometric determination of esterase activity of

---

[10] C. Huggins and J. Lapides, *J. Biol. Chem.* **170**, 467 (1947).

trypsin, which was first described by Schwert *et al.*,[11] is applied with small modification for pronase. The principle of the method is a continuous titratiton of liberated carboxyl groups from an appropriate ester, using a potentiometer. In our laboratory, Radiometer Model TTTlc in conjunction with a TTA31 titration assembly and a type SBR2 recorder (Radiometer, Copenhagen) and $0.1\ N$ NaOH as titrating agent have been used.

Reagents

Substrates. $0.1\ M$ BAEE in distilled water or $0.1\ M$ ATEE in ethanol.

Buffer. $0.005\ M$ Tris(hydroxymethyl)aminomethane-hydrochloric acid buffer containing $0.04\ M$ sodium chloride and $0.02\ M$ calcium chloride, pH 8.0.

Enzyme solution. 1% (for hydrolysis of ATEE) or 0.2% (for hydrolysis of BAEE) pronase dissolved in 1 mM hydrochloric acid.

Titrating agent. $0.1\ M$ sodium hydroxide.

Procedure. All the measurements are made at pH 8.0 and at 37°. Three milliliters of the buffer (pH 8.0) and 0.3 ml of $0.1\ M$ BAEE or 0.3 ml of $0.1\ M$ ATEE are pipetted into a 5-ml reaction vessel with a jacket for circulating thermostatted water and the solution is stirred with a small mechanical stirrer. A slow stream of nitrogen that had been washed in a medium similar to that in the reaction vessel is used to expel carbon dioxide. After standing for 5 minutes, the solution is adjusted to pH 8.0 by the addition of $0.1\ M$ sodium hydroxide, and 0.1 ml of a suitably diluted enzyme solution is added to the reaction vessel.

Esterase activity by this method is expressed as the amount of reacted BAEE or ATEE in micromoles per minute per milliliter of enzyme solution.

Other substrates used for BAEE are tosyl-L-arginine methyl ester, benzoyl-L-arginine methyl ester, and tosyl-L-lysine methyl ester.

Since the hydrolysis of ATEE is apparently of first order, the results are calculated on the basis of the initial slope.

Esterase activity with BAEE may be determined by both potentiometric[12] and the spectrophotometric methods.[13] However, the other esterase activity toward ATEE should not be determined by the spectro-

[11] G. W. Schwert, H. Neurath, S. Kaufman, and J. E. Snoke, *J. Biol. Chem.* **172**, 221 (1948).

[12] M. Laskowski, Vol. II, p. 36.

[13] B. C. W. Hummel, *Can. J. Biochem. Physiol.* **37**, 1393 (1959).

photometric method of Schwert and Takenaka[14] because of the high blank value resulting from low affinity of the enzyme toward ATEE and of inactivation of enzyme by ethanol or methanol used for dissolving the substrate. When the concentration of ethanol or methanol in the reaction mixture is above 10%, the enzyme activity is inhibited. Therefore, it is impossible to use benzoyl-L-tyrosine ethyl ester for ATEE as a substrate since 50% methanol has to be used for dissolving the substrate.[13]

Determination of Aminopeptidase Activity of Pronase

Principle. It appears that aminopeptidases present in pronase have a substrate specificity similar to that of the leucine aminopeptidase from swine kidney,[7,15] and hydrolyze amino acid amides and peptide bonds of N-terminal amino acids of di-, tri-, and polypeptides whose amino group must be free.

The sensitive and convenient substrate for this enzyme is L-leucylglycine. Hydrolysis of the substrate is measured by the ninhydrin method of Yemm and Cocking[16] for the determination of amino groups liberated.

Reagents

Substrate. 4.8 mM L-leucylglycine in distilled water, adjusted to pH 8.0 with sodium hydroxide.

Enzyme contained 2 to 10 μg pronase per milliliter in 0.05 M Veronal sodium–0.05 N HCl buffer containing 0.01 M calcium chloride and 0.001 M cobaltous chloride, pH 8.0.

Hydrochloric acid, 0.05 N

Citrate buffer, 0.2 M, pH 5.0

Citric acid ($C_6H_8O_7 \cdot H_2O$, 21.008 g) is dissolved in 200 ml of distilled water, mixed with 200 ml of 1 N sodium hydroxide, and diluted to 500 ml. Store in the cold with a little thymol.

Potassium cyanide, 0.01 M

Ethanol in water, 60% by volume

Potassium cyanide–methyl Cellosolve solution. 5 ml of 0.01 M potassium cyanide are diluted to 250 ml with methyl Cellosolve.

Methyl Cellosolve–ninhydrin solution. 5% (w/v) solution of ninhydrin in methyl Cellosolve is prepared.

Potassium cyanide–methyl Cellosolve–ninhydrin solution. 50 ml of the methyl Cellosolve–ninhydrin solution are mixed with 250

[14] G. W. Schwert and Y. Takenaka, *Biochim. Biophys. Acta* **26**, 570 (1955).
[15] E. L. Smith and R. L. Hill, *in* "The Enzymes" (P. D. Boyer, H. Lardy, and K. Myrbäck, eds.), Vol. 4, p. 37. Academic Press, New York, 1960.
[16] E. W. Yemm and E. C. Cocking, *Analyst* **80**, 209 (1955).

ml of the potassium cyanide—methyl Cellosolve solution. The resulting solution is at first red, but soon becomes yellow. It should be stored overnight before use.

Procedure. One milliliter of suitably diluted enzyme solution is in-cubated at 40° with 1 ml of L-leucylglycine soultion and the enzyme reaction is terminated exactly 10 minutes later by the addition of 1 ml of 0.05 N hydrochloric acid. After 8-fold dilution of the reaction mixture with distilled water, 1 ml of it is mixed with 0.5 ml of citrate buffer and 1.2 ml of the potassium cyanide–methyl Cellosolve–ninhydrin solution is added. The mixed solution is heated at 100° for 15 minutes, cooled for 5 minutes in running tap water, 3 ml of ethanol are added, and the color intensity is read at 570 nm. The reading is corrected for the value of blank which is mixed with enzyme solution and 0.05 N hydro-chloric acid before the addition of substrate.

One aminopeptidase unit is defined as the amount of enzyme which catalyzes the hydrolysis of 1 micromole of L-leucylglycine (LG) per minute under the conditions described.

As shown in Fig. 2, 1 g of pronase-P, research grade, has 25,000 aminopeptidase units.

Aminopeptidase activity of pronase can also be determined by a method of Goldbarg and Rutenburg, using L-leucyl-β-naphthylamide as a substrate.[17]

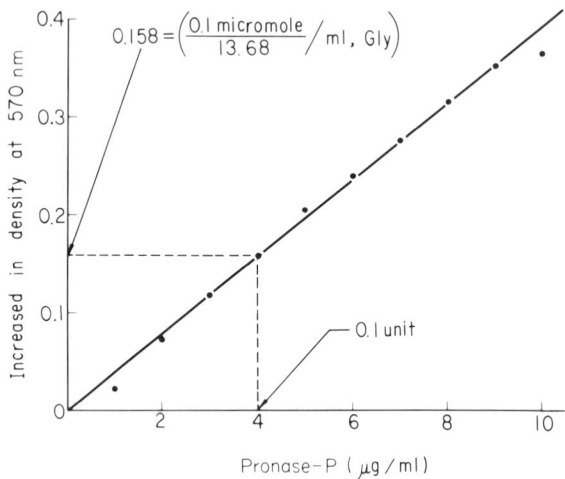

Fig. 2. Standard curve of aminopeptidase activity of pronase-P.

[17] J. A. Goldbarg and A. M. Rutenburg, *Cancer* **11**, 283 (1958).

DETERMINATION OF CARBOXYPEPTIDASE ACTIVITY OF PRONASE

Principle. This enzyme hydrolyzes carbobenzoxy-glycyl-L-leucine at the glycylleucyl linkage.[2,7] Assay is conducted by the ninhydrin method of Yemm and Cocking[16] for liberated amino groups.

Reagents

   Substrate. 0.04 M carbobenzoxy-glycyl-L-leucine (CGL). In 10 ml of 0.1 M sodium borate containing 0.01 M calcium chloride, pH 7.6, is dissolved CGL (322.3 mg), adjusted to pH 7.6 with 1 N sodium hydroxide, and diluted to 25 ml with distilled water.

   Enzyme sample contained 10 to 100 μg of pronase per milliliter of 0.1 M sodium borate–0.05 M hydrochloric acid buffer containing 0.01 M calcium chloride, pH 7.6.

Procedure. Five-tenths milliliter of 0.2 M citrate buffer, pH 5.0, and 1.2 ml of potassium cyanide–methyl Cellosolve–ninhydrin solution are pipetted into the test tube provided with an aluminum cap. To 1 ml of 0.04 M CGL preincubated at 40° 1 ml of enzyme solution is added at zero time. At zero time and at 5-minute intervals, a 0.5 ml aliquot of the reaction mixture is removed and pipetted into the test tube containing ninhydrin solution (prepared previously). The procedure of color development by the ninhydrin reagent is the same as that for aminopeptidase assay. A blank value is estimated by the reaction mixture pipetted at zero time.

   One carboxypeptidase unit is defined as the amount of enzyme which hydrolyzed 1 micromole of CGL per minute under the conditions described. Figure 3 shows the standard curve of carboxypeptidase activ-

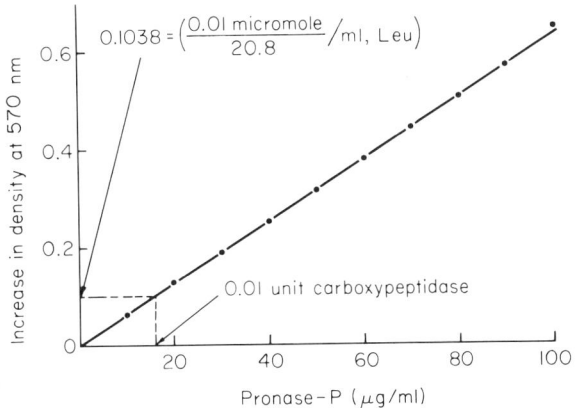

FIG. 3. Standard curve of carboxypeptidase activity of pronase-P.

ity of pronase, and usually pronase-P has 610 units carboxypeptidase activity per gram.

Purification Procedure

Pronase is a partially purified preparation obtained from the culture broth of *Streptomyces griseus* K-1, purified on an industrial scale at Kaken Chemical Co., Ltd. The isolation and purification of each enzyme, neutral and alkaline proteinase, aminopeptidase, and carboxypeptidase, from pronase will be described below.

Step 1. Initial Group Separation.[7] The best initial purification procedure developed for group separation of these enzymes is chromatography of pronase on CM-cellulose column. A column (4.1 × 40 cm) of CM-cellulose (100 g by dry weight, Serva, 0.57 meq/g) is equilibrated with 0.01 M sodium acetate buffer containing 5 mM calcium chloride, pH 5.2. In the pH range of 5.0 to 5.5, chromatographic behavior of pronase is independent of pH. Unless the equilibrium of CM-cellulose with the buffer is sufficient with respect to both pH and calcium ion, neutral proteinases and aminopeptidases become inactive during the chromatography. After application of 200 ml of 10% pronase solution dialyzed against the same buffer, the column is washed with about 500 ml of the starting buffer until the absorption of the effluent at 280 nm will return nearly to a base line value. After appearance of the breakthrough peak, the remaining enzymes are eluted from the column by linear gradient developed by use of 0.25 M sodium chloride in the starting buffer, as shown in Fig. 4. All procedures should be performed at 4°–5°, and a flow rate is 12 ml/cm²/hour.

Highly reproducible results are obtained under the conditions described and four protein peaks having enzyme activity are resolved. The first peak which is not adsorbed on CM-cellulose consists of neutral proteinase and an aminopeptidase. The aminopeptidase overlaps the protein peak of neutral proteinase but is chromatographically distinct from it. The second peak appears immediately after the beginning of sodium chloride gradient and contains neutral proteinase together with a small amount of aminopeptidase and carboxypeptidase. The third peak includes carboxypeptidase and two different kinds of proteinase, one of which hydrolyzes acetyl-L-tyrosine ethyl ester and the other hydrolyzes benzoyl-L-arginine ethyl ester. The fourth peak contains mainly a proteinase which hydrolyzes acetyl-L-tyrosine ethyl ester.

Proteinases present in peaks I and II are neutral proteinase in contrast to peaks III and IV in which alkaline proteinases are eluted. Three alkaline proteinases with esterolytic activity are designated in the order of their elution from the column of CM-cellulose as alkaline pro-

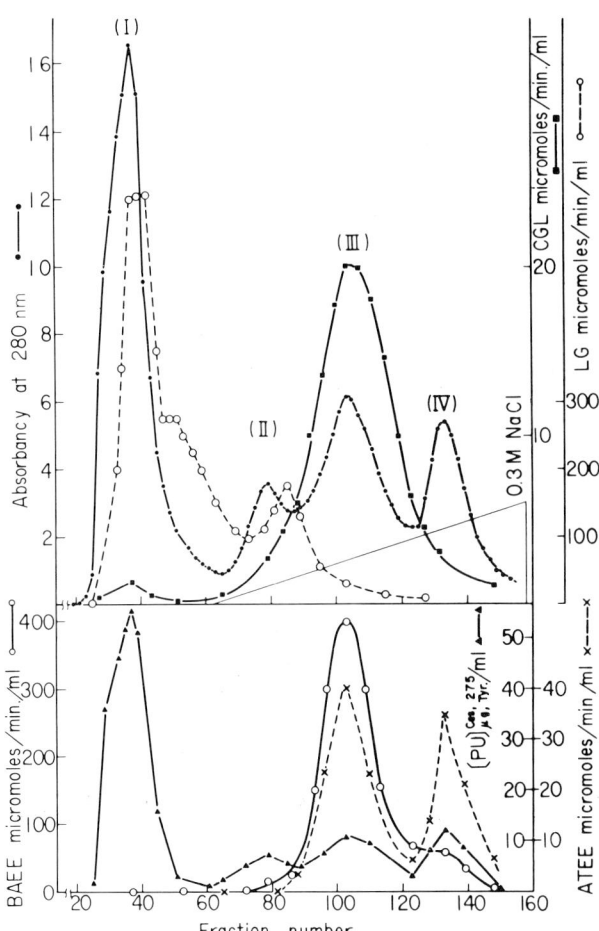

FIG. 4. Chromatography on CM-cellulose of pronase-P. Two hundred milliliters of 10% pronase dialyzed against 0.01 M sodium acetate buffer containing 5 mM calcium chloride, pH 5.2, were applied on a column (4.1 × 40 cm) of CM-cellulose equilibrated previously with the same buffer. After washing the column with 500 ml of the same buffer, a linear gradient was developed using 0.25 M sodium chloride in the starting buffer, pH 5.2. Effluent of 20 ml was collected in one tube.

The top graph gives the locations of the protein, CGL and LG hydrolyzing activities. The bottom chart gives the caseinolytic activity and the esterase activities toward BAEE and ATEE. The symbols are as follows: (●) absorbance at 280 nm. (-○-) activity toward LG, (■) activity toward CGL, (▲) caseinolytic activity, (-○-) activity toward BAEE, (×) activity toward ATEE.

teinase a, b, and c; proteinase a is an enzyme which hydrolyzes ATEE and appears in peak III, proteinase b hydrolyzes BAEE, and c is present in peak IV and hydrolyzes ATEE. Not only the separation of neutral and alkaline proteinases but also that of aminopeptidase and carboxypeptidase is achieved by this first group separation. The recoveries of protein, caseinolytic, BAEE-hydrolyzing, ATEE-hydrolyzing, aminopeptidase, and carboxypeptidase activities are about 90, 80, 75, 75, 90, and 75%, respectively.

Step 2. Isolation of Carboxypeptidase from Alkaline Proteinases.[7] The fractions of peak III in the previous step are combined and dialyzed against 0.05 M sodium acetate buffer containing 0.01 M calcium chloride, pH 5.2. A column of Amberlite CG-50 resin (Rohm & Haas) is equilibrated with 0.05 M sodium acetate buffer containing 0.01 M calcium chloride, pH 5.2, and washed with distilled water. The dialyzed enzyme

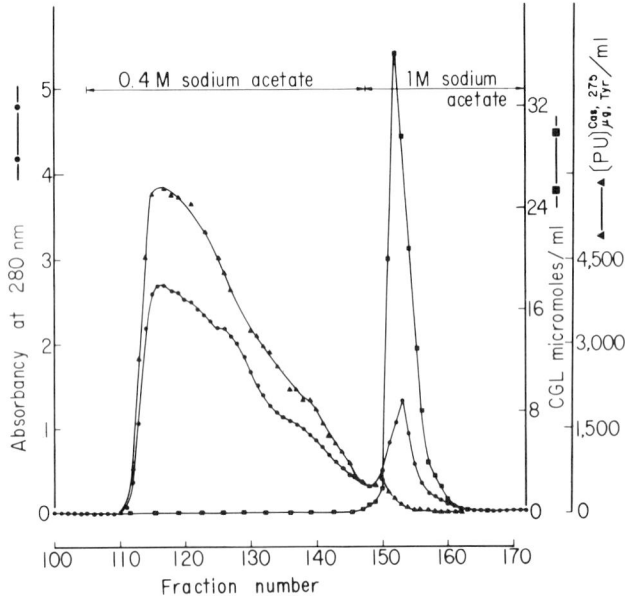

FIG. 5. Chromatographic separation of carboxypeptidase from alkaline proteinases on Amberlite CG-50 resin. One-sixth of the combined solution obtained from peak III in the previous step was applied on the column (1.3 × 25 cm) of Amberlite CG-50 resin. After washing the column with 1 liter of the starting buffer, the elution was developed by a stepwise increase of sodium acetate (0.4 M, 1 M) at flow rate of 10 ml/hour. Effluent of 5 ml was collected in one tube.

Only the section of the elution pattern where the enzyme activities appeared is shown in the figure. (●) Absorbance at 280 nm, (■) carboxypeptidase activity, (▲) caseinolytic activity.

solution is applied to the column, the column is first washed with the same buffer, and the enzymes adsorbed are eluted by a stepwise increase in sodium acetate concentration. Proteinases are first eluted at the concentration of 0.4 M sodium acetate, carboxypeptidase is eluted at the concentration of 1 M sodium acetate.

Step 3. Separation and Purification of Alkaline Proteinase.[7] By the methods of the first and second steps, alkaline proteinases are freed from aminopeptidase, neutral proteinases (step 1), and carboxypeptidase (step 2). However, two (a, b) of the three alkaline proteinases still remain inseparable from each other in step 2. Complete separation of these alkaline proteinases, a and b, is achieved by subsequent CM-Sephadex chromatography of the proteinase sample obtained from the previous step, and proteinase c is further purified by rechromatography on CM-cellulose under the same conditions as step 1.

The fractions with proteinases obtained by step 2 are combined and

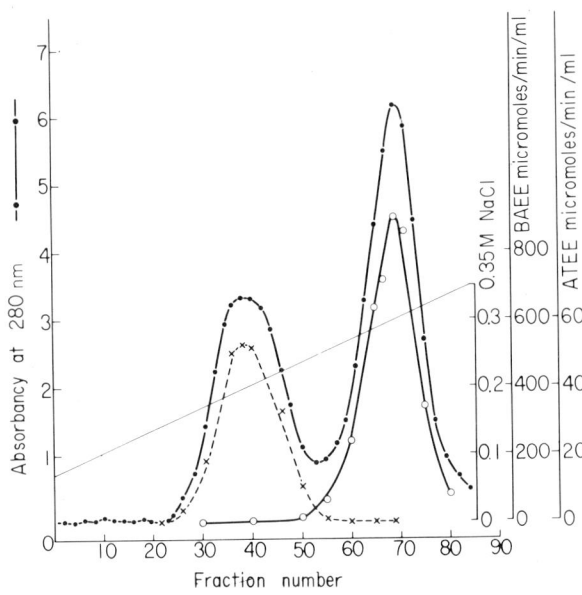

Fig. 6. Chromatographic separation of alkaline proteinases a and b on CM-Sephadex. The combined enzyme solution obtained from peak III on CM-cellulose (610 ml) was dialyzed against 0.01 M sodium acetate buffer containing 0.07 M sodium chloride, pH 4.2. Dialyzed enzyme solution was applied on the column (2.8 × 40.5 cm). After washing the column with 250 ml of the same buffer, a linear gradient was developed with sodium chloride. Effluent of 10 ml was collected in one tube at flow rate of 50 ml/hour. (●) Absorbancy at 280 nm, (○) activity toward BAEE, (×) activity toward ATEE.

dialyzed against 0.01 M sodium acetate buffer containing 0.07 M NaCl, pH 4.2.

The fractions of peak III obtained from step 1 may also be used, and carboxypeptidase which coexists with alkaline proteinases in peak III is denatured and removed by extensive dialysis of the combined solution against 0.01 M sodium acetate buffer, pH 4.2. The dialyzed enzyme solution is applied to a column of CM-Sephadex (C-50, capacity; 4.5 ± 0.5 meq/g, Pharmacia) previously equilibrated with the same buffer. After being washed with the same buffer, the column is eluted with a linear gradient of 0.07 M to 0.35 M sodium chloride in 0.01 M sodium acetate buffer, pH 4.2. By using this chromatographic step, alkaline proteinase a which hydrolyzes acetyl-L-tyrosine ethyl ester is first eluted and is completely separated from the other alkaline proteinase b which hydrolyzes benzoyl-L-arginine ethyl ester and is eluted at a higher concentration of sodium chloride.

Recoveries of alkaline proteinase a and b in this step are about 70 and 80%, respectively, and by the use of the above steps, purified alkaline proteinase b is obtained in approximate yield of 50–60% of pronase and is purified 12-fold over pronase-P.

Wählby[5] reported purification of three alkaline proteinases of pronase by consecutive chromatography on CM-cellulose and phosphorylated cellulose. Our effort for separation of these alkaline proteinases by using phosphorylated cellulose, sulfoethylcellulose, or rechromatography on CM-cellulose did not effect complete separation but chromatography on CM-Sephadex did. For further purification, molecular sieving with Sephadex G-100 or G-75 is recommended for each of the separated alkaline proteinases.

Step 4. Separation of Neutral Proteinases from Aminopeptidases.[7] In the CM-cellulose chromatography of pronase (step 1), a greater part of neutral proteinase was found together with aminopeptidase in the non-adsorbed fractions. For separation of aminopeptidases from neutral proteinases, the fractions with neutral proteinase in peak I are combined and the pH of the enzyme solution is adjusted to 7.0 with sodium hydroxide. The enzymes are salted out with ammonium sulfate (0.5 saturation). The precipitate is collected, dissolved in 0.005 M Veronal sodium–hydrochloric acid buffer containing 1 mM calcium chloride, pH 7.6, and dialyzed against the same buffer. The dialyzed enzyme solution is then concentrated by lyophilization. The lyophilized enzyme is dissolved in the same buffer (about 5%) and is applied to a column of DEAE-Sephadex A-50 (capacity; 3.5 ± 0.5 meq/g, size; 40–120 μ, Pharmacia) which was equilibrated with the same buffer. The equilibrium should be checked by determining both the pH and the counter-ion concentration of the washings. Otherwise, reproducible results may not

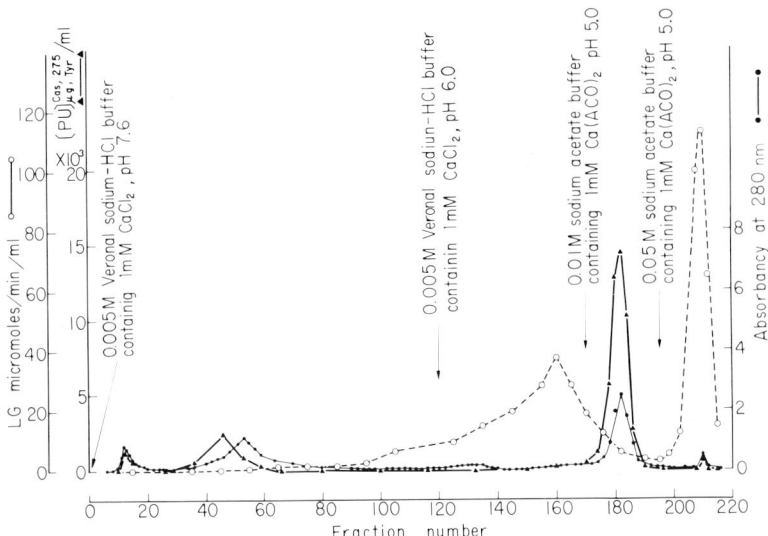

Fig. 7. Chromatographic separation of neutral proteinases from aminopeptidases. The lyophilized enzyme obtained from peak I (200 mg) was dissolved in 10 ml of 0.005 M Veronal sodium–hydrochloric acid buffer containing 1 mM calcium chloride, pH 7.6, and the enzyme solution was applied on the column (2.0 × 55 cm) of DEAE-Sephadex equilibrated with the same buffer. After washing the column with 1 liter of the same buffer, enzymes were eluted by a stepwise method. An effluent of 10 ml was collected in one tube at flow rate of 25 ml/hour. (●) Absorbancy at 280 nm, (▲) caseinolytic activity, (○) LG-hydrolyzing activity.

be expected. After washing the column with the same buffer, the elution is carried out by a stepwise increase of concentration of the buffer and change of pH as shown in Fig. 7. The neutral proteinase found in the nonadsorbed fractions (peak I) on CM-cellulose chromatography is separated from amino peptidase and is shown to be further fractionable into three proteinase components. The amino peptidase is also separated into two fractions. The recoveries of protein, caseinolytic, and amino peptidase activities are about 90, 70, and 100%, respectively. The use of DEAE-cellulose instead of DEAE-Sephadex to separate amino peptidase and neutral proteinase was not successful.

Properties[2,7]

Purity of each enzyme has not yet been established. Recently, in our laboratory, alkaline proteinase a which hydrolyzes acteyl-L-tyrosine ethyl ester was crystallized in columns,[18] and its homogeneity was confirmed in free-boundary electrophoresis and column chromatography.

[18] Y. Narahashi, M. Yanagita, and K. Sibuya, *Sym. Enzyme Chemistry Japan*, p. 168 (1968).

The neutral proteinases are stable in a narrow pH range of 7.0–7.5 in the absence of calcium ion, but highly stable at pH of 5.0–9.0 in the presence of calcium ion. Addition of calcium ion to the solution in which neutral proteinase is dissolved protects the enzyme from denaturation during purification procedures such as dialysis, chromatography, and salting out, etc. The alkaline proteinase b which hydrolyzes benzoyl-L-arginine ethyl ester is the most stable in 1 mM hydrochloric acid and is also stable in a buffer of pH 4.0–5.0, and becomes unstable above pH 5.5. Alkaline proteinases a and c are stable at a pH range of 4.0 to 6.5 and become unstable at pH below 4.0 and above 7.0. Aminopeptidase and carboxypeptidase are stable at pH 5.0–8.0 in the presence of calcium ion.

Aminopeptidase and carboxypeptidases are both metalloenzymes, and the binding of activating metal with EDTA renders the enzyme inactive. Upon reactivation by calcium and cobalt ions, both activities are restored. Aminopeptidase is heat-stable below 80° but is very labile on dialysis against distilled water. Carboxypeptidase is less heat-stable but is stable on dialysis against distilled water. Neutral proteinases are completely inhibited by EDTA but not by diisopropylphosphorofluoridate (DFP), while three alkaline proteinases are inhibited by DFP but are not affected by EDTA. The alkaline proteinase which hydrolyzes benzoyl arginine ethyl ester is inhibited by 1-chloro-3-tosylamido-7-amino-2-heptanone (TLCK) but other two alkaline proteinases which hydrolyze acetyl-L-tyrosine ethyl ester are not inhibited by either tosylyhenyl-alanine chloromethyl ketone (TPCK) or TLCK, which are specific inhibitors for chymotrypsin and trypsin, respectively.[19, 20]

Reporting the substrate specificity of one of the neutral proteinases by using synthetic substrates, Morihara[21] indicated that the enzyme hydrolyzes the peptide bond containing the amino group of leucine, phenylalanine, or tyrosine. Independently, Narahashi found that the enzyme is capable of hydrolyzing the peptide bond containing the leucine[7] amino group.

Some kinetic paramters of alkaline proteinase b which hydrolyze benzoyl-L-arginine ethyl ester have been studied. The enzyme hydrolyzes, in general, substrates similar to trypsin in pancreas. The K_m values are $9.0 \times 10^{-6} M$ for BAEE, $7.7 \times 10^{-6} M$ for TAME, and $1.4 \times 10^{-5} M$ for BAPNA.

[19] G. Schoellmann and E. Shaw, *Biochemistry* **2**, 252 (1963).
[20] E. Shaw, M. Marse-Guia, and W. Cohen, *Biochemistry* **4**, 2219 (1965).
[21] K. Morihara, H. Tsuzuki, and T. Oka, *Arch. Biochem. Biophys.* **123**, 572 (1968).

[48] Urinary Plasminogen Activator (Urokinase)[1]

By WILFRID F. WHITE and GRANT H. BARLOW

Assay Methods

Principle. Two general types of assay are available, based on (1) the specific activator of plasminogen and (2) the esterolytic activity of urokinase toward synthetic substrates. We have used an adaptation of the former method to monitor the purification process and have used the latter method for a precise evaluation of the finished products.

First Method (Fibrin Tube)

Reagents

Sodium barbital buffer, $0.05\,M$, pH 7.6
Bovine thrombin (17 NIH units/ml)
Fibrinogen (bovine fraction I), 16 mg/ml in barbital buffer

Procedure. Transfer 0.2 ml of ten different dilutions (made with 0.1% human serum albumin, Fr. V, Pentex, Inc., Kankakee, Ill.) of the urokinase to be assayed to separate tubes. Add to each, in order, 0.3 ml bovine thrombin and 1.0 ml of fibrinogen solution. Incubate the tubes at 37° for 16 hours. If the proper dilutions have been used, a graded series of clots will remain, ranging from unchanged to complete lysis. The level of urokinase resulting in 50% clot lysis is estimated and the activity of the original sample is calculated against a standard of assigned potency, which is run simultaneously.

Second Method (Esterolytic)

Reagents

Phosphate $(0.06\,M)$ NaCl $(0.09\,M)$ buffer, pH 7.5
Perchloric acid $(0.75\,M)$
Methanol-spectroanalyzed $0.1\,M$
Potassium permanganate 2%
Sodium sulfite 10%
Sulfuric acid 67% (v/v)
N-α-Acetyl-L-lysine methyl ester hydrochloride (ALME) (Cyclo Chem. Corp., Los Angeles, Calif.) 0.24%

---

[1] W. F. White, G. H. Barlow, and M. M. Mozen, *Biochemistry* 5, 2160 (1966).

Chromotropic acid reagents: 200 mg of 4,5-dihydroxy-2,7-naph-thalene-disulfonic acid disodium (Eastman Kodak Co., Rochester, N.Y.) dissolved in 10 ml of distilled water and then mixed with 90 ml of sulfuric acid reagent.

Procedure. A standard methanol curve is made by diluting the methanol stock with phosphate buffer to give samples in the range 0.25 to 1.5 micromoles/ml. A 1-ml aliquot of each dilution is transferred to a tube containing 0.5 ml of 0.75 M perchloric acid; a blank is prepared in the same manner using phosphate buffer.

The contents of each tube are mixed and 1 ml from each is trans-ferred to a test tube suitable for boiling. To each tube is added 0.1 ml $KMnO_4$ close to the fluid surface. The contents are mixed well, and exactly 1 minute is allowed for the oxidation of methanol to HCHO; this is followed by the addition of 0.1 ml sodium sulfite to each tube which is immediately shaken vigorously to insure thorough mixing and decolorizing of the solution. Four milliliters of chromotropic acid reagent is added to each tube, and the contents mixed well, preferably on a mechanical mixer because of the corrosive and viscous nature of the acid. The tubes, covered with marbles or small beakers, are placed in a boiling water bath for 15 minutes, then cooled to room temperature and the absorbancy read in a spectrophotometer at 580 nm.

With the standard urokinase a dilution series is made containing from 200 to 1000 CTA units per milliliter. An 0.2 ml aliquot of each dilution is mixed with 2 ml ALME, and the mixtures are incubated at 37°. After 2 and 32 minutes incubation, respectively, a 1-ml aliquot is removed and added to 0.5 ml of 0.75 M perchloric acid and kept cold until all tubes are ready for centrifugation. From each supernatant, 1-ml aliquots are transferred to boiling tubes and treated in the same manner as described above for methanol. The absorbancy of the 32-minute sample is read using the corresponding 2-minute sample as a blank. An unknown urokinase preparation is assayed in the manner described above and the activity is found by direct extrapolation to the control curve and multiplication of the dilution factor.

Definition of Unit. The unit of urokinase activity, referred to as the CTA unit, was adopted in 1964 by the Committee on Trombolytic Agents, National Heart Institute. This unit is based on the activity of a working standard urokinase preparation[2] which was independently assayed in several laboratories. The standard urokinase preparation was also as-

---

[2] The working standard urokinase is available from WHO International Laboratory for Biological Standards, National Institute for Medical Research, London, England.

sayed by its ability to split the synthetic substrate N-acetyl-L-lysine methyl ester (ALME). One unit of urokinase activity releases 5×10^{-4} micromoles/CTA unit/minute at $37°$.[3]

Purification Procedure

Freshly collected urine from healthy young males is transported rapidly to the processing area and cooled quickly to less than $10°$ by pumping through a heat exchanger into a tank[4] where the temperature is maintained at $10°$ until enough urine is collected for foaming. This is conveniently done in batches of 900–1400 liters by the use of a Model N33 G-300, 3 hp, Lightning Fixed Mounted Agitator. With the agitator blades 15–30 cm below the urine surface a vortex will be created when the agitator is turned on. After 20 minutes agitation, the foam layer is allowed to collect on the liquid surface for 10 minutes. The urine is rapidly drained from the bottom of the tank through a 2-inch pipe fitted with a Teflon diaphragm valve and a sight glass. When the foam level reaches the sight glass, the valve is quickly closed. The foam is then allowed to break down further until an additional 25 minutes has elapsed, when the urine layer is again drawn off. The foam is then liquefied by the addition of a minimal amount of octyl alcohol. Foam concentrates are frozen and stored at $-20°$ unless further processing is carried out immediately.

Step 1. Isolation of Crude Urokinase.[5] The activity is precipitated from fresh or newly thawed foam concentrate by adding $(NH_4)_2SO_4$ to 65% saturation $(2°)$. The precipitate is recovered by centrifugation or filtration, dissolved in 3% NaCl in a volume 1/200 of the original urine volume, and dialyzed overnight against 12 volumes (two changes) of 0.1 M phosphate buffer, pH 6.5, containing 0.1% EDTA. The bag contents are clarified by centrifugation (30,000 g for 30 minutes). This solution, designated crude urokinase, has a specific activity of about 250 units/A_{280}.

Step 2. Purification on IRC-50 Resin. Amberlite IRC-50 resin (previously equilibrated against 0.1 M phosphate buffer, pH 6.5, containing 0.1 M NaCl and 0.1% disodium versenate) is poured as a slurry into a column tall enough to give a bed height of 100 cm and having a cross-

---

[3] This value appears in article by A. J. Johnson, D. L. Kline, and N. Alkjaersig, *Thromb. Diath. Haemorrhag.* **21,** 259 (1969). However, other values ranging from 6.9 to 8.8×10^{-4} have been reported. It is essential that a CTA unit be defined in each laboratory in terms of the standard preparation.

[4] Equipment used during the processing of urokinase is either of stainless steel, polyethylene, or ceramic.

[5] Unless otherwise stated, all purification procedures are carried out at $4°$.

sectional area of about 1 cm⁻² per 600,000 CTA units of crude urokinase. The crude urokinase is introduced into the column, which is then washed with distilled water until the optical absorbancy of the effluent 280 nm is about 0.06. The adsorbed urokinase is eluted with 3% NaCl solution containing 0.1% EDTA. Suitable fractions are collected and a plot is made of the optical absorbancies at 280 nm. The major portion of the peak usually contains 50 to 60% of the starting crude urokinase activity at a specific activity of about 10,000 CTA units/A_{280}. It is usually considered unrewarding to attempt to work up the tail of the peak, which contains only about 10% of the starting activity. A pool is made of the fractions from the main part of the elution peak and it is either lyophilized directly or first concentrated by ultrafiltration and then lyophilized.

Step 3. Gel Filtration on Sephadex G-100 Columns. Step 2 material is further purified by gel filtration in a column of Sephadex G-100. The column is poured to a height of 100–110 cm in a column with a cross-sectional area of about 20 cm⁻² per gram of step 2 material. Figure 1 shows a typical large-scale run made in a standard 4-foot length of

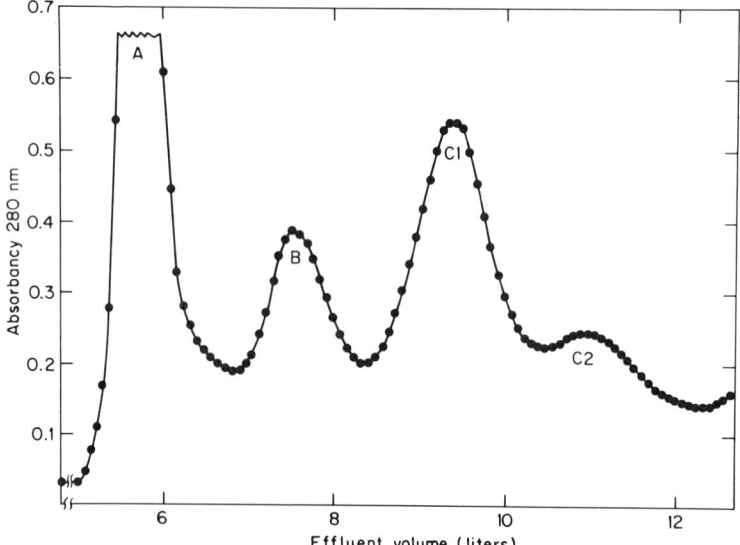

Fig. 1. Gel filtration of step 2 urokinase. Column size: 15.2 × 110 cm. Equilibrated to pH buffer (0.1 M NaCl; 0.1 M phosphate). Sample (5 g step 2 fraction) applied in volume of 73 ml. Flow rate: 200 ml/hour. Fractions cut at low points between peaks except on left side of Cl, where 280 ml was rejected to reduce contamination by peak B. An additional peak, designated D, emerges beyond the right-hand limit of this figure. Reprinted from *Biochemistry* **5**, 2160 (1966). Copyright (1966) by the American Chemical Society. Reprinted by permission of the copyright owner.

6-inch Pyrex pipe, using a 325 mesh stainless steel screen supported by a perforated Teflon plate at the bottom. The clarified sample of step 2 material (having been equilibrated by dialysis against the pH 6.5 buffer, is layered carefully on top of the column, the surface of which is stabilized by means of a Whatman No. 1 paper disk held in place by a lightweight perforated Lucite plate. Filtration is done by gravity alone with a buffer head of about 40 cm above the top of the Sephadex bed. In addition to the peaks shown in Fig. 1, continued elution results in the emergence of a fifth peak, the beginning of which is apparent in the righthand edge of the figure. Bioactivity is found only in the peaks labeled C-1 and C-2. The total yield of activity is in the range of 80 to 85%. The cutting of fractions from the column depends on the use to which the product will be put. It is likely that the smaller molecular weight material (C-2) is produced by enzymatic degradation[1] of the larger material either during the storage of the urine in the bladder or during processing, perhaps both. Therefore, the C-2 material is usually disregarded in the further workup of the product even though it is active in lysing clots as well as offering an interesting subject for study by protein chemists desiring to work with the smallest molecule possessing activator activity. In working up to the main peak (C-1), a few tubes on either edge of the peak are rejected in order to restrict contamination by adjacent peaks. After pooling the tubes from the central region, the C-1 product is concentrated by ultrafiltration. Ordinarily, C-1 type material constitutes at least 70% of the total biological activity contained in the parent step 2A fraction and exhibits a specific activity in the range of 30,000 to 35,000 units/A_{280}.

Step 4. Refractionation of Sephadex G-100. For final removal of contaminants (especially the inert proteins of peak B), fraction C-1 is usually subjected to refiltration on Sephadex G-100. This is done either by combining four or five batches of step 3 material and using the same column as employed in step 3 or by running a single batch of step 3 material in a column having a cross-sectional area one-fifth that of the step 3 column. In either case a large, symmetrical peak of 280 nm absorbancy emerges with the only evidence of impurities being a small peak or shoulder at the leading edge of the main peak. Ninety percent or more of the optical density applied to the column is contained in the main peak, which emerges at an effluent volume equal to 50% ± 1% of the total volume of the column. Expressed in another way, the ratio of the emergence volume for urokinase to that of egg albumin is 0.92. The material from the main peak is concentrated by ultrafiltration, dialyzed against 0.05 M sodium chloride containing 0.1% EDTA, and lyophilized. This step 4 product has a potency in the range of 50,000 to 75,000 CTA units/A_{280}. It has been found to be suitable for clinical use as well as for laboratory experiments in the activation of plasminogen.

Physical Properties

Purity. Despite the symmetrical appearance of C-1 peak after re-chromatography on Sephadex G-100, examination of the step 4 product by polyacrylamide disc electrophoresis and by immunochemical methods suggested the presence of two subtypes of urokinase. A separation into two active types, designated S1 and S2, has been achieved by chromatography on hydroxylapatite.[1] A list of physical constants of these two types is given in the table.[6] Despite the difference in the molecular weights

PHYSICAL CONSTANTS ON UROKINASE

	S1	S2
Sedimentation coefficient ($s_{20,w}^0$), S		
pH 2.6	—	3.18
pH 4.5	2.75	3.33
pH 6.5	2.66	3.27
pH 8.5	2.71	—
Heterogeneity constant (p)		
pH 4.5	0.28×10^{-13}	—
pH 6.5	0.35	—
pH 8.5	0.32	—
Diffusion coefficient ($D_{20,w}^0$), cm²/sec		
pH 6.5	7.41×10^{-7}	—
Partial specific volume,[a] V, ml/g	0.724	0.728
Frictional ratio	1.35	—
Molecular weight		
S and D	31,700	—
Equilibrium	31,300	54,700
Electrophoretic mobility $\times 10^5$ cm²/V sec (pH 4.8, acetate buffer, U 0.1)	+3.5	+2.2
$E_{1\,cm}^{1\%}$ 280 nm, pH 6.5	13.2	13.6

[a] Calculated from amino acid composition data. Reprinted from *Biochemistry* **5**, 2160 (1966). Copyright (1966) by the American Chemical Society. Reprinted by permission of the copyright owner.

[6] An alternative method for the purification of urokinase has been given by A. Lesuk (U.S. Patent 3,355,361), whose crystalline product has been described [*Science* **147**, 880 (1965)]. Ultracentrifugal methods have given a molecular weight of approximately 54,000 and the product shows an activity of about 104,000 CTA units per milligram protein. These values agree closely with our values for type S2 material, leaving unexplained the origin of our S1 species. Lesuk *et al. Thromb. Diath. Haemorrhag.* **18**, 293 (1967) have shown that enzymes such as trypsin can effectively reduce the molecular size of urokinase. However, Burges *et al., Nature* **208**, 894 (1965) have shown by chromatography in calibrated Sephadex columns that the activator activity of urine and of crude concentrates from urine shows a molecular weight of 34,500 ± 2000, which agrees with our value for type S1.

obtained by ultracentrifugation as shown in the table, the two forms do not show a separation on Sephadex G-100 on which separately their effluent volumes differ by only 6%. The fibrinolytic activity for S1 is 120,000 CTA units per A_{280} or about 200,000 CTA units per milligram protein,[7] while for S2 the values are 60,000 CTA units per unit A_{280} and 100,000 units per milligram protein.

Stability. Urokinase is a moderately stable enzyme, showing no appreciable loss in activity over years in lyophilized form or over months in sterile solutions at 1 mg/ml or more at refrigerator temperatures. However, in dilute solution it has been found advisable to maintain a concentration of 0.1% EDTA when dialyzing or holding for more than a few hours. Stability is decreased at salt concentration below 0.03 M sodium chloride, and precipitation with loss in activity occurs at very low salt concentrations. In diluting to the levels of activity measured in the fibrinolytic assay, it is advisable to add a protein such as human serum albumin fraction V or gelatin to prevent surface denaturation.

Activators and Inhibitors

Unlike numerous analogous systems in the field of enzymes, no pro-enzyme has been found for urokinase. Apparently urokinase is excreted into the urine in fully activated form.

Human plasma (and serum) contains an inhibitor to urokinase which, in certain pathological conditions, reaches levels ten to twenty times normal.[8] The inhibitor is destroyed by heating at 56° for 30 minutes and is contained in Cohn's fraction IV-I.

Synthetic ε-aminocaproic acid inhibits the action of urokinase on plasminogen[9] as well as on synthetic ester substrates.[10]

Specificity

Urokinase resembles trypsin and plasmin in its activity against simple synthetic substrates, although there are differences in the relative rates for the three enzymes, depending on the substrate. However, urokinase is much more fastidious in its action on naturally occurring proteins, plasminogen being the only protein substrate against which it has been shown to have any activity. By contrast, both trypsin and plasmin attack a variety of proteins including casein, the plasma proteins, ACTH and

[7] Protein determination made by the Folin-Ciocalteau method [O. H. Lowry, N. J. Rosebrough, A. L. Farr, and R. J. Randall, *J. Biol. Chem.* **193**, 265 (1951)].

[8] I. M. Nilsson, H. Krook, N. H. Sternby, E. Soederberg, and N. Soderstrom, *Acta Med. Scand.* **169**, 323 (1961).

[9] N. Alkjaersig, A. P. Fletcher, and S. Sherry, *J. Biol. Chem.* **234**, 832 (1959).

[10] L. Lorand and E. V. Condit, *Biochemistry* **4**, 265 (1965).

β-MSH, all of which resist the action of urokinase. Robbins *et al.*[11] have shown that the activation of plasminogen to plasmin by urokinase is due to the cleavage of a single arginyl-valine bond.

Distribution

Activator activity is found in most tissues and in plasma. However, despite circumstantial evidence that the kidney is primarily involved in regulating the level of circulating activator, Kucinski *et al.*[12] were unable to detect the presence of urokinase in renal venous blood by immunochemical means and also established lack of identity between urokinase and tissue activator. Since the activator produced in culture of human kidney cells is immunochemically identical to urokinase[13] and shows molecular size contrasts indistinguishable from urokinase,[14] it appears reasonable to postulate at this time that urokinase is synthesized in the kidney and is secreted only in the urine.

[11] K. C. Robbins, L. Summaria, B. Hsieh, and R. J. Shah, *J. Biol. Chem.* **242**, 2333 (1967).
[12] C. S. Kucinski, A. P. Fletcher, and S. Sherry, *J. Clin. Invest.* **47**, 1238 (1968).
[13] M. Bernik and H. C. Kwaan, *J. Lab. Clin. Med.* **70**, 650 (1967).
[14] G. H. Barlow and W. F. White, unpublished observations.

[49] Mouse Submaxillary Gland Proteases[1]

By Milton Levy, L. Fishman, and I. Schenkein

Mouse submaxillary glands contain at least four enzymes capable of hydrolyzing tosylarginine methyl ester (TAM), a substrate hydrolyzed by several proteolytic enzymes, including trypsin and thrombin. We will describe a method for the isolation of two of the mouse enzymes, "A" and "D," in pure form from the glands of mature male mice.[1a] The size of the glands and their enzyme contents are controlled by sex hormones. Our preparations were primarily directed toward the isolation of "nerve growth factor"[2] and as a consequence there may be some steps not needed when only the enzymes are wanted.

The enzymes A and D are related but not identical by immunological tests. Using TAM hydrolysis as the criterion, A has much the higher specific activity. On the same basis both are more active than trypsin.

[1] EC 3.4.4—.
[1a] I. Schenkein, M. Boesman, E. Tokarsky, L. Fishman, and M. Levy, *Biochem. Biophys. Res. Commun.* **36**, 156 (1969).
[2] I. Schenkein, M. Levy, E. D. Bueker, and E. Tokarsky, *Science* **159**, 640 (1968).

D is strongly activated by all amino acids (both *d* and *l*) as well as by tosylarginine.[1] The enzymes act on protein substrates such as protamines, histones, and lysozyme but not in proportion to their "tamase" activity as compared with trypsin. They seem to have a strong preference for arginyl bonds.

Assay of Enzymes

The assay reaction is

$$C_7H_7SO_2NH—CH[—(CH_2)_3—NH—C(=NH)—NH_3^+]—COOCH_3 + OH^- \rightarrow$$
$$C_7H_7SO_2NH—CH[—(CH_2)_3—NH—C(=NH)—NH_3^+]—COO^- + CH_3OH$$

It is followed by measurement of the alkali uptake in the pH-stat technique.[3] A manual method is described in Vol. II and spectrophotometric[4] and fluorometric[5] methods have been described for the same reaction with trypsin. With the addition of glycine to the substrate solution (insuring the activation of D) these can be used for the assay of submaxillary enzymes. A unit of enzyme catalyses the hydrolysis of 1 micromole of TAM per minute at 37° and pH 8.0.

The apparatus used consists of the Radiometer (Copenhagen) Titrator TTTlc, which controls the Titragraph (SBR 2c) and associated syringe buret (SBUla). The syringe delivers alkali into the reaction mixture to maintain the pH at 8.0 on signal from the Titrator. The Titragraph chart moves at uniform speed and makes a record of the alkali delivered as a function of time. We have found Hamilton syringes[6] less subject to leakage than glass plunger syringes. They have the nominal disadvantage of not delivering volumes with whole number ratios to the chart paper divisions. We standardize the solutions on the Titragraph scale and label them in terms of micromoles delivered per chart division. This is called the D value. Thus a 0.2 D alkali delivers 0.2 micromoles of reagent per chart division. The actual volume delivery is about 0.44 ml/100 chart divisions with the syringe in use.

A thermostat operated at 37° is used for temperature control. As suggested by Jacobsen et al.[3] the calomel electrode is also thermostated by immersion in this bath. A cross section of this part of the apparatus is shown in Fig. 1. The reaction vessel is a 12-ml weighing bottle with

[3] C. F. Jacobsen, J. Leonis, K. Linderstrøm-Lang, and M. Ottesen *in* "Methods of Biochemical Analysis" (D. Glick, ed.), Vol. 4, p. 171. Wiley (Interscience), New York, 1957; M. Laskowski, Vol. II [3].
[4] M. A. Seligman, A. S. Carlson, and T. Robertson, *Arch. Biochem. Biophys.* **97**, 159 (1962).
[5] V. M. Saresai and H. S. Provido, *J. Lab. Clin. Med.* **64**, 1023 (1965).
[6] Hamilton Company, Whittier, Calif.

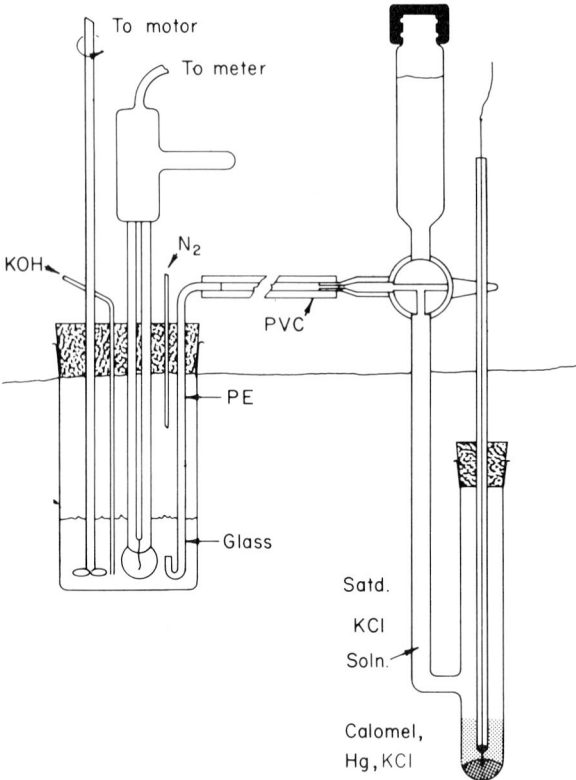

Fig. 1. Cross section of titration vessel and calomel electrode. Not to scale. PE, Polyethylene tubing; PVC, polyvinylchloride tubing. The rubber stopper has a hole (not shown) about 5 mm in diameter for the introduction of sample. N_2 is passed through a saturator immersed in the bath before entering the vessel. The calomel and glass electrode parts are supported independently. The wavy line indicates the bath water level.

a 24/12 internal grinding[7] fitted onto a rubber stopper carrying the items shown and having an addition hole through which the enzyme solution may be added. It is mounted on a clamp and rod by which it may be lifted and rotated out of the bath for the manipulation of renewing the liquid junction and the substrate solution. The flexible polyvinylchloride KCl bridge allows the calomel electrode to remain in the thermostat. The liquid junction is made at the small U-shaped glass tubing as shown. Only the end of this tube is glass. The main part passing through the stopper is polyethylene tubing. It has sufficient

[7] Kimble Products Cat. #15145 25 × 40 mm.

flexibility to avoid breakage. The openings in the liquid junction and diameter of tubing are about 1 mm ($\frac{1}{8}$ in. in the PVC section). Openings that are too small may result in too great resistance in the circuit. For the same reason even partial bubble blockage is not tolerable in the tubings.

The calomel electrode is constructed from a three-way T-bore Teflon plug stopcock, a screw cap test tube, a rimmed test tube, and a tube carrying a sealed wire to the mercury, calomel, KCl section. The latter is held by a tightly fitting rubber stopper which is waxed in place with the calomel electrode partly filled. The filling is completed with saturated KCl. Air is removed through the stopcock by appropriate manipulation.

The substrate solution contains 2 mg of TAM HCl and 0.2 mg of glycine per milliliter in 0.1 M KCl. Three milliliters of this solution is measured into one of the weighing bottles. A few drops of KCl are permitted to flow through the system from the reservoir and the stopcock turned to the position shown. External KCl solution is wiped from the U-tube with a bit of absorbent paper and the weighing bottle put on the stopper. The rod carrrying the glass electrode, motor, etc., is raised, rotated, and the reaction vessel lowered into the thermostat. The stirrer is started, the tip of the buret placed in the solution, the equipment set for pH-stat operation at a pH of 8.0 and after a few minutes for temperature equilibration, the addition of alkali begun by the use of the appropriate switch. The pH will come to 8.0 and the titragraph trace then becomes practically vertical. After 1 or 2 minutes to demonstrate this stability, the enzyme solution is added, usually in a volume of 1 to 200 μl, using a constriction pipette.[8] In the presence of enzyme, alkali uptake will begin and continue at a linear rate for 30–70% of complete hydrolysis. If the alkali is 0.2 D and the chart speed 0.5 cm per minute, then the units of enzyme added are equal to 0.1 the number of vertical lines crossed on the chart between two of the horizontal lines. Runs of 4–16 minutes are usual and the rates can be measured to about 1% accuracy. The zero enzyme rate should be determined and subtracted. It is about 0.015 micromoles per minute. At the end of the run the vessel is lifted and rotated out of the bath, the electrode and fittings rinsed with distilled water, blotted with a cellulose wipe, the liquid junction renewed, and a new weighing bottle with substrate solution put in place. Rates are proportional to enzyme concentration from 0.2 to 1.5 micromoles per minute.

Qualitative Test. A useful qualitative test for activity consists of a solution of TAM, 1 mg/ml, and glycine, 0.1 mg/ml, containing phenol

[8] M. Levy, *Compt. Rend. Trav. Lab. Carlsberg, Ser. Chim.* pp. 21–101 (1936).

red and adjusted to a clear red by addition of alkali. A drop of enzyme solution added to 0.5 ml of this solution soon changes the color from red to yellow. The time required is roughly a measure of enzyme concentration. The test is used to test the effluents from chromatographs for activity.

Preparation of Enzymes

Raw Material. The submaxillary glands of sexually mature male mice (6 weeks and older) are dissected and placed in a minimal volume of cold 0.1 M phosphate buffer, pH 7.4. The mixture is then frozen with dry ice and may be held for some months in this condition. Several commercial sources (Pel-Freez Inc., St. Louis Serum Co.) can supply this material.

Extraction. Our standard preparation uses 2000 submaxillary glands (1000 mice). The frozen glands are partly thawed and then homogenized for 3 minutes at 3°–5° in a Waring blendor run at high speed with enough water to make a liter. The resultant mixture is centrifuged at 10,000 g for 10 minutes. The supernatant fluid is collected.

Purification. A 0.2 M solution of streptomycin sulfate is adjusted to pH 7.4 with sodium hydroxide. To each volume of the supernatant liquid one-ninth volume of the streptomycin solution is added with stirring. After storage for 3 hours at 3°–5° the cold mixture is centrifuged at 10,000 g for 10 minutes. The supernatant fluid is collected. Seven milliliters of absolute alcohol at −15° is added slowly with stirring to each 100 ml of the fluid. After storage at −15° for 15 minutes, it is centrifuged at 10,000 g for 10 minutes. The supernatant solution is collected. Forty-eight milliliters of ethanol at −15° is added, with stirring, per 100 ml of the supernatant solution obtained from the streptomycin precipitate. The mixture is stored for 2 hours at −15° and the fluid decanted from the gummy red precipitate. The supernatant contains considerable enzymatic activity and appears to be richer in A than is the precipitate. The precipitate is dissolved in 500 ml of water, and 24.5 g of ammonium sulfate is added per 100 ml (35% saturation). The pH is kept at 7.4 by addition of Tris buffer. After 30 minutes at 3°–5° the mixture is centrifuged for 10 minutes at 10,000 g. The supernatant solution is brought to 80% saturation with ammonium sulfate by addition of 31.5 g per 100 ml of the original solution. After 2–4 hours the mixture is centrifuged for 15 minutes at 10,000 g. The precipitate is dissolved in 200 ml of water and the solution dialyzed in regenerated cellulose casing[9] with changes of external water every 2 hours until no sulfate is detectable in the external solution by addition of 1% $BaCl_2$ solution.

[9] A. H. Thomas Cat. #4465-A ⅝ inch.

Chromatography on CM-Cellulose. Twenty grams of Whatman CM23 carboxymethyl cellulose is stirred with 0.5 M NaOH containing 0.5 M NaCl and filtered on a fritted glass Büchner funnel. This is washed with water until the washings are free of alkali and finally with 0.005 M NaCl. A jacketed column kept at 3°–4° and 2 cm in diameter is provided. The cellulose derivative is poured to form a column 60 cm long and the 0.005 M NaCl pumped through the column at about 4 ml per minute until in and out flow have the same pH and conductivity. The entire dialyzed solution is applied to the column and eluted with distilled water at 3–4 ml per minute. The first three column volumes contain nerve growth factor and enzymes. The collected fluid is lyophilized. The white material, designated CMI, is stable for some months at −50°.

Chromatography on DEAE-Sephadex. A jacketed column 5 cm in diameter and capable of holding a column 89 cm long is loaded with DEAE-Sephadex A-25 (Pharmacia Inc.). The material is prepared by washing 500 g in succession with 0.5 M NaOH, water, 2 hours with 0.5 M HCl and finally with water until free of acid. To the suspension in water solid Tris is added until the pH is stable at 7.45 for an hour while stirring. The material is then suspended in buffer (0.05 M Tris-HCl, pH 7.45) and decantation used to remove fines. The suspension is cooled to 3°–4° and poured to form the column. Buffer is pumped overnight at 4 ml per minute with the column kept at 3°–4°. The column should now be at equilibrium with the buffer in that the in and out flow are at the same pH.

Two grams of CMI is dissolved in 20 ml of the buffer and dialyzed against it overnight. The resultant volume of solution (about 21 ml) is layered on the column and the elution begun with the buffer at 780 ml per hour. Twenty-six milliliter fractions are collected. The elution diagram is given in Fig. 2. The location of enzyme is highly dependent on elution rate but the qualitative test enables the enzyme to be detected quickly. Disc electrophoresis on this material shows 4 or 5 bands. The appropriate fractions are combined, dialyzed against water, and lyophilized.

Electrophoretic Separation. The final step of purification involves the use of a preparative acrylamide gel apparatus[10] (Fractophorator, Joseph Buchler, Inc.). The gel tube of the apparatus is extended by a piece of polyvinylchloride (Tygon) tubing about 6 cm long which in turn is closed by a rubber stopper (Fig. 3). The tube is clamped in a vertical position. The gelation mixture is made in a buffer containing 0.0125 M Tris and 0.096 M glycine which is also used as collecting fluid and

---

[10] I. Schenkein, M. Levy, and P. Weis, *Anal. Biochem.* **25**, 387 (1968).

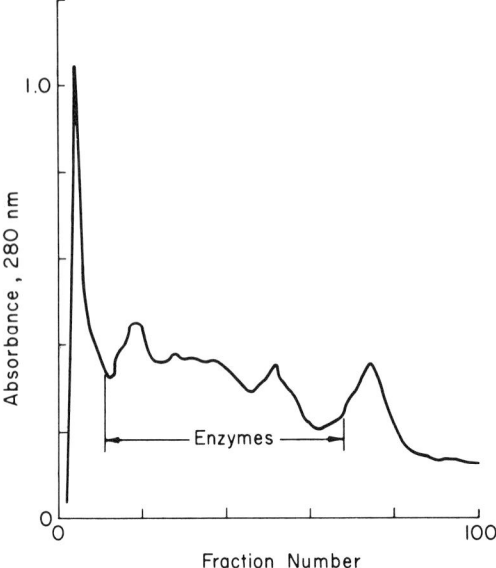

FIG. 2. Elution from DEAE-Sephadex A-25. See text for details. Fractions are 26 ml each.

electrode wash. To 100 ml of buffer 7.5 g of acrylamide, 0.2 g of bisacrylamide, and 0.06 ml of TEMED ($NNN'N'$-tetramethylethylene diamine) are added. This solution is deaerated. Ammonium persulfate (0.7 mg/ml) is added to a sufficient volume of the gel mixture and poured into the closed gel tube to form a meniscus 3.5 cm above the end of the glass. A layer of buffer with a tracer dye is carefully placed above the heavy solution to flatten the meniscus. After the gel has set, the extender tube and gel are cut through at the end of the glass with a razor to produce a cut surface for the exit of constituents. The apparatus is kept at about

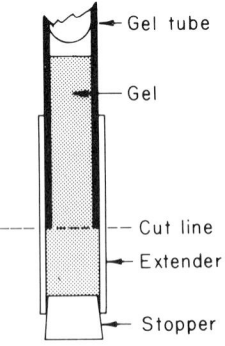

FIG. 3. Preparation of gel tube for Fractophorator operation. The extender is polyvinylchloride tubing.

10°. The gel tube is placed in the apparatus and prerun for an hour at 15 mA (about 400 V).

The sample is prepared for electrophoresis by dissolving about 100 mg of lyophilized DS II in 1.75 ml of 0.00125 M Tris and 0.0096 M glycine buffer (1–10 dilution of standard buffer). About 200 mg of sucrose is dissolved in the solution. The protein solution is layered above the gel through the supernatant buffer. For the initial 15–20 minutes of run the current is held to 5 mA and the electrode wash buffer is not circulated. A zone of protein forms and enters the gel. When this has happened, the electrode buffer circulation is begun and the current raised to 15 mA. The volume of each collection is kept at about 3 ml and the time per fraction is 3 minutes. A typical result is shown in Fig. 4. In this run some dye and other impurities emerged before the enzymes. Appropriate combining of fractions as indicated in Fig. 4 gives pure A and D as shown by analytical disc electrophoresis on acrylamide gels.

The combined fractions are freed of buffer using Sephadex G-25 and lyophilized. For testing, the solids are dissolved in 0.001 M HCl at about 0.2 mg/ml. The preparations are pure white powders which on disc electrophoresis show only one band each and which when tested by immunoelectrophoresis and by double diffusion against an antiserum to DS

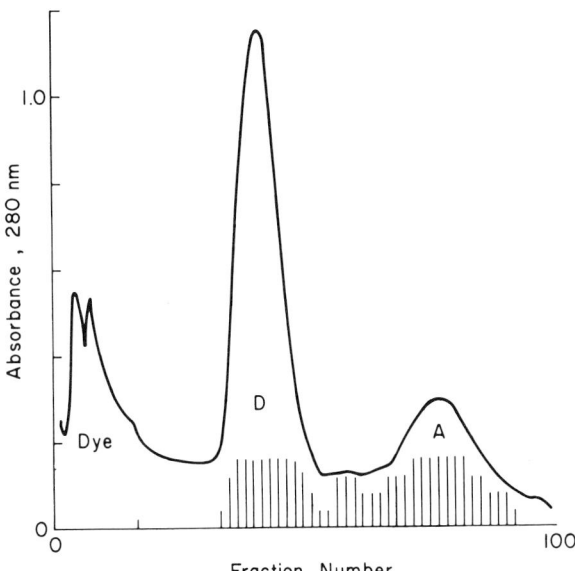

Fig. 4. Record of electrophoretic separation on "Fractophorator." Fractions at 3-minute intervals. Total load 100 mg fraction DS II. The vertical bars indicate enzyme tests on a 0–4 scale by the qualitative test described in the text.

TABLE I

PREPARATION OF MOUSE SUBMAXILLARY PROTEASE

Step	Protein (mg)	Enzymes (units × 10⁻⁶)	Specific activity (μ/mg)	Yield (%)[a]
Homogenate	24,800	3.65	150	100
85% Ammonium sulfate ppt	3,680	2.66	725	73
CM I	2,362	2.32	1,000	64
DS II	400	0.72	1,800	20
"D"	174	0.10–0.07	400–600	10
Fractophorator "A"	86	0.26	3,000	

[a] Several enzymes with the same qualitative activity are involved. The yields are therefore recoveries of activity, not of substances.

II show only a single precipitation line for each preparation. When tested against DS II itself five precipitation lines appear with the antiserum. Table I is a record of a preparation made with 2000 glands.

Enzymatic Properties

The optimum pH is 8.0–8.5 for either enzyme. The acid branches of the activity-pH curve show a midpoint at 6.5 for A and 7.1 for D. All measurements are made in the presence of glycine. In the absence of activator (glycine in the assay system) the hydrolysis curve is sigmoid with D. This results from the activating effect on D of the tosylarginine liberated. The initial rate is increased 2- to 4-fold by the activators. The K_m for TAM is $6.5 \times 10^{-4} M$ for A and $2 \times 10^{-4} M$ for D. K_m for the hydrolysis of protamine sulfate is 0.63 g/liter.

Chemical Properties

The ultraviolet absorbance curves for the two enzymes show rather wide flat peaks with maxima at 276 and 275 nm. At 280 nm a solution containing 0.16 mg of nitrogen (presumed 1 mg of protein) has an optical density of 2.49 for A and 2.59 for D.

The enzymes are inhibited by diisopropylphosphofluoridate and by 1-chloro-3-tolsylamido-7-amino-2-heptanone hydrochloride (TLCK) at second-order rates considerably less than those for trypsin. D is denatured by urea, reversible at low $(4 M)$ and high $(10 M)$ concentrations but irreversibly at $6 M$ urea. The irreversible loss at $6 M$ is second order and thus appears to be self-hydrolysis of a denatured form by a still-active form.

The sedimentation patterns are symmetrical and lead to $S = 3.1$ for A and $S = 2.8$ for D. These are consistent with the summation of residues from the amino acid analyses given in Table II. Analytically significant differences in numbers of residues are present only for Pro, Val, Met, Ile,

TABLE II
COMPOSITIONS OF MOUSE SUBMAXILLARY GLAND PROTEASES

TABLE II
COMPOSITIONS OF MOUSE SUBMAXILLARY GLAND PROTEASES

Amino acid	Residues[a]/molecule	
	A	D
Asp	31	30
Thr	16	16
Ser	18	17
Glu	22	21
Pro	21	16
Gly	30	29
Ala	15	16
Cys	9	9
Val	13	9
Met	6	(4)
Ile	11	6
Leu	30	26
Tyr	8	9
Phe	9	7
Lys	20	22
His	6	7
Arg	(4)	5
Trp[b]	12	7
$\sum$ residue wt.	29,700	27,900

[a] Rounded off to nearest whole numbers, averages of seven analyses. The calculation base is in parenthesis in each case.
[b] Calculated from tyrosine contents and ratio of tryptophan to tyrosine from ultraviolet absorption.

Leu, and Trp. These are all less in D than in A. A few amino acids are present in (doubtfully) greater amounts in D than in A (Ala, Tyr, Lys, His, Arg).

[50] Glandular Kallikreins from Horse and Human Urine and from Hog Pancreas[1]

By MARION E. WEBSTER and ELINE S. PRADO

kallikrein + kininogen → kinin

The kallikreins (EC 3.4.4.21) have been defined as endogenous enzymes which rapidly and specifically liberate the polypeptides called kinins from the substrates in plasma called kininogens.[1a] Perhaps more comprehensively they might be defined as those proteolytic enzymes of animal origin which release a kinin from kininogen but do not readily cleave proteolytic bonds in other proteins. The kallikreins, due to their different properties, have been separated into two classes—those derived from glandular sources and those derived from plasma. Each kallikrein derived from different sources or different species varies not only in structure[2] but in the proteins from which it must be separated and, therefore, purification techniques for individual kallikreins are somewhat different. As methods for purification of plasma kallikreins are only now being developed,[3] this chapter will be restricted to the preparation and properties of three of the glandular kallikreins—that is, those derived from horse and human urine, and from hog pancreas.

Assay Methods

The enzyme activity of the kallikreins can be estimated either by measuring the quantity of kinin formed from kininogen or by measuring their ability to cleave synthetic N-substituted arginine esters. The former method, once the technique has been mastered, provides a rapid method for the identification of the active fractions. The latter method, although more precise, is not as specific since other proteolytic enzymes cleave these ester bonds. Either method can be used for following the purification of horse or human urinary kallikreins, but the method based on formation of kinins should be employed when purifying hog pancreatic kallikrein, particularly during the early steps of the purification.

Formation of Kinins

Principle. Cleavage of the kininogens in plasma by the kallikreins releases the biologically active polypeptides called kinins. The amount of kinin generated from excess kininogen is a reflection of the quantity of enzyme and may be quantitated by direct bioassay on a number of biological preparations or by radioimmunoassay.[4] As the latter technique has not as yet been developed to the point at which it can be applied to biological material, direct bioassay of the kinins utilizing one or more of the available pharmacological methods is currently the method of

[1] Supported in part by Fundação de Amparo à Pesquisa do Estado de São Paulo, Grant No. 68/638; National Institutes of Health, Special Fellowship, Grant No. 1 FO3 AM 42717-01 and Bôlsa Chefe de Pesquisas, Conselho Nacional de Pesquisas.
[1a] M. E. Webster, in "Hypotensive Peptides" (E. G. Erdös, N. Back, F. Sicuteri, and A. F. Wilde, eds.), p. 648. Springer Verlag, New York, 1966.
[2] H. Moriya, J. V. Pierce, and M. E. Webster, *Ann. N.Y. Acad. Sci.* **104**, 172 (1963).
[3] E. Habermann and W. Klett, *Biochem. Z.* **346**, 133 (1966).
[4] R. C. Talamo, E. Haber, and K. F. Austen, *J. Immunol.* **101**, 332 (1968).

choice. Direct measurement of the decrease in the carotid blood pressure of the anesthetized dog following injections of varying quantities of kallikrein as initially proposed[5] is a useful technique but has the disadvantage that some heterologous kallikreins, such as horse urinary kallikrein,[6] do not readily liberate kinins from dog kininogens. Recording the height of contraction *in vitro* of various isolated smooth muscles such as the rat uterus or guinea pig ileum provides an alternate method for measuring the kinin which is formed. Although the rat uterus will detect smaller amounts of kinin, the guinea pig ileum has a more quantitative response and this method has been chosen to be described in detail. In order to minimize the destructive effects of the kininases, enzymes which cleave the kinins at various peptide bonds thus destroying their biological activity, varying amounts of kallikrein are added to a constant amount of plasma directly in the tissue bath. Whenever possible, it is recommended that the plasma from the same species as the kallikrein be used as substrate. However, dog plasma is a much better substrate for hog pancreatic kallikrein than is hog plasma.

Reagents

Tyrode's Fluid
Solution I   NaCl, 200 g/liter
                        KCl, 5 g/liter
                        $CaCl_2$, 5 g/liter
                        $MgCl_2 \cdot 6\ H_2O$, 2.5 g/liter
Solution II $NaH_2PO_4 \cdot H_2O$, 2.5 g/liter
                        $NaHCO_3$, 50 g/liter

All solutions are made with recently glass-distilled water. The stock solutions may be stored at 4° for at least 2 months. On the day of assay, 40 ml solution I, 20 ml solution II, 1 g glucose, and 500 μg atropine sulfate are added to 400–500 ml water and diluted to a final volume of 1 liter.

Plasma. Citrated blood (0.2 ml 20% sodium citrate/10 ml blood) is taken in silicone-coated (Siliclad, Clay-Adams, Inc., New York) containers, the plasma separated and either used the same day or frozen in silicone-coated tubes. The plasma should be frozen and thawed only once as repeated freezing and thawing will cause loss of substrate. The use of silicone-coated glassware,

[5] E. K. Frey, H. Kraut, and E. Werle, "Das Kallikrein-Kinin-System und seine Inhibitoren." Ferdinand Enke Verlag, Stuttgart, 1968.
[6] E. S. Prado and J. L. Prado, *Experientia* **17**, 31 (1961); E. S. Prado, J. L. Prado, and C. M. W. Brandi, *Arch. Intern. Pharmacodyn.* **137**, 358 (1962).

although preferable, is not essential if sufficient kininogen remains in the plasma following its contact with glass. Also, the plasma may be heated to 60° for ½ hour. Following this treatment, it can usually be dialyzed against running tap water at 4° without loss of substrate. Heating plasma to this temperature is thought to destroy the plasma prekallikrein and, while eliminating the possibility that activators of prekallikrein will be identified as kallikreins, also prevents their discovery.

Procedure. Guinea pigs (150–250 g) are stunned by a blow on the head and exsanguinated by cutting the blood vessels on the neck. The guinea pig is washed thoroughly with water, the abdomen opened and a 10–15 cm portion of the terminal ileum is removed, care being taken to prevent contact of the intestine with the outer surface of the guinea pig. The lumen is washed several times with approximately 20 ml Tyrode's solution warmed to 37°. The washed ileum is allowed to equilibrate for 15–30 minutes in warm Tyrode's solution and a portion (2–3 cm) is tied at both ends with surgical needle and thread as shown in Fig. 1a and cut as shown in Fig. 1b so that the bath fluid will circulate within the lumen freely. Forceps should be used in handling the tissue and stretching should be avoided. One end of the tissue is tied to the hook at the bottom of the muscle chamber and the other is suspended from the muscle lever as shown in Fig. 2. Tension is applied and the tissue allowed to equilibrate for approximately ½ hour before starting the assay. During this period both the sensitivity of the tissue and the proportionality of its response to varying amounts of a standard kallikrein preparation may be determined. For these determinations, a known quantity of kallikrein (e.g., 0.2 KL units of horse urinary kallikrein) is added to the bath for 1 minute, followed by 0.4 ml plasma and the contraction is recorded for 2 minutes. The Tyrode's solution in the muscle chamber is changed twice at 1-minute intervals and 1 minute allowed for the tissue to relax. Each new addition to the bath occurs in exactly

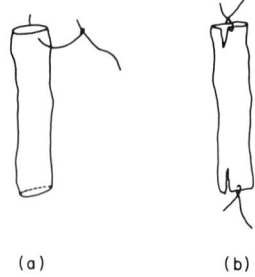

(a)   (b)

Fig. 1. Method for tying and cutting guinea pig ileum.

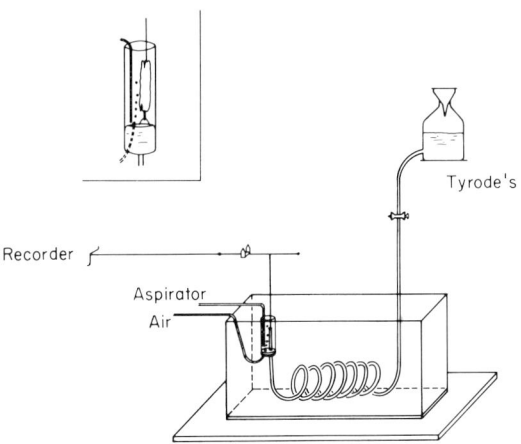

Fɪɢ. 2. The tissue bath (7–8 ml capacity) can be purchased from various suppliers of pharmacological equipment such as Phipps and Bird, Inc., Richmond, Va., or C. F. Palmer, Ltd., London, England, or may be made by simply inserting a rubber stopper in a glass cylinder (12 × 70 mm). Air is bubbled slowly through the chamber and can be supplied by an aquarium air pump (vibrator-type). It can be admitted to the muscle chamber by inserting a hypodermic needle through the rubber stopper. Removing the head from a hypodermic needle and bending the shaft to insert it into the rubber stopper provides a method for supplying the hook to which the tissue is tied at the bottom of the muscle chamber. For direct recording, a magnification of 6–7 times is usually employed and can be achieved by tying the muscle 3–4 cm from the axle and having the recording tip, preferably equipped with an all-metal frontal writing point, at 21–24 cm. Tension is applied by adding weight (e.g., a piece of modeling clay) to the shorter arm to exactly balance the longer arm of the rod and placing the required additional weight on the longer arm at a point equidistant from the place the muscle is tied. Usually 1 g of tension is employed and this weight is increased somewhat (0.2–0.3 g) in those tissues which show normal contraction. Recording speed, 6 mm/minute. Bath temperature, 37°.

5 minutes and this rhythm is maintained throughout the day. As individual tissues can vary in sensitivity to the kinin over a 4-fold range, increasing or decreasing concentrations of kallikrein (e.g., 0.4 or 0.1 unit) are added to the bath with the same amount of plasma until the height of contraction of the larger dose corresponds to less than 80% of the maximum muscle contraction. After several such contractions, when it has been shown that the height of contraction of at least two concentrations of kallikrein is reproducible and related to the quantity of kallikrein added to the bath, unknown solutions may be bioassayed. A typical dose-response curve is shown in Fig. 3 as given by 0.1 and 0.2 units of horse urinary kallikrein when added to 0.4 ml horse plasma. The plasma by itself should cause no contraction of the ileum and this should

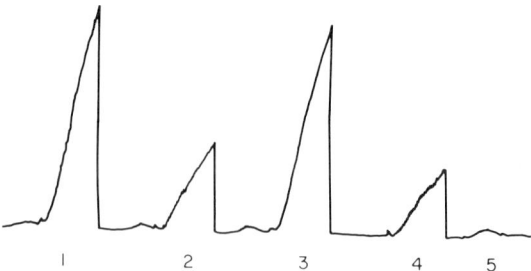

Fig. 3. Record showing the contractions of the isolated guinea pig ileum induced by the addition of 0.4 ml horse plasma to horse urinary kallikrein. 1 and 3, 0.2 units of enzyme. 2 and 4, 0.1 units of enzyme. 5, 0.4 ml horse plasma. Height of contraction given by 0.2 units = 56, 52 mm, respectively.

be demonstrated with each tissue both initially and throughout the day. If contraction should occur, the volume of plasma can sometimes be reduced, the plasma may be heated to 60° and dialyzed, or another plasma or tissue substituted. The height of contraction given by the various concentrations of unknown and standard are usually measured in millimeters and have been found to be linearly related to the logarithm of the dose. A four-point assay is to be preferred for those samples where precise quantitation is desired and details of the statistical treatment of the data for bradykinin have previously been published.[7] However, for routine assay of a large number of samples, the addition of two concentrations of unknown solutions (e.g., 0.1 and 0.2 ml) in a volume up to 0.5 ml, one of which has a height of contraction which falls between that of the standards, usually furnishes a sufficiently accurate estimate to permit location of the active fractions during purification.

Definition of Unit. During the purification of hog pancreatic and human urinary kallikreins as reported in this paper, the original kallikrein unit (KU) was employed. This unit was defined as the decrease in the blood pressure of the dog following intravenous injection of 5.0 ml of dialyzed normal urine which had been taken from a pool of 50 liters of human urine. Varying pools of urine of this magnitude, however, have not always contained one kallikrein unit per 5.0 ml and the standard unit has been maintained by exchange of biologically active material. This unit of activity can only be measured using the dog's blood pressure as the method of bioassay and reflects the ability of the various kallikreins to release kinin only from dog kininogen. As this substrate was unsuitable for horse urinary kallikrein, one KL unit of horse urinary kallikrein was arbitrarily defined as that amount of activity per milligram of a crude

[7] M. Rocha e Silva, *Acta Physiol. Latinoam.* **2**, 238 (1952).

preparation (Table I) and gives a calculated value of 0.25–0.75 KL units/ml horse urine. Recently, a new unit (KaU) has been proposed[8] wherein one milliunit of kallikrein is arbitrarily defined as that amount of enzyme which will release in 1.0 minute at 30° 1 nanomole of bradykinin equivalent from an amount of citrated plasma which has been heated for ½ hour at 60° and which contains an amount of substrate to the enzyme being assayed which is equivalent to 5 nanomoles of bradykinin. This provisional unit, however, has only recently been proposed and cannot be determined by direct bioassay as detailed above. However, it is recommended that in future publications, wherever possible, the units used for purification be standardized in terms of KaU.

Cleavage of Synthetic Esters

Principle. Many methods are available for following the enzymatic cleavage of ester bonds. The method which has been chosen to be presented in detail is a modification of the Hestrin method[9] and is based on the determination of residual ester by formation of a ferric complex with the hydroxamic ester resulting from the reaction of an ester with alkaline hydroxylamine.

Reagents

Buffer. 0.1 M Tris, pH 8.5
Substrate. 0.12 M p-toluenesulfonyl-L-arginine ethyl ester (TAME). Dissolve 45.4 mg/ml distilled water and store at 4° when not in use.
Alkaline hydroxylamine solution. 2 M (13.9%) aqueous hydroxylamine hydrochloride is mixed with an equal volume of 3.5 M NaOH just before use. Store 2 M hydroxylamine at 4°.
Hydrochloric acid, 3 M. Dilute 1 volume of concentrated hydrochloric acid with 3 volumes distilled water.
Ferric chloride, 10%. Dissolve 100 g $FeCl_3 \cdot 6 H_2O$ in 1 liter 0.1 M HCl.
Trichloroacetic acid (TCA), 7.5%

Procedure. Tubes containing 0.1 ml substrate and 1.5–1.8 ml buffer are placed in the water bath at 30° for exactly 5 minutes followed by the addition of 0.4–0.1 ml enzyme solution. The enzyme-substrate mixture

[8] M. E. Webster, *in* "Bradykinin, Kallidin und Kallikrein," Handbook of Experimental Pharmacology (E. G. Erdös, ed.), Vol. XXV, p. 659. Springer Verlag, Heidelberg, 1970.
[9] P. S. Roberts, *J. Biol. Chem.* **232**, 285 (1958); M. E. Brown, *J. Lab. Clin. Med.* **55**, 616 (1960).

is incubated at 30° for varying periods of time at one concentration of enzyme or for 1 hour with at least three dilutions of enzyme. The reaction is stopped by the addition of 2.0 ml 7.5% TCA, allowed to stand at room temperature for at least 30 minutes, and the precipitate removed by filtration or centrifugation. To 2.0 ml of the supernatant is added 2.0 ml of the alkaline hydroxylamine solution. After standing at room temperature for at least 25 minutes, 1.0 ml 3 M HCl and 2.0 ml 10% ferric chloride are added to the mixture and the color which develops is measured in exactly 15 minutes at 540 nm. Formation of gas bubbles in the colorimeter cell is largely prevented if the solution is thoroughly mixed and transferred to the cuvette just prior to reading. The samples are read against a reagent blank prepared by substituting buffer for enzyme and TAME. A blank of TAME should also be included which is prepared by substituting buffer for enzyme and appropriate blanks of the enzyme solution can be prepared either by adding the enzyme to one series of tubes after the addition of the TCA or by substituting buffer for TAME. As the kallikreins are purified, little or no precipitate is obtained on the addition of the TCA. The procedure, therefore, can be simplified and the sensitivity increased by reducing the concentration of substrate to 0.06 M and stopping the enzyme reaction by the addition of the alkaline hydroxylamine. In fact, even with crude solutions, provided no protein precipitates when the acid and iron additions are made, the step involving precipitation with TCA can be omitted with consequent economy in substrate and effort.

Definition of Unit. Various authors have employed a number of different units for measurement of this enzyme activity. In future studies it is suggested that a unit based on the recommendations of the International Union of Biochemistry[10] be employed. One unit would therefore be defined as that amount which will catalyze the transformation of 1 micromole of substrate per minute at 30° under the above standard conditions.

Purification Procedures

Horse Urinary Kallikrein[11]

Step 1. Ammonium Sulfate Fractionation. Freshly collected horse urine (20–30 liters) can be stored at 4° and is fractionated with ammonium sulfate at this same temperature. The pH is adjusted to 7–7.5 with 10 M

[10] "Enzyme Nomenclature," p. 6. Elsevier, New York, 1965.
[11] E. S. Prado, J. L. Prado, and C. M. W. Brandi, *Arch. Intern. Pharmacodyn.* **137**, 358 (1962); J. L. Prado, E. S. Prado, C. M. W. Brandi, and A V. Katchburian, *Ann. N.Y. Acad. Sci.* **104**, 186 (1963).

TABLE I
PURIFICATION OF HORSE URINARY KALLIKREIN

Purification steps	KLU[a]/liter urine	Specific activity	
		KLU/mg protein[b]	EU[c]/mg protein
Horse urine, dialyzed	520	0.37	—
1. Ammonium sulfate fractionation, 0.4–0.6 saturation	315	1.3	0.12
2. Acetone fractionation, pH 6.5, 45–60%	220	25	2.3
3. DEAE-cellulose chromatography, peak of activity	114	260	24
4. Acetone fractionation, pH 5.9, 60–75%	80	330	30
5. Sephadex G-75 gel filtration, peak of activity	24	800	65

[a] KLU (kallikrein liberating units) are defined as that amount of activity per milligram found in a crude 0.4–0.7 saturated ammonium sulfate fraction. The kinin liberated was measured by direct bioassay on the guinea pig ileum. A recent determination would suggest that one of these units is equivalent to approximately 1.0 mKaU as determined on the guinea pig ileum.

[b] Protein measured by method of Lowry et al. (see text footnote 13).

[c] EU (esterase units) are defined as that amount of kallikrein which will hydrolyze 1 micromole of TAME per minute at 30° and pH 8.5.

NaOH, solid ammonium sulfate (223 g/liter) is added with stirring and the solution stored overnight. Most of the clear supernatant is removed by siphoning and the remaining solution clarified by centrifugation. The precipitate (0.4 saturation), which contains less than 3% of the kallikrein activity of the urine, is discarded and additional ammonium sulfate (132 g/liter of supernatant) is added and the mixture again allowed to stand overnight. The precipitate is collected, suspended in the smallest possible volume of water, and dialyzed overnight against running tap water. The dialyzed solution is centrifuged to remove insoluble material and the supernatant frozen until several batches have been accumulated. The total mixture (150 liters of urine) contains about 60% of the kallikrein activity of the starting urine with a specific activity of 1.3 (Table I).

Step 2. Acetone Fractionation, pH 6.5. The combined solution from step 1 is diluted to approximately 30 mg/ml protein, solid sodium chloride is added to make the solution 0.1 M, and the pH is adjusted to 6.5. Acetone is added with stirring at room temperature to a final concentration of around 45% (v/v). After standing for 30 minutes at room temperature, the precipitate, which should contain about 15% of the enzyme activity and 75% of the protein, is removed by centrifugation and discarded. The concentration of acetone in the supernatant is raised to around 60% (v/v) and after 1 hour the precipitate, which should contain 70% of the enzyme and 3% of the protein, is separated by centrifugation, allowed to drain briefly at room temperature to remove traces of acetone, and dissolved in the smallest possible volume of 0.075 M sodium phosphate buffer, pH 5.9. The concentration of acetone necessary to achieve this purification may vary somewhat from preparation to preparation and preliminary small-scale trials (in increments of 5% acetone) should be conducted.

Step 3. DEAE-Cellulose Chromatography. The enzyme solution is dialyzed for 16 hours at 4° against 20 times its volume of 0.075 M sodium phosphate buffer, pH 5.9, and added at room temperature to a column (2.7 × 20 cm) of DEAE-cellulose which had been previously equilibrated with the same buffer.[12] The column is washed with the same buffer (about 600 ml) until the E_{280} (extinction coefficient in a 1-cm cell at 280 nm) becomes relatively constant (0.2 E_{280}) and 46% of the protein and 0.8% of the enzyme activity has been removed. The column is eluted stepwise with about 250 ml 0.1 M sodium phosphate buffer, pH 5.9, only until the E_{280} again comes to a value of 0.2. This eluate should contain about 7.8% of the protein and 15% of the enzyme activity. The major portion of the enzyme activity is eluted from the column with 300 ml 0.125 M sodium phosphate buffer, pH 5.9, collected in 50-ml fractions and the enzyme activity and specific activity of the various fractions determined. Pooling the fractions with the highest specific activity should result in the recovery of about 50% of the starting activity with a specific activity of 260.

Step 4. Acetone Fractionation, pH 5.9. The enzyme activity in the pooled fractions from step 3 is rapidly concentrated by fractionation with acetone. In this procedure solid sodium chloride is added to the solution to make a final concentration of 0.03 M and 1½ volumes of acetone added at room temperature. After 60 minutes, the mixture is centrifuged and the precipitate discarded. The concentration of acetone in the supernatant is increased to 75% (v/v), and the mixture allowed to stand 1 hour. The precipitate is removed by centrifugation and dissolved in the smallest possible volume of water. The solution is dialyzed overnight against several changes of distilled water at 4° and freeze dried to yield 75 mg of dry powder which can be stored in a dessicator at 4°. Horse urinary kallikrein prepared in this manner is quite stable. However, further purification or, for example, preparation of horse urinary kalli-

---

[12] E. A. Peterson and H. A. Sober, Vol. V, p. 3.

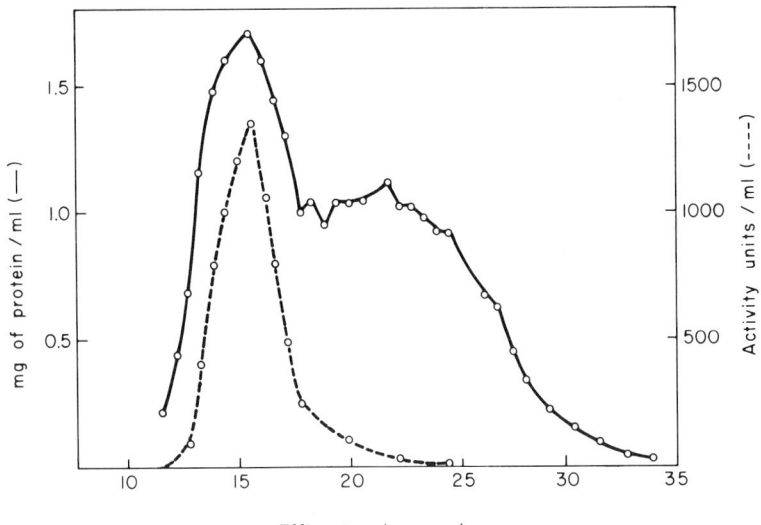

Effluent volume , ml

Fig. 4. Sephadex G-75 chromatogram of partially purified horse urinary kallikrein. Protein measured by method of Lowry *et al.*[13]

krein from a DEAE-cellulose column employing a gradient rather than the stepwise procedure recommended above, yields a product which rapidly loses activity. Therefore, until methods are devised to stabilize this kallikrein, it is recommended that when more highly purified kallikrein is required for a particular experiment, it should be prepared as detailed in step 5 only in the amounts necessary and on the same day as the experiment.

Step 5. Sephadex G-75 Gel Filtration. The freeze-dried material from step 4 (17 mg) is dissolved in 1.0 ml distilled water and added at 4° to a 1.2 × 29 cm column of Sephadex G-75. It is washed through the column with distilled water (30 ml) at a flow rate of 4 ml/hour, collecting fractions of 0.5 ml. The enzyme activity is associated with the initial protein peak as shown in Fig. 4 and 55% of the activity can be recovered by pooling the main fractions. Those fractions with the highest specific activity (800) can be usually recovered in a 30% yield of the starting activity.

Human Urinary Kallikrein[14]

Step 1. Amberlite IRC-50 Adsorption. Normal male human urine is collected daily and stored at 4° until about 180 liters is accumulated.

[13] O. H. Lowry, N. J. Rosebrough, A. L. Farr, and R. J. Randall, *J. Biol. Chem.* **193**, 265 (1951).
[14] H. Moriya, J. V. Pierce, and M. E. Webster, *Ann. N.Y. Acad. Sci.* **104**, 172 (1963).

To 174 liters of urine in a 200-liter tank is added 1.8 kg Amberlite IRC-50 (150–325 mesh) in the acid form[15] and the pH adjusted to 4.0 with approximately 2 liters of concentrated HCl. The mixture is stirred vigorously for 2 hours at 16°, the resin recovered by filtration, washed on the funnel with 5 liters 0.1 M sodium acetate buffer, pH 4.0, suspended in 10 liters with distilled water and made 0.1 M phosphate and pH 6.0 with tribasic potassium phosphate and 10 M NaOH. The suspension is stored overnight at 4°, stirred for 1 hour at 15°, and filtered. The resin is washed on the funnel with several liters of buffer and the combined eluates contain 80–100% of the enzyme activity of the urine (Table II).

TABLE II

PURIFICATION OF HUMAN URINARY KALLIKREIN

Purification steps	EU[a]/liter urine	Specific activity EU/E$_{280}$
Human urine, dialyzed	120[b]	0.04[b]
1. IRC-50 adsorption	100[b]	0.1[b]
2. DEAE-cellulose batchwise adsorption	75	2.2
3. Acetone precipitation, pH 5.4, 50–60%	56	15
4. Ammonium sulfate fractionation, 69–83%	21	29
5. Sephadex G-75 gel filtration	16	60
6. DEAE-Sephadex A-50 chromatography, peak of activity	7.2	230

[a] EU (esterase units) were determined by measuring the hydrolysis of TAME at, pH 8.5, 37° and defined as that amount of activity given by one KU. A recent determination on a crude human kallikrein preparation would indicate that 1 EU will hydrolyze 0.02 micromoles of TAME per minute at 30°, pH 8.5. The ratio of activity of EU/KU, where KU was determined by measuring the increase in arterial blood flow of the dog, was approximately 1.0 at all levels of purification.

[b] These results were averages obtained on earlier pools of urine which had been dialyzed, freeze dried, and concentrated prior to assay. The bioassay employing the guinea pig ileum more readily determines the activity of these dialyzed fractions.

Step 2. DEAE-Cellulose Batchwise Adsorption. The eluates from step 1 are diluted to 40 liters with tap water, 70 g DEAE-cellulose (0.45 meq/g) added, the pH adjusted to 6.7, and the mixture stirred for 1 hour at room temperature. The adsorbent is recovered by filtration, washed on the funnel with 600 ml 0.1 M sodium phosphate buffer, pH 5.8, containing 0.1 M NaCl to remove inactive materials, and eluted on the funnel with 700 ml 0.5 M sodium phosphate buffer, pH 6.3. The latter eluate contained 55–85% of the enzyme activity in the urine and

---

[15] C. H. W. Hirs, Vol. I, p. 113.

is dialyzed overnight against running tap water and freeze dried to afford 3.84 g of material which had been purified 55 times.

Step 3. Acetone Precipitation. The dried material from step 2 is dissolved at 5 mg/ml in 0.065 M sodium acetate containing 0.035 M NaCl and the pH adjusted to 5.4 with 5 M acetic acid. The solution is cooled in an ice bath and an equal volume of acetone added dropwise while stirring. After standing overnight at 0°–4°, the precipitate which contained 12% of the activity and 50% E_{280} is removed by centrifugation at 4°, and the concentration of acetone in the supernatant increased to 60% (v/v). The following day the precipitate is removed by centrifugation, allowed to drain briefly at room temperature to remove traces of acetone, dissolved in 200 ml distilled water, and the pH adjusted to 7.0.

Step 4. Ammonium Sulfate Fractionation. The above solution (E_{280}/ ml = 3) is cooled in an ice bath and a saturated solution of ammonium sulfate at 25°, previously adjusted to pH 7.0 with ammonium hydroxide, is added dropwise with stirring to make 55% saturation. The mixture is allowed to stand overnight at 4° and the precipitate is removed by centrifugation at 4°. In this manner, precipitates are removed at 60%, 65%, 69%, and 83% saturation. The fraction which precipitated between 69% and 83% contained 37% of the activity with a 2-fold increase in purity. The earlier fractions contain 53% of the starting activity at concentrations from 1.3–12 EU/liter and a specific activity of 2.7–19, respectively. The volume of the 69–83% fraction (10 ml) is reduced to 2.5 ml by placing the fraction in a dialysis bag and surrounding the bag with Sephadex G-75 in a suitable container. The level upper surface is covered with a flat plate, and weights are applied to increase the rate of water removal.[16] After 4 hours at 4° the volume is usually reduced and, if not, the damp Sephadex can be peeled from the dialysis bag and replaced with fresh. The Sephadex is recovered by suspending in water, filtering, washing slowly on the funnel with distilled water and absolute ethanol (carefully layered on top of the gel), removing excess ethanol by suction, and air drying.

Step 5. Sephadex G-75 Gel Filtration. The concentrated solution from step 4 (E_{280}/ml = 40) is added to a 1 × 46 cm column of Sephadex G-75 and washed through the column with approximately 100 ml of distilled water at 4° (flow rate, 4 ml/hour). Two E_{280} peaks are obtained, the second of which contained the enzyme activity. The three 4-ml fractions (48–60 ml effluent) comprised 75% of the starting activity with a 2-fold purification and are eluted from the column just ahead of the ammonium sulfate.

---

[16] H. Palmstierna, *Biochem. Biophys. Res. Commun.* **2**, 53 (1960).

Step 6. DEAE-Sephadex A-50 Chromatography. The pooled fractions from step 5 are concentrated to 2.5 ml (E_{280}/ml = 17) by dialysis against dry Sephadex G-75 as described in step 4 and ammonium formate, pH 6.7, is added to make a final concentration of 0.2 M. The solution is added to a 1 × 25 cm column of DEAE-Sephadex A-50 which has been previously equilibrated with the same buffer. The column is washed with 40 ml of the same buffer and a gradient of 0.2 to 0.8 M ammonium formate and of pH 6.7 to 6.85 (125 ml constant-volume mixing chamber) is applied (flow rate, 5 ml/hour). The enzyme activity is eluted just slightly after the largest protein peak and pooling of the four best 2.5-ml fractions (70–80 ml of gradient effluent) gives a 45% recovery of enzyme activity with a 4-fold increase in purity. Based on the nondialyzable solids found in human urine, a 650-fold purification is realized.

Hog Pancreatic Kallikrein[17]

Step 1. Extraction. Hog pancreas, which had been obtained as fresh as possible from the slaughterhouse, is stored at −20°. The glands (500 g) are thawed, cleaned of extraneous fat and minced in a meat grinder. The suspension is allowed to autolyze at 15°–25° for 18–24 hours, distilled water (2500 ml) is added, and the pH of the mixture adjusted to 6.0 with 6 M HCl. The mixture is stirred at room temperature for 5 hours, heated at 55° for 5 minutes, allowed to cool to room temperature, and the pH adjusted to 4.5 with 10 M HCl. The suspension is cooled in an ice bath and one-third the volume of acetone (30% v/v) added dropwise while stirring. Addition of the acetone directly to the suspension greatly improved the rate of filtration when compared to the aqueous extract and the precipitate is rapidly removed by filtration at 4° through filter paper placed on a Büchner funnel. The clear supernatant contains 50–75,000 KU at a purity of 0.4–1 KU/mg (Table III).

Step 2. DEAE-Cellulose Batchwise Adsorption. A portion of the clear supernatant from step 1, corresponding to the extract from 100 g, is diluted 20-fold with tap water and 3.0 g DEAE-cellulose added. The pH of the suspension is adjusted to 7.0 with 1 M NaOH and the mixture stirred vigorously at 10°–20° for 2 hours, the pH being readjusted if necessary. The adsorbent is recovered by filtration on a large glass funnel, transferred to a small glass funnel, and eluted on this funnel with about 60 ml 0.6 M ammonium formate, pH 6.0. The filtrate from the DEAE-cellulose adsorption contains no detectable enzyme activity and 80% or

[17] H. Moriya, K. Yamazaki, H. Fukushima, and C. Moriwaki, *J. Biochem. (Tokyo)* **58**, 208 (1965); H. Moriya, A. Kayto, and H. Fukushima, *Biochem. Pharmacol.* **18**, 549 (1969).

TABLE III
PURIFICATION OF HOG PANCREATIC KALLIKREIN

Purification steps	KU^a/g tissue	Specific activity	
		KU/mg^b	EU^c/mg
1. Extraction	125	0.7	—
2. DEAE-cellulose batchwise adsorption	110	10	3
3. Acrinol precipitation	70	45	15
4. Precipitation at pH 4.2	30	75	54
5. Acetone precipitation, 0–67%	30	140^d	58^d
6. Hydroxyapatite chromatography, peak of activity	15	900^d	220^d
7. Sephadex G-100 gel filtration, peak of activity	10	1200^d	260^d

[a] KU (kallikrein units) were determined by measuring the decrease in the blood pressure or increase in arterial blood flow of the dog.

[b] Dry weight.

[c] EU (esterase units) were determined by measuring the hydrolysis of TAME at pH 8.5, 37° and defined as that amount of activity given by one KU of human urinary kallikrein. These units will hydrolyze approximately 0.02 micromoles of TAME per minute at 30° and pH 8.5.

[d] Protein estimated by measuring E_{280}, purest material 1 E_{280} = 0.5 mg protein as estimated by Lowry et al.,[13] bovine serum albumin used as the standard.

more of the enzyme is recovered in the eluate together with 4% of the nondialyzable solids. This step is repeated until all of the supernatant from step 1 has been adsorbed and eluted from DEAE-cellulose.

Step 3. Acrinol Precipitation. The eluates from the DEAE-cellulose adsorption are pooled and the pH adjusted to 5.0 with 1 M formic acid. The solution is cooled in an ice bath and 30 ml 3% (w/v) aqueous acrinol (2-ethoxy-6,9-diaminoacridinium lactate) (Pfaltz & Bauer, Inc., Flushing, N.Y.) is added while stirring vigorously. The suspension is allowed to equilibrate for 4 hours at 4° and the supernatant clarified by centrifugation (10,000 rpm, 15 minutes, 4°). The pH of the supernatant is adjusted to pH 7.8 with 1 M ammonium hydroxide and additional acrinol (45 ml) added and the mixture allowed to stand overnight at 4°. The precipitate, which contains the activity, is washed with 0.01 M ammonium acetate, pH 8.0, and extracted twice by stirring for 30 minutes with 125 and 50-ml portions of 0.6 M ammonium acetate, pH 4.5, in an ice bath. After centrifugation, the yellow supernatants are pooled, cooled in an ice bath, and precipitated by the addition of 3 volumes of acetone added dropwise with stirring. The suspension is stored overnight at 4° and the precipitate is collected by centrifugation, washed with acetone, allowed to drain briefly at room temperature to remove traces of acetone, and dissolved in distilled water. The slightly yellow solution contains 50–80% of the starting activity and only 10–15% of the nondialyzable solids.

Step 4. Precipitation at pH 4.2. The pH of the aqueous solution from step 3 is adjusted to pH 4.2 with 1 M acetic acid and the suspension allowed to stand for 2 hours at 4°. The voluminous inactive precipitate which forms is removed by centrifugation for 15 minutes at 10,000 rpm, 4°. The clear supernatant is immediately adjusted to pH 6.5 with 1 M NaOH and contains 50–70% of the enzyme activity and 15–40% of the nondialyzable solids to give an average increase in purity of 2-fold or more.

Step 5. Acetone Precipitation. Sodium acetate and sodium chloride are added to the supernatant from step 4 to give a final concentration of 0.065 M and 0.035 M, respectively. The pH of the solution is adjusted to 7.5 with 5 M NH$_4$OH and precipitated by the addition of 2 volumes of acetone under the conditions described in step 3. The precipitate is suspended in the smallest possible volume of distilled water and kept frozen or freeze dried, if necessary, to give an average yield of around 60 mg (E$_{280}$/mg = 0.6). In subsequent purification steps the material can no longer be dried as it is unstable under these conditions.

Step 6. Hydroxyapatite Chromatography. The concentrated solution from step 5, obtained from three 500-g batches of hog pancreas, is added to a 1.6 × 25 cm column of hydroxyapatite which has been previously equilibrated with 0.01 M sodium phosphate buffer, pH 6.8. The column is washed with 60 ml of the same buffer and a gradient of 0.01 to 0.15 M phosphate buffer (150 ml constant-volume mixing chamber) is applied. A flow rate of 6.0 ml/hour is used and fractions of 5.0 ml volume are collected. The enzyme activity is eluted in one peak (5 fractions) between 0.08 and 0.1 M buffer with 60–80% of activity recovered at a purity of 700–1000 KU/E$_{280}$.

Step 7. Sephadex G-100 Gel Filtration. The active fractions from step 6 are pooled and concentrated to a volume of 1–3 ml by dialysis against dry Sephadex G-100 as described in step 4 of the purification of human urinary kallikrein or by pressure dialysis at 4°. The solution is added at 4° to a 2.5 × 40 cm column of Sephadex G-100, washed through the column with distilled water (180 ml) at a flow rate of 12 ml/hour, collecting fractions of 4 ml. The enzyme activity is eluted in one main peak (50–70 ml effluent) and can be concentrated by the methods described above. The final product, which represents 80% of the starting activity, had a purity of 1,200 KU/E$_{280}$ and is obtained in a 5–8% yield having been purified approximately 2000-fold as calculated from the initial extract.

Properties

Stability. Highly purified horse urinary kallikrein is unstable in solution, an instability which can be detected only if the kinin-forming or esterase activity is determined immediately. Storage of the enzyme in water or phosphate buffer, pH 8.0, even though the solution is maintained most of the time in the frozen state, does not prevent the gradual loss of activity to a value around 500 KL units/mg protein. Human urinary kallikrein, at the level of purity so far attained, appears to be quite stable if maintained in solution at neutral pH (6–8). The hog pancreatic kallikrein prepared by this procedure is unstable in the dry form and is maintained in Tris buffer in the frozen state. Other highly purified preparations of this same enzyme have resulted in material which can be maintained in the dry form at −20° for several months.[18] These latter preparations are stable for at least 5 hours at 20°, pH 7–9.5, and even at pH 4.4 the enzyme shows little loss of activity in 1 hour.

Purity and Physical Properties. Human urinary kallikrein migrated as a single band in electrophoresis on polyacrylamide gel at pH 8.7 and a molecular weight of 40,500 was found by short column sedimentation equilibrium. The hog pancreatic kallikrein prepared by this procedure migrated as a single sharp band in disc electrophoresis and immunoelectrophoresis. A molecular weight for this enzyme of 25,200 was estimated by sedimentation equilibrium. Other preparations,[18, 19] have had molecular weights of 25,000 or 33,000 and at least two different kallikreins (A and B) have been separated by electrophoresis near pH 8.0 or chromatography on DEAE-cellulose. Even these highly purified kallikreins could again be separated into several active fractions which differed in their content of sialic acid. Whether the formation of some of these multiple forms of kallikrein depends on the degree of autolysis or on contamination of the autolyzates with a neuraminidase-containing bacteria from the intestine has not been clearly established. However, preliminary evidence suggests that two different prekallikreins, which may correspond to forms A and B, can be detected in pancreatic

[18] H. Fritz, J. Eckert, and E. Werle, *Hoppe-Seylers Z. Physiol. Chem.* **348**, 1120 (1967); F. Fiedler and E. Werle, *Hoppe-Seylers Z. Physiol. Chem.* **348**, 1087 (1967); F. Fiedler, *Hoppe-Seylers Z. Physiol. Chem.* **349**, 926 (1968); F. Fiedler and E. Werle, *European J. Biochem.* **7**, 27 (1968).

[19] H. Moriya, *J. Pharm. Soc. Japan* **79**, 1451 (1959); H. Moriya, J. V. Pierce, and M. E. Webster, *Ann. N.Y. Acad. Sci.* **104**, 172 (1963). E. Habermann, *Hoppe-Seylers Z. Physiol. Chem.* **328**, 15 (1962); I. Trautschold, H. Fritz, and E. Werle, *in* "Hypotensive Peptides" (E. G. Erdös, N. Back, F. Sicuteri, and A. F. Wilde, eds.), p. 221. Springer Verlag, New York, 1966.

homogenates[18] and Moriya, in recent unpublished studies, has found that the preparation described in this paper can also be separated by isoelectric focusing into two or more active components.

Activators and Inactivators. Following exhaustive dialysis, horse urinary kallikrein can no longer form kinins from dialyzed horse plasma and it has been shown that Na+, K+, Li+, Rb+, or Ca+ are required for its activation.[20] The velocity of the reaction of a preparation of hog pancreatic kallikrein with BAEE has been shown to increase up to a sodium chloride concentration of 0.05 M.[18] The activation reported for this latter enzyme by sulfydryl compounds and chelating agents has been interpreted as due to a reversal of heavy metal inhibition.

The esterolytic activity of both of the above kallikreins, as well as the formation of kinins by horse urinary kallikrein, is inhibited by heavy metal ions such as Co^{2+}, Hg^{2+}, Cu^{2+}, Ni^{2+}, and Mn^{2+} while Ca^{2+} and Mg^{2+} are without effect. All of the kallikreins are inhibited by certain trypsin inhibitors [diisopropylphosphofluoridate, benzamidine, pancreatic inhibitor (Kunitz), and Trasylol] but not by others [soybean, ovomucoid, and TLCK (1-chloro-3-tosylamide-7-amino-2-heptanone)]. Egg white inhibits human urinary kallikrein but not hog pancreatic kallikrein. Antibody to human urinary kallikrein prevents the formation of kinins by human urinary and pancreatic kallikreins but not hog pancreatic kallikrein. Other inhibitors have been found in plasma, various animal tissues, and plants and recently a number of synthetic trypsin inhibitors have been tested for their ability to inhibit some of the glandular kallikreins.

Specificity. All three glandular kallikreins release kallidin from the kininogens in plasma. All are capable of cleaving various N-substituted arginine and lysine esters, the velocity of the lysine esters lower than the corresponding arginine ester. Although these enzymes do not usually cleave protein substrates such as casein and hemoglobin, horse urinary kallikrein at high concentrations has been shown to cleave the arginine bonds in salmine and in poly-L-arginine but it does not hydrolyze poly-L-lysine.

Kinetics. The rate at which the kallikreins cleave the various synthetic esters varies. For example, horse urinary kallikrein is only slightly more active on BAEE than on TAME, human urinary kallikrein is less active while a hog pancreatic kallikrein preparation was 30 times more active. The purest preparations of horse urinary kallikrein, human urinary kallikrein, and hog pancreatic kallikrein will hydrolyze, respec-

[20] J. L. Prado, A. V. Katchburian, J. Mendes, and E. S. Prado, *Acta Physiol. Latinoam.* **15**, 386 (1965); C. Kramer, E. S. Prado, and J. L. Prado, *Med. Pharmacol. Exp.* **15**, 389 (1966).

tively, 65, 9, and 5 micromoles of TAME per milligram per minute at 30°, pH 8.5. A K_m of 1.3×10^{-3} and 6×10^{-5} on this same substrate has been reported for horse urinary kallikrein and hog pancreatic kallikrein. Other kinetic values on synthetic substrates for a hog pancreatic kallikrein preparation have been given by Fiedler.[18] Calculated values for the initial rate of kinin formation by horse urinary kallikrein under conditions of zero-order kinetics are 800 nanomoles equivalent of bradykinin per minute per milligram at 37° from heated and dialyzed horse plasma[20] and 140 nanomoles from a partially purified bovine kininogen.[21]

Chemical Composition and Active Centers. Hog pancreatic kallikrein is a glycoprotein containing glucosamine, galactose, fucose, mannose, glucose, and sialic acid.[18] This same preparation (a mixture of A and B forms) contained 283 amino acid residues and 19–21 sugar residues to give a calculated molecular weight of 33,400–33,800. Both a seryl and a histidyl residue are thought to be present in the active center of the kallikreins.

Distribution. All exocrine glands, with the notable exception of the liver and gastric glands, contain a kallikrein which is excreted by all of the glands except the pancreas in a free active form. A more comprehensive discussion and bibliography on the kallikreins can be found in following reviews.[22, 23]

[21] C. R. Diniz, A. A. Pereira, J. Barroso, and M. Mares-Guia, *Biochem. Biophys. Res. Commun.* **5**, 448 (1965).
[22] E. K. Frey, H. Kraut, and E. Werle, "Das Kallikrein-Kinin-System und seine Inhibitoren," Ferdinand Enke Verlag, Stuttgard, 1968.
[23] "Bradykinin, Kallidin and Kallikrein," Handbook of Experimental Pharmacology (E. G. Erdös, ed.), Vol. 25. Springer Verlag, Heidelberg, 1970.

[51] Renin

By R. R. SMEBY and F. M. BUMPUS

$$\text{Renin substrate} \xrightarrow{\text{[Renin]}} \text{angiotensin I}$$
$$\text{(angiotensinogen)}$$

Assay

Principle. Renin has been assayed directly by its pressor effect on test animals or indirectly by measuring the amount of angiotensin generated by incubation *in vitro.* The direct assay has previously been described in this series.[1] Recently, major interest has been shown in meas-

[1] E. Haas, Vol. II, p. 124.

uring renin activity in plasma of patients with hypertensive disease and most work has centered on indirect measurement of the plasma enzyme, since the amount of enzyme in plasma is too small to measure by direct assay. Many indirect assay procedures have been used, but most work has been done with the method reported by Helmer and Judson[2] or modifications of their procedure. Therefore, this assay and the modified methods of Pickens et al.[3] and Boucher et al.[4] will be presented. Although these methods were developed for measurement of the enzyme in plasma, they can be used for the measurement of renin in kidney extracts if suitable substrate is supplied.

Assay of Helmer and Judson[2]

Reagents

NaCl, saturated

Crude human renin prepared as described earlier in this series[1] by the first procedure

Procedure. Venous or arterial blood is collected using heparin as an anticoagulant and chilled immediately with ice. The sample is centrifuged, the plasma separated and adjusted to pH 5.5. The plasma is transferred to dialysis bags and dialyzed against cold running tap water (13°) for 18–20 hours. After dialysis, the pH is again adjusted to 5.5 and the precipitated proteins removed by centrifugation. The supernatant is made isotonic by the addition of 0.26 ml of saturated NaCl per 10 ml plasma. Renin substrate is determined by adding 3 ml saline containing 0.2 Goldblatt units[1] of human renin per milliliter of dialyzed plasma. This amount of renin provides an excess so that all substrate is converted to angiotensin. The samples are incubated for 1 or 2 hours at pH 5.5 and 37° and the reaction stopped by heating at 100°. The amount of angiotensin produced can be measured by assay with a spirally cut strip of rabbit aorta,[5,6] or a 48-hour bilaterally nephrectomized, pithed cat,[7] or the ganglion-blocked, vagotomized rat.[3]

---

[2] O. M. Helmer and W. E. Judson, *Circulation* **27**, 1050 (1963).

[3] P. T. Pickens, F. M. Bumpus, A. M. Lloyd, R. R. Smeby, and I. H. Page, *Circulation Res.* **17**, 438 (1965).

[4] R. Boucher, R. Veyrat, J. deChamplain, and J. Genest, *Can. Med. Assoc. J.* **90**, 194 (1964).

[5] O. M. Helmer, *Am. J. Physiol.* **188**, 571 (1957).

[6] R. F. Furchgott and S. Bhadrakom, *J. Pharmacol. Exptl. Therap.* **108**, 129 (1953).

[7] R. E. Shipley, O. M. Helmer, and K. G. Kohlstaedt, *Am. J. Physiol.* **149**, 708 (1947).

Calculation of Activity. Activity is expressed in terms of the first-order reaction constant calculated as:

$$K = \left[\frac{2 - \log 100 \ (M - X/M)}{t} \right] 2.303 \qquad (1)$$

Where M is the amount of angiotensin produced by incubation with excess renin, X is the amount of angiotensin produced by renin present in the plasma, and t is the time of incubation in minutes.

Assay of Pickens *et al.*[3]

Reagents

Disodium ethylenediaminetetraacetate (EDTA), 2.2 g/liter
Diisopropylphosphofluoridate (DFP), 1 g dissolved in 19 ml isopropanol
NaCl, 0.9%
NaOH, 0.1 N
0.1% acetic acid
HCl, 0.1 N

Heparinized plasma (10 ml) is adjusted to pH 6.5, dialyzed against EDTA solution for 24 hours, and then against distilled water for 24 hours at 4°. The precipitate is removed by centrifugation, 1 drop of DFP solution is added to the supernatant, and the pH adjusted to 5.5 with 0.1 N HCl. It is incubated for 4 hours at 37°, then 1 ml 0.9% saline added, the volume adjusted to 20 ml with distilled water, the pH adjusted to 5.0, and the sample heated in a boiling water bath for 10 minutes. The protein is removed by centrifugation, the pH adjusted to 7.2 with 0.1 N NaOH, and back to 6.8 with 0.1% acetic acid and the solution taken to dryness *in vacuo.* The residue is redissolved in 1 ml of distilled water and the amount of angiotensin produced assayed with a ganglion-blocked, vagotomized rat.

Expression of Renin Activity. The amount of renin in the plasma is expressed as nanograms angiotensin generated per ml plasma per 4-hour incubation period.

Assay of Boucher *et al.*[4]

Reagents

Ammonium EDTA prepared by stirring 3.8 g of EDTA in 100 ml water and adjusting the pH to 6.5 with ammonium hydroxide

Dowex resin 50W-X2, 100–200 mesh. Five hundred grams of resin are washed with 2 liters 4 N NaOH, then 1 liter water, 1 liter 2 N HCl, 2 liters water, and then with 0.2 N ammonium acetate, pH 6, until the eluate reaches pH 6.

Ammonium acetate, 0.2 N, pH 6

Diethylamine, 0.1 N

Ammonium hydroxide, 0.2 N

Procedure. Blood is collected with ammonium EDTA as anticoagulant, chilled immediately to 0.5° and centrifuged. The pH of plasma is adjusted to 5.5 with 1 N HCl and the plasma filtered through glass wool. To 10 ml plasma in a 50-ml Erlenmeyer flask is added 4 ml of moist, suspended Dowex 50W-X2(NH_4^+) resin. The mixture is incubated for 3 hours at 37° with vigorous shaking. Following incubation, the mixture is transferred to a glass column (1×10 cm) containing 1 ml resin. The column is washed with 15 ml ammonium acetate (pH 6) and 15 ml water. These washes are discarded. Angiotensin is eluted with 15 ml of 0.1 N diethylamine followed by 15 ml of 0.2 N ammonium hydroxide into a flask containing a small amount of concentrated acetic acid. The solution is taken to dryness below 50° *in vacuo*. The residue is dissolved in 2 ml of 20% aqueous ethanol and evaporated to dryness *in vacuo*. This is repeated 5 times to remove ammonium acetate. The final residue is dissolved in 1 ml of 20% aqueous ethanol for assay of angiotensin. Renin activity is expressed as nanograms angiotensin generated per 100 ml plasma per 3 hours of incubation.

Comments on Renin Assay. In these procedures, renin activity found may be modified by the presence of renin inhibitors[8] or activators present in plasma. The result has been termed "effective renin activity." To avoid this, Lever *et al.*[9] isolated the renin from plasma before assay and Skinner[10] attempted to destroy these factors. The results of these assays have been mistakenly referred to as "renin concentration." Both types of assay have led to the same clinical conclusions and it is not known which may prove to be the most useful type of assay for the clinician.

Preparation of Hog Renin

The procedure which yielded renin of 780 U/mg protein has previously been given in this series.[1] A simpler, chromatographic procedure

[8] S. Sen, R. R. Smeby, and F. M. Bumpus, *Biochemistry* **6**, 1572 (1967).

[9] A. F. Lever, J. I. S. Robertson, and M. Tree, *Biochem. J.* **91**, 346 (1964).

[10] S. L. Skinner, *Circulation Res.* **20**, 391 (1967).

has recently been described by Skeggs *et al.*[11] which gave renin of 454 U/mg protein. During this purification four forms of the enzyme were observed and termed renins I, II, III, and IV from the order of elution from a DEAE-cellulose column. Renin I seems to be the major renin present in kidney with the other forms being formed during dialysis at pH 5. The procedure of Haas yields a mixture of renins II, III, and IV while the procedure of Skeggs *et al.*[11] given below yields renin II.

Procedure. The entire procedure is conducted at temperatures of 5° or less. Fresh, frozen hog kidneys (4.55 kg) are partially thawed and passed through an electric meat grinder into 5 liters of cold distilled water. The mixture is stirred for 1 hour and then filtered through several layers of cheesecloth. The kidney pulp is stirred an additional hour with a second 5-liter portion of cold water. The mixture is strained as before and the pulp discarded. The two filtrates are combined and shaken thoroughly with 2 liters of toluene. After standing for 1 hour the major portion of the upper layer is discarded and the lower layer centrifuged at 2500 rpm in a large centrifuge for 2 hours. The upper jellylike toluene layer and the gray bottom sediment are also discarded. The red, turbid, middle layer is diluted with 6 volumes of cold distilled water and the protein content determined. For each gram of protein in the solution, 15 g of a well-squeezed but moist filter cake of the free base form of DEAE-cellulose is added. The mixture is stirred for 3 minutes, then filtered on a large vacuum funnel and the DEAE pads squeezed by means of a rubber dam. The inactive filtrate is discarded. The DEAE pad is stirred with an original volume of cold water and pH adjusted to 4.6 with 1 N acetic acid. The pH is measured on a small portion of filtrate to which one-tenth volume of 1 M sodium chloride has been added. After adjusting the pH, the batch is stirred an additional 10 minutes and then filtered without further delay. The inactive filtrate is discarded. The DEAE pad is immediately suspended in a volume of cold 0.1 M sodium chloride which is one-third of the volume originally treated with DEAE cellulose. The pH is adjusted to 4.6 with 1 N acetic acid and the mixture filtered after stirring for 10 minutes. The DEAE pad is extracted once more in an identical fashion with 0.1 M sodium chloride. The two eluates are combined and solid ammonium sulfate added to obtain 2.5 M concentration. After stirring, the batch is allowed to stand in the refrigerator for 48 hours and the precipitate recovered by centrifugation and dissolved in 400 ml of cold water. At this point a typical preparation contains 14,000 Goldblatt units with a specific activity of about 1 U/mg protein.

[11] L. T. Skeggs, K. E. Lentz, J. R. Kahn, and H. Hochstrasser, *Circulation Res.* Suppl. II **21**, 91 (1967).

The enzyme is reprecipitated with ammonium sulfate by adjusting this solution to pH 7.5. Ammonium sulfate is then added to 1 M concentration, the mixture stirred thoroughly, and the precipitate removed by centrifugation and discarded. Sufficient ammonium sulfate is added to the supernatant to make the concentration 2.0 M. The mixture is stirred for a few minutes, centrifuged, and the supernatant discarded. The precipitate is dissolved in cold distilled water, dialyzed against cold distilled

TABLE I
Summary of Reported Michaelis Constants for Various Renins

Source of renin	Source of substrate	pH	Ionic environment	Michaelis constant (ng/ml)	Ref.
Hog kidney	Hog	7.0	Tris, 0.066 M	0.71	a
Hog kidney	Rat	7.0	Tris, 0.066 M	2.5	a
Human kidney	Human	5.5	Salt free	142	b
Human kidney	Human	6.9	NaCl, 0.9%	185	c
Human kidney	Hog	7.4	Phosphate, 0.1 M	413	d
Hog (I) kidney	Hog (A)	7.5	Phosphate, 0.05 M	1724	e
Hog (I) kidney	Hog (B₁)	7.5	NaCl 0.1 M	742	e
Hog (I) kidney	Hog (B₂)	7.5	NaCl 0.1 M	742	e
Hog (I) kidney	Hog (C₁)	7.5	NaCl 0.1 M	1380	e
Hog (I) kidney	Hog (C₂)	7.5	NaCl 0.1 M	1254	e
Hog (I) kidney	Synthetic tetradeca	7.5	NaCl 0.1 M	1724	e
Hog (II) kidney	Hog	7.5	NaCl 0.1 M	1066	e
Hog (III) kidney	Hog	7.5	NaCl 0.1 M	1066	e
Hog (IV) kidney	Hog	7.5	NaCl 0.1 M	1200	e
Rabbit kidney	Ox	5.7	0.15 M phosphate/saline	620	f
Rabbit plasma	Ox	5.7	0.15 M phosphate/saline	550	f
Dog kidney	Ox	5.7	0.15 M phosphate/saline	550	f
Dog plasma	Ox	5.7	0.15 M phosphate/saline	530	f
Human kidney	Ox	5.7	0.15 M phosphate/saline	590	f
Human plasma	Ox	5.7	0.15 M phosphate/saline	530	f
Human urine	Ox	5.7	0.15 M phosphate/saline	510	f
Human amniotic fluid	Ox	5.7	0.15 M phosphate/saline	530	f

[a] P. Blaquier, *Am. J. Physiol.* **208**, 1083 (1965).

[b] P. T. Pickens, F. M. Bumpus, A. M. Lloyd, R. R. Smeby, and I. H. Page, *Circulation Res.* **17**, 438 (1965).

[c] E. Haas and H. Goldblatt, *Circulation Res.* **20**, 45 (1967).

[d] A. B. Gould, L. T. Skeggs, and J. R. Kahn, *Lab. Invest.* **15**, 1802 (1966).

[e] L. T. Skeggs, K. E. Lentz, J. R. Kahn, and H. Hochstrasser, *Circulation Res.* Suppl. II **21**, 91 (1967).

[f] R. R. Smeby and F. M. Bumpus, *in* "Renal Hypertension" (I. H. Page and J. W. McCubbin, eds.), p. 26. Year Book Medical Publishers, Chicago, Illinois, 1968.

water containing 0.5 ml of 1 M sodium bicarbonate and 0.5 ml of 1 M sodium acetate per liter and having a pH of 7. Excess toluene is added to the buffer to act as a preservative. In this way the pH of the dialyzate was maintained between pH 7.0 and 7.5. Following dialysis, the solution is adjusted to pH 5 and the precipitate which formed is quickly removed by centrifugation. The pH of the supernatant is immediately adjusted to 7 and stored in the frozen state until used. A typical preparation contains about 3800 U of activity with a specific activity of 2.33 U/mg protein.

This preparation is purified further by chromatography on a DEAE-cellulose column. The column is 1.15 cm in diameter by 55 cm high and equilibrated with 0.025 M sodium acetate buffer, pH 5.35. The pH of the enzyme sample is adjusted to 5.35 and diluted if necessary to achieve an ionic strength no greater than that of the 0.025 M sodium acetate buffer. After application of the sample, the 0.025 M buffer, pH 5.35, is pumped through the column at a rate of 0.48 ml/minute until ten 5-ml fractions are collected. Buffer was then pumped into the column from a cylindrical gradient mixing chamber which initially contained 500 ml. At the same time, 0.025 M acetic acid was added to the chamber with mixing at 0.24 ml/minute. Thus, a descending pH gradient was formed which usually reads 3.8 after about 200 5-ml fractions are collected. Renin I is eluted from such a column at pH 5.16, renin II at pH 4.72, renin III at pH 4.53, and renin IV at pH 4.25. Renin I represents the largest renin activity peak. Renin I peak is dialyzed at pH 5.0 for 45 hours at 5° during which renin I is converted to renin II. This is rechromatographed on a column as described above and renin II represents the main renin

TABLE II

pH OPTIMA FOR RENIN FROM DIFFERENT SOURCES

Source of renin	Substrate	pH Optimum	Ref.
Rabbit	Ox	5.4–6.2	a
Human	Human	5.5	b
Human	Ox	6.0	c
Human	Hog	7–8	d
Hog	Synthetic	6.5–7	e

a A. F. Lever, J. I. S. Robertson, and M. Tree, *Biochem. J.* **91,** 346 (1964).

b P. T. Pickens, F. M. Bumpus, A. M. Lloyd, R. R. Smeby, and I. H. Page, *Circulation Res.* **17,** 438 (1965).

c J. J. Brown, D. L. Davies, A. F. Lever, J. I. S. Robertson, *Can. Med. Assoc.* **90,** 201 (1964).

d A. B. Gould, L. T. Skeggs, and J. R. Kahn, *Lab. Invest.* **15,** 1802 (1966).

e D. Montague, B. Riniker, and F. Gross, *Am. J. Physiol.* **210,** 595 (1966).

activity. The specific activity of this fraction is usually about 450 U/mg protein.

Properties. The Michaelis constants for renins from a number of sources and under a variety of conditions have been reported. These are summarized in Table I.

The molecular weight of hog renin has been determined by Sephadex chromatography to be between 40,000 and 50,000.[12, 13] Highly purified hog renin is highly unstable in the cold.[12]

The pH optima for renin from several species are given in Table II.

[12] W. S. Peart, A. M. Lloyd, G. N. Thatcher, A. F. Lever, N. Payne, and N. Stone, *Biochem. J.* **99**, 708 (1966).
[13] E. Kemp and I. Rubin, *Acta Chem. Scand.* **18**, 2403 (1964).

[52] Staphylokinase

By CHARLES H. LACK and KITTY L. A. GLANVILLE

Staphylokinase is an enzyme, produced by many strains of M. *Staphylococcus aureus*, which converts plasminogen to plasmin.

Assay Method

Principle. All methods available for staphylokinase assay depend on the initial activation of plasminogen to plasmin by staphylokinase under standard conditions followed by measurement of the fibrinolytic or proteolytic activity of the plasmin formed.

Most fibrinolytic methods are based on visual observation of clot lysis time and suffer from poor reproducibility. The fibrin plate method[1] is a convenient, simple procedure for screening activator activity. For a fuller discussion of assay methods, see Chapter by F. B. Taylor and R. H. Tomar on streptokinase [63] in this volume.

Casein is a less sensitive substrate for plasmin than fibrin, but the results obtained with casein are more reproducible and lend themselves to rate determination. The degree of casein digestion can be easily determined from the extinction at 280 nm of the perchloric acid-soluble fraction of the digest. The method described was developed by Davidson[2] and is a modification of a technique reported by Norman[3] for the assay of plasminogen.

[1] T. Astrup and S. Mullertz, *Arch. Biochem. Biophys.* **40**, 346 (1952).
[2] F. M. Davidson, *Biochem. J.* **76**, 56 (1960).
[3] P. S. Norman, *J. Exptl. Med.* **106**, 423 (1957).

Reagents

(1) Human plasminogen, prepared by the method of Kline.[4] This is assayed by Davidson's[2] casein method which is carried out under identical conditions to the activator assay described below except that various dilutions of plasminogen are activated by 60 Christensen units of streptokinase for 5 minutes at 37° before casein solution is added. To determine plasmin contamination in the preparation of plasminogen, the assay is carried out using buffer in place of activator.

Activity is expressed in terms of an "extinction" unit which gives, under the conditions of the test, an increase in E at 280 nm of 1.00/minute of digestion. Therefore the activity of a solution of plasminogen assayed will be $R/(V \times 30)$ units, where R is the extinction reading, after subtraction of the appropriate blank value, for the particular volume V of plasminogen assayed.

(2) Palitzsch's borate saline buffer, pH 7.4.[3] Boric acid 0.2 M containing 50 mM NaCl is mixed with sufficient 50 mM sodium borate to give the required pH.

(3) Casein solution. 4% Light white soluble casein (B.D.H. Poole, Dorset) is used without further purification. The casein is dissolved in Palitzsch's buffer, pH 7.4, by stirring at 37° or 56°. The solution is stored at 4° and is made up fresh about every 2 weeks.

(4) Perchloric Acid. A one in ten aqueous dilution of the 60% solution obtained commercially is used.

Procedure. METHOD. Several volumes of the staphylokinase solution to be assayed are incubated in duplicate in a 37° water bath in the presence of 50 milliunits of human plasminogen (about 0.1 ml) in a total volume of 1.1 ml, the volume being made up with Palitzsch's buffer, pH 7.4. After exactly 30 minutes, 1 ml of casein solution is added to each tube as substrate for the plasmin formed, and incubation at 37° is continued for exactly 30 minutes. Undigested casein is then precipitated by the addition of 3 ml of perchloric acid. The precipitate is allowed to form for 1 hour or more before it is removed by filtration on Whatman No. 1 filter paper (5.5 cm). The extinction coefficient (E) at 280 nm of the PCA-soluble digestion products contained in the filtrates is measured in a spectrophotometer. Readings are made against a "blank" filtrate prepared by adding 1 ml of casein and 3 ml of perchloric acid to 1.1 ml of buffer. The assay includes a plasminogen blank and staphylokinase blanks in which the activator or plasminogen respectively is replaced by Palitzsch's buffer. Both blank values are subtracted from the corresponding assay reading. Davidson determined the activity by interpolation. It

[4] D. L. Kline, *J. Biol. Chem.* **204**, 949 (1953).

was preferred, however, to read the activity off a standard graph prepared previously, since the curve is sigmoid in shape and it is thus easier to assess within which range of staphylokinase concentration used, the readings are reliable.

Definition of Unit and Specific Activity. A unit of staphylokinase activity is defined as that activity which gives rise, under the conditions of the assay described, to an increase in E. of 0.300. This unit is found to be approximately equal to one third of a Christensen unit.[2] The specific activity of a preparation is expressed as units per milligram protein.[5]

Production and Purification Procedure

A strain of *Staphylococcus aureus*, which has been selected for its high yield of staphylokinase production,[6] is stored in the freeze-dried state and fresh cultures are prepared as required. These are used to inoculate a casein hydrolyzate medium[7] which is shaken for 7 hours at 37°.[2]

Two-liter batches of broth culture containing staphylokinase were obtained by this process.

Step 1. Preparation of Culture Supernatant Fluid. Cells are removed from the culture by centrifugation and the supernatant fluid is heated at 75° in a water bath for 30 minutes to kill any that remain. This treatment has little effect on staphylokinase activity.[8]

Step 2. Ammonium Sulfate Precipitation of Active Material.[9] The culture supernatant fluid is brought to 75% saturation with solid $(NH_4)_2SO_4$ (i.e., 3.07 M) and allowed to stand at 4° overnight. The precipitate is separated by centrifuging, and washed with a saturated $(NH_4)_2SO_4$ solution. It is then suspended in a suitable volume of water and dialyzed at 4° against running tapwater for 3 days, followed by distilled water, until free from sulfate ions.

This step serves mainly to concentrate staphylokinase from a large volume without considerably increasing the specific activity. Loss of activity at this stage is variable (26–80%).

Step 3. Chromatography on a CM-Cellulose Column.[9] The staphylokinase concentrate is purified by chromatography on a CM-cellulose column[10] at room temperature in one of two ways:

[5] O. H. Lowry, N. J. Rosebrough, A. L. Farr, and R. J. Randall, *J. Biol. Chem.* **193**, 265 (1951).
[6] C. H. Lack, *J. Clin. Pathol.* **10**, 208 (1957).
[7] E. S. Duthie and G. Haughton, *Biochem. J.* **70**, 125 (1958).
[8] E. B. Gerheim, J. H. Ferguson, B. L. Travis, C. L. Johnston, and P. W. Boyles, *Proc. Soc. Exptl. Biol. N.Y.* **68**, 246 (1948).
[9] K. L. A. Glanville, *Biochem. J.* **88**, 11 (1963).
[10] E. A. Peterson and H. A. Sober, Vol. V, p. 3.

(1) When a concentrated staphylokinase solution of only moderate purity is required, the material is adsorbed on a column at pH 4 and eluted at pH 7 as follows:

The dialyzed staphylokinase suspension is dissolved at pH 4 by the addition of an equal volume of 0.2 M citric acid-0.4 M phosphate buffer at pH 4, the alkaline component of the buffer system being added first to aid solution at the higher pH value. After addition of the citric acid component, the pH is checked and adjusted to 4 if necessary. Any undissolved material is removed by centrifugation. The volume of the supernatant fluid is usually about one-tenth that of the broth culture from which it was derived. The supernatant fluid at pH 4 is applied to a CM-cellulose column (e.g., 1.7 × 11 cm) equilibrated with 0.1 M citric acid-0.2 M phosphate buffer at pH 4 at the maximum flow rate obtained by gravity. Under these conditions all staphylokinase activity is adsorbed on the column; the highly pigmented effluent contains 50% or more of the protein applied. The column is washed with pH 4 buffer until the effluent is free of protein, as indicated by measurement of $E_{280\,nm}$. Elution is carried out at pH 7 using the same buffer system. Most of the pigment has been removed and recovery of activity from the column is high, purification with respect to protein being 5- to 9-fold. Staphylokinase cannot be separated satisfactorily from hemolysin by this method. As the kinase moves with the pH 7 solvent front together with a faint pigment, the active fraction can be easily collected and only a minimum of control assays have to be carried out.

(2) To obtain a more highly purified staphylokinase preparation free from hemolysin, a technique is used similar to that above, but employing stepwise elution with pH 6 sodium phosphate buffers of increasing molarity as follows:

The staphylokinase concentrate is adsorbed on the column, equilibrated with the starting buffer either at pH 4 as described for (1) or at pH 6 from 15 mM sodium phosphate buffer. The column is developed by stepwise increase of molarity of the pH 6 phosphate buffer solution, i.e., 15 mM, 45 mM, and 0.25 M. Hemolysin, being more strongly adsorbed than staphylokinase, can thus be completely removed from the preparation. Staphylokinase is eluted by 45 mM buffer solution whereas hemolysin is collected either by prolonged elution with the same buffer strength, or by increasing the molarity of the buffer to 0.25 M. Forty-four-fold purification of staphylokinase with respect to protein could be achieved by this method.

Step 4. Precipitation by Ethanol.[11] Staphylokinase is purified further

---

[11] K. L. A. Glanville, Ph.D. thesis, London University, England, 1966.

TABLE I

PURIFICATION OF STAPHYLOKINASE

Step	Fraction	Volume (ml)	Total activity (units)	Total protein (mg)	Specific activity (units/mg protein)	Yield (%)	Purification
1	Heated culture supernatant	2180	50,140	359	139	100	1.0
2	Dissolved (NH₄)₂SO₄ precipitate	50	30,300	—	—	60	—
	After dialysis, nondiffusate	105	22,400	168	133	45	0.96
2ᵃ	Dialyzed (NH₄)₂SO₄ precipitate in pH 6 buffer	85	12,665	167	70	100	1.0
3	Active eluate from column	25	5,000	2.0	2,500	40	36
4	Dissolved EtOH precipitate	25	2,312	0.31	7,400	18	106

ᵃ The concentrate was supplied by the Microbiological Research Establishment, Porton, England.

by precipitation with an equal volume of absolute ethanol. The precipitate that is allowed to develop for about 20 hours at 4° is separated by centrifugation, washed once with 50% ethanol and dissolved in a suitable volume of distilled water. Recovery ranged from 50 to 100% and purification of staphylokinase, with respect to protein, varied from 3-to 7-fold.

Freeze Drying. The purified staphylokinase solution is distributed in ampoules in 1 or 2 ml-aliquots and freeze dried. Loss on freeze-drying was negligible under these conditions and no activity was lost on storage of the freeze dried material for 2 years at 4°.

Degree of Purification. The highest degree of purification achieved by this method, using elution method (2) in step 3, was 100-fold-purification of the $(NH_4)_2SO_4$ precipitated concentrate (Table I).

Modification. This method has been modified by Dr. K. Sargeant of the Microbiological Research Establishment, Porton, England, who carried out purification of 100-liter batches of culture supernatant fluid. The dialysis step to remove $(NH_4)_2SO_4$ was replaced by deionization on Sephadex G-50 in 25 mM phosphate buffer at pH 6. For the chromatographic separation, CM-cellulose was replaced by CM-Sephadex G-50 equilibrated with 50 mM citric acid-phosphate buffer at pH 5. The active fraction, adjusted to pH 5 with McIlwaine's buffer at pH 4.4, was filtered and the filtrate applied to the column which was washed with 50 mM citric acid-phosphate buffer at pH 5. Elution was carried out using a gradient obtained by means of a two-chamber system containing 50 mM citric acid-phosphate buffer at pH 5 and 0.2 M citric acid-phosphate buffer at pH 6. The bulk of staphylokinase activity was eluted after the main protein peak and had a specific activity of 1000–3000 U/mg.

A different method for the purification of staphylokinase involves fractional precipitation with $(NH_4)_2SO_4$, dialysis, precipitation at pH 5.5, and chromatography on Amberlite XE-64.[12]

Properties

Stability. Staphylokinase is relatively heat stable.[8] Maximum stability of the purified solution is at pH 6 to 8. At this pH no activity of the semisterile solution was lost in 24 hours at 37°; when heated to 100° for 45 minutes the solution lost about 40% of its activity. Stability is greater in concentrated than in diluted solutions, particularly in the presence of neutral buffer. Solutions in Palitzsch's buffer at pH 7.4 were stored for over 2 weeks at 4° without loss of activity. At −20° solutions in buffer containing above 40 U/ml were stable for more than 4 months.

[12] E. Soru and M. Sternberg, *Enzymologia* 25, 231 (1963).

Dilute solutions of purified staphylokinase usually lost some activity when being pipetted. This could not be remedied by silicone treatment of the glassware but was prevented by addition of 0.2% (v/v) of Triton X-100, a nonionic detergent, to the staphylokinase solution. As this treatment reduced the stability of purified staphylokinase on storage, it seemed advisable to add Triton X-100 to the solution only shortly before it was pipetted out for an experiment. When assaying staphylokinase containing 0.2% (v/v) Triton X-100, the activator activity appeared to be enhanced.[11]

Purity and Physical Properties.[11] A solution of staphylokinase (specific activity 900 U/mg), purified as described in the text using method (2) in step 3 still contains a protein impurity as shown by: (1) Starch gel electrophoresis. Two protein bands could be observed, the stronger band being associated only with activator activity. (2) Gel filtration on Sephadex G-100. The main protein peak which contained the activator activity was closely followed by an inactive smaller protein peak. This method could be used for further purification of staphylokinase. (3) Immunodiffusion on agar gel. Using rabbit antiserum prepared against an impure staphylokinase preparation, one main line of precipitation was obtained together with an extremely faint line.

No coagulase, leukocidin, α-, β-, or δ-hemolysin were detected in the preparation.

The freeze-dried material is a white powder soluble in water near or above neutral pH. Buffer solution enhances the solubility. The ultraviolet absorption spectrum of the solution is that of a protein.

The approximate molecular weight of staphylokinase as determined by gel filtration on a calibrated column of Sephadex G-100 is 22,500, while that of streptokinase determined on the same column is 53,000.

The isoelectric point was found by paper electrophoresis to be near pH 6.7. This value is only approximate because of the limitations imposed by electrophoresis on a supporting medium.

Inactivators. STAPHYLOKINASE ANTIBODY. Staphylokinase is antigenic.[13] A wide range of inhibitor level in the population has been demonstrated[14] and a rise in antistaphylokinase titer in nine out of thirty-five cases of staphylococcal osteomyelitis was observed.[15] Antibody can be obtained in serum of rabbits sensitized experimentally to staphylokinase over a period of 35 days.

ϵ-AMINOCAPROIC ACID (ϵ-ACA). ϵ-ACA is an inhibitor of plasmin

[13] F. Aoi, Kitasoto, *Arch. Exptl. Med.* **9,** 171 (1932).
[14] B. Sweet, G. P. McNicol, and A. S. Douglas, *Clin. Sci.* **29,** 375 (1965).
[15] C. H. Lack and A. G. Towers, *Brit. Med. J.* **2,** 1227 (1962).

activity[16] but its effect on plasminogen activation by streptokinase as a competitive inhibitor is much greater.[17] Activation of human plasminogen by staphylokinase is also inhibited by ε-ACA, but to obtain comparable inhibition of plasminogen activation about ten times higher ε-ACA concentrations are required for staphylokinase than for streptokinase activation. At very low ε-ACA concentration (e.g., 20 mM) its stabilizing effect on plasmin may be more pronounced than its inhibitory action.[11]

CASEIN. Activation by staphylokinase is inhibited in the presence of 2% casein, or an impurity in the casein used.[2]

Specificity. The only known action of staphylokinase is that of plasminogen activation. It does not itself hydrolyze casein and has no esterolytic effect using TAME as substrate.[14]

Unlike streptokinase, staphylokinase can activate the plasminogen of a number of animal species without the addition of human globulin as source of proactivator, but in contrast to urokinase, it does not activate ox plasminogen. Table II shows the results obtained by a number of workers when activating plasminogen prepared from various animal species. The resistance of bovine plasminogen to activation by staphylokinase is not due to the presence of an inhibitor of staphylokinase since the activation of human plasminogen by staphylokinase is not affected by the addition of ox euglobulin.[2]

Cliffton and Cannamela[18] noted a difference in the fibrinolytic and proteolytic activities of the plasmin derived from staphylokinase activation of certain animal serum plasminogens.

TABLE II
STAPHYLOKINASE ACTIVATION OF DIFFERENT PLASMINOGENS

Activated	References	Not activated	References
Human	19,20,21	Rat	18,20,21
Monkey	18	Ox	18,19,20,21
Dog	18,20,21	Chicken	18,21
Cat	20	Horse	18,21
Rabbit	18,19,20,21	Sheep	21
Guinea pig	18,19,20,21	Pig	18

[16] S. Okamoto and F. Nagasawa, Patent Specification 770, 693, London, England. The Patent Office, Oct. 21st (1954). p. 9 Granted, 1957.
[17] N. Alkjaersig, A. P. Fletcher, and S. Sherry, J. Biol. Chem. 234, 832 (1959).
[18] E. E. Cliffton and D. A. Cannamela, Blood 8, 554 (1953).
[19] F. M. Davidson, Nature 185, 626 (1960).
[20] J. H. Lewis and J. H. Ferguson, Am. J. Physiol. 166, 594 (1951).
[21] E. B. Gerheim and J. H. Ferguson, Proc. Soc. Exptl. Biol. Med. 71, 261 (1949).

Kinetic Properties. The maximum rate of activation of human plasminogen by staphylokinase is between $37°–45°$[14] and at pH 8.6.[11] The resulting plasmin, however, is not stable under these conditions unless a stabilizing agent, e.g., glycerol,[22] is added.

The enzymatic nature of staphylokinase has been demonstrated for the activation of dog[20] and human[2] plasminogens. A free α-amino group in the staphylokinase molecule is essential for its activator activity.[12] Human plasminogen can be activated as completely by staphylokinase as by streptokinase,[2] yielding a plasmin whose molecular weight and amino acid composition are of the same order as those of streptokinase or urokinase-activated plasmin.[23]

Distribution

Staphylokinase is usually produced by coagulase-positive strains of staphylococcus of human origin. Christie and Wilson[24] found that 92 out of 99 coagulase positive but none of 12 coagulase-negative strains produced fibrinolysis. The coagulase-positive strains which failed to produce lysis were strong producers of β-hemolysin. The production of β-hemolysin is stated to exclude the production of staphylokinase.[25] Strains have been isolated which produce staphylokinase but no coagulase or hemolysins.[26] Of the strains isolated from human infections 96.4%,[27] 70%,[28] 60%,[29] and 16%[26] were reported to be fibrinolytic, whereas 100%[30] and 50%[26] isolated from the dog and only 5%[26,31] isolated from the ox were found to be fibrinolytic.

[22] S. Shulman, N. Alkjaersig, and S. Sherry, *J. Biol. Chem.* **233**, 91 (1958).

[23] E. Soru, *Rev. Roum. Biochim.* **5**, 17 (1968).

[24] R. Christie and H. Wilson, *Australian J. Exptl. Biol. Med. Sci.* **19**, 329 (1941).

[25] P. M. Rountree, *Australian J. Exptl. Biol. Med. Sci.* **25**, 359 (1947).

[26] G. Hentschel and H. Blobel, *Zentr. Bakteriol. Parasitenk, Abt. I, Orig.* **206**, 193 (1968).

[27] A. Kaffka, *Zentr. Bakteriol. Parasitenk. Abt. I, Orig.* **168**, 381 (1957).

[28] C. H. Lack and D. G. Wailling, *J. Pathol. Bacteriol.* **68**, 431 (1954).

[29] G. H. Chapman, *J. Bacteriol.* **43**, 313 (1941).

[30] H. W. Smith, *J. Comp. Pathol.* **57**, 98 (1947).

[31] St. H. George, K. E. Russell, and J. B. Wilson, *J. Infect. Diseases* **110**, 75 (1962).

[53] Snake Venom Proteases Which Coagulate Blood

By M. P. ESNOUF

Malayan Pit Viper (*Ancistrodon rhodastoma*) Venom

Assay Method[1]

The coagulant fraction of this venom, while acting almost exclusively on fibrinogen, has in addition a slight factor X converting activity.[2] The thrombinlike activity of the fraction is conveniently assayed by adding different dilutions of the fraction to a standard solution of fibrinogen and recording the clotting times. The clotting times are then compared with those obtained with dilutions of the crude venom.

Reagents

Fibrinogen: A bottle of human fibrinogen (grade L from A.B. Kabi, Stockholm, Sweden) was dissolved in 100 ml distilled water and then diluted with an equal volume of 0.15 M Tris and the pH adjusted to 7.5 with hydrochloric acid.

Method. The fibrinogen solution (0.3 ml) is incubated at 37° with 0.1 ml of the buffer in 2½ inch × ⅜ inch glass tubes. The venom solution (0.1 ml) is added, and the time from the addition of the venom to formation of the clot is recorded. Under these conditions a venom solution (30 μg/ml) gives a clot in about 20 seconds.

The factor X converting activity of the venom fraction is readily demonstrated by first incubating the venom fraction with *A. rhodastoma* antivenom until the thrombinlike activity has been lost and then adding the venom to factor X-deficient plasma which will give a clotting time of at least 3 minutes compared with a clotting time of about 20 seconds when factor X and calcium ions are added to the factor X-deficient plasma.

Assay Method[2]

In addition to the coagulation assay, the purified fraction can be assayed by its esterase activity on various synthetic substrates, such as the esters of arginine or the p-nitrophenyl esters of carbobenzoxy derivatives

[1] M. P. Esnouf and G. W. Tunnah, *Brit. J. Haematol.* **13**, 581 (1967).
[2] L. Nahas, K. W. E. Denson, and R. G. Macfarlane, *Thromb. Diath. Haemorrhag.* **12**, 355 (1964).

of the aromatic amino acids. The method using benzoyl arginine ethyl ester[3] (BAEE) will be described.

Reagents

> Benzoyl arginine ethyl ester is dissolved in 0.05 M Tris adjusted to pH 8.0 with hydrochloric acid, to give a final concentration of ester of 1.5 mM.
>
> Venom solution. Only the purified coagulant fraction can be assayed by this procedure, because there are other components in the venom which will hydrolyze these synthetic substrates but which have no activity on fibrinogen.

Method. Two milliliters of the 1.5 mM benzoyl arginine ethyl ester solution was incubated at 37° in a 1-cm spectrophotometer silica cuvette; after a few minutes a sample (1 ml) of the venom fraction in 0.05 M Tris-hydrochloric acid buffer, pH 8.0, was added. The change in the optical density of the mixture was measured at 253 nm. If p-toluenesulfonyl arginine ester is to be used as the substrate instead of BAEE, the reaction is followed at 247 nm. The rate of hydrolysis obtained in this way is corrected for the rate of spontaneous hydrolysis of these esters at pH 8.0. Under these conditions the rates of hydrolysis of benzoyl arginine ethyl ester and p-toluenesulfonyl arginine are about 100 millimicromoles/minute/μg and 40 millimicromoles/minute/μg, respectively.

Units. Since the venom contains other esterases, this latter type of assay can only be used to relate the different activities of purified preparations. Because of this, the clotting assay, though less precise, is more specific, and can be used in computation of specific clotting activity relative to the crude venom. However, since different batches of venom vary in their content of thrombinlike activity, the most convenient unit to use is based on equivalent thrombin units.

Purification

The purification consists of a two-step chromatography process. In the first stage the crude venom is chromatographed on triethylaminoethyl (TEAE) cellulose, and the coagulant fraction rechromatographed on Sephadex G-100.

Fractionation of the Crude Venom on TEAE-Cellulose. The TEAE-cellulose (Serva Entwicklungs Labor., Heidelberg, Germany) is suspended in 2 M sodium chloride buffered with 0.1 M Tris adjusted to pH 6.0 with phosphoric acid. The slurry is poured into a glass column 3.6 cm in diameter, until the packed height of the column is 20 cm. The column

[3] G. W. Schwert and Y. Takenaka, *Biochim. Biophys. Acta* **16,** 570 (1955).

is then washed with a further 2 liters of the same solvent and then equilibrated with 0.01 M Tris adjusted to pH 8.5 with phosphoric acid. The crude venom (350 mg) is dissolved in the 20 ml of 0.01 M Tris-phosphate, pH 8.5, and centrifuged to remove the insoluble material. The clear supernatant is applied to the column. The venom is washed onto the column with a further 50 ml of the same buffer. After the venom has been washed onto the column, the buffer is changed to 0.01 M Tris-phosphate, pH 7.0, and the fractionation is carried out at room temperature at a flow rate of 90–100 ml/hour. After 1200 ml of this solvent, the eluting buffer is changed to 0.02 M Tris-phosphate, pH 6.0, and the fractionation continued until the pH of the effluent has reached pH 6.0, which occurs after about 600 ml of this solvent has been passed through the column. The solvent is now changed to 0.04 M Tris-phosphoric acid buffer at the same pH and the column eluted with about 600 ml of this solvent, when no further protein will be eluted with this solvent. The buffer is then changed to 0.1 M Tris-phosphoric acid buffer, pH 6.0, and it is with this solvent that the coagulant fraction is eluted. The recovery of the thrombinlike activity from the column is about 60% of the applied activity.

Fractionation of the Thrombinlike Activity on Sephadex G-100. The Sephadex G-100 after standing in 0.1 M sodium chloride buffer with 0.1 M sodium phosphate, pH 7.0, for 48 hours is packed into a column 5.8 cm in diameter until the Sephadex reaches a packed height of 97 cm. The void volume of the column is about 700 ml. The fraction from the TEAE-cellulose is concentrated about 20-fold and dialyzed against 0.1 M sodium chloride buffered with 0.1 M sodium phosphate and applied to the column, which is eluted with the same buffer at a flow rate of 80 ml/hour. The coagulant activity is eluted after 940 ml of this solvent. The venom fraction is sufficiently stable for it to be concentrated by rotary evaporation. Some slight losses in activity are observed when the fraction is concentrated by ultrafiltration, although this may depend on the nature of the membrane used.

Properties

Stability. Dilute solutions (100 μg/ml) of the venom dissolved in 0.1 M Tris-phosphate, pH 7.0, have remained stable for many months at −25° or at +5°.

Purity and Physical Properties. The coagulant fraction obtained as described above gives only a single precipitin arc on immunoelectrophoresis[4] against the specific horse. *A. rhodastoma* antivenin; it also gives

[4] J.-J. Scheidegger, *Intern. Arch. Allergy Appl. Immunol.* **7**, 103 (1955).

a single band after electrophoresis in 7% polyacrylamide gel.[5] The electrophoretic mobility of this protein at pH 7.0 is 3.9×10^{-5} V/cm²/second in a moving-boundary electrophoresis apparatus.

The partial specific volume $\bar{V}$ of the protein is 0.69, which suggests that the protein contains carbohydrate; this has been confirmed by chemical analysis.[1]

The sedimentation coefficient $s_{20,w}^0$ of 3.35 S was found for the protein at a concentration of 4.86 mg/ml, and the diffusion coefficient D_{20}^0 is 4.81×10^{-7} cm²/second at the same concentration.

The approach to equilibrium method for the determination of the molecular weight gives a range of values from 37,000 to 44,000; the curvature of the plots obtained indicates that the material is polydisperse.

The same conclusion is reached from low speed equilibrium experiments, in which plots of log c/r^2 (see footnote 6) have a distinct upward curvature. The weight average molecular weight for the total material present in the centrifuge cell is 45,000. The value of the Z average molecular weight M_Z calculated from schlieren photographs taken during the same experiment[7] is 51,000. High-speed equilibrium experiments for the calculation of the number average molecular weight M_n[8] give a value of 41,500. Assuming that the protein is biologically homogeneous, these results suggest that under the conditions of these experiments in which the protein was dissolved in 0.1 M sodium chloride buffered with 0.04 M sodium phosphate, the protein is a mixture of monomers and dimers, the monomer having a molecular weight of 30,000.

Activators and Inactivators

Biological. Unlike thrombin, this venom protein is unaffected by inhibitors such as heparin, hirudin, soya bean trypsin inhibitor, or by the various plasma antithrombins. The enzymatic activity is inhibited by the specific antivenin.

Chemical. The esterase and coagulant activity of this protein are inhibited by $5 \times 10^{-3} M$ diisopropylphosphofluoridate and by $5 \times 10^{-3} M$ toluenesulfonyl fluoride, the latter causing complete inhibition after 10 minutes of incubation at room temperature, whereas diisopropylphosphofluoridate gave only 68% inhibition after 40 minutes.

Specificity. The specificity of this venom protein for macromolecular substrates seems to be confined to fibrinogen, although it does have slight

[5] B. J. Davis, *Ann. N.Y. Acad. Sci.* **121**, 404 (1964).
[6] W. D. Lansing and E. O. Kraemer, *J. Am. Chem. Soc.* **57**, 1369 (1935).
[7] O. Lamm, *Z. Physik. Chem.* **143**, 177 (1929).
[8] D. A. Yphantis, *Biochemistry* **3**, 297 (1964).

factor X converting activity. It appears to be without activity on the other clotting factors, and does not activate plasminogen directly.

The enzyme, like trypsin, hydrolyzes the two arginine esters, benzoyl arginine ethyl ester, and p-toluene arginine methyl ester. It will not hydrolyze acetyl tyrosine ethyl ester, the synthetic substrate for chymotrypsin. The venom protein will hydrolyze p-nitrophenyl esters of the carbobenzoxy derivatives of both aromatic and aliphatic amino acids.

Russell's Viper (*Vipera russellii*) Venom[1]

Assay Method

Principle. The coagulant protein of this snake venom is specific for one coagulation factor, namely factor X, and can be readily assayed using normal citrated plasma as the indicator. The clotting time of the plasma after the addition of the venom is proportional to the rate of factor X activation and hence the amount of venom added to the plasma. Phospholipid is also added to the plasma at the same time as the venom, to ensure that the rate of thrombin formation is not a rate-determining reaction.

Reagents

Normal citrated plasma. This is obtained by collecting either human or bovine blood into a one-tenth volume of 0.1 M sodium citrate. The blood is then centrifuged to remove the formed elements and the supernatant plasma is stored in plastic tubes in a deep freeze.

Phospholipid emulsions. A dry powder of the acetone-insoluble fraction of human brain is extracted with 15 volumes of chloroform-methanol (1:1 v/v). The extract is washed twice with water and the solvent evaporated in a rotary evaporator at 30°. The residue is redissolved in benzene and stored at −20°. The emulsion is prepared from the phospholipid solution by first evaporating the benzene, by blowing a stream of nitrogen over the solution, and redissolving the lipid residue in peroxide-free ether. 0.02 M Tris-hydrochloride acid buffer pH 7.5 is then added and the ether is removed by bubbling nitrogen through the mixture for 30 minutes at room temperature.

Calcium chloride solution. This is a 0.05 M solution of calcium chloride buffered with 0.02 M Tris-hydrochloric acid, pH 7.5.

Venom. Crude or purified coagulant protein is dissolved in 0.02 M Tris-hydrochloric acid, pH 7.5.

Procedure. It should be stressed that none of the solutions used in this assay should contain phosphate, because factor X or activated factor X is readily adsorbed by very small amounts of precipitated calcium phosphate.

The citrated plasma (0.1 ml) is incubated at 37°, with 0.1 ml of the venom solution and 0.1 ml of the phospholipid emulsion (50 mg phospholipid/ml). To this mixture is added 0.1 ml 0.05 M calcium chloride and the clotting time of the plasma is recorded from the addition of the calcium chloride. In this assay crude venom solutions of 10 μg/ml give a clotting time of about 11 seconds. The same clotting time is given the purified fraction at a concentration of 0.7 μg/ml. A plot of the clotting time against decreasing concentrations of the venom solution gives a straight line on double logarithmic paper.

Williams and Esnouf[9] found that the coagulant protein of this venom possessed esterase activity, and that synthetic substrates such as *p*-toluenesulfonyl arginine methyl ester would competitively inhibit the coagulant activity of this. However, recently D. J. Hanahan (private communication) has been able to separate the esterase activity from the coagulant activity of the venom. At present the only explanation for these apparently conflicting results is the possibility that the venom of the same species of snake, living in different regions, may be different.

In view of this uncertainty of the esterolytic activity of this, an assay based on the esterase activity of the protein will be omitted.

Purification Procedure

In the original method[9] the venom was fractionated on Whatman DEAE-cellulose (DE 50). Similar results have been obtained on the more recent version of this cellulose (DE 52). The description of the method is based on Whatman DE 50. Twenty-five grams of the dry cellulose powder is suspended in 2 M sodium chloride buffered with 0.01 M Tris and the pH adjusted to 6.0 with phosphoric acid. The slurry is poured into a glass column 3.4 cm in diameter and the column is packed under pressure until the height of the packed column reaches about 30 cm. The column is then washed with a further 2 liters of the same solution, and equilibrated with 0.01 M Tris-phosphoric acid buffer, pH 8.5. The crude venom is dissolved in this buffer (10 mg/ml). The insoluble material is removed by centrifugation, and 30 ml of the supernatant is applied to the column. The column is then eluted with 500 ml of the pH 8.5 buffer. The buffer is then changed to 0.01 M Tris-phosphoric acid pH, 7.0, and the column eluted with this buffer until the pH of the effluent has reached

[9] W. J. Williams and M. P. Esnouf, *Biochem. J.* **84,** 52 (1962).

7.0. The column is then eluted with 0.01 M Tris-phosphoric acid buffer, pH 6.0, until the column is equilibrated at pH 6.0.

The column is now eluted with 0.15 M sodium chloride dissolved in the same buffer, and about 90% of the applied coagulant activity is recovered from the column with this solvent. The chromatography is carried out at room temperature at a flow rate of 75 ml/hour.

The fractions with coagulant activity can be concentrated by ultrafiltration.

Examination of the concentrated protein in the ultracentrifuge shows that two components are present. These are readily separated by centrifugation into a linear density gradient of sucrose (5–20% w/v) dissolved in 0.1 M sodium phosphate buffer, pH 7.0. This gradient (4.2 ml) is formed in each of three 2 inch by ½ inch cellulose nitrate centrifuge tubes, and 0.6 ml of a solution containing not more than 30 mg/ml protein in 0.1 M phosphate buffer, pH 7.0, is carefully layered onto the gradient. The centrifuge tubes are then placed into a previously cooled swing-out rotor (Beckman SW 39) and centrifuged for 40 hours at 36,000 rpm at about 3°. At the end of this period the rotor is allowed to decelerate without the brake, so as not to disturb the gradient. The contents of the tubes are then fractionated by piercing the bottom of the centrifuge tube with a pin and collecting 2 drops of solution into a series of test tubes. The optical density of the samples are measured at 280 nm after adding 2.5 ml of 0.15 M sodium chloride to each tube. The fractions containing the faster component (coagulant protein) are pooled and dialyzed against distilled water and freeze dried.

Properties

Stability. The freeze-dried protein kept in the deep freeze remains fully active for many years. Solutions of the protein in buffered saline at concentrations of 100 g/ml, or greater, are stable for many weeks when deep frozen or kept at 1°. Dilute solutions of the protein 10 g/ml, or less, remain fully active for only a few hours.

Purity and Physical Properties. The coagulant protein obtained by the method described above does not possess any of the several enzymatic activities known to be present in the crude venom.

Electrophoresis into 7% acrylamide gel[10] gives only one component and similar observations have been made in free boundary electrophoresis between pH 5.5 and pH 7.0. The coagulant protein has an $s^0_{20,w}$ of 5.05 S and a $D^0_{20,w}$ of 4.20×10^{-7} cm²/second. The molecular weight, as calculated by the procedure of Archibald,[11] is 105,000.

[10] B. J. Davis, *Ann. N.Y. Acad. Sci.* **121**, 404 (1964).
[11] W. T. Archibald, *J. Phys. Chem.* **51**, 204 (1947).

Activators and Inhibitors

Activators. The rate at which factor X is converted to the activated form by the venom protein is influenced by the concentration of calcium ions. The optimum rate of activation is obtained in the presence of 7 mM calcium chloride, while in the absence of calcium ions there is no reaction. Calcium can be replaced by other divalent metal ions, although they do not activate to the same extent as calcium. They can be written in a descending order of activity as Ca > Mn > Sr > Ba > Mg.

Inhibitors. Apart from the specific antiserum, and reagents which bind calcium ions, the venom is uninfluenced by inhibitors such as the various trypsin inhibitors, although competitive inhibition of the coagulant activity[9] has been obtained in the presence of *p*-toluenesulfonyl arginine.

Specificity. The purified coagulant protein is specific for factor X present in human, bovine, and rabbit plasma, and it appears to be without any activity on the other coagulation factors. The venom fraction has a slight hydrolytic activity on gelatin in the presence of calcium ions.

In addition to the esterolytic activity on TAME, for which it has a K_m of 5.9×10^{-4} moles, it also slowly hydrolyzes lysine ethyl ester, although, as mentioned earlier, it has to be resolved whether the coagulant protein is always associated with esterolytic activity.

[54] Renal Dipeptidase

By BENEDICT J. CAMPBELL

Glycylphenylalanine + H$_2$O → glycine + phenylalanine
Glycyldehydrophenylalanine + H$_2$O → glycine + phenylpyruvic acid + NH$_3$

The above equations demonstrate the hydrolysis of a saturated dipeptide and an unsaturated dipeptide, of which glycylphenylalanine and glycyldehydrophenylalanine are representative.

Assay Method

Principle. Renal dipeptidase isolated from fresh hog kidney exhibits activity against a variety of dipeptides.[1] By means of using the unsaturated dipeptide, glycyldehydrophenylalanine, enzymatic hydrolysis may be conveniently followed by observing the fall in optical density at 275 nm. This assay can be modified for application to particulate enzyme

---

[1] B. J. Campbell, Y. C. Lin, R. V. Davis, and E. Ballew, *Biochim. Biophys. Acta* **118,** 371 (1966).

fractions.[2] For saturated dipeptides the colorimetric ninhydrin assay[1,3] and a titrimetric assay[4] have been employed. In the isolation procedure to be described below, the spectrophotometric assay using glycyldehydrophenylalanine as substrate was followed.

Reagents

  0.025 M Tris-HCl buffer, pH 8.0
  Glycyldehydrophenylalanine (0.052 mM)

Synthesis of Substrate. A solution of 20 g of phenylserine is prepared in 80 ml of 1 M NaOH. This solution is treated alternatively with 8 g of chloroacetylchloride and 180 ml of 1 M NaOH. Each is added in 20 equal portions with shaking after each addition. The temperature is maintained at 0° in an ice bath during reaction. After 1 hour the reaction mixture is acidified with 5 N HCl. Crystallization of chloroacetylphenylserine occurs overnight. The yield is 19.8 g (70%), and the melting point of the crystals is 152°–155°.

A reaction mixture of 5 g of chloroacetylphenylserine in 50 ml of acetic anhydride is heated on the steam bath for 1 hour. The solution is then filtered and vacuum distilled to concentrate the reaction product. Crystallization occurs after most of the liquid has been removed. Evaporation is again carried out after the addition of 13 ml of CCl$_4$. The solution is then allowed to stand overnight at −20°. The azlactone is filtered off the next morning as yellow crystals. The yield is 1.2 g (24%), and the product melts at 102°–107°.

A mixture of 1.8 g of the azlactone of chloroacetylphenylserine, 25 ml of acetone, and 13 ml of water is refluxed for 2 hours. After cooling, the mixture is shaken twice with petroleum ether, and crystals of chloroacetyldehydrophenylalanine form in the aqueous portion. The crude product is recrystallized from distilled water. The weight of the purified crystals is 1.1 g (60%), and the crystals melt at 185°–188°.

A saturated aqueous solution of NH$_3$ is prepared by bubbling NH$_3$ through 4 ml of water maintained at 0° in an ice and salt bath. To the NH$_3$ solution is added 80 mg of chloroacetyldehydrophenylalanine, and NH$_3$ is bubbled through the solution at 0° for an additional 30 minutes. The mixture is allowed to come to room temperature and shaken at 5-minute intervals until all of the chloroacetyldehydrophenylalanine dissolves. The solution is kept at 4° for 4 days and then concentrated by directing an air stream over its surface. A crude precipitate is collected

---

[2] A. M. René and B. J. Campbell, *J. Biol. Chem.* **244**, 1445 (1969).
[3] A. T. Matheson and B. L. Tattrie, *Can. J. Biochem. Physiol.* **42**, 95 (1964).
[4] G. F. Bryce and R. B. Rabin, *Biochem. J.* **90**, 509 (1964).

and recrystallized from water. A yield of 48 mg is obtained (52%) and the final product, glycyldehydrophenylalanine, melts at 250°–257°. The unsaturated peptide exhibits a molecular extinction coefficient of 1.53 × 10^4 at 275 nm.

Procedure. To a cuvette with a 1-cm light path is added a solution of 2.40 ml of 0.052 mM glycyldehydrophenylalanine in 0.025 M Tris-HCl buffer at pH 8.0. Then 0.10 ml of enzyme solution is added by means of an adder-mixer.[5] A cuvette employed as a spectrophotometric blank contains 2.40 ml of 0.025 M Tris-HCl buffer at pH 8.0 and 0.10 ml of enzyme solution. The initial rate is measured as the fall in optical density at 275 nm as the reaction proceeds. The temperature of reaction is controlled, usually at 35°, by means of a temperature-controlled jacket through which water is circulated from a Brinkmann-Haake ultra thermostat type F.

Definition of Unit and Specific Activity. One unit of renal dipeptidase activity is defined as the amount that catalyzes the hydrolysis of 1 micromole of glycyldehydrophenylalanine per minute when the substrate is at a concentration of 0.050 mM. Specific activity is expressed as units per milligram of protein. The protein concentration is determined by the Folin-Lowry method,[6] and the protein standard employed is 3 × crystallized bovine serum albumin.

Purification Procedure

Renal dipeptidase was first partially purified from swine kidney by Robinson, Birnbaum, and Greenstein.[7] The method described below includes additional steps and modification of earlier techniques.[2]

Step 1. Homogenization. Fresh kidney cortex (1.5 kg) are cut into pieces and homogenized in a Waring blendor with 2 volumes of ice water. The homogenate is centrifuged at 1200 g for 20 minutes to remove cellular debris.

Step 2. Precipitation at pH 5. The supernatant is chilled to 0° in a cold bath and then adjusted to pH 5 by the careful addition of 1 N HCl. The resulting thick suspension is centrifuged immediately at 0° and 3000 g for 30 minutes. The supernatant fluid is discarded.

Step 3. Washing. The pH 5 precipitate is washed from the tubes with an equal volume of 0.066 M phosphate buffer at pH 7. The last traces of

[5] P. D. Boyer and H. L. Segal, *in* "The Mechanism of Enzyme Action" (W. D. McElroy and B. Glass, eds.), p. 520. Johns Hopkins Press, Baltimore, Maryland, 1954.

[6] O. H. Lowry, N. J. Rosebrough, A. L. Farr, and R. J. Randall, *J. Biol. Chem.* **193**, 265 (1951).

[7] D. S. Robinson, S. M. Birnbaum, and J. P. Greenstein, *J. Biol. Chem.* **202**, 1 (1953).

protein in the centrifuge tubes are removed with the same volume of
0.1 M KCl solution and added to the first washings. The lumpy suspension is homogenized briefly in the blender and frozen at $-10°$. After
thawing, the suspension is again adjusted to pH 5 with 1 M HCl and
centrifuged for 30 minutes at $0°$ and 3000 g. The sediment is taken up
with 0.066 M phosphate buffer at pH 7 and briefly homogenized. Cold
acetone ($0°$) is added to the suspension and stirred to a concentration of
70% (v/v). The reaction mixture is centrifuged at $0°$ and 8000 g for 40
minutes. The supernatant is carefully decanted and discarded. The sediment is washed from the tubes with an equal volume of 0.066 M phosphate buffer at pH 7. Precipitation at pH 5 and washing with phosphate
buffer is repeated until the supernatant fluid is proved to be protein-free
by testing it with 10% trichloroacetic acid solution.

   Step 4. Solubilization with n-Butanol. The washed pH 5 precipitate
is suspended in 1 liter of distilled, demineralized water. n-Butanol, chilled
to $0°$, is added to a concentration of 20% by volume with rapid stirring.
The suspension is stirred with a magnetic stirrer for 1 hour at $2°$. The
resulting emulsion is then dialyzed for 18 hours against several changes
of distilled water, until all butanol is removed. The dialyzate is adjusted
to pH 5 by dropwise addition of 1 M HCl at $0°$, and the resulting mixture centrifuged at 3000 g for 30 minutes at $0°$. The sediment is discarded. A solution of 0.1 M Tris-HCl buffer at pH 7 is added to the
supernatant in the amount of 1% by volume, and the mixture is adjusted
to pH 7. The volume of the preparation at this stage is approximately 2
liters. Lyophilization is employed to reduce this volume to about 400 ml.
The insoluble material which appears following lyophilization is centrifuged and discarded. The solution is then dialyzed against 0.002 M
Tris-HCl buffer at pH 7.

   Step 5. Fractionation with Ammonium Sulfate. To the aqueous solution containing solubilized protein is added solid ammonium sulfate to
50% saturation at $0°$. The precipitate which is formed after standing
overnight at $0°$ is collected by centrifugation at 4500 g for 1 hour and
discarded. Solid ammonium sulfate is then added to the supernatant to
75% saturation, and after standing for 48 hours in the cold, the suspension is centrifuged at 10,000 g for 1 hour. The sediment is dissolved in 8
ml of 0.002 M Tris-HCl buffer at pH 8.0 and dialyzed against the same
buffer for 18 hours with four changes of dialyzing buffer.

   Step 6. Gel-Filtration with Sephadex G-150. Pharmacia Sephadex
G-150 (Lot No. 8511, water regain, 15 ± 1.5 g/g, particle size 40–120 μ)
is equilibrated in 0.002 M Tris-HCl containing 0.01 mM ZnCl$_2$ at pH 8.0.
The equilibrated gel is poured to form a column of dimensions 2.5 cm $\times$
90 cm. In a typical preparation 10 mg of the dialyzed ammonium sulfate

fraction in 2 ml of 2 mM Tris-HCl buffer at pH 8.0 are applied to the column. Elution is carried out with the initial buffer at a flow rate of 25 ml/hour, and fractions are collected as 5 ml of effluent per tube. The absorbance of each tube at 280 nm and 260 nm is measured. The enzyme activity is measured by the spectrophotometric assay. The elution pattern from the column exhibits four absorption peaks. The second peak which is eluted before the first peak reaches base line and which also forms a shoulder for the third peak contains the enzyme activity. It is necessary to assay all tubes in this region to obtain a complete enzyme yield from the column. The tubes containing enzyme activity are pooled, and the resulting solution lyophilized to a concentration appropriate for step 7.

Step 7. Carboxymethylcellulose Chromatography. CM-Cellulose (Schleicher and Schuell Co., Lot No. 1310, Capacity 0.8 meq/g) is treated as outlined by Peterson and Sober.[8] The water-washed CM-cellulose is suspended in 5 mM ammonium acetate buffer containing 0.25 mM ZnCl$_2$ at pH 4.6 and poured to form a column of dimensions 3.9 $\times$ 9 cm. A sample containing 64 mg of protein in 21.7 ml of Tris-HCl buffer from the Sephadex G-150 column is dialyzed against distilled, demineralized water for 3 hours with three changes of water. One volume of 0.01 M ammonium acetate at pH 4.6, containing 0.5 mM ZnCl$_2$ is added and mixed with the sample immediately before application to the column. Starting buffer, 5 mM ammonium acetate at pH 4.6, containing 0.25 mM ZnCl$_2$ is allowed to pass through the column at a rate of 4 ml per minute. As soon as the first protein peak is eluted, the second buffer, 0.05 M ammonium acetate at pH 6.0, containing 0.25 nM ZnCl$_2$ is applied. As the tubes containing enzyme activity are collected, 0.1 M Tris is added dropwise immediately to bring the pH of the eluted material to 8. The tubes in the active peak are pooled and neutralized to pH 8. All of the peptidase activity is eluted in the second peak. The resulting solution is concentrated by lyophilization prior to the final purification step.

Step 8. Gel Filtration with Sephadex G-200. Pharmacia Sephadex G-200 (Lot No. 3177, water regain, 20 $\pm$ 2 g/g, particle size 40-20 μ) is equilibrated in 2 mM Tris-HCl buffer at pH 8.0 containing 0.01 mM ZnCl$_2$. The equilibrated gel is poured to form a column of dimensions 2.5 $\times$ 92 cm. In a typical experiment 6 mg of CM-cellulose purified enzyme in 4 ml of 2 mM Tris-HCl buffer at pH 8.0 containing 0.01 mM ZnCl$_2$ are applied to the column. Elution is carried out with the same buffer at a flow rate of 8 ml per hour, and 3 ml of effluent are collected per tube. In the first gel filtration purification three absorbance peaks are eluted from the column. The second peak which contains the enzyme

---

[8] E. A. Peterson and H. A. Sober, see Vol. V [1], p. 3.

activity is collected and concentrated by lyophilization. Refiltration of
this material on the same column is carried out using the same conditions
after prewashing the column overnight with the same buffer. The absorbance and activity peaks coincide in this final purification step. The purified renal dipeptidase is concentrated to approximately 1 mg/ml and
stored frozen in a Tris-HCl buffer at pH 8.0. The overall purification
produced an enzyme approximately 500 times more active than source
material. From 1.5 kg of kidney cortex, 132 g of protein homogenate
were obtained, and the complete purification from this homogenate
yielded 5.8 mg of peptidase. A summary of the solubilization and purification of renal dipeptidase is shown in Table I.

TABLE I
PURIFICATION OF RENAL DIPEPTIDASE

Fraction	Volume (ml)	Total activity (units)	Total protein (mg)	Specific activity × 100	Yield (%)
Kidney homogenate	3000	62,000	132,000	0.47	100
First pH 5 precipitate	2000	100,000	100,000	1.0	161
Washed precipitate	1000	66,000	55,000	1.2	106
Solubilized enzyme	1400	12,600	2,100	6.0	20
Ammonium sulfate fraction	40	5,800	200	29.0	9
Sephadex G-150	40	1,600	38.4	41.7	2.6
Carboxymethylcellulose chromatography	30	1,360	24.0	56.6	2.2
Sephadex G-200	40	1,440	6.0	240	2.3
Sephadex G-200 refiltration	40	1,410	5.8	243	2.3

Properties

Stability. Renal dipeptidase may be concentrated by lyophilization
without loss in activity if the lyophilizing flasks are kept cold in an ice
bath and if the material is shell frozen in Tris-HCl buffer at pH 8. Extended dialysis against nonchelating buffers does not result in loss in
activity. The enzyme is rapidly inactivated below pH 5 in ammonium
acetate buffer but is stable at pH 8 in Tris or phosphate buffers. If the
enzyme is stored frozen in dilute solution (0.13 mg/ml Tris-HCl buffer,
pH 8), it loses approximately two-thirds of its activity over a period of
6 months.

Purity and Physical Properties. The purified enzyme exhibits only one
symmetrical peak after ultracentrifugation at 59,780 rpm for 1 hour, and
migration of the enzyme by acrylamide-gel electrophoresis at pH 8.0 indicates that the purified peptidase is monophoretic. It has been estimated

from approach to equilibrium measurements that the molecular weight is approximately 47,200. But more recent results obtained by sedimentation equilibrium and by gel filtration indicate that the enzyme exists in a form with molecular weight 87,000–93,000.

Activators. An increase in zinc content is observed with increasing purification of the peptidase until a 1:1 mole ratio of zinc to enzyme is reached if the molecular weight is assumed to be 47,200. Peptidase activity can be reduced by removal of zinc using o-phenanthroline dialysis and can be restored by dialysis against zinc-containing buffers. It is suggested that renal dipeptidase is a metalloenzyme with the zinc atom strongly bound to the protein portion of the enzyme. The peptidase does not exhibit absolute specificity for zinc ion activation. Preincubation of the purified native enzyme with Mg^{2+} or Co^{2+} produced a 24% activation, whereas the same treatment with Ni^{2+} or Cd^{2+} leads to a 29% inhibition in each case.

Inhibitors. A time-dependent inhibition of the enzyme is produced upon incubation with o-phenanthroline. This inhibition can only be reversed by dialysis to remove o-phenanthroline followed by dialysis against zinc-containing buffers. The peptidase is inhibited by inorganic phosphate as well as nucleotides. This inhibition is not time dependent and is completely reversible upon dilution or dialysis. The degree of inhibition for the adenine series is in the order ATP > ADP > AMP > inorganic phosphate. Inhibition is also produced by treatment with monovalent anions. This inhibition is competitive and is in the order $CN^- >$ $SCN^- > N_3^- > I^- > NO_3^- > HCO_3^- > Br^- > OAc^- > Cl^- > F^-$.

Specificity. Renal dipeptidase acts upon a variety of dipeptides to produce the constituent amino acid products. Early work using the colorimetric ninhydrin assay suggested that glycylglycine was the most readily hydrolyzed substrate,[1] but recent measurements using the titrimetric assay[4] indicate that L-alanylglycine breaks down most rapidly at a rate of 229 ± 19 micromoles/minute/mg enzyme under conditions of substrate-saturated enzyme. The terminal amino group of the dipeptide substrate cannot be blocked by groups such as the carbobenzoxy group, but methylation of the amino group yields derivatives that are hydrolyzed at reduced rates. The enzyme requires that the N-terminal amino acid of the dipeptide be in the L-configuration. However, activity is exhibited against dipeptides in which the C-terminal amino acid is in the D- as well as the L-configuration. The enzyme has no esterase activity, nor does it act against leucinamide, tripeptides, or proteins (casein, hemoglobin).

Kinetic Properties and pH Dependence. The kinetic properties and the effect of pH on enzyme activity were determined by measurements of peptidase-catalyzed hydrolysis of glycyldehydrophenyalanine. The K_m for this substrate was 1 mM at the optimum pH of 7.60 and 35° Proton dissociations from the enzyme-substrate complex occur during peptidase catalysis at pK's of 6.9 and 8.5.

Amino Acid Composition. The amino acid composition of renal dipeptidase is given in Table II.

TABLE II

AMINO ACID COMPOSITION OF RENAL DIPEPTIDASE

Amino acid	Amino acid residues	Nitrogen (g/100 g protein)	Residues[a] per molecule	Assumed residues per molecule
Lysine	7.88	1.70	28.65	29
Histidine	2.91	0.88	9.93	10
Arginine	5.63	2.03	17.08	17
Aspartic acid	9.51	1.15	38.83	39
Threonine	4.72	0.65	21.92	22
Serine	5.17	0.85	28.41	28
Glutamic acid	11.77	1.27	42.60	43
Proline	4.32	0.62	20.85	21
Glycine	3.03	0.74	25.04	25
Alanine	4.82	0.95	31.93	32
Half-cystine	3.72	0.51	17.24	17
Valine	5.67	0.79	26.52	27
Methionine	1.95	0.22	7.31	7
Isoleucine	3.60	0.45	15.10	15
Leucine	9.83	1.22	40.96	41
Tyrosine	4.15	0.35	11.90	12
Phenylalanine	5.62	0.54	17.98	18
Tryptophan	4.34	0.68	11.22	11
Ammonia	0.40	0.33	10.99	11
Zinc	0.14		1.01	1
Total	99.18	15.93		414
Recovery %		100.82		

[a] Calculated on the basis of a molecular weight of 47,200.

Distribution. The complete purification of renal dipeptidase as reported above has only been applied to hog kidney tissue; however, enzymatic activity against glycyldehydrophenylalanine by crude extracts of other mammalian tissues has been reported.[9]

[9] J. P. Greenstein, *Adv. Enzymol.* **8**, 117 (1948).

[55] β-Aspartyl Peptidase[1]

By Edward E. Haley

Assay Method A

Principle. The amino acids produced in the reaction were estimated by a photometric ninhydrin method. A ninhydrin-cyanide reagent was used which gives quantitative color yields with free amino acids, but much lower yields with most di- and tripeptides.[2] This reagent was used in a more sensitive assay for peptidase activity. The assay method was used for all experiments, except those to determine the effect of substrate concentration on hydrolysis.

Reagents

Tris-HCl buffer, 0.05 M, pH 8.1

β-Aspartylleucine, 0.01 M, in 0.05 M Tris-HCl buffer, pH 8.1. The peptide was dissolved in water and the solution adjusted to pH 8.1 with NaOH, then to 0.02 M with water. An equal volume of 0.1 M buffer was added to the peptide solution. This solution was used as the substrate.

β-Aspartylleucine, 1.25 mM

Aspartic acid, 2.5 mM

Leucine, 2.5 mM

Trichloroacetic acid, 1.2 M

Sodium citrate buffer, 0.067 M, pH 5.0

Ninhydrin–potassium cyanide–methyl Cellosolve reagent. The reagent was prepared by mixing 1 volume of ninhydrin dissolved in methyl Cellosolve, 5 g per 100 ml, with 5 volumes of potassium cyanide in methyl Cellosolve. The latter solution was made by diluting 2 ml of 0.01 M aqueous potassium cyanide to 100 ml with methyl Cellosolve. The final reagent should be stored overnight before use to ensure low blank readings.[3]

Ethyl alcohol, 60%

Procedure. The reaction mixtures were made in 12-ml conical centrifuge tubes and had a total volume of 0.20 ml. The standard substrate,

---

[1] E. E. Haley, *J. Biol. Chem.* **243**, 5748 (1968).

[2] A. T. Matheson and B. L. Tattrie, *Can. J. Biochem.* **42**, 95 (1964); E. W. Yemm and E. C. Cocking, *Analyst* **80**, 209 (1955).

[3] Room temperature stabilities are: ninhydrin-potassium cyanide–methyl Cellosolve, 1 week; ninhydrin–methyl Cellosolve, 1 month; potassium cyanide–methyl Cellosolve, 1 week; and aqueous potassium cyanide, 1 week.

0.05 ml of β-aspartylleucine solution, was evaporated to dryness under vacuum in the tube to enable the entire 0.20-ml reaction volume to be made up of enzyme and other reagent solutions. The total buffer content was 6.5 micromoles of Tris-HCl, pH 8.1. The reaction was initiated by the addition of the enzyme solution, and incubation was at 37° for 1 hour. Controls without substrate and without enzyme were included. The reaction was stopped by the addition of 10 μl of trichloroacetic acid solution (TCA) and the tubes were centrifuged when required. The extent of hydrolysis was estimated by mixing 0.05-ml aliquots of the reaction mixture with 1.5 ml of citrate buffer and 1.2 ml of ninhydrin–potassium cyanide–methyl Cellosolve reagent, and heating in a boiling water bath for 7½ minutes. The tubes were immediately chilled in an ice bath for 5 minutes and the contents diluted by the addition of 7.3 ml of 60% ethyl alcohol. Absorbance at 570 nm was measured in the Coleman Junior Spectrophotometer. A standard curve was obtained by preparing mixtures of β-aspartylleucine, aspartic acid, and leucine solutions to simulate 0, 10, 20, 40, and 80% hydrolysis. Controls containing samples of Tris buffer were included. The response was linear with enzyme concentration and with time for the first 60 minutes, when the extent of hydrolysis did not exceed 50%.

Assay Method B

Principle. The leucine formed in the enzymatic reaction was estimated by the ninhydrin method after separation of the leucine from the peptide and aspartic acid by ion exchange. This method was used in the tests to determine the effect of substrate concentration on hydrolysis. At high substrate concentrations the micromoles of peptide hydrolyzed could be determined more accurately by this method than by Method A, which determines percent hydrolysis of the peptide. Method B could be used generally, but is more lengthy.

Reagents

    Tris-HCl buffer, 0.05 M, pH 8.1
    β-Aspartylleucine, 0.01 M, in 0.05 M Tris-HCl buffer, pH 8.1
    Trichloroacetic acid, 1.2 M
    Dowex 50W-X4, 100–200 mesh, hydrogen form
    Dowex 2-X8, 100–200 mesh, acetate form
    Pyridine, 2 M
    Citrate buffer, 0.067 M, pH 5.0
    Ninhydrin–potassium cyanide–methyl Cellosolve reagent
    Ethyl alcohol, 60%

Procedure. The reaction mixtures were made in conical centrifuge tubes and had a total volume of 1.0 ml. The mixture consisted of the

enzyme sample, 20 micromoles of Tris-HCl buffer and 0.2 to 7.5 micromoles of peptide in the tests to determine the effect of substrate concentration, or would be 2.5 micromoles of peptide in other tests. Incubation
was at 37° for 30 minutes, and the reaction was stopped by the addition
of 0.05 ml of TCA. All of the reaction mixture, or an aliquot of the
supernatant if centrifugation was required, was added to a 1 × 2.5 cm
column of Dowex 50W-X4. The column was washed with 8.5 ml of water
and eluted with 10 ml of pyridine solution. The pyridine eluate was
evaporated to dryness under vacuum and the residue dissolved in 5 ml
of water. This solution was added to a 1 × 2.5 cm column of Dowex
2-X8, and the column was washed with 15 ml of water. The effluent was
evaporated to dryness, and the residue dissolved in 0.2 ml of water. An
aliquot, 0.1 ml, was analyzed by the ninhydrin method described in
Method A, except that the experimental samples and samples of leucine
for a standard curve were heated at 100° for 5 minutes rather than 7½
minutes.

Definition of Unit of Enzyme Activity. One unit of enzyme activity is
defined as that amount required to hydrolyze 1 micromole of β-aspartyl-
leucine per minute under the conditions of the assay.

Purification Procedure

Step 1. Extraction and Ammonium Sulfate Fractionation. All the
operations in this step were carried out in the cold. *Escherichia coli* strain
B, midlogarithmic harvest, was obtained commercially.[4] The cells, 67 g
of frozen paste, were thawed and homogenized in a Waring blendor with
900 ml of 0.05 M Tris-HCl buffer, pH 8.1, for 30 seconds. The suspended
cells were disrupted in a French pressure cell at 15,000 to 20,000 psi, and
the suspension separated in a Spinco model L ultracentrifuge at 23,000 ×
g for 30 minutes. Manganese chloride, 0.05 volume of 1 M concentration,
was added to the supernatant with stirring in an ice bath.[5] The mixture
was stirred for 40 minutes, then centrifuged as described above. Solid
$(NH_4)_2SO_4$ was added to the supernatant to 90% saturation at 0°, and
10 ml of 7.4 M NH_4OH was added to readjust the pH to 8.1. The mixture was stirred in an ice bath for 3 hours and centrifuged as described
above. The precipitate was dissolved in 0.05 M Tris-HCl buffer, pH 8.1,
to a volume of 90 ml and dialyzed against the same buffer until free of
$(NH_4)_2SO_4$. The enzyme solution was concentrated to about two-thirds
of its volume by dialysis against 500 ml of 15% polyvinylpyrrolidone
(PVP) in 0.05 M Tris-HCl buffer, pH 8.1, with no loss of activity.

[4] The enzymatic activity in the supernatant of disrupted cells harvested in late
logarithmic phase was almost identical to that for the midlogarithmic phase.
[5] L. T. Mashburn and J. C. Wriston, Jr., *Arch. Biochem. Biophys.* **105**, 450 (1964).

Step 2. Sephadex G-200 Chromatography. Due to the stability of the enzyme and the relatively short time the enzyme was on the column, this step could be carried out at room temperature. The $(NH_4)_2SO_4$ precipitate concentrate above, 59 ml, was passed through a 4.7×29 cm column of Sephadex G-200, 40 to 120 μ particle size, prepared from gel swelled in 0.05 M Tris buffer, pH 8.1, at 100° for 5 hours. The column was equilibrated and eluted with the same buffer at a flow rate of 30 ml per hour. The effluent passed through the ultraviolet absorption meter, and 3-ml fractions were collected and refrigerated. Two peaks of enzymatic activity were detected. Fractions representing the center of the principal, earlier peak were combined and dialyzed against 15% PVP in 0.02 M Tris buffer, pH 8.1. The final volume was 20 ml.

Step 3. DEAE-Cellulose Chromatography. The most effective purification step was chromatography on DEAE-cellulose, which served also to separate asparaginase and other peptidases. The Sephadex enzyme described above, 19 ml, was applied to a 0.6×137 cm column of Whatman DE-32 equilibrated with 0.02 M Tris-HCl buffer, pH 8.1, at 4°. The column was eluted at approximately 13 ml per hour, first with 40 ml of the above buffer, then with a 300-ml linear gradient of NaCl in the buffer to 0.4 M. Four milliliter fractions were collected. Elution was continued with 190 ml of 0.02 M buffer and 0.4 M NaCl, then with 80 ml of 0.02 M buffer and 0.8 M NaCl. Protein concentration was measured by absorbance at 280 nm. Two peaks of enzymatic activity were detected in this fractionation step also. The principal one began from the point at which the eluting buffer reached 0.3 M in NaCl. The activity reached its peak just as the gradient was completed. The active fractions were combined and concentrated to 10 ml with 15% PVP in 0.02 M buffer. The other enzyme activity was eluted with 0.02 M buffer and 0.8 M NaCl. Further purification and testing was with the principal enzyme, fractions 80 through 96.

Step 4. Polyacrylamide Gel Electrophoresis. Enzymatic activity for α-aspartyl dipeptides was removed from the β-aspartyl peptidase by electrophoresis in polyacrylamide gels. Additional purification by this method was only about 2-fold. Separation of the proteins in the DEAE-cellulose enzyme in limited quantity was accomplished in 5% polyacrylamide gel in a pH 8.9 to 9.1 buffer composed of 10.3 g of Tris, 1.30 g of disodium EDTA, and 0.76 g of boric acid per liter in a vertical gel apparatus.[6] Solid sucrose was added to the 0.3-ml sample of DEAE enzyme to give a 10% solution, and bromophenol blue was added as a marker. The sample was introduced beneath the buffer into a 3×78 mm gel slot.

[6] Apparatus for vertical gel and horizontal electrophoresis was obtained from E-C Apparatus Corp., Philadelphia, Pa.

TABLE I
PURIFICATION OF PEPTIDASE

Fraction	Volume (ml)	Protein (mg)	Total enzyme (units)	Specific activity (units/mg)	Yield[a] (%)	Purification (-fold)
Disruption supernatant	910	6000	42.0	0.007	100	1.0
(NH₄)₂SO₄ ppt. soln.	121	1920	44.2	0.023	105	3.1
Sephadex G-200	129	1070	38.5	0.036	95	5.1
DEAE-cellulose	63	28.9	21.7	0.75	57	107

[a] Corrected for sampling.

Electrophoresis was carried out at 300 V (20 V/cm) and 140 to 160 mA for 2–3 hours. A sample of the electrophoretogram was stained with amido black, and the protein pattern was used as a guide for the extraction of sections of the untreated portion of the gel. This portion had been kept frozen during the staining and destaining procedure.

The enzyme was removed from the gel by one of two methods: (1) homogenization of sections in 0.05 M Tris buffer, pH 8.1, followed by dialysis of the homogenate at room temperature for 2 hours and then overnight in the cold against the same buffer, then separation by centrifugation; or (2) electrophoretic migration out of the gel sections, which had been put into dialysis sacs with the Tris-EDTA-borate buffer. The sacs were placed side by side in a trough in the horizontal apparatus, and each sac was separated by sections of polyurethane foam saturated with the buffer. Migration into the buffer required 3 hours at 400 V (13 V/cm)

TABLE II
ELECTROPHORETIC PURIFICATION OF DEAE-CELLULOSE ENZYME

Fraction	Volume (ml)	Protein (mg)	Total enzyme (units)	Specific activity (units/mg)	Yield[a] (%)	Purification (-fold)
DEAE-cellulose PVP conc.	10	28.4	21.3	0.45	34	64
DEAE-cellulose PVP conc.[b]	0.6	1.70	0.78	0.45	34	64
Polyacrylamide gel conc.[b]	4.5	1.03	0.79	0.77	34	110

[a] Corrected for sampling.
[b] Results from purification of two 0.3-ml portions of DEAE-cellulose enzyme whose active gel fractions were combined.

and 8 mA. The sacs were dialyzed as above, and active fractions were concentrated by dialysis against PVP solution. The quantities of protein were so low in the gel fractions that protein estimation was made by amino acid analysis of an acid hydrolyzate of a sample. The total amino acid residue value was related to that of the DEAE-cellulose enzyme hydrolyzate, which had been determined also by the method of Lowry et al.[7]

A summary of the purification data is given in Tables I and II.

Properties

Stability. The enzyme lost no detectable activity in Tris-HCl buffer, pH 8.1, stored in the freezer for several months, nor at pH 8.1 for 3 days at room temperature. Some instability was observed closer to the pH optimum, at pH 7.5, where 17% of the activity was lost during 24 hours at room temperature. Recovery of activity was good throughout the purification procedure.

Purity and Physical Properties. The preparation represented a 110-fold purification over the disrupted cell supernatant, and was, of course, not pure. The enzyme was extracted with several protein bands from the polyacrylamide gel electrophoretogram. Insufficient amounts of this material were available for analytical studies, but the material was free of asparaginase activity and peptidase activity for substrates other than β-aspartyl dipeptides. Molecular weight estimation on the cruder DEAE-cellulose enzyme on a calibrated Sephadex G-200 column was 120,000.

Specificity. The results of tests on the limited types of compounds tested as substrates for the enzyme showed that the activity was restricted to β-aspartyl dipeptides. The action of the enzyme on various types of substrates is shown in Table III. Not all β-aspartyl dipeptides are substrates, and the tripeptides β-aspartylleucylglycine, β-aspartyl-glycylalanine, and β-aspartylglycylglycine were not hydrolyzed. γ-Glutamylleucine, a homolog of β-aspartylleucine, was unaffected as were other leucine dipeptides, α-aspartylleucine, asparaginylleucine, glycyl-leucine, and leucylglycine.

Activators and Inhibitors. The peptidase does not require metal ions for activation. Of the metals tested Mn^{2+}, Co^{2+}, and Zn^{2+} were inhibitory, and Ca^{2+} and Mg^{2+} seemed slightly stimulatory. Sulfhydryl reagents p-hydroxymercuribenzoate, iodoacetamide, and o-iodosobenzoate were not inhibitory.

pH Optimum and Buffer Effects. The pH optimum is 7.8 to 8.0 in

[7] O. H. Lowry, N. J. Rosebrough, A. L. Farr, and R. J. Randall, *J. Biol. Chem.* **193**, 265 (1951).

TABLE III
Action of Peptidase on Various Types of Substrates[a]

Substrate	Relative hydrolysis	Substrate	Relative hydrolysis
β-Asp-Leu	100	α-Asp-Leu	0
β-Asp-Ser	82	α-Asp-Ser	0
β-Asp-Met	68	α-Asp-Phe	0
β-Asp-Val	56	α-Asp-Ala	0
β-Asp-Gln	48	α-Asp-Ile	0[b]
β-Asp-Phe	38	α-Asp-Asn	0
β-Asp-Ala	33	α-Asp-Gly-Ala	0[b]
β-Asp-Ile	19	α-Asp-Leu-Gly	0
β-Asp-Thr	18	N-Acetyl Met	0[c]
β-Asp-Asn	10	Asn	0[d]
β-Asp-Gly	0	β-Asp-glucosylamine	0[e]
β-Asp-His	0	γ-Glu-Leu	0
β-Asp-Gly-Ala	0[b]	Gly-Leu	0
β-Asp-Gly-Gly	0	Leu-Gly	0[f]
β-Asp-Leu-Gly	0	Carnosine	0
		Asn-Leu	0

[a] Extent of hydrolysis of the substrate with reference to β-aspartylleucine as 100. Assay Method A unless otherwise indicated. The enzymes were the PVP concentrates of the gel electrophoresis fractions unless otherwise indicated.

[b] With DEAE-cellulose enzyme, PVP concentrate.

[c] With DEAE and gel enzymes at pH 8.1 and pH 7.2.

[d] Asparaginase assay method of L. T. Mashburn and J. C. Wriston, Jr., *Biochem. Biophys. Res. Commun.* **12**, 50 (1963).

[e] Two enzyme levels tested, one which hydrolyzed 0.11 micromole of β-Asp-Leu and 5 times that level which hydrolyzed 0.40 micromole. Assay Method A.

[f] Tested in buffer with and without magnesium acetate at 0.01 M. See S. Simmonds, *J. Biol. Chem.* **241**, 2502 (1966).

three 0.05 M buffers: Tris-HCl, Tris-maleate, and sodium phosphate. The enzyme activity at the pH optimum was greatest in Tris-maleate and least in phosphate, the ratio between them being about 2 to 1.

Substrate Affinity. The Michaelis constant for β-aspartylleucine is $8.1 \times 10^{-4}\ M$.

Distribution. Other microorganisms have not been tested for β-aspartyl peptidase activity. This activity in low level has been reported also in rat tissues,[8] but its specificity is markedly different than that of *Escherichia coli* B. For example, β-aspartylglycine is the substrate most readily hydrolyzed by the rat enzyme, and the tripeptides β-aspartylglycylglycine and β-aspartylglycylalanine are hydrolyzed at the aspartyl-glycyl bond. Both enzymes lack metal ion requirements and have the same pH opti-

[8] F. E. Dorer, E. E. Haley, and D. L. Buchanan, *Arch. Biochem. Biophys.* **127**, 490 (1968).

mum, but the rat enzyme is more active in phosphate than in Tris buffers. Enzyme activity was not detected in human serum.

[56] β-Aspartyl Peptidase from Rat Liver[1]

By Edward E. Haley

Assay Method

Principle. The glycine formed in the enzymatic reaction was estimated by a photometric ninhydrin method.[2] Assay Method B for β-aspartyl peptidase from *Escherichia coli* was adapted from this assay method with slight modifications.

Reagents

Sodium phosphate buffer, a molar mixture of 3 parts Na_2HPO_4 and 1 part NaH_2PO_4. The pH depends on the dilution.

β-Aspartylglycine, 0.02 M

Dowex 50W-X4, 100–200 mesh, hydrogen form

Dowex 2-X8, 100–200 mesh, acetate form

Pyridine, 2 M

Citrate buffer, 0.067 M, pH 5.0

Ninhydrin–potassium cyanide–methyl Cellosolve reagent[3]

Ethyl alcohol, 60%

Procedure. The reaction mixtures had a total volume of 2.0 ml and consisted of the enzyme sample, 100 micromoles of sodium phosphate buffer, and 5 micromoles of β-aspartylglycine. The peptide was omitted from the blank. The final pH was 7.2. Incubation was at 37° for 2 hours and the reaction was stopped by heating in a boiling water bath for 5 minutes. After centrifugation 1.5 ml of the supernatant was added to a 1 × 2.5 cm column of Dowex 50W-X4. The column was washed with 8.5 ml of water and eluted with 10 ml of pyridine solution. The pyridine eluate was evaporated to dryness under vacuum and the residue dissolved in 5 ml of water. This solution was added to a 1 × 2.5 cm column of Dowex 2-X8, and the column was washed with 15 ml of water. The effluent was evaporated to dryness, and the residue analyzed for glycine by the ninhydrin method described in Method A for the *E. coli* enzyme.

---

[1] F. E. Dorer, E. E. Haley, and D. L. Buchanan, *Arch. Biochem. Biophys.* **127**, 490 (1968).

[2] A. T. Matheson and B. L. Tattrie, *Can. J. Biochem.* **42**, 95 (1964); E. W. Yemm and E. C. Cocking, *Analyst* **80**, 209 (1955).

[3] Prepared as described in the chapter on β-aspartyl peptidase from *E. coli*.

Definition of Unit of Enzyme Activity. One unit of enzyme activity is defined as that amount required to hydrolyze 1 micromole of β-aspartylglycine in 1 minute at 37° under the conditions of the assay.

Purification Procedure

All operations except CM-cellulose and hydroxylapatite chromatography and the heat treatment were carried out in the cold.

Step 1. Extraction. Albino male rats, 300–400 g, which had been maintained on Purina chow pellets, were killed by decapitation, and the livers quickly removed. About 50 g of liver was blended with 3 volumes of cold 0.1 M sodium phosphate buffer (a molar mixture of 3 parts Na_2HPO_4 and 1 part NaH_2PO_4) for 30 seconds in a prechilled stainless steel Waring blendor. The homogenate was centrifuged to remove cell debris.[4] The supernatant (fraction A, Table I) was centrifuged for 1 hour at 105,000 g in a Spinco Model L ultracentrifuge.

Step 2. Ammonium Sulfate Fractionation. Solid ammonium sulfate was added to the supernatant (fraction B, 18 mg of protein/ml) to 40% saturation (0.243 g/ml). The mixture was stirred for another 30 minutes and the suspension was centrifuged. More ammonium sulfate was added to the supernatant to 55% saturation (0.097 g/ml). The mixture was stirred an additional 30 minutes and the suspension was centrifuged. The precipitate was dissolved in 20 ml of 0.1 M sodium phosphate buffer and dialyzed against 5 liters of this buffer for 16 hours.

Step 3. Heat Treatment. The dialyzed solution (fraction C) was immersed in a water bath at 60° for 5 minutes, cooled, centrifuged, and the supernatant dialyzed against 5 liters of 0.005 M sodium phosphate buffer for 16 hours. The turbid solution was centrifuged.

Step 4. CM-Cellulose Chromatography. The supernatant from above (fraction D) was applied to a 2.3 × 23 cm column of Whatman CM 32 equilibrated with 0.005 M sodium phosphate buffer. The column was eluted with the same buffer. The fractions from a wide yellow band were combined and adjusted to 0.01 M sodium phosphate buffer concentration.

Step 5. Hydroxylapatite Chromatography. The above eluate solution (fraction E) was applied to a 2.3 × 11 cm column of Bio-Gel HTP[5] equilibrated with 0.01 M sodium phosphate buffer. The column was eluted with 0.05 M sodium phosphate buffer and the first 120 ml of effluent discarded. The fractions from a narrow yellow band, which was eluted with 1 M potassium phosphate buffer, pH 7.5, were combined and dialyzed against 5 liters of 0.01 M sodium phosphate buffer for 16 hours (fraction F).

[4] Centrifugation was at 600 g for 1 hour in the cold unless otherwise stated.
[5] Purchased from Bio-Rad Laboratories, New York, New York.

TABLE I

PARTIAL PURIFICATION OF β-ASPARTYL PEPTIDASE FROM RAT LIVER[a]

Fraction	Protein[b] (mg)	Activity (units)	Specific activity (units/mg)	Yield (%)	Purifica-tion (-fold)
A. "Low-speed" supernatant	3300	1.55	0.00047	(100)	1.0
B. 105,000 g Supernatant	2370	0.55	0.00065	100	1.4
C. Ammonium sulfate (40–55% satn)	570	0.70	0.0012	45	2.6
D. Heat at 60°	216	0.60	0.0028	39	6.0
E. CM-cellulose	147	0.55	0.0037	35	8.0
F. Hydroxylapatite	72	0.50	0.0070	32	14.9

[a] Starting with 50 g (wet weight) of liver.

[b] Protein was determined by the method of O. H. Lowry, N. J. Rosebrough, A. L. Farr, and R. J. Randall, *J. Biol. Chem.* **193**, 265 (1951).

A summary of the purification data is given in Table I.

Properties

Stability. The enzyme activity is stable for as long as 5 days at room temperature, and is stable to freezing and thawing in 0.01 M sodium phosphate buffer. Stability is dependent upon the presence of sodium ions. The activity is unstable to dialysis against water or 0.01 M Tris-HCl buffer, pH 8.0. Activity cannot be restored by the addition of NaCl, but if NaCl, 0.01 M, is included in the dialysis solution, 80% of the activity is recovered.

Purity. The enzyme preparation, fraction F, represents a 15-fold purification over the liver homogenate. The extent of purification with regard to separation from α-aspartylglycine and leucylglycine hydrolyzing activities, however, was 500- and 40-fold, respectively. Attempts to purify the β-aspartyl peptidase activity further by DEAE-cellulose chromatography were unsuccessful due to loss of activity on the column.

Specificity. The hydrolysis of peptides other than β-aspartyl peptides appears to be due to contaminating enzymes present in the preparation, because of the relative enrichment toward β-aspartylglycine activity in the course of the purification. In Table II is shown the action of the enzyme on various types of substrates, and the relative difference in activities between fractions B and F toward β-aspartyl peptides and other peptides. The β-aspartyl tripeptides, β-aspartylglycylglycine, -glycylalanine, and -glycylvaline, were cleaved at the aspartyl bond.

Activators and Inhibitors. The peptidase appears to require sodium or potassium for maximal activity, but not alkaline earth or transition

TABLE II
Hydrolysis of Peptides by Rat Liver Preparations

| Peptide | Relative hydrolysis[a] (% of β-aspartylglycine) | | Method of analysis |
	Fraction B[b]	Fraction F[b]	
β-Aspartylglycine	(100)	(100)	e
β-Aspartylalanine		51	e
β-Aspartylvaline		28	e
β-Aspartylleucine	59	65	e
β-Aspartylisoleucine		37	e
β-Aspartylserine	47	56	e
β-Aspartylthreonine		29	e
β-Aspartylmethionine		82	e
β-Aspartylglycylglycine[c]		95	e
β-Aspartylglycylalanine[d]		55	e
β-Aspartylglycylvaline		13	e
α-Aspartylglycine	1500	3	e
α-Aspartylalanine		10	e
α-Aspartylleucine		9	e
α-Aspartylserine		10	e
α-Aspartylglycylglycine		3	e
α-Aspartylglycylalanine		3	e
γ-Glutamylleucine		0	e
α-Glutamylleucine		5	e
Asparagine		0	f
Glycylglycine		40	g
Glycylalanine		40	g
Glycylvaline		Trace	g
Glycylleucine		Trace	g
Leucylglycine	2000	50	g
Glycylglycylglycine	1000	30	g
β-Alanylglycine		0	g
N-Acetylglycine		0	g
Carnosine		0	g

[a] All incubations were carried out as described for β-aspartylglycine under assay method and included a substrate blank for each peptide to correct for any free amino acid present in or formed from substrate in the absence of enzyme.

[b] Enzyme fractions are described in Table I.

[c] At the end of the incubation of β-aspartylglycylglycine, glycylglycine was identified by paper chromatography [F. E. Dorer, E. E. Haley, and D. L. Buchanan, *Anal. Biochem.* **19**, 366 (1967)] in a desalted aliquot of the incubation mixture.

[d] Glycylalanine from β-aspartylglycylalanine was identified as above.

[e] The extent of hydrolysis was determined by ninhydrin reaction of the Dowex 2 effluent as described for β-aspartylglycine under assay method. The results were corrected for differences in the ninhydrin color values of the various substrate C-terminal amino acids or dipeptides.

metal ions. There was a slight activation by mercaptoethanol, while inhibition was caused by p-hydroxymercuribenzoate but not by iodoacetamide.

pH Optimum and Buffer Effects. The pH optimum is 7.5–8.0 in sodium phosphate buffer. There is a broad pH optimum between 7 and 9 in Tris-HCl buffer, and the activity is about half of that in the sodium phosphate buffer.

Distribution. Homogenates of rat kidney, brain, lung, skeletal muscle, and heart muscle, in addition to liver, all catalyze a slow but significant hydrolysis of β-aspartylglycine. Rat kidney enzyme, prepared under the same conditions as for fraction B of liver, showed a specificity similar to that of the liver enzyme. The specific activity of the kidney enzyme is about twice that of the liver enzyme at the same stage of purification.[6]

[6] See also the discussion of distribution of β-aspartyl peptidase in Chapter [55] on the *E. coli* enzyme.

---

[f] Ammonia was determined by Nesslerization of an aliquot of the incubation mixture.

[g] The extent of hydrolysis was estimated by semiquantitative determination of hydrolysis products by paper chromatography in a desalted aliquot of the incubation mixture.

---

[57] Peptide Hydrolases in Mammalian Connective Tissue

By CHRISTIAN SCHWABE

Connective tissue constitutes about 20% of the body proteins of a macroorganism. Its function is as varied at its appearance in topographically different areas such as skin, bone, or as the supporting tissue of various organs. The cellular element of the connective tissue, the fibroblasts, are responsible for the production and maintenance of the intercellular components, collagen, and acid mucopolysaccharides.

Peculiarly, in spite of its substantial contribution to the total body weight, large portions of homogeneous cellular connective tissue are rare. A notable exception is the pulp of teeth which consists of fibroblasts, collagen, and mucopolysaccharides with a negligible amount of endothelium (capillaries) and nerve fibers interwoven. This tissue provides an unambiguous source of fibroblasts needed for the study of connective tissue peptide hydrolases.

The peptide bond-hydrolyzing enzymes found in connective tissue are divided into two groups, the neutral peptide hydrolases and the acid proteases, according to the hydrogen ion concentration at which they are effective. With the exception of a leucine aminopeptidase, the neutral peptide hydrolases do not attack proteins but hydrolyze amino acid amides, dipeptides, and a variety of oligopeptides. The leucine aminopeptidases hydrolyze protein substrates, although very slowly. The enzymes which hydrolyze peptide bonds at low pH values all appear to be typical proteases, commonly referred to as "cathepsins." It should be emphasized that the name cathepsin is a collective term given to intracellular proteolytic enzymes and does not necessarily imply similarity within this group. Some cathepsins have been characterized, such as beef spleen, cathepsin A, B, C,[1] and D.[2]

Hydrogen-Ion Activated Peptide Peptidohydrolases

Assay Method

Principle. The main enzyme of this group in mammalian fibroblasts catalyzes the hydrolysis of internal peptide bonds of denatured proteins and some peptides (peptide peptidohydrolase 3.4.4). The activity is most conveniently assayed with hemoglobin as substrate but other proteins, such as oxidized ribonuclease, or peptides like the insulin A or B chain, are also hydrolyzed.

Reagents

  4% Hemoglobin solution in pH 4.0 acetate buffer (0.1 M). [The hemoglobin substrate powder (Worthington) is stirred at pH 2.0 for 1/2 hour at room temperature, then readjusted to pH 4.0 and made up to volume.]
  10% Trichloroacetic acid (TCA). This solution should be stored at 4° and prepared new every 2 weeks.
  2.7% Solution of sodium potassium tartrate
  1% Solution of copper sulfate
  Sodium carbonate–sodium bicarbonate buffer, 0.5 M, pH 10.5, made up in 0.05 M in sodium hydroxide to provide base reserve.
  Phenol reagent (Folin-Ciocalteau),[3] 1 N diluted from the commercially available 2 N reagent with water just prior to use.

Procedure. The total volume of the assay system should be adjusted according to the number of points desired to plot ΔOD vs. time. Usually 2 ml of 4% hemoglobin is placed into a test tube and incubated at 40°

---

[1] J. S. Fruton, *in* "The Enzymes" (P. D. Boyer, H. Lardy, and K. Myrbäck, eds.), Vol. 4, p. 233. Academic Press, New York, 1960.
[2] E. M. Press, R. Porter, and J. Cebra, *Biochem. J.* **74**, 501 (1960).
[3] O. Folin and V. Ciocalteau, *J. Biol. Chem.* **73**, 627 (1927).

with 400–800 μl of crude tissue extract and sufficient pH 4.0 acetate buffer (0.1 N) to bring the volume up to 4 ml. After 5 minutes, 500 μl are pipetted into a tube containing 1.5 ml of 10% TCA. The TCA tube is vigorously shaken for a few seconds (test tube shaker) and after 15 minutes the precipitate is filtered through Whatman No. 4 paper.

Plastic funnels, 35 mm in diameter, and 5-cm filter circles (Whatman No. 4) are most convenient, particularly when many assays are performed at the same time. The withdrawal of 500 μl aliquots from the digestion mixture is repeated at any desired time interval.

The color is developed using an adaptation of the Lowry method for proteins.[4] From every tube of filtrate 2 × 500 μl are placed into a duplicate set of clean test tubes. Using a syringe pipette, 2 ml of a freshly prepared solution (1 ml of 2.7% sodium potassium tartrate and 1 ml of 1% copper sulfate made to 100 ml with the carbonate–bicarbonate buffer) are added to the 500 μl of filtrate and the contents mixed. After 10 minutes at room temperature 0.2 ml of 1 N phenol reagent is added from a graduated pipette and every tube immediately and vigorously agitated on a test tube shaker. The test tubes may be left at room temperature for 1 hour, or at 40° for 10 minutes to fully develop the color. The readings are made on any suitable colorimeter or spectrophotometer at 700 nm. The units given in this description are defined for a 1-cm light path. The results are plotted as OD (700 nm) vs. time and the zero time is obtained by extrapolation from the linear portion of the plot. Since hemoglobin and extract do not precipitate completely immediately upon mixing, the direct zero time reading is not reliable when crude extracts are assayed. With purified samples this precaution is not necessary.

Definition of Activity Unit and Specific Activity. The enzyme unit is defined as the amount of enzyme giving rise to a change in optical density (700 nm) of 0.1 per minute. The reaction is essentially linear if the enzyme concentration is chosen to produce a ΔOD of 0.1 to 0.15 in 30 minutes. Slopes obtained during this period are used for computation. The specific activity is defined as units per milligram of protein. For protein determination it has been assumed that a concentration of 1 mg/ml gives rise to an optical density of 1.0 at 280 nm in a 1-cm cell.

Source and Sample Collection. Only the lower two molar teeth of 1 to 1½-year-old commercially slaughtered cattle (the usual age of animals used for meat production) are used since they are easy to recover and contain most of the desired tissue. The last tooth is usually buried under a thin plate of bone and the preceding tooth is about one-half to three-fourths erupted. An unskilled laborer in a meat-packing plant can be

[4] O. H. Lowry, N. J. Rosebrough, A. L. Farr, and R. J. Randall, *J. Biol. Chem.* 193, 265 (1951).

trained to chisel these teeth out of the demeated mandible and to crack the teeth and extract the pulp. The tissue (about 1 g per tooth) is washed briefly in running cold water to remove gross contamination and then immediately placed on dry ice for shipment. In about 2 months 100 kg of tissue can be obtained. In order to retain the protease activity indefinitely, the whole sample is lyophilized and stored in the dry state at −20°. Lyophilization facilitates mechanical disruption of the tissue required for the extraction of enzymes. Collagenous tissue resists usual degradation methods in the hydrated state.

While it is as yet unknown whether the enzymes described here occur in all fibroblasts, it is very likely that they do not occur in ectodermally derived tissues. Only cathepsin D is similar in molecular size but is less restricted in its activity.[2] (See [19], this volume.)

Purification of the Acid Protease. The dry lyophilized tissue is broken up in a Waring blendor and subsequently ground to a 60-mesh powder in a bench type Wiley mill. An extract is prepared by stirring 200 g of pulp powder in 3 liters of distilled water at room temperature. The suspension is centrifuged at 5000 rpm for 10 minutes and the supernatant collected. The pellet is resuspended in 1 liter of distilled water and centrifuged as before. The combined supernatants constitute about 4000 ml of original extract.

Purification on DEAE-Cellulose. In order to effect a group separation of the acid and neutral peptide hydrolases, earlier purification methods were replaced by an initial chromatographic step on a DEAE-cellulose column.[5] A 9×100 cm jacketed column (made of acrylic resin) is packed with 700 g of DEAE-cellulose (a 1:1 mixture of Schleicher and Schuell type 40 and 70) suspended in 0.1 M NaOH, and subsequently washed with 16 liters of $5 \times 10^{-2} M$ HCl and 16 liters of pH 8.0 Tris buffer $(5 \times 10^{-2} M)$, containing $10^{-2} M$ calcium chloride, $10^{-2} M$ sucrose, and $10^{-3} M$ magnesium chloride. With an average liquid head of 120 cm this column should flow at 250 ml per hour, if maintained at 4°. The extract (4000 ml) is permitted to run onto the column. The vessel containing the extract should be cooled in an ice bath since the application to a column takes approximately 16 hours. The first 2 liters of column effluent are discarded and subsequent effluent is collected in 25-ml fractions in a refrigerated fraction collector while the protein sample is still being applied to the top of the column. Following the application of the 4 liters of extract, elution is begun with 4 liters of the same buffer used to equilibrate the column initially, followed by the additional 4-liter portions of the same basic buffer but 0.1 M, 0.2 M, and 0.5 M in NaCl, respec-

[5] C. Schwabe and G. Kalnitsky, *Arch. Biochem. Biophys.* **109**, 68 (1965).

tively. This schedule leads to the elution of a protease (referred to as DEAE I)[6] in the first 4 liters of eluent. The next 2 liters contain no activity, followed by 2.5 liters containing a second protease (DEAE II) and an aminopeptidase activity.

Purification of DEAE I by Ammonium Sulfate Fractionation. The 4 liters of fraction 1 obtained from the DEAE column are 40% saturated with solid ammonium sulfate.[7] The heavy precipitate is centrifuged off and discarded, the remaining supernatant is saturated to 85% with ammonium sulfate and stirred for about 15 minutes. The second precipitate is collected by centrifugation at 5000 rpm and the supernatant discarded. This fractionation is performed at room temperature.

Adsorption on Calcium Phosphate. The pellet obtained from the 85% ammonium sulfate fraction is redissolved in 250 ml of distilled water and dialyzed for 16 hours against a $10^{-2} M$ solution of $CaCl_2$. Calcium phosphate slurry (Bio-Rad)[8] is added to the dialyzed solution in 20-g portions, followed by centrifugation at 5000 rpm between each addition. The protein contents of the original solution as well as each of the supernatants obtained has to be checked before the next addition of calcium phosphate. This procedure should be carried on until 70% of the protein has been adsorbed on the calcium phosphate gel. The remaining supernatant contains 70–80% of the original activity.

Chromatography on Hydroxylapatite. About 45 g of wet hydroxyapatite slurry (prepared according to Jenkins)[9] is mechanically mixed with 50 g of a wet slurry of Whatman cellulose for 15 minutes. The adsorbent is then poured into a 2×25-cm jacketed column and permitted to settle under a gravity flow of $0.02 M$ phosphate buffer (pH 6.8). With a liquid head of 60 cm (between the outlet of the column and the buffer reservoir) a fairly constant flow rate of 16 ml per hour is attained. The total 260 ml supernatant from the calcium phosphate fractionation procedure is applied to the column in about 18 hours. A plateau of nonadsorbed protein is observed to elute from the column long before the protein sample has been applied *in toto*. The adsorbed protein is eluted by a simple gradient produced by 250 ml of $0.02 M$ phosphate buffer (pH 6.8) in vessel I and 400 ml of $0.15 M$ phosphate buffer in vessel II (vessel I is a 250-ml Erlenmeyer flask, fitted with airtight inlet and outlet). The volume in vessel I (which is magnetically stirred) remains constant until depletion of the limiting buffer. After about 400 ml of

---

[6] Naming of the connective tissue enzymes will be postponed until specificities are established beyond doubt.

[7] See Vol. I, p. 76.

[8] Bio-Rad Laboratories, Richmond, California.

[9] W. T. Jenkins, *Biochem. Prep.* **9**, 84 (1962).

eluant (including the 250 ml from the protein peak) have been collected, a sharp peak of proteolytic activity should be observed, usually followed by a second, less pronounced, activity peak. The first larger fraction is pooled for further purification. These fractions are referred to as hydroxylapatite fractions I and II.

Chromatography on Bio-Gel P-60. A 2 × 100 cm column with upward flow adapter is packed with Bio-Gel P-60 (Calbiochem, molecular exclusion value ∼60,000) and equilibrated with an acetic acid triethylamine buffer. This buffer solution is prepared by the addition of 2 ml of trimethylamine to each liter of 0.1 N acetic acid. The flow rate through the column should be maintained at 4 ml/cm^2/hour by means of a peristaltic pump. The inlet tubing leading to this pump is fitted with a Y-piece to permit the application of a sample through the buffer feeding line. Hydroxylapatite fraction I should be concentrated to 7 ml on an Amicon UM I[10] filter and pumped into the Bio-Gel column. The activity elutes in a symmetrical peak in front of a larger inactive protein component. The enzyme, which should be better than 90% pure, is pooled and sucrose added to a final concentration of 0.05 M before lyophilization. The summary of the purification steps given in Table I represents one typical experiment.

TABLE I

PURIFICATION OF AN ACID PEPTIDE PEPTIDOHYDROLASE FROM BOVINE FIBROBLASTS

Fraction	Volume (ml)	Protein (mg/ml)	Total protein (ml)	Total units	Specific activity (units/ mg × 10³)	Percent[a] yield
Crude extract	4,000	19.3	77,200	318	4.4	—
DEAE	4,200	2.0	8,400	185	22.0	58
Ammonium sulfate (40–85%)	230	6.7	1,540	109	71.0	35
CaPO₄	230	2.1	483	77	160.0	24
Hydroxylapatite	72	0.9	65	35	540.0	11
Bio-Gel P-60	40	0.2	8	39	4,900.0	12

[a] Pertaining to the main fraction only.

In general the best results will be achieved if freezing and lyophilization can be avoided between procedures. Should the enzyme be kept in solution for a longer time CaCl₂ (0.02 M) must always be present. Sucrose (0.05 M) protects the enzyme largely against denaturation during freezing.

---

[10] Amicon Corporation, Lexington, Massachusetts.

Properties of the Acid Hydrolase

Specificity. The highly purified preparation produces only very few peptides after 24 hours of incubation at pH 4.0 (40°) with oxidized bovine ribonuclease A. The peptide containing residues 1-52 (N-terminal lysine, C-terminal alanine), 53-120 (N-terminal aspartic acid, C-terminal phenylalanine), 121-124. (N-terminal aspartic acid, C-terminal valine) have been positively identified. Several minor peptides have not yet been analyzed. The A-chain of oxidized insulin is hydrolyzed between residues 14 and 15 (-Tyr-Gln-) and 16 and 17 (-Leu-Glu-). At high enzyme concentration (100 μg/ml) the tetrapeptide Leu-Phe-Glu-Ala is hydrolyzed between Phe and Glu. When the latter reaction is followed by thin-layer chromatography, some evidence for transamidation becomes apparent. It is not clear whether the transamidation at this unusual pH (4.0) is due to a trace impurity. The peptides (or derivatives) Tyr-Meth-Asp-Phe NH$_2$, Gly-Phe-Phe, Z-Gly-Phe, N-Benz DL-Phe-β naphthyl ester and Leu-Tyr NH$_2$ were not hydrolyzed.

Activators and Inhibitors. The enzyme is insensitive to diisopropylphosphofluoridate (DFP) and slightly inhibited by HgCl$_2$. No activator other than hydrogen ions has been found.

Purity. The enzyme obtained following the above procedure should show a single band upon discontinuous electrophoresis at pH 8.0 and 4.0.[11] Occasionally a minor impurity above and below the active band may be observed which can be removed by repeating the Bio-Gel column step.

Stability. The purified enzyme is unstable in dilute solutions unless 0.02 M CaCl$_2$ is added. Denaturation during freezing and lyophilization is prevented by 0.05 M sucrose. About 50% of its activity is lost in 15 minutes if heated to 50° and total inactivation occurs after 15 minutes at 60°. At pH values of 2.0 or less the enzyme will be irreversibly denaturated, while it retains activity well at higher pH values. The purification is performed at pH 8.0 without undue reduction of enzymatic activity.

pH Optimum. The activity of the enzyme with hemoglobin as substrate is maximal at pH 3.5 to 4.0. At pH 2.0 the enzyme activity ceases abruptly. The catalytic activity slopes down more gently toward the neutral pH range with 50% activity remaining at pH 5.0.

Molecular Weight. The molecular weight determined by exclusion chromatography is 40,000 $\pm$ 2000.[11] According to preliminary data, the protein is formed by a single polypeptide chain.

Distribution. The acid protease appears to be a lysosomal enzyme al-

[11] C. Schwabe and G. Kalnitsky, *Biochemistry* **5**, 158 (1966).

though it has not been established that this is its exclusive location in the cell.

Other Acid Proteases. The second active peaks obtained from the DEAE column and a second peak or shoulder observed during hydroxylapatite chromatography remain to be characterized. It seems certain that enzymes different from spleen cathepsins are present and that a carboxypeptidaselike activity does not occur in connective tissue cells.

Neutral Peptide Hydrolases

This group includes leucine aminopeptidase as a major enzyme. In addition an aminotripeptidase, glycylglycine dipeptidase, (gly)$_3$ tripep-

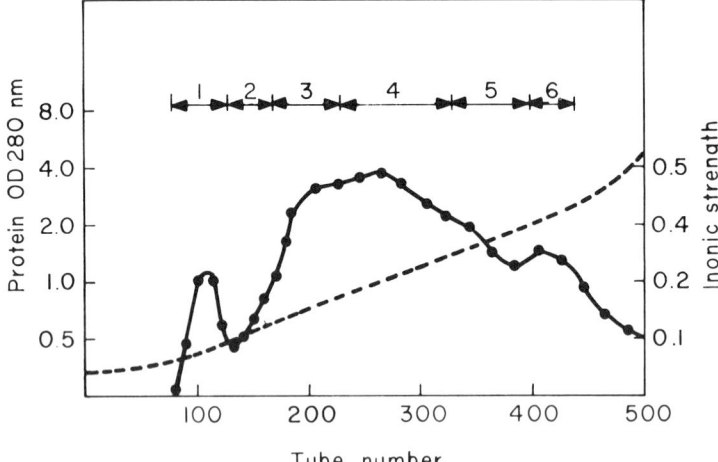

FIG. 1. Distribution of various peptide hydrolases of mammalian fibroblasts after chromatography on DEAE-cellulose. The fractions as indicated above the protein line ●————● contain the following enzymes: (1) Acid peptide peptidohydrolase (cathepsin). (2) Acid peptide hydrolase, imidodipeptidase I. (3) Aminotripeptidase, iminodipeptidase, triglycine peptidase I, glycylalanine dipeptidase. (4) Glycylleucine dipeptidase, imidodipeptidase II, leucine aminopeptidase. (5) Formylleucine hydrolyzing enzyme. (6) Triglycine peptidase II.

tidase, glycylleucine dipeptidase, imino and imidodipeptidase, and a formylleucine hydrolyzing enzyme are found in a crude neutral extract of the fibroblasts of bovine dental connective tissue. Charging the large DEAE column described in the previous section with only about 50 g of protein and eluting by a buffer gradient rather than a stepwise procedure, the enzyme distribution as depicted in Fig. 1 is observed.[12] Fractions of the effluent of this column might be used to further purify a certain

[12] C. Schwabe, *Biochemistry* **8**, 783 (1969).

enzyme of this group of neutral peptide hydrolases. Below the purification of the Mn^{2+} ion-dependent soluble leucine aminopeptidase (an α-aminopeptide aminoacidohydrolase 3.4.1) is described. Other enzymes will be considered briefly at the end of this chapter.

Principle. The enzyme hydrolyzes leucinamide and a variety of peptides with N-terminal residues other than glycine, proline, or hydroxyproline. Denatured proteins with susceptible N-terminals are hydrolyzed much slower than dipeptides. Leucylleucine is a better substrate than leucinamide and is routinely used to follow purification steps. The activity is measured by a ninhydrin reaction essentially as described by Matheson and Tattri.[13] The reagent reacts only slightly with dipeptides but yields normal color values with amino acids.

Reagents

A 10 mM solution of leucylleucine in 0.1 M Tris buffer at pH 8.0
Acetic acid, 0.1 M
$MnCl_2$ solution, 0.1 M
Methyl Cellosolve, 50% in distilled water
Stock ninhydrin solution: 50 ml 5% ninhydrin in Cellosolve, 250 ml methyl Cellosolve, 5 ml KCN 0.01 M. This reagent is stored at −20°.
Working ninhydrin solution. Dilute stock ninhydrin 1:1 with 50% Cellosolve just prior to use.
Tris buffer 0.1 M, pH 8.0
Citrate buffer, 0.25 M, pH 5.0

Procedure. An appropriate amount of crude extract (25 μl) or partially purified enzyme preparation is made to 200 μl with 0.1 M Tris HCl buffer (pH 8.0, 2 × 10^{-3} M in $MnCl_2$) and placed into a water bath (40°) for about 5 minutes. The prewarmed Leu-Leu solution is added (200 μl) and three 25-μl samples of the reaction mixture are immediately withdrawn and placed in three different tubes, each containing 0.5 ml of 0.1 N acetic acid. In 5, 10, 15, and 20-minute intervals again triplicate samples of 25 μl each are withdrawn and pipetted into acetic acid-containing tubes (acid stops the reaction). All volume transfers, except the acetic acid, must be made with micropipettes to be sufficiently accurate. The whole contents of the acetic acid-containing tubes are used for color development as described below.

To assay the column effluent for leucine aminopeptidase activity 2 × 25 μl may be withdrawn from every third or fifth tube (depending upon the size of the column and volume collected) and placed into a duplicate

[13] A. T. Matheson and B. L. Tattri, *Can. J. Biochem.* **42**, 95 (1964).

set of test tubes. Each member of one set receives 25 μl of Tris buffer (pH 8.0) while the second set receives 25 μl Leu-Leu substrate. Incubation for 15 minutes at 40° is usually sufficient to locate the enzyme in the effluent. The reaction is stopped by the addition of 0.5 ml of 0.1 N acetic acid from a pipette-syringe.

Color Development. This procedure is identical for both the regular assay and the column assay.

The tube racks are placed in an ice bath and 1 ml of citrate buffer (pH 5.0) is added (speedy addition of the reagents for color development is best achieved with pipette-syringes), followed by 1 ml of the working ninhydrin solution. Each test tube is removed from the rack, shaken vigorously, and placed back into the ice bath. After all tubes have been shaken, the whole rack is transferred to a vigorously boiling water bath for exactly 8½ minutes. Thereafter the rack is rapidly transferred back into the ice bath and 2 ml of 50% aqueous isopropanol added. The test tubes are shaken again and left in the cold water for 5 minutes. Subsequently the tubes may be kept at room temperature for 30 minutes or may be warmed up briefly at 40° prior to color determination at 570 nm. The absorbance should be measured within 1 hour.

The ΔOD is plotted as a function of time. Enzyme concentrations should be adjusted to give a linear reaction for at least 10 minutes. The column assay is evaluated by plotting the difference between blank and experimental sets of tubes as a function of tube number.

Assay of Tissue Slices or Whole Homogenates. Tissue slices may be prepared from either fresh or thawed connective tissue and suspended in Ringer's solution at pH 7.5 in double-sidearm Warburg flasks.[14] A single 15-ml respirometer flask can be loaded with several grams of wet tissue slices. One sidearm of the flask receives 200 μl of a 0.2% solution of snake venom[15] or correspondingly less purified L-alpha amino acid oxidase, while the second sidearm is filled with a substrate peptide (about 1 millimole). At least one, preferably both amino acids of a dipeptide must be a substrate for the amino acid oxidase.

The amino acid oxidase is added first to the tissue slices, and the immediate oxygen uptake, which is due to free amino acids in the tissue, is measured. When the rate drops to near zero the peptide substrate is added and the rate of peptide hydrolysis measured as O_2 uptake. Appropriate controls can be used to compensate for respiratory activity. This interference will be low since no CO_2 absorbing base is included in the system.

[14] W. W. Umbreit, R. H. Burns, and J. F. Stauffer, "Manometric Techniques," Burgess, Minneapolis, 1964.
[15] Ross Allan's Snake Farm, Florida.

The result should be expressed in micromoles oxygen used per minute per milligram of wet or dry weight of tissue. The oxidation of every mole of amino acid causes the uptake of 1 mole of O_2 if no peroxidase is present. The efficiency of the system must be tested by the addition of standard amounts of leucine to the tissue slices.

Definition of Activity Unit and Specific Activity. The activity unit is defined as the amount of enzyme hydrolyzing 1 micromole of leucylleucine per minute in Tris-HCl buffer at pH 8.0 (buffer effects are significant). Specific activity designates the number of units per milligram of protein, assuming that 1 mg of protein in 1 ml of water gives rise to an optical density of 1 in a 10-mm light path at OD = 280 nm. The conversion of ninhydrin color to micromoles is possible if the ϵ-values for products and reactants are determined in a system duplicating assay conditions. Considering the ϵ_r is the molar extinction of reactants and ϵ_p represents the molar extinction of products and that 1 mole reactant (Leu-Leu) gives rise to 2 moles of product, the following relation is obtained.

$$\frac{\text{moles hydrolyzed}}{\text{liter minutes}} = \frac{\Delta OD_{570}}{2\epsilon_p - \epsilon_r \text{ min}} = \frac{\Delta OD_{570}}{\Delta\epsilon \text{ min}}$$

(In case of a mixed dipeptide the term $2\epsilon_p$ is replaced by $\epsilon_{p1} + \epsilon_{p2}$.) The ϵ-values will have to be determined by each investigator. The data may also be expressed as percent hydrolysis since the limiting ΔOD is the value of the denominator and any ΔOD obtained divided by the limiting value and multiplied by 100 gives percent hydrolysis. Knowing the assay conditions, it is possible to calculate C_0 values from specific activities obtained from initial velocities (zero order) for comparison. The ninhydrin method used here permits good measurements with as little as 2% hydrolysis if a 5 mM leucylleucine solution is used as substrate. In contrast to the discontinuous titration method of Grassman and Hyde[16]; the assay can always be restricted in time to observe the apparent zero-order period of reaction.

Purification. The purification of the leucine aminopeptidase of bovine fibroblasts begins with the DEAE column described under Purification of Acid Proteases. After 8.5 liters have been collected from the column, the leucine aminopeptidase begins to elute (this must be checked since batches of DEAE may vary). The following 6 liters of eluant are collected and lyophilized to provide a convenient, stable starting material for subsequent steps.

Benzalkonium Chloride (Zephiran) Precipitation. This quaternary

---

[16] W. Grassman and W. Hyde, *Z. Physiol. Chem.* **183**, 32 (1929).

ammonium compound has proven useful for the purification of the connective tissue LAP although its mechanism of action and limiting conditions are not well understood. The procedure should be followed as closely as possible as deviation might yield unpredictable results.

The lyophilized LAP fraction (10 g) is redissolved in 500 ml of distilled water and the solution magnetically stirred in an ice bath. A 1.7% solution of Zephiran chloride in 0.1 M Tris at pH 8.0, 10^{-3} in $MnCl_2$, is prepared and added to the enzyme solution in 20-ml increments. Slow addition is essential and is best achieved by siphoning the Zephiran chloride through an intravenous tubing into the gently stirred protein solution. Following each addition of 20 ml Zephiran, the solution is centrifuged at 5000 rpm, the pellet suspended in 100 ml of 0.1 M Tris buffer (pH 8.0, 10^{-3} M in $MnCl_2$, 10^{-2} M in $MgCl_2$, and 10^{-1} M in $NaCl_2$) and stirred gently at 4°. Up to 60 ml of Zephiran chloride solution may be added without precipitating a significant portion of the activity, although it is advisable to retain all the pellets obtained until final tests have been made. The addition of the next 20 ml of Zephiran chloride precipitates the LAP activity. All pellets are stirred overnight at 4° and tested for activity. If a substantial number of enzyme units are missing after this procedure, the pellet of the enzyme-containing fraction is resuspended in 100 ml of Tris buffer. The combined supernatants should contain 70% of the activity present prior to Zephiran fractionation.

Generally, if the ionic strength is too high the enzyme will not precipitate at pH 8.0. The column effluent may be fractionated with Zephiran without prior lyophilization, in which case the buffers present should be removed by a short dialysis or the pooled effluent diluted with about 50% of its volume of distilled water.

Acetone Fractionation. The enzyme solution obtained from the previous step is cooled to 0° in an ice bath and ice-cold acetone is added slowly. Any precipitate formed between 0–20% acetone concentration is centrifuged off and discarded. The acetone concentration in the supernatant is then increased to 40%, the pellet collected by centrifugation at 4000 rpm and redissolved in 10 ml of 0.05 M Tris buffer (pH 8.0, 10^{-3} M in $MgCl_2$).

Ammonium Sulfate Precipitation. In order to remove most of the solvent, the enzyme solution obtained from the acetone fractionation is diluted to 300 ml and reconcentrated to 100 ml with an Amicon UM I filter. This solution, containing about 1 mg protein per milliliter, is cooled in an ice bath and 50 ml of saturated ammonium sulfate containing 1 g of Tris base per liter, are added. No precipitate forms at this time. Adding solid ammonium sulfate the salt concentration is increased to 40% saturation, and the precipitate centrifuged off at 5000 rpm and

TABLE II

THE PURIFICATION OF CONNECTIVE TISSUE LEUCINE AMINO PEPTIDASE

Procedure	Volume (ml)	Protein mg/ml OD 280 nm	Total protein OD 280 mg	Specific activity $\times 10^2$	Total activity units	Yield (%)	Purity (integers)
First extract	1,000	30.0	30,000	1.3	400	—	1
DEAE fraction	500	6.4	3,200	9.3	300	75	7
Benzalkonium chloride	125	0.6	90	244.0	220	55	188
Acetone (30–40%)	10	5.0	50[a]	260.0	130	32	244
Ammonium sulfate (41–58%)	10	2.7	27	370.0	100	25	285
Agarose column	50	0.25	12.5	800.0[b]	100	25	615

[a] Probably too high due to residual acetone bound to the protein.
[b] An average of several forms of the enzyme possessing different specific activities.

discarded. The ammonium sulfate in the supernatant solution is increased to 60% saturation and the precipitate, collected by centrifugation, is redissolved in a small amount of pH 8.0 Tris buffer.

Agarose Chromatography. A 150 × 2.5 cm jacketed column maintained at 4° is used for exclusion chromatography with Agarose (Bio-Gel A 1.5, exclusion limit 1.5 × 10⁶ g/mole). The sample is concentrated to 5–10 ml, applied to the top of the column, and eluted with a 0.025 M Tris buffer (pH 8.0) containing 10⁻³ moles of $MgCl_2$ per liter. A broad protein peak elutes in tubes 100 to 300 (if 3-ml fractions are collected) which contain nearly all of the activity placed onto the column. The peak broadening is due to aggregates varying in size from 4 × 10⁵ to 5 × 10⁴ g/mole. An acrylamide gel run at pH 8.0 in a glycine buffer system[17] of the pooled active fraction should show one band at the origin and four equally spaced bands traveling toward the anode.[12] The purification procedure is summarized in Table II.

Properties of the Purified Connective Tissue Leucine Aminopeptidase

Specificity. The specificity of the fibroblast enzyme is similar to that of other leucine aminopeptidases.[18] An absolute requirement exists for the L-configuration of the N-terminal residue possessing a free primary amine function. The penultimate residue must be an L-amino acid other than proline or hydroxyproline but could also be glycine. Peptides with glycine or imino acids at the N-terminal position are not hydrolyzed. The amides of leucine or phenylalanine are hydrolyzed one order of magnitude slower than dipeptides. In contrast to the kidney enzyme, no activity toward glycinamide has been detected.

The rate of hydrolysis of dipeptides is greatest if norleucine, isoleucine, leucine, methionine, or phenylalanine occupies the N-terminal position. The activity decreases with increasing length of the peptide substrate chain and becomes very slow with denatured proteins. N-Acetylated peptides, denatured cytochrome c (horse heart), and mercuripapain are not hydrolyzed.

Activators and Inhibitors. In contrast to other leucine aminopeptidases, the connective tissue enzyme is activated solely by Mn^{2+} ions. EDTA removes the ions from the enzyme only at elevated temperatures (40°). The inhibition can be reversed by Mn^{2+} ions. In contrast, removal of Mn^{2+} ions by dialysis against distilled water inhibits the enzyme irreversibly. Copper ions reduce the rate of hydrolysis of Leu-Tyr (K_i 5 × 10⁻³ M) at pH 8.0 (the K_m for this substrate at pH 8.0 is 8 × 10⁻⁴ M). Some inhibition is also observed with Zn^{2+}, Cd^{2+}, and Ca^{2+} ions. Exposure to N-ethylmaleimide and p-mercuribenzoate abolishes the

[17] L. Ornstein and B. J. Davis, *Ann. N.Y. Acad. Sci.* **121**, 428 (1964).
[18] E. L. Smith, *Advan. Enzymol.* **12**, 191 (1951); see also Vol. II, p. 88.

aminopeptidase activity. The enzyme can be reacted with iodoacetamide or 1-dimethylaminonaphthalene-5-sulfonyl chloride without influencing the activity or the rate of migration in an electric field at pH 8.0.

Purity. Due to the multiple forms of this enzyme, it is difficult to ascertain absolute purity of a preparation of connective tissue leucine aminopeptidase. Upon acrylamide electrophoresis in a discontinuous buffer system (pH 8.0), four bands are observed to travel toward the anode while a fifth band remains at the origin. Each band has been demonstrated to be active.[12]

Stability. The enzyme is relatively stable in solution and may be frozen and thawed repeatedly without causing inactivation. In frozen samples (−20°) the activity is retained for at least 1 year. Lyophilization, in contrast, irreversibly inactivates the enzyme up to 80%. In 8.0 M urea (40°) the LAP remains active for 16 hours while the addition of 0.1 M LiCl to this solution destroys the activity as well as the banding pattern on acrylamide electrophoresis. The enzyme also remains active if kept at 55° for 30 minutes in 0.1 M Tris buffer (pH 8.0) or at 50° in the same buffer containing 30% acetone or ethanol. The presence of 30% dioxane in the Tris buffer deactivates the enzyme completely within 30 minutes.[19] The pH stability curve of connective LAP shows a maximum at pH 8.0 and a steep decline within 2 pH units above and below this value. At the higher pH the enzyme is protected by substrate.

pH Optimum. The pH optimum of the connective tissue leucine aminopeptidase is substrate concentration dependent. At pH 7.5 the affinity for Leu-Leu is greatest but the maximal catalysis rate is obtained at pH 9.0 and high substrate concentration.[20] Below pH 7.0 the activity disappears.

Molecular Weight. In the crude extract aggregates of connective tissue leucine aminopeptidase as large as 1.5×10^6 g/mole have been observed during exclusion chromatography. The main form of the enzyme with a molecular weight of 4×10^5 dissociates into functional units, each 10^5 g/mole. Preliminary experiments suggest that smaller units (5×10^4 g/mole) exist and that these are possibly formed by two kinds of polypeptide chains. A leucine aminopeptidase isolated from bovine lenses by Kretschmer and Hansen[21] shows a similar pattern.

Other Neutral Peptide Hydrolases from Bovine Fibroblasts. The enzymes described here have not been purified. The fractions in which they are contained are shown in Fig. 1. Molecular weights were determined by exclusion chromatography.

[19] C. Schwabe, *Biochemistry* **8**, 795 (1969).
[20] C. Schwabe, *Biochemistry* **8**, 771 (1969).
[21] K. Kretschmer and H. Hansen, *Z. Physiol. Chem.* **349**, 831 (1968).

Imidopeptidase. (Molecular weight ~150,000.) The existence of two forms of this enzyme is suggested by its appearance in two widely separated fractions of the DEAE column (2 and 4). The imidodipeptidase differs in some respects from the intestinal enzyme. In the presence of Tris buffer (pH 8.0) Mn^{2+} ions have no effect but stimulate in the presence of phosphate buffer. Zinc ion in Tris buffer inhibits the activity but stimulates in phosphate buffer.

Glycylglycine Dipeptidase. The Co^{2+} activated peptidase is observed in the crude extract but appears to denature upon DEAE chromatography. Differences in glycylglycine dipeptidase from other tissues have not been observed.

Aminotripeptidase. (Molecular weight ~100,000.) The activity is independent of Mn^{2+} ion, active in the presence of EDTA, and slightly depressed by Cd^{2+} ions (DEAE fraction 2). Tripeptides of L-amino acids are hydrolyzed to a dipeptide and the free N-terminal amino acid. Triglycine is not a substrate.

Triglycine Peptidase. (Molecular weight ~400,000, possibly 100,000 subunits.) The peptidase catalyzes the hydrolysis of triglycine to glycine and glycylglycine and does not require Co^{2+} ions (DEAE fraction 2).

Imino Dipeptidase. (Molecular weight ~250,000.) Dipeptides possessing an N-terminal proline are hydrolyzed by this enzyme. The activity is stimulated by Mn^{2+} ions and inhibited by EDTA and inorganic pyrophosphate (DEAE fraction 2).

Glycylalanine Dipeptidase. (Molecular weight ~50,000.) A small amount of 200,000 molecular weight activity was observed. The ion requirement is not known (DEAE fraction 2).

Formylleucine Hydrolyzing Enzyme. This weak activity has not been further characterized (DEAE fraction 5).

Triglycine Peptidase II. It is not known whether the enzyme differs from the triglycine peptidase I (DEAE fraction 6).

[58] ε-Lysine Acylase from *Achromobacter pestifer*

By ICHIRO CHIBATA, TSUTOMU ISHIKAWA, and TETSUYA TOSA

$$\epsilon\text{-}N\text{-Acyl-L-lysine} + H_2O \to \text{fatty acid} + \text{L-lysine}$$

Assay Method[1]

Principle. The assay of ε-lysine acylase (ε-N-acyl-L-lysine amidohydrolase, EC 3.5.1.17) is based on the determination of liberated

[1] T. Ishikawa, T. Tosa, and I. Chibata, *Agr. Biol. Chem.* **26**, 581 (1962).

L-lysine from ε-*N*-acyl-L-lysines by the acidic ninhydrin colorimetric method.[2]

Reagents

Sodium acetate-acetic acid buffer solution, 1×10^{-1} *M*, pH 5.0

ε-*N*-benzoyl-L-lysine solution, 1.2×10^{-2} *M*, in water

Enzyme. Prior to assay, dilute in 5×10^{-2} *M* sodium acetate solution (1×10^{-2} *M*) so that 1 ml will contain approximately 1–5 units.

Acidic ninhydrin solution. One gram of ninhydrin is dissolved in 64 ml of acetic acid and 16 ml of 6×10^{-1} *M* phosphoric acid.

50% Ethanol–water solution (v/v)

Procedure. A mixture of 0.5 ml of buffer solution, 0.5 ml of ε-*N*-benzoyl-L-lysine solution, and 0.5 ml of enzyme solution is incubated at 42° for 10 minutes. After completion of the enzyme reaction, 0.5 ml of the above enzyme reaction mixture is placed into 1.0 ml of acidic ninhydrin solution. The mixture is heated at 100° in a boiling water bath for exactly 5 minutes, and cooled rapidly to room temperature. To the resulting mixture 3.5 ml of 50% ethanol–water solution (v/v) is added. The optical density of the solution is read at 340 nm. A reaction blank containing water instead of ε-*N*-benzoyl-L-lysine solution is treated identically.

Definition of Unit and Specific Activity. One unit of activity is the amount of enzyme liberating 1 micromole of lysine per hour under the above conditions. Protein content is determined by the method of Lowry *et al.*[3] Specific activity is expressed as units per milligram of protein.

Purification Procedure[4,5]

Preparation of Cells[6]

Five liters sterilized medium containing 1% peptone, 5% glucose, 0.2% KH_2PO_4, and 0.1% $MgSO_4 \cdot 7 H_2O$ in a 10-liter jar fermentor are inoculated with 150 ml of the seed culture of *Achromobacter pestifer* EA ATCC 23584, which was incubated with reciprocal shaking (110 cycles/minute, stroke 8 cm) at 30° for 48 hours in 500-ml flasks containing 100 ml of the same medium described above. Fermentation in the jar is

[2] E. Work, *Biochem. J.* **67**, 416 (1957).

[3] O. H. Lowry, N. J. Rosebrough, A. L. Farr, and R. J. Randall, *J. Biol. Chem.* **193**, 265 (1951).

[4] T. Ishikawa, T. Tosa, and I. Chibata, *Agr. Biol. Chem.* **26**, 412 (1962).

[5] I. Chibata, T. Ishikawa, and T. Tosa, *Nature* **195**, 80 (1962).

[6] T. Ishikawa, T. Tosa, and I. Chibata, *Agr. Biol. Chem.* **26**, 193 (1962).

carried out at 30° for 24 hours with aeration (5 liters air/minute/5 liters medium) and agitation (400 rpm). The cells are harvested by centrifugation and washed with 0.9% sodium chloride solution.

Step 1. Preparation of Cell-Free Extract. One kilogram of wet cells from 75 liters of culture broth are mixed with 500 ml of toluene to initiate autolysis. After being liquefied, 12 liters of $6.67 \times 10^{-2} M$ sodium sulfate solution is added, and the suspension is kept for 24 hours at 30° with occasional stirring. The suspension is centrifuged at 50,000 g and 0° for 30 minutes. A yellowish, clear, viscous solution (ca. 12 liters) is obtained.

Step 2. Fractionation with Ammonium Sulfate. The crude extract is brought to 40% saturation by addition of ammonium sulfate at 5° and adjusted to pH 6.5. After standing for 30–60 minutes at 5°, the precipitate is removed by centrifugation at 20,000 g and 5°. The resulting supernatant is brought to 70% saturation by further addition of ammonium sulfate at 5° and is adjusted to pH 6.5. After standing for 30–60 minutes at 5°, the precipitate is collected by centrifugation at 20,000 g and 5°. The precipitate obtained is dissolved in 350 ml of $5 \times 10^{-2} M$ sodium acetate solution and dialyzed against the same solution at 5° for 24 hours. After dialysis of the solution, any insoluble material is removed by centrifugation at 20,000 g and 5°.

Step 3. Rivanol Treatment. To the dialyzed solution (ca. 380 ml) a 15% volume of 2% Rivanol solution is added dropwise at 5°. After allowing the solution to stand at 5° for 1–2 hours, the precipitated acidic protein is removed by centrifugation at 20,000 g and 5°. To remove excess Rivanol, the supernatant is passed through a column packed with Amberlite IRC-50 buffered with $5 \times 10^{-1} M$ sodium acetate-acetic acid buffer solution, pH 6.0. The effluent containing ϵ-lysine acylase activity is collected.

Step 4. Fractionation with Acetone. To the effluent (ca. 600 ml) cold acetone is added dropwise at −10° to the final concentration of 45%. The precipitate is immediately removed by centrifugation at 15,000 g and −10°. The resulting supernatant is brought to 60% volume by further dropwise addition of cold acetone at −10°. The precipitate is immediately collected by centrifugation at 15,000 g and −10°. The precipitate obtained is dissolved in a small volume (ca. 50 ml) of $5 \times 10^{-2} M$ sodium acetate solution, and any insoluble material is removed by centrifugation at 20,000 g and 5°.

Step 5. Fractionation with Ammonium Sulfate. The supernatant is brought to 48% saturation by dropwise addition of saturated ammonium sulfate solution at 5° and adjusted to pH 7.0. After standing for 60 minutes at 5°, the precipitate is removed by centrifugation at 20,000 g and 5°. The supernatant is brought to 60% saturation by further dropwise addition of saturated ammonium sulfate solution at 5° and adjusted to pH 7.0. After standing for 60 minutes at 5°, the precipitate is collected by centrifugation at 20,000 g and 5°. The precipitate obtained is dissolved in a small quantity of Veronal buffer solution (pH 8.8, $\mu =$ 0.0505) and dialyzed against the same buffer for 24 hours at 5°.

Step 6. Vertical Zone Electrophoresis. The dialyzed solution is used for vertical zone electrophoresis to obtain the pure enzyme. The apparatus and technique are the same as those employed by H. Chiba *et al.*[7] To allow the electrophoresis to be performed for a long time, the apparatus is furnished with Hg—Hg_2Cl_2 reversible electrodes, and is so designed to prevent electroosmosis and pH changes. The cellulose powder used as a supporting material is prepared by refluxing clean cotton of good quality in absolute methanol containing 1 M hydrogen chloride for 24 hours. The cellulose powder is packed in a column using a Veronal buffer solution (pH 8.8, $\mu = 0.0505$). Nine milliliters of the dialyzed enzyme solution (step 5) containing 200 mg of protein (specific activity 10,800) is placed on the electrophoresis column (50 cm length) as described by H. Chiba *et al.* A potential difference of 300 V (20 mA) between the two electrodes is applied for 160 hours in the Veronal buffer solution (pH 8.8, $\mu = 0.0505$) at 6°. When the run is finished, the column is detached from the apparatus and mounted on a fraction collector. The elution is carried out with the same buffer solution at 6° and the elution rate is controlled to 12 ml per hour by a pump. Fractions of 3.2 ml are collected, and are assayed for protein content and enzyme activities. The fractions of higher specific activity (14,300–14,600 units/mg of protein) are collected.

To the pure ε-lysine acylase solution of higher specific activity, saturated ammonium sulfate solution is added to give 60% saturation; thus the enzyme is precipitated. The final preparation is stored in a refrigerator.

The purification procedure is summarized in Table I.

Padayatty and Van Kley[8,9] reported that the enzyme was effectively purified by the addition of gel filtration technique in 0.1 M sodium acetate buffer solution (pH 6.0) on Sephadex G-200 before column zone electrophoresis.

Properties

Stability. A suspension of the pure enzyme in 60% ammonium sulfate may be stored at 0° for several months without decrease of enzyme activity.

[7] H. Chiba, F. Sugimoto, and M. Kito, *Agr. Biol. Chem.* **24**, 428 (1960).

[8] J. D. Padayatty and H. Van Kley, *Biochemistry* **5**, 1394 (1966).

[9] J. D. Padayatty and H. Van Kley, *Arch. Biochem. Biophys.* **120**, 296 (1967).

TABLE I
Purification of ε-Lysine Acylase

Step	Volume (ml)	Total protein (mg)	Total enzyme activity (unit)	Cumulative yield (%)	Enzyme specific activity (unit/mg of protein)	Cumulative purification (-fold)
1. Cell-free extract	12,000	26,560	$6,365 \times 10^3$	100	240	1
2. Fractionation with $(NH_4)_2SO_4$	380	6,570	$5,130 \times 10^3$	80.6	780	3.3
3. Rivanol treatment	600	1,350	$4,525 \times 10^3$	71.1	3,360	14.0
4. Fractionation with acetone	50	480	$3,410 \times 10^3$	53.6	7,140	29.8
5. Fractionation with $(NH_4)_2SO_4$	9	200	$2,163 \times 10^3$	34.0	10,800	45.0
6. Vertical zone electrophoresis	—	130	$1,885 \times 10^3$	29.6	14,500	60.4

Optimum pH. Optimum pH activity of the ε-lysine acylase from *A. pestifer* EA ATCC 23584 is found to be around 4.8–5.2. This value is different from those of the other reported ε-lysine acylases, that is, the pH optima for the enzymes from rat kidney, *Pseudomonas* species and *Aspergillus oryzae* No. 10 are pH 7.0, 6.0, and 8.2–8.4, respectively.

Purity and Physical Properties. In the vertical zone electrophoresis, specific activities of all fractions in the main component are apparently identical. Thus, the enzyme having a specific activity of 14,500 units per milligram of protein appears homogeneous on electrophoresis.

In the absorption of ultraviolet light, the extinction coefficients of 1% solution in a 1-cm cell at 280 nm is 12.00 and the ratio of optical density 280 nm/260 nm is 1.84.

Inhibitors and Activators. ε-Lysine acylase activity is inhibited by both the heavy metal ions (Ag^+, Hg^+ and Hg^{2+}) and the sulfhydryl reagents (monoiodoacetic acid, *p*-chloromercuribenzoate and ω-chloroacetophenon). The inhibition by *p*-chloromercuribenzoate is completely recovered by the addition of cysteine. Among the chelating agents, the inhibitory effect of ethylenediaminetetraacetate and citrate are slight, but oxalate and pyrophosphate inhibit markedly the enzyme activity. Potassium cyanide, azide, and *o*-phenanthroline show no inhibitory effect.

The slight activation of the enzyme is observed by the addition of metal ions or salts in higher concentration ($1 \times 10^{-3} M$ to $10^{-1} M$). A specific and more effective activator is not found.

TABLE II
RATES OF HYDROLYSIS OF ε-*N*-ACYLATED L-LYSINES BY ε-LYSINE ACYLASE

ε-Acyl-L-lysines	Rate of hydrolysis[a]	
	at pH 5.0	at pH 7.2
Benzoyl	15,000	6,100
Acetyl	2,700	1,400
Monochloroacetyl	5,300	2,600
Dichloroacetyl	34,800	7,700
Trichloroacetyl	7,500	2,000
Monoiodoacetyl	14,500	5,900
n-Butyryl	16,000	4,500
iso-Butyryl	6,400	3,500
Methacrylyl	4,500	3,200
n-Caproyl	15,700	5,200
n-Caprylyl	12,500	5,000
Glycyl (HCl salt)	4,100	3,700
Carbobenzoxy	5,200	2,600
Phenylacetyl	1,500	1,100
p-Toluoyl	8,700	3,500
o-Nitrobenzoyl	0	0
m-Nitrobenzoyl	3,100	3,600
p-Nitrobenzoyl	5,900	2,700
α-Naphthoyl	1,200	1,100
Cyclohexanecarboxyl	4,200	3,300

[a] Liberated lysine micromoles/hour/mg of the enzyme.

Specificity.[10] Among the various *N*-acyl derivatives of lysine and its analogs, the enzyme is unsusceptible to the following compounds: ε-*N*-benzoyl-D-lysine, α-*N*-acyl, and α,ε-*N*-diacyl-L-lysines, *N*-acetyl-DL-methionine, ε-*N*-benzoyl-L-lysine ethyl ester, ε-*N*-benzoyl-L-lysine amide, monobenzoyl cadaverine, ε-*N*-benzoylaminocaproic acid, ε-*N*-acetyl-aminocaproic acid, α-chloro-ε-*N*-benzoylamino-DL-caproic acid, α-bromo-ε-*N*-benzoylamino-DL-caproic acid, δ-*N*-benzoyl-L-ornithine, ζ-*N*-ben-zoyl-DL-homolysine, ε-*N*-benzoyl-ε-*N*-methyl-DL-lysine, and ε-*N*-tosyl-L-lysine.

From these results, it is shown that the enzyme has a greater substrate specificity than α-aminoacylase, and susceptible substrates are limited to ε-*N*-acyl-L-lysines. Table II shows the rates of hydrolysis of various ε-*N*-acyl-L-lysines by the enzyme.

Kinetics. Lineweaver-Burk plot yields a K_m of 2.9×10^{-4} M for ε-*N*-chloroacetyl-L-lysine and 5.0×10^{-4} M for ε-*N*-acetyl-L-lysine. The K_m for ε-*N*-benzoyl-L-lysine is below 1.0×10^{-4} M, but a correct measurement is practically impossible due to the low concentration of liberated lysine.

[10] I. Chibata, T. Tosa, and T. Ishikawa, *Arch. Biochem. Biophys.* **104**, 231 (1964).

The activation energy of the enzyme on ϵ-N-benzoyl-L-lysine is 9280 calories per mole by means of an Arrhenius' plot.

Distribution. The enzyme is widely distributed in animal tissues[11-13] and in microorganisms.[14-18] Among the former, kidney (pigeon, chicken, pig, rat, etc.) is the richest source, but the enzyme is also found in brain, liver, spleen, pancreas, and heart of a few animals.[12] Of the microorganisms, besides *Achromobacter pestifer* EA, *Pseudomonas* species,[14, 15] a number of soil bacteria,[16, 17] *Aspergillus oryzae*,[18] and mushrooms[12] have been reported to contain the enzyme.

ϵ-Lysine acylase activities of various enzyme sources are summarized in Table III.

TABLE III

ϵ-Lysine Acylase Activity of Various Enzyme Sources

		Enzyme activity[a]			
Enzyme sources	Substrates	Crude extract	Purified preparation	Optimum pH	Ref.
Rat kidney	ϵ-N-Acetyl-L-lysine	0.023	2.59	7.0	11
Hog kidney	ϵ-N-Acetyl-L-lysine	0.05	8.21	8.0	13
Pseudomonas (KT 83)	ϵ-N-Benzoyl-L-lysine	0.32	1,190	6.0	15
Aspergillus oryzae No. 10	ϵ-N-Benzoyl-L-lysine	2.90	31.6	8.2–8.4	18
Achromobacter pestifer EA	ϵ-N-Benzoyl-L-lysine	240	14,500	5.0	4

[a] Liberated lysine micromoles/hour/mg of the protein.

The animal enzyme (rat kidney and hog kidney) cleaves the benzoyl derivative at a lower rate than the acetyl one, and is also susceptible to ϵ-N-α-N-disubstituted derivatives of L-lysine. The enzyme in *Aspergillus oryzae* is inducibly formed by ϵ-N-benzoyl-L-, -D-, or -DL- lysine, and does not have the optical specificity on the hydrolysis of ϵ-N-acyl-DL-lysine. These major differences indicate that the ϵ-lysine acylases from animal, mold, and bacterial sources are quite different enzymes.

[11] W. K. Paik, L. Bloch-Frankenthal, S. M. Birnbaum, M. Winitz, and J. P. Greenstein, *Arch. Biochem. Biophys.* **69**, 56 (1957).

[12] W. K. Paik, *Comp. Biochem. Physiol.* **9**, 13 (1963).

[13] W. K. Paik and L. Benoiton, *Can. J. Biochem. Physiol.* **41**, 1643 (1963).

[14] Y. Kameda, E. Toyoura, Y. Kimura, and K. Matsui, *Chem. Pharm. Bull.* **6**, 394 (1958).

[15] S. Wada, *J. Biochem.* (*Tokyo*) **46**, 445 (1959).

[16] T. Ishikawa, T. Tosa, and I. Chibata, *Agr. Biol. Chem.* **26**, 43 (1962).

[17] I. Chibata, T. Ishikawa, and T. Tosa, *Bull. Agr. Chem. Soc. Japan* **24**, 31 (1960).

[18] I. Chibata, T. Ishikawa, and T. Tosa, *Bull. Agr. Chem. Soc. Japan* **24**, 37 (1960).

Section III

Enzymes Primarily Considered as Transpeptidases

[59] Lobster Muscle Transpeptidase

By Rolf Myhrman and Joyce Bruner-Lorand

Introduction

Tissue coagulin,[1,2] an enzyme found in lobster muscle, induces clot formation when added to the plasma of the same species. The reaction is calcium dependent and may be inhibited by the same amines which prevent the cross-linking of vertebrate fibrin.[3,4] Sulfhydryl reagents such as mercurials and iodoacetate[5] are also inhibitory. The enzyme catalyzes the incorporation of amines such as the fluorescent N-(5-aminopentyl)-5-dimethylamino-1-naphthalene sulfonamide into α-casein.[6] In this respect it is similar to guinea pig liver transglutaminase and the fibrinoligase (activated factor XIII) cross-linking enzyme of vertebrates (see this volume, [59a]).[7-9] Guinea pig liver transglutaminase can substitute for the lobster enzyme in the clotting test described.[10] We thus consider the enzyme to be a sulfhydryl-dependent transpeptidase, and propose the name "lobster muscle transpeptidase." As with the other two related enzymes, it is also capable of hydrolyzing nitrophenyl esters.

Assay Methods

Activity is measured either by the ability of the enzyme to clot plasma or by its effect on the incorporation of N-(5-aminopentyl)-5-dimethylamino-1-naphthalene sulfonamide into α-casein.

Clotting Assay

The standard assay is performed in a 10×75 mm glass tube at room temperature. The enzyme, 0.1 ml, is mixed with 0.1 ml of $0.2\,M$ calcium

[1] J. Glavind, *in* "Studies on the Coagulation of Crustacean Blood" (A. Busuk, ed.). Nyt Nordisk Forlag, Kjobenhavn, 1948.
[2] G. Duchateau and M. Florkin, *Bull. Soc. Chim. Biol.* **36**, 295 (1954).
[3] L. Lorand, R. F. Doolittle, K. Konishi, and S. K. Riggs, *Arch. Biochem. Biophys.* **102**, 171 (1963).
[4] L. Lorand, N. G. Rule, H. H. Ong, R. Furlanetto, A. Jacobsen, J. Downey, N. Oñer, and J. Bruner-Lorand, *Biochemistry* **7**, 1214 (1968).
[5] R. F. Doolittle and L. Lorand, *Biol. Bull.* **123**, 481 (1962).
[6] R. V. Myhrman and L. Lorand, *Biol. Bull.* **135**, 430 (1968).
[7] L. Lorand, T. Urayama, J. W. C. deKiewiet, and H. L. Nossel, *J. Clin. Invest.* **48**, 1054 (1969).
[8] O. Lockridge and L. Lorand, unpublished.
[9] "Report of the Committee on Nomenclature of Blood Clotting Factors." *Thromb. Diath. Haemorrhag. Suppl.* **13**, 428 (1963).
[10] J. Bruner-Lorand, T. Urayama, and L. Lorand, *Biochem. Biophys. Res. Commun.* **23**, 828 (1966).

chloride, and the reaction is initiated 45 seconds later by the addition of 0.2 ml of citrated lobster plasma containing approximately 12 mg of protein. Plasma is added rapidly and the tube is immediately shaken several times by hand to thoroughly mix the contents. Then the tube is gently tipped and inspected at short intervals to detect the first sign of gelation. The clotting time is taken as the interval between addition of plasma and first appearance of gel.

Homarus plasma is obtained by entering the pericardial sinus with a siliconized Pasteur pipette through the first abdominal segment and it is collected by swirling into one fourth volume cold 0.1 M sodium citrate. The blood is then clarified by centrifuging at approximately 2000 g for 5 minutes. The supernatant plasma is decanted and may be frozen and stored in small aliquots (5 ml) at $-30°$ for at least 10 months without loss of activity. Alternately, it may be freeze dried and stored at $-30°$, without loss of activity, for a similar period of time. It is reconstituted by adding water. For duration of short-term experiments, the citrated plasma is simply kept in ice where it retains its ability to clot for about 5 days.

Definition of Unit. A unit of activity is defined as that amount of enzyme which will produce gelation in 100 seconds under the conditions described.

Incorporation of N-(5-Aminopentyl)-5-dimethylamino-1-naphthalene Sulfonamide (or Monodansylcadaverine) into α-Casein

Principle. As this fluorescent amine is incorporated enzymatically into casein, presumably by an amide exchange with some of the γ-glutaminyl residues of the protein, a marked enhancement of fluorescence of the naphthalene residue is observed, as if the chromophore were placed in a more hydrophobic environment. The initial rate of enhancement is proportional to enzyme concentration over a wide range of dilutions for a given enzyme fraction.

The simpler clotting assay is used to monitor the specific activity of preparations during purification of the enzyme.

Reagents

Tris-chloride, 0.05 M, 0.001 M EDTA, pH 7.5 (buffer I)
Calcium chloride, 40 mM
N-(5-Aminopentyl)-5-dimethylamino-1-naphthalene sulfonamide, 1 mM, in 0.05 M Tris-chloride, pH 7.5
Substrate: α-casein (Mann) 7 mg dissolved in 1.0 ml 0.05 M Tris-chloride, pH 7.5 (care must be taken to see that all the casein is dissolved).

Enzyme: Lobster muscle transpeptidase is contained in 0.05 M Tris-chloride, 0.001 M EDTA, pH 7.5, and kept in ice until use.

Procedure. The Turner Model 111 fluorometer is used with $\lambda_A = 360$ nm, $\lambda_F = 546$ nm. The reaction mixture contains: 1.0 ml Tris-chloride; 0.1 ml calcium chloride; 25λ fluorescent amine; 0.1 ml α-casein; 0.1 ml enzyme. The reaction is initiated by the addition of enzyme, and run at 24°C. Control tubes, lacking enzyme, show no incorporation of the fluorescent amine into casein.

Purification Procedure

Principle. The purification of lobster muscle transpeptidase is based on differential centrifugation and DEAE chromatography.

Step 1. Homogenization. Fresh lobster tails (*Homarus americanus*) are cut into 5-g pieces and frozen at −30° for at least 2 weeks prior to use. To prepare the enzyme, frozen 5-g portions are minced in a small beaker with 2 ml cold 0.25 M sucrose, and homogenized for 2 minutes in a total of 10 ml cold sucrose, using a Teflon-glass homogenizer. (Kontes, size D, motorized.)

Step 2. Differential Centrifugation. The homogenate is clarified at 9000 g for 20 minutes (Beckman Model L, No. 30 rotor, 10,000 rpm, 4°). The supernatant is respun at 90,000 g for 4 hours (No. 30 rotor, 30,000 rpm, 4°), and the resulting pelleted enzyme is suspended in cold 0.05 M Tris-chloride, 0.001 M EDTA, pH 7.5 (buffer I) to an optical density of about 60 at 280 nm.

Step 3. Sonication. Five-milliliter aliquots of the above enzyme solution are sonicated in three bursts of 15 seconds each, allowing 15 seconds between bursts for cooling. (Bronwill Biosonik III, with probe, 19 mm diameter, and an intensity setting of 60.)

Step 4. DEAE-Cellulose Chromatography. The sonicate (about 12 ml) is applied to a 19 mm × 20 cm DEAE column (Whatman DE-52) previously equilibrated at 4° with buffer I, and the column is washed with 250 ml of buffer I at an average flow rate of 35 ml/hour. Eight-milliliter fractions are collected. The enzyme is eluted with a linear sodium chloride gradient to 0.4 M in 600 ml, which releases the activity at a sodium chloride concentration of about 0.03 M, as shown in Fig. 1. The fractions of highest specific activity are pooled and concentrated 4-fold at 0°, by the addition of polyacrylamide gel (Lyphogel, Gelman Instrument Co.).

The purification scheme is outlined in the table. Yields and purification factors vary from one preparation to another, but the concentrated DEAE pool represents a 200 to 400-fold purification relative to the 9000-g supernatant at a recovery of 20–40%. The DEAE pool shows two

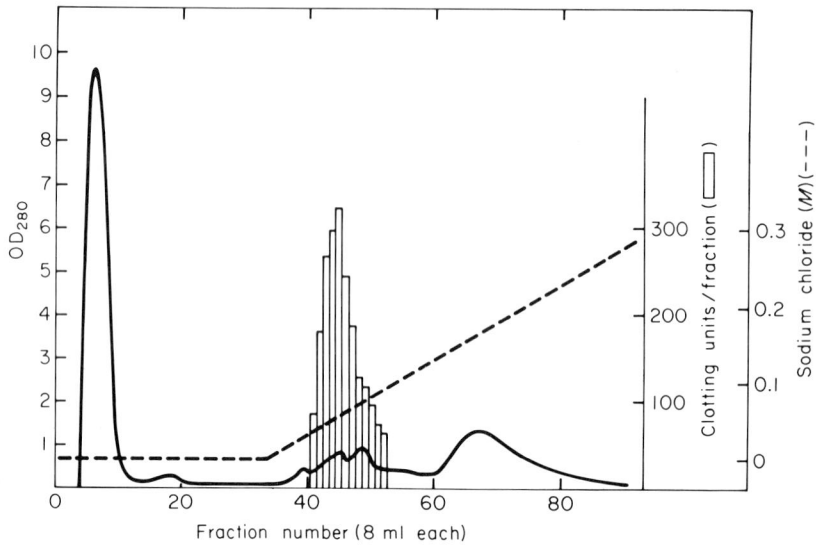

FIG. 1. DEAE-cellulose chromatography of lobster muscle transpeptidase.

PURIFICATION OF LOBSTER MUSCLE TRANSPEPTIDASE

Fraction	Volume (ml)	Clotting (units/ml)	Total units	Protein (mg/ml)	Specific activity	Purification factor	% Yield
9000 g Supernatant	135.0	62.5	8430	53.4	1.17	1	100
90,000 g Supernatant	121.0	27.8	3360	43.1	0.65	—	—
90,000 g Ppt. in buffer I	9.2	2000	16500	109.0	18.4	16	195
Sonicate	7.2	2000	14400	103.0	19.4	17	171
DEAE pool	34.0	133	4520	0.717	186	159	54
DEAE pool concentrated	10.4	315	3280	1.26	250	214	39

major bands in polyacrylamide disc gel electrophoresis at pH 9.3, one of which is essentially eliminated by subjecting the material to sedimentation in a cold sucrose gradient. The pool shows no loss of activity in 3 months if adjusted to 20% glycerol and stored at —8°.

Polyacrylamide Disc Gel Electrophoresis. Gel solutions are made to 100 ml with deionized water and contain the following:

A: Acrylamide	30	g
N,N'-Methylenebisacrylamide	0.8	g
B: Tris (hydroxymethylaminomethane)	18.15	g
1 N Hydrochloric acid	24	ml
N,N,N',N'-Tetramethylethylenediamine	0.24	ml
C: Ammonium persulfate	0.14	g

Gels 6 mm diameter $\times$ 60 mm are poured from a solution containing one part each of A and B and two parts of C. They are allowed to polymerize for 40 minutes at room temperature, and are then equilibrated with buffer III (28.8 g Tris, 6.0 g glycine in a total of 1 liter, pH 9.3) for at least 20 minutes at a current of 2.5 mA per gel. Samples are layered onto the gels, and a current of 2.5 mA per gel is applied for 80 minutes. Gels are then stained in 1% amido black in 7% acetic acid for 30 minutes, and destained electrophoretically in 7% acetic acid.

Protein concentration is estimated at OD_{280} using bovine serum albumin as a standard.

Physical Properties

Isoelectric Point. A gradient spanning pH 4 to 7 is established in the 110 ml LKB electrofocusing column, with the less dense solution in some of the middle fractions replaced by the concentrated DEAE pool. The voltage is raised stepwise from an initial 300 V to a final 700 V, and the column run for a total of 43 hours at 3°–4°. The resulting gradient is collected in 2-ml fractions for the determination of pH and clotting activity. The isoelectric point, as indicated by the peak in clotting activity, is 5.9.

Molecular Weight. A linear sucrose gradient, 10% to 35% in 4.5 ml, is established in a ½ $\times$ 2 inch cellulose nitrate tube using a two-chamber Buchler mixer loaded with 2.6 ml of 8% sucrose in buffer I, and 2.4 ml of 40% sucrose in buffer I. The gradients are cooled to 4°, and a mixture of 0.26 ml concentrated DEAE pool, 10λ beef liver catalase [Mann, 16 mg/ml in 0.05 M Tris-chloride, pH 7.5 (buffer II)], and 10λ rabbit muscle lactic dehydrogenase (Calbiochem, 5 mg/ml in 2.2 M ammonium sulfate) is layered on top. The gradient is spun at 40,000 rpm $\times$ 13 hours in the Beckman SW 50L rotor, and eluted in 28 fractions of 13 drops (approx. 0.17 ml) each. Clotting is assayed as above except that 0.05 ml enzyme fraction and 0.05 ml buffer I are substituted for the usual 0.1 ml enzyme. Catalase is assayed by the loss of OD_{240} of 2.0 ml of a 0.06% solution of hydrogen peroxide in buffer II, in 6 minutes following the addition of 10λ of gradient fraction. Lactic dehydrogenase is assayed by the loss of OD_{340} in the minute following the addition of 0.05 ml of sodium pyruvate (2.5 mg/ml in buffer II) to a solution of 2.0 ml of buffer

II, 0.05 ml of reduced β-diphosphopyridine nucleotide (1.25 mg/ml in buffer II), and 10λ of gradient fraction.

For the further purification of the concentrated DEAE pool by sucrose gradient sedimentation, 0.3 ml of the pool is substituted for the enzyme mix in the above procedure.

A comparison of the distance migrated by lobster muscle transpeptidase and each of the other enzymes yields a sedimentation coefficient of about 10 and a molecular weight of about 200,000 assuming that all three proteins are spherical and have the same partial specific volume.

Ester Hydrolysis

Lobster muscle transpeptidase will catalyze the hydrolysis of p-nitrophenyl acetate in a mixture containing 0.2 ml of buffer I, 0.1 ml of 0.2 M calcium chloride, 0.4 ml of enzyme solution, and 0.1 ml of a 1 mM solution of the ester in acetone. Reactions are initiated by the addition of enzyme, and the increase in nitrophenol is followed at 400 nm. The reaction is inhibited by a 1-hour preincubation of the enzyme with iodoacetamide or iodoacetate at a concentration of 1 mM.

[59a] Fibrinoligase[1]

The Fibrin Stabilizing Factor System

By L. LORAND and T. GOTOH

Introduction

Fibrinoligase is a physiological transpeptidating (i.e., transamidating) enzyme[1a] which catalyzes the formation of selective γ-glutamyl-ϵ-lysine cross links between fibrin molecules.[2-7] In terms of the $(\alpha\beta\gamma)_2$ three-chain fibrin structure[8,9] (α and β contain N-termini of glycine, represent-

[1] This work was aided by a USPHS Research Career Award and by grants from the National Institutes of Health (HE-02212) and the American Heart Association.
[1a] L. Lorand, K. Konishi, and A. Jacobsen, *Nature* **194**, 1148 (1962).
[2] L. Lorand and H. H. Ong, *Biochem. Biophys. Res. Commun.* **23**, 188 (1966).
[3] L. Lorand, H. H. Ong, B. Lipinski, N. G. Rule, J. Downey, and A. Jacobsen, *Biochem. Biophys. Res. Commun.* **25**, 629 (1966).
[4] L. Lorand, J. Downey, T. Gotoh, A. Jacobsen, and S. Tokura, *Biochem. Biophys. Res. Commun.* **31**, 222 (1968).
[5] L. Lorand, N. G. Rule, H. H. Ong, R. Furlanetto, A. Jacobsen, J. Downey, N. Oñer, and J. Bruner-Lorand, *Biochemistry* **7**, 1214 (1968).
[6] S. Matacic and A. G. Loewy, *Biochem. Biophys. Res. Commun.* **30**, 356 (1968).
[7] J. J. Pisano, J. S. Finlayson, and M. P. Peyton, *Science* **160**, 892 (1968).
[8] L. Lorand and W. R. Middlebrook, *Biochem. J.* **52**, 196 (1952).
[9] B. Blomback and I. Yamashina, *Arkiv Kemi* **12**, 299 (1958).

ing the end groups created by the hydrolytic removal of fibrinopeptides A and B from fibrinogen, γ contains an N-terminal tyrosine), cross-linking consists essentially of the pairing of α and γ chains.[3, 10-13] The β chains do not seem to be directly involved.

The cross-linking reaction constitutes the last step in the enzymatic events of normal blood clotting. The enzyme itself appears only in a transient manner during the course of coagulation; no doubt, the continued functioning in the circulation of such an enzyme, which could secondarily act on proteins other than fibrin, cannot be safely tolerated. Normally only the zymogen form [fibrinoligase precursor; fibrin stabilizing factor (FSF) or, in relation to other coagulation components, also called factor XIII[14]] is present in plasma. This constitutes a significant difference which distinguishes fibrinoligase from two other related and somewhat interchangeable enzymes,[15] liver transglutaminase (Vol. XVIIA [127]) and the muscle transpeptidase (this volume [59]), both of which occur as fully active enzymes in their physiological states. In blood, conversion of fibrinoligase precursor to the active enzyme is brought about by a limited proteolysis with thrombin.[4, 16, 17] Thus, the latter hydrolytic enzyme (it itself being only transiently produced in the cascading appearance of enzymes during coagulation) exercises a unique control function not only over the rate of production of the fibrin substrate which is to be cross-linked but also over the timing of the appearance of the enzyme which catalyzes cross-linking. Coordination of these biosynthetic steps, which represent the final enzymatic phase of clotting in vertebrates, can be summarized as follows:

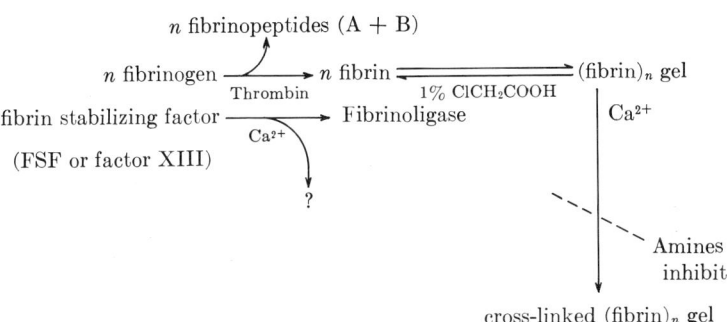

[10] L. Lorand and D. Chenoweth, *Proc. Natl. Acad. Sci. U.S.* **63**, 1247 (1969).

[11] L. Lorand, D. Chenoweth, and R. A. Domanik, *Biochem. Biophys. Res. Commun.* **37**, 219 (1969).

[12] R. Chen and R. F. Doolittle, *Proc. Natl. Acad. Sci. U.S.* **63**, 420 (1969).

[13] T. Takagi and S. Iwanaga, *Biochem. Biophys. Res. Commun.* **38**, 129 (1970).

[14] *Thromb. Diath. Haemorrhag.* Suppl. 13, p. 428 (1963).

We have succeeded in separating the thrombin-catalyzed conversion

$$\text{fibrin stabilizing factor} \xrightarrow[\text{Ca}^{2+}]{\text{thrombin}} \text{fibrinoligase} \qquad (1)$$

of the zymogen, from the later cross-linking reaction which takes place between fibrin (*not* fibrinogen!) and fibrinoligase:

$$(\text{fibrin})_n \text{ gel} \xrightarrow[\text{Ca}^{2+}]{\text{fibrinoligase}} \text{cross-linked (fibrin)}_n \text{ gel} \qquad (2)$$

soluble in 1% ClCH₂COOH   *insoluble* in 1% ClCH₂COOH

Both of these steps may now be studied individually in purified systems.

Reaction (2) serves as the basis of bioassays for fibrinoligase activity. The "cross-linked" fibrin gel is characterized by a greatly increased elastic modulus[18] and it is also very much more resistant both to lytic enzymes[19-22] and to dispersion by added solutes, such as weak acids and bases[23]; urea[24,25] or some neutral salts.[26,27] Most routine assays make use of the latter property and 1% monochloroacetic acid proved to be the best choice[28] for the purpose. The "extent of cross-linking" of a given amount of fibrin (4–5 mg is used in most tests) is estimated by solubilization of the clot in monochloroacetic acid, following a fixed cross-linking period with (varying dilutions of) the enzyme. Complete or partial solubilization of the clot can be seen with the naked eye or may be readily quantitated either by measuring the protein content of the acid-insoluble clot residue or that of the acid-solubilized supernatant. The enzyme required to convert half of the fibrin substrate into a monochloroacetic acid-insoluble gel form is defined as possessing "one cross-linking unit

[15] L. Lorand, J. Bruner-Lorand, and T. Urayama, *Biochem. Biophys. Res. Commun.* **23**, 828 (1966).

[16] L. Lorand and K. Konishi, *Arch. Biochem. Biophys.* **105**, 58 (1964).

[17] K. Konishi and L. Lorand, *Biochim. Biophys. Acta* **121**, 177 (1966).

[18] W. Roberts, L. Mockros, and L. Lorand, *Proc. Intern. Conf. Med. Biol. Eng. 8th* No. 27-1 (1969).

[19] L. Lorand and A. Jacobsen, *Nature* **195**, 911 (1962).

[20] L. Lorand, J. Bruner-Lorand, and T. R. E. Pilkington, *Nature* **210**, 1273 (1966).

[21] L. Lorand, in "Dynamics of Thrombus Formation and Dissolution" (S. A. Johnson and M. M. Guest, eds.), pp. 212–244. Lippincott, Philadelphia, 1969.

[22] L. Lorand, *Thromb. Diath. Haemorrhag., Suppl.* XXXIX, p. 75, 1970.

[23] K. C. Robbins, *Am. J. Physiol.* **142**, 581 (1944).

[24] L. Lorand, *Hung. Acta Physiol.* **1**, 192 (1948).

[25] L. Lorand and K. Laki, *Science* **108**, 280 (1948).

[26] S. Shulman, S. Katz, and J. D. Ferry, *J. Gen. Physiol.* **36**, 759 (1953).

[27] T. H. Donnelly, M. Laskowski, Jr., N. Notley, and H. A. Scheraga, *Arch. Biochem. Biophys.* **56**, 369 (1955).

[28] L. Lorand, *Nature* **166**, 694 (1950).

(XLU)" of activity.[28a] Perhaps it should be pointed out that "cross-linking," as judged by the criterion of clot solubility has a rather complex meaning and that, quantification and reproducibility notwithstanding, bioassays based on this property must be regarded as empirical at best. A fibrin clot appears to be totally insoluble in monochloroacetic acid when only about 60% of the protein is "cross-linked" and the rest is still monomeric.[4, 22] Furthermore, the cross-linked portion itself consists of a spectrum of oligomeric clusters of fibrin, in which the dimeric species predominates. Nevertheless, the clot solubility test proved to be of great use, particularly in the evaluation of specific inhibitors of fibrin cross-linking.[5]

We have also developed (and are still in the process of developing) various rate assays for measuring the activity of fibrinoligase. These fall into two categories: (a) those based on the hydrolytic activity; and (b) those on the transpeptidating property of the enzyme. p-Nitrophenyl esters (e.g., acetate or trimethylacetate) have been used as hydrolytic substrates[29]; obviously, however, measurements have significance only in conjunction with the enzyme in its highest purity. The transpeptidating reaction utilizes synthetic amines of various types which were shown to be specific inhibitors in the enzyme-catalyzed cross-linking of fibrin and were also shown to act so by virtue of becoming incorporated into the acceptor function of fibrin itself. Various types of amine tracers could be used: hydroxylamine, hydrazine[30] as chemical markers; glycine ethyl ester,[30, 31] histamine as radioisotopes[32]; N-(5-aminopentyl)-5-dimethyl-aminonaphthalene-1-sulfonamide (or "monodansylcadaverine") as a fluorescent label.[3, 5] On account of the favorable $K_{m,\text{app}}$ (ca. $2 \times 10^{-4} M$) of the latter substrate, and because of the great sensitivity and the ease of fluorescent measurements, most quantitative data so far are based on the incorporation of monodansylcadaverine. This is particularly so in regard to the clinical diagnostic tests in plasma relating to the hereditary disease of fibrin stabilizing factor deficiency.[32, 33] The scope of the test for other clinical applications is being enlarged.

[28a] L. Lorand, in "Blood Coagulation, Hemorrhage and Thrombosis, Methods of Study" (L. M. Tocantins and L. A. Kazal, eds.), p. 239. Grune & Stratton, New York, 1964.

[29] O. Lockridge, Ph.D. dissertation, Northwestern University, Evanston, Illinois, 1970.

[30] L. Lorand and H. H. Ong, Biochemistry 5, 1747 (1966).

[31] L. Lorand and A. Jacobsen, Biochemistry 3, 1939 (1964).

[32] L. Lorand, T. Urayama, J. deKiewiet, and H. A. Nossel, J. Clin. Invest. 48, 1054 (1969).

[33] L. Lorand, T. Urayama, A. C. Atencio, and D. Y. Y. Hsia, Am. J. Human Genetics 22, 89, 1970.

Both fibrin and casein may be used as amine acceptors; the test devised for clinical application makes use of casein.[32, 33] Interestingly, fibrinogen is a much poorer acceptor substrate than fibrin. The hydrolytic removal of the small fibrinopeptide A fragment is the prerequisite for unmasking the incorporating sites of the protein.[3, 21, 22, 30]

Assays for the precursor (i.e., fibrin stabilizing factor) are the same as for the ligase, except that the thrombin-catalyzed conversion step (reaction 1) has to be carried out prior to measuring the generated enzyme activity either by the fibrin cross-linking, ester hydrolysis or the amine incorporation test. In the present description, two different methods are given for the assay of the zymogen. Its potency is expressed in the same units [i.e., "cross-linking units" (XLU) and "amine incorporating units" (AIU)] as used for the fibrinoligase enzyme itself. If the latter alone is to be assayed, the activation step with thrombin is simply omitted.

Assay Methods

Bioassay, Based on Clot Solubility, for the Fibrin Stabilizing Factor

The fibrin substrate is prepared essentially as given by Donnelly *et al.*,[27] an adaptation of the method of Lorand and Middlebrook.[8] The starting fibrinogen may be of bovine (e.g., Armour fraction I precipitated with 0.25 saturated ammonium sulfate, as prescribed by Laki[34]) or of human origin (e.g., Kabi). A 0.1% solution of fibrinogen in 0.05 M Tris-0.1 M NaCl solution at pH 7.5 (with HCl) is allowed to stand with 1 mM iodoacetic acid, and clotting is initiated by the addition of thrombin (ca. 20 NIH units per 100 ml of solution). The fibrin clot formed over a 30-minute period is collected on a glass rod, gently squeezed of excess liquid and is taken up in 1/30 of the volume of the original fibrinogen in 2 M NaBr adjusted to pH 5.3 with acetic acid. When fully dissolved, water is added to lower the NaBr concentration to 1 M. The protein solution is then poured into 20 times its volume of a phosphate buffer of pH 6.5 (0.03 M KH$_2$PO$_4$, 0.015 M Na$_2$HPO$_4$, and 0.09 M KCl) where the change in pH and ionic strength causes it to precipitate. The fibrin precipitating over a 2-hour period is wound out on a glass rod as before and is dissolved in 1/50 of the volume of starting fibrinogen in the 2 M NaBr (pH 5.3) solvent. It is then dialyzed overnight in the cold ($\sim 4°$) against 1 M NaBr of pH 5.3. (All the earlier steps are carried out at room temperature.) Any precipitate present after dialysis should be removed by centrifugation. The protein concentration is adjusted by dilution with the 1 M NaBr to about 1.5% (absorbancy at 280 nm for

---

[34] K. Laki, *Arch. Biochem.* **32**, 317 (1951).

1% protein is 15.12[35]). The fibrin solution may be stored at 0°. We arbitrarily set an expiration date of about 3 weeks.

Different (chromatographically purified) thrombin preparations (bovine or human[36]) may be used for preparing fibrin (as above) as well as for activating the fibrin stabilizing factor in the test. Thrombin is diluted to the required strength with the 0.05 M Tris–0.1 M NaCl buffer of pH 7.5. (For definition of NIH units of thrombin activity see this volume, [8], [9], and [9a].)

Cysteine and tosylarginine methyl ester (TAME) are each freshly made up to 0.1 M in the 0.05 M Tris–0.1 M NaCl and the pH of the solutions is adjusted to pH 7.5 (with a few drops of 1 N NaOH if necessary). One millimolar aqueous $CaCl_2$ is used. Purification of the fibrin stabilizing factor zymogen is described separately. It is made up in the Tris-NaCl solution of pH 7.5. Monochloroacetic acid is prepared as a 2% aqueous solution.

In its most useful version the multistage assay consists of the following main steps:

(1) Activation of the zymogen by thrombin, in the presence of calcium and cysteine.

(2) Quenching of thrombin by the addition of the competing substrate, TAME. (For less precise measurements, this step is often omitted.)

(3) Cross-linking of fibrin by the enzyme generated in step (1).

(4) Testing of the solubility of the fibrin, after the cross-linking period, in 1% monochloroacetic acid.

A set of mixtures, containing various amounts of the fibrin stabilizing factor, is prepared in 18-mm wide test tubes at room temperature. Into each tube is pipetted 0.5 ml of $CaCl_2$, 0.5 ml of cysteine, 0.6 ml of the fibrin stabilizing factor, and 0.1 ml of thrombin. After 10 minutes of activation, 0.5 ml of TAME is added and is followed by the admixing of 0.3 ml of fibrin solution, whereupon uniform gelation ensues. Half an hour later, 2.5 ml of 2% monochloroacetic acid is introduced in order to test the solubility of the fibrin clot.[28, 28a] Swirling of the tubes aids the penetration of the acid; however, breaking up of the clot is to be avoided if the extent of solubility is to be judged visually. Various concentrations of fibrin stabilizing factor are selected such that the test tubes would yield a range of clots from the fully soluble to the completely monochloroacetic acid-insoluble type. The "threshold" activity producing half an acid-insoluble clot represents 1 cross-linking unit (XLU).

[35] P. Johnson and E. Mihalyi, *Biochim. Biophys. Acta* **102**, 476 (1965).
[36] P. S. Rasmussen, *Biochim. Biophys. Acta* **16**, 157 (1955).

Tubes for residual clots are read 2 hours after the addition of mono-chloroacetic acid and are checked again some 16 hours later. If quantitation is required, the remaining clot residues are sedimented by centrifugation. Optical density of the supernatants may be read at 280 nm and expressed in terms of either fibrin or standardized crystalline bovine serum albumin quantities. Alternately, the sediments are washed twice with 0.15 M NaCl, centrifuged, and digested in NaOH for colorimetric measurement by the Folin-Ciocalteau[37] method.

Amine Incorporation Test for the Fibrin Stabilizing Factor

An outline of the coupled rate assay may be given as:

$$\text{casein} + \text{H}_2\text{NR} \xrightarrow[\text{Ca}^{2+}]{\text{fibrinoligase}} \text{casein-NHR} + \text{(H)}$$

with the overhead notation:
fibrin stabilizing factor (factor XIII)
Ca^{2+} | Thrombin

As mentioned above, the amine (H_2NR) in the particular test to be described was chosen as the fluorescent monodansylcadaverine. When concentrations of Ca^{2+}, thrombin, casein, and that of the amine are optimized together with pH and time of conversion of the factor to the ligase enzyme, the rate of formation of casein-bound amine reflects the original concentration of fibrin stabilizing factor. If the activity of the preactivated fibrinoligase is to be measured, the simplified

$$\text{casein} + \text{H}_2\text{NR} \xrightarrow[\text{Ca}^{2+}]{\text{fibrinoligase}} \text{casein-NHR} + \text{(H)}$$

system is applicable.

For highly purified fibrin stabilizing factor and fibrinoligase preparations, the continuous rate assay (described in this volume, [59]) may be used to good advantage. However, for measuring the fibrin stabilizing factor content of biological fluids, such as human plasma, a sampling procedure has been worked out with the following details.[32]

Desensitization of the Test Plasma. (This step is unnecessary when testing purified fibrin stabilizing factor.) Mix 0.2 ml of citrated plasma with 0.05 ml of 50% (v/v) aqueous glycerol in a 16 × 100 mm (Sorvall Pyrex) test tube. Bring rapidly to 56° and keep at this temperature for 2.5 minutes. Then by immersion in ice, allow it to cool to room temperature (24°).

Activation of Fibrin Stabilizing Factor with Thrombin. Add 0.05 ml of 0.2 M glutathione solution [dissolved in 50% (v/v) aqueous glycerol

---

[37] A. Hirsch and C. Cattaneo, *Z. Physiol. Chem.* **304,** 53 (1956).

and adjusted to pH 7.5 with $3 N$ NaOH]. Activation of factor XIII is initiated by admixing 0.2 ml of thrombin [125 NIH units/ml; dissolved in 25% (v/v) aqueous glycerol and $0.02 M$ $CaCl_2$; pH 7.5] and it is allowed to proceed for 20 minutes.

Amine Incorporation. Add 0.5 ml of the synthetic amine substrate, 2 mM monodansylcadaverine[5] solution in $0.05 M$ Tris-HCl of pH 7.5 and 3 mM $CaCl_2$. The incoporation reaction is started with the addition of 1.0 ml of 0.4% casein solution; made up in a mixture containing 10% glycerol, 3 mM $CaCl_2$, $0.05 M$ Tris-HCl at pH 7.5 [in preparing the casein (Hammarsten) solution, it is clarified after dialysis by centrifuging at 20,000 rpm for 60 minutes in a Spinco L No. 30 rotor].

Incorporation Measurement. Amine incorporation is stopped 30 minutes later by adding 2 ml of 10% trichloroacetic acid. The protein precipitate is washed successively with 6×10 ml of ethanol–ether (1:1) and dried. Finally, it is taken up in a 2 ml solution of $8 M$ urea, 0.5% sodium dodecylsulfate in $0.05 M$ Tris-HCl at pH 8, for measuring the protein-bound fluorescent amine.

Fluorescent intensities are obtained with excitation at approximately 355 nm; emission at approximately 525 nm. All fluorescence measurements are carried out in the mixture containing $8 M$ urea and 0.5% sodium dodecyl sulfate in $0.05 M$ Tris buffer at pH 8. Monodansylcadaverine at a concentration of 1 micromole/liter in the same solvent system serves as reference.

A given rate of amine incorporation is conveniently expressed in terms of units. It is of further advantage to normalize these unit values for 1 ml of citrated plasma.

Using 0.2 ml of citrated plasma in a 2-ml test mixture, units are defined in the following manner:

$$\text{AIU per ml of citrated plasma} = \frac{2}{0.2} \times \frac{1}{3} \times \frac{1}{0.9} \left[\frac{i_T - i_B}{i_R - i_S} \right]$$

$$= 3.7 \left[\frac{i_T - i_B}{i_R - i_S} \right]$$

The first term on the left accounts for the 10-fold dilution of plasma in the test; the second simply serves to reduce the numerical values into a convenient range and may be considered to express units in terms of a 10-minute incorporation instead of the measured 30 minutes; the third is to correct for the estimated 10% loss of factor during heat desensitization plasma. If units are to be expressed for original plasma and not the citrated one, another correction for dilution wth the anticoagulant should be included.

The symbols within the brackets represent the measured fluorescent intensities (i) of the test sample (T) after 30 minutes of amine incorporation; of the blank (B) control mixture with only the monodansyl-cadaverine missing, but otherwise similarly treated; of a 1 μM reference (R) solution of free monodansylcadaverine and finally of the mixed (urea-sodium dodecyl sulfate) solvent (S) system itself.

Purification Procedure

Fibrin Stabilizing Factor (Factor XIII) from Plasma

Plasma, rather than serum serves as the source for isolating the fibrin stabilizing factor zymogen.[38,39] (Again, it should be recalled that no trace of the active fibrinoligase enzyme is present in normal plasma.) Both oxalated bovine plasma (Pentex, Kankakee, Ill.) or citrated human plasma from a blood bank can be readily used as starting materials. The factor appears to be quite stable in freshly frozen plasma over extended periods (months) of storage.

Purification procedures consist of two phases: (1) preparation of crude concentrates by one (or a combination) or several precipitation steps; and (2) chromatography on DEAE-cellulose.

Precipitation procedures are based on the use of ether,[38,39] alcohol,[40] ammonium sulfate,[40,41] polyethyleneglycol[42] or isoelectric fractionation at pH 5.4.[38-40] Since the factor combines readily with fibrinogen,[43] a mild heat treatment was developed[41] for the differential precipitation of the latter protein. This does not seem to affect the potency of the factor; however, it is not known whether its fine structure is modified. In any case, it is a useful step in preparing concentrates.

The following is a well-proven procedure for production on a laboratory scale:

To 8 liters of thawed plasma, 2 liters of saturated ammonium sulfate solution is added (4°) and the precipitate formed in an hour is removed by centrifugation (International head No. 858; 8000 rpm × 20 minutes, 0°). The sediment is taken up in 0.15 M KCl and the 1250 ml solution is adjusted to pH 5.4 with 1 N acetic acid. Then 240 ml of saturated ammonium sulfate is added again and the precipitate is removed as before. It is dissolved in 0.05 M Tris–0.1 M NaCl buffer, with the total volume being 450 ml and the pH of the solution set to 7. (This concentrate may be stored for several days in the cold room which is particularly conve-

[38] L. Lorand, Doctoral thesis, Leeds University, England, 1951.
[39] L. Lorand, Physiol. Rev. 34, 742 (1954).
[40] L. Lorand and A. Jacobsen, J. Biol. Chem. 230, 421 (1958).
[41] A. G. Loewy, C. Veneziale, and M. Forman, Biochim. Biophys. Acta 26, 670 (1957).
[42] M. Kazama and R. D. Langdell, Federation Proc. 28, 746, No. 2725 (1969).
[43] L. Lorand, in "Anticoagulants and Fibrinolysins" (R. L. MacMillan and J. F. Mustard, eds.), p. 333. Macmillan, Canada, 1961.

nient if several batches of the above preparation are to be pooled for chromatography.)

The solution is diluted 2-fold with the $0.05\,M$ Tris–$0.1\,M$ NaCl buffer and (in 225-ml aliquots) it is heated to $56°$ for 3 minutes (by rapid immersion in an $80°$ water bath). While at $56°$, the denatured fibrinogen is removed with the aid of a glass rod and the supernatant is squeezed out as completely as possible and is cooled to $15°$. Small floating fragments are removed by centrifugation.

The supernatant (ca. 460 ml) is brought to $4°$ and is mixed with 260 ml of saturated ammonium sulfate. The precipitate is removed by centrifuging, as in earlier steps, and it is dissolved by adding 40 ml of an $0.05\,M$ Tris–$0.001\,M$ EDTA buffer set to pH 7.5 with HCl. The 42-ml solution is dialyzed against the buffer (2×1 liter) overnight in the cold room.

Fibrin Stabilizing Factor Zymogen on DEAE-Cellulose

Both stepwise[44] and gradient elution[4,43] procedures have been reported. On account of the good reproducibility of eluting fibrin stabilizing factor with a salt gradient at pH 7.5 and because of our extensive experience with this methodology over the past 5 years, only this technique is recommended.

DEAE-cellulose (Whatman DE-52, microgranular, preswollen) is equilibrated with $0.05\,M$ Tris–$0.001\,M$ EDTA set to pH 7.5 with HCl and made into a column of approximately 3×24 cm. Washing with the buffer is stopped when the pH and conductivity of the effluent is identical to that of the original. The chromatographic operations are carried out at $4°$. Flow rate is adjusted to about 32 ml per hour and approximately 8-ml fractions are to be collected. The fibrin stabilizing factor concentrate isolated earlier is then applied. Protein contents of these crude preparations vary appreciably. As much as 400 mg in 80 ml can be applied. (For larger quantities, we have successfully used a 4×60 cm column.) About 500 ml of buffer is allowed to pass so as to elute the protein impurities not retained by the DEAE-cellulose. Then a linear gradient is applied with the proximal chamber containing 600 ml of the $0.05\,M$ Tris–$0.001\,M$ EDTA at pH 7.5 and the distal one the same volume of $0.3\,N$ NaCl in the $0.05\,M$ Tris–$0.001\,M$ EDTA buffer. (With the larger column 1300 ml solution is placed in each mixing chamber.)

Figure 1 shows an elution pattern which is typical only in the position of the fibrin stabilizing factor activity on the gradient. The size of the impurities emerging before the gradient is applied as well as those eluting at higher salt concentration varies appreciably from one preparation to

[44] A. G. Loewy, K. Dunathan, R. Kriel, and H. L. Wolfinger, Jr., *J. Biol. Chem.* **236**, 2625 (1961).

another. The active peak is pooled and concentrated with 0.36 saturation of ammonium sulfate. The precipitate is dissolved in minimal (2–3 ml) volume of 0.05 M Tris–0.1 M NaCl–0.001 M EDTA, pH 7.5 buffer (with HCl) and is dialyzed against the same. The solution is stored at −10° after adding glycerol (20%).

Zonal Centrifugation and Disc Gel Electrophoresis of Fibrin Stabilizing Factor

Routinely, sucrose-density centrifugation[4,22] of the chromatographed material (Fig. 2) reveals only minor impurities which sediment slower than the fibrin stabilizing factor activity. Tentatively these are identified as byproducts of the spontaneous activation of the factor to fibrinoligase. The latter sediments indistinguishably from the zymogen and does occasionally arise to some extent.

The fractions with uniform specific activity obtained by zonal centrifugation give a single band in disc gel electrophoresis,[4] unless some of the spontaneously activated fibrinoligase is present.

Sucrose density centrifugation proved to be so useful that it is often used as a final preparative step of purifying fibrin stabilizing factor. A linear gradient (4.5 ml) to 40% sucrose is formed in 0.05 M Tris–0.1 M NaCl–0.001 M EDTA buffer at pH 7.5 and 0.5 ml of chromatographed fibrin stabilizing factor is applied. Centrifugation (Spinco, SW 50 or 65 rotors) is carried out at 44,000 rpm for 18.5 hours at 0°.

Properties of Fibrin Stabilizing Factor

In both chromatographic behavior on DEAE-cellulose and in sedimentation properties, the zymogen as well as the fibrinoligase enzyme itself are strikingly different from liver transglutaminase, an enzyme of related activity in many respects.[15] Transglutaminase elutes at much higher salt concentration[45] and sediments slower in sucrose density centrifugation.[29] With catalase and lactic dehydrogenase, fibrin stabilizing factor (and fibrinoligase, too), sediments between the two markers, whereas transglutaminase is considerably slower than lactic dehydrogenase.[22,29] In chromatography and sedimentation, fibrinoligase and its precursor show great similarity, however, with the transpeptidase (see this volume [59]) isolated from lobster muscle. It must be recalled that the latter is not known to occur in the form of a zymogen, but only as the active enzyme.

There are a number of kinetic dissimilarities[29] among the three transpeptidating enzymes mentioned.

[45] J. E. Folk and P. W. Cole, *J. Biol. Chem.* **241**, 5518 (1966).

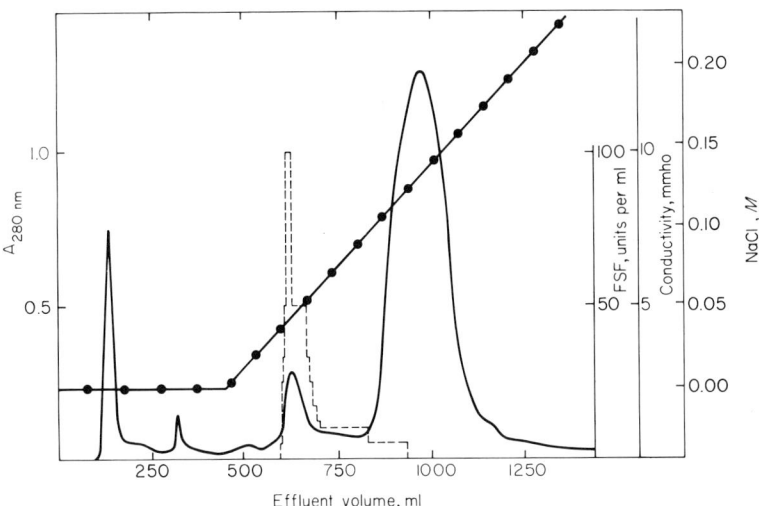

Fig. 1. Salt gradient-elution chromatography of fibrin stabilizing factor (bovine or human) from DEAE-cellulose. ———, Absorbancy at 280 nm; - - - - fibrin stabilizing factor activity in terms of cross-linking units (XLU); ●———●, conductivity or NaCl concentration.

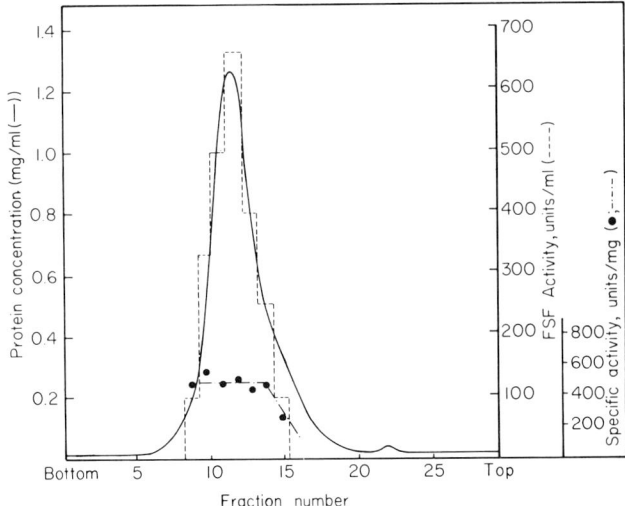

Fig. 2. Sucrose-density centrifugation of chromatographed fibrin stabilizing factor (bovine or human). Activity is given in terms of cross-linking units.

Conversion of the Fibrin Stabilizing Factor Zymogen
to the Fibrinoligase Enzyme

Thrombin, as a hydrolytic enzyme, serves the purpose of physiologically converting the zymogen into the active transpeptidase.[1,16] Disc gel electrophoresis[4] as well as electrofocusing[22] may be used to monitor the conversion and to analyze for fragments which seem to be released as byproducts of activation.

The activating thrombin can be removed by gel filtration.[17] Trypsin[46] and reptilase[47] also seem capable of converting the zymogen to active fibrinoligase. Solid-phase enzymes such as water-insoluble trypsin have been used.[4,47]

[46] K. Konishi and T. Takagi, *Abstr. 7th Intern. Congr. Biochem., Tokyo, 1967*, J. 381.
[47] M. Kopec, Z. S. Latallo, M. Stahl, and Z. Wegrzynowicz, *Biochim. Biophys. Acta* **181**, 437 (1969).

[60] γ-Glutamyl Transpeptidase

By G. E. CONNELL and E. D. ADAMSON

Glutathione + gly-gly → γ-glu-gly-gly + cySH-gly
2 Glutathione → γ-glu-γ-glu-cySH-gly + cySH-gly
Glutathione + H_2O → glutamic acid + cySH-gly

γ-Glutamyl transpeptidase from most sources catalyzes the transfer of the γ-glutamyl group from glutamine or a variety of substituted γ-amides of glutamic acid to many primary amines, most notably, the α-amino acids and small peptides.[1,2] γ-Glutamyl amides and peptides may be both acceptors and donors in the reaction.

Assay

Principle. The method described below is a rapid spectrophotometric method devised by Orlowski and Meister.[3,4] When a preparation of γ-glutamyl transpeptidase is incubated with γ-L-glutamyl-*p*-nitroanilide, *p*-nitroaniline is liberated. The colorless substrate exhibits a maximum absorbancy at 315 nm and no absorbancy at 410 nm, while *p*-nitroaniline exhibits high absorbancy in the range 350 to 420 nm. The release of *p*-nitroaniline in reaction mixtures may be followed directly by continuous photometric recording and is linear for several minutes at moderate enzyme concentrations.

[1] C. S. Hanes, F. J. R. Hird, and F. A. Isherwood, *Nature* **166**, 288 (1950).
[2] C. S. Hanes, F. J. R. Hird, and F. A. Isherwood, *Biochem. J.* **51**, 25 (1952).
[3] M. Orlowski and A. Meister, *Biochim. Biophys. Acta* **73**, 679 (1963).
[4] M. Orlowski and A. Meister, *J. Biol. Chem.* **240**, 338 (1965).

Reagents. The substrate is prepared so that 0.90 ml contains 5 micromoles γ-L-glutamyl-*p*-nitroanilide, 10 micromoles MgCl₂, 100 micromoles Tris-HCl buffer, pH 9.0. The substrate is sparingly soluble, and heating to 50°–60° may be required to achieve complete solution.

Synthesis of L-γ-*glutamyl-p-nitroanilide.* This compound is prepared by a modification of the general procedure of King and Kidd[5]; 13.2 g (0.051 mole) phthaloyl-L-glutamic anhydride[6,7] is suspended in 35 ml glacial acetic acid, and 6.9 g (0.05 mole) *p*-nitroaniline is added. The mixture is heated at 60°–65° for 30 minutes and then concentrated under reduced pressure at 25° to yield a crude product which is dried in a vacuum desiccator over CaCl₂ and NaOH. The product is ground to a fine powder and dissolved in 250 ml methanol. Hydrazine hydrate (80%, 0.1 mole) is added, and the solution filtered to remove a small amount of insoluble material. It is allowed to stand 48 hours at 26° while a crystalline product separates. The crystals are washed on a Büchner funnel with cold distilled water and with ethanol, then dried in a vacuum over CaCl₂. The material is suspended with vigorous agitation in 10 parts of cold 0.5 *N* HCl and the mixture filtered into a separatory funnel. Free *p*-nitroaniline is removed by extraction with 5 volumes of ethyl acetate, after which the aqueous layer is quickly filtered and brought to pH 7 by addition of 1 *M* Na₂CO₃. A crystalline product forms immediately; this is washed with water and with ethanol and then dried in a vacuum over CaCl₂. The yield is 8.2 g (61%), m.p. 183°–186°, ε at 315 nm = 13,000.

Procedure. To 0.90 ml buffered substrate solution in a cuvette of 1-cm light path, held at 37°, 0.10 ml enzyme solution is added with rapid stirring. Using a photometer with or without a recording device, the increase in absorbance (Δ*A*) at 410 nm is determined for 3–5 minutes and the average Δ*A* per minute calculated. This figure is converted to micromoles of *p*-nitroaniline/minute using the molar extinction coefficient of *p*-nitroaniline at 410 nm of 8800 M^{-1} cm^{-1}.

In an alternative procedure, the enzyme and buffered substrate may be incubated for a fixed time, e.g., 3–5 minutes; 2 ml 1.5 *N* acetic acid is added to terminate the reaction and precipitate the protein. The absorbance is then measured at 410 nm against a reference solution containing the same components, except that the enzyme is added after addition of acetic acid. This procedure may be useful in working with crude enzyme preparations which would give turbid reaction mixtures. Szasz[8] has suggested minor modifications of the *p*-nitroanilide method for use in clinical

[5] F. E. King and D. A. A. Kidd, *J. Chem. Soc.* p. 3315 (1949).
[6] J. C. Sheehan and W. A. Bolhofer, *J. Am. Chem. Soc.* **72**, 2469 (1950).
[7] F. E. King, J. W. Clark-Lewis, and R. Wade, *J. Chem. Soc.* p. 886 (1957).
[8] G. Szasz, *Clin. Chem.* **15**, 124 (1969).

laboratories. The quantitative aspects of several assay procedures have been compared by Dimov and Kulhanek.[9]

Definition of Unit and Specific Activity. One unit of γ-glutamyl transpeptidase activity is defined as the amount of enzyme required to release 1 micromole *p*-nitroaniline per minute under the conditions of the assay.

Specific activity is expressed in units per milligram protein as determined by the method of Lowry *et al.*[10]

Preparation of Enzyme[11]

Beef kidneys (4 kg) free of connective tissue, which have been stored frozen at $-15°$, are homogenized in a Waring blendor with 12 liters 0.9% NaCl solution at 4°. The homogenate is then centrifuged at 900 g and the supernatant retained.

Step 1. Acetone and Sodium Deoxycholate Treatment. The supernatant is treated with half its volume of cold acetone at 4°, the precipitate spun off at 900 g and washed with 2 volumes 0.9% NaCl mixed with 1 volume acetone. The precipitate is suspended in 2 volumes 0.9% NaCl and for each liter of suspension 50 ml 20% sodium deoxycholate solution is added. The mixture is homogenized for 5 minutes and left overnight at 4°. After centrifugation at 10,000 g the precipitate is discarded.

Step 2. Ammonium Sulfate Precipitation. To the clear supernatant, 472 g solid ammonium sulfate per liter is added to give 70% saturation. After standing overnight the precipitate is spun off, suspended in a small volume of water, and dialyzed for 48 hours against tap water. The residual precipitate is removed by centrifugation.

Step 3. Butanol, Heat, and Acetone Treatment. To the supernatant 0.4 volume cold *n*-butanol is added slowly over 10 minutes and the mixture is warmed for 15 minutes at 38°. The biphasic mixture is centrifuged and the lower water layer is collected and cooled. The protein dissolved in the water layer is precipitated by the addition of 1.5 volumes of cold acetone, the precipitate is spun off, suspended in a small volume of water and dialyzed overnight against distilled water. The solution is clarified by centrifugation and the precipitate discarded.

Step 4. Second Ammonium Sulfate Precipitation. The protein which salts out between 52 and 70% saturation in fractionation with solid ammonium sulfate is collected, suspended in a small amount of water, and dialyzed for 48 hours at 4° against 0.01 M Tris-citric acid buffer, pH 8.9.

[9] D. M. Dimov and V. Kulhanek, *Clin. Chim. Acta* **16,** 271 (1967).
[10] O. H. Lowry, N. J. Rosebrough, A. L. Farr, and R. J. Randall, *J. Biol. Chem.* **193,** 265 (1951).
[11] A. Szewczuk and T. Baranowski, *Biochem. Z.* **338,** 317 (1963).

Step 5. DEAE-Cellulose Treatment. The clear supernatant is applied, at 4°, to a column (0.9 × 30 cm) of DEAE-cellulose (0.8 meq/g) previously equilibrated with 0.01 M Tris-citric acid, pH 8.9, at a flow rate of 5 ml per hour. The same buffer is used for elution, and the active fractions are combined and dialyzed against 0.01 M sodium acetate buffer, pH 5.5.

Step 6. Second DEAE-Cellulose Column. The product of the previous column (55 mg) is applied to a DEAE-cellulose column (1.5 × 30 cm) previously equilibrated with 0.01 M sodium acetate buffer, pH 5.5. The column is eluted with a stepwise gradient of (a) 0.01 M buffer, (b) 0.1 M buffer, (c) 0.2 M buffer, (d) 0.1 M NaCl in 0.2 M buffer, (e) 0.2 M NaCl in 0.2 M buffer, and (f) 0.5 M NaCl in 0.2 M buffer. A flow rate of 30 ml/hour is used and fractions of 10 ml are collected. One active fraction of the enzyme A is eluted with 0.2 M buffer and a second active fraction B with 0.1 M NaCl in 0.2 M buffer. Fraction A has 34% and fraction B has 52% of the activity of the preparation applied to the final column. The properties of the preparations at various stages of purification are summarized in the table.

SUMMARY OF PURIFICATION OF γ-GLUTAMYL TRANSPEPTIDASE

Step	Preparation	Volume (ml)	Total protein	Specific activity[a] (units/mg protein)	Yield (%)
	Centrifuged homogenate	11,000	390 g	0.078	100
1	Deoxycholate extract	4,700	27.7 g	0.69	63
2	1st Ammonium sulfate precipitation	550	8.6 g	1.55	44
3	Butanol treatment	53	1.22 g	9.6	38
4	2nd Ammonium sulfate precipitation	37	470 mg	17.6	27.2
5	1st DEAE-cellulose treatment	43	77 mg	66.5	16.8
6	2nd DEAE-cellulose column				
	Fraction A	—	21 mg	81.5	14.3
	Fraction B	—	31 mg	85.5	

[a] The specific activities quoted in this table are based upon an assay using γ-glutamyl-α-naphthylamide as substrate rather than γ-L-glutamyl-p-nitroanilide. The values obtained are slightly lower than those which would be obtained using the standard procedure.

Properties

Specificity. γ-Glutamyl transpeptidase has absolute specificity with respect to the donor substrate for the glutamyl moiety in γ-linkage. It will not, for example, act upon L-asparagine, L-homoglutamine, or the *p*-nitroanilide of β-aminoglutaric acid.[4] Among the γ-glutamyl peptides which have been shown to be donor substrates are glutathione, in both the oxidized and reduced form,[2] and a number of dipeptides, for example, γ-glutamylglycine, -phenylalanine, -tyrosine, -glutamic acid, -aspartic acid. Several γ-glutamylarylamides also are donor substrates, namely, the derivatives of α- and β-naphthylamine, aniline, and *p*-nitroaniline.[3, 12–14] Glutamine has been shown to be a γ-glutamyl donor using highly purified transpeptidase preparations from beef[11] or hog[4] kidney. Glutamine is not a donor substrate for the sheep kidney enzyme in partly purified preparations.[2, 15] Peptides containing D-glutamic acid in γ-linkage are substrates but the reaction rates are much lower than with the analogous L-glutamyl peptides.

The transpeptidase has broad specificity with respect to the nature of the acceptor amine. All of the naturally occurring L-amino acids which have been tested proved to be active.[2, 16–18] A substantial number of dipeptides and several tripeptides are also known to be active.[2, 17] Glutathione itself will act as an acceptor as well as a donor substrate.[19] Evidence has been obtained for the synthesis by the enzyme of tetra-, penta-, or higher peptides of glutamic acid in γ-linkage beginning with a simple γ-glutamyl substrate.[4, 11] Orlowski and Meister have shown that hydroxylamine is an acceptor substrate of transpeptidase and that the product is the γ-hydroxamate of glutamic acid.[4]

Most preparations of γ-glutamyl transpeptidase catalyze hydrolytic as well as transfer reactions.[2, 4, 16, 17] After prolonged incubation all γ-glutamyl peptides in a reaction mixture may be hydrolyzed completely to free glutamate.[11] The pH optimum for the hydrolytic reaction is lower than that for the transfer reaction. The values for the pig kidney enzyme acting on γ-glutamyl-*p*-nitroanilide are approximately pH 8.1 and pH 8.8 for hydrolysis and transfer, respectively. Leibach and Binkley have described a preparation from pig kidney which catalyzes γ-glutamyl transpeptidation but not hydrolysis.[20] The relationship of this enzyme to γ-glutamyl transpeptidase is not clear; it differs from the transpeptidase with respect to specificity, heat lability, and carbohydrate content.

[12] M. Orlowski and A. Szewczuk, *Clin. Chim. Acta* **7**, 755 (1962).
[13] G. G. Glenner, J. E. Folk, and P. J. McMillan, *J. Histochem.* **10**, 481 (1962).
[14] J. A. Goldbarg, E. P. Pineda, E. E. Smith, O. M. Friedman, and A. M. Rutenberg, *Gastroenterology* **44**, 127 (1963).
[15] G. E. Connell, Ph.D. Thesis, University of Toronto, Canada, 1955.
[16] P. J. Fodor, A. Miller, and H. Waelsch, *J. Biol. Chem.* **202**, 551 (1953).
[17] P. J. Fodor, A. Miller, and H. Waelsch, *J. Biol. Chem.* **203**, 991 (1953).
[18] F. J. R. Hird and P. H. Springell, *Biochem. J.* **56**, 417 (1954).
[19] J. P. Revel and E. G. Ball, *J. Biol. Chem.* **234**, 577 (1959).
[20] F. H. Leibach and F. Binkley, *Arch. Biochem. Biophys.* **127**, 292 (1968).

Activators and Inhibitors

Revel and Ball[19] demonstrated that γ-glutamyl transpeptidase is inhibited by serine in the presence of borate buffer. Kinetic analysis[21] suggested that the mixture of serine and borate forms an inhibitor competitive with the γ-glutamyl substrate.

Iodoacetamide ($0.05\ M$) inactivates the enzyme rapidly and irreversibly.[21] Iodoacetate and other sulfhydryl reagents are much less effective. Iodoacetamide may be reacting with a carboxyl group in the active center to form a glycolyl ester, in a reaction analogous to that of ribonuclease T_1 with iodoacetate.[22] The enzyme is protected from iodoacetamide inhibition by serine in the presence of borate.[21]

Goldbarg[23] demonstrated complete inhibition of γ-glutamyl transpeptidase activity in rat kidney homogenate by sulfobromophthalein at $2 \times 10^{-2}\ M$. Greenberg[24] also observed sulfobromophthalein inhibition of the enzyme in homogenates of human jejunal mucosa. Sulfobromophthalein inhibits weakly (about 50% at $2 \times 10^{-2}\ M$) the purified beef kidney enzyme.[11]

Orlowski and Meister have reported slight activation of the pig kidney enzyme by Mg^{2+}.[4] No activation by Mg^{2+} was observed with the beef kidney,[11] serum,[8, 14] or bean[25] enzymes. The kidney enzyme of Leibach and Binkley[20] is absolutely dependent upon the presence of Mg^{2+}. In general the activity of the enzyme is not significantly affected by the presence of traces of other metallic ions, nor of ethylenediaminetetraacetate.[8, 11]

pH Dependence. The pH optimum for transpeptidation is 8.8–9.0 for both the beef kidney[11] and the pig kidney[4] enzymes. A value of 8.2 has been reported for human serum transpeptidase by Szasz.[8] The bean enzyme[26] has a higher pH optimum, 9.5 for transpeptidation, while for hydrolysis the optimum is between pH 6.0 and 6.5.

Stability. γ-Glutamyl transpeptidase of beef kidney is stable for at least 10 months at $-20°$, whether in the frozen or lyophilized state.[11]

Physical and Chemical Properties. As highly purified preparations of γ-glutamyl transpeptidase from mammalian sources are heterogeneous, no meaningful physicochemical characterization has been possible. The heterogeneity may be explained by the membrane-bound state of the

[21] A. Szewczuk and G. E. Connell, *Biochim. Biophys. Acta* **105**, 352 (1965).

[22] K. Takahashi, W. H. Stein, and S. Moore, *J. Biol. Chem.* **242**, 4682 (1967).

[23] J. A. Goldbarg, O. M. Friedman, E. P. Pineda, E. E. Smith, R. Chatterji, E. H. Stein, and A. M. Rutenberg, *Arch. Biochem. Biophys.* **91**, 61 (1960).

[24] E. Greenberg, E. E. Wollaeger, G. A. Fleisher, and G. W. Engstrom, *Clin. Chim. Acta* **16**, 79 (1967).

[25] J. F. Thompson, D. H. Turner, and R. K. Gering, *Phytochemistry* **3**, 33 (1964).

[26] M. Y. Goore and J. F. Thompson, *Biochim. Biophys. Acta* **132**, 15 (1967).

enzyme in tissues, and the solvation procedures which are required in the purification.

Both the beef and pig kidney enzymes appear to be glycoproteins containing approximately 36% carbohydrate by weight. Enzymatic removal of the sialic acid results in no loss of activity.[27] Indirect evidence suggests that a soluble form of the enzyme from mammalian tissues is also a glycoprotein.[28] The pig kidney enzyme preparation contains a small proportion of fatty acid.[4]

Distribution

γ-Glutamyl transpeptidase was discovered by Hanes, Hird, and Isherwood[1,2] in sheep kidney and it has been highly purified both from beef[11] and pig[4] kidney. Although the enzyme is found in most tissues of man and other mammals, the level is far higher in kidney than in any other tissues.[15,23]

Histochemical analysis and centrifugal fractionation of tissue have shown that the transpeptidase is a membrane-bound enzyme,[13,15,23,29,30] Glenner, Folk, and McMillan demonstrated,[30] by a specific histochemical method, that in the kidney the enzyme is located on the brush border of the proximal convoluted tubules and loops of Henle. This pattern of distribution has suggested the possibility that the enzyme may play a role in kidney function, for example in the tubular reabsorption of amino acids. Histochemical observations on the jejunal mucosa of man[24] which in some respects parallel those of Glenner, Folk, and McMillan on kidney, have stimulated speculation regarding a general role for the enzyme in amino acid transport.

Szewczuk[28] has shown that tissues of man contain, in addition to the membrane-bound enzyme, a small amount of γ-glutamyl transpeptidase which is recovered in the high-speed supernatant after centrifugal fractionation of homogenates. Evidence was presented that the serum enzyme was closely related in properties to the soluble enzyme of liver. The urinary enzyme differed from these in some respects but resembled, on the other hand, the soluble transpeptidase of kidney. Other workers[28,31-33] have provided evidence by electrophoresis and gel filtration for heterogeneity of the serum and urine enzymes.

Thompson et al.[25] and Goore and Thompson[26] have purified a γ-

[27] A. Szewczuk and G. E. Connell, Biochim. Biophys. Acta 83, 218 (1964).
[28] A. Szewczuk, Clin. Chim. Acta 14, 608 (1966).
[29] G. G. Glenner and J. E. Folk, Nature 192, 338 (1961).
[30] G. G. Glenner, J. E. Folk, and P. J. McMillan, J. Histochem. Cytochem. 10, 481 (1962).
[31] F. Kokot and J. Kuska, Clin. Chim. Acta 11, 118 (1965).
[32] M. Orlowski and A. Szczeklik, Clin. Chim. Acta 15, 387 (1967).
[33] K. Jacyszyn and T. Laursen, Clin. Chim. Acta 19, 345 (1968).

glutamyl transpeptidase from kidney bean fruit which resembles the mammalian enzyme in many respects. The plant enzyme is not membrane-bound but is recovered in the high-speed supernatant after centrifugal fractionation of homogenates. γ-Glutamyl peptides accumulate in dormant bulbs and seeds in several species[34] which suggests that the enzyme may have a special role in the metabolism of plants. An enzyme from *Agaricus bisporus* (mushroom)[35] differs from the bean enzyme; although it will catalyze hydrolysis and transfer of the γ-glutamyl radical to hydroxylamine, it will not transfer to α-amino acids.

γ-Glutamyl transpeptidase has been found in several species of bacteria, for example, *Escherichia coli*, *Proteus vulgaris*, *Proteus morganii*, and *Pseudomonas aeruginosa*. A specialized form of transpeptidase from *Bacillus subtilis* has been characterized by Williams and Thorne.[36] This enzyme is involved in the synthesis of an extracellular soluble poly-γ-glutamyl peptide by the organism. A similar polypeptide appears to be present in a bound form as a component of the capsule of anthrax bacillus and other strains of the mesentericus group of bacilli.[37, 38]

[34] P. M. Dunnill and L. Fowden, *Biochem. J.* **86**, 388 (1963).
[35] H. J. Gigliotti and B. Levenberg, *J. Biol. Chem.* **239**, 2274 (1964).
[36] W. J. Williams and C. B. Thorne, *J. Biol. Chem.* **210**, 203 (1954).
[37] M. Bovarnick, *J. Biol. Chem.* **145**, 415 (1942).
[38] V. Bruckner, J. Kovacs, and H. Nagy, *J. Chem. Soc.* p. 148 (1953).

[61] γ-Glutamyl Cyclotransferase (γ-Glutamyl Lactamase)

By E. D. Adamson, G. E. Connell, and A. Szewczuk

γ-L-Glu-NH—R   2 Pyrrolidone-5-carboxylic acid + H₂N—R

γ-Glutamyl cyclotransferase catalyzes the transfer of the γ-carboxyl group of terminal L-glutamyl residues from linkage with α-amino groups of amino acids and peptides to linkage with the amino group of the same glutamyl residue. The products are the cyclic derivative of glutamic acid, L-pyrrolidone carboxylic acid (PCA), and a free amino acid or peptide.

Assay

Principle. γ-Glutamyl-γ-glutamylnaphthylamide labeled with ^{14}C in the terminal glutamyl group yields ^{14}C-labeled pyrrolidone carboxylic acid in the presence of the enzyme. The other product, γ-glutamyl-naphthylamide is not further degraded by the enzyme. The PCA which is formed is separated from other substances in the reaction mixture by paper electrophoresis, and the quantity is determined by liquid scintillation counting.

Reagents

γ-L-^{14}C-glutamyl-γ-L-glutamyl-α-naphthylamide, 0.032 M, adjusted to pH 8.0 with 0.05 M Tris-HCl buffer. Substrate solutions should be prepared freshly at least monthly.[1]

Tris-HCl buffer, 0.80 M, pH 8.0

Synthesis of the Substrate

γ-L-GLUTAMYL-α-NAPHTHYLAMIDE. Phthaloyl-L-glutamic anhydride (12.95 g, 0.05 mole) obtained by the method of Sheehan and Bolhofer,[1a] is dissolved with heating in 30 ml dry dioxane, and after cooling, 7.15 g (0.05 mole) α-naphthylamine is added. The solution is heated to 45° for 30 minutes with occasional stirring and then evaporated at reduced pressure. The remaining oily product is dissolved in 50 ml 1 M sodium carbonate, filtered, diluted to 300 ml with water, and then acidified with 70 ml 2 N hydrochloric acid. The precipitate is filtered off, washed with distilled water, and dried in a vacuum desiccator over calcium chloride. The dried product is dissolved in 250 ml ethyl alcohol and 7.45 g (0.05 mole) triethanolamine, and 8.1 g (0.075 mole) pure phenylhydrazine is added. The mixture is boiled 2 hours with refluxing, cooled, and 3 g (0.05 mole) glacial acetic acid is added. The material is left overnight at 4°, filtered on a Büchner funnel, washed three times with alcohol and once with ether, and then allowed to dry in air. The crude product is dissolved in 400 ml water with heating to 90°–95°, 0.5 g charcoal is stirred in, and the mixture filtered while hot. After 12 hours the crystals are collected on a funnel, washed with alcohol, and dried in a vacuum desiccator over calcium chloride. The yield is 6 g (44% of theoretical) pure γ-L-glutamyl-α-naphthylamide (m.p. = 185°, $[\alpha]_D^{20} = 28°$, 3% in 1 N HCl).

---

[1] If the substrate is radiochemically impure and contains ^{14}C-PCA (i.e., if an aged solution is used) acceptable results may be obtained by running a blank reaction mixture containing no enzyme. The radioactivity of PCA in the blank is determined and the excess PCA formed in the presence of enzyme is calculated.

[1a] J. C. Sheehan and W. A. Bolhofer, *J. Am. Chem. Soc.* **72**, 2469 (1950).

γ-L-¹⁴C-GLUTAMYL-γ-L-GLUTAMYL-α-NAPHTHYLAMIDE. This compound is synthesized by an adaptation of the method of Orlowski and Szewczuk[2] from phthaloyl-L-¹⁴-C-glutamic anhydride and unlabeled γ-L-glutamyl-α-naphthylamide. For this synthesis 100 μCi L-¹⁴C-glutamic acid is diluted with 1.50 g unlabeled glutamic acid and phthaloyl-L-¹⁴C-glutamic anhydride is prepared according to the method of Sheehan and Bolhofer.[1a] Phthaloyl-L-¹⁴C-glutamic anhydride (780 mg) is dissolved in a hot solution of 816 mg γ-L-glutamyl-α-naphthylamide in 3 ml glacial acetic acid. The solution is heated for 30 minutes at 60°. Acetic acid is removed *in vacuo*, and the remaining powder is dissolved in 15 ml dry dioxane. After 3 hours the solution is filtered, concentrated at reduced pressure, and the residue is dissolved in 60 ml 0.1 M sodium carbonate. The material is precipitated by the addition of 30 ml 0.5 N hydrochloric acid. The precipitate is filtered, washed with water, and dried in a desiccator over calcium chloride. Of this material, 1.06 g is dissolved in 9 ml 0.5 M sodium carbonate and 1 ml 2 M hydrazine is added. After 2 days the mixture is acidified to pH 3 with fresh distilled hydroiodic acid, filtered, and concentrated under reduced pressure. Ethanol (50 ml, 99%) is added to the concentrate, the product is recrystallized from 99% ethanol and then dried (m.p. = 183°–184° with decomposition).

Procedure. Incubation tubes are prepared by sealing one end of 2 mm inside diameter glass tubing (about 4 cm lengths). To 5 μl 0.80 M Tris-HCl buffer in an incubation tube, 10 μl enzyme solution of appropriate dilution is added. Buffered substrate solution (25 μl) is then added with thorough mixing. The mixture is incubated at 37° for 60 minutes, and the reaction is stopped by heating to 100° for 3 minutes. The reaction mixture is transferred quantitatively to Whatman 3 MM paper for electrophoresis. The sample is applied in a band 1.2 cm long and is subjected to drying in a current of warm air during application. After high-voltage electrophoresis, pH 3.6, at 55 V/cm for 45 minutes, the paper is dried and a strip of the paper containing a marker spot of authentic PCA (0.1 micromole) is cut off. The strip is treated to locate the PCA by the method of Rydon and Smith,[3] as follows: The strip is placed in an atmosphere of chlorine for 7–10 minutes, then the excess gas is removed by placing the strip in a ventilated oven at 60° for about ½ to 1 hour. When the paper strip is sprayed lightly with 1% starch–1% potassium iodide solution, PCA is located immediately as a blue-black spot on a fainter colored background. The radioactive spots of PCA from the reaction mixtures are located by reference to the guide strip, cut out, and

[2] M. Orlowski and A. Szewczuk, *Acta Biochim. Polon.* **8**, 189 (1961).
[3] H. N. Rydon and P. W. G. Smith, *Nature* **196**, 922 (1952).

placed in scintillation fluid for counting. The scintillation fluid is prepared by dissolving 5 g 2,5-diphenyloxazole and 300 mg 1,4-bis-2-(5-phenyloxazolyl)-benzene in 1000 ml toluene. Counts are corrected for quenching by the channels ratio method.

The yield of PCA is calculated as follows:

Number of micromoles of PCA produced per minute

$$= \frac{\text{PCA disintegrations/minute} \times 0.80}{\text{substrate disintegrations/minute} \times 60}$$

The relationship between enzyme concentration and PCA formation is linear up to 30% production of the theoretical amounts; with pure enzyme preparations this figure reaches 50%.

Alternative assays have been used by Orlowski, Richman, and Meister.[4]

Definition of Unit and Specific Activity. One unit of enzyme activity is defined as that amount of enzyme which catalyzes the production of 1 micromole of pyrrolidone carboxylic acid per minute under the conditions of the assay. Specific activity is expressed as the number of units of enzyme per milligram protein (determined by the method of Lowry et al.[5]).

Adaptation of the Assay for Crude Enzyme Preparations. The presence of γ-glutamyl transpeptidase in crude γ-glutamyl cyclotransferase preparations may deplete the theoretical amount of substrate added. The addition of L-serine and sodium tetraborate to the incubation mixture so that the final concentration of each is $0.05\,M$ will completely inhibit transpeptidase activity.

Rapid, Semiquantitative Estimation of Enzyme Activity. Enzyme solution (10 μl) is added to 10 μl $0.80\,M$ Tris-HCl buffer, pH 8.0, in an incubation tube. The substrate, 20 μl $0.05\,M$ γ-glutamylalanine, is added rapidly so that thorough mixing is achieved. The mixture is incubated for a standard time (between 15 and 60 minutes may be chosen) at 37° and the reaction is stopped by heating to 100° for 3 minutes. A suitable aliquot (30 μl) is then transferred to Whatman 3 MM paper for electrophoresis. The paper is subjected to high-voltage electrophoresis at pH 3.6, 55 V/cm for 40 minutes, and dried and stained by the method of Rydon and Smith[3] (see above). Spots of PCA produced during incubation may then be visualized and the quantity estimated by the intensity and size of the colored spot by reference to a marker spot of PCA. In addition, the paper can then be sprayed with ninhydrin solution so as to con-

[4] M. Orlowski, P. G. Richman, and A. Meister, *Biochemistry* 8, 1048 (1969).
[5] O. H. Lowry, N. J. Rosebrough, A. L. Farr, and R. J. Randall, *J. Biol. Chem.* 193, 265 (1951).

firm these observations by estimation of the amount of alanine produced by enzymatic action.

Preparation of Enyme

About 5 pig livers (7 kg) is a quantity conveniently handled in the preliminary steps. The tissue is obtained from freshly killed animals and processed immediately in the cold. The liver is first minced in a meat grinder. For every 100 g mince, 200 ml 0.05 M Tris-HCl, pH 7.4, is added and the mixture is homogenized for 3 minutes in a Waring blendor. The homogenate is centrifuged at 12,000–13,000 g for 30 minutes; the supernatant is retained and diluted with an equal volume of 0.05 M Tris-HCl buffer, pH 7.4.

Step 1. Ammonium Sulfate Precipitation. The protein which precipitates at room temperature from the diluted supernatant between 60 and 90% saturation with solid ammonium sulfate is suspended in a small volume of 0.01 M potassium phosphate buffer, pH 6.5, and dialyzed in the cold against frequent changes of this buffer. The red solution is clarified by centrifugation.

Step 2. CM-Cellulose Treatment. CM-Cellulose (microgranular) is equilibrated with 0.01 M potassium phosphate buffer, pH 6.5, and placed in a Büchner funnel. For every 100 g CM-cellulose about 200 ml 10% protein solution may be processed. The solution is allowed to pass at room temperature through the CM-cellulose and is eluted with an equal volume of phosphate buffer. The eluate is conveniently concentrated to 1/3 its volume at this point, by pervaporation in dialysis tubing. The concentrated protein solution is brought to pH 5.4 by dialysis against 0.15 M sodium acetate buffer, and then clarified by centrifugation.

Step 3. DEAE-Sephadex Chromatography. The protein solution (about 600 ml 10%) is passed at 4°, through a column (10 cm $\times$ 45 cm) of DEAE-Sephadex previously equilibrated with 0.15 M sodium acetate buffer, pH 5.4, and is eluted with the same buffer. A great deal of colored inactive protein is eluted immediately from the column and when the protein concentration has dropped to a low level, the enzyme emerges. The active fractions are pooled and concentrated to about 2% protein concentration. To avoid assaying many fractions of eluate by the standard procedure, the more rapid, semiquantitative method may be employed (see above).

Step 4. Preparative Polyacrylamide Gel Electrophoresis.[6] The concentrated product (500 mg protein) of the previous column is dialyzed against 0.1 M Tris-glycine, pH 9.5, and is subjected to electrophoresis

---

[6] T. Jovin, A. Chrambach, and M. A. Naughton, *Anal. Biochem.* **9**, 351 (1964).

in 18% polyacrylamide gel (containing 1 g bisacrylamide per 100 g monomer) at pH 9.5. The gel height is 10 cm and the current is maintained at 40 mA during the run. The enzyme migrates ahead of other proteins emerging after 16 hours and is eluted with 0.1 M Tris-glycine buffer, pH 9.35. Active fractions are detected by the "rough assay," pooled, and immediately concentrated by ultrafiltration.

Step 5. Preparative Isoelectric Focusing.[7-11] The product (100 mg protein) from electrophoresis is dialyzed overnight against 1% glycine solution and is applied in a sucrose gradient to an isoelectric focusing column (440 ml) maintained at 15° with tap water cooling. Ampholine[11a] of pH range 4.6 to 5.2 (previously prepared from Ampholine of pH range 4–6) at a final concentration of 1% is also incorporated into the gradient. A potential of 350 V is applied initially to the column, and as the ampholytes become focused this is raised gradually until a value of 600 V is reached after 48 hours. The column is emptied at a rate of 2.5 ml/min and fractions of 3–5 ml are collected. The pH and absorbance at 280 nm are measured, and fractions are assayed for activity. At least two proteins with different isoelectric points are obtained. Protein is separated from sucrose and Ampholine after prolonged dialysis against 0.05 M Tris-HCl buffer, pH 8.0, or by passage of the mixture through a column of Sephadex G-50.

The purification process is summarized in Table I.

Properties

Specificity. The best substrates so far described for γ-glutamyl cyclotransferase are those with two γ-glutamyl residues, namely γ-glutamyl-γ-glutamyl-α-naphthylamide and the analogous p-nitroanilide. Neither γ-glutamyl-α-naphthylamide nor γ-glutamyl-p-nitroanilide is a substrate. Di-γ-glutamylnaphthylamide is acted upon much more rapidly than the triglutamyl analog. Among the compounds of glutamic acid with other amino acids, the best substrates are γ-glutamylalanine, γ-glutamylglutamine, γ-glutamylglycine, and γ-glutamyl-α-aminobutyric acid. The enzyme shows only slight activity toward the γ-glutamyl derivatives of the hydrophobic amino acids phenylalanine, tyrosine, leucine, valine, and toward glutathione and ophthalmic acid. The enzyme is inactive toward glutamine.

[7] H. Svensson, *Acta Chem. Scand.* **15**, 325 (1961).

[8] H. Svensson, *Acta Chem. Scand.* **16**, 356 (1962).

[9] H. Svensson, *Arch. Biochem. Biophys. Suppl.* **1**, 132 (1962).

[10] O. Vesterberg and H. Svensson, *Acta Chem. Scand.* **20**, 820 (1966).

[11] O. Vesterberg, *Acta Chem. Scand.* **21**, 206 (1967).

[11a] Ampholine from LKB-produkter AB, Stockholm-Bromma 1, Sweden.

TABLE I

SUMMARY OF THE PURIFICATION OF γ-GLUTAMYL CYCLOTRANSFERASE
FROM 7 KG PIG LIVER

Step	Preparation	Volume	Total protein	Specific activity (units/mg protein)	Yield
	Homogenate	20.33 liters	1.31 kg	0.0416	100
1	Ammonium sulfate extract	2 liters	93.5 g	0.174	30
2	CM-Cellulose treatment	670 ml	21.5 g	0.61	24
3	DEAE-Sephadex chromatography	63 ml	587 mg	10.81	11.6
4	Preparative polyacrylamide gel electrophoresis	18 ml	56.8 mg	67	7
5	Isoelectric focusing			84–147	4.9

All substrates so far identified have a γ-glutamyl residue with free α-amino and α-carboxyl groups, linked to the α-amino group of an adjacent amino acid residue with an α-carboxyl group either free or bound in amide, ester, or peptide linkage.

No compounds containing D-glutamyl residues have been identified as substrates.

In crude enzyme preparations containing γ-glutamyl transpeptidase[12] in addition to cyclotransferase, some substances which are not substrates of cyclotransferase may be converted to pyrrolidone carboxylic acid by sequential action of the two enzymes.[4, 13]

pH Dependence. In the presence of Tris buffers the activity of the enzyme from human liver[13] and brain[4] and sheep brain[4] is practically independent of pH between pH 5 and 9. In phosphate and borate buffers the brain enzyme is less active over most of the range but a distinct optimum is observed in borate buffer near pH 8.0.

Stability. Both the pig liver and human brain[4] enzymes are stable in solution near 0° for at least 1 month except in extremely dilute solutions. The enzyme is also stable in the frozen state at —20°.

Physical and Chemical Properties. The sedimentation velocity of pig liver cyclotransferase in phosphate buffer-NaCl (pH 6.5; ionic strength 0.20) is 2.1 Svedberg units, corrected to water at 20°. The sedimentation velocity does not vary with concentration between 0.2 mg and 1.2 mg/ml. The human brain enzyme has a sedimentation coefficient of 1.2 Svedberg units ($s_{20, w}^0$) in 0.05 M Tris-citrate, pH 8.8.

[12] G. E. Connell and E. D. Adamson, this volume [60].

[13] G. E. Connell and A. Szewczuk, *Clin. Chim. Acta* **17**, 423 (1967).

TABLE II
AMINO ACID ANALYSIS OF γ-GLUTAMYL CYCLOTRANSFERASE

Amino acid	Micromoles per 100 micromoles recovered	Number of residues per 22,400 MW
Glycine	7.4	14.9
Alanine	5.9	11.9
Leucine	8.1	16.3
Isoleucine	5.7	11.4
Valine	5.7	11.6
Half-cystine	2.0	4.2
Proline	4.9	9.9
Methionine	1.9	3.7
Serine	7.1	14.4
Threonine	4.4	8.9
Aspartic	9.1	18.4
Glutamic	15.2	30.6
Lysine	8.6	17.3
Histidine	1.5	3.0
Arginine	3.3	6.7
Tyrosine	3.4	6.8
Phenylalanine	4.0	8.1
Tryptophan	1.0	2.0

The amino acid composition of γ-glutamyl cyclotransferase of pig liver is presented in Table II.

The molecular weight of cyclotransferase from pig liver, determined by sedimentation equilibrium in phosphate buffer-NaCl (pH 6.5, ionic strength 0.20) is 22,400.

Different Forms of the Enzyme. γ-Glutamyl cyclotransferase from pig liver occurs in at least two forms with isoelectric points between 4.82 and 5.05; the enzyme from human brain[4] is also reported to occur in at least two forms with isoelectric points at 4.06 and 4.25 while that of sheep brain[4] occurs in five forms with isoelectric points ranging from 6.20 to 4.65. The relationship between the different forms remains to be elucidated; they appear to have similar sedimentation coefficients and molecular weights but different isoelectric points.

Distribution

The highest level of γ-glutamyl cyclotransferase so far known occurs in human brain[13] and the enzyme has been purified 1000-fold from this source by Orlowski, Richman, and Meister.[4] Appreciable levels are found in other organs of mammals, namely, kidney, liver, spleen, pancreas, lung, and muscle of man,[13] mouse,[4] and guinea pig,[4] in liver of rat,[14]

[14] G. E. Connell and C. S. Hanes, *Nature* **177**, 377 (1956).

rabbit,[15] and pig,[4] and in sheep brain,[4] A smaller amount occurs in calf and rabbit lens,[15] and serum of man.[13]

In all cases so far recorded the enzyme has been recovered entirely in the supernatant fraction after high-speed centrifugation of homogenates or extracts of the above tissues.

[15] E. E. Cliffe and S. G. Waley, *Biochem. J.* **79**, 118 (1961).

[62] Ribosomal Peptidyl Transferase (*E. coli*)

By ROBIN E. MONRO

$$\underset{\overset{|}{R''_n}}{\overset{\overset{|}{OR'''_n}}{R'CHCO}} + \underset{\overset{|}{R''_{n+1}}}{\overset{\overset{|}{OR'''_{n+1}}}{NH_2CHCO}} \rightarrow \underset{\overset{|}{R''_n}}{\overset{\overset{|}{OR'''_{n+1}}}{R'CHCO-NHCHCO}} + HOR'''_n$$

where R' is an acylamido group; R'' is the side chain of one of the 20 protein amino acids; HOR''' is tRNA (HO is the 2' or 3' hydroxyl group of the terminal adenosine); and n corresponds to the number of amino acid residues in the growing peptide chain.

Assay Method[1,2]

Principle. Ribosomal peptidyl transferase, the catalyst of peptide bond formation in protein synthesis, is not a typical enzyme. It exists as a catalytic center on the 50 S ribosomal subunit, and its expression is interlinked (largely through the tRNA substrates) with that of other functional centers on the ribosome (Fig. 1). The peptidyl transferase activity of the 50 S subunit normally requires, for its expression, the presence of the 30 S subunit[3] and mRNA.[4,5] However, it is possible to induce isolated 50 S subunits to catalyze peptidyl transfer by addition of methanol or ethanol.[1,2] Under such conditions the normal substrates can also be replaced by fragments thereof (or other small analogs such as puromycin), and by this means interactions with the ribosome in the immediate vicinity of the catalytic site can be investigated. This reaction

[1] R. E. Monro and K. A. Marker, *J. Mol. Biol.* **25**, 347 (1967).
[2] R. E. Monro, T. Staehelin, M. L. Celma, and D. Vazquez, *Cold Spring Harbor Symp. Quant. Biol.*, in press, 1969.
[3] M. B. Hille, M. J. Miller, K. Iwasaki, and A. J. Wahba, *Proc. Natl. Acad. Sci. U.S.* **58**, 1652 (1967).
[4] I. Rychilik, *Biochim. Biophys. Acta* **114**, 425 (1966).
[5] M. S. Bretscher and K. A. Marcker, *Nature* **211**, 380 (1966).

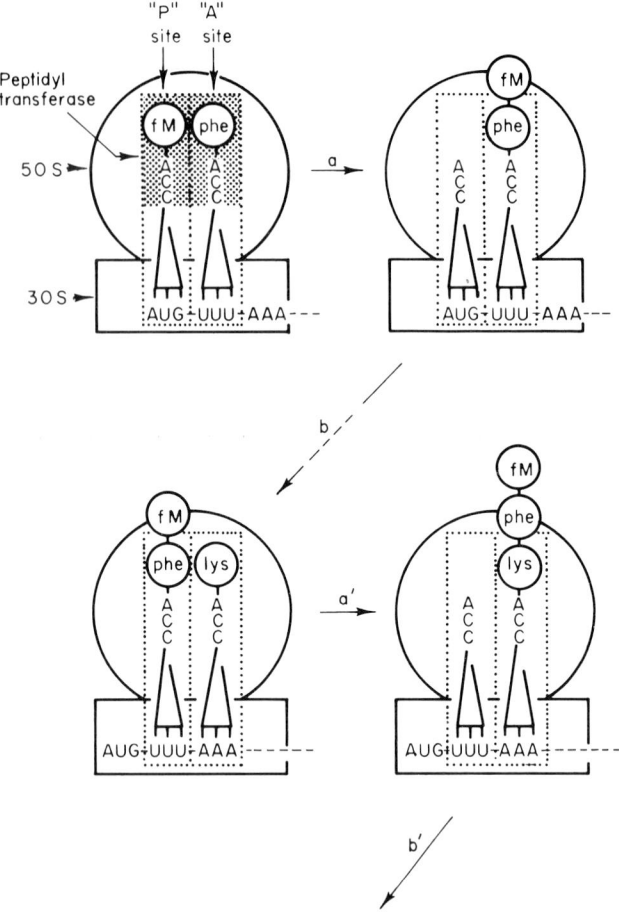

Fɪɢ. 1. Diagrammatic representation of interactions occurring on the ribosome during two cycles of chain elongation. (a) Peptidyl transfer, (b) other reactions of protein synthesis. The "P-site" is defined as the collection of sites with which the peptidyl-tRNA interacts at the moment prior to peptide bond formation, and the "A-site" as the corresponding collection of sites at which the aminoacyl-tRNA interacts. The P- and A-sites of the peptidyl transferase center are depicted as interacting with the CCA-peptide and CCA-amino acid moieties of the two substrates. [Reproduced from *Nature* **223**, 903 (1969) by kind permission of the editor.]

(Fig. 2) is termed the "fragment reaction." Other assay systems are also available[3-6] but are less resolved since they require the presence of the 30 S subunit, mRNA, and intact peptidyl donor substrate.

[6] P. Leder and H. Bursztyn, *Biochem. Biophys. Res. Commun.* **25**, 233 (1966).

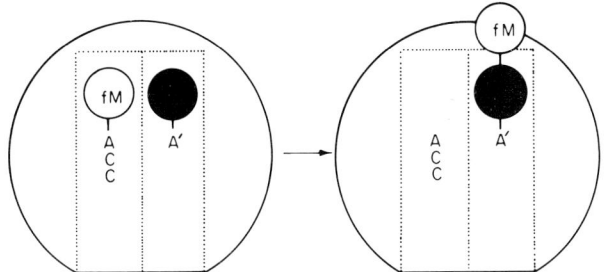

Fig. 2. Diagrammatic representation of the "fragment reaction." Met-f is transferred from CCA to puromycin to give CCA and f-met-puromycin. CACCA-Leu-Ac and other related fragments can also serve as substrates.

The assay method is based on the reaction in Eq. (1) (footnotes 2

$$CACCA\text{-}(^3H)leu\text{-}Ac + puromycin \rightarrow Ac\text{-}(^3H)leu\text{-}puromycin + CACCA \qquad (1)$$

and 7). The product, Ac-(^{3}H)leu-puromycin, is selectively extracted into ethyl acetate, and the radioactivity estimated in a scintillation counter. The original ethyl acetate method[6] has been modified[8] by raising the ionic strength and lowering the volume of the aqueous phase, in order to increase the extent and rapidity of extraction. Alternative methods for estimation of the extent of reaction include alcohol precipitation or paper ionphoresis.[1,7]

Reagents

Buffered salts mix:
    Tris(hydroxymethyl) aminomethane (Tris)-HCl buffer, pH 7.4, at 20° (giving pH 7.8 in the assay mixture at 0°), 0.25 M in Tris
    KCl, 2 M
    Mg acetate, 0.25 M
Puromycin dihydrochloride, 10 mM
CACCA-(^{3}H)Leu-Ac (10–30 Ci/mM),[7,9] 200–600 cpm/μl
Ribosomes
Methanol or ethanol (absolute)
Buffered MgSO$_4$ solution:
    Na acetate, pH 5.5, 0.3 M in acetate saturated with solid MgSO$_4$ at room temperature

[7] R. E. Monro, J. Černá, and K. A. Marcker, *Proc. Natl. Acad. Sci. U.S.* **61**, 1042 (1968).
[8] B. E. H. Maden and R. E. Monro, *European J. Biochem.* **6**, 309 (1968).
[9] R. E. Monro, Vol. XX [50].

Scintillation fluid (modified from footnote 10):
Toluene, 1 liter
2-Methoxyethanol, 250 ml
Butyl PBD (from Ciba), 5 g

Procedure.[8] The incubation mixture (150 μl, assuming volumes of water and alcohol to be additive) is placed in small conical tubes at 0°, and contains 20 μl of buffered salts mix, 10 μl of puromycin, water, ribosomes (1 OD_{260} unit), 50 μl of methanol or ethanol (precooled to 0°), and 10 μl of CACCA-(^{3}H)leu-Ac. The components are added in the order indicated, and the reaction is initiated by addition of CACCA-leu-Ac. (If ethanol is used, a precipitate of ribosomes is formed, but this does not affect the kinetics of the reaction. If aliquots are taken from a common mix, care should be taken to disperse the precipitate before pipetting, in order to avoid uneven distribution of ribosomes.) After 2–30 minutes of incubation at 0°, the reaction is stopped by addition of 0.1 ml of buffered $MgSO_4$. Ethyl acetate, 1.5 ml, is then added, the mixture agitated for 5 seconds at room temperature, and centrifuged briefly at low speed. One milliliter of the top layer (ethyl acetate) is transferred to a 4-ml vial, mixed with 2.5 ml of scintillation fluid, and the radioactivity estimated in a scintillation counter (efficiency of counting about 20%). Over 95% of the Ac-leu-puromycin is recovered in the ethyl acetate phase after one extraction. Blank values obtained by addition of buffered $MgSO_4$ before alcohol, or by incubation without ribosomes, puromycin, or alcohol amount to less than 3% of the added radioactivity. Estimates are reproducible to within ±5%. The extent of reaction is expressed as percentage of CACCA-leu-Ac converted to Ac-leu-puromycin. Values can be corrected for the presence of free Ac-(^{3}H)leu in CACCA-(^{3}H)-leu-Ac preparations, as determined by paper ionphoresis. Initial rates are estimated by extrapolation of progress curves to zero time, or, less accurately, from single incubations in which less than 20% of the added CACCA-leu-Ac has reacted.

Definition of Unit and Specific Activity. Because of technical difficulties in the preparation of large amounts of suitable substrate,[9] the CACCA-leu-Ac is labeled with very high specific activity (^{3}H)leu, and is used at a very low concentration (in the order of 5 mM). This concentration is more than an order of magnitude less than the K_m of CACCA-leu-Ac, and the other substrate, puromycin, is in large excess, so that (a) the rate of reaction is proportional to the concentration of CACCA-leu-Ac, (b) the reaction is first order with respect to CACCA-leu-Ac, and (c) the time taken to convert a given percentage of the

[10] D. J. Prockop and P. S. Ebert, *Anal. Biochem.* **6,** 263 (1963).

added CACCA-leu-Ac to Ac-leu-puromycin is independent of the initial concentration of CACCA-leu-Ac. Until a substrate becomes available which can be used at saturating concentrations, it is proposed that a unit of peptidyl transferase activity be defined as that amount of ribosomes which catalyzes the conversion of CACCA-leu-Ac to Ac-leu-puromycin at an initial rate of 10% per minute in the above assay. The unit defined in this way is independent of the concentration of CACCA-leu-Ac (in the range under study), so that the concentration can be adjusted, with substrate preparations of various specific activities, to have the appropriate amount of radioactivity. The definition also obviates technical difficulties in obtaining CACCA-(^{3}H)leu-Ac of accurately known specific activity.[9]

The specific activity of peptidyl transferase is expressed as the number of units per milligram of ribosomes; ribosome concentration is determined by use of an $E_{260}^{1\%}$ of 160. It is premature to define an expression for specific activity of components of the peptidyl transferase center derived from 50 S subunits.

Range of Application. The assay technique is applicable to ribosomes from a variety of species (see below), but not to preparations containing soluble proteins (due to the formation of bulky precipitates in the presence of alcohol). The inapplicability to crude extracts is not a serious handicap, since the purification of ribosomes is followed by physical criteria. The peptidyl transferase center is firmly integrated into the structure of the 50 S subunit and there is no tendency for it to be lost during the purification of ribosomes.[11,12,13]

Purification Procedure

The purification of 70 S ribosomes and their 50 S subunits from crude cell extracts is effected by differential and zonal centrifugation under controlled ionic conditions. The method for *Escherichia coli* ribosomes is described elsewhere.[11,14] Active preparations of 70 S ribosomes and 50 S subunits from *E. coli* have specific activities of about 10 and 15, respectively.

It is possible to remove about 30% of the proteins from the 50 S subunit without loss of activity, thus giving a further purification of peptidyl transferase.[2,11] Removal of more proteins leads to inactivation.

[11] T. Staehelin, D. Maglott, and R. E. Monro, *Cold Spring Harbor Symp. Quant. Biol.* **34** (1969).

[12] B. E. H. Maden, R. R. Traut, and R. E. Monro, *J. Mol. Biol.* **35**, 333 (1968).

[13] R. E. Monro, *J. Mol. Biol.* **26**, 147 (1967).

[14] T. Staehelin, Vol. XX [47].

Recombination of the resultant fractions under suitable conditions leads to restoration of activity. Work is in progress to purify and identify components of the 50 S subunit which constitute the peptidyl transferase center.

Properties

Stability. Ribosomes may be stored without loss of activity for several days at $0°$, or for several months at temperatures below $-70°$. Frozen ribosomes should be distributed in small aliquots and used only once, since repeated freezing and thawing leads to inactivation.

Peptidyl transferase activity varies with the conformational state of the ribosome. Salt-washed ribosomes are more active than unwashed ribosomes.[12, 13] Exposure to absence of monovalent cations leads to inactivation, which can be reversed by suitable treatment.[15]

Purity and Physical Properties.[16] Subunits (50 S) prepared by the recommended procedure have less than 3% contamination with 30 S subunits, and are free from other particulate components and supernatant proteins. It is possible that a small fraction of the ribosomal protein is lost during preparation, but evidence that the peptidyl transferase center is recovered intact is provided by the demonstration that typical 50 S preparations have approximately one such center per 50 S particle.[2]

Subunits with 50 S have a sedimentation coefficient of 50 S in the presence of Mg^{2+} and monovalent cation. Their molecular weight is about 1.8×10^6. Each 50 S subunit is a highly organized structure containing 1 molecule of 23 S RNA (molecular weight about 1.2×10^6) and 1 molecule of 5 S RNA (molecular weight 4.1×10^4), and in the order of 30 different proteins, some of which are present as 1 mole per mole of ribosomes, others in lesser amounts. The proteins range in molecular weight from about 10,000 to 50,000. As already noted, work is in progress to determine which component(s) of the 50 S subunit is present at the peptidyl transferase center.[2, 11]

Specificity. Evidence from the use of various types of substrate analog[2, 7] suggests that effective interaction at the P-site of the peptidyl transferase center (Fig. 1) involves CpCpA, the 3' terminal grouping of nucleotides common to all species of tRNA. Interaction is also favored by acylation (at the α-amino group) of the aminoacyl residue attached to the adenosine, and is influenced by the nature of the amino acid side group.

[15] R. Miskin, A. Zamir, and D. Elson, *Biochem. Biophys. Res. Commun.* 33, 551 (1968).
[16] For details and references see footnote 11.

Analogous studies with peptidyl acceptor substrates[17, 18] indicate that there is specific interaction of the bases, "C" and "A," in the terminal dinucleotide of tRNA, at the A-site of the peptidyl transferase center. Evidence is not available as to whether the third base (C) also takes part in the interaction. The aminoacyl residue must be attached to the 3' position of the terminal adenosine, and must be of the L isomeric form.[19] Activity is influenced by the nature of the amino acid side group.[18, 19]

Activators and Inhibitors. Catalysis of peptidyl transfer takes place most rapidly in the region of pH 8–9.[8] Below pH 8 the reaction is progressively inhibited. Above pH 8 it is difficult to assay the reaction, due to base-catalyzed hydrolysis of the substrates (for this reason the reaction is assayed at pH 7.8). The reaction shows an absolute dependence upon monovalent and divalent cations. Specificity toward cations, and effects of cation concentration have been reported.[8] The reaction is inhibited by Be^{2+}, Zn^{2+}, and Hg^{2+},[8] but sulfhydryl group reagents[13] and serine inhibitors (R. E. Monro and P. Siegler, unpublished data) are without effect. A number of antibiotic inhibitors of protein synthesis act specifically on the peptidyl transferase center.[20]

Distribution. Peptidyl transferase is integrated into the structure of the larger ribosomal subunit in all species which have so far been examined, i.e., *E. coli, B. subtilis*,[21] *Anacystis montana*,[21] yeast,[22] protozoa,[23] rat liver,[24] and human tonsils.[22] The ribosomal peptidyl transferases in these various species are similar to one another in many respects, but a difference between the fine structures at the catalytic centers of 70 S and 80 S ribosomes (from procaryotic and eucaryotic organisms, respectively) is indicated by differences in sensitivity to antibiotics.[22]

[17] I. Rychlik, S. Chladek, and J. Zemlička, *Biochim. Biophys. Acta* **138**, 640 (1967).
[18] R. H. Symons, R. J. Harris, L. P. Clarke, J. F. Wheldrake, and W. H. Elliott, *Biochim. Biophys. Acta* **179**, 248 (1969).
[19] D. Nathans and A. Neidle, *Nature* **197**, 1076 (1963).
[20] R. E. Monro and D. Vazquez, *J. Mol. Biol.* **28**, 161 (1967).
[21] D. Vazquez, M. L. Celma, and R. E. Monro, unpublished data.
[22] D. Vazquez, E. Battaner, R. Neth, G. Heller, and R. E. Monro, *Cold Spring Harbor Symp. Quant. Biol.,* in press, 1969.
[23] G. Cross, personal communication.
[24] T. Staehelin and A. Falvey, personal communication.

Section IV

Naturally Occurring Activators and Inhibitors of Proteolytic Enzymes

[63] Streptokinase

By FLETCHER B. TAYLOR, JR., and RUSSELL H. TOMAR

Introduction

Streptokinase (SK), an extracellular protein produced by various strains of streptococci, is capable of converting human plasminogen to plasmin. Its capacity to cause lysis of blood clots was first described by Tillet and Garner in 1933.[1] This effect was thought to be due to direct enzymatic action on the fibrin of these clots. However, Milstone in 1941 demonstrated that streptokinase achieved its effect through activation of a plasma protein.[2] Thus, the proenzyme and enzyme forms of this plasma protein were given the names plasminogen (profibrinolysin), and plasmin (fibrinolysin), respectively, and the bacterial extract was given the name streptokinase.[3] As of now plasminogen and plasmin are the only substrates with which SK is known to to react. However, it is not known whether this interaction involves enzymatic hydrolysis of plasminogen by: (1) SK itself or (2) by an activator complex composed initially of SK and either plasminogen and/or plasmin or (3) by an activator complex composed of SK and a proactivator which is distinct from plasminogen. While there is little doubt that a complex of SK and plasminogen is formed, the first and third possibilities have not been excluded as being important in the conversion of plasminogen to plasmin by SK. In any case, however, it is still not known whether streptokinase is a proteolytic enzyme in the classic sense (hydrolysis of covalent bonds) or a physicochemical modifier of a preexisting enzyme. Since the interaction of SK with plasminogen is apparently unique, we will describe: (1) SK activity in terms of its effect on the activity of plasminogen and expand this description to include data on the nature of this interaction (under Assay Procedures), (2) purification of both SK and plasminogen (under Preparative Procedures) and (3) the known physical and chemical characteristics of SK (under Physical Properties).

Assay of Streptokinase Activity and Discussion of Its Mode of Activity

All of the many assays described detect the presence of streptokinase indirectly through its capacity to activate plasminogen to plasmin.

---

[1] W. S. Tillet and R. L. Garner, *J. Exptl. Med.* **58**, 485 (1933).

[2] H. Milstone, *J. Immunol.* **42**, 109 (1941).

[3] At first streptokinase was called fibrinolysin by Tillet and Milstone until it was found that this material induced fibrinolysis indirectly through activation of a plasma protein. The term streptokinase was then coined by Christensen[4] to describe

Plasmin then hydrolyzes an indicator substrate such as a fibrin clot, casein, or synthetic substrate. The rate of hydrolysis of the substrate is a function of the amount of plasmin formed from the proenzyme, and, within the limits described, this plasmin formation in turn is directly related to the amount of SK present.

Clot Lysis Assay

The most commonly used assay for SK is that originally described by Christensen.[4] It involves the determination of the smallest amount of streptokinase that will cause lysis of a standard fibrin clot in 10 minutes. One-tenth milliliter of serial dilutions of streptokinase in gelatin buffer, 0.8 ml of standard human "fibrinogen" solution (which also contains plasminogen), and 0.1 ml of thrombin solution (described below) are mixed and placed in a water bath at 37°. The fibrinogen clots in 1 minute or less. The clots then lyse because the SK activates the plasminogen "contaminating" the original fibrinogen reagent. The lysis times of the tubes of the series are followed up to 20 minutes, and the lysis time of each dilution of SK is plotted against the reciprocal of the dilution on log-log paper (Fig. 1). A straight line can be drawn through the points beginning with a lysis time of 1.5 to 2 minutes and extending to about 20 to 30 minutes. Above and below these points the curve is not linear and therefore cannot be used for purposes of calculation. The upper limit

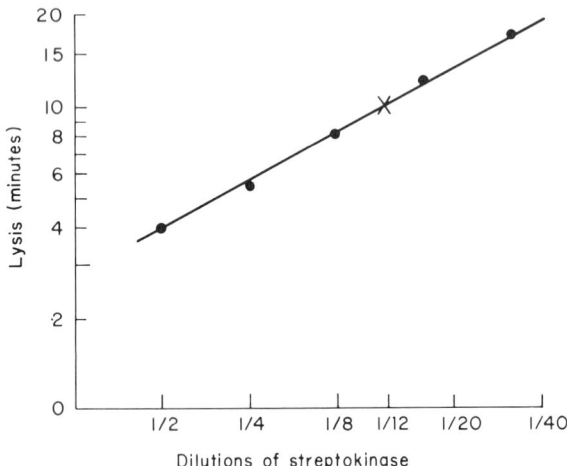

Fig. 1. Plot of the logs of lysis time of the clot (minutes) vs. the reciprocal of the dilution of the SK solution being tested.

the bacterial extract, and the terms profibrinolysin and fibrinolysin to describe the inactive (proenzyme) and active (enzyme) form of the plasma protein.

[4] L. R. Christensen, J. Clin. Invest. **28**, 163 (1949).

of the curve is of course determined by the amount of substrate (fibrin) relative to active enzyme (plasmin). However, it also should be noted that the upper limit is influenced by the fact that high concentrations of SK combine with plasminogen or plasmin to form an activator complex (to be defined below) which has a lower affinity for fibrin than does plasmin. This assay of clot lysis can detect the presence of nanogram quantities of streptokinase.

Under the above conditions, one unit is that amount of SK which will lyse a standard clot in 10 minutes as determined by interpolation of the lysis time curve. A sample assay and calculation is as follows: One-tenth milliliter aliquots of undiluted, 1:2, 1:4, 1:8, and 1:16, 1:32 dilutions of streptokinase are added to the system as described above and lysis time recorded. Figure 1 shows the log plot of the lysis time vs. the dilutions. Employing the above definition of 1 unit of SK activity together with interpolation of this plot, the units/ml or unit/mg sample can be calculated by recording the lysis time for a given dilution or concentration of SK and using the following equation:

$$\left(\frac{\text{Lysis time (minutes)}}{10 \text{ minutes}}\right) \times \text{(dilution factor)}$$
$$\times \text{ (10, or correction to 1 ml of SK sample being assayed)}$$
$$= \text{units/ml of sample being assayed} \quad (1)$$

Interpolating along the plotted graph in Fig. 1, one sees that a 1:12 dilution of the starting material should lyse a standard clot in 10 minutes. By definition then, 0.1 ml of a 1:12 dilution contains one SK unit. Therefore, the starting material contained 120 SK units/ml.

$$\frac{10 \text{ minutes}}{10 \text{ minutes}} \times \text{(12-fold dilution) (10)} = 120 \text{ units/ml} \quad (2)$$

The WHO Expert Committee on Biological Standardization used the Christensen assay to establish an international standard for SK based on assay of aliquots of a 6-g lot (#48035-154) of an impure but stable preparation of SK provided by Lederle Laboratories.[5] One international unit (that amount of the SK standard given above which would lyse a clot in 10 minutes) was found to be equivalent to 0.002090 mg of this material.

Reagents

Fibrinogen (Containing Human Plasminogen). Cohn fraction I lyophilized and stored at 4° is made into a 0.25% solution w/v with borate saline buffer on the day of the study.

[5] Bull. World Health Organ. **33**, 235 (1965).

Thrombin. Lyophilized bovine thrombin (10,000 NIH units, Parke-Davis) is made into a solution containing I NIH unit thrombin activity per 0.1 ml with borate saline buffer on the day of the study.

Lyophilized Standard and Test Sample. SK diluted in gelatin buffer on the day of the study.

Buffers

Borate Saline

Sodium borate hydrated, 7 g

Sodium chloride, 9 g

Boric acid, 11.1 g

Distilled water, 1000 ml

HCl, 5 N, to pH 7.4

Gelatin Buffer

Gelatin, 5 g dissolved in a small amount of warm water

NaCl, 10 g

KH_2PO_4, 13.6 g

make up to 1000 with distilled water

NaOH, 10 N, to pH 7.4

Individual aliquots of these reagents can be prepared beforehand and stored at −60°. However, neither the fibrinogen nor thrombin can be frozen and thawed more than once without affecting solubility and activity.

Alternative reagents which are used less often are as follows: 0.7 ml of a 0.25% solution of highly purified fibrinogen (95% clottable) in phosphate buffer (pH 7.4, 0.1 M) 0.1 ml containing one Remmert and Cohen casein unit of highly purified plasminogen, 0.1 ml of 10 units/ml of bovine thrombin and 0.1 ml of the SK to be tested all in the same buffer. These reagents, though more difficult to prepare, can yield results which are more consistent than those obtained when fibrinogen (Cohn I) containing plasminogen is used.

The Christensen assay is useful because it allows: (1) a comparatively simple determination of relative SK activity within a given laboratory; and (2) an approximation of specific activity of SK for biological purposes.

However, the difficulty with such an assay for use in more rigorous biochemical studies is alluded to in the report on the establishment of the international unit of SK where a range of 2150 to 3886 Christensen units or 970 to 5093 British Units was recorded by different laboratories.[5] This wide range is due to the fact that this assay requires subjective determination of the end point and the use of three biological products, fibrinogen, plasminogen, and SK. As with most biologicals, lot-to-lot variability, instability, and contamination by inhibitors multiply problems of standardization. This difficulty has been obviated in part by preparation of large batches of crude fibrinogen (plus plasminogen) for use solely in standardizing successive batches of SK. However, unless all laboratories have access to the same batch of SK or crude fibrinogen, more rigorous direct comparison of the various SK preparations is not possible. Other assays which have been employed for determinations of SK activity include the fibrin plate assay[6] and casein assay.[7]

Synthetic Substrate (LMe) Assay

Synthetic esters instead of fibrinogen have been employed as substrates to define more rigorously the specific activity of SK.[8,9] The following is a description of an assay for SK activity on lysine methyl ester (LMe) which utilizes plasminogen of known specific activity. It involves the addition of streptokinase to highly purified human plasminogen of known specific activity and assaying the rate of hydrolysis of a stable synthetic substrate (LMe).

In this assay 0.200 mg of plasminogen (20 casein units/mg protein) in 0.25 ml 0.001 N HCl is activated by 0.25 ml of the SK test sample in Tris buffer with 0.5 ml of 0.08 M LMe (obtained from Mann Research Co., N.Y., N.Y.) in Tris buffer (final pH 7.4) at 37°. At the end of the desired time interval (usually 30 minutes), 2 ml of a 1:1 mixture of reagents I and II (described below) are added to the above reaction mixture. The samples are then allowed to stand 2 minutes, followed by the addition of 1 ml reagent III. After at least 30 minutes and not more than 24 hours, the samples are centrifuged at 1000 g for 10 minutes to remove the precipitate. One milliliter of the supernatant is added to 4 ml of reagent IV and the color of the samples is read at 525 nm within 2 minutes. Control samples include LMe alone (0.5 ml) plus buffer (0.5 ml), and LMe (0.5 ml) plus plasminogen (0.25 ml) plus buffer (0.25 ml).

In defining the activity of SK by esterolysis, it is necessary to do so indirectly in terms of the amount of plasmin activity generated by the SK conversion of plasminogen to plasmin. Therefore, for each batch or lot of SK a standard curve is drawn of the change in optical density

[6] T. Astrup and S. Mullertz, *Arch. Biochem. Biophys.* **40**, 346 (1952).
[7] A. J. Johnson, W. R. McCarty, W. S. Tillet, A-O. Tse, L. Skoza, J. Newman, and M. Semar, *in* "Blood Coagulation Hemorrhage and Thrombosis," p. 449. Grune & Stratton, New York, 1964.
[8] F. B. Taylor and J. Botts, *Biochemistry* **7**, 232 (1968).
[9] P. R. Roberts, *J. Biol. Chem.* **235**, 2262 (1960).

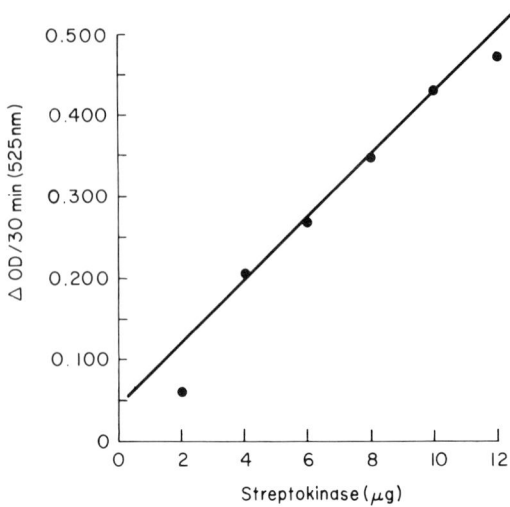

Fig. 2. Plot of the change in OD_{525} in 30 minutes due to hydrolysis of LMe by plasmin versus the amount of SK in assay system (μg).

(moles of LMe hydrolyzed) per minute by the activated plasminogen, (plasmin, 4 casein units/0.2 mg) vs. the concentration of SK in micrograms per milliliter of assay solution (Fig. 2). The explanation of the upper and lower limits of such a plot is the same as that given previously for the clot lysis assay. The specific activity of the plasmin used must be constant from assay to assay and this should be approximately 20 casein units/mg protein as determined by casein assay (to be described below). In order that the activity of different lots or batches of SK can be compared under the conditions described above, SK activity is expressed in units. One unit of SK activity is arbitrarily defined as 1×10^{-9} moles LMe hydrolyzed min^{-1}, mg N^{-1} SK by 0.200 mg of plasmin (4 casein units/0.2 mg)

$$\frac{\text{units of SK activity}}{\text{mg N of SK}} = \frac{(\Delta OD \text{ during 30 minutes})(2.79 \times 10^{-8})(10^3)}{(30 \text{ minutes})(\text{no. mg N})(1 \times 10^{-9})} \quad (3)$$

where: ΔOD is the difference in optical density at 525 nm between the test sample and control buffer and control plasminogen samples after the desired period of incubation. In this way, spontaneous hydrolysis of LMe as well as hydrolysis by plasmin contamination of plasminogen are taken into account; ϵ-2.79 $\times$ 10^{-8} is the molar extinction coefficient for assay of LMe products; 10^3 is the factor which converts the OD reading to OD units; 30 minutes is the time interval during which the reaction took place; No. mg N is the total number of mg of nitrogen in the SK being assayed; 1×10^{-9} is the moles of LMe hydrolyzed chosen arbitrarily to represent one unit of SK activity. A sample assay and calculations are as follows: 2, 4, 6, 8, 10, and 12 μg of SK (per 0.25 ml) are added to the plasminogen-LMe system as described above and the change in OD recorded. Figure 2 shows a plot of ΔOD (LMe hydrolyzed) vs. micrograms of SK. Given the above conditions and definition of 1 unit of SK activity, the number of units of SK activity in 10 μg of SK is given by recording the ΔOD which corresponds to 10 and performing the following calculations:

$$\frac{\text{units of SK activity}}{\text{mg N of SK}} = \frac{(0.433)(2.79 \times 10^{-8})(10^3)}{(30)(0.0015)(1 \times 10^{-9})} = 2.6 \times 10^5 \text{ SK}^{(10)} \quad (4)$$

Reagents

 I

   $NH_2OH \cdot HCl$, 4 M
   Distilled H_2O, 139.5 g/500 ml, filtered and stored at 4°
 II
   NaOH, 7 M
   Distilled H_2O, 280 g/1000 ml
 III
   12% TCA in distilled H_2O. The TCA is hydrophilic. Therefore, it
     must be dried under vacuum with P_2O_5 and then weighed.
 IV
   $FeCl_3$, 0.11 M, in approx. 0.04 M HCl
   29 g $FeCl_3$ in 1000 ml of H_2O, pH adjusted to 1.2 with HCl

To check that these reagents when mixed give the proper pH, add 1 ml of reagents I, II, and III to 1 ml of water or buffer in which the sample is to be dissolved, and determine the pH. It should be 1.2 or slightly less. Then add 1 ml of the I, II, and III mixture to 1 ml of reagent IV. The pH of this should be 1.2 ± 0.1. The pH is critical in production of color. If it is too high, a precipitate forms. If it is too low, the color fades too rapidly.

Plasminogen is prepared as described in the next section. After its preparation its specific activity must be determined in order to assure uniformity in all assays of SK activity over an extended period of time or between different laboratories. The assay for plasminogen activity is as follows.

[10] 10 μg = original amount of protein (SK) used in assay. 14.8 = grams nitrogen per 100 g of SK. Therefore, 0.148×10.0 μg = 1.48 μg protein nitrogen in 10.00 μg of SK or 0.014 mg protein nitrogen in 0.100 mg of SK.

To 1 ml of 4% casein prepared as described by Derechin,[11] add 1 ml of plasminogen solution (borate saline buffer, pH 7.4). Then add 0.1 ml of SK (200 units) in the same buffer. Control samples consist of casein (1 ml) plus buffer (1 ml), and casein (1 ml) plus plasminogen (1 ml). Allow to incubate at 37° for 30 minutes and then add 3 ml of 1 M PCA (or 12% TCA). Allow the system to sit in the cold for 60 minutes. Filter through Whatman No. 40 filter paper. Read the filtrate at 280 nm. Employing this assay, a standard curve relating ΔOD to the amount of plasmin (in units) in the system can be constructed in the same manner as has been described in the assays of SK activity. Given the limits of the system described by this curve, a unit of plasmin activity has been defined. One unit (casein units) of plasmin activity has been defined as that activity which releases 450 μg of TCA-soluble tyrosine (from casein) per milliliter of sample per hour.[12]

Casein units/ml

$$= \frac{(\Delta OD)(150 \ \mu g \ \text{tyrosine})(2)(5)(\text{correction to 1 ml of plasmin})}{450 \ \mu g \ \text{of tyrosine}} \quad (5)$$

where ΔOD is the difference in absorbances at 280 nm between the sample and the control plasminogen sample incubated for the same length of time, 150 μg of tyrosine is equal to 0.001 OD reading; 2 is the correction from 30 minutes to 60 minutes; and 5 represents the final volume in ml of the solution, including the PCA. The correction to 1 ml in this case is 1 since 1 ml of plasmin (plgn) as described here is used in the assay. From this the specific activity of a plasmin preparation may be obtained.

$$(\text{Remmert and Cohen}^{12}) \text{ or casein units/mg N} = \frac{\text{casein units/ml}}{\text{mg N/ml}} \quad (6)$$

Another unit of plasmin activity is that very recently proposed by the Committee on Thrombolytic Agents as the CTA unit.[13] The assay is similar to that described above and is described in the chapter on plasminogen by Robbins and Summaria [10].

The LMe assay may represent a useful more quantitative alternative to the Christensen or fibrin clot assays because: (1) the end point is *not* based on visualization of lysis of a clot which is subject to error or variation from laboratory to laboratory; (2) multiple points in a single determination may be made; (3) standardization of the substrate LMe is far easier than for fibrin or fibrinogen; (4) the plasminogen must

[11] M. Derechin, *Biochem. J.* **82,** 42 (1962).
[12] F. L. Remmert and P. O. Cohen, *J. Biol. Chem.* **181,** 431 (1949).
[13] A. J. Johnson, D. L. Kline, and N. Alkjaersig, *Thromb. Diath. Haemorrhag.* **30,** 259 (1969).

be added as a separate entity with less possibility of plasminogen or plasmin or their inhibitors contaminating the substrate as is the case when fibrinogen is used.

Mode of SK Interaction with Plasminogen

The mode of action of SK on plasminogen and/or plasmin as assayed above remains to be fully elucidated. However, the following facts are known.

1. In 1961 working from the original observations on "activator" by Mullertz,[14] evidence was presented by Kline and Fishman[15] that suggested that increasing amounts of SK react directly with performed plasmin to form a complex with increased affinity for certain synthetic substrates (i.e., LMe) and decreased affinity for certain protein substrates (i.e., casein). This plasmin–SK mixture (complex) was termed "activator" because: (1) it was an efficient activator of bovine plasminogen to plasmin whereas SK or plasmin alone was not; and (2) it was an effective activator of human plasminogen whereas plasmin alone was not.

They also observed that the original plasmin activity could be recovered from this SK-plasmin mixture (activator complex) by exposure to pH 2.5 for 1 hour. Upon readjusting the pH to 7.4 and reassay of this mixture on casein and LMe, the original affinity of plasmin for casein was restored along with the corresponding decrease of affinity of the mixture for LMe. Thus, the activity of the SK-plasmin "activator" complex had been transformed back to the activity of plasmin alone. From these studies in which large amounts of SK and plasmin were used, it was assumed that activation of plasminogen by small amounts of SK was mediated through its reaction with the small amount of plasmin (1% or more of which is always associated with plasminogen) to form the activator mentioned which, in turn, initiates further conversion of plasminogen to plasmin.

2. In 1964, the actual physical existence of such a complex was demonstrated by Davis *et al.*,[16] who showed that as one approached a 1:1 to 1:2 molar ratio of streptokinase relative to plasmin (or plasminogen), a complex was formed which sedimented faster in the ultracentrifuge than either of the two components alone. It was postulated but not shown that this was activator. Later DeRenzo *et al.*,[17] of this same group, demon-

[14] S. Mullertz, *Biochem. J.* **61**, 424 (1955).

[15] D. L. Kline and J. B. Fishman, *J. Biol. Chem.* **236**, 2807 (1963).

[16] M. C. Davis, E. Boggiano, W. F. Barg, and F. F. Buck, *J. Biol. Chem.* **239**, 2651 (1964).

[17] E. C. DeRenzo, E. Boggiano, W. F. Barg, and F. F. Buck, *J. Biol. Chem.* **242**, 2428 (1967).

strated this same complex by starch gel electrophoresis and went on to show that it had activator activity as defined by its capacity to convert bovine plasminogen to plasmin.

3. Recently, in our laboratory a series of studies has been made of the ratios of SK and plasminogen in the "activator complex"; the split products which arise from such an interaction; and of the active fragment which can be split off from the complex upon reduction with mercapto-ethanol (MSH). These studies have been made in an attempt to define more completely in macromolecular terms what happens to these molecules when they interact. These observations fall into two categories, those SK-plasminogen reactions leading to plasmin formation and those SK-plasminogen reactions leading to activator formation.

(a) When only 1 mole or less of SK is reacted with 40 moles of plasminogen, plasmin is produced. Under these conditions, a split product was identified with an approximate molecular weight of 6200 which originated from plasminogen and which was clearly associated with the conversion of plasminogen to plasmin. We also demonstrated that between one and two hydrogen ions per mole of plasminogen were released during its conversion to plasmin under these conditions.[8] Barlow et al., later confirmed this finding by showing that the molecular weight as determined by ultracentrifugation with photoelectric scanning of plasminogen differed from that of plasmin by 5600.[18] More recently we have shown (by employing SK and plasminogen labeled with ^{125}I and ^{131}I, respectively, and analysis and separation of the labeled reactants on discontinuous acrylamide gel electrophoresis) that activation of plasminogen to plasmin resulted in: (1) splitting off of one or more pieces from plasminogen; (2) movement of the plasmin fraction further into the gel; (3) association of the small amount of SK used with the plasmin or plasminogen peaks. As the ratio of SK was increased relative to plasminogen (toward conditions favoring activator formation) SK appeared in three forms: that which was bound to plasminogen or plasmin to form "activator," that which was free (native SK), and that which has been hydrolyzed (split products). Under these conditions the split products from plasminogen are no longer visible.[19]

The chemistry of this conversion of plasminogen to plasmin by small amounts of urokinase or SK in glycerol involves hydrolysis of arginyl-valine peptide bonds.[20] This is described in detail by Robbins [10].

[18] G. H. Barlow, L. Summaria, and K. C. Robbins, J. Biol. Chem. 244, 1138, (1969).
[19] R. H. Tomar and F. B. Taylor, Federation Proc. 28, No. 2, 322 (1969).
[20] K. C. Robbins, L. Summaria, B. Hsieh, and R. J. Shah, J. Biol. Chem. 242, 2333 (1967).

(b) When approximately 1–4 moles of SK are reacted with 1 mole of plasminogen, activator complex, defined previously, is formed.

Under these conditions bovine plasminogen activator activity, not plasmin activity, is generated. Again employing radiolabeling and gel techniques, we have shown that: (1) a complex with activator activity consisting of plasminogen or plasmin and SK is formed with the molar ratio of these components being dependent on the conditions of reaction; (2) that upon reduction with MSH, a fragment retaining activator activity is generated. Since the molecular weight of the fragment is in the range of 70–100,000 and radio-label from both SK and plasminogen were present, it is conceivable that the piece consists of some combination of intact SK and a portion of the plasmin molecule.

Two additional points are of importance in this study: (1) the same results are obtained when plasmin instead of plasminogen is used. (2) The two key questions regarding the SK-plasminogen interaction remain unresolved: Is SK which is actually in the complex modified? Where is the active site of "activator" (modified SK, plasmin, or both)?

In regard to this last question, studies of the effect of urea on this complex have been recently conducted by Summaria et al.[21] Urea is said to dissociate the complex into modified plasmin which can be used again to make activator with fresh SK, and into an SK which cannot be used again. These studies together with the electrophoretic studies of DeRenzo et al.[17] suggest that SK might be modified irreversibly upon interaction with plasmin to form activator.

Finally, the possibility of a direct action (without complex formation with plasmin) of streptokinase on plasminogen has been suggested by the studies of Kline and Ts'ao[22] and Summaria et al.[23]

Preparation of Streptokinase and Reagents for Streptokinase Assay

Preparation of Crude Streptokinase. Crude streptokinase–streptodornase (SK-SD) is usually prepared from cultures of β-hemolytic streptococci of Group C, in a manner similar to that described by Christensen.[24] Large amounts of Crude SK-SD are distributed under the trade name Varidase by the Lederle Co. Although the crude SK-SD can be and is prepared in individual laboratories, the above preparation has been used as a starting material for preparation of purified SK.

[21] L. Summaria, W. R. Groskoff, and K. C. Robbins, *Federation Proc.* **28**, No. 2, 322 (1969).
[22] D. Kline and C. H. Ts'ao, *Thromb. Diath. Haemorrhag.* **18**, 288 (1967).
[23] L. Summaria, B. Hsieh, W. R. Groskoff, and K. C. Robbins, *Proc. Soc. Exptl. Biol. Med.* **130**, 737 (1969).
[24] L. R. Christensen, *J. Gen. Physiol.* **30**, 465 (1947).

Preparation of High Purity Streptokinase. Several satisfactory methods of SK purification have been published by DeRenzo *et al.*; Taylor and Botts; and Tomar.

1. The first of these methods was described by DeRenzo *et al.* for the Lederle Co., under U.S. patent 3,226,304.[25] They describe in detail four different procedures for preparing the crude starting material. They then described further purification procedures employing either DEAE-cellulose chromatography or density-gradient electrophoresis. Both these later procedures yield SK of high purity with an approximate specific activity of 600–700 Christensen units/μg N.

2. The second of these methods was developed simultaneously in our laboratories.[8] The details are as follows:

Approximately 100 mg of SK is dissolved in 3 ml of 0.10 M NaCl–0.01 M Tris buffer (pH 8.5) and dialyzed for 48 hours at 4° against two changes of 2-liter volumes of the same buffer. This material is then adsorbed and fractionated on DEAE-Sephadex A-50 at 4° employing a linear NaCl-Tris gradient. The DEAE-Sephadex A-50 is obtained from Pharmacia Fine Co., Uppsala, Sweden. This material has a binding capacity of 3.5 ± 0.5 meq/g. Lots (5–10 g) of the DEAE-Sephadex are washed 8 times with 1-liter volumes of 0.1 M NaCl–0.01 M Tris buffer (pH 8.5) and allowed to equilibrate with this buffer over 3 days after a final adjustment of the pH of the slurry to pH 8.5 with 0.1 N HCl. This is packed into a 45 × 3 cm column by gravity (at a flow rate of 20 ml/ hour). The total bed volume is 250 ml. The linear gradient is formed by using two 300-ml Erlenmeyer flasks (level with each other and open to the atmosphere) that are connected in series to each other and the column. These two flasks are connected at their bases by tubing and hence to the column by a ⅛-inch diameter polyethylene tube and plug. The first flask contains 250 ml of a 0.5 M NaCl–0.01 M Tris solution that runs into the second flask (mixing flask) which contains 0.1 M NaCl– 0.01 M Tris. The thoroughly mixed solutions are then delivered into the top of the column by gravity feed at a rate of 20 ml/hour. The total elution of volume is 500 ml collected in 2–3-ml aliquots by a GME Model T 15² fraction collector. The protein concentration is determined by absorbance measurements in 1-cm quartz cells at 280 nm with a Beckman DB spectrophotometer. The effluent containing SK is pooled and dialyzed for 48 hours at 4° against three changes of 3-liter volumes of 0.001 M phosphate and 0.3 M NaCl at pH 7.4. This dialyzate is then filtered on G-100 Sephadex gel at 4° at a flow rate of 20 ml/hour. The

[25] E. C. DeRenzo, P. K. Siiteri, B. L. Hutching, and P. H. Bell, *J. Biol. Chem.* **242**, 533 (1967).

Sephadex G-100 is also obtained from Pharmacia Fine Co. The dry powder is suspended in a sufficient amount of phosphate buffer (0.001 M phosphate 0.30 M NaCl at pH 7.4) and stored for 72 hours to assure complete swelling. The fines were removed by several decantations and the gel is packed into a 45 × 3 cm column. The total bed volume is 250 ml and the void volume is 32 ml as determined with blue dextran. The streptokinase is pooled, concentrated by ultrafiltration, and stored in 2-ml volumes at −60°.

Rechromatography of the effluent from the G-100 (containing SK) on DEAE (Sephadex) is frequently necessary in order to complete the separation of α-globulins which are closely associated with streptokinase. The final product has an approximate specific activity of 600–800 Christensen units per microgram nitrogen or 2.5 × 10⁵ SK LMe units. This represents a 10- to 11-fold increase in specific activity and is close to the results obtained by others.[25-27]

The third of these methods described by Tomar involves a precipitation of SK from the starting material with 40–50% ammonium sulfate resulting in a 2- to 3-fold increase in specific activity together with a concentration of the material for subsequent variable gradient chromatography on DEAE-cellulose.[28] The simple $(NH_4)_2SO_4$ step originally described by Fletcher and Johnson[26] with high increase in specific activity represents the most simple single batch procedure for preparing SK from Varidase described to date.

Dillon and Wanamaker[29] in a comparative study have described the separation and partial purification of SK from both Group C and Group A, β-hemolytic streptococci employing methods similar to those described above. Finally, it should be noted that Blatt et al.[27] have also reported preparations of SK of the same order of purity.

Preparation of Plasminogen. Since plasminogen is the key reagent in the LMe assay of streptokinase activity and since it is a major source of variability in all assays of SK activity, we will describe our method of purification.

Either Cohn fraction III or Cutter pseudoglobulin plasminogen is used as starting material. Approximately 150 mg of starting material are dissolved in 0.0075 M sodium acetate buffer, pH 4.5. This material is adsorbed at 4° on a 30 × 2.5 cm column of carboxymethyl cellulose with a binding capacity of 0.61 meq/g. The carboxymethyl cellulose obtained

[26] A. P. Fletcher and A. J. Johnson, *Proc. Soc. Exptl. Biol. Med.* **94**, 233 (1957).
[27] W. F. Blatt, H. Segal, and J. L. Gray, *Thromb. Diath. Haemorrhag.* **11**, 393 (1966).
[28] R. H. Tomar, *Proc. Soc. Exptl. Biol. Med.* **127**, 239 (1968).
[29] H. C. Dillon and L. W. Wanamaker, *J. Exptl. Med.* **121**, 351 (1965).

from Schleicher and Schuell (Keene, N.H.) is prepared by treating 50 g of CM-cellulose with 3 liters of 0.1 HCl twice, followed by 1 liter of H_2O twice, followed by 3 liters 0.5 N NaOH and 0.5 M NaCl, followed by 3 liters of 10% ethanol twice, followed by 2 liters of H_2O twice, followed by equilibration in 2 liters of 0.0075 M sodium acetate buffer. This material is then packed into the 30 × 2.5 cm column by gravity. The adsorbed plasminogen is eluted from the column in 3-ml aliquots with a linear NaCl gradient (0.2–0.5 N NaCl) employing the same volumes and two-vessel system described for SK. The eluates are assayed for plasminogen activity on casein as described above. The trailing half of the major peak of activity is then pooled, adsorbed, and rechromatographed on the same system. The specific activity of final material is assayed and then stored at −60° in approximately 20 casein unit aliquots (approx. 1 mg). This method of preparation has two major advantages over other methods we have utilized. Since the chromatography and storage is done at pH 4.5, plasminogen remains both stable and soluble at a pH which is not lower than 4.5. Other techniques in which the chromatography is done at a higher pH require the use of urea[30] or EACA or lysine[31] in order to prevent autodigestion and ensure maximum solubilization of the plasminogen.

Physical Properties

The physical and chemical properties of streptokinase as reported in 1967 by DeRenzo et al.[25] and our laboratories,[8] are almost identical. Its molecular weight as determined by equilibrium sedimentation is between 47,000[25] and 49,000.[8] Its partial specific volume as determined by pycnometry and calculation from amino acid analysis data is between 0.75[25] and 0.73[8] and its specific viscosity as determined by measurement with an Ostwald type viscometer with an outflow time for water of 70 seconds at 37° is 0.10.[8] The isoelectric point of SK is 4.7,[25] and it migrates on gel electrophoresis as an α-globulin.[8] Chemical analysis of streptokinase reveals a hexose and hexoseamine content of less than 0.2 g/100 liters and 0.1 g 100 liters respectively.[8] No lipids are present.[8] The nitrogen content is 14.5%.[8] From this, an extinction coefficient $E_{1cm}^{1\%}$ of 9.49 was determined.[8] Optical rotatory dispersion studies suggest that the helix content is 10–12% as determined from the depth of the trough of the Cotton effect.[8]

The table shows the results of amino acid analysis.[8]

[30] F. B. Taylor and I. Staprans, *Arch. Biochem. Biophys.* **114**, 26 (1966).

[31] K. C. Robbins, L. Summaria, D. Elwyn, and G. H. Barlow, *J. Biol. Chem.* **240**, 541 (1965).

RECOVERY OF AMINO ACIDS AND AMMONIA IN ACID HYDROLYZATES
OF PURIFIED STREPTOKINASE

Amino acid residue	Amount recovered in hydrolyzates at	
	23 hours	48 hours
	(millimicromoles/μg of nitrogen)	
Asp	8.16	8.09
Thr	3.48	3.36
Ser	3.72	2.34
Glu	5.54	5.76
Pro	2.45	2.53
Gly	2.55	2.50
Ala	2.83	2.80
Val	2.80	2.82
Met	0.28	0.28
Ile	2.86	2.66
Leu	5.01	4.97
Tyr	2.35	2.43
Phe	1.90	1.84
NH$_3$	6.50	8.63
Lys	4.08	4.01
His	1.08	1.00
Arg	2.51	2.57

[64] Assay and Preparation of a Tissue Plasminogen Activator[1]

By TAGE ASTRUP and PREBEN KOK

Introduction

The fibrinolytic activity in tissues is caused by an agent which converts a zymogen in blood or serous fluids (plasminogen, profibrinolysin) into a trypsinlike proteinase (plasmin, fibrinolysin; EC 3.4.4.14).[1a] Work leading to the demonstration of this tissue plasminogen activator (fibrinokinase or cytofibrinokinase) was reviewed in 1956.[2] The development of a method for its quantitative assay was delayed because the activator is strongly attached to structural cellular proteins from which it cannot

[1] Supported by Grant HE-05020 from the U.S. Public Health Service, National Institutes of Health, National Heart Institute.
[1a] T. Astrup and P. M. Permin, Nature 159, 681 (1947).
[2] T. Astrup, Blood 11, 781 (1956).

be separated by solvents commonly used for such purposes. When it was observed that the activator could be brought into solution by molar potassium thiocyanate[3] and isolated by acid precipitation,[4] an assay method was worked out[5] using the fibrin plate method[6] for activity determinations. An activator preparation made from a thiocyanate extract of pig heart served as a reference standard. Later, Bachmann *et al.*[7] observed that activator could be extracted from pig hearts with dilute acetate buffer at pH 4.2. However, in the systematic assay of activator concentrations in tissues it has not been possible to replace the use of molar solutions of thiocyanate salts in the quantitative extraction with other procedures. Hence, the presence of thiocyanate in the solutions puts a limitation to the choice of methods for activity determinations. Thus, the determination of plasminogen activation in a caseinolytic system is influenced by thiocyanate, as is the clotting of fibrinogen by thrombin required in assays by the lysis time method. These theoretically better methods are applicable only when highly purified and concentrated activator preparations free from thiocyanate salts are available. In the fibrin plate method clot formation has taken place before application of the thiocyanate extracts. In addition, the fibrin plate is about 100 times more sensitive than the lysis time assay, an important advantage because most tissues, in particular from animal species, contain the activator in low concentrations.[8,9]

Assay Method

Principle. Tissue samples are extracted with 1 or 2 M potassium thiocyanate, the solutions precipitated with acid and the sediment redissolved at neutral reaction. Serially diluted solutions are applied to fibrin plates prepared with bovine plasminogen-rich fibrinogen. The activities, recorded as diameter products of the lysed zones (mean of triplicate determinations) after incubation at 37° overnight (16–18 hours), are plotted double logarithmically with relative concentrations of the dilutions on the abscissa and diameter products on the ordinate (see Fig. 1). If inhibitory agents are absent, parallel linear dilution curves are obtained. The actual activator concentrations in units per milliliter are then determined by appropriate interpolation on a dilution curve simultaneously obtained

---

[3] T. Astrup and A. Stage, *Nature* **170**, 929 (1952).

[4] T. Astrup and I. Sterndorff, *Acta Physiol. Scand.* **36**, 250 (1956).

[5] T. Astrup and O. K. Albrechtsen, *Scand. J. Clin. Lab. Invest.* **9**, 233 (1957).

[6] T. Astrup and S. Müllertz, *Arch. Biochem. Biophys.* **40**, 346 (1952).

[7] F. Bachmann, A. P. Fletcher, N. Alkjaersig, and S. Sherry, *Biochemistry* **3**, 15 (1964).

[8] O. K. Albrechtsen, *Acta Physiol. Scand. Suppl.* **47**, No. 165 (1959).

[9] T. Astrup, *Federation Proc.* **25**, 42 (1966) (Review).

from a reference standard. Determinations of activator concentrations in tissues require an intermediate acid precipitation in order to remove contaminating inhibitory material or destroy labile activators probably originating in the blood. In some tissues or species the quantitative recovery of activator after acidification presents difficulties requiring additional precautions (see below).

Applied over many years and under a wide variety of conditions, several improvements have been introduced in the fibrin plate method and important details of technique have been described.[10-12] Those essential to the assay of tissue activator will be included in the following discussion. The use of fibrinogen preparations low in plasminogen content decreases the sensitivity to plasminogen activators. Presence of citrate makes the fibrin susceptible to liquefaction by the strong solutions of KSCN required in the extraction of activator. The addition of agar recommended by some authors is unfortunate in the assay of tissue activator because the thiocyanate produces zones of clarification which can be mistaken for zones of lysis since there is no true liquefaction of the agar-containing substrate. The need of agar addition to stabilize the fibrin clot indicates that a fibrinogen preparation is unsuitable since solid and stable clots are easily obtained with good fibrinogen preparations free from citrate. Since commercial fibrinogen preparations do not satisfy fully the requirements of the fibrin plate method,[10] the preparation of an ammonium sulfate-precipitated bovine fibrinogen will be described in greater detail.

Reagents

Reagent grade chemicals and distilled water are used throughout when not otherwise stated.

Potassium thiocyanate, 2 M. KSCN (194.4 g) is dissolved in water and diluted to 1000 ml. An appropriate amount is adjusted each day to pH 7.75 with solid sodium bicarbonate.

Potassium thiocyanate, 1 M, with 0.25% gelatin. Prepared fresh each week from 100 ml of 2 M KSCN to which is added a solution of 0.5 g gelatin (bacteriological) in 100 ml water (dissolved by heating to 70°–80° and stirring). An appropriate amount is adjusted each day to pH 7.75 with solid sodium bicarbonate.

[10] P. Brakman, "Fibrinolysis. A Standardized Fibrin Plate Method and a Fibrinolytic Assay of Plasminogen." Scheltema and Holkema NV, Amsterdam, 1967.
[11] P. Brakman and T. Astrup, in "Thrombosis and Bleeding Disorders" (N. Bang, F. K. Beller, E. Deutsch, and E. F. Mammen, eds.). Academic Press, New York.
[12] T. Astrup, P. Glas, and P. Kok, in "Thrombosis and Bleeding Disorders" (N. Bang, F. K. Beller, E. Deutsch and E. F. Mammen, eds.). Academic Press, New York.

Fibrin Plate Buffer. Consists of 0.05 M sodium barbital (N.F. grade; Merck) with 0.09 M sodium chloride, 0.0017 M calcium chloride, and 0.0007 M magnesium chloride, and is adjusted to pH 7.75. Total ionic strength is 0.15. The presence of magnesium and calcium ions improves clot stability and opacity, making the lysed zones more demarcated and visible.[10]

Saline Barbital Buffer. Consists of 0.05 M sodium barbital and 0.10 M NaCl at pH 7.75.

Fibrinogen. Bovine fibrinogen is prepared by ammonium sulfate precipitation of bovine oxalate plasma following methods outlined in the original descriptions and including later improvements.[6, 10, 11, 13]

Cattle blood is collected in 1 volume of 0.1 M potassium oxalate per 9 volumes of blood, avoiding contamination with tissue fluids. Plasma is separated the same day, preferably by high speed centrifugation. The separated plasma is then stirred for 20 minutes at room temperature with 60 g calcium phosphate per liter followed by centrifugation in a refrigerated centrifuge with horizontal head for 30 minutes at about 1400 g. All subsequent centrifugations are performed in a similar manner when not otherwise mentioned. The separated plasma is centrifuged at high speed to remove traces of calcium phosphate and stored overnight in the cold room. The next day, after stirring at room temperature until 10°–15° is reached, a sediment is removed by filtration through glass wool or by centrifugation. One liter of the adsorbed plasma is then diluted with 500 ml cold water and 600 ml of saturated ammonium sulfate is added dropwise through a glass tube with an inverted tip placed under the surface of the mixture and with slow stirring in the cold room. Foaming should be avoided. The fibrinogen is collected by centrifugation for 5 minutes and redissolved in 500 ml of 0.15 M NaCl, the solution diluted with 1000 ml cold water and again precipitated in the cold with 600 ml saturated ammonium sulfate. After separation by centrifugation (10 minutes), the walls of the beakers are wiped with filter paper to remove excess of ammonium sulfate solution and the fibrinogen dissolved at room temperature in 200 ml H_2O adjusting pH to 7.5 with 0.1 N NaOH. Insoluble material is removed by filtration or centrifugation and the solution stored overnight in the cold room. The sediment which has formed is again removed by filtration or centrifugation. The fibrinogen concentration is determined and the electrical conductivity of the solution measured.[6, 10] The apparent ionic strength of the solution is obtained by interpolation on a conductivity curve obtained from measurements of appropriate dilutions of ammonium sulfate at the same temperature. It is convenient to

---

[13] T. Astrup and S. Darling, *Acta Physiol. Scand.* **4**, 45 (1942).

adjust the ionic strength to 0.45 by addition of solid NaCl. The fibrinogen concentration is usually about 1.5%. It has a clottability between 80 and 90%. Solutions are stored in plastic containers at −20° at which temperature they are stable for years.

From each batch of fibrinogen, fibrin plates are prepared and tested against a standard solution of tissue activator to determine the sensitivity. Blood collected from individual animals may yield fibrinogen preparations of varying sensitivity. Batches with low sensitivity, suggesting the presence of excess of inhibitor or lack of plasminogen, are discarded. The fibrinogen stock solution may be kept in the refrigerator for a few weeks. Solutions for the fibrin plate assay are prepared daily and adjusted to 0.1% fibrinogen, pH 7.75, and ionic strength 0.15. As an example, 10 ml of a solution with 1.5% fibrinogen and of ionic strength 0.45 yield 150 ml of 0.1% fibrinogen by addition of 20 ml water (to adjust the ionic strength) and 120 ml fibrin plate buffer. To avoid precipitation and denaturation of fibrinogen the water should be added last. The final pH is adjusted if required.

Fibrin plates are made in disposable plastic petri dishes (Optilux, Falcon Plastics) selected for flatness and having an inside diameter of 84 mm. In each dish 9 ml of the 0.1% fibrinogen solution is clotted with 0.3 ml of thrombin solution on a horizontal support and left for not less than an hour for clot stabilization before solutions of samples are applied.

Thrombin. Bovine thrombin, practically free from contaminating fibrinolytic agents (Parke-Davis, Detroit; "Thrombin Reagent," Leo Pharmaceuticals, Copenhagen), is dissolved in 0.15 M NaCl to 20 NIH units per milliliter.

Tissue Activator Standard. Results are calculated in units of a reference standard prepared from pig heart.[5] Freshly collected pig hearts are rinsed, trimmed for fat and connective tissue, and the myocardium cut into pieces, passed several times through a meat grinder, and the ground tissue stored overnight at −20°. The next day after thawing, 1 kg is stirred at room temperature for 15 minutes in 10 liters of 0.15 M NaCl with 50 ml of toluene to remove soluble proteins. After separation by centrifugation and suspension in 5 liters of 0.15 M NaCl and 25 ml of toluene, stirring is continued overnight in the cold room. The ground tissue is then separated by brief centrifugation, strained through a double layer of gauze, and again similarly treated twice with 5 liters of 0.15 M NaCl, and 3 times with distilled water, followed by four treatments with acetone. The dehydrated powder is suspended in 0.5 liters of dry ether, filtered through coarse filter paper, washed with dry ether, and cautiously dried on filter paper in the air (avoid a humid atmosphere). The light grayish powder is ground thoroughly in a mortar. It weighs about 100 g and is stable for years when stored dry in the cold. From this acetone-dried tissue a water-soluble lyophilized preparation for use in the assays is prepared. Ten grams tissue powder is stirred for 2 hours with 100 ml 2 M KSCN adjusting pH to 7.0. After separation by centrifugation the supernatant (approximately 80 ml) is diluted with 7 volumes of distilled water and pH adjusted to 1.0 with 1 N HCl. After 30 minutes the sediment is collected by centrifugation, cautiously rinsed with 10 ml water, the remaining fluid is wiped from the walls of the beaker with filter paper, and the sediment is redissolved in the original volume ($\sim$80 ml) of saline-barbital buffer. Insoluble material is removed by centrifugation and the clear solution lyophilized and stored in airtight containers at $-20°$. The yield is about 2 g with an activity between 1 and 5 tissue activator units per milligram (see below). The product is stable for long periods when stored dry in the freezer. Its potency is determined by comparison with a primary reference standard (kept in our laboratory) as described below. Solutions prepared for assays should preferably contain between 10 and 20 tissue activator units per milliliter in 1 M KSCN with gelatin.

Assay Procedure

Fresh tissue samples are rinsed and excess water removed with blotting paper. Samples are weighed and either immediately extracted and assayed, or placed in small tightly closed containers at $-20°$. Brief storage causes no significant decrease in activator content in most tissues. Prolonged storage should be avoided. The samples are treated in a Potter-Elvehjem homogenizer with 3 ml of 2 M KSCN per 100 mg. Disintegration requires variable periods of time, depending upon tissue structure. Overheating should be prevented by external cooling of the homogenizer in ice water. The tissue suspension is then shaken slowly for 1 hour at room temperature. In the original assay three extractions of tissue were made instead of the single extraction with a three times larger volume now in use. The slightly higher accuracy provided by three extractions is of no significance considering the large spreading of the activity of individual samples. With small samples of tissues (20–50 mg) the relative extraction volume can be increased without seriously affecting the results.[14]

After centrifugation (10 minutes at 2900 g) an aliquot (1.00 ml) of the clear supernatant is separated and diluted with 7 volumes (7 ml) of distilled water and 1 N HCl added to pH 1.0. The sediment is separated by centrifugation after standing for 30 minutes at room temperature and

[14] T. Astrup, F. K. Beller, P. Glas, and J. Rasmussen, *Obstet. Gynecol.* **25**, 853 (1965).

redissolved in 1 M KSCN containing gelatin using the original volume of the aliquot (1 ml) and adding solid sodium bicarbonate until neutral reaction (checked easily and with sufficient accuracy on neutral litmus paper). Some tissues leave turbid, fibrinolytically active supernatants after acid precipitation. In such cases a crude extract prepared from an inactive tissue, such as rabbit skeletal muscle, can be added before acidification in order to complete precipitation of plasminogen activator.[15] For assays on the fibrin plates serial dilutions in 1 M KSCN containing gelatin are prepared. A solution of a standard preparation of tissue activator is similarly diluted to obtain a reference curve. Preparations available as active solutions are diluted directly in 1 M KSCN and assayed.

Using a 0.1-ml long-tipped measuring pipette, Corex No. 7064-A, 1/10 in 1/100, exactly 30 μl of each dilution are applied in triplicate on the fibrin plates which are then incubated at 37° overnight (usually from 16 to 18 hours) on shelves exactly level. The activity is recorded arbitrarily as the product in mm² of two perpendicular diameters of each lysed zone and the mean of triplicate estimations calculated for each dilution. These are then plotted in a double logarithmic graph as in Fig. 1. The dilution curves should be linear and parallel with the reference curve simultaneously obtained. Deviations indicate presence of inhibitory material. Diameter products should be between 100 and 800 mm.² The concentration of activator in the test solution is obtained by interpolation on the reference curve. The concentration of activator per gram fresh tissue may then be calculated.

To exemplify calculations, the data on Fig. 1 are used. The reference curve was from a solution containing 8.0 tissue activator units per milliliter. Because of the high activator concentrations in the tissues selected, the original stock solutions were appropriately diluted before serial dilutions were prepared. A thiocyanate extract (470 ml) was prepared from pregnant hog ovaries (51.6 g) and a sample diluted 200 times with 1 M KSCN containing gelatin. Serial dilutions were prepared and assayed as described, yielding the upper curve in Fig. 1. Interpolation shows the standard solution to correspond to 35% of the test solution. The activator concentration was then $(8 \times 100 \times 200 \times 470)/(35 \times 51.6) = 41,600$ units per gram ovary.

Similarly, an extract (5000 ml) was prepared from normal hog ovaries (447 g) and a sample diluted 100 times and assayed after serial dilution (bottom curve in Fig. 1). Interpolation shows the test solution to correspond to 28% of the standard solution. Hence, the activator concentra-

---

[15] P. Glas and T. Astrup, *Am. Heart J.* **76**, 504 (1968).

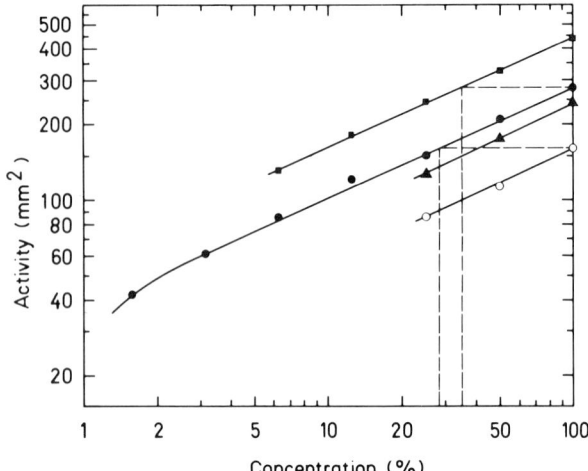

Fig. 1. Assay curves for tissue plasminogen activator. Serial dilutions of stock solutions assayed as follows: (●————●) reference standard containing 8.0 tissue activator units per milliliter, (■————■) solution prepared from pregnant hog ovaries (sample diluted 200 times), (○————○) solution prepared from normal hog ovaries (sample diluted 100 times), (▲————▲) solution prepared from intestinal serosa from hogs (sample diluted 10 times). Abscissa (logarithmic): concentrations of serial dilutions in percentages of the test solution. Ordinate (logarithmic): diameter products in mm² of lyzed zones (each the mean of triplicate).

tion was $(8 \times 28 \times 100 \times 5000)/(100 \times 447) = 2510$ units per gram ovary.

An extract (160 ml) made from 18 g of intestinal serosa from hog was diluted 10 times for assay (curve marked by triangles in Fig. 1). The test solution corresponded to 73% of the standard solution. The activator concentration was $(8 \times 73 \times 10 \times 160)/(100 \times 18) = 520$ units per gram fresh tissue.

The accuracy of the fibrin plate assay of tissue activator is about ±7% (coefficient of variation). In the assay of tissue samples this is an insignificant variation because of the large individual differences. Higher accuracy, about 1%, is provided by the lysis time method (see below) but this can be used only for concentrated activator solutions free from KSCN.

The tissue activator unit was originally defined as the amount of activator in 1 mg of an arbitrarily selected pig heart preparation.[5] All data from our systematic studies of the content of plasminogen activator in tissues have been reported in these units.[9] One tissue activator unit (A and A unit) equals approximately 0.1 CTA unit of urokinase (as defined by the Committee on Thrombolytic Agents, advisory to the National

Heart Institute) but, since tissue plasminogen activator differs chemically from urokinase, comparisons in the same unitage will be valid only with limitations. It is important to remember that the relationship between activator concentration and activity is linear only in a double logarithmic graph, so that, e.g., a 2-fold increase in activity might reflect a 5-fold increase in activator concentration.

Purification of Tissue Activator

Source of Activator. The high specific activity of purified tissue plasminogen activator preparations, in terms of units of activator per milligram of protein, indicates that the molecular concentration in even the most active tissues is low. Hence, it is important, first, to select a highly active tissue for purification, and, second, to develop simple chemical procedures in the early steps of the purification amenable to the treatment of larger quantities of crude material. When the bulk of inert material has been removed, more sophisticated methods of purification can be applied. Unfortunately, nearly all animal tissues are low in activity and the few known exceptions are not easily available in large quantities. After trying several different tissues, we selected ovaries from pregnant hogs because of their extremely high activity. Normal hog ovaries contain from 200 to 5000 A and A units per gram fresh tissue. In late pregnancy they regularly reach 30,000 units, and occasionally even 50,000 units per gram, the highest concentrations yet observed in any organ or species. The selection of such potent material saves many preliminary steps aimed at concentrating the active material. The purification follows with minor modifications the procedure recently described.[16]

Extraction of Activator. One kilogram of frozen ovaries is partially thawed, passed twice through a meat grinder in the cold room, and disintegrated briefly in a rotating knife blender with 1.5 liters of cold 0.05 M KCl adjusting the pH to 6.0 with 1 N HCl. After dilution with 8.5 liters of 0.05 M cold KCl, the suspension is stirred slowly for 2 hours in the cold room keeping pH at 6.0 by addition of HCl. The sediment is collected by centrifugation in the cold for $\frac{1}{2}$ hour at about 1400 g, and the supernatant, containing only 3–5% of the total activity, is discarded. The sediment is then stirred overnight at room temperature with 10 liters of 2 M ammonium thiocyanate adjusting pH to 7.3–7.4 with 1 N KOH. Centrifugation at room temperature yields about 10 liters of an extract with a potency regularly reaching 3000 units per milliliter, corresponding to a total yield of 30 $\times$ 10^6 units per kilogram of ovaries. The protein nitrogen, determined by micro-Kjeldahl after removal of thiocyanate by

[16] P. Kok and T. Astrup, *Biochemistry* **8**, 79 (1969).

dialysis, amounts to about 0.6 mg/ml, yielding a specific activity of about 5000 units per milligram nitrogen.

Separation of Activator. The thiocyanate extract is diluted with 1 volume of water (10 liters) and pH adjusted to 4.5 with 1 N HCl. After standing for 15 minutes, addition of acid is continued until pH 2.5. The sediment is collected after ½ hour by centrifugation and kept overnight at $-20°$ in order to destroy lipoproteins. The next day the material is thawed and extracted for 1 hour at room temperature with 1000 ml 1 M NH$_4$SCN, adjusting pH to 5.9 with 1 N KOH, followed, after centrifugation, by a second similar treatment with 500 ml NH$_4$SCN. The combined extracts are acidified to pH 2.5 with 1 N HCl, left for ½ hour at room temperature and centrifuged. The sediment is suspended in 800 ml of cold ethyl ether and kept overnight at $-20°$ or below. The next day the mixture is thawed and stirred briefly at 4°. The sediment is collected by centrifugation in the cold and again treated with 800 ml of ether followed by freezing overnight. The sediment is again collected by centrifugation and then stirred for 1 hour with 1000 ml of 1 M NH$_4$SCN adjusting pH to 5.9. After centrifugation the treatment is repeated and the two extracts combined, yielding approximately 2000 ml containing about 10,000 units per milliliter and 0.5–0.6 mg protein nitrogen per milliliter. The yield is about 20×10^6 units per kilogram of ovaries with a specific activity between 17,000 and 20,000 units per milligram protein nitrogen.[16a]

Fractionation with Zn^{2+}. A solution containing 0.2 M ZnCl$_2$ and 0.6 M glycine is prepared and adjusted to pH 8 with 7 M NH$_4$OH. Equal volumes (2000 ml) of this solution and the activator extract, similarly adjusted to pH 8, are mixed and the pH carefully changed to 5.8 with 1 N HCl. After ½ hour the precipitate is separated by centrifugation, suspended in 1000 ml of a solution containing 0.02 M ZnCl$_2$ and 0.06 M glycine at pH 5.8 and again collected by centrifugation. It is then extracted for 30 minutes with 500 ml of a solution containing 0.01 M ZnCl$_2$ and 0.03 M glycine adjusting pH to 4.3. The solution is separated by centrifugation and the extraction is repeated. The combined extract (about 1000 ml) contains about 18,000 units/ml and 0.3–0.5 mg protein nitrogen per milliliter. The yield is about 18×10^6 units per kilogram ovaries with a specific activity between 35–60,000 units per milligram nitrogen.

Activator Product I. The acidity is now adjusted to pH 5.8 with 0.5 N

[16a] Utmost care must be applied in the ether treatment in the separation step. We have tried a number of other solvents in order to replace the ethyl ether with a less dangerous fluid. We have met with little success because of great losses of activity, except in the case of *n*-butanol which, preliminary experiments indicate, might be applicable in a modification of the method.

KOH and the active precipitate collected by centrifugation after 1 hour in the cold room. After treatment in the cold for ½ hour at pH 5.8 with 200 ml 0.05 M glycine and centrifugation, followed by a similar treatment with 0.05 M KCl, the precipitate is treated with cold acetone and cold ether, each three times, and the product dried in the air. The resulting dry powder from 1 kg of ovaries weighs about 1.5 g. It contains about 10,000 units per milligram, equal to 70,000 units per milligram nitrogen. The yield is about 15×10^6 units per kilogram of ovaries. The activity of product I is stable when kept in a dry place and is soluble at acid reaction in aqueous solutions of low ionic strength.

Activator Product II. Four hundred milligrams of product I is stirred for 1 hour with 16 ml of 0.1 M glycine-HCl buffer at pH 2.35 containing 0.1 M KCl and the solution clarified by high-speed centrifugation (10,000 g, ½ hour). The solution, 15 ml, is applied to a Sephadex G-200 column, 90×2.5 cm, prepared with a 0.1 M glycine-HCl buffer, pH 2.35, containing 0.2 M KCl. Effluents are collected at 30-minute intervals at a rate of $2–2.5$ ml/cm² per hour using the same buffer and activities and ultraviolet absorbancies determined. Gel filtration is preferably performed in the cold room with glycine buffer half saturated with chloroform. (When processing a whole batch of product I, prepared from 1 kg of ovaries, a 100×5 cm column is used.)

The most active fractions (Fig. 2) are combined (40–50 ml) and

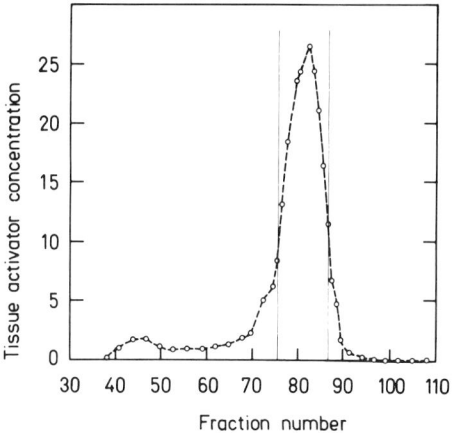

FIG. 2. Collection of active fractions of product I by gel filtration on a Sephadex G-200 column (90×2.5 cm) by ascending flow in a 0.1 M glycine-HCl buffer at pH 2.35 and containing 0.2 M KCl. Effluents collected at intervals of 30 minutes. Abscissa: Fraction number. Ordinate: Activator concentration in units $\times$ 10^3 per milliliter. For preparation of product II, fractions within the thin vertical lines are combined.

dialyzed in the cold room against a 0.01 M glycine-HCl buffer at pH 2.35 containing 0.1 M KCl. The solution is adjusted to pH 7.1 with dilute KOH, left in the cold for 15 minutes, and the sediment collected by centrifugation (in the cold). After treatment three times with cold acetone followed by anhydrous ethyl ether (3 times), it is dried in the air. The resulting powder (about 20 mg from 400 mg of product I) contains from about 100,000 to 175,000 units per milligram protein, and reaching about 1.2×10^6 units per milligram protein nitrogen. The yield is between 50 and 65% of product I corresponding to $8-10 \times 10^6$ units per kilogram of ovaries.

Properties

The acid stability of the tissue activator greatly facilitates its isolation and purification. The products are stable for long periods when stored dry and cold. Crude extracts frequently contain additional amounts of activator activity of a heat-labile nature. Whether these labile activators originate in the blood or interstitial fluids or whether they represent a different type of tissue activator is not known. In the purification procedure described here they are destroyed by the acid treatment, leaving only acid-stable activator. Extraction of plasminogen activator from hog ovaries with $2 M$ thiocyanate gives higher yields than any of many other procedures tested. The purified activator is soluble in acid solutions at low ionic strength. At neutral pH a concentration about 30,000 units per milliliter can be obtained in saline barbital buffer using product I.

The yield and purity of activator preparations may vary in each step and from batch to batch. Both depend much upon the activity of the crude material. The data given represent those obtained in batches prepared from ovaries with activities around 30,000 units per gram. The preparations obtained are of uniform molecular size in gel filtration experiments, suggesting a molecular weight around 60,000. However, zone electrophoresis has shown them to be amenable to further purification, probably 3–5 times. The tissue activator preparations described here compare well in activity with those obtained from pig heart by Bachmann et al.[7] but the selection of a different source and method makes the purification procedure less cumbersome. The porcine tissue activator differs from human urokinase in molecular size[16] and in its interaction with ε-aminocaproic acid,[17] or with inhibitors in certain pathological blood samples.[18] Human tissue activator also differs from human urokinase in

[17] S. Thorsen and T. Astrup, *Proc. Soc. Exptl. Biol. Med.* **130**, 811 (1969).
[18] P. Brakman, E. R. Mohler, and T. Astrup, *Scand. J. Haematol.* **3**, 389 (1966).

its interaction with ε-aminocaproic acid.[19] Tissue activator, though stable at acid reaction, is easily destroyed by pepsin.[20,21] Purified preparations of tissue activator from pig heart were found to activate plasminogen by an enzymatic reaction,[7,22] reportedly through the cleavage of an arginine-valine bond.[23]

Comments

The fibrin plate method is applicable to the assay of crude tissue activator preparations and is little influenced by the presence of various salts (such as thiocyanate) and contaminants in the test samples introduced during purification procedures. It often gives more reliable results than the otherwise more accurate lysis time method[17,24] which requires 100 times higher concentrations of activator. The tissue activator acts specifically with plasminogen to form plasmin; it does not digest casein or plasminogen-free fibrin. The presence of enzymes, which can digest fibrin directly, can be tested on plasminogen-free fibrin plates.[10] The histochemical fibrin slide technique,[25,26] useful for the cellular localization of plasminogen activator activity, is highly sensitive but does not yield quantitative results.

The tissue plasminogen activator is widely distributed in the animal organism.[9,27] Its cellular localization in human venous endothelium was first reported by Todd.[25] In animals, arteries may also be active.[28] Budding capillaries in granulation tissue are particularly rich in tissue activator.[29] Recently, additional sources of tissue plasminogen activator have been revealed, namely corneal epithelial cells,[30] vaginal epithelial cells,[31,32] endometrial epithelial cells,[33,34] and serosal cells.[35] It is quite

[19] P. Kok and T. Astrup, *Federation Proc.* **28**, No. 2, 441 (1969). (Abstract #1037).
[20] K. Alkjaer and T. Astrup, *Trans. 6th Congr. European Soc. Haematol. Copenhagen*, p. 42. Karger, Basel, 1957.
[21] J. Rasmussen and O. K. Albrechtsen, *Fertility Sterility* **11**, 264 (1960).
[22] D. C. McCall and D. L. Kline, *Thromb. Diath. Haemorrhag.* **14**, 116 (1965).
[23] L. Summaria, B. Hsieh, and K. C. Robbins, *J. Biol. Chem.* **242**, 4279 (1967).
[24] M. Lassen, *Scand. J. Clin. Lab. Invest.* **10**, 384 (1958).
[25] A. S. Todd, *J. Pathol. Bacteriol.* **78**, 281 (1959).
[26] H. C. Kwaan and T. Astrup, *Lab. Invest.* **17**, 140 (1967).
[27] T. Astrup, *in* "Dynamics of Thrombus Formation and Dissolution" (S. A. Johnson and M. M. Guest, eds.), p. 275. Lippincott, Philadelphia, 1969.
[28] B. A. Warren, *Brit. J. Exptl. Pathol.* **44**, 365 (1963).
[29] H. C. Kwaan and T. Astrup, *J. Pathol. Bacteriol.* **87**, 409 (1964).
[30] M. Pandolfi and T. Astrup, *Arch. Ophthalmol.* **77**, 258 (1967).
[31] J. Henrichsen and T. Astrup, *J. Pathol. Bacteriol.* **93**, 706 (1967).
[32] K. Tympanidis, A. E. King, and T. Astrup, *Am. J. Obstet. Gynecol.* **100**, 185 (1968).
[33] K. Tympanidis and T. Astrup, *Acta Endocrinol.* **60**, 69 (1969).
[34] K. Tympanidis and T. Astrup, *J. Clin. Pathol.* **22**, 36 (1969).
[35] O. M. Jensen, S. B. Larsen, and T. Astrup, *Arch. Pathol.* **88**, 623 (1969).

possible that more than a single species of tissue activator exists in the organism. Since the activity of endometrial and vaginal epithelial cells is influenced by hormones, it could represent a type of activator different from that present in vascular endothelial cells.

[65] Tissue Activator of Plasminogen (Cytokinase)

By S. Y. ALI

Mammalian tissues contain an activator which converts the blood proenzyme plasminogen to the active protease plasmin.[1] The purified tissue activator of plasminogen is referred to as cytokinase to be consistent with the terminology of plasminogen activators. Tissue activator has been purified from pig ovaries[2] and from pig heart.[3] The present paper will deal with the extraction and purification of tissue activator or cytokinase from rabbit kidney.[4] The preparation of plasminogen and casein and the fibrin-lysis assay method have been dealt with in earlier chapters and they will not be described here.

Assay Method

Principle. The activator is incubated with its substrate, the proenzyme plasminogen, in the first stage and the amount of plasmin so formed is measured by the extent of proteolysis of casein. The trichloroacetic acid-soluble aromatic amino acids and peptides released from casein are measured spectrophotometrically.

Reagents

0.1 M Tris buffer, pH 7.4
4% (w/v) Casein (light white soluble casein, British Drug Houses Ltd.)
10% (w/v) Trichloroacetic acid
Human plasminogen[5]
Rabbit kidney cytokinase[4]

[1] T. Astrup, *Federation Proc.* **25**, 42 (1966).
[2] P. Kok and T. Astrup, *Biochemistry* **8**, 79 (1969).
[3] F. Bachmann, A. P. Fletcher, N. Alkjaersig, and S. Sherry, *Biochemistry* **3**, 1578 (1964).
[4] S. Y. Ali and L. Evans, *Biochem. J.* **107**, 293 (1968).
[5] D. L. Kline, *J. Biol. Chem.* **204**, 949 (1953). Activity of plasminogen is expressed in milliunits as described by F. M. Davidson, *Biochem. J.* **76**, 56 (1960).

Procedure. Nine test tubes are placed in a water bath at 37°. The first set of three test tubes contains (a) 0.2 ml of plasminogen (50 milliunits[5]) the second three (b) 0.2 ml of cytokinase (about 200 units[4]). The third set (c) consists of 0.2 ml of cytokinase and 0.2 ml of plasminogen. The volume in all the tubes is made up to 2 ml with Tris buffer and the tubes are incubated at 37° for exactly 10 minutes. Casein (1 ml) is then added to all the tubes. Trichloroacetic acid (3 ml) is added to one tube from each set. The remaining two test tubes from each set are incubated at 37° for another 30 minutes. Trichloroacetic acid is now added to these tubes as well.

The precipitated protein is then removed by filtration through Whatman No. 3 filter paper. Extinction values of the filtrates are determined in a spectrophotometer at 280 nm. From the average extinction value obtained from the duplicate tubes in each set is subtracted the value for the first blank tube. The value thus obtained from set (a) provides the amount of plasmin contamination in the plasminogen preparation and when this is subtracted from (c) the difference indicates the plasmin formed by the action of cytokinase. The value for set (b) is nil when dealing with purified preparations of cytokinase, but it is positive when crude tissue activator is used and this value should be subtracted from (c).

Definition of Unit and Specific Activity. The E_{280} value, obtained after subtraction of the plasmin and protease activity, is expressed in terms of an arbitrary unit of cytokinase activity. It is defined as the amount of cytokinase that, when allowed to activate 50 milliunits[5] of human plasminogen for 10 minutes and the resulting plasmin is allowed to digest casein for 30 minutes at pH 7.4 and at 37°, produces an E_{280} difference of 0.001. Specific activity is expressed in terms of this unit per milligram of protein[6] in the cytokinase preparation.

Purification Procedure

Kidneys, obtained from young rabbits, are weighed, minced with scissors, and introduced into a Potter-type homogenizer (smooth glass-walled tube and Teflon pestle) and homogenized at 4° at 5000 rpm in sufficient 0.25 M-sucrose to make a 10% (w/v) suspension. The homogenate is subjected to differential centrifugation.[7] The sediment obtained after centrifugation at 5000 g for 10 minutes is rejected and the supernatant fluid is then centrifuged at 140,000 g for 1 hour. The supernatant fluid is decanted and rejected. The sediment, which consists of the lysosome-

[6] O. H. Lowry, N. J. Rosebrough, A. L. Farr, and R. J. Randall, *J. Biol. Chem.* **193**, 265 (1951).
[7] S. Y. Ali and C. H. Lack, *Biochem. J.* **96**, 63 (1965).

microsome fraction,[7] is suspended in water to give a concentration equivalent to 0.5 g of the tissue/ml on a wet weight basis.

The lysosome-microsome fraction, which contains about 90% of the cytokinase activity in the kidney homogenate,[7] is then adjusted to pH 10.5 with a few drops of 1 N NaOH, and centrifuged at 20,000 g for 10 minutes at 4°. The supernatant fluid is neutralized with 1 N HCl and then dialyzed against 5 mM phosphate buffer, pH 7.6, for 18 hours. The nondiffusate (12 ml) is then layered over a DEAE-cellulose column (2.2 $\times$ 35 cm) equilibrated with 5 mM phosphate buffer, pH 7.6. A linear gradient of the same buffer, 0.005 M (300 ml) to 0.4 M (300 ml), is then used for elution of cytokinase at room temperature. Cytokinase[4] and protein[6] assays are performed on those fractions which produce lysis when spotted on fibrin plates.[4,8] Cytokinase appears in the eluate when the molarity of phosphate buffer reaches about 50 mM. The fractions that give the highest specific activity for cytokinase are pooled, dialyzed overnight against distilled water at 4°, and freeze dried. The freeze-dried material is stable at 4° almost indefinitely. The table shows a summary

PURIFICATION OF RABBIT KIDNEY CYTOKINASE

Fraction	Volume (ml)	Total protein (mg)	Total activity (units)	Specific activity (units/mg of protein)	Yield (%)	Purification
Rabbit kidney homogenate	66	416	27,300	66	100	1.0
Lysosome-microsome fraction	13	74	21,800	295	80	4.5
Soluble supernatant fluid after alkaline extraction	12	53	11,700	222	43	3.4
Column fractions 6, 7, and 8	99	3	7,750	2,600	28	40.0
Column fraction 7	33	1	3,070	2,910	11	44.3

of the purification steps. An 86-fold purification can be achieved by stepwise increase of phosphate buffer molarity but there is a consequent fall in percentage recovery of cytokinase activity.

Properties

Stability. Cytokinase is quite stable in neutral buffer solution when stored at 4°. When kept at 22° for 30 minutes cytokinase was stable over the range pH 3–11. It becomes unstable at acid pH values when the

[8] T. Astrup and S. Mullertz, *Arch. Biochem. Biophys.* **40**, 346 (1952).

temperature is raised. It retains all its activity at pH 8–9 even when treated at 37° or 56° for 30 minutes. It is inactivated completely when heated at 80° for 15 minutes.

Purity and Physical Properties. From its behavior on free electrophoresis, and judging from ultracentrifugal sedimentation and immunological studies, it appeared that a cytokinase preparation with a specific activity of 3000 units/mg protein was about 70% pure. Cytokinase has an absorption maximum at 280 nm ($E_{1\,cm}^{0.1\%} = 0.87$ at 280 nm) and does not contain enough hexose or sialic acid to be a glycoprotein. It has a $s_{20,\,w}^{0}$ 2.9–3.1 S and a molecular weight of about 50,000 when measured on a calibrated Sephadex G-100 column. It has an isoelectric point at pH 8.6.[4]

Activators and Inactivators. Although Triton X-100 appears to enhance the activity of crude tissue activator, presumably due to the disruption of subcellular membranes,[9] it does not have any enhancing effect on the activity of purified cytokinase.[4,7] Like urokinase, cytokinase is also inhibited in a competitive manner by ϵ-aminocaproic acid ($0.2\,M$) and aminocyclohexane-carboxylic acid ($50\ mM$). At high inhibitor-to-substrate ratio some nonspecific side effects are apparent and a partial competitive type of inhibition may be taking place. Cytokinase is also inhibited by arginine ($50\ mM$) and by cysteine ($10\ mM$) but is unaffected by p-chloromercuribenzoate ($0.1\ mM$) and iodoacetamide ($5\ mM$).

Specificity. Cytokinase is specific for activation of plasminogen to plasmin although the mode of action is not completely understood. Like urokinase, and unlike streptokinase and staphylokinase, cytokinase can activate plasminogen from various mammalian species. It does not have any proteolytic activity against casein or when examined on a heated-fibrin plate in which the plasminogen has been inactivated. Its action on synthetic amino acid substrates has not been examined but, if its enzymatic nature is similar to urokinase, it should have an esterase activity toward lysine substrates.

Kinetic Properties. Cytokinase activates plasminogen maximally at pH 8.5 and at 37°. The kinetics of activation are those of a zero-order reaction. When the concentration of cytokinase is adjusted with streptokinase and urokinase, in terms of their individual units of activity, it appears to have the same reaction velocity. On this basis 1 Christensen unit of streptokinase was approximately equal to 4 Ploug units of urokinase or to 40 units of cytokinase.

Cytokinase appears to follow typical Michaelis-Menten kinetics when activating plasminogen. Experimental points fit the Lineweaver and

[9] C. H. Lack and S. Y. Ali, *Nature* **201**, 1030 (1964).

Burk plots and cytokinase and urokinase possessed the same Michaelis constant toward human plasminogen with a $K_m = 3.1 \times 10^{-5}\,M$.

Distribution. Tissue activator or cytokinase is present in most mammalian tissue[1,4] and the content in each varies from species to species. Kidney, lung, and spleen are found consistently to have a high tissue activator content. Its presence in the liver can be demonstrated only if the inhibitors present in a homogenate are removed in the supernatant fluid after differential centrifugation. Subcellular fractionation shows that tissue activator is distributed in the lysosome and microsome fractions and the proportion in either varies in different tissues.[4,10] The presence of a high content of cytokinase in the microsome fraction may imply that the tissue is synthesizing the activator. Its predominance in the lysosomes may indicate the storage capacity of that tissue and the release of activator can then be controlled by labilizers and stabilizers of lysosomes. It should be noted that cytokinase is a basic protein, with a dominant positive charge at physiological pH, and will therefore tend to form electrostatic complexes with negatively charged polymers and structural components such as ribonucleic acids, acid glycoproteins, and mucopolysaccharides.

The relationship of plasminogen activators in tissues, in blood, and in other body fluids has not been determined satisfactorily. The vascular endothelium of veins and venules is now considered to be a major source of tissue activator as well as of blood activator.[1,11,12] Urokinase which was once considered to be filtered blood activator, has been shown to originate in the kidney and urinary tract by tissue culture and other experiments.[4,13,14] Rabbit kidney cytokinase has been shown to be similar to human urokinase[4] but its relationship to blood activator is not yet clear.

[10] E. L. Beard, M. H. Montuori, G. J. Danos, and R. W. Busuttil, *Proc. Soc. Exptl. Biol. Med.* **129**, 804 (1968).

[11] A. S. Todd, *J. Pathol. Bacteriol.* **78**, 281 (1959).

[12] R. Holemans, D. McConnell, and J. G. Johnstone, *Thromb. Diath. Haemorrhag.* **25**, 192 (1966).

[13] R. H. Painter and A. F. Charles, *Am. J. Physiol.* **202**, 1125 (1962).

[14] K. Buluk and M. Malofiejew, *Acta Physiol. Polon.* **14**, 351 (1963).

[66] Naturally Occurring Inhibitors of Proteolytic Enzymes

By Beatrice Kassell

Proteinase inhibitors are widely distributed in plants and animals, and a very large number of them have been reported. These chapters are restricted to some representative examples of well-characterized inhibitors from different sources. Other chapters in this volume cover inhibitors present in blood [68] and the secretory inhibitors of pancreas [67]. The book by Vogel, Trautschold, and Werle[1] is an excellent source of further information and includes data on all known inhibitors. Useful reviews are also available.[2,3]

The assays and units in this series of articles are those used by the individual investigators whose purification procedures are described. Because so many different assays and variations of the same assay are used, it has not been possible to express the units of inhibiting activity in any uniform manner. This problem (previously discussed by Vogel et al.[1]) is a serious one. Although methods and units have been recommended as international standards for trypsin and chymotrypsin,[4] it is unfortunate that these have not been generally adopted.

Most of the inhibitors have been isolated as *trypsin* inhibitors. However, almost without exception, they inhibit other enzymes. Their function in plant and animal tissues is a subject of considerable interest, although speculative at present.

I am grateful to many colleagues in this field for personal communications that aided substantially in the preparation of these articles. Thanks are due particularly to Drs. Y. Birk, D. Čechová, J. W. Donovan, R. E. Feeney, H. Fritz, M. Laskowski, Jr., J. Mikola, S. Moore, R. J. Peanasky, P. Portmann, J. Pudles, J. J. Rackis, E. Sach, and E. Werle.

[1] R. Vogel, I. Trautschold, and E. Werle, "Natürliche Proteinasen-Inhibitoren." Thieme, Stuttgart, 1966; also available updated in English translation, Academic Press, New York, 1968.

[2] M. Laskowski and M. Laskowski, Jr., *Advan. Protein Chem.* **9**, 203 (1954).

[3] P. Desnuelle, in "The Enzymes" (P. D. Boyer, H. Lardy, and K. Myrbäck, eds.), Vol. 4, p. 128. Academic Press, New York, 1960.

[4] International Commission for the Standardization of Pharmaceutical Enzymes, First Report, *J. Mondial de Pharmacie* **1**, 6 (1965).

[66a] A Trypsin Inhibitor from Barley

By BEATRICE KASSELL

Introduction

Trypsin inhibitors have been isolated from several grains, e.g., whole wheat flour,[1,2] corn seed,[3] rye and wheat germ,[4] and barley.[5] The inhibitor from barley, prepared by Mikola and Suolinna,[5] has been selected for detailed description. The presence of proteolytic inhibitors in barley had been noted previously.[6,7] Some properties of the other inhibitors are included in Table II.

Assay[5,8]

Principle. Inhibition of the tryptic hydrolysis of benzoylarginine *p*-nitroanilide (BAPA) is measured by the change in absorbance at 410 nm.

Reagents

Buffer 1: Tris chloride, 50 mM, pH 8.2. Just before use, add 1 ml of 2 M CaCl$_2$ to 100 ml of buffer.
Substrate: Dissolve (by gently warming) 43.5 mg of DL-BAPA·HCl (Nutritional Biochemicals Corp., Cleveland, Ohio) in 1 ml of dimethyl sulfoxide. Be sure no crystals remain. Dilute to 100 ml with warmed buffer 1. Use within 1 hour.
Buffer 2: Sodium acetate, 50 mM, pH 5.4 containing 20 mM CaCl$_2$
Enzyme solution: 0.1 unit (about 100 μg) per milliliter in buffer 2
Inhibitor solution: Dilute with buffer 2 to a concentration to give about 50% inhibition of the trypsin (about 30 μg/ml of purified inhibitor).
Acetic acid: 30% in water (v/v)

Procedure. Pipette 5 ml of substrate and 0.5 ml of water into a series of 18 × 150 mm test tubes in a 25° bath. In a second series of small tubes in the bath, place 1 ml of trypsin plus 1 ml of buffer or 1 ml of

[1] G. Shyamala, B. M. Kennedy, and R. L. Lyman, *Nature* **192**, 360 (1961).
[2] G. Shyamala and R. L. Lyman, *Can. J. Biochem.* **42**, 1825 (1964).
[3] K. Hochstrasser, M. Muss, and E. Werle, *Z. Physiol. Chem.* **348**, 1337 (1967).
[4] K. Hochstrasser and E. Werle, *Z. Physiol. Chem.* **350**, 249 (1969).
[5] J. Mikola and E.-M. Suolinna, *European J. Biochem.* **9**, 555 (1969).
[6] J. Laporte and J. Trémolières, *Compt. Rend. Soc. Biol.* **156**, 1261 (1962).
[7] W. C. Burger and H. W. Siegelman, *Plant Physiol.* **19**, 1089 (1966).
[8] B. F. Erlanger, N. Kokowsky, and W. Cohen, *Arch. Biochem. Biophys.* **95**, 271 (1961).

trypsin plus 1 ml of inhibitor solution for standards and unknowns, respectively. Incubate 10 minutes. Timing with a stopwatch, add 0.5 ml of the enzyme (or enzyme and inhibitor) to the substrate and mix well. Exactly 10 minutes later, stop the reaction by adding 1 ml of 30% acetic acid. Measure the absorbance at 410 nm against a blank without enzyme or inhibitor. Subtract the absorbance of the unknowns from the absorbance of the trypsin standard.

Definition of Unit.[5] One unit of enzyme is the amount that liberates 1 micromole of p-nitroaniline per minute.

One unit of inhibitor decreases the reaction rate by one trypsin unit. Specific activity is expressed as inhibitor units per A_{280}.

Purification Procedure[5]

The procedure is summarized in Table I. Operations are at room temperature except as indicated.

Step 1. Extraction. The starting material is Pirkka barley (a Finnish six-row variety, see Distribution). Suspend 1 kg of finely ground barley in 2 liters of water; acidify to pH 4.9 with 1 M acetic acid. Extract for 2 hours at 5° without agitation and centrifuge.

Step 2. Heat Precipitation. Divide the extract into 250-ml portions in 500-ml conical flasks. Heat 15 minutes in a boiling water bath. Cool to room temperature and centrifuge.

Step 3. Ammonium Sulfate Precipitation. Precipitate the active material by adding dry ammonium sulfate to 40% saturation at 5°. Stir for 1 hour at 5° and centrifuge. Resuspend the precipitate in 200 ml of water.

TABLE I
PURIFICATION OF BARLEY TRYPSIN INHIBITOR[a]

Fraction	Volume (ml)	Total activity (units)	Specific activity (units/A_{280})	Recovery (%)
Extract	1,430	738	—	100
After treatment at 100°	1,300	650	—	89
Ammonium sulfate precipitate	265	597	—	81
Same after dialysis	330	525	0.37	71
DEAE-cellulose peak	120	412	0.97	56
20% Ethanol precipitate	9.8	400	1.33	54
Sephadex G-75 peak	91	355	1.83	48
2nd Ethanol precipitate	10.5	345	1.90	47
Final product	130 mg	333	1.90	45

[a] J. Mikola and E.-M. Suolinna, *European J. Biochem.* **9,** 555 (1969) (reproduced by permission of Springer-Verlag, Heidelberg and of the author).

Stir for 30 minutes. Separate the inactive precipitate by centrifuging and wash it once with 70 ml of 0.1 M sodium acetate buffer of pH 4.9. Combine the two active supernatants, and dialyze against 2×10 liters of 10 mM Tris chloride buffer of pH 7.5 at 5°. If a precipitate forms during dialysis, centrifuge and extract the solid (containing 5–15% of the activity) with 25 mM acetic acid. Adjust the extract to pH 7.5 and add it to the main solution.

Step 4. Chromatography on DEAE-Cellulose. Prepare a 4×14 cm column of DEAE-cellulose, previously recycled with 0.5 M HCl and 0.5 M NaOH, adjusted to pH 7.5 and equilibrated with 10 mM Trischloride of pH 7.5, the starting buffer. Apply the solution of step 3. Elute inert proteins first with starting buffer and then elute the active material with 60 mM Tris chloride of pH 7.5. Pool the active peak and precipitate the inhibitor by adding 20 volumes percent of ethanol at −6°. Centrifuge after 30 minutes and dissolve the precipitate in 5–15 ml of 0.1 M sodium acetate buffer of pH 4.4.

Step 5. Gel Filtration on Sephadex G-75. Prepare a 2.5×95 cm column of the Sephadex equilibrated with sodium acetate buffer pH 4.9, ionic strength 0.1. Elute with the same buffer, collecting 5-ml fractions at the rate of 20 ml/hour. Some inactive material is eluted before the active peaks, which appears approximately in tubes 55 to 70. Pool the active peak and dialyze overnight at 5° against 1 liter of water. Adjust the dialyzed solution to pH 7.5 with NaOH and precipitate the inhibitor by adding 20 volumes percent of ethanol at −6°. Dissolve the precipitate in a few milliliters of 15 mM acetic acid and lyophilize. The yield is about 130 mg.

Properties[5]

Stability. The inhibitor is stable to exposure to 100° for 15 minutes at pH 4.9, and to 1 hour's exposure to pH 2.1 at 25°. It is completely destroyed by incubation with pepsin under the latter conditions.

Purity. The purified inhibitor is homogeneous on disc electrophoresis at pH 3.5 and on ultracentrifugation at pH 5.4.

Physical Properties. The molecular weight is 14,400 by equilibrium centrifugation. The $s_{20,w}^0$ is 1.79 S. The partial specific volume is 0.718 calculated from amino acid composition. The frictional ratio (f/f_{min}) is 1.26. A solution of 1 mg of inhibitor per milliliter has an absorbance of 1.27 at 280 nm.

Specificity. The barley inhibitor has equal effect on the tryptic digestion of casein, hemoglobin, and BAPA.

The inhibitor is totally inactive against the proteolytic enzymes of germinated barley: green malt endopeptidase (gelatin substrate), green malt peptidase (BAPA substrate), and barley carboxypeptidase (*N*-carbobenzoxyphenylalanyl-alanine substrate). It does not affect the hydrolysis of casein by chymotrypsin, papain, subtilopeptidase A, *Aspergillus oryzae* proteinases, pH 6.5 and 10.3, *Streptomyces griseus* proteinases or *Pseudomonas fluorescens* proteinases. It does not inhibit the hydrolysis of hemoglobin by pepsin at pH 2.

Kinetic Properties. Inhibition of tryptic digestion of casein or BAPA is linear up to 65% inhibition. The molar ratios of trypsin to inhibitor are 0.94 and 1.12, respectively, for the two substrates at 50% inhibition.

TABLE II

AMINO ACID COMPOSITION AND OTHER PROPERTIES OF TRYPSIN INHIBITORS FROM SEVERAL GRAINS

Amino acid	Barley[a]	Corn seed[b]	Rye germ[c]	Wheat germ[c]
	\multicolumn{4}{c}{Residues per molecule}			
Alanine	10	21	12	9
Arginine	9	17	15	15
Aspartic acid	10	12	15	15
Half-cystine	10	9	24	24
Glutamic acid	14	20	12	10
Glycine	10	22	7	9
Histidine	3	2	1	0
Isoleucine	5	12	5	5
Leucine	9	20	2	3
Lysine	2	2	10	10
Methionine	2	1	2	2
Phenylalanine	3	2	5	5
Proline	11	25	25[d]	23[d]
Serine	8	13	17	15
Threonine	7	14	11	11
Tryptophan	3	—	—	—
Tyrosine	5	3	3	3
Valine	6	9	7	9
Amide groups	(12)	—	—	—
Carbohydrate	0	0		
Molecular weight	14,300	21,190	17,000[d]	17,000[d]
Amino terminus	—	Serine	Serine	Serine
Carboxyl terminus	—	Isoleucine	—	—

[a] J. Mikola and E.-M. Suolinna, *European J. Biochem.* **9**, 555 (1969).
[b] K. Hochstrasser, M. Muss, and E. Werle, *Z. Physiol. Chem.* **348**, 1337 (1967).
[c] K. Hochstrasser and E. Werle, *Z. Physiol. Chem.* **350**, 249 (1969).
[d] Approximate value.

Distribution. Similar trypsin-inhibiting activity is present in 2-row varieties of barley tested (e.g., Balder, Ingrid, Zephyr).

Trypsin inhibitor activity has been detected in several tissues of barley.[9] The inhibitor of endospermal tissues (aleurone layer and starchy endosperm) is immunologically identical to the inhibitor obtained by the isolation procedure described. The inhibitors of embryos, coleoptiles, and rootlets are different from this inhibitor.

Amino Acid Composition. A comparison of the barley inhibitor with some other grains is shown in Table II. The lack of resemblance of the barley inhibitor to the others may be explained by its derivation from a different tissue (see Distribution).

[9] J. Mikola and M. Kirsi, personal communication (1970).

[66b] Bovine Trypsin-Kallikrein Inhibitor (Kunitz Inhibitor, Basic Pancreatic Trypsin Inhibitor, Polyvalent Inhibitor from Bovine Organs)

By BEATRICE KASSELL

Introduction

This protein was first isolated by Kunitz and Northrop[1] from bovine pancreas as a trypsin inhibitor. Later it was found that a kallikrein inactivator widely distributed in bovine tissues is identical in its properties[2] and in amino acid sequence[3] to the Kunitz trypsin inhibitor.[4,5] It differs in structure and other properties from the specific trypsin inhibitor of the pancreatic juice[6] (this volume, [67]).

The inhibitor is commercially available in highly purified form, e.g., as Iniprol (Laboratoire Choay, Paris), as Trasylol (Farbenfabriken Bayer AG, Elberfeld, Germany), as Basic Pancreatic Trypsin Inhibitor (Worthington Biochemical Corp., Freehold, N.J.).

A related inhibitor has been isolated from bovine colostrum.[6a, 6b] A comparison of the two structures is included in this article.

[1] M. Kunitz and J. H. Northrop, *J. Gen. Physiol.* **19**, 991 (1936).
[2] H. Kraut, N. Bhargava, F. Schultz, and H. Zimmermann, *Z. Physiol. Chem.* **334**, 230 (1964).
[3] F. A. Anderer, *Z. Naturforsch.* **20b**, 462 (1965).
[4] B. Kassell, M. Radicevic, S. Berlow, R. J. Peanasky, and M. Laskowski, Sr., *J. Biol. Chem.* **238**, 3274 (1963).
[5] B. Kassell and M. Laskowski, Sr., *Biochem. Biophys. Res. Commun.* **20**, 463 (1965).
[6] L. J. Greene and D. C. Bartelt, *J. Biol. Chem.* **244**, 2646 (1969).

Assay Method[7]

Principle. Inhibition of tryptic hydrolysis of N-α-benzoyl-DL-arginine-p-nitroanilide (BAPA) is measured by following the change in absorbance at 405 nm.

Reagents

Substrate: Dissolve 100 mg of BAPA with warming in 100 ml of water.
Buffer: $0.2\,M$ Triethanolamine-HCl, pH 7.8
Trypsin: (Trypure,[8] Novo Industri A/S, Copenhagen). Dissolve 10 mg in 100 ml of $0.001\,M$ HCl. Store in the cold.
Diluted inhibitor solution in buffer: 0.2 ml approximately equivalent to 10 μg of trypsin.

Procedure. Place 3-ml, 1-cm cuvettes in a spectrophotometer connected to a circulating water bath at $25°$. Equilibrate the solutions in the water bath. Incubate 0.2 ml of trypsin solution, 0.2 ml of inhibitor solution, and 1.6 ml of buffer for 3 minutes in the cuvette. Start the enzymatic reaction by adding 1 ml of substrate. Mix with a plastic spatula. Record the increase in absorbance at 405 nm for 5 minutes. Repeat the determination without the inhibitor.

Definition of Unit and Specific Activity. One enzyme unit corresponds to the hydrolysis of 1 micromole of substrate per minute (ΔA_{405}/min = 3.32 for 3 ml.

One inhibitor unit decreases the activity of two enzyme units by 50%, thus decreasing the absorbance by 3.32/minute for 3 ml.

The specific activity equals inhibitor units per milligram protein.

Purification Procedure

The purification is based on the selective adsorption of the inhibitor from a crude extract by an insoluble polyanionic derivative of trypsin (trypsin-resin) originated by Katchalski and co-workers.[9] The trypsin-inhibitor complex, attached to the resin, is washed free of contaminating material. The inhibitor is then removed by dissociating the complex under

---

[6a] D. Pospíšilová, V. Dlouhá, and F. Šorm, *Fifth Meeting, Federation European Biochem. Soc.* Abstr. No. 971 (1968).

[6b] D. Čechová-Pospíšilová, V. Svestková, and F. Šorm, *Sixth Meeting, Federation European Biochem. Soc.* Abstr. No. 225 (1969).

[7] H. Fritz, G. Hartwich, and E. Werle, *Z. Physiol. Chem.* **345**, 150 (1966).

[8] This is a stabilized trypsin preparation. If another kind of trypsin is used, add $0.02\,M$ CaCl$_2$ to the buffer.

[9] Y. Levin, M. Pecht, L. Goldstein, and E. Katchalski, *Biochemistry* **3**, 1905 (1964).

acidic conditions. The procedure is that of Fritz *et al.*[10-12] This method is particularly suited to the Kunitz inhibitor. Because of the extreme resistance of this protein to enzymatic hydrolysis, it is recovered entirely in its native form from the complex.[12a]

Lung has been selected as starting material, but liver or parotid gland may be used. The amount of tissue processed may be increased easily. All operations are carried out at $0°-4°$.

Preparation of Trypsin-Resin

Titration of Maleic Anhydride-Ethylene Copolymer (EMA).[11] The product, obtained from Monsanto Chemical Co., Inorganic Chemicals Division, Des Plaines, Ill., varies from type to type and from lot to lot. Determine both the free carboxyl groups and the rate of saponification by titrating with a pH-stat as follows. Place a weighed amount of resin (about 1 mg) in a titration vessel of 2.5 ml volume. Introduce the electrodes and a magnetic stirrer. Add 1.2 ml of water. Record the uptake of 10 mM NaOH, added from a microburet. Calculate the content of free carboxyl groups from the initial rapid uptake of NaOH, i.e., from the beginning of the titration to the break in the curve. Resins with 1–5% of free carboxyl groups are suitable. To convert an excessive number of free carboxyl groups back to the anhydride, heat the resin at $105°-110°$ for 24 hours *in vacuo*,[9] then repeat the determination. The time for complete saponification, determined with different types and lots of EMA, varied from 0.17 to 43 hours.

Binding Capacity and Binding Rate of the Resin for Trypsin.[11] Add 20 mg of EMA to an ice-cooled $(0°-4°)$ mixture of 2 ml of 0.1% hexamethylene diamine solution and 7.5 ml of 0.2 M triethanolamine buffer of pH 7.8. Homogenize the suspension briefly, then stir briskly with cooling. Two minutes after adding the resin, add a cold solution of 100 mg of trypsin in 7.5 ml of the same buffer. Remove aliquots every few minutes for rapidly saponifying resins and at longer intervals for slowly saponifying resins. Filter the aliquots at once and determine the trypsin in the filtrate. This is the amount not bound to the resin.

For resins with 1–5% of free carboxyl groups, trypsin binding reaches

[10] H. Fritz, H. Schult, M. Hutzel, M. Wiedemann, and E. Werle, *Z. Physiol. Chem.* **348**, 308 (1967).

[11] H. Fritz, M. Gebhardt, E. Fink, W. Schramm, and E. Werle, *Z. Physiol. Chem.* **350**, 129 (1969).

[12] H. Fritz, personal communication (1969).

[12a] A simpler method for purification of the inhibitor by affinity chromatography is being developed in the author's laboratory, and will be ready for publication in the latter part of 1970.

its maximum of 80–95% within a few minutes. As the free carboxyl groups rise above 5%, the *amount* of trypsin bound falls rapidly. With resins of too low free carboxyl content, the binding of the enzyme is complete only after 6–10 hours.

Trypsin-Resin Reaction.[10] Suspend 500 mg of copolymer in 50 ml of 0.2 M potassium phosphate buffer, pH 7.5, homogenize briefly (cooling in ice), and dilute to 500 ml with the same buffer. Add 50 ml of 0.1% hexamethylene diamine in water and stir for 3 minutes. Add 2.5 g of trypsin (Novo Industries) dissolved in 250 ml of the phosphate buffer and stir for 12 hours with the pH maintained at 7.5. Dilute with 2 volumes of water and centrifuge. Calculate the trypsin content of the resin by determining the activity of the supernatant solution.

To obtain a suitable flow rate, mix the solid enzyme-resin with a 2- to 4-fold volume of cellulose powder (No. 123, Schleicher and Schull, Keene, N.H.), previously suspended and decanted several times. Pour the resin into a column of 4 cm diameter. Wash with 0.1 M triethanolamine buffer, pH 7.8, containing 0.3 M NaCl and 0.01 M CaCl$_2$ until no more active trypsin can be detected in the effluent.

The column may be stored for a long period at 0° in the buffer, and may be reequilibrated and used at least 100 times without loss of activity. Each gram of trypsin attached to the resin will retain about 70 mg of inhibitor.

Inhibitor Purification

Step 1. Crude Extract.[10, 12, 13] Homogenize 1 kg of lung (containing approximately an amount of inhibitor equivalent to 600 mg of the Novo Trypure or 600 units) with 1 liter of 0.1 M triethanolamine buffer, pH 7.8, containing 0.3 M NaCl and 0.01 M CaCl$_2$. Stir the mixture for 1 hour. Add 100 g of Celite 545 and centrifuge for 1 hour at 13,000 g in a refrigerated centrifuge. Resuspend the solid in an equivalent volume of the same buffer and centrifuge again. Combine the two supernatant solutions. Add 50% trichloroacetic acid (w/v) to a final concentration of 2.5%. Allow the solution to stand for 30 minutes at room temperature. Centrifuge as before and adjust the supernatant solution to pH 7.8 with 1 N NaOH. Determine the amount of inhibitor in the supernatant solution.

Step 2. Affinity Chromatography. Pass the inhibitor solution through the column at a rate of 2 ml/min, collecting 15-ml fractions of effluent, until all the solution has passed through or until excess inhibitor appears in the effluent. Wash the column with the same buffer until the effluent is entirely protein-free. Elute the inhibitor with 0.25 M KCl-HCl buffer

---

[13] A few details have been added by the present author.

of pH 1.7–2.0 until the effluent is free from inhibitor. Reequilibrate the column with the pH 7.8 buffer.

Step 3. Concentration and Desalting. Adjust the pH of the combined inhibitor fractions to 5–6 with 1 M KOH and concentrate the solution on a rotary evaporator at 30°. Remove the precipitated salts by centrifuging and wash the precipitate with a small amount of water. Desalt by passing through a column of Sephadex G-25, 2 × 50 cm, equilibrated with 0.1 M ammonium bicarbonate buffer of pH 7.0. Lyophilize the fractions containing inhibitor.

At this stage, the inhibitor is 70–80% pure with 85% yield.

Step 4. Final Purification. The final purification may be carried out by a second passage through the trypsin-resin.[12] Alternate final purification steps are chromatography on CM-cellulose[14,15] or on Amberlite IRC-50.[4]

Properties

Stability. The inhibitor shows remarkable stability to high temperature, acid, alkali, and enzymes. It may be heated in dilute acid[16] at 100° and in 2.5% trichloroacetic acid[1] at 80° without loss of activity. It may be kept for 18 months at room temperature in 0.14 M NaCl without loss of activity.[14] The activity remains constant for 24 hours at room temperature up to pH 12.6, but begins to decrease at pH 12.8.[17] The inhibitor is not inactivated by pepsin,[18,19] trypsin,[18,20] chymotrypsin α[18,20] or B,[21] Pronase,[18,20] kallikrein,[20] plasmin,[20] elastase,[20] collagenase,[20] carboxypeptidase A[21] or B,[20] cobra venom,[20] Russell's viper venom,[20] ficin,[20] papain,[20] bromelain,[20] *Aspergillus oryzae* protease,[20] *Bacillus subtilis* proteases[20] A and B, *Leucostoma* peptidase A[21] (from venom of the western cottonmouth moccasin[22]), *Streptomyces griseus* proteases[21] VI, VII, and VIII, silkworm digestive juice alkaline proteinase,[21,22a] or crude

[14] E. Sach, M. Thély, and J. Choay, *Compt. Rend. Acad. Sci., Paris* **260**, 3491 (1965).

[15] R. Avineri-Goldman, I. Snir, G. Blauer, and M. Rigbi, *Arch. Biochem. Biophys.* **121**, 107 (1967).

[16] N. M. Green and E. Work, *Biochem. J.* **54**, 257, 347 (1953).

[17] M. P. Sherman and B. Kassell, *Biochemistry* **7**, 3634 (1968).

[18] B. Kassell and M. Laskowski, Sr., *J. Biol. Chem.* **219**, 203 (1956); *Federation Proc.* **24**, 593 (1965).

[19] H. Kraut and N. Bhargava, *Z. Physiol. Chem.* **334**, 236 (1963).

[20] H. Kraut and N. Bhargava, *Z. Physiol. Chem.* **348**, 1498 (1967).

[21] B. Kassell and T.-W. Wang, unpublished experiments.

[22] F. W. Wagner, A. M. Spiekerman, and J. M. Prescott, *J. Biol. Chem.* **243**, 4486 (1968). A gift of the enzyme is gratefully acknowledged.

[22a] We are grateful to Dr. J. I. Mukai, Kyushu University, Japan, for a gift of this enzyme.

Penicillium notatum proteinase.[21, 22b] It is, however, digested by thermolysin at 60°–80°.[22c]

Purity. Inhibitor preparations purified by earlier chromatographic methods[4, 14, 15, 23] were homogeneous by electrophoresis, sedimentation, chromatography, amino acid analysis, and end group determination. Inhibitor purified as described in this article gives the correct amino acid analysis[12] after step 4.

Physical Properties. The molecular weight by osmotic pressure[1] is 6000 and by amino acid analysis[4] is 6513. In the ultracentrifuge, the inhibitor shows a monomer–dimer equilibrium, with the dimer predominating in neutral solution. The molecular weights are 11,600–11,700 in 0.9% $NaCl$[24] and 10,300–11,800 in 0.8% $NaCl$ at pH 7.3.[25] At the extremes of the pH range, the monomer predominates. The molecular weights are 6500 in 8 M urea[25] at pH 10.5, 6700 in 0.1 M acetic acid,[25] 6500 ± 200[26] at pH 1.1 and 10.5. The sedimentation constant ($s_{20, w}^{0}$) is 0.91[26] to 1.0.[4] The frictional coefficient[26] (f/f_o) is 1.387.

The isoelectric point is at pH 10[16] to 10.5.[27] The absorbance[4, 16] ($A_{1 cm}^{1\%}$ at 280 nm) is 8.3. The inhibitor is soluble[16] in water, 70% methanol, 70% ethanol, and 50% acetone. The optical rotation[1] ($[\alpha]_D$ per milligram of nitrogen) is −65°. The dispersion constant[28] of the one-term Drude equation ($\lambda_c = 234$ nm) and the Moffitt constant[28] ($b_0 = 169$) are independent of pH and are not altered by the addition of urea or 2-chloroethanol.

Specificity. The following enzymes are inhibited: Trypsins of cow,[1] pig,[20] man,[29] and turkey,[30] acetyl trypsin,[15] bovine chymotrypsins[31] α and B (the latter only slightly), chicken chymotrypsin,[30] bovine[32] and porcine[20, 33] kallikreins, rabbit,[33a] human[29, 34, 34a] and porcine[20] plasmin, and

[22b] W. E. Marshall and J. Porath, *J. Biol. Chem.* **240**, 209 (1965). We are indebted to Dr. Marshall for a sample of this proteinase.

[22c] T.-W. Wang and B. Kassell, *Biochem. Biophys. Res. Commun.,* in press.

[23] V. Dlouhá, J. Neuwirthová, B. Meloun, and F. Šorm, *Collect. Czech. Chem. Commun.* **30**, 1705 (1965).

[24] H. Kraut, W. Körbel, W. Scholtan, and F. Schultz, *Z. Physiol. Chem.* **321**, 90 (1960).

[25] F. A. Anderer and S. Hörnle, *Z. Naturforsch.* **20b**, 457 (1965).

[26] W. Scholtan and S. Y. Lie, *Makromol. Chem.* **98**, 204 (1966).

[27] J. Chauvet, G. Nouvel, and R. Acher, *Biochim. Biophys. Acta* **92**, 200 (1964).

[28] W. Scholtan and H. Rosenkranz, *Makromol. Chem.* **99**, 254 (1966).

[29] R. E. Feeney, G. E. Means, and J. C. Bigler, *J. Biol. Chem.* **244**, 1957 (1969).

[30] C. A. Ryan, J. J. Clary, and Y. Tomimatsu, *Arch. Biochem. Biophys.* **110**, 175 (1965).

[31] F. C. Wu and M. Laskowski, Sr., *J. Biol. Chem.* **213**, 609 (1955).

[32] H. Kraut, E. K. Frey, and E. Bauer, *Z. Physiol. Chem.* **175**, 97 (1928).

[33] E. Sach and M. Thély, *Compt. Rend. Acad. Sci., Paris* **266**, 1200 (1968).

a trypsinlike component of Pronase.[35] Numerous other enzymes have been tested, but are *not* inhibited: pepsin,[18] elastase,[15, 20] liver esterase,[36] thrombin,[29, 37] carboxypeptidase A[15] and B,[20] renin,[38] angiotensin converting enzyme,[39] angiotensinase,[40] enterokinase,[41] ribonuclease,[15] lyzozyme,[15] ribosomal polypeptide chain elongation factors,[42] *Bacillus subtilis* proteases[20] A and B, *Aspergillus oryzae* protease,[20] clostripain[43] (from *Clostridium histolyticum*), ficin,[20] papain,[20] or thermolysin.[22b] It does not react with diisopropyl phosphoryl trypsin,[44] but does form a complex with trypsinogen.[44a]

Kinetic Properties. The inhibition of trypsin is stoichiometric[1] and is considered to be competitive[44, 45] or pseudononcompetitive.[41] Values for the dissociation constant (K_I) between 10^{-9} M and 3×10^{-11} M have been reported.[16, 41, 44, 46, 47] The latter value,[41] determined at pH 7.8, is probably the most accurate. The change in K_I with pH is shown in Table I for complexes with trypsin and kallikrein. The values for trypsin are in good agreement with those obtained earlier by Green[36] below pH 4. The reaction with kallikrein is noncompetitive, and complete inhibition is not achieved even with a large excess of inhibitor.[48]

The reaction of the inhibitor with trypsin is not instantaneous. The rate of the reaction increases[16, 41] with pH and is at its maximum[16] (about 1 minute) between pH 7 and pH 10.5 at 25°. The bimolecular reaction rate constant[41] at 25° and pH 7.8 is 3.05×10^5 liters/mole/second.

[33a] E. G. Vairel and M. Thély, *Ann. Biol. Clin. (Paris)* **18**, 363 (1960).

[34] V. Mansfeld, M. Rybák, Z. Horáková, and J. Hladovek, *Z. Physiol. Chem.* **318**, 6 (1960).

[34a] A. DeBarbieri, M. E. Scevola, and M. Franchini, *Boll. Chim. Farm.* **103**, 37 (1964).

[35] M. Trop and Y. Birk, *Biochem. J.* **109**, 475 (1968).

[36] N. M. Green, *Biochem. J.* **66**, 407 (1957).

[37] C. J. Amris, *Scand. J. Haematol.* **3**, 19 (1965).

[38] L. T. Skeggs, personal communication (1968).

[39] Y. S. Bakhle, *Nature* **220**, 919 (1968).

[40] R. Ripa and P. Gilli, *Boll. Soc. Ital. Biol. Sper.* **44**, 1297 (1968).

[41] J. Pütter, *Z. Physiol. Chem.* **348**, 1197 (1967).

[42] K. Moldave, personal communication (1968).

[43] B. Labouesse and P. Gros, *Bull. Soc. Chim. Biol.* **42**, 543 (1960), and personal communication (1963).

[44] N. M. Green, *J. Biol. Chem.* **205**, 535 (1963).

[44a] V. Dlouhá and B. Keil, *Federation European Biochem. Soc. Letters* **3**, 137 (1969).

[45] P. Jeanteur and E. Fournier, *Compt. Rend. Acad. Sci., Paris* **260**, 722 (1965).

[46] D. Grob, *J. Gen. Physiol.* **33**, 103 (1949).

[47] H. Kraut and R. Körbel-Enkhardt, *Z. Physiol. Chem.* **312**, 161 (1958).

[48] I. Trautschold and E. Werle, *Z. Physiol. Chem.* **325**, 48 (1961).

TABLE I
pH-Dependence of the Dissociation Constants of the Complexes of Trypsin and Kallikrein with Bovine Trypsin-Kallikrein Inhibitor[a]

pH	Trypsin-inhibitor complex K_I (mole/liter)	Kallikrein-inhibitor complex K_I (mole/liter)
2.0	4.5×10^{-4}	—
3.0	2.4×10^{-6}	—
4.0	2.6×10^{-9}	Fully dissociated
5.0	—	1.9×10^{-5}
6.0	1.3×10^{-10}	3×10^{-6}
7.8	2×10^{-11}[b]	1.2×10^{-8}

[a] From R. Vogel, I. Trautschold, and E. Werle, "Natural Proteinase Inhibitors," p. 91. Academic Press, New York, 1968.
[b] Interpolated value.

The rate of reaction with kallikrein is also pH-dependent and not instantaneous.[49]

Distribution.[49] The inhibitor occurs in numerous organs of cattle and goats: pancreas, lung, parotid gland, lymph nodes, spleen, and liver.

Amino Acid Composition and Sequence. The composition is shown in Table II. The structure shown in Fig. 1, independently determined for

TABLE II
Amino Acid Composition of the Bovine Trypsin-Kallikrein Inhibitor[a,b]

Amino acid	Residues per molecule	Amino acid	Residues per molecule
Alanine	6	Methionine	1
Arginine	6	Phenylalanine	4
Aspartic acid	5	Proline	4
Half-cystine	6	Serine	1
Glutamic acid	3	Threonine	3
Glycine	6	Tryptophan	0
Histidine	0	Tyrosine	4
Isoleucine	2	Valine	1
Leucine	2	Amide groups	4
Lysine	4		
		Total N	17.65%

[a] B. Kassell, M. Radicevic, S. Berlow, R. J. Peanasky, and M. Laskowski, Sr., *J. Biol. Chem.* **238**, 3274 (1963); determined with trypsin inhibitor isolated from pancreas.
[b] F. A. Anderer and S. Hörnle, *Z. Naturforschung* **206**, 457 (1965), found the same composition for the lung kallikrein inhibitor.

[49] R. Vogel, I. Trautschold, and E. Werle, "Natural Proteinase Inhibitors," p. 91. Academic Press, New York, 1968.

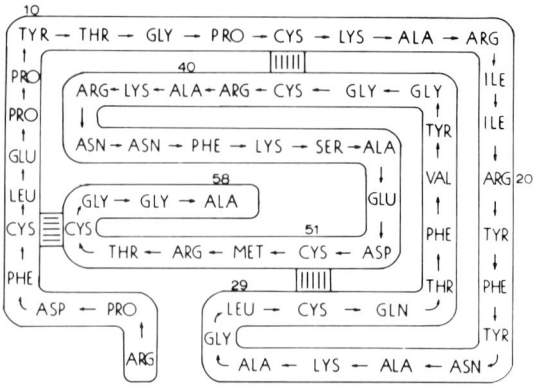

FIG. 1. A schematic representation of the structure of the trypsin inhibitor (based on footnotes 3 and 50).

the proteins isolated as trypsin[50] and kallikrein[3] inhibitors, has been confirmed by two other groups.[51, 52]

Figure 2 compares the basic pancreatic inhibitor with the bovine colostrum inhibitor.[53] The location of at least 21 amino acid residues is identical. In contrast to the pancreatic inhibitor, the colostrum inhibitor is a glycoprotein.

A   Arg-Pro-Asp-Phe-Cys-Leu-Gly-Pro-Pro-Tyr-Thr-Gly-Pro-Cys-Lys-Ala-Arg-Ile -Ile -Arg-

B   Phe-Gln-Thr-Pro-Pro-Asp-Leu-Cys-Gln-Pro-Pro-Gln-Ala-Arg-Gly-Pro-Cys-Lys-Ala-Ala -Leu-Leu-Arg-

A   -Tyr-Phe-Tyr-Asn-Ala-Lys-Ala-Gly-Leu-Cys-Gln-Tyr-Phe-Val-Tyr-Gly-Gly-Cys-Arg-Ala-Lys-Arg-Asn-

B   -Tyr-Phe-Tyr-Asx-Ser-Thr-Ser-Asn-Ala -Cys-Glu-Pro-Phe-Thr-Tyr-Gly-Gly-Cys(asx,asx,asx,asx,glx,

A   -Asn-Phe-Lys-Ser-Ala-Gly-Asp-Cys-Met-Arg-Tyr-Cys-Gly -Gly-Ala

B   ,glx,glx,thr,thr,gly,met,phe)Cys-Leu-Arg-Ile -Cys-Asx-Pro-Pro-Glx-Glx-Thr-Glx-Lys-Ser

FIG. 2. Comparison of the amino acid sequences of the basic pancreatic inhibitor (A) and of the main component of the cow colostrum inhibitor (B). Residues occupying identical positions in both proteins are outlined. (Reproduced with permission of North Holland Publishing Company, Amsterdam.)

[50] B. Kassell and M. Laskowski, Sr., *Biochem. Biophys. Res. Commun.* **20**, 463 (1965).

[51] R. Acher and J. Chauvet, *Bull. Soc. Chim. France* p. 3954 (1967).

[52] V. Dlouhá, D. Pospíšilová, B. Meloun, and F. Šorm, *Collect. Czech. Chem. Commun.* **33**, 1363 (1968).

[53] D. Čechová, V. Svestková, B. Keil, and F. Šorm, *Federation European Biochem. Soc. Letters* **4**, 155 (1969).

[66c] Trypsin and Chymotrypsin Inhibitors from Soybeans

By Beatrice Kassell

Introduction

In 1946, Kunitz[1] crystallized soybean trypsin inhibitor and Bowman[2] discovered a second inhibitor. Since then, other inhibitors have been purified from soybeans,[3, 4, 4a-7] but these have not been studied as much as the first two inhibitors and are not included in this article.

For purification of the Kunitz inhibitor, the procedure described is that of Rackis et al.,[3, 8] based on chromatography on DEAE-cellulose. Several other purification methods have been used: similar DEAE chromatographies,[5-7] preparative electrophoresis,[8a] and isoelectric focusing.[9, 9a]

For the second inhibitor, the purification procedure is derived mainly from the work of Birk and co-workers,[10, 11] but includes modifications made by Frattali[12] and in the author's laboratory. The name Bowman-Birk inhibitor has been used previously.[12, 13] The 1.93 S inhibitor isolated by Yamamoto and Ikenaka[6] is the same or very similar.

Both inhibitors are available commercially, the Kunitz inhibitor from several sources (e.g., California Corporation for Biochemical Research, Los Angeles, Calif.; Gallard-Schlesinger Chemical Manufacturing Co., Garden City, New York; Novo Industri A/S, Copenhagen, Denmark;

[1] M. Kunitz, *J. Gen. Physiol.* **29**, 149 (1946).
[2] D. E. Bowman, *Proc. Soc. Exptl. Biol. Med.* **63**, 547 (1946).
[3] J. J. Rackis, H. A. Sasame, R. L. Anderson, and A. K. Smith, *J. Am. Chem. Soc.* **81**, 6265 (1959).
[4] Y. Birk and A. Gertler, *Bull. Res. Council Israel, Sect. A* **11**, 48 (1962).
[4a] Y. Birk, A. Gertler, and S. Khalef, *Biochim. Biophys. Acta* **67**, 326 (1963).
[5] J. J. Rackis and R. L. Anderson, *Biochem. Biophys. Res. Commun.* **15**, 230 (1964).
[6] M. Yamamoto and T. Ikenaka, *J. Biochem. (Tokyo)* **62**, 141 (1967).
[7] V. Frattali and R. F. Steiner, *Biochemistry* **7**, 521 (1968).
[8] J. J. Rackis, H. A. Sasame, R. K. Mann, R. L. Anderson, and A. K. Smith, *Arch. Biochem. Biophys.* **98**, 471 (1962).
[8a] V. Frattali and R. F. Steiner, *Anal. Biochem.* **27**, 285 (1969).
[9] N. Catsimpoolas, C. Ekenstam, and E. W. Meyer, *Biochim. Biophys. Acta* **175**, 76 (1969).
[9a] N. Catsimpoolas, *Separation Sci.* **4**, 483 (1969).
[10] Y. Birk, *Biochim. Biophys. Acta* **54**, 378 (1961).
[11] Y. Birk, A. Gertler, and S. Khalef, *Biochem. J.* **87**, 281 (1963).
[12] V. Frattali, *J. Biol. Chem.* **244**, 274 (1969).
[13] R. Vogel, I. Trautschold, and E. Werle, "Natürliche Proteinasen-Inhibitoren," pp. 8–21. Thieme, Stuttgart, 1966.

Nutritional Biochemical Corp., Cleveland, Ohio; Sigma Chemical Corp., St. Louis, Mo.; Worthington Biochemical Corp., Freehold, N.J.; and the Bowman-Birk inhibitor from Miles Laboratories, Elkhart, Indiana). None of these commercial preparations is electrophoretically homogeneous.[12,14] Isoelectric focusing has been used[9a] for further purification of both inhibitors from commercial sources.

Assay Method

The casein digestion method of Kunitz[15] has already been described in a previous volume of this treatise.[16]

Definition of Unit. Kunitz[15] defined one tryptic unit as the enzyme activity which causes an increase of one unit of absorbance at 280 nm per minute of digestion of casein, under the standard conditions. The specific activity is expressed as units per microgram of trypsin protein. Trypsin-inhibiting activity is expressed as units of trypsin inhibited, and specific activity as units of trypsin inhibited per microgram of inhibitor.

The inhibitor unit, defined by Rackis,[8] is the amount of trypsin inhibited in micrograms by 1 μg of a standard commercial preparation of the Kunitz soybean inhibitor. Birk[11] uses the original Kunitz definition of the unit (see above).

Kunitz Soybean Inhibitor (SBTIA$_2$)

Purification Procedure[3,8,17,18]

Step 1. Extraction. The starting material is 100 g of defatted soybean meal, a commercial product made from seed-grade dehulled beans of the Adams or Hawkeye variety. Extract the meal at pH 7.2 by stirring slowly, to avoid foaming, for 1 hour at room temperature with 1 liter of water. Centrifuge and reextract the solid with 500 ml of water for ½ hour. Combine the supernatant solutions and adjust to pH 4.4 with 1 M HCl. Discard the precipitate. Bring the solution (the whey) to pH 8.0 with 1 N NaOH and allow to stand at 4° for 1 hour to precipitate the phytate salts. Centrifuge and dialyze the whey at 4° against two changes of distilled water. Lyophilize. The product (4 to 5 g of crude whey protein,

[14] A. C. Eldridge, R. L. Anderson, and W. J. Wolf, *Arch. Biochem. Biophys.* **115**, 495 (1966).
[15] M. Kunitz, *J. Gen. Physiol.* **30**, 291 (1947).
[16] M. Laskowski, Sr., Vol. II [3].
[17] J. J. Rackis, *Proc. Seed Protein Conference, New Orleans,* p. 98, Southern Utilization and Development Division, Agricultural Research Service, U.S. Department of Agriculture (1963).
[18] J. J. Rackis, personal communication (1969).

depending on the source of the beans) contains 0.8–1.0 g of total trypsin inhibitor, of which about half is the Kunitz inhibitor.[18]

Step 2. Chromatography on DEAE-Cellulose. Cycle DEAE-cellulose[19] (Type 40, Brown Co., Berlin, N.H.) with 0.5 M HCl and 0.5 M KOH; neutralize with a concentrated solution of KH_2PO_4. Wash several times on the Büchner funnel with 10 mM potassium phosphate buffer of pH 7.6, the starting buffer, and resuspend in the same buffer. Remove the fines by settling. Prepare a 2.25 × 39 cm column by allowing the

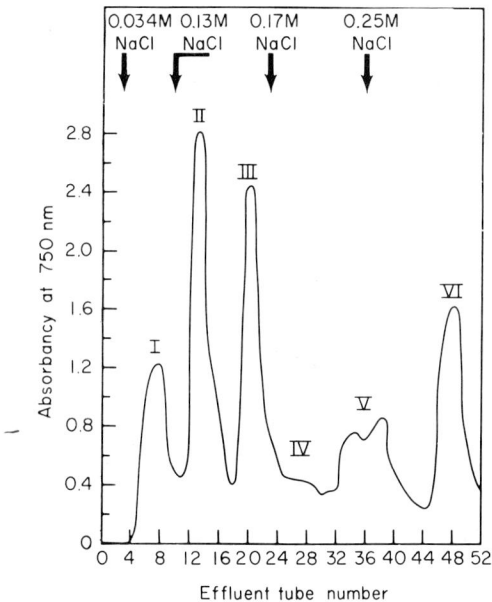

Fig. 1. Stepwise elution of whey proteins from DEAE-cellulose. Sample: 600–800 mg of dialyzed and lyophilized protein in 30 ml of 0.01 M potassium phosphate buffer, pH 7.6. Column: 2.25 × 39 cm; fractions 10 ml. Absorbancy at 750 nm: optical density of 0.1 ml aliquots with Folin-Lowry reagent. The arrows indicate points of change of sodium chloride concentration in buffer. [J. J. Rackis, *Proc. Seed Protein Conference*, p. 98 (1963). Reproduced with permission.[17]]

DEAE to settle to a height of 43 cm and compressing to 39 cm. Dissolve about 600 mg of the whey protein in 30 ml of buffer. Clarify by centrifuging at 12,000 rpm for 5 minutes and apply to the column. Elute at room temperature with stepwise increments of NaCl in starting buffer, as shown in Fig. 1. Inhibiting activity is present in peaks V and VI, the latter being inhibitor $SBTIA_2$. Discard part of the leading edge of peak

---

[19] E. A. Peterson and H. A. Sober, Vol. V [1].

VI: pool the remainder. Dialyze and lyophilize. The yield is nearly 90% of the SBTIA$_2$ in the whey. The purified inhibitor has 1.05 units of trypsin-inhibiting activity per microgram.[8]

Properties

Stability. Below 30° the inhibitor retains its activity[15,20] from pH 1 to 12, although the conformation shows some changes with pH.[21,22] It is reversibly[15] denatured by short heating to 80°, and irreversibly[15] denatured by heating at 90° (50% in 33 minutes, using a 0.5% solution in 2.5 mM HCl). It is digested by pepsin at pH 2.0[15] or 1.5,[23] but not at pH 3.0.[15]

Purity. The inhibitor prepared by the method described gives a single peak[8] on rechromatography, on moving boundary electrophoresis at pH 7.6, and in the ultracentrifuge. However, polyacrylamide gel electrophoresis[14] at pH 9.2 shows several extra bands in very small amounts. Commercial preparations vary in the intensity of the extra bands[14] appearing under these conditions.

The inhibitor prepared by isoelectric focusing[9] is 99% pure by microdensitometry of a disc electrophoresis gel, and shows only one component upon immunoelectrophoresis.

Physical Properties. The molecular weight is 24,000 by osmotic pressure,[15] 21,600[8] and 21,500[22] by ultracentrifugal analysis, 20,000 by light scattering,[24] and 21,700[22] and 22,000[6] by amino acid analysis. The partial specific volume by density determinations is 0.735[8] or 0.698[22] (the latter in 1 M KCl) and by amino acid analysis is 0.74.[6] The sedimentation coefficient, $s_{20,w}^0$ is 2.29.[8]

The electrophoretic mobility at pH 7.6, 2° and 0.1 ionic strength[8] is -8.0×10^{-5}, and at pH 5.9 and 0.1 ionic strength[6] is -7.14×10^{-5}. The isoelectric point[15,21] is 4.5–4.6.

The absorbance of a solution containing 1 mg/ml at pH 7.6 is 0.994[8]–1.013[6] at 280 nm.

Optical rotatory dispersion[25] values are $a_0 = -552°$ and $b_0 = -10°$ (Moffitt plots). The ORD and circular dichroism measurements indicate a nonhelical structure[25,26] in the native state. Some α-helix is formed in the presence of n-propanol,[26] anionic detergents,[26] or 2-chloroethanol.[25] The CD spectrum of the native inhibitor has a strong negative ellipticity

[20] M. Kunitz, *J. Gen. Physiol.* **32**, 241 (1948).
[21] R. F. Steiner and H. Edelhoch, *J. Biol. Chem.* **238**, 925 (1963).
[22] Y. V. Wu and H. A. Scheraga, *Biochemistry* **1**, 698 (1962).
[23] B. Kassell and M. Laskowski, Sr., *J. Biol. Chem.* **219**, 203 (1956).
[24] R. F. Steiner, *Arch. Biochem. Biophys.* **49**, 71 (1954).
[25] K. Ikeda, K. Hamaguchi, M. Yamamoto, and T. Ikenaka, *J. Biochem.* (*Tokyo*) **63**, 521 (1968).
[26] B. Jirgensons, *J. Biol. Chem.* **242**, 912 (1967).

band at 200 nm and a weaker positive band at 226 nm.[26a] Changes in the CD spectrum with pH[26b] are indicative of conformational changes below pH 7 and above pH 11.5. At pH 1.8–2.0, the amplitude of the positive Cotton effect at 226 nm decreases and the position shifts from 226 to 230 nm.[26a]

Specificity. The trypsins of various species are inhibited, e.g., cow,[1] human,[27, 28] salmon,[29] sting ray,[30] barracuda,[30] and turkey.[31] Of other enzymes, bovine chymotrypsin[32] α and B, chicken chymotrypsin,[31] human plasmin,[27, 33, 34] cocoonase,[34a, 34b] plasma kallikrein,[35] and a serum trypsin-like enzyme[36] are inhibited. It blocks the conversion of prothrombin to thrombin.[37] Kallikreins from bovine organs,[13] porcine kallikrein,[38, 39] human[28] and bovine[37] thrombin, liver esterase,[40] and β-subtilisin[41] BPN′ are not inhibited, although complexes are formed with both thrombins.[37] The Kunitz inhibitor does not form a complex with the tosyllysine chloromethyl ketone[41a] (TLCK) derivative of trypsin[41b] or with diisopropylphosphoryl trypsin.[41c]

Kinetic Properties. The inhibition of trypsin is stoichiometric. With benzoylarginine ethyl ester as substrate, the inhibition is competitive.[42, 43]

[26a] B. Jirgensons, M. Kawabata, and S. Capetillo, *Makromol. Chem.* **125**, 126 (1969).

[26b] M. Baba, K. Hamaguchi, and T. Ikenaka, *J. Biochem. (Tokyo)* **65**, 113 (1969).

[27] F. F. Buck, M. Bier, and F. F. Nord, *Arch. Biochem. Biophys.* **98**, 528 (1962).

[28] R. E. Feeney, G. E. Means, and J. C. Bigler, *J. Biol. Chem.* **244**, 1957 (1969).

[29] C. B. Croston, *Arch. Biochem. Biophys.* **112**, 218 (1965).

[30] E. N. Zendzian and E. A. Barnard, *Arch. Biochem. Biophys.* **122**, 714 (1967).

[31] C. A. Ryan, J. J. Clary, and Y. Tomimatsu, *Arch. Biochem. Biophys.* **110**, 175 (1965).

[32] F. C. Wu and M. Laskowski, Sr., *J. Biol. Chem.* **213**, 609 (1955).

[33] L. B. Nanning and M. M. Guest, *Arch. Biochem. Biophys.* **108**, 542 (1964).

[34] V. Mansfeld, M. Rybák, Z. Horáková, and J. Hladovec, *Z. Physiol. Chem.* **318**, 6 (1960).

[34a] F. C. Kafatos, J. H. Law, and A. M. Tarkakoff, *J. Biol. Chem.* **242**, 1488 (1967).

[34b] H. F. Hixson, Jr. and M. Laskowski, Jr., *Biochemistry* **9**, 166 (1970).

[35] E. Werle and L. Mayer, *Biochem. Z.* **323**, 279 (1952).

[36] A. M. Siegelman, A. S. Carlson, and T. Robertson, *Arch. Biochem. Biophys.* **97**, 159 (1962).

[37] G. F. Lanchantin, J. A. Friedmann, and D. W. Hart, *J. Biol. Chem.* **244**, 865 (1969).

[38] E. Werle and B. Kaufmann-Boetsch, *Naturwissenschaften* **19**, 559 (1959).

[39] E. Sach and M. Thély, *Compt. Rend. Acad. Sci. Paris, Ser D,* **266**, 1200 (1968).

[40] N. M. Green, *Biochem. J.* **66**, 416 (1957).

[41] H. Matsubara and S. Nishimura, *J. Biochem. (Tokyo)* **45**, 413, 503 (1958).

[41a] E. Shaw, Vol. XI [80].

[41b] G. Feinstein and R. E. Feeney, *J. Biol. Chem.* **241**, 5183 (1966).

[41c] L. W. Cunningham, Jr., F. Tietze, N. M. Green, and H. Neurath, Faraday Soc. Symposium, Aug. 1952, p. 58.

[42] N. M. Green, *J. Biol. Chem.* **205**, 535 (1953).

[43] P. Métais, H. Schirardin, and J. Warter, *Compt. Rend. Soc. Biol.* **57**, 2307 (1963).

TABLE I
VARIATION OF THE ASSOCIATION CONSTANT^{a,b} OF THE
KUNITZ SOYBEAN INHIBITOR WITH pH

pH	q average number of protons released on association	Log K_{app} $\left(K_{app} = \dfrac{[TI]}{[T][I]}\right)$
3.75	3.35	4.578^g
4.00	2.92	5.349
4.25	2.53	6.004^g
4.50	2.34	6.639
4.75	2.00	7.182
5.00	1.80	7.657
5.25	1.50	8.070
5.50	1.33	8.424
5.75	1.13	8.732
8.30	0^h	9.832^h

^{a} From J. Lebowitz and M. Laskowski, Jr., *Biochemistry* **1**, 1044 (1962), determined by potentiometric measurement at 20° in 0.5 M KCl, 0.05 M CaCl$_2$. (Copyright by the American Chemical Society. Reprinted by permission.)
^{b} The values are in good agreement with partial data obtained by an enzymatic method,^{c} by light scattering,^{d} by fluorescence polarization,^{d} by sedimentation measurement,^{e} and by gel filtration.^{f}
^{c} N. M. Green, *J. Biol. Chem.* **205**, 535 (1953).
^{d} R. F. Steiner, *Arch. Biochem. Biophys.* **49**, 71 (1954).
^{e} E. Sheppard and A. D. McLaren, *J. Am. Chem. Soc.* **75**, 2587 (1953).
^{f} G. A. Gilbert, *Nature* **210**, 299 (1966).
^{g} Average of two methods of calculation.
^{h} Interpolated value.

With protein substrates, the type of inhibition is uncertain.[42] The association constants of the inhibitor-trypsin complex at different pH values are given in Table I. When the inhibitor reacts with trypsin at low pH (3.75), an arginyl-isoleucine bond in the inhibitor is hydrolyzed.[44] Resynthesis of the cleaved bond by trypsin has been achieved,[44a] and an equilibrium mixture of 14% native and 86% cleaved form is established[44b] in the presence of trypsin at pH 4.0.

Amino Acid Composition. The amino acid composition is shown in Table II.

[44] K. Ozawa and M. Laskowski, Jr., *J. Biol. Chem.* **241**, 3955 (1966).
[44a] W. R. Finkenstadt and M. Laskowski, Jr., *J. Biol. Chem.* **242**, 771 (1967).
[44b] C. W. Niekamp, H. F. Hixson, Jr., and M. Laskowski, Jr., *Biochemistry* **8**, 16 (1969).

TABLE II
AMINO ACID COMPOSITION OF TWO SOYBEAN INHIBITORS

Amino acid	Residues per molecule	
	Kunitz inhibitor[a]	Bowman-Birk[b] inhibitor
Alanine	9	4
Arginine	10	2
Aspartic acid	29	11–12
Half-cystine	4	14
Glutamic acid	19	7
Glycine	17	0
Histidine	2	1
Isoleucine	15	2
Leucine	15	2
Lysine	12	5
Methionine	3	1
Phenylalanine	10	2
Proline	11	6
Serine	12	8–9
Threonine	8	2
Tryptophan	2	0
Tyrosine	4	2
Valine	15	1
Amide groups	13	6
Total N	16.34%	15.9%
Molecular weight	22,000	7,975

[a] M. Yamamoto and T. Ikenaka, *J. Biochem. (Tokyo)* **62**, 141 (1967).

[b] Based on the data from three laboratories: Y. Birk, *Ann. N.Y. Acad. Sci.* **146**, 388 (1968); M. Yamamoto and T. Ikenaka, *J. Biochem. (Tokyo)* **62**, 141 (1967); V. Fratteli, *J. Biol. Chem.* **244**, 274 (1969).

Partial Sequences. The amino terminus[45] is Asp-Phe-Val-Leu-Asp. The carboxyl terminus is leucine.[46] The sequences surrounding the two disulfide bridges[47] are:

(1)   Glu-Arg-Cys-Pro-Leu-Thr
              |
      Trp-Leu-Cys-Val-Gly-Ile-Pro-Thr-Glu

(2)   Val-Phe-Cys-Pro-Gln-Gln-Ala
              | |
      Gly-Ile-Asp-Gly-Cys-Lys-Asp-Asp-Glu

[45] T. Ikenaka, K. Shimada, and Y. Matsushima, *J. Biochem. (Tokyo)* **54**, 193 (1963).

[46] E. W. Davie and H. Neurath, *J. Biol. Chem.* **212**, 507 (1955).

[47] J. R. Brown, N. Lerman, and Z. Bohak, *Biochem. Biophys. Res. Commun.* **23**, 561 (1966).

Bowman-Birk Inhibitor from Soybeans (Inhibitor AA)

Purification Procedure[11, 12, 48]

Starting Material. Defatted soybean flour is used, obtained by 48-hour ether extraction (in a Soxhlet apparatus) of finely ground soybeans, Harasoy,[48] Lee,[12] Hawkeye,[12] or any other variety. "Hizyme" (Central Soya, Chicago, Ill.), a hexane-treated flour from mixed varieties of soybeans, proved satisfactory in the author's laboratory without further extraction.

Step 1. Preparation of Crude Inhibitor. Warm 6 liters of 60% ethanol to 60° and add 600 g of the starting material. Stir 1 hour at 55°–60°, cool to room temperature, and filter through Whatman No. 50 paper on a 24-cm Büchner funnel, using a flattened circle of cloth of 21-cm diameter under the filter paper to prevent clogging. Discard the precipitate. Adjust the pH of the filtrate to 5.3 with 1 M HCl (with some increase in cloudiness). Add 2 volumes of acetone with stirring and allow the precipitate to settle. Filter (as above) by decanting most of the liquid through the Büchner funnel before adding the rather sticky precipitate. Dissolve the air-dried precipitate in 500 ml of water and dialyze against water at 5°, using Visking 18/100 cellulose casing.

Step 2. Chromatography on CM-Cellulose.[11,48] Cycle CM-cellulose[19] with 0.5 N NaOH and 0.5 N HCl, adjust to pH 4.0 with acetic acid, and settle several times in the starting buffer (5 mM sodium acetate of pH 4.0). Prepare a 4 × 40 cm column and equilibrate with the starting buffer. Adjust the dialyzed inhibitor solution to pH 4.0 with 1 M acetic acid, centrifuge to remove a small amount of precipitate if present, and apply the clear supernatant solution to the column. Elute the column (a) with 1.5 liters of the starting buffer (including the volume of the applied sample); (b) a gradient of NaCl concentration formed by passing 1 liter of 0.12 M NaCl in starting buffer through 1 liter of the starting buffer in a constant volume mixing flask; (c) an increased gradient of NaCl concentration formed by passing 1 liter of 0.25 M NaCl in starting buffer into the mixing flask. A flow rate of 150 ml per hour with 8 ml fractions is satisfactory. Following elution of an inactive peak, the trypsin inhibitor appears in a peak extending from about tubes 250 to 350. Combine only the tubes of highest and constant activity (approximately the center half of the peak). Dialyze the solution against water at 5° and lyophilize. This gives an almost homogeneous inhibitor, but there is considerable sacrifice in yield. Almost the entire peak can be used if step 3 is to be

[48] Y. Birk and A. Gertler, *in* "Biochemical Preparations," (W. E. M. Lands, ed.) Vol. 12, p. 25. Wiley, New York, 1968. (Material from this article is reproduced by permission of the publisher.)

carried out. In this case, concentrate the pooled peak (e.g., in an ultrafilter).

Step 3. Chromatography on DEAE-Cellulose.[12] Adjust the solution to 140 ml with 0.05 M ammonium acetate solution, pH 6.5, the starting buffer, and dialyze at 5° against the same solution. Warm the protein solution to room temperature and apply to a column of DEAE-cellulose,[19] 2 × 28 cm, equilibrated with the same buffer. Elute with a salt and pH gradient, prepared in a rectangular Varigrad, by adding to compartments 1 through 9, respectively, (a) 0.20 M ammonium acetate, pH 5.0: 0, 5, 10, 20, 40, 60, 80, 100 and 100 ml; (b) 0.05 M ammonium acetate, pH 6.5, as needed to bring all compartments to 100 ml. A flow rate of 40 ml/hour and 5 ml fractions are satisfactory. The active peak appears approximately in tubes 175 to 195. Again, pool tubes of the highest and constant activity, dialyze against water in the cold, and lyophilize.

The yield is about 800 mg (40%). The specific activity of the final product is 10 trypsin inhibitor units per milligram.

Properties

Stability.[10,48] There is no loss in activity when the inhibitor is heated in the dry state at 105° or in 0.02% aqueous solution for 10 minutes at 100°. Autoclaving for 20 minutes at 15 lb pressure destroys the activity. The inhibitor is stable to acid (pH 1.5, 2 hours, 37°) and to peptic and Pronase digestion. Reduction with 2-mercaptoethanol in the presence of urea inactivates the inhibitor and the activity is not recovered upon reoxidation.

Purity. The product of Birk et al.[11,48] (completed by rechromatography on CM-cellulose) behaves as a single substance by rechromatography on CM- and DEAE-cellulose, by sedimentation velocity, by paper electrophoresis at pH 4.9 and 6.9, and by acrylamide gel electrophoresis at pH 4.7 and 8.9. Amino acid analysis, however, shows a fraction of a residue of glycine (Table II).

The preparation of Frattali[12] (for which the final step is the DEAE chromatography above) is almost free of glycine, with ratios of amino acid residues very close to whole numbers. Polyacrylamide electrophoresis at pH 8.35 indicates a single component.

Physical Properties. The molecular weight is 8,000[49]–16,400[6] by ultracentrifugation (monomer–dimer–trimer equilibrium[49]). It is 7975 by amino acid analysis.[12] The $s_{20,w}^0$ value is 1.9[6] to 2.3.[11] The diffusion constant[11] ($D_{20,w}^0$) is 9.03 × 10⁻⁷ cm²/sec. The partial specific volume is 0.69 from amino acid composition.[6,49]

[49] D. B. S. Millar, G. E. Willick, R. F. Steiner, and V. Frattali, *J. Biol. Chem.* **244**, 281 (1969).

The isoelectric point is 4.0^6 to $4.2.^{11}$ The absorbance, $A_{1\,cm}^{1\%}$ at 280 nm is $4.4^{6,\,49}$ to $4.8.^{11}$

The electrophoretic mobility[6] is 8.45×10^{-5} cm^2/V/cm at pH 5.9.

Specificity.[50] The Bowman-Birk preparation inhibits both the proteolytic and esterolytic activities of trypsin and chymotrypsin, when assayed on casein, or on benzoylarginine ethyl ester (BAEE) and acetyltyrosine ethyl ester (ATEE), respectively. The inhibitor-trypsin complex inhibits chymotrypsin (with casein or ATEE assay). The inhibitor-chymotrypsin complex inhibits trypsin (with BAEE, but not with casein assay). It also inhibits a trypsinlike component of Pronase.[51] It has a slight inhibiting effect on human trypsin and on human plasmin, but not on human thrombin.[28]

Kinetic Properties. The inhibition of trypsin and chymotrypsin is noncompetitive[50] and nonstoichiometric.[12] The K_I for trypsin[50] (casein) is 5.6×10^{-8}. The K_I for α-chymotrypsin[50] (casein) is 5.0×10^{-7}.

When the inhibitor reacts with a catalytic amount of trypsin at pH 3.75, a Lys-X bond is hydrolyzed.[52,53] The inhibitor retains activity against both trypsin and chymotrypsin,[52] but becomes a less efficient trypsin inhibitor.[52,54] When the inhibitor reacts with a small amount of chymotrypsin at pH 3.75, its activity toward chymotrypsin is markedly decreased, but that toward trypsin is unaffected.[52] Both modified inhibitors are reconverted in part to the native form by incubation with equimolar amounts of the original enzyme.[54]

Distribution. The inhibitor is found in all varieties of soybeans.

Amino Acid Composition. This is shown in Table II. There is no resemblance to the Kunitz inhibitor.

Terminal group. The amino terminal group is aspartic acid.[6,53]

[50] Y. Birk, *Ann. N.Y. Acad. Sci.* **146**, 388 (1968).
[51] M. Trop and Y. Birk, *Biochem. J.* **109**, 457 (1968).
[52] Y. Birk, A. Gertler, and S. Khalef, *Biochim. Biophys. Acta* **147**, 402 (1967).
[53] Y. Birk, A. Gertler, and S. Khalef, personal communication (1969).
[54] V. Frattali and R. F. Steiner, *Biochem. Biophys. Res. Commun.* **34**, 480 (1969).

[66d] Trypsin Inhibitors from Other Legumes

By Beatrice Kassell

Introduction

In the years following the isolation of inhibitors from soybeans, a large number of other vegetable food proteins have been shown to contain trypsin inhibitors. This article describes the lima bean trypsin inhibitors in some detail, and includes some of the information available about inhibitors from other legumes.

Lima bean (*Phaseolus lunatus*) inhibitors were purified and studied by Tauber and co-workers[1] and then by Fraenkel-Conrat *et al.*[2] Further purification and separation into several active fractions was later achieved using chromatographic methods by Jirgensons and co-workers,[3] by Jones, Moore, and Stein,[4] and by Haynes and Feeney.[5] The method of purification described by Jones *et al.* is used in the present article.

The lima bean inhibitor is commercially available in partially purified form (e.g., Worthington Biochemical Corp., Freehold, N.J., and Sigma Chemical Co., St. Louis, Mo.). The purification procedure described below, starting with step 2, has also been used for further purification of commercial preparations.[4]

Inhibitors have been highly purified from the following sources: red gram[6] (*Cajanus cajan* L.), mung bean[7] (*Phaseolus aureus* Roxb.), kidney bean[8] (*Phaseolus vulgaris*), navy bean[9] (*Phaseolus vulgaris*), black-eyed pea[10] (*Vigna sinensis*), peanut[11] (*Arachis hypogaea* L.), field bean[12] (*Dolichos lablab*), French bean[13] (*Phaseolus coccineus*), and sweet pea[13] (*Lathyrus odoratus*).

Assay

The casein digestion method of Kunitz has been described previously in this treatise.[14]

Definition of Unit.[4] The specific inhibiting activity is expressed as milligrams of *active* trypsin (determined by the rate of hydrolysis of benzoyl-L-arginine ethyl ester) inhibited per milligram of inhibitor.

[1] H. Tauber, B. B. Kershaw, and R. D. Wright, *J. Biol. Chem.* **179**, 1155 (1949).

[2] H. Fraenkel-Conrat, R. C. Bean, E. D. Ducay, and H. S. Olcott, *Arch. Biochem. Biophys.* **37**, 393 (1952).

[3] B. Jirgensons, T. Ikenaka, and V. Gorguraki, *Makromol. Chem.* **39**, 149 (1960).

[4] G. Jones, S. Moore, and W. H. Stein, *Biochemistry* **2**, 66 (1963).

[5] R. Haynes and R. E. Feeney, *J. Biol. Chem.* **242**, 5378 (1967).

[6] S. Tawde, *Ann. Biochem. Exptl. Med (Calcutta)* **21**, 359 (1961).

[7] H.-M. Chü and C.-W. Chi, *Sci. Sinica (Peking)* **14**, 1441 (1965).

[8] A. Pusztai, *Biochem. J.* **101**, 379 (1966).

[9] L. P. Wagner and J. P. Riehm, *Arch. Biochem. Biophys.* **121**, 672 (1967).

[10] M. M. Ventura and J. X. Filho, *Anales Acad. Brasil Cienc.* **38**, 553 (1967).

[11] R. Tixier, *Compt. Rend. Acad. Sci. Paris* **266**, 2498 (1968).

[12] A. P. Banerji and K. Sohonie, *Enzymologia* **36**, 137 (1969).

[13] J. Weder and H.-D. Belitz, *Deut. Lebensm. Rundschau* **65**, 78 (1969).

[14] M. Laskowski, Sr., Vol. II [3].

Purification Procedure[4,15]

Operations are at 4°–5° except as stated.

Step 1. Extraction. The starting material is 100 g of finely ground lima beans (var. Fordhook, certified seed). Extract for 30 minutes with 500 ml of 80% ethanol. Filter by suction. Stir the semidry meal with 500 ml of 0.25 M H_2SO_4 for 1 hour. Centrifuge for 30 minutes at 1200 g. Reserve the supernatant solution and reextract the solid in the same manner.

Combine the extracts and adjust to pH 5.0–5.1 with 5 M NH_4OH. Precipitate the inhibitor by adding solid ammonium sulfate with stirring to 50% saturation. After 2 hours, remove the precipitated protein by centrifugation and take it up to 50 ml of 0.1 M ammonium formate buffer of pH 3.2. Remove insoluble material by centrifugation. The extract contains about 90% of the original inhibiting activity. Based upon the specific activities of the purified inhibitors, 100 g of beans contains about 0.5 g of inhibitors. The extract may be frozen at this stage.

Step 2. Gel Filtration on Sephadex G-75. Prepare a 2 × 150-cm column of the Sephadex, after decanting fines and deaerating in the ammonium formate buffer. Apply 10 ml of the solution of step 1, and elute with the same buffer, at the rate of 16–20 ml per hour. Measure absorbance at 280 nm and inhibiting activity in the effluent. The peak containing the inhibiting activity appears when about 200 ml of effluent have been collected. Pool the active material and lyophilize. The specific activity is 1.6–1.8, and the recovery is 80–90%. Repeat this step to bring all the material to this stage.

Step 3. Chromatography on DEAE-Cellulose. Dissolve the material of step 2 in 50–100 ml of the starting buffer (0.01 M sodium phosphate of pH 7.60 ± 0.02). Equilibrate the solution by passage through a 4 × 40 cm column of Sephadex G-25, medium, at a flow rate of 30–40 ml per hour, collecting 10-ml fractions. Pool the protein peak.

Recycle DEAE-cellulose[16] with HCl and NaOH and equilibrate with the starting buffer. Prepare a 2 × 55 cm column at 25°. Apply up to 150 mg of the protein from the Sephadex column. Elute the column at 25° with a gradient solution prepared by passing 0.4 M NaCl in starting buffer through a constant-volume, 1-liter mixing chamber containing the starting buffer. Collect 5-ml fractions at the rate of 20 to 25 ml per hour. Detect the protein in the effluent fractions by measuring the ninhydrin color of 0.1 ml aliquots after alkaline hydrolysis.[17] Some inactive material

---

[15] S. Moore, personal communication, 1969.
[16] E. A. Peterson and H. A. Sober, Vol. V [1].
[17] C. H. W. Hirs, Vol. XI [35].

is eluted first, then four or more peaks containing inhibitor appear between about 480 and 800 ml of effluent solution. Pool separately the main portion of each peak, sacrificing the material in the valleys between the peaks, and lyophilize.

Step 4. Rechromatography on DEAE-Cellulose. Dissolve each fraction in 10–20 ml of water and equilibrate as before by passing through the Sephadex column. By carrying out the rechromatography under the same conditions as before, each peak appears in the expected position, with small amounts of contaminating neighboring peaks now separated. Lyophilize the pooled fractions from each peak, dissolve the product in 10–20 ml of 0.1 M acetic acid, and desalt by passage through a 2 $\times$ 35 cm column of Sephadex G-25 equilibrated with 0.1 M acetic acid. Lyophilize the pooled fractions. The activities of components 1 to 4 were 2.69, 2.77, 2.28, and 2.44. Calculated on a molar basis (see below), this corresponds to 0.95, 0.99, 0.95, and 0.97 moles of trypsin inhibited by 1 mole of inhibitor.

Variability with Different Batches of Lima Beans. Different commercial preparations and different samples of fresh-frozen lima beans give peaks that vary in amount, in position, and in amino acid composition.[18] As many as six active peaks have been found even with highly inbred varieties of lima beans.[5]

Properties

Stability. Most of the legume inhibitors show considerable stability to heat. There is no loss of activity when lima bean inhibitors[2] are heated for 15 minutes at 90° at pH 5 or pH 7. Kidney bean inhibitor[19] does not lose activity in 90 minutes at 90° in 0.15 M NaCl. At 100° and pH 7.6, red gram inhibitor[6] retains 50% of its activity after 80 minutes and guar meal inhibitor[20] retains 50% after 10 minutes. At 100° and pH 7.2, field bean inhibitor[12] loses 59% of its activity in 30 minutes and 86% in 1 hour; 30 minutes of autoclaving at 15 lb pressure destroys 93% of the activity.

These inhibitors are also relatively stable to extremes of pH. Lima bean inhibitors[2] can be kept for 24 hours at 23° in 10 mM NaOH or for 24 hours at 40° in 10 mM HCl. Red gram inhibitor[6] retains partial activity when exposed for 60 minutes to pH 2.5 and to pH 10. Field bean inhibitor[12] is stable for a week at 15° between pH 3 and pH 10. Kidney bean inhibitor[19] does not lose activity in 2 hours at pH 2 and 37°.

[18] W. Ferdinand, S. Moore, and W. H. Stein, *Biochim. Biophys. Acta* **96**, 524 (1965).
[19] A. Pusztai, *European J. Biochem.* **5**, 252 (1968).
[20] J. R. Couch, C. R. Creger, and Y. K. Bakshi, *Proc. Soc. Exptl. Biol. Med.* **123**, 263 (1966).

TABLE I
PURITY OF SOME LEGUME INHIBITORS

Inhibitors from	Criteria of homogeneity		
	Single peak on rechromatography	Single component on electrophoresis	Single component in ultracentrifuge
Lima bean[a]	Yes (peaks 1–4)	—	—
Lima bean[b]	—	No	Yes (fraction 6)
Mung bean,[c,d] component B	Yes	Yes	Yes
Kidney bean[e]	—	Yes	Yes
Navy bean[f]	Yes	—	Yes
Black-eyed pea[g]	Yes	Yes	Yes
Peanut[h]	—	Yes	—
Field bean[i]	—	Yes	—
Sweet pea[j]	—	Yes	—

[a] G. Jones, S. Moore, and W. H. Stein, *Biochemistry* **2**, 66 (1963).

[b] R. Haynes and R. E. Feeney, *J. Biol. Chem.* **242**, 5378 (1967).

[c] H.-M. Chü and C.-W. Chi, *Sci. Sinica (Peking)* **14**, 1441 (1965).

[d] H.-M. Chü, S.-S. Lo, M.-H. Jen, C.-W. Chi, and T.-C. Tsao, *Sci. Sinica (Peking)* **14**, 1454 (1965).

[e] A. Pusztai, *European J. Biochem.* **5**, 252 (1968).

[f] L. P. Wagner and J. P. Riehm, *Arch. Biochem. Biophys.* **121**, 672 (1967).

[g] M. M. Ventura and J. X. Filho, *Anales Acad. Brasil. Cienc.* **38**, 553 (1967).

[h] R. Tixier, *Compt. Rend. Acad. Sci. Paris* **266**, 2498 (1968).

[i] A. P. Banerji and K. Sohonie, *Enzymologia* **36**, 137 (1969).

[j] J. Weder and H.-D. Belitz, *Deut. Lebensm. Rundschau* **65**, 78 (1969).

Two of the inhibitors are known to resist inactivation by pepsin; they are the inhibitors from lima bean (24 hours at pH 2[2] or pH 1.5[4]) and kidney bean[19] (2 hours at pH 2 and 37°). Lima bean inhibitor is not inactivated by papain.[2]

Purity. From the criteria shown in Table I, several of the inhibitors have now been prepared in a high degree of purity.

Physical Properties. The molecular weights of some inhibitors are shown in Table II. The lima bean inhibitors range from 8300 to 10,000, while the other inhibitors vary up to 24,000.

Other physical constants are known for a few of the inhibitors. For lima bean inhibitors the $s_{20,w}^0$ is 1.5 for the mixture[2] and 1.81 for fraction 6[5]; the calculated partial specific volume ($\bar{v}$) for fraction 6 is 0.725.[5] For navy bean inhibitor,[9] $\bar{v}$ is 0.693. For black-eyed pea inhibitor,[10] $s_{20,w}^0$ is 1.6, $D_{20,w}$ is 8.6×10^{-7} cm²/sec, $\bar{v}$ is 0.69 and the ratio f/f_0 is 1.47. Some values of $A_{280}^{1\%}$ are: 8.23 for black-eyed pea,[10] 7.58 for field bean,[12] and 7.78 for red gram[6] inhibitors. The isoelectric points are: kidney bean[19] inhibitor, 5; field bean[12] inhibitor, 4.7; and peanut[11] in-

TABLE II

MOLECULAR WEIGHTS OF SEVERAL TRYPSIN INHIBITORS OF LEGUMES

Inhibitors from	Amino acid composition	1:1 Ratio to trypsin	Ultracentrifugal measurement	Osmotic pressure
Lima bean[a]				
Peak 1	8,408	8,800	—	—
Peak 2	8,291	8,400	—	—
Peak 3	9,892	10,400	—	—
Peak 4	9,423	9,700	—	—
Lima bean[b]				
Fraction 4	9,100	8,950	—	—
Fraction 6	9,667	9,050	16,200	—
Lima bean[c]	—	—	10,000	9,400
Mung bean[d]	8,113	—	9,000	9,200
Kidney bean[e]	—	9,600	10,000[f]	—
Navy bean[g]	—	11,500	23,000	—
Black-eyed pea[h]	16,923	15,300	17,100	—
Peanut[i]	17,000[j]	—	—	—
Field bean[k]	—	23,700	—	—

[a] G. Jones, S. Moore, and W. H. Stein, *Biochemistry* **2**, 66 (1963).

[b] R. Haynes and R. E. Feeney, *J. Biol. Chem.* **242**, 5378 (1967).

[c] H. Fraenkel-Conrat, R. C. Bean, E. D. Ducay, and H. S. Olcott, *Arch. Biochem. Biophys.* **37**, 393 (1952).

[d] K. Wang, C.-W. Chi, and T.-C. Tsao, *Acta Biochim. Biophys. Sinica* **5**, 510 (1965); *Chem. Abstr.* **64**, 13,000b (1966).

[e] A. Pusztai, *European J. Biochem.* **5**, 252 (1968).

[f] The inhibitor shows a tendency to aggregate.

[g] L. P. Wagner and J. P. Riehm, *Arch. Biochem. Biophys.* **121**, 672 (1967).

[h] M. M. Ventura and J. X. Filho, *Anales Acad. Brasil. Cienc.* **38**, 553 (1967).

[i] R. Tixier, *Compt. Rend. Acad. Sci. Paris* **266**, 2498 (1968).

[j] Calculated by the author. Two peanut inhibitors isolated by Hochstrasser *et al.*[l] had molecular weights of 17.000 by gel filtration, but are believed to be made up of 4 subunits.

[k] A. P. Banerji and K. Sohonie, *Enzymologia* **36**, 137 (1969).

[l] K. Hochstrasser, K. Illchmann, and E. Werle, *Z. Physiol. Chem.* **350**, 929 (1969).

hibitor, 6.75. Optical rotatory dispersion measurements have been made for mung bean inhibitor,[21] component A: the specific rotation is $-16°$, and λ_c is 298 nm; for components III, IV, V, and VI of lima bean inhibitors[3]: the values are, respectively, $-32.0°$, $-32.4°$, $-20.5°$, and $-20.3°$ for $[\alpha]_{546}$ and 258, 258, 280, and 274 for λ_c. Circular dichroism studies of three purified lima bean inhibitors were made by Ikeda *et al.*[22] From pH 7.4 to 11.8, the positive band at 248 nm increases, while there

[21] K. Wang, C.-W. Chi, and T.-C. Tsao, *Acta Biochim. Biophys. Sinica* **5**, 510 (1965); *Chem. Abstr.* **64**, 13,000 b (1966).

[22] K. Ikeda, K. Hamaguchi, M. Yamamoto, and T. Ikenaka, *J. Biochem. (Tokyo)* **63**, 521 (1968).

TABLE III
Amino Acid Composition of Inhibitors from the Seeds of Some Legumes

Amino acid	Lima beans[a]				Peanuts[f]	Navy beans[g]	Black-eyed peas[h]	Mung beans[i]
	Peak 1	Peak 2	Peak 3	Peak 4	Residues per mole			
Alanine	3	3	4	3	7	8	10	3
Arginine	2	2	2	2	14	7	5	4
Aspartic acid	12	14	13	13	14	30	21	10
Half-cystine	12	14	16	14	22	30	13	8
Glutamic acid	6	5	7	7	16	17	14	7
Glycine	1	0	1	1	12	5	4	2
Histidine	5	3	6	6	3	10	7	4
Isoleucine	4	4	5	4	1	9	5	2
Leucine	3	3	3	3	4	6	2	2
Lysine	4	4	4	4	7	11	8	6
Methionine	0	0	0	0	1	1	0	2
Phenylalanine	1	1	2	2	4	4	6	1
Proline	6	6	7	7	13	16	8	6

Serine	12	12	15	13	10	35	28	10
Threonine	4	3	5	5	14	14	5	3
Tryptophan	0	0	0	0	0	0	2	0
Tyrosine	1	1	2	1	3	4	10	1
Valine	1	1	1	1	9	2	2	1
Amide groups	4	5	5	5	—	17	—	5
Total residues	77	76	93	86	154[j]	209	150	72
Total Nitrogen (%)						16.18		15.9
Carbohydrate						2 (hexose)		almost carbohydrate free

[a] The data for the four inhibitors isolated by Jones et al.[b] are given as examples. The fractions isolated from lima beans by Jirgensons et al.[c] and by Haynes and Feeney[d] have similar, but not identical, compositions. Different lots of lima beans subjected to the same purification procedure also yield inhibitors that vary slightly in amino acid composition.[e]

[b] G. Jones, S. Moore, and W. H. Stein, Biochemistry 2, 66 (1963).

[c] B. Jirgensons, T. Ikenaka, and V. Gorguraki, Makromol. Chem. 39, 149 (1960).

[d] R. Haynes and R. E. Feeney, J. Biol. Chem. 242, 5378 (1967).

[e] W. Ferdinand, S. Moore, and W. H. Stein, Biochim. Biophys. Acta 96, 524 (1965).

[f] R. Tixier, Compt. Rend. Acad. Sci. Paris 266, 2498 (1968).

[g] L. P. Wagner and J. P. Riehm, Arch. Biochem. Biophys. 121, 672 (1967).

[h] M. M. Ventura and J. X. Filho, Anales. Acad. Brasil. Cienc. 38, 553 (1967).

[i] H.-M. Chü, S.-S. Lo, M.-H. Jen, C.-W. Chi, and T.-C. Tsao, Sci. Sinica (Peking) 14, 1454 (1965).

[j] The two peanut inhibitors isolated by Hochstrasser et al. [K. Hochstrasser, K. Illchmann and E. Werle, Z. Physiol. Chem. 350, 929 (1969)] contained 156 and 164 amino acid residues.

is little or no change in the negative band near 280 nm. These proteins are nonhelical.

Specificity. Most of the purified inhibitors and many other unpurified extracts of legumes[13, 23] combine with both bovine trypsin and bovine chymotrypsin. However, this double action has been shown to exist at different sites in the same molecule only for Bowman-Birk inhibitor from soybeans (this volume [66c]), for one of the lima bean inhibitors,[5] for black-eyed pea inhibitor,[10] and for kidney bean inhibitor.[19]

Bovine trypsin is inhibited stoichiometrically by lima bean,[4, 5] black-eyed pea,[10] kidney bean,[19] mung bean,[24] green gram[25] (*Phaseolus mungo*), red gram,[6] field bean,[12] and double bean[26] (*Faba vulgaris*) inhibitors. Navy bean inhibitor[9] appears to react with 2 molecules of trypsin; mung bean inhibitor[24] forms a second complex with 2 molecules of trypsin. Peanut inhibitor[11] reacts with different amounts of trypsin, depending on the substrate.

Stoichiometric inhibition of bovine chymotrypsin occurs with kidney bean[19] and field bean[12] inhibitors, while black-eyed pea inhibitor[10] reacts with 2 molecules of chymotrypsin. In the presence of high chymotrypsin concentrations, the kidney bean inhibitor[19] combines with more than 1 mole in a nonstoichiometric manner. Other inhibitors show varying degrees of action against chymotrypsin, e.g., lima bean,[5] peanut,[11] mung bean,[27] double bean,[25] and navy bean.[28]

Human trypsin is inhibited by lima bean, navy bean, kidney bean, and black-eyed pea inhibitors.[28] Lima bean inhibitor is effective against turkey trypsin and chicken chymotrypsin.[29] Bovine acetyltrypsin is inhibited by lima bean[2] and field bean[12] inhibitors.

A few examples of tests with other enzymes can be given, and the reader is referred to the book by Vogel, Trautschold, and Werle[30] for further data. Field bean inhibitor[12] has antithrombin activity. Kidney bean inhibitor reacts stoichiometrically with elastase[19] and with human plasmin,[19, 28] does not inhibit[19] pepsin, papain, β-subtilis protease, or

[23] E. P. Abramova and M. P. Chernikov, *Federation Proc., Transl. Suppl.* **24,** 635 (1965).
[24] H.-M. Chü and C.-W. Chi, *Acta Biochim. Biophys. Sinica* **6,** 22 (1966); *Chem. Abstr.* **65,** 4432h (1966).
[25] P. M. Honawar and K. Sohonie, *J. Sci. Ind. Res.* (*India*) **18C,** 202 (1959).
[26] K. Sohonie and K. S. Ambe, *Nature* **175,** 508 (1955).
[27] H.-M. Chu and C.-W. Chi, *Acta Biochim. Biophys. Sinica* **3,** 229 (1963); *Chem. Abstr.* **59,** 10405 f (1963).
[28] R. E. Feeney, G. E. Means, and J. C. Bigler, *J. Biol. Chem.* **244,** 1957 (1969).
[29] C. A. Ryan and J. J. Clary, *Arch. Biochem. Biophys.* **108,** 169 (1964).
[30] R. Vogel, I. Trautschold, and E. Werle, "Natural Proteinase Inhibitors." Academic Press, New York, 1968.

carboxypeptidase A, has weak[19] action on carboxypeptidase B, and has little[19] or no[28] action on thrombin. Lima bean inhibitor acts on a trypsin-like component of Pronase (fraction C),[31] but does not inactivate Nagarse[5] (subtilisin BPN'), Pronase[5] (type VI from *Streptomyces griseus*), *Aspergillus oryzae* proteinase,[5] human thrombin,[28] or *Clostridium histolyticum* collagenase.[32] Navy bean and black-eyed pea inhibitors are weakly effective against human plasmin,[28] but are ineffective against human thrombin.[28] The peanut inhibitors act on chymotrypsin, plasmin, and serum kallikrein, but not on kallikreins from pancreas, sub-maxillary gland, or urine.[32a]

Kinetic Properties. The known trypsin inhibition constants for the legume inhibitors fall in a narrow range: lima bean,[33] approximately $4 \times 10^{-10} M$; mung bean,[27] 10^{-9} to $10^{-10} M$; field bean,[12] $3.3 \times 10^{10} M$; green gram,[25] $6.1 \times 10^{-10} M$. The reaction with trypsin is instantaneous for field bean[12] and red gram[6] inhibitors.

Distribution. The inhibitors are widely distributed in the seeds of leguminous plants. In double bean plants,[34] the inhibitor is present in seeds, leaves, stems, and roots of the plant, with the highest concentration in the seeds and the lowest in the roots.

Amino Acid Composition. The compositions of several inhibitors are shown in Table III. A few common features are: high cystine content, little or no methionine and tryptophan, and in most cases low tyrosine and phenylalanine content.

Mung bean inhibitor has amino terminal serine.[7] The two peanut inhibitors studied by Hochstrasser *et al.*[32a] differ in their amino terminal residues, one having serine and one valine.

[31] M. Trop and Y. Birk, *Biochem. J.* **109**, 475 (1968).
[32] I. Mandl, H. Zipper, and L. T. Ferguson, *Arch. Biochem. Biophys.* **74**, 465 (1958).
[32a] K. Hochstrasser, K. Illchmann, and E. Werle, *Z. Physiol. Chem.* **350**, 929 (1969).
[33] D. Grob, *J. Gen. Physiol.* **33**, 103 (1949).
[34] K. S. Ambe and K. Sohonie, *Experientia* **12**, 302 (1956).

[66e] Proteolytic Enzyme Inhibitors from *Ascaris lumbricoides*

By BEATRICE KASSELL

Introduction

Early in this century,[1,2] it was discovered that *Ascaris lumbricoides* extracts inhibit trypsin, chymotrypsin, and pepsin. Several groups have since succeeded in purifying chymotrypsin and trypsin inhibitors from *Ascaris*.

The *Ascaris* trypsin inhibitor was partially purified by Collier[3] and was further studied by Green.[4] Rhodes *et al.*[5] isolated two trypsin inhibitors, one from body walls and one from perienteric fluid. Portmann and Fraefel[6] obtained several trypsin inhibitors from whole ascarids. The inhibitor isolated by Pudles *et al.*,[7] which at first appeared to differ in amino acid composition, proved to be identical to the main inhibitor (CM-1 below) of Portmann and Fraefel when a Sephadex G-50 filtration was substituted for dialysis.[8] Recently, Kucich and Peanasky[9] reported two inhibitors from *Ascaris* body walls. It is clear that there are several inhibitors that show some resemblances in amino acid composition. The best characterized is the main inhibitor of Portmann and Fraefel.[6] The purification described is their procedure, but the properties of the other inhibitors are included.

The chymotrypsin inhibitors were also studied by Collier[3] and by Green.[4] The predominant inhibitor of body walls was first obtained pure by Peanasky and co-workers[10,11] and their procedure is described below. Rhodes *et al.*[5] found one inhibitor predominant in perienteric fluid and another (probably the same as Peanasky's) predominant in body walls. Rola and Pudles,[12] using whole ascarids, obtained three inhibitors.

[1] E. Weinland, *Z. Biol.* **44**, 1 (1903).
[2] L. B. Mendel and A. F. Blood, *J. Biol. Chem.* **8**, 177 (1910).
[3] H. B. Collier, *Can. J. Res.* **B19**, 90 (1941).
[4] N. M. Green, *Biochem. J.* **66**, 416 (1957).
[5] M. B. Rhodes, C. L. Marsh, and G. W. Kelley, Jr., *Exptl. Parasitol.* **13**, 266 (1963).
[6] P. Portmann and W. Fraefel, *Helv. Chim. Acta* **50**, 2078 (1967).
[7] J. Pudles, F. H. Rola, and A. K. Matida, *Arch. Biochem. Biophys.* **120**, 594 (1967).
[8] P. Portmann, personal communication, 1969.
[9] U. Kucich and R. J. Peanasky, *Biochim. Biophys. Acta* **200**, 47 (1970).
[10] R. J. Peanasky and M. Laskowski, Sr., *Biochim. Biophys. Acta* **37**, 167 (1960).
[11] R. J. Peanasky and M. M. Szucs, *J. Biol. Chem.* **239**, 2525 (1964).
[12] F. H. Rola and J. Pudles, *Arch. Biochem. Biophys.* **113**, 134 (1966).

Trypsin Inhibitors

Assay Method[8,13]

Principle. Inhibition of tryptic hydrolysis of N-α-benzoyl-DL-arginine-p-nitroanilide (BAPA) is measured by the change in absorbance at 405 nm.

Reagents

Buffer: Tris chloride, 0.1 M, pH 8.0, containing 0.025 M $CaCl_2$

Substrate: Dissolve 47.7 mg of BAPA with warming in 100 ml of buffer. The final concentration in the reaction mixture is 0.844 mM.

Trypsin: 10 mg/100 ml of 0.1 N HCl, containing 0.1 ml of chloroform and 3 drops of saturated $CaCl_2$. This solution keeps 2–3 weeks at 4°. Before using, dilute to 40 μg/ml with water.

Diluted inhibitor solution in buffer, 0.1 ml approximately equivalent to 10 μg of trypsin

Procedure. Place 3-ml, 1-cm cuvettes in a spectrophotometer connected to a circulating water bath at 25°. Equilibrate the solutions in the water bath. Add 2 ml of substrate, 0.5 ml of trypsin (20 μg) and 0.1 ml of inhibitor solution in rapid succession. Mix with a plastic spatula. After 5 minutes, begin to record the increase in absorbance at 405 nm for an additional 5 minutes. Repeat the determination without the inhibitor to determine the total amount of trypsin present.

Definition of Unit.[8] One unit is equivalent to 0.26 micromole of p-nitroaniline not liberated by 20 μg of trypsin in 30 minutes.

Specific activity = Units/mg N.

Purification Procedure[6]

Step 1. Crude Extract. Collect 25 kg of swine ascarids and preserve in 90% ethanol until use. Grind the ascarids in a meat grinder and mix with 5–6 liters of 1% NaCl solution. Homogenize the suspension in a blender in suitable portions. Add 1 g of Diastase and homogenize again until the particles are very fine. Cover with toluene and autolyze for 7 days. Add 25% trichloroacetic acid (TCA) (about 1.5 liters) until the pH of the mixture is 2.3–2.5. Heat rapidly to 80° and cool at once. Centrifuge, resuspend the precipitate in 3 liters of 2.5% TCA, and centrifuge again. The yellow, slightly turbid, supernatant solution contains almost 100% of the inhibitor originally present in the ascarids.

[13] A modification of the method of W. Nagel, F. Willig, W. Peschke, and F. H. Schmidt, *Z. Physiol. Chem.* **340**, 1 (1965).

Step 2. Magnesium Sulfate Precipitation. Adjust the crude extract (13–14 liters) to pH 3.0 with 1 N NaOH and heat to 70°. Saturate with $MgSO_4$ and maintain at 70° until an oily, semisolid layer (containing the inhibitor) forms on the surface and can be easily skimmed off. Discard the lower layer, which contains little inhibitor.

Step 3. Alcohol Fractionation (to reduce the sugar content). Dissolve the upper layer of step 2 in 3 liters of water and adjust to pH 5.0 with 1 N NaOH. Mix with an equal volume of absolute ethanol and cool to 4°. After 24 hours, filter off the precipitated $MgSO_4$ and wash with 50% (v/v) alcohol until the solid is pure white. Adjust the filtrate, containing all the inhibitor, to pH 6.0 with NaOH and concentrate it to 200 ml in a rotary evaporator. Carefully adjust the very viscous solution to pH 7.5 with NaOH and add an equal volume of absolute ethanol. Centrifuge and discard the almost inactive precipitate. Raise the alcohol concentration to 70% by volume and again discard the precipitate, which contains only about 1% of the inhibiting activity. Add alcohol to 90% by volume. Separate the precipitate containing 90% of the inhibitor by centrifuging and dissolve it in 200 ml of water.

Step 4. Ammonium Sulfate Precipitation (further reduction of sugar content). Allow the solution of step 3 to stand 24 hours at 4°. Centrifuge and discard a small amount of inactive precipitate. Add solid ammonium sulfate to the clear supernatant solution to 80% saturation and centrifuge. Dissolve the sticky yellow precipitate in 200 ml water. Repeat the precipitation four times, and each time remove the precipitate by centrifuging. The precipitate formed in the fifth precipitation is yellowish and granular and contains 97% of the starting activity with 95% of the carbohydrate removed.

Step 5. Picric Acid Precipitation. Dissolve the final precipitate of step 4 in 200 ml of water and add an equal volume of absolute alcohol. Keep overnight at 4°. Filter off the precipitated ammonium sulfate on a sintered glass funnel. Evaporate the filtrate to 150 ml in a rotary evaporator. Add 450 ml of saturated (1.2%) picric acid. The inhibitor precipitates as the picrate. Centrifuge, wash the precipitate three times in the centrifuge bottles with 50 ml of absolute alcohol, and suspend it in 150 ml of water. Dissolve the precipitate by adding Dowex 1-X8 (bicarbonate form) with stirring until CO_2 evolution stops. The pH of the solution should not go above 7.5 or at most 8.0. Filter off the yellow-colored Dowex 1 and wash it three times with 25–30 ml of water. Evaporate the combined filtrates *in vacuo* to 30 ml.

Step 6. Gel Filtration on Sephadex G-50. Swell 60 g of Sephadex G-50 overnight in 5 liters of water and transfer it with water to a 3.5 × 55 cm column. Apply the solution of step 5 and make a quantitative transfer with 2 ml of water. Elute the column with water at the rate of 20 ml per hour, collecting 5-ml fractions. Run ninhydrin determinations and trypsin-inhibiting activity on the fractions. The first three peaks contain relatively little activity, while the fourth (peak D) contains 70% of the inhibitor. Pool peak D (approximately tubes 87–103) and evaporate to dryness in a rotary evaporator. Dissolve in 15 ml of Tris acetate buffer [per liter: 1.21 g Tris (0.01 M) plus 0.525 g magnesium acetate (0.0025 M) adjusted to pH 8.0 with acetic acid].

Step 7. Chromatography on DEAE-Sephadex A-25. Swell 60 g of the resin in 3 liters of water, decant the water and cycle with 0.5 N HCl and 0.5 N NaOH. Wash free of alkali with water and suspend in the Tris buffer above. Transfer to a 3.5 × 55 cm chromatography tube and equilibrate with the buffer until the pH of the effluent is 8.0. Apply the inhibitor solution to the column and begin elution with the same buffer. After the first peak has appeared, elute with a gradient prepared by passing 0.1 M magnesium acetate in 0.01 M Tris acetate into 500 ml of the starting buffer in a constant volume mixing bottle. Collect 5-ml fractions at the rate of 20 ml per hour. Locate the peaks by ninhydrin assay and trypsin-inhibiting activity. The active material is distributed among three or four peaks. If the last peaks do not come off the column, change the gradient buffer to 0.2 M magnesium acetate and continue the elution. Pool each peak separately and evaporate to 5 ml on a rotary evaporator. Adjust each fraction to pH 7.5 and cool at 4°. Add ethanol to a concentration of 95%. Centrifuge in the cold and wash the precipitated inhibitor with 20 ml of 95% ethanol. After centrifuging again, dissolve in doubly distilled water and lyophilize.

The highest activity attained in the main (first) fraction is about 1200 units/mg N. The inhibitor should be kept a minimum time at pH 8 because there is some loss of activity. This pH is necessary, however, for sufficient retention of the inhibitor on the resin. At this stage, the hexose content is not greater than 31 μg/mg N. Amino acid analysis and paper electrophoresis at pH 2 show that the fractions are not yet homogeneous.

Step 8. Chromatography on CM-Cellulose. Dissolve the main fraction of step 7 in 3 to 5 ml of 0.01 M sodium acetate buffer, pH 5.5, and apply the solution to a 1.2 × 30 cm column of CM-cellulose,[14] equilibrated with the same buffer. Elute the column with a sodium acetate gradient, prepared as above, 0.01 to 0.1 M, at pH 5.5. The activity appears in up to seven peaks,[8] CM-1, CM-2, etc. Evaporate each pooled fraction *in vacuo* to 5 ml, and precipitate the inhibitor by adding absolute ethanol to a concentration of 95%. Centrifuge in the cold and wash the precipitate twice with 95% alcohol. Dissolve in 0.1% acetic acid and lyophilize.

The activities of fractions CM-1 and CM-2 are the highest, 2600–

---

[14] E. A. Peterson and H. A. Sober, Vol. V [1].

TABLE I

TEMPERATURE AND pH DEPENDENCE[a] OF THE STABILITY OF INHIBITOR[b] CM-1

	Percent activity remaining at pH						
	5	6	7	8	9	10	11
Native inhibitor at 25° (control)	100	100	100	100	100	100	100
10 minutes at 60°	100	100	95	100	96	91	107
10 minutes at 100°	100	100	59	31	6	—	—
24 hours at room temperature	100	86	89	89	89	79	56

[a] U. R. Hänggi, thesis, University of Freiburg, Switzerland, 1969 (reprinted with permission of the author).

[b] P. Portmann and W. Fraefel, Helv. Chim. Acta 50, 2078 (1967).

3200 units/mg N.[8] The other fractions are less active. CM-1 and CM-2 show only one component by paper electrophoresis at pH 2.0, but high-voltage electrophoresis at pH 7 shows that CM-1 still contains about 2% of CM-2. This does not affect the amino acid composition, since the two fractions are similar. If too large a sample is put on the column, incomplete separation results and this chromatography should be repeated.

Properties

Stability. Inhibitor CM-1[6] is stable in acid and neutral solution (Table I). It is not affected by $8 M$ urea.[7] It is not hydrolyzed by

TABLE II

MOLECULAR WEIGHTS OF VARIOUS *Ascaris* TRYPSIN INHIBITORS

Source	Inhibitor	Molecular weight
Whole ascarids[a]	CM-1	$7,192,$[b] $7,000$[c]
Whole ascarids[a]	CM-2	$6,500$[b]
Whole ascarids[a]	CM-3	$15,000-20,000$[c]
Body walls[d]	—	$4,650$[e]
Perienteric fluid[d]	—	$7,100$[e]
Body walls[f]	Peak 1	$5,520$[b]
Whole ascarids[g]	—	$s^0_{20,w} = 1.1$

[a] P. Portmann and W. Fraefel, Helv. Chim. Acta 50, 2078 (1967).

[b] From amino acid analysis.

[c] By Sephadex chromatography.

[d] M. B. Rhodes, C. L. Marsh, and G. W. Kelley, Jr., Exptl. Parasitology 13, 266 (1963).

[e] Derived from 1:1 ratio to trypsin.

[f] U. Kucich and R. J. Peanasky, Biochim. Biophys. Acta, 200, 47 (1970).

[g] J. Pudles, F. H. Rola, and A. K. Matida, Arch. Biochem. Biophys. 120, 594 (1967).

chymotrypsin.[15] It is relatively stable to digestion by pepsin; the activity falls to 80% of the control after 3 days of digestion and no further on longer digestion.[8]

Physical Properties. Table II summarizes the molecular weights of *Ascaris* inhibitors.

Purity. Inhibitor CM-1[6] is almost homogeneous (2% of CM-2), and is free from carbohydrate.

Specificity. Inhibitor CM-1 of Portmann and Fraefel[8,15] has a small amount of activity against chymotrypsin. It does not inhibit pepsin, papain, or Pronase. The two inhibitors isolated by Kucich and Peanasky[9] do not inhibit chymotrypsin.

Kinetic Properties. Green[4] reported the rate constant for inhibitor-trypsin reaction (0.007 ionic strength, Tris buffer, pH 7.8) to be 2.1 ×

TABLE III
AMINO ACID COMPOSITION OF *Ascaris* TRYPSIN INHIBITORS

Amino acid (residues/mole)	Portmann and Fraefel[a]		Kucich and Peanasky[b] peak 1
	CM-1	CM-2	
Lysine	7	6	5
Histidine	0	0	0
Arginine	3	3[c]	3
Aspartic acid	5	4	3
Threonine	4	4	3
Serine	1	1	1
Glutamic acid	11	9–10	8
Proline	6	5	4
Glycine	6	6	4
Alanine	5	5	4
Half-cystine	10	8	8
Valine	2	2	2
Methionine	0	0	0
Isoleucine	3	3	2
Leucine	0	0	0
Tyrosine	0	0	0
Phenylalanine	2	2	2
Tryptophan	1	1	1
NH₃	6–7[d]	—	9

[a] P. Portmann and W. Fraefel, *Helv. Chim. Acta* **50**, 2078 (1967).
[b] U. Kucich and R. J. Peanasky, *Biochim. Biophys. Acta* **200**, 47 (1970).
[c] P. Portmann, personal communication.
[d] W. Fraefel and R. Acher, *Biochim. Biophys. Acta* **154**, 615 (1968).

[15] U. R. Hänggi, thesis, University of Freiburg, Switzerland, 1969.

10^6 liter/mole/second. Kucich and Peanasky[9] report instantaneous reaction between their two inhibitors and trypsin.

The K_I values found by Green,[4] using bovine trypsin at pH 7.8 with BAEE or casein substrate, are in the range of 0.5 to $3 \times 10^{-9} M$. The values found by Kucich and Peanasky,[9] using porcine trypsin at pH 7.5 with hemoglobin substrate, are: for peak 1, $K_I = 9.0 \times 10^{-8} M$; for peak 2, $K_I = 1.3 \times 10^{-8} M$.

Amino Acid Composition. This is shown in Table III. The absence of histidine, leucine, methionine, and tyrosine from all of the inhibitors is notable.

Terminal Groups. All of the Portmann and Fraefel[6] inhibitors have amino terminal glutamic acid.[15]

Amino Acid Sequence.[16] The sequence of inhibitor[6] CM-1 is:

Glu-Ala-Glu-Lys-Cys-(Asx, Glx, Glx, Pro, Gly, Trp)-Thr-Lys-
Gly-Gly-Cys-Glu-Thr-Cys-Gly-Cys-Ala-Gln-Lys-Ileu-Val-
Pro-Cys-Thr-Arg-Glu-Thr-Lys-Pro-Asn-Pro-Gln-Cys-Pro-
Arg-Lys-Gln-Cys-Cys-Ileu-Ala-Ser-Ala-Gly-Phe-Val-Arg-
Asp-Ala-Gln-Gly-Asn-Cys-Ileu-Lys-Phe-Glu-Asp-Cys-Pro-Lys

The positions of the disulfide bridges are shown in Fig. 1.

Chymotrypsin Inhibitors

Assay Method[11, 17]

Reagents

Buffer: sodium phosphate, 0.1 M, pH 7.6
Globin[18] substrate solution: Dissolve 3.6 g of globin in 116 ml of
    a 33% (w/v) aqueous solution of urea (recrystallized from 95%
    alcohol). Add 1 M NaOH with stirring and allow the globin to
    denature at 25° for 30–60 minutes. Adjust the pH to 7.5 with
    18.5 ml of a solution containing 40 g of urea plus 100 ml of
    1 M KH$_2$PO$_4$. The solution may be kept for several weeks in the
    refrigerator.
α-Chymotrypsin solution: Dissolve a few milligrams in 0.0025 M
    HCl. Measure the absorbance at 280 nm. $A_{280} \times 0.5 = \mu g/ml$.

---

[16] W. Fraefel and R. Acher, *Biochim. Biophys. Acta* **154**, 615 (1968).
[17] N. M. Green and E. Work, *Biochem. J.* **54**, 257 (1953).
[18] Peanasky's globin was prepared from hemoglobin crystallized from fresh, citrated human blood.[19] Globin may be obtained commercially from Pentex Biochemicals, Kankakee, Illinois.
[19] A. Rossi-Fanelli, E. Antonini, and A. Caputo, *Biochim. Biophys. Acta* **30**, 608 (1958); *J. Biol. Chem.* **236**, 391 (1961).

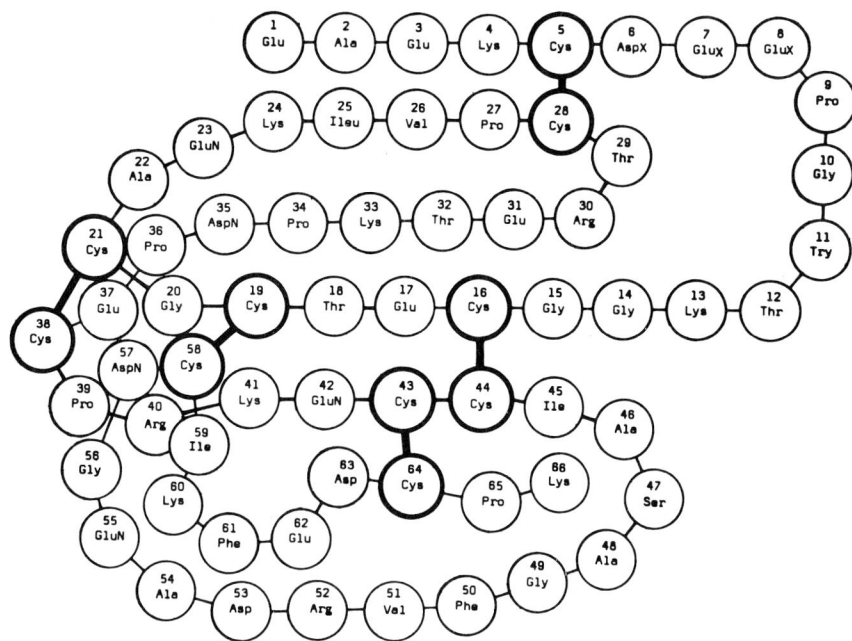

FIG. 1. The complete structure of trypsin inhibitor CM-1, showing the position of the disulfide bridges. (From A. Induni, thesis, University of Freiburg, Switzerland, 1969; reproduced with permission of the author.)

Adjust the solution to contain about 180 μg of protein per milliliter. Chymotrypsin preparations are not 100% "active"; the chymotrypsin used by Peanasky[11] was 71% active, determined by titration with *N-trans*-cinnamoyl imidazole.[19a]

Inhibitor solution: Dilute the solution with water to about 20 μg/ml so that it gives about 50% inhibition of the chymotrypsin.

Trichloroacetic acid (TCA); 5% solution

Procedure. Place a series of polycarbonate centrifuge tubes (16 × 100 mm) in a bath at 37°. Add 0.1 ml (2 μg) of inhibitor solution, 0.8 ml of phosphate buffer and 0.1 ml (18 μg) of chymotrypsin solution. Mix and equilibrate for 15 minutes. With timing, add 1 ml of preequilibrated globin solution, mix, and stop the reaction exactly 5 minutes later by adding 2 ml of TCA. The timed additions are best made with automatic pipettes. Place in ice for 30 minutes and centrifuge for 30 minutes at 20,000 *g* in a Sorvall centrifuge. Use cellulose nitrate tubes (17 × 95 mm) as adapters for the polycarbonate tubes in the SM-24 rotor.

[19a] A. R. Schonbaum, B. Zerner, and M. L. Bender, *J. Biol. Chem.* **236**, 2930 (1961).

Prepare a standard curve by using chymotrypsin without inhibitor in a range of concentrations. Include in each set of assays chymotrypsin standards and blanks prepared by adding equivalent amounts of inhibitor and chymotrypsin after the TCA.

Definition of Unit.[11] One unit inhibits 1 μg of "active" chymotrypsin. Specific activity = units per milligram of protein.

Purification Procedure[10, 11]

Step 1. Crude Extract. Collect *Ascaris lumbricoides*, var. *suis* in the salt medium of Baldwin and Moyle.[20] Wash the ascarids and separate the body walls by dissection. Carry out subsequent operations at 5°, except as indicated.

Homogenize 100-g portions of body walls in 4 volumes of water in a blender for 2 minutes and extract for an additional 15 minutes. Centrifuge at 25,000 g for 20 minutes and then at 57,000 g for 3.5 hours.

Step 2. Acid Precipitation. Adjust the clear red supernatant solution to pH 1.9 with 5 N H_2SO_4 and incubate at 37° for 75 minutes. Cool to 5° and adjust to pH 5.7 with 5 M NaOH. Filter off the heavy precipitate with the aid of Celite.

Step 3. Ammonium Sulfate Fractionation. Adjust the filtrate to pH 4.7 and collect the fraction that precipitates between 50 and 80% saturation with ammonium sulfate. Dissolve the precipitate in water (0.1 the volume of step 2) and adjust the concentration to 10 mg/ml (Biuret reaction,[21] $A_{540} \times 31.5$ = mg protein per 10 ml of Biuret solution). The yields in steps 1, 2, and 3 are almost quantitative.

Step 4. TCA Precipitation. Add 90% (w/v) TCA to bring the solution to 7.5% concentration. After 5 minutes, centrifuge at 25,000 g for 5 minutes. Precipitate[22] the inhibitor from the supernatant solution by adding ammonium sulfate to 30% saturation. Centrifuge as before and suspend the precipitate in 0.2 M sodium phosphate buffer of pH 7.2 (0.2 the volume of step 3). Dialyze the suspension against 0.07 M sodium acetate buffer of pH 5.15 using acetylated[23] 18/32 tubing. The yield at this step is about 50%. At this point repeat the procedure to accumulate about 500 mg of protein; this requires six to seven 100-g portions of ascarids.[24]

Step 5. Continuous-Flow Electrophoresis. The protein concentration of

[20] E. Baldwin and V. Moyle, *J. Exptl. Biol.* **23**, 277 (1947).

[21] A. G. Gornall, C. J. Bardawill, and M. M. David, *J. Biol. Chem.* **177**, 751 (1949).

[22] Modified procedure, R. J. Peanasky, personal communication, 1969.

[23] L. C. Craig, in "Methods of Protein Chemistry" (P. Alexander and R. J. Block, eds.) Vol. 1, p. 116. Macmillan (Pergamon), New York, 1960.

[24] R. J. Peanasky, personal communication, 1969.

the dialyzed solution should be about 15 mg/ml. Equilibrate a continuous-flow electrophoresis apparatus (Beckman-Spinco, Palo Alto, Calif., model CP) with 0.07 M acetate buffer, pH 5.15. Introduce the solution onto the center tab at the rate of 1.5 ml per hour, with an applied potential of 500 V. Determine the absorbance at 280 nm and the chymotrypsin and trypsin-inhibiting activity of each tube. The main chymotrypsin-inhibiting peak is located from tubes 2 to 4 and is free of trypsin-inhibiting activity. Collect fraction 3, containing about 20% of this peak. Lyophilize and dialyze against water, using the acetylated dialysis tubing.

Step 6. Crystallization. Concentrate the dialyzed solution to a 10% solution by lyophilization. Add an equal volume of saturated ammonium sulfate solution. Adjust the pH to 7.2 with 5 M NH$_4$OH plus a drop of 1 M phosphate, pH 7.2. Then add saturated ammonium sulfate dropwise to the first sign of silkiness. The inhibitor crystallizes in the form of fine needles after 8 hours at room temperature. The crystalline material has 100 times the activity of the solution of step 2.

Step 7. Gel Filtration. Prepare two columns equilibrated with 0.05 M Tris chloride, pH 8.5. The first (2.4 $\times$ 34 cm) contains Sephadex G-75; the second (2.4 $\times$ 37 cm) contains Sephadex G-25. Apply about 75 mg of the inhibitor of step 6 to the first column and elute with the same buffer. Collecting 3.0 ml fractions, the most active material appears in fractions 25 to 33. Pool these tubes and lyophilize. Apply about 100 mg of the lyophilized material (from 2 to 3 runs) to the second column and again elute with the same buffer. Collecting 3-ml fractions, the most active material appears in fractions 22 to 29. Pool and lyophilize. Repeat these two gel filtrations in sequence. The product then gives a symmetrical peak with uniform specific activity (about 2000), when chromatographed once more on Sephadex G-75. Finally, dialyze against water in the acetylated bags and lyophilize. The final yield[24] is about 10%. The losses occur mainly by selection of only the purest material in the last steps. If the less pure fractions are reworked, the yield is improved.

Properties

Stability. The inhibitors of Rola and Pudles[12] retain activity when kept at 70° and pH 7.2 for 2 hours, or in 7 M urea at room temperature and pH 7.2 for 1 hour.

Peanasky's inhibitor retains full activity[24] after digestion with leucine aminopeptidase, carboxypeptidases A and B, and trypsin, but is digested[24] by ficin and papain.

Purity. Peanasky's inhibitor[11] appears pure by two criteria: constant

TABLE IV

MOLECULAR WEIGHTS OF CHYMOTRYPSIN INHIBITORS

Inhibitor	Source	Molecular weight
I	Perienteric fluid[a]	12,400[b]
II	Body walls[a]	8,760[b]
	Body walls[c]	7,955[d]
	Body walls[c]	8,000[e]
Pooled	Whole ascarids[f]	8,600[g]

[a] M. B. Rhodes, C. L. Marsh, and G. W. Kelley, Jr., *Exptl. Parasitol.* **13**, 266 (1963).
[b] Derived from 1:1 ratio to chymotrypsin.
[c] R. J. Peanasky, personal communication, 1969.
[d] From amino acid analysis.
[e] By sedimentation equilibrium, using $\bar{v} = 0.749$ determined by pycnometry ($\bar{v} = 0.709$ by amino acid analysis).
[f] F. H. Rola and J. Pudles, *Arch. Biochem. Biophys.* **113**, 134 (1966).
[g] By Sephadex chromatography.

specific activity across the final chromatography peak and absence of more than traces of extra amino terminal groups.

Rola and Pudles[12] inhibitor II behaves as a single component upon electrophoresis at pH 8.6; inhibitor I separates into two active components.

Physical Properties. The isoelectric point[25] is over 10.75. The absorbance at 280 nm ($A_{1\,cm}^{1\%}$) is 7.14.[11] See Table IV for a summary of molecular weights of chymotrypsin inhibitors.

Specificity. Peanasky's preparation inhibits α-chymotrypsin,[11] chymotrypsin B,[11] and subtilisin.[24] It does not inhibit[24] trypsin, leucine aminopeptidase, carboxypeptidase A or B, ficin, or papain. It also does not inhibit[24] the chymotryptic activity of trypsin.

Kinetic Properties. The inhibitor forms a 1:1 complex with α chymotrypsin and chymotrypsin B.

The K_I values are as follows: complex with α-chymotrypsin[11]: 1.1 to $3 \times 10^{-8}\,M$, at pH 5.0–9.0 and 31°; complex with chymotrypsin B[11]: 2.1 to $3.2 \times 10^{-8}\,M$ at pH 5.0 to 9.0 and 31°; complex with subtilisin[24]: $1.2 \times 10^{-6}\,M$.

The reactions[11] with chymotrypsin α and B are not instantaneous; both reactions require 12–15 minutes at pH 7.0 at 0° or 28°.

Distribution.[5] The inhibitors of *Ascaris* occur in body wall, perienteric fluid, intestines, ovaries, and uterus.

Amino Acid Composition. The composition is given in Table V. The

[25] R. J. Peanasky, *Federation Proc.* **22**, 246 (1963).

TABLE V

AMINO ACID COMPOSITION OF THE CHYMOTRYPSIN INHIBITOR[a]

Amino acid	Residues/mole	Amino acid	Residues/mole
Lysine	6	Half-cystine	10
Histidine	1	Valine	4
Arginine	7	Methionine	3
Aspartic acid	4	Isoleucine	1
Threonine	5	Leucine	2
Serine	3	Tyrosine	0
Glutamic acid	8	Phenylalanine	0
Proline	9	Tryptophan	1
Glycine	7	Amide groups	4
Alanine	1		
		Total N	16.59%

[a] R. J. Peanasky, personal communication, 1969.

absence of tyrosine and phenylalanine and the large amounts of cystine and proline are unusual.

Terminal Groups. Peanasky's inhibitor has arginine in the amino[11] and carboxyl[24] terminal positions.

[66f] Chymotrypsin Inhibitor I from Potatoes

By C. A. RYAN and BEATRICE KASSELL

Introduction

Numerous proteinase inhibitors have been isolated from potatoes[1-8a] and sweet potatoes.[9] The present article describes chymotrypsin inhibitor I, first isolated[3] by Ryan and Balls in 1962 and crystallized[4] in 1963.

[1] E. Werle, L. Maier, and F. Löffler, *Biochem. Z.* **321**, 372 (1951).

[2] K. Sohonie and K. S. Ambe, *Nature* **176**, 972 (1955).

[3] C. A. Ryan and A. K. Balls, *Proc. Natl. Acad. Sci. U.S.* **48**, 1839 (1962).

[4] A. K. Balls and C. A. Ryan, *J. Biol. Chem.* **238**, 2976 (1963).

[5] M. Yoshikawa, T. Kiyohara, and K. Ito, *Hyogo Noka Daigaku Kenkyu Hokoku, Nogei-kagaku Hen* **6**, 35 (1963).

[6] V. Rábek and V. Mansfeld, *Experientia* **19**, 151 (1963).

[7] F. H. Rola and C. Correia-Aguiar, *Anales Acad. Brasil Cienc.* **39**, 521 (1967).

[8] J. M. Rancour and C. A. Ryan, *Arch. Biochem. Biophys.* **125**, 380 (1968).

[8a] K. Hochstrasser, E. Werle, R. Siegelmann, and S. Schwarz, *Z. Physiol. Chem.* **350**, 897 (1969).

[9] K. Sohonie and P. M. Honavar, *Sci. Cult. (Calcutta)* **21**, 538 (1956).

Assay Method[3, 10–12]

Principle. The hydrolysis of tyrosine ethyl ester (TEE) by chymotrypsin is measured by potentiometric titration at constant pH.

Reagents

Substrate: TEE (2.09 mg of tyrosine ethyl ester/ml), 10 mM, in 25 mM CaCl$_2$, adjusted to pH 6.3

Stock Enzyme: 1.00 mg/ml of α-chymotrypsin in 1 mM HCl. Determine and adjust the concentration by the absorbance, $A_{280} \times 500 = \mu$g/ml.

Inhibitor: Prepare the inhibitor solution in 10 mM Tris buffer of pH 7.0. Mix suitable amounts of this solution with the chymotrypsin in small vials so that about 50% of the chymotrypsin is inhibited. (5 to 50-μl aliquots of the mixture should contain 0 to 40 μg of active chymotrypsin for accurate titrations.) Allow to stand 3–4 minutes.

NaOH: Standard 0.100 M, free from carbonate

Procedure. The apparatus for manual titration consists of a small thermostated cup connected to a circulating water bath at 25° and mounted on a magnetic stirrer. Place the combination microelectrode of a pH meter having an expanded scale at pH 6–7, a fine tube to deliver nitrogen, and the tip of a microburet containing the NaOH in position so that they will be beneath the surface of the solution. Equilibrate the solutions in the bath. Pipette 2.0 ml of substrate into the cup. Adjust to pH 6.4 with NaOH from the buret. Start the reaction by adding 50 μl of the solution prepared by mixing enzyme and inhibitor. When the pH reaches 6.3, start timing the reaction. Add 2 to 4 μl of NaOH and note the time that the pH again reaches 6.3. Continue the addition of increments of NaOH and timing over a period of 3 to 5 minutes (5 to 6 readings). Repeat the entire procedure, substituting 50 μl of a solution containing the same amount of enzyme, but with buffer in place of inhibitor. A pH-stat may be used.

Calculate the activity of the chymotrypsin from the slope of the reaction curve without inhibitor and the inhibition from the difference between the two values. The reaction is first order, so the tangent to the initial rate curve is used.

Definition of Unit and Specific Activity. An esterase unit is expressed[12]

---

[10] G. W. Schwert, H. Neurath, S. Kaufman, and J. E. Snoke, *J. Biol. Chem.* **172**, 221 (1948).

[11] F. L. Aldrich, Jr. and A. K. Balls, *J. Biol. Chem.* **233**, 1355 (1958).

[12] C. A. Ryan, *Biochemistry* **5**, 1592 (1966).

as micromoles of substrate hydrolyzed per minute per milligram of enzyme in 2 ml.

The reaction between chymotrypsin and inhibitor I is stoichiometric and therefore the milligrams of chymotrypsin neutralized can be used to calculate the milligram of inhibitor in each preparation. The best values for milligrams of α-chymotrypsin neutralized per milligram inhibitor I approach 3.0.[12]

Purification Procedure[13]

The procedure is based on the property of the inhibitor to dissociate into subunits in a solution at least 3 M in guanidine concentration. Inhibitor I is separated on Sephadex G-75, first as a molecule of 39,000 molecular weight, and then as subunits of 9000 to 10,000 molecular weight. Finally, the native protein is reformed and isolated.

The starting material is 100 lb of new Russet Burbank potatoes (*Solanum tuberosum*). The procedure can be scaled up or down as desired. Gram quantities can be prepared in a relatively short time.

Step 1. Extraction. Cut the potatoes into small pieces, with peels intact, and soak in 5-gallon buckets containing a sodium dithionite solution (7 g per liter) until the process of cutting is complete. Pour off the sodium dithionite solution and discard it. Homogenize the potato pieces to a mushy pulp in a 4-liter capacity Waring-type blendor. Add about 500 ml of the sodium dithionite solution to facilitate the blending. Collect the pulp in buckets and express the juice through style A nylon cloth (W. G. Runkles Machinery Co., Trenton, N.J.) using an apple-type press at approximately 1000 psi. For a laboratory scale, use four layers of cheesecloth and express the juice by hand. Adjust the expressed juice (about 15 liters) to pH 3.0, using about 22 ml of 6 N HCl per liter, and then centrifuge for 15 minutes at 5°. Carefully pour off the clear supernatant solution and collect it in large glass cylinders.

Step 2. Ammonium Sulfate Precipitation. Add solid ammonium sulfate slowly at 5° to give 70% saturation (472 g/liter). Stir the mixture gently for 1 hour. Centrifuge as above and discard the supernatant solution. Suspend the precipitate in 2 liters of 70% ammonium sulfate. Filter at room temperature through Whatman No. 1 filter paper on a 25-cm Büchner funnel using a 1/4-in. pad of Celite No. 545 filter aid. Wash the precipitate, while on the pad, with 1 liter of 70% ammonium sulfate. This removes residual sodium dithionite.

Dissolve the precipitate in 2 liters of water by stirring in a beaker for 1 hour with a magnetic stirrer at room temperature. Remove the

[13] C. A. Ryan and J. C. Melville, manuscript in preparation.

Celite by filtering by suction through Whatman No. 1 filter paper. The yield of inhibitor I is 74% based on the original potato juice.

Step 3. Heat Treatment. Divide the clear filtrate into 500-ml portions and heat in 2-liter Erlenmeyer flasks in a steam bath, with stirring, to a temperature of 80°. The heating should not exceed 5–6 minutes. At approximately 60° precipitation begins and by 80° a voluminous precipitate results. Inhibitor I survives this step almost quantitatively. Filter the hot solution through Whatman No. 1 filter paper, using Celite No. 545 as a filter aid. Pool the clear filtrates from each heated fraction and lyophilize.

Step 4. Dialysis. Disperse the dry lyophilized powder in 500 ml of water and place in dialysis tubing (4.5 cm flat width). Dialyze against distilled water for 48 hours using several changes. The mixture in the dialysis tubing first clears as salts are removed and then precipitates as salt-free globulins come out of solution. Remove the mixture from the tubing and filter as described above through Whatman No. 1 filter paper using Celite as a filter aid. Lyophilize the filtrate. The yield of inhibitor I is 90% from the ammonium sulfate precipitate.

Step 5. Gel Filtration on Sephadex G-75 in Buffer. Dissolve 1–5 g of dry material from step 4 in 30 to 60 ml of 0.05 M Tris chloride, 0.1 M KCl, pH 8.2, and apply to a 100 × 10.0 cm column of Sephadex G-75 equilibrated with the same buffer. Apply the sample to the column from the bottom and use an upward flow of about 450 ml per hour for elution, regulating the flow with a peristaltic pump (e.g., Sigma Motor, Inc., Middleport, N.Y.). Monitor the effluent at 280 nm. Collect 50-ml fractions. Begin collection of the effluent just preceding the void volume of the column (previously determined with blue dextran). Figure 1 shows the elution pattern. The second peak to elute ($V/V_0 = 1.33$) contains inhibitor I. Pool fractions 27 through 50. Reduce the volume to 300 ml using a rotary vacuum evaporator and desalt the pooled solution by passage through a 100 × 5.0-cm column of Sephadex G-25 with distilled water as eluent. Collect the breakthrough peak (the only peak) in its entirety, and lyophilize. The resulting salt-free protein is approximately 90% inhibitor I and represents a 76% yield of inhibitor I based on the heated, dialyzed, lyophilized material from step 4.

Step 6. Gel Filtration on Sephadex G-75 in 4 M Guanidine. Dissolve 100 mg of the product of step 4 in 5 ml of 4 M guanidine, pH 8.0. Apply the solution to an upward flow Sephadex G-75 column (100 × 2.5 cm) equilibrated with 4 M guanidine, pH 8.5. Elute the column with 4 M guanidine, pH 8.5, collecting 3.5-ml fractions at a flow rate of 60 ml per hour. The major peak appears at V/V_0 of 1.82 (Fig. 2). Pool fractions 36 through 59. Dilute the combined fractions 8-fold with water and dialyze overnight against 12 liters of distilled water. Reduce the volume

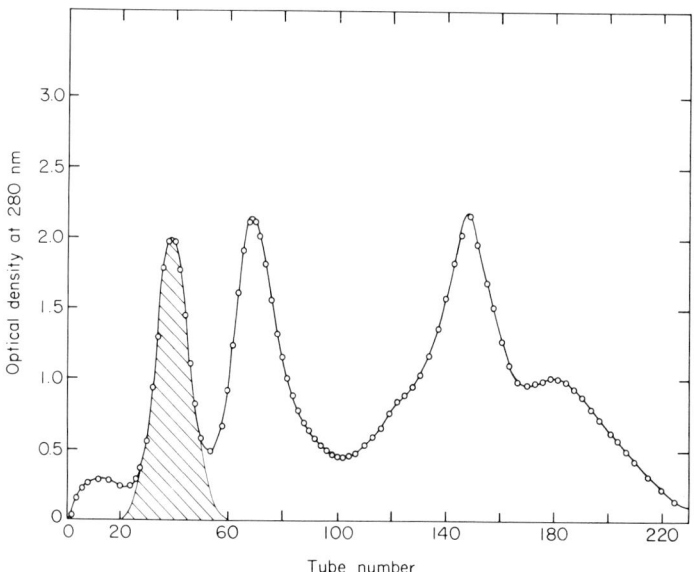

Fig. 1. Gel filtration of potato Inhibitor I on Sephadex G-75 in 0.05 M Tris chloride buffer, pH 8.2, containing 0.1 M KCl. Column 100×10 cm, flow rate 450 ml/hour, fractions 50 ml. See text, step 5. (From C. A. Ryan and J. C. Melville, manuscript in preparation.)

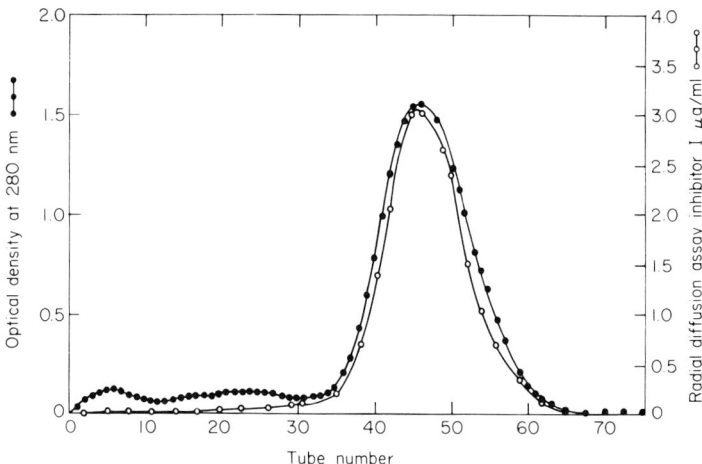

Fig. 2. Gel filtration of potato Inhibitor I on Sephadex G-75 in 4 M guanidine solution, pH 8.5. Column 100×2.5 cm, flow rate 60 ml/hour, fractions 3.5 ml. See text, step 6. (From C. A. Ryan and J. C. Melville, manuscript in preparation.)

to 100 ml in a rotary evaporator and dialyze for 24 hours against several changes of distilled water. Filter through a medium sintered glass filter. Lyophilize and store at —20°, although storage at room temperature is not detrimental. The yield of inhibitor I in this step is 82%. The overall recovery of inhibitor I from the crude juice is 41%.

Properties

Stability. Inhibitor I is stable at pH 1 to 10 at room temperature. Above pH 10 it is denatured; for example, its half-life of inhibitory activity toward chymotrypsin at pH 11.5 is 48 minutes and at 12.2 is 25 minutes.[14] It is stable at 80° for 10 minutes from pH 2 to 10.[13]

Full activity can be restored after dissolving inhibitor 1 in 8 M urea or in 6 M guanidine and diluting 10-fold in a pH range of 4 to 9.[14] It is stable in 5% TCA in the cold[13] and in 2.5% TCA at room temperature. The denatured precipitate formed in 10% TCA, when dissolved in 6 M urea and diluted 10-fold regains native characteristics.[13]

Pepsin destroys the activity below pH 6.[12]

Purity. Inhibitor I gives a single peak upon gel filtration[15] and electrophoresis in the presence of sodium dodecyl sulfate and mercaptoethanol. It shows a single peak in the ultracentrifuge, both in buffer and in 4 M guanidine solutions (see the table and Fig. 2).

MOLECULAR WEIGHT DETERMINATIONS OF INHIBITOR I AND ITS
COMPLEX WITH CHYMOTRYPSIN[a]

Protein	Method	Molecular weight or equivalent weight
Inhibitor I	Ultracentrifuge-Archibald (0.5–1.0% solution)	38,800 ± 2,400
	Ultracentrifuge-Velocity (0.25–1.0% solution)	37,500 ± 2,800
	Sephadex G-75 in buffer, pH 8.2	37,000 ± 2,000
	Sephadex G-75 in 4 M guanidine, pH 8.5	9,800 ± 200
	Sephadex G-75 in 0.1% sodium dodecyl sulfate[a]	9,300 ± 1,100
	α-Chymotrypsin inhibition (synthetic substrate)[b]	9,500 ± 500
	α-Chymotrypsin inhibition (casein substrate)[c]	10,000
Inhibitor I-chymotrypsin complex	Ultracentrifuge-Velocity (1.0% solution)	123,000

[a] J. C. Melville and C. A. Ryan, *Arch. Biochem. Biophys.* **138**, 700 (1970).
[b] Substrate: either L-tyrosine ethyl ester or benzoyl-L-tyrosine ethyl ester.
[c] C. A. Ryan, *Biochemistry* **5**, 1592 (1966).

[14] J. C. Melville and C. A. Ryan, *Arch. Biochem. Biophys.* **138**, 700 (1970).
[15] C. A. Ryan, unpublished observations.

Physical Properties. The molecular weight is between 37,000 and 39,000 and the molecule consists of four subunits (see the table).

The absorption maximum of the protein is at 280 nm with a second, sharp maximum at 290 nm and a shoulder at 275 nm. The optical factor (OD $\times$ optical factor = mg protein) is 1.27 $\pm$ 0.01 at pH 3.0 and 1.3 $\pm$ 0.04 at pH 7.0.

Specificity. The following enzymes are potently inhibited by inhibitor I: Bovine α-chymotrypsin,[4] bovine chymotrypsin B,[15] chicken chymotrypsin,[16] subtilisin,[12] and one of the pronase enzymes.[12] Trypsin is inhibited, but only its proteinase activity and not esterolytic activity.[12]

Inhibitor I does not inhibit pepsin,[12] rennin,[15] carboxypeptidases A and B,[12,17] bromelain,[12] ficin,[12] and papain.[12] A milk-clotting enzyme from *Endothia parasitica*, called "Sure Curd" is also not inhibited.[15] Human trypsin, plasmin, and thrombin are not inhibited.[18]

Inhibitor I forms a complex[19] with inactive, anhydrochymotrypsin[20] (serine-195 converted to dehydroalanine) and also interacts[21] with TPCK-chymotrypsin.[22]

Kinetic Properties. As shown in the table, 1 molecule of inhibitor I inactivates 4 molecules of chymotrypsin, i.e., each subunit is equivalent to a molecule of the enzyme.[14] The association constant[23] (K_A) of the inhibitor α-chymotrypsin reaction at 30° is 2.9 $\pm$ 0.9 $\times$ 10^{10} M^{-1}. The thermodynamic parameters for the reaction are: $\Delta F = -14.5 \pm 0.2$ kcal/mole enzyme; $\Delta H = -18 \pm 4$ kcal/mole and $\Delta S = -11 \pm 4$ entropy units.

Distribution.[24] Inhibitor I has a transitory occurrence in all tissues of potato plants, except xylem and seeds, where it is absent.

[16] C. A. Ryan, J. J. Clary, and Y. Tomimatsu, *Arch. Biochem. Biophys.* **110**, 175 (1965).
[17] Impure preparations of inhibitor I contain contaminants of potato carboxypeptidase B inhibitor.
[18] R. E. Feeney, G. E. Means, and J. C. Bigler, *J. Biol. Chem.* **244**, 1957 (1969).
[19] R. J. Foster and C. A. Ryan, *Federation Proc.* **24**, 473 (1965).
[20] D. H. Strumeyer, W. N. White, and D. E. Koshland, Jr., *Proc. Natl. Acad. Sci. U.S.* **50**, 931 (1963).
[21] G. Feinstein and R. E. Feeney, *J. Biol. Chem.* **241**, 5183 (1966).
[22] E. Shaw, Vol. XI [80].
[23] J. De Moura and R. J. Foster, personal communication, 1970.
[24] C. A. Ryan, O. C. Huisman, and R. W. Van Denburgh, *Plant Physiol.* **43**, 589 (1968).

[66g] Proteinase Inhibitors from Egg White

By Beatrice Kassell

Introduction

Egg white was one of the earliest known trypsin inhibitors[1] and this inhibiting property was shown to be associated with the ovomucoid fraction by Meyer et al.[2] and by Lineweaver and Murray.[3] This fraction consists of several closely related glycoproteins,[4,5] all with inhibiting activity, and of a second type of inhibitor, ovoinhibitor.[6]

Trypsin and/or chymotrypsin inhibitors have been purified from the egg whites of numerous avian species[7] and have been detected in still more species.[8] These homologous inhibitors are of great interest, since they differ in their reactions with trypsin and chymotrypsin, while resembling each other in chemical and physical properties.

Chicken ovomucoid is commercially available in partially purified form from several sources (e.g., Worthington Biochemical Corp., Freehold, N.J.; Sigma Chemical Corp., St. Louis, Mo.; and L. Light and Co., Great Britain).

The present article describes the chicken ovomucoids and ovoinhibitors in detail and includes properties of the trypsin and chymotrypsin inhibitors of other species. Several recent[9–11] and older[12–15] reviews are

---

[1] C. Delezenne and E. Pozerski, *Compt. Rend. Soc. Biol.* **55**, 935 (1904).

[2] K. Meyer, J. W. Palmer, R. Thompson, and D. Khorazo, *J. Biol. Chem.* **113**, 479 (1936).

[3] H. Lineweaver and C. W. Murray, *J. Biol. Chem.* **171**, 565 (1947).

[4] R. E. Feeney, F. C. Stevens, and D. T. Osuga, *J. Biol. Chem.* **238**, 1415 (1963).

[5] R. E. Feeney, D. T. Osuga, and H. Maeda, *Arch. Biochem. Biophys.* **119**, 124 (1967).

[6] K. Matsushima, *Science* **127**, 1178 (1958).

[7] M. B. Rhodes, N. Bennett, and R. E. Feeney, *J. Biol. Chem.* **235**, 1686 (1960).

[8] S. Nakamura, M. Nagao, and R. Suzuno, *Comp. Biochem. Physiol.* **18**, 937 (1966).

[9] M. D. Melamed, in "Glycoproteins" (A. Gottschalk, ed.), Vol. V, Chap. 11, Section 2. BBA Library, Elsevier, Amsterdam, 1966.

[10] R. Vogel, I. Trautschold, and E. Werle, "Natural Proteinase Inhibitors," p. 45. Academic Press, New York, 1968.

[11] R. E. Feeney and R. G. Allison, "Evolutionary Biochemistry of Proteins," pp. 83, 208ff. Wiley, New York, 1969.

[12] K. Meyer, *Advan. Protein Chem.* **2**, 264 (1945).

[13] H. L. Fevold, *Advan. Protein Chem.* **6**, 188 (1951).

[14] R. C. Warner, in "The Proteins" (H. Neurath and K. Bailey, eds.), Vol. IIA, p. 465. Academic Press, New York, 1954.

[15] M. Laskowski and M. Laskowski, Jr., *Advan. Protein Chem.* **9**, 203 (1954).

available. A ficin and papain inhibitor (a protein of 12,700 molecular weight) has also been purified from chicken egg white.[16] It does not inhibit trypsin or chymotrypsin.

The term "ovomucoid" used below without mention of species refers to chicken ovomucoid.

Assay[7, 17, 18]

Principle. Change in the absorbance at 395 nm of the indicator, *m*-nitrophenol, is a measure of acid liberated during the hydrolysis of ester substrates by trypsin and chymotrypsin.

INHIBITION OF TRYPSIN

Reagents

Substrate–buffer–indicator solution. *p*-toluenesulfonyl arginine methyl ester (TAME), 0.01 M, in 0.006 M Tris-chloride buffer containing 0.02% *m*-nitrophenol, with a final pH of 8.2.

Trypsin solution. 65–80 μg/ml in 0.004 M acetic acid containing 0.02 M CaCl$_2$.

Inhibitor solution. 15–20 μg/ml in 0.006 M Tris-chloride buffer of pH 8.95.

Procedure. A recording spectrophotometer connected to a thermostated bath at 37° is used. The speed of the chart is 4 in./minute.

Equilibrate all the solutions in the bath. Pipette into the cuvette 0.3 ml of enzyme solution and 0.7 ml of inhibitor solution. The final pH is 8.2 ± 0.05. Add 2 ml of substrate–buffer–indicator solution by blowing it into the cuvette from a fine-tipped pipette to insure mixing. Start the recorder within 2 to 3 seconds, and record the change in transmittance at 395 nm for 1 minute.

Repeat the determination with 0.7 ml of 0.006 M buffer of pH 8.95 substituted for the inhibitor solution. The amount of trypsin inhibited is the difference between the two determinations.

The slope has a linear relationship to the amount of trypsin. The conversion factor is predetermined with variable amounts of a trypsin preparation of known "active enzyme" content, e.g., titrated with *p*-nitrophenyl *p*-guanidinobenzoate.[19]

Unit. Based on a standard preparation of chicken ovomucoid, 1 μg of

[16] K. Fossum and J. R. Whitaker, *Arch. Biochem. Biophys.* **125**, 367 (1968).
[17] M. B. Rhodes, R. M. Hill, and R. E. Feeney, *Anal. Chem.* **29**, 376 (1957).
[18] M. M. Simlot and R. E. Feeney, *Arch. Biochem. Biophys.* **113**, 64 (1966).
[19] T. Chase, Jr., and E. Shaw, *Biochem. Biophys. Res. Commun.* **29**, 508 (1967).

inhibitor is equivalent to 0.86 μg of trypsin;[17] this is a 1:1 molar ratio.

For ovoinhibitor,[20] equivalence was determined with three fractions of similar activity. The average value was 1 μg of inhibitor equivalent to 0.85 μg of trypsin, or a molar enzyme to inhibitor ratio of 1.74.

INHIBITION OF CHYMOTRYPSIN

Reagents

Substrate–buffer–inhibitor solution: benzoyl-L-tyrosine ethyl ester (BTE), 0.01 M, in 0.006 M Tris-chloride buffer containing 30% methanol and 0.02% m-nitrophenol, with a final pH of 8.2. Acetyl-L-tyrosine ethyl ester may be substituted[21, 22] for BTE; in this case the methanol is omitted.

Chymotrypsin solution. 65–80 μg/ml in 0.004 M acetic acid, containing 0.02 M CaCl$_2$.

Inhibitor solution. Same as for trypsin assay.

Procedure. The method is the same as for the trypsin assay. The standardization should be carried out with an α-chymotrypsin preparation of known "active enzyme" content, e.g., titrated with N-*trans*-cinnamoyl imidazole.[23]

Unit. Chicken ovomucoid does not inhibit chymotrypsin;[24, 25] measurement of the inhibition of chymotrypsin is a measure of contamination with ovoinhibitor.[25]

For ovoinhibitor,[20] 1 μg inhibits 0.84 μg of chymotrypsin; this is a molar enzyme:inhibitor ratio of 1.65 (average value of three fractions).

Chicken Ovomucoid

Purification Procedure

The ovomucoid is precipitated with trichloroacetic acid (TCA)-acetone by the method of Lineweaver and Murray[3] and further purified by chromatography according to Feeney, Osuga, and Maeda.[5] Other ovomucoids are prepared by similar methods.[7]

Step 1. Separation from Egg White.[3] Adjust 750 ml of fresh egg white

[20] J. G. Davis, J. C. Zahnley, and J. W. Donovan, *Biochemistry* **8**, 2044 (1969).
[21] C. A. Ryan, J. J. Clary, and Y. Tomimatsu, *Arch. Biochem. Biophys.* **110**, 175 (1965).
[22] Y. Tomimatsu, J. J. Clary, and J. J. Bartulovich, *Arch. Biochem. Biophys.* **115**, 536 (1966).
[23] G. R. Schonbaum, B. Zerner, and M. L. Bender, *J. Biol. Chem.* **236**, 2930 (1961).
[24] F. C. Wu and M. Laskowski, Sr., *J. Biol. Chem.* **213**, 609 (1955).
[25] R. E. Feeney, F. C. Stevens, and D. T. Osuga, *J. Biol. Chem.* **238**, 1415 (1963).

at 25°–30° to pH 3.5 by the slow addition, with stirring, of about 750 ml of TCA-acetone solution (1 volume of aqueous 0.5 M TCA plus 2 volumes of acetone). Continue stirring for 30 minutes, then store overnight at 4°. Centrifuge for 30 minutes at 13,000 g in a refrigerated centrifuge. Discard the precipitate. A sample of the supernatant solution should remain almost clear when heated to 80° for 5 minutes.

Precipitate the inhibitor at 4° by adding 3 to 3.5 liters of cold acetone. After settling, filter on a Büchner funnel, using Celite as a filter aid. Suspend the filter cake in about 400 ml of water. Remove the Celite by filtration and precipitate the inhibitor with 2½ volumes of acetone. Filter or centrifuge the suspension and dissolve the precipitate in a minimum amount of water adjusted to pH 4.5. Dialyze the inhibitor solution against water to remove residual TCA, then against 0.02 M sodium phosphate buffer of pH 6.5. The yield is about 5 g, and the activity is increased 8.5-fold over the original egg white.

Step 2. Chromatography on DEAE-Cellulose.[5] Cycle DEAE-cellulose[26] (Whatman DE-50, Reeve Angel and Co., Clifton, N.J.) with 0.5 M HCl and 0.5 M NaOH. Adjust to pH 6.5 with a concentrated solution of NaH_2PO_4. Wash several times on the Büchner funnel with 0.02 M sodium phosphate buffer, pH 6.5, the starting buffer, and resuspend in the same buffer. Remove the fines by settling. Prepare a 35 × 3.2 cm column by allowing the resin to settle to a height of 39 cm and compressing to 35 cm. Equilibrate the column overnight. Apply the entire sample[27] of step 1 (or 5 g), and elute the column with two successive linear gradients of NaCl in the starting buffer, as shown in Fig. 1. The details of the procedure are given in the legend. Table I indicates that the inhibiting activity is mainly in peaks II and III; peak I contains many constituents, including lysozyme and ovoinhibitor. Pool each peak, lyophilize, and dialyze against 0.1 M sodium acetate buffer of pH 4.2.

Step 3. Chromatography on CM-Cellulose.[5, 7, 25] Recycle CM-cellulose[26] (Whatman CM-70) with 0.5 M NaOH and 0.5 M HCl, and adjust to pH 4.2 with 4 M sodium acetate. Wash the resin on the Büchner funnel with 0.1 M sodium acetate pH 4.2, the starting buffer. Resuspend in this buffer and remove fines. Prepare two 35 × 1.8 cm columns as described for DEAE, and equilibrate with the starting buffer. Apply the material from peak II and peak III to the separate columns. Elute with 150 ml of starting buffer, collecting 10-ml fractions. Change to a linear gradient prepared from 500 ml of starting buffer and 500 ml of 1 M NaCl in starting buffer in interconnected open bottles of the same size.

---

[26] E. A. Peterson and H. A. Sober, Vol. V [1].

[27] The present author recommends applying a smaller sample and using a slower flow rate to achieve greater resolution than that shown in Fig. 1.

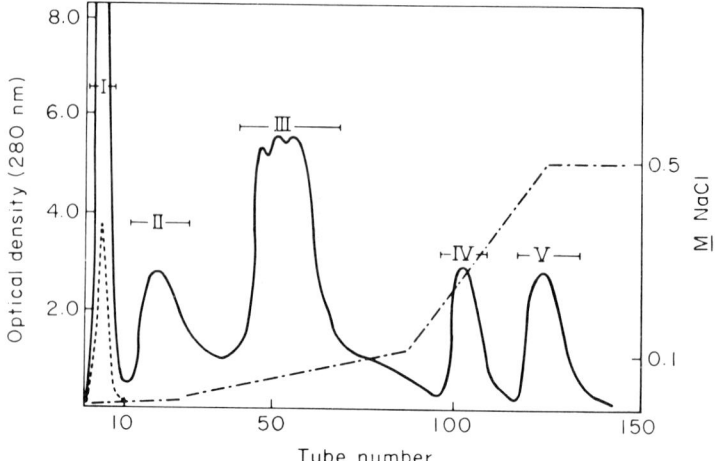

Fig. 1. Chromatography of trichloroacetic acid-acetone preparation of chicken ovomucoid on DEAE-cellulose. Sample 5 g, column 35×3.2 cm, flow rate 3.6 ml/minute, fractions 12.5 ml. (———) Optical density; (·—·—) salt gradient; (· · · · ·) 10-fold scale reduction of peak 1. Reproduced from Feeney, Osuga, and Maeda, *Arch. Biochem. Biophys.* **119**, 124 (1967).

DEAE fractions II and III both give several peaks,[5] but in each case there is one predominant peak. Pool the most active fractions from the predominant peaks, lyophilize, dialyze extensively against water, and lyophilize again.

TABLE I

DISTRIBUTION OF CHICKEN OVOMUCOID FRACTIONS
FROM DEAE-CELLULOSE[a] (Fig. 1)

Fraction no.	% (by weight)	Trypsin-inhibiting activity		Sialic acid content[c] (%)
		Specific activity[b]	% of total activity	
I	11.0	0.3	4	
II	21.6	1.0	24	0.50
III	48.0	1.0	54	0.63
M[d]	14.4	1.0	16	
IV	2.2	0.58	1	
V	3.2	0	0	

[a] R. E. Feeney, D. T. Osuga, and H. Maeda, *Arch. Biochem. Biophys.* **119**, 124 (1967).

[b] With chromatographically pure turkey ovomucoid taken as 1.00.

[c] After rechromatography on CM-cellulose (see text, step 3).

[d] Mixture of fractions between peaks II and III and between peaks III and IV.

Other Methods of Purification. A procedure[28] involving precipitation with alcohol instead of acetone may be used for the preliminary separation of the ovomucoid from egg white. Similar DEAE-cellulose[29] and TEAE-cellulose[30, 31] methods have been suggested for further purification, as well as gel filtration,[23] chromatography on sulfoethyl cellulose,[32] chromatography on hydroxylapatite,[32] and electrophoresis.[33, 33a] However, a thorough testing of these products[32] by electrophoretic and immunoelectrophoretic methods has shown that chicken ovomucoid has not yet been prepared as a single entity.

Properties

Stability. Chicken ovomucoid shows unusual stability to heat and to high concentrations of urea in neutral or acid solutions; activity is rapidly lost in alkaline solutions. The data have been summarized by Melamed.[9]

Between pH 3 and 7, after 30 minutes at 80° more than 90% of the activity remained[3]; at pH 9 only 6% was retained.[34] After 15 minutes[35] at 90°, 80% of the activity remained at pH 5 and 62% at pH 7. At pH 6 and 100°, 30% of the activity was still present after 15 minutes.[34] Turkey and pheasant ovomucoids are even more stable to heat than chicken ovomucoids.[34]

Jirgensons and co-workers[36] have shown that the denaturation in alkali, first detected by loss of activity,[37] is also reflected in changes in viscosity and optical rotation (see Physical Properties).

In 9 M urea at pH 4, more than 90% of the activity was retained[3] after 18 hours at 25° or after 30 minutes at 80°; at 100°, 86% remained after 15 minutes and 66% after 30 minutes.[3] At pH 7.4, also in 9 M urea, 80% of the activity was still present[34] after 90 minutes at 80°. Ovomucoid in 8 M urea showed a small increase in viscosity.[28]

In 9 M urea, turkey ovomucoid lost its activity against trypsin and

[28] E. Fredericq and H. F. Deutsch, *J. Biol. Chem.* **181**, 499 (1949).
[29] A. K. Chatterjee and R. Montgomery, *Arch. Biochem. Biophys.* **99**, 426 (1962).
[30] F. R. Jevons, *Biochim. Biophys. Acta* **45**, 384 (1960).
[31] J. G. Beeley and F. R. Jevons, *Biochim. Biophys. Acta* **101**, 133 (1965).
[32] J. Montreuil, B. Castiglioni, A. Adam-Chosson, F. Caner, and J. Queval, *J. Biochem.* (*Tokyo*) **57**, 514 (1965).
[33] M. Jutisz, M. Kaminski, and J. Legault-Démare, *Biochim. Biophys. Acta* **23**, 173 (1957).
[33a] M. D. Melamed, *Biochem. J.* **103**, 805 (1967).
[34] F. C. Stevens and R. E. Feeney, *Biochemistry* **2**, 1346 (1963).
[35] H. Fraenkel-Conrat, R. C. Bean, E. D. Ducay, and H. S. Olcott, *Arch. Biochem. Biophys.* **37**, 393 (1952).
[36] B. Jirgensons, T. Ikenaka, and V. Gorguraki, *Makromol. Chem.* **39**, 149 (1960).
[37] A. K. Balls and T. L. Swenson, *J. Biol. Chem.* **106**, 409 (1934).

chymotrypsin[34] at different rates. After 90 minutes at pH 7.4 and 80°, less than 5% of the activity against trypsin remained, while 85% of the chymotrypsin inhibiting ability was still present. Pheasant ovomucoid was stable under these conditions.[34]

Chicken ovomucoid is readily digested by pepsin.[35, 38]

Purity. The electrophoretic heterogeneity of the earlier preparations[3, 28] of ovomucoid has been recognized for a long time.[28, 39, 40] Present methods have reduced, but have not eliminated, the heterogeneity. Even ovomucoids from the eggs of single hens[5] give chromatographic patterns similar to Fig. 1. The distinct components do not differ significantly in antitryptic activity or in amino acid composition.[5, 34] The isolation,[40a] from four components of ovomucoid, of four glycopeptides with the same amino acid composition, but with definite quantitative differences in carbohydrate content, makes it likely that differences in the size of the carbohydrate moieties account for most of the heterogeneity. However, the possibility[11] of microheterogeneity in amino acid composition or in amide content cannot yet be excluded.

The ovomucoids of ten other avian species are also electrophoretically heterogeneous.[5]

Physical Properties. Molecular weight determinations[3, 7, 28, 41] by osmotic pressure and by sedimentation and diffusion methods are in the range from $27,000^{7, 28}$ to $31,500^{41}$; 28,000 is usually taken as the average value. The sedimentation constants for chicken ovomucoid range from 2.3 to 2.8, with diffusion constants from 6.0 to 8.1×10^{-7} cm²/sec at 20°. The partial specific volume is 0.685^{28} to $0.71.^{7}$ The frictional ratio[28] from sedimentation and diffusion is 1.35, giving an axial ratio of 6.3 for the unhydrated particle. However, ovomucoid is probably not compact and highly asymmetric, but highly hydrated, with an axial ratio near unity.[42] Jirgensons[36] found the intrinsic kinematic viscosity in 0.25 M NaBr solution to be 5.1 ml/g. In 0.1 M NaBr + 0.1 M NaOH, it changed to 18. Donovan[42] reported 6.31 ml/g at pH 2.1, 5.51 at pH 3.8, and 5.46 at pH 4.6.

Sedimentation constants of 2.1 to 2.3 have been found for other ovomucoids (Table IV), but a molecular weight (28,000) has been reported[7] only for duck ovomucoid. Presumably, they are all quite similar.

[38] B. Kassell and M. Laskowski, Sr., *J. Biol. Chem.* 219, 203 (1956).

[39] L. G. Longsworth, R. K. Cannan, and D. A. MacInnes, *J. Am. Chem. Soc.* 62, 2580 (1940).

[40] M. Bier, L. Terminiello, J. A. Duke, P. J. Gibbs, and F. F. Nord, *Arch. Biochem. Biophys.* 47, 465 (1953).

[40a] M. Kana Mori and M. Kawabata, *Agr. Biol. Chem.* 33, 75, 220 (1969).

[41] H. F. Deutsch and J. I. Morton, *Arch. Biochem. Biophys.* 93, 654 (1961).

[42] J. W. Donovan, *Biochemistry* 6, 3918 (1967).

The optical rotation $[\alpha]_D$ was reported as $-56°$ (5% in water) by Lineweaver and Murray,[3] while Chatterjee and Montgomery[29] reported $-78°$ (0.79% in water); the latter preparation was chromatographically purified.

The optical rotatory dispersion has been studied in the visible[36, 43] and in the ultraviolet[43, 44] ranges and has led to the conclusion[43] that ovomucoid has 22–27% of helical structure. The effect of pH on the rotatory

TABLE II

OPTICAL ROTATORY DISPERSION (ORD) AND CIRCULAR DICHROISM (CD)
DATA FOR CHICKEN OVOMUCOID

Type	pH	$[\alpha]_{546}$ (degrees)	$K \times 10^{-8}$	$\lambda_c(\pm 3)$ (nm)
ORD[a]	1.1	-84.1	20.8	228.5
	6.1	-91.5	22.2	237.3
	9.7	-83.6	21.0	228.9
	12.0	-65.9	16.5	220.0

	pH	λ_{max} (nm)	$[\theta]_{max}$ (degrees decimole^{-1} cm^2)
CD[b,c]	7.1	210	$-16,000$
	7.1	222	$-10,500$
	7.1	243	$+390$
	7.1	263	-210
	7.1	271	-190
	7.1	292	-100

[a] B. Jirgensons, T. Ikenaka, and V. Gorguraki, *Makromol. Chem.* **39**, 149 (1960). (Reproduced with permission of Hüthig and Wepf Verlag, Basel.)
[b] K. Ikeda, K. Hamaguchi, M. Yamamoto, and T. Ikenaka, *J. Biochem.* (*Tokyo*) **63**, 521 (1968). (Reproduced with permission of the Japanese Biochemical Society.)
[c] Ionic strength 0.1; see text for changes with pH.

properties in visible light is shown in Table II. Denaturation in alkali decreases the dispersion constant,[44a] λ_c, in the normal way, while the specific rotation becomes less negative than for the native protein. This is interpreted as indicating an unusual conformation.[36] In the ultraviolet,[44] in 0.1 M NaCl, there is a negative Cotton effect with a trough at 230 nm. This trough decreases when the pH is raised to 11.5 and is greatly diminished at pH 12.8. The data given by Tomimatsu and Gaffield[43] for

[43] Y. Tomimatsu and W. Gaffield, *Biopolymers* **3**, 509 (1965).
[44] K. Ikeda, K. Hamaguchi, M. Yamamoto, and T. Ikenaka, *J. Biochem.* (*Tokyo*) **63**, 521 (1968).
[44a] The dispersion constant is derived from the one-term Drude equation.

TABLE III

REACTION OF INHIBITORS WITH VARIOUS TRYPSINS AND CHYMOTRYPSINS[a,b]

| | Molar complexing ratio (enzyme/inhibitor) | | | | |
| | Trypsins | | | Chymotrypsins | |
Ovomucoids	Bovine	Human	Turkey	Bovine-α	Chicken
Chicken (*Gallus gallus*)	1	0	1	0	0
Peking duck (*Anas platyrhynckos*)	2			1	
Khaki Campbell duck (*Anas platyrhynckos*)	2			1	
Turkey (*Meleagris gallopavo*)	1	0	1	1	1
Guinea fowl (*Numida meleagris*)	1			1	
Goose (*Anser anser*)	2			0	
Emu (*Dromiceius novae-hollandiae*)	2			1	
GAX pheasant (a genetic cross between golden pheasant and Lady Amherst pheasant)	0			1	
Lady Amherst's pheasant (*Chrysolophus amherstiae*)	0			1	
Ring-necked pheasant) (*Phasianus colchicus*)	1			1	
Golden pheasant (*Chrysolophus pictus*)	0			1	
Ostrich (*Struthio camelus*)	Positive[c]			Weak[c]	
Cassowary (*Casuarius casuarius*)	1			0	
Red jungle fowl (*Gallus gallus*)	1			0	
California valley quail (*Lophortyx californica*)	1–2			1	
Painted quail (*Oreortyx picti*)	1			0	
Japanese quail (*Coturnus coturnus japonica*)	Positive	Weak		0	
Tinamou (*Eudromia elegans*)	Weak	0		Positive	
Penguin (*Pygoscelis adeliae*)	Positive			Weak	
Rhea (*Rhea americana*)	Positive			Weak	
Chicken ovoinhibitor	2		1	2	1

[a] The data are compiled from the following sources:

F. C. Wu and M. Laskowski, Sr., *J. Biol. Chem.* **213**, 609 (1955).

M. B. Rhodes, N. Bennett, and R. E. Feeney, *J. Biol. Chem.* **235**, 1686 (1960).

A. J. Vithayathil, F. Buck, M. Bier, and F. F. Nord, *Arch. Biochem. Biophys.* **92**, 532 (1961).

C. A. Ryan, J. J. Clary, and Y. Tomimatsu, *Arch. Biochem. Biophys.* **110**, 175 (1965).

pH 4.5 are as follows: A_{193} 49°, A_{225} —503°, b_0 —86°, a_0—524°, λ_c 231 nm.

The circular dichroism spectrum[44] of ovomucoid at pH 7.1 shows minima at 292 and 264 nm and a maximum at 243 nm (Table II). The helical content was estimated to be 26% from this data. At pH 11.6 the 243 maximum is greatly enlarged and the intensity of the band at 264 decreases; the band at 292 is almost unaltered. At pH 12.8, all the bands are greatly diminished, showing disorganization.

Between pH 3.5 and 4.5, there is[45] a small decrease in specific rotation at both 578 and 313 nm accompanied by a decrease in solvent perturbation effect on the absorption spectrum; these observations indicate a limited conformational change. Spectrophotometric and spectrofluorometric methods[42] associate the protonation of a carboxyl group of pK 2.8 with this change.

Isoelectric points ranging from 3.9 to 4.5 have been reported[28, 39, 46] for chicken ovomucoid. Bier *et al.*,[40, 47] in agreement with Fredericq and Deutsch,[28] separated five components with isoelectric points from 3.83 to 4.41; subsequent electrophoretic separations have confirmed the differences among the components. Osuga and Feeney[48] have compared the isoionic points of ovomucoids of several species with the results shown in Table IV. The electrophoretic mobility[41] at pH 8.6 and 0.1 ionic strength is —3.5 × 10⁵.

The absorbance [$A_{1cm}^{1\%}$] is 4.13[29] at 280 nm and 4.10[42] at 277.5 nm.

Specificity. From the extensive work of Feeney and his co-workers with ovomucoids derived from the egg whites of many avian species, different patterns of specificity have emerged (Table III). Ovomucoids may

[45] T. T. Herskovits and M. Laskowski, Jr., *J. Biol. Chem.* **237**, 3418 (1962).

[46] Z. Hesselvik, *Z. Physiol. Chem.* **254**, 144 (1938).

[47] M. Bier, J. A. Duke, R. J. Gibbs, and F. F. Nord, *Arch. Biochem. Biophys.* **37**, 491 (1952).

[48] D. T. Osuga and R. E. Feeney, *Arch. Biochem. Biophys.* **124**, 560 (1968).

Y. Tomimatsu, J. J. Clary, and J. J. Bartulovich, *Arch. Biochem. Biophys.* **115**, 536 (1966).

D. T. Osuga and R. E. Feeney, *Arch. Biochem. Biophys.* **118**, 340 (1967).

R. E. Feeney, G. E. Means, and J. C. Bigler, *J. Biol. Chem.* **244**, 1957 (1969).

R. E. Feeney and R. G. Allison, "Evolutionary Biochemistry of Proteins," p. 85. Wiley, New York, 1969.

b For reactions with other enzymes, see the text.

c The terms "weak" and "positive" are used when quantitative data were not available.

specifically inhibit either trypsin or chymotrypsin or they may react with both enzymes, separately or at the same time ("double-headed" inhibitors). Some of the ovomucoids react with 2 molecules of trypsin. The reactions of chicken and turkey ovomucoids with turkey trypsin[21] and chicken chymotrypsin[21] are the same as with the bovine enzymes (Table III).

Human trypsin is not significantly inhibited by any of the ovomucoids tested.[49,50] Chicken ovomucoid inhibits porcine and ovine trypsin in a 1:1 molar ratio,[51] and also the acetyl derivative of bovine trypsin.[35,51] Some of the ovomucoids show varying degrees of inhibition of subtilisin.[11]

Several ovomucoids form complexes[52] with inactive anhydrochymotrypsin (active center serine replaced by dehydroalanine[53]) and with trypsin and chymotrypsin inactivated at the active center histidine residue (TLCK-trypsin[54] and TPCK-chymotrypsin[55]), but not with tosylchymotrypsin[52] or DFP-trypsin.[56]

Numerous enzymes in addition to those in Table III are not significantly inhibited by chicken ovomucoid: bovine chymotrypsin B,[24] *Clostridium histolyticum* collagenase,[57] kallikreins from several sources,[58-60] human plasmin,[50,58] human thrombin,[50] and liver esterase.[61]

Kinetic Properties. The dissociation constant (K_I) of the complex of chicken ovomucoid with bovine trypsin is $5 \times 10^{-9}\ M$[56] to $5.8 \times 10^{-9}\ M$.[51] K_I values of $6 \times 10^{-9}\ M$ were found for the complexes of chicken ovomucoid with porcine and ovine trypsins.[51] The inhibition has been characterized as noncompetitive[62,63] by classical methods. Green[56] suggested that the usual criteria cannot be applied and regarded the inhibition as competitive. However, the formation of complexes between ovomucoids

[49] F. F. Buck, M. Bier, and F. F. Nord, *Arch. Biochem. Biophys.* **98**, 528 (1962).

[50] R. E. Feeney, G. F. Means, and J. C. Bigler, *J. Biol. Chem.* **244**, 1957 (1969).

[51] A. J. Vithayathil, F. Buck, M. Bier, and F. F. Nord, *Arch. Biochem. Biophys.* **92**, 532 (1961).

[52] G. Feinstein and R. E. Feeney, *J. Biol. Chem.* **241**, 5183 (1966).

[53] D. H. Strumeyer, W. N. White, and D. E. Koshland, Jr., *Proc. Natl. Acad. Sci. U.S.* **50**, 931 (1963).

[54] E. Shaw, M. Mares-Guia, and W. Cohen, *Biochemistry* **4**, 2219 (1965).

[55] G. Schoellman and E. Shaw, *Biochemistry* **2**, 252 (1963).

[56] N. M. Green, *J. Biol. Chem.* **205**, 535 (1953).

[57] I. Mandl, H. Zipper, and L. T. Ferguson, *Arch. Biochem. Biophys.* **74**, 465 (1958).

[58] E. Werle, *Ärztliche Forschung* **22**, 41 (1968).

[59] N. Back and R. Steger, *Federation Proc.* **27**, 96 (1968).

[60] E. Sach and M. Thély, *Compt. Rend. Acad. Sci., Paris, Ser. D* **266**, 1200 (1968).

[61] N. M. Green, *Biochem. J.* **66**, 416 (1957).

[62] H. Fraenkel-Conrat, R. S. Bean, and H. Lineweaver, *J. Biol. Chem.* **177**, 385 (1949).

[63] P. Métais, H. Schirardin, and J. Warter, *Compt. Rend. Soc. Biol.* **157**, 2307 (1963).

and inactive derivatives of trypsin and chymotrypsin (see Specificity) makes it doubtful that the inhibition is competitive.

The rate of reaction of ovomucoids with trypsin and chymotrypsin has been studied[18] at pH 8.2 and 37°. Turkey and duck ovomucoids react rapidly with trypsin (in less than 1 minute), but about 4 minutes are required to complete the reaction with chymotrypsin. Ovomucoid from the ring-necked pheasant reacts rapidly with both enzymes. Chicken ovomucoid reacts rapidly with trypsin and golden pheasant ovomucoid reacts rapidly with chymotrypsin.

Chicken ovomucoid is a temporary[64] inhibitor of trypsin, i.e., it is gradually hydrolyzed in the presence of trypsin.[65] At pH 3.75, upon reaction with trypsin, an Arg-Ala bond is hydrolyzed.[66]

Distribution. The occurrence of ovomucoid-type inhibitors in egg white appears to be common to all the avian species tested.

Composition and Terminal Groups. The data for nine ovomucoids are given in Table IV. There are some general similarities: large amounts of hydroxy, acidic, and branched chain amino acids, as well as lysine and cystine, and little or no methionine. Arginine varies from zero to six residues. Ovomucoid resembles other trypsin inhibitors in its high cystine and low methionine content, and in the absence of tryptophan.

The nitrogen content of chicken ovomucoid[3, 34] is 13.3%. Turkey and pheasant ovomucoids have 12.7 and 13.3%, respectively.[34]

Chicken ovomucoid has amino terminal alanine[67] and carboxyl terminal phenylalanine.[68, 69] The other ovomucoids in Table IV have amino terminal valine.[48]

The ovomucoids are glycoproteins, unlike most other trypsin inhibitors. Studies on the nature and linkage of the carbohydrate moiety of chicken ovomucoid have been reviewed by Melamed.[9] Some data on carbohydrate composition are given in Table IV; the hexoses have been identified as D-mannose and D-galactose. Several investigations of the type of attachment of the carbohydrate to the protein have led to diverse conclusions (cf. footnote 9). However, the most recent work of Montreuil and co-workers[70] appears to establish linkages between glucosamine and asparagine of the type N-(β-aspartyl)-N-acetylglucosaminylamine.

[64] M. Laskowski, Sr., and F. C. Wu, *J. Biol. Chem.* 204, 797 (1953).
[65] L. Gorini and L. Audrain, *Biochim. Biophys. Acta* 8, 702 (1952).
[66] K. Ozawa and M. Laskowski, Jr., *J. Biol. Chem.* 241, 3955 (1966).
[67] H. Fraenkel-Conrat and R. R. Porter, *Biochim. Biophys. Acta* 9, 557 (1952).
[68] L. Pénasse, M. Jutisz, C. Fromageot, and H. Fraenkel-Conrat, *Biochim. Biophys. Acta* 9, 551 (1952).
[69] R. A. Turner and G. Schmerzler, *Biochim. Biophys. Acta* 11, 586 (1953).
[70] M. Monsigny, A. Adam-Chosson, and J. Montreuil, *Bull. Soc. Chim. Biol.* 50, 857 (1968).

TABLE IV

COMPOSITION AND PROPERTIES OF OVOMUCOIDS[a]

Residues per mole[b]

	Chicken	Turkey	Cassowary	Emu	Ostrich	Rhea	Tinamou	Duck	Penguin
Alanine	11.7	8.6	7.3	7.0	5.6	5.2	8.5	8.4	5.4
Arginine	6.3	5.8	1.0	0.0	3.4	0.0	3.5	1.2	3.1
Aspartic acid	31.9	27.2	25.8	27.1	27.3	25.7	30.8	33.0	28.0
Half-cystine	17.5	16.7	19.9	16.7	19.9	17.4	19.6	20.2	18.8
Glutamic acid	14.9	19.0	17.7	16.6	18.0	20.1	18.1	20.1	15.7
Glycine	16.1	17.3	14.3	14.9	17.0	14.6	17.6	19.2	14.7
Histidine	4.3	5.2	3.1	3.0	2.2	4.1	3.4	3.5	2.2
Isoleucine	3.2	4.4	6.0	6.0	4.5	3.7	4.7	2.6	4.1
Leucine	12.2	13.5	12.5	13.4	15.3	12.6	9.4	13.5	12.6
Lysine	13.6	11.2	16.5	16.2	14.8	15.8	17.1	17.2	13.5
Methionine	1.9	1.8	1.0	0.9	1.0	0.9	0.0	7.9	2.0
Phenylalanine	5.3	3.2	3.1	3.1	3.3	5.0	4.7	4.7	2.5
Proline	7.7	8.8	10.4	8.9	11.4	9.7	14.0	10.4	9.5
Serine	12.5	10.0	14.2	13.4	18.3	16.7	12.5	13.3	13.4
Threonine	14.6	14.2	11.8	12.2	17.0	16.2	14.0	19.9	16.0
Tryptophan	0.0	0.0	0.0	0.0	0.0	0.0	0.0	0.0	0.0
Tyrosine	6.7	6.8	10.6	6.4	11.1	8.2	10.4	11.5	10.1
Valine	16.0	15.7	15.4	15.1	16.9	17.9	17.6	17.4	19.4
Sialic acid	0.3	2.2	4.7	8.9	2.9	5.3	0.1	0.0	6.8
Hexose	16.7	18.0	16.4	15.6	15.2	16.9	15.9	12.8	14.8
Glucosamine	21.0	18.5	16.2	15.1	16.0	14.5	16.4	10.2	12.8
N-Terminus	Alanine	Valine	Valine	Valine	Valine	Valine	Valine	Valine	Valine
$s_{20,w}^{0}$[c]	2.3	2.2	2.1	2.2	2.2	2.1	2.2	2.1	2.1
Isoionic pH	4.48	4.28	3.90	3.78	3.97	3.82	4.73	4.28	3.62

[a] From: R. E. Feeney and R. G. Allison, "Evolutionary Biochemistry of Proteins," pp. 85–86. Wiley, New York, 1969. (Reprinted by permission.)

[b] The value 28,000 was used for molecular weight of chicken ovomucoid and was assumed for molecular weights of other proteins.

[c] 1.0% solutions in 0.05 M phosphate buffer at pH 7.0.

Chicken Ovoinhibitor

Ovoinhibitor differs from chicken ovomucoid in several properties, including solubility,[6] molecular weight,[22] and ability to inhibit chymotrypsin and bacterial proteinases.[6] It is therefore considered a separate inhibitor.

Purification Procedure[22]

The starting material is 17 g of ovomucoid prepared by the Lineweaver-Murray method[3] (step 1, above) and dialyzed only against water. Commercially prepared ovomucoid may be used. Operations are carried out at 23° except for the dialysis step (4°).[70a]

Separation of Ovoinhibitor from Ovomucoid. Titrate a 5% solution of the ovomucoid in water to pH 6.0 with 1 N NaOH. Add solid ammonium sulfate to 50% saturation. Collect the precipitate by centrifugation, dissolve it in a minimum amount of water, and dialyze the solution overnight against 0.25% NaCl. Centrifuge the bag contents. Bring the supernatant solution to 35% saturation with ammonium sulfate. Centrifuge and dissolve the precipitate in a minimum volume of water. Dialyze the solution against water, and lyophilize. The yield is 250 mg of ovoinhibitor.

Other Methods of Purification. There are several alternate methods.[6,22] A more elaborate method has been developed by Davis et al.[20]; it consists of separation from egg white by ammonium sulfate fractionation, followed by gel filtration and DEAE-cellulose chromatography. This procedure results in the separation of five ovoinhibitor fractions, identical in amino acid composition, but differing in carbohydrate content (Table VI). These fractions, however, are still heterogeneous.

Properties

Stability.[71] In acid solution, ovoinhibitor shows considerable stability: 93–95% of the activity is retained when the inhibitor is kept for 15 minutes at 90° at pH 3 and 5, for 24 hours at 40° at pH 2, or for 3 hours at 40° at pH 1. In 0.01 N NaOH, the activity is stable at 23° for 24 hours, but the activity is lost in 3 hours at 40° in 0.1 N NaOH and in 15 minutes at 90° at pH 7 and 9. Pepsin at pH 2 destroys the activity, but papain at pH 5 does not.

Purity. Ovoinhibitor preparations are electrophoretically[20,22] and chromatographically[20] heterogeneous, but give a single boundary in sedimentation experiments.[20]

Physical Properties. The data are summarized in Table V. The high

[70a] Y. Tomimatsu, personal communication, 1969.
[71] K. Matsushima, *J. Agr. Chem. Soc. Japan* **32**, 211 (1958).

TABLE V
PHYSICAL PROPERTIES OF OVOINHIBITOR

Property	Method	Value
Molecular weight	Light scattering[a]	49,000
	Sedimentation (Archibald)[a]	44,000
	Sedimentation equilibrium[b,c]	48,700
	Sedimentation equilibrium[c,d]	48,600
	Composition[c,e]	46,000–49,000
Partial specific volume	Amino acid composition[a]	0.693
($\bar{v}$)	Amino acid composition[c]	0.708–0.712
	Density[c]	0.707
Refractive increment[a]		0.185
dn/dc, $\lambda = 436$ nm		
Absorbance, $[A_{1\,cm}^{1\%}]$		$7.40^a, 6.5$–6.9^c
at 278 nm		

[a] Y. Tomimatsu, J. J. Clary, and J. J. Bartulovich, *Arch. Biochem. Biophys.* **115**, 536 (1966).

[b] In H_2O and D_2O.

[c] J. G. Davis, J. C. Zahnley, and J. W. Donovan, *Biochemistry* **8**, 2044 (1969).

[d] In 6 M guanidinium chloride plus 2 mM dithiothreitol, pH 8.

[e] Molecular weight range for five fractions calculated as follows: amino acids 44,000 + hexose 800–1800 + glucosamine 1500–3000.

molecular weight compared to most other inhibitors is not due to dimer formation,[20] since it remains unchanged in a denaturing, reducing solvent (Table V, footnote d).

Specificity. Ovoinhibitor reacts with trypsins and chymotrypsins of both bovine[7] and avian[21] origin (Table III). It also inhibits *Aspergillus protease*[6,71] and *Bacillus subtilis*[6] var. *biotecus* (Nagarse). With pronase, the esterase activity[20] and part of the protease activity[72] are inhibited.

Kinetic Properties. Enzyme-to-inhibitor molar combining ratios of 1.74 to 1.90 and 1.65 to 1.70 have been determined[20,22] for the reactions with trypsin and chymotrypsin, respectively. Ovoinhibitor reacts with both trypsin and chymotrypsin simultaneously; the inhibition of one enzyme is unaffected by the presence of the other, indicating distinct inhibition sites for the two enzymes.[4,6,22] The two active sites differ in their stability toward heat and photooxidation.[73] Subtilisin and trypsin are independently inhibited[22] in mixtures. With mixtures of chymotrypsin and subtilisin, the inhibition of one enzyme is affected[22] by the presence of the

[72] R. Haynes and R. E. Feeney, *J. Biol. Chem.* **242**, 5378 (1967).

[73] T.-S. Chu and C.-L. Hsu, *Acta Biochim. Biophys. Sinica* **5**, 434 (1965); *Chem. Abstr.* **64**, 5373g.

other; this suggests that the two enzymes compete for the same or closely adjacent sites.

Inhibition constants (K_I) for both enzymes[20,73,74] are between 10^{-8} and $10^{-9}\,M$. The inhibition of trypsin and chymotrypsin appears to be intermediate between competitive and noncompetitive inhibition, from Lineweaver-Burk plots.[20] When the inhibition constant is calculated[20] for 1:1 binding for three separate ovoinhibitor fractions as competitive inhibition, K_I values of 1.4 to 1.6 $\times$ $10^8\,M$ for trypsin and 0.96 to 1.2 $\times$ $10^{-8}\,M$ for chymotrypsin are found. Calculated[20] for noncompetitive inhibition, K_I values are 4.4 to 5.0 $\times 10^{-8}\,M$ and 4.2 to 4.8 $\times 10^{-8}\,M$, respectively.

Composition.[20] The amino acid composition is given in Table VI. Al-

TABLE VI

COMPOSITION OF OVOINHIBITOR[a]

Amino acid	Residues per mole[b]	Amino acid	Residues per mole[b]
Alanine	20	Methionine	4
Arginine	21	Phenylalanine	6
Aspartic acid	47	Proline	18
Half-cystine	34	Serine	27
Glutamic acid	39	Threonine	33
Glycine	32	Tryptophan[c]	<1
Histidine	14	Tyrosine	17
Isoleucine	17	Valine	27
Leucine	22	Amide groups	37
Lysine	23	Total N (%)	16.3–16.8

Component	Residues per mole				
Fraction[d]	A	B	C	D	E
Glucosamine	14.0	15.2	11.0	7.6	9.0
Hexose	9.7	10.2	7.7	5.6	
Sialic acid	0.2		0.4	0.5	

[a] J. G. Davis, J. C. Zahnley, and J. W. Donovan, *Biochemistry* **8**, 2044 (1969). [Copyright by the American Chemical Society. Reprinted (condensed) by permission of the copyright owner.]

[b] Average of five chromatographically separated fractions; there were no significant differences among the fractions in amino acid composition.

[c] No tryptophan was detected by ultraviolet absorbance of neutral and alkaline solutions and only a small fraction of a residue by the method of Spies and Chambers [*Anal. Chem.* **21**, 1249 (1949)].

[d] Five fractions which were separated on DEAE-cellulose and rechromatographed.

[74] J. C. Zahnley and J. G. Davis, *Biochemistry* **9**, 1428 (1970).

though the different fractions have markedly different carbohydrate contents, they do not differ significantly in amino acid composition. There is considerable resemblance to the proportions of amino acids in the ovomucoids (Table IV); the content of carbohydrate is smaller than in the ovomucoids.

[67] Pancreatic Secretory Trypsin Inhibitors

By Philip J. Burck

Introduction

A pancreatic trypsin inhibitor distinct from the basic Kunitz inhibitor, first crystallized in 1936,[1] was isolated from bovine pancreas by Kazal et al. in 1948.[2] Using a side fraction of a commercial insulin process, they purified this inhibitor by fractional precipitation with alcohol and then crystallized it from a trichloroacetic acid solution. Unlike the basic trypsin inhibitor, which is found only in some ruminants and is not secreted by the pancreas, the Kazal-type inhibitor, which we shall refer to as the pancreatic secretory trypsin inhibitor (PSTI), has been found in the pancreas of the cow,[3] pig,[3] rat,[4] dog,[5] and human.[3,6,7] Bovine PSTI has been isolated as a homogeneous polypeptide by chromatographic methods from pancreatic juice,[8] from acid–ethanol extracts of pancreas,[9,10] and from pancreas extracts by adsorption to water-insoluble enzyme resins.[11] Purified preparations of the analogous porcine inhibitor have been obtained from the pancreas[11-13] and pancreatic juice.[14]

[1] M. Kunitz and J. H. Northrop, J. Gen. Physiol. 19, 991 (1936).

[2] L. A. Kazal, D. S. Spicer, and R. A. Brahinsky, J. Am. Chem. Soc. 70, 3034 (1948).

[3] H. Fritz, F. Woitinas, and E. Werle, Hoppe-Seylers Z. Physiol. Chem. 345, 168 (1966).

[4] M. I. Grossman, Proc. Soc. Exptl. Biol. Med. 99, 304 (1958).

[5] H. Fritz, G. Hartwick, and E. Werle, Hoppe-Seylers Z. Physiol. Chem. 345, 150 (1966).

[6] B. J. Haverback, B. Dyce, H. Bundy, and H. A. Edmondson, Am. J. Med. 29, 424 (1960).

[7] P. J. Keller and B. J. Allen, J. Biol. Chem. 242, 281 (1967).

[8] L. J. Greene, M. Rigbi, and D. S. Fackre, J. Biol. Chem. 241, 5610 (1966).

[9] E. W. Cerwinsky, P. J. Burck, and E. L. Grinnan, Biochemistry 6, 3175 (1967).

[10] P. J. Burck, R. L. Hamill, E. W. Cerwinsky, and E. L. Grinnan, Biochemistry 6, 3180 (1967).

[11] H. Fritz, I. Huller, M. Wiedemann, and E. Werle, Hoppe-Seylers Z. Physiol. Chem. 348, 405 (1967).

Assay Method

Principle. The inhibition of the hydrolysis of esters of L-arginine by trypsin is used as the quantitative assay for the pancreatic secretory trypsin inhibitors (see Vol. II [3,4] for previous descriptions of this method).

Inhibition of the Hydrolysis of Benzoyl Arginine Ethyl Ester (BAEE)

The most convenient assay for PSTI is the inhibition of trypsin esterase activity using the spectrophotometric method of Schwert and Takenaka[15] modified by Kassell *et al.*[16]

Reagents

Crystalline bovine or porcine trypsin (80 μg/ml) in 0.0025 N HCl–0.02 M $CaCl_2$

BAEE, 0.00025 M, in 0.05 M Tris, pH 8.0. Alternatively, the Determatube TRY (Worthington) may be used.[9]

PSTI in 0.05 M Tris, pH 8.0. The concentration is adjusted so that 25–75% of the trypsin is inhibited. Normally solutions containing 5–20 μg/ml of PSTI are used.

Procedure. One milliliter of the PSTI solution is added to 1.0 ml of the trypsin solution and then incubated at 25° for 5 minutes. This solution (0.2 ml) is then pipetted into 2.8 ml of the BAEE solution in a cuvette. After quickly mixing the solution, the increase in optical density at 253 nm is recorded using a spectrophotometer.

Definition of Unit and Specific Activity. One inhibition unit is the amount of inhibition that has caused the reduction of BAEE hydrolysis by 1.0 OD_{253} units per minute. Specific activity is defined as inhibitor units per OD_{280}.

Inhibition of the Hydrolysis of p-Toluenesulfonyl-L-arginine Methyl Ester (TAME)

This assay can be used in laboratories equipped with a recording pH-stat. This method is best used for the assay of crude solutions of PSTI

[12] P. J. Burck, E. W. Cerwinsky, and E. L. Grinnan, *Abstr. Am. Chem. Soc. Meeting, September 1967,* p. C-22.

[13] H. Fritz, H. Schult, M. Neudecker, and E. Werle, *Angew. Chem.* **78,** 775 (1966).

[14] L. J. Greene, J. J. DiCarlo, A. J. Sussman, and D. C. Bartelt, *J. Biol. Chem.* **243,** 1804 (1968).

[15] G. W. Schwert and Y. Takenaka, *Biochim. Biophys. Acta* **16,** 570 (1955).

[16] B. Kassell, M. Radicevic, S. Berlow, R. J. Peanasky, and M. J. Laskowski, Sr., *J. Biol. Chem.* **238,** 3274 (1963).

where possible nucleotide absorption may interfere with the spectro-photometric determination of the hydrolysis of BAEE at 253 nm.

Reagents

> Stock trypsin solution containing crystalline bovine or porcine tryp-sin (1 mg/ml) in 0.01 M KCl-HCl, pH 2.9. This is used to pre-pare 1.0 ml of solution containing 25 μg trypsin in 0.1 M KCl, 0.015 M Tris, 0.02 M CaCl$_2$.
>
> TAME, 0.01 M, in 0.005 M Tris-HCl, 0.1 M KCl, 0.02 M CaCl$_2$
>
> PSTI in 0.1 M KCl, 0.015 M Tris, 0.02 M CaCl$_2$, pH 7.8. The con-centration is adjusted so that 25–75% of the trypsin is inhibited. Normally solutions containing 2–6 μg/ml of PSTI are used.

Procedure. One milliliter of the PSTI solution is added to 1.0 ml of the trypsin solution and then incubated at 25° for 5 minutes. One milli-liter of this solution is then pipetted into 4.0 ml of the TAME solution. The reaction is followed using a pH-stat.

Definition of Unit and Specific Activity. One inhibition unit is the amount of inhibition that has caused the reduction of TAME hydrolysis by 1 micromole per minute. Specific activity is defined as inhibitor units per OD$_{280}$.

Purification Procedure

Procedures are described for the purification of the bovine and porcine pancreatic secretory trypsin inhibitors from both pancreas and pancreatic juice. Since the porcine inhibitor exists in two forms, a chromatographic procedure employing SE-Sephadex for their separation is also described. The yields of bovine and porcine PSTI from pancreas and pancreatic juice are summarized in Tables I and II, respectively.

Isolation of PSTI from Pancreas[9,12]

Step 1. Extraction of the Pancreas. One kilogram ground, frozen pan-creas is placed into the following solution: 2080 ml 95% ethanol, 286 ml distilled water, 44 ml 85% phosphoric acid. After stirring for 5 hours at 5°, the alcohol extract is squeezed from the pancreas using cheesecloth. The pancreas residue is then resuspended in the following solution: 1592 ml 95% ethanol, 728 ml distilled water, 10 ml 85% phosphoric acid. After stirring for 2 hours at 5°, the alcohol extract is recovered as before. The pH of the combined extracts is then adjusted to 8.2 with NH$_4$OH. Hyflo (1% w/v) is then added to this solution, and the precipitate is removed by filtration on a Büchner funnel. The pH of the filtrate is ad-justed to 3.6 with 5 N H$_2$SO$_4$. This solution is then concentrated to one-third volume using a rotary evaporator at 25°. The concentrated extract is decanted from the fat layer which forms. The pH of the decanted solution is adjusted to 2.2 with 85% H_3PO_4, and Hyflo (1.5% w/v) is added. After stirring for 15 minutes, the precipitate is removed by filtration on a Büchner funnel. The filtrate is then further concentrated to one-sixth of the original volume. Hyflo (0.5% w/v) is added to the concentrated extract, and after stirring for 15 minutes the precipitate is removed by filtration on a Büchner funnel. The filtrate is cooled to 5°. One kilogram pancreas should yield 1.0 liter concentrated extract.

Step 2. Salt Fractionation. Sodium chloride (150 g) is added to 1.0 liter of the cold, concentrated extract and stirred for 30 minutes. Hyflo (1% w/v) is added, and stirring is continued for another 15 minutes. The precipitate is then removed by filtration on a Büchner funnel. Sodium chloride (160 g) is then added to the 15% NaCl filtrate. After stirring for 30 minutes, Hyflo (1% w/v) is added. Stirring is continued for another 15 minutes, and the 31% NaCl precipitate is then collected by filtration on a Büchner funnel. This filter cake is then extracted three times for 30 minutes with cold, distilled water. The final volume of the combined extracts is one-tenth the volume of the concentrated pancreas extract. At this point it is most convenient to pool five lots before continuing purification. The water extract (500 ml) is brought to 0.60 saturated ammonium sulfate and is stirred for 30 minutes. The precipitate is collected by centrifugation at 14,000 *g* at 5° and is dissolved in 50 ml 0.01 *M* ammonium acetate, pH 5.0.

Step 3. Desalting on Sephadex G-25C. The solution of the 0.60 saturated ammonium sulfate precipitate is desalted at room temperature on a G-25C Sephadex (Pharmacia) column equilibrated with 0.01 *M* ammonium acetate, pH 5.0. A ratio of column feed solution volume to column bed volume of 1:10 and a height:diameter ratio of the column of at least 10:1 are most satisfactory. To desalt 50 ml of a solution of the 0.60 saturated ammonium sulfate precipitate, a column 3.7 × 45 cm is used. Fractions of 10 ml are collected at a flow rate of 200 ml per hour (0.4 bed volumes per hour). The eluted fractions are tested for sulfate ions with 1–2 drops of 1.0 *N* $BaCl_2$. The crude PSTI is eluted with the front-running, salt-free, protein-rich peak. The active fractions within this peak are pooled and lyophilized. Several lots may be conveniently pooled at this point.

Step 4. Chromatography on CM-Cellulose. The crude, salt-free inhibitor is dissolved in 0.005 *M* ammonium acetate, pH 5.0, at a concentration of 1 g per 10 ml buffer. Crude PSTI (7.5 g) can be conveniently chromatographed on a 3.7 × 45 cm column (15 g crude PSTI per 1.0 liter bed volume CM-cellulose). The height:diameter ratio of the column should be at least 10:1. Prior to chromatography the solution should be dialyzed for at least 4 hours against 10 volumes 0.005 M ammonium acetate, pH 5.0. The dialyzed solution is then adsorbed onto the CM-cellulose column equilibrated with 0.005 M ammonium acetate, pH 5.0. Fractions of 10 ml are collected at a flow rate of 200 ml per hour (0.4 bed volumes/hour). Unadsorbed protein is eluted through the column with the solvent buffer. When the OD_{280} of the column effluent has returned to a point near zero, elution of the column with 0.05 M ammonium acetate, pH 5.0, is begun. The purified PSTI is contained in the second effluent protein peak following the buffer change. The fractions comprising this peak are pooled and lyophilized.

Step 5. Chromatography on DEAE-Cellulose.[8] Purified PSTI from the CM-cellulose column is dissolved in 0.028 M Tris, pH 9.0 at a concentration of 10 mg/ml of buffer. Four hundred milligrams of purified PSTI can be conveniently chromatographed on a 3.7 × 45 cm column (800 mg PSTI per 1.0 liter bed volume). The height:diameter ratio of the column should be at least 10:1. Prior to chromatography the solution should be dialyzed for at least 4 hours against 10 volumes of 0.028 M Tris, pH 9.0. The dialyzed solution is then adsorbed onto a DEAE-cellulose column equilibrated with 0.028 M Tris, pH 9.0 at room temperature. Fractions of 10 ml are collected at a flow rate of 200 ml per hour (0.4 bed volumes per hour). The purified bovine PSTI is recovered in the column effluent after the column has been eluted with five to six bed volumes of the buffer. The porcine PSTI is not bound to the resin and appears in the breakthrough fraction of the column effluent. The active

TABLE I
SUMMARY OF PURIFICATION PROCEDURE FOR PSTI FROM PANCREAS

Fraction	Bovine			Porcine		
	Total[a] units	Specific activity	Overall yield (%)	Total[a] units	Specific activity	Overall yield (%)
Step 1	2200	—	—	2400	—	—
Step 2	1870	5.0	85	2160	8.0	90
Step 3	1720	7.6	78	1920	10.0	80
Step 4	1250	215	57	1440	225	60
Step 5	1050	370	48	1320	375	55
SE-Sephadex[b]	—	—	—	1000	410	42
				250	390	10.5

[a] Units are defined using the BAEE assay method.
[b] The two values under each heading for porcine PSTI are for forms I and II of the inhibitor.

fractions are pooled, dialyzed overnight in NoJax Visking dialysis casing against distilled water at 5°, and then lyophilized. Porcine PSTI, which exists as a combination of its two forms after chromatography on DEAE-cellulose, is further purified by chromatography on SE-Sephadex as described in step 4 of the following section.

See Table I for a summary of this purification procedure.

Purification of PSTI from Pancreatic Juice[8, 14]

Step 1. Collection of Pancreatic Juice. The pancreatic juice is collected into bottles at 4° by direct cannulation of the main pancreatic duct. The juice is treated with diisopropylphosphofluoridate (DFP) (0.1 M in isopropyl alcohol, 10 ml per liter of juice) for 1 hour with continuous stirring and is then lyophilized.

Step 2. Chromatography on Sephadex G-75. Lyophilized pancreatic juice is dissolved in distilled water containing $10^{-4} M$ DFP to a final concentration of 1.5–2.0% protein. This solution is applied to a column of Sephadex G-75 (Pharmacia) equilibrated with 0.5 M KCl, 0.01 M Tris, $10^{-4} M$ DFP, pH 8.0. A ratio of column feed solution volume to column bed volume of 1:10 and a height:diameter ratio of the column of about 25:1 are most satisfactory. To desalt 900 ml of a solution of the lyophilized pancreatic juice, a column 7.6 × 175 cm is used. Fractions of 60 ml are collected at a flow rate of 60 ml/hour with the column outflow limited by a peristaltic pump. The active fractions of bovine PSTI having a specific activity above 5.0 (TAME assay) and of porcine PSTI having a specific activity above 2.4 (BAEE assay) are pooled and lyophilized.

Step 3. Chromatography on DEAE-Cellulose. The crude PSTI from the Sephadex G-75 column is dissolved in distilled water in a volume one-seventh that of the Sephadex G-75 effluent volume. This solution is then dialyzed using NoJax Visking dialysis casing against distilled water until it has the same conductivity as 0.028 M Tris, pH 9.0. After clarification by centrifugation at 13,000 g, this solution is adsorbed onto a column of DEAE-cellulose equilibrated with 0.028 M Tris, pH 9.0, at room temperature. A sample solution (75 ml) having an OD_{280} of 14 can be adsorbed to a 1.8 × 37 cm column of DEAE-cellulose. Fractions of 20 ml are collected at a flow rate of 120 ml/hour. The purified bovine PSTI is recovered in the column effluent after the column has been eluted with five to six bed volumes of the buffer. The porcine PSTI is not bound to the resin and appears in the breakthrough fraction of the column effluent. The active fractions are pooled, dialyzed overnight in NoJax Visking dialysis casing against distilled water at 5°, and then lyophilized. Porcine PSTI is further purified as described in the next step.

Step 4. Chromatography on SE-Sephadex. Lyophilized, salt-free porcine PSTI from the DEAE-cellulose column is dissolved in 0.02 M ammonium acetate, pH 5.0, to a final concentration of 0.2% protein. This solution is then adsorbed onto a column of SE-Sephadex equilibrated with 0.05 M ammonium acetate, pH 5.4, at room temperature. Five milliliters of sample solution having an OD_{280} of 1.6 can be adsorbed to a 0.9 × 143 cm column of SE-Sephadex. Five-milliliter fractions are collected at a flow rate of 5–8 ml/hour using a pump or air pressure. The two forms of the porcine inhibitor, PSTI I and II, are separated by elution of the column with the starting buffer and are recovered in the column effluent after the column has been eluted with 2.5 and 3.0 bed volumes respectively. The active fractions of each inhibitor are pooled and lyophilized.

See Table II for a summary of the purification of PSTI from pancreatic juice.

TABLE II
SUMMARY OF PURIFICATION PROCEDURE FOR PSTI FROM PANCREATIC JUICE

	Bovine		Porcine	
Fraction	Specific[a] activity	Overall yield (%)	Specific[b] activity	Overall yield (%)
Step 1	0.4–0.6	—	0.16–0.20	—
Step 2	8.6	72	6.4	85
Step 3	2750	60	320	68
Step 4[c]	—	—	410	52
			390	13

[a] Units are defined using the TAME assay method.

[b] Units are defined using the BAEE assay method.

[c] The two values under each heading for porcine PSTI are for forms I and II of the inhibitor.

Properties of the Inhibitors

Specificity. The pancreatic secretory trypsin inhibitors are much more specific in their spectrum of inhibition than the Kunitz inhibitor. Like KPTI, the secretory inhibitors form an equimolar complex with trypsin.[8,10] However, unlike KPTI, the inhibition observed is the temporary inhibition first observed with the original Kazal inhibitor by Laskowski and Wu[17] (see Vol. II [4]). Bovine and porcine PSTI do not inhibit the activity of chymotrypsin, kallikrein, or urokinase. Bovine PSTI does weakly inhibit plasmin fibrinolytic and esterase activity and inhibits the

[17] M. Laskowski and F. C. Wu, *J. Biol. Chem.* **204**, 797 (1953).

clotting of recalcified plasma and thrombin-clotting activity. However, the inhibition of thrombin does not appear to be stoichiometric, since a large excess of inhibitor is required to effect a delay in clotting time. Porcine PSTI I and II have the same spectrum of inhibition. Both inhibit plasmin esterase activity, but do not inhibit the fibrinolytic activity of this enzyme. Porcine PSTI I and II are also weaker inhibitors of the clotting of recalcified plasma and thrombin clotting activity than bovine PSTI. The inhibitory specificity of the secretory inhibitors is summarized in Table III.

TABLE III
ACTIVITY OF PANCREATIC TRYPSIN INHIBITORS

Type of inhibition	Porcine PSTI	Bovine PSTI	Kunitz PTI
Trypsin protease	+	+	+
Trypsin esterase	+	+	+
Chymotrypsin protease	−	−	+
Chymotrypsin esterase	−	−	+
Thrombin clotting	+	+	+
Thrombin esterase	−	−	+
Plasmin fibrinolysis	−	+	+
Plasmin esterase	+	+	
Kallikrein esterase	−	−	+
Recalcified plasma clotting	+	+	+
Urokinase fibrinolysis	−	−	+

Stability. Both bovine and porcine PSTI are stable to lyophilization, dialysis against distilled water, and storage at 5° as dry powders or solutions. The inhibitors are stable over the pH range of 2.2 to 9.0 used during purification. A crystalline preparation of the Kazal inhibitor, which had a specific activity 90% of that of chromatography purified bovine PSTI, was active after 18 years of storage.[8]

Physical Properties. The bovine inhibitor contains 56 amino acid residues[8,9,11] and has a molecular weight of 6155. The linear sequence of these amino acids has been determined.[18] The porcine inhibitor exists in two forms. One contains 56 amino acid residues; the other contains 52, having 4 less residues on the amino terminal end of the peptide chain. These are present in a 4:1 ratio in the pancreas[12] and pancreatic juice[14] and have molecular weights of 6024 and 5609, respectively.

The absorption spectrum of bovine PSTI in 0.01 M Tris–0.1 M KCl, pH 7.8, has a maximum at 276 nm, and the ratio of the absorbance at 280 and 260 nm is 1.42. In 0.1 N NaOH the UV-absorption maximum for bovine PSTI is 294 nm.[8] The porcine inhibitors, like bovine PSTI,

[18] L. J. Greene and D. C. Bartelt, *J. Biol. Chem.* **244**, 2646 (1969).

contain 2 residues of tyrosine and 3 of cystine per molecule as the only chromophores absorbing to an appreciable extent over 250 nm. Porcine inhibitor I (56 amino acids) also has an absorption maximum at 276 nm which shifts to 294 nm in 0.1 N NaOH.[14, 19]

Equilibrium sedimentation studies have been conducted on both inhibitors. Using a calculated partial specific volume of 0.71, an average molecular weight of 6500 ± 300 was obtained for the inhibitor from bovine pancreas[9] while a molecular weight of 6100 ± 260 was obtained for the inhibitor from pancreatic juice[8] using a calculated partial specific volume of 0.719. Porcine inhibitors I and II from porcine pancreatic juice have molecular weights of 6040 ± 300 and 5400 ± 300, respectively, as determined by sedimentation equilibrium studies.[14]

Electrophoretic examination of the pancreatic secretory trypsin inhibitors is most conveniently carried out using disc electrophoresis. Either the 15% acrylamide gel system at pH 4.5[20] or 8.9[21, 22] may be used at 6 mA per tube for 60 minutes. The gels are stained with amido black in 7% acetic acid. Excess stain can be removed electrophoretically or by washing in 7% acetic acid. These systems do not separate porcine PSTI I and II.

Reactive Site. Both bovine and porcine PSTI undergo a limited proteolysis at their "reactive sites." Proteolysis by catalytic amounts of trypsin hydrolyzes a specific arginine bond in bovine PSTI[23] and a specific lysine bond in porcine PSTI.[24, 25] The specific bond hydrolyzed in bovine PSTI is the arginyl-isoleucine bond at positions 18 and 19. Removal by carboxypeptidase B of the newly formed COOH-terminal arginine results in inactivation of the bovine inhibitor. The "reactive site" of bovine and porcine PSTI probably occurs within a disulfide loop in the molecule as suggested by Ozawa and Laskowski[26] for other trypsin inhibitors.

[19] H. Tschesche, *Hoppe-Seylers Z. Physiol. Chem.* **348**, 1653 (1967).
[20] R. A. Reisfeld, U. J. Lewis, and D. E. Williams, *Nature* **195**, 281 (1962).
[21] B. J. Davis, *Ann. N.Y. Acad. Sci.* **121**, 404 (1964).
[22] L. Ornstein, *Ann. N.Y. Acad. Sci.* **121**, 321 (1964).
[23] M. Rigbi and L. J. Greene, *J. Biol. Chem.* **243**, 5457 (1968).
[24] H. Tschesche, *Hoppe-Seylers Z. Physiol. Chem.* **348**, 1216 (1967).
[25] H. Tschesche and H. Klein, *Hoppe-Seylers Z. Physiol. Chem.* **349**, 1645 (1968).
[26] K. Ozawa and M. Laskowski, Jr., *J. Biol. Chem.* **241**, 3955 (1966).

[68] Preparation and Assay of Plasma Antithrombin

By Frank C. Monkhouse

Introduction

While antithrombin activity can be neutralized by a number of non-specific agents and surfaces, the main progressive antithrombin of plasma is associated with the alpha globulins.[1-3] This protein fraction brings about the irreversible destruction of thrombin in a progressive manner typical of enzymatic degradation.[4,5]

Assay Method

Any method designed to estimate antithrombin activity should involve measurements over a time period. Since thrombin is known to be adsorbed on both dry glass and fibrinogen,[6,7] measurements should be carried out in silicone-treated glassware or plastic, and if the measurements are done on plasma, this should first be defibrinated. Since heating to 54° for 3–4 minutes has no measurable effect on antithrombin, this is the preferable way to remove fibrinogen. In general, with small concentrations of thrombin relative to antithrombin, practically all thrombin is destroyed and the reaction is of the first order. With high concentrations of thrombin in relation to antithrombin, equilibrium is reached more slowly, with some thrombin always remaining in the active state. Two basic types of assay are recommended therefore, one (Method A) based on the use of small concentrations of thrombin and the other (Method B) on large concentrations relative to antithrombin concentration.

Method A

This is a first-order reaction and since clotting times are inversely proportional to the thrombin concentration, a straight line results when log clotting times are plotted against thrombin–antithrombin incubation

[1] F. C. Monkhouse and S. Milojevic, *Can. J. Physiol. Pharmacol.* **46**, 347 (1968).

[2] P. Porter, M. C. Porter, and J. N. Shanberge, *Clin. Chim. Acta* **17**, 189 (1967).

[3] V. Abilgaard, *Scand. J. Clin. Invest.* **21**, 89 (1968).

[4] E. Mihalyi, *J. Gen. Physiol.* **37**, 139 (1953).

[5] F. C. Monkhouse, "Blood Clotting Enzymology," p. 323. Academic Press, New York, 1967.

[6] W. H. Seegers, M. Nieft, and E. C. Loomis, *Science* **101**, 520 (1945).

[7] W. H. Seegers, K. D. Miller, E. B. Andrews, and R. C. Murphy, *Ann. J. Physiol.* **169**, 700 (1952).

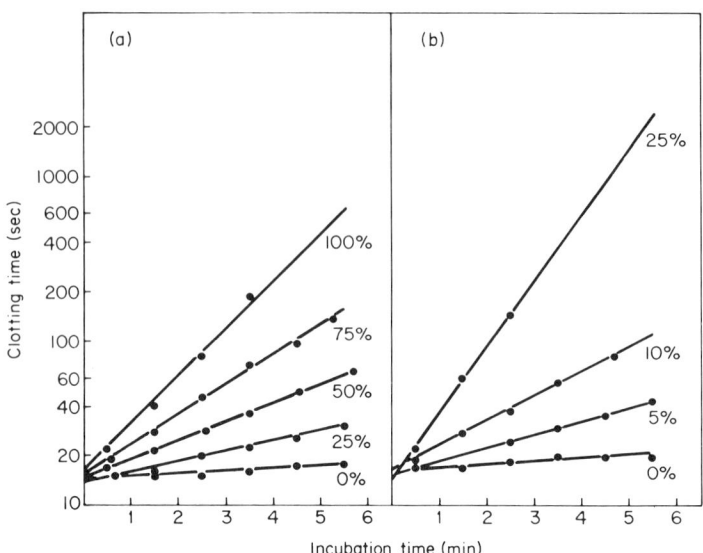

Fig. 1. The straight-line relation between the logarithm of the clotting time and the thrombin–antithrombin incubation time. (a) With varying dilutions of defibrinated plasma as the source of antithrombin, and (b) dilution of purified antithrombin.

times. The slope of this line is the rate constant K, and varies directly with the antithrombin concentration. Validity of this statement is illustrated by Fig. 1. The simplest and best method was introduced by Gerendas.[8] Porcelain spot plates are used in place of test tubes and a fine glass hook is used to detect the first signs of coagulation. To 0.3 ml of plasma (or antithrombin fraction), 0.2 ml of $0.05\,M$ phosphate buffer pH 7.8 and 0.3 ml of thrombin solution (10 μ/ml) are added. At 30-second and minute intervals thereafter, 0.1 ml of this incubation mixture is added to 0.2 ml of a fibrinogen solution and the clotting time noted. Clotting times over a period of 6 minutes are plotted and the best-fitting straight line is drawn. Antithrombin activity is expressed in terms of frozen pooled defibrinated beef plasma stored in small aliquots at $-20°$, and prepared fresh at 6-month intervals. Specific activity refers to the activity per milligram of protein as compared to this standard beef plasma. When it is desirable to use units to give a quantitative comparison rather than percent of standard, 1 ml of standard beef plasma is considered to contain 100 units of antithrombin. In our hands this has proven to be much more satisfactory than attempting to express anti-

[8] M. Gerendas, *Nature* **157**, 837 (1946).

thrombin in terms of units of thrombin destroyed, since a thrombin unit is itself a somewhat arbitrary figure.

Method B

If one wishes to use larger quantities of thrombin and to avoid clotting time as a measure, then the following method is recommended.

Two hundred units of thrombin are placed in each of a number of siliconized test tubes. Varying amounts of the antithrombin solution to be assayed are then added. At least one tube has no antithrombin added, and serves as a control. The volumes are then adjusted to 1 ml by the addition of Tris buffer and the tubes are allowed to incubate at 28° for 1 hour. The amount of residual thrombin is then estimated by adding 0.2-ml aliquots of the incubated thrombin-antithrombin mixture to 0.2 ml of a $0.4\,M$ TAME (p-toluenesulfonyl arginine methyl ester) solution. The volumes are adjusted to 2 ml with Tris buffer, pH 8.5, the tubes are incubated at 37° for 30 minutes and the extent of TAME hydrolysis determined. A curve is then made by plotting volume of antithrombin solution against units of thrombin neutralized. For further details of the method, consult the publication by Monkhouse et al.[9]

Purification

The first stage in the purification of antithrombin is adsorption on aluminum hydroxide, and since the preparation of the aluminum hydroxide is the key to successful purification, details are given.

Preparation of Aluminum Hydroxide Gel for Adsorption of Plasma Antithrombin

Note: This procedure should be carried out as rapidly as possible, never over more than 2½ hours.

1. Weigh out 22 g of ammonium sulfate. Dissolve in approximately 750 ml of distilled water. Bring to 63°.

2. Weigh out 76.5 g of aluminum ammonium sulfate. Dissolve in approximately 2¼ liters of distilled water. Bring to 58°.

3. Prepare a 100-ml quantity of a 50% solution of absolutely fresh ammonium hydroxide (ammonia analysis—minimum 28%, maximum 30%) from a previously unopened bottle. This is important.

4. Quickly pour 50% ammonium hydroxide in one lot into ammonium sulfate solution. The temperature will drop from 63° to approximately 60°. Immediately pour this solution in one lot into the aluminum ammonium sulfate solution. A precipitate will form immediately. Begin

[9] F. C. Monkhouse, E. S. France, and W. H. Seegers, *Circulation Res.* **3**, 397 (1955).

stirring the gel very vigorously, maintaining the temperature between 58° and 60°, and continue for 10 minutes.

5. Centrifuge the gel rapidly (2000 rpm for 5 minutes sufficient). Use the centrifuge brake to speed up the process, and pour off the supernatant.

6. Suspend the precipitated gel by stirring or shaking in 1 liter of distilled water to which 0.44 ml of concentrated ammonium hydroxide has been added.

7. Centrifuge as before, suspending the precipitated gel in 1 liter of distilled water to which has been added 0.88 ml of ammonium hydroxide.

8. Centrifuge three times more, suspending the precipitated gel each time in 1 liter of distilled water (no ammonium hydroxide added).

9. Centrifuge and suspend gel in the least amount of distilled water which still allows the gel to be pipetted. Store at 4°.

Adsorption of Antithrombin

Defibrinate the plasma by heating to 53°, maintaining that temperature for 3 minutes and cooling quickly to room temperature. Remove the denatured fibrinogen by filtering the plasma through several layers of gauze. To remove the prothrombin, add 50 mg of $BaCO_3$ powder for each milliliter of plasma, stir gently for 10 minutes, and centrifuge. To each milliliter of supernatant add 0.2 ml of aluminum hydroxide, stir gently for 10 minutes, centrifuge, and discard the supernatant. Elute the antithrombin from the precipitate with 0.3 ml of 0.05 M sodium phosphate buffer (pH 7.8) per milliliter or original plasma.

Preparation for Chromatography

Each liter of original plasma yields approximately 300 ml of eluate. For further purification, this eluate is reduced in volume to approximately 40 ml. Lyophilization frequently results in loss of specific activity. In our hands, pervaporation in cellophane bags has proven to give the most consistent results. Following the reduction in volume, dialyze the material against a constantly stirred 0.05 M phosphate buffer for 5 hours, with a change of buffer every hour. The material is now referred to as crude antithrombin and usually contains a concentration of antithrombin activity 9–12 times that of normal defibrinated plasma. It can be maintained without loss of activity at this stage for an indefinite period of time if stored at −20°. Before applying to a chromatographic column for further purification, dialyze the crude antithrombin for a further 2 hours against 0.065 M Tris buffer, pH 8.6.

Preparation of Chromatographic Columns

Suspend 120 g of N,N-diethylaminoethyl ether (DEAE) cellulose in 3.5 liters of distilled water to which is added approximately 5 g of sodium hydroxide pellets. Stir thoroughly with a magnetic stirrer (pH will rise to 11 or higher) and filter. Add concentrated HCl to the cellulose suspended in water until the pH is reduced to 2. Following this, add sufficient 5 N sodium hydroxide to raise the pH to 3. Filter and wash with distilled water and reconstitute in Tris buffer at pH 8.6. When the cellulose has been equilibrated to the buffer (this takes three to four changes of 2-liter quantities of buffer with 1 hour of equilibration per change), run the column for 3 hours with 0.065 M tris buffer, pH 8.6. The above amount of cellulose will normally prepare a column 40 mm in diameter and 120 cm in length.

Separation by a DEAE-Cellulose Column

With the stopcock at the bottom of the column closed, pipette 25 ml of crude antithrombin (concentrated eluate) carefully onto the surface of the column. When the eluate has covered the column with an even layer, partially open the stopcock and allow the eluate to flow into the cellulose. Allow a head of 30 to 40 mm of the first eluting solution, Tris buffer containing 0.075 M sodium chloride, to build up, and seal the attached head. Adjust the flow rate to 3 ml per minute. Apply a one-bed volume of each of the following solutions in the order given: (1) 0.075 M, (2) 0.125 M, and (3) 0.200 M sodium chloride, all in Tris buffer at pH

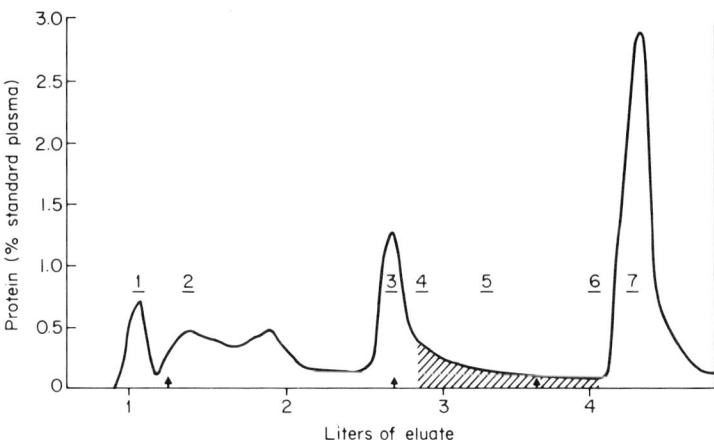

Fig. 2. Chromatography on DEAE-cellulose of crude antithrombin from beef plasma. The column was equilibrated with Tris buffer, 0.065 M, pH 8.6. The proteins were eluted from the column with gradients of 0.075 M, 0.125 M, 0.200 M, and 2.0 M sodium chloride in Tris buffer. Arrows indicate where changes of concentration at the top of the column began. The shaded area indicates the area where antithrombin activity was recovered. The small numbers over the bars indicate specific fractions which were subjected to electrophoresis on cellulose acetate strips shown in Fig. 3.

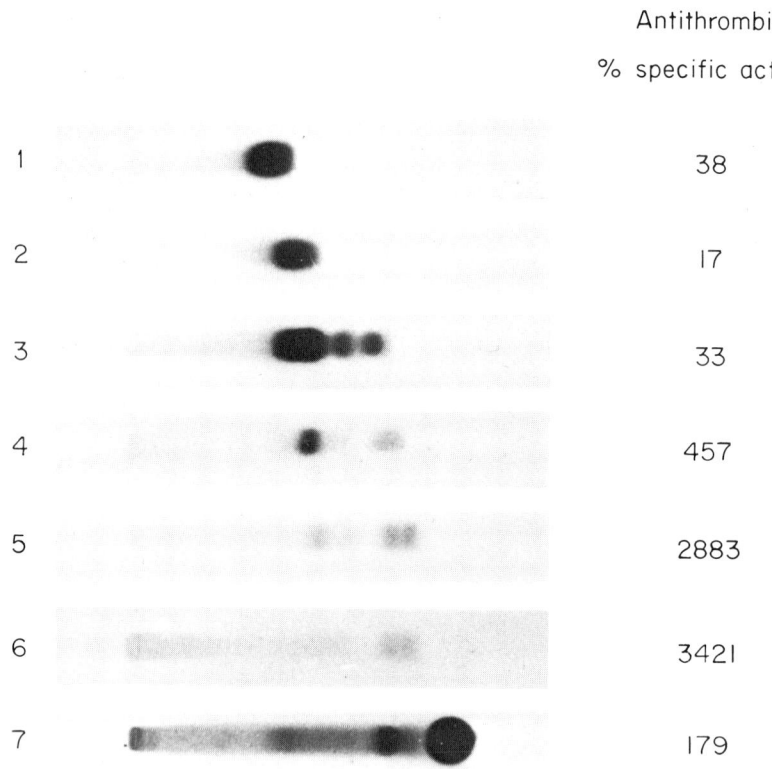

Antithrombin

% specific activity

1	38
2	17
3	33
4	457
5	2883
6	3421
7	179

Fig. 3. Electrophoretic patterns on cellulose acetate strips run 75 minutes at 200 V in 0.03 M phosphate buffer, pH 7.8. Numbers 1–7 refer to the samples from the DEAE column in the regions indicated by the numbers over the bars in Fig. 2.

8.6. Figure 2 illustrates the pattern of fractionation and the shaded area indicates where the antithrombin activity is eluted. In Fig. 3, electrophoretic patterns of aliquots from this area are shown along with the antithrombin activity.

Further Purification with Starch Gel Electrophoresis

Small quantities of antithrombin with high specific activity can be obtained by subjecting the best material from the chromatographic column to starch gel electrophoresis according to the method of Smithies.[10] By making the gel double the standard thickness, and using a slot double the thickness used by Smithies, up to 2 ml of eluate can be processed at one time. The buffer used for both starch suspension and the bridge solu-

[10] O. Smithies, *Biochem. J.* **61**, 629 (1955).

tion should be $0.006 M$ in respect to phosphate and contain 1.85 g of NaH_2PO_4 per liter. Best results have been obtained at pH 8.5 with runs for 17 hours at 280 V. During this time the current gradually increases from approximately 20 to 50 mA. Following the electrophoresis, a strip of gel cut longitudinally is stained with amido black and the remainder cut in sections perpendicular to the direction of flow, as indicated by bands on the stained portion. The sections are frozen and thawed and the fluid expressed by pressure. They are washed once with $0.05 M$ phosphate buffer. The extracts from a number of runs can be pooled and reduced in

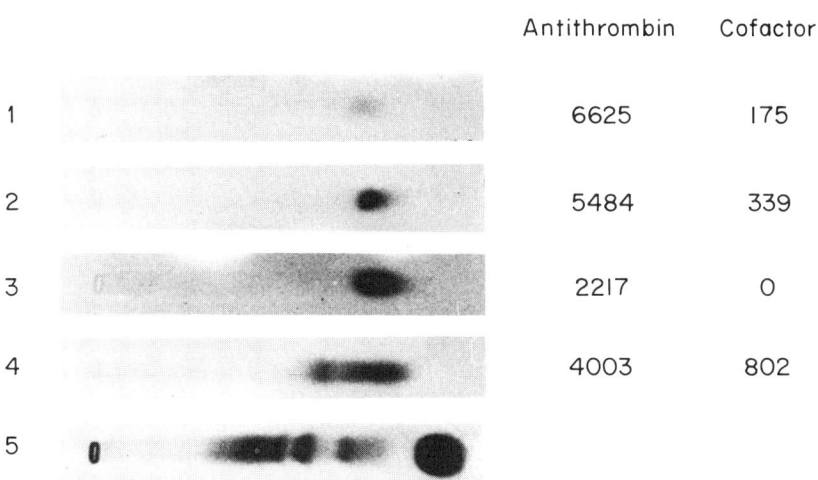

| | % specific activity | |
	Antithrombin	Cofactor
1	6625	175
2	5484	339
3	2217	0
4	4003	802
5		

Fig. 4. Electrophoretic patterns of material recovered from starch gel after electrophoresis. (1) Trailing edge of protein band after elution. (2) Center of band. (3) Leading edge of band. (4) Material applied to starch gel. (5) Standard plasma. Cofactor refers to heparin cofactor activity. Percent specific activity refers to the activity per milligram of protein compared to a standard plasma.

volume. In Fig. 4, electrophoretic patterns of material prepared by the starch gel method are illustrated as they appear on cellulose acetate strips.

Properties of Antithrombin

Optimal activity of antithrombin occurs between pH 8 and 8.5.[11] At pH's below 5.7, antithrombin loses its activity and if maintained at this pH for 1 hour or more, this loss of activity is irreversible. Antithrombin

[11] F. C. Monkhouse, *Thromb. Diath. Hemorrhag.* 9, 387 (1963).

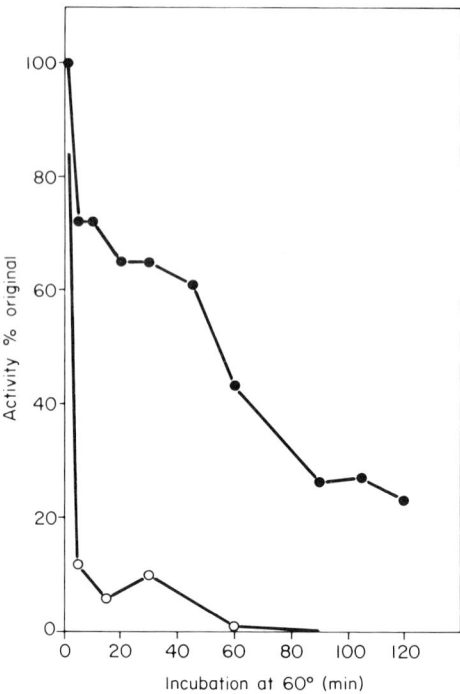

Fɪɢ. 5. The decrease in antithrombin activity of plasma when incubated at 60°. (●———●) Dog plasma; (○———○) human plasma.

loses activity at low ionic concentration.[11] Full activity can be restored by additions of di- or trivalent anions, with phosphate ion the most active. Loss of activity occurs if antithrombin is kept in a low ionic solution for any length of time, thus it should be stored in a $0.05\ M$ phosphate solution.

Antithrombin is relatively stable below 60° but loses activity rapidly above 60°. There is some species difference, with human antithrombin being most labile (Fig. 5). Human antithrombin is more labile than that of dog, rabbit, beef, or rat.

Many organic solvents destroy antithrombin activity. Ethyl alcohol, anesthetic ether, and chloroform are particularly destructive, but petroleum ether has little effect. If anything, it tends to increase activity.

At least three washes with 3 volumes of anesthetic ether are necessary to remove all antithrombin activity from plasma. When this is accomplished, the thrombin formed from the prothrombin in the ether-extracted plasma remains stable for several hours.[12]

[12] F. C. Monkhouse, Can. J. Physiol. Pharmacol. **43,** 819 (1965).

One of the oddities of the antithrombin–thrombin relationship is that they mutually destroy each other.[13] This accounts for the somewhat lower antithrombin activity in serum than in plasma, and the greater the amount of thrombin generated during coagulation, the lower the antithrombin activity in the resulting serum.[13] Similarly, the addition of thrombin to plasma lowers the antithrombin level in proportion to the amount of thrombin added.

Highly purified fractions of antithrombin have been found by Seegers *et al.* to quantitatively inactivate their purified autoprothrombin C[14] and purified autoprothrombin I (probably equivalent to factor VII).[15]

More recently, work in our laboratory has shown that highly purified antithrombin inactivates plasmin (Fig. 6). However, while the antithrombin fractions lose their ability to inactivate plasmin on freezing and storage at −20°, they retain full antithrombin activity. No change in

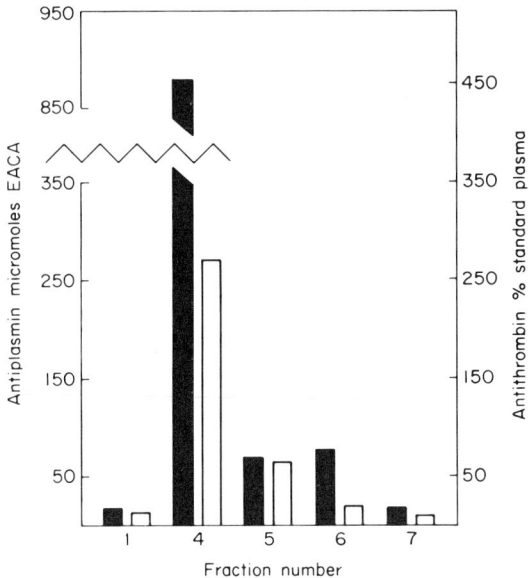

FIG. 6. The antithrombin and antiplasmin activity of fractions of crude antithrombin subjected to purification by DEAE-cellulose chromatography. The antiplasmin is expressed in terms of micromoles of ε-amino-N-caproic acid which would give an equivalent degree of inhibition. Solid bars represent antiplasmin activity, open bars antithrombin activity.

[13] F. C. Monkhouse, *Am. J. Physiol.* **197**, 984 (1959).

[14] W. H. Seegers, E. R. Cole, C. R. Harmison, and F. C. Monkhouse, *Can. J. Biochem.* **42**, 359 (1964).

[15] W. H. Seegers, H. Schroer, and K. Mitsyasu, *Can. J. Biochem.* **42**, 1425 (1964).

protein pattern following the freezing has been demonstrable so far. It may be, therefore, that the two activities are functions of different parts of the same molecule. An association between antithrombin and antiplasmin has been reported by other workers.[16, 17]

While the main antithrombin activity is associated with the alpha globulins, it has been shown that some antithrombin activity is associated with preparations of alpha macroglobulins.[18] This antithrombin apparently does not inhibit the esterase activity of antithrombin.[19] The main antithrombin, that associated with the alpha globulins, inactivates both the clotting activity and the esterase activity of thrombin.

The exact relationship of these closely associated proteolytic activities cannot be settled until more chemically pure fractions are available. To obtain this purity without loss of activity is proving extremely difficult.

[16] N. Heimburger, *First International Symposium on Tissue Factors in the Homeostasis of the Coagulation-Fibrinolysis System. Florence, Italy, May, 1967* p. 353.
[17] H. Gans and B. H. Tan, *Clin. Chim. Acta* 17, 111 (1967).
[18] M. Steinbuch, C. Blatrix, and F. Josso, *Rev. Franc. Etudes Clin. Biol.* 13, 179 (1968).
[19] M. Steinbuch, C. Blatrix, and F. Josso, *Nature* 216, 500 (1967).

[69] Hirudin as an Inhibitor of Thrombin

By Fritz Markwardt

Introduction

The salivary glands (also called neck glands or pharyngeal glands) of the leech *Hirudo medicinalis* contain a substance with anticoagulant properties which has been named hirudin. During leeching, the blood sucker secretes this anticoagulant into the wound in order to keep the blood from clotting. Hirudin was first isolated in 1955 and identified as a polypeptide.[1, 2] The following amino acid composition was found (number of residues per mole): 10 Asp, 13 Glu, 6 Cys, 4 Ser, 9 Gly, 4 Thr, 1 Ala, 3 Val, 4 Leu, 2 Ile, 3 Pro, 2 Phe, 2 Tyr, 1 His, and 4 Lys.

The amino acid sequence from the C-terminal end of the molecule is -Ala-Gly-Ser-Glu-Leu.[3] The molecular weight based on the amino acid composition was 10,800. This is in agreement with a value of 9060 obtained by measurements with the analytical ultracentrifuge. The sedi-

[1] F. Markwardt, *Naturwissenschaften* 42, 587 (1955).
[2] F. Markwardt, *Hoppe-Seylers Z. Physiol. Chem.* 308, 147 (1957).
[3] P. de la Llosa, C. Tertrin, and M. Jutisz, *Biochim. Biophys. Acta* 93, 40 (1964).

mentation constant of hirudin ($s_{20,w}^0$) was 0.98, the diffusion constant ($D_{20,w}^0$) was 10.8. The electrophoretic mobility at pH 5.0 (ionic strength 0.1) was 6.6×10^{-5} cm²/V/second. The isoelectric point was found to be pH 3.9.[4] The partial specific volume ($\bar{V}$) was 0.741 ml/g, the sedimentation coefficient at zero concentration being 0.98 S, the diffusion coefficient 10.8 F, and the molar frictional ratio 1.42.[5]

Hirudin is a specific inhibitor of thrombin.[6-8] For its effect, it does not require the presence of other coagulation factors or plasma constituents. Hirudin inhibits blood coagulation by blocking the end product of the first stage of clotting, thrombin, and thereby prevents the conversion of fibrinogen to fibrin.

The reaction of thrombin with hirudin is faster than the reaction of thrombin with fibrinogen. Like an ionic reaction, the time interval of the thrombin–hirudin reaction is too fast to be measured accurately. When thrombin reacts with hirudin, a complex is formed which can be identified by means of electrophoresis and chromatography. Hirudin can be dissociated from this inactive complex by denaturation with heat or acids. The inactive thrombin–hirudin complex is only poorly dissociable, as seen in Fig. 1. The noticeable deviation from the stoichiometric reaction indicates the point of dissociation. From these data a dissociation constant (K_m) of 0.8×10^{-10} (pH 7.4, 20°) was calculated. This dissociation constant is so small that for all practical purposes the complex can be regarded as nondissociable.

Investigations into which groups of the hirudin molecule might be important for the binding of thrombin revealed that esterification of the carboxyl groups of hirudin destroyed its activity. Besides free carboxyl groups, hydroxyl groups of phenol or imidazole groups also could be important for the binding. However, chemicals that reacted with both of these groups did not display an effect on the biological activity of hirudin. In contrast, oxidation of the disulfide bridges by performic acid destroyed the capability of hirudin to bind thrombin.

In order to further elucidate the mode of action of hirudin, its reaction with a chemically modified form of thrombin was of special interest. The acetylation of the free amino groups of thrombin results in esterase–thrombin which has lost its specific proteolytic action on the conversion

[4] F. Markwardt and P. Walsmann, *Hoppe-Seylers Z. Physiol. Chem.* **348**, 1381 (1967).
[5] H. Triebel and P. Walsmann, *Biochim. Biophys. Acta* **120**, 137 (1966).
[6] F. Markwardt, *Arch. Exptl. Pathol. Pharmakol.* **229**, 389 (1956).
[7] F. Markwardt and P. Walsmann, *Hoppe-Seylers Z. Physiol. Chem.* **312**, 85 (1958).
[8] F. Markwardt, "Blutgerinnungshemmende Wirkstoffe aus blutsaugenden Tieren." Fischer, Jena, 1963.

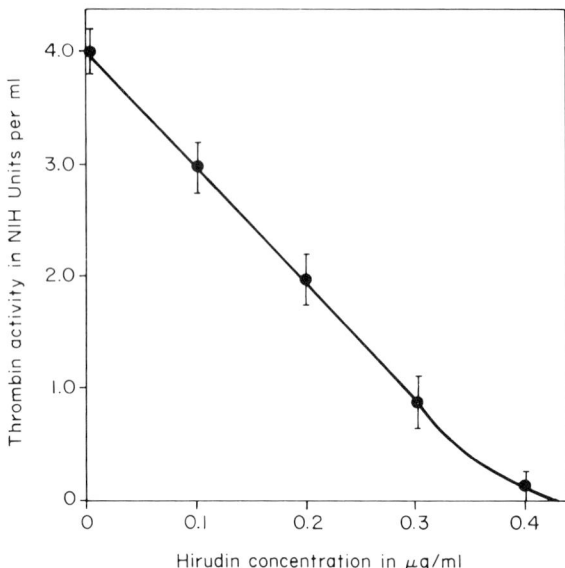

FIG. 1. Inhibition of the clotting activity of thrombin by hirudin.

of fibrinogen to fibrin. Such an esterase–thrombin did not bind hirudin, and therefore, the esterolytic activity of this enzyme cannot be inhibited by hirudin. In contrast, diisopropylphosphoryl-thrombin (DIP-thrombin) in which the serin residue in the active enzyme center has been phosphorylated by diisopropylphosphofluoridate (DFP), is like the native form of thrombin bound by hirudin.

When thrombin reacts with hirudin, which leads to the immediate neutralization of the enzymatic activity of thrombin, the active surface of the enzyme is apparently occupied by hirudin. It has recently been suggested[9] that when thrombin reacts with fibrinogen, the fibrinopeptides, due to their polyanionic characteristics, are bound to positively charged sites on the surface of the enzyme. This brings the fibrinogen molecule into a position where its fibrinopeptides can be cleaved. From the finding that DIP-thrombin still binds hirudin, it can be concluded that the hydrolytically active groups of thrombin are not involved in the binding of hirudin. Apparently the binding sites for hirudin can be found in the cationic groups of the active surface of the enzyme which binds the fibrinopeptide portion of the fibrinogen substrate. Acetylation of these

[9] L. Lorand, "Anticoagulants and Fibrinolysis" (R. L. MacMillan and J. F. Mustard, eds.), p. 333. Pitman, London, 1961.

groups, as in esterase-thrombin, blocks the binding of hirudin and thrombin.

Methods of Assay

For the standardization and control of hirudin activity, its inhibitory effect can be determined in biochemical test systems. A known amount of thrombin is added to a hirudin solution with unknown activity and the remaining thrombin activity is measured. Thrombin activity can be tested on fibrinogen (clotting activity) or on synthetic substrates (esterolytic activity) (Fig. 1).

Since thrombin activity is standardized in National Institute of Health units, the activity of hirudin can easily be related to this international standard. The hirudin activity is expressed in antithrombin units (AT-U), whereby one antithrombin unit is the amount of hirudin which neutralizes one NIH unit of thrombin (fibrinogen assay). One antithrombin unit (AT-U) corresponds to 0.1 μg of pure hirudin.[9]

Hirudin Titration with Thrombin

Based on the specific, rapid, and stoichiometric reaction between hirudin and thrombin, hirudin activity can be quantitatively determined by titration with a standardized thrombin solution.[6] The following principle is involved: A fibrinogen solution to which hirudin has been added will not clot until enough thrombin is added to neutralize all of the hirudin present. An excess amount of thrombin can be noted within seconds by the appearance of a fibrin clot.

Procedure. Aliquots of 0.01–0.1 ml of a hirudin solution with unknown activity are added to 0.2 ml of a 0.5% fibrinogen solution in Tris-NaCl buffer, pH 7.4. After mixing, a standardized thrombin solution (100 NIH units/ml) is gradually added at room temperature. Since the formation of fibrin takes some time, the thrombin cannot be added as quickly as reagents are added when performing chemical titrations. Therefore, thrombin is added in 0.005-ml aliquots (corresponding to 0.5 NIH units) in minute intervals. The thrombin is added with micropipettes and immediately mixed with the fibrinogen–hirudin mixture. Only if the approximate amount of thrombin to be added is already known (repeat titration or after preliminary titration), can thrombin initially be added faster in larger quantities, but as the expected final concentration is approached, small amounts of thrombin are added in minute intervals. The end point of titration is reached when the fibrinogen coagulates within 1 minute. Instead of purified fibrinogen solutions, citrated plasma also can be used as the source of fibrinogen. The standard deviation for this method is ±0.5 AT-U.

Use of Hirudin in Coagulation Studies

The effect which hirudin displays on thrombin can be employed for the determination of the activity of this enzyme. The practical application of this procedure has revealed certain advantages over other techniques. The determination of thrombin and those reactions which are closely related to the formation of thrombin can be easily handled in a rather simple procedure. The developed techniques require, however, pure or standardized hirudin which can either be obtained commercially (VEB Arzneimittelwerk, Dresden, German Democratic Republic) or purified by the technique outlined below.

Determination of Thrombin[10]

Principle. After knowing that thrombin activity is neutralized by hirudin, hirudin can be employed for the determination of thrombin activity by titrating a thrombin solution of unknown activity against a known concentration of pure or standardized hirudin. In principle, the procedure is similar to the one already described for the determination of hirudin activity.

Procedure. An aliquot of 0.1 ml hirudin solution (containing 10 AT-U hirudin) is mixed with 0.1 ml of a 0.5% fibrinogen solution or citrated plasma. After mixing, small quantities of thrombin (from 50 to 200 NIH units) are added at room temperature and the clotting of fibrinogen is recorded. The clotting of the substrate indicates that 10 NIH units of thrombin have been added. The standard deviation is ± 0.2 NIH units of thrombin.

Determination of Prothrombin[11]

Prothrombin can be assayed only after it is completely converted to thrombin. One unit of prothrombin is defined as the amount of prothrombin which gives rise to 1 NIH unit of thrombin. In order to correctly determine prothrombin, one must be certain that its conversion to thrombin is complete.

Principle. Freshly drawn blood is mixed with an excess and known amount of hirudin. By adding tissue thromboplastin, a complete conversion of prothrombin to thrombin is achieved, but the generated thrombin will be immediately neutralized by the hirudin. The remaining amount of hirudin in the blood sample is next titrated by adding known quantities of standardized thrombin, until the blood sample clots. The

[10] F. Markwardt, *Arch. Pharm.* **290/62**, 281 (1957).
[11] F. Markwardt, *Arch. Exptl. Pathol. Pharmakol.* **232**, 487 (1958).

prothrombin content is calculated from the consumption of hirudin during the coagulation of the blood.

Procedure. In a test tube, 0.05 ml hirudin (containing 30 AT-U) is mixed with exactly 0.2 ml freshly drawn blood and 0.1 ml thromboplastin solution. After 30 minutes of incubation, a thrombin solution of known activity is slowly added using a micropipette, as described for the assay of hirudin. The coagulation of the blood sample is recorded. The amount of thrombin that had to be added to facilitate clotting is substracted from the amount of hirudin added (30 AT-U), the difference being the amount of thrombin generated from the prothrombin in the blood sample by tissue thromboplastin. Since one unit of prothrombin equals 1 NIH unit of thrombin, the exact amount of prothrombin can be calculated.

The use of hirudin in other coagulation procedures, such as the hirudin tolerance test, has been described by Markwardt.[8, 12, 13]

Isolation of Hirudin[3, 14]

Preparation of the Animals and Extraction of the Inhibitor

The glands containing hirudin are located in the region of the body of the leeches which is next to the head region or suctorial disc or, more specifically, in body segments VII, VIII, and IX. These comprise the so-called neck region of the leeches. Since a clean anatomical dissection of the glands for the purpose of obtaining hirudin is not practical, the entire corresponding body parts are separated. For this purpose, leeches of more than 1.5 g body weight, who have not been leeching for at least 3 months, are sacrified by placing them in a 96% ethanol solution. After 24 hours, the frontal portion of the leeches is separated by dissecting this portion from the body approximately 5 mm before the anterior (male) genital orifice. The sections are once more placed into 96% ethanol and dehydrated for an additional 24 hours. The still-wet head sections are next chopped into small pieces and twice extracted for 30 minutes under stirring with a 10-fold volume of a 40% acetone–water solution. The extracts are combined and diluted with ½ volume of an 80% acetone–water solution. By adding glacial acetic acid, the pH is lowered to 4.3–4.5 and the resulting precipitate removed by centrifugation. The supernatant is mixed with a diluted ammonia solution until a pH of 6.0 is obtained. The volume is reduced to 1/10 of the original volume by placing the solution in a vacuum at 40°. Next, the pH is adjusted to 1.8 by adding 10% trichloroacetic acid. The raw hirudin

[12] F. Markwardt, *Klin. Wochschr.* 37, 1142 (1959).
[13] R. Schmutzler and F. Markwardt, *Klin. Wochschr.* 40, 796 (1962).
[14] F. Markwardt, G. Schäfer, H. Töpfer, and P. Walsmann, *Pharmazie* 22, 239 (1967).

PURIFICATION OF HIRUDIN

Procedure	Dry weight (mg)	Total activity (10^3 AT-U)	Specific activity (AT-U/mg)
1. 1000 leech heads dehydrated in 96% ethanol, chopped to small pieces, extracted with 40% acetone, separation of impurities	900	450	500
2. Fractionation with ethanol	270	405	1,500
3. Adsorption on cation-exchange resin	80	360	4,500
4. Gel filtration	35	295	8,400
5. Chromatography on anion-exchange resin	20	208	10,400

product is precipitated from the 10-fold dilution with acetone, washed with acetone, and the solvents are removed in the vacuum. The raw hirudin preparations have a specific activity of approximately 500 AT-U/mg. (See the table.)

Fractionation with Ethanol

Approximately 1.0 g of raw hirudin is dissolved in 30 ml distilled water and chilled to 0°–5°. Over a period of 30 minutes, 54 ml of 96% ethanol are slowly added. The resulting precipitate is removed by centrifugation. The main bulk of hirudin remains in solution. In order to extract additional hirudin, the precipitate is extracted with ethanol two more times as described above. The combined supernatants are cooled to 0° and cold (−10°) ethanol is slowly added until a concentration of 85% (v/v) is obtained. For more complete precipitation, 0.5% ammonium acetate was added to the ethanol. The precipitate is collected by centrifugation and after washing with ethanol, it is dried in a dessicator under vacuum.

These hirudin preparations contain 10–15% pure hirudin. In addition, inert material and small amounts of a trypsin inhibitor are present.[15] These preparations are useful for experimentally inhibiting coagulation and for certain blood clotting tests.

Adsorption on Cation-Exchange Resins

About 0.3 g of the ethanol-precipitated hirudin is dissolved in 90 ml of a 0.01 M ammonium acetate buffer, pH 4.6. Amberlite IRC-50 (80 ml) in its hydrogen form is added to this solution under stirring until the inhibitor is completely adsorbed into the resin. The resin is collected

[15] H. Fritz, K.-H. Oppitz, M. Gebhardt, I. Oppitz, and E. Werle, *Hoppe-Seylers Z. Physiol. Chem.* **350**, 91 (1969).

by filtration and washed three times with 100 ml of distilled water. For elution of the hirudin, the resin is suspended in 50 ml of a 1 M ammonium acetate solution. The pH is adjusted to pH 7.0 by adding a 5% ammonia solution. The eluate is removed by filtration. The resin is once more suspended in 50 ml 1 M ammonium acetate solution and the pH adjusted to 8.0 by adding a 5% ammonia solution. The eluates are combined and concentrated to a volume of 50 ml by placing them in a vacuum at 20°–30°. Next, the hirudin is precipitated in the cold from the eluate by adding a 10-fold volume of 95% ethanol containing 0.5% ammonium acetate. After storage for several hours, the hirudin-containing precipitate is collected by centrifugation, washed with cold acetone, and dried in a vacuum dessicator.

Instead of bulk adsorption column chromatography can be employed and Amberlite IRC-50, equilibrated with 0.05 M ammonium formate, pH 4.9, or CM-Sephadex C-50 can be used. First eluation is performed with the equilibration buffer, followed by a 1 M ammonium formate solution, pH 7.0. The obtained preparations contain 40–50% pure hirudin.

Gel Filtration on Sephadex G-50

Approximately 100 mg hirudin (40–50% pure) are dissolved in 1 ml of a 0.1 M NaCl solution and placed on a Sephadex G-50 column (1 × 130 cm), previously equilibrated with 0.1 M NaCl, and eluated with 0.1 M NaCl. The eluate is collected in 3-ml aliquots. The tubes containing active material are combined, concentrated in a vacuum, desalted on Sephadex G-25 and lyophilized. These preparations contain about 15–20% impurities.

Chromatography on Anion-Exchange Resins

About 40 mg hirudin (80–90% pure) are dissolved in water and placed on a DEAE-Sephadex A-25 column (0.7 × 20 cm) which has been equilibrated with a 0.05 M pyridine acetate solution, pH 7.4. Hirudin is eluted by establishing a linear gradient with 0.05 M pyridine acetate buffer, pH 7.4 and 1 M NaCl in 0.5 M pyridine acetate buffer, pH 6.9. The tubes containing active material are combined, concentrated in a vacuum over P_2O_5 and potassium hydroxide at 4° and desalted by filtering on Sephadex G-25. The salt-free solution contains pure hirudin and is dried from the frozen state.

Criteria of Purity

Hirudin is stable in dried form. In water (preservative added) it is stable for 6 months at room temperature. It is also stable when heated for 15 minutes at 80°. The heat stability decreases with increasing pH values of the solvent, and in contrast, decreasing the pH increases the heat stability. Hirudin is also stable for 15 minutes at 20° in 0.1 N HCl or 0.1 N NaOH. Trypsin and α-chymotrypsin do not destroy the hirudin activity. The resistance against these proteolytic enzymes is probably due to the tertiary structure of the peptide. This assumption is supported by the finding that oxidized hirudin is readily destroyed by trypsin. Papain, pepsin, and subtilopeptidase A destroy hirudin completely. Using paper electrophoresis, several ninhydrin-positive spots could be identified in the digestion products. Since hirudin does not contain arginine residues, the presence of arginine will identify proteins that contaminate hirudin. Also interesting is the absence of tryptophan and methionine residues in hirudin.

The purity of the hirudin preparations was investigated by means of ultracentrifugation and electrophoresis, using a Tiselius electrophoretic apparatus. With both techniques hirudin was homogeneous. Homogeneity was also observed with other electrophoretic techniques. The staining properties of the spots with ninhydrin or amido black 10 b coincided with the antithrombin activity.

Section V

Water-Insoluble Proteolytic Enyzmes

[70] Water-Insoluble Derivatives of Proteolytic Enzymes

By Leon Goldstein

Enzyme derivatives in which the biologically active protein is covalently bound to a water-insoluble polymeric carrier may serve as easily removable reagents of considerably improved shelf stability; they are well suited for repeated or continuous use and provide means for a more adequate control of enzyme reactions.[1-3] Immobilized enzyme derivatives can be used as specific adsorbents for the isolation and purification of enzyme inhibitors.[4-6] Moreover, new properties may sometimes be imposed on the immobilized enzyme by the chemical nature of the polymeric carrier.[1-3]

The methods available for the immobilization of proteins have been recently summarized in several reviews.[1-3] Some of the more important methods involving the covalent binding of proteins to water-insoluble polymeric carriers are listed below.

The amino groups on proteins have been utilized to effect covalent linking to several carboxylic polymers via the corresponding azides,[7-10] or by activation of the polymer carboxyls by carbodiimides[11-13] or by Woodward's Reagent K (N-ethyl-5-phenylisoxazolium-3'-sulfonate).[14] More recently, cellulose activated by s-trichlorotriazine (cyanuric chloride)[15-17] or by a dichloro-sym-triazinyl dyestuff (Procion brilliant

[1] L. Goldstein and E. Katchalski, Z. Anal. Chem. **243**, 375 (1968).

[2] I. H. Silman and E. Katchalski, Ann. Rev. Biochem. **35**, 873 (1966).

[3] L. Goldstein, in "Fermentation Advances" (D. Perlman, ed.), p. 391. Academic Press, New York, 1969.

[4] H. Fritz, H. Schult, M. Hutzel, M. Wiedermann, and E. Werle, Z. Physiol. Chem. **348**, 308 (1967).

[5] H. Fritz, M. Neudecker, and E. Werle, Angew. Chem. **78**, 775 (1966); Angew. Chem. (Intern. Ed.) **5**, 735 (1966).

[6] H. Fritz, K. Hochstrasser, E. Werle, E. Brey, and B. M. Gebhardt, Z. Anal. Chem. **243**, 452 (1968).

[7] F. Micheel and J. Evers, Makromol. Chem. **3**, 200 (1949).

[8] M. A. Mitz and L. J. Summaria, Nature **189**, 576 (1961).

[9] C. W. Wharton, E. M. Crook, and K. Brocklehurst, European J. Biochem. **6**, 565 (1968).

[10] W. E. Hornby, M. D. Lilly, and E. M. Crook, Biochem. J. **98**, 420 (1966).

[11] N. Weliky and H. H. Weetall, Immunochemistry **2**, 293 (1965).

[12] D. G. Hoave and D. E. Koshland, J. Biol. Chem. **242**, 2447 (1967).

[13] N. Weliky, F. S. Brown, and E. C. Dale, Arch. Biochem. Biophys. **131**, 1 (1969).

[14] R. P. Patel, D. V. Lopiekes, S. R. Brown, and S. Price, Biopolymers **5**, 577 (1967).

[15] B. P. Surinov and S. E. Manoilov, Biokhimiya **31**, 387 (1966).

[16] G. Kay and E. M. Crook, Nature **216**, 514 (1967).

orange MGS),[18] and Sephadex or Sepharose, activated by cyanogen bromide,[19–21] have been successfully applied to the immobilization of several enzymes. A polymeric acylating reagent, ethylene-maleic anhydride (1:1) copolymer (EMA), has been successfully used for the preparation of polyanionic water-insoluble derivatives of enzymes, antigens, and protein enzyme inhibitors.[4–6, 22–24]

In the case of acidic proteins containing large excess of carboxyl groups, immobilization could be achieved by coupling the protein to aminoethyl cellulose through soluble carbodiimide activation of the protein carboxyls.[25]

Proteins containing relatively large amounts of aromatic amino acids have been coupled to the polydiazonium salts derived from p-aminobenzyl cellulose,[26, 27] poly-p-aminostyrene,[24, 28] the m-aminobenzyloxymethyl ether of cellulose,[11, 15, 29] a p-amino-DL-phenylalanine copolymer[30, 31] and, more recently, a synthetic diazotizeable resin, S-MDA.[31] (See Scheme 2.)

Kinetic Behavior of Immobilized Enzymes

The kinetic behavior of immobilized enzyme systems is dominated by several factors not encountered in the kinetics of free enzymes: (a) effects of the chemical nature of the carrier, stemming from the modified environment within which the immobilized enzyme is located; (b) steric restrictions imposed by the carrier; and (c) diffusion control on the rate of substrate penetration.

[17] G. Kay, M. D. Lilly, A. K. Sharp, and R. J. H. Wilson, *Nature* **217**, 641 (1968).

[18] R. J. H. Wilson, G. Kay, and M. D. Lilly, *Biochem. J.* **108**, 845 (1968).

[19] R. Axen, J. Porath, and S. Ernback, *Nature* **214**, 1302 (1967).

[20] J. Porath, R. Axen, and S. Ernback, *Nature* **215**, 1491 (1967).

[21] J. Porath, in "Gamma Globulins" (J. Killander, ed.), Nobel Symp. No. 3 (June 1967), p. 287. Wiley (Interscience), New York, 1967.

[22] Y. Levin, M. Pecht, L. Goldstein, and E. Katchalski, *Biochemistry* **3**, 1905 (1964).

[23] B. Alexander, A. Rimon, and E. Katchalski, *Biochemistry* **5**, 792 (1966).

[24] A. H. Sehon, in "International Symposium on Immunological Methods of Biological Standardization" Royaumont, 1965, *Symp. Series Immunobiol. Stand.* **4**, 51, Karger, Basel, 1967.

[25] L. Goldstein, unpublished data.

[26] D. H. Campbell, E. Leuscher, and L. S. Lerman, *Proc. Natl. Acad. Sci. U.S.* **37**, 575 (1951).

[27] L. Goldstein, M. Pecht, S. Blumberg, D. Atlas, and Y. Levin, *Biochemistry* **9**, 2322 (1970).

[28] N. Grubhofer and L. Z. Schleith, *Z. Physiol. Chem.* **297**, 108 (1954).

[29] A. E. Gurvich, *Biokhimiya* (*Engl. transl.*) **22**, 977 (1957).

[30] A. Bar Eli and E. Katchalski, *J. Biol. Chem.* **238**, 1690 (1963).

[31] I. H. Silman, M. Albu-Weissenberg, and E. Katchalski, *Biopolymers* **4**, 441 (1966).

Low Molecular Weight Substrates

The pH-activity profiles of the polyanionic derivatives of several proteolytic enzymes acting on their specific low molecular weight substrates have been shown to be displaced toward more alkaline pH values by 1–2.5 pH units, at low ionic strength ($\Gamma/2 \approx 0.01$) as compared to the native enzymes.[1,32] Polycationic derivatives of the same enzymes exhibit the reverse effect, i.e., displacement of the pH-activity profile toward more acidic pH values.[1] These anomalies are abolished at high ionic strength ($\Gamma/2 \geqslant 1$). To illustrate this phenomenon, the pH-activity profiles of a polyanionic derivative of trypsin (EMA-trypsin) and of the polyanionic and polycationic derivatives of chymotrypsin (EMA-chymotrypsin and polyornithyl chymotrypsin) are shown in Figs. 1 and 2. Charged derivatives of papain,[25] ficin,[10] and subtilopeptidase A (subtilisin Carlsberg)[25] exhibit similar behavior. Furthermore, the apparent Michaelis constant (K_m) of a polyanionic derivative of trypsin (EMA-trypsin), with the positively charged substrate benzoyl-L-arginine amide

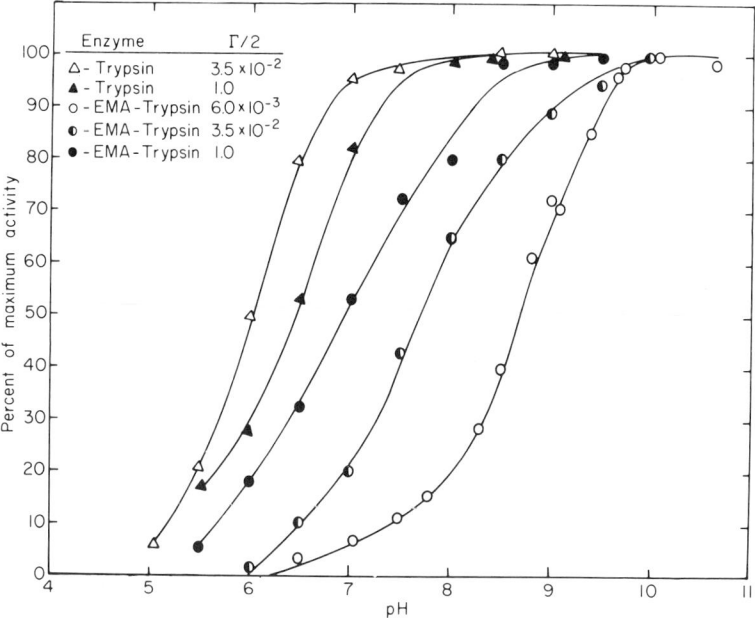

FIG. 1. pH-Activity curves for trypsin and a polyanionic, ethylene-maleic acid copolymer derivative of trypsin (EMA-trypsin), at different ionic strengths, using benzoyl-L-arginine ethyl ester as substrate (redrawn from the data of footnote 32).

[32] L. Goldstein, Y. Levin, and E. Katchalski, *Biochemistry* **3**, 1913 (1964).

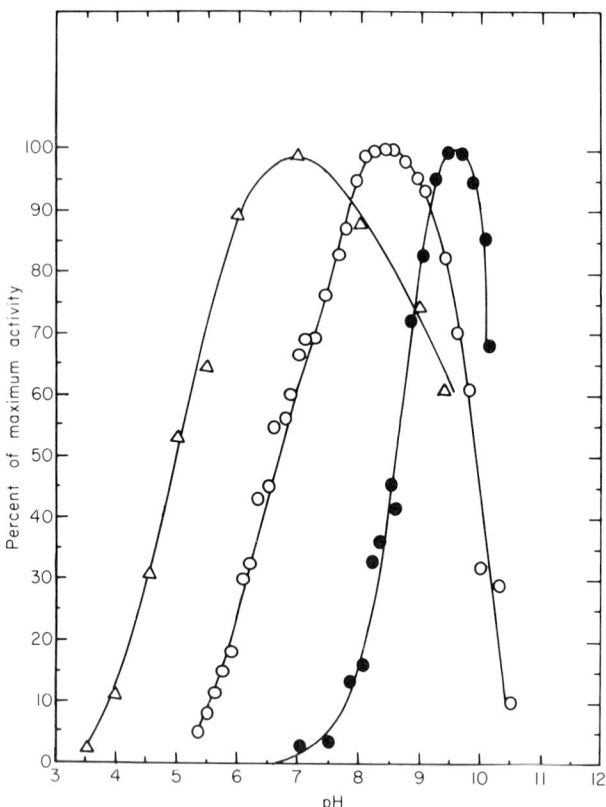

FIG. 2. pH-Activity curves at low ionic strength ($\Gamma/2 = 0.008$) for chymotrypsin, a polyanionic derivative of chymotrypsin (EMA-chymotrypsin) and a polycationic, polyornithyl derivative of chymotrypsin, using acetyl-L-tyrosine ethyl ester as substrate.[1, 25] (○) chymotrypsin, (●) EMA-chymotrypsin, (△) polyornithyl-chymotrypsin.

(BAA) has been found to be markedly lower, at low ionic strength, than that of the native enzyme[32] (Figs. 3 and 4). Similar effects have been reported for the polyanionic derivatives of papain (EMA-papain),[25] ficin[10] (CM-cellulose-ficin) and bromelain[34] (CM-cellulose-bromelain) using benzoyl-L-arginine ethyl ester and substrate. Again, the perturbation of the apparent Michaelis constant is abolished at high ionic strength[32, 33, 34] (see Fig. 4). K_m values closely similar to those of the native enzyme have been demonstrated with uncharged substrates for the poly-

[33] M. D. Lilly, W. E. Hornby, and E. M. Crook, *Biochem. J.* **100**, 718 (1966).
[34] C. W. Wharton, E. M. Crook, and K. Brocklehurst, *European J. Biochem.* **6**, 572 (1968).

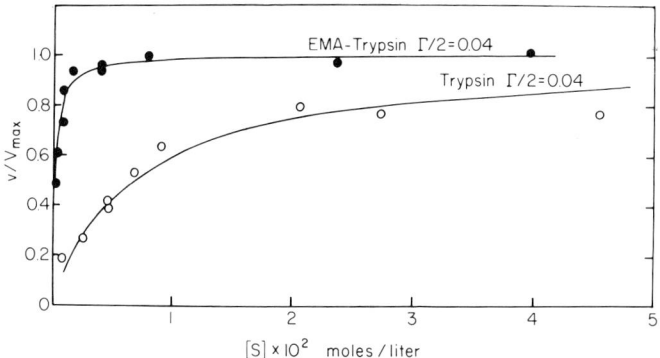

FIG. 3. Normalized Michaelis-Menten plots for trypsin and a polyanionic derivative of trypsin (EMA-trypsin), acting on benzoyl-L-arginine amide.[32]

anionic and polycationic derivatives of chymotrypsin and for a polyanionic derivative of papain.[25] These phenomena have been explained as resulting from the unequal distribution of ionic species between the charged "solid-phase," the polyelectrolyte-enzyme particle, and the surrounding solution.[32]

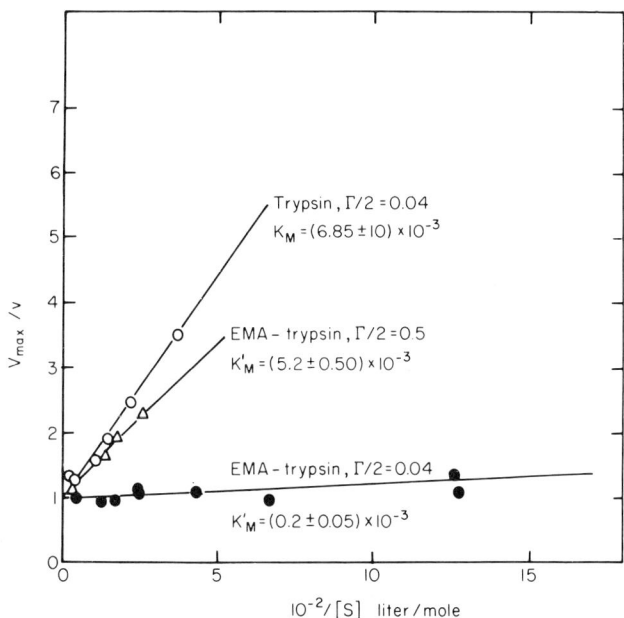

FIG. 4. Normalized Lineweaver-Burk plots for trypsin and a polyanionic derivative of trypsin (EMA-trypsin) acting on benzoyl-L-arginine amide.[32]

Thus the local hydrogen ion concentration in the domain of the charged enzyme derivative could be described by Eq. (1)

$$a_{H^+}^i = a_{H^+}^0 e^{z\epsilon\psi/kT} \tag{1}$$

where $a_{H^+}^i$ and $a_{H^+}^0$ are the hydrogen ion activities in the polyelectrolyte-enzyme derivative phase and the external solution, $\psi =$ the electrostatic potential in the domain of the charged immobilized enzyme particle, $\epsilon =$ the positive electron charge, $z =$ a positive or negative integer, of value unity in the case of hydrogen ions, $k =$ the Boltzmann constant, and $T =$ the absolute temperature.

It follows from Eq. (1) that the local pH in the domain of a poly-anionic enzyme derivative will be lower than that measured in the external solution. The reverse will be true for a polycationic enzyme derivative. Consequently the pH-activity profile of an enzyme immobilized onto a charged carrier will be displaced toward more alkaline or toward more acid pH values, for a negatively or positively charged carrier respectively, i.e.,

$$\Delta pH = pH^i - pH^0 = 0.43 \frac{z\epsilon\psi}{kT} \tag{2}$$

where ΔpH is the difference between the local pH within the polyelectrolyte-enzyme particle (pH^i) and the pH of the external solution (pH^0). pH^i is deduced from the pH-activity profile of the native enzyme (see Figs. 1 and 2).

The dependence of enzymatic activity on pH is related to the dissociation of ionizing groups participating in the enzymatic catalysis mechanism[35,36] (e.g., the acidic limb of the pH-activity curve of chymotrypsin associated with the ionization of histidine-57[35,37]). The displaced pH-activity profiles of a polyelectrolyte enzyme derivative can therefore be alternatively represented in terms of changes in the values of the apparent acidic dissociation constants [pK_a (app)] of the "active-site" ionizing group effected by the polyelectrolyte "microenvironment"[38] of the enzyme derivative, i.e.,

$$pK_a' = pK + 0.43 \frac{z\epsilon\psi}{kT} \tag{3}$$

or

[35] H. Gutfreund, *Trans. Faraday Soc.* **51**, 441 (1955); M. L. Bender and F. J. Kézdy, *Ann. Rev. Biochem.* **34**, 49 (1965).
[36] M. Dixon and E. C. Webb, "Enzymes" (2nd ed.). Longmans, London, 1964.
[37] P. B. Sigler, D. M. Blow, B. W. Matthews, and R. Henderson, *J. Mol. Biol.* **35**, 143 (1968).
[38] J. T. Edsall and J. Wyman, "Biophysical Chemistry," Vol. I, Chap. 9. Academic Press, New York, 1958; C. Tanford, "Physical Chemistry of Macromolecules" Chaps. 7 and 8. Wiley, New York, 1961.

$$\Delta pK_a = pK'_a - pK_a = 0.43 \frac{z\epsilon\psi}{kT} \tag{4}$$

pK_a and pK'_a are the apparent dissociation constants for the native enzyme and polyelectrolyte enzyme derivative, calculated from the appropriate pH-activity profile.

Equations (1–2) and (3–4) are of course identical.

The model summarized in Eqs. (1) to (4) explains satisfactorily most of the known data on the pH dependence of activity of chemically modified enzymes: The pH-activity profiles of both acetylated and succinylated, enzymatically active derivatives of trypsin and chymotrypsin are displaced toward higher pH's.[1, 3, 32, 38, 39] A similar alkaline shift in the pH optimum has been reported for insoluble preparations of chymotrypsin coupled to Sepharose[19, 20] (by means of cyanogen bromide activation of the polysaccharide carrier). These effects are most probably due to the increase in the net negative charge on the protein, resulting from the acylation of amino groups, and thus to an increase in the values of pK_a (app) of the active-site histidines of these enzymes [see Eqs. (3) and (4) and footnotes 35–37]. Water-insoluble derivatives of enzymes bound to electrically neutral carriers such as a p-amino-DL-phenylalanine-leucine copolymer,[29, 30] S-MDA resin (see Schemes 2 and 3), and p-aminobenzyl cellulose via the polydiazonium salts derived from these resins, often show pH-activity profiles displaced toward more alkaline pH's.[31, 39a] This anomaly is somewhat unexpected in view of the assumed chemistry of binding, i.e., coupling of the polymeric diazonium salt to the tyrosyl residues on the protein, via azo bonds. Amino acid analysis of acid hydrolyzates of several such derivatives has, however, revealed among the missing amino acids not only tyrosine but also lysine and arginine.[31] It seems, therefore, that in this case as well, the increase in pK_a (app) is due to an increase in the net negative charge on the protein, effected by the disappearance of lysyl-ϵ-amino and guanido groups at some stage of the coupling reaction. In addition, the "acid-strengthening" effect of the azo-bond[40] on the hydroxyl ionization of "coupled" tyrosyl residues might also have contributed to the increase of net negative charge on the enzymes, and thus to the displacement of the pH-activity curves.

---

[39] J. Sri Ram, L. Terminiello, M. Bier, and F. F. Nord, *Arch. Biochem. Biophys.* **52**, 464 (1954); J. Sri Ram, N. Bier, and P. H. Maurer, *Advan. Enzymol.* **24**, 105 (1962).

[39a] The water-insoluble derivatives of papain (with the above resins as carriers) are the only well-characterized case known to the author where both the pH-activity profiles and the values of K_m of the immobilized enzyme preparations (with BAEE as substrate) are closely similar to those of the crystalline enzyme.

[40] H. Zollinger, "Azo and Diazo Chemistry," Wiley (Interscience), New York, 1961; M. Sokolovsky and B. L. Vallee, *Biochemistry* **5**, 3574 (1966); *ibid.*, **6**, 700 (1967).

The changes in the values of the apparent Michaelis constants (K_m) of polyelectrolyte enzyme derivatives acting on charged low-molecular weight substrates (cf. Figs. 3 and 4) can be related to the unequal distribution [cf. Eq. (1)] of substrate between the charged enzyme particle and the external solution [Eq. (5)].

$$[S]^i = [S]^0 e^{z\epsilon\psi/kT} \tag{5}$$

$[S]^i$ and $[S]^0$ are the substrate concentrations in the domain of the polyelectrolyte enzyme particle and the external solution. It follows from Eq. (5) that $[S]^i > [S]^0$ when polyelectrolyte enzyme derivative and substrate are of opposite charge; the enzyme derivative will attain the limiting rate, V_{max}, at lower bulk concentration of substrate as compared to the native enzyme, and the value of the Michaelis constant for the immobilized enzyme (K'_m) will be correspondingly lower. For substrate and polyelectrolyte enzyme particles of the same charge, the opposite will be true, i.e., $[S]^i < [S]^0$ and K'_m of the enzyme derivative will be higher. This can be seen by introducing Eq. (5) into the Michaelis-Menten rate equation

$$v = \frac{V_{max}[S]}{K_m + [S]}$$

to obtain Eq. (6).

$$v' = \frac{V_{max}[S]^0 e^{z\epsilon\psi/kT}}{K_m + [S]^0 e^{z\epsilon\psi/kT}} \tag{6}$$

Primed symbols indicate quantities pertaining to the polyelectrolyte enzyme derivative. From Eq. (6) $v' = \frac{1}{2}V_{max}$ when $[S]^0 = K_m e^{-z\epsilon\psi/kT}$. Thus the value of external substrate concentration $[S]^0$, at which half-maximum velocity is attained, leads to an apparent Michaelis constant K'_m related to the Michaelis constant of the native enzyme by the expression

$$K'_m = K_m e^{-z\epsilon\psi/kT} \tag{7}$$

Eq. (7) can be rewritten as

$$\Delta pK_m = pK'_m - pK_m = \log \frac{K_m}{K'_m} = 0.43 \frac{z\epsilon\psi}{kT} \tag{8}$$

High Molecular Weight Substrates

The activities of immobilized proteases toward protein substrates, as determined by the standard procedures (Kunitz's casein digestion method, appearance of nonprotein nitrogen, potentiometric determination of the number of peptide bonds split, ninhydrin, etc.) are usually lower than those of the crystalline enzymes, when compared on a weight basis. Water-insoluble derivatives of polytyrosyl trypsin[29] and papain[30] hydrolyzed casein at 15 and 30% of the rates which would have been expected from the amount of active bound enzyme (determined by a rate assay with low molecular weight substrate). Similar results have been obtained with carboxymethyl cellulose conjugates of ficin,[10,33] bromelain,[9] and trypsin,[8] and with polyanionic polycations and polyalcohol derivatives of papain[25] and subtilopeptidase A.[25] In the majority of cases the lowering in proteolytic activity can be attributed to steric hindrance induced by the carrier.[1-3]

An additional effect might become prominent in the case of polyelectrolyte derivatives of enzymes, i.e., the electrostatic interaction between charged carrier and charged, high molecular weight substrate: In the digestion of casein by polyanionic derivatives of trypsin (see EMA-trypsin, p. 950) the rate of hydrolysis has been found to depend on the carrier-to-enzyme ratio.[22] Preparations of high enzyme content (high enzyme:carrier ratio) showed caseinolytic activities close to those of native trypsin while preparations of low enzyme content had about 20% of the expected proteolytic activity.[22] Raising the pH and thus increasing the charge density of the carrier led to a marked decrease in the caseinolytic activity of both EMA-trypsin preparations.[22] The data conveyed that the number of peptide bonds split could be controlled by varying the charge density on the polyelectrolyte enzyme derivative. Furthermore, indications that the sites of attack as well as the rates of cleavage of a high molecular weight substrate might be affected by the chemical nature of the polymeric carrier could be found in the literature.[1-3,22,23,41-46] For example, both water-insoluble polytyrosyl trypsin and EMA-trypsin attacked prothrombin. The polytyrosyl trypsin derivative, like native trypsin, converted prothrombin to thrombin which was not digested any further. In contrast, EMA-trypsin rapidly degraded the newly formed thrombin.[23] It was also found that while the clotting factor VII was activated by insoluble polytyrosyl trypsin, it was not affected by EMA-trypsin.[2]

These observations have been substantiated in studies on the EMA-trypsin digestion of myosin and the meromyosins.[41,42] It was found that

[41] S. Lowey, L. Goldstein, and S. Luck, *Biochem. Z.* **345**, 248 (1966).
[42] S. Lowey, L. Goldstein, C. Cohen, and S. M. Luck, *J. Mol. Biol.* **23**, 287 (1967).
[43] H. S. Slayter and S. Lowey, *Proc. Natl. Acad. Sci. U.S.* **58**, 1611 (1967).
[44] S. Lowey, *in* "Symposium on Fibrous Proteins, Australia 1967" (W. A. Crewther, ed.). Butterworth, London, 1968; S. Lowey, H. S. Slayter, A. G. Weeds, and H. Baker, *J. Mol. Biol.* **42**, 1 (1969).
[45] E. B. Ong, Y. Tsang, and G. E. Perlmann, *J. Biol. Chem.* **241**, 5661 (1966).
[46] T. L. Westman, *Biochem. Biophys. Res. Commun.* **35**, 313 (1969).

the first-order rate constants estimated for the EMA-trypsin digestion of myosin were about 50-fold lower as compared to those of native trypsin, and about half as many peptide bonds were split by the poly-anionic trypsin derivative. Moreover, different protein fragments were obtained on limited EMA-trypsin digestion of both heavy-meromyosin (HMM) and myosin. In these investigations, a new helical subfragment (HMM-subfragment-2) was isolated from EMA-trypsin digests of heavy meromyosin.[41,42] The separation of the globular part of the myosin molecule HMM-subfragment-1 as well as the isolation of its intact helical rod portion have been recently achieved by the controlled digestion of myosin with a water-insoluble papain derivative.[43,44]

It has been recently established[45] that whereas trypsin hydrolyzed the 15 lysyl peptide bonds in pepsinogen, the maximal number of bonds cleaved by EMA-trypsin never exceeded 10; the same difference was observed when reduced carboxymethylated pepsinogen was used as substate. These findings were confirmed by peptide mapping.

The data indicate that the chemical nature of the carrier, at least in the case of polyelectrolyte enzyme derivatives, may impose additional restrictions on the specificity of the bound enzyme. These restrictions probably result from charge interactions between carrier and different regions, or different sequences on the high molecular weight substrate molecule. Hence, in studies of protein sequence, the restricted specificity displayed by polyelectrolyte enzyme derivatives could be utilized to obtain overlapping peptides by a series of derivatives of one enzyme only.[1,3]

Enzyme Columns

Columns of immobilized enzymes have been employed for the continuous preparation of product, for regulation of the extent of conversion of substrate to product, and in automated analytical procedures.[1-3,29,33,47,48]

A modified form of the integrated Michaelis-Menten rate equation has been used by Lilly et al.[33] to describe the kinetics of substrate hydrolysis in an enzyme column.

Defining the residence time of substrate in the enzyme column t, by

$$t = V_l/Q \tag{9}$$

where V_l is the void volume of the column, Q the flow rate, and substituting Eq. (9) into the integrated Michaelis-Menten equation

[47] R. Goldman, O. Kedem, I. H. Silman, S. R. Caplan, and E. Katchalski, *Biochemistry* **7**, 486 (1968).
[48] R. Goldman, O. Kedem, and E. Katchalski, *Biochemistry* **7**, 4518 (1968).

$$S_0 - S_t = k_3[E]t + K_m \ln(S_t/S_0)$$

Eq. (10) is obtained:

$$PS_0 - K'_m \ln(1 - P) = k_3 E\beta/Q = C/Q \tag{10}$$

where $\beta = V_l/V_t$ is voidage of the column; $P = (S_0 - S_t)/S_0$, fraction of substrate reacted in column; $C = k_3 E\beta$, reaction capacity of column; V_t = total volume, and S_0, S_t, k_3, E, and K_m have their usual significance.

Eq. (10) may be rearranged

$$PS_0 = K'_m \ln(1 - P) + C/Q \tag{11}$$

If values of P are measured when various initial concentrations of substrate are perfused through the same column at an identical flow rate (i.e., Q = const.), PS_0 plotted against $(1 - P)$ will give a straight line if K_m and C are constant at this flow rate. The slope of the line will be equal to K_m and the intercept to C/Q. Thus K_m and C can be determined for any flow rate through the column. Kinetic data on the hydrolysis of benzoyl-L-arginine ethyl ester in packed columns of carboxymethyl cellulose-ficin when plotted according to Eq. (11) indicated that the apparent Michaelis constant (K'_m) decreased with increasing flow rate and asymptoted toward a minimal value at high flow rates.[33] In general, the value of the Michaelis constant was higher than that observed under comparable conditions in stirred suspensions of the carboxymethyl cellulose ficin derivative.[10,33] The data also suggested that at very low values of Q there was a tendency for C to increase. These deviations from the behavior expected on the basis of Eq. (11) could be qualitatively explained as due to diffusion-limited transport of substrate into the enzyme particles. A more general study of diffusion-controlled processes in immobilized enzyme systems has been carried out by Goldman et al.[47,48] using papain-collodion membranes.

Carriers

The water-insoluble derivatives of trypsin, chymotrypsin, papain, and subtilopeptidase A (subtilisin Carlsberg), the preparation and properties of which are described below, can be divided into two groups according to the type of carrier used:

1. Polyanionic, water-insoluble derivatives, in which the enzyme is bound to a carboxylic polymer (an ethylene-maleic acid copolymer) via amide linkages.
2. Water-insoluble derivatives in which the enzyme is bound to a neutral, synthetic resin (see S-MDA, p. 946) mainly via azo linkages.

The polyanionic, water-insoluble enzyme derivatives are prepared by coupling the enzyme protein to a polyfunctional acylating reagent, a

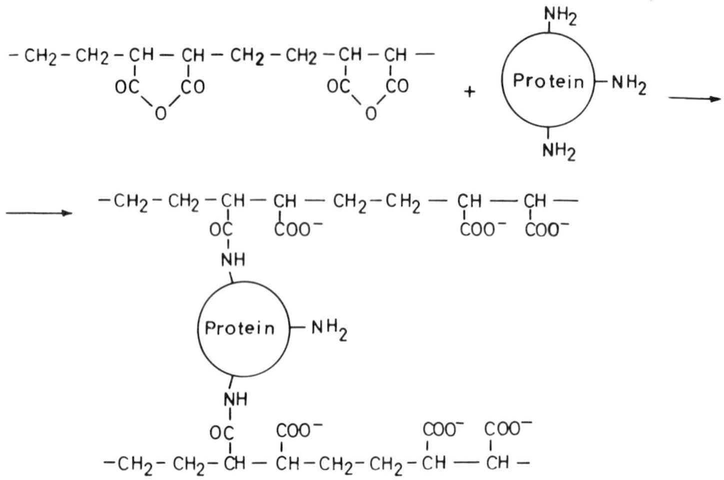

$$- CH_2 - CH_2 - CH - CH - CH_2 - CH_2 - CH - CH - \quad + \quad \text{(Protein)} \overset{NH_2}{\underset{NH_2}{-NH_2}} \longrightarrow$$

$$\longrightarrow -CH_2 - CH_2 - CH - CH - CH_2 - CH_2 - CH - CH -$$

$$\text{(Protein)} - NH_2$$

$$-CH_2 - CH_2 - CH - CH - CH_2 - CH_2 - CH - CH -$$

SCHEME 1. Coupling of proteins to an ethylene–maleic anhydride (1:1) copolymer (EMA).

1:1 copolymer of ethylene and maleic anhydride, EMA (p. 948).[22,32] EMA reacts mainly with the ε-amino groups of the lysyl residues on the protein (Scheme 1). Additional cross-linking with hexamethylene diamine (1,6-diaminoethane) enhances the insolubility of the materials. The unreacted maleic anhydride residues on the polymeric carrier subsequently undergo hydrolysis in the aqueous medium. The ratio of enzyme protein to carrier in the reaction mixture can be varied within fairly wide limits. Samples containing up to 80% by weight of bound protein have been prepared by this procedure.[22] The physical properties of the EMA-protein derivatives are dependent on their protein-to-carrier ratio.[1,22,32] Preparations of low protein content are gel-like in appearance. The swollen EMA-protein gels contract significantly on increasing the ionic strength of the suspending medium. These materials can be separated from the medium by centrifugation. They cannot, however, be filtered except at very high ionic strength ($\Gamma/2 \geqslant 1.0$). Preparations of high protein content (60–80% protein by weight) have flaky texture, are easily separated from the suspending medium by centrifugation, and in most cases by filtration, at moderate ionic strengths. High protein-content EMA derivatives can be used in columns.

The second type of water-insoluble enzyme derivatives is obtained by coupling the protein, mainly through its aromatic amino acid residues, to the polydiazonium salt derived from a synthetic resin, S-MDA[48a] (Schemes 2 and 3).[31] The S-MDA resin is prepared by condensation of

[48a] The name of the resin S-MDA is derived from the names of its chemical components: (dialdehyde)-starch methylene-dianiline (see Scheme 2).

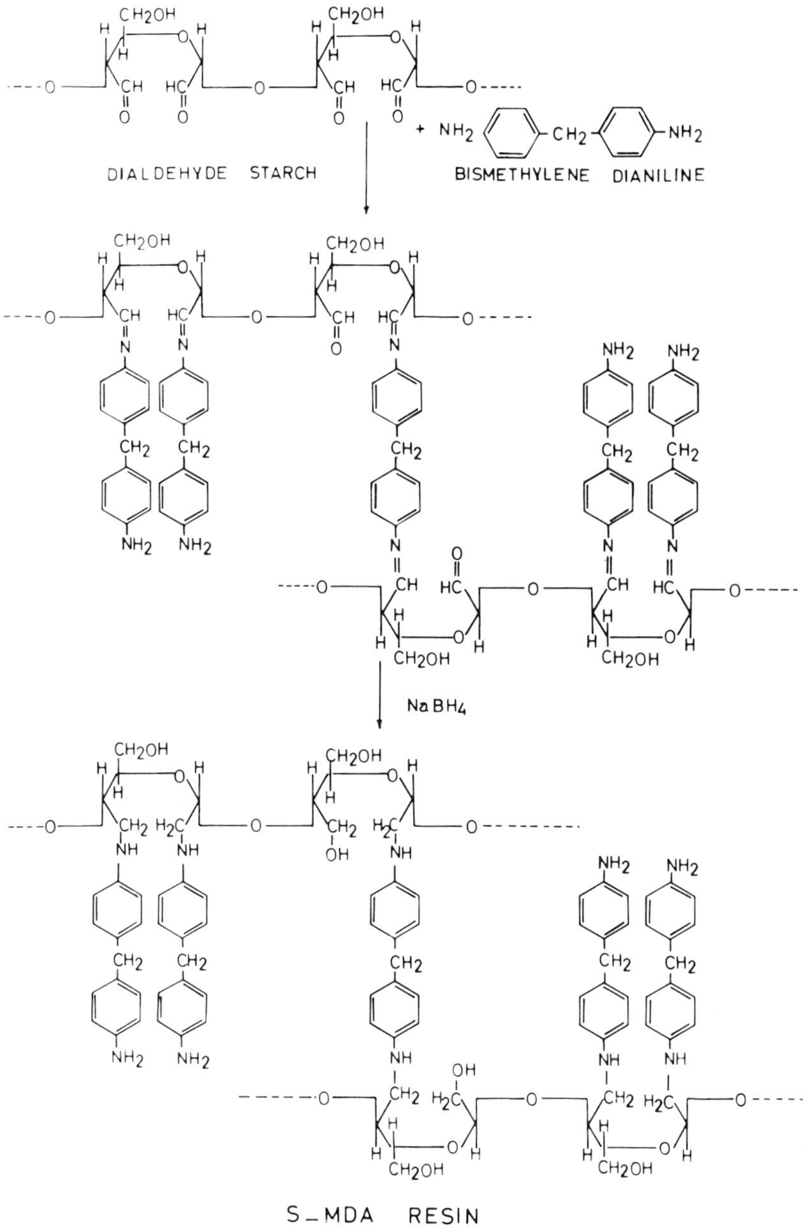

SCHEME 2. Synthesis of S-MDA resin.

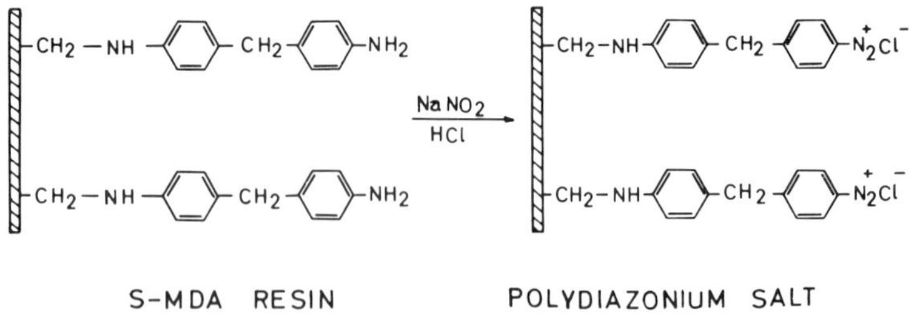

S-MDA RESIN   POLYDIAZONIUM SALT

SCHEME 3. Preparation of a polydiazonium salt derived from S-MDA resin.

dialdehyde starch (a commercially available periodate-oxidation product of starch) with bismethylene dianiline, and the subsequent reduction of the Schiff's base highly cross-linked polymeric product (Scheme 2). Preparations containing up to about 10% by weight of bound protein can be prepared using the S-MDA resin as carrier.[48b] The water-insoluble, S-MDA enzyme conjugates have particulate form, are easily filterable, and can be conveniently used in columns.[31]

Ethylene-Maleic Anhydride (1:1) Copolymer (EMA)

EMA samples of average molecular weight in the range of 20,000–100,000 are available commercially (Monsanto Chemical Co., Inorganic Chemical Division, St. Louis, Mo.). EMA undergoes, on standing, partial hydrolysis as a result of absorption of moisture. The carboxyl groups liberated in this process can be reconverted into the anhydride form on heating at 105°–110° for 24 hours over phosphorus pentoxide. Experience has shown, however, that when the coupling reaction is carried out in an aqueous medium, both the binding of protein and the recovery of enzymatic activity in the insoluble derivative are consistently higher with EMA samples in which only 60–80% of the carboxylic groups are in the anhydride form. When, on the other hand, the coupling reaction is carried out in a mixed solvent system, where a solution of the EMA polymer in an organic solvent (e.g., dimethylsulfoxide, dimethylformamide, acetone, etc.) is added to an aqueous solution of the protein, oven-dried EMA is to be preferred.

[48b] As an alternative choice for a diazotizable resin, commercial p-aminobenzyl cellulose (PAB-cellulose) could be used. This resin has, however, distinctly inferior properties as carrier, in comparison to the S-MDA resin. Both its protein binding capacity and the recovery of enzymatic activity in the insoluble form are considerably lower.[31] Moreover, insoluble enzyme preparations utilizing p-aminobenzyl cellulose as carrier have a tendency to liberate color into aqueous buffers. This deficiency of PAB cellulose can be partly overcome by refluxing the resin with methanol for about 30 minutes and air drying prior to diazotization.[31]

Characterization. The anhydride content of an ethylene-maleic anhydride copolymer sample is determined as follows[22]: The total amount of carboxyl plus anhydride groups in an EMA sample is determined after hydrolysis of the sample (50–100 mg) in boiling water for 10–15 minutes; the hydrolyzed copolymer in aqueous solution, 0.5 M in NaCl, is titrated with 0.1 N NaOH using phenolphthalein as indicator. Two moles of titrant are consumed per base mole of maleic acid or maleic anhydride. In a parallel experiment the intact copolymer is dissolved in anhydrous dimethylformamide and the solution is titrated with 0.1 N sodium methoxide in methanol–benzene, using thymol blue as indicator.[49] One mole of titrant is consumed per base mole of anhydride, whereas 2 moles are consumed per base mole of maleic acid. The amount of anhydride and free carboxyl groups initially present in the ethylene-maleic anhydride copolymer sample is calculated from the two titration values.

Preparation of S-MDA Resin[31, 49a]

Reagents. Dialdehyde starch, the periodate-oxidation product of starch (90% oxidized), can be obtained under the trade name Sumstar-190, from Miles Chemical Co., Elkhart, Ind. Bismethylene dianiline (MDA, p,p'-diaminodiphenyl methane) and sodium borohydride are standard commercially available chemicals.

Procedure. Dialdehyde starch, Sumstar-190 (10 g) is suspended in water (200 ml), stirred at room temperature for 10–15 minutes to obtain a fine slurry, and 2 M carbonate buffer, pH 10.5 (40 ml) is then added. The suspension is poured slowly into a vigorously stirred, 10% solution of bismethylene dianiline in methanol (300 ml). The reaction mixture is stirred at room temperature for 2–3 days. The insoluble, polymeric Schiff's base thus obtained is separated on a funnel, washed with methanol, and then suspended in water and reduced with sodium borohydride (40 g). The reaction mixture is brought to neutral pH with acetic acid and the resin is washed with water and methanol and then refluxed with methanol (3–4 changes of solvent, 300 ml each) to remove methanol-soluble aromatic amines. Refluxing is continued until a negative test is obtained with N-dimethylaminobenzaldehyde (Ehrlich's reagent).[50, 51] The material is then filtered and air dried. The total weight of dry material is 14–15 g.

The nitrogen content of the S-MDA resin is 6.5–6.8%. The diazotization capacity is 0.24–0.26 meq/g, which comprises about 10% of the

---

[49] A. Patchornik and S. Ehrlich-Rogozinki, *Anal. Chem.* **33**, 803 (1961).
[49a] S-MDA (dialdehyde)-starch-*methylene-dianiline*.
[50] C. Menzie, *Anal. Chem.* **28**, 1321 (1956).
[51] M. M. Sprung, *Chem. Rev.* **26**, 297 (1940).

total MDA content of the resin.[31] The maximal protein binding capacity of the S-MDA resin (following diazotization) is 8–10 mg protein per 100 mg resin.[31]

Characterization of the S-MDA Resin. The diazotization capacity of the S-MDA resins is estimated by coupling its polydiazonium salt with *p*-bromophenol and determination of the nitrogen and bromine contents of the reaction product: S-MDA resin (50 mg) is suspended in 50% acetic acid (4 ml) and stirred for 1 hour over ice. An aqueous solution of sodium nitrite (10 mg in 1 ml) is then added dropwise to the chilled suspension. The diazotization mixture is stirred for 1 hour over ice and then brought to pH 8.5 by the dropwise addition of 5 N NaOH. The precipitate is separated on a funnel, washed with cold 0.2 M phosphate buffer, pH 7.8, and suspended in the same buffer (5 ml). A solution of *p*-bromophenol (50 mg, dissolved in water by the dropwise addition of 2 N NaOH) is then added to the chilled suspension. The reaction mixture is stirred overnight at 4°. The precipitate is separated by filtration, washed with 0.1 M carbonate buffer, pH 10.5, water and methanol, and dried *in vacuo* over phosphorus pentoxide. The nitrogen and bromine contents of the dry resin are determined by the Dumas and Schöniger combustion methods,[52] respectively.

The protein binding capacity of the S-MDA resin can be estimated from the saturation curves obtained by coupling the polydiazonium salt, derived from S-MDA (50 mg) with varying amounts of protein (1–10 mg) and determining the enzymatic activity of the reaction mixture supernatants and the insoluble precipitates,[31] or by determining the amino acid composition of acid hydrolyzates of the latter.[31]

Water-Insoluble Trypsin Derivatives

Trypsin (EC 3.4.4.4), × 2 crystallized, salt free and lyophilized is commercially available. Trypsin is assayed titrimetrically,[53, 54] at 25°, pH 8, using benzoyl-L-arginine ethyl ester (BAEE) as substrate (5 ml; 1.16×10^{-2} M BAEE, 0.002 M in phosphate) and 0.1 N NaOH as titrant.

Polyanionic, Water-Insoluble Trypsin Derivative[22, 32]

EMA-Trypsin

Carrier. Ethylene-maleic anhydride (1:1) copolymer (EMA) (see p. 948).

[52] A. Steyermark, "Quantitative Organic Microanalysis" (2nd ed.). Academic Press, New York, 1961.
[53] C. F. Jacobsen, J. Leonis, K. Lindestrøm-Lang, and M. Ottesen, *in* "Methods of Biochemical Analysis" (D. Glick, ed.), Vol. 4, p. 171. Wiley (Interscience), 1957.
[54] M. Laskowski, Vol. II [3].

Procedure. Ethylene-maleic anhydride copolymer, EMA (100 mg) is suspended in ice-cooled, 0.2 M potassium phosphate buffer, pH 7.5 (5–10 ml). The suspension is homogenized and the cross-linking agent, hexamethylene diamine (1 ml of a 1% aqueous solution), is added with stirring. The appropriate amount of trypsin (10–400 mg) dissolved in the same buffer (5–30 ml) is added and the reaction mixture left overnight at 4° with magnetic stirring. The precipitate is separated by centrifugation and washed by suspending in 1 M KCl (100 ml) and recentrifuging (about twenty times) at 4°, until no residual enzymatic activity toward benzoyl-L-arginine ethyl ester hydrochloride can be detected in the supernatant after filtration through a Millipore filter.

The recovery of enzyme activity (tested with BAEE) in the water-insoluble EMA-trypsin derivatives is 30–40%.[22, 32]

Notes. It should be stressed that the binding of enzymatically active protein is consistently higher when the EMA carrier is partially hydrolyzed. Best results are obtained when 60–80% only of the carboxyl groups are in the anhydride form.[22]

In the preparation of high protein content samples (300–400 mg protein per 100 mg EMA in the coupling mixture), the cross-linking agent, hexamethylene diamine, can be omitted.[22]

Determination of Activity. The EMA-trypsin derivatives are assayed as described for trypsin, at pH 9.5 (the optimal pH of the EMA-trypsin derivatives is in the pH range 9.5–10[22, 32]). The amount of active enzyme is calculated from the rate of substrate hydrolysis.

The protease activity is determined at pH 7.6 by the casein digestion method.[54, 55]

Determination of Bound Protein. The EMA-trypsin preparation is hydrolyzed in an evacuated, sealed tube using 6 N HCl (48 hours, 110°). The acid is evaporated and the residue suspended in 0.2 M citrate buffer, pH 2.2 (4 ml); insoluble material is removed by centrifugation and amino acid analysis is carried out employing an automatic amino acid analyzer.[56] The amount of protein is calculated from the amounts of alanine, leucine, glycine, and valine.[57]

In the case of EMA-trypsin preparations containing 60–80% protein by weight, an estimate of the protein content can be obtained from the nitrogen content (Kjeldahl) of the sample.

Properties. EMA-trypsin derivatives are considerably more stable in the alkaline pH range (pH 7.0–10.7) than the crystalline enzyme.[22] EMA-trypsin suspensions in distilled water or in 0.1 M phosphate buffer,

[55] M. Kunitz, *J. Gen. Physiol.* **30**, 291 (1947).
[56] D. H. Spackman, Vol. XI [1].
[57] M. O. Dayhoff and R. V. Eck, "Atlas of Protein Sequence and Structure," National Biomedical Research Foundation, Silver Spring, Maryland, 1967–68.

pH 7, can be stored in the cold (4°) for several months without significant loss in enzyme activity.[22] They can be lyophilized without significant
decrease in activity. The dried powders can be stored at room temperature or at 4° for long periods of time.[22]

The pH-activity profiles of EMA-trypsin derivatives using benzoyl-
L-arginine ethyl ester as substrate are displaced by about 2.5 pH units
at low ionic strength ($\Gamma/2 \simeq 0.005$) toward more alkaline pH values as
compared with native trypsin under similar conditions.[22, 32] The pH
optimum for the polyanionic trypsin derivatives lies in the region of pH
9.5–10.0[22, 32] (see Fig. 1). This effect is reversed on increasing the ionic
strength. In the case of EMA-trypsin preparations of high protein content (60–80% by weight), the displaced pH-activity profile is essentially
unaffected by increasing the ionic strength.[22, 32] The apparent Michaelis
constant (K_m) of the EMA-trypsin preparations, with benzoyl-L-argi-
nine amide (BAA) as substrate, at low ionic strength ($K'_m = 0.2 \times 10^{-3} M$), is lower than that of the native enzyme ($K_m = 6.85 \times 10^{-3} M$)[32]
(see Figs. 3 and 4).

The activity of EMA-trypsin preparations toward protein substrates
is considerably lower than that of the native enzyme as calculated on
the basis of the respective esterase activities.[22] Moreover, it has been
shown in several cases[1-3, 41, 42, 45-46] that the sites of attack as well as the
rates of cleavage of protein substrates such as myosin,[41, 42] pepsinogen,[45]
and oxidized ribonuclease,[46] are affected by the polyelectrolyte nature of
the EMA-trypsin derivative (p. 944).

Water-Insoluble Polytyrosyl Trypsin Derivative

The direct coupling of trypsin to the polydiazonium salts derived from
the S-MDA resin and other similar resins leads to excessive inactivation
of the bound protein.[29, 31] To circumvent this limitation, polytyrosyl side
chains are grown onto the protein by initiating the polymerization of
N-carboxyl-L-tyrosine anhydride[58] with the enzyme in aqueous solution.
Polytyrosyl trypsin,[59] subsequently coupled with the insoluble S-MDA
polydiazonium salt, yields highly active water-insoluble derivatives.[31]

Polytyrosyl Trypsin[59]

Materials. N-Carboxyl-L-tyrosine anhydride[58] is available commercially (Miles Chemical Co.).

Procedure. Trypsin (1 g) is dissolved in 0.0025 N HCl (36 ml) and
0.1 M phosphate buffer, pH 7.6 (36 ml) is then added. The final pH of

[58] A. Berger, J. Kurtz, T. Sadeh, A. Yaron, R. Arnon, and Y. Lapidoth, *Bull. Res.
Council Israel* **7A**, 98 (1958).
[59] A. N. Glazer, A. Bar Eli, and E. Katchalski, *J. Biol. Chem.* **237**, 1832 (1962).

the solution is 7.2. The protein solution is chilled in an ice bath and a solution of N-carboxy-L-tyrosine anhydride (800 mg) in anhydrous dioxane (16 ml) is added dropwise with stirring. The milky reaction mixture is stirred magnetically at 4° for 16 hours and then exhaustively dialyzed against 0.0025 N HCl (6 liters changed daily for about 7 days) until a clear solution is obtained. The solution is clarified by centrifugation, if traces of insoluble material are still present at the end of the dialysis, then lyophilized and stored at 4°. The yield of protein and the recovery of enzymatic activity are almost quantitative.

Characterization and Properties. Polytyrosyl trypsin is assayed as described for trypsin. The protease activity is determined by the casein digestion method.[54, 55]

The enrichment of the polytyrosyl trypsin preparations in tyrosine can be estimated spectrophotometrically by determining the absorbance at 295 nm of a polytyrosyl trypsin solution in 0.1 N NaOH and correcting for the absorbance of the unmodified enzyme under the same conditions. The molar extinction coefficient of tyrosine at 295 nm, pH 13, is 2300.[59]

Polytyrosyl trypsin dissolves readily in aqueous solution of pH's lower than 5 or higher than 9.5.[59] It shows limited solubility between pH 5 and 9.5.

Water-Insoluble Polytyrosyl Trypsin Derivative[31] (*S-MDA Polytyrosyl Trypsin Conjugate*)

Carrier. S-MDA resin (see p. 949).

Procedure. S-MDA resin (100 mg) is suspended in 50% acetic acid (8 ml) and stirred for 1 hour over ice. An aqueous solution of sodium nitrite (20 mg in 1 ml) is then added dropwise to the chilled suspension. The diazotization mixture is stirred for 1 hour over ice and then brought to pH 8.5 by the dropwise addition of 5 N NaOH. Crushed ice is added, as needed, to keep the temperature down. The polydiazonium salt separates as a dark-brown lumpy precipitate. The precipitate is separated on a Büchner funnel, washed with cold 0.2 M phosphate buffer, pH 7.8, and suspended in the same buffer (10 ml). A solution of polytyrosyl trypsin (30 mg) in 0.001 N HCl (8 ml) is then added to the chilled suspension and the coupling reaction allowed to proceed overnight with stirring, in the cold room. The water-insoluble S-MDA-polytyrosyl trypsin conjugate is separated by filtration on a Büchner funnel, washed with 1 M KCl (100–200 ml) and then with water. The precipitate is resuspended in water, or in 0.1 M phosphate buffer, pH 7.0.

The recovery of enzymatic activity (tested with BAEE) in the water-insoluble polytyrosyl trypsin derivative is about 40%.

Determination of Activity. The S-MDA polytyrosyl trypsin derivatives are assayed as described for trypsin at pH 10 (the optimal pH of the S-MDA-polytyrosyl trypsin derivatives[31]). The amount of active enzyme is calculated from the rate of substrate hydrolysis.

The protease activity is determined at pH 7.6 by the casein digestion method.[54, 55]

Determination of Bound Protein. The S-MDA-polytyrosyl trypsin preparation is hydrolyzed in an evacuated sealed tube using $6\,N$ HCl (48 hours, 110°). The acid is evaporated and the residue suspended in $0.2\,M$ citrate buffer, pH 2.2 (4 ml). Insoluble colored material is removed by strong centrifugation (Sorval 12,000 rpm) and amino acid analysis is carried out employing an automatic amino acid analyzer.[56] The amount of protein is calculated from the amounts of alanine, leucine, glycine, and valine.[57]

Note. After analysis of hydrolyzates of insoluble azo derivatives of proteins, the amino acid analyzer column is regenerated with five times the volume of $0.2\,N$ NaOH normally employed, in order to remove colored diazotization products adsorbed on the column.

Properties. S-MDA-polytyrosyl trypsin samples can be stored in aqueous suspensions at 4° for long periods of time without significant loss of enzyme activity.[31] At room temperature, suspensions of these materials lose 15–20% of their activity after 1 week.[31] S-MDA-polytyrosyl trypsin preparations are almost completely inactivated on lyophilization or air drying.[31]

The pH-activity profile of the S-MDA-polytyrosyl trypsin derivatives acting on benzoyl-L-arginine ethyl ester as substrate is displaced by 2 pH units toward more alkaline pH values as compared to the native enzyme under similar conditions. The pH optimum for this insoluble derivative lies in the region of pH 10–10.5.[31]

The caseinolytic activity of S-MDA-polytyrosyl trypsin is about 30% that of crystalline trypsin or polytyrosyl trypsin.[31]

Other Methods of Preparation

Water-insoluble derivatives of trypsin have been obtained by coupling the enzyme with carboxymethyl cellulose azide.[8, 60] Water-insoluble polytyrosyl trypsin derivatives have been prepared by coupling polytyrosyl trypsin[58] with the polydiazonium salt derived from a p-amino-DL-phenylalanine-leucine copolymer.[29] Cross-linking trypsin with glutaraldehyde has led to an enzymatically active, insoluble preparation.[61] The im-

[60] C. J. Epstein and C. B. Anfinsen, *J. Biol. Chem.* **237**, 2175 (1962).
[61] A. F. S. A. Habeeb, *Arch. Biochem. Biophys.* **119**, 264 (1967).

mobilization of trypsin adsorbed onto silica gel particles by glutaraldehyde cross-linking of the adsorbed protein has been recently described.[62]

Water-Insoluble Chymotrypsin Derivatives

Chymotrypsin (EC 3.4.4.5) crystallized three times, lyophilized is
available commercially (Worthington Biochemicals). Chymotrypsin is
assayed titrimetrically[53,54] at 25°, pH 8.2, using acetyl-L-tyrosine ethyl
ester (ATEE) as substrate (5 ml; 0.018 M ATEE, 0.01 M KCl) and
0.1 N NaOH as titrant.

Polyanionic Water-Insoluble Chymotrypsin Derivative[1,25]

EMA-Chymotrypsin

Carrier. Ethylene-maleic anhydride (1:1) copolymer (EMA).

Procedure. EMA-chymotrypsin derivatives are prepared by the procedure described for EMA-trypsin derivatives.

The recovery of enzymatic activity (tested with ATEE) in the water-
insoluble EMA-chymotrypsin derivatives is about 30%.[25]

Determination of Activity. The EMA-chymotrypsin derivatives are
assayed as described for chymotrypsin, at pH 9.5 (the optimal pH for
these derivatives[25]). The amount of active enzyme is calculated from the
rate of substrate hydrolysis.

The protease activity is determined at pH 7.6 by the casein digestion
method.[54,55]

Determination of Bound Protein. The protein content of EMA-chymotrypsin derivatives is determined as described for EMA-trypsin.

Properties. Aqueous suspensions of EMA-chymotrypsin preparations
can be stored at 4° without significant loss of activity for long periods of
time. The EMA-chymotrypsin preparations can be lyophilized, retaining
most of their activity. The dried powders can be stored at 4° with no
significant loss of activity.[25]

The pH-activity profile of the EMA-chymotrypsin derivatives, using
acetyl-L-tyrosine ethyl ester (ATEE) as substrate is displaced at low
ionic strength ($\Gamma/2 = 0.01$) by about 1.5 pH units toward more alkaline
pH values as compared to crystalline chymotrypsin under similar conditions.[1,63] This effect is partially reversed on increasing the ionic strength.
The pH optimum of the EMA-chymotrypsin derivatives lies in the region
of pH 9.3–9.6.[1,25] The apparent Michaelis constant of the EMA-chymo-

[62] R. Haynes and K. A. Walsh, *Federation Proc.* **28**, 534 (1969); *ibid., Biochem.
Biophys. Res. Commun.* **36**, 235 (1969).
[63] E. L. Smith and M. J. Parker, *J. Biol. Chem.* **233**, 1387 (1958).

trypsin derivatives with ATEE as substrate $(K'_m = 1.2 \times 10^{-3} M)$ is similar to that of the crystalline enzyme $(K'_m = 0.8 \times 10^{-3} M)$.[1,25]

The proteolytic activity of the EMA-chymotrypsin derivatives is considerably lower than that of the native enzyme as calculated on the basis of the respective esterase activities.[1,25]

Other Methods of Preparation

Sephadex and Sepharose conjugates of chymotrypsin have been prepared by cyanogen bromide activation of the polysaccharide carriers.[19–21] Polyanionic, water-insoluble derivatives of chymotrypsin have been obtained by coupling the enzyme to CM-cellulose, polyglutamic acid, and polyacrylic acid through activation of the carrier carboxyls with Woodward's Reagent K (N-ethyl-5-phenylisoxazolium-3'-sulfonate)[14] or to carboxymethyl cellulose azide.[8,64] A water-insoluble derivative of polytyrosyl chymotrypsin (with a p-amino-DL-phenylalanine leucine copolymer as carrier) has been reported.[65,65a]

Water-Insoluble Papain Derivatives

It has been the experience of several investigators, as well as of the author, that in the preparation of immobilized derivatives of papain, the recoveries of enzyme activity in the insoluble form vary considerably with different batches of enzyme.[30,31] In order to increase both the reproducibility of the methods as well as the recovery of "insolubilized" enzyme activity, the procedures described below employ the mercury-blocked enzyme, mercuripapain,[66,67] as starting material.

Papain (EC 3.4.4.10), crystallized twice, in suspension in aqueous acetate buffer pH 4.5 is available commercially (Worthington Biochemicals). Papain is assayed titrimetrically,[53] at 25°, pH 6.3 using benzoyl-L-arginine ethyl ester (BAEE) as substrate (5 ml; 0.05 M BAEE, 0.005 M cysteine, 0.002 M EDTA), and 0.1 N NaOH as titrant.[65] One unit of esterase activity is defined as that amount of enzyme which catalyzes the hydrolysis of 1 micromole of substrate per minute in the above assay.

Mercuripapain. Mercuripapain is prepared by a modification of the procedure of Kimmel and Smith.[66,67] A suspension of crystalline papain (1 ml; 20–30 mg/ml; 13–18 esterase units per milligram) is added to a

[64] M. Lilly, C. Money, W. Hornby, and E. M. Crook, *Biochem. J.* **95**, 45 P (1965).

[65] E. Katchalski, *in* "Polyamino Acids, Polypeptides and Proteins" (M. A. Stahmann, ed.), p. 283. University of Wisconsin Press, Madison, Wisconsin, 1962; I. H. Silman, Ph.D. thesis, Hebrew University, Jerusalem, 1964.

[65a] The direct coupling of chymotrypsin to the polydiazonium salt derived from the S-MDA resin leads to excessive inactivation of the bound protein.

[66] J. R. Kimmel and E. L. Smith, *J. Biol. Chem.* **207**, 515 (1954); E. L. Smith, B. J. Finkle, and A. Stockell, *Discussion Faraday Soc.* **20**, 96 (1955).

[67] J. R. Kimmel and E. L. Smith, *Biochem. Prep.* **6**, 61 (1958).

solution $0.005\,M$ in cysteine, $0.002\,M$ in EDTA, pH 7 (20 ml) and the mixture is incubated at 37° for 15 minutes. The solution of activated papain is brought to pH 5.5 with 0.1 N HCl and a solution of mercuric chloride (33 mg $HgCl_2$, dissolved in 1 ml distilled water and adjusted to pH 5.5) is added dropwise with stirring. The reaction mixture is stirred for a few minutes, transferred to a dialysis tube (Visking Cellophane Tubing, size 23/32) and dialyzed exhaustively against distilled water (5 liters) at 4°. In the course of dialysis varying amounts of insoluble material at very low enzymatic activity separate out. The dialyzed mercuripapain solution is clarified by centrifugation and used directly in the binding experiments.

Characterization and Properties. Mercuripapain is assayed as described under papain.[66,67,67a] The protein content of the mercuripapain solution is estimated from its absorbance at 280 nm using the value $E_{280\ nm}^{1\%} = 24.6$.[68,69] The recovery of protein varies with the papain sample used and is in the range 60–85%. The recovery of enzymatic activity is 70–95%. The specific activity of the mercuripapain preparations after activation with cysteine[66,67] is 18–24 esterase units per milligram of protein.[25]

Polyanionic Water-Insoluble Papain Derivative[25]

EMA-Papain

Carrier. Ethylene-maleic anhydride (1:1) copolymer.

Procedure. The ethylene-maleic anhydride copolymer (EMA) is dried before use, at 105°–110° for 24 hours, over phosphorus pentoxide.

The mercuripapain solution is brought to pH 7.8 by the addition of $1\,M$ phosphate buffer, pH 8; the final concentration of phosphate in the protein solution should be $0.1\,M$.

A solution of dried EMA (50 mg) in redistilled dimethyl sulfoxide, DMSO (5 ml) is added dropwise to the ice-cooled, strongly stirred solution ($0.1\,M$ in phosphate, pH 7.8) of mercuripapain (10–200 mg). The cross-linking agent hexamethylene diamine (0.5 ml of a 1% aqueous solution) is then added and the reaction mixture stirred overnight at 4°. The insoluble precipitate is separated by centrifugation (Sorval; 12,000 rpm; 15 minutes) and washed by suspending and recentrifuging with $0.5\,M$ KCl, and then with water, until no residual enzymatic activity

---

[67a] The mercuripapain preparations are devoid of enzyme activity, as tested on BAEE, in the absence of reducing agent. The cysteine and EDTA present in the assay mixture restore their full activity.

[68] A. N. Glazer and E. L. Smith, *J. Biol. Chem.* **236**, 2948 (1961).

[69] J. R. Whitaker and M. L. Bender, *J. Am. Chem. Soc.* **87**, 2728 (1965).

toward benzoyl-L-arginine ethyl ester can be detected in the supernatant, after filtration through a Millipore filter. The washed precipitate is then suspended in water.

The recovery of enzymatic activity in the water-insoluble EMA-mercuripapain derivative is 65–90%.[25] When crystalline papain is coupled to EMA using the same procedure, the recovery of enzymatic activity in the EMA-papain derivative is about 30%.[25]

Notes

1. In the preparation of high protein content samples (300–400 mg protein per 100 mg EMA in the coupling mixture) the cross-linking agent, hexamethylene diamine, can be omitted.[22, 25]

2. The final concentration of dimethylsulfoxide in the coupling mixture should not exceed 20% (v/v).

3. Mixing dimethylsulfoxide with water is accompanied by evolution of heat. During the addition of the EMA-dimethylsulfoxide solution care should be taken to keep down the temperature of the reaction mixture (e.g., by addition of crushed ice).

Determination of Activity. The EMA-mercuripapain derivatives are assayed as described for papain and mercuripapain at pH 7.5[25] (the optimal pH for these derivatives). The amount of active enzyme is calculated from the rate of substrate hydrolysis.

The protease activity is determined, at pH 7, by the casein digestion method.[54, 55]

Determination of Bound Protein. The protein content of EMA-mercuripapain derivative is determined as described for EMA trypsin.

Note. After analysis of mercuripapain derivatives the amino acid analyzer column is washed with 50 ml of a 1% cysteine solution, to remove mercuric salts adsorbed on the column.

Properties. Aqueous suspensions of EMA-mercuripapain preparations can be stored at 4° with no loss in activity for long periods of time. The EMA-mercuripapain derivatives can be lyophilized, retaining most of their enzymatic activity. The dried powders can be stored at 4° or at room temperature in a dessicator for extended periods of time with no significant loss of activity.[25] EMA-mercuripapain derivatives are stable in the pH range 4–10.[25]

The pH-activity profile of EMA-papain derivatives, using benzoyl-L-arginine ethyl ester (BAEE) as substrate is displaced at low ionic strength ($\Gamma/2 = 0.05$), by about 1–1.5 pH units toward more alkaline pH values as compared to crystalline papain under similar conditions.[25] The pH optimum of the polyanionic derivatives at $\Gamma/2 = 0.05$ lies in the region of pH 7.2–7.9. At high ionic strengths ($\Gamma/2 = 1$–2), this effect is partly reversed. The Michaelis constant (K_m) of EMA-papain preparations acting on BAEE is considerably lower at low ionic strength than that of the crystalline enzyme.[25]

The activity of EMA-papain derivatives toward protein substrates is 40–50% of that of the native enzyme, as calculated on the basis of the respective esterase activities.[25]

EMA-mercuripapain preparations of high protein content are well suited for the digestion of protein substrates. Because of their flaky texture they are easy to remove by centrifugation (and in certain cases by filtration through Millipore filters) from the reaction mixture.[25] They can also be preactivated, the activator (cysteine-EDTA) washed off, and the activated enzyme preparation added to the protein substrate solution without the added complication of reducing agent being present in the reaction mixture.[70, 71]

Water-Insoluble Papain Derivative[31]

S-MDA-Mercuripapain Conjugate

Carrier. S-MDA resin.

Procedure. S-MDA resin (100 mg), is diazotized as described in the procedure given for the preparation of water-insoluble polytyrosyl trypsin. The washed polydiazonium salt precipitate is suspended in $0.2 M$ phosphate buffer, pH 7.8 (6 ml). A solution of mercuripapain (8–10 mg) is added dropwise. The reaction mixture is stirred overnight at 4°. The water-insoluble S-MDA-mercuripapain conjugate is separated by filtration on a Büchner funnel, washed with $1 M$ KCl (200 ml), then with water (100 ml) and resuspended in water.

The recovery of enzymatic activity in the water-insoluble mercuripapain derivative is about 60%.[31] When crystalline papain is coupled with diazotized S-MDA resin using the same procedure, the recovery of enzymatic activity in the insoluble precipitate is only 20–30%.[31]

Determination of Activity. The S-MDA-mercuripapain conjugates are assayed as described for papain and mercuripapain.[31]

The protease activity is determined at pH 7 by the casein digestion method.[54, 55]

Determination of Protein Content. The protein content of the S-MDA mercuripapain conjugates is determined as described for S-MDA-polytyrosyl trypsin.

Properties. Aqueous suspensions of S-MDA-mercuripapain conjugates can be stored at 4° for long periods without significant loss of activity.[31] On lyophilization, the S-MDA-mercuripapain conjugates retain about

[70] J. J. Cebra, D. Givol, I. H. Silman, and E. Katchalski, *J. Biol. Chem.* **236,** 1720 (1961).

[71] J. J. Cebra, D. Givol, and E. Katchalski, *J. Biol. Chem.* **237,** 751 (1962).

30% of their enzymatic activity.[31] Dry powders of S-MDA-mercuri-papain can also be obtained by washing with 70% methanol and air-drying on a filter. The material thus obtained retains 50–60% of the enzymatic activity. Moreover, the methanol washing removes traces of physically adsorbed enzyme.

The pH dependence of the esterase activity of the S-MDA-papain preparations with benzoyl-L-arginine ethyl ester (BAEE) resembles closely that of the native enzyme. The pH optimum is in the region of pH 6–6.5. These derivatives also have the same apparent Michaelis constant $[K_m(\text{app}) = 1.9 \times 10^{-2} M]$ as crystalline papain.[31]

The caseinolytic activity of S-MDA-mercuripapain derivatives is similar to that of the native enzyme, as calculated on the basis of the respective esterase activities.[31] The S-MDA-papain derivatives can be easily removed from the digestion mixture by filtration or centrifugation; they have good flow properties[31] and are suitable for column work (see footnotes 70 and 71).

Other Methods of Preparation

Water-insoluble derivatives of papain have been obtained by coupling the enzyme with the polydiazonium salts derived from a p-amino-DL-phenylalanine-leucine copolymer[30] or p-aminobenzyl cellulose[31,43,44] and by cross-linking papain with bisdiazobenzidine-2,2′-disulfonic acid.[30] A papain membrane has been prepared by adsorbing papain onto a synthetic collodion membrane and cross-linking the adsorbed protein with bisdiazobenzidine-2,2′-disulfonic acid.[47]

Water-Insoluble Derivatives of Subtilopeptidase A
(Subtilisin Carlsberg)

Subtilopeptidase A (Subtilisin Carlsberg, EC 3.4.4.16), crystalline, is available commercially (Novo Industries, Copenhagen, Denmark). Subtilopeptidase A is assayed titrimetrically,[53] at 25°, pH 8.6, using acetyl-L-tyrosine ethyl ester (ATEE) as substrate (5 ml; 0.018 M ATEE, 0.01 M in KCl) and 0.1 N NaOH as titrant.[72,73]

Polyanionic Water-Insoluble Subtilopeptidase A Derivative[25]

EMA-Subtilopeptidase A

Carrier. Ethylene-maleic anhydride (1:1) copolymer (EMA).

Procedure. EMA-subtilopeptidase A derivatives are prepared by the procedure described for EMA-trypsin derivatives.

[72] A. N. Glazer, *J. Biol. Chem.* **242**, 433 (1967).
[73] A. O. Barel and A. N. Glazer, *J. Biol. Chem.* **243**, 1344 (1968).

The recovery of enzymatic activity (tested with ATEE) in the water-insoluble EMA-subtilopeptidase A derivative is about 20%.[25]

Determination of Activity. The EMA-subtilopeptidase A derivatives are assayed as described for subtilopeptidase A, at pH 9.4 (the optimal pH for these derivatives).[25] The amount of active enzyme is estimated from the rate of substrate hydrolysis. The protease activity is determined, at pH 7.6, by the casein digestion method.[54, 55]

Determination of Bound Protein. The protein content of EMA-subtilopeptidase A derivatives is determined as described for EMA-trypsin.

Properties. Aqueous suspensions of EMA-subtilopeptidase A preparations can be stored at 4° without significant loss of enzymatic activity for long periods of time. The EMA-subtilopeptidase A derivative can be lyophilized retaining 50–60% of their activity. The dried powders can be stored at 4° with no significant loss of activity. The EMA-subtilopeptidase derivatives are stable in the pH range 5 to 10.[25]

The pH-activity profile of EMA-subtilopeptidase A derivatives, with acetyl-L-tyrosine ethyl ester (ATEE) as substrate, is displaced, by 1–1.5 pH units at low ionic strength ($\Gamma/2 = 0.01$) toward more alkaline pH values as compared to crystalline subtilisin under similar conditions. The pH optimum of the polyanionic subtilopeptidase A derivatives, at low ionic strengths ($\Gamma/ = 0.01$), lies in the region of pH 9.5–9.6. This effect is partially reversed on increasing the ionic strength. The EMA-subtilopeptidase A derivatives have the same apparent Michaelis constant ($K_m = 0.9 \times 10^{-2}\,M$), with ATEE as substrate, as the crystalline enzyme[25] (see footnotes 72 and 73).

The proteolytic activity of the EMA-subtilopeptidase A derivatives is considerably lower than that of the native enzyme as calculated on the basis of the respective esterase activities. The initial rates of hydrolysis have been found to vary considerably with the nature of the protein substrate used.[25]

Water-Insoluble Subtilopeptidase A Derivative[31]

S-MDA-Subtilopeptidase A Conjugate

Carrier. S-MDA resin.

Procedure. The carrier, S-MDA resin (100 mg), is diazotized as described in the procedure given for the preparation of water-insoluble polytyrosyl trypsin.

The washed diazonium salt precipitate is suspended in 0.2 M phosphate, pH 7.8 (4 ml), and a solution of subtilopeptidase A (15 mg) in the same buffer (3 ml) is added dropwise. The reaction mixture is left overnight at 4° with magnetic stirring. The water-insoluble S-MDA-subtilopeptidase A conjugate is separated by filtration, washed with 1 M KCl (200 ml), then with water (100 ml), and resuspended in water or in 0.1 M phosphate buffer, pH 7.0.

The recovery of enzymatic activity (tested with ATEE) in the water-insoluble subtilopeptidase A derivative is 15–20%.

Determination of Activity. The S-MDA-subtilopeptidase A derivatives are assayed as described for subtilopeptidase A, at pH 9.4 (the optimal pH of these derivatives).[31] The amount of active enzyme is calculated from the rates of substrate hydrolysis.

The protease activity is determined at pH 7.6, by the casein digestion method.[54, 55]

Determination of Bound Protein. The protein content of the S-MDA subtilopeptidase A conjugates is determined as described for S-MDA-polytyrosyl trypsin.

Properties. S-MDA-subtilopeptidase A samples can be stored in aqueous or dilute buffer (pH 7) suspensions at 4° for long periods of time without significant loss of enzymatic activity. S-MDA-subtilopeptidase A derivatives are almost completely inactivated on lyophilization or air drying.[31]

The pH-activity profile of the S-MDA-subtilopeptidase A derivatives, with acetyl-L-tyrosine ethyl ester as substrate, is displaced by about one pH unit at low ionic strength ($\Gamma/2 = 0.01$) toward more alkaline pH values, as compared to the crystalline enzyme. The pH optimum of the S-MDA-subtilopeptidase A derivatives lies in the region of pH 9.2–9.6.[31]

The apparent Michaelis constant of the S-MDA-subtilopeptidase A derivatives, with ATEE as substrate $[K_m(\text{app}) = 1.7 \times 10^{-2} M]$ is slightly higher than that of the crystalline enzyme $[K_m(\text{app}) = 0.9 \times 10^{-2} M]$.[31]

The proteolytic activity of S-MDA-subtilopeptidase A derivatives is slightly lower than that of the native enzyme.[31]

Other Methods of Preparation

A water-insoluble subtilisin Novo preparation has been prepared by cross-linking the enzyme with glutaraldehyde.[74] More recently, subtilopeptidase A (subtilisin Carlsberg) has been coupled to Sepharose activated by cyanogen bromide.[75]

---

[74] K. Ogata, M. Ottesen, and I. Svendsen, *Biochim. Biophys. Acta* **159**, 403 (1968).
[75] C. B. Anfinsen, unpublished data.

[71] Cellulose-Insolubilized Enzymes

By E. M. Crook, K. Brocklehurst, and C. W. Wharton

In general, assay techniques applicable to soluble enzyme systems may be used for the estimation of the activity of. cellulose-insolubilized (CI) enzymes. Direct spectrophotometric assays are not readily applicable, however, since the particulate nature of the support causes a high degree of scattering, particularly in the ultraviolet region of the spectrum. Diffuse reflectance spectroscopy has not been applied to CI-enzyme systems, but it is this technique which holds the greatest promise. The titrimetric method has general applicability as a relatively simple continuous rate assay. Almost any available analytical technique may be used for discontinuous assays since the insoluble enzyme may be removed from the assay solution by either centrifugation or filtration. The catalytic behavior of *O*-(carboxymethyl) cellulose-insolubilized enzymes in columns has been examined (see p. 977) and this represents a possible method of assay. In this case the eluate is devoid of particulate material and continuous spectrophotometric measurement is possible. A number of parameters characteristic of the column and the CM-cellulose involved need to be determined before a meaningful estimate of the activity of the preparation may be obtained. Continuous titrimetric rate assays have been used exclusively in this laboratory for the determination of the specific activity of CI-enzymes in stirred suspensions. In order to assess the specific activity, it is necessary to obtain an estimate of the protein content of the CI-enzyme. Methods for determining both the enzyme protein content and in the case of the sulfhydryl enzymes, the quantity of attached enzyme possessing a reactive active center thiol group are given.

Assay Methods

Determination of the Dry Weight of a Standard Suspension

Procedure. A standard aliquot of suspension is placed in a preweighed Gooch filter, the water is removed by suction and the solid is washed with a little water. The Gooch filter is dried in the oven, allowed to cool to room temperature in a desiccator, and reweighed.

Protein Content of CI-Enzymes

Method I. Mass Balance. A known mass of enzyme is used in the reaction in which the enzyme is coupled to the activated support material. The amount of enzyme attached is determined by difference after spectrophotometric measurement (A_{280}) of the total enzyme recovered from the coupling solution and washings. The spectrophotometric absorption parameters of the proteases which have been coupled to cellulose derivatives are given in Table I.

TABLE I
EXTINCTION COEFFICIENTS AND MOLECULAR WEIGHTS OF PROTEOLYTIC ENZYMES[a]

Enzyme	Molecular weight	$10^{-4} \epsilon_{280} M^{-1} cm^{-1}$	$E_{280}^{1\%}$
Bromelain	33,200[1]	6.33[1]	19
Ficin	25,000[2]	5.25	21[2]
Papain	20,700[3]	5.10[4]	25
α-Chymotrypsin	24,800[5]	4.96	20[6]
Trypsin	23,800[7]	3.40	14.3[8]

[a] Superscript numbers refer to text footnotes.

Method II. Amino Acid Analysis. This method is considerably more sensitive than method (I) and is used when only small quantities of enzyme and/or support are available. The method involves acid hydrolysis of the enzyme to amino acids followed by estimation of the amino acid amino groups using ninhydrin.[9]

Reagents

Solid CO_2
Acetone
Citrate buffer, 0.2 M, pH 5.0
5% w/v Ninhydrin in methyl Cellosolve
Aq. KCN, 0.01 M
Aq. KCN in methyl Cellosolve, 2% v/v 0.01 M

Procedure. CI-enzyme (1–10 mg) is placed in a thick-walled Pyrex tube with a neck prepared for vacuum sealing. A similar quantity of the CM-cellulose, from which the hydrazide was prepared, and a sample of enzyme (0.2–2.0 mg) in H_2O (up to 1 ml) is placed in another tube. Constant boiling HCl (5 ml) is added to both tubes which are then sealed

[1] T. Murachi, M. Yasui, and Y. Yasuda, *Biochemistry* **3**, 48 (1964).
[2] P. T. Englund, T. P. King, L. C. Craig, and A. Walti, *Biochemistry* **7**, 163 (1968).
[3] E. L. Smith, B. J. Finkle, and A. Stockell, *Discussions Faraday Soc.* **20**, 96 (1955).
[4] A. N. Glazer and E. L. Smith, *J. Biol. Chem.* **236**, 2948 (1961).
[5] M. L. Bender, G. R. Schonbaum, and B. Zerner, *J. Am. Chem. Soc.* **84**, 2540 (1962).
[6] M. Laskowski, Vol. II, p. 8.
[7] P. Desnuelle, *in* "The Enzymes" (P. D. Boyer, H. Lardy, and K. Myrbäck, eds.), 2nd ed., Vol. 4, p. 124. Academic Press, New York, 1960.
[8] E. W. Davie and H. Neurath, *J. Biol. Chem.* **212**, 507 (1955).
[9] E. C. Cocking and E. W. Yemm, *Biochem. J.* **58**, xii (1958).

in vacuo after repeatedly freezing in solid CO_2/acetone, evacuating, and thawing to remove dissolved air. The tubes are incubated for 48–72 hours at $110°$ in an oven. The tubes are allowed to cool to room temperature and carefully opened. Solid material is removed by filtration and the filtrates transferred to two 25-ml round-bottom flasks, the tubes and filters being washed with a little distilled water. The combined filtrates and washings in the flasks are evaporated to dryness *in vacuo*. The residues are transferred to 10 ml volumetric flasks and made up to the mark with water. Aliquots (0.01–1.0 ml) are taken for amino acid estimation, a standard curve being constructed for the hydrolyzate from the tube which contained uncoupled CM-cellulose and enzyme. The aliquots are placed in test tubes and made up to 1 ml with water. To each tube is added citrate buffer (0.5 ml) followed by ninhydrin solution (0.2 ml) and KCN solution (1 ml). The solutions are mixed well and heated in a boiling water bath for at least 15 minutes and then cooled for 15 minutes in running tap water. The resulting solution is made up to a convenient volume (10–100 ml) with 60% ethanol and the absorbance at 570 nm determined. The repeatability of the method is $\pm 0.2\%$ and Methods I and II have been found to agree to within $\pm 0.25\%$. The quantity of enzyme attached to the support is calculated using the standard curve.

Determination of Total Nitrogen Content. The total nitrogen content of the CI-enzyme preparations may be determined, e.g., by the Kjeldahl method.[10]

Determination of the Reactive Thiol Content of CI-Sulfhydryl Proteases. The reactive thiol per mole attached enzyme is estimated by titration of the enzyme with 5,5'-dithiobis-(2-nitrobenzoic acid) (DTNB) according to the scheme:

$$ES^- + RSSR \rightleftharpoons ESSR + RS^-$$

At pH 8 RS^- absorbs strongly at 412 nm ($\epsilon_{412} = 1.36 \times 10^4$).[11]

Reagents

Tris buffer, $0.2\,M$, pH 8.0 ($I = 0.1$)
DTNB, $0.05\,M$ neutralized to pH 7.0

Procedure.[12] An aliquot of CI-thiol enzyme (ca. 50 mg in ca. 1 ml H_2O) is placed in a small beaker and stirred magnetically. Tris buffer (3 ml) is added, followed by DTNB solution (0.05 ml). The suspension

---

[10] R. Ballentine, Vol. III, p. 984.
[11] G. L. Ellman, *Arch. Biochem. Biophys.* **82**, 70 (1959).
[12] C. W. Wharton, E. M. Crook, and K. Brocklehurst, *European. J. Biochem.* **6**, 565 (1968).

is stirred for 30 minutes at room temperature. The solid material is removed by filtration, using a sintered glass filter under slight suction. The filtrate is made up to 5 ml with ca. 1 ml of buffer which is used to wash the filter. The absorbance of the solution at 412 nm (A_1) is then measured. A similar quantity of CM-cellulose is treated as above and the absorbance at 412 nm (A_2) subtracted from the value for the CI-thiol enzyme. The molarity of thiol (M_t), when distributed evenly throughout 5 ml is given by:

$$M_t = \frac{A_1 - A_2}{1.36 \times 10^{-4}} M$$

The thiol per mole of enzyme (TPM) is given by:

$$\text{TPM} = \frac{W \times P \times 2 \times 10^2}{\text{MW} \times M_t}$$

where W is the weight of CI-enzyme present in the assay, P is the percent composition of the CI-enzyme [i.e., (wt. enzyme/total wt.) $\times$ 100] and MW is the molecular weight of the enzyme (see Table I).

Titrimetric Rate Assays. All the proteolytic enzymes which have been attached to cellulose catalyze the hydrolysis of N-acyl-α-amino acid esters. At pH values near neutrality the products of the hydrolysis:

$$\text{RCOOR}' \xrightarrow{\text{H}_2\text{O}} \text{RCOO}^- + \text{R}'\text{OH} + \text{H}^+$$

are a carboxylate anion, a proton, and an alcohol. Accordingly, in order to maintain a constant pH value, a hydroxyl anion must be added to the assay system for each substrate molecule hydrolyzed. The reaction is followed using an automatic recording pH-stat, which maintains constant pH, recording the amount of alkali added as a function of time.

Reagents

NaOH, 1 M ⎫
NaOH, 0.1 M ⎬ volumetric
KCl, 0.5 M ⎭
EDTA (di-Na salt), 0.1 M

Substrates

Bromelain ⎫
Ficin   ⎬ 0.25 M α-N-Benzoyl-L-arginine ethyl ester (BAEE)
Papain   ⎬
Trypsin   ⎭
α-Chymotrypsin 0.05 M N-acetyl-L-tyrosine ethyl ester (ATEE) in 50% aq. methanol

Procedure. The reaction is followed using a Radiometer TTT1 automatic titrator, SBU1- 0.5 ml buret and SBR2 recorder. The micro reaction vessel (TTA 31) is maintained at 25° by means of a water jacket and water bath. A stream of O_2-free nitrogen is passed through a wash bottle containing water at 25° and thence over the surface of the fluid in the reaction vessel to exclude CO_2 and O_2. The total reaction volume at the start of the reaction is 5 ml and the titrant is 0.1 M NaOH. With the exception of the CI-bromelain catalyses, the substrate concentrations given above are sufficient to saturate the enzyme and thus approximate estimates of V_{max} may be obtained directly.

Assays involving BAEE. One milliliter 0.5 M KCl, 0.05 ml 0.1 M EDTA, x ml standard suspension (1 ml in this laboratory, containing ca. 50 mg CI-enzyme in the case of CI-bromelain, 5–10 mg in the case of CI-trypsin) and $(1.95 - x)$ ml deionized water are placed in the reaction vessel and equilibrated to pH 7 using 1 M NaOH added from an Agla micrometer syringe. The equilibration takes about 5 minutes and must be complete before the substrate (2.0 ml) is added and the pH rapidly readjusted to 7.0 using the Agla syringe. The pH-stat is now switched on and the reaction followed for a time sufficient to obtain a zero-order trace (a degree of curvature may occur in the early stages of the reaction due to final reequilibration of the CI-enzyme, particularly in the case of CM-cellulose derivatives).

Assays involving ATEE. The procedure is the same as that given for assays involving BAEE except that 1.0 ml ATEE and $(2.95 - x)$ ml water are used.

Convenient chart speeds for the titrimeter assays are 1–4 divs/minute and the rate of the reaction is given by:

$$v_i = \text{chart divisions per minute} \times 10^{-4} \, M \, \text{min}^{-1}$$

Specific Activity. The unit of specific activity of CI-enzymes is defined as the number of micromoles of substrate hydrolyzed per minute per milligram CI-enzyme. The units of activity are derived from v_i by

$$\text{Units of activity} = 5 \times 10^3 \, v_i/W \text{ micromoles/minute/mg}$$

where W is the weight of CI-enzyme in the assay.

It is important to note that in general the activity obtained by the above technique will not be strictly comparable with a similar estimate for the free enzyme catalysis. A number of factors which perturb the kinetic properties of CI-enzymes have to be taken into account (see p. 973).

Preparation of CI-Enzymes

The principal methods which have been used to attach enzymes covalently to cellulose are summarized below. In all cases the attachment depends upon the attack of a nucleophilic center in the enzyme on an electrophilic center in a cellulose derivative. Except in Method 4 the nucleophilic center in the enzyme is represented here as an amino group. While this group is the most probable attachment site on the enzyme, in no case has this been verified experimentally.

1. The Curtius Azide method was first used to attach enzymes to cellulose by Micheel and Ewers[13] and was modified subsequently by Mitz and Summaria.[14]

Cellulose—OH + ClCH$_2$CO$_2$H $\xrightarrow{\text{NaOH}}$ Cellulose—O—CH$_2$—CO$_2$H

Cellulose—O—CH$_2$—CO$_2$H $\xrightarrow{\text{CH}_3\text{OH}, \text{HCl}}$ Cellulose—O—CH$_2$—CO$_2$CH$_3$

Cellulose—O—CH$_2$—CO—NHNH$_2$ $\xleftarrow{\text{NH}_2\text{NH}_2}$ Cellulose—O—CH$_2$—CO$_2$CH$_3$

Cellulose—O—CH$_2$—CO—NHNH$_2$ $\xrightarrow{\text{NaNO}_2, \text{HCl}}$ Cellulose—O—CH$_2$—CON$_3$

Cellulose—O—CH$_2$—CON$_3$ $\xrightarrow[\text{protein-NH}_2]{\text{mild alkali}}$ Cellulose—O—CH$_2$—CONH—Protein

2. Acylation of cellulose hydroxyl groups by bromoacetyl bromide followed by alkylation of the protein amino group[15] or presumably reactive thiol groups if they are available.

Cellulose—OH + Br—CO—CH$_2$Br $\longrightarrow$ Cellulose—O—CO—CH$_2$Br

Cellulose—O—CO—CH$_2$Br $\xrightarrow{\text{Protein}—\text{NH}_2}$ Cellulose—O—CO—CH$_2$—NH—Protein

This method suffers from the lability, even at neutral pH, of the ester linkage produced in the first reaction.

3. Reaction of both cellulose and the protein with a reactive triazine.[16]

[13] F. Micheel and J. Ewers, *Makromol. Chem.* **3,** 200 (1949).

This method has great potential; the third reactive chlorine atom may be replaced by a group whose function would be to perturb the environment of the matrix in a given manner, e.g., a charged group.

4. Attachment to cellulose of an aromatic amine, diazotization of the amino group and subsequent coupling with the protein, presumably with one of the activated positions on the ring of a tyrosine residue.[17]

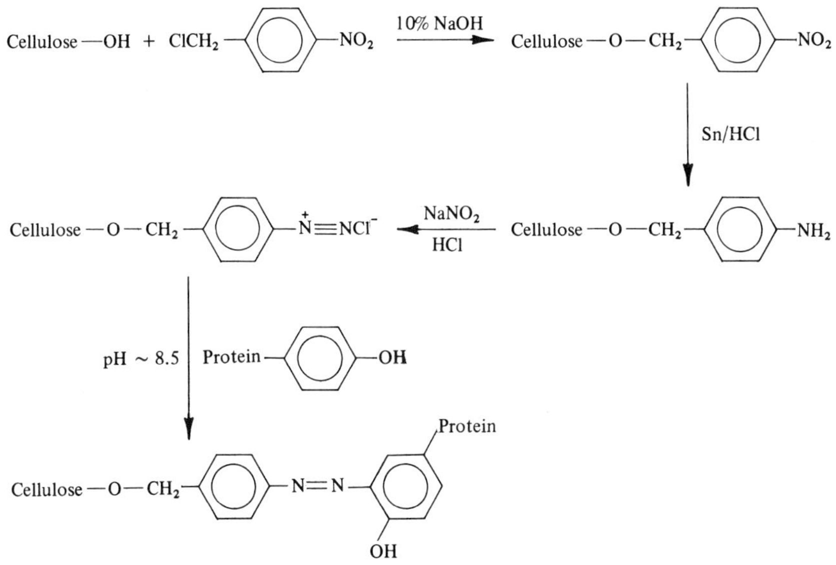

Cellulose—OH + ClCH$_2$—〈O〉—NO$_2$ $\xrightarrow{\text{10\% NaOH}}$ Cellulose—O—CH$_2$—〈O〉—NO$_2$

↓ Sn/HCl

Cellulose—O—CH$_2$—〈O〉—$\overset{+}{\text{N}}\equiv\text{NCl}^-$ $\xleftarrow[\text{HCl}]{\text{NaNO}_2}$ Cellulose—O—CH$_2$—〈O〉—NH$_2$

pH ~ 8.5 │ Protein—〈O〉—OH

↓

Cellulose—O—CH$_2$—〈O〉—N=N—〈O〉—Protein
                                          │
                                          OH

5. Synthesis of amide bonds between CM-cellulose and the protein using isoxazolium salts, developed by Woodward, Olofson, and Mayer[18] for peptide synthesis. This method has been used to attach proteases to a copolymer of alanine and glutamic acid.[19]

Details of Method I, the method used most extensively to date, are given below. A great deal of research in the authors' laboratory has led to adaptations and improvements of the original procedure of Micheel and Ewers and the CI-enzymes discussed in the section on properties have been prepared using this method.

[14] M. A. Mitz and J. Summaria, *Nature* **189,** 576 (1961).
[15] A. T. Jagendorf, A. Potchornik, and M. P. Sela, *Biochim. Biophys. Acta* **78,** 516 (1963).
[16] G. Kay and E. M. Crook, *Nature* **216,** 514 (1967).
[17] D. H. Campbell, E. Luescher, and L. S. Lerman, *Proc. Natl. Acad. Sci. U.S.* **37,** 575 (1951).
[18] T. Wagner, C. J. Hsu, and G. Kelleher, *Biochem. J.* **108,** 892 (1968).
[19] R. B. Woodward, R. A. Olofson, and H. Mayer, *J. Am. Chem. Soc.* **83,** 1010 (1961).

Preparation of CM-Cellulose Bromelain (CMCB): Experimental Details

This version of the Curtius azide method for attaching enzymes to CM-cellulose is illustrated by the preparation of CMCB.[12] This procedure has been used also to attach papain, ficin, trypsin, and α-chymotrypsin to CM-cellulose to produce CMCP, CMCF, CMCT, and CMCC, respectively.

Preparation of CM-Cellulose Hydrazide

CM-11 (Whatman; exchange capacity 0.6 meq/g) (100 g) is suspended in dry methanol (1.5 liter) by stirring the mixture vigorously with a powerful magnetic stirrer in a 2-liter conical flask fitted with a reflux condenser. Concentrated HCl (10 ml) is added and the mixture is heated carefully until the methanol is just refluxing. The mixture is stirred and heated under reflux for 4 hours. The CM-cellulose methyl ester is then collected by vacuum filtration and washed thoroughly on a Büchner funnel with dry methanol (3 × 500 ml).

The conversion of the CM-cellulose methyl ester to the hydrazide is effected by stirring a mixture of the methyl ester (100 g), methanol (1 liter), and hydrazine hydrate (100 ml) in a 2-liter flask fitted with a mechanical stirrer and immersed in a water bath at 40° for 72 hours. The CM-cellulose hydrazide is collected by vacuum filtration and washed successively with 2 × 250 ml of methanol, methanol saturated with CO_2, 90% aqueous methanol, and dry methanol. This procedure removes unreacted hydrazine as hydrazine carbonate.

Preparation of CM-Cellulose Azide

CM-cellulose hydrazide (5 g) is suspended in 0.6 N HCl (200 ml) cooled to 0° in an ice bath. Aqueous sodium nitrite (25 ml of a 5% solution) is added slowly with vigorous magnetic stirring over 20 minutes. The suspension is stirred at 0° for a further 20 minutes and then the CM-cellulose azide is collected by vacuum filtration using a 250-ml sintered filter. The azide is washed with ice-cold water (2 × 200 ml) ice-cold 1 M NaCl (200 ml) and again with ice-cold water (2 × 200 ml) and dried on the filter by suction.

Preparation of CMCB

To a solution of purified bromelain (1 g) in water (100 ml) immersed in an ice bath and stirred magnetically is added the freshly prepared CM-cellulose azide (5 g) all at once. The pH is rapidly adjusted to 8.7 with saturated sodium tetraborate solution. The suspension is stirred in the ice bath for 20 minutes after which the pH is readjusted to 8.7 if necessary. The suspension is then stirred for a further 40 minutes. The solid is then collected on a 250-ml sintered filter by vacuum filtration and washed on the filter as follows: H_2O (1 × 200 ml), 1 M NaCl (2 × 200 ml), 0.05 M acetic acid (2 × 200 ml), 1 M NaCl (2 × 200 ml), 0.5 M $NaHCO_3$ (2 × 200 ml), 1 M NaCl (2 × 200 ml), and H_2O (4 × 200 ml). The solid CMCB is then suspended in a known volume of water to give a standard suspension which is stored at 4° until required. For long periods of storage 1 drop of purified chloroform or sodium azide (0.1% w/v) is added to the suspension to prevent bacterial contamination.

Reactivation of CMCB[12]

A partly oxidized sample of CMCB may be reactivated by stirring a suspension of it in 5 mM aqueous dithiothreitol (DTT) at pH 7.0 and room temperature for 15 minutes. The CMCB is then filtered off, washed with water, 1 M NaCl, and 0.5 M $NaHCO_3$ until the filtrate is free from DTT as shown by lack of reaction with DTNB.

Properties

Stability. CI-enzymes are at least as stable as the corresponding enzymes in free solution. In most cases stability to heat, extremes of pH, and oxidation is enhanced. The fact that attachment to CM-cellulose of ficin[20] but not of bromelain[12] stabilizes the enzyme toward heat denaturation may be a consequence of different pH-dependencies of the thermal stabilities of the two systems (see Levin et al.[21]). CI-proteases are not subject to self-digestion as are the corresponding enzymes in solution. The CM-cellulose enzymes may be stored for long periods at 4° either as suspensions in deionized water or as freeze-dried preparations. In general about 50% of the activity of a CM-cellulose enzyme is lost during the freeze-drying process, although CMCF has been freeze dried without loss of activity.[20]

1. CMCB. CMCB prepared from purified bromelain (untreated with cysteine) stored at 4° in air as a suspension in deionized water (50–100 mg/ml) maintains its activity toward BAEE for a period of 6 weeks. The activity toward BAEE in the absence of cysteine is about 50% of that obtained by preincubation of the CMCB with 5 mM cysteine. CMCB activated with DTT and washed free of activator has the same activity toward BAEE as the unactivated CMCB when this is assayed in the presence of 5 mM cysteine. The DTT activated CMCB maintains its activity also for about 6 weeks when stored at 4°. After this time, in both cases, the activity may fall off rapidly. This is due to bacterial contamination and may be prevented by including 0.1% v/v chloroform or

[20] W. E. Hornby, M. D. Lilly, and E. M. Crook, *Biochem. J.* **98**, 420 (1966).
[21] Y. Levin, M. Pecht, L. Goldstein, and E. Katchalski, *Biochemistry* **3**, 1905 (1964).

0.1% w/v sodium azide in the stock suspension. Attachment of bromelain to CM-cellulose reduces both the rate of loss of activity and the rate of loss of the reactive thiol group when the enzyme preparations are stored at 4°. The stabilities of bromelain and CMCB are comparable both at 60° and at 70°. At the latter temperature the rate of loss of activity is considerably higher than the rate of loss of the reactive thiol group.

2. CMCF.[20] A suspension of CMCF in water at pH 5.0 stored at 2° for 4 months retains > 85% of its activity toward BAEE. Attachment of ficin to CM-cellulose increases the stability of the enzyme toward thermal denaturation.

Purity and Physical Properties. Most CI-enzyme preparations contain about 5–10% protein on a dry weight basis and most of their physical properties such as particle size, density, solubility, charge, etc. are thus determined by the nature of the cellulose matrix. Cellulose is stable over a wide range of pH, possesses great ease of wetting, is available in readily activatable forms, and can be prepared as powders and flocs with excellent filtration and flow characteristics.

In the case of the CM-cellulose enzymes, the enzyme protein is attached to the cellulose backbone by stable, covalent linkages (probably amide bonds between the CM-cellulose carboxyl groups and protein amino groups) and cannot be removed by any process other than hydrolysis, e.g., by hot concentrated acid or alkali. The washing procedure described on p. 971 ensures that all protein adsorbed or ion-exchanged onto the cellulose is removed and only covalently linked enzyme remains. This enzyme remains firmly bound to the cellulose matrix even under conditions which lead to its complete denaturation. Inadequate washing of the preparations leads to products containing noncovalently linked protein, much of which may be brought into solution by buffer salts and other constituents of a reaction mixture with consequent loss of the advantages of an insolubilized enzyme.

The magnitude of the apparent catalytic constant (see p. 977) for CI-enzymes shows that they are not homogeneous preparations; some enzyme molecules are attached to the cellulose in modes which are either unreactive or less reactive than the free solution enzyme. This loss of activity at least in the case of CMCF, does not seem to be due to covalent attachment of the enzyme by its active center thiol group.[20]

Activators and Inactivators. There is no reason to suppose that substances which either activate or inhibit catalyses in free solution by a given enzyme will not act in an analogous manner on CI-enzymes, although the rate of reaction of the modifier with the enzyme may be changed as a result of the attachment. For example, the rates of inactivation of both ficin[20] and bromelain[22] by reaction of their essential thiol

[22] C. W. Wharton, E. M. Crook, and K. Brocklehurst, unpublished results, 1968.

groups with DTNB at low ionic strength are markedly decreased by attachment of the enzyme to CM-cellulose. This decrease in rate, which is due to the unfavorable electrostatic interactions of the CM-cellulose and the DTNB, is diminished at high ionic strength when the negative charges on the matrix and the reagent are shielded from each other.

Reactivation of CMCB, CMCF, and CMCP. When these CI thiol enzymes are wholly or partly in a reversibly oxidized state, they may be reactivated by stirring a suspension of the CM-cellulose enzyme in an excess of a solution of a sulfhydryl compound, e.g., DTT at pH 6–7. The excess of reducing agent may be readily removed by filtration and washing the activated CM-cellulose enzyme with O_2-free $NaHCO_3$ solution and then with O_2-free water.

The use of cysteine may be undesirable for reactivating CM-cellulose enzymes because although it gives full activation, its use sometimes results in preparations which possess abnormally high SH/protein ratios. This is due probably to reaction of the cysteine thiol group with residual CM-cellulose azide and subsequent intramolecular transfer of the CM-cellulose acyl moiety to the cysteine amino group.

Specificity. There has been no demonstration of a fundamental change in the specificity of an enzyme consequent upon its attachment to cellulose as a result of matrix-imposed conformational changes in the protein. Such specificity changes as have been observed are due to changes in the apparent kinetic parameters which control the specificity as a result of such factors as charged groups on the cellulose matrix, steric hindrance by the matrix, diffusion limitation of substrates in and of products out of the enzyme matrix system.

Kinetic Properties. REACTIONS IN STIRRED SUSPENSIONS. The reactions catalyzed by CMCB, CMCC, CMCF, CMCP, and CMCT follow Michaelis-Menten kinetics (see, for example, footnotes 12 and 20). In addition to the factors which influence the rate of the free solution enzyme catalysis, the rates of the catalyses by CI-enzymes are determined also by (1) the nature of the support, particularly its charge when the substrate also is charged and (2) the rate of stirring of the suspension. As a result of these additional factors the ratios of reactivities of a series of substrates for a given free solution enzyme may not be the same for the corresponding CI-enzyme. For example, CMCF preparations exhibit, relative to free ficin, a greater loss in activity toward casein than toward BAEE.[20] This differential effect presumably reflects steric hindrance and diffusion limitation by the support. CI-enzymes may thus be used for the preferential removal of low molecular weight substrates in the presence of susceptible macromolecules.

Michaelis Constants. The measured Michaelis constant, K_m, for a CI-enzyme catalysis is generally similar (within a factor of ca. 2) to

TABLE II

KINETIC PARAMETERS FOR CATALYSES BY CI-ENZYMES AND BY THE CORRESPONDING FREE SOLUTION ENZYMES AT pH 7.0 AND 25.0°

Enzyme preparation	Exchange capacity of CM-cellulose (meq/g)	Substrate	Reaction conditions	I	K_m (mM)	k_{cat}(app) (sec^{-1})^a	Reference
Bromelain	0.7	BAEE		0.1	110	0.63	
Bromelain	0.7	BAEE		0.2	110	0.73	
Bromelain	0.7	BAEE	Free solution; 1 mM EDTA; 5 mM cysteine	0.5	110	0.88	
Bromelain	0.7	BAEE		1.0	100	1.10	
Bromelain	0.7	BAEE		1.6	140	1.40	24
CMCB	0.7	BAEE		0.023	7	0.21	
CMCB	0.7	BAEE		0.046	19	0.30	
CMCB	0.7	BAEE		0.10	30	0.33	
CMCB	0.7	BAEE	Stirred suspension, 1 mM EDTA; 5 mM cysteine	0.20	37^b	0.38(0.56)	12, 24
CMCB	0.7	BAEE		0.50	55	0.48	
CMCB	0.7	BAEE		1.00	60	0.58	
CMCB	0.7	BAEE		1.30	56	0.54	24
CMCB	0.7	BAEE		1.60	63	0.64	

Enzyme		Substrate	Condition				Ref.
Ficin	0.7	BAEE	Free solution; 1 mM EDTA; 5 mM cysteine	0.15	20	1.83	
CMCF	0.7	BAEE	Stirred suspension	0.15	2.5	0.16	
CMCF	0.9	BAEE	Stirred suspension	0.15	2.5	0.18	
CMCF	0.9	BAEE	Stirred suspension	0.55	7.0	—	
CMCF	0.7	BAEE	Packed-bed; flow; 30 ml/hour	0.40	10.0	0.14	
CMCF	0.7	BAEE	Packed-bed; flow: 140 ml/hour	0.40	5.4	0.14	
CMCF	0.7	BAEE	Packed-bed; flow: very fast	0.40	ca. 5.2	0.14	26
CMCF	0.9	BAEE	Packed-bed; flow: 30 ml/hour	0.40	15.9	0.15	
CMCF	0.9	BAEE	Packed-bed; flow: 30 ml/hour	0.10	9.0	0.16	
CMCF	0.9	BAEE	Packed-bed; flow: 140 ml/hour	0.40	4.9	0.15	
CMCF	0.9	BAEE	Packed-bed; flow: 140 ml/hour	0.10	4.5	0.16	
CMCF	0.9	BAEE	Packed-bed; flow: very fast	0.40	ca. 3.5	0.15	
CMCF	0.9	BAEE	Packed-bed; flow: very fast	0.10	3.5	0.16	
α-Chymotrypsin	0.7	ATEE	Free solution	0.05	0.27	—	27
CMCC	0.7	ATEE	Stirred suspension	0.05	0.56	—	

a k_{cat}(app) = $V_{max}/[E]$, $[E]$ calculated from A_{280} in the case of the free solution enzymes and from protein content in the case of the CMC-enzymes except when $[E]$ calculated from thiol content estimated by reaction with DTNB.

b Value misprinted in footnote 12.

that for the corresponding free enzyme catalysis when (1) the supporting matrix and/or the substrate is uncharged and (2) irrespective of the charges on the components, the assays are carried out at high ionic strength. When the matrix is negatively charged and the substrate is positively charged, the substrate concentration is higher in the interstices of the matrix than in the bulk medium and the K_m for the CI-enzyme catalysis is considerably smaller at low ionic strength than the K_m for the free enzyme catalysis. When the substrate is negatively charged this effect is reversed and the K_m for the CI-enzyme catalysis is larger at low ionic strength than the K_m for the free enzyme catalysis. This effect has not been demonstrated for a CM-cellulose protease catalyzing the conversion of a negatively charged substrate but has been demonstrated with CM-cellulose-ATP-creatine phosphotransferase.[23] At high ionic strength the electrostatic interactions of the matrix and the substrate are shielded from each other and the K_m's for the CI and free solution systems are similar (see Table II).

The value of K_m for the CMCB-catalyzed hydrolysis of BAEE at pH 7.0 and 25.0° at a given ionic strength, I, is predicted satisfactorily[24] by the equation:

$$K_m = \frac{\gamma K_m(\text{lim})I}{(\gamma Z m_c/2 + I)}$$

in which $\gamma K_m(\text{lim}) = 0.066$ and $\gamma Z m_c = 0.26$; γ is the ratio of the mean ion activity coefficients of the CM-cellulose matrix and bulk phases, $K_m(\text{lim})$ is the limiting value of K_m when the net charge on the matrix is zero, Z is the modulus of the number of charges on the matrix and m_c is the concentration of the matrix in its own hydrated volume. When appropriate values of $\gamma K_m(\text{lim})$ and $\gamma Z m_c$ are used, this equation should have general applicability for an enzyme attached to a negatively charged matrix catalyzing the conversion of a positively charged substrate. Other equations[20, 25] have been derived to describe the equilibrium distribution of a charged substrate between the internal microenvironment of the enzyme on its matrix and the bulk solution, but they do not permit an assessment of the perturbing effect of the matrix on K_m in terms of readily measurable quantities.

Catalytic Constants. A comparison in a meaningful way of the intrinsic activity of an enzyme attached to a solid matrix with that of an enzyme in free solution is difficult. A comparison of the reaction rate per

[23] W. E. Hornby, M. D. Lilly, and E. M. Crook, *Biochem. J.* **107**, 669 (1968).
[24] C. W. Wharton, E. M. Crook, and K. Brocklehurst, *European. J. Biochem.* **6**, 572 (1968).
[25] L. Goldstein, Y. Levin, and E. Katchalski, *Biochemistry* **3**, 1913 (1964).

milligram enzyme protein at a fixed substrate concentration can be misleading because of the marked change in K_m at a given pH which sometimes results from attachment of the enzyme to the matrix (see above) and also because of shifts in pH-activity curves consequent upon attachment (see below). The latter may sometimes be obviated by working in a pH-independent region and the former by computing V_{max} for both the supported and free solution enzyme catalyses and comparing the values of the catalytic constant, k_{cat}. The calculation of k_{cat} defined as $V_{max}/$ enzyme active site concentration, demands a knowledge of the concentration of catalytically active enzyme both in solution and attached to the matrix. This constitutes one of the major problems besetting the study of enzymes attached to solid matrices. In the case of all the CI-enzymes examined to date, the apparent catalytic constant $[k_{cat}(app)]$ defined as $V_{max}/$"enzyme concentration" is significantly lower than the value for the corresponding enzymatic catalysis in free solution. The usual method of measuring the amount of enzyme attached to the matrix, estimation of the total protein content of the preparation, does not distinguish active from inactive enzyme. When the total protein content of CMCB is used as a measure of the enzyme concentration, $k_{cat}(app)$ for the hydrolysis of BAEE at pH 7.0, I = 0.2 and 25° is 52% of the corresponding value for the catalysis by bromelain.[12] A better estimate than the protein concentration of active bromelain on the CM-cellulose matrix should be the concentration of thiol groups which are reactive toward a thiol reagent such as DTNB. When this is used as a measure of the concentration of active sites on CMCB, $k_{cat}(app)$ is 77% of the value for the bromelain catalysis.[12] Thus some CMCB molecules which display a reactive thiol group are inactive or less active in catalyzing the hydrolysis of BAEE than are free bromelain molecules. The value of $k_{cat}(app)$ for the CMCB-catalyzed hydrolysis of BAEE increases with increase in ionic strength as does $k_{cat}(app)$ for the corresponding bromelain catalysis[24] (see Table II).

pH Characteristics. Interaction of charges on the cellulose matrix with hydrogen ions may cause a change in the rate-pH profile of the enzyme catalysis when the enzyme is attached to the matrix. In the case of CMCF the accumulation of hydrogen ions in the negatively charged matrix results in the pH in the matrix being lower than the pH in the bulk solution (in the range of pH where the CM-cellulose-carboxyl groups are ionized) with consequent shift of the alkaline limb of the rate-pH profile to higher pH's.[20]

Reactions in Packed Beds (Column Assays). Passage of a solution of substrate at a known rate through a column of a CI-enzyme leads to a readily measurable difference in the inflowing and emerging substrate concentrations. The characterizing kinetic parameters vary from column to column and must be established for each set of experimental conditions.

The catalysis of the hydrolysis of both BAEE and casein by a column of CMCF is described by the modified Michaelis-Menten equation,[26]

$$P[S_0] - K_m \ln(1 - P) = k_{cat}(\text{app}) \, E\beta/Q$$

where P is the fraction of the substrate reacted in the column $= ([S_0] - [S_t])/[S_0]$; $[S_0]$ = concentration of substrate entering the column; $[S_t]$ = concentration of substrate leaving the column; β, the voidage of the column, is the ratio of the void volume and the total volume of the cylindrical packed bed; E = total amount of enzyme (in moles) in the packed bed; Q = flow rate. The kinetic parameters for the catalysis of the hydrolysis of BAEE by CMCF in packed beds are compared with the parameters for catalysis by CMCF in stirred suspensions and by free solution ficin in Table II.

[26] M. D. Lilly, W. E. Hornby, and E. M. Crook, *Biochem. J.* **100**, 718 (1966).
[27] C. Money and E. M. Crook, unpublished results.

[72] Water-Insoluble Thrombin[1, 1a]

By BENJAMIN ALEXANDER and ARACELI M. ENGEL

Water-insoluble forms of some enzymes can now be readily prepared from soluble, preferably highly purified material. The procedures are based upon two principles: The first involves the preparation of *derivatives* of certain enzymes or proenzymes, notably trypsin, chymotrypsin, papain, plasminogen, and prothrombin (factor II) which are supposedly linked to macromolecular inert synthetic backbone, thereby being converted into insoluble products.[2-7] The second consists of embedding the soluble enzyme in an insoluble, highly cross-linked synthetic polymer

[1] Supported by U.S.P.H.S. Grants No. HE 09011 and No. HE 11447.
[1a] A preliminary report of this work was presented at the 49th Annual Meeting of the Federation of the American Societies for Experimental Biology.[1b]
[1b] A. M. Engel and B. Alexander, *Federation Proc.* **24**, 512 (1965).
[2] J. J. Cebra, D. Givol, H. Silman, and E. Katchalski, *J. Biol. Chem.* **236**, 1720 (1961).
[3] A. Bar-Eli and E. Katchalski, *J. Biol. Chem.* **238**, 1832 (1963).
[4] Y. Levin, M. Pecht, L. Goldstein, and E. Katchalski, *Biochemistry* **5**, 1905 (1964).
[5] Q. Z. Hussain and T. F. Newcomb, *Proc. Soc. Exptl. Biol. Med.* **115**, 301 (1964).
[6] I. H. Silman and E. Katchalski, *Ann. Rev. Biochem.* **35**, 873 (1966).
[7] S. Rimon, Y. Stupp, and A. Rimon, *Can. J. Biochem.* **44**, 415 (1966).

without involving chemical bonding between enzyme and insoluble carrier.[8] In the procedure described below for the preparation of insoluble thrombin, the first principle was employed.

Method of Preparation

The method consists in coupling purified thrombin (EC 3.4.4.13) to a diazotized copolymer of p-amino-DL-phenylalanine L-leucine, similar to the procedure used for the preparation of one form of water-insoluble trypsin.[3]

Materials

"Citrate" bovine thrombin is derived from purified factor II obtained from oxalated plasma, according to the method of Goldstein et al.[9] Factor II is then converted to thrombin in 25% sodium citrate,[10] which is subsequently removed by dialysis. The specific activity of the thrombin is 800–1200 NIH units/mg protein. Some of this material is further purified by starch gel electrophoresis, according to Tishkoff et al.[11] The material does not contain detectable factors VII (proconvertin) and X (Stuart factor).

p-Amino-DL-phenylalanine L-leucine copolymer, MW ~ 5000 (Miles Laboratories, Elkhart, Ind.)

p-Tosyl-L-arginyl methyl ester (TAME) and benzoyl arginyl ethyl ester (BAEE) (K and K Laboratories)

Trypsinogen, 1 × crystallized and chymotrypsinogen A, chromatographically homogeneous (Worthington)

Plasminogen (Lot No. 2259-164-28), derived from human Cohn plasma fraction 3, was kindly provided by Dr. A. J. Johnson of New York University

Hammarsten casein (General Biochemical) was purified according to Norman[12]

Acetyl-L-tyrosine-ethyl ester (ATEE) (Calbiochem)

Procedure. Water-insoluble thrombin is prepared as follows: the copolymer of p-amino-DL-phenylalanine L-leucine (1:2 ratio) is diazotized by dissolving 50 mg of the copolymer in a mixture of 0.65 ml of

[8] P. Bernfeld and J. Wan, *Science,* 142, 578 (1963).
[9] R. Goldstein, A. Le Bolloc'h, B. Alexander, and E. Zonderman, *J. Biol. Chem.* 234, 2857 (1959).
[10] W. H. Seegers, *Proc. Soc. Exptl. Biol. Med.* 72, 677 (1949).
[11] G. H. Tishkoff, L. Pechet, and B. Alexander, *Blood* 15, 778 (1960).
[12] D. Norman, *J. Exptl. Med.* 106, 423 (1957).

50% acetic acid and 0.65 ml 2 N HCl, cooling to 0°, and adding drop-wise (while stirring) 0.30 ml of 0.5 N NaNO$_2$. After 1½ hours at 2°, the pH is raised to 7.5 with 5 N NaOH. The evolved coarse brown-orange precipitate is centrifuged, and washed at least 4 times with 7 ml of phosphate buffer (0.1 M, pH 7.4) until the last wash, after separation by centrifugation and filtration from the precipitate, had a pH of 7.4.

The precipitate, finely homogenized in an electric homogenizer (Tri-R Instrument, Inc.) and resuspended in 7.0 ml of the buffer, is mixed with a solution of purified thrombin in saline (6–7 mg/ml), and interaction is allowed to proceed for 40 hours at 4° with constant stirring.

The mixture is then centrifuged, the supernatant is filtered through a Millipore (45 μ) pad (sieve appropriate for the size particle of the material under preparation), and the filtrate is tested for clotting activity. The filtered material is resuspended in the phosphate buffer, and washed (with the buffer) and separated by centrifugation, as many times as needed (usually around 20 times) until the supernatant of the last centrifugation, when Millipore filtered and tested, exhibits no clotting activity, as indicated below for a clotting test with the insoluble material.

The precipitate, resuspended in the desired volume of buffer or saline, is kept at 2°–4°. The p-amino-DL-phenylalanine L-leucine insoluble thrombin (PLITh) thus obtained is a reddish-brown fine suspension which constitutes the stock solution.

Assay Methods

The clotting activity of PLITh is determined with 0.2 ml of a suitable suspension-dilution in saline of the stock solution, to which 0.2 ml of a 1.5% factor I (fibrinogen, 90% clottable) is added. Conglomeration of the dispersed PLITh heralds the first appearance of fibrin, which is recorded as the clotting end point. The thrombic activity (in NIH units) is determined by interpolation of the observed clotting time on a standardization curve derived with dilutions of a preparation of NIH thrombin (Lot No. B-3), obtained from the Division of Biologic Standards of the National Institutes of Health.

The esterase activity is measured by the catalyzed hydrolysis of 0.04 M TAME or 0.01 M BAEE, calculated from the continuously recorded uptake of 0.10 N NaOH in the pH-stat with attached automatic buret (Radiometer, Copenhagen).

The caseinolytic activity is determined with Hammarsten casein, according to the procedure of Engel et al.[13]

The protein of the insoluble thrombin is determined by measuring the valine content, as has been applied to water-insoluble trypsin.[3]

[13] A. M. Engel, B. Alexander, and L. Pechet, *Biochemistry* **5**, 1543 (1966).

Definition of Units. One thrombin clotting unit of PLITh is defined as that amount which clots 0.2 ml of a 1.5% factor I solution in 15 seconds at room temperature.

Units of thrombin esterase activity are defined as the number of micromoles H^+ liberated from TAME or BAEE by the sample in question at 22°, titrated at pH 7.8 (± 0.05) with $0.10\,N$ NaOH/minute/ml of TAME or BAEE/ml of sample.

One caseinolytic unit is defined as the activity which increases the optical density, at 280 nm, by 1 unit of the optical reading/minute, after 30 minutes of incubation at 37° of the enzyme with a 1% solution of Hammarsten casein.[13]

Properties

Some properties of PLITh are summarized in Table I.

TABLE I
SOME PROPERTIES OF PLITh

Thrombin (mg)[a] per 100 mg carrier[b]	62
Bound protein (mg)[c] per 100 mg insoluble enzyme	34
Clotting activity of bound protein (percent of original native thrombin)	3
TAME esterolytic activity of bound protein (percent of original native thrombin)	15

[a] Specific activity, 300 NIH units/mg protein.

[b] p-Amino-DL-phenylalanine L-leucine copolymer (MW $\sim$ 5000).

[c] Calculated from valine content, assuming valine/protein N ratio of 0.32 which was found for the parent native thrombin.

The optimum pH for esterolytic activity (TAME) over a range of 6.5 to 10.0 is 9.0, as compared with 8.0 for the soluble parent thrombin.

Stability. Stability of both esterase and clotting activities at 4° compared with soluble thrombin is shown in Fig. 1. No bacterial contamination was detected during the period of testing. In the frozen state the material maintained full activity for at least 6 months.

Purity. PLITh prepared as above does not contain measurable amounts of factors VII and X as determined by the method of Owren[14] and Bachman *et al.,*[15] respectively. These procedures are capable of detecting as little as 1% factors VII and X.

Specific Clotting Activity. The average specific activity of seven

[14] P. A. Owren, *in* "Transactions of the 5th Conference on Blood Clotting and Allied Problems" (J. E. Flynn, ed.), p. 98. Josiah Macy, Jr. Foundation, New York, 1952.

[15] F. Bachman, F. Duckert, and F. Koller, *Thromb. Diath. Haemorrhag.* **2**, 24 (1958).

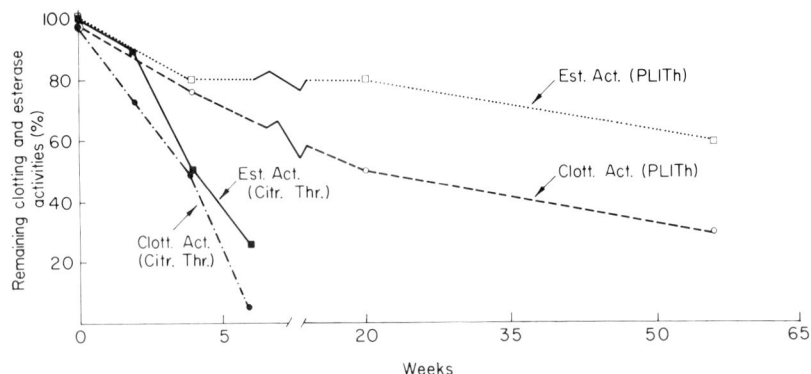

FIG. 1. Comparative stability of PLITh and the parent soluble thrombin. The two preparations were kept at 4° (thrombin, 500 U/mg protein; PLITh, 10 U/mg protein), and their clotting and esterase activity measured at the intervals indicated.

unselected preparations was 3.2 NIH units/mg of protein (range 0.7–10.0) with an average yield of 3.0% of the original thrombin activity.

Inhibitors. Clotting activity is markedly retarded by TAME, as has also been reported for the native soluble enzyme.[16] Esterase activity was inhibited by 0.10 M to 0.20 M Na+, as we observed with the soluble thrombin, and as has been confirmed more recently by Roberts and Burkat.[17]

Soybean trypsin inhibitor (Worthington) had no effect on the clotting or esterase activities of PLITh, in accord with its well-known inertness in these respects on soluble thrombin.[18]

Specificity. As is true with soluble thrombin,[13, 19] besides clotting factor I and catalyzing TAME and BAEE hydrolysis, PLITh cleaves casein, and activates trypsinogen, plasminogen, and chymotrypsinogen. Relevant comparative data with PLITh and parent thrombin are shown in Table II.

Kinetic Properties. Kinetic enzyme data on PLITh have not yet been obtained.

Reversibility. As already indicated, during its preparation the water-insoluble material had been washed extensively with buffer until the washings failed to exhibit thrombin-clotting activity. Over an extended interval (~6 months) samples of the stock solution were Millipore filtered, and the filtrates were examined for clotting activity. None was

[16] S. Sherry, N. Alkjaersig, and A. Fletcher, *Am. J. Physiol.* **209**, 577 (1965).
[17] P. S. Roberts and R. K. Burkat, *Proc. Soc. Exptl. Biol. Med.* **127**, 447 (1968).
[18] G. F. Lanchantin, J. A. Friedmann, and D. W. Hart, *J. Biol. Chem.* **244**, 865 (1969).
[19] A. M. Engel and B. Alexander, *Biochemistry* **5**, 3590 (1966).

TABLE II

SPECIFIC ACTIVITIES OF PLITh AND THE PARENT SOLUBLE THROMBIN

	Clotting U per mg protein	TAME U per mg protein	BAEE U per mg protein	Caseinolytic U per mg protein	$m\mu M$ Trypsinogen activated per mg protein[a]	TAME U Plasminogen activated per mg protein[a]	$m\mu M$ Chymotrypsinogen activated per mg protein[b]
Thrombin	200	21.6	5.3	7.2	0.2	5.1	2.6
PLITh	23	4.0	1.0	2.0	0.1	2.0	0.7

[a] As measured by TAME esterase activity of evolved trypsin or plasmin, respectively.
[b] As measured by ATEE esterase activity of evolved chymotrypsin.

detected, indicating that no freeing or "leaking" of soluble thrombin had occurred. Nevertheless, a cautionary note should be sounded. Extensive washing with buffer does not preclude the possibility that *soluble parent material* may not be released by washing with a protein acting as a possible substrate.

In collaboration with Y. Shamash,[20] we have found that insoluble trypsin previously washed extensively, suspended in buffer, and yielding a trypsin-free Millipore filtrate after long storage, when washed with a solution of albumin, exhibited trace esterase activity in the albumin-containing filtrate. Several albumin washings thereafter could remove the last traces of soluble tryptic activity. This has also been the experience of Ephraim Katchalski of the Weizmann Institute, Israel (personal communication).

Similar observations[21] with PLITh are shown in Table III. The data indicate that the copresence of PLITh and albumin or fibrin monomer resulted in liberation of a trace amount of thrombic clotting activity into

TABLE III
THROMBIN "LEAKAGE" FROM INSOLUBLE THROMBIN[a]

| | Clotting time | |
PLITh suspended in	Nonfiltered suspension (minutes)	Millipore filtered (hours)
0.15 N NaCl	11	24 hours
1.0 N NaCl	12	24
1.0 N NaBr	12	24
1.0% Albumin in 0.15 N NaCl	12	Firm clot in 2 hours
0.5% Fibrin monomer[b] in 1.0 N NaBr	14	Firm clot in 2 hours

[a] *Procedure:* To 0.1 ml of a 1:10 saline dilution of stock PLITh suspension was added 0.9 ml of each of the following: 0.15 N NaCl; 1.09% purified human albumin (Courtland) in 0.15 N NaCl; 1.0 N NaCl; 1.0 N NaBr; fibrin monomer. An aliquot of each solution was Millipore filtered (0.45 μ), and 0.15 ml of the filtrate was added to 1.0 ml of a 0.5% solution of factor I (Pentex, bovine, 95% clottable) in 0.3 M phosphate–0.075 N NaCl buffer, pH 6.8, followed by enough H_2O or 0.15 N NaCl to bring the salt concentration to 0.15 ionic strength, thereby obtaining the same ionic strength, volumes, and factor I concentrations throughout. The clotting time (room temperature) was then recorded.

[b] Prepared according to T. H. Donnelly, M. Laskowski, N. Notley, and H. A. Scheraga, *Arch. Biochem. Biophys.* **56**, 369 (1955).

[20] Research Associate, New York Blood Center, N.Y. Present address: Marcus Center, Magen David Adom, P.O. Box 15025, Jaffa, Israel.
[21] B. Alexander, Y. Shamash, and A. M. Engel, unpublished observations.

the solution, albeit immeasurable, in contrast to suspensions of the insoluble thrombin in the respective electrolyte solutions alone. This could be progressively reduced by successive washing of the insoluble enzyme with albumin, as indicated by the following experiment: 1.5 ml of PLITh in 0.15 NaCl (clotting activity, 0.5 NIH units/ml) was washed repeatedly with equal volumes of saline, and the washings set aside for clotting determination. The final centrifuged pellet was resuspended in 3.0 ml of the albumin solution, the mixture was centrifuged, the supernatant Millipore filtered, and the filtrate set aside for clotting test. This procedure was repeated five more times. The observed clotting times were as follows: all saline washes, >24 hours; PLITh in albumin, 4 minutes; first albumin extract, 12 minutes; second albumin extract, 15 minutes; third extract, 24 minutes; fourth, 27 minutes; fifth, 29 minutes; sixth, 29 minutes; PLITh resuspended in albumin after 6 albumin extractions, 6½ minutes.

Thus, although repeated washing with albumin substantially decreased thrombin "leakage," it could not be totally eliminated, in contrast to our experience with water-insoluble trypsin.

PLITh, and a water-insoluble derivative of factor II (prothrombin) prepared by the same procedure,[22] have been particularly useful in our laboratory in the investigation of fibrinogen-fibrin conversion,[23] and in the activation of factor II[22] via several pathways.

[22] A. M. Engel and B. Alexander, *Intern. Congr. Biochem.* (*Tokyo*) Abstract IV, F-177, 793 (1967).
[23] B. Alexander, "Symposium on Fluid Replacement in the Surgical Patient." Columbia Univ. Coll. of Physicians & Surgeons, Depts. of Surgery and Anaesthesiology, in association with Division of Medical Sciences of the National Research Council, May 26 and 27, 1969, in press.

Author Index

F

Fabre, C., 42, 47(2)

Fackre, D., 52(k), 53, 56(g), 57, 613

Fackre, D. S., 906, 910(8), 911(8), 912(8), 913(8), 914(8)

Fahey, J. L., 168, 170(34)

Fahrney, D., 317, 335(15), 590

Fahrney, D. E., 8

Falla, F., 335, 336(62)

Falvey, A., 803

Fambrough, R. M., 648, 649

Fantes, K. H., 622

Farr, A. L., 387, 511, 526(9), 528, 547, 583, 592, 596(6), 627(61, a), 671, 689(b), 691, 695(d), 708, 724, 735, 739, 743, 757, 784, 792, 835

Fasman, G. D., 79, 80(17)

Fasold, H., 373, 386(11)

Fawwal, I., 373, 375(16), 440, 442(9), 443 (9), 444(9)

Feder, J., 20, 21, 25(2), 26(2), 61, 119, 120 (26), 131(26), 136(26), 137(26), 227 (7), 228, 234(7), 377, 573, 650, 651

Feeney, R. E., 197, 211, 849, 850(29), 857, 862(28), 863, 865(5), 866, 867, 868(a), 869, 870, 871(5, 28), 889, 890, 891, 892, 893(5, 7, 25), 894, 895, 896(5, 11, 34), 898(a), 899, 900, 901(18, 34, 48), 902, 904

Feinstein, G., 280, 281(14), 857, 889, 900

Felber, J.-P., 496

Feldsted, E. T., 163, 171(17)

Femfert, U., 515, 518, 521

Ferdinand, W., 865, 869

Ferguson, J. H., 708, 711(8), 713, 714(20)

Ferguson, L. T., 871, 900

Ferold, H. L., 890

Ferrari, C., 139

Ferry, G., 446, 447(16)

Ferry, J. D., 772

Fidlar, E., 163, 171(17)

Fiedler, F., 697, 698(18), 699(18)

Filho, J. X., 863, 866, 867, 868(h), 869, 870(10)

Fink, E., 846

Finkenstadt, W. R., 858

Finkenstaedt, J. T., 300

Finkle, B. J., 233, 235(14), 236(14), 956, 957(66), 964

Finlayson, J. S., 770

Fischer, E. H., 482

Fischer, F. G., 544, 545(3)

Fish, J. C., 435

Fisher, Jr., E., 616(14), 619

Fishman, J. B., 815

Fishman, L., 672

Fleisher, G. A., 787

Fletcher, A., 196, 198, 199(51), 671, 672, 713, 819, 822, 832(7), 833(7), 834, 982

Florkin, M., 47, 48, 49(24, 25), 52(c), 53, 765

Fodor, P. J., 786

Foldes, I., 139

Folin, O., 398, 742

Folk, J. E., 91, 94(32), 96(32), 97(32), 98 (32), 99, 100, 109, 110, 111, 112, 463, 476, 499, 500, 504, 505, 507, 508, 780, 786, 788(13)

Foltmann, B., 422, 423, 424, 425, 427(8, 13), 429, 430, 431(4, 5, 8), 432, 433, 434, 435, 436, 455, 457

Forman, M., 778

Fortmann, B., 421

Fossum, K., 262, 271(7), 272(7), 891

Foster, R. J., 889

Fournier, E., 850

Fowden, L., 789

Fox, S. W., 272

Fraefel, W., 872, 873(6), 876, 877, 878

Fraenkel-Conrat, H., 405, 444, 649, 863, 865(2), 866(2), 867, 870(2), 895, 896 (35), 900, 901

France, E. S., 917

Franchini, M., 849(34a), 850

Frankel, S., 113, 114(13), 135(13), 609

Frater, R., 237, 239(37), 240(37), 241(37), 271, 510, 513(3), 533, 534(18)

Frattali, V., 853, 854(12), 859, 860, 861, 862

Fredericq, E., 895, 896(28), 899

Freisheim, J. H., 460, 467, 468, 471, 472(7)

Frey, E. K., 683, 699, 849

Fried, M., 307, 308(42)

Friedenson, B., 263, 265, 266, 271, 272(8)

Friedman, L., 433

Friedman, O. M., 786, 787(14), 788(23)

Friedmann, J. A., 145, 168, 169, 857, 982

Subject Index